The Elements

Element	Symbol	Atomic number	Atomic mass (amu)	Element	Symbol	Atomic number	Atomic mass (amu)
actinium	Ac	89	(227)	manganese	Mn	25	54.94
aluminum	Al	13	26.98	meitnerium	Mt	109	(268)
americium	Am	95	(243)	mendelevium	Md	101	(258)
antimony	Sb	51	121.8	mercury	Hg	80	200.6
argon	Ar	18	39.95	molybdenum	Mo	42	95.94
arsenic	As	33	74.92	neodymium	Nd	60	144.2
astatine	At	85	(210)	neon	Ne	10	20.18
barium	Ba	56	137.3	neptunium	Np	93	237.0
berkelium	Bk	97	(247)	nickel	Ni	28	58.69
beryllium	Be	4	9.012	niobium	Nb	41	92.91
bismuth	Bi	83	209.0	nitrogen	N	7	14.01
bohrium	Bh	107	(267)	nobelium	No	102	(259)
boron	B	5	10.81	osmium	Os	76	190.2
bromine	Br	35	79.90	oxygen	O	8	16.00
cadmium	Cd	48	112.4	palladium	Pd	46	106.4
calcium	Ca	20	40.08	phosphorus	P	15	30.97
californium	Cf	98	(251)	platinum	Pt	78	195.1
carbon	C	6	12.01	plutonium	Pu	94	(244)
cerium	Ce	58	140.1	polonium	Po	84	(209)
cesium	Cs	55	132.9	potassium	K	19	39.10
chlorine	Cl	17	35.45	praseodymium	Pr	59	140.9
chromium	Cr	24	52.00	promethium	Pm	61	(145)
cobalt	Co	27	58.93	protactinium	Pa	91	231.0
copernicium	Cn	112	285	radium	Ra	88	(226)
copper	Cu	29	63.55	radon	Rn	86	(222)
curium	Cm	96	(247)	rhenium	Re	75	186.2
darmstadtium	Ds	110	(269)	rhodium	Rh	45	102.9
dubnium	Db	105	(262)	roentgenium	Rg	111	(272)
dysprosium	Dy	66	162.5	rubidium	Rb	37	85.47
einsteinium	Es	99	(252)	ruthenium	Ru	44	101.1
erbium	Er	68	167.3	rutherfordium	Rf	104	(261)
europium	Eu	63	152.0	samarium	Sm	62	150.4
fermium	Fm	100	(257)	scandium	Sc	21	44.96
flerovium	Fl	114	289	seaborgium	Sg	106	(266)
fluorine	F	9	19.00	selenium	Se	34	78.96
francium	Fr	87	(223)	silicon	Si	14	28.09
gadolinium	Gd	64	157.3	silver	Ag	47	107.9
gallium	Ga	31	69.72	sodium	Na	11	22.99
germanium	Ge	32	72.40	strontium	Sr	38	87.62
gold	Au	79	197.0	sulfur	S	16	32.07
hafnium	Hf	72	178.5	tantalum	Ta	73	180.9
hassium	Hs	108	(277)	technetium	Tc	43	(98)
helium	He	2	4.003	tellurium	Te	52	127.6
holmium	Ho	67	164.9	terbium	Tb	65	158.9
hydrogen	H	1	1.008	thallium	Tl	81	204.4
indium	In	49	114.8	thorium	Th	90	232.0
iodine	I	53	126.9	thulium	Tm	69	168.9
iridium	Ir	77	192.2	tin	Sn	50	118.7
iron	Fe	26	55.85	titanium	Ti	22	47.87
krypton	Kr	36	83.80	tungsten	W	74	183.9
lanthanum	La	57	138.9	uranium	U	92	238.0
lawrencium	Lr	103	(260)	vanadium	V	23	50.94
lead	Pb	82	207.2	xenon	Xe	54	131.3
lithium	Li	3	6.941	ytterbium	Yb	70	173.0
livermorium	Lv	116	293	yttrium	Y	39	88.91
lutetium	Lu	71	175.0	zinc	Zn	30	65.41
magnesium	Mg	12	24.31	zirconium	Zr	40	91.22

Note: Parentheses ( ) denote the most stable isotope of a radioactive element.

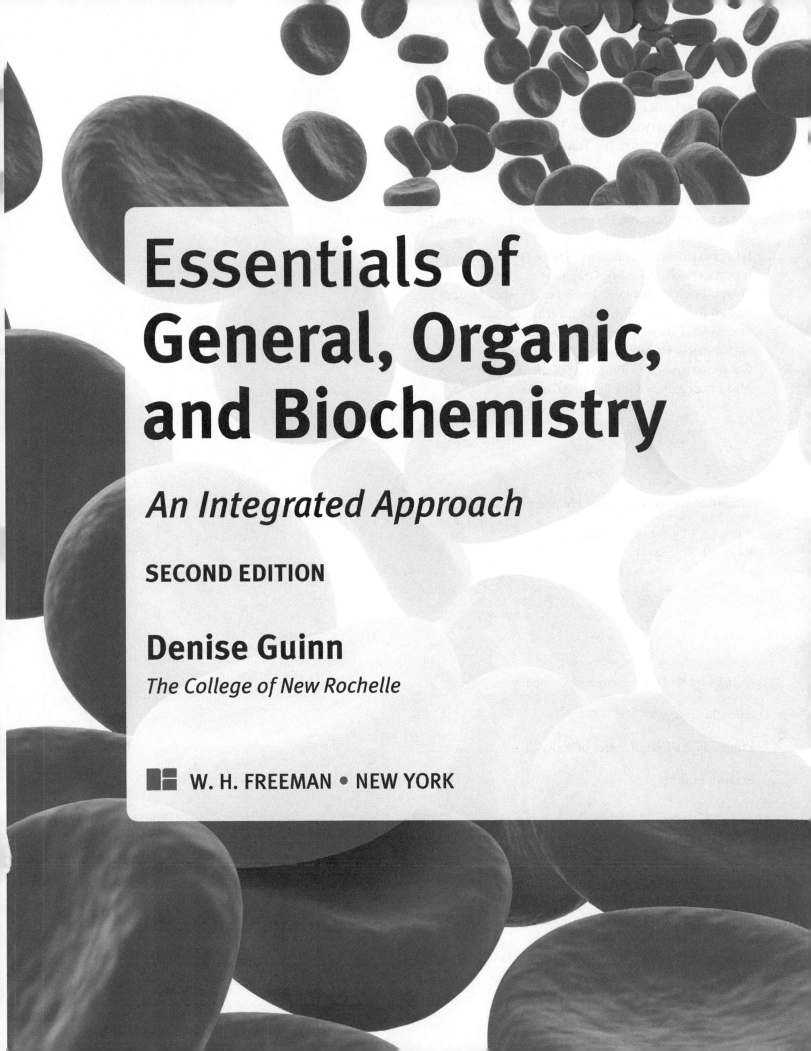

Essentials of General, Organic, and Biochemistry

An Integrated Approach

SECOND EDITION

Denise Guinn
The College of New Rochelle

W. H. FREEMAN • NEW YORK

SENIOR DEVELOPMENT EDITOR: Randi Blatt Rossignol
PUBLISHER: Jessica Fiorillo
SENIOR ACQUISITIONS EDITOR: Bill Minick
EDITORIAL ASSISTANT: Tue Tran
ASSOCIATE DIRECTOR OF MARKETING: Debbie Clare
MARKETING ASSISTANT: Samantha Zimbler
MEDIA EDITOR: Dave Quinn
SUPPLEMENTS EDITOR: Courtney Lyons, Tue Tran and Dave Quinn
ART DIRECTOR: Diana Blume
ILLUSTRATION COORDINATOR: Janice Donnola
ILLUSTRATIONS: Network Graphics
PHOTO EDITOR: Robin C. Fadool and Bianca Moscatelli
PHOTO ASSISTANT: Nicholas H. Fernandez
PHOTO RESEARCHER: Dena Diglio Betz
PRODUCTION COORDINATOR: Paul Rohloff
COMPOSITION AND LAYOUT: Aptara®, Inc.
MANUFACTURING: King Printing Co., Inc.

Library of Congress Control Number:

ISBN-13: 978-1-4292-3124-4
ISBN-10: 1-4292-3124-6

W. H. Freeman and Company
41 Madison Avenue
New York, NY 10010
Houndmills, Basingstoke RG21 6XS, England
www.whfreeman.com

To my sons, Charles and Scott,
and to all the Vogels for their continued inspiration and support
–Denise

About the Author

Jorge Madrigal/Madrigal Studios

Denise Guinn received her B.A. in chemistry from the University of California at San Diego and her Ph.D. in organic chemistry from the University of Texas at Austin. She was a National Institutes of Health postdoctoral fellow at Harvard University before joining Abbott Laboratories as a Research Scientist in the Pharmaceutical Products Discovery Group. In 1992, Dr. Guinn joined the faculty at Regis University, in Denver, Colorado, as Clare Boothe Luce Professor of chemistry, where she taught courses in general chemistry, organic chemistry, and the general, organic, and biochemistry course for nursing and allied health majors. In 2008, she joined the chemistry department at The College of New Rochelle, New York where she teaches organic chemistry, biochemistry, and the one-semester GOB course for nursing students. She has published in the Journal of Organic Chemistry, the Journal of the American Chemical Society, and the Journal of Medicinal Chemistry. She currently resides in Nyack, New York.

Brief Contents

Contents

Chapter 11 | Carbohydrates: Structure and Function   493

Chapter 12 | Lipids: Structure and Function   527

Chapter 13 | Proteins: Structure and Function   561

A Letter from The Author

In teaching the general, organic, and biochemistry course for the past 22 years, it has been a great pleasure and opportunity to be at the forefront of the integrated approach to teaching general, organic, and biochemistry. In writing the first edition of *Essentials of General, Organic, and Biochemistry,* our goal was to make it obvious why chemistry is a cornerstone in the education of today's health care professionals by using health and medicine as the framework for learning the fundamentals of chemistry. The second edition of this text has been further shaped by the hundreds of instructors and students who shared with us their experience using the first edition. There is a consensus that the integrated approach effectively engages students in the course early-on, while at the same time making it feasible to learn the fundamental concepts of organic chemistry and biochemistry in a one-semester course. I hope that you feel, as I do, that the second edition has retained the elements of the first edition that worked so well, while incorporating some organizational changes and new material that better support student learning in chemistry.

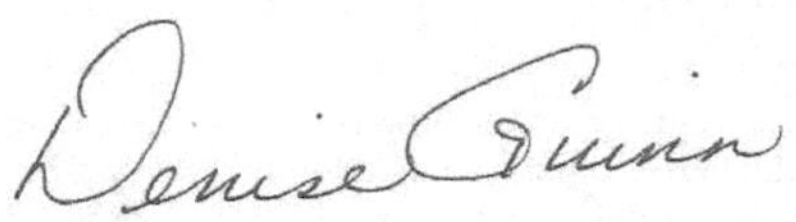

Denise Quinn

Preface

This text takes an integrated approach to general, organic, and biochemistry as applications of chemistry in health and medicine are used to illustrate the key concepts. To achieve this goal, we approach the course differently in several important ways.

Integration of Organic and Biochemistry in Every Chapter

The GOB course has traditionally been taught sequentially covering the three major areas of chemistry—general chemistry, organic chemistry, and biochemistry—in that order. By relegating much of the interesting, relevant content to the end, some students may lose interest in the course.

In this textbook, organic and biochemistry concepts are included in every chapter, so that during every week of the course, students are engaged in topics directly related to their field of study. To make this integration even more effective, organic chemistry is introduced relatively early in the text (Chapters 6 and 7).

Historically, when **organic compounds** were first studied, it was believed that they could *not* be prepared in the laboratory and that only a living plant or animal could produce an organic compound. This is the origin of the term *organic*, which means "from living things." Compounds that do not contain carbon were known as **inorganic compounds**. Today, almost any organic compound can be prepared in the laboratory, even extremely complex organic molecules. Nevertheless, the terms *organic* and *inorganic* remain with us today to distinguish these two basic classes of compounds.

Organic compound:	Contains carbon
Inorganic compound:	Does not contain carbon

Many compounds produced in nature are synthesized in the laboratory. Some are used as **pharmaceuticals**—drugs used for therapeutic purposes. Plants and animals are rich sources of medicinally valuable organic compounds, such as Taxol, the lifesaving anti-cancer drug first isolated from the yew tree (Figure 6-4). Most pharmaceuticals, however, are synthesized in the laboratory. For example, Lipitor (atorvastatin), the best-selling pharmaceutical in the history of medicine, used for the treatment of high cholesterol, is entirely synthetic; it is not produced naturally by any plant or animal (Figure 6-5). You will see examples of pharmaceuticals throughout this chapter and the next as you are introduced to the fundamental principles of organic chemistry.

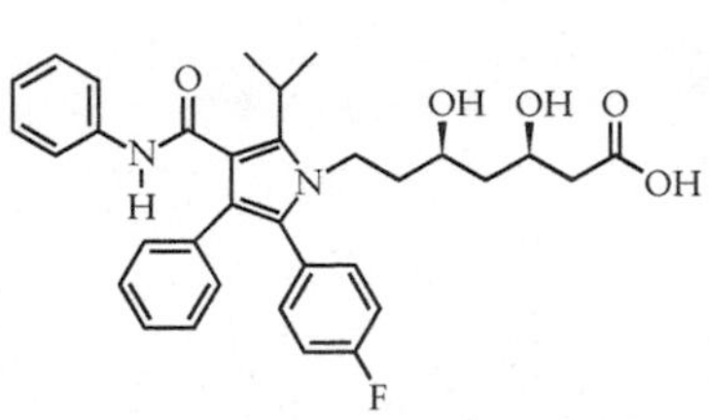

Lipitor

Figure 6-5 Lipitor (atorvastatin), a synthetic drug used for the treatment of high cholesterol.

NEW: Improved Organization

While retaining the integration of general, organic, and biochemistry, some topics have been rearranged to make the book an even better fit for most courses. Here are some of the organizational changes in the second edition:

- Chapter 1 from the first edition has been divided into two chapters. The benefits of this are that Chapter 1, Measuring Matter and Energy, is now shorter and focused on concepts relating to the measurement of matter and energy, thus allowing the introduction of energy—a central theme throughout the text—earlier in the course.

- Chapter 2, Atomic Structure and Nuclear Radiation, now covers atomic structure, including a streamlined section on nuclear radiation (previously in Chapter 16).

- Chemical reactions are now covered much earlier in the text. Chapter 4, Chemical Quantities and Chemical Reactions, introduces the concept of the mole, along with balancing equations, enthalpy, kinetics, and a new section on chemical equilibrium.

- The chapter on functional groups, Chapter 7, Organic Chemistry and Biomolecules, has a new section on stereochemistry and integrates the structure of the important biomolecules when introducing the functional groups.

- The chapter on solutions (Chapter 8, Mixtures, Solution Concentrations, and Diffusion) and the chapter on acids and bases (Chapter 9, Acids and Bases) are now next to each other while still following the chapters on organic chemistry (Chapters 6 and 7), thereby retaining the examples of organic compounds.

- The last five chapters (11–15) are presented in a new order: Carbohydrates, Lipids, Proteins, Nucleic Acids, and a capstone chapter on Energy and Metabolism. The structure and function of the biomolecules are described in Chapters 11–14, while the catabolic biochemical pathways and energy implications now appear in Chapter 15.

Content Tailored to Prepare Students for Their Careers in Health Care

As scientists, we already know that chemistry is the central science, and as such is an important foundation for understanding health and medicine. For students to be motivated to learn chemistry, they need to see how the concepts they are learning are relevant to their chosen field of study. Studies show that consistently motivated students are much more likely to succeed than those searching for relevance throughout the course, especially early in the course. While engineering examples are best for engineering students, medical examples, and other health- and consumer-based examples are most effective for teaching the fundamental concepts of chemistry to nursing students and allied health majors.

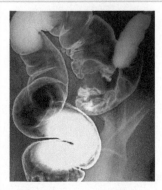

Figure 8-7 X ray image of a patient with colon cancer, after receiving a barium enema—a suspension of radioactive barium sulfate, $BaSO_4$, in water. [ZEPHYR/Getty Images]

Saturated and Unsaturated Solutions Soluble ionic compounds dissolve in water because the ion-dipole forces of attraction are stronger than the ionic bonds of the intact solid lattice in water. Some ionic compounds, however, do not dissolve in water. For example, magnesium hydroxide, $Mg(OH)_2$, the active ingredient in milk of magnesia, used to relieve the symptoms of heartburn, is insoluble in water. Radioactive barium sulfate, $BaSO_4$, a suspension taken orally to image the gastrointestinal tract or as a "barium enema," used to screen for colon cancer (Figure 8-7), is another example of an insoluble ionic compound. Insoluble ionic compounds have stronger electrostatic inter-

We make the connections between chemical concepts and health care in each chapter beginning with an intriguing opening story; throughout the main

text as chemical concepts are covered; in all of the exercises; and at the end of each chapter in the Chemistry in Medicine feature, which utilizes the concepts of the chapter to explain a particular medical condition or disease.

- For example, when discussing units of concentration in Chapter 8, we include units commonly used in medicine. As a consequence, we are able to write exercises that ask students to calculate concentrations, dosages, and flow rates, all while providing practice with dimensional analysis and the metric system.

- When naming ionic compounds in Chapter 3, we include some that are commonly found in health and consumer products. When discussing the conversion of units in Chapter 4, we use actual blood test results so that students are performing calculations using real data.

- When describing organic functional groups, many prescription and over the counter medicines are used as examples.

- When discussing the gas laws in Chapter 5, we discuss the volume of air in the lungs during an asthmatic attack.

- Rather than introducing the traditional reactions such as displacement reactions between salts and redox reactions of metals, in Chapter 10 we use the five basic organic reactions catalyzed by enzymes in the cell as examples.

WORKED EXERCISE Applying Gas Laws to the Human Body

5-13 When an asthmatic has an asthma attack, a smaller amount of air enters the lungs, and therefore a lower oxygen partial pressure exists in the lungs and consequently the concentration of oxygen in the blood is lower. How is the concentration of oxygen in the blood affected during an asthma attack? Explain using Henry's law.

Solution

5-13 When an asthmatic has an asthma attack, the partial pressure of oxygen in the lungs is less because of the decreased amount of air inhaled, and therefore, according to Henry's law, the concentration of dissolved oxygen in the blood is less.

NEW: Many More Authentic Images. A liberal use of photos related to actual clinical practice, consumer health care products, and other natural products reinforce the chemical concepts, while showing an application.

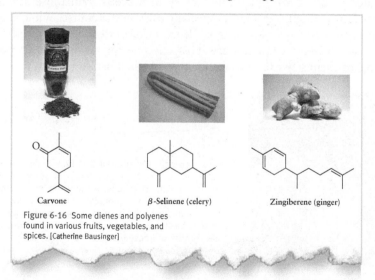

Carvone   β-Selinene (celery)   Zingiberene (ginger)

Figure 6-16 Some dienes and polyenes found in various fruits, vegetables, and spices. [Catherine Bausinger]

Chapter Walkthrough

Chapter Opening Vignettes. Because context and connection are crucial to motivation and learning chemistry, the first encounter with new concepts arrives in the form of a short, practical, and real-life example of a health-related topic connected to the concepts in the chapter. These stories immediately immerse the student in a high-interest topic related to health and medicine.

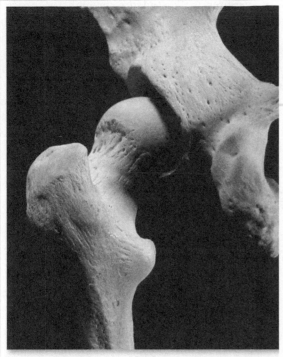

1 | Measuring Matter and Energy

Osteoporosis and Bone Density

One in five American women over the age of 50 is estimated to have osteoporosis, and over half of these women will break a bone at some point in their lives as a result of the condition. Osteoporosis is a condition characterized by the progressive loss of bone density, resulting in an increased risk of bone fractures. For example, even a cough or stumble is often enough to trigger a fracture. Most bone fractures occur in the hip, wrist, and spine.

Bone density is the measurement most commonly used to assess bone strength. Density is a physical property of a substance, which can be calculated from the mass of a sample of the substance divided by its volume: $d = m/V$. Mass is similar to weight and is a measure of the amount of material, in this case, bone mineral. Volume is a measure of the three-dimensional space occupied by a material, in this case, a segment of bone. The greater the bone density is, the greater the bone mineral per volume and the stronger the bone. Stronger bone is better able to withstand stress and less likely to fracture. The average bone density for a person is 1.5 g/cm³, read as "one point five grams per centimeter cubed."

To screen for osteoporosis, several techniques have been developed for measuring or estimating the bone density of the hip, wrist, and lumbar spine. Currently, the most common technique is the DEXA scan (*dual energy x-ray absorptiometry*), which

The hip bone is often used for a bone mineral density (BMD) measurement because it is a good indicator of whether or not a person has osteoporosis. Shown is a healthy hip bone.
[James Stevenson/Science Source]

Detailed Worked Exercises, paired with Practice Exercises. Throughout each chapter, Worked Exercises reinforce the text's explanations. They give students a helpful roadmap for solving problems as well as the opportunity to practice the concepts they are learning in the text. Practice Exercises follow each set of Worked Exercises, offering students an immediate check on their understanding of the concepts. Answers to the Practice Exercises appear at the end of each chapter.

Figure 4-5 A can of Red Bull contains 80 mg of caffeine. An eight-ounce cup of coffee contains between 100 and 200 mg of caffeine.
[Catherine Bausinger]

WORKED EXERCISES | Interconverting between Mass and Moles

4-7 Sulfur reacts with oxygen, producing a blue flame as it forms sulfur dioxide, a compound with a noxious odor. How many grams of sulfur are there in 5.600 mol of sulfur?

4-8 In a 250 mL can of Red Bull there are 80. mg of caffeine (Figure 4-5). How many moles of caffeine, $C_8H_{10}N_4O_2$, are in 80. mg of caffeine?

Solutions

4-7 First determine the molar mass of sulfur because this will be the conversion between moles and mass. Remember, molar mass has the same numerical value as atomic mass, molecular mass, and formula mass but has units of g/mol instead of amu. According to the periodic table, the molar mass of sulfur is 32.07 g/mol.

Next, express the molar mass of S as two possible conversion factors:

$$\frac{32.07 \text{ g S}}{1 \text{ mol S}} \quad \text{or} \quad \frac{1 \text{ mol S}}{32.07 \text{ g S}}$$

Using dimensional analysis, multiply the supplied unit by the appropriate form of the conversion factor that allows the supplied units to cancel, leaving the requested unit. Note that the supplied unit here is moles and the requested unit is grams, so we use the conversion factor shown in red above:

$$5.600 \text{ mol S} \times \frac{32.07 \text{ g S}}{1 \text{ mol S}} = 179.6 \text{ g S} \quad \text{(four significant figures)}$$

Using Molecular Models. These exercises walk students through the process of building and examining a ball-and-stick model to illustrate an important concept in the chapter. We have found that this type of tactile exercise is effective, and students begin to use the model kit on their own when solving problems. A small inexpensive model kit, specifically designed to be used for the modeling exercises, is available with the book. Building models is a superb way for students to experience and readily understand the three-dimensional aspects of chemistry. The exercises are accompanied by Inquiry Questions that encourage students to build a deeper understanding of molecular structure and relate it to the more abstract structures seen in print.

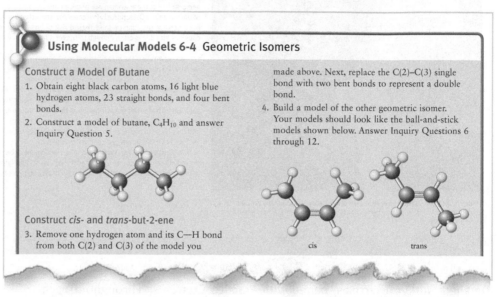

Using Molecular Models 6-4 Geometric Isomers

Construct a Model of Butane

1. Obtain eight black carbon atoms, 16 light blue hydrogen atoms, 23 straight bonds, and four bent bonds.
2. Construct a model of butane, C_4H_{10} and answer Inquiry Question 5.

Construct *cis*- and *trans*-but-2-ene

3. Remove one hydrogen atom and its C—H bond from both C(2) and C(3) of the model you

made above. Next, replace the C(2)–C(3) single bond with two bent bonds to represent a double bond.

4. Build a model of the other geometric isomer. Your models should look like the ball-and-stick models shown below. Answer Inquiry Questions 6 through 12.

cis   trans

Guidelines. Step-by-step instructions, such as naming organic compounds, can be used as both explanations and a quick reference when doing homework or other exercises. They are clearly set off in the text to emphasize their importance and to help students find them easily.

Guidelines for Solving a Dilution Calculation

Step 1. Begin by determining the three supplied variables and the one variable that needs to be calculated. Remember from algebra, that if you have one equation, you can solve for at most one unknown. Applied to a dilution calculation, any one of these variables can be determined if the other three are known. In this example:

$C_1 = 10\%$	concentration of stock solution
$C_2 = 1\%$	desired concentration of dilute solution
$V_1 = ?$	the variable being solved: how much of the stock solution we need to dilute
$V_2 = 100$ mL	the requested final volume of the dilute solution

Step 2. Rearrange the dilution equation so that the variable you are solving for—the unknown—is isolated on one side of the equality, by dividing both sides of the equation by the appropriate variable(s). In this example, we want to solve for V_1, so we isolate it on one side of the equality by dividing both sides by C_1:

$$V_1 = \frac{C_2 \times V_2}{C_1}$$

Step 3. Substitute the supplied variables listed in step 1 into the algebraically rearranged equation from step 2 and solve for the unknown variable. In this example:

$$V_1 = \frac{1\% \times 100 \text{ mL}}{10\%} = 10 \text{ mL}$$

Step 4. Prepare the dilute solution by transferring the calculated volume of stock solution, from step 3, to a volumetric flask with volume V_2. Then add water to the mark, with mixing. In this example, you would transfer 10 mL of the 10% stock saline solution to a 100. mL volumetric flask and then add water, the solvent, to the mark on the 100. mL volumetric flask, along with mixing. You now have a 1% saline solution.

   Note that the amount of solute removed from the concentrated solution is the same amount of solute that is in present in the dilute solution since only solvent was added to the solution.

Chemistry in Medicine. A Chemistry in Medicine feature concludes each chapter by providing an in-depth look at how the chemical principles described in the chapter can be directly applied to a problem or issue in health care.

Chemistry in Medicine The Chemistry of Vision

In memory of Ingeborg Vogel (1935–2011), my mother, who suffered from macular degeneration in the last decade of her life.

The leading cause of blindness in people over the age of 60 is age-related macular degeneration (ARMD). It is estimated that 11 million people in the United States have some form of ARMD, and that this number is expected to double by the year 2050. Macular degeneration is a condition in which a person loses his or her central vision, preventing them from being able to read, recognize faces, drive, and see detail. Figure 6-25 shows what a person with ARMD sees compared to someone with normal vision.

We can better understand this disease if we consider the chemistry of vision, a process that involves an important chemical reaction initiated by light.

Figure 6-25 Image of what a person with age-related macular degeneration sees compared to someone with normal vision, showing loss of central vision. [National Eye Institute/National Institutes of Health]

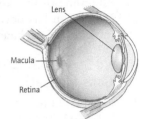

Figure 6-26 Parts of the eye: lens, retina, and macula. The macula is located at the center of the retina right behind the lens.

The retina, located at the back of the eye, contains millions of special photoreceptor cells known as rods and cones. In the center of the retina is a small area known as the macula, which contains a high concentration of cones. Light focused on the macula enables us to see small detail and color. ARMD is caused by a deterioration of the tissue that supports the macula (Figure 6-26).

from one geometric *isomer*—the cis—to the other geometric isomer—the trans. As a result of this cis-to-trans isomerization reaction, the overall shape of the protein-retinal complex changes significantly. The change in shape of the protein-retinal complex initiates a nerve impulse that travels along the optic nerve to the brain. Nerve impulses from the rods and cones are then interpreted in

Chapter Summaries. The chapter's key concepts are presented as a bulleted list at the end of each chapter, offering students a quick reference guide.

Key Words. All boldfaced terms in every chapter are defined in the Key Words section at the end of every chapter, as well as the Glossary at the end of the text.

Summary

Alcohols and Ethers

- An alcohol has the general structure R—O—H, where the —OH group is referred to as a hydroxyl group. The carbon atom bearing the hydroxyl group is an alkyl carbon.
- An ether has the general structure: R—O—R′. The carbon atoms bonded to the oxygen atom can be either alkyl or aromatic carbons.
- An R group is a carbon atom or chain of carbons of undefined length and composition.

have a bent molecular shape around the .5° bond angles.
1°, 2°, and 3° alcohols, depending on the the carbon atom bonded to the —OH as two R groups, and 3° has three R

groups are known as diols; three hydroxyl yl groups, polyols.
des are carbohydrates, a class of yl groups—polyols.
ains an —OH group bonded to an

Key Words

Achiral An object or a molecule that is not chiral. It is superimposable on its mirror image and so identical to its mirror image.

Alcohol A functional group derived from water where one of the H atoms has been replaced with an —R group: R—O—H. The carbon atom bearing the hydroxyl group is an alkyl carbon.

Aldehyde A carbonyl containing functional group characterized by a hydrogen atom bonded to the carbonyl carbon. Also includes formaldehyde which contains two —H atoms bonded to the carbonyl carbon.

Alkaloid A compound containing an amine that is found in nature.

Amide A carbonyl compound with a nitrogen atom bonded to the carbonyl carbon. The nitrogen atom has two additional bonds to either —H or —R.

Amine A functional group derived from ammonia in which one, two, or all three of the hydrogen atoms have been replaced by R groups: RNH_2, R_2NH, or R_3N. There is no carbonyl group bonded to the nitrogen.

Amino acid Biological compounds used to build proteins, characterized by an amine and a carboxylic acid functional group bonded to the same carbon atom.

Analgesic A substance that reduces pain.

Carbohydrate A type of biomolecule that includes monosaccharides—simple sugars—and disaccharides. Carbohydrates are a source of energy for cells.

Carbonyl group A carbon-oxygen double bond, C=O.

Carboxylic acid A carbonyl containing functional group that contains an —OH group bonded directly to the carbonyl carbon.

Additional Exercises. At the end of each chapter, additional exercises reinforce the concepts and skills presented in the text. Clearly labeled by section to help both students and instructors, there are on average 100 exercises in each chapter, and they are also available in the textbook's accompanying online homework system. Answers to the odd-numbered exercises are available at the back of the book in Appendix B, and detailed solutions for all the exercises are available in the student solutions manual.

Student Ancillary Support

Supplemental learning materials allow students to interact with concepts in a variety of scenarios. By analyzing figures, reinforcing problem-solving methods, reviewing chapter objectives, and viewing podcasts of an instructor work out solutions, students obtain a practical understanding of the core concepts. With that in mind, W. H. Freeman has developed the most comprehensive student learning package available.

Printed Student Study Guide and Solutions Manual

by Rachel C. Lum,
ISBN: 1-4641-2506-6

The combined Student Study Guide and Solutions Manual provides students with a manual designed to help them avoid common mistakes and understand key concepts. After a brief review of each section's critical ideas, students are taken through step-by-step worked examples, try-it-yourself examples, and chapter quizzes, all structured to reinforce chapter objectives and build problem-solving techniques. The Solutions Manual includes detailed solutions to all odd-numbered exercises in the text.

Molecular Model Kit

ISBN: 1-4292-2687-0

Molecular models help students understand the physical and chemical properties of molecules by providing a way to visualize the three-dimensional arrangement of atoms. This model set, created specifically for *Essentials of General, Organic, and Biochemistry*, uses different color polyhedra to represent atoms and plastic connectors to represent bonds.

Electronic Student Resources

The *Essentials of General, Organic, and Biochemistry, Second Edition* Book Companion Website, accessed at www.whfreeman.com/guinn2e, provides a range of tools for problem solving and understanding concepts, from self-quizzes and pre-drawn molecules for use in homework problems to an interactive periodic table of the elements and an English/Spanish Glossary. In addition, the Book Companion Website includes these interactive activities:

Problem-Solving Tutorials reinforce the concepts presented in the classroom, providing students with self-paced explanations and practice. Each tutorial relates to a worked example from the text, and consists of the following four components:

- Conceptual Explanation: An easy-to-follow, thorough explanation of the topic.
- Worked Example: A step-by-step walkthrough of the problem-solving technique.
- Try It Yourself: An interactive version of the Worked Example that prompts students to complete the problem and supplies answer-specific feedback.
- Practice Problems: A series of problems designed to check understanding.

ChemCasts are videos that replicate the face-to-face experience of watching an instructor solve a problem. Using a virtual whiteboard, the ChemCast tutors show students the steps involved in solving key worked examples, while explaining the concepts along the way. The worked examples were chosen with the input of general, organic, and biochemistry students. ChemCasts can be viewed online or downloaded to a portable media device, such as an iPod.

The Virtual Model Kit allows students to build molecular models on the computer screen. Students select elements and bonds according to instructions in the text's Using Molecular Models features, as well as from additional exercises available within the associated Website and homework system. After constructing the models, students are presented with automatically graded questions to evaluate their understanding.

Electronic Textbooks

The CourseSmart eTextbook provides the full digital text, along with tools to take notes, search, and highlight passages. A free app allows access to CourseSmart eTextbooks on Android and Apple devices, such as the iPad. CourseSmart eTextbooks can be downloaded to your computer and accessed without an Internet connection, removing any limitations for students when it comes to reading a digital text. The CourseSmart eTextbook can be purchased at www.coursesmart.com

Online Tutoring

W. H. Freeman knows that learning chemistry can be difficult, and that sometimes students need some extra support. That's why we have partnered with NetTutor to bring students tutoring they can access anytime, anywhere.

NetTutor®

www.nettutor.com

NetTutor® is an online tutoring service that specializes in a customized tutoring experience. Using the Socratic

Method, tutors guide students through problems or concepts but never present the answer. This approach develops critical thinking skills and encourages students to persevere. Powered by an easy-to-use interactive interface, NetTutor® provides the most student-friendly online tutoring environment available, conducted by local tutors who understand how to help today's students learn.

With NetTutor®, students get:

- Expert tutors ready to help students with problems from the textbook and assignments.
- Tutors who use questions and familiar concepts to guide students through the learning process.
- A virtual whiteboard, which enables students to interact with tutors as they work to solve problems, share information, and ask questions.
- The ability to print or save copies of each tutoring session for future reference.
- The option for students to extend access via direct online purchase.

Assistance is always available!

- Online tutoring seven days a week.
- Offline question-and-answer 24-hours a days.

For more information about packaging tutoring access with your *Essentials of General, Organic, and Biochemistry* text, please contact your W. H. Freeman Publisher's Representative.

For the Laboratory Lab Manual

by Julie Klare, Fortis College
ISBN: 1-4641-2507-4

The lab manual provides a wide variety of classic and innovative experiments covering the basic topics of general, organic, and biochemistry. These experiments emphasize biological applications of chemical concepts, often within the context of the health sciences. Each experiment can easily be completed within a 3-hour time frame and is accompanied by data sheets and questions that guide students through the analysis of their data.

LabPartner® Chemistry
www.whfreeman.com/labpartner

W. H. Freeman's latest offering in custom lab manuals provides instructors with a diverse and extensive database of experiments published by W. H. Freeman and Hayden-McNeil Publishing—all in an easy-to-use, searchable online system. With the click of a button, instructors can choose from a variety of traditional and inquiry-based labs. LabPartner Chemistry sorts labs in a number of ways, from topic, title, and author, to page count, estimated completion time, and prerequisite knowledge level. You can add content on lab techniques and safety, reorder the labs to fit your syllabus, and include your original experiments with ease. Wrap it all up in an array of bindings, formats, and designs. It's the next step in lab publishing—the perfect partner for your course.

Instructor Resources

For instructors using *Essentials of General, Organic, and Biochemistry, Second Edition*, W. H. Freeman provides a complete suite of assessment tools and course materials.

Computerized Test Bank

by Rachel Jameton, Lewis-Clark State College
ISBN: 1-4641-2511-2

The Computerized Test Bank offers over 1600 multiple-choice questions. It is designed to assess student knowledge at all levels of learning, from basic definitions to application and synthesis. Utilizing diagrams, figures, and structures, the test bank emphasizes visual understanding of an array of concepts, including applications to the health sciences. While the test bank is also available as a printed manual, the easy-to-use CD-ROM includes Windows and Macintosh versions of the widely used test generation software, which allows instructors to add, edit, and re-sequence questions to suit their testing needs. The Test Bank is also available in Word Format.

Electronic Instructor Resources

Instructors can access valuable teaching tools through www.whfreeman.com/guinn2e. These password-protected resources are designed to enhance lecture presentations, and include textbook images (available in .jpeg and PowerPoint format), Clicker Questions, Lab Information, and more. Enhanced Lecture PowerPoints, which contain complete class-lecture content, highlight key chapter ideas and include worked examples, pre-class questions, and textbook figures and diagrams. There are approximately 60 slides per chapter.

Course Management System Cartridges

W. H. Freeman provides seamless integration of resources in your Course Management Systems. Four cartridges are available (Blackboard, Canvas, Desire2Learn, and Angel) and other system cartridges (Moodle, Sakai, etc.) can be produced upon request.

Online Homework Systems

W. H. Freeman offers the widest variety of online homework options on the market.

sapling learning with integrated eText

www.saplinglearning.com

Sapling Learning provides highly effective interactive homework and instruction that improve student learning outcomes for the problem-solving disciplines. They offer an enjoyable teaching and effective learning experience that is distinctive in three important ways:

- **Targeted Instructional Content:** Sapling Learning increases student engagement and comprehension by delivering immediate feedback and targeted instructional content.

- **Performance Tracking:** Sapling Learning grades assignments, tracks student participation and progress, and compiles performance analytics—helping instructors save time and tailor assignments to address student needs.

- **Proven Results:** Independent university studies have shown Sapling Learning improves student performance by three-fourths to a full letter grade.

- **Unsurpassed Service and Support:** Sapling Learning makes teaching more enjoyable by providing a dedicated Master's- and Ph.D.-level colleague who provides software, course development, and consulting support throughout the semester.

Our Tech TAs help instructors to:

- Customize assignments by editing our existing questions or creating new questions that support the course, not just the text.

- Choose point values and grading policies to help achieve educational goals, analyze class statistics, and apply the results to help students learn more efficiently.

- Provide instructors with one-on-one training on use of Sapling Learning's online homework system and customization tools.

- Resolve any issues that may arise with the software or online assignments so that instructors spend less time managing technology and more time with their students.

Sapling Learning is not tied to a specific textbook or edition, giving instructors freedom to customize Sapling to their syllabus and students the flexibility to choose more affordable used or rental textbooks.

For budget-conscious students, a lower-priced, homework-only option is also available. This version does not include an integrated eText.

WebAssign Premium (includes integrated eBook)

www.webassign.com

For instructors interested in online homework management, WebAssign Premium features a time-tested secure online environment already used by millions of students worldwide. Featuring algorithmic problem generation and supported by a wealth of learning tools, WebAssign Premium for Essentials of General, Organic, and Biochemistry presents instructors with a powerful assignment manager and student environment. WebAssign Premium provides the following resources:

- Algorithmically generated problems: Students receive homework problems containing unique values for computation, encouraging them to work out the problems on their own.

- Complete access to the Multimedia-Enhanced e-Book is available from a live table of contents, as well as from relevant problem statements.

- Links to select ChemCasts and Problem-Solving Tutorials are provided as hints and feedback to ensure a clearer understanding of the problems and the concepts they reinforce.

- Personal Study Plan allowing students to review key prerequisite algebra concepts at their own pace.

For budget-conscious students, a lower-priced, homework-only option is also available. This version does not include an integrated eBook.

Acknowledgments

We are grateful to the many instructors who throughout the development of this text contributed their expertise and experience by providing thoughtful reviews of individual chapters. We are especially grateful to Valerie Keller of the University of Chicago, who reviewed the entire final manuscript for accuracy.

A special debt of gratitude goes to the many instructors who gave their time and expertise to reviewing drafts of chapters in the second edition, especially Vahan Ghazarian, Martin Brock, Mikhail Goldin, Angela Allen, and John Singer, who all reviewed many chapters of the book:

Angela M. Allen, *Lenoir Community College*
Ricardo Azpiroz, *Richland College*
Juan M. Barbarin-Castillo, *Tarrant County College*
Lori A. Bolyard, *University of Indianapolis*
Martin Brock, *Eastern Kentucky University*
Krys Bronk, *Blue Mountain Community College*
John Bumpus, *University of Northern Iowa*
Gerald J. Buonopane, *Seton Hall University*
David Cartrette, *South Dakota State University*
Rosemarie Chinni, *Alvernia University*
Douglas S. Cody, *Farmingdale State*
Milagros Delgado, *Florida International University*
David L. Gallaher, *Carlow University*
Vahan Ghazarian, *East Los Angeles College*

Louis Giacinti, *Milwaukee Area Technical College*
Mike Goldin, *Liberty University*
Bonnie L. Hall, *Grand View University*
Melanie Harvey, *Johnson County Community College*
Ryan C. Jeske, *Ball State University*
Regis Komperda, *Catholic University of America*
Ed Kremer, *Kansas City Kansas Community College*
Joseph Kremer, *Alvernia University*
Mathangi Krishnamurthy, *Fitchburg State University*
Allison Lamanna, *Boston University*
Mary Lamar, *Eastern Kentucky University*
Andrea D. Leonard, *University of Louisiana at Lafayette*
Nicholas Madhiri, *Southwestern Adventist University*
Christopher Massone, *Molloy College*
David F. Maynard, *California State University*

Phil McBride, *Eastern Arizona College*
Peter P Mullen, *Florida State College*
Justin P'Pool, *University of Indianapolis*
Tanea T. Reed, *Eastern Kentucky University*
Linda Roberts, *California State University Sacramento*
Robert Shapiro, *Becker College*
John Singer, *Jackson Community College*
Julie Smist, *Springfield College*
Allison S. Soult, *University of Kentucky*
Donald Spencer, *East Texas Baptist University*
Barbara Stallman, *Lourdes University*
Daniel J. Stasko, *University of Southern Maine, Lewiston-Auburn College*
Eric R. Taylor, *University of Louisiana at Lafayette*
William Wagener, *West Liberty University*
Linda Waldman, *Cerritos College*

I remain indebted to the instructors listed below who shared our vision for the book and agreed to class test the manuscript for the first edition, prior to its publication. Together with the comments and support offered by instructors at focus groups, this feedback has been extremely valuable.

Class Testers

Mamta Agarwal, *Chaffey College*
George Bandik, *University of Pittsburgh*
Lois Bartsch, *Graceland University*
Martin Brock, *Eastern Kentucky University*
Stephen Dunham, *Moravian College*
Michelle Hatley, *Sandhills Community College*
Andrea Martin, *Widener University*
Phil McBride, *Eastern Arizona University*
Edmond O'Connell, *Fairfield University*
Anuhadha Pattanayak, *Skyline College*
Matthew Saderholm, *Berea College*
Jeffrey Sigman, *Saint Mary's College*
Tara Sirvent, *Vanguard University*
Carnetta Skipworth, *Bowling Green Community College of Western Kentucky University*
Lorraine Stetzel, *Lehigh Carbon Community College*
Christy Wheeler, *College of Saint Catherine*

Focus Group Participants

Loyd Bastin, *Widener University*
Scott Carr, *Anderson University*
Rosemarie Chinni, *Alvernia College*
Ana Ciereszko, *Miami-Dade College*

Mian Jiang, *University of Downtown Houston*
Booker Juma, *Fayetteville State University*
Annie Lee, *Rockhurst University*
Carol Libby, *Moravian College*
Samar Makhlouf, *Lewis University*
Andrea Martin, *Widener University*
Bryan May, *Central Carolina Technical College*
Elizabeth Pulliam, *Tallahassee Community College*
Rita Rhodes, *University of Tulsa*
Trineshia Sellars, *Palm Beach Community College*
Carnetta Skipworth, *Bowling Green Community College of Western Kentucky University*
Lee Ann Smith, *Western Kentucky University*

Reviewers

Ronald T. Amel, *Viterbo University*
Laura Anna, *Millersville University*
Tasneem Ashraf, *Cochise College*
Theodore C. Baldwin, *Olympic College*
George Bandik, *University of Pittsburgh*
Thomas Barnard, *Cardinal Stritch University*
Bal Barot, *Lake Michigan College*
Lois M. Bartsch, *Graceland University*

Nick Benfaremo, *St. Joseph's College of Maine*
Jerry Bergman, *Northwest State Community College*
Chirag Bhagat, *William Rainey College*
John Blaha, *Columbus State Community College*
Carol E. Bonham, *Pratt Community College*
Martin Brock, *Eastern Kentucky University*
Diane M. Bunce, *The Catholic University of America*
Joe C. Burnell, *University of Indianapolis*
N. J. Calvanico, *Lehigh Carbon Community College*
David Canoy, *Chemeketa Community College*
Scott R. Carr, *Anderson University*
Stephen Cartier, *Warren Wilson College*
Lynne C. Cary, *Bethel College*
Amber Flynn Charlebois, *Fairleigh Dickinson University*
Ana A. Ciereszko, *Miami-Dade College*
Joana Ciurash, *College of the Desert*
Stuart C. Cohen, *Horry-Georgetown Technical College*
Jeannie T. B. Collins, *University of Southern Indiana*
Felicia Corsaro-Barbieri, *Gwynedd-Mercy College*

Brian Cox, *Cochise College-Sierra Vista Campus*

Milagros Delgado, *Florida International University*

Anthony B. Dribben, *Tallahassee Community College*

Stephen U. Dunham, *Moravian College*

Robert G. Dyer, *Arkansas State University, Mountain Home*

Eric Elisabeth, *Johnson County Community College*

Anne Felder, *Centenary College*

Francisco Fernandez, *City University of New York, Hostos Community College*

K. Thomas Finley, *The College at Brockport*

Hao Fong, *South Dakota School of Mines and Technology*

Karen Frindell, *Santa Rosa Junior College*

Priscilla J. Gannicott, *Lynchburg College*

Andreas Gebauer, *California State University, Bakersfield*

Louis A. Giacinti, *Milwaukee Area Technical College*

Eric Goll, *Brookdale Community College*

Nalin Goonesekere, *University of Northern Iowa*

Maralea Gourley, *Henderson State University*

Ernest Grisdale, *Lord Fairfax Community College*

Mehdi H. Hajiyani, *University of the District of Columbia*

Melanie Harvey, *Johnson County Community College*

Michelle L. Hatley, *Sandhills Community College*

Michael A. Hauser, *St. Louis Community College–Meramec*

John W. Havrilla, *University of Pittsburgh–Johnstown*

Jonathan Heath, *Horry-Georgetown Technical College*

Sherry Heidary, *Union County College*

Sara Hein, *Winona State University*

Steven R. Higgins, *Wright State University*

Jason A. Holland, *University of Central Missouri*

Byron Howell, *Tyler Junior College*

Michael O. Hurst, *Georgia Southern University*

T. G. Jackson, *University of South Alabama*

Rachel A. Jameton, *Lewis-Clark State College*

Mike Jezercak, *University of Central Oklahoma*

Matthew Johnston, *Lewis-Clark State College*

Booker Juma, *Fayetteville State University*

Cathie Keenan, *Chaffey College*

Mushtaq Khan, *Union County College*

Edward A. Kremer, *Kansas City Kansas Community College*

Bette A. Kreuz, *The University of Michigan-Dearborn*

Peter J. Krieger, *Palm Beach Community College*

Lida Latifzadeh Masoudipour, *El Camino College*

Annie Lee, *Rockhurst University*

Scott Luaders, *Quincy University*

Julie Lukesh, *University of Wisconsin, Green Bay*

Riham Mahfouz, *Thomas Nelson Community College*

Samar Makhlouf, *Lewis University*

Karen Marshall, *Bridgewater College*

Andrea Martin, *Widener University*

Christopher Massone, *Molloy College*

Bryan May, *Central Carolina Technical College*

Phil McBride, *Eastern Arizona College*

Ann H. McDonald, *Concordia University Wisconsin*

Patrick McKay, *San Mateo County Community College*

S. Ann Melber, *Molloy College*

Stephen Milczanowski, *Florida Community College at Jacksonville*

Michael N. Mimnaugh, *Chicago State University*

Luis D. Montes, *University of Central Oklahoma*

Peter P. Mullen, *Florida Community College*

Michael P. Myers, *California State University, Long Beach*

Grace M. Ndip, *Shaw University*

Nelson Nunez-Rodriguez, *City University of New York, Hostos Community College*

E. J. O'Connell, *Fairfield University*

Janice J. O'Donnell, *Henderson State University*

Elijah Okegbile, *Pima Community College*

C. Edward Osborne, *Northeast State Community College*

John D. Patton, *Southwest Baptist University*

Lynda R Peebles, *Texas Woman's University*

Paul Popieniek, *Sullivan County Community College of SUNY*

Jerry Poteat, *Georgia Perimeter College*

Ramin Radfar, *Wofford College*

Christina Ragain, *University of Texas, Tyler*

S. Ramaswamy, *University of Iowa*

Rita T. Rhodes, *The University of Tulsa*

Rosalie Richards, *Georgia College and State University*

Shashi Rishi, *Greenville Technical College*

Ghassan Saed, *Oakland University*

Steve P. Samuel, *SUNY College at Old Westbury*

Karen Sanchez, *Florida Community College at Jacksonville*

Shaun E. Schmidt, *Washburn University*

William Seagroves, *New Hampshire Technical Institute*

Sara Selfe, *Edmonds Community College*

Trineshia N. Sellars, *Palm Beach Community College*

Paul Seybold, *Wright State University*

Sonja Siewert, *West Shore Community College*

Jeffrey A. Sigman, *Saint Mary's College of California*

Nancy C. Simet, *University of Northern Iowa*

John W. Singer, *Jackson Community College*

Joseph F. Sinski, *Bellarmine University*

Carnetta Skipworth, *Western Kentucky University*

Robert Smith, *Metropolitan Community College*

Koni Stone, *California State University, Stanislaus*

Mary W. Stroud, *Xavier University*

K. Summerlin, *Troy University, Montgomery Campus*

Erach R. Talaty, *Wichita State University*

Eric R. Taylor, *University of Louisiana–Lafayette*

Jason R. Taylor, *Roberts Wesleyan College*

Rod Tracey, *College of the Desert*

Robert C. Vallari, *St. Anselm University*

Sarah Villa, *University of California, Los Angeles*

Maria Vogt, *Bloomfield College*

Linda Waldman, *Cerritos College*

Karen T. Welch, *Georgia Southern University*

Christy Wheeler West, *College of St. Catherine*

Ryan S. Winburn, *Minot State University*

Corbin Zea, *Grand View College*

This textbook would not have been possible were it not for the exceptional dedication and talent provided by the editorial team at W. H. Freeman. My deepest gratitude goes to Randi Rossignol, the senior developmental editor on the second edition, and Susan Moran, the senior developmental editor for the first edition. Their commitment and considerable editorial talents were instrumental in achieving the text you see before you today. It has been a great pleasure undertaking such a creative endeavor in collaboration with individuals with such considerable talent and insight. I also appreciate the considerable experience brought to the project by publisher Jessica Fiorillo and acquisitions editors Anthony Palmiotto and Bill Minick, who deftly managed the many facets of producing a modern textbook: ancillaries, media package, and all the associated activities that must be in place for a textbook to come to fruition.

This textbook benefits from the dedicated efforts of our supplements authors, to whom we extend our most heartfelt gratitude. Among them is an experienced laboratory text author, Julie Klare of Fortis College, who has created an impressive array of laboratory experiments designed to accompany this textbook, as well as the comprehensive PowerPoint slides and online quizzes for instructors. Dr. Rachel Lum prepared the solutions guide that accompanies the more than 1500 questions that appear in the book, and who wrote and revised the new exercises that appear at the end of each chapter. Thanks also go to Jonathan Bergmann and Aaron Sams, Woodland Park High School, Colorado, who created the ChemCast problem-solving videos.

We extend our appreciation to Dennis Free and the staff at Aptara Corporation, and to the many essential W. H. Freeman production staff members including project editor Jane O'Neill, who carefully shepherded the book through the proof stages; Beth Rosario and Margaret Comaskey, the copyeditors; Diana Blume, who created the elegant design; Bill Page, who guided the production of an entirely new illustration program; Paul Rohloff, who arranged the typesetting and printing; Donna Ranieri, who researched many of the photographs, and Bianca Moscatelli, photo editor.

We owe special thanks to Dave Quinn, Courtney Lyons, and Tue Tran for ably guiding the development of the impressive print and online set of resources available on our book's Website. In addition, Mark Santee managed the WebAssign integration. Finally, we thank Debbie Clare, Associate Director of Marketing, and the entire sales force for all of their enthusiasm and support.

The pedagogy of this book is greatly enhanced by the artwork produced by Network Graphics. The text is further enhanced by the number and quality of ball-and-stick models and space-filling models rendered by Alex Panov. Authentic protein structures and electron density models were rendered by Gregory Williams.

A special debt of gratitude goes to Catherine Bausinger who shot the more than 50 new photographs that appear in the second edition, and Charles Guinn who created the much improved math appendix for the second edition. I would also like to express my deepest gratitude to Rebecca Brewer for collaborating with me on the development and writing of the first edition. Finally, I want to thank my colleagues at various academic institutions who supported my efforts in the writing of the second edition with their encouragement and moral support. They are Rachel Jameton (Lewis and Clarke State College), Rachel Lum, Stephen Cartier (Warren Wilson College), Madeline Mignone (Dominican College), and my colleages at the College of New Rochelle: Dorothy Escribano, Richard Thompson, Lee Warren, Elvira Longordo, Arlene Rosen, Terry Colarusso, Melanie Harasym, Lynn Petrullo, and Faith Kostel-Hughes.

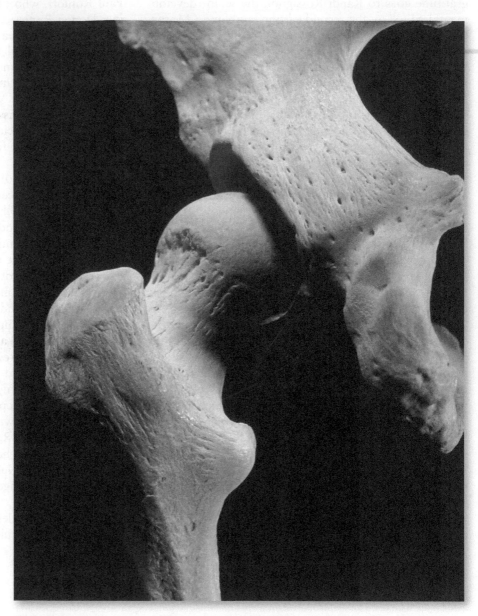

The hip bone is often used for a bone mineral density (BMD) measurement because it is a good indicator of whether or not a person has osteoporosis. Shown is a healthy hip bone.
[James Stevenson/Science Source]

1 Measuring Matter and Energy

Osteoporosis and Bone Density

One in five American women over the age of 50 is estimated to have osteoporosis, and over half of these women will break a bone at some point in their lives as a result of the condition. Osteoporosis is a condition characterized by the progressive loss of bone density, resulting in an increased risk of bone fractures. For example, even a cough or stumble is often enough to trigger a fracture. Most bone fractures occur in the hip, wrist, and spine.

Bone density is the measurement most commonly used to assess bone strength. Density is a physical property of a substance, which can be calculated from the mass of a sample of the substance divided by its volume: $d = m/V$. Mass is similar to weight and is a measure of the amount of material, in this case, bone mineral. Volume is a measure of the three-dimensional space occupied by a material, in this case, a segment of bone. The greater the bone density is, the greater the bone mineral per volume and the stronger the bone. Stronger bone is better able to withstand stress and less likely to fracture. The average bone density for a person is 1.5 g/cm^3, read as "one point five grams per centimeter cubed."

To screen for osteoporosis, several techniques have been developed for measuring or estimating the bone density of the hip, wrist, and lumbar spine. Currently, the most common technique is the DEXA scan (*d*ual *e*nergy *x*-ray *a*bsorptiometry), which is a type of low dose x-ray that measures the *mass* and *area* of a section of bone and then estimates the third dimension to obtain a *volume* measurement.

From a DEXA scan, a bone density or BMD (*b*one *m*ineral *d*ensity) measurement is obtained. The result is then compared with the average BMD for healthy young adults of the same gender and ethnicity. A person is given a T-score that corresponds to how far they are from the average BMD of a similar population (Figure 1-1). A T-score of +1 to −1 means they fall within the "normal" range. A person with a T-score of −1 to −2.5 is classified as having osteopenia, a condition characterized by lower-than-average bone density, which may eventually develop into osteoporosis. People with a T-score less than or equal to −2.5 are classified as having osteoporosis. Basically, a T-score compares a patient's bone density to a similar segment of the population and classifies that patient as having normal bone density, osteopenia, or osteoporosis. ●

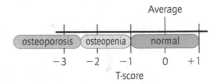

Figure 1-1 When a patient's BMD (bone mineral density) measurement is obtained, they are given a T-score that corresponds to how far they are from the average BMD of a similar population.

In this chapter, you will see how important measurements are in medicine and science. You will learn that there are limitations to any measurement, how measurements are reported, common units of measurement, and how to calculate density and other physical properties. While working in the health care field, you will often be called on to take measurements and to perform critical calculations, such as the dosage of medicine to administer to a patient. These skills are a foundational and critical part of your training, so it is with measurement that we start your venture into the fascinating field of chemistry.

■ 1.1 Matter and Energy

Chemistry is the study of matter and changes in matter. ***Matter is defined as anything that has mass and occupies volume.*** Hence, matter is all the "stuff" around you and in you. It includes matter that you can see—the macroscopic—as well as matter that is too small for you to see—the microscopic and atomic scales.

Matter is found in three **states** or **phases**: solid, liquid, and gas. Examples of all three states of matter are found in the body. The oxygen you breathe is in the gas state, making it possible to quickly fill the lungs with each breath you take. Blood is in the liquid state, so it can be pumped throughout the circulatory system transporting important nutrients to cells. Skin and bone are in the solid state, providing structural integrity to the body.

The macroscopic differences between solids, liquids, and gases can be described by their shape and volume relative to their container, as illustrated in the top part of Figure 1-2:

- A **solid** (s) has a definite *volume* and *shape*, which are independent of its container.
- A **liquid** (l) has a definite *volume* but does not have a definite *shape*, as it conforms to the shape of its container.
- A **gas** (g) has neither a definite *volume* nor a definite *shape*. A gas conforms to the volume and shape of its container.

What makes a gas take on the shape of its container while a solid has a shape independent of its container? To understand these macroscopic properties, we must examine matter on the atomic scale—the incredibly small particles of matter that cannot be seen even with a light microscope, as illustrated in the bottom part of Figure 1-2. Experiments have shown that matter is made up of particles known as atoms, which is the subject of chapter 2. In the solid phase, atomic particles are very close to one another in an ordered lattice. In

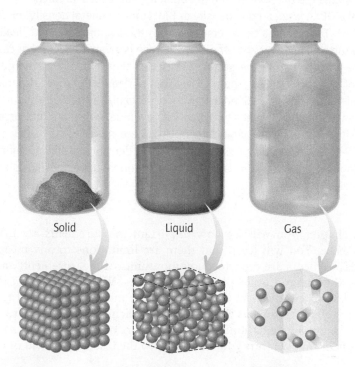

Figure 1-2 The differences in macroscopic properties of solids, liquids, and gases are a result of differences at the atomic scale.

the liquid phase, atomic particles are farther apart but still interacting. In the gas phase, atomic particles are far apart from one another and have little if any interaction.

As a science, chemistry endeavors to explain the macroscopic world—that which we can see—by understanding events on the atomic scale—that which we cannot see. Our understanding of the macroscopic world through knowledge of the atomic world has been achieved through centuries of experiments, measurements, and careful observation. Medicine has seen advances as a result, since medicine is a science with a foundation in chemistry and biology. As a person entering the health care professions, it is important for you to understand the fundamental principles of chemistry at both the macroscopic and the atomic level.

Kinetic and Potential Energy

Particles of matter are not stationary but instead are vibrating and moving about, to a degree that depends on the amount of energy they possess. To understand the physical properties of matter described above, we must first examine the relationship between matter and energy. Central to the physical properties of all matter is energy, one of the most important concepts in science. *Energy, broadly defined, is the capacity to do work, where* **work** *is defined as the act of moving an object against an opposing force.* Energy determines the state of matter. For a given substance, the gas phase has more energy than the liquid phase, which has more energy than the solid phase.

Kinetic Energy There are two basic forms of energy: *kinetic energy* and *potential energy. Kinetic energy is the energy of motion, including the energy a substance possesses as a result of the motion of its particles.* A skier has kinetic energy as she skies down a mountain (Figure 1-3). Similarly, the particles of matter in any state of matter have kinetic energy. The kinetic energy (KE) of an object or particle depends on its mass, m, and its velocity, v (speed), which can be expressed by the mathematical equation:

$$KE = \tfrac{1}{2}mv^2$$

Thus, faster-moving objects have greater kinetic energy than slower-moving objects with the same mass, and heavier objects have greater kinetic energy than lighter objects moving at the same velocity. For example, a high-speed car collision results in more damage than a low-speed car collision (velocity) if the mass of the cars is the same, and colliding with an SUV causes more damage than colliding with a compact car (mass) if they are each traveling at the same velocity.

Heat is a form of kinetic energy because it involves the *motion* of the particles of matter. *Heat energy always flows from the hotter object to the colder object.* Our bodies are able to perceive heat transfer. Thus, if you touch ice, heat transfers from your body to the ice, causing the particles of ice to increase in kinetic energy and the particles on the surface of your hand to lose kinetic energy. Your nerves detect this transfer of heat energy out of your hand, and your brain interprets it as interacting with something cold. If you touch a hot stove, a similar thing happens except in the opposite direction. Your brain knows it has touched something with more kinetic energy. You interpret it as something hot.

It is important to note that temperature is *not* the same as heat. Heat is kinetic energy—motion of particles of matter—while **temperature** is a *measure* of the particles' kinetic energy. For example, the temperature of the air is

Figure 1-3 This skier has kinetic energy by virtue of her motion. [Jupiterimages/ Getty Images]

a measure of the average kinetic energy (heat energy) of the oxygen and nitrogen particles that constitute air. States of matter in which particles have a greater average kinetic energy will have a higher temperature than states of matter whose particles have a lower average kinetic energy, for a given substance. As particles move faster, their kinetic energy increases, and the temperature rises. Rub your hands together quickly and vigorously. Do you feel them getting warmer? Your hands become warmer because the skin particles on the top layer of skin are moving faster, resulting in greater kinetic energy.

Potential Energy *Potential energy is stored energy, the energy a substance possesses as a result of its position or composition* (Figure 1-4). A rock poised at the top of a precipice, for example, possesses potential energy as a result of its position. When the rock falls, its potential energy is converted into kinetic energy. A gallon of gasoline has potential energy as a result of its composition—the gasoline particles. When the gasoline is burned in a car engine, the potential energy of these gasoline particles is turned into kinetic energy, which is used to perform the work of moving your car. Food also contains potential energy, which is transferred to your cells after you eat and digest the food so that your cells can do the "work" they need for living.

Kinetic Molecular View of the States of Matter

Consider the physical states of matter from a **kinetic molecular view** (Figure 1-2). In the *gas* phase, particles of matter have the greatest kinetic energy. Particles in the gas phase are moving faster than when they are in the liquid or solid phase. Particles in the gas phase are far apart from one another, interacting only when they collide. In the *liquid* phase, these same particles are moving slower than in the gas phase but faster than in the solid phase. In the liquid state, the particles are much closer together, moving randomly and tumbling over one another. This is why liquids flow when poured. In the *solid* phase, the particles exist in a regular ordered pattern with much less kinetic energy than when in the liquid or gas phases, so they are closer together with mainly vibrational motion.

Figure 1-4 Left: The balanced rock has potential energy by virtue of its position. [iStockphoto/Thinkstock] Right: The gasoline in this container has potential energy by virtue of its composition. [© Ron Chapple/Corbis]

> **WORKED EXERCISES** | Matter and Energy |
>
> **1-1** In which of the following states of matter is the velocity of the particles of a given substance the slowest?
>
> **a.** solid **b.** liquid **c.** gas
>
> **1-2** Indicate whether each of the following is an example of kinetic energy or potential energy.
>
> **a.** a ball rolling down a hill
> **b.** standing on the edge of a diving board
> **c.** a piece of bread
>
> **1-3** Is heat energy kinetic energy or potential energy?
>
> **1-4** Which direction does heat flow when you place your hand over a pot of steam: from the pot to your hand or from your hand to the pot?
>
> **1-5** Using the kinetic molecular view of matter, offer an explanation for why cooking odors travel quickly from the kitchen to neighboring rooms.
>
> Solutions
>
> **1-1** **a.** Particles in the solid phase are the slowest because they have the least amount of kinetic energy.
>
> **1-2** **a.** Kinetic energy because the ball is moving
> **b.** Potential energy because of the diver's position at the edge of the diving board. When he jumps, this potential energy will be converted to kinetic energy.
> **c.** Potential energy because of the composition of bread, a food that can be turned into energy.
>
> **1-3** Heat is a form of kinetic energy because it represents the motion of particles of matter.
>
> **1-4** Heat flows from the pot to your hand because the pot is hotter than your hand and heat transfer always occurs in the direction from hot to cold.
>
> **1-5** Cooking odors are gas particles detected by your nose, and as a gas, they fill the volume of their container—the rooms, in this example.

> **PRACTICE EXERCISES***
>
> **1** Describe the macroscopic differences between the *solid*, *liquid*, and *gas states* by comparing their shape and volume relative to the container they occupy.
>
> **2** Indicate whether each of the following examples illustrates *potential energy* or *kinetic energy*:
>
> **a.** a compressed spring
> **b.** a windmill turning
> **c.** particles in the gas phase colliding with the walls of their container
> **d.** a skier skiing down a mountain
> **e.** the breakfast you eat for sustenance during the day
>
> *You can find the answers to the Practice Exercises at the end of each chapter.

Physical and Chemical Changes

Matter can undergo both physical changes and chemical changes. A **physical change** is a process that does *not* affect the composition of the substance. A **change of state** is an example of a physical change. For example, when water in the solid phase (ice) changes to a liquid (s → l, melting), the composition of the water molecules has not changed. The particles simply have greater kinetic energy, so they are moving more rapidly and are farther apart. One way to tell that the composition has not changed is that liquid water can be turned back into ice by removing heat. Thus, the observable properties of the substance have not changed physically.

A **chemical change**, also known as a **chemical reaction**, involves a change in the composition of the substance. For example, cooking food is a chemical change, because the food's properties have changed as a result of cooking. Cooling the food back to room temperature does not bring back the uncooked food, because a chemical change has occurred. Since all matter possesses energy, any changes in matter will have an associated change in energy. The energy associated with chemical changes is described in Chapter 4 and for physical changes in Chapter 5.

WORKED EXERCISE   Physical and Chemical Changes

1-6   Indicate whether each of the following changes illustrates a *physical* or a *chemical change*.

    a. water vapor condensing on the outside of a cold glass on a humid summer day
    b. a shiny piece of iron rusting to form brown-colored iron(II) oxide
    c. a sample of liquid water freezing
    d. cooking a steak over hot coals

Solution

1-6   **a.** a physical change because water is undergoing a phase change from gas to liquid, but its composition has not changed—it is still water
    b. a chemical change because the composition of matter changes, as observed in a color change
    c. a physical change because water is undergoing a phase change from liquid to solid, but it is still water
    d. a chemical change because the appearance of the steak changes, and it cannot return to uncooked steak on cooling

---

PRACTICE EXERCISES

3   Indicate whether each of the following changes is a *physical* or a *chemical change*:
    a. silver tarnishing to form silver sulfide
    b. sugar granules dissolving in water
    c. combustion of gasoline in a car engine producing carbon dioxide and water
    d. boiling water to make water vapor

■ 1.2 Measurement in Science and Medicine

To understand matter and energy, we must be able to measure them. Thus, we turn our attention to measurement, a critical skill for anyone working in a field of science. Consider a drop of blood on the head of a pin, which you can *see* with the naked eye (Figure 1-5). If you look at this droplet of blood through a microscope, you will see that it is composed of millions of red blood cells, each with a diameter 1,000 times smaller than the droplet of blood. Now imagine—because you can't see it with a microscope—peering inside one of these red blood cells. Among other things, you will "see" millions of hemoglobin molecules, each composed of many atoms. Hemoglobin is the substance that carries oxygen from the lungs to the tissues throughout the body. A hemoglobin *molecule* has a diameter 1,000 times smaller than a red blood cell and a million times smaller than the droplet of blood on the head of a pin. If you imagine zooming in further, you will "see" that a single hemoglobin molecule is composed of about 10,000 atoms, including four iron atoms, about 100 times smaller than a hemoglobin molecule.

From the above exercise, we see that matter can be described on different scales, illustrated in Figure 1-5. The **macroscopic scale** includes all matter that

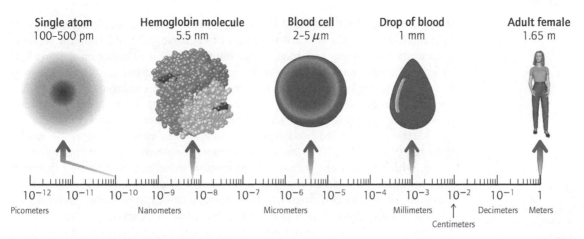

Figure 1-5 Metric lengths ranging from the incredibly small 100-pm diameter of a single iron atom—the atomic scale—to the 1.65-m height of the average adult female—the macroscopic scale.

you can see, such as the drop of blood and the human body, containing on average 5 liters of blood. The **microscopic scale** includes matter such as red blood cells that cannot be seen with the naked eye but can be seen with magnification through a microscope. The **atomic scale** describes matter such as a hemoglobin molecule or an atom that is far too small to be seen except with a very specialized type of microscope, the scanning electron microscope (SEM).

Throughout this text you will see that an understanding of the *atomic scale* brings with it a better understanding of the *macroscopic*—the things you can see—especially as the understanding applies to health and disease. You will learn that many disease processes are caused by some malfunction at the atomic level.

English and Metric Units

Medical professionals make measurements every day, whether to obtain a patient's weight, take a patient's temperature or blood pressure, or administer a particular dose of medication. *A measurement consists of two parts: a number and a unit.* For example, a baby might weigh 10 *lb*, not simply 10. Every measurement also has a margin of error associated with it, which is conveyed by the number of digits (figures) reported in the measurement. Thus, the baby's weight may be recorded as 10. lb, 10.0 lb, or 10.00 lb, depending on the precision of the balance used to weigh the baby, and each of these measurements conveys a different degree of uncertainty in the measurement.

The two most common *systems* of measurement that you will encounter in medicine are the metric system and the English system. The **English system** is used only in the United States and a few other countries. The **metric system** is the most widely used system of measurement in the world and the system of units used in the sciences. The metric system is convenient because it involves units that are multiples of 10 of the base unit.

The metric system employs various **base units** that measure a particular quantity, such as

- the *meter* (m) for measuring length
- the *gram* (g) for measuring mass
- the *second* (s) for measuring time
- the *calorie* (cal) for measuring energy

The *i*nternational *s*ystem of units (SI) was established by an international group of scientists for the purpose of setting a uniform set of units in the sciences. SI units are the preferred unit for science and commerce.

Prefixes in the Metric System If you have ever looked at the label on a bottle of multivitamins, you will see the mass of each vitamin and mineral reported with units such as mg and mcg. Large bags of flour and sugar have units such as kg, beverages have volume labels such as mL, and the thickness of a rope or string is given in units of cm or mm. These are all examples of metric units containing a *prefix* in front of a base unit (m, L, g). As you can see, metric prefixes are used not only in science but are part of our everyday world.

Metric *prefixes* are used when a measurement is much larger or smaller than the base unit so as to avoid the need to use many zeros in the numerical value, which makes the measurement cumbersome and difficult to interpret. For example, the head of a pin is 0.001 meter in diameter, a blood cell has a diameter of 0.000 001 meter, and there are 500,000,000,000 bytes in my computer hard drive. It is much simpler to report and interpret that the head of a pin is 1 millimeter in diameter, a blood cell has a diameter of 1 micrometer, and my hard drive has 500 gigabytes. To avoid writing numbers with many zeros, the metric system employs **prefixes**, which when placed in front of the base unit represent a multiplier or divider that makes the prefixed unit larger or smaller by some multiple of 10 or 1/10 (Table 1-1). For example, the prefix *milli* always represents the multiple 1,000 (10^3); thus, 10^3 millimeters (mm) is equal to 1 m. Similarly, 10^3 mg is equal to 1 gram (g), and 10^3 mL is equal to 1 liter (L). *Each prefix represents a specific multiplier independent of the base unit.* Other prefixes represent different multipliers. For example, the prefix *micro* represents the multiple 1,000,000 (10^6); thus, 10^6 micrometers (μm) is equal to 1 meter (m).

Some prefixes represent negative exponents. For example, the prefix *kilo* represents 1/1,000 or (10^{-3}), so 10^{-3} kilometers (km) is equal to 1 meter (m) and 10^{-3} kilogram (kg) is equal to 1 gram (g). The prefix *mega* represents 10^{-6} and the prefix *giga* represents 10^{-9}. Thus, my hard drive with 500,000,000,000 bytes can be written as 500 gigabytes, a number much easier to interpret. The common prefixes in the metric system, the multiplier/divider they represent, and their abbreviations are shown in Table 1-1.

The metric system with its prefixes is easy to learn because you need only memorize the prefixes (Table 1-1, column 2) and the associated multiple of 10 (columns 3 and 4) represented by each prefix. *Moreover, use Table 1-1 to create a mathematical equality, known as a* **conversion**, *between the base unit and the prefixed unit of interest when performing a metric conversion.* For example, 10^9 nm = 1 m. It is critical when writing a metric conversion that the exponent, 10^9 in this case, is always placed in front of the *prefixed* unit (nm) and 1 is

> Students often do a great job memorizing the metric prefixes and their associated exponents but then forget whether the exponent should be placed in front of the prefixed unit or the base unit. Remember that the exponent goes in front of the prefixed unit and a 1 goes in front of the base unit. For example, 10^3 mm = 1 m, not 1 mm = 10^3 m. One way to help you remember this is to note which is larger, the base unit or the prefixed unit. Clearly, more smaller units are required to equal a larger unit.

TABLE 1-1 Common Metric Prefixes and the Multipliers They Represent*

Prefix	Prefix Symbol	Multiplier	Multiplier in Scientific Notation
giga	G	0.000 000 001	10^{-9}
mega	M	0.000 001	10^{-6}
kilo	k	0.001	10^{-3}
deci	d	10	10
centi	c	100	10^2
milli	m	1000	10^3
micro	μ	1 000 000	10^6
nano	n	1 000 000 000	10^9
pico	p	1 000 000 000 000	10^{12}
femto	f	1 000 000 000 000 000	10^{15}

*Examples of how to use the table to create a metric conversion:
1 g = 10^3 mg; 1 g = 10^6 μg; 1 g = 10^{-3} kg

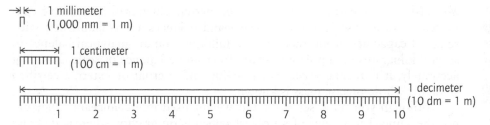

Figure 1-6 Actual size of 1 millimeter, 1 centimeter, and 1 decimeter. There are 10 millimeters in 1 centimeter, 10 centimeters in 1 decimeter, and 10 decimeters in 1 meter. With respect to the base unit, the meter, there are 10^3 millimeters in 1 meter, 10^2 centimeters in 1 meter, and 10 decimeters in 1 meter.

always placed in front of the *base* unit (m), not the other way around, when using this table.

You will notice that most of the multipliers in Table 1-1 are written in scientific notation. Scientific notation allows us to express numbers containing many zeros without writing the zeros. If you need a review of scientific notation, refer to Appendix A: Mathematics Review with Tips on How to Use a Calculator, where you will also find instructions for how to input a number that is given in scientific notation into a calculator. Let us now consider some common measurements using the metric system and the English system.

Length The meter is the base unit of **length** and distance in the metric system. There are 10 decimeters (dm) in 1 meter (m), 10^2 centimeters (cm) in 1 meter (m), and 10^3 millimeter (mm) in 1 meter (m) as shown in actual size in **Figure 1-6**. As you can see from Table 1-1, there are 10^6 micrometers (μm) in 1 meter, 10^9 nanometer (nm) in 1 meter, 10^{12} picometers (pm) in 1 meter, and so forth. Notice these units represent lengths smaller than the base unit, the meter. Table 1-1 also shows lengths and distances larger than the base unit, represented by the prefixes *kilo*, *mega*, and *giga*; however, the kilometer is the only commonly used prefix for distance measurements.

If you consider 1 mm the smallest length that you can reasonably see with the naked eye, then every 1,000-fold decrease takes you into the range of another scale: macroscale (greater than mm) → microscale (μm range) → atomic scale (smaller than a nm). Figure 1-5 shows a range of metric lengths from an iron atom with a diameter of 126 pm (the atomic scale) to a woman measuring 1.65 m tall (the macroscale).

The common English units of length are the inch (in), the foot (ft), and the mile (mi). The conversions between some common English and metric units of length are as follows:

$$1.00 \text{ in} = 2.54 \text{ cm (exact)}$$
$$39.37 \text{ in} = 1.00 \text{ m}$$
$$1.00 \text{ mi} = 1.61 \text{ km}$$

Measurements of length are routinely used to assess a developing fetus. An ultrasound scan of a developing fetus (**Figure 1-7**) is a noninvasive technique used to measure the size of various parts of the fetus, including the crown-rump length, the biparietal diameter (distance between the sides of the head), the femur length (thigh bone), and the abdominal circumference. These measurements of length assess gestational age, size, and growth of the fetus. For example, the biparietal diameter in a healthy fetus increases from about 2.4 cm at 13 weeks to about 9.5 cm at term. Structural abnormalities of the fetus such as spina bifida and cleft palate can be reliably diagnosed using ultrasound measurements taken before 20 weeks.

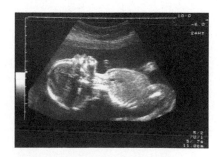

Figure 1-7 An ultrasound scan of a human fetus at 20 weeks. This fetus has a crown–rump length of 11.8 cm. Ultrasound scans are a noninvasive way to monitor the growth of the fetus. [© Craig Holmes Premium/Alamy]

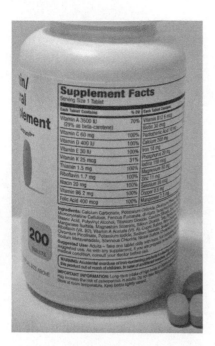

Figure 1-8 A nutritional label for a multivitamin showing the mass of each vitamin in 1 tablet. [Catherine Bausinger]

Mass Mass is a measure of the amount of matter, which is measured on a calibrated balance or scale. Precision scales and balances are available in a wide range of capacities, from fractions of a milligram on an analytical balance to several kilograms on a precision scale. To ensure a balance or scale is reading accurately, it is regularly *calibrated* against an internal or external certified standard whose precise mass is known.

Although you often see mass and weight used interchangeably, they are not the same. The weight of an object depends on gravity, whereas its mass does not. For example, the weight of an object will be different on the earth than on the moon, but its mass will be the same. Since you will be taking your measurements on earth, you can use either term.

The base unit of mass in the metric system is the gram (g). If you look at a multivitamin label, such as the one shown in Figure 1-8, you will see the mass of each vitamin and mineral in one tablet printed on the label. You may recognize several metric prefixes on the label since the amount of each vitamin in a tablet is significantly less than a gram, the base unit. For example, the vitamin tablet in the label shown in Figure 1-8 contains 60 mg of vitamin C.

The symbol for the prefix *micro* is the Greek letter μ, pronounced "myou." Thus, 10^6 μg is equal to 1 g. The microgram is also sometimes abbreviated "mcg" to avoid some of the confusion associated with a Greek letter. The vitamin label in Figure 1-8 shows that each tablet contains 25 mcg of vitamin K.

$$10^6 \text{ mcg} = 10^6 \text{ μg} = 1 \text{ g}$$

The most common English unit of mass is the pound (lb). The relationship between the pound and the kilogram (kg) is

$$1.000 \text{ kg} = 2.205 \text{ lb}$$

Volume Volume is a measure of how much three-dimensional space a substance occupies. For example, 1 L of gasoline and 10 mL of cough syrup both represent volume measurements. Volume units are derived from units of length. For example, the volume of a box can be obtained by multiplying the width, length, and depth of the box. A cube measuring 1 cm along each side, therefore, has a volume of 1 cm × 1 cm × 1 cm = 1 cm³, as shown in Figure 1-9. The volume

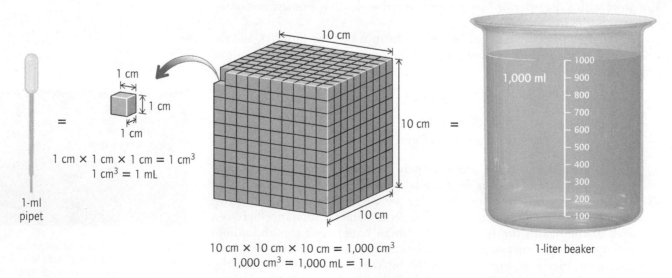

10 cm × 10 cm × 10 cm = 1,000 cm³
1,000 cm³ = 1,000 mL = 1 L

1-ml pipet

1 cm × 1 cm × 1 cm = 1 cm³
1 cm³ = 1 mL

1-liter beaker

Figure 1-9 Volume is derived from length measurements. A cube measuring 1 cm along each side has a volume of 1 cm³, equal to 1 mL. A cube measuring 10 cm along each side has a volume of 1,000 cm³, equal to 1,000 mL, equal to 1 L.

of the box is 1 cm^3 or 1 cc, read "one cubic centimeter" or "one centimeter cubed." Notice centimeters cubed (cm^3) arises from the multiplication of three units of length in cm. *Units are treated the same as numbers for mathematical operations such as multiplication, division, addition, and subtraction.*

The base unit of volume in the metric system is the liter (L). One liter is the volume occupied by a cube measuring 10 cm a side: 10 cm × 10 cm × 10 cm = 1,000 cm^3, as shown in Figure 1-9. Since 1,000 milliliters = 1 liter and 1,000 cm^3 = 1 L, then it must also be true that 1 cm^3 is equal to 1 milliliter (mL):

$$1 \text{ cm}^3 = 1 \text{ mL}$$

Special marked glassware is used to measure the volume of a liquid, including the pipet, the syringe, and the graduated cylinder (Figure 1-10). Beakers also contain volume markings, but they should not be used for precise volume measurements.

The volume of a solid object cannot be measured with a pipette or graduated cylinder. If the solid object has a regular shape, its length or diameter can be measured and the volume calculated mathematically, as in the case of the cube shown in Figure 1-9. Alternatively, and often the simplest way to measure the volume of a solid, is to determine how much water it displaces. For example, if a block of copper were fully submerged in a graduated cylinder filled with a known initial volume of water ($V_{initial}$), shown in Figure 1-11a, the level of the water would rise (V_{final}), shown in Figure 1-11b, by an amount corresponding to the volume of the block of copper. Thus, the volume of copper could be calculated by subtracting the initial volume of water from the final volume of water:

$$V_{object} = V_{final} - V_{initial}$$

You could, for example, determine the volume of your body by measuring the volume of water that you displace when fully submerged in a bathtub.

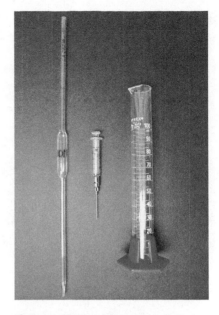

Figure 1-10 Examples of laboratory glassware routinely used to measure the volume of a liquid: pipet, syringe, and graduated cylinder (left to right). [Catherine Bausinger]

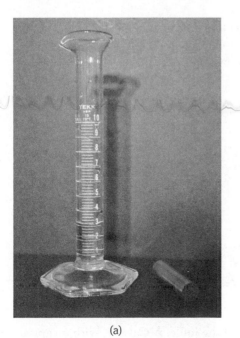 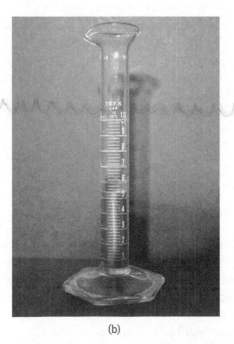

(a)　　　　　　　　　　　　　　　　(b)

Figure 1-11 Measuring the volume of a copper cylinder by displacement: (a) 3.0 mL of water is added to a graduated cylinder. (b) The volume rises to 6.1 mL after a copper cylinder has been carefully added to the graduated cylinder. The volume of the copper cylinder, obtained by displacement, is 6.1 cm^3 − 3.0 cm^3 = 3.1 cm^3. [Catherine Bausinger]

Figure 1-12 The Caloric content listed on nutritional labels, written with a capital *C*, are actually kilocalories (kcal). [Catherine Bausinger]

There are several English units of volume in common use, including the gallon (gal), the quart (qt), the pint, the tablespoon (tbsp), and the teaspoon (tsp). The conversion between the gallon, an English unit, and the liter, the metric base unit of volume, is

$$1 \text{ gal} = 3.785 \text{ L}$$

Energy Like matter, energy can be measured. The common units of energy used in science are the calorie and the joule. *A calorie (cal) is the amount of heat energy required to raise the temperature of 1 gram of water by 1 °C (Celsius).* As expected, there are 10^{-3} kilocalorie (kcal) in 1 calorie (cal). The Calorie listings that you see on nutritional labels, spelled with a capital C, are actually kilocalories (kcal) (**Figure 1-12**). Therefore, 1 **Calorie** is equal to 1 kilocalorie and 10^{-3} Calories are equal to 1 calorie (cal). Generally the Calorie, with a capital C, is only used in nutritional applications.

$$1 \text{ Calorie} = 1 \text{ kcal}$$
$$10^{-3} \text{ Calories} = 1 \text{ cal}$$

The standard unit of energy used in science is the joule, J, pronounced "jewel." *A joule is defined as the amount of energy required to lift a 1 kilogram weight a height of 10 centimeters.* The conversion between a calorie and a joule is

$$1 \text{ cal} = 4.184 \text{ J (exact)}$$

WORKED EXERCISES | Units

1-7   Using Table 1-1, write a conversion that gives the mathematical relationship between the gram and the microgram.

1-8   What is the volume, in units of cubic centimeters, of a cube measuring 5.00 cm a side? What does this volume correspond to in units of milliliters?

1-9   When a chunk of gold is placed in a graduated cylinder containing 100.0 mL of water, the water level rises to 120.0 mL. What is the volume of the chunk of gold in mL? What is the volume in cm^3?

1-10 What are two common units for energy in science? Which represents more energy, 1 calorie or 1 Calorie?

1-11 Rank the following lengths from longest to shortest:
    a. 1 nm   **b.** 1 mm   **c.** 1 cm   **d.** 1 μm   **e.** 1 m

Solutions

1-7   From Table 1-1: 10^6 μg = 1 g.

1-8   The volume of the cube is 5.00 cm × 5.00 cm × 5.00 cm = 125 cm^3. Remember to multiply the units as well as the numerical values. Since 1 mL = 1 cm^3, 125 cm^3 is equal to 125 mL.

1-9   The volume of the gold is equal to the difference in the volume before and after the gold chunk is submerged: V_{gold} = 120.0 mL − 100.0 mL = 20.0 mL, which corresponds to 20.0 cm^3.

1-10 Two common units of energy are the calorie and the joule. A Calorie, with a capital C, is 1,000 times larger than a calorie, with a lowercase *c*, and is typically used only in nutritional applications.

1-11 Longest to shortest: 1 m > 1 cm > 1 mm > 1 μm > 1 nm

PRACTICE EXERCISES

4 How many picometers are there in 1 meter? How many picograms are there in 1 gram? What does the prefix *pico* stand for?

5 What is the length, in centimeters, of one side of a cube that has a volume of 27 cm^3?

6 For each pair of measurements below, indicate which represents the larger amount. If they are the same, state so.
   a. 1 μm or 1 mm
   b. 10^{-3} kg or 1 g
   c. 1 Calorie or 1,000 calories
   d. 1 Calorie or 1 cal

7 How many of the following are in 1 meter?
   a. millimeter   b. decimeter   c. kilometer

8 Which of the objects with the diameters shown below can be seen with the naked eye?
   a. 1 km   b. 1 m   c. 1 nm

9 For each pair, indicate which represents the greater mass. If they are the same, state so.
   a. 1 mg or 1 μg   b. 1 ng or 1 μg   c. 10 mcg or 10 μg

10 What is the volume of an object, in cubic centimeters, if it is placed in 10 mL of water and causes the level of the water to rise to 11 mL?

1.3 Significant Figures and Measurement

Every measurement contains some degree of uncertainty. The uncertainty in a measurement depends on the accuracy and precision of the measuring device as well as the human error associated with reading any measuring device—balance, thermometer, ruler, and so on. *The number of digits (figures) reported by the person taking the measurement conveys information about the uncertainty of the measurement.*

Precision and Accuracy

Precision is an indicator of how close repeated measurements are to one another, while **accuracy** is an indicator of how close repeated measurements are to the "true" value. Ideally, a measurement is both accurate and precise. However, measurements can be inaccurate and/or imprecise. Consider, for example, the weight of an infant measured three times on four different balances (Table 1-2). Assume the "true" weight of the infant is 7.5 lb.

TABLE 1-2 Three Mass Readings of a 7.5-lb Infant Taken on Four Different Balances, A–D

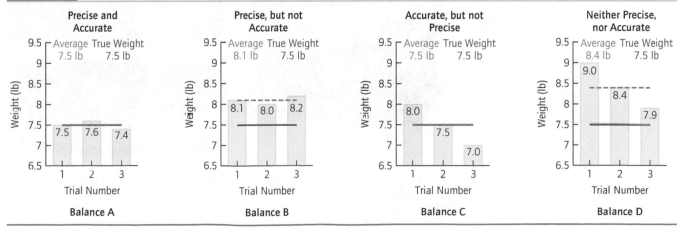

The average weight reported on balance A is 7.5 lb, the same as the "true" value, and all measurements are ±0.1 lb of the average weight; therefore, the measurements obtained on balance A are both accurate and precise. The measurements obtained using balance B are also within 0.1 lb of the average weight, but the average—8.1 lb—is not close to the "true" value; therefore, balance B is precise but not accurate. Balance C shows an average close to the "true" value, but the individual measurements differ from one another by as much as 1.0 lb; therefore, this balance is accurate but not precise. The measurements taken on balance D are neither accurate nor precise because the average is not near the "true" value and each measurement differs significantly from the others.

Significant Figures in Measured Values

Even using balance A, you cannot know the infant's weight exactly because of imperfections in the balance, whether the balance had been recently calibrated, and the inherent human error in taking any measurement. If you report a measurement of 7.5 lb, it suggests the infant weighed between 7.4 and 7.6 lb. *When reporting measurements, the convention is to record all of the certain digits plus one uncertain digit to a best approximation. The certain digits as well as the one uncertain digit are known as **significant figures**.* In the example of the baby weighing 7.5 lb, 7 is the certain digit, and 5 is the uncertain digit (it could be 4, 5, or 6), so the correct way to report the baby's weight would be to write two significant figures: 7.5 lb. A scientist interpreting this measurement understands it to mean that the infant's weight is 7.5 ± 0.1 lb, or somewhere between 7.4 and 7.6 lb.

The greater the precision of the measuring device, the more significant figures you can report. For example, a bathroom scale can be read to the tenths place (one place to the right of the decimal place), as in 5.0 lb. A top-loading balance (Figure 1-13 left), such as you might find in the chemical laboratory, can be read to the hundredths place, as in 10.4 g, and an analytical balance (Figure 1-13 right) can be read to the ten-thousandths place, as in 10.4977 g. Hence, more significant digits should be reported

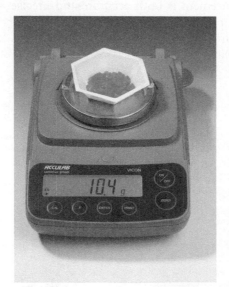

Figure 1-13 A top-loading balance (left) and an analytical balance (right) are designed to weigh to different levels of precision: the digital display on the top-loading balance shows only three significant figures for this sample, whereas the display on the analytical balance shows six significant figures for the same sample. [© Richard Megna/Fundamental Photographs]

for a measurement made on an analytical balance than on a top-loading balance or a bathroom scale.

By definition, the last digit is always understood to be the uncertain significant figure. For example, the uncertain significant figure is shown in red in the measurements:

$$5.0$$
$$5.00$$
$$5.000$$
$$5.0000$$

These measurements have 2, 3, 4, and 5 significant figures respectively and therefore indicate measurements made on increasingly more precise instruments. Note that these measurements would normally also be followed by a unit.

In the laboratory and in the clinic you will routinely be taking measurements and recording them. You should record the number of significant figures appropriate for the measuring device you are using, including zeros when applicable. For example, is the weight of the infant 7 lb, 7.0 lb, or 7.00 lb, measurements representing one, two, and three significant figures, respectively? Students often forget to include the measured zeros after a decimal. They should be included when they represent a certain digit or the uncertain digit.

Follow the guidelines below to learn how to count the number of significant figures in a measurement. Pay particular attention to zeros since sometime they are significant, whereas at other times they are merely place holders.

Guidelines for Determining the Number of Significant Figures in a Measured Number

- All nonzero digits are significant.

  3.45   3 significant figures

- Zeros between nonzero digits are always significant regardless of whether or not there is a decimal point.

  3.05   3 significant figures

- Zeros following a nonzero digit and to the right of the decimal point are significant.

  0.400   3 significant figures
  4.0   2 significant figures

- If there is no decimal point, zeros following a nonzero digit are not significant; they are merely placeholders.

  6,000   1 significant figure

- A decimal point placed after zeros indicates the zeros are significant:

  6,000.   4 significant figures

- Zeros that appear *before* nonzero digits, whether or not there is a decimal point, are not significant. The zeros merely serve as placeholders.

  0.00040   2 significant figures

- All digits in a number expressed in scientific notation are significant.

  2.30×10^4 3 significant figures

Exact Numbers

If you counted the number of students in a classroom carefully, there would be no uncertainty in the measurement. Indeed, numbers *obtained by counting are known as* **exact numbers,** *which have no uncertainty and therefore have an infinite number of significant figures.* Some definitions contain exact numbers because they are defined, not measured. All the metric conversions in Table 1-1, for example, are exact numbers. Other definitions such as "60 minutes = 1 hour" contain the exact numbers 60 and 1 and therefore have an infinite number of significant figures—no uncertainty. In this text, consider all metric conversions and conversions followed by the word *exact* to be exact numbers and therefore to have an infinite number of significant figures.

WORKED EXERCISE | Counting Significant Figures

1-12 Indicate the number of significant figures in the following measured values:

- **a.** 4.507 cm
- **b.** 0.00550 g
- **c.** 2.0×10^5 m
- **d.** 53,000. seconds
- **e.** 189 students

Solution

1-12
- **a.** There are four significant figures in 4.507. Zeros between nonzero digits are significant, and all nonzero digits are significant.
- **b.** There are three significant figures in 0.00550. The zeros before the 5 are not significant; they are simply placeholders. The zero after the 5 is significant.
- **c.** There are two significant figures in 2.0×10^5. Scientific notation is the only way to convey two significant figures in this measurement because writing 20,000 indicates only one significant figure, while writing 20,000. indicates five significant figures.
- **d.** There are five significant figures in 53,000. The decimal point after the zeros indicates that the three zeros are significant.
- **e.** There are an infinite number of significant figures in this exact number because it was obtained by counting. There is no uncertainty in an exact number.

PRACTICE EXERCISES

11 How many significant figures are there in the measured values shown below? Explain your reasoning.
- **a.** 0.007 m
- **b.** 50 people
- **c.** 23,000. seconds
- **d.** 0.004050 mg

12 The measurements below were made on three different balances:

i. 5.5 g   **ii.** 5.51 g   **iii.** 5.5093 g

- **a.** Which measurement was made on the most precise balance?
- **b.** Which is the uncertain digit in each measurement?
- **c.** A student takes a reading from a balance where the display fluctuates between 5.50 g and 5.53 g. Which of the three measurements above should the student report?

Significant Figures in Calculations

For calculations involving measured values, it is important to report the final answer with the correct number of significant figures. This is done by determining the correct number of significant figures that the final answer is allowed to have and rounding the last significant figure. It is incorrect to simply report all the digits displayed by the calculator! The correct number of significant figures depends on the number of significant figures in the measured value(s) and the type of calculation: multiplication and division or addition and subtraction. *In calculations involving several mathematical steps, round only the final answer.* If you round after each step, your answer might have "rounding errors."

Calculations Involving Multiplication and Division *When multiplying or dividing measured values, the final calculated answer cannot have more significant figures than the measurement with the fewest number of significant figures.* For example, if you multiplied the measured lengths 3.55 m and 11.65 m to determine the area of a room, the answer displayed on the calculator would read many more digits than either of the measured values in the calculation. If you reported all of the digits from the calculator, it would suggest the area of the room was more precise than either of the measurements that went into calculating the area, which is impossible. Instead, the answer must be rounded to three significant figures because the measurement with the *fewest* significant figures, 3.55, has three significant figures:

3.55 m	×	11.65 m =	<41.3575 m²> =	**41.4 m²**
measured value		measured value	calculator answer	final answer after
3 significant		4 significant	6 significant	rounding to
figures		figures	figures	3 significant
				figures

The correct answer to report would be 41.4 m². According to rounding rules, the third significant figure, 3, is rounded *up* to 4 because the next digit, 5, is equal to or greater than 5.

Calculations Involving Addition and Subtraction *When adding or subtracting measured values, the final calculated answer should not have more places past the decimal than the measurement with the fewest places past the decimal.* For example, the sum of the measurements shown below should be reported as 19.3 cm, not 19.294 cm as displayed on the calculator, because the measurement with the fewest decimal places, 5.4, has its last digit in the tenths place.

$$\begin{array}{r} 5.4 \quad \text{cm} \\ 6.55 \quad \text{cm} \\ +7.344\ \text{cm} \\ \hline \end{array}$$

<19.294 cm>   Calculator answer
19.3   cm   Final answer after rounding

According to rounding rules, the 2 in the tenths place should be rounded up to 3 because the next digit, 9, is equal to or greater than 5, and this digit and the others that follow it should be dropped.

> **Review of rounding rules**
>
> Once you determine how many significant digits to include in your final answer:
>
> - Round up the last significant figure if the next digit is greater than or equal to 5, then drop this digit as well as all digits that come after it.
>
>   0.846862 becomes 0.847 because 8 is greater than 5.
>
> - Leave the last significant figure as is if the next digit is less than 5, then drop this digit as well as all digits that come after it.
>
>   0.846324 becomes 0.846 because 3 is less than 5.

WORKED EXERCISE | Significant Figures in Calculations

1-13 Perform the following calculations, assuming the numbers represent measurements, by rounding the final answer to the correct number of significant figures:

a. $0.0022 \times 58.88 =$

b. $7.0 + 8.55 + 233 =$

Solution

1-13 a. The calculation involves multiplication, so the final answer cannot have more significant figures than the measurement with the fewest significant figures. The number 0.0022 has two significant figures and the number 58.88 has four significant figures; therefore, the final answer should have two significant figures. The calculator gives the following result: $0.0022 \times 58.88 = 0.129536$. Round up the last of two significant figures to 3, giving a final answer of 0.13 because the third digit, 9, is greater than or equal to 5.

b. The calculation involves addition, so the final answer cannot have more digits past the decimal place than the measurement with the fewest digits past the decimal place, the value 233, which has its last digit in the ones place. The calculator gives the sum as 248.55, but the final answer cannot have any digits past the ones place. Thus, 248.55 should be rounded up to 249 because the next digit is 5—which is a number greater than or equal to 5.

$$
\begin{array}{r}
7.0 \\
8.55 \\
+233. \\
\hline
\end{array}
$$

$<248.55>$ Calculator answer

$\phantom{<}249$ Final answer after rounding

PRACTICE EXERCISES

13 Perform the following calculations assuming all values represent measurements. Report the answer to the correct number of significant figures.

  a. $56.50 \times 37.99 =$

  b. $5.999 + 6.001 + 3.2222 =$

  c. $57.200 \times \dfrac{67.55}{1.220} =$

■ 1.4 Using Dimensional Analysis

Medical professionals are routinely called on to convert one unit of measurement into another unit of measurement as part of their daily routine. For example, a patient's weight might be taken in the English unit of pounds (lb), but the dosage of his medicine might be prescribed in terms of his weight in kilograms, a metric unit. To determine the correct dosage, you will need to perform a unit conversion from English to metric units. *A **unit conversion** is a type of calculation in which a measurement in one unit (the supplied unit) is converted into the equivalent value in another unit (the requested unit).*

In the sciences, conversions are performed using a method known as dimensional analysis, which will be used throughout this text. **Dimensional analysis,** also known as the factor-label method, is a process that evaluates the *units*—dimensions—to properly set up the calculation. Using dimensional analysis ensures that values aren't multiplied when they should be divided and vice versa. As you can imagine, giving the incorrect dosage to a patient as a result of a calculation error could be catastrophic. Unfortunately, these types of mistakes occur, but if you learn dimensional analysis, they won't happen to you!

The three basic steps for using dimensional analysis are illustrated in the guidelines on the next page. Take the time to learn this technique now while the calculations are still relatively straightforward, and then the technique will serve you well for the more complex calculations seen in later chapters.

As a final step, always double-check that you have shown the correct number of significant figures. Since multiplication and division are involved in metric conversions, the final answer should have no more significant figures than the measured value with the fewest number of significant figures.

The beauty of dimensional analysis is that more than one conversion factor can be used, by multiplying conversion factors sequentially and setting

Guidelines for Using Dimensional Analysis

1. **Step 1: Identify the conversion needed.** Write the mathematical expression that equates the supplied unit—the one given to you—and the requested unit—the one you are being asked to determine. When there is no single conversion, more than one conversion may be required. *For example, to convert 0.500 gram into milligrams, you need the metric-metric conversion that equates grams, the supplied unit, and milligrams, the requested unit. From Table 1-1:*

$$10^3 \text{ mg} = 1 \text{ g}$$

2. **Step 2: Express the conversion as two possible conversion factors.** A conversion can be turned into a **conversion factor** by expressing the equality in the form of a ratio, or fraction. Turn the conversion into a conversion factor by placing one side of the equality in the numerator and the other side of the equality in the denominator. Then invert the fraction to get the other conversion factor. Remember, the numerical value and its associated unit must remain together. For example, the conversion factors for the conversion in step 1 are

$$\frac{1 \text{ g}}{10^3 \text{ mg}} \quad \text{and} \quad \frac{10^3 \text{ mg}}{1 \text{ g}}$$

3. **Step 3: Set up the calculation by multiplying the supplied unit by the correct conversion** factor. The correct conversion factor is the one that has the supplied *unit* in the denominator so that it will cancel, leaving only the requested unit in the final answer. As with identical *numbers*, identical *units* cancel algebraically when they appear in both the numerator and the denominator of a fraction. The result is an answer in the requested unit.

$$\cancel{\text{supplied unit}} \times \frac{\text{requested unit}}{\cancel{\text{supplied unit}}} = \text{requested unit}$$

For example, to convert 0.500 g into mg, dimensional analysis directs us to set up the calculation using the form of the conversion factor shown, because it allows the supplied units, grams, to cancel, leaving a final answer in mg, the requested unit:

$$0.500 \cancel{\text{ g}} \times \frac{10^3 \text{ mg}}{1 \cancel{\text{ g}}} = 500. \text{ mg}$$

In this problem, the measured value is 0.500 g, which has three significant figures, so the final answer, 500. mg, should also have three significant figures. Metric conversion factors are exact numbers, and therefore the conversion factor itself will not limit the number of significant figures you can report.

them up so that units cancel, until only the requested unit remains. The worked exercise below shows how to convert a prefixed metric unit to another prefixed metric unit by using *two* known metric conversions involving the metric base unit.

WORKED EXERCISE   Metric Conversions

1-14 Convert 15 mm into units of cm using dimensional analysis and the conversion factors shown in Table 1-1.

Solution

1-14 The supplied unit is millimeters, mm, and the requested unit is centimeters, cm. There is no direct conversion between mm and cm in Table 1-1 or between any two prefixed units. The easiest way to convert one metric prefixed unit to another metric prefixed unit is to go through the *base unit*, by multiplying by two conversion factors, one that goes from the *supplied* unit to the *base* unit and another that goes from the *base* unit to the *requested* unit: mm → m and m → cm.

Step 1: Identify the conversions needed. From Table 1-1:

$$10^3 \text{ mm} = 1 \text{ m} \quad \text{and} \quad 10^2 \text{ cm} = 1 \text{ m}$$

Step 2: Express each conversion as two possible conversion factors.

$$\frac{10^3 \text{ mm}}{1 \text{ m}} \quad \text{or} \quad \frac{1 \text{ m}}{10^3 \text{ mm}} \quad \text{and} \quad \frac{10^2 \text{ cm}}{1 \text{ m}} \quad \text{or} \quad \frac{1 \text{ m}}{10^2 \text{ cm}}$$

Step 3: Set up the calculation by multiplying the supplied unit by the correct conversion factors. The correct conversion factor always has the unit that needs to be canceled in the denominator. Here, mm is canceled with the first conversion factor, the mm → m conversion, and then m is canceled with the second conversion factor, the m → cm conversion. Note the correct form of each conversion factor is the one that has the unit that needs to be canceled in the denominator.

$$15 \text{ mm} \times \frac{1 \text{ m}}{10^3 \text{ mm}} \times \frac{10^2 \text{ cm}}{1 \text{ m}} = 1.5 \text{ cm}$$

Finally, check that the answer has the correct number of significant figures. Here, the only measured value is 15 mm, which has two significant figures; thus, the final answer, 1.5 cm, can have only two significant figures.

PRACTICE EXERCISES

14 How many centimeters are there in 0.65 m?

15 How many grams are there in 7.7 kg?

16 How many liters are there in 561 mL of water?

17 How many picometers are there in 56 nm?

18 Hemoglobin has a spherical shape with a diameter of 5.5 nm. What is the diameter of hemoglobin in meters? What is it in millimeters?

19 One patient has a tumor 150. μm in diameter, while another has a tumor 2.00 mm in diameter. Which patient has the larger tumor?

20 How many calories are there in 0.177 kcal?

Converting between English and Metric Units

In the United States, English units are still in common use, so you will find it necessary at times to convert from an English unit to a metric unit and vice versa. This type of conversion can also be done using dimensional analysis by using known metric-English conversions. Table 1-3 lists some of the common conversions between metric and English units for length, mass, and volume.

TABLE 1-3 Some Common English-Metric Conversions

Type of Measurement	Conversions
Length	1 in = 2.54 cm (exact)
	1 mi = 1.609 km
	39.37 in = 1 m
Mass	2.205 lb = 1 kg
Volume	1 gal = 3.785 L

WORKED EXERCISES   English-Metric Conversions

1-15 A patient weighs 125 lb. What is her weight in kilograms?

1-16 The length of the femur bone in a fetus at 20 weeks gestation is on average 32 mm. What is this length in inches?

Solutions

1-15 In this exercise, the supplied unit is pounds, lb, and the requested unit is kilograms, kg.

Step 1: Identify the conversion needed. Table 1-3 shows that the conversion between pounds and kilograms is 1 kg = 2.205 lb.

Step 2: Express the conversion as two possible conversion factors:

$$\frac{1 \text{ kg}}{2.205 \text{ lb}} \quad \text{or} \quad \frac{2.205 \text{ lb}}{1 \text{ kg}}$$

Step 3: Set up the calculation by multiplying the supplied unit by the conversion factor that allows the supplied unit to cancel:

$$125 \text{ lb} \times \frac{1 \text{ kg}}{2.205 \text{ lb}} = 56.7 \text{ kg}$$

The measurement, 125 lb, has three significant figures, and the conversion factor has four significant figures, so the final answer should also have three significant figures: 56.7 kg.

1-16 In this exercise, the supplied unit is millimeters, mm, and the requested unit is inches, in.

Step 1: Identify the conversion(s) needed. Table 1-3 does not show a direct conversion between millimeters and inches; however, it lists a conversion between inches and meters, and you know the metric conversion between mm and m, so multiply by two conversion factors: millimeters to meters and inches to meters.

Step 2: Express each conversion as two possible conversion factors:

$$\frac{10^3 \text{ mm}}{1 \text{ m}} \quad \text{or} \quad \frac{1 \text{ m}}{10^3 \text{ mm}} \quad \text{and} \quad \frac{1 \text{ m}}{39.37 \text{ in}} \quad \text{or} \quad \frac{39.37 \text{ in}}{1 \text{ m}}$$

Step 3: Set up the calculation by multiplying the supplied unit by the appropriate conversion factors so that units cancel in a way that leaves only the requested unit. Here, the supplied value is 32 mm, which, when multiplied by the millimeter-to-meter conversion with mm in the denominator, causes millimeters to cancel. When this is multiplied by the inches-to-meter conversion with meters in the denominator, it causes meters to cancel, leaving inches, the requested unit:

$$32 \text{ mm} \times \frac{1 \text{ m}}{10^3 \text{ mm}} \times \frac{39.37 \text{ in}}{1 \text{ m}} = 1.3 \text{ in (2 significant figures)}$$

The measurement, 32 mm, has two significant figures, so the final answer must have two significant figures.

PRACTICE EXERCISES

21 If you wanted to fill your empty 15-gal gas tank in Europe, how many liters of gasoline would you need to purchase?

22 A premature baby weighs 906 g. What is the weight of the baby in pounds?

Energy Conversions

Unit conversions involving energy can also be performed using the common energy conversions given on page 12. To convert between nutritional Calories and joules, use two conversion factors, one from the Calorie to the calorie and then a second conversion factor from the calorie to the joule since we have not provided a direct conversion between the nutritional Calorie and the joule.

WORKED EXERCISES   Energy Conversions

1-17 How many calories (cal) of energy are there in 1.56×10^4 J?

1-18 One avocado contains 305 Calories. How many joules does an avocado contain?

Solutions

1-17 In this exercise you are asked to convert between two energy units.

Step 1: Identify the conversion. Write the expression that equates the supplied unit (joules) and the requested unit (calories) from page 12:

$$1 \text{ cal} = 4.184 \text{ J (exact)}$$

Step 2: Express the conversion as two conversion factors:

$$\frac{1 \text{ cal}}{4.184 \text{ J}} \quad \text{or} \quad \frac{4.184 \text{ J}}{1 \text{ cal}}$$

Step 3: Set up the calculation by multiplying the supplied value by the conversion factor that allows the supplied unit to cancel. Multiply the supplied unit by the form of the conversion factor in step 2 that has joules in the denominator so that the supplied unit (J) cancels, leaving the requested unit of calories:

$$1.56 \times 10^4 \text{ J} \times \frac{1 \text{ cal}}{4.184 \text{ J}} = 3.73 \times 10^3 \text{ cal (3 significant figures)}$$

1-18 In this exercise you are asked to convert between nutritional Calories and joules, two energy units, but there is no direct conversion between the two given. Therefore, use two conversion factors, one that converts nutritional Calories to calories and one that converts calories to joules. The plan: Calories $\rightarrow$ cal $\rightarrow$ joule.

Step 1: Identify the conversions needed. Write the conversions needed from page 12.

$$10^{-3} \text{ Calories} = 1 \text{ cal}$$
$$1 \text{ cal} = 4.184 \text{ J (exact)}$$

Step 2: Express each conversion as two conversion factors. Write each conversion as two possible conversion factors:

$$\frac{10^{-3} \text{ Cal}}{1 \text{ cal}} \quad \text{or} \quad \frac{1 \text{ cal}}{10^{-3} \text{ Cal}} \quad \text{and} \quad \frac{1 \text{ cal}}{4.184 \text{ J}} \quad \text{or} \quad \frac{4.184 \text{ J}}{1 \text{ cal}}$$

Step 3: Set up the calculation so that the supplied units cancel. Multiply 305 Cal by the form of the first conversion factor from step 2 that has Calories in the denominator so that Calories cancel. Then multiply by the second conversion factor in step 2 that has calories in the denominator so cal cancel, leaving only the requested unit (J):

$$305 \text{ Cal} \times \frac{1 \text{ cal}}{10^{-3} \text{ Cal}} \times \frac{4.184 \text{ J}}{1 \text{ cal}} = 1.28 \times 10^6 \text{ J (3 significant figures)}$$

PRACTICE EXERCISES

23 Using the conversions on page 12, convert the following units into calories:

    a. 5.79 kcal   **b.** 48.8 J

24 How many joules are there in 2.45 cal?

25 How many joules are there in 2,720 Calories, the amount of energy the average person consumes in a day?

Dosage Calculations

For some medicines prescribed for patients, the dosage must be adjusted according to the patient's weight. This is especially true when administering medicine to children. For example, a dosage of "8.0 mg of tetracycline per kilogram body weight daily" is a dosage based on the weight of the patient. A patient's weight is often given in pounds, yet many drug handbooks give the dosage per kilogram body weight of the patient. Therefore, to calculate the correct amount of medicine to give the patient, you must first convert the patient's weight from pounds into kilograms with an English-metric conversion, using Table 1-3.

It is important to recognize that the dosage is itself a conversion factor between the mass or volume of the medicine and the weight of the patient. Whenever you see the word *per*, it means *in every* and can be expressed as a ratio or fraction where *per* represents a division operation (divided by). For example, 60 miles *per* hour can be written as the ratio 60 mi/1 hr. Similarly, a dosage of 8.0 mg *per* kg body weight can be expressed as the fraction 8.0 mg/1 kg. Hence, dosage *is* a conversion factor:

$$\frac{8 \text{ mg}}{1 \text{ kg}} \quad \text{or} \quad \frac{1 \text{ kg}}{8 \text{ mg}}$$

Dimensional analysis is used to solve dosage calculations by multiplying the patient's weight by the appropriate English-metric conversion factor and then multiplying by the dosage conversion factor, as shown in the following worked exercise.

> Some common abbreviations indicating the frequency with which a medication should be administered include *q.d.* and *b.i.d.*, derived from the Latin meaning administered "daily" and "twice daily," respectively. If the medicine is prescribed for two times daily or four times daily, divide your final answer by two or four to determine how much to give the patient at each administration.

WORKED EXERCISE | Dosage Calculations

1-19 Tetracycline elixir, an antibiotic, is ordered at a dosage of 8.0 mg per kilogram of body weight q.d. for a child weighing 52 lb. How many milligrams of tetracycline elixir should be given to this child daily?

Solution

Step 1: Identify the conversions. Since the dosage is given based on a patient's weight in kilograms, an English-to-metric conversion must be performed. From Table 1-3 this is 1.000 kg = 2.205 lb. The dosage itself is already a conversion factor.

Step 2: Express each conversion as two possible conversion factors. The English-to-metric conversion factors for the patient's weight are

$$\frac{1 \text{ kg}}{2.205 \text{ lb}} \quad \text{or} \quad \frac{2.205 \text{ lb}}{1 \text{ kg}}$$

The dosage *is* a conversion factor between the mass of medicine in milligrams and the weight of the patient in kilograms:

$$\frac{8.0 \text{ mg}}{1 \text{ kg}} \quad \text{or} \quad \frac{1 \text{ kg}}{8.0 \text{ mg}}$$

Since the dosage is marked q.d., the dose should be given once a day.

Step 3: Set up the calculation by multiplying the supplied unit by the appropriate conversion factors so that supplied units cancel. Write the patient's weight in pounds (the supplied unit) and multiply by the correct metric-English conversion so that pounds cancel, leaving kg. Then multiply by the dosage conversion factor so that units of kg cancel. The result is the mass, in mg, of the medicine to give the patient daily:

$$52 \text{ lb} \times \frac{1 \text{ kg}}{2.205 \text{ lb}} \times \frac{8.0 \text{ mg}}{\text{kg}} = 190 \text{ mg given once a day (2 significant figures)}$$

Sometimes this type of calculation is confusing because units of mass appear in two places. The key to solving this type of problem is recognizing that the mass of the patient is different from the mass of the medication.

Finally, check that your final answer has no more significant figures than the supplied values. In this example, 52 lb and 8.0 mg/kg both have two significant figures, so the final answer should have two significant figures.

If the medicine had been prescribed for two times daily or four times daily, you would have divided your final answer by two or four to determine how much medicine to give the patient at each administration.

PRACTICE EXERCISES

26 Quinidine is an antiarrhythmic agent. It is prescribed for an adult patient weighing 110.0 lb at a dosage of 25.00 mg per kilogram of body weight per day. Express the dosage as a fraction. How many milligrams of quinidine should be given to the patient daily?

27 Ampicillin, an antibiotic, is prescribed for a child weighing 63.0 lb at a dosage of 20.0 mg per kilogram of body weight per day, in four equally divided doses. How many milligrams should be given at each administration?

Density Calculations

Density is a physical property of matter often used to characterize a substance or material, as seen in the opening vignette for bone density. The **density** (d) of a substance is defined as its mass (m) per unit volume (V):

$$d = \frac{m}{V}$$

Density is derived from two measured values: *mass* and *volume*; therefore it typically has units of g/mL or g/cm³. The density of a substance is independent of the amount of the sample. For example, the density of one drop of water is the same as a bathtub full of water. The density of some common substances is shown in Table 1-4. Substances less dense than water, which do not dissolve in water, such as oil, will float on top of it.

Density (m/V) is a conversion factor and so can be used to calculate the *mass* of a substance whose volume and density are known or, in its inverted

TABLE 1-4 Density of Some Elements and Compounds at 25 °C

Element or Compound	Density (g/mL or g/cm³)
Water	1.00
Iron	7.9
Gold	19.32
Ethanol	0.79

form (V/m), can be used to calculate the *volume* of a substance whose mass and density are known, as demonstrated in the worked exercises below.

Specific gravity is the ratio of the density of a substance to the density of water at 4 °C:

$$\text{specific gravity} = \frac{\text{density of substance} \left(\dfrac{g}{mL}\right)}{\text{density of water} \left(\dfrac{g}{mL}\right)}$$

Since both the numerator and the denominator have the same units, specific gravity is a *unitless* quantity. Since the density of water at 4 °C is 1.00 g/mL, the value in the denominator, *the specific gravity of a substance is the same as its density but without any units*. Substances with a density greater than water will have a specific gravity greater than water, and substances with a density less than water will have a specific gravity less than water. Specific-gravity measurements are often used to determine the ratio of various solvent-water mixtures, such as an ethanol-water mixture. Indeed, beer brewers and wineries use specific gravity to determine the alcohol (ethanol) content of their beer or wine.

In medicine, the specific gravity of urine is measured to screen for certain medical conditions. The normal specific gravity of urine is 1.002 to 1.030. Values higher than the normal range may indicate diabetes, kidney failure, or kidney infection. Specific gravity values lower than normal may indicate a urinary tract infection, high or low sodium levels, or a condition that leads to excessive urination.

WORKED EXERCISES   Density and Specific Gravity Calculations

1-20 What is the density of a substance with a mass of 0.90 g and a volume of 1.2 mL?

1-21 Using Table 1-4, calculate the mass, in grams, of a 5.0-cm³ block of gold.

1-22 Using Table 1-4, calculate the volume, in milliliters, of a 22-g block of gold.

1-23 A patient supplies a urine sample that has a specific gravity of 1.040. What is the density of this patient's urine? Is the specific gravity of the patient's urine within the normal range?

Solutions

1-20 Density is defined as $d = m/V$. To calculate density, substitute the values supplied for mass and volume into this equation:

$$d = \frac{0.90 \text{ g}}{1.2 \text{ mL}} = 0.75 \text{ g/mL (2 significant figures)}$$

No units cancel, so the final answer has units of g/mL, which are typical units of density.

1-21 **Step 1: Identify the conversions.** From Table 1-4, we find the density of gold is 19.32 g/cm³. The density of gold *is* the conversion factor.

Step 2: Express the conversion as two possible conversion factors:

$$\frac{19.32 \text{ g}}{1 \text{ cm}^3} \quad \text{or} \quad \frac{1 \text{ cm}^3}{19.32 \text{ g}}$$

Step 3: Set up the calculation by multiplying the supplied unit by the correct conversion factor that allows the supplied units to cancel. Volume is the supplied unit, so use the first form of the density conversion factor shown in step 2 because it will allow volume to cancel, leaving the requested units of mass:

$$5.0 \; \cancel{cm^3} \times \frac{19.32 \; g}{1 \; \cancel{cm^3}} = 97 \; g \; of \; gold$$

The final answer is rounded to two significant figures because the supplied measurement, 5.0 cm^3, has two significant figures.

1-22 **Steps 1 and 2 are the same as the previous question.**
Step 3: Set up the calculation by multiplying the supplied unit by the conversion factor that allows the supplied units to cancel. Mass is the supplied unit, so use the inverted form of density as a conversion factor so that mass will cancel, leaving only the requested unit of volume:

$$22 \; \cancel{g} \times \frac{1 \; cm^3}{19.32 \; \cancel{g}} = 1.1 \; cm^3$$

The final answer is rounded to two significant figures because the supplied value, 22 g, has two significant figures.

1-23 Since the specific gravity of the patient's urine is 1.040 and the density of water is 1 g/mL, the density of the patient's urine must be 1.040 g/mL. The specific gravity of the patient's urine is above the normal range of 1.002–1.030.

PRACTICE EXERCISES

28 What is the density of a liquid that has a mass of 5.5 g and a volume of 5.0 mL?

29 Which of the following pairs is more dense?
  a. a soccer ball or a bowling ball
  b. a beaker of water or a swimming pool full of water
  c. ethanol or water (Use Table 1-4.)

30 Would a gold sphere with a mass of 5 g have a greater diameter or a smaller diameter than a gold sphere with a mass of 10. g? *Hint:* No calculation is required.

31 An object has a mass of 10.5 g. When it is submerged in a graduated cylinder initially containing 82.5 mL of water, the water level rises to 95.0 mL. What is the density of the object?

32 Using Table 1-4, what is the mass, in grams, of 25 mL of water?

33 Using Table 1-4, what volume of water, in milliliters, would a 65-g piece of iron displace?

34 A piece of unknown metal was found to have a mass of 71.1 g and a volume of 9.0 cm^3. Using Table 1-4, is this unknown metal gold or iron? Explain.

35 Would you expect a woman with osteoporosis to have a bone density greater than or less than normal? Explain.

36 Ice floats on liquid water. Does this mean ice is more or less dense than liquid water?

Temperature

In section 1.1 you learned that temperature is a measure of the average kinetic energy of particles of matter. For example, the molecules of water in a beaker of boiling water, with a temperature of 100 °C, have more kinetic energy than the molecules of water in a beaker of frozen water, with a temperature of 0 °C.

Temperature is measured with a thermometer using one of three temperature scales: Celsius (°C), Fahrenheit (°F), or kelvin (K). The United States is one of few countries that still use the Fahrenheit scale, although only for nonscientific applications. The Celsius scale is used in science, medicine, and everyday applications in the rest of the world. The **kelvin scale** is used primarily in chemistry and physics because it represents an **absolute temperature scale**, where 0 K is the temperature that corresponds to zero kinetic energy; hence, the kelvin scale does not use a degree sign (°). Although absolute zero has never been reached, temperatures as low as 0.1 K have been attained in specialized experiments.

In contrast, the Celsius and Fahrenheit scales are relative scales, which are based on the freezing and boiling points of water: assigned the values 0 °C and 100 °C, respectively, in the Celsius scale, and 32 °F and 212 °F, respectively, in the Fahrenheit scale. The Celsius scale is divided into 100-degree increments, whereas the Fahrenheit scale is divided into 180-degree increments. Figure 1-14 compares the three temperature scales. For example, normal body temperature on each of these scales is 98.6 °F, 37 °C, and 310 K.

To convert a temperature in one scale to a temperature in another scale is not simply a matter of multiplying conversion factors as you have done up to this point, because the size of a degree Celsius is larger than the size of a degree Fahrenheit, and the two scales are offset from each other by 32 degrees. A degree Celsius is the same increment as a kelvin, but the two scales are offset by 273 degrees. To perform a temperature conversion, first determine which temperature scales you are converting between:

$$°F \leftrightarrow °C \leftrightarrow K$$

Conversions between °F ↔ °C or °C ↔ K are direct and require using only *one* of the conversion equations in Table 1-5. Conversions between °F ↔ K require two steps, where the first step requires converting to °C.

Select the equation from Table 1-5 that shows the temperature scale you are trying to convert to isolated on one side of the equality. For example,

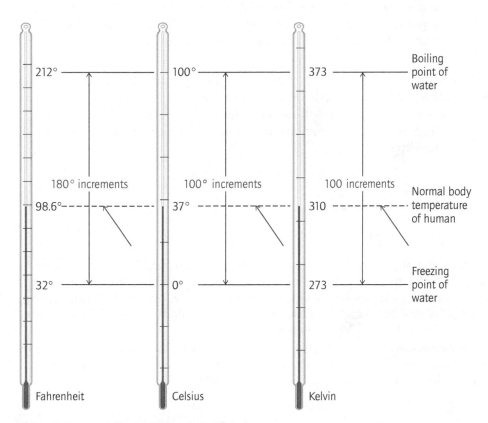

Figure 1-14 Comparison of Fahrenheit, Celsius, and kelvin temperature scales.

TABLE 1-5 Equations for Temperature Scale Conversions*

From °C to °F	$°F = (1.8 \times °C) + 32.0$
From °F to °C	$°C = \dfrac{(°F - 32.0)}{1.8}$
From °C to K	$K = °C + 273.15$
From K to °C	$°C = K - 273.15$

*Note that a degree sign (°) is not used in the kelvin scale because it is an absolute scale.

if the supplied value is in °C and you are asked to convert to °F, use the equation reading "°F = ..."

To convert between °F and K, first convert from °F to °C, as in the previous example, and then convert from °C to K, using the equation from Table 1-5 reading "K = . . ." In temperature conversions, remember to follow the rules for order of operations in mathematics: parentheses first, followed by multiplication and division, then addition and subtraction. Consult Appendix A for a math review on these topics.

WORKED EXERCISES Temperature Conversions

1-24 Name the three temperature scales. What temperature scale is used most in medicine and for everyday purposes all over the globe except for the United States?

1-25 While traveling in Europe, you feel ill and visit a health clinic. A nurse tells you that your temperature is 38.3 °C. Convert this temperature into °F. Do you have a fever?

1-26 It is a comfortable summer day on the beach in southern California, where the air temperature is 72 °F. Your friend who is visiting from Europe would like to know what this temperature is in °C since they don't use the Fahrenheit scale in Europe. What would you tell her? Being a student of science, you decide to also inform her of the air temperature in kelvins. What is the temperature in kelvins?

Solutions

1-24 The three temperature scales are the Celsius, Fahrenheit, and kelvin scales. The Celsius scale is used in most of the world and in medicine.

1-25 Yes, you have a fever of 100.9 °F. First determine that you are asked to convert from °C to °F. Next, find the equation from Table 1-5 that has °F isolated on one side of the equality:

$$°F = (1.8 \times °C) + 32.0$$

Then substitute the supplied temperature, **38.3 °C,** into the equation:

$$°F = (1.8 \times \mathbf{38.3 \ °C}) + 32.0 = 100.9 \ °F$$

Solve the equation, by performing the multiplication in parentheses first and then adding this value to 32.0. Follow the rules that apply to addition and subtraction when determining the number of significant figures. Here, the supplied value given has one place past the decimal, so the final answer should be reported to one place past the decimal (1.8 is an exact number).

1-26 You are supplied with the temperature in °F and first asked to convert to °C. Identify the equation from Table 1-5 that shows °C isolated on one side of the equality, substitute the value 72 °F into the equation, and solve for °C:

$$°C = \frac{(°F - 32.0)}{1.8} = \frac{(72 - 32.0)}{1.8} = 22 \ °C$$

Remember to perform the operation in parentheses first and then divide by 1.8. The measured value has its last digit in the ones place and so too must the answer.

Next you are basically asked to convert from °C to kelvins. Identify the equation from Table 1-5 that has K isolated on one side of the equality:

$$K = °C + 273.15$$

Using the answer from the previous calculation, 22 °C, substitute this value into the equation and solve for K:

$$K = 22 \, °C + 273.15 = 295 \, K$$

PRACTICE EXERCISES

37 Normal body temperature is 37.0 °C. Convert this temperature into °F and kelvins. Show your work.

38 Many birds have a normal body temperature of 106 °F. Convert this temperature into kelvins.

39 The boiling point of helium is 4 K. What is the boiling point of helium in °C and °F?

Specific Heat

You have seen that when energy is added to a substance, its kinetic energy increases and its temperature rises. The extent of the temperature rise depends on the amount of heat added, the amount of the substance, and the identity of the substance. *Specific heat is a measure of the amount of heat required to raise the temperature of 1 g of a particular substance by 1 °C.* Table 1-6 lists the specific heat of some common substances. For example, the specific heat of water is 1 cal/g · °C. As you can see, specific heat is a physical property of a substance in a given state of matter. The specific heat of a substance can be calculated using the following equation:

$$\text{specific heat} = \frac{\text{Heat(cal)}}{\text{mass(g)} \times \Delta T(°C \text{ or } K)}$$

This equation can be used to calculate any one of these variables if the other three are known. "Heat" refers to the amount of energy transferred to or from the

TABLE 1-6 Specific Heat of Some Common Substances

Substance	Specific Heat $\left(\dfrac{cal}{g \cdot °C}\right)$
Water (liquid)	1.00
Paraffin wax	0.60
Ethanol	0.58
Water (gas)	0.497
Water (solid)	0.490
Ambient air	0.24
Aluminum	0.22
Brick	0.20
Sand	0.16
Iron	0.11
Wood	0.10
Copper	0.093
Lead	0.031

substance, mass is the amount of sample, and ΔT is the change in temperature—before and after heat has been transferred. Heat is measured in calories, mass is measured in grams, and temperature can be measured in either the Celsius or kelvin scales. Therefore, specific heat usually has units of cal/g · °C.

This equation tells us that a substance with a high specific heat requires a greater input of heat energy to achieve the same increase in temperature as a substance with a lower specific heat. Imagine walking on the beach on a day when the air temperature is hot. The sand under your feet feels quite warm, but when you step into the water, the water is cool. This is largely because the heat energy transferred to water causes a much smaller temperature rise than the same amount of heat transferred to sand because sand has a much lower specific heat than water: 0.16 cal/g °C for sand compared to 1.00 cal/g °C for water. Indeed, water has one of the highest specific heats of any substance. It is the high specific heat of water that allows our bodies to maintain a relatively constant temperature.

WORKED EXERCISE | Specific Heat Calculations |

1-27 Using Table 1-6, calculate the number of calories that must be added to raise the temperature of 22.1 g of each of the substances shown from 25.0 °C to 42.0 °C. Which substance requires the most heat to raise its temperature?

    a. water   **b.** paraffin wax   **c.** copper

Solution

1-27 You have been asked to determine the heat, so rearrange the specific heat equation algebraically so that *heat* is isolated on one side of the equality:

$$\text{Heat} = \text{specific heat} \left(\frac{\text{cal}}{\text{g} \times \text{°C}} \right) \times \text{mass(g)} \times \Delta T(\text{°C})$$

First, calculate the change in temperature, ΔT:

$$\Delta T = 42.0° = 25.0° = 17.0\,\text{°C}$$

Obtain the value for the specific heat of each substance from Table 1-6. Then substitute this and the other supplied values into the equation:

a. For water: $\text{Heat} = 1.00\,\dfrac{\text{cal}}{\text{g °C}} \times 22.1\,\text{g} \times 17.0\,\text{°C} = 376\,\text{cal}$

b. For paraffin wax: $\text{Heat} = 0.60\,\dfrac{\text{cal}}{\text{g °C}} \times 22.1\,\text{g} \times 17.0\,\text{°C} = 225\,\text{cal}$

c. For copper: $\text{Heat} = 0.093\,\dfrac{\text{cal}}{\text{g °C}} \times 22.1\,\text{g} \times 17.0\,\text{°C} = 34.9\,\text{cal}$

Water requires 376 calories, the greatest amount of heat to raise the temperature 17.0 °C, which is more than the 34.9 calories that it takes to raise the temperature of copper by the same amount.

PRACTICE EXERCISES

40 Use Table 1-6 to calculate the number of calories required to raise 45.0 g of each of the following substances from 23.0 °C to 32.0 °C.

    a. sand   **b.** liquid water   **c.** ethanol

41 An unknown piece of metal weighing 5.3 g could be iron, aluminum, or lead. In going from a temperature of 26 °C to 38 °C, the metal absorbed 14 cal of heat. Use Table 1-6 to determine the identity of the metal.

42 How is the high specific heat of water important to the human body?

43 Two houses are identical in all aspects except that one is made of brick and the other is made of wood. Which house will be cooler on the same hot summer day? Explain.

Chemistry in Medicine Matter and Energy in Malnutrition

The world currently produces enough food to provide each person on the planet with 2,720 Calories of energy per day, which is more than enough for the average person. Yet one-third of the world is well fed, one-third of the world is underfed, and one-third of the world is starving, according to estimates from the World Health Organization (Figure 1-15).

Food, energy, and chemistry are all interconnected. *Energy is defined in science as the ability to do work.* Your body does work as it performs basic involuntary physiological functions, such as breathing, keeping the heart beating, repairing damaged cells, and so forth. Your body obtains energy from the food you eat, which allows this cellular work to occur, as well as tasks such as walking, talking, and studying.

Food is composed of three types of chemical substances known as carbohydrates, proteins, and triglycerides (fats). Digestion of carbohydrates, for example, produces glucose, an important source of energy for the body. Glucose circulating in your blood is called "blood sugar." When glucose enters a cell, several chemical changes occur. Chemical changes that occur in living systems are referred to as metabolism. The metabolism of glucose requires oxygen, which is why you breathe. During the metabolism of glucose, potential energy from glucose is transferred to the cell, where it is used for a variety of energy-demanding functions. The overall chemical reaction is

Glucose + oxygen → carbon dioxide + water + energy

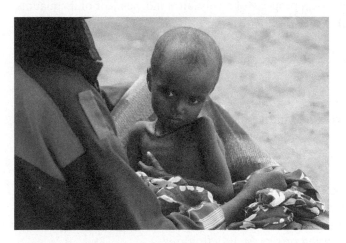

Figure 1-15 A child needs approximately *1,000 to 2,000* calories of energy a day depending on his or her age. Yet every year 15 million children around the world die from starvation.
[The Times/Gallo Images/Getty Images]

In the cell, oxygen comes from the air you breathe and most of the glucose comes from the carbohydrates you eat. Most of the carbon dioxide produced is exhaled, and water either becomes a part of the fluid systems of your body or is eliminated by sweating or through the urine.

The most important product of the reaction of glucose with oxygen, as far as your body is concerned, is energy. Indeed, you probably already know, from reading nutrition labels, that food contains Calories. The Calorie is a unit for measuring energy. When glucose is metabolized, the amount of energy released can be measured in units of Calories. In order to maintain his or her weight, a person must consume enough Calories to obtain the energy that he or she needs for daily living.

Although fewer than 2.5 percent of Americans suffer from malnutrition, health care professionals can and do encounter underfeeding in the United States and Canada. The elderly, the poor, and those suffering from anorexia, bulimia, and some mental illnesses are especially susceptible to malnutrition. Starvation occurs when the body does not take in enough food to provide the Calories it needs to maintain basic involuntary physiological functions. Because glucose is the sole source of energy for the brain, without a steady supply of glucose, hypoglycemic coma will eventually occur. Therefore, when the body is starving, it starts to convert muscle into glucose to provide a steady supply of glucose for the brain. However, the metabolism of muscle eventually damages the heart muscle. Ironically, the process of converting muscle into glucose consumes as much energy as it produces. Thus, there is no net gain in energy, and the body withers away trying to keep itself alive. Starvation for one or more months ultimately results in death.

While starvation remains a world crisis, the number of overweight people in the world has actually overtaken the number of malnourished people. There are 1 billion overweight people compared to 800 million undernourished. A study done at the University of North Carolina found that this transition from a starving world to an obese world is actually accelerating. Nutrition management is indeed an important aspect of health care today.

In this chapter, you have learned how measurement is important in chemistry and how energy and other quantities can be measured. You will see throughout this text that matter, energy, and life are all interconnected.

Summary

Matter and Energy

- Matter exists in three physical states: solid, liquid, and gas.
- Energy is defined as the capacity to do work.
- There are two forms of energy: kinetic energy and potential energy.
- Kinetic energy is the energy of motion.
- Potential energy is stored energy.
- Heat is a form of kinetic energy that is transferred from a hot to a cold object.
- The temperature of a substance reflects the average kinetic energy of the particles of that substance.
- In the gas phase, the particles are far apart and moving rapidly and randomly. In the liquid phase, they are close together and move randomly. In the solid phase, they are close together and are arranged in a regular ordered pattern with only vibrational motion.
- A physical change, such as a change of state, does not affect the composition of a substance, whereas a chemical change (a chemical reaction) alters the composition of a substance.

Measurement in Science and Medicine

- Every measurement has a numerical value and a unit.
- The metric system uses prefixes that represent a multiple of 10 of the base unit: *giga, kilo, deci, centi, milli, micro, nano, pico*.
- In the metric system, the base unit of length is the meter, for mass it is the gram, and for volume it is the liter.
- Energy is a measurable quantity. The units used to describe energy are the joule and the kilocalorie. The Calorie (with a capital C) is equal to a kilocalorie and used in nutritional applications only.
- Density, m/V, is a physical property of a substance independent of its quantity.
- Temperature is a measure of kinetic energy and is measured using a thermometer.
- The three temperature scales are the Celsius, Fahrenheit, and kelvin scales.

Significant Figures and Measurement

- The degree of uncertainty in a measurement is indicated by the number of significant figures reported in the measured value.
- Precision is an indicator of how close repeated measurements are to each other, whereas accuracy is a measure of how close the measurements are to the true value.
- Significant figures are all the nonzero digits and zeros in a measurement that are not placeholders.
- Exact numbers have an infinite number of significant figures because they are obtained by counting or represent definitions.
- Solutions to calculations that require multiplying and dividing measured numbers must be rounded so that there are no more significant figures than in the measured value with the least number of significant figures.
- Solutions to calculations that require addition and subtraction may not have more places past the decimal than the measurement with the fewest decimal places.

Using Dimensional Analysis

- Unit conversions are performed using dimensional analysis, a technique for solving problems based on units. The supplied unit is multiplied by one or more conversion factors that relate the supplied and requested units, allowing the supplied units to cancel algebraically, leaving the requested unit.

- Dosage calculations typically involve an English metric conversion of the patient's weight multiplied by the dosage expressed as a conversion factor between the mass of the medicine and the weight of the patient.

- Density is a physical property of a substance defined as its mass per volume: $d = m/V$. Density can be used as a conversion factor to calculate mass or volume, if density and either mass or volume is known.

- Temperature conversions between Celsius and Fahrenheit and Celsius and kelvin require the use of one of the temperature equations.

Key Words

Absolute temperature scale A temperature scale, like the kelvin scale, where zero represents a temperature in which particles have no kinetic energy.

Accuracy An indicator of how close repeated measurements are to the "true" value.

Atom The smallest intact particle of matter. All matter is made up of atoms.

Atomic scale Matter that is too small to be seen by a light microscope. The scale of atoms.

Base unit In the metric system, a base unit is the standard unit of measurement for a particular quantity, such as the gram for mass, the meter for length, the liter for volume, and the calorie for energy.

calorie The amount of heat energy required to raise the temperature of one gram of water by 1 °C.

Calorie A nutritional calorie, equivalent to 1,000 calories, or 1 kcal.

Change of state The process of going from one state of matter (s, l, g) to another.

Chemical change A process where the composition of the substance changes. Also known as a chemical reaction.

Chemical reaction *See* chemical change.

Conversion A mathematical expression equating two different units.

Conversion factor A mathematical relationship between two units expressed as a ratio or fraction.

Density A physical property of a substance defined as its mass per unit volume: $d = m/V$.

Dimensional analysis A method for converting between units and solving other problems that contain units. Conversion factors are used to cancel units.

Energy The capacity to do work, where work is the act of moving an object.

English system A system of measure used in the United States in nonscientific everyday applications.

Exact number Numbers with an infinite number of significant figures. They are obtained by counting and apply to definitions.

Gas A state of matter that on the macroscopic scale has neither a definite shape nor a definite volume and takes on the shape of its container. Particles in the gas phase are moving rapidly and in random motion, with mainly empty space between them. The kinetic energy of particles is higher in this state than in the other states.

Heat Kinetic energy (molecular motion) that is transferred from a hot object to a cooler object due to a difference in kinetic energy.

Joule A unit of measure for energy.

Kelvin scale The absolute temperature scale that assigns a temperature of zero to the theoretical condition in which molecular motion has stopped.

Kinetic energy The energy of motion; the energy a substance possesses as a result of the motion of its particles.

Kinetic molecular view A model of the arrangement and motions of particles from the perspective of the particles.

Length A measurement of distance. The base unit of length in the metric system is the meter. In the English system, the common units of length are the mile, the yard, the foot, and the inch.

Liquid A state of matter that occupies a definite volume but does not have a definite shape. Particles in the liquid state are interacting and have more kinetic energy than in the solid phase but less than in the gas phase.

Mass A measure of the amount of matter, measured on a calibrated balance or scale. The metric base unit of mass is the gram. The common English units of mass are the pound (lb) and the ounce (oz).

Macroscopic scale Matter that you can see with the naked eye.

Matter Anything that has mass and occupies volume.

Metric prefix When placed before a metric base unit, it signifies a multiple or fraction of 10. For example, *milli* is the multiplier 1,000.

Metric system The system of measurement used in science and medicine and in most of the world.

Microscopic scale Matter that requires a microscope to see.

Phases of matter *See* states of matter.

Physical change A process that does not affect the composition of the substance. Changes of state and dissolving are physical changes.

Potential energy Stored energy. The energy a substance possesses as a result of its position or composition.

Precision An indicator of how close repeated measurements are to each other.

Significant figures The number of certain digits plus one uncertain digit in a measurement. The number of significant figures reflects the degree of uncertainty in a measurement.

Solid A state of matter that has a definite shape and volume independent of its container. Particles in the solid phase are in a lattice with only vibrational motion, the least kinetic energy of the three states of matter.

Specific gravity The ratio of the density of a substance to the density of water at 4 °C. Specific gravity is a unitless quantity.

Specific heat A measure of the amount of heat required to raise the temperature of 1 g of a substance by 1 °C.

State of matter Matter is found in three states: solid, liquid, and gas.

Temperature A measure of the average kinetic energy of a substance. The three temperature scales are Celsius, Fahrenheit, and kelvin.

Unit conversion A type of calculation in which a measurement in one unit is converted into the equivalent value in another unit.

Volume A unit of measure that describes the three-dimensional space occupied by a substance. The metric base unit of volume is the liter. Common English units of volume include the gallon, quart, cup, tablespoon, and teaspoon.

Work The act of moving an object against an opposing force.

Additional Exercises

Osteoporosis and Bone Density

44 Does a strong bone have more or less bone mineral per volume than a weaker bone?

45 What two physical properties of a bone does a bone density scan measure? What physical properties does the bone density scan estimate?

46 A 50-year-old woman had a DEXA scan. Her T-score is −1.3. What is the condition of her bone?

47 A 65-year-old woman had a DEXA scan, and her T-score is +0.9. What is the condition of her bone?

48 A 62-year-old woman stumbles and breaks her hip. Should she be screened for osteoporosis?

49 Does a person with osteoporosis have more or less bone density than a person with osteopenia?

Matter and Energy

50 What are the three physical states of matter?

51 What two states of matter are influenced by the shape of the container?

52 What two states of matter have fixed volumes compared to the container?

53 Is heat, temperature, or work a measure of the average kinetic energy of particles?

54 Is heat, temperature, or work involved in moving an object?

55 Indicate whether each of the following examples is a demonstration of work or heat:
a. feeling the warmth from a fire
b. jumping rope
c. using an ice pack to treat a sprain
d. moving a patient on a gurney

56 Indicate whether each of the following examples is a demonstration of temperature or heat:
a. determining if a patient has a fever
b. preparing a vein for an IV with a hot compress

57 Does a rock sitting at the top of a hill have potential energy or kinetic energy?

58 Does a rock rolling down a hill have potential energy or kinetic energy?

59 Indicate whether each of the following examples is a demonstration of potential energy or kinetic energy:
a. water flowing over a dam
b. a skier standing at the top of a hill
c. water in a reservoir behind a dam
d. atoms bonded together within a crystal lattice

60 Indicate whether each of the following examples is a demonstration of potential energy or kinetic energy:
a. a biker pedaling up a hill
b. a hiker standing at the top of a mountain
c. helium atoms in a balloon
d. the wax in a candle

61 Which molecules with similar mass have more kinetic energy, faster-moving ones or slower-moving ones?

62 Which molecules with similar velocity have more kinetic energy, heavier ones or lighter ones?

63 In which physical state do molecules have the least amount of kinetic energy?
a. solid phase
b. liquid phase
c. gas phase

64 In which physical state do molecules have the greatest amount of kinetic energy?
a. solid phase
b. liquid phase
c. gas phase

65 In which state of matter do water molecules have the most kinetic energy: liquid water, steam, or ice?

66 Indicate whether each of the following examples is a physical change or a chemical change:
a. snow melting
b. toast burning
c. food spoiling
d. mowing the grass

67 Indicate whether each of the following examples is a physical change or a chemical change:
a. melting butter on popcorn
b. bleaching your hair
c. glass breaking
d. burning paper

Measurement in Science and Medicine

68 Classify the size of the following items as macroscopic, microscopic, or on the atomic scale:
a. a hospital   c. DNA
b. a skin cell   d. a red blood cell

69 Classify the size of the following items as being macroscopic, microscopic, or on the atomic scale:
a. a chromium atom   c. a grain of sand
b. the human body   d. a virus

70 Arrange the following metric prefixes in order of increasing size:
a. *nano*   c. *pico*
b. *kilo*   d. *micro*

71 Arrange the following metric prefixes in order of increasing size:
a. *femto*   c. *milli*
b. *centi*   d. *deci*

72 Express the following numbers in scientific notation:
a. 1,000,000,000,000,000,000
b. 2,305,000,000
c. 0.0000000000015
d. 0.0208

73 Express the following numbers in scientific notation:
a. 0.0000076   c. 10,000
b. 0.001   d. 1400

74 Write the following numbers, written in scientific notation, in conventional form:
a. 1×10^5 c. 1.65×10^2
b. 2.4×10^{-3}

75 Write the following numbers, written in scientific notation, in conventional form:
a. 4.2×10^4 c. 1.3×10^{-2}
b. 1.37×10^7

76 Which is the larger number: 4.5×10^{-2} or 4.5×10^2?

77 Which is the smaller number: 6.3×10^{-3} or 6.3×10^3?

78 For each of the following pairs, indicate which is the smaller number. If the numbers are the same, state so.
a. 10^4 or 10^8
b. 10^{-3} or 10^{-6}
c. 3.7×10^4 or 3.7×10^{-4}
d. 0.46 or 4.6×10^{-1}
e. 54,000 or 5.4×10^3

79 For each of the following pairs, indicate which is the smaller number. If the numbers are the same, state so.
a. 10^3 or 1.5×10^3
b. 10^2 or 10^3
c. 2.5×10^{-3} or 2.5×10^3
d. 800,000 or 8×10^5
e. 0.00078 or 7.8×10^{-3}

80 For each of the following pairs, which length is shorter? If the lengths are the same, state so.
a. 1 m or 10 mm   c. 1 cm or 1 dm
b. 1 cm or 10 mm   d. 15 cm or 1 dm

81 For each of the following pairs, which length is shorter? If they are the same, state so.
a. 10 m or 1 km   c. 10^{-3} m or 1 mm
b. 1 nm or 10^{-9} m   d. 1 nm or 1 μm

82 For each of the following pairs, state which measurement represents the smaller mass. If they have the same mass, state so.
a. 1 ng or 1 mg
b. 100 mg or 1 g
c. 1,000 mg or 1 g
d. 50 mcg or 100 μg

83 For each of the following pairs, state which measurement represents the smaller mass. If they have the same mass, state so.
a. 1 g or 1 kg   c. 1 g or 1×10^{-3} kg
b. 1 μg or 1 kg   d. 10 μg or 100 mcg

84 Which of the following is equivalent to 1 m?
a. 10 cm   c. 1,000 mm
b. 100 dm   d. 10^{12} μm

85 Which of the following is equivalent to 150 μg?
a. 1.5×10^{-5} g
b. 1.5×10^{-2} mg
c. 1.5×10^{-7} kg
d. 150 g

86 A lead ball is dropped into a graduated cylinder containing 15.0 mL of water, causing the level of the water to rise to 16.5 mL. What is the volume of the lead ball?

87 An irregular-shaped metal object is placed in a graduated cylinder containing 200. mL of water. The water level increases to 203.5 mL. What is the volume of the metal sample?

88 Which is the larger volume? If they are the same, state so.
a. 5 cm^3 or 5 mL
b. 500 mL or 0.5 L
c. 25 mL or 250 cm^3
d. 0.5 L or 500 cm^3

89 Which is the larger volume? If they are the same, state so.
a. 0.25 cm^3 or 25 μL
b. 80 mL or 80 L
c. 0.75 mL or 0.75 cm^3
d. 0.05 μL or 5×10^{-3} cm^3

90 What is the volume of a cube measuring 24 cm a side?

91 What is the length of the side of a cube with a volume of 8 cm^3?

92 How many "small *c*" calories are found in one "capital *C*" Calorie? Are "small *c*" calories or "capital *C*" calories normally reported on nutritional labels for foods?

93 List two of the most common units of energy in science.

Significant Figures and Measurement

94 Answer TRUE or FALSE:
a. _____ Only sloppy work leads to uncertainty in a measurement.
b. _____ A balance must be calibrated in order to give accurate readings.
c. _____ In experimental work, if a series of repeated measurements gives values very close to one another, they are considered precise.
d. _____ In experimental work, if a series of repeated measurements gives values very close to the "true value," they are considered accurate.

95 Answer TRUE or FALSE:
a. _____ Environmental factors can affect a measurement.
b. _____ Measurements do not contain uncertainty.
c. _____ In experimental work, if a series of repeated measurements gives values close to one another and close to the "true value," they are considered precise and accurate.
d. _____ In experimental work, if a series of repeated measurements gives values not close to one another but close to the "true value," they are considered precise but not accurate.

96 A nurse is testing out a new balance. She weighs a newborn baby who weighs 3.5 kg four different times on the new balance and obtains the following weights: 2.83 kg, 2.83 kg, 2.81 kg, and 2.82 kg. Is the new balance accurate? Is it precise?

97 A chef weighs a 2,272-g bag of flour on a balance four times and obtains the following weights: 2,372 g, 2,072 g, 2,172 g, 2,472 g. Is the balance accurate? Is it precise?

98 Indicate the number of significant figures in each of the following measurements:
  a. 57,000
  b. 4.60
  c. 0.00011
  d. 23,304.60

99 Indicate the number of significant figures in each of the following measurements:
  a. 304
  b. 5,000
  c. 5,110
  d. 0.000330

100 Which of the following contain exact numbers and therefore contain an infinite number of significant figures?
  a. 2.54 cm = 1 in
  b. a room with an area measuring 360 cm × 256 cm
  c. 55 students in a classroom

101 Which of the following represent exact numbers and therefore contain an infinite number of significant figures?
  a. 1 mm = 10^{-3} m
  b. a mass of 56.7 kg
  c. a density of 1.2 g/mL

102 Round the following numbers to 3 significant figures:
  a. 2.306
  b. 9,312
  c. 1.555

103 Round the following numbers to 2 significant figures:
  a. 1.7777
  b. 4.25
  c. 28.1

104 Perform the following calculation: $\dfrac{3.27}{5.2}$. Assume these are measured values. The answer is
  a. 0.6288
  b. 0.629
  c. 0.63
  d. 0.6

105 Perform the following calculation 124.893 − 45.01. Assume these are measured values. The answer is
  a. 79.9
  b. 79.88
  c. 79.883
  d. 79

106 Perform the following calculations. Assume these are measured values.
  a. 3.2 × 8.54 =
  b. 3.2 + 8.54 =
  How many significant figures were in the answer to question **a**? How many significant figures were in the answer to question **b**?

107 Perform the following calculations:
  a. 2.26 + 8.1 =   b. 2.26 × 8.1 =
  How many significant figures were in the answer to question **a**? How many significant figures were in the answer to question **b**?

108 Perform the following calculations. Show the correct number of significant figures in your answer, assuming each number is a measured value. Include units in your answer.
  a. 56.33 cm × 2.5 cm =
  b. 3.4 cm + 2.2 cm + 5.11 cm + 8.777 cm =
  c. $\dfrac{34.22 \text{ g}}{39.0 \text{ mL}} =$

109 Perform the following calculations. Show the correct number of significant figures in your answer, assuming each number is a measured value.
  a. 33,000. + 910. =
  b. 0.333 × 0.22 =
  c. (37.55 mL + 22.2 mL) × 56.66 =

Using Dimensional Analysis

110 Carry out the following metric conversions. Remember to report your final answer to the correct number of significant figures.
  a. Convert 50,000. m into kilometers.
  b. How many micrograms are there in 0.66 g?

111 Carry out the following metric conversions. Remember to report your final answer to the correct number of significant figures.
  a. How many milliliters are there in 6.000 L?
  b. How many kilometers are there in 2.0×10^6 centimeters?

112 What conversion factors would you use to convert 61,000. mm into picometers?

113 What conversion factor would you use to convert 4,000 m into kilometers?

114 How many milliliters are there in 75.6 μL?

115 Ibuprofen can be found in 200.-mg doses in over-the-counter analgesics such as Advil and Motrin. How many grams of ibuprofen does such an Advil tablet contain?

116 Convert 500. mg of vitamin C into grams and micrograms (μg).

117 How many seconds are there in 2.000 minutes?

118 Using dimensional analysis, calculate the number of seconds in 5.2 years. Express the final answer in scientific notation.

119 What is the mass in pounds of an animal with a mass of 150. kg?

120 There are 5,280 ft in a mile. How many kilometers are there in 68.2 miles?

121 A doctor must make an incision 2.5 cm long. How long will the incision be in meters?

122 There are 4 quarts in a gallon. How many liters are there in 86 gal?

123 An ultrasound tech measures the humerus bone in an 18-week-old fetus. The bone is 26.7 mm long. How long is the bone in inches?
  a. 6.78×10^{-3} in   c. 1.05×10^3 in
  b. 1.05 in

124 A phlebotomist draws two tubes of blood with a volume of 14.3 cm^3 to run an HIV test. How many liters of blood are in the sample?
   a. 14.3 L
   b. 1.43 L
   c. 1.43 × 10^3 L
   d. 1.43 × 10^{-2} L

125 There are 8 oz in a cup and 4 cups in a quart and 4 quarts in a gallon. How many milliliters are there in 2 oz?

126 How many calories are there in
   a. 0.234 Cal
   b. 0.0991 kcal
   c. 20.7 kcal
   d. 352 Cal

127 How many calories are there in
   a. 4.14 J
   b. 36.2 kJ
   c. 0.0587 kJ
   d. 367 J

128 Rank from largest to smallest volume: 50.00 mL; 5,000 μL; 0.5000 L; 8.000 cm^3.

129 The recommended dose of Ceclor, an antibiotic, is 20 mg/kg a day, in three equally divided doses, one given every 8 hours. How many milligrams should a baby weighing 12 lb receive at each administration?

130 Tylenol is ordered for a child with a fever at a dose of 25 mg/kg a day. If the child weighs 34 lb, how much Tylenol should be given to the child every day?

131 Which object in the following pairs has the greater density? Explain why.
   a. a loaf of bread or a brick
   b. a bowling ball or a soccer ball
   c. a bucket full of water or a bucket full of concrete
   d. a bone from a woman with osteoporosis or a normal bone
   e. a *high-density lipoprotein* (HDL) circulating in the blood or a *low-density lipoprotein* (LDL) circulating in the blood?

132 A sample of muscle tissue has volume of 8.7 mL and a mass of 9.22 g. What is the density of the sample?

133 A sample of compact bone has a mass of 3.8 g and a volume of 2.0 cm^3. What is the density of the sample?

134 What substance has a density of 1.0 g/mL?

135 If a liquid has a density greater than 1.0 g/mL, will it float or sink in water?

136 What is the mass of a gold sphere that displaces 2.3 mL of water? (Use Table 1-5.)

137 Using Table 1-4, calculate the mass of a gold cube having sides 2.20 cm in length.

138 Would a gold sphere with a mass of 15 g have a greater diameter or a smaller diameter than a gold sphere with a mass of 6 g? (*Hint:* No calculation required.)

139 The density of a patient's urine is 1.025 g/mL. What is the specific gravity of the urine sample? Is the specific gravity within the normal range?

140 A patient provides a urine sample. The density of the patient's urine is 1.037 g/mL. What is the specific gravity of the urine sample? Is the specific gravity within the normal range?

141 A patient provides a urine sample. The specific gravity of the patient's urine is 1.014. What is the density of the urine sample? Is the specific gravity within the normal range?

142 A patient provides a urine sample. The specific gravity of the patient's urine is 0.997. What is the density of the patient's urine? Is the specific gravity within the normal range?

143 What is the freezing point of water in Celsius and Fahrenheit?

144 What is normal body temperature in Celsius and Fahrenheit?

145 Benzamycin is a topical antibiotic gel. The gel needs to be stored between 2 °C and 8 °C. Where is this gel stored?
   a. in the freezer
   b. in the refrigerator
   c. in a cabinet

146 Accutane is used to treat acne. These gelatin capsules need to be stored between 15 °C and 30 °C. Where should these capsules be stored?
   a. in the freezer
   b. in the refrigerator
   c. in a cabinet

147 While traveling in Europe, you notice that a thermometer reads 20. °C. Are you wearing winter clothes (coat, hat, scarf, etc.) or summer clothes? What is this temperature in kelvins and in degrees Fahrenheit?

148 The thermometer in Christchurch, New Zealand, reads 31 °C. Are you wearing winter clothes (coat, hat, scarf, etc.) or summer clothes? What is this temperature in degrees Fahrenheit and kelvins?

149 The thermometer in Tokyo, Japan, reads 11 °C. Are you wearing summer clothes or winter clothes? What is this temperature in degrees Fahrenheit and kelvins?

150 A patient has moderate hypothermia. His body temperature is 87.4 °F. What is this temperature reading in degrees Celsius?

151 While on vacation in Europe, you took your temperature and it was 38 °C. Do you have a fever?

152 The temperature in outer space is 2.7 K. What is this temperature in degrees Celsius and in degrees Fahrenheit?

153 Medical couriers use dry ice to keep perishable medical materials cold while they are transported. Dry ice has a temperature of −78 °C. What is this temperature in kelvins and degrees Fahrenheit?

154 Liquid nitrogen is often used as a treatment to remove warts by freezing them off. Liquid nitrogen has a temperature of 77 K. What is this temperature in degrees Celsius and degrees Fahrenheit?

155 Using Table 1-6, calculate in calories the amount of heat that must be added to warm 13.4 g of each of the substances from 29 °C to 51 °C. Which substance requires the greatest input of heat energy and why?
   a. water
   b. copper
   c. sand

156 Using Table 1-6, calculate in calories the amount of heat that must be added to warm 10.9 g of each of the substances from 22 °C to 49 °C. Which substance requires the greatest input of heat energy and why?
   a. brick
   b. ethanol
   c. wood

157 It takes 412 cal to raise the temperature of a sample of water from 30 °C to 45 °C. What is the mass of the sample in grams?

158 It takes 504 cal to raise the temperature of a sample of ethanol from 16 °C to 43 °C. What is the mass of the sample in grams?

159 It takes 137 cal to raise the temperature of 56.8 g of iron. How much did the temperature of iron change in °C?

160 It takes 2,820 cal to raise the temperature of 127.2 g of water. How much did the temperature of water change in °C?

161 A metal object with a mass of 23 g is heated to 97 °C and then transferred to an insulated container containing 81 g of water at 19 °C. The water temperature rises and the temperature of the metal object falls until they both reach a final temperature of 24 °C. What is the specific heat of this metal object? *Hint:* Assume that the heat lost by the metal object is equal to the heat gained by the water.

Chemistry in Medicine: Matter and Energy in Malnutrition

162 What are the three types of substances in our food that provide us with energy?

163 What substance is the most important source of energy for the body? What type of substance is it?

164 What two products are produced when glucose is metabolized in the presence of oxygen?

165 In order for a person to maintain his or her weight, how many Calories of food do they need to consume compared to the amount of energy they expend in daily living?

166 When the body is starving, how does it produce energy?

167 Is glucose a form of potential energy or kinetic energy?

168 What is energy?

Answers to Practice Exercises

1 The solid state has a definite volume and shape independent of its container. The liquid state has a definite volume but no definite shape and conforms to the shape of its container. The gas state has no definite volume or shape. It conforms to the shape and volume of its container.

2 a. potential energy
   b. kinetic energy
   c. kinetic energy
   d. kinetic energy
   e. potential energy

3 a. chemical change
   b. physical change
   c. chemical change
   d. physical change

4 There are 1×10^{12} pm in 1 m. There are 1×10^{12} pg in 1 g. The prefix *pico* is the multiplier 1×10^{12}.

5 The length of each side of the cube is 3.0 cm because the volume is 3.0 cm $\times$ 3.0 cm $\times$ 3.0 cm = 27 cm^3.

6 a. 1 mm is larger than 1 μm.
   b. 10^{-3} kg is the same as 1 g.
   c. 1 Calorie and 1,000 calories are the same amount of heat energy.
   d. 1 Calorie is larger than 1 cal.

7 a. 10^3 mm = 1 m
   b. 10 dm = 1 m
   c. 10^{-3} km = 1 m

8 a. 1 km and b. 1 m can be seen with the naked eye.

9 a. 1 mg
   b. 1 μg
   c. 10 mcg is the same as 10 μg.

10 11 mL − 10 mL = 1 mL = 1 cm^3

11 a. One; the zeros are placeholders.
   b. Infinite. It is an exact number because it is a value obtained by counting.
   c. 5; the zeros are significant because there is a decimal following the zeros.
   d. 4; the zeros before the 4 are placeholders, whereas those after are significant.

12 a. iii
   b. The last digit in each number is uncertain.
   c. ii

13 a. 2,146
   b. 15.222
   c. 3,167

14 65 cm

15 7,700 g or 7.7×10^3 g

16 0.561 L

17 56,000 pm or 5.6×10^4 pm

18 5.5×10^{-9} m; 5.5×10^{-6} mm

19 The patient with the 2.00-mm tumor has the larger tumor.

20 177 calories

21 57 L

22 2.00 lb

23 a. 5,790 cal or 5.79×10^3 cal
   b. 11.7 cal

24 10.3 J

25 1.14×10^7 J

26 $\dfrac{25 \text{ mg}}{1 \text{ kg}}$
   1247 mg daily

27 143 mg every 6 hours (4 times a day)

28 1.1 g/mL

29 **a.** a bowling ball   **b.** the same   **c.** water

30 smaller diameter

31 0.840 g/mL

32 25 g (The density of water is 1.0 g/mL.)

33 8.2 mL of water would be displaced.

34 The unknown metal has a density of 7.9 g/cm^3, which corresponds to the density of iron, so the unknown metal is iron.

35 Someone with osteoporosis has lost bone mass; therefore, their bone density (m/V) would be less than normal.

36 For ice to float on water, it must be less dense than liquid water.

37 98.6 °F.   310.2 K.

38 314 K (First convert from °F to °C and then from °C to K.)

39 −269 °C

   −452 °F

40 **a.** sand: 65 cal

   b. water: 400 cal

   c. ethanol: 230 cal

41 The metal is aluminum with a specific heat of 0.22 cal/g · °C.

42 Because the human body is composed of mainly water, the high specific heat of water means that a significant amount of energy is required to raise or lower the temperature of water, which ensures that the body maintains a constant temperature.

43 The brick house will be cooler on a hot summer day because, according to Table 1–6, brick has a higher specific heat than wood and therefore requires more heat to increase its temperature.

Scanning electron micrograph (SEM) of human red blood cells. Red blood cells carry oxygen to tissues throughout the body. The red color is due to the iron atoms in hemoglobin, which bind oxygen. [Kenneth Eward/Science Source]

2 | Atomic Structure and Nuclear Radiation

Anemia and Iron

An iron deficiency is the most common nutritional deficiency in the world. When a person's iron reserves become depleted, a condition known as iron deficiency anemia (IDA) develops. Anemia is a general term for any condition in which there is a shortage of red blood cells. People with IDA experience fatigue as well as some of the other symptoms listed in Table 2-1. IDA can also slow motor and mental development in children.

Every cell in the body requires iron, especially red blood cells. Red blood cells serve the important function of transporting oxygen to cells throughout the body. The characteristic biconcave shape of a red blood cell gives it more surface area, maximizing its ability to bind oxygen.

Red blood cells constitute 40 percent of the volume of the blood. A single red blood cell contains about 250 million hemoglobin molecules, the oxygen-carrying protein molecules in these cells. A model of a hemoglobin molecule is shown in Figure 2-1. A molecule is made up of two or more *atoms* held together by chemical bonds, and atoms are the smallest building blocks of all matter. Hemoglobin is a large, complex molecule composed of 10,000 atoms, and four of these atoms are iron (Fe) atoms that bind to oxygen molecules. It is the iron atoms in hemoglobin that give red blood cells their distinctive red color. The other atoms in a hemoglobin molecule are carbon (C), hydrogen (H), oxygen (O), nitrogen (N), and sulfur (S). In this chapter you will learn about *atoms* and in the next chapter you will learn how *atoms* bond together to form *molecules*. In chapter 13 you will learn about proteins, where you will revisit the hemoglobin molecule in greater detail.

Every cell in the body requires oxygen to produce energy, and this is why we must continuously breathe air into our lungs. Air is a mixture of mostly oxygen (O_2) and nitrogen (N_2). Oxygen is a molecule composed of only two oxygen atoms—O_2, so it is a much smaller molecule than hemoglobin. Each of the four iron atoms in a hemoglobin molecule will bind one oxygen molecule, carrying it through the bloodstream and delivering it to cells, as illustrated in Figure 2-2.

IDA can be caused by poor absorption of iron, lack of iron in the diet, chronic blood loss, or pregnancy. Also, some medications reduce the body's ability to absorb iron. When the body's supply of iron is depleted, as in IDA, hemoglobin production is reduced, so fewer red blood cells are made, and tissues become starved of oxygen. Not surprisingly, fatigue is one of the symptoms of anemia.

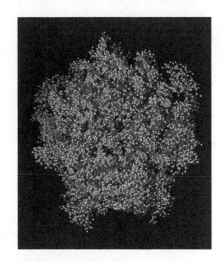

Figure 2-1 A ball-and-stick model of hemoglobin, a large molecule composed of many atoms: Most of the atoms in hemoglobin are carbon (C), hydrogen (H), oxygen (O), and nitrogen (N), with some sulfur (S). The four iron (Fe) atoms bind molecular oxygen (O_2). [Iculig/Shutterstock]

TABLE 2-1 Symptoms of Iron Deficiency Anemia	
Inflammation of the tongue (glossitis)	Shortness of breath
Brittle nails	Pale skin
Frequent headaches	Taste disturbances
Lethargy	Dizziness

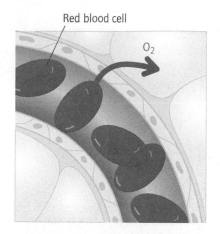

Red blood cell

O₂

Figure 2-2 Red blood cells travel through the circulatory system delivering oxygen to cells.

The binding of oxygen to iron in hemoglobin is an example of how atoms and molecules interact—chemistry. You will continue to see the central role chemistry plays in our understanding of health and disease in the coming chapters as you build your knowledge of chemistry. ●

All matter is composed of atoms, including the lifesaving drug given to a patient, the food you eat, and the hemoglobin in your blood cells. Matter is composed of *elements* and *compounds*. An **element** is a substance composed of only *one* type of atom. Elements cannot be broken down into anything simpler by chemical means. Thus, nickel is an element composed of only nickel atoms, and gold is an element composed of only gold atoms.

A **compound** is a substance composed of *two* or *more* different atoms held together by chemical bonds. A compound *can* be broken down into its elements by chemical means. For example, water (H_2O) is a compound composed of molecules containing *two* hydrogen atoms and *one* oxygen atom. Water can be broken down into its elements, oxygen (O_2) and hydrogen (H_2), by a chemical reaction. In this chapter you will learn about the different types of atoms that make up the elements. In the next chapter, chapter 3, you will learn how atoms come together to form compounds. In chapter 4 you will learn how elements and compounds are changed in a chemical reaction.

■ 2.1 Elements and the Structure of the Atom

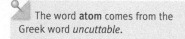
The word **atom** comes from the Greek word *uncuttable*.

We as a society have applied our knowledge of the atom to build nuclear weapons, to generate nuclear power, and to treat and diagnose disease (nuclear medicine).

It was the Greek philosophers who first hypothesized that matter could not be cut into smaller and smaller pieces indefinitely, that a point would be reached when you could not cut any further. We now know this point to be the **atom**, the smallest stable component of all matter. Thus, if you were to take a sample of graphite, such as that found on the tip of your pencil, and cut it into smaller and smaller pieces, the smallest indivisible piece of graphite would be a single carbon atom, with the incredibly small diameter of 140 pm (140×10^{-12} m), shown in the colored scanning tunneling micrograph (STM) image in Figure 2-3. The graphite we see on a pencil—the macroscopic—is made up of an enormous number of carbon atoms, the atomic scale.

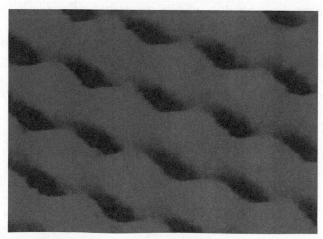

Figure 2-3 The tip of a pencil is composed of an enormous number of carbon atoms as shown in the colored scanning tunneling micrograph (STM) of graphite (carbon). [Left: Ron Chapple Studios/iStockphoto/Thinkstock. Right: Colin Cuthbert/Science Source]

TABLE 2-2 Charge, Mass, and Location of the Subatomic Particles in an Atom

Subatomic Particle	Charge	Mass	Location within the Atom
Proton	+1	1 amu	Nucleus
Neutron	0	1 amu	Nucleus
Electron	−1	0.00055 amu	Electron orbitals

The Parts of an Atom

All atoms contain

- **protons,**
- **neutrons,** and
- **electrons**

Each of these *subatomic particles* is characterized by its *charge, mass,* and *location* within the atom, as summarized in Table 2-2. A proton has a positive charge, +1; an electron has a negative charge, −1; and a neutron has no charge, 0. An atom contains an equal number of protons and electrons; therefore, atoms are electrically neutral, with a net charge of zero.

Protons and neutrons are approximately equal in mass. Since their mass is so small, a unit known as the **atomic mass unit** (amu) is used. The conversion between the amu and the metric unit for mass, the gram, is

$$1 \text{ amu} = 1.66 \times 10^{-24} \text{ g}$$

Thus, the mass of a proton and the mass of a neutron is approximately 1 amu. By comparison, the electron is so much lighter that we can ignore its contribution to the mass of an atom, and so the mass of an atom depends on the number of protons and neutrons that it contains. Atoms with a greater number of protons and neutrons have a greater mass (they are heavier) than atoms with fewer neutrons and protons.

The Structure of the Atom

Discovering the structure of the atom was one of the greatest scientific discoveries of the early twentieth century. Experiments showed that protons and neutrons are concentrated in the center of the atom, known as the **nucleus,** which has a diameter of approximately 1×10^{-15} m. The nucleus of the atom is extremely dense since all its mass is concentrated in this incredibly small volume. In contrast, electrons, with their significantly smaller mass, are found in a much larger volume of space around the nucleus, defining the volume of the atom. Electrons are often represented as an electron cloud, a type of cross section of the atom, as shown in Figure 2-4 for a hydrogen atom. An electron cloud is like a time-lapse photograph of an atom's electrons, showing their positions over time. The darker regions represent areas where the probability of finding the electron is greater.

The diameter of an atom ranges from 74 to 300 pm. To appreciate the size of the nucleus compared to the entire atom, imagine scaling up the nucleus of the atom to the size of a football. The size of the atom would scale up to the size of the football stadium (Figure 2-5). Most of the volume of the atom is composed of empty space. While there is no single model that perfectly represents the atom, Figure 2-6 shows the protons and neutrons concentrated in the nucleus surrounded by an electron cloud that occupies a much larger volume of space.

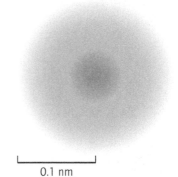

0.1 nm

Figure 2-4 The electron cloud for a hydrogen atom. The darker areas show where the electron is more likely to be found. The radius of the atom is defined by the electron cloud, about 0.1 nm for a hydrogen atom.

Figure 2-5 The Georgia Dome. A football is to a football stadium as the nucleus of an atom is to the atom.
[© John Korduner/Icon SMI/Corbis]

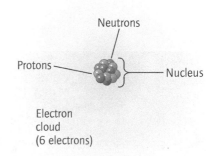

Neutrons

Protons

Nucleus

Electron cloud (6 electrons)

Figure 2-6 Parts of an atom: protons and neutrons are in a small volume at the center of the atom, known as the nucleus, and the electrons are in a much greater volume of space surrounding the nucleus.

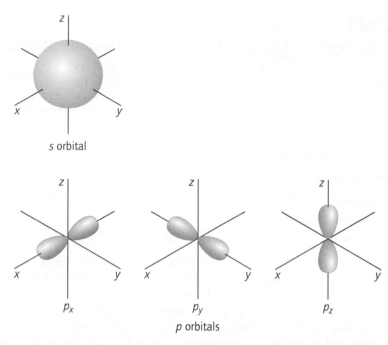

Figure 2-7 The shapes of some common electron orbitals: an *s* orbital and the three perpendicular p_x, p_y, and p_z orbitals. The *s* orbital has a spherical shape. Each *p* orbital has the same dumbbell shape, just oriented perpendicular to each other.

We know that electrons do not orbit the nucleus, as scientists initially hypothesized. Although the *path* traveled by the fast-moving electron cannot be determined, the *probability* of finding the electron in a particular volume can be described and is known as an **electron orbital**. Figure 2-7 illustrates the shapes of some electron orbitals, such as the spherically shaped *s orbital* and the three perpendicular dumbbell-shaped *p orbitals*. Electrons play a central role in the physical and chemical properties of elements, which is described in greater detail in section 2.3.

Elements and the Periodic Table

*An **element** is defined by the number of protons that its atoms contain, known as the **atomic number** of the element.* For example, the element *helium* is composed of helium atoms, which all contain *two* protons, so the atomic number for helium is 2. The element iron is composed of iron atoms containing 26 protons, so *iron* has atomic number 26. To date, 114 different elements have been discovered and named, some of which were prepared artificially and existed only briefly in a laboratory experiment. All but two of the elements with atomic numbers 1 through 92 are found naturally on the earth. You will no doubt be familiar with some elements, whereas others will be new to you.

Every element has a one- or two-letter **atomic symbol**, in which the first letter is capitalized. For example, the atomic symbol for helium is He. Element symbols are abbreviations derived from the English or Latin name for the element. Latin names account for the unexpected element symbols such as iron, Fe, and gold, Au (Figure 2-8). The atomic symbol for every element, its atomic number, and its atomic mass are given in the **periodic table of the elements**, shown in Figure 2-9. For example, the element gold, highlighted in Figure 2-9, and photographed in Figure 2-8, is shown on the next page with its atomic symbol (Au) atomic number (79), and average atomic mass (197.0).

Figure 2-8 Gold (Au), the element with atomic number 79. [Stockbyte/iStockphoto/Thinkstock]

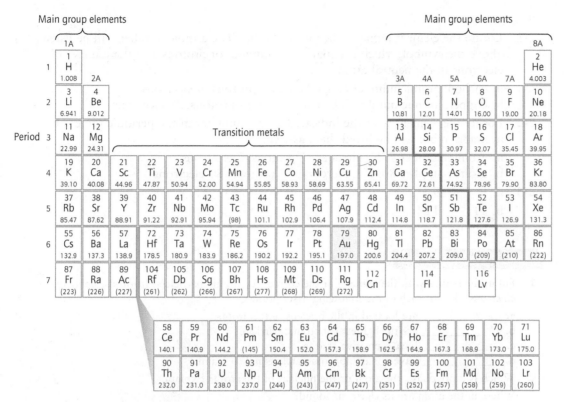

Figure 2-9 The periodic table of the elements shows the atomic symbol, atomic number, and average atomic mass of each element.

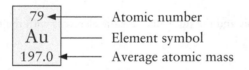

79	◀──	Atomic number
Au	───	Element symbol
197.0	◀──	Average atomic mass

The periodic table of the elements displays all the elements sequentially by atomic number, from lowest atomic number to highest atomic number, from left to right and from top to bottom, across the table. The periodic table on the inside front cover of the text also includes the full name of each element. You will make frequent reference to this table throughout this text.

WORKED EXERCISES   Using the Periodic Table to Obtain Information about an Element

2-1   Using the periodic table on the inside cover, which includes the full name of each element, state the atomic number, element symbol, and the number of protons for each of the following elements:

a. carbon   b. silver

2-2   Provide the element symbol, full name, number of protons, and number of electrons for elements with the following atomic numbers?

a. atomic number 7   b. atomic number 80

2-3   How many protons and electrons do the following elements contain?

	number of protons	number of electrons
a. fluorine	_____	_____
b. sodium	_____	_____

Solutions

2-1   Locate the element name in the periodic table. The atomic number will be listed above the symbol, which is equal to the number of protons and the number of electrons in the neutral atom.

   a. carbon: atomic number 6; symbol, C; 6 protons, 6 electrons
   b. copper: atomic number 29; symbol, Cu; 29 protons, 29 electrons

2-2   Locate the element with the indicated atomic number in the periodic table. The atomic symbol will be listed there as well.

   a. atomic number 7: N, nitrogen; 7 protons, 7 electrons
   b. atomic number 80: Hg, mercury; 80 protons, 80 electrons

2-3   a. 9   9
      b. 11   11

PRACTICE EXERCISES

1   Fill in the blanks with the name(s) of the correct subatomic particle(s) (*protons*, *neutrons*, *electrons*) that fit each of the following descriptions:

   a. _____ are located in the nucleus of the atom.
   b. _____ are the lightest subatomic particle.
   c. _____ have a positive charge.
   d. _____ have a mass approximately equal to 1 amu.

2   Which is more dense, the atom or the nucleus of the atom? Explain.

3   Where in the atom are its electrons found?

4   Calcium is a major constituent of bone and teeth. The atomic number for calcium is 20. How many protons and electrons does a calcium atom contain?

5   What is the atomic number for each of the following elements?

   a. krypton   b. silver

6   What are the element symbols and the full names for the elements with the following atomic numbers?

   a. 10   b. 8

7   Which element has the fewest number of protons? What is its atomic number and element symbol?

8   Vanadium is found in some mineral supplements. How many protons does vanadium contain?

9   Which of the following substances are *not* elements? How can you tell?

   a. copper   b. bronze   c. gold   d. silver
   e. steel   f. uranium   g. tin   h. tungsten

Isotopes of an Element

Although the number of protons is the same, the number of neutrons varies among the atoms of a particular element. *Atoms with the same number of protons, but a different number of neutrons, are known as* **isotopes**. For example, there are three naturally occurring isotopes of carbon. All three isotopes of carbon have 6 protons, by definition (atomic number 6), but they differ in the number of neutrons they contain: 6, 7, or 8 neutrons, as shown in Table 2-3.

TABLE 2-3 The Naturally Occurring Isotopes of Carbon

Isotope	Atomic Number	Number of Protons	Number of Neutrons	Mass Number
Carbon-12	6	6	6	12
Carbon-13	6	6	7	13
Carbon-14	6	6	8	14

Since neutrons are neutral particles, the number of neutrons does not affect the net charge of an isotope, but it does affect the mass of the isotope.

Most elements have more than one naturally occurring isotope, and additional isotopes may be prepared artificially. For example, silicon (Si) has three naturally occurring isotopes and 17 man-made isotopes. An isotope is identified by its **mass number**, which is the sum of its protons and neutrons:

Mass number = number of protons + number of neutrons

As you can see from the last column in Table 2-3, the three naturally occurring isotopes of carbon have mass numbers 12, 13, and 14, which means they have 6, 7, and 8 neutrons, respectively.

The nuclear symbol for an isotope is written with the mass number as a *super*script to the left of the element symbol and the atomic number written as a *sub*script to the left of the element symbol, as shown for iodine-125, an isotope used to treat prostate cancer and brain tumors:

Mass number: $^{125}_{53}I$
Atomic number:

mass number = number of protons + number of neutrons
atomic number = number of protons

Alternatively, the mass number can be written following the full name or symbol of the element, separated by a hyphen, as shown in Table 2-4 for the three isotopes of carbon.

TABLE 2-4 Ways to Write the Isotopes of Carbon

Isotope Symbol	Alternative Representations
$^{12}_{6}C$	C-12 or carbon-12
$^{13}_{6}C$	C-13 or carbon-13
$^{14}_{6}C$	C-14 or carbon-14

WORKED EXERCISES   Isotopes and Mass Number

2-4   How many neutrons are there in strontium-90, a radioactive isotope with atomic number 38?

2-5   What are the three acceptable ways to represent an isotope with 17 protons and 20 neutrons?

Solutions

2-4   Calculate the number of neutrons in an isotope by subtracting the number of protons (the atomic number) from the mass number:

number of neutrons = mass number − atomic number
number of neutrons = 90 − 38 = 52 neutrons

2-5   Chlorine-37, or Cl-37, or $^{37}_{17}Cl$. The sum of the number of protons and neutrons is equal to the mass number: 17 + 20 = 37.

PRACTICE EXERCISES

10   Fill in the blanks in the table below:

Isotope	Mass Number	Atomic Number	Number of Protons	Number of Neutrons
Chlorine-35				
		80		117
	125		53	

11 Silicon has three naturally occurring isotopes: silicon-28, silicon-29, and silicon-30. Complete the table below:

Isotope	Mass Number	Atomic Number	Number of Protons	Number of Neutrons
Silicon-28				
Silicon-29				
Silicon-30				

    a. What do all silicon isotopes have in common?
    b. How are the isotopes of silicon different?
    c. Which isotope of silicon has the greatest mass?

12 Only one stable isotope of gold exists, and it contains 116 neutrons and 79 protons. Write the symbol for this isotope with the atomic number as a subscript and the mass number as a superscript to the left of the symbol. What are two alternate ways of writing this isotope?

Average Atomic Mass

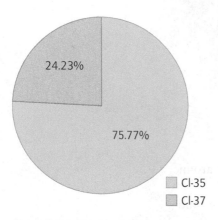

Figure 2-10 The pie chart shows the natural abundance of chlorine isotopes on earth.

We have seen that the mass of an atom depends on the number of protons and neutrons that it contains; therefore, isotopes with more neutrons will have a greater mass. For example, a chlorine-35 isotope has an approximate mass of 35 amu because it contains 17 protons and 18 neutrons, each with a mass of approximately 1 amu. Chlorine-37 has two more neutrons than chlorine-35, so its mass is approximately 37 amu. The pie chart in **Figure 2-10** shows the natural abundance of chlorine on earth: 75 percent chlorine-35 (shown in blue) and 25 percent chlorine-37 (shown in orange). Therefore, the *average* atomic mass of chlorine is 35.45 amu—a value that represents a *weighted average* of Cl-35 and Cl-37. In other words, chlorine-35 makes a greater contribution to the average mass than chlorine-37 because there are three times as many chlorine-35 atoms as there are chlorine-37 atoms in a naturally occurring sample of chlorine (Cl). *The average atomic mass of an element is a weighted average of the mass of all its isotopes based on their natural abundance.* The term *average atomic mass* is often simply referred to as *atomic mass*. The atomic mass of each element is given along with its element symbol and atomic number in the periodic table.

PRACTICE EXERCISE

13 Calcium has four naturally occurring isotopes: calcium-40, calcium-42, calcium-43, and calcium-44. The following table shows the natural abundance of each of these isotopes as a percentage of the total.

Calcium Isotope	Natural Abundance (%)
Calcium-40	96.95
Calcium-42	0.76
Calcium-43	0.19
Calcium-44	2.09

    a. Which calcium has the greatest mass?
    b. Which calcium isotope is the most abundant on earth?
    c. Which isotope has the greatest number of neutrons?
    d. How many more neutrons does calcium-44 have than calcium-40?
    e. What are the typical units for average atomic mass?

The average atomic mass of carbon is 12.007 although no single carbon atom has this mass, just as the average number of children per couple in the United States is 2.5, even though there is no such thing as a couple with half a child! It's just the nature of averages.

2.2 Navigating the Periodic Table of the Elements

You have seen that the periodic table of the elements shows all the elements, but it is also organized in a unique array of 18 columns and seven rows, which conveys additional information about patterns observed among the elements known as **periodicity**; hence the name *periodic* table. In this section we will consider the significance of the organization of the periodic table and learn how to navigate the periodic table and obtain additional information about the elements.

Groups and Periods

Elements in the same column of the periodic table are known as a **group** or **family** because they exhibit similar physical and chemical properties. Each group (column) is identified by a *group number*, given at the top of each column in the periodic table. Elements in the first two columns and the last six columns, Groups 1A to 8A, are known as **main group elements** (Figure 2-11). The groups in the middle section of the periodic table, numbered 1B to 8B, are known as **transition metal elements**. In addition to their group *number*, some groups are identified by common names:

Group 1A   **alkali metals**
Group 2A   **alkaline earth metals**
Group 7A   **halogens**
Group 8A   **noble gases**

Elements within the same group exhibit similar chemical and physical properties. For example, the alkaline earth metals (group 2A) all have a shiny silvery-white color with a soft, malleable consistency, a physical property

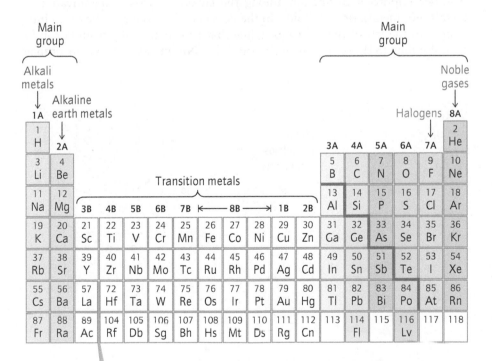

Figure 2-11 Groups are represented in 18 columns in the periodic table. Groups 1A through 8A, representing the main group elements, are shown here each in a different color.

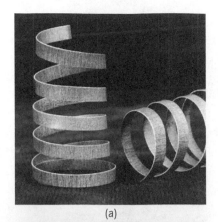

(a)

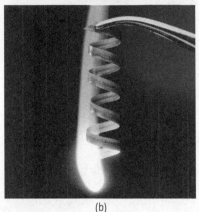

(b)

Figure 2-12 (a) The element magnesium (Group 2A) is a malleable, shiny, silvery-white alkaline earth metal. [Paul Silverman/Fundamental Photographs] (b) Magnesium undergoes a chemical reaction with oxygen when heated to produce magnesium oxide, released in the form of a white smoke. [Richard Megna/Fundamental Photographs]

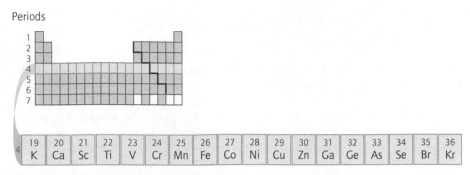

Figure 2-13 The periodic table of the elements contains seven rows, known as periods. Here, the elements in period 4 are called out and shown in orange.

(Figure 2-12a). One readily observable chemical property that the alkaline earth elements share is reactivity in oxygen (Figure 2-12b) when heated. If you were to examine another group of elements, you would observe a different set of physical and chemical properties but similarities among members of the group.

Each row of the periodic table is known as a **period**. There are seven periods in the periodic table, as shown in **Figure 2-13**, which features the elements in period 4—the fourth row. Sections of periods 6, the **lanthanides**, and 7, the **actinides**, are called out from the rest of the periodic table (Figure 2-11) and appear as two rows below the main table, each containing 14 elements, to signify that they belong in the gap between the elements La and Hf, in period 6, and between the elements Ac and Rf in period 7.

Metals, Nonmetals, and Metalloids

Another important distinction among the elements is their classification as metals, nonmetals, or metalloids. In the periodic table, **metals** (shown in blue) appear to the left of the bold zigzag line that runs diagonally from boron (5) to polonium (84), as shown in **Figure 2-14**. **Nonmetals** (shown in orange)

1 H																	2 He
3 Li	4 Be			☐ Metals								5 B	6 C	7 N	8 O	9 F	10 Ne
11 Na	12 Mg			☐ Metalloids ☐ Nonmetals								13 Al	14 Si	15 P	16 S	17 Cl	18 Ar
19 K	20 Ca	21 Sc	22 Ti	23 V	24 Cr	25 Mn	26 Fe	27 Co	28 Ni	29 Cu	30 Zn	31 Ga	32 Ge	33 As	34 Se	35 Br	36 Kr
37 Rb	38 Sr	39 Y	40 Zr	41 Nb	42 Mo	43 Tc	44 Ru	45 Rh	46 Pd	47 Ag	48 Cd	49 In	50 Sn	51 Sb	52 Te	53 I	54 Xe
55 Cs	56 Ba	57 La	72 Hf	73 Ta	74 W	75 Re	76 Os	77 Ir	78 Pt	79 Au	80 Hg	81 Tl	82 Pb	83 Bi	84 Po	85 At	86 Rn
87 Fr	88 Ra	89 Ac	104 Rf	105 Db	106 Sg	107 Bh	108 Hs	109 Mt	110 Ds	111 Rg	112 Cn		114 Fl		116 Lv		

58 Ce	59 Pr	60 Nd	61 Pm	62 Sm	63 Eu	64 Gd	65 Tb	66 Dy	67 Ho	68 Er	69 Tm	70 Yb	71 Lu
90 Th	91 Pa	92 U	93 Np	94 Pu	95 Am	96 Cm	97 Bk	98 Cf	99 Es	100 Fm	101 Md	102 No	103 Lr

Figure 2-14 Metals (blue) and nonmetals (orange) are separated by a bold zigzag line running diagonally from boron (5) to polonium (84) in the periodic table. Metalloids (pink) appear one on either side of the bold zigzag line, except for aluminum (Al), which is a metal.

TABLE 2-5 Physical Properties Distinguishing Metals and Nonmetals

Metals	Nonmetals
Shiny	Dull
Exist as solids at room temperature (except mercury)	Exist as solids, liquids, and gases at room temperature
Good conductors of electricity	Poor conductors of electricity
Malleable (can be shaped)	Brittle, hard, or soft

appear to the right of this bold zigzag line. **Metalloids** (shown in pink) are found one on each side of the bold zigzag line except aluminum (Al), which is a metal. Although hydrogen is located at the top of the alkali metals group, it is actually classified as a nonmetal. It is located in group 1A because it has similar electron characteristics to the alkali metals (group 1A).

Generally, metals and nonmetals have very different physical and chemical properties, whereas metalloids display characteristics of both metals and nonmetals. Table 2-5 compares some of the physical properties of metals and nonmetals. Later in the chapter you will learn that metal atoms tend to *lose* electrons and nonmetal atoms tend to *gain* electrons, which is an important chemical difference between metals and nonmetals.

Important Elements in Biochemistry and Medicine

Elements with an important role in biochemistry and medicine are highlighted in the color-coded medical periodic table of the elements shown in Figure 2-15:

- The building block elements used to construct biological compounds (C, H, N, O, P, and S) are shown in blue.
- Macronutrients (Na, K, Mg, Ca, P, S, Cl) are shown in orange.
- Micronutrients (V, Cr, Mn, Fe, Co, Cu, Zn, Mo, Si, Se, F, I) are shown in yellow.

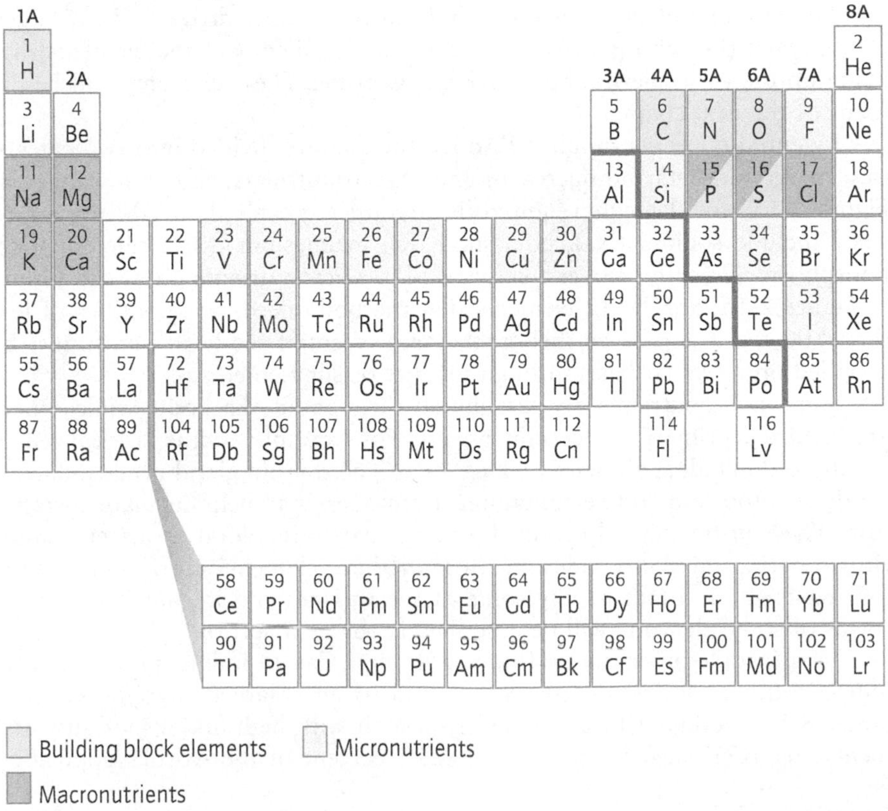

☐ Building block elements   ☐ Micronutrients

☐ Macronutrients

Figure 2-15 A periodic table of the elements showing the building block elements in blue, the macronutrients in orange, and the micronutrients in yellow.

Figure 2-16 The organs and cells in which the micronutrients zinc, fluoride, chromium, iodine, and iron are concentrated.

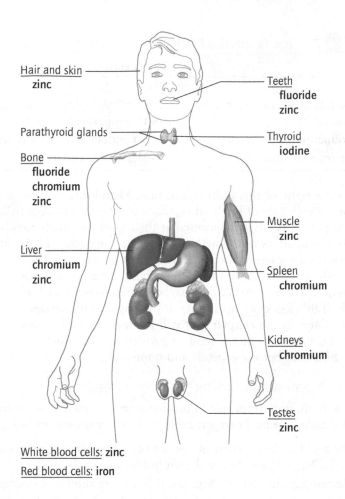

Hair and skin
zinc

Teeth
fluoride
zinc

Parathyroid glands

Thyroid
iodine

Bone
fluoride
chromium
zinc

Muscle
zinc

Liver
chromium
zinc

Spleen
chromium

Kidneys
chromium

Testes
zinc

White blood cells: zinc
Red blood cells: iron

Micronutrients are also called **trace minerals.**

The building block elements include carbon (C), hydrogen (H), nitrogen (N), oxygen (O), phosphorus (P), and sulfur (S). They are the elements that make up the molecules found in living organisms. These elements will be the focus of the next chapter.

Essential nutrients supplied through the diet are divided into two categories: macronutrients and micronutrients. **Macronutrients,** shown in orange in Figure 2-15, are elements required in quantities greater than 100 mg a day. They include sodium (Na), potassium (K), magnesium (Mg), calcium (Ca), phosphorus (P), sulfur (S), and chlorine (Cl). **Micronutrients,** shown in yellow in Figure 2-15, are the 12 elements that the body needs in quantities of less than 100 mg a day. Nevertheless, they are essential for health and must be supplied on a regular basis through the diet or supplements.

Some micronutrients are distributed throughout the body, while others are localized primarily in certain organs. For example, iodine is localized in the thyroid gland, as shown in Figure 2-16. The thyroid gland needs iodine to produce important iodine-containing hormones that help regulate metabolism. These hormones help control weight, heart rate, blood cholesterol, muscle strength, and skin condition. Most of these elements are found in their ionized form, and that is why some of their names end in *-ide*. The ionized forms of some elements will be described in the next section.

Most people ingest about 100 to 200 mg a day of iodine, primarily from iodized salt. An iodine deficiency can lead to an enlarged thyroid—called a goiter. Severe iodine deficiencies are associated with high infant mortality and mental retardation. It is estimated that 29 percent of the world's population lives in areas where the soils are depleted in iodine, causing locally grown foods to be deficient in this essential micronutrient. Iodine supplements and the fortification of food with iodine have been used to fight iodine deficiency disorders worldwide. A total of one teaspoon of iodide is all a person requires in a lifetime!

Some micronutrients, such as fluoride, serve a structural role in the body. Fluoride provides part of the structure of bone and teeth. Optimal fluoride intake has been shown to reduce dental cavities because fluoride helps produce stronger tooth enamel. Fluoride is found in drinking water and tea. However, excessive intake of fluoride can cause yellowing of teeth, an enlarged thyroid gland, and brittle bones and teeth. The acute toxic dose of fluoride is 2–8 mg per kg of body weight.

Many municipalities add 0.7–1.2 parts per million (ppm) of fluoride to the local water supply to increase the availability of this micronutrient. One ppm of fluoride means there is 1 mg of fluoride in every 1 L of water. Municipal water fluoridation is controversial. Some fear possible harmful health effects from adding fluoride to municipal water systems. Advocates of fluoridation argue that it is analogous to fortifying salt with iodine or adding vitamin D to fortify milk.

Zinc is important in the immune system and is often found in over-the-counter throat lozenges. The concentration of zinc is highest in bone, teeth, hair, skin, liver, muscle, white blood cells, and the testes, as shown in Figure 2-16. Adults usually require 8–12 mg a day of zinc. Good sources of zinc include oysters, breakfast cereal, beef, pork, chicken, yogurt, baked beans, and nuts. Zinc deficiencies can result in growth retardation, hair loss, diarrhea, skin lesions, and slow healing of wounds.

In the chapter-opening vignette you learned that iron is a part of the structure of hemoglobin, the molecule that transports oxygen in our red blood cells. Iron is also found in many other oxygen transport molecules and enzymes involved in extracting energy from the foods we eat. Iron-rich foods include clams, pork, beef liver, iron-fortified cereal, and pumpkin seeds. Iron is believed to be absorbed more efficiently from the diet when meat, fish, and vitamin C are part of the diet; while coffee and tea hinder the absorption of iron.

WORKED EXERCISES Groups and Periods in the Periodic Table

2-6   Answer the following questions for the element potassium:
   a. To what group does this element belong?
   b. What is the common name of this group?
   c. What period does this element belong to?
   d. Is this element a metal, nonmetal, or metalloid?

2-7   What element is located in group 5A and period 3?

2-8   Is hydrogen a metal, a nonmetal, or a metalloid?

Solutions

2-6   a. Potassium belongs to group 1A because it is found in the first column in the periodic table.
   b. Potassium is an alkali metal because that is the common name for group 1A.
   c. Potassium belongs to period 4 because it is located in the fourth row of the periodic table.
   d. Potassium is a metal because it is located left of the bold zigzag line in the periodic table, indicative of metals.

2-7   Phosphorus, P.

2-8   Hydrogen is a nonmetal even though it is located left of the bold zigzag line.

14 State the group number and family name, if one exists, for the following elements:

    a. beryllium   **b.** iodine   **c.** argon   **d.** nitrogen

15 Which of the following elements are metals?

    a. sulfur   **b.** mercury   **c.** neon

    d. sodium   **e.** uranium   **f.** molybdenum

16 Provide the name of the element that fits the description listed in each case:

    a. A group 2A metal in period 5

    b. A halogen in period 3

    c. A transition metal in period 4 with two more protons than calcium

17 Complete the table below for the six building block elements.

Element	Metal or Nonmetal	Group Number	Period Number	Atomic Number/ Number of Protons
C				
H				
N				
O				
P				
S				

18 What distinguishes a micronutrient from a macronutrient? How does your body obtain these nutrients?

19 How many of the seven macronutrients are also main group elements?

20 How many of the building block elements are nonmetals?

2.3 Electrons

Our current model of the electrons in an atom is based on **quantum mechanics**, a theory developed in the early twentieth century by such now famous scientists as Einstein, Heisenberg, and Bohr. Quantum mechanics explains phenomena observed on the atomic scale, such as the behavior of an electron in a hydrogen atom. An electron is so incredibly small and light, it behaves like nothing we have ever seen or encountered in our macroscopic world. Indeed, in some experiments an electron acts like a particle, and in other experiments it behaves more like a *wave*. Think of the electron as a standing wave like the oscillating head of a drum or guitar string but in three dimensions. Quantum mechanics helps us understand the peculiar nature of the electron. In this section we will examine electrons, especially as they pertain to the atoms in the first 20 main group elements.

Like all moving objects, electrons have *energy*. Indeed, in many ways an electron is defined by its energy. Quantum mechanics tells us that an electron in an atom can exist in only certain allowed energy levels and no energies in between. This would be like a car that could only be driven at 10, 35, 50, and 65 miles per hour and no speeds in between! Figure 2-17 illustrates how the energy levels for the electrons in an atom increase in steps (*quanta*) rather than continuously. Think of electron energy levels as unevenly spaced steps rather than a ramp. Electron energy levels are designated $n = 1, 2, 3, \ldots$ where the greater the value of n, the higher the energy of the electron. Electrons in higher energy levels occupy orbitals with a greater volume, so the electron can be found farther from the nucleus. These atoms are, therefore, larger. In the lowest energy

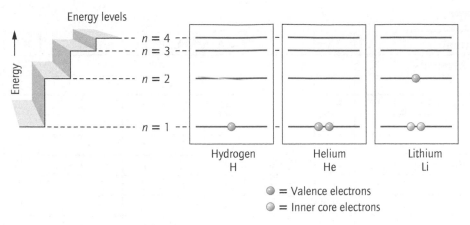

Figure 2-17 Electron energy levels, $n = 1, 2, 3, 4 \ldots$, occupied in the ground state for the first three elements: H, He, Li.

state of an atom, known as the **ground state**, electrons are found in their lowest allowable energy levels. When an atom absorbs energy, an electron is promoted to a higher energy level, known as an **excited state** of the atom.

Ground-State Electron Arrangements

Quantum mechanics tells us that a given energy level, n, can accommodate a maximum of $2n^2$ electrons, where $n = 1, 2, 3$, etc. Thus, the $n = 1$ level can accommodate only two electrons, the $n = 2$ level can accommodate up to eight electrons, and the $n = 3$ level can accommodate up to 18 electrons, and so forth, as shown in Table 2-6.

We know that every element differs in the number of protons (the atomic number) that its atoms contain, which is equal to the number of electrons in the atom. The periodic table of the elements shows that the first 20 elements, hydrogen through calcium, have from 1 to 20 electrons. Atoms in their ground state will have their electrons in the lowest energy levels but not exceeding the maximum allowed for any given energy level. Thus, when determining the *electron arrangement* for a particular element, fill its $n = 1$ energy level first, up to two electrons, before filling the $n = 2$ energy level, up to eight electrons, and so forth.

For example, a hydrogen atom has *one* electron, which will be in the $n = 1$, and lowest energy level, when in the ground state. Similarly, helium has its *two* electrons also in the $n = 1$ energy level. Lithium has two electrons in the $n = 1$ energy level, the maximum allowed in this energy level, and one electron in the $n = 2$ energy level, the next-highest energy level, as illustrated in Figure 2-17. Here we see that helium has a full $n = 1$ energy level and no partially filled energy levels, a particularly stable electron arrangement and characteristic of all the noble gases. The stable electron arrangements for the noble gases account for their lack of chemical reactivity (they are inert).

Hydrogen and helium are the elements in period 1 of the periodic table because they have their only electrons in the $n = 1$ energy level, as shown in Table 2-7. The elements lithium through neon have atoms with 3–10 electrons, whose electron arrangements are also shown in Table 2-7. These elements have atoms with a full $n = 1$ energy level (two electrons), and one to eight electrons in the $n = 2$ energy level. Thus, these elements make up the elements in period 2 of the periodic table. Since eight electrons is the maximum number of electrons in the $n = 2$ energy level, neon is the element with a full $n = 2$ outermost level. Like helium, neon has a full outermost energy level, and therefore is in the same group (group 8A) as helium.

TABLE 2-6 Maximum Number of Electrons Allowed in Each Energy Level

Energy Level $n =$	Maximum Number of Electrons, $2n^2$
1	2
2	8
3	18
4	32

TABLE 2-7 Electron Arrangements by Energy Level for the First 20 Elements*

Group Number	1A	2A			3A	4A	5A	6A	7A	8A
Period 1 Elements	H									He
$n = 1$	1									2
Period 2 Elements	Li	Be			B	C	N	O	F	Ne
$n = 1$	2	2			2	2	2	2	2	2
$n = 2$	1	2			3	4	5	6	7	8
Period 3 Elements	Na	Mg			Al	Si	P	S	Cl	Ar
$n = 1$	2	2			2	2	2	2	2	2
$n = 2$	8	8			8	8	8	8	8	8
$n = 3$	1	2			3	4	5	6	7	8
Period 4 Elements	K	Ca			Ga	Ge	As	Se	Br	Kr
$n = 1$	2	2			2	2	2	2	2	2
$n = 2$	8	8			8	8	8	8	8	8
$n = 3$	8	8			8	8	8	8	8	8
†$n = 4$	1	2			3	4	5	6	7	8

*Valence electrons indicated in blue. All other electrons indicated in black are inner core electrons.
†Fourth energy level begins to fill before the third energy level is completely full.

The elements sodium (Na) through argon (Ar) have atoms containing 11 to 18 electrons, and each of their electron arrangements is shown in Table 2-7. These elements have a full $n = 1$ and $n = 2$ energy level, and the outermost one to eight electrons occupy the $n = 3$ energy level. Thus, these elements make up the elements in period 3 of the periodic table. Although the maximum number of electrons that can be accommodated in the $n = 3$ energy level is 18, the $n = 4$ energy level begins to fill before the $n = 3$ level is completely filled. Thus, eight electrons in the $n = 3$ energy level is also a particularly stable electron arrangement, so argon, like neon and helium, is a group 8A noble gas element. Although the formula $2n^2$ tells us the maximum number of electrons allowed in an energy level, the actual filling of energy levels for the first 20 elements follows a pattern of 2, 8, 8, 2, for the energy levels $n = 1, 2, 3$, and 4, respectively.

Valence Electrons and Periodicity

The electrons in the highest energy level for a ground state atom—the outermost electrons—are known as **valence** electrons, whereas electrons in the lower energy levels are known as **inner core electrons**. The valence electrons are farthest from the nucleus and determine the bonding characteristics of an element as well as its physical and chemical properties.

One of the most important trends represented in the organization of the periodic table is that main group elements in the same group have the same number of valence electrons, a number equal to the group number. For example, the alkali metal (group 1A) elements hydrogen (H), lithium (Li), sodium (Na), potassium (K), rubidium (Rb), cesium (Cs), and francium (Fr) all have one valence electron. These elements differ only in the energy level in which this one valence electron is found: for hydrogen, $n = 1$; lithium, $n = 2$; sodium, $n = 3$; potassium, $n = 4$, and so forth as shown in Table 2-7 and Figure 2-18.

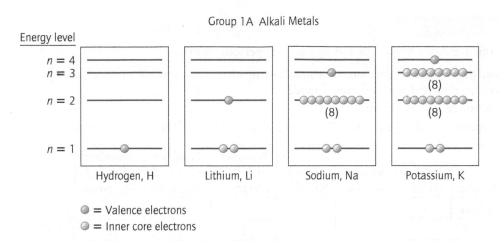

= Valence electrons

= Inner core electrons

Figure 2-18 Electron arrangements for the first four group 1A elements. Each group 1A element has one valence electron, shown in green, in successively higher energy levels: hydrogen (H) in $n = 1$, lithium (Li) in $n = 2$, sodium (Na) in $n = 3$, and potassium (K) in $n = 4$.

A similar pattern is seen with the alkaline earth (group 2A) elements, beryllium (Be), magnesium (Mg), calcium (Ca), strontium (Sr), barium (Ba), and radium (Ra), except they each have *two* valence electrons, as shown in Table 2-7 and Figure 2-19. As you can see, the valence electrons are found in successively higher energy levels as you go down a column as shown in Table 2-7.

Similar trends exist for groups 3A–8A. Group 8A represents the noble gases—helium (He), neon (Ne), argon (Ar), krypton (Kr), xenon (Xe), and radon (Rn)—the elements with a full valence energy level. In practice, you can determine the number of valence electrons for an atom simply by finding which group it belongs to, and you can determine which energy level the valence electrons are in by determining in which period the element is located. For example, locate arsenic (As) in the periodic table. Since it is in group 5A and in period 4, we know that it has five valence electrons and they are located in the $n = 4$ energy level.

In general, atoms with their valence electrons in higher energy levels are larger because the electrons can be found farther from the nucleus. For example, a lithium atom with its one valence electron in an $n = 2$ energy level is smaller than a potassium atom with its one valence electron in an $n = 4$ energy level.

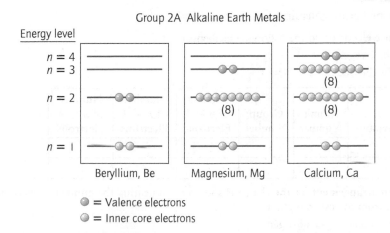

= Valence electrons

= Inner core electrons

Figure 2-19 Electron arrangements for the first three group 2A elements. Each group 2A element has two valence electrons, shown in green, in successively higher energy levels: beryllium (Be) in $n = 2$, magnesium (Mg) in $n = 3$, and calcium (Ca) in $n = 4$.

WORKED EXERCISES   Valence Electrons and Energy Levels

Answer the questions below consulting only the periodic table.

2-9   How many valence electrons do the following elements contain? In which energy level are their valence electrons found?

   a. bromine   **b.** strontium

2-10 Complete the table below by filling in the empty cells with the information requested.

Element	Symbol	Atomic Number	Group Number	Number of $n = 1$ Electrons	Number of $n = 2$ Electrons	Number of $n = 3$ Electrons
Carbon	C					
				2	8	2

2-11 Give the electron arrangement for the elements shown below from lowest to highest.

   a. oxygen   **b.** calcium   **c.** silicon

Solutions

2-9   **a.** Bromine is a group 7A element (a halogen), so it has seven valence electrons. We know they are in the $n = 4$ energy level because Br is in period 4.

   b. Strontium is a group 2A element (an alkaline earth metal), so it has two valence electrons. They are in the $n = 5$ energy level because Sr is a period 5 element.

2-10

Element	Symbol	Atomic Number	Group Number	Number of $n = 1$ Electrons	Number of $n = 2$ Electrons	Number of $n = 3$ Electrons
Carbon	C	6	4A	2	4	0
Magnesium	Mg	12	2A	2	8	2

2-11 **a.** oxygen   **b.** calcium   **c.** silicon
   $n = 1$: 2 electrons $n = 1$: 2 electrons $n = 1$: 2 electrons
   $n = 2$: 6 electrons $n = 2$: 8 electrons $n = 2$: 8 electrons
                                 $n = 3$: 8 electrons $n = 3$: 4 electrons
                                 $n = 4$: 2 electrons

PRACTICE EXERCISES

Answer the questions below consulting only the periodic table.

21 How many valence electrons do the following elements contain?
   a. barium   **b.** oxygen

22 Complete the table below.

Element	Symbol	Atomic Number	Group Number	Number of $n = 1$ Electrons	Number of $n = 2$ Electrons	Number of $n = 3$ Electrons
Sulfur	S					
				2	8	8

23 Show the electron arrangement for the elements shown by indicating the number of electrons in each energy level from lowest to highest.
   a. boron   **b.** neon   **c.** nitrogen

24 What is the maximum number of electrons that can be accommodated in the $n = 2$ energy level?

25 What is characteristic of the electron arrangement for the group 8A elements?

■ 2.4 Ions

Ions are atoms that have *lost* or *gained* one or more valence electrons, giving them a net charge. Positively charged ions are known as **cations**, and negatively charged ions are known as **anions**. Main group *metals* tend to *lose* their valence electrons in a process known as **ionization** to become *cations* with a *full* outermost energy level. The main group *nonmetals* can *gain* electrons to become *anions* with a *full* valence energy level. The unequal number of protons and electrons in an ion creates a positive (+) or a negative (–) charge. *The magnitude of the charge is equal to the difference between the number of protons and electrons in the ion.* When there are more electrons than protons, the charge on the ion is negative, making it an anion. When there are fewer electrons than protons, the charge on the ion is positive, making it a cation.

> **Cation:** an ion with a *positive* charge: *fewer* electrons than protons
>
> **Anion:** an ion with a *negative* charge: *more* electrons than protons

Consider sodium (Na), a group 1A metal (Figure 2-20a), with one valence electron in the $n = 3$ energy level. Sodium can lose its one valence electron to become a sodium cation, Na^+ (Figure 2-20b), with a full outermost $n = 2$ energy level, a stable electron arrangement the same as neon (Figure 2-20c). An elemental sodium atom has 11 protons and 11 electrons. After losing one electron, the sodium cation, Na^+, has 11 protons and 10 electrons and so a net +1 charge, as illustrated in Figure 2-20 and Table 2-8.

Similarly, a group 2A metal can lose its two valence electrons. For example, magnesium (Mg) can lose its two valence electrons in the $n = 3$ energy level to become a magnesium cation (Mg^{2+}), with a full outermost $n = 2$ energy level, as illustrated in Figure 2-21. Similarly, aluminum loses its three valence

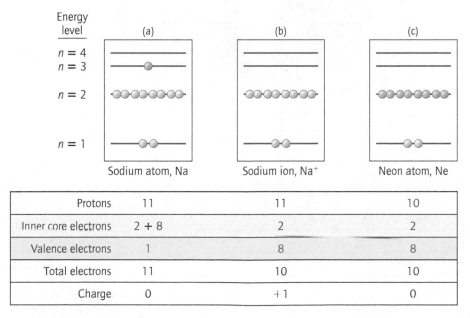

	Sodium atom, Na	Sodium ion, Na⁺	Neon atom, Ne
Protons	11	11	10
Inner core electrons	2 + 8	2	2
Valence electrons	1	8	8
Total electrons	11	10	10
Charge	0	+1	0

Figure 2-20 Electron arrangements for (a) sodium, Na, (b) sodium cation, Na^+, and (c) neon, Ne. Sodium cation, Na^+, has the same electron arrangement as neon, Ne.

TABLE 2-8 Number of Protons and Electrons in Na^+, F^-, and Ne

Element/Ion	Number of Protons	Number of Electrons	Charge
Na^+ (sodium cation)	11	10	+1
F^- (fluoride anion)	9	10	−1
Ne (neon)	10	10	0

electrons to become Al^{3+}. These ionization patterns are another example of the periodicity of the elements.

The symbol for an ion includes a superscript to the right of the symbol that shows the magnitude and sign of the charge on the ion. For example, the symbol for a sodium ion is Na^+ and the symbol for a magnesium ion is Mg^{2+}. The convention when writing the symbol for an ion is to place the sign of the charge (+ or −) *after* the number. If the charge is 1+ or 1−, the number "1" is implied and a + or − alone is used. The name of a cation formed from a main group element is the same as the element from which it is derived.

In contrast, the nonmetals in group 5A to 7A, on the right side of the periodic table, can *gain* valence electrons to become *anions*, giving them the same electron arrangement as the noble gas in the same period. These elements gain the number of electrons they need to fill their valence energy level (8 − group number). For example, fluorine (F), a group 7A nonmetal, gains one valence electron (8 − 7 = 1) to become a fluoride ion (F^-) with a full $n = 2$ energy level, a stable electron arrangement that is the same as that of neon, the noble gas in the same period, period 2 (Figure 2-22). An elemental fluorine atom has nine protons and nine electrons. On gaining one electron, the fluoride ion has nine protons and 10 electrons, creating a net −1 charge (Table 2-8). Indeed, all the group 7A elements form ions with a −1 charge: F^-, Cl^-, Br^-, and I^-. The name of an anion is the same as the name of the

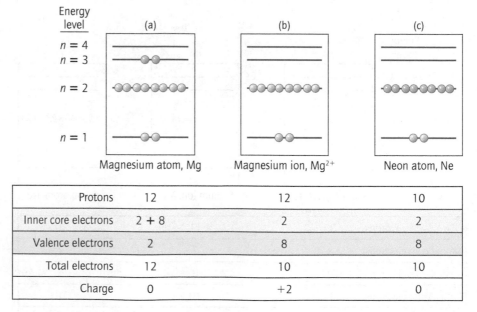

Protons	12	12	10
Inner core electrons	2 + 8	2	2
Valence electrons	2	8	8
Total electrons	12	10	10
Charge	0	+2	0

Figure 2-21 Electron arrangements for (a) magnesium, Mg, (b) the magnesium ion, Mg^{2+}, and (c) neon, Ne. A magnesium atom loses its two valence electrons to become a magnesium cation, Mg^{2+}, which has the same electron arrangement as Ne.

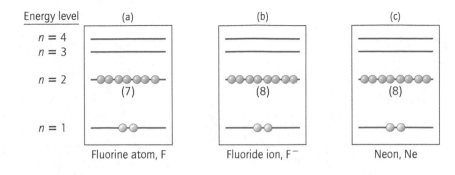

Energy level	(a)	(b)	(c)

	Fluorine atom, F	Fluoride ion, F⁻	Neon, Ne
Protons	9	9	10
Inner core electrons	2	2	2
Valence electrons	7	8	8
Total electrons	9	10	10
Charge	0	−1	0

Figure 2-22 Electron arrangements for (a) fluorine, F, (b) fluoride ion, F⁻, and (c) neon, Ne. A fluorine atom gains one electron in its valence energy level to become a fluoride ion, F⁻, which has the same electron arrangement as Ne.

element from which it is derived, but the ending is changed to *-ide*, as in fluor*ide*.

Similarly, the nonmetal elements in group 6A gain two electrons to fill their highest energy level, creating anions with a −2 charge. For example, the sulfide ion, S^{2-}, is formed when sulfur, S, in period 3, gains two electrons, creating an electron arrangement the same as that of the noble gas argon—the period 3 noble gas. Similarly, the group 5A nonmetals form −3 ions, such as nitride, N^{3-}. Most of the important metal and nonmetal macronutrients and micronutrients are actually found in the body as ions.

The group 4A nonmetal elements have a half-filled valence shell, but they do not gain or lose four electrons to become ions because a +4 or −4 charge is unstable for a nonmetal ion. Carbon, the most important group 4A element, achieves stability instead by *sharing* electrons with other nonmetal atoms. Shared electrons are the subject of the next chapter.

Keep in mind that ions are always formed by the gain or loss of *electrons*; an atom cannot gain or lose protons. Only in nuclear reactions do we see the number of protons or neutrons in an atom change. Nuclear reactions will be described in the next section. *As a general rule, metals lose electrons to become cations, whereas nonmetals gain electrons to become anions or share electrons to become molecules.*

An ion has significantly different physical and chemical properties than the element from which it was formed. For example, elemental sodium, Na, is a shiny metal that is extremely flammable, exploding into a fireball on contact with water (Figure 2-23). In contrast, the sodium cation, Na^+, is part of many ionic compounds, such as sodium chloride (NaCl), the familiar and unreactive white crystalline substance used as table salt.

Transition metals and heavy metals also form ions. However, we cannot readily predict the number of electrons that these elements lose simply from its group number. Moreover, many transition metals and heavy metals can lose a variable number of electrons. For example, iron (Fe) can lose either two or three electrons to become Fe^{2+} or Fe^{3+}, respectively. All iron ions contain

Figure 2-23 Ions have very different chemical and physical properties from the elements from which they are derived. For example, sodium metal (Na) reacts violently with water (top), whereas the sodium ion (Na^+) in sodium chloride (NaCl), table salt, is unreactive (bottom). [(Top) Andrew Lambert Photography/Science Source; (bottom) Timothy Lozinski/Alamy]

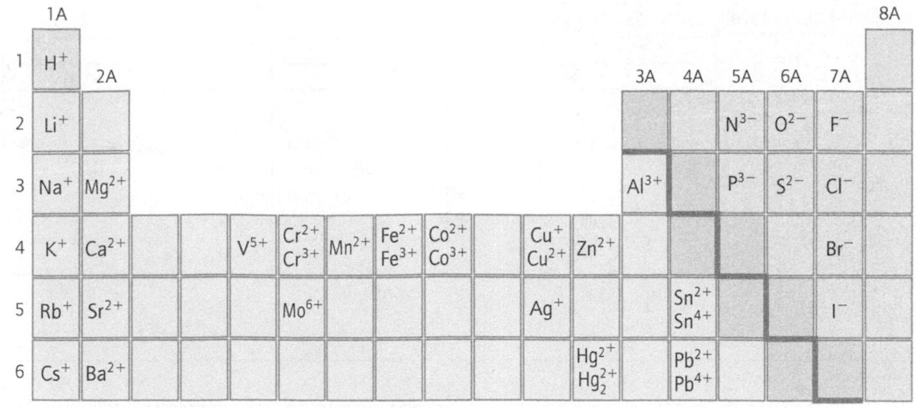

	1A													8A
1	H$^+$	2A								3A	4A	5A	6A	7A
2	Li$^+$											N^{3-}	O^{2-}	F$^-$
3	Na$^+$	Mg^{2+}								Al^{3+}		P^{3-}	S^{2-}	Cl$^-$
4	K$^+$	Ca^{2+}		V^{5+}	Cr^{2+} Cr^{3+}	Mn^{2+}	Fe^{2+} Fe^{3+}	Co^{2+} Co^{3+}	Cu$^+$ Cu^{2+} Zn^{2+}					Br$^-$
5	Rb$^+$	Sr^{2+}		Mo^{6+}				Ag$^+$			Sn^{2+} Sn^{4+}			I$^-$
6	Cs$^+$	Ba^{2+}					Hg^{2+} Hg$_2^{2+}$	Pb^{2+} Pb^{4+}						

Figure 2-24 Main group ions and some common transition metal ions.

Figure 2-25 Heavy metal ions are poisonous, such as the lead ion (Pb^{2+}) found in paints prior to 1978. [© Jeff Albertson/Corbis]

26 protons, but Fe^{2+} has 24 electrons and Fe^{3+} has 23 electrons. The ionic forms of some common transition metals are shown in Figure 2-24. Metal ions that exist in more than one form are named after the element from which they are derived, followed by parentheses containing a Roman numeral that represents the charge on the ion. For example, iron(II) and iron(III) are iron atoms that have lost two and three electrons, respectively.

Lead (Pb), mercury (Hg), and other "heavy metals" form ions such as Hg$_2^{2+}$ (always found as a dimer—a pair) and Pb^{2+} that are poisonous (Figure 2-25). Lead ions are especially harmful to the developing brain of a child. Lead poisoning in children has been shown to occur when they are exposed to the dust from old lead-based paint, which was used in homes prior to 1978. Lead and mercury ions bind to the sulfur atoms in important proteins in the body, preventing them from performing their necessary function.

WORKED EXERCISES   Ions

2-12 Write the symbol for the ion(s) formed from the following elements. Indicate whether the element is a metal or a nonmetal:

    a. oxygen   **b.** barium   **c.** copper

2-13 Consider the ions Br$^-$, Sr^{2+}, and Co^{2+}

    a. Indicate the number of protons and electrons in each ion.

    b. State whether the ion was formed from a metal or a nonmetal.

    c. Explain why the charge on the ion is as indicated based on its location in the periodic table.

Solutions

2-12 a. O^{2-}, nonmetal because oxygen is a group 6A nonmetal and by gaining two electrons ($8 - 6 = 2$), it will attain a full $n = 2$ valence energy level, an electron arrangement like that of neon.

    b. Ba^{2+}, metal because barium is a group 2A metal and by losing its two valence electrons in the $n = 6$ valence shell, it will have a full $n = 5$ valence energy level, an electron arrangement like that of the period 5 noble gas xenon.

    c. Cu$^+$ and Cu^{2+}, metal. Copper is a transition metal that can lose either one or two electrons, as shown in Figure 2-24.

2-13 a. Br^- has 35 protons and 36 electrons
Sr^{2+} has 38 protons and 36 electrons
Co^{2+} has 27 protons and 25 electrons
 b. Br^- derived from a non-metal
Sr^{2+} derived from a main group metal
Co^{2+} derived from a transition metal
 c. Br^- is −1 because Br has 7 valence electrons and needs one electron to fill its valence energy level (8 − 7 = −1).
Sr^{2+} is +2 because Sr acquires two valence electrons and by losing all valence electrons has a full valence energy level.
Co^{2+} charge cannot be predicted from its group number because it is a transition metal.

PRACTICE EXERCISES

26 Write the symbol and the name of the ions formed from the following elements, based on their position in the periodic table:
 a. selenium   **b.** potassium   **c.** phosphorus

27 Consider the ions I^-, N^{3-}, and Cr^{3+}
 a. Indicate the number of protons and electrons in each.
 b. State whether the ion was formed from a metal or a nonmetal.
 c. Explain why the charge on the ion is as indicated based on its location in the periodic table.

28 Why do group 6A elements gain two electrons? What is the charge on these ions? Provide an example of a group 6A anion.

29 In the movie *Erin Brockovitch*, based on a true story, Brokovitch learns that the high incidence of cancer in a small town is the result of groundwater contamination from Cr(VI). However, another chromium ion, Cr(III), is an important micronutrient. Write the symbol for each of these ions. What is the difference between these two forms of chromium?

■ 2.5 Radioisotopes

In section 2.1 you learned that isotopes of an element have the same number of protons and electrons but differ in the number of neutrons they contain. Some isotopes are unstable due to an imbalance in the ratio of neutrons to protons or because they contain too many protons and neutrons in the nucleus. Unstable isotopes are known as radioactive isotopes or **radioisotopes**. There are more than 300 naturally occurring isotopes, and 36 of these are radioactive. All isotopes with an atomic number greater than 82 are radioisotopes. In addition to naturally occurring radioisotopes, there are the more than 2,000 man-made isotopes known as **artificial radioisotopes**. All artificial isotopes are radioactive. Smoke detectors, computer monitors, and instruments used for some medical imaging contain artificial isotopes.

The nucleus of a radioactive isotope undergoes a natural process of **radioactive decay** to become a more stable nucleus. *Radioactive decay usually produces an isotope with a different atomic number and hence a different element.* Radioactive decay is also accompanied by the release of **radiation**, a form of energy. About 80 percent of the radiation you are exposed to comes from natural sources, known as **background radiation**. These natural sources include cosmic rays from the sun and radon gas filtering up from the ground. Your body also contains some naturally radioactive substances, such as K-40 and organic molecules containing trace amounts of C-14.

Air travel increases your exposure to radiation because cosmic radiation is 100 times greater at cruising altitudes than it is on the ground.

TABLE 2-9 Types of Radiation and Their Nuclear Symbols

Radiation Type	Nuclear Symbol	Energy Type
α particle	$_{2}^{4}\alpha$ or $_{2}^{4}\text{He}$	High-energy particles
β particle	$_{-1}^{0}\beta$ or $_{-1}^{0}e$	
x-ray	—	Electromagnetic radiation
γ-ray	γ	

Radiation

Radiation from the decay of radioisotopes takes the form of either high-energy *electromagnetic radiation*, such as x-rays and γ-rays (pronounced "gamma rays"), or *high-energy particles*, such as α particles and β particles (Table 2-9).

Electromagnetic Radiation Electromagnetic radiation, or "light," is a form of energy that travels through space as a wave at the speed of light (3×10^8 meters per second). Electromagnetic radiation varies in terms of its **wavelength**, the distance between wave crests as shown in Figure 2-26, which is related to its **frequency**, the number of times a full wave cycle passes a given point. One wave cycle per second is known as a **hertz (Hz)**, a unit of frequency. The longer the wavelength of light, the lower its frequency:

Longer *wavelength* corresponds to lower *frequency*.

Shorter *wavelength* corresponds to higher *frequency*.

Electromagnetic radiation is divided into regions according to its frequency and wavelength. For example, if you tune your car radio to a radio station broadcasting at 92.8 megahertz (MHz), the antenna of your car is receiving electromagnetic radiation in the radio frequency region of the electromagnetic spectrum, which is traveling at the speed of light, at a frequency of 92.8×10^6 cycles per second (hertz), which corresponds to electromagnetic radiation with a wavelength of 3.25 meters.

Electromagnetic radiation ranges from the very long wavelength radio waves to the short wavelength x-rays and γ-rays. The entire range of wavelengths is known as the **electromagnetic spectrum** (Figure 2-27). Visible light is the narrow region in the middle of the electromagnetic spectrum that we can see with our eyes, ranging from 400 to 700 nm. Other regions of the electromagnetic spectrum include x-ray, ultraviolet (UV), infrared (IR), microwave, and radio

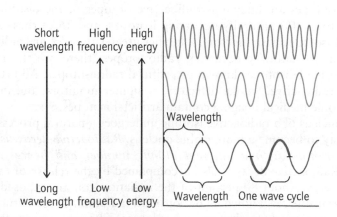

Figure 2-26 Electromagnetic radiation is defined by its wavelength—the distance between wave crests, which is inversely proportional to its frequency—the number of wave cycles that pass a given point every second. Longer wavelengths correspond to lower frequencies; shorter wavelengths correspond to higher frequencies.

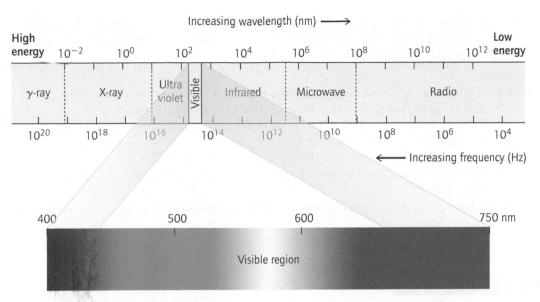

Figure 2-27 The electromagnetic spectrum is divided into regions ranging from high-energy/ high-frequency/short wavelengths (left side of illustration) to low-energy/low-frequency/ long wavelengths (right of illustration). The narrow region of the electromagnetic spectrum between 400 and 750 nm is known as visible light, expanded here to show the wavelengths of the different colors our eyes are able to detect.

waves (Figure 2-28). The sun produces the entire electromagnetic spectrum. Special instruments, such as an x-ray machine or a microwave, generate a particular wavelength of electromagnetic radiation.

The energy of a particular type of electromagnetic radiation depends on its wavelength or frequency. The shorter the wavelength (higher frequency), the higher in energy the electromagnetic radiation is. Thus, at one end of the electromagnetic spectrum are the low-energy radio waves and at the other end of the spectrum are the high-energy γ-rays. Electromagnetic radiation with wavelengths shorter than visible light has higher energy and can damage biological tissue. You know this if you have ever had a sunburn. Exposure to the ultraviolet light from the sun is high enough in energy to burn your skin. The more damaging forms of electromagnetic radiation are the very short, and therefore higher energy, γ-rays and x-rays, both of which are emitted during certain radioactive decay processes and certain types of medical imaging techniques.

Figure 2-28 Cell phones emit electromagnetic radiation in the microwave region of the electromagnetic spectrum. [Catherine Bausinger]

Longer wavelength, *lower* frequency corresponds to *lower* energy.

Shorter wavelength, *higher* frequency corresponds to *higher* energy.

WORKED EXERCISES Electromagnetic Radiation

2-14 Referring to Figure 2-27, answer the questions below by choosing from among the following forms of electromagnetic radiation:

    i. x-ray   ii. visible   iii. ultraviolet

    a. Which form causes the greatest damage to biological tissue?

    b. Which form is the lowest in energy?

    c. Which has the shortest wavelength?

    d. Which travels at the fastest speed?

2-15 Mice cannot see red light, but they do see blue light and green light. Red light has a wavelength of 700 nm and blue light has a wavelength of 450 nm. Which type of light has more energy, red light or blue light?

Solutions

2-14 **a.** Of the three forms of electromagnetic radiation listed, x-rays are the most damaging to biological tissue because they have the highest energy.

b. Visible light is the lowest in energy of the three listed because it has the longest wavelength and lowest frequency.

c. X-rays have the shortest wavelength of the three listed.

d. All forms of electromagnetic radiation travel at the *same* speed, the speed of light: 3×10^8 m/s.

2-15 Blue light has a shorter wavelength and higher frequency than red light; therefore it has more energy.

PRACTICE EXERCISES

30 Referring to Figure 2-27, answer the questions below by choosing from among the following forms of electromagnetic radiation:

   i. γ-rays   ii. microwaves   iii. radio waves

  a. Which form causes the greatest damage to biological tissue?

  b. Which form is the lowest in energy?

  c. Which contains less energy than visible light but more energy than radio waves?

31 Bumblebees can "see" UV light. How is UV light different from visible light?

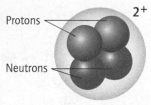

Figure 2-29 An α particle is a slow-moving high-energy particle consisting of two protons and two neutrons with a +2 charge.

α Decay We have learned that radioisotopes emit radiation to become more stable—lower energy—isotopes. Some radioisotopes undergo radioactive decay by emitting an **α particle**. An α particle (pronounced "alpha particle") is a slow moving, high-energy particle consisting of two protons and two neutrons as shown in Figure 2-29. The nuclear symbol for an α particle is $^4_2\alpha$ or $^4_2$He. The symbol for helium is sometimes used because an α particle has the same composition as a helium nucleus. Keep in mind, however, that there are no corresponding electrons, so an α particle is not the same as a helium atom; instead, it is an extremely dense particle with a +2 charge. Emission of an α particle always results in a daughter nuclide with a mass number that is four less and an atomic number that is two less than the parent nuclide. For α decay, most of the energy released in the nuclear decay process appears in the form of kinetic energy of the α particle.

The radioactive decay process is typically shown by writing a **nuclear equation**, as seen in the example below for the α decay of Th-232:

$$^{232}_{90}\text{Th} \longrightarrow {}^{228}_{88}\text{Ra} + \boxed{^4_2\alpha}$$

Parent   **Daughter**   α particle
nuclide   **nuclide**

The radioisotope undergoing radioactive decay, also known as the **parent nuclide**, is written on the left side of the arrow using the convention for representing an isotope: the mass number as a superscript and the atomic number as a subscript. The new isotope produced, known as a **daughter nuclide**, and the radioactive particle emitted are shown on the right side of the arrow, also with nuclear symbols.

In a nuclear equation, the sum of the atomic numbers (subscripts) of all nuclides on the left side of the arrow must equal the sum of the atomic numbers for all nuclides on the right side of the arrow; likewise for the mass numbers (superscripts). Indeed, this arithmetic can be used to determine the identity of the daughter nuclide if you know the radioisotope and the type of radiation it emits, as shown in the worked exercise.

Superscripts: $\quad$ 232 $\quad = \quad$ 228 $\quad + \quad$ 4

$$^{232}_{90}\text{Th} \longrightarrow {}^{228}_{88}\text{Ra} + {}^{4}_{2}\alpha$$

Subscripts: $\quad$ 90 $\quad = \quad$ 88 $\quad + \quad$ 2

WORKED EXERCISE $\quad$ α Decay

2-16 Predict the daughter nuclide produced in the α decay of U-238. Write the nuclear equation.

Solution

2-16 Begin by writing the parent nuclide on the left side of the arrow. Include its atomic number and mass number as a subscript and a superscript, respectively, using the periodic table to determine its atomic number. On the right side of the arrow write the nuclear symbol for an α particle:

$$^{238}_{92}\text{U} \rightarrow {}^{4}_{2}\alpha + X$$

Next, determine the identity of the daughter nuclide, X, by subtracting 4 from the mass number and subtracting 2 from the atomic number of the parent nuclide:

$$\text{Mass number} = 238 - 4 = 234$$
$$\text{Atomic number} = 92 - 2 = 90$$

This tells us the daughter nuclide is $^{234}_{90}X$. Look up the atomic symbol for the element with atomic number 90 in the periodic table: Th; hence $^{234}_{90}\text{Th}$ is the daughter nuclide. The complete nuclear equation is

$$^{238}_{92}\text{U} \longrightarrow {}^{234}_{90}\text{Th} + \boxed{{}^{4}_{2}\alpha}$$

Parent $\qquad$ **Daughter** $\qquad$ α particle
nuclide $\qquad$ **nuclide**

The sum of the mass numbers is equal on both sides of the arrow, and the sum of the atomic numbers is equal on both sides of the arrow. Note that it does not matter whether the α particle or the daughter nuclide is written first.

PRACTICE EXERCISES

32 List four types of radioactive decay. Which of the four types of decay are forms of electromagnetic radiation?

33 Predict the daughter nuclide formed when polonium-210 undergoes α decay. Write the nuclear equation.

34 Radon-222 often builds up in the basements of homes located in areas with a high
concentration of this nuclide in the surrounding earth. A radioisotope produces radon-222 by
α decay. What is the parent nuclide that produces radon-222? Write the nuclear equation for
this radioactive decay process.

β Decay Some radioisotopes undergo radioactive decay by emitting a **β** particle. A β particle is *a high-energy electron whose nuclear symbol is* $_{-1}^{0}\beta$, *indicating it contains no protons or neutrons.* The daughter nuclide formed when
a β particle is emitted will therefore have the same mass number as the parent
nuclide, but an atomic number one greater than the parent nuclide; hence,
the daughter nuclide is the element with the next-higher atomic number in the
periodic table. A high-energy electron traveling at about 90 percent of the
speed of light, the **β particle**, is emitted from the nucleus into the surrounding
area.

Consider, for example, the radioactive decay of P-32, used in the detection
of breast cancer and eye tumors. P-32 undergoes β decay according to the
nuclear equation shown:

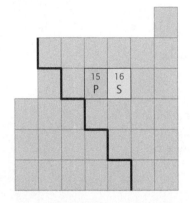

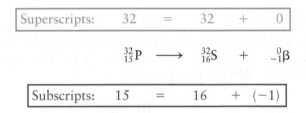

Superscripts: 32   =   32   +   0

$$_{15}^{32}\text{P} \longrightarrow \,_{16}^{32}\text{S} + \,_{-1}^{0}\beta$$

Subscripts: 15   =   16   +   (−1)

Figure 2-30 Radioactive decay of P-32,
atomic number 15, produces S-32,
atomic number 16, as the daughter
nuclide and a β particle. The daughter
nuclide formed from β decay is the
element with the next-higher atomic
number in the periodic table.

The nuclear equation shows that the parent nuclide P-32, an unstable isotope
of phosphorus, decays to S-32, a stable isotope of sulfur and the next element
in the periodic table (Figure 2-30). A β particle is emitted in the process. The
subscripts on both sides of the arrow are equal: 15 = 16 + (−1), as are the
superscripts: 32 = 32 + 0, as required in a nuclear equation.

WORKED EXERCISE   β Decay

2-17 The radioisotope cerium-141 is known to undergo β decay. It is often used to assess
blood flow through the heart.
  a. How many protons and neutrons does the parent nuclide contain?
  b. Write the nuclear equation for the β decay of cerium-141.
  c. What daughter nuclide is produced from the β decay of Ce-141?
  d. What form of radiation is produced in this radioactive decay? Is it
  electromagnetic radiation or a high-energy particle?

Solution

2-17 **a.** There are 58 protons and 83 neutrons in a Ce-141 radioisotope. Use the periodic
table to obtain the atomic number for Ce, and subtract this value from the mass
number to obtain the number of neutrons in the parent nuclide.
  b. To write the nuclear equation, locate the next element in the periodic table to
determine the daughter nuclide, and give the daughter nuclide the same mass
number (superscript) as the parent nuclide: Pr-141. Then write the nuclear
equation by writing the nuclear symbol for the parent nuclide on the left of the

arrow and the nuclear symbol for the daughter nuclide and the β particle on the right side of the arrow:

$$\text{Superscripts:} \quad 141 \quad = \quad 141 \quad + \quad 0$$

$$^{141}_{58}\text{Ce} \longrightarrow {}^{141}_{59}\text{Pr} + {}^{0}_{-1}\beta$$

$$\text{Subscripts:} \quad 58 \quad = \quad 59 \quad + \quad (-1)$$

c. Pr -141, the next element in the periodic table, has the same mass number, 141, as the parent nuclide.
d. A β particle is emitted, which is a type of high-energy particle.

PRACTICE EXERCISES

35 The radioisotope I-131 is known to undergo β decay. It is often used to diagnose thyroid conditions.
   a. How many protons and neutrons does this parent nuclide contain?
   b. Write the nuclear equation.
   c. What is the daughter nuclide produced from the β decay of I-131?
   d. What form of radiation is produced in this nuclear reaction? Is it electromagnetic radiation or a high-energy particle?
36 Gold-198 is used in the diagnosis of kidney disease. Gold-198 undergoes β decay. Write the nuclear equation and label the parent and daughter nuclides.

X-rays and γ Radiation You learned earlier that x-rays and γ radiation are short-wavelength, high-energy forms of electromagnetic radiation. X-rays accompany some forms of radioactive decay and when focused on bone or other tissue can be used to create valuable diagnostic images such as x-ray and CT scans. To learn more about x-rays as a diagnostic tool, see Chemistry in Medicine: Nuclear Radiation in Medical Imaging, at the end of this chapter.

Gamma-ray (γ) emission, higher in energy than x-rays, accompanies almost all forms of radioactive decay (see page 76). However, γ emission is typically not shown in a nuclear equation because it does not affect the atomic number or mass number of the nuclides involved. Few radioisotopes are pure γ emitters and typically emit some other type of radiation as well.

After radioactive decay, the daughter nuclide is often in an **excited state**, a condition in which the *nucleus* contains excess energy. An isotope in an excited state is referred to as a **metastable** isotope and is notated by the abbreviation *m* following the mass number of the isotope. A metastable daughter nuclide eventually releases its excess energy as it relaxes back to the ground state by releasing a pulse of γ radiation.

The most common γ emitter used in medicine today is metastable technetium, Tc-99m. When Tc-99m relaxes to its ground state, Tc-99, γ radiation is emitted:

$$^{99m}_{43}\text{Tc} \rightarrow {}^{99}_{43}\text{Tc} + {}^{0}_{0}\gamma$$

Since no particle is emitted, the atomic number, mass number, and element symbol for the daughter nuclide are the same as for the parent nuclide; the nuclides differ only in the amount of energy their nuclei contains.

The food irradiation process uses γ-rays to destroy disease-causing microorganisms such as *E. coli* and *Salmonella* on fruits and vegetables. It also reduces spoilage caused by bacteria, giving food a longer shelf life. The food itself does *not* become radioactive because only the electromagnetic radiation from the radioisotope, not the radioisotope itself, comes in contact with the food.

WORKED EXERCISE γ Radiation

2-18 Cesium-137 undergoes β decay to barium-137m, which then undergoes γ emission to a stable nuclide.

  a. Write the two nuclear equations for this two-step nuclear decay sequence.
  b. What is the stable nuclide formed at the end of this two-step process?
  c. What does the *m* in barium-137m stand for?

Solution

2-18 a.

$$^{137}_{55}\text{Cs} \rightarrow\ ^{137m}_{56}\text{Ba} +\ ^{0}_{-1}\beta$$
$$^{137m}_{56}\text{Ba} \rightarrow\ ^{137}_{56}\text{Ba} +\ ^{0}_{0}\gamma$$

  b. Ba-137.
  c. The *m* in Ba-137m indicates a metastable isotope, one whose nucleus has excess energy.

Half-Life

You have seen that when some radioisotopes decay, they produce a new daughter nuclide and some form of radiation. As a sample of radioisotope decays, the amount of the parent nuclide diminishes over time.

The *time* that it takes a sample of a radioisotope to decay to one-half of its original mass is known as its **half-life**. The half-life of radioactive decay varies widely for the different radioisotopes, ranging from mere seconds to billions of years. Consider iodine-131 which has a half-life of eight days. This means that a 100-g sample of I-131 will decay to 50.0 g after eight days, 25.0 g after 16 days (two half-lives), 12.5 g after 24 days (three half-lives), and so forth. After five half-lives the original sample is practically gone: only 3 percent of the original sample remains. Mathematically, this rate of decay is known as **exponential decay**, as illustrated in the graph in **Figure 2-31**.

> The fraction of the radioisotope remaining after *n* half-lives can also be calculated using the equation
>
> $$N = \left(\tfrac{1}{2}\right)^n$$
>
> where *N* is the fraction remaining and *n* is the number of half-lives. Since the number of half-lives, *n*, appears as an exponent, we call this *exponential decay.*

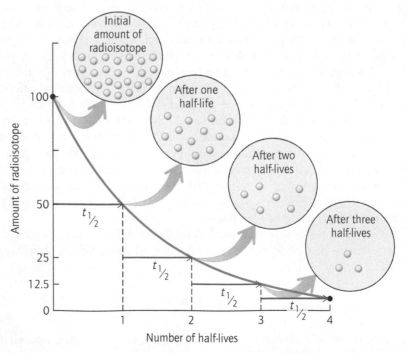

Figure 2-31 Radioisotopes decay exponentially. One half-life ($t_{1/2}$) is the time it takes the radioisotope to decay to one-half its original amount. Two half-lives is the time it takes the radioisotope to decay to one-half of one-half, or one-fourth, of the original amount and so forth.

TABLE 2-10 Radioisotopes, Radiation Type, and Half-Life

Radioisotope	Type of Emission	Half-Life
*Barium-131	γ	11.6 days
Carbon-14	β	5,730 years
*Cerium-141	β	32.5 days
*Chromium-51	γ	27.8 days
*Gallium-67	γ	78 hours
*Gold-198	β	64.8 hours
*Iodine-131	β and γ	8 days
*Iron-59	β and γ	44 days
*Krypton-79	γ	34.5 hours
*Krypton-81m	γ	13.3 seconds
*Phosphorus-32	β	14.3 days
†Technetium-99m	γ	6 hours
Uranium-238	α	4×10^9 years

*Isotope used in medicine.

†Most widely used radioisotope in medicine.

Disposal and storage of nuclear waste is problematic because some radioisotopes have extremely long half-lives. For example, Pu-239 has a half-life of 24,100 years! That means it takes 24,100 years for 10 grams of Pu-239 to decay to 5 g and another 24,100 years to decay to 2.5 g.

Krypton-81, which is used in medicine to diagnose lung function, has an extremely short half-life of 13 seconds. At the other extreme, carbon-14 has a half-life of 5,730 years. Carbon-14 is used in carbon dating to determine the age of archeological objects. For medical applications, short half-lives are usually preferred in order to minimize a patient's exposure to radiation. Table 2-10 lists some radioisotopes along with their half-lives and the type of radioactive emission produced.

WORKED EXERCISES   Half-Life

2-19 Iron-59 has a half-life of 44 days. How many grams of iron-59 would remain from an initial 100.-g sample after 132 days?

2-20 Radon-222 has a half-life of 3.8 days. How long would it take for a 20.-g sample of Rn-222 to decay to 1.3 g?

Solutions

2-19 First, determine how many half-lives have elapsed in the course of 132 days, given that *one* half-life is 44 days:

$$132 \text{ days} \times \frac{1 \text{ half-life}}{44 \text{ days}} = 3 \text{ half-lives}$$

Next, multiply the original mass of material, 100. g, by $\frac{1}{2}$ three consecutive times (three half-lives, shown above the arrows):

$$100. \xrightarrow{1} 50.0 \text{ g} \xrightarrow{2} 25.0 \text{ g} \xrightarrow{3} 12.5 \text{ g}$$

Thus, 12.5 g of the original sample of Fe-59 would remain after 132 days.

2-20 First, determine how many half-lives need to elapse for the sample to decay to 1.3 g by counting the number of times that you have to multiply the sample by $\frac{1}{2}$ to get 1.3 g:

$$20. \xrightarrow{1} 10. \xrightarrow{2} 5.0 \xrightarrow{3} 2.5 \xrightarrow{4} 1.25$$

Next, determine the total time elapsed in four half-lives of this radioisotope, given that one half-life is 3.8 days:

$$4 \text{ half-lives} \times \frac{3.8 \text{ days}}{1 \text{ half-life}} = 15 \text{ days}$$

PRACTICE EXERCISES

37 Using Table 2-10, determine how many hours it would take a 1.0-g sample of gold-198 to decay to 0.50 g. What has happened to the other 0.50 g of Au-198?

38 For a 10.0-g sample of cerium-141,
   a. How much Ce-141 is left after three half-lives?
   b. How much time has elapsed in three half-lives?

39 A 50.0-g sample of C-14 has decayed, to 3.13 g. How much time has elapsed since there were 50.0 g of C-14?

Biological Effects of Nuclear Radiation

Radiation emitted from certain radioisotopes is classified as **ionizing radiation** because it has sufficient energy to dislodge a valence electron from an atom, forming a cation:

$$\text{Radiation} \quad + \quad \underbrace{X}_{\substack{\text{Neutral} \\ \text{atom or} \\ \text{molecule}}} \quad \longrightarrow \quad \underbrace{X^+}_{\text{Cation}} \quad + \quad \underbrace{e^-}_{\substack{\text{Valence} \\ \text{electron}}}$$

Your body is composed of an enormous number of atoms. We have seen earlier in the chapter that when an atom or a molecule is ionized (loses an electron), it changes in a significant way. In living organisms, this change can be quite destructive, causing cell death or gene mutations. Radiation, therefore, can damage DNA, the genetic material of our cells. Mutations—changes in the DNA—are passed on when a cell reproduces, which can be the beginning of a cancer. The effect of ionizing radiation is greatest on rapidly reproducing cells such as lymphocytes (white blood cells), blood-producing cells, and cancer cells. In contrast, nerve and muscle cells reproduce slowly and therefore are the least sensitive to ionizing radiation.

The biological effects of nuclear radiation depend in large part on the specific type of radiation. Two characteristics of radiation determine its biological effects:

- the energy of the radiation, and
- the penetrating power of the radiation.

The **penetrating power** of a particular type of radiation is a measure of the extent to which it passes through matter, as illustrated in Figure 2-32. An α particle is relatively large and slow moving but high in energy; therefore, it is very destructive to human tissue. However, due to its greater size and slow speed, an α particle has little penetrating power. Light clothing, skin, or even a piece of paper is sufficient protection against α particles. However, ingestion

γ-rays kill bacteria and other microorganisms, so they can be used to sterilize hospital instruments.

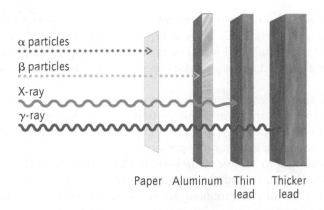

Paper Aluminum Thin Thicker
lead lead

Figure 2-32 The penetrating power of different types of radiation: an α particle has the least penetrating power and γ-rays have the most penetrating power.

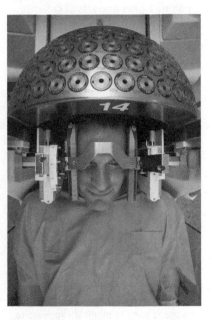

Figure 2-33 The gamma knife uses γ-rays to destroy tumors, avoiding the risks associated with traditional surgery. [© BSIP SA/Alamy]

or inhalation of an α emitter can cause major damage to delicate internal organs due to the high energy of α particles. Few α emitters are used in medicine for this reason.

Radioisotopes that emit β particles have significantly less energy than α particles but much more penetrating power because they are substantially lighter (8,000 times lighter) than α particles. Specialized heavy clothing or a thick piece of aluminum is required for protection against β particles. β particles can penetrate through the skin. In many respects, β particles are more damaging than α particles because of their penetrating power.

The energy of γ-rays and x-rays is less than or equal to the energy of β particles, but they have more penetrating power, because they are forms of electromagnetic radiation. A thin sheet of lead is sufficient protection against x-rays. This is the reason you are asked to wear a lead apron when your dentist takes dental x-rays. γ-rays have the greatest penetrating power of all forms of radiation and so they can be very damaging. Several inches of lead are required to adequately protect against γ radiation.

Carefully focused γ radiation can be used effectively to destroy tumors or cancer cells in a noninvasive type of surgery known as radiosurgery, also referred to as the **gamma knife**. A gamma knife uses γ radiation to destroy tumors, avoiding the risks associated with traditional open surgery (Figure 2-33).

PRACTICE EXERCISES

40 What type(s) of radiation can be stopped by the following barriers:
  a. a piece of paper?
  b. a sheet of aluminum?
  c. a thin sheet of lead?
  d. a thick slab of lead?
  e. skin?

41 α particles have the least penetrating power, so why would it be dangerous to swallow a radioisotope that is an α emitter?

42 What subatomic particle is lost from an atom that has been subjected to ionizing radiation? How has the atom changed?

43 Does γ radiation have more or less penetrating power than high-energy particles?

44 Which is more damaging to biological tissue, x-rays or γ-rays? Explain in terms of energy and penetrating power. What type(s) of radioactive decay produces γ-rays?

Figure 2-34 A Geiger counter is a handheld instrument used to measure radiation. More frequent audible clicks indicate higher amounts of radiation. [Scientifics Direct, LLC. www.scientificsonline.com]

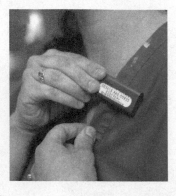

Figure 2-35 A radiation badge is worn by medical professionals who routinely work around radioisotopes and other sources of radiation. The radiation badge is used to monitor exposure to radiation. [Cliff Moore/Science Source]

Measuring Radiation Like other forms of energy, radiation can be measured. A **Geiger counter** (Figure 2-34) is a simple handheld instrument used in the field for the detection of all forms of radiation. A Geiger counter produces an audible series of clicks that increases as you move closer to areas with more radiation.

A radiation badge (Figure 2-35) is typically worn by personnel who regularly work in areas or use instruments that produce ionizing radiation. These badges consist of a special type of photographic film that is regularly processed to monitor radiation exposure by personnel.

It is important to be able to measure both the amount of radiation emitted by a radioactive source and the amount of radiation absorbed by the individual. Several units of measurement are therefore in use to measure radiation. The units differ in the type of information conveyed and include:

- the number of radioactive *emissions* per second emitted from the radioisotope,
- the amount of energy absorbed (*absorbed dose*) by an individual from a radiation source, and
- the biological effectiveness of the energy absorbed (*effective dose*) by an individual.

The number of radioactive emissions is the most basic unit of radiation and indicates the number of emissions per second from a radioactive source. A Geiger counter can be used to obtain this measurement. Two common units are the *becquerel* (**Bq**) and the *curie* (**Ci**), as shown in Table 2-11. These units of measurement include *all* radioactive emissions from a sample and do not distinguish between different types of radioactive decay, such as α particles and β particles.

Units of absorbed dose more adequately convey the amount of energy absorbed by an individual as a result of exposure to different forms of radiation (α particle, β particle, γ-ray, x-ray). An **absorbed dose** measurement indicates the energy of the radiation absorbed per mass of tissue. There are two units of absorbed dose in use (Table 2-11), but the **gray** (**Gy**) is the most common in medical applications. The **rad** is the other common unit of absorbed dose.

Absorbed dose measurements still do not convey the differences in *penetrating power* of different forms of radiation. The unit that encompasses both penetrating power and amount of energy to give the actual biological effect of a particular type of radiation is the human **effective dose**, shown in Table 2-11.

TABLE 2-11 Radiation Units and Their Abbreviations

Unit	Symbol
Amount of Radioactive Decay	
Becquerel	Bq
Curie	Ci
Absorbed Dose	
Gray	Gy
Radiation absorbed dose	Rad
Effective Dose	
Sievert	Sv
Roentgen equivalent man	Rem

The human effective dose is calculated by multiplying the absorbed dose by a quality factor, Q, which varies for the different types of radiation. For example, $Q = 20$ for an α particle, and $Q = 1$ for γ-rays and x-rays.

$$\text{Effective dose} = \text{absorbed dose} \times Q$$

When the unit of absorbed dose is the *gray*, the unit of effective dose is the **sievert** (Sv). For example, natural background radiation is approximately 0.0024 Sv per year or 2.4 mSv (millisieverts) per year, although it varies with geographical location. Another common unit of effective dose is the **rem**. There are 100 rem in 1 Sv. For example, a CT scan of the head and body has an effective dose of 110 mrem (millirem).

WORKED EXERCISES   Measuring Radiation

2-21 If you live within 50 miles of a nuclear power plant, you receive 0.01 mrem of additional radiation per year. Dental x-rays produce 1 mrem per procedure. What type of information does the rem convey? How many years living next to a nuclear power plant is equivalent to one dental x-ray?

2-22 International health and safety authorities endorse the safety of irradiated foods up to a dose of 10,000 Gy. What type of information does this unit convey?

Solutions

2-21 A rem is a unit for measuring the effective dose of radiation exposure. The rem takes into account both the energy of the radiation and its penetrating power. Living next to a nuclear power plant for 100 years is the equivalent to the radiation from one dental x-ray: 1 ~~mrem~~ × 1 year/.01 ~~mrem~~ = 100 years.

2-22 It indicates that the irradiated foods may absorb 10,000 gray. The gray is a unit of radiation measurement that conveys the absorbed dose, a measure of energy exposure. This unit takes into account the different energies for α particles, β particles, x-rays, and γ-rays, but it does not take into account the penetrating power of the radiation.

PRACTICE EXERCISES

45 Explain the difference between the sievert and the gray as a unit of measure for radiation.

46 Match each unit in the column on the left with the type of information it conveys in the column on the right.

Bq	absorbed dose (energy exposure)
Gy	
Sv	
Ci	effective dose
rem	
rad	number of radioactive emissions

47 Would the effective dose of a 5-Gy α emitter be greater than or less than a 5-Gy β emitter? Explain.

48 The amount of exposure to cosmic radiation that an individual receives per year depends on the altitude at which he or she lives. The value is 26 mrem at sea level, 35 mrem at 3,500 ft, 52 mrem at 6,000 ft, and 96 mrem at 9,000 ft. Estimate approximately how much cosmic radiation you receive per year where you live. Why is the value greater at higher altitudes?

TABLE 2-12 Dose Related Symptoms of Acute Radiation Sickness

Effective Dose (Sv)	Symptoms
0.05–0.2	None
0.2–0.5	Temporary decrease in white blood cell count
0.5–1.0	Headache and increased risk of infection; possible temporary male sterility
1.0–2.0	LD_{10}; nausea, hair loss, fatigue; loss of white blood cells; temporary male sterility
2–3	LD_{35}; hair loss, fatigue, and general illness; high risk of infection
3–4	LD_{50}; uncontrollable bleeding in the mouth; permanent sterility in women
4–6	LD_{60}; death resulting from internal bleeding and infection; permanent female sterility
6–10	LD_{100}; death after 14 days

Radiation Sickness Exposure to radiation can occur either as a single large dose (acute exposure) or as many smaller doses over a longer period of time (chronic exposure). Moreover, exposure to radiation can occur either by accident or intentionally as part of radiation therapy. *Radiation sickness results from acute exposure to radiation.* The severity of the symptoms is directly proportional to the effective dose received, as shown in Table 2-12. The LD_x values shown in this table refer to the *lethal dose* of the radiation in x percent of the population after 30 days. An LD_{50}, for example, is the level of exposure that would result in death in 50 percent of the population after 30 days. Symptoms of radiation sickness often do not occur immediately after exposure but follow a latent phase, during which time the individual shows no symptoms.

Although excessive exposure to radiation can have serious health consequences, exposure to specific forms of radiation in a controlled manner can have benefits that outweigh the risks, such as the use of radiation in the diagnosis and treatment of disease.

PRACTICE EXERCISES

49 Using Table 2-12, what are the visible symptoms of acute exposure to 1.2 Sv of radiation?

50 What does it mean if the LD_{35} is 2–3 Sv?

51 What is the difference between acute and chronic exposure to radiation?

52 What level of radiation causes no overt symptoms of radiation sickness but results in a temporary decrease in white blood cell count?

Chemistry in Medicine Nuclear Radiation in Medical Imaging

You already know that visible light—electromagnetic radiation—makes it possible for us to see our surroundings. Can other forms of electromagnetic radiation also be used to produce images? The answer is yes: many imaging techniques have been developed that allow us to "see" internal organs and systems in the body using electromagnetic radiation. These imaging techniques are based on radioisotopes emitting radiation as they decay to a more stable isotope. For example, Figure 2-36 shows images of a hand with radiation from various regions of the electromagnetic spectrum.

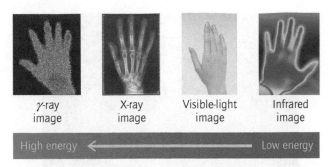

Figure 2-36 Images of a hand, using various regions of the electromagnetic spectrum. Each type of radiation reveals a different feature of the hand. X-rays and γ-rays show bones, visible light shows the surface of the skin, and infrared shows the temperature on the surface of the hand. The orange areas in the γ-ray image are bone. [From left to right: Alfred Pasieka/Science Source; BSIP/Science Source; Yashuhide Fumoto/Digital Vision/Getty Images; Ted Kinsman/Science Source.]

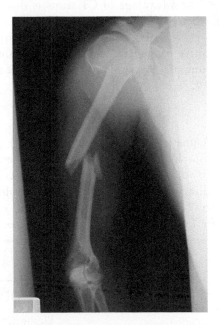

Figure 2-37 Chemistry plays an important role in therapeutic and diagnostic medicine. Here a patient is undergoing MRI, used to assess damage to soft tissue. [© Pete Saloutos/Corbis]

Traditional diagnostic techniques in medicine include symptom appraisal, blood tests, and even exploratory surgery. Often these tests do not lead to a definitive diagnosis. Radiation-based diagnostic techniques provide a more definitive way to diagnose certain medical conditions that often eliminate the need for exploratory surgery.

X-rays and radio waves are used to image anatomical structures of the body. These techniques include x-ray imaging, computed tomography (CT) scans, and magnetic resonance imaging (MRI). These imaging techniques are noninvasive, relatively comfortable procedures and are often used as complementary techniques. The instruments used for these techniques often look the same from the outside (Figure 2-37).

X-rays

Chances are you have had an x-ray taken at the dentist's office, at the emergency room for a broken bone, or at a medical laboratory for some other health-related issue. X-ray imaging is by far the most widely used imaging technique in medicine. X-rays are a form of electromagnetic radiation that when focused on a specific area of the body produces a 2-D image (Figure 2-38). As x-rays pass through the body (high penetrating power), they encounter structures of different densities, such as bone and muscle, which absorb the radiation to a different extent. High-density anatomical structures such as bone will absorb more x-rays, whereas lower-density anatomical structures such as muscle will absorb less. Some x-rays will pass completely through the body and eventually collide with the **x-ray detector**—the part of the x-ray instrument that measures the amount of x-ray radiation that has passed through the tissue. The detector provides an image showing lighter and darker areas: lighter areas indicate higher-density tissues, which have absorbed more

x-rays, and darker areas indicate lower-density tissues, which have absorbed fewer x-rays. Thus, a break in a bone will show up as a darker, discontinuous area (low-density space) in the middle of a lighter, continuous area (the high-density bone). Exposure to x-rays is kept as

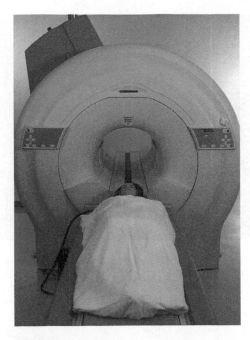

Figure 2-38 An x-ray image of a broken bone. The break will appear as a darker, discontinuous area (low-density space) in the middle of a lighter, continuous area (the high-density bone). [© David Frazier/Corbis]

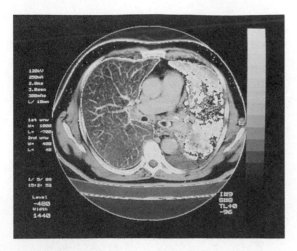

Figure 2-39 A computed x-ray tomography (CT) scan of the chest shows cancer in the left lung. Notice the difference in appearance between the healthy right lung (left on image) and the cancerous left lung (right on image). The heart is in the center. [Simon Fraser/Science Source]

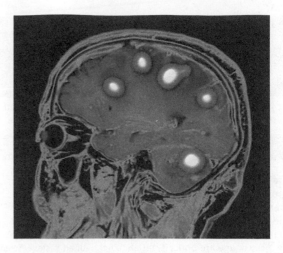

Figure 2-40 A color-enhanced MRI scan of a brain shows multiple malignant tumors (yellow-red). MRI is an imaging technique that uses strong magnetic fields and radio frequency pulses to generate 3-D images. [Medical Body Scans/Science Source]

brief as possible to avoid damaging surrounding tissue. The typical effective dose of radiation from a single chest x-ray is approximately 0.1 mSv (millisieverts).

CT Scans

Like conventional x-ray instruments, **computed *tomography* (CT)** scanning instruments use x-rays to create images but with a circular array of detectors that surround the body. The information from these detectors is downloaded to a computer for processing to create a 3-D image. The advantage of CT scans is that they can distinguish between two separate, similar structures adjacent to each other. Computed tomography is used as the definitive diagnostic technique for brain hemorrhages, pneumonia, appendicitis, and complex fractures (**Figure 2-39**).

MRI

Like CT, *magnetic *resonance* imaging* (**MRI**) produces three-dimensional images of organs and tissues. MRI is valuable for diagnosing conditions such as tumors, edema (swelling of organs and tissues), and multiple sclerosis (Figure 2-40). *MRI is best for imaging soft-tissue areas of the body such as the brain and liver, whereas x-ray techniques (including CT) are best for imaging denser tissue such as bones and joints.*

MRI is a technique that employs a magnetic field and pulses (short bursts) of electromagnetic radiation in the radio frequency region of the electromagnetic spectrum. Your body is composed of mostly water, and water contains hydrogen atoms. When the protons in the hydrogen atoms of water molecules in your body are exposed to the strong magnetic field of the MRI instrument, they behave like tiny magnets and align themselves with the magnetic field of the MRI, similar to a compass needle aligning itself with magnetic north. As the MRI scan progresses, radio wave pulses of various energies are applied. Protons in different environments of the body (muscle, nerve tissue, tumor tissue, and so on) will resonate at different radio frequencies, allowing the computer to produce an image that distinguishes different tissues in the body using color images.

A major advantage of MRI is that it is harmless to the patient. The radio frequency region of the electromagnetic spectrum, as you know, is low-energy electromagnetic radiation and does not pose any risk of damaging tissue. The strong magnetic field also does not pose a risk to the patient unless the patient is wearing a pacemaker or other similar electromagnetic device.

Chapter Summary

Elements and the Structure of the Atom

- The atom is the smallest stable component of matter. An atom is composed of protons, neutrons, and electrons.
- Protons have a positive charge, electrons have a negative charge, and neutrons have no charge.

- The mass of an atom depends on the number of protons and neutrons it contains.
- Protons and neutrons are located in the dense nucleus at the center of the atom.
- Electrons are located in electron orbitals and have a negligible mass.
- An *s* orbital has a spherical shape.
- The three *p* orbitals are perpendicular and have a dumbbell-like shape.
- An element is composed of atoms that contain the same number of protons. An element cannot be broken down into a simpler form of matter.
- The atomic number for an element corresponds to the number of protons its atoms contain.
- Isotopes of an element are atoms with the same number of protons but a different number of neutrons.
- The mass number for an isotope is the sum of its protons and neutrons.
- The average atomic mass of an element is the weighted average of its isotopes based on the natural abundance of each isotope and the mass of each isotope.
- Every element has a one- or two-letter atomic symbol.

Navigating the Periodic Table of the Elements

- The periodic table of the elements shows the 114 elements from lowest to highest atomic number in a unique organization of seven rows—periods—and columns—groups.
- Elements in the same column in the periodic table belong to the same group or family. Elements in the same row in the periodic table belong to the same period.
- The main group elements are in groups 1A through 8A.
- Elements in the same group contain the same number of valence electrons, accounting for their similar chemical and physical properties.
- Metals and nonmetals are separated by the bold diagonal zigzag line on the right side of the periodic table. Metalloids are the elements along both sides of this line.
- Many elements have an important role in biochemistry as either building block elements, micronutrients, or macronutrients.

Electrons

- Electrons occupy different energy levels, n, where $n = 1, 2, 3 \ldots$
- Our current model of the electron comes from quantum mechanics, which states that the energy of an electron is fixed in one of certain allowed energy levels.
- Atoms in the ground state, the lowest energy state of the atom, have their electrons in the lowest energy levels, up to the maximum allowed in any given energy level.
- Valence electrons are the outermost electrons of an atom, in the energy level with the highest value of n. The other electrons, in lower energy levels, are inner core electrons.
- Elements in the same group have the same number of valence electrons, which corresponds to the group number.

Ions

- A cation is formed when a metal atom loses electrons, and an anion is formed when a nonmetal atom gains electrons. The formation of a cation is known as ionization.
- Main group metals lose their valence electrons and main group nonmetals can gain valence electrons until they have a full valence shell, like the noble gas nearest them in the periodic table. For cations, it is the noble gas in the period beneath it, and for nonmetals, it is the noble gas in the same period.
- Many transition metals and some main group metals lose a variable number of electrons and therefore exist in more than one ionic form.

Radioisotopes

- Radioactive isotopes—radioisotopes—are unstable isotopes that achieve stability by radioactive decay, a process in which electromagnetic radiation and/or high-energy particles are emitted from the radioisotope.
- Electromagnetic radiation is a form of energy that travels through space as a wave at the speed of light.
- Electromagnetic radiation is described by its wavelength, which is related to its frequency. Different wavelengths of light contain different amounts of energy. The longer the wavelength, the lower the frequency and the lower its energy.
- Radioactive decay can be described by a nuclear equation. The sum of the subscripts on both sides of the equation must be equal, and similarly with the superscripts. Subscripts represent the atomic number of the nuclides, and superscripts represent the mass number.
- There are four major types of radioactive decay: α-particle emission, β-particle emission, x-ray emission, and γ-ray emission.
- An α particle ($^{4}_{2}\alpha$) is a helium nucleus: two protons and two neutrons. It carries a 2+ charge, and it is a high-energy, relatively slow-moving particle.
- A β particle is a high-energy electron ($^{0}_{-1}\beta$).
- β decay yields a daughter nuclide with an atomic number one greater than the parent nuclide and the same mass number.
- Most radioactive decay is accompanied by γ-rays, high-energy electromagnetic radiation.
- A half-life is the time it takes a particular radioisotope to decay to one-half its mass—exponential decay.
- Radioactive decay produces ionizing radiation that can damage or destroy cells through ionization, the loss of an orbital electron, producing an ion.
- The biological effects of nuclear radiation depend on the energy and penetrating power of the radiation.
- α particles have the highest energy but least penetrating power.
- β particles and x-rays have less energy than α particles but greater penetrating power.
- Gamma rays have more energy and greater penetrating power than x-rays.
- The units of measurement that indicate the number of radioactive emissions produced by a radioisotope are the becquerel (Bq) and the curie (Ci).

- The absorbed dose is a unit of radiation measurement that takes into account the energy of the radiation absorbed. Common units are the gray (Gy) and the rad.
- The effective dose is a unit of radiation that takes into account the energy of the radiation absorbed together with the penetrating power of the radiation. Common units of effective dose are the sievert (Sv) and the rem.
- Radiation sickness occurs with acute—a single large dose—exposure to radiation.

Key Words

α particle A slow-moving, high-energy particle consisting of two protons and two neutrons emitted as a result of nuclear decay. Its nuclear symbol is $_2^4\alpha$ or $_2^4He$.

Absorbed dose The energy of radiation absorbed per mass of tissue.

Actinides The elements Ac through Lr in period 7, offset from the main body of the periodic table.

Alkali metals The metal main group elements in group 1A.

Alkaline earth metals The metal main group elements in group 2A.

Anion A negatively charged ion resulting from a nonmetal atom gaining enough electrons (8-group number) to fill its valence energy level and obtain an electron arrangement like that of the noble gas in the same period.

Artificial radioisotopes Man-made radioactive isotopes.

Atom The smallest indivisible stable component of an element.

Atomic mass unit A unit of mass. Protons and neutrons have a mass of 1 amu.

Atomic number The number of protons in an atom; the atomic number defines the element.

Atomic symbol The one- or two-letter symbol used to identify an element.

Average atomic mass A weighted average of the mass of all the isotopes of an element based on the natural abundance of each isotope.

β particle A high-energy electron emitted as a result of nuclear decay. Its nuclear symbol is $_{-1}^0\beta$.

Background radiation Radiation emitted from natural sources such as the sun and the earth.

Becquerel (Bq) A unit of measurement indicating the number of radioactive emissions from a sample.

Cation A positively charged ion resulting from the loss of valence electrons from a metal.

Compound A substance composed of two or more different atoms held together by chemical bonds.

Curie (Ci) A unit of measurement indicating the number of radioactive emissions from a sample.

Daughter nuclide The new isotope formed after radioactive decay.

Effective dose A measure of radiation that includes both the penetrating power and the amount of energy of a particular type of radiation. It conveys the actual biological effect of the radiation.

Electromagnetic radiation A form of energy that travels through space as a wave at the speed of light.

Electromagnetic spectrum The entire range of wavelengths ranging from γ-rays to radio waves.

Electron A subatomic particle of the atom with a negative charge and negligible mass, found in an electron orbital.

Electron orbital A region of space describing where the electron is most likely to be found.

Element A substance composed of only one type of atom, which cannot be broken down into a simpler form of matter. There are 114 elements that have been identified and named.

Excited state A nucleus with excess energy after undergoing radioactive decay.

Exponential decay A rate of decay expressed by the equation: $N = (\frac{1}{2})^n$, where the number of half-lives, n, is an exponent in the equation. Every half-life, the sample diminishes to half its mass.

Family Elements in the same column of the periodic table; also known as a group.

Frequency The number of times a full wave cycle passes a given point; used to describe a type of electromagnetic radiation. It is inversely related to the wavelength of light and related to the energy of the light.

γ radiation A form of electromagnetic radiation having the shortest wavelengths and greatest energy. Accompanies radioactive decay.

Gamma knife A type of noninvasive radiosurgery in which carefully focused γ radiation is used to destroy tumors or cancer cells.

Geiger counter An inexpensive instrument used in the field to detect all forms of radiation.

Gray (Gy) A unit of radiation measurement indicating the energy of the radiation absorbed per mass of tissue.

Ground state The lowest energy state of an atom wherein the electrons occupy the lowest allowable energy levels.

Group A column in the periodic table; also known as a family.

Half-life The time that it takes a sample of a radioisotope to decay to one-half of its mass.

Halogens The elements in group 7A.

Hertz (Hz) A unit of frequency that equals one wave cycle per second.

Inner core electrons The electrons in energy levels lower than the energy level containing the valence electrons.

Ion A positive or negatively charged atom. The charge results from an unequal number of protons and electrons due to the loss or gain of electrons.

Ionization A process whereby a metal atom loses electrons to become a cation.

Ionizing radiation High-energy radiation with sufficient energy to dislodge an orbital electron from an atom or a molecule, creating a cation.

Isotopes Atoms having the same number of protons but a different number of neutrons. Isotopes are distinguished by their mass number.

Lanthanides The elements La through Lu in period 6, seen offset from the main body of the periodic table.

Lethal dose (LD_x) A level of radiation exposure that would result in death in x percent of the population exposed in 30 days.

Macronutrients Elements that must be supplied through the diet in a quantity greater than 100 mg/day.

Main group elements The elements in groups 1A through 8A.

Mass number The sum of the number of protons and neutrons in an isotope.

Metalloids Elements found one on each side of the dark zigzag line (excluding aluminum, which is a metal) running diagonally on the right side of the periodic table, separating metals from nonmetals.

Metals The elements on the left side of the periodic table (left of the dark zigzag line).

Metastable An isotope whose nucleus is in an excited state.

Micronutrients Elements that must be obtained through the diet in quantities of less than 100 mg/day.

Molecule Two or more atoms held together by chemical bonds.

Neutron A subatomic particle with no charge and a mass of approximately 1 amu; located in the nucleus of the atom.

Noble gases The elements in group 8A. They are uniquely stable because they have a full valence energy level.

Nonmetals Elements on the right side of the periodic table (right of the dark zigzag line).

Nuclear equation A common way to represent radioactive decay that shows the parent nuclide on the left side of the arrow and the daughter nuclide and the particle of radiation on the right side of the arrow. The sum of the superscripts on both sides of the arrow must be equal. The sum of the subscripts on both sides of the arrow must be equal.

Nuclear medicine The use of radioisotopes in the diagnosis and treatment of disease.

Nucleus The small, dense center of the atom that contains its protons and neutrons.

Parent nuclide A radioisotope that underdoes radioactive decay.

Penetrating power A measure of the extent to which a particular type of radiation passes through matter.

Period A row in the periodic table.

Periodicity The repeating trends in chemical and physical properties of the elements, reflected in the layout of the periodic table.

Periodic table of the elements The table showing the 114 elements by their atomic symbol, atomic number, and atomic mass, displayed in characteristic rows and columns.

Proton Subatomic particle with a positive charge and a mass of 1 amu; located in the nucleus of the atom.

Quantum mechanics A theory developed in the early twentieth century that explains phenomena observed on the atomic scale, such as the energy of an electron is quantized: it can only exist in certain fixed energy levels.

Rad A common unit of absorbed dose of radiation.

Radiation The energy released by a radioisotope.

Radiation sickness Illness resulting from acute exposure to radiation.

Radioactive decay The release of radiation by an unstable nucleus in order to become a more stable nucleus.

Radioisotope A radioactive isotope, unstable due to an imbalance in the number of neutrons and protons or because it contains too many neutrons and protons in the nucleus.

Rem A common unit of effective dose of radiation.

Sievert The unit of effective dose when the unit of absorbed dose is the gray.

Subatomic particles The parts of an atom: protons, neutrons, and electrons.

Trace minerals Another term for *micronutrients*.

Transition metal elements The metals in groups 1B through 8B, positioned between groups 2A and 3A in the periodic table.

Valence electrons The outermost electrons of an atom; those with the highest n value; equal in number to the group number.

Wavelength The distance between wave crests used to describe a type of electromagnetic radiation. Inversely related to frequency and energy.

X-ray detector The part of the x-ray instrument that measures the amount of x-ray radiation that has passed through the tissue.

X-rays Electromagnetic radiation with short wavelengths, high frequency, and high energy.

Additional Exercises

Anemia and Iron

53 What is anemia? What is the main function of red blood cells?

54 What is hemoglobin? Why is it so important in the body?

55 What causes iron deficiency anemia?

56 What are some of the symptoms of iron deficiency anemia?

Elements and the Structure of the Atom

57 The fundamental component of all matter is the _____.

58 Compounds are made up of _____ held together by _____.

59 What is the difference between an element and a compound?

60 What are the three types of subatomic particles in an atom? Where are these subatomic particles located within an atom?

61 What is the charge on a proton, an electron, and a neutron?

62 What subatomic particles are located in the nucleus of the atom?

63 Why is the nucleus of the atom so dense?

64 Which of the following is an *s* orbital and which is a *p* orbital? What does an orbital represent?

(a)   (b)   (c)

65 Which is the lightest subatomic particle: the proton, the electron, or the neutron?

66 How many protons and electrons are there in the following elements?
a. oxygen, a building block element
b. chromium, a micronutrient
c. phosphorus, a macronutrient

67 How many protons and electrons are there in the following elements?
a. vanadium, a micronutrient
b. sulfur, a building block element and a macronutrient
c. magnesium, a macronutrient

68 Provide the element name and atomic symbol for the element that has
a. 51 protons.
b. 33 protons.
c. 56 electrons.
d. 88 electrons.

69 What does the atomic number of an element represent?

70 What does the mass number of an isotope represent?

71 What is the mass of a proton in amu? What is the mass of a proton in grams?

72 How do isotopes of an element differ from each other? How are they similar?

73 Oxygen has three stable isotopes: oxygen-16, oxygen-17, and oxygen-18. Write each of these isotopes using the symbol for the element along with the appropriate subscript and superscript.

74 Technetium-99 is a radioactive isotope frequently used in medicine. How many protons, electrons, and neutrons does Tc-99 have? Write this isotope using the element symbol with the atomic number and mass number in the conventional format.

75 Sulfur has four naturally occurring isotopes: sulfur-32, sulfur-33, sulfur-34, and sulfur-36. Complete the table below, which gives information about each of these isotopes.

Isotope	Mass Number	Atomic Number	Number of Protons	Number of Neutrons
Sulfur-32				
Sulfur-33				
Sulfur-34				
Sulfur-36				

76 What do all sulfur isotopes have in common?

77 Which sulfur isotope is the lightest? Explain.

78 In nature, 51 percent of bromine atoms are bromine-79 and the other 49 percent are bromine-81.
a. What is the atomic number for each of these isotopes?
b. What is the mass number for each of these isotopes?
c. What is the difference between these two isotopes?
d. Using the periodic table, look up the average atomic mass of bromine. Does this value make sense given the relative abundance of bromine isotopes?
e. Why is the average atomic mass of bromine not 79 or 81?

79 Iron has four natural isotopes: iron-54, iron-56, iron-57, and iron-58. The natural abundance of the isotopes are as follows:

Isotope	Natural Abundance (%)
Iron-54	5.80
Iron-56	91.72
Iron-57	2.20
Iron-58	0.28

a. Which isotope is the lightest?
b. Which isotope is the least abundant?
c. Which isotope is the most abundant?
d. Which isotope has the fewest neutrons?
e. Why is the average atomic mass of iron, 55.845 amu, closest to the mass of iron-56, yet not exactly 56?

Navigating the Periodic Table of the Elements

80 What is the full name and atomic number of the following elements?
a. O   b. Na   c. Cu   d. Sn
e. Ru   f. W   g. Eu

81 What is the full name and atomic number of the following elements?
a. B   b. Mg   c. Os   d. Ag
e. Hg   f. Am   g. Cs

82 What is the full name and element symbol for the elements with the following atomic numbers?
a. 6   b. 13   c. 92
d. 78   e. 27

83   What is the full name and element symbol for the elements with the following atomic numbers?
a. 4   b. 25   c. 46
d. 90   e. 106

84   Lead is used in doctors' and dentists' offices as a radiation shield. Between which two elements in the periodic table is lead located and why?

85   Titanium is used in medical prostheses and orthopedic implants. Between which two elements in the periodic table is titanium located?

86   How are elements with similar physical and chemical properties arranged on the periodic table?

87   What is a family or group of elements?

88   What groups constitute the main group elements? Where in the periodic table are the transition metal elements?

89   What is a row of elements in the periodic table called?

90   State the group number and the name of the family, if one exists, for the following elements:
a. calcium, which plays a vital role in signaling for many cellular processes
b. krypton, which is used in fluorescent lights
c. bromine, which is used in fire retardants
d. rubidium, which is used in imaging heart function
e. chlorine, which is a macronutrient

91   State the group number and the name of the family, if one exists, for the following elements:
a. sodium, which is a macronutrient and regulates blood volume, blood pressure, and pH
b. germanium, which is used in semiconductors in electronics
c. palladium, which is used in catalytic converters
d. thallium, which is used in cardiac stress tests
e. neon, which is used in lighted advertising signs

92   What are the differences between nonmetals and metals?

93   Classify the following elements as nonmetal, metal, or metalloid:
a. oxygen   b. germanium   c. carbon

94   Classify the following elements as nonmetal, metal, or metalloid:
a. tin
b. beryllium
c. helium
d. silicon
e. cerium
f. phosphorus

95   Provide the name of the element that fits the following description:
a. A group 1A metal in period 4
b. A noble gas in period 6

96   Provide the name of the element that fits the following description:
a. A transition metal in period 5 with three more protons than technetium
b. An actinide with two fewer protons than plutonium

97   Name three building block elements. Write the number of valence electrons for each of these elements. Why are they called building block elements?

98   Name two macronutrients and two micronutrients. What distinguishes micronutrients from macronutrients?

99   Are the building block elements metals or nonmetals?

100   What organ in the body requires the most iodine? What is the source of iodine for most people?

101   Why is iron an essential micronutrient?

102   What role does fluorine play in the body?

103   What role does zinc play in the body? What foods are good sources of zinc?

Electrons

104   What subatomic particle is most influential in determining the physical and chemical properties of an element?

105   Gold has 79 electrons, whereas helium has 2. Which atom would you expect to have the smaller diameter? Explain.

106   Which atom do you expect to have a larger diameter, oxygen or selenium? Explain.

107   What is the relationship between group number and the number of valence electrons for main group elements?

108   Complete the table below for the ground state of the elements indicated.

Element Name	Atomic Symbol	Atomic Number	Group Number	Number of $n = 1$ Electrons	Number of $n = 2$ Electrons	Number of $n = 3$ Electrons
Boron			3A			
	P					
		11				
				2	8	6

109   Complete the table below for the ground state of the elements indicated.

Element	Symbol	Atomic Number	Group Number	Number of $n = 1$ Electrons	Number of $n = 2$ Electrons	Number of $n = 3$ Electrons
				2	6	0
		4				
Argon						
	F					

110   Which family of elements has seven valence electrons?

111   Which family of elements has four valence electrons?

112   What does it mean for an atom to be in an "excited state"?

113   Is an atom more likely to be in its ground state or an excited state?

114   Which family of elements has a full outermost energy level?

115   What is the difference between the electron arrangement for boron and aluminum?

116 What is the maximum number of electrons allowed in the $n = 3$ energy level?

117 What is the maximum number of electrons allowed in the $n = 1$ energy level?

118 What is the highest occupied energy level in the following atoms? How many electrons does this energy level contain?
a. Ca
b. Na
c. Xe
d. O
e. H

Ions

119 How is the magnitude of charge determined for an ion?

120 What are the two types of ions? What type of element forms anions?

121 What is the difference between a cation and an anion? What type of element forms cations?

122 Write the common ions formed from the following elements:
a. lithium
b. phosphorus
c. iodine
d. vanadium

123 Write the common ions formed from the following elements:
a. calcium
b. tin
c. nitrogen
d. silver

124 Why are ions not formed from group 8A elements?

125 Why are ions not formed from carbon?

126 Hydrogen can gain or lose one electron to form an ion. Write the symbol for these two ions derived from hydrogen. Offer a plausible explanation for why hydrogen can form both an anion and a cation.

127 What is the difference between Cu^+ and Cu^{2+}? How are these two ions similar?

128 Indicate the number of protons and electrons in the following ions. Explain why the charge is as indicated, based on the location of the element in the periodic table. Name the ion.
a. Cs^+ b. Ag^+ c. Cr^{3+}
d. Se^{2-} e. Br^-

129 Indicate the number of protons and electrons in the following ions. Explain why the charge is as indicated, based on the location of the element in the periodic table. Name the ion.
a. Mg^{2+} b. Fe^{2+} c. Cl^-
d. F^- e. O^{2-}

130 When mercury loses one electron to form an ion, it forms a dimer between two mercury ions, Hg_2^{2+}. The charge on each mercury ion is +1. How many protons and electrons are on each mercury ion in the dimer? Mercury also forms another ion; what is its symbol?

Radioisotopes

(Refer to Table 2-10 for the half-lives of the radioisotopes.)

131 What are radioisotopes?

132 What is radioactive decay?

133 What is background radiation?

134 Co-59 is a natural radioisotope, whereas the other 28 isotopes of cobalt are artificial radioisotopes. Co-57 is used to estimate organ size. What is the difference between Co-59 and Co-57 in terms of subatomic particles?

135 When the nucleus of a radioisotope undergoes radioactive decay, what is released in the process?

136 Define electromagnetic radiation.

137 Why is electromagnetic radiation often referred to as "light"?

138 List in order of decreasing energy the following types of electromagnetic radiation:
a. microwave
b. ultraviolet
c. radio waves
d. visible
e. γ-rays
f. x-rays
g. infrared

139 Which form of electromagnetic radiation has the longer wavelength in each pair?
a. radio wave or microwave
b. x-ray or γ-ray
c. visible or ultraviolet

140 Which form of electromagnetic radiation is higher in energy in each pair?
a. visible or infrared
b. visible or ultraviolet
c. x-ray or γ-ray

141 Which form of electromagnetic radiation is more damaging to biological tissue in each pair? Explain why.
a. x-ray or γ-ray
b. ultraviolet or visible

142 Which of the following forms of radiation are not considered electromagnetic radiation?
a. x-ray
b. α particle
c. neutron
d. microwave
e. β particle
f. γ-ray

143 Bismuth-213 is used in targeted α therapy to treat cancer. The α particles that are released from the α decay of bismuth-213 go directly to a cancer cell. Write a nuclear equation for this decay.

144 Lutetium-177 is an ideal therapeutic radioisotope because it is a strong β emitter and has just enough γ radiation to enable imaging.
a. Write the nuclear equation for this β decay.
b. What is the daughter nuclide produced from β decay of this radioisotope?
c. What types of radiation (electromagnetic or high-energy particle) are produced in the radioactive decay of lutetium-177?

145 Iridium-192 is used as a wire inserted directly into a cancer tumor, where it undergoes β decay. After the radiation treatment, the wire is removed. What is the daughter nuclide produced?

146 Which type of radioactive decay produces a daughter nuclide that is a different element from the parent nuclide? More than one correct answer is possible.
a. α　　b. β　　c. γ

147 Write nuclear equations for the following:
a. β decay of Na-26
b. formation of Po-206 through α decay
(Refer to Table 2-10 for the half-lives of the radioisotopes.)

148 Fluorine-18 is used in imaging cells. How much of a 16-g sample of fluorine-18 is left after two half-lives? How much time has elapsed in two half-lives?

149 Iodine-131 is used to treat thyroid cancer. How much of an 18.0-g sample of iodine-131 is left after 32 days? How many half-lives does 32 days represent?

150 Iron-59 is used to study iron metabolism in the spleen. How long would it take a 28-g sample of iron-59 to decay to 1.75 g?

151 Why is Tc-99m an ideal radioisotope for use in medicine? (*Hint:* Consider only the half-life.)

152 If 25.0 mg of I-131 is given to a patient, how much I-131 remains in the patient after 24 days?

153 Tc-99m is produced from the radioactive decay of Mo-99. Mo-99 has a half-life of 66 hours. If you start with 100 g of Mo-99, how long would it take to form 50.0 g of Tc-99m?

154 Why is it not dangerous to eat fruit and vegetables that have been irradiated? What is the benefit of irradiating fruits and vegetables?

155 Show how 80.0 g of a radioisotope decays exponentially by showing how much is present after each half-life. Go through eight half-lives. How many half-lives have elapsed when about 3 percent of the material remains?

156 What are the dangers of nuclear waste? Do hospitals produce nuclear waste?

157 How does ionizing radiation damage biological tissue?

158 What two characteristics of radiation determine its biological effect?

159 What does ionizing radiation do to a molecule to cause it to become an ion?

160 How can ionizing radiation cause cancer?

161 You are continually exposed to radio waves in the environment, and need no protection. However, you need to wear a protective lead apron when you have your teeth x-rayed briefly at the dentist. Explain.

162 It is extremely dangerous to swallow an α emitter; however, you can stop an α emitter with just a piece of paper. Explain the apparent contradiction.

163 Why does a lead apron give your neck and chest sufficient protection against x-rays when you have a dental x-ray taken?

164 Indicate which form of radiation has more energy:
a. x-ray or γ-ray
b. α particle or β particle
c. γ-ray or β particle

165 Indicate which form of radiation has greater penetrating power:
a. α particle or β particle
b. β particle or γ-ray
c. γ-ray or x-ray

166 What type of radiation simultaneously has the highest energy but the least penetrating power?

167 What types of radiation are stopped by a thin piece of lead?

168 How many sieverts are there in 1 rem?

169 Do the becquerel and the curie measure the same or different properties of radiation? Explain. What are the abbreviations for these radiation units? How would you abbreviate a millicurie?

170 Do the becquerel and the rad measure the same or different properties of radiation? Explain.

171 What is the difference between an absorbed dose and an effective dose measurement of radiation? What unit(s) are used for absorbed dose? What unit(s) are used for effective dose?

172 Using Table 2.12, what symptoms can you expect from acute exposure to 2.4 Sv of radiation?

173 What does LD_{50} mean?

174 Identify the following situations as either an acute or a chronic exposure to radiation.
a. A hospital worker carries out CT scans regularly on patients.
b. An uninformed person picks up an α emitter with his unprotected hand.
c. A patient receives radiation therapy.
d. A farmer is standing in his field near Fukushima, Japan, at the time of the Fukushima nuclear reactor accident.

175 On a one-hour flight a man was exposed to 0.000003 Sv of radiation from cosmic rays. A CT scan of the head of the same man exposed him to 0.0015 Sv of radiation. Did the flight or the CT scan expose him to a higher dose of radiation?

176 Thyroid cancer is the most common form of cancer among survivors of the Chernobyl nuclear power plant accident in 1986. What radioisotope do you expect is responsible for this type of cancer?

177 Tc-99m has been used to detect infection in knee prostheses. A patient with a knee prosthesis was injected with 20 mCi of Tc-99m, then monitored with a radiation-based technique. The detection technique monitored the blood flow through the prosthesis.
a. Medical personnel will often use the curie to describe the amount of Tc-99m used rather than the mass in grams. In the example above, how much Tc-99m is left after one half-life?
b. What type of radiation is emitted from Tc-99m?

178 High-dose-rate temporary brachytherapy (another name for sealed source radiotherapy) is used to treat some head and neck cancers. In temporary brachytherapy, the radioactive source is removed from the patient after a short period of time. Suppose that iridium-192 is applied to the tumor for about 6.5 minutes.
a. Iridium-192 undergoes β decay and release of γ radiation. Identify the daughter nuclide formed in this process. Write the nuclear equation for the decay of iridium-192.
b. What kind of material should be used to protect the healthy tissues (i.e., the torso, arms, and legs) of the patient from the β and γ radiation?

Chemistry in Medicine: Nuclear Radiation in Medical Imaging

179 A small child swallowed a quarter. An x-ray showed the quarter clearly as a light spot in a dark area of the esophagus. Explain how the radiologist was able to see where in the child's body the quarter was located.

180 A patient came into the ER with a possible head injury. He was given a CT scan and an MRI. Which one of these techniques was most likely used to determine if he had a skull fracture? Which one of these techniques was most likely used to determine if he had any brain trauma?

181 What are the advantages of performing an MRI instead of a CT scan? What are the disadvantages?

182 What are the advantages of performing an MRI or a CT scan instead of taking an x-ray? What are the disadvantages?

Answers to Practice Exercises

1 a. protons and neutrons;
   b. electrons;
   c. protons;
   d. protons and neutrons.

2 The nucleus is more dense than the atom because it has the same mass in a much smaller volume than the atom.

3 Electrons are found in orbitals.

4 Calcium, Ca, has 20 protons and 20 electrons.

5 a. Kr, 36;   b. Ag, 47.

6 a. Ne, neon;
   b. O, oxygen.

7 Hydrogen, H, atomic number 1.

8 Vanadium, V, has 23 protons.

9 b. Bronze is not an element (it is a mixture of the elements copper and tin). You can tell it is not an element because it is not in the periodic table of the elements.
   e. Steel is not an element (it is a mixture of the elements iron and carbon). You can tell it is not an element because it is not in the periodic table.

10

Isotope Symbol	Mass Number	Atomic Number	Number of Protons	Number of Neutrons
Chlorine-35	35	17	17	18
Mercury-197	197	80	80	117
Iodine-125	125	53	53	72

11

Isotope	Mass Number	Atomic Number	Number of Protons	Number of Neutrons
Silicon-28	28	14	14	14
Silicon-29	29	14	14	15
Silicon-30	30	14	14	16

a. All silicon isotopes have 14 protons and 14 electrons.
b. The number of neutrons is different: 14, 15, and 16.
c. Silicon-30 because it has more neutrons and each neutron has a mass of 1 amu.

12 $^{195}_{79}Au$; alternate ways of writing the isotope: Au-195 or gold-195.

13 a. Ca-44 has the greatest mass.
   b. Ca-40 is the most abundant isotope of calcium.
   c. Ca-44 has the most neutrons (24).
   d. Ca-44 has four more neutrons than Ca-40.
   e. The units for average atomic mass are amu.

14 a. group 2A, alkaline earth metals
   b. group 7A, halogens
   c. group 8A, noble gases
   d. group 5A

15 b. mercury
   d. sodium
   e. uranium
   f. molybdenum

16 a. strontium, Sr
   b. chlorine, Cl
   c. titanium, Ti

17

Element	Metal or Nonmetal	Group Number	Period Number	Atomic Number/ Number of Protons
C	Nonmetal	4A	2	6
H	Nonmetal	1A	1	1
N	Nonmetal	5A	2	7
O	Nonmetal	6A	2	8
P	Nonmetal	5A	3	15
S	Nonmetal	6A	3	16

18 A micronutrient is required in amounts of less than 100 mg/day, whereas a macronutrient is required in amounts of greater than 100 mg/day. We obtain micronutrients and macronutrients from our diet.

19 All the macronutrients are main group elements.

20 All of the building block elements are nonmetals.

21 a. 2;   b. 6.

22

Element	Symbol	Atomic Number	Group Number	$n = 1$	$n = 2$	$n = 3$
Sulfur	S	16	6A	2	8	6
Argon	Ar	18	8A	2	8	8

23 a. B: $n = 1$ has has two electrons, $n = 2$ has three electrons.
b. Ne: $n = 1$ has two electrons, $n = 2$ has eight electrons.
c. N: $n = 1$ has two electrons, $n = 2$ has five electrons.

24 Maximum number of electrons in the $n = 2$ energy level is $2n^2 = 2(2)^2 = 8$.

25 The group 8A elements have their valence energy level full: eight electrons except for $n = 1$, where two electrons is a full energy level.

26 a. Selenide, Se^{2-};
b. potassium, K^+;
c. phosphide, P^{3-}.

27 a. I^- has 53 protons and 54 electrons. N^{3-} has seven protons and 10 electrons; Cr^{3+} has 24 protons and 21 electrons.
b. I^- is formed from a nonmetal. N^{3-} is formed from a nonmetal. Cr^{3+} is formed from a transition metal.
c. I^- has a -1 charge because it is a group 7A element and needs one more electron to fill its valence energy level $(8 - 7 = 1)$. N^{3-} has a -3 charge because it is a group 5A element, which needs three more electrons to fill its valence energy level $(8 - 5 = 3)$. Cr^{3+} has a $+3$ charge because that is one of the variable forms of this transition metal (see Figure 2–24).

28 Group 6A elements gain two electrons because they are short of a full valence energy level by two electrons: they have six valence electrons and need eight. The charge on these ions is -2. For example: O^{2-}, S^{2-}, Se^{2-}.

29 The symbol for Cr(VI) is Cr^{6+} and for Cr(III) is Cr^{3+}. Both chromium ions have 24 protons, but Cr^{3+} has 21 electrons, whereas Cr^{6+} has 18 electrons.

30 a. γ-ray radiation causes the greatest damage to biological tissue.
b. Radio waves have the lowest energy.
c. Microwaves have more energy than radio waves but less than visible light.

31 UV light has a shorter wavelength and a higher frequency than visible light, so it is higher in energy than visible light.

32 Types of radioactive decay: α particles, β particles, γ-ray, and x-ray. Electromagnetic radiation: γ-ray and x-ray.

33 $^{210}_{84}Po \rightarrow ^{4}_{2}\alpha + ^{206}_{82}Pb$ Daughter nuclide: $^{206}_{82}Pb$

34 $^{226}_{88}Ra \rightarrow ^{4}_{2}\alpha + ^{222}_{86}Rn$ Parent nuclide: $^{226}_{88}Ra$

35 a. 53 protons, 78 neutrons
b. $^{131}_{53}I \rightarrow ^{0}_{-1}\beta + ^{131}_{54}Xe$
c. $^{131}_{54}Xe$ d. β particle, a high-energy particle

36 $^{198}_{79}Au \rightarrow ^{0}_{-1}\beta + ^{198}_{80}Hg$
   Parent   Daughter
   nuclide   nuclide

37 For a 1.0-g sample to decay to half that amount, 0.50 g, requires one half-life. Since the half-life of Au-198 is 64.8 hours, the decay would take 64.8 hours. The other 0.50 gram decayed to mercury-198.

38 a. After three half-lives there is 1.25 g of the sample remaining.
b. The half-life of Ce-141 is 32.5 days, so three half-lives is 3×32.5 days $= 97.5$ days.

39 First determine how many half-lives it takes for the C-14 to decay from 50.0 g to 3.13 g:

$$50.0\text{ g} \xrightarrow{1} 25.0\text{ g} \xrightarrow{2} 12.5\text{ g} \xrightarrow{3} 6.25\text{ g} \xrightarrow{4} 3.13\text{ g}$$

Each arrow represents one half-life; therefore, four half-lives have elapsed. The half-life of C-14 is 5,730 years, so the total number of years elapsed is

$$4\text{ half-lives} \times \frac{5{,}730\text{ years}}{\text{half-life}} = 22{,}920\text{ years}$$

40 a. an α particle;
b. an α particle and a β particle;
c. an α particle, a β particle, and an x-ray;
d. all particles;
e. an α particle.

41 Since the α particle was swallowed, it is delivered directly into the body. The α particle is one of the most energetic particles, so it can cause a significant amount of tissue damage to whatever it encounters first.

42 Ionizing radiation removes a valence electron, forming a cation. The cation has one less electron than it has protons, and it has a $+1$ charge.

43 γ radiation has more penetrating power than high-energy particles.

44 A γ-ray is more damaging than an x-ray because it has both more energy and more penetrating power. Almost all forms of radioactive decay involve the emission of γ-rays.

45 The γ-ray is a measure of absorbed dose, an amount of energy, whereas the sievert is a measure of effective does. The effective dose is the absorbed dose times a quality factor, which factors in the type of radiation.

46 Bq ↔ number of radioactive emissions

Gy ↔ absorbed dose

Sv ↔ effective dose

Ci ↔ number of radioactive emissions

rem ↔ effective dose

rad ↔ absorbed dose

47 The 5-Gy α emitter would have a greater effective dose because it is a higher energy particle that can cause more biological damage for the same amount of absorbed dose.

48 When I lived in Denver, the mile-high city, my exposure was around 52 mrem a year. Now that I live in New York, at sea level, my exposure is 26 mrem a year.

49 Nausea, hair loss, and fatigue.

50 An LD_{35} of 2–3 Sv means that 35 percent of those exposed to 2–3 Sv would die within 30 days.

51 Acute is a single large dose, whereas chronic is smaller doses over a longer period of time.

52 0.2–0.5 Sv.

Caffeine

Caffeine is just one of many molecules in the mixture we call coffee. The ball-and-stick model shows the atoms, bonds, and three-dimensional shape of a caffeine molecule. Each sphere represents an atom, and different colors represent different types of atoms: carbon is black or gray, oxygen is red, nitrogen is blue, and hydrogen is light blue or white. The "sticks" joining the atoms represent bonds. [Catherine Bausinger]

3 Compounds and Molecules

What Is In Your Morning Coffee?

Do you start your day with a cup of coffee or tea? Do you drink soft drinks through-out the day to give yourself a lift? If so, you may be among the 90 percent of Americans who consume more than 300 mg of caffeine a day. Caffeine, the central nervous system stimulant in these beverages, is a *compound*. In chapter 2, you learned about elements, and in this chapter you will learn about compounds, the most abundant type of matter. Compounds are composed of two or more different types of atoms that are held together by chemical bonds. Most beverages are actually composed of several different compounds and elements, known as a *mixture*. You will learn about mixtures in chapter 8. A cup of coffee is a mixture composed of mainly water, a compound composed of two hydrogen atoms and one oxygen atom held together by chemical bonds. A cup of coffee also contains caffeine, a compound more complex than water, composed of eight carbon atoms, 10 hydrogen atoms, four nitrogen atoms, and two oxygen atoms. Chemical bonds hold this assemblage of atoms together, creating what is known as a *molecule*.

What happens when caffeine enters your body and how does it keep you awake? Within 15 minutes of ingestion, caffeine molecules make their way through the circulatory system to your brain, where they interact with brain cells. Caffeine blocks an important signal between brain cells that prepares the brain for sleep. That signal begins with the release of adenosine, another molecule, which on binding to the adenosine receptor, a huge protein molecule on the surface of brain cells, signals the cell to slow its activity and prepare for sleep, as shown in Figure 3-1a. When caffeine binds to the adenosine receptor instead, it prevents adenosine from binding to its receptor, so the message is never sent, and you remain awake, as illustrated in Figure 3-1b.

Eventually, all this extra brain activity stimulates your pituitary and adrenal glands, which interpret this as a crisis, thereby releasing additional hormones—the fight-or-flight response. Epinephrine, also known as adrenaline, is produced so that you will be more alert during the perceived crisis. Epinephrine in turn stimulates the release of dopamine, the "feel good" compound that activates the pleasure centers of the brain. Thus, caffeine makes you feel more alert as a result of the release of epinephrine, which in turn makes you feel good as a result of the release of dopamine. Eventually, the extra epinephrine and dopamine wear off, and you begin to feel tired. At that point, you are probably headed for your next cup of coffee. Understanding some of the biochemical events in a healthy individual as well as those involved in disease processes requires an understanding of compounds, the subject of this chapter. ●

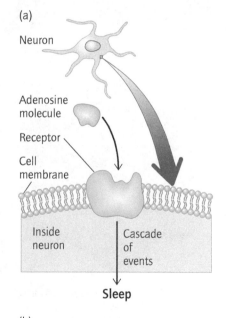

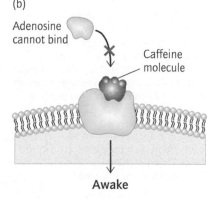

Figure 3-1 Biochemical events in a neuron leading to or preventing sleep. (a) When a molecule of adenosine binds to the adenosine receptor, it initiates a cascade of biochemical events that leads to sleep. (b) When caffeine, instead, binds to the adenosine receptor, it blocks adenosine from binding and prevents the biochemical events leading to sleep.

(a)

(b)

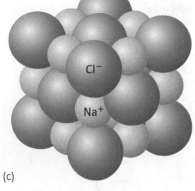

(c)

Figure 3-2 The ionic compound sodium chloride (NaCl) viewed on different scales. (a) On the macroscopic scale, sodium chloride is the familiar crystals we see in table salt. (b) Viewed with a microscope, sodium chloride has a distinctive crystalline appearance. (c) On the atomic scale, sodium chloride is an ionic lattice composed of an equal number of sodium, Na^+, ions (yellow spheres) and chloride, Cl^-, ions (green spheres), arranged so that each cation is completely surrounded by anions and each anion is completely surrounded by cations. [(a) Catherine Bausinger; (b) Charles D. Winters/Science Source]

Matter is composed of elements and compounds. In chapter 2 you learned about elements, and in this chapter you will learn about compounds. Elements are the building blocks for all compounds. *A compound is composed of two or more different atoms in a definite whole-number proportion.* For example, water is a compound composed of two hydrogen atoms and one oxygen atom. The atoms in a compound are held together by *chemical bonds*. There are two basic types of compounds, which differ in the types of bonds they contain:

- Ionic compounds contain ionic bonds.
- Covalent compounds contain covalent bonds.

We first consider ionic compounds and then turn our attention to the covalent bonds that hold together the atoms in molecules.

■ 3.1 Ionic Compounds

An **ionic compound** is composed of metal **cations** and nonmetal **anions** held together by the strong attraction between oppositely charged ions, known as an **ionic bond**, a type of **electrostatic attraction**. Ionic bonds are like a "glue" that holds ions together. *Recall that a central theme in science is that like charges repel and opposite charges attract.*

Most ionic compounds are also called **salts**. Sodium chloride, commonly known as table salt, is perhaps the most familiar ionic compound. Table salt is composed of a 1:1 ratio of sodium (Na^+) and chloride (Cl^-) ions held together by the electrostatic attraction between the positive charge on the sodium cations and the negative charge on the chloride anions.

Like sodium chloride, shown in Figure 3-2, most ionic compounds are brittle solids at room temperature (Figure 3-2a). When viewed under the scanning electron microscope, sodium chloride has the crystalline appearance shown in Figure 3-2b. On the atomic scale, sodium chloride exists in a crystalline lattice structure made up of an equal number of sodium, Na^+, and chloride, Cl^-, ions, as shown in Figure 3-2c. In a **lattice** structure, every cation is surrounded by anions, and every anion is surrounded by cations. An extended lattice of ions goes on until the edge of the crystal. In fact, the only way to disrupt the lattice is to dissolve the compound in water or to heat the salt to extremely high temperatures until it melts.

An ion formed from a single atom, like those described in chapter 2, is a **monatomic ion**. Recall that a monatomic cation is formed when a metal loses one or more valence electrons and a monatomic anion is formed when a nonmetal gains one or more valence electrons. A **polyatomic ion** (*poly-* means "many") is an ion formed when a *molecule*, rather than a single *atom*, gains or loses one, two, or three electrons. In a polyatomic ion, the bonds between atoms are primarily covalent bonds, but an imbalance in the number of protons and electrons overall creates a net charge, and hence it is an ion. The charge on a polyatomic ion can be *localized* on one atom or it can be *spread* over several atoms in the polyatomic ion. The charge on a polyatomic ion is indicated as a superscript following the formula, much like for a monatomic ion. For example, the ammonium ion, NH_4^+, and the carbonate ion, CO_3^{2-}, are polyatomic ions with a +1 and −2 charge, respectively. Other common polyatomic ions are listed in Table 3-1. Each polyatomic ion has a unique *name*, *formula*, and *charge*, which you should learn to associate together. For example, the carbonate ion has the formula unit CO_3^{2-}, a polyatomic ion composed of one carbon and three oxygen atoms, and a −2 charge.

The Formula Unit

An ionic compound is identified by its formula unit, the *ratio* of cations and anions, indicated by subscripts in the formula unit. *The lowest whole number ratio of ions that gives an electrically neutral ionic compound determines the subscripts.* The **formula unit** of an ionic compound composed of monatomic ions is written with the atomic symbol for the cation followed by the atomic symbol for the anion, *excluding* their charges. The ratio of the cations to anions is indicated with subscripts following the symbols. A subscript is understood to be 1 when none is shown. The *lowest* whole-number ratio is used for the subscripts. For example the 2 in Li_2O means there are two lithium cations for every one oxide anion, a 2:1 ratio of Li^+ to O^{2-}. A 1 following oxygen is inferred since no subscript is shown.

In chapter 2 you learned that the charge for a cation derived from a main group metal is the same as its group number and the charge for an anion derived from a main group nonmetal is determined by subtracting 8 from its group number. *The sum of all the positive and negative charges in the formula unit of an ionic compound must add up to zero because an ionic compound is electrically neutral.* Thus, ionic compounds in which the cation and anion have the same magnitude of charge exist in a 1:1 ratio so no subscripts are needed. For example, sodium chloride and magnesium oxide are electrically neutral as a 1:1 ratio of cation to anion:

NaCl: Na^+ Cl^-

              $(+1)$ $+$ (-1) $=$ 0

MgO: Mg^{2+} O^{2-}

              $(+2)$ $+$ (-2) $=$ 0

When the magnitude of the charge on the cation and the anion are different, the ratio of ions is not 1:1 and subscripts must be included to indicate the ratio that gives a neutral ionic compound. For example, the ionic compound formed from Li^+ and O^{2-} has the formula unit Li_2O because *two* lithium cations, Li^+, are necessary to cancel the charge on *one* oxide anion, O^{2-}

Li_2O: $2\,Li^+$ O^{2-}

             $2(+1)$ $+$ (-2) $=$ 0

             $(+2)$ $+$ (-2) $=$ 0

Writing the formula unit for an ionic compound containing a polyatomic ion is similar to the process described for monatomic ions except that if there is a subscript following the polyatomic ion, the entire polyatomic ion formula

> Do not include the charges on the ions when writing the formula unit of an ionic compound.

TABLE 3-1 Common Polyatomic Ions

Name	Formula
Anions	
Acetate	$CH_3CO_2^-$
Carbonate	CO_3^{2-}
Cyanide	CN^-
Dihydrogen phosphate	$H_2PO_4^-$
Hydrogen carbonate (bicarbonate)	HCO_3^-
Hydrogen phosphate	HPO_4^{2-}
Hydroxide	OH^-
Hypochlorite	OCl^-
Nitrate	NO_3^-
Nitrite	NO_2^-
Phosphate	PO_4^{3-}
Sulfate	SO_4^{2-}
Sulfite	SO_3^{2-}
Cations	
Ammonium	NH_4^+
Hydronium	H_3O^+

must be enclosed in parentheses. For example, magnesium hydroxide has the formula unit $Mg(OH)_2$. Parentheses around the hydroxide ion, OH^-, followed by the subscript 2 indicate that there are two hydroxide polyatomic ions (OH^-) for every one magnesium monatomic ion, Mg^{2+}:

$$
\begin{array}{lccc}
\mathbf{Mg(OH)_2} & Mg^{2+} & 2\,OH^- & \\
& +2 & +\ 2(-1) & =\ 0 \\
& +2 & +\ -2 & =\ 0
\end{array}
$$

The general guidelines for determining the formula unit for an ionic compound are listed in the Guidelines box below.

Guidelines for Determining the Formula Unit for an Ionic Compound

Step 1: Determine the charge on the cation.

a. Main group metal elements. For monatomic cations derived from main group elements in group 1A and 2A, the charge on the cation corresponds to its group number.

b. Transition metal elements. If the monatomic cation is derived from a transition metal (or a main group metal in group 3A-5A), there is often more than one possible charge for the cation. You will need to determine which form it is in from other information provided, such as a Roman numeral in parentheses following the name of the cation. For example, if the cation is iron(II), its charge is +2, whereas if the cation is iron(III), its charge is +3.

c. Polyatomic cations. Either memorize the charge associated with a particular polyatomic ion or look it up in Table 3-1.

Step 2: Determine the charge on the anion.

a. The magnitude of the charge on a monatomic anion corresponds to the group number minus 8.

b. For a polyatomic anion, memorize the charge associated with a particular ion or look it up in Table 3-1.

Step 3: Determine the subscripts. Write the cation followed by the anion and insert subscripts so that the sum of all the charges on the ions in the formula unit is equal to zero. If the subscript is 1, it is implied and should not be written in. One aid for determining subscripts is demonstrated below: The subscript on each ion is equal to the numerical value of the magnitude of charge on the *other* ion.

For example, the formula unit for calcium chloride can be obtained by performing the exercise shown:

$$Ca^{2+}\diagdown\!\!\!\!\diagup Cl^- \longrightarrow CaCl_2$$

The sum of the charges is equal to zero when the subscript 2 is placed after Cl:

$$
\begin{array}{lc}
CaCl_2 & Ca^{2+}\quad 2\,Cl^- \\
& (+2)+2(-1)=0
\end{array}
$$

If a subscript is needed for a polyatomic ion, parentheses must be used to enclose the formula of the polyatomic ion, and the subscript placed after the parentheses:

For example,

$$Ca(HCO_3)_2 \qquad \begin{array}{cc} Ca^{2+} & 2\ HCO_3^- \\ (+2) & + & 2(-1) = 0 \end{array}$$

Step 4: If both subscripts can be divided by a common divisor, do so.
The subscripts in a formula unit must be the *lowest* whole-number ratio of cations to anions.

For example, in Mg_2O_2 both subscripts can be divided by 2, so the correct formula unit is MgO and not Mg_2O_2.

Naming Ionic Compounds

To name an ionic compound given its formula unit, write the name of the cation followed by the name of the anion. For example, the compound with the formula unit $CaBr_2$ has the name calcium bromide. A cation derived from a main group element has the same name as the main group metal. For metals that have variable charged forms, the charge on the cation is indicated by including a Roman numeral, corresponding to the magnitude of the charge, in parentheses immediately following the name of the cation. To name a monatomic anion, change the ending on the element name to *-ide*. For a polyatomic ion, the name given in Table 3-1 is used, and no change in the ending is made. For example, Na_2CO_3 is named sodium carbonate, because the compound is composed of sodium ions, Na^+, and the polyatomic carbonate anions, CO_3^{2-}.

Another example, $FeCl_2$ is named iron(II) chloride, where the (II) indicates that the transition metal monatomic iron cation is in its +2 state and not in its alternative +3 state. Roman numerals should not be used for metals that exist in only one form, such as silver (Ag^+) and all main group metal ions.

An alternate naming system for ions with variable charge is shown below. The ending *-ous* is used for the ion with the lower charge form and the ending *-ic* is used for the ion with the higher charge form. For example:

Fe^{2+}	ferrous ion	Cu^+	cuprous ion	Pb^{2+}	plumbous
Fe^{3+}	ferric ion	Cu^{2+}	cupric ion	Pb^{4+}	plumbic

Applying the alternate naming system, $FeCl_3$ would be called ferric chloride rather than iron(III) chloride. Both are acceptable names for this ionic compound.

Ions in Health and Consumer Products

Ionic compounds can be found in many foods and other consumer products (Figure 3-3). For example, calamine lotion, used to soothe skin irritation, is a mixture of zinc oxide and iron(III) oxide. Sodium fluoride, NaF, a common additive in some toothpastes, helps fortify enamel and prevent cavities. Zinc oxide, ZnO, the active ingredient in many sunscreens, effectively blocks the most damaging forms of ultraviolet light.

Tooth enamel (Figure 3-4) is composed of the ionic compound calcium hydroxyapatite, which has the formula unit $Ca_5(PO_4)_3OH$. As you can tell from the formula unit, this ionic compound is made up of two different polyatomic anions: three phosphate ions (PO_4^{3-}) and one hydroxide ion (OH^-). One of the reasons fluoride ion, F^-, is added to municipal water systems, toothpastes, and mouthwashes is that it replaces the hydroxide ion in calcium hydroxyapatite to form $Ca_5(PO_4)_3F$, which is a much stronger substance than hydroxyapatite and therefore less prone to decay.

Note that the numerical subscripts of the formula unit do not appear anywhere in the name of an ionic compound.

Figure 3-3 Some common consumer products that contain ionic compounds: calamine lotion contains zinc oxide (ZnO) and iron(III) oxide (Fe_2O_3), toothpaste contains sodium fluoride (NaF), and sunscreen contains zinc oxide. [Catherine Bausinger]

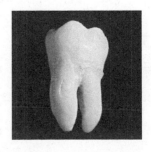

Figure 3-4 Tooth enamel, the top outer part of the molar, is composed of calcium hydroxyapatite, $Ca_5(PO_4)_3OH$. [Anna Blume, Alamy]

The active ingredient in bleach is the hypochlorite ion (OCl^-). This ion is a bactericide—a substance that kills bacteria. The preservatives sodium nitrate, $NaNO_3$, and sodium nitrite, $NaNO_2$, are compounds containing the polyatomic ions NO_3^- and NO_2^-, respectively, which are used to inhibit the growth of microorganisms in food.

WORKED EXERCISES   Formula Units and Names of Ionic Compounds

3-1   What is the formula unit for aluminum sulfide, which is formed when removing tarnish, Ag_2S, from silver using aluminum? Show that aluminum sulfide is electrically neutral.

3-2   What is the name of the compound with the formula unit Fe_2O_3? Show that this ionic compound is electrically neutral.

3-3   What is the name of the compound with the formula unit $(NH_4)_2S$?

3-4   The active ingredient in bleach is sodium hypochlorite. What is the formula unit for this ionic compound?

Solutions

3-1 Al_2S_3

Step 1: Determine the charge on the cation. The cation is listed first, and the charge on aluminum is +3 because it is a cation derived from a group 3A metal: $Al \rightarrow Al^{3+}$

Step 2: Determine the charge on the anion. The charge on sulfur is −2 because it is an anion derived from a group 6A nonmetal: $S \rightarrow S^{2-}$

Step 3: Determine the subscripts. Write the cation followed by the anion and insert subscripts such that the total of all the positive charges cancels the total of all the negative charges. Alternatively, use the method shown below:

$$Al^{3+} \diagdown\!\!\!\!\diagup S^{2-} \longrightarrow Al_2S_3$$

(cation subscript × cation charge) + (anion subscript × anion charge) = 0
$$2(+3) + 3(-2) = (+6) + (-6) = 0$$

Step 4: If the subscripts can be divided by a common divisor, do so. No common divisor exists for the subscripts 2 and 3, so the formula unit is correct as is.

3-2   **Iron(III) oxide.** Iron is a transition metal that exists in more than one form: +2 or +3. To determine the unknown charge on iron (x) in Fe_2O_3, use a trial-and-error method or create a formula equating cations and anions and solve for the unknown cation charge:

(cation subscript × cation charge) + (anion subscript × anion charge) = 0
$$2(x) + 3(-2) = 0$$
$$2x - 6 = 0$$
$$2x = 6$$
$$x = +3$$

Construct the name of the ionic compound by listing the name of the cation first, including parentheses enclosing the form of the transition metal cation that was calculated, as a Roman numeral, followed by the anion name: iron(III) oxide (or ferric oxide in the alternate naming system). Fe_2O_3 is electrically neutral because $2(+3) + 3(-2) = 0$.

3-3   **Ammonium sulfide.** Use Table 3-1 to determine that the name of the polyatomic ion NH_4^+ is ammonium. The anion is a monatomic ion formed from the element sulfur, so change the ending to -ide—sulfide. List the cation name followed by the anion name: ammonium sulfide.

3-4 **NaOCl.** The cation, sodium, is listed first, Na, but understood to be Na^+. The anion, hypochlorite, does not end in -ide, so we know it is not a monatomic ion, but instead a polyatomic anion, whose formula we determine from Table 3-1 is ClO^-. Since the cation has a $+1$ charge and the anion has a -1 charge, we know they exist in a 1:1 ratio in the lattice, so no subscripts are needed in the formula unit. We then write the formula unit listing the cation, followed by the anion, excluding charges: NaOCl.

PRACTICE EXERCISES

1. Answer the following questions for the lattice structures of NaCl and CsCl illustrated in Figure 3-5:
   a. How do the lattice structures shown support the formula units given?
   b. Why is the ratio of cations to anions 1:1 in these two salts?
   c. Why does the lattice show more ions than the formula unit does?
   d. What is holding the ions together in these lattice structures?
   e. What color are the spheres representing the chloride ion in each lattice?
   f. Which ion, Na^+ or Cl^-, has the smaller diameter, based on the figure?

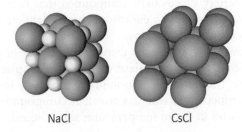

NaCl   CsCl

Figure 3-5 Lattice structure of sodium chloride, NaCl, and cesium chloride, CsCl.

2. Write the formula unit for the following ionic compounds:
   a. sodium sulfide
   b. magnesium nitride
   c. copper(I) chloride
   d. copper(II) chloride

3. Name the following ionic compounds:
   a. Na_2O b. MgI_2 c. KCl

4. From the formula unit PbO_2, deduce whether the lead ion is in its $+2$ or $+4$ form. Explain your answer.

5. Using Table 3-1, name the following ionic compounds:
   a. KNO_3
   b. Na_2HPO_4
   c. NaOH
   d. $AgNO_3$
   e. $(NH_4)_2CO_3$

6. Using Table 3-1, write the formula unit for the following ionic compounds:
   a. lithium hydroxide
   b. strontium carbonate
   c. magnesium phosphate

7. Potassium cyanide is a poison. Write the formula unit for potassium cyanide.

8. Lithium carbonate is an antipsychotic drug used to treat bipolar disorder. What is the formula unit for lithium carbonate?

9. Most gems are finely polished minerals. A mineral is an ionic compound. Turquoise, used in jewelry, has the formula unit $CuAl_6(PO_4)_4(OH)_8$. Notice that there is more than one type of cation and anion; nevertheless, the ionic compound must be electrically neutral. Based on the known charges on the polyatomic anions and the monatomic cations, determine what form of copper is present in turquoise, Cu^{2+} or Cu^+?

■ 3.2 Covalent Compounds

The greatest variety of compounds is seen among covalent compounds. Some of the most important substances found in a biological cell are covalent compounds, including proteins, carbohydrates, lipids, DNA, hormones, and vitamins. A **covalent compound** is composed of identical molecules in the same way an element is composed of the same atoms. *A **molecule** is a discrete entity—not a lattice—composed of two or more nonmetal atoms held together by covalent bonds.* Although there are some elements that exist as molecules, such as hydrogen (H_2), oxygen (O_2), nitrogen (N_2), and the halogens F_2, Cl_2, Br_2, and I_2, most molecules are composed of two or more different atoms, as you saw in the opening vignette for caffeine, which has the molecular formula $C_8H_{10}N_4O_2$. Most covalent compounds are composed of molecules assembled from the building block elements: carbon, hydrogen, nitrogen, oxygen, phosphorus, and sulfur. Molecules come in many different sizes depending on the number of atoms that they contain. Some molecules are quite large, as for example proteins, which can have over 10,000 atoms in their structure, all composed of carbon, hydrogen, oxygen, nitrogen, and sulfur atoms. The variety of covalent compounds that can be constructed from the building block elements is enormous, and scientists create and discover new covalent compounds every day.

Ionic and covalent compounds display very different physical and chemical properties. For example, covalent compounds require less heat energy to melt or vaporize than ionic compounds. Moreover, ionic compounds are solids at room temperature, whereas covalent compounds can be found in all three states of matter at room temperature: solid, liquid, and gas.

The Covalent Bond

In chapter 2 you learned that metals tend to lose electrons and nonmetals tend to gain electrons to become ions so that they achieve full valence energy levels. The resulting metal cations and nonmetal anions are attracted by their opposite charges and form ionic bonds. Alternatively, nonmetal atoms can achieve a full valence energy level by sharing some or all of their valence electrons with another nonmetal. *A shared pair of electrons, between two nonmetal atoms, is known as a single **covalent bond**.* For example, two hydrogen atoms, each with one valence electron, can come together and share their valence electrons to form a hydrogen molecule, H_2, which has two electrons associated with *both* hydrogen nuclei—a covalent bond. Sharing a pair of electrons in a hydrogen molecule effectively gives each hydrogen atom two valence electrons, creating a full $n = 1$ energy level for both hydrogen atoms. The two shared electrons in H_2 are in an *orbital* that encompasses both hydrogen nuclei, as shown in **Figure 3-6**.

In a hydrogen molecule both electrons are attracted to the positive charge of each nucleus. Although the two electrons repel and the two nuclei likewise repel, these repulsions are more than offset by the strong attraction between the negatively charged electrons and the positively charged nuclei when the two atoms are at their optimal separation between the two nuclei, known as the **bond length**. A covalent bond has lower potential energy than two separate atoms. We say the molecule is "more stable" and two separate atoms are "less stable."

The Molecular Formula

A covalent compound is composed of molecules with the same unique composition, described in part by its molecular formula. A **molecular formula** shows the number of each atom type in the molecule by listing the atomic symbols in alphabetical order, followed by a subscript indicating how many atoms of each type there are in the molecule. For example, a water molecule

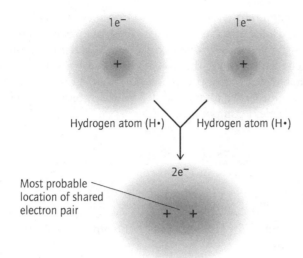

Figure 3-6 Two isolated hydrogen atoms (H), each containing a single electron, come together to form a hydrogen molecule (H_2) where both electrons spend time around both nuclei.

contains two hydrogen atoms and one oxygen atom, so its molecular formula is H_2O. As with formula units, a 1 is assumed if no subscript is written. Molecular formulas differ from formula units in that they do not represent a ratio but the actual number of each type of atom in the molecule. For example, the molecular formula for caffeine, $C_8H_{10}N_4O_2$, indicates there are eight carbon atoms, 10 hydrogen atoms, four nitrogen atoms, and two oxygen atoms in a single discrete caffeine molecule. Trillions and trillions of caffeine molecules constitute the compound caffeine, one of many compounds in the mixture we call coffee.

It is important to recognize that a molecular formula cannot be altered without also changing the identity of the compound. For example, hydrogen peroxide has the molecular formula H_2O_2, so it is composed of the same type of atoms as water (H_2O) but in different numbers, and therefore it is a different compound (Figure 3-7). Although a molecular formula indicates the number and type of atoms in a molecule, it usually does not give sufficient information about the arrangement of the atoms and the type of covalent bonds present. For this, one needs a Lewis structure, described on the next page.

Naming Simple Binary Covalent Compounds

Covalent compounds containing only *two* different types of nonmetal elements—**binary compounds**—can be named using a simple set of rules. Construct the name of the binary compound by naming the first atom in the formula according to its element name followed by the element name of the second atom in the formula, with a change to the ending of its name to *-ide*. For example, NO is named nitrogen oxide. Note that while the ending *-ide* is also used to name ionic compounds, these are covalent compounds.

If more than one atom of a given type is present, as indicated by subscripts, insert a *prefix* before the element name to indicate the number of atoms of that type in the molecule. For example, N_2O has the name dinitrogen oxide, where the prefix *di-* before *nitrogen* indicates that there are two nitrogen atoms (N) in the molecule. Dinitrogen oxide, more familiar as "laughing gas," is widely used in dentistry and other areas of medicine. Table 3-2 is a list of the common prefixes used to name binary compounds.

Figure 3-7 Although water and hydrogen peroxide are both composed of hydrogen (H) and oxygen (O) atoms, they are molecules with a different number of these atoms and therefore entirely different substances. We drink water, H_2O, whereas hydrogen peroxide, H_2O_2, is a disinfectant. [Catherine Bausinger]

TABLE 3-2 Prefixes Used in Naming a Binary Covalent Compounds

Number of Atoms	Prefix
1	*mono*
2	*di*
3	*tri*
4	*tetra*
5	*penta*
6	*hexa*

Be careful not to mix up the rules for naming binary covalent compounds, where prefixes coincide with subscripts, and the rules for naming ionic compounds, where prefixes are *not* included as part of the name. In other words, Al_2O_3 is aluminum oxide, not dialuminum trioxide, because it is an ionic compound, but CCl_4 is carbon tetrachloride because it is a binary covalent compound.

Sometimes, the prefix *mono-* is used (*mono* means "one"), but normally it is assumed that if an atom has no prefix, only one such atom is present. For example, CO is named carbon monoxide, but NO, with a similar formula, is named nitrogen oxide. Note that when the two vowels *ao* or *oo* appear together, the first vowel is dropped. Thus, the binary compound N_2O_5 is named dinitrogen pentoxide. In chapters 6 and 7 you will learn the rules for naming more complex covalent compounds.

WORKED EXERCISES · Naming Binary Compounds

3-5 Name the binary compound PF_5. How can you tell from the molecular formula that this is a covalent compound?

3-6 What is the molecular formula for sulfur hexafluoride? Is this a covalent or ionic compound?

3-7 What is the difference between sulfur dioxide and sodium oxide?

Solutions

3-5 Phosphorus pentafluoride. There is one phosphorus atom named after the element: phosphorus (P). There are five fluorine (F) atoms, so the prefix *penta-* is inserted before the element name, fluorine, and the ending is changed to *-ide*. This is a covalent compound because it is composed of only nonmetal atoms.

3-6 SF_6. Since there is no prefix before sulfur, we know there is only one sulfur atom. The prefix *hexa-* before *fluoride* indicates that there are six fluorine atoms. This is a binary covalent compound because it is composed of only nonmetal atoms.

3-7 Sulfur dioxide is a binary covalent compound with the formula SO_2. Since it is a covalent compound, the number of oxygen atoms, indicated by the subscript in the formula, also appears in the name. The presence of the metal "sodium" is a clue that sodium oxide, Na_2O, is an ionic compound, so no prefixes are used in its name. The ending in both compounds is changed to *-ide* according to the convention.

PRACTICE EXERCISES

10 Name the following binary compounds:
   a. CO_2 b. N_2O_3

11 What are the molecular formulas for the following binary compounds?
   a. nitrogen trifluoride
   b. sulfur trioxide

12 What are the names of the following compounds? Indicate whether each compound is ionic or covalent.
   a. $AlCl_3$ b. CS_2 c. ZnO   d. N_2O_4

Writing Lewis Dot Structures

While a molecular formula tells us the total number of each type of atom in a molecule, it does not tell us how the atoms are arranged and where the bonds are. For this, we use **Lewis dot structures**. Writing and interpreting Lewis dot structures is the foundation for understanding organic chemistry and biochemistry.

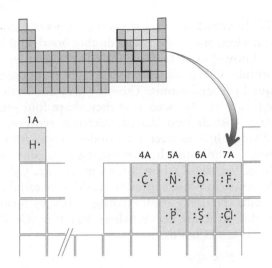

Figure 3-8 Lewis dot symbols for the building block elements.

The **Lewis dot symbol** for an element is a way of representing an atom and its valence electrons by writing its atomic symbol surrounded on up to four sides with its valence electrons represented as *dots*. The Lewis dot symbol for the important nonmetal atoms is shown in Figure 3-8. Remember that the number of valence electrons for an atom corresponds to its group number.

The building block elements in period 2 share one to four valence electrons in order to achieve a full $n = 2$ valence energy level (eight electrons), similar to the noble gas neon (Ne) in period 2. Since the most common building block elements are in period 2, eight electrons is the usual number of electrons we find surrounding atoms in a Lewis dot structure. Hydrogen, in period 1, needs two electrons to fill its $n = 1$ energy level. *To write a Lewis dot structure, arrange the atoms so that each atom in period 2 and higher is surrounded by eight electrons, known as the **octet rule**.* For example, CH_4 has the Lewis dot structure shown in Figure 3-9, where the carbon atom shares each of its four valence electrons with four hydrogen atoms and each hydrogen atom shares its one valence electron with carbon, so that carbon has a full $n = 2$ energy level with eight electrons and each hydrogen has a full $n = 1$ energy level with two electrons. A *pair* of shared electrons is represented by a single line drawn between the two atoms sharing electrons, and known as a single bond. Thus, there are four C—H single bonds in the Lewis structure of CH_4, a molecule known as methane.

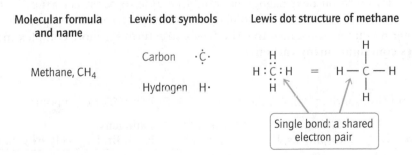

Figure 3-9 Lewis dot structure for methane. Two shared electrons are represented by a line, known as a single bond.

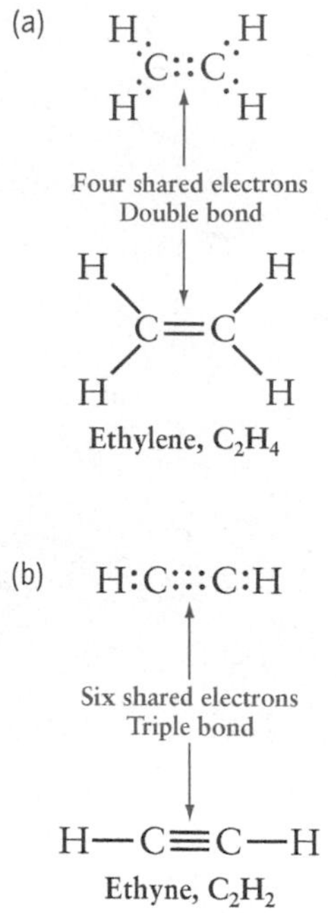

(a)

Four shared electrons
Double bond

Ethylene, C_2H_4

(b) H:C:::C:H

Six shared electrons
Triple bond

H—C≡C—H

Ethyne, C_2H_2

Figure 3-10 The two carbon atoms in (a) a molecule of ethylene (C_2H_4) sharing four electrons—a double bond. (b) A molecule of ethyne (C_2H_2) sharing six electrons—a triple bond.

Octet
Red: Blue:
nonbonding   H—Ö—H   bonding
electrons   electrons
Water, H_2O

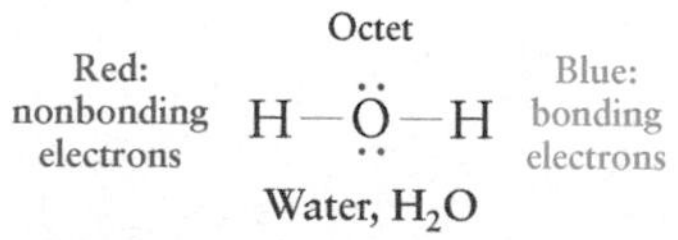

Figure 3-11 Lewis dot structure for water, H_2O: oxygen has an octet with two nonbonding pairs of electrons (red) and two bonding pairs of electrons (blue).

Double and Triple Bonds In order for some atoms to achieve an octet, they must share four electrons, known as a **double bond**, and others must share six electrons, known as a **triple bond**. For example, ethylene has the molecular formula C_2H_4, which means it is composed of two carbon atoms and four hydrogen atoms. The only way for both carbon atoms in this molecule to achieve an octet is if they share four electrons, as shown in Figure 3-10a. Just as two shared electrons represents a single bond, indicated by a single line between the bonded atoms, four shared electrons, a double bond, is indicated by writing two parallel lines between the atoms. Similarly, the two carbon atoms in ethyne, C_2H_2, share six electrons, a triple bond, indicated by writing three parallel lines between the two carbon atoms (Figure 3-10b). Hydrogen always forms only one bond, so the Lewis dot structure for ethylene has four C—H bonds and for ethyne it has two C—H bonds.

Nonbonding Electrons The group 5A, 6A, and 7A elements do not share *all* of their valence electrons. For example, oxygen has six valence electrons, so an oxygen atom needs to share only two electrons (from one or two other atoms) to achieve an octet. The other four valence electrons on oxygen are not shared, hence these electrons are known as **nonbonding electrons** or **lone pair electrons** because they spend their time around only one nucleus. Nonbonding electrons are always shown as pairs of dots in a Lewis dot structure, written beside the atom they belong to but without a second atom attached. The nonbonding electrons on oxygen are shown in red in the Lewis dot structure of water in Figure 3-11. The oxygen has an octet: two nonbonding pairs ($2 \times 2 = 4$ electrons) plus two bonding pairs ($2 \times 2 = 4$) of electrons. Both hydrogen atoms share two electrons, as required to complete the $n = 1$ energy level of hydrogen.

Periodicity: Bonding and Nonbonding Electrons Another trend in periodicity of the nonmetal elements is seen in the number of covalent bonds that elements within a group form. Halogens typically form one covalent bond because they have seven valence electrons—one short of an octet. The group 6A elements form two bonds because they have six valence electrons—two short of an octet. Similarly, group 5A elements form three bonds and group 4A elements form four bonds. Table 3-3 compares the number of bonding and nonbonding pairs of electrons typically seen in the nonmetal elements when they are part of a Lewis structure. For example, according to Table 3-3, oxygen usually has two bonding and two nonbonding pairs of electrons, which means that oxygen is seen with either one double bond, =O, or two single bonds, —O—, in a Lewis structure. Using Table 3-3 together with the guidelines given on next page, you should be able to construct a Lewis dot structure from the molecular formula of most simple covalent compounds. In chapter 6 you will learn how to write Lewis structures for more complex molecules containing many carbon atoms.

TABLE 3-3 Typical Number of Bonding and Nonbonding Pairs of Electrons in the Building Block Elements

Electrons	Carbon, C	Nitrogen, N Phosphorus, P	Oxygen, O Sulfur, S	Halogens: F, Cl, Br, I	Hydrogen, H
Bonding pairs (covalent bonds)	4	3	2	1	1
Nonbonding pairs	0	1	2	3	0

Guidelines for Writing Lewis Dot Structures

1. **Add up the total number of valence electrons from all the atoms in the molecule.** Remember that the number of valence electrons for an element is equal to its group number.

   In CCl_4 the total number of valence electrons is $4(7) + 4 = 32$ electrons: 7 for each of the four chlorine atoms (group 7A) and 4 from the carbon atom (group 4A).

2. **Determine which is the central atom (or two central atoms) and which atoms are on the periphery of the molecule and bonded to the central atom. Initially place a single bond between each pair of atoms.** The central atom is usually an atom found closer to the center of the periodic table because those elements form more bonds. Hydrogen is always on the periphery, and halogens are usually on the periphery. Use Table 3-3 to guide you. Except in the case of hydrogen peroxide, H—O—O—H, avoid forming a bond between two oxygen atoms.

   For example, in CCl_4, carbon is closer to the center of the periodic table than chlorine, so it is the central atom, and the four Cl atoms are on the periphery, each with a single bond to the carbon atom. This is supported by Table 3-3: carbon forms four bonds and chlorine forms one bond.

$$Cl-\underset{\underset{Cl}{|}}{\overset{\overset{Cl}{|}}{C}}-Cl$$

3. **Determine the number of valence electrons remaining after step 2 and distribute these remaining electrons as nonbonding pairs by placing *pairs* of dots around each atom until it has an octet (8), except hydrogen (2). Never place nonbonding electrons on hydrogen. If you run out of electrons before every atom has an octet, proceed to step 4.**

   In CCl_4, we see that out of the 32 valence electrons calculated in step 1, eight were used to form the four C—Cl bonds in step 2. Thus, 24 valence electrons remain to be distributed. The carbon atom already has an octet from step 2,

   so place three pairs (6) of nonbonding electrons around each Cl atom to give each Cl atom an octet. This uses the remaining 24 valence electrons.

$$:\overset{..}{\underset{..}{Cl}}-C-\overset{..}{\underset{..}{Cl}}:$$

4. **If any atoms are short of an octet after all the valence electrons have been distributed in step 3, turn a nonbonding pair of electrons from an adjacent atom into a double bond or turn two nonbonding pairs of electrons into a triple bond.**

   No multiple bonds are required in the example of CCl_4. In a molecule of C_2H_4, there are $2(4) + 4(1) = 12$ valence electrons, which are initially distributed as shown in the partial structure at bottom left, after steps 2 and 3. As you can see, all 12 electrons have been distributed, but one of the carbon atoms is still short of an octet. Resolve this by turning the lone pair of electrons on the adjacent carbon atom into a covalent bond between the two carbon atoms, forming a double bond, which allows both carbon atoms to achieve an octet:

   $$H-C \overset{\frown}{\underset{\underset{H}{|}}{C}}-H \dashrightarrow \overset{H}{\underset{H}{\diagdown}}C=C\overset{H}{\underset{H}{\diagup}}$$

5. **Double-check each atom in the molecule against Table 3-3 to make sure that each atom is surrounded by the expected number of bonding and nonbonding electrons.**

   For CCl_4, according to Table 3-3, the carbon atom should have four bonds and zero lone pairs and each chlorine atom should have one bond and three lone pairs. In the example of C_2H_4, each carbon should have four bonds and zero lone pairs and each hydrogen atom should have one bond and zero lone pairs. Both examples meet these criteria.

WORKED EXERCISE | Writing Lewis Dot Structures |

3-8   Using the five steps in the guidelines on page 103, write the Lewis dot structure for hydrogen cyanide, HCN.

Solution

3-8   **Step 1:** H is in group 1A and has one valence electron, C is in group 4A and has four valence electrons, and N is in group 5A and has five valence electrons. The total number of valence electrons in HCN is therefore $1 + 4 + 5 = 10$ valence electrons.

Step 2: Carbon is the central atom in this molecule because it is the atom closest to the center of the periodic table. H is on the periphery because it forms only one bond. That leaves nitrogen, which must form a bond to carbon because hydrogen cannot form two bonds. Initially, place a single bond between H and C and a single bond between C and N:

$$H—C—N$$

Step 3: Out of the 10 total valence electrons in the molecule calculated in step 1, four electrons have been distributed in step 2. Temporarily place the remaining six electrons as three pairs of electrons around nitrogen to give it an octet. All electrons have been distributed, but carbon still does not have an octet, so proceed to step 4.

$$H—C—\ddot{N}:$$

Step 4: Convert two of the nonbonding pairs of electrons initially placed on nitrogen into carbon-nitrogen bonds, forming a triple bond. In order for both C and N to have an octet, a triple bond is required—six electrons shared between carbon and nitrogen. Every atom has an octet except H:

$$H—C≡N:$$

Step 5: In accordance with Table 3-3, carbon has four bonds (one single and one triple bond) and no lone pairs—the usual arrangement for carbon. The nitrogen atom has three bonds (one triple bond) and one nonbonding pair—the usual arrangement for nitrogen. The hydrogen atom has one single bond and no lone pairs, also as expected.

PRACTICE EXERCISES

13 Answer the following questions for the Lewis dot structure of carbon dioxide:

$$:\ddot{O}=C=\ddot{O}:$$

   a. How many nonbonding pairs of electrons and how many bonding pairs of electrons does each oxygen atom have?
   b. How many nonbonding pairs of electrons and how many bonding pairs of electrons does the carbon atom have?
   c. Does each atom have an octet?
   d. What types of covalent bonds are present in this molecule: single, double, or triple? How many electrons are shared between the carbon and oxygen atoms in this molecule?

14 Write Lewis dot structures for the following compounds using the guidelines and Table 3-3. Do all the period 2 elements have an octet? Which of these molecules are elements?
   a. ammonia, NH_3
   b. oxygen, O_2
   c. hydrogen sulfide, H_2S
   d. chlorine, Cl_2

Extension Topic 3-1 Expanded Octets

The building block elements in period 2—carbon, nitrogen, oxygen, and fluorine—always have an octet when they are part of a neutral molecule. Atoms in period 3—phosphorus, sulfur, and chlorine—also typically have an octet, but they are also capable of being surrounded by *more* than eight electrons, known as an **expanded octet**. Several important biomolecules contain a phosphorus atom with an expanded octet, including adenosine triphosphate (ATP), the energy carrier molecule of the cell, and DNA, the molecule containing a cell's genetic information.

> Expanded octets are available to nonmetals in period 3 or higher because they have available *d* orbitals, a type of atomic orbital different from the s and p orbitals described in chapter 2.

Biomolecules that contain phosphorus are **derivatives** of phosphoric acid, H_3PO_4, whose Lewis dot structure is shown in **Figure 3-12**. Notice that the central phosphorus atom is surrounded by 10 electrons rather than eight. Each oxygen atom has an octet, as expected for a period 2 element.

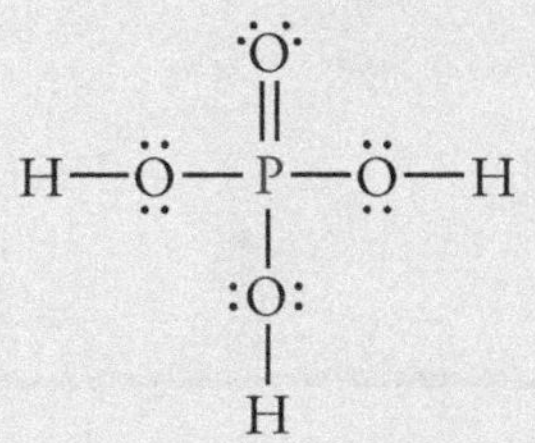

$$H-\overset{..}{\underset{..}{O}}-\overset{\overset{..}{O}}{\underset{\underset{\underset{H}{|}}{:O:}}{P}}-\overset{..}{\underset{..}{O}}-H$$

Figure 3-12 The Lewis dot structure of phosphoric acid, H_3PO_4, showing an expanded octet (10 electrons) for phosphorus, P.

WORKED EXERCISE Identifying Expanded Octets in Molecules

E3.1 For each of the two phosphorus-containing molecules shown below, indicate whether or not it has an expanded octet and how many electrons surround the central phosphorus atom. In which molecule does phosphorus have the normal arrangement of electrons according to Table 3-3?

a.

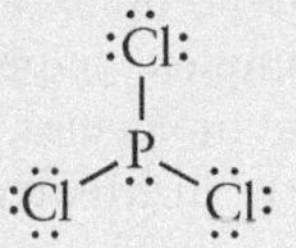

$$:\overset{..}{\underset{..}{Cl}} - \overset{\overset{:\overset{..}{Cl}:}{|}}{\underset{:\underset{..}{Cl}:}{P}} \cdot\cdot \overset{..}{\underset{..}{Cl}}:$$

PCl₃

b.

$$:\overset{..}{Cl}: \quad :\overset{..}{Cl}: \quad :\overset{..}{Cl}:$$

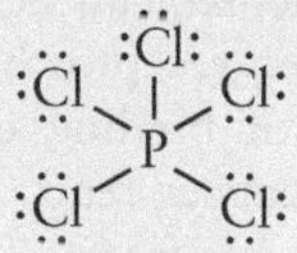

PCl₅

Solution

a. PCl_3 contains a phosphorus atom with an octet (one nonbonding pair plus three bonding pairs). This is the normal arrangement of electrons for a group 5A element such as phosphorus.

b. PCl_5 has an expanded octet with 10 electrons around phosphorus (five bonding pairs and zero nonbonding pairs). This is not the normal arrangement of electrons for a group 5A element, but because it is in period 3, it is allowed to have an expanded octet. Each chlorine atom in both compounds has an octet (three nonbonding pairs and one bonding pair) as expected.

PRACTICE EXERCISES

E3.1 Indicate which atoms, if any, have an expanded octet in the molecules shown. How many electrons are on the atoms with an expanded octet?

a.

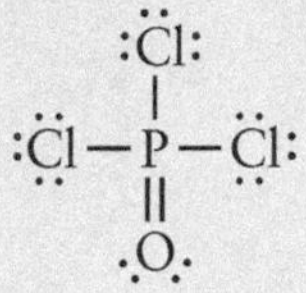

$$:\overset{..}{\underset{..}{Cl}} - \overset{\overset{:\overset{..}{Cl}:}{|}}{\underset{\underset{..}{\overset{..}{O}}:}{P}} - \overset{..}{\underset{..}{Cl}}:$$

b. 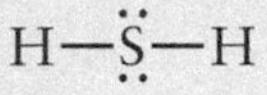 $H-\overset{..}{\underset{..}{S}}-H$

c.

$$H-\overset{..}{\underset{..}{O}}-\overset{\overset{:\overset{.}{O}\cdot}{||}}{\underset{\underset{..}{\overset{.}{O}}:}{S}}-\overset{..}{\underset{..}{O}}-H$$

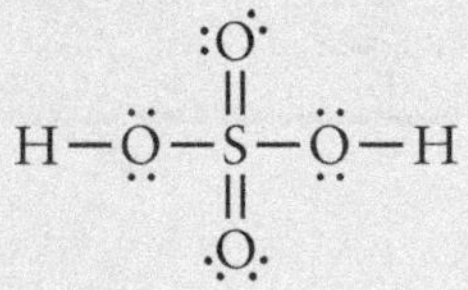

E3.2 Taurine, found in some high-energy drinks and cat food, has the structure shown below. It is believed to enhance the effects of caffeine in high-energy drinks.

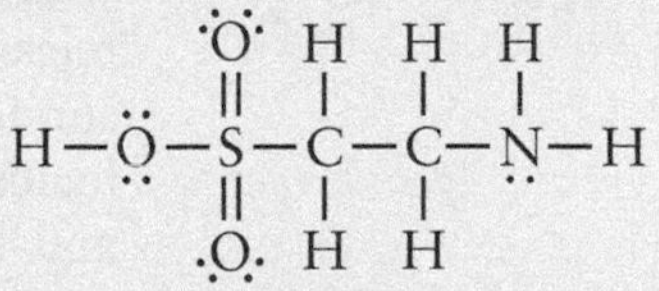

$$H-\overset{..}{\underset{..}{O}}-\overset{\overset{\cdot\overset{..}{O}}{||}}{\underset{\underset{..}{\overset{.}{O}}:}{S}}-\overset{\overset{H}{|}}{\underset{\underset{H}{|}}{C}}-\overset{\overset{H}{|}}{\underset{\underset{H}{|}}{C}}-\overset{\overset{H}{|}}{N}-H$$

Taurine

a. Which element in taurine has an expanded octet? Is the element with an expanded octet in period 3? How many electrons are on this atom?

b. How many bonding electrons does each of the carbon atoms have? How many nonbonding electrons does carbon have?

c. How many bonding and nonbonding electrons does the nitrogen atom have?

Using Molecular Models 3-1 Covalent Bonds in Molecules

Construct H_2, O_2, HCN, and CH_4

1. Obtain two black carbon atoms, seven light blue hydrogen atoms, two red oxygen atoms, and one blue nitrogen atom.

2. Your model kit has two types of bonds: straight and bent. Use one straight bond for any single bond between atoms. Use two bent bonds whenever you need to construct a double bond and three bent bonds whenever you need to construct a triple bond.

3. Make a model of a hydrogen molecule, H_2.

4. Make a model of O_2 similar to the ball-and-stick model shown. You will need to use two bent bonds. Write a Lewis dot structure for O_2.

Oxygen, O_2

5. Make a model of CH_4 as shown. Write a Lewis dot structure for CH_4.

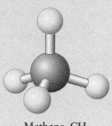

Methane, CH_4

6. Make a model of HCN. You will need three bent bonds to create the triple bond. Write a Lewis dot structure for HCN.

Hydrogen cyanide, HCN

Inquiry Questions

1. Which of the four models represent elements?

2. Which of the models has a double bond? How many shared electrons does a double bond have?

3. Does the carbon atom in CH_4 have an octet? What type of bonds does the model have?

4. Why does HCN have a triple bond? How many shared electrons does a triple bond have?

5. Which models contain only single bonds? How many shared electrons does a single bond have?

3.3 Three-Dimensional Shapes of Molecules

Cells recognize molecules in large part by their three-dimensional shapes, which are important in cellular communication. For example, a hormone released by a cell in one part of the body can interact with its receptor on the surface of another cell in another part of the body, signaling a series of biological events. In this section you will learn about the shapes of molecules, and later in the chapter you will learn about the intermolecular forces of attraction between these molecules.

Molecules are three-dimensional structures; in other words, they are not necessarily two dimensional (flat), as their Lewis dot structures might lead you to believe. Large molecules, in particular, can have quite complex and elaborate shapes. For example, the estrogen receptor has a complex shape, which is necessary for its specific function of binding estrogen molecules. To learn about the role of estrogen in breast cancer, read Chemistry in Medicine: Treating Breast Cancer, at the end of this chapter.

Although Lewis dot structures are two dimensional and are not intended to represent the actual shape of a molecule, they do contain all the information necessary for predicting the shape of a molecule. The shape of a simple molecule containing a central atom surrounded by two or more atoms is defined by the spatial arrangement of all the *atoms*, which in turn is determined by the number of and type (bonding or nonbonding) of *electrons* on the *central* atom. The arrangement of electrons around the central atom is referred to as the **electron geometry,** and the arrangement of atoms around the central atom is known as the **molecular shape** of the molecule.

Molecular Models

Molecular models are frequently used in chemistry to visualize the three-dimensional shapes of molecules (Figure 3-13). Two types of molecular models are routinely used and will be seen throughout this text: the ball-and-stick model and the space-filling model.

Ball-and-stick models represent atoms as *balls* and covalent bonds as *sticks*. Note that the balls are color coded to represent the various building block elements. Ball-and-stick models are used primarily to show molecular shape, although they ignore the amount of actual space (volume) occupied by the atoms in a molecule.

A **space-filling model** is often used to show the amount of space occupied by the atoms in a molecule, but in so doing, it obscures the geometry. Thus, the model one chooses to use depends on what structural aspect of the molecule needs to be visualized.

Occasionally, a tube model will be used in place of a ball-and-stick model, particularly when the molecule is large or complex. In a **tube model,** bonds and atoms are part of a tube, where each end represents an atom, color coded the same as ball-and-stick models. Each half of the bond (a tube) bears the color of the atoms to which it is bonded.

Using Lewis Dot Structures to Predict Electron Geometry

A three-step process is used to determine the *shape* of a molecule:

Step 1: Write the Lewis dot structure for the molecule.
Step 2: From the Lewis dot structure determine the electron geometry.
Step 3: From the electron geometry determine the molecular shape.

In section 3.2 you learned how to do step 1: write the Lewis dot structures. Here we describe the remaining two steps for determining the shape of a simple molecule from its Lewis dot structure.

A Lewis dot structure provides all the information needed to determine the electron geometry of a simple molecule. Predicting electron geometry is based on **v**alence **s**hell **e**lectron **p**air **r**epulsion (**VSEPR**) theory. This theory holds that groups of valence electrons (bonding and nonbonding) surrounding a central atom repel because electrons have like charges. Therefore, bonding and nonbonding groups of electrons around a central atom will assume a geometry that places them as far apart from one another as is geometrically possible while still being bonded to the central atom.

To determine electron geometry, we simply need to count the number of "groups" of electrons surrounding the central atom in the Lewis dot structure. A *group* of electrons for this purpose refers to the six electrons of a triple bond, the four electrons of a double bond, the two electrons of a single bond, or the two electrons of a nonbonding pair of electrons. An electron geometry that places the electron *groups* as far apart from one another as possible is adopted, which depends only on the number of groups involved. These electron geometries, based on the number of groups surrounding the central atom, are

- Four groups of electrons: tetrahedral electron geometry
- Three groups of electrons: trigonal planar electron geometry
- Two groups of electrons: linear electron geometry

**Ethanol, C_2H_5OH
(drinking alcohol)**

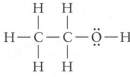

Lewis dot structure

Ball-and-stick model

Space-filling model

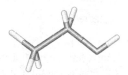

Tube model

● Carbon   ● Oxygen
○ Hydrogen   ● Nitrogen

Figure 3-13 A comparison of the common models used to represent a molecule, showing ethanol (C_2H_5OH): Lewis dot structure, ball-and-stick model, space-filling model, and tube model.

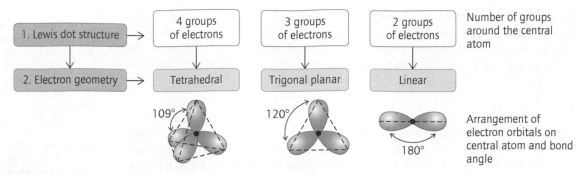

Figure 3-14 Determining the electron geometry of a molecule.

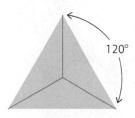

Recall from geometry that an angle is defined by three points. Point *b* is the central atom in a *bond angle*.

Angle (0–360°)

Since most of the molecules you will encounter in biological systems have two-, three-, or four electron *groups* surrounding the central atom, these are the three basic *electron geometries* you will see throughout this text.

The three common electron geometries are illustrated in Figure 3-14. The *one-dimensional* **linear** geometry looks like a straight line, with two groups of electrons separated by 180°. The **trigonal planar** geometry is *two dimensional* with each of the three groups of electrons directed toward one of the three corners of an equilateral triangle, separated by 120° (Figure 3-15a). The **tetrahedral** geometry is *three dimensional* with each of the four groups of electrons directed toward one of the four corners of a tetrahedron, separated by 109.5° (Figure 3-15b). As you can see, the fewer the number of *groups* of electrons around the central atom, the farther apart the electron groups can be placed and the greater the angle between electron groups.

Using Electron Geometry to Determine Molecular Shape

In the third step of the three-step process for determining molecular shape, the electron *geometry* is used to predict the *shape* of the molecule. *When the central atom has only bonding electrons and no nonbonding electrons, its molecular shape is the* same *as its electron geometry.* Thus, four bonding groups around a central atom produce a molecule with a tetrahedral shape, three bonding groups around a central atom produce a molecule with a trigonal planar shape, and two bonding groups of electrons around a central atom produce a molecule with a linear shape, as shown in the ball-and stick models in Figure 3-16 for carbon dioxide (linear), formaldehyde (trigonal planar), and methane (tetrahedral), which each have a central atom surrounded by bonding groups of electrons and no nonbonding electrons.

When the central atom in a molecule has one or two nonbonding pairs of electrons, the molecular shape is different from its electron geometry because

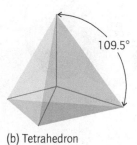

(a) Equilateral triangle

(b) Tetrahedron

Figure 3-15 (a) The trigonal planar electron geometry has the central atom at the center of an equilateral triangle with electron groups directed at each of the three corners, 120° apart from each other. (b) The tetrahedral electron geometry has the central atom at the center of a tetrahedron with electron groups directed at each of the four corners, 109.5° apart from one another.

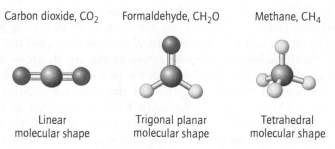

Figure 3-16 Ball-and-stick models of molecules with a molecular shape the same as their electron geometry (a) carbon dioxide, linear; (b) formaldehyde, trigonal planar; (c) methane, tetrahedral.

shape is defined by the relative position of the atoms only. For example, you would not describe a water molecule as having a tetrahedral shape, even though it has a tetrahedral electron geometry (four electron groups), because the relative position of the two hydrogen atoms and the one oxygen atom describes a **bent shape** not a tetrahedral shape (Figure 3-17). However, it is the nonbonding electrons on the oxygen atom that prevent the water molecule from having a linear shape, because that would place the nonbonding and bonding electrons too close to one another.

Thus, the shape of a molecule depends on both its electron geometry *and* the relative number of bonding and nonbonding groups of electrons on the central atom. Let us first consider the three different shapes of molecules derived from a tetrahedral electron geometry and then consider the two different shapes of molecules derived from a trigonal planar geometry. We have already seen that molecules with a linear electron geometry always have a linear molecular shape because they have only bonding electrons.

Water, H₂O

Figure 3-17 Water has a tetrahedral electron geometry (four electron groups) and a bent molecular shape.

Shapes of Molecules with a Tetrahedral Electron Geometry There are three *molecular shapes* derived from a tetrahedral *electron geometry*, which is determined by the number of *nonbonding* groups of electrons on the central atom, as summarized in Table 3-4.

A tetrahedral molecular shape arises when a central atom is surrounded by four bonding groups of electrons and zero nonbonding electrons. For example, methane (CH₄), which has the Lewis dot structure shown in the first row in Table 3-4, has a carbon atom surrounded by four single bonds, so the *electron geometry* is tetrahedral. Since every group is a bonding group, the *molecular shape* is also tetrahedral.

The **trigonal pyramidal** *molecular shape* arises when there are three bonding groups and one nonbonding group of electrons around a central atom with a tetrahedral electron geometry. For example, the Lewis dot structure for ammonia, shown in the second row in Table 3-4, has three N—H single bonds and one nonbonding pair of electrons on the central nitrogen atom. Since there are four groups of electrons around the central nitrogen atom, ammonia

TABLE 3-4 Molecular Shapes Resulting from a Tetrahedral Electron Geometry

Total Number of Electron Groups	Number of Bonding Groups	Number of Nonbonding Groups (Lone Pairs)	Electron Geometry	Molecular Shape	Approximate Bond Angle	Example Lewis Dot Structure	Example Ball-and-Stick Model
4	4	0	Tetrahedral 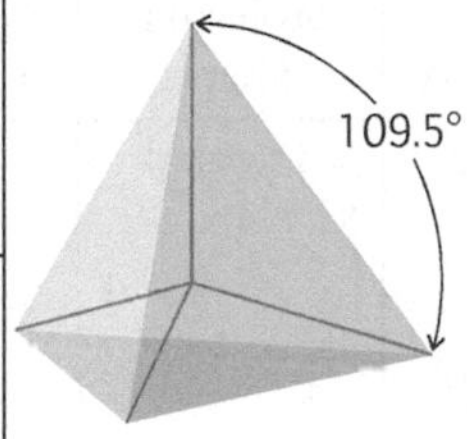	Tetrahedral	109.5°	H—C—H (Methane)	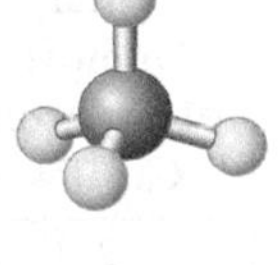
	3	1		Trigonal pyramidal		H—N̈—H (Ammonia)	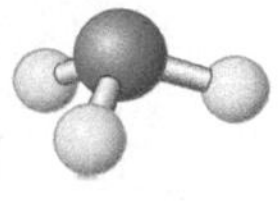
	2	2		Bent		H—Ö—H (Water)	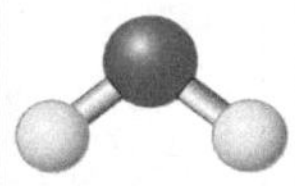

has a tetrahedral *electron geometry*, but the *molecular shape* described by the nitrogen and three hydrogen atoms is trigonal pyramidal.

A **bent** *molecular shape* arises when there are two bonding and two nonbonding groups of electrons around a central atom with a tetrahedral electron geometry. The Lewis dot structure for water, for example, has two O—H bonds and two nonbonding pairs of electrons on oxygen, as shown in the bottom row in Table 3-4. The four groups of electrons on oxygen give it tetrahedral *electron geometry*, but the molecular shape is defined by the two hydrogen atoms and the oxygen atom, a *molecular shape* known as bent.

A **bond angle** describes the angle between the central atom and any two of the atoms to which it is bonded. Since the tetrahedral, trigonal pyramidal, and bent molecular shapes are all derived from the same electron geometry (tetrahedral), they have approximately the same bond angles: 109.5°, as indicated in Table 3-4.

Shapes of Molecules with a Trigonal Planar Electron Geometry

Two molecular shapes are derived from a trigonal planar electron geometry, and both give rise to molecules with bond angles of approximately 120° (Table 3-5). A trigonal planar shape arises when a central atom is surrounded by three bonding groups of electrons and zero nonbonding electrons.

For example, formaldehyde has a trigonal planar *electron geometry* and a trigonal planar *molecular shape*, as shown in the first row in Table 3-5, because all three groups of electrons is bonding.

If one of the electron groups in a molecule with a trigonal planar *electron geometry* is a nonbonding pair of electrons, the *molecular shape* is bent. An example is seen in sulfur dioxide (SO_2), shown in the last row in Table 3-5, which has two bonding groups (the two S—O double bonds) and a nonbonding pair of electrons on the central sulfur atom. The bent shape here is similar in appearance to the bent shape described for water except that the bond angle is approximately 120°. Note also that in sulfur dioxide the *sulfur* atom has an expanded octet, which does not affect the way we count bonding and nonbonding groups to determine *electron geometry* and *molecular shape*.

Shapes of Molecules with a Linear Electron Geometry

When a molecule has a central atom surrounded by two groups of electrons, these groups will

> Remember to count a double bond or a triple bond as one group of electrons. Hence, there are three groups of electrons around the carbon atom in formaldehyde.

TABLE 3-5 Molecular Shapes Resulting from a Trigonal Planar Electron Geometry

Total Number of Electron Groups	Number of Bonding Groups	Number of Nonbonding Groups (Lone Pairs)	Electron Geometry	Molecular Shape	Approximate Bond Angle	Example Lewis Dot Structure	Example Ball-and-Stick Model
3	3	0	Trigonal planar 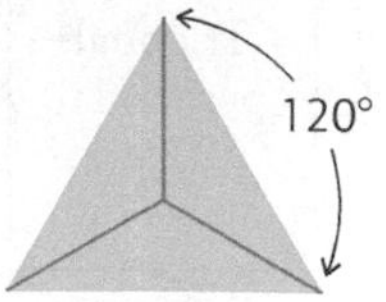	Trigonal planar	120°	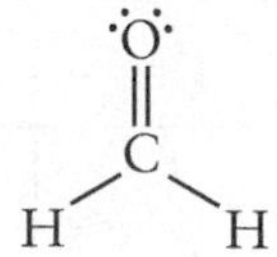 Formaldehyde	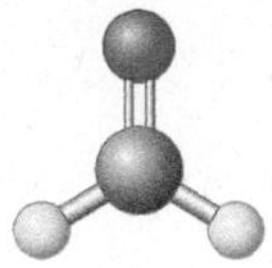
	2	1		Bent		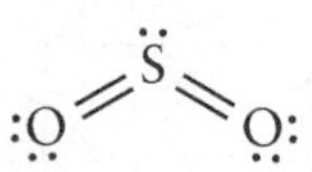 Sulfur dioxide	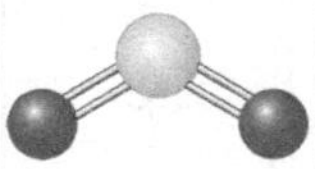

TABLE 3-6 The Molecular Shape Resulting from a Linear Electron Geometry

Total Number of Electron Groups	Number of Bonding Groups	Number of Nonbonding Groups (Lone Pairs)	Electron Geometry	Molecular Shape	Approximate Bond Angle	Example Lewis Dot Structure	Example Ball-and-Stick Model
2	2	0	Linear	Linear	180°	H—C≡N: Hydrogen cyanide	

always be bonding groups; therefore, the molecule will have a linear *electron geometry* and a linear *molecular shape* (Table 3-6). The bond angle in a linear molecule is 180°. For example, hydrogen cyanide, HCN, has two groups of electrons around the central atom, one triple bond and one single bond, so its *electron geometry* and its *molecular shape* is linear and the H—C—N bond angle is 180°.

Techniques for Drawing a Three-Dimensional Representation of a Tetrahedral Molecular Shape Most molecules are easily drawn in a way that accurately represents their molecular shape. However, the tetrahedral shape is a little more difficult to draw because it is three dimensional, while paper is two dimensional. Nevertheless, it is important to learn how to draw and interpret the tetrahedral molecular shape because it is the most common molecular shape encountered in biological molecules. Chemists have adopted a technique used by artists to draw the structure of a tetrahedral center in three dimensions: write the central atom and two of its four bonds in the plane of the paper, using a normal line to represent the bond. The bonds should form an angle greater than 90° to represent the 109.5° bond angle. Next, draw the third bond attached to the central atom projecting toward you by writing a *wedged* line (simulates the appearance of getting larger as it gets closer to you). Finally, write the fourth bond attached to the central atom projecting away from you as a *dashed* line (simulates the appearance of getting smaller), as illustrated in Figure 3-18 for methane. It will help if you have a model beside you the first time you attempt to draw this molecular shape. Using Molecular Models 3-2: The Tetrahedral Molecular Shape is a modeling exercise that will give you greater familiarity with these basic molecular shapes encountered in organic chemistry.

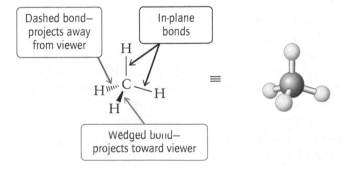

Figure 3-18 Technique for drawing a tetrahedral shape in three-dimensions. A dashed line is used to show the atom projecting away from the viewer and a solid wedged line is used to show the atom projecting toward the viewer. Regular lines represent bonds in the same plane as the paper.

Using Molecular Models 3-2 The Tetrahedral Electron Geometry

Construct Chloromethane, CH₃Cl

1. Obtain one black carbon atom, three light blue hydrogen atoms, one green chlorine atom, and four straight bonds.
2. Using the Lewis dot structure as a guide, make a model of chloromethane.

$$
\begin{array}{c}
\text{H} \\
| \\
\text{H}-\text{C}-\ddot{\text{C}}\text{l}: \\
| \\
\text{H}
\end{array}
$$

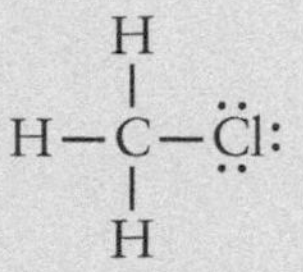

Construction of Ammonia, NH₃

1. Obtain one blue nitrogen atom, three light blue hydrogen atoms, and three straight bonds.
2. Using the Lewis dot structure as a guide, make a model of ammonia.

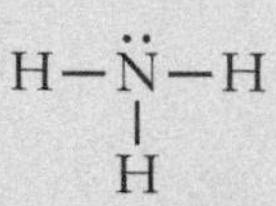

$$
\begin{array}{c}
\text{H}-\ddot{\text{N}}-\text{H} \\
| \\
\text{H}
\end{array}
$$

Construction of Water, H₂O

1. Obtain one red oxygen atom, two light blue hydrogen atoms, and two straight bonds.
2. Using the Lewis dot structure as a guide, make a model of water.

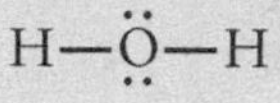

$$
\text{H}-\ddot{\underset{..}{\text{O}}}-\text{H}
$$

Inquiry Questions

1. Examine your models. What are the approximate bond angles in all three of these models? Why are the bond angles the same in all three models?

2. Chloromethane: Is there any way to arrange the hydrogen and chlorine atoms around the central carbon atom in a way that would place the hydrogen and chlorine atoms farther apart from one another? What is the advantage of adopting a tetrahedral geometry with 109.5° bond angles rather than a flat planar geometry with 90° bond angles?

3. Chloromethane: Carefully examine your model of chloromethane. Situate the carbon and two of the C—H bonds in the same plane. Do you see that the other H atom and the Cl atom are behind and in front of the H—C—H plane? Draw a 3-D representation of the model using the dashes and wedges convention.

4. Ammonia: Compare your ammonia model to the model of methane. How is the electron geometry of ammonia similar to that of methane? How is the molecular shape of ammonia different from that of methane?

5. Ammonia: What prevents the H—N—H bond angles from being 120° in this model?

6. Water: Compare your model of water to your models of methane and ammonia. How is the electron geometry of your water model similar to that of methane and ammonia? How is the molecular shape different?

7. Water: What prevents the H—O—H bond angle from being 180° in this model?

WORKED EXERCISES Predicting Molecular Shape

3-9 Answer the following questions for carbon dioxide, CO_2.
  a. Write the Lewis dot structure.
  b. How many electron groups surround the central carbon atom? What are these groups: nonbonding electrons, single bonds, double bonds, or triple bonds?
  c. What is the electron geometry: linear, trigonal planar, or tetrahedral?
  d. What is the molecular shape?
  e. What is the bond angle?

3-10 Ozone, O_3, is a form of elemental oxygen that in the stratosphere protects us from harmful UV radiation. The Lewis dot structure for ozone is shown below (ignore the fact that two of the oxygen atoms do not have the usual two bonding and two nonbonding electron pairs). What is the electron geometry? What is the molecular shape of this molecule? What is the O—O—O bond angle?

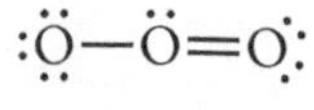

$$
:\ddot{\text{O}}-\ddot{\text{O}}=\ddot{\text{O}}:
$$

3-11 Answer the following questions for a molecule of PBr_3.

    a. Write the Lewis dot structure.

    b. How many electron groups surround the central phosphorus atom?

    c. What is the electron geometry? Explain.

    d. What is the molecular shape? Explain.

    e. What is the approximate Br—P—Br bond angle?

Solutions

3-9

    a. $\ddot{\text{O}}=\text{C}=\ddot{\text{O}}$

    b. The central carbon atom is surrounded by *two* groups of electrons. Both groups are double bonds (C=O double bonds).

    c. The electron geometry is linear because according to VSEPR theory, a linear geometry maximizes the distance between two groups of electrons around a central atom.

    d. The molecular shape is linear because both groups of electrons are bonding, which is always the case for two groups of electrons around a central atom.

    e. The O—C—O bond angle is 180°, the bond angle for a linear geometry.

3-10 The electron geometry for ozone is trigonal planar because three groups of electrons surround the central oxygen atom: one double bond, one single bond, and one nonbonding pair. The molecular shape is bent because one of the three groups of electrons is nonbonding. The O—O—O bond angle is approximately 120° because the electron geometry is trigonal planar.

3-11

    a. 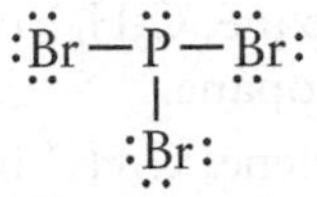 $:\!\ddot{\text{B}}\text{r}-\overset{\displaystyle ..}{\underset{\displaystyle |}{\text{P}}}-\ddot{\text{B}}\text{r}\!:$
            :Br:

    b. There are four electron groups surrounding the central P atom: three single bonds and a nonbonding pair of electrons.

    c. The electron geometry is tetrahedral because there are four groups of electrons around the central atom.

    d. The molecular shape is trigonal pyramidal because there are three bonding groups and one nonbonding group of electrons around the central atom.

    e. The Br—P—Br bond angle is approximate to 109.5° because the electron geometry is tetrahedral.

PRACTICE EXERCISES

15 Fill in the table below for the compounds listed. Note that boron is an exception to the octet rule, surrounded by only six electrons.

	Formula	Lewis structure	Electron geometry	Molecular shape	Bond angles	3-D drawing
a.	CCl_4					
b.	HCN					
c.	BF_3					

16 Write the Lewis dot structure for hydrogen sulfide, H_2S, the substance that gives rotten eggs their bad smell. What is the electron geometry of this molecule? What is the molecular shape? What is the H—S—H bond angle? Why is this molecule not linear?

17 Write the Lewis dot structure and then indicate the electron geometry, molecular shape, and bond angles for each of the following compounds:

    a. CH_2Cl_2 **b.** OF_2 **c.** $SiCl_4$ **d.** PF_3

18 What do the central atoms in all the compounds in exercise 17 have in common?

Methanol, CH₃OH

Figure 3-19 Lewis dot structure and ball-and-stick model of methanol, CH₄O, showing the tetrahedral shape of the carbon center and the bent shape of the oxygen center.

Shapes of Larger Molecules

The six basic molecular shapes described for simple molecules can also be used to evaluate the shapes of larger molecules with many more atoms. The shapes of larger molecules are not described by a unique shape; instead each of the individual atom centers in a complex molecule is evaluated independently, as though it were a central atom surrounded by two, three, or four atoms, in the same manner as shown previously. Then, the shapes of the individual atom centers are spliced together. For example, Figure 3-19 shows a Lewis structure and a ball-and-stick model of methanol (CH_4O). In a molecule of methanol, the carbon atom is surrounded by four bonding groups of electrons, so it has a tetrahedral shape. The oxygen atom is surrounded by four groups of electrons, two bonding pairs and two nonbonding pairs, so according to Table 3-4 it has a bent molecular shape. Splicing the two shapes together yields the overall shape illustrated by the ball-and-stick model shown. We say the carbon atom center is tetrahedral and the oxygen center is bent. Model kits are extremely valuable in helping to visualize the overall shape of larger molecules since the holes for the bonds are predrilled in the balls so that they produce the correct shapes.

WORKED EXERCISES   Shapes of Larger Molecules

3-12 What is the molecular shape around each carbon atom in propane, C_3H_8? In your own words, describe the overall appearance of the shape of propane.

3-13 What is the molecular shape around each carbon atom in ethylene, C_2H_4? In your own words describe the overall shape of ethylene.

Solutions

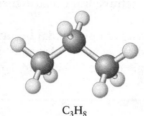

C_3H_8

3-12 Propane is composed of three tetrahedral carbon centers, giving it an overall zigzag appearance which can be seen in the model.

3-13 To determine the shape of ethylene, begin by writing its Lewis dot structure:

$$\underset{H}{\overset{H}{\diagdown}} C = C \underset{H}{\overset{H}{\diagup}}$$

Next, determine the electron geometry and molecular shape around each carbon atom. Both carbon atoms are surrounded by three groups of electrons and no nonbonding pairs, so both carbon atoms have a trigonal planar electron geometry and molecular shape, as shown in the ball-and-stick models below. The overall shape of the molecule can be described as flat—or two-dimensional.

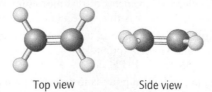

Top view   Side view

PRACTICE EXERCISE

19 Describe the overall shape of acetylene (C_2H_2) by writing the Lewis dot structure and analyzing the electron geometry and molecular shape of each carbon center.

■ 3.4 Polarity

Have you ever wondered why raindrops bead up on a leaf (Figure 3-20) or saran wrap sticks to itself? These properties are a result of molecular *polarity*, a characteristic of molecules that determines how they interact with other molecules. In this section you will learn what makes a molecule polar or nonpolar, and then in the next section you will learn how polarity determines how molecules interact with one another.

Polarity is an important concept in biology. It accounts for the formation of cell membranes, which separate the inside of the cell from the outside of the cell. Polarity also determines whether drug molecules can cross the blood-brain barrier. You will see applications of polarity throughout this text.

In section 3.2 you learned that a covalent bond is formed when two atoms share some of their valence electrons. You also learned that the shared electrons of a covalent bond spend their time around both atomic nuclei. While the word *sharing* may imply that the electrons spend their time *equally* around both nuclei, this is not necessarily the case. Some atoms are better able to draw electrons toward their nuclei than other atoms. *Electronegativity is a measure of an atom's ability to draw electrons toward itself in a covalent bond.*

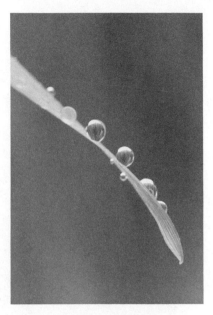

Figure 3-20 Water droplets form beads on a leaf as the result of strong intermolecular forces of attraction between water molecules. [Creatas Images/jupiterimages/Getty]

Electronegativity

Figure 3-21 shows electronegativity values for the main group elements. The height of the bars in this three-dimensional periodic table corresponds to electronegativity; hence, taller bars and darker colors correspond to more electronegative

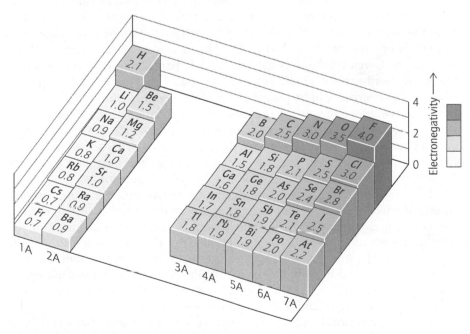

Figure 3-21 Electronegativity values for the main group elements. Darker shades and taller bars represent more electronegative atoms.

Electronegativity is a unitless quantity that ranges from 0 to 4.

elements. Not surprisingly, a trend in electronegativity is observed across the periodic table: elements toward the *bottom left* part of the periodic table: are the *least* electronegative (light green) and elements toward the *top right* part of the periodic table are the *most* electronegative (dark green). Hence, the most electronegative element is fluorine, with an electronegativity value of 4.0. Note that the noble gases are not included on the electronegativity table because these elements do not usually form covalent bonds.

Several factors affect the electronegativity of an element. One is the proximity of the valence electrons to the nucleus. The closer the valence electrons are to the positively charged nucleus, the more electronegative the atom. Thus, within a group of elements, electronegativity increases as you go from the bottom to the top of a group. For example, fluorine is more electronegative than chlorine, which is more electronegative than bromine, and so forth:

$$\text{Trend within a } group: \boxed{F > Cl > Br > I}$$

Within a period, electronegativity increases from left to right. For example, fluorine is more electronegative than oxygen, which is more electronegative than nitrogen, and so forth:

$$\text{Trend within a } period: \boxed{F > O > N > C}$$

Bond Dipoles

A **polar covalent bond** exists between two atoms with a significant difference in electronegativity, such that the bonding electrons spend more time around the more electronegative atom. Since electrons carry a negative charge, this creates a partial negative charge (δ^-) on the more electronegative atom and a partial positive charge (δ^+) on the less electronegative atom. This separation of opposite charges in a polar covalent bond is known as a **bond dipole**—two poles. Specifically, a polar covalent bond is defined as a bond between two atoms with an electronegativity value that differs by more than 0.5 but less than 2.

As shown in **Figure 3-22a**, the covalent bond in hydrogen fluoride, H—F, is a polar covalent bond because the electronegativity difference between fluorine (the most electronegative element EN = 4) and hydrogen (the least electronegative nonmetal EN = 2.1) is 1.9. This is evident in the electron cloud shown. The Greek symbol δ is used to indicate partial charges on atoms in a molecule, as illustrated in the space-filling model in Figure 3-22a. Alternatively, a **dipole arrow** can be placed alongside the bond, written with the head of the arrow pointing toward the more electronegative atom. Figure 3-22a also shows an electron density diagram of the HF molecule. An **electron density diagram** is a space-filling model that employs color to show regions of high electron density (red) and low electron density (blue) within a molecule. Neutral regions of a molecule appear green to yellow in color.

In a covalent bond formed between two atoms with the same or very similar electronegativity values (electronegativity difference is less than 0.5), the bonding electrons are evenly distributed between both atoms, and the bond is known as a **nonpolar covalent bond**. This is illustrated for a molecule of hydrogen, H_2, in **Figure 3-22b**. A carbon-hydrogen bond is another common nonpolar bond because carbon and hydrogen have very similar electronegativity values (2.5 − 2.1 = 0.4). *Hence, all hydrocarbons—compounds containing only carbon and hydrogen atoms—are nonpolar.*

Ionic and covalent bonds are part of a polarity continuum that spans from nonpolar covalent bonds on one end of the continuum to ionic bonds on the

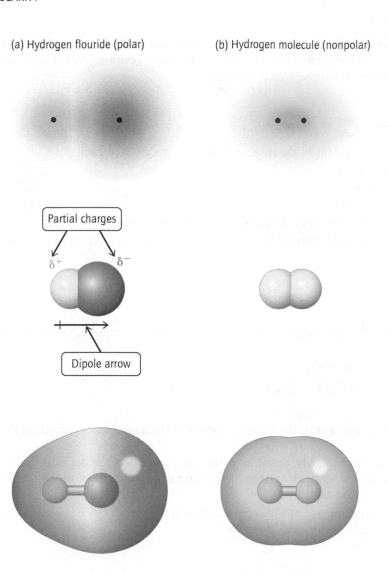

(a) Hydrogen flouride (polar)   (b) Hydrogen molecule (nonpolar)

Partial charges

δ^+ δ^-

Dipole arrow

Figure 3-22 (a) Hydrogen fluoride, HF, a polar molecule and (b) Hydrogen, H_2, a nonpolar molecule. Electron cloud (top), space-filling model (middle), electron density model (bottom).

other end of the continuum, as shown in Figure 3-23. Polar covalent bonds are in the middle of the continuum. The figure shows nonpolar bonds (on the left) have equally shared electrons because the difference in electronegativity is close to zero (less than 0.5). Polar covalent bonds (middle) have electrons that spend more of their time around the more electronegative atom, creating separated partial charges, because the difference in electronegativity of the atoms is between 0.5 and 2.0, as shown on the scale in Figure 3-23. Ionic bonds occur

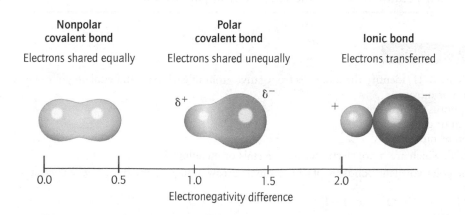

Nonpolar covalent bond	Polar covalent bond	Ionic bond
Electrons shared equally	Electrons shared unequally	Electrons transferred

δ^+ δ^- $+$ $-$

0.0   0.5   1.0   1.5   2.0

Electronegativity difference

Figure 3-23 The bond polarity continuum ranges from nonpolar bonds (far left), with their equally shared electrons, to ionic bonds (far right), with full separate charges. Polar covalent bonds are in the middle, with electrons spending more time on the more electronegative atom, creating separated partial charges.

between atoms that have actually transferred their electrons (from metal to nonmetal) to become cations and anions with separated full charges (electronegativity difference greater than 2.0).

WORKED EXERCISES | Distinguishing Polar and Nonpolar Covalent Bonds

3-14 Using Figure 3-21, identify the more electronegative atom in each pair:

    a. carbon or oxygen

    b. sulfur or oxygen

    c. calcium or iodine

3-15 As a group, which are more electronegative: metals or nonmetals? Explain.

3-16 Label the polar covalent bonds by placing a properly drawn dipole arrow alongside the bond

    a. C—Cl

    b. N—H

    c. C—H

3-17 Using Figure 3-21, label each of the following bonds as nonpolar, polar covalent, or ionic:

    a. NaCl   **c.** C—H

    b. O—H   **d.** C—S

Solutions

3-14 a. Oxygen, because it is to the right of carbon in the same period on the periodic table.

    b. Oxygen, because it is above sulfur in the same group on the periodic table.

    c. Iodine, because it is much further to the right of calcium, even though iodine is in a higher period than Ca.

3-15 Nonmetals as a group are more electronegative than metals since nonmetals are to the right of metals on the periodic table.

3-16 a. C—Cl

    b. N—H

    c. The bond is nonpolar because C and H have approximately equal electronegativity values (EN difference less than 0.5).

3-17 a. ionic because Na^+ is a metal ion and Cl^- is a nonmetal anion (difference in EN >2)

    b. polar covalent because oxygen is much more electronegative than hydrogen (difference in EN $= 3.5 - 2.1 = 1.4$)

    c. nonpolar because the electronegativity values for C and H are approximately the same (difference in EN <0.5)

    d. nonpolar because they have the same electronegativity values.

PRACTICE EXERCISES

20 Using Figure 3-21, identify the more electronegative atom in each pair and explain your choice:

    a. carbon or nitrogen

    b. phosphorus or nitrogen

    c. lithium or chlorine

    d. silicon or nitrogen

21 As a group, which are more electronegative: metals or metalloids?

22 Label the polar covalent bonds by placing a dipole arrow alongside the bond in the correct direction:

    a. O—H   **b.** C—O   **c.** F—F

Polar and Nonpolar Molecules

A molecule with more than one polar bond can be either polar or nonpolar, depending on its molecular shape. A **polar molecule** has a separation of charge—a positive end and a negative end of the molecule—like a north pole and a south pole. In contrast, a **nonpolar molecule** has an even distribution of electrons throughout the molecule. Polar and nonpolar molecules have very different physical properties, which are very important in living cells. For example, fats do not dissolve in blood because fats are composed of mainly carbon and hydrogen atoms, nonpolar molecules, while blood is composed primarily of water, a polar molecule.

If a molecule has only nonpolar covalent bonds, the molecule is nonpolar. Hydrocarbons, for example, are nonpolar molecules because they contain only C—C and C—H bonds, both nonpolar bonds. For a molecule with two or more polar covalent bonds, the shape of the molecule determines whether the molecule is polar or nonpolar. For example, carbon dioxide is a nonpolar molecule even though it has two polar bonds. The linear shape of the molecule causes the two identical bond dipoles to be directed 180° apart from each other and hence cancel each other.

$$\overleftrightarrow{O}\!=\!\overrightarrow{C}\!=\!\overleftrightarrow{O}$$
$$\underset{\delta^-}{}\ \underset{\delta^+}{}\ \underset{\delta^-}{}$$

Carbon dioxide
Nonpolar

Bond dipoles also cancel when there are three identical bond dipoles in a molecule with a trigonal planar molecular shape or four identical bond dipoles in a molecule with a tetrahedral molecular shape. For example, CF_4, BF_3, and CO_2 (Figure 3-24) are all nonpolar molecules, even though they contain polar covalent bonds. They are nonpolar molecules because their bond dipoles cancel. Note that in order for bond dipoles to cancel, the bonds must be identical, as in CF_4, which has four identical C—F bonds, and BF_3 which has three identical B—F bonds, and CO_2 which has two identical C—O bonds. In contrast, the dipoles in a molecule of CF_3Cl do not cancel because the amount of charge separation in a C—F bond is different from that in a C—Cl bond.

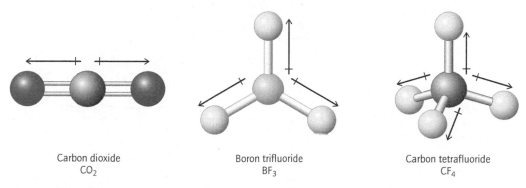

Carbon dioxide
CO_2

Boron trifluoride
BF_3

Carbon tetrafluoride
CF_4

Figure 3-24 Examples of nonpolar molecules with bond dipoles that cancel. In the linear molecule, CO_2, the equivalent C=O bond dipoles are directed 180° apart. In the trigonal planar molecule, BF_3, the equivalent B—F bond dipoles are directed 120° apart. In the tetrahedral molecule, CF_4, the equivalent C—F bond dipoles are directed 109.5° apart.

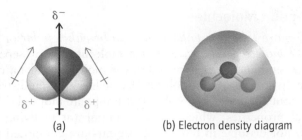

(a) (b) Electron density diagram

Figure 3-25 Models showing water as a polar molecule. (a) Space-filling model showing bond dipoles and molecular dipole, and partial charges. (b) Electron density diagram showing that the oxygen atom has a partial negative charge, indicated in red, and the hydrogen atoms have a partial positive charge, indicated in blue.

Molecules with one or more bond dipoles and a trigonal pyramidal or bent molecular shape are polar because the bond dipoles *cannot cancel* as a result of the shape of the molecule. Consider, for example, the two O—H bonds in a water molecule, H_2O, shown in **Figure 3-25a**. Since water has a bent molecular shape rather than a linear shape, the two identical O—H bond dipoles do not cancel. The center of partial negative charge is on the more electronegative oxygen atom and the center of partial positive charge is between the two less electronegative hydrogen atoms, creating a net molecular dipole, indicated by the large black dipole arrow. If you add the two bond dipoles (blue arrows), the result is a net molecular dipole, shown by the black arrow. The electron density diagram in Figure 3-25b also illustrates that oxygen has a partial negative charge (red) and the hydrogen atoms have partial positive charges (blue). To help you determine whether a molecule is polar or nonpolar, follow the flowchart given in **Figure 3-26**.

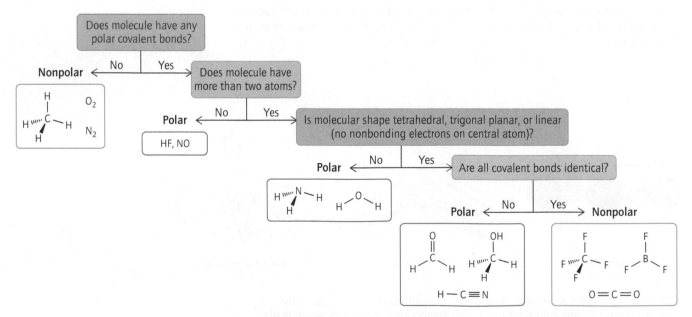

Figure 3-26 Flowchart for determining whether a molecule is polar or nonpolar based on its shape and bond dipoles.

WORKED EXERCISES   Identifying Polar and Nonpolar Molecules

3-18 Carbon tetrafluoride, CF_4, has four polar C—F bonds, yet the molecule is nonpolar. Explain.

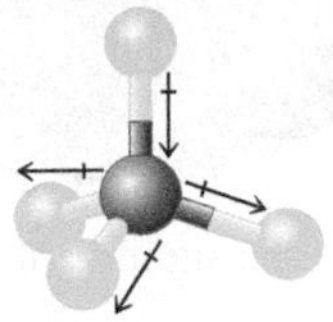

3-19 Explain why BF_3 is nonpolar but NF_3 is polar. Note that boron typically does not have an octet and no lone pairs.

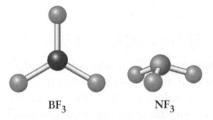

BF_3 NF_3

Solutions

3-18 Each C—F bond dipole is equivalent, and because the molecule has a tetrahedral shape, these bond dipoles are directed to the four opposite corners of a tetrahedron, causing the bond dipoles to cancel; hence the molecule is nonpolar.

3-19 Both BF_3 and NF_3 have three equivalent polar covalent bonds (B—F or N—F), because fluorine is more electronegative than boron or nitrogen. However, BF_3 has a trigonal *planar* molecular shape, whereas NF_3 has a trigonal *pyramidal* molecular shape. Thus, the three B—F bond dipoles cancel in BF_3 because they point to the three opposite corners of an equilateral triangle, making the molecule nonpolar. In contrast, NF_3 has a trigonal pyramidal molecular shape, and so the three N—F bond dipoles do not cancel, making NF_3 a polar molecule with a net dipole as shown below:

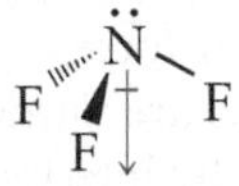

PRACTICE EXERCISES

23 Indicate whether the following molecules are polar or nonpolar. Explain.
   a. Cl_2
   b. C_3H_8
   c. SO_2 (trigonal planar electron geometry; bent molecular shape)
   d. HCl
   e. CCl_4

24 A Lewis dot structure and an electron density diagram for formaldehyde, CH_2O, are shown below.

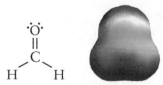

   a. What is the electron geometry? What is the molecular shape? What are the bond angles?
   b. Is this molecule polar or nonpolar? If it is polar, label the positive and negative ends of the molecule with a dipole arrow.
   c. What does the blue region in the electron density diagram signify? What does the red region signify?

25 An electron density diagram for estradiol, $C_{18}H_{24}O_2$, an estrogen, is shown below:

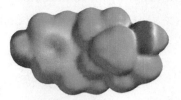

a. What information is conveyed by the significant amount of green color in this electron density diagram?

b. There are two O—H groups in the molecule. Where are they and how can you spot them so readily?

3.5 Intermolecular Forces of Attraction

At room temperature (25°C), why is water (H_2O) a liquid and methane (CH_4) a gas? This question asks about a macroscopic property—a property that you can observe with the naked eye. Yet, the answer depends in large part on how the molecules of the substance interact with one another at the atomic level. You have already seen that atoms interact to form covalent bonds, creating molecules. In this section, you will see how molecules interact with one another through *inter*molecular forces of attraction—those forces that exist *between* molecules rather than *within* molecules. Intermolecular forces of attraction determine the physical properties of a substance, such as boiling point, melting point, and solubility.

Intermolecular forces are generally much weaker forces of attraction than covalent bonding forces. Typically the strongest intermolecular forces of attraction are 5 to 10 percent of the strength of the average covalent bond because the charges involved are much smaller—partial charges rather than full charges—and the distances between charges are greater. Although intermolecular forces of attraction are weaker than covalent bonds, they are extremely important. Covalent bonds are found where relatively *permanent* connections between atoms are required, and intermolecular forces of attraction are found where *temporary* connections between molecules are needed. For example, the two strands of a DNA molecule are held together by intermolecular forces of attraction, and so they can be pulled apart—like a zipper—when the information on the two strands must be read by the cell, such as during DNA replication or transcription.

As with ionic bonds, intermolecular forces arise from electrostatic interactions: the attraction between opposite charges and the repulsion between like charges. Partial charges resulting from dipoles in one molecule are attracted to the opposite partial charges in another molecule. A dipole can be either a *permanent dipole*, such as those found in the polar molecules you learned about in section 3.4, or a *temporary dipole*, those induced in nonpolar molecules. The three basic types of intermolecular forces of attraction, in order of increasing strength, are

- dispersion forces
- dipole–dipole forces
- hydrogen-bonding forces

We can determine the intermolecular force of attraction present by evaluating the strength of the *individual* bond dipoles in the molecules. Figure 3-27 shows a flowchart that summarizes this evaluation process.

Dispersion Forces

All molecules and elements interact through dispersion forces, also known as London forces, named after the German physicist who was the first to explain

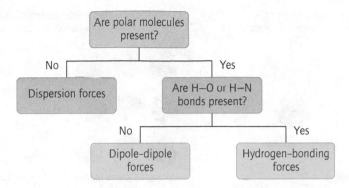

Figure 3-27 Flowchart for determining the type of intermolecular forces of attraction present in a covalent compound.

these forces. Dispersion forces are the only intermolecular force of attraction available to nonpolar molecules. Since nonpolar molecules do not have a permanent dipole, they have a uniform distribution of electrons. The *average* distribution of electrons over time in a nonpolar molecule is illustrated in Figure 3-28a. However, since electrons are in constant random motion, at any given *instant* in time they may temporarily shift toward one end of the molecule, creating a *temporary* molecular dipole—a momentary separation of charge (Figure 3-28b). A temporary molecular dipole in one molecule *induces* a corresponding temporary molecular dipole in neighboring molecules, causing them to shift their electrons in a way that brings opposite partial charges together and like charges apart, thereby drawing the molecules together (Figure 3-28c). Because *induced* dipoles are short lived, dispersion forces are the weakest of the three types of intermolecular forces of attraction.

Since electrons are distributed throughout a molecule, there are many opportunities for temporary dipoles to form. It follows, therefore, that dispersion forces of attraction are greater in molecules that have

- more electrons
- more surface area

For example, hexane (C_6H_{14}) has more electrons and more surface area than ethane (C_2H_6), so there are more dispersion forces in a sample of hexane than in a sample of ethane. Indeed, this is why hexane is a liquid at room temperature and ethane is a gas at the same temperature. Fewer intermolecular forces of attraction are holding the molecules of ethane together, so it is a gas at this temperature.

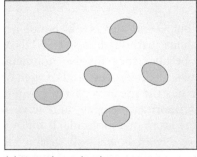

(a) Nonpolar molecules

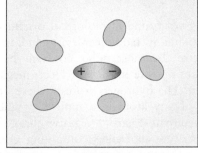

(b) Temporary dipole in one molecule

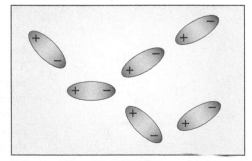

(c) Induced dipoles in neighboring molecules

Figure 3-28 Dispersion forces: (a) The distribution of electrons, on average, is uniform in nonpolar molecules. (b) At any given instant, electrons may temporarily shift toward one end of a molecule, creating a temporary molecular dipole. (c) A temporary molecular dipole in one molecule induces temporary molecular dipoles in neighboring molecules, which align so that opposite charges come together.

Figure 3-29 Dipole–dipole intermolecular forces of attraction in formaldehyde. (a) A formaldehyde molecule showing a permanent dipole. (b) Formaldehyde molecules arrange themselves so that the permanent dipoles orient themselves so that opposite partial charges interact.

(a) Permanent dipole in a single formaldehyde molecule, three views

(b) Dipole-dipole interactions of many formaldehyde molecules, two views

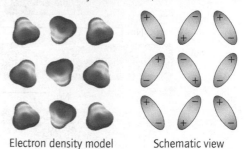

Electron density model   Schematic view

Dipole–Dipole Forces

In addition to dispersion forces, polar molecules have **dipole–dipole** intermolecular forces of attraction, created by the electrostatic attraction that occurs between the opposite partial charges of the permanent dipoles in these molecules. Polar molecules orient themselves in a way that brings opposite charges together and keeps like charges apart. Because the dipoles in these molecules are *permanent*, dipole–dipole intermolecular forces of attraction are stronger than the *temporary* dipoles that are responsible for dispersion forces.

For example, formaldehyde has a permanent dipole as a result of its C=O bond (Figure 3-29a). Therefore, the partial positively charged end of one formaldehyde molecule will orient itself toward the partial negatively charged end of another formaldehyde molecule and so forth throughout the sample, as shown in Figure 3-29b.

Hydrogen Bonding Forces

If you have ever placed a water bottle in the freezer and seen it burst, you have experienced the effects of hydrogen-bonding intermolecular forces of attraction. As water freezes, it forms more and more hydrogen bonds, which creates a solid lattice of water molecules, which has a greater volume than liquid water.

Hydrogen bonding is the strongest intermolecular force of attraction between molecules. A hydrogen bond is *not* a covalent bond as the term *bond* might suggest but instead is a type of dipole–dipole interaction that exists in molecules that have one of the three most polar covalent bonds and hence the strongest dipoles:

- H—F
- H—O
- H—N

H—F is not found in biological organisms, but O—H and N—H bonds are found in many biomolecules, including proteins, carbohydrates, and DNA. These are the most polar covalent bonds because they involve nonmetal atoms at opposite ends of the electronegativity scale: hydrogen, and oxygen or nitrogen. The hydrogen atom in these bonds has the greatest partial positive charge of any atom in a molecule, whereas fluorine, nitrogen, and oxygen have the greatest partial negative charge of any atom in a molecule. It follows therefore that molecules containing O—H or N—H bonds will exhibit the strongest force of attraction between the positive pole—hydrogen—of one molecule and the negative pole—a nitrogen or an oxygen atom—in another molecule. A dashed line between the oppositely charged atoms is the convention used to represent a hydrogen bond formed between the hydrogen atom (δ^+) of one molecule and the nitrogen or oxygen atom (δ^-) of another molecule, as shown in Figure 3-30a for a pair of space-filling models of water.

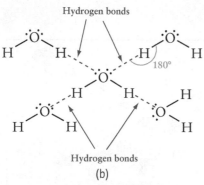

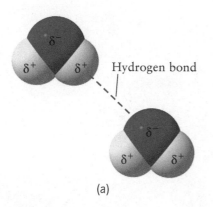

Figure 3-30 (a) Hydrogen bonding is a type of dipole–dipole interaction. (b) Hydrogen bonding in ice. Each water molecule forms hydrogen bonds to four other water molecules, causing ice to occupy a greater volume, and thus making it less dense, than liquid water.

Water in the solid phase (ice) has four hydrogen bonds per molecule, as shown in Figure 3-30b. Hydrogen bonding causes ice to occupy a greater volume than water in the liquid phase. As ice is heated, hydrogen bonds break, creating the liquid phase, with one to three hydrogen bonds per water molecule. It is only in the gas phase (steam) that there are no hydrogen bonds between water molecules because molecules are too far apart.

A network of hydrogen bonds extends throughout a liquid or solid sample of water, acting like a glue that holds water molecules together. Indeed, it is hydrogen bonding that causes water to a be a liquid at room temperature and the absence of hydrogen bonding that causes methane to be a gas at the same temperature. In order for liquid water to become a gas (steam), all the hydrogen bonds must be broken. Because breaking hydrogen bonds requires energy, sufficient energy must be added (by heat or microwaves) before the molecules can become separated and enter the gas phase. In contrast, methane has no hydrogen bonds, only dispersion forces, so it becomes a gas at well below room temperature. In chapter 5, you will learn more about the role of intermolecular forces of attraction and energy in phase changes.

Hydrogen Bonding in DNA Deoxyribonucleic acid (DNA) contains an organism's genetic information: stored within the atoms and bonds of its chemical structure (see chapter 14). DNA is composed of two extremely large molecules—strands—wound together in the shape of a double helix. Hydrogen bonds hold the two strands of DNA together throughout the entire length of the helix. Figure 3-31 shows a portion of the DNA double helix. Projecting perpendicular to the two strands are groups of atoms that form hydrogen bonds between an N—H or O—H bond on one strand and either a nitrogen or an oxygen atom on the other strand. Hydrogen bonding connects these ladderlike rungs to create the three-dimensional structure of DNA (Figure 3-31).

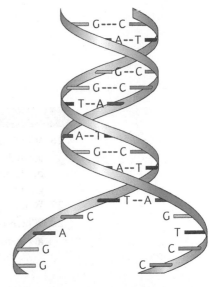

Figure 3-31 DNA, showing its two strands twisted in the shape of a double helix. The two strands are held together by hydrogen bonds formed between atoms on adjacent strands, indicated by the dashed red lines along the entire length of the helix.

WORKED EXERCISES Intermolecular Forces of Attraction

3-20 Which of the following molecules has a permanent dipole? Explain.
   a. C_2H_6 **b.** H_2O

3-21 Which molecule in each pair exhibits stronger intermolecular forces of attraction? Explain your choice.
   a. C_3H_8 or C_6H_{14}
   b. H_2O or H_2S

3-22 Provide an illustration showing how a single water molecule would hydrogen bond with four adjacent water molecules in a sample of ice. Label the relevant partial charges.

Solutions

3-20 a. Water, H_2O, is a bent molecule containing two polar O—H bonds, and therefore it has a permanent dipole. Ethane is a hydrocarbon and therefore contains only nonpolar covalent bonds.

3-21 a. C_6H_{14} has stronger dispersion forces than C_3H_8 because there are more electrons in C_6H_{14}. The strength of dispersion forces increases with the number of electrons in a molecule.

   b. H_2O has stronger intermolecular forces of attraction because it forms hydrogen bonds, which are the strongest intermolecular force of attraction. Sulfur is less electronegative than oxygen (it is located below oxygen on the periodic table), so S—H is not as strong a dipole as O—H.

3-22

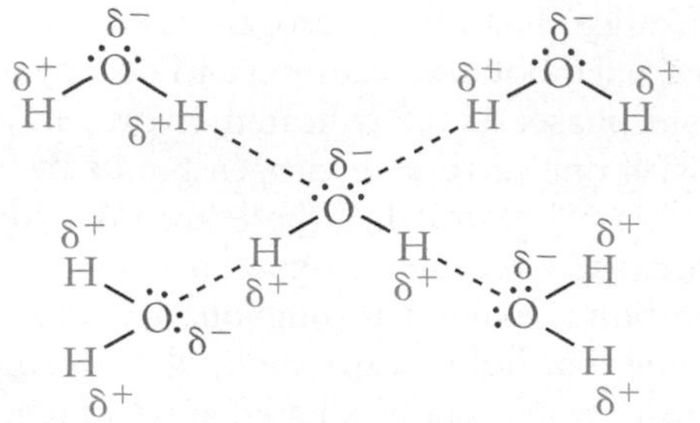

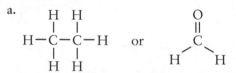

PRACTICE EXERCISES

26 Which of the intermolecular force of attraction listed is the weakest? Explain why.

   a. hydrogen bonding
   b. dispersion forces
   c. dipole–dipole forces

27 What type of molecules exhibit only dispersion forces and why?

28 Which of the following molecules contains a permanent dipole?

   a. acetone (fingernail polish remover)

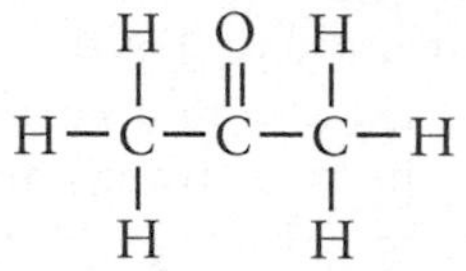

$$\begin{array}{c} \text{H} \quad \text{O} \quad \text{H} \\ | \quad\ \ || \quad\ \ | \\ \text{H}-\text{C}-\text{C}-\text{C}-\text{H} \\ | \quad\quad\quad | \\ \text{H} \quad\quad\ \ \text{H} \end{array}$$

   b. C_3H_8

29 Which molecule in each pair exhibits stronger intermolecular forces of attraction and what type is it? Explain your choice.

   a.

$$\begin{array}{c} \text{H} \quad \text{H} \\ | \quad\ \ | \\ \text{H}-\text{C}-\text{C}-\text{H} \\ | \quad\ \ | \\ \text{H} \quad \text{H} \end{array} \quad \text{or} \quad \begin{array}{c} \text{O} \\ || \\ \text{C} \\ \diagup\ \ \diagdown \\ \text{H} \quad\quad \text{H} \end{array}$$

   b. HCl or HF
   c. F_2 or HF
   d. CH_2Cl_2 or CCl_4

30 Provide an illustration of two ammonia molecules hydrogen bonding. Label the partial charges in the hydrogen bond.

31 What type of intermolecular force of attraction holds the two strands of a DNA molecule together?

Chemistry in Medicine Treating Breast Cancer

A woman born today has a one in eight chance of developing breast cancer in her lifetime. Throughout a woman's life, estrogens cause normal breast, liver, bone, and uterine cells to grow and divide. In the liver, estrogens promote the production of good cholesterol, and in bone, estrogens promote strong bones. These are some of the normal and beneficial effects of estrogen.

Scientists now understand how estrogens work on a molecular level. Estrogens are molecules that act as female

hormones. The most important estrogen is *estradiol*, whose ball-and-stick model is shown in **Figure 3-32** (an electron density diagram for estradiol is shown in Practice Exercise 25). Estradiol is the main estrogen produced in women of child-bearing age as well as the main estrogen involved in breast cancer. *Estradiol promotes cell growth in cells that contain estrogen receptors.* When estradiol binds to an estrogen receptor, it signals the cell to grow and reproduce. Since 75 percent of breast cancer cells also

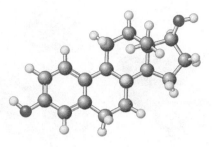

Figure 3-32 Ball-and-stick model of Estradiol, the most important estrogen.

contain estrogen receptors, known as ER+ breast cancer cells, estradiol stimulates their growth as well. Breast cancer cells thus thrive on estradiol.

Estrogen receptors are located in the nucleus of the cell, where the genetic information for the cell is contained. An estrogen receptor is an extremely large protein molecule. Its three-dimensional shape (space-filling model) with two bound estradiol molecules is shown in **Figure 3-33**. The receptor appears symmetrical because it is actually a pair of identical molecules, like a molecular twin, known as a *dimer*.

Estradiol binds to a cavity within the receptor, known as the estrogen binding site. Estradiol fits into the cavity because it has a shape complementary to that of the binding site. In addition, estradiol is a relatively nonpolar molecule, which creates dispersion forces of attraction between estradiol and nonpolar regions in complementary regions within the estrogen receptor binding site.

When estradiol binds to the estrogen receptor, the overall shape of the receptor changes. The change in shape in turn allows other specialized proteins necessary for DNA activation to bind to the estrogen receptor, signaling cell growth. This biochemical process is illustrated in **Figure 3-34a**.

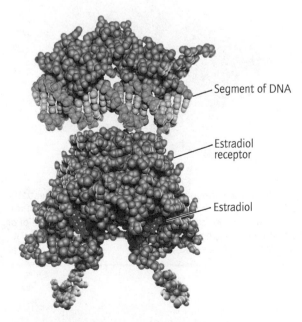

Figure 3-33 Space-filling model of the estrogen receptor (blue), a protein, showing a segment of DNA (in green) and two bound estradiol molecules (in fuscia).

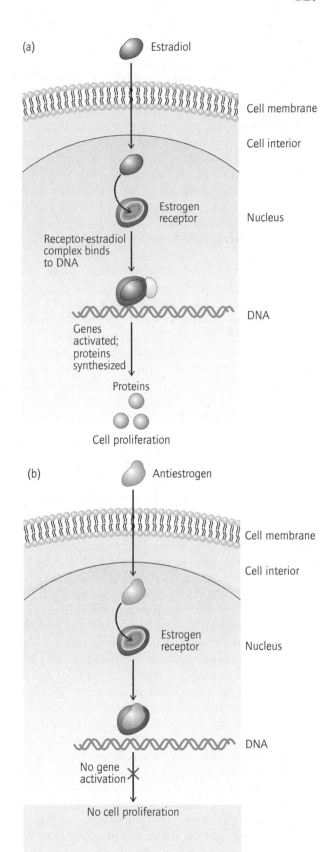

Figure 3-34 The action of estrogens and antiestrogens. (a) Estradiol binds to the estrogen receptor, causing gene activation. Proteins required for cell growth are therefore produced. (b) Antiestrogens bind to the estrogen receptor, preventing gene activation. Proteins required for cell growth are therefore not produced.

Unfortunately, estradiol also stimulates the growth and production of breast cancer cells in breast tissue and uterine cancer cells in the uterus because these cells contain estrogen receptors. The challenge in designing a drug for the treatment of breast cancer is to create one that selectively blocks estradiol from binding to the estrogen receptor in breast and uterine cells but not in liver and bone cells. In other words, an effective treatment for breast cancer is one that doesn't have the side effect of causing osteoporosis or uterine cancer. Hormone therapy for the treatment of breast cancer involves the long-term administration of a drug that selectively blocks the effects of estrogen—a drug known as an *antiestrogen*.

In the case of the estrogen receptor, its natural ligand is estradiol, but similarly shaped synthetic molecules can also bind to the estrogen receptor. Indeed, some of these synthetic drugs are better than estradiol at binding to the receptor. But because they are slightly different in shape and polarity, they can't activate the receptor—a characteristic of an antiestrogen. By occupying the estrogen receptor binding site, antiestrogens block estradiol from binding to the receptor and thereby slow the growth of breast cancer cells, as illustrated in Figure 3-34b.

Tamoxifen, introduced in 1977, is one of the oldest antiestrogens still in use today. Tamoxifen is typically prescribed as a hormone therapy for many years following the surgical removal of breast cancer. Tamoxifen has been shown to reduce the risk of breast cancer by 49 percent. Tamoxifen is transformed in the body into *hydroxytamoxifen*, which is the active drug molecule that actually binds to the estrogen receptor. Estradiol and hydroxytamoxifen have noticeably different shapes (**Figure 3-35**), yet they both fit the estrogen receptor binding site. However, they bind at slightly different places within the binding site, causing the receptor to change its shape in different ways, as shown in **Figure 3-36**. Consequently, the binding of hydroxytamoxifen prevents the activation of DNA in breast tissue.

Unfortunately, one of the side effects of Tamoxifen is that it stimulates the proliferation of uterine cells, increasing the likelihood of developing uterine cancer. For reasons not understood, it blocks the estrogen receptor only in breast tissue. Instead of blocking the effects of estrogen, as it does in breast cells, Tamoxifen acts like estrogen in uterine cells.

As you can see, shape and polarity are important characteristics of compounds. The binding of Tamoxifen to the estrogen receptor is just one of many examples of

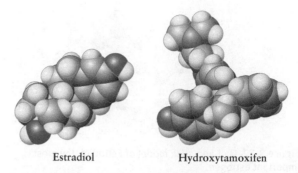

Estradiol   **Hydroxytamoxifen**

Figure 3-35 Space-filling models of estradiol and hydroxytamoxifen showing their different overall molecular shape.

a drug molecule interacting with a receptor. It is estimated that there are hundreds of different receptors that bind substances ranging from the thyroid hormones to vitamin D. Indeed, receptors are a major target for drug therapy. Efforts continue today to find better drugs for the treatment of breast cancer.

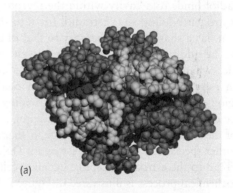

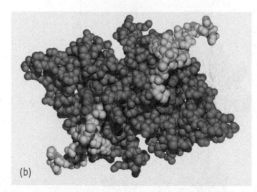

Figure 3-36 The estrogen receptor (blue) bound to (a) estradiol (fuscia) and to (b) hydroxytamoxifen (orange). The shape of the receptor changes noticeably depending on whether estradiol or hydroxytamoxifen is bound, which consequently affects how it interacts with DNA (green).

Summary

Ionic Compounds

- An ionic compound is composed of a lattice structure containing cations and anions formed as a result of the electrostatic attraction between opposite charges.

- Monatomic cations are derived from metals that lose one or more electrons to create a full valence energy level. Their name is the same as the metal element from which they are derived.

- Monatomic anions are derived from nonmetals that gain one or more electrons to achieve a full valence energy level. Their name is the same as the nonmetal element from which they are derived, except the ending is changed to "ide."

- Polyatomic ions are derived from molecules that have gained or lost electrons, creating a localized or delocalized charge in the polyatomic ion. These ions have a unique name and charge associated with them.

- The formula unit of an ionic compound indicates the lowest whole-number ratio of cations to anions that gives an electrically neutral compound.

- An ionic compound is named by writing the name of the cation first, followed by the name of the anion. If the cation has multiple forms, a Roman numeral is placed in parentheses immediately following the cation name.

Covalent Compounds

- Covalent compounds are composed of molecules. Molecules are discrete entities composed of nonmetal atoms that share electrons—covalent bonds.

- A shared pair of electrons between two atoms is known as a single covalent bond and written as a single line between the two atoms. Double bonds occur when four electrons are shared, written as two parallel lines, and triple bonds occur when six electrons are shared, written as three parallel lines.

- Nonbonding electrons are valence electrons that are not shared and belong to only one atom in the molecule. They are indicated as a pair of dots in a Lewis dot structure.

- A molecular formula shows the number of each atom type in the molecule by listing the atomic symbols in alphabetical order, followed by a subscript indicating how many atoms of each type there are in the molecule.

- The Lewis dot symbol for an element is a way of representing an atom and its valence electrons, by writing the atomic symbol surrounded by its valence electrons as dots.

- Lewis dot structures are created by combining Lewis dot symbols so that every atom in period 2 or higher has eight electrons, known as the octet rule.

- Hydrogen always forms one bond and shares two electrons.

- Elements in period 3 or higher are capable of forming expanded octets, wherein they have more than 8 electrons.

Three-Dimensional Shapes of Molecules

- Ball-and-stick models are used primarily to show molecular shape, whereas space-filling models are used to show the amount of space occupied by the atoms in a molecule.

- Valence shell electron pair repulsion (VSEPR) theory can be used to predict the electron geometry of a molecule from its Lewis dot structure.

- VSEPR is based on the premise that the central atom in a molecule and the atoms surrounding it will adopt a molecular shape that places all groups of electrons on the central atom, bonding and nonbonding, as far apart from each other as possible while maintaining bonds to the central atom.
- The three electron geometries are tetrahedral for four electron groups, trigonal planar for three electron groups, and linear for two electron groups.
- A molecule with no nonbonding electrons on the central atom will have the same molecular shape as its electron geometry.
- Molecules with nonbonding electrons will have a shape that depends on the electron geometry as well as the number of bonding versus nonbonding electrons around the central atom: three bonding groups and one nonbonding group adopt a trigonal pyramidal shape. Two bonding groups and two nonbonding groups adopt a bent shape. Two bonding groups and one nonbonding group also adopt a bent shape.
- A tetrahedral electron geometry results in 109.5° bond angles, a trigonal planar electron geometry results in 120° bond angles, and a linear electron geometry results in a 180° bond angle, regardless of the molecular shape.
- The shapes of large molecules are determined from a combination of the shapes of the individual atom centers in the molecule.
- Wedges and dashes are used to illustrate the three-dimensional structure of a tetrahedral molecule.

Polarity

- Electronegativity is a unitless measure of an atom's ability to attract electrons toward itself in a molecule.
- Electronegativity increases as you go from the bottom to the top of a group in the periodic table. It also increases from left to right within a period.
- A nonpolar covalent bond is formed between identical atoms and atoms with similar electronegativities (an EN difference of less than 0.5).
- A polar covalent bond is formed between nonmetal atoms with different electronegativities (an EN greater than 0.5 but less than 2.0). A polar covalent bond has a bond dipole—a separation of charge. A partial negative charge exists on the more electronegative atom in the bond and a partial positive charge exists on the less electronegative atom.
- Polarity is a continuum that ranges from nonpolar (the same or similar electronegativities) to polar covalent (partial charge separation) to ionic (full transferred charges).
- Diatomic molecules are nonpolar if they are elements or have a nonpolar bond. They are polar if they have a polar bond.
- A molecule is nonpolar if it contains no bond dipoles. Hydrocarbons are a class of nonpolar molecules.
- A molecule is nonpolar, despite the presence of bond dipoles, if it has a tetrahedral shape and contains four equivalent bond dipoles, or trigonal planar with three equivalent bond dipoles, or linear with two equivalent bond dipoles. In these cases the bond dipoles cancel because they are directed in opposite and equal directions.
- A molecule is polar if it contains polar bonds that do not cancel, as in the case of a trigonal pyramidal or bent shape or in the case of a diatomic molecule with a bond dipole.

- An electron density diagram is used to show charge separation in the form of color in a space-filling model.
- A dipole arrow is used to show the direction of bond dipoles in a molecule. The head of the arrow points toward the more electronegative atom.

Intermolecular Forces of Attraction

- There are three basic types of intermolecular forces of attraction: dispersion forces, dipole–dipole forces, and hydrogen-bonding forces.
- Dispersion forces are present in all molecules and are the only forces of attraction in nonpolar molecules.
- Dispersion forces are electrostatic attractions between induced temporary dipoles.
- Dipole–dipole forces are electrostatic forces of attraction between the permanent dipoles of polar molecules. Polar molecules will arrange themselves so that like charges avoid each other and opposite charges attract.
- Hydrogen bonding is the strongest intermolecular force of attraction and occurs between the hydrogen atom in any of the bonds H—F, H—O, or H—N and an electronegative atom such as N or O.
- Intermolecular forces of attraction are weaker than covalent bonds.
- Hydrogen bonding holds the two strands of DNA together.

Key Words

Ball-and-stick model A model of a molecule that represents atoms as colored spheres and covalent bonds as sticks. It is a tool for visualizing the shape of a molecule.

Bent shape A molecular shape that arises either from a tetrahedral electron geometry when there are two nonbonding pairs and two bonding groups of electrons or from a trigonal planar electron geometry when there is one nonbonding pair of electrons and two bonding groups. The bond angle is 109.5° and 120°, respectively.

Binary compound A covalent compound composed of only two different types of atoms.

Bond angle The angle generated by a central atom and any two atoms bonded to the central atom.

Bond dipole The charge separation in a polar covalent bond, created when two atoms have different electronegativities (EN >0.5 and <2.0).

Bond length The distance between two nuclei in a covalent bond.

Covalent bond Electrons shared between two nonmetal atoms in a molecule.

Covalent compound A molecule composed of more than one type of nonmetal atom held together by covalent bonds.

Dipole arrow An arrow symbol, with a hatch mark on the end opposite the arrow, that indicates the direction of a bond dipole. The head of the arrow points toward the more electronegative atom.

Electron density diagram A space-filling model color coded to show the relative charge accumulation on different atoms in a molecule. Red is used for δ^- and blue for δ^+.

Electron geometry The relative position of the bonding and nonbonding electrons around a central atom that determines the molecular shape of the molecule.

Electronegativity A measure of an atom's ability to attract electrons toward itself in a molecule.

Electrostatic attraction The force of attraction between oppositely charged entities such as a proton and an electron or a cation and an anion.

Formula unit For an ionic compound the ratio of cations to anions, shown as subscripts, and atomic symbols without charges.

Hydrogen bonding The strongest type of intermolecular forces of attraction between molecules. It occurs in molecules that contain an H—F, H—O, or H—N bond, the three strongest bond dipoles.

Ionic bond The strong electrostatic attraction that holds anions and cations together, such as in the lattice structure of an ionic compound.

Ionic compound A compound composed of ions held together by electrostatic attractions.

Lattice A three-dimensional array of cations and anions that is an ionic compound. Cations are surrounded by anions and anions by cations.

Lewis dot structure The structure of a molecule showing how valence electrons are distributed around atoms and shared between atoms in a molecule. Constructed by combining Lewis dot symbols. Covalent bonds are drawn as lines, representing shared pairs of electrons, and nonbonding electrons are drawn as pairs of dots.

Lewis dot symbol A way of representing an atom and its valence electrons by writing the atomic symbol surrounded by its valence electrons represented as dots.

Linear An electron geometry and molecular shape resembling a straight line, formed by two groups of electrons surrounding a central atom with a 180° bond angle.

Mixture A combination of two or more elements and/or compounds.

Molecule Two or more nonmetal atoms held together by covalent bonds.

Molecular dipole A charge separation in a molecule created by bond dipoles together with a molecular shape that doesn't allow cancellation of the bond dipoles.

Molecular shape The geometry of a molecule based on the relative positions of the atoms in the molecule. Molecular shape is determined from the electron geometry and the relative number of bonding and nonbonding groups around the central atom.

Monatomic ions: Ions formed from an atom, such as Na^+ and Cl^-.

Nonpolar covalent bond A bond formed between two atoms with the same or comparable electronegativity (EN <0.5 electronegativity difference).

Nonpolar molecule A molecule with an even distribution of electrons, no separation of charge. It is the result of either all nonpolar bonds in the molecule or polar bonds that cancel because of a symmetrical molecular shape.

Nonbonding electrons or lone pair electrons Valence electrons on an atom in a molecule that belong solely to that atom and are not shared; represented as a pair of dots in a Lewis dot structure.

Octet rule The tendency for most atoms in a molecule to share electrons—form covalent bonds—so that they have eight valence electrons.

Polar covalent bond A covalent bond that contains a bond dipole. A polar covalent bond is formed when two atoms

with significantly different electronegativities share electrons (EN >0.5 and <2.0).

Polar molecule A molecule that has a separation of charge. A polar molecule is formed when one or more bond dipoles in the molecule do not cancel, creating a charge separation within the molecule.

Polyatomic ion: An ion formed from the loss or gain of electrons from a molecule, creating a positive or negative charge.

Space-filling model A model of a molecule that shows the relative amount of space occupied by the atoms in the molecule.

Tetrahedral An electron geometry or molecular shape resembling a tetrahedron. The electron groups around the central atom point to the four corners of a tetrahedron, and bond angles are 109.5°. Occurs when there are four bonding groups of electrons around a central atom.

Trigonal planar An electron geometry or molecular shape resembling an equilateral triangle. The electron groups around the central atom point to the three corners of the triangle, and bond angles are 120°. Requires three bonding groups of electrons around a central atom.

Trigonal pyramidal shape The shape of a central atom with a tetrahedral electron geometry in which there is one nonbonding pair of electrons and three bonding groups. Bond angles are approximately 109.5°.

Tube model A model used to represent complex molecules wherein the bonds and atoms appear as part of a tube. Each end of the tube represents the color-coded atoms.

Valence shell electron pair repulsion theory (VSEPR) A theory used to predict the shapes of simple molecules. VSEPR predicts the electron geometry from the number of electron groups (usually two to four) around the central atom. Groups of electrons are positioned to achieve the maximum distance between them while maintaining the bond to the central atom.

Additional Exercises

What Is in Your Morning Coffee?

32 Is caffeine an element, a molecule, or an ion?

33 Caffeine works by blocking the _____ receptor on neurons and preventing _____.

34 What chemical is first produced in the fight-or-flight response? How does this chemical prepare the body for fight or flight?

35 Epinephrine stimulates the production of what "feel-good" compound?

36 What is the difference between an element and a compound?

37 What are the two types of compounds? What kinds of bonds distinguish each type of compound?

Ionic Compounds

38 What is an ionic compound?

39 Is an ionic bond a strong or weak force of attraction? What is another name for this force of attraction?

40 An ionic bond forms between _____ and _____.

41 What is a salt?

42 What is a lattice structure?

43 What is the difference between monatomic ions and polyatomic ions?

44 Polyatomic ions have which of the following (There can more than one correct answer):
a. a single atom that has lost or gained electrons
b. a molecule that has lost or gained electrons
c. ionic bonds between the atoms of a polyatomic ion
d. covalent bonds between most of the atoms

45 Sodium ions are macronutrients that play an important role during nerve impulses. Write the symbol for the sodium ion. Is the sodium ion a monatomic ion or a polyatomic ion?

46 Characterize the following ions as monatomic or polyatomic:
a. Br^- c. Mg^{2+} e. HO^-
b. HCO_3^- d. $CH_3CO_2^-$

47 Identify the following ions as monatomic or polyatomic:
a. HPO_4^{2-} d. Fe^{2+}
b. H_3O^+ e. NH_4^+
c. Cl^-

48 What ions are present in the following compounds? Provide the name, the symbol, and the charge on the ions:
a. KBr, often used in veterinary medicine to treat epilepsy in dogs
b. $MgCl_2$, used to treat and prevent low levels of magnesium in patients
c. KI, used to treat an overactive thyroid and to protect the thyroid gland from inhaled or swallowed radioactive iodine
d. $BaCl_2$
e. NaF, used to prevent cavities in children when the water supply is not fluoridated

49 What ions are present in the following compounds? Provide the name, the symbol, and the charge on the ions:
a. Na_2O
b. CuO
c. AgCl
d. ZnS
e. GaAs

50 What is the charge on each of the ions in the following ionic compounds?
a. $Zn(OH)_2$
b. $CuCH_3CO_2$
c. $SnCl_4$
d. V_2O_5
e. CrO_3

51 What is the charge on each of the ions in the following ionic compounds?
a. $BaSO_4$, a contrast media used in x-rays and CT scans of the esophagus and intestines
b. Li_2CO_3, used in the treatment of bipolar disorder
c. $MgSO_4$, used as an anticonvulsant and a laxative
d. KH_2PO_4, used to help control the amount of calcium in the body
e. $NaNO_2$, used to prevent botulism and to treat cyanide poisoning

52 Identify the cations and anions present in the following compounds. What is the ratio of cations to anions for each compound and why?
a. NaOH
b. KCH_3CO_2
c. NH_4OH
d. $(NH_4)_2HPO_4$

53 Write the formula unit for the ionic compound derived from each of the following ions. Indicate which ion is the cation and which ion is the anion.
a. lithium and iodide
b. rubidium and fluoride
c. calcium and bromide
d. barium and iodide
e. iron(II) and sulfide
f. aluminum and oxide

54 Write the formula unit for the ionic compound formed from each of the following ions. Indicate which ion is the cation and which ion is the anion.
a. strontium and chloride
b. chromium(III) and oxide
c. iron(II) and oxide
d. cobalt(II) and chloride
e. platinum(IV) and fluoride
f. iron(II) and selenide

55 Name the following compounds:
a. SrO
b. KI
c. NaI
d. LiF
e. Ga_2O_3

56 Name the following compounds:
a. $CaSO_4$, used in plaster of Paris to make casts
b. NH_4HCO_3, a component of smelling salts
c. $Al_2(CO_3)_3$, used as an antacid and to manage abnormally high levels of phosphate in the body
d. $AgNO_3$, used to cauterize superficial blood vessels in the nose to prevent nosebleeds
e. Fe_2O_3, a component of rust

57 What is the correct name of K_2HPO_4?
a. potassium phosphate
b. potassium hydrogen phosphate
c. potassium dihydrogen phosphate
d. potassium phosphide

58 What is the correct formula unit for sodium carbonate?
a. $NaCO_3$ c. $Na(HCO_3)_2$
b. $NaHCO_3$ d. Na_2CO_3

59 Write the formula unit for the following compounds:
a. sodium phosphate, which can be used as a laxative in preparing a patient for a colonoscopy
b. ammonium chloride, an expectorant in cough syrup
c. magnesium hydroxide (milk of magnesia)
d. sodium bicarbonate, an antacid

60 Write the formula unit for the following compounds:
a. calcium carbonate, an antacid
b. potassium hydrogen phosphate, used to help control phosphate levels in the body
c. calcium hydrogen carbonate
d. magnesium hydrogen phosphate

61 Lead accumulates in bones as lead(II) phosphate. Write the formula unit for lead(II) phosphate.

62 Calcium phosphate is one of the compounds found in kidney stones. What is the formula unit for calcium phosphate?

Covalent Compounds

63 What is a molecule?

64 How is a molecule different from an ionic lattice?

65 What is the difference between an ionic compound and a covalent compound?

66 Bond length is the optimal distance between two nuclei sharing electrons, such that the_____ between nuclei and electrons exceeds the_____ between two nuclei and two electrons. Select *attraction* or *repulsion*.

67 Characterize each of the following compounds as either ionic or covalent:
a. $CH_3CH_2NH_2$
b. NH_4OH
c. $CaCO_3$

68 What is the difference between a molecular formula and a formula unit?

69 A covalent bond is formed between
a. two metals
b. a metal and a nonmetal
c. two nonmetals
d. all of the above

70 Which elements exist as diatomic molecules?

71 What does it mean when two atoms share electrons?

72 Identify the following statements as True or False:
____a. a covalent bond between two atoms has a lower potential energy than the two separate atoms
____b. a covalent bond between two atoms is more stable than the two separate atoms
____c. atoms in a molecule achieve a stable electron arrangement by sharing electrons with other atoms
____d. ionic compounds share electrons

73 What is the octet rule?

74 How many electrons are shared in the following covalent bonds?
a. single bond
b. double bond
c. triple bond

75 Write the Lewis dot symbol for the following atoms:
a. carbon
b. hydrogen
c. oxygen
d. phosphorus

76 Write a Lewis dot symbol for the following atoms:
a. nitrogen
b. chlorine
c. sulfur
d. silicon

77 Answer the following questions for the Lewis dot structure of Ampyra shown below. Ampyra is a drug used to improve walking in patients with multiple sclerosis (MS).

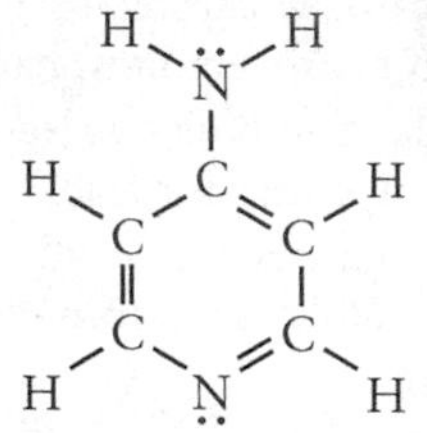

a. How many bonding pairs of electrons and how many nonbonding pairs of electrons do each of the carbon atoms contain?
b. How many bonding pairs of electrons and how many nonbonding pairs of electrons does the nitrogen atom at the top of the structure contain?
c. Does every atom, except hydrogen, have an octet of electrons?
d. What types of covalent bonds are present in this molecule: single, double, or triple bonds?

78 Answer the following questions for the Lewis dot structure of hydrogen fluoride, shown below:

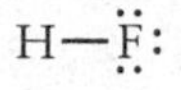

$$H - \ddot{\underset{..}{F}}:$$

a. How many bonding electrons does the fluorine atom contain?
b. How many nonbonding electrons does the fluorine atom contain?
c. Does the fluorine atom have an octet?
d. How many electrons are shared between the fluorine atom and the hydrogen atom?
e. Is hydrogen fluoride a covalent or an ionic compound?

79 Write a Lewis dot structure for Br_2. Is this molecule an element or a compound?

80 Write the typical number of bonds and nonbonding pairs of electrons for the following atoms when they are part of a molecule:
a. carbon
b. oxygen
c. sulfur
d. halogens (F, Cl, Br, I)
e. nitrogen
f. phosphorus
g. hydrogen

81 Predict how many bonds and nonbonding pairs of electrons would be found on a selenium atom? Explain.

82 Write the Lewis dot structure for the following compounds:
a. ethane, C_2H_6
b. PH_3
c. chloroform, CH_3Cl
d. carbon dioxide, CO_2
e. methanol, CH_4O

83 Write the Lewis dot structure for the following compounds:
a. PF_3
b. SH_2
c. formaldehyde, CH_2O
d. HCl
e. Carbonic acid, H_2CO_3. (*Hint:* The hydrogen atoms are bonded to the oxygen atoms and all the oxygen atoms are bonded to the carbon atom.)

84 Part of the structure of ATP is shown below. Indicate which atoms have an expanded octet. How many electrons do these atoms have?

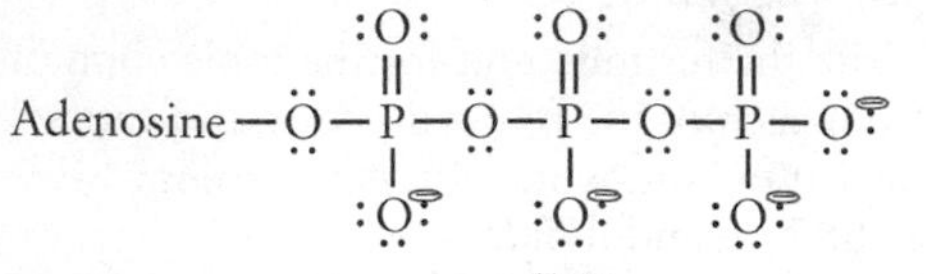

ATP

85 Draw the Lewis dot structure for acetonitrile, CH_3CN. (*Hint:* The atoms are arranged in the order shown in the molecular formula.)

86 Which of the following is the correct Lewis dot structure for propane, C_3H_8?

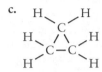

87 Name the compound SO_3.

88 What is the difference in the name for PCl_3 and the name for PCl_5?

89 Provide the names for the following binary compounds that contain nitrogen and oxygen:
a. NO, a signaling molecule in cells and fast-acting vasodilator
b. NO_2, an air pollutant
c. N_2O, an anesthetic and analgesic also known as "laughing gas." It is also a greenhouse gas.
d. N_2O_3
e. N_2O_4, used as a rocket propellant
f. N_2O_5

Three-Dimensional Shapes of Molecules

90 What structural information do ball-and-stick models convey? When the size of atoms is important, what type of model is often more suitable than a ball-and-stick model?

91 Write the Lewis dot structure represented by each model below. Indicate whether each model is a ball-and-stick model or a space-filling model. For the ball-and-stick model indicate the bond angle(s).

92 What is the basic premise of valence shell electron pair repulsion (VSEPR) theory?

93 If a molecule has only two groups of electrons on the central atom, what is the shape of the molecule?

94 What is the electron geometry for a molecule with three groups of electrons on the central atom? What are the possible molecular shapes of the molecule? What is/are the bond angle(s)?

95 What is the electron geometry for a molecule with four groups of electrons on the central atom? What are the possible molecular shapes of the molecule? What is/are the bond angle(s)?

96 Which of the molecular shapes listed below are three dimensional (not flat)?
a. linear   b. trigonal planar   c. tetrahedral

97 Which of the molecular shapes listed below are three dimensional (not flat)?
a. trigonal pyramidal   c. trigonal planar   b. bent

98 What are the bond angles in a molecule that has:
a. a trigonal planar electron geometry
b. a linear electron geometry
c. a trigonal pyramidal molecular shape
d. a tetrahedral molecular shape

99 What are the two possible bond angles for a molecule with a bent shape? Why do molecules with a bent geometry not necessarily have the same bond angle?

100 Fill in the blanks in the table below for the six common molecular shapes:

1. Total number of electron groups		3				
2. Number of bonding groups	2	3			3	
3. Number of nonbonding groups	0		1	0		2
4. Electron geometry		Trigonal planar				
5. Molecular shape					Trigonal pyramidal	
6. Bond angle					109.5°	

101 Write the Lewis dot structure for the following molecules and then determine the electron geometry and the molecular shape. Indicate the bond angles.
a. CH_2Cl_2 b. NI_3

102 Write the Lewis dot structure for the following molecules and then determine the electron geometry and the molecular shape. Indicate the bond angles.
a. PF_3
b. SCl_2
c. $SiCl_4$

103 What are the bond angles and what is the molecular shape of

$$:\ddot{S}=C=\ddot{O}:$$

a. 120°, trigonal planar
b. 109.5°, tetrahedral
c. 180°, linear
d. 109.5°, bent

104 What are the bond angles and what is the molecular shape of

$$:\ddot{Cl}-\overset{\overset{\displaystyle \cdot\cdot\,O\,\cdot\cdot}{\|}}{C}-\ddot{Cl}:$$

a. 120°, trigonal planar
b. 109.5°, tetrahedral
c. 180°, linear
d. 109.5°, bent

105 What are the bond angles and the molecular shape of the following molecules?
a. HCN
b.

$$:\ddot{Cl}-\overset{\overset{\displaystyle \cdot\cdot\,O\,\cdot\cdot}{\|}}{\underset{\underset{\displaystyle :\ddot{Cl}:}{|}}{P}}-\ddot{Cl}:$$

106 The Lewis dot structure for the sulfate ion, SO_4^{2-}, is shown below. What is the electron geometry of this polyatomic ion? What is the O—S—O bond angle?

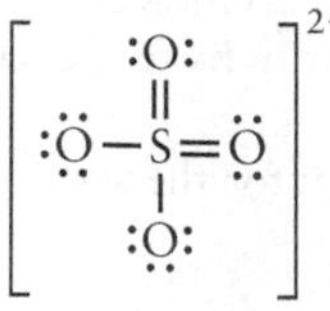

107 The Lewis dot structure for the nitrate ion, NO_3^-, is shown below. What is the geometry of this polyatomic ion? What is the O—N—O bond angle?

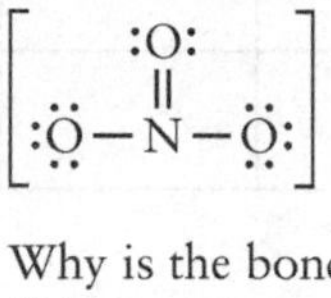

108 Why is the bond angle around the central atom smaller if the number of groups of electrons on the central atom is greater? What is the greatest bond angle? What is the smallest bond angle seen in organic molecules?

109 Write the Lewis dot structures of the following molecules and then indicate the molecular shape and the bond angles for each central atom.
   a. PH_3
   b. CCl_4
   c. BF_3 (boron is an exception to the octet rule and has six electrons)
   d. SF_2
   e. OCS

110 Write the Lewis dot structures for the following molecules and then indicate the molecular shape and the bond angles for each central atom.
   a. C_2H_4 c. C_2H_6
   b. SiF_4 d. C_2H_2

111 Isoflurane is used as an inhaled anesthesic. It has the following Lewis dot structure:

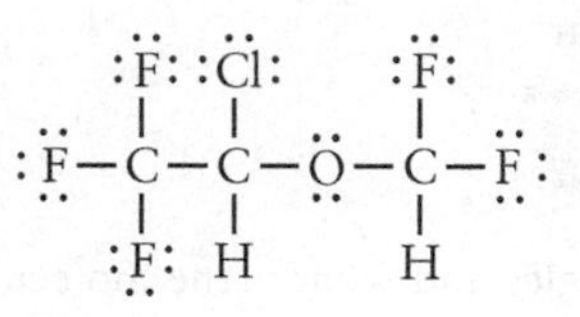

   a. Provide a three-dimensional representation of isoflurane using dashed and wedged bonds.
   b. What is the bond angle around each carbon atom?
   c. What is the bond angle around the oxygen atom?
   d. What is the molecular shape at each carbon center?
   e. What is the molecular shape around the oxygen atom?

112 Acetic acid is found in vinegar and is also used to treat ear infections. Its Lewis dot structure is shown below:

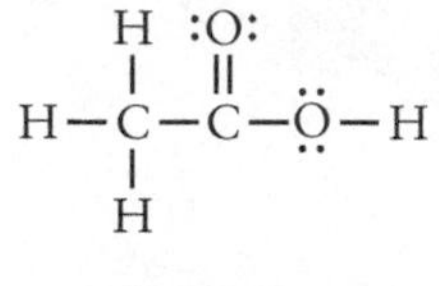

   a. What is the molecular shape at each carbon center? Do both carbon atoms have the same molecular shape? Explain.

   b. What is the H—C—H bond angle?
   c. What is the H—C—C bond angle?
   d. What is the O—C—O bond angle?

113 What is the difference between a trigonal planar molecular shape and a trigonal pyramidal molecular shape? How are the bond angles different? Why are they different?

114 Why are diatomic molecules always linear?

115 The Lewis dot structure of an epoxide is shown below. Based on VSEPR, what is the optimal bond angle around each carbon atom? Since this molecule has the shape of an equilateral triangle, what bond angle does each carbon atom actually have? Why do you think this molecule is not commonly found in nature?

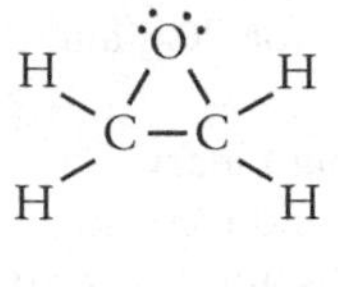

Modeling Exercises

Exercises 116 and 117 require the use of a molecular modeling kit.

116 Construct Formaldehyde, H_2CO
   • Obtain one black carbon atom, two light blue hydrogen atoms, one oxygen atom, two bent bonds, and two straight bonds.
   • Make a model of H_2CO. *Hint:* Both hydrogen atoms are attached to the carbon atom.

Inquiry Questions

1. How many different groups of electrons surround the central carbon atom in your model: two, three, or four?

2. How many groups of electrons are bonding electrons: two, three, or four?

3. Are there any nonbonding electrons on the central atom?

4. Determine the electron geometry for formaldehyde.

5. What is the molecular shape of formaldehyde?

6. What is the H—C—O bond angle? Are all the bond angles in the model approximately the same?

7. Why is the molecular shape of formaldehyde not a T shape? In other words, what advantage does a trigonal planar geometry offer over a T shape?

117 Construct Propene, C_3H_6
   • Obtain three black carbon atoms, six light blue hydrogen atoms, seven straight bonds, and two bent bonds.
   • Using the Lewis dot structure as a guide, make a model of propene.

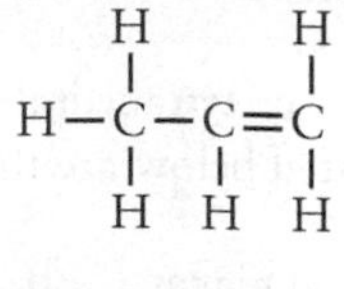

Inquiry Questions

1. How many different groups of electrons surround the central carbon atom in your model: two, three, or four?

2. Are these groups of electrons bonding or nonbonding?

3. Are there any nonbonding electrons on the central atom?

4. Determine the electron geometry around the central carbon atom in propene.

5. Is the electron geometry of the two carbon atoms on each end of the molecule the same?

6. What is the C—C—C bond angle around the central carbon atom?

Polarity

118 Define the term *electronegativity*.

119 Indicate which element in each pair is more electronegative.
   a. carbon or oxygen
   b. sulfur or oxygen
   c. lithium or fluorine

120 Indicate which element in each pair is more electronegative:
   a. phosphorus or oxygen
   b. iodine or chlorine
   c. sodium or chlorine

121 Does electronegativity increase or decrease as you move from bottom to top in a group of elements?

122 Does electronegativity increase or decrease as you move from left to right in a period of elements?

123 Excepting the noble gas elements, which family of elements is the most electronegative?
   a. alkali metal
   b. halogens
   c. alkali earth metals

124 When is a covalent bond polar?

125 How does the electronegativity of the atoms in a covalent bond determine whether it is a polar covalent bond or a nonpolar covalent bond?

126 Answer the following questions for a molecule of chloroform, $CHCl_3$, which was used as an anesthetic in the mid-1800s.
   a. Write the Lewis dot structure.
   b. Provide a three-dimensional representation of the molecule using dashed and wedged bonds.
   c. Are there any bond dipoles in the molecule? If so, which atom in each bond is more electronegative? Show all bond dipoles with a properly drawn dipole arrow.
   d. Is chloroform a polar molecule? Explain.

127 Which of the following molecules are polar? For polar molecules write a three-dimensional representation of the molecule along with dipole arrows.
   a. H_2O
   b. ethanol

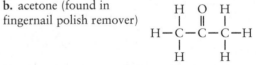

   c. C_2H_4

128 Which of the following molecules are polar? For polar molecules write a three-dimensional representation of the molecule along with dipole arrows. For nonpolar molecules explain why they are nonpolar.
   a. HF   b. Br_2 c. CO_2

129 Which of the following molecules are nonpolar? Explain.
   a. I_2 b. CH_4 c. HBr

130 Which of the following molecules are nonpolar? Explain.
   a. HCN   b. CS_2 c. CH_2F_2

Intermolecular Forces of Attraction

131 What is the difference between a covalent bond and an intermolecular force of attraction?

132 Which is stronger: a covalent bond or an intermolecular force of attraction?

133 What are the three types of intermolecular forces? Which one is the strongest? Which one is the weakest?

134 Which type of intermolecular force of attraction do all molecules exhibit?

135 Nonpolar molecules are attracted to each other by what kind of intermolecular force of attraction?

136 Explain why dispersion forces are the weakest of the intermolecular forces of attraction.

137 What makes dipole–dipole interactions stronger than dispersion forces?

138 Draw a representation of how you might expect a sample of HBr molecules to arrange themselves. What type of intermolecular force of attraction exists between HBr molecules?

139 What three covalent bonds are involved in hydrogen bonding?

140 What is the strongest type of intermolecular forces present in butane, C_4H_{10}?
   a. hydrogen bonding
   b. dipole–dipole interactions
   c. dispersion forces

141 What is the strongest type of intermolecular forces present in the following compounds?
   a. C_5H_{12}
   b. acetone (found in fingernail polish remover)

```
        H   O   H
        |   ||  |
    H — C — C — C — H
        |       |
        H       H
```

   c. water

142 What is the strongest type of intermolecular force present in each of the following compounds?
   a.
```
        H
        |
    H — C — Ö — H
        |   ··
        H
```

   b. HF
   c. C_3H_8

143 What intermolecular force of attraction supports the double helical shape of DNA?

144 DNA sequencing is used in crime scene investigations and also used to diagnose genetic disorders. The first step in sequencing DNA is to heat the DNA strand so that it separates into two strands. Explain on a molecular level why heat is needed to separate the two strands of DNA.

Chemistry in Medicine: Treating Breast Cancer

145 How does estrogen affect the growth of breast cancer cells?

146 How do antiestrogens prevent the growth of breast cancer cells?

147 What type of molecule is the estrogen receptor? What allows estradiol to bind to the estrogen binding site on an estrogen receptor?

148 Besides hydrogen bonding, what other intermolecular forces of attraction bind estradiol to the estrogen receptor?

149 When Tamoxifen binds to the estrogen receptor in breast cancer cells, does gene activation occur? Why or why not?

150 Why can a decrease in circulating estrogen contribute to osteoporosis?

Answers to Practice Exercises

1 **a.** Both lattice structures show a one-to-one ratio of cations to anions, which agrees with the formula units NaCl and CsCl.

b. The ratio is 1:1 because the charge on the cation is the same as that on the anion, and this is the lowest whole-number ratio.

c. The formula unit represents the lowest whole-number ratio of ions in the lattice, whereas the lattice structure is a continuous extended array of ions that goes on until the edge of the crystal.

d. The electrostatic force of attraction between oppositely charged ions, also known as ionic bonds.

e. Chloride ions are shown in green.

f. The sodium cation has a smaller diameter than a chloride anion, as seen in the smaller spheres.

2 **a.** Na_2S **b.** Mg_3N_2 **c.** $CuCl$ **d.** $CuCl_2$

3 **a.** sodium oxide

b. magnesium iodide

c. potassium chloride

4 Pb^{4+}. Since oxide is a -2 ion and there are two of them, lead must be $+4$ in order to have an electrically neutral compound with one lead ion.

5 **a.** potassium nitrate

b. sodium hydrogen phosphate

c. sodium hydroxide

d. silver nitrate

e. ammonium carbonate

6 **a.** $LiOH$ **b.** $SrCO_3$ **c.** $Mg_3(PO_4)_2$

7 KCN

8 Li_2CO_3

9 Cu^{2+}. Calculated $x + 6(+3) + 4(-3) + 8(-1) = 0$. Therefore, $x = +2$.

10 **a.** carbon dioxide **b.** dinitrogen trioxide

11 **a.** NF_3 **b.** SO_3

12 **a.** aluminum chloride; ionic

b. carbon disulfide; covalent

c. zinc oxide; ionic

d. dinitrogen tetroxide; covalent (the vowels *ao* appear together, so drop the first vowel)

13 **a.** Each oxygen has two bonding and two nonbonding pairs of electrons.

b. Carbon has four bonding pairs of electrons and zero nonbonding pairs.

c. yes

d. double bonds; sharing four electrons

14 All period 2 elements have an octet. (b) and (d) are elements because they are composed of molecules containing only one type of atom.

a.
$$H-\overset{\displaystyle H}{\underset{\displaystyle |}{\overset{\displaystyle |}{N}}}-H$$

b. $\ddot{O}=\ddot{O}$

c. $H-\ddot{S}-H$

d. $:\!\ddot{Cl}-\ddot{Cl}\!:$

Extension Topic 3-1: Expanded Octets

E3.1 **a.** phosphorus, P, has 10 electrons

b. no expanded octet

c. sulfur, S, has 12 electrons

E3.2 **a.** Sulfur, S, has an expanded octet. Yes, sulfur is in period 3. The sulfur atom has 12 electrons.

b. Each carbon has eight bonding electrons (four single bonds). Carbon has zero nonbonding electrons.

c. The nitrogen atom in taurine has six bonding electrons (three single bonds) and two nonbonding electrons for a total of eight.

15

	Lewis structure	Electron geometry	Molecular shape	Bond angles	3-D drawing
a.	$:\!\ddot{Cl}\!:$ bonded C with four Cl	Tetrahedral	Tetrahedral	109.5°	Cl–C with four Cl
b.	$H-C\equiv N:$	Linear	Linear	180°	$H-C\equiv N$
c.	$:\!\ddot{F}\!:$ B with three F	Trigonal planar	Trigonal planar	120°	F–B with three F

16   H—S̈—H. Tetrahedral electron geometry. Bent molecular shape and 109.5° bond angle. H_2S is not linear because the sulfur atom has four groups of electrons that according to VSEPR must be spaced as far apart as possible, which is a tetrahedral electron geometry; a linear geometry would put the nonbonding electrons too close to the bonding electrons.

$$\begin{array}{c} :\ddot{C}l: \\ | \\ H-C-\ddot{C}l: \\ | \\ \end{array}$$

17 **a.**   H   Tetrahedral electron geometry; tetrahedral molecular shape; 109.5° bond angles

b. :F̈—Ö—F̈: Tetrahedral electron geometry; bent molecular shape; 109.5° bond angles

$$\begin{array}{c} :\ddot{C}l: \\ | \\ :\ddot{C}l-Si-\ddot{C}l: \\ | \\ :\ddot{C}l: \end{array}$$

c.   Tetrahedral electron geometry; tetrahedral shape; 109.5° bond angles

$$\begin{array}{c} :\ddot{F}-\ddot{P}-\ddot{F}: \\ | \\ :\ddot{F}: \end{array}$$

d.   Tetrahedral electron geometry; trigonal pyramidal molecular shape; 109.5° bond angles

18   They all have a tetrahedral electron geometry and 109.5° bond angles because the central atoms all have four groups of electrons.

19   H—C≡C—H. The electron geometry and the molecular shape is linear at both carbon atom centers. Therefore, overall shape of acetylene is linear.

20 **a.** Nitrogen because it is in the same period as but to the right of carbon

b. Nitrogen because it is in the same group as but above phosphorus in the periodic table

c. Chlorine because even though it is in period 3 and lithium is in period 2, Cl is on the far right and lithium on the far left of the periodic table

d. Nitrogen because it is both above and to the right of silicon on the periodic table

21   Metalloids are more electronegative than metals because they are to the right of the metals in the periodic table.

22 **a.** O⃖—H

b. C⃗—O

c. Nonpolar bond, no bond dipole because identical atoms have the same electronegativity

23 **a.** Nonpolar because it is an element, so both atoms by definition have the same electronegativity

b. Nonpolar because it is a hydrocarbon and carbon and hydrogen have very similar electronegativity values

c. Polar because the two S—O bond dipoles do not cancel because the molecule has a bent shape

d. Polar because there are only two atoms and the bond between them is polar. The bond is polar because Cl is more electronegative than H.

e. Nonpolar because the C—Cl bond dipoles cancel as a result of the tetrahedral shape and identical bond dipoles

24 **a.** The electron geometry is trigonal planar. The molecular shape is trigonal planar. The bond angles are 120°.

$$\begin{array}{c} O \quad \uparrow \\ || \\ C \\ \end{array}$$

b. H   H Formaldehyde is polar because there is one polar bond.

c. The blue region shows the part of the molecule that has a partial positive charge and the red region shows the part of the molecule that has a partial negative charge. The electron density diagram shows there is a charge separation in the molecule: red and blue are at two opposite ends (poles) of the molecule.

25 **a.** The significant amount of green color in this electron density diagram indicates there is a large nonpolar region in the molecule, carbon and hydrogen atoms.

b. An O—H group is polar, so look for an area where there is red or blue: on the far left and right of the molecule as drawn.

26 **b.** Dispersion forces because they arise from temporary dipoles

27   Hydrocarbons and molecular elements because they contain only nonpolar bonds

28 **a.** Acetone because of the C═O bond dipole

29 **a.** CH_2O has dipole-dipole interactions, whereas C_2H_6 has only dispersion forces because the former has a permanent dipole and the latter does not.

b. H—F has hydrogen bonding, which is stronger than the dipole–dipole forces of attraction in H—Cl.

c. H—F can form hydrogen bonds, which is a stronger intermolecular force of attraction than the dispersion forces seen in the element F_2.

d. CH_2Cl_2 has dipole-dipole forces, which are stronger than the dispersion forces in CCl_4 (nonpolar molecule because the bond dipoles cancel).

30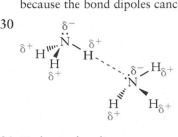

31   Hydrogen bonding

The ethanol in alcoholic beverages can be produced from the fermentation of grapes. [Purestock/Thinkstock.com]

4 Chemical Quantities and Chemical Reactions

How Does the Body Handle Alcoholic Beverages?

Why does drinking too many alcoholic beverages leave a person feeling nauseated and light-headed, and maybe hungover into the next day? The answer lies in the chemical reactions that ethanol undergoes in liver cells. Ethanol is the "alcohol" found in champagne, beer, wine, and other alcoholic beverages. The Lewis dot structure for ethanol is shown below and a ball-and-stick model is shown on the previous page.

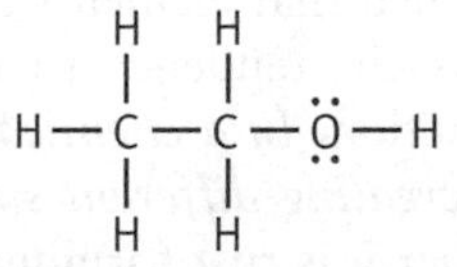

$$H-\overset{\displaystyle H}{\underset{\displaystyle H}{C}}-\overset{\displaystyle H}{\underset{\displaystyle H}{C}}-\ddot{O}-H$$

Since our bodies cannot store ethanol, as they do sugars and fats, they must metabolize it in order to eliminate it from the body. *Metabolism* is a general term for the chemical reactions carried out by a biological cell. In a chemical reaction, bonds are broken and new bonds are formed to produce different chemical substances. Ethanol is metabolized in a series of chemical reactions that ultimately turn ethanol into carbon dioxide (CO_2) and water (H_2O). This metabolism occurs in the liver, which is why alcoholics often suffer from liver damage.

When ethanol is consumed, liver cells carry out the sequence of reactions known as a *biochemical pathway*, shown below.

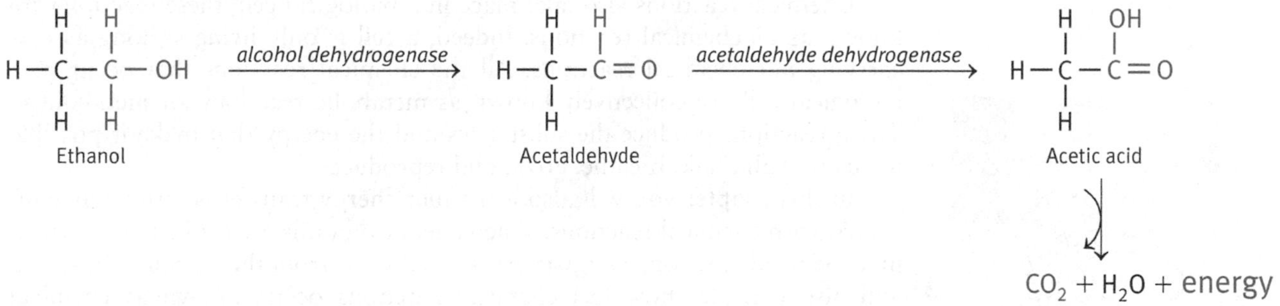

Ethanol → (alcohol dehydrogenase) → Acetaldehyde → (acetaldehyde dehydrogenase) → Acetic acid

$$CO_2 + H_2O + energy$$

The first compound produced in this biochemical pathway is acetaldehyde. Acetaldehyde then undergoes a chemical reaction that transforms it into acetic acid, which is eventually converted into carbon dioxide and water in yet another biochemical pathway. Most of the carbon dioxide formed is exhaled, and the rest remains dissolved in the blood.

The reactions involved in the metabolism of ethanol, like all biological reactions, proceed with the help of biological catalysts known as enzymes. An enzyme has the specific structure and function necessary to catalyze one particular chemical reaction. In the first reaction of ethanol metabolism, the enzyme *alcohol dehydrogenase* catalyzes the reaction that converts ethanol into acetaldehyde. In the second reaction of the biochemical pathway, *acetaldehyde dehydrogenase* catalyzes the reaction that converts acetaldehyde into acetic acid. If a person consumes too much alcohol in too short a period of time, the enzyme catalyzing the second reaction, *acetaldehyde*

Outline

dehydrogenase, cannot keep up with the production of acetaldehyde, causing acetaldehyde to build up in the cell. Elevated levels of acetaldehyde are believed to be responsible, in part, for the headache, nausea, dizziness, and "hangover" felt by the person consuming too much ethanol in too short a time.

Genetically, some people have an ineffective form of the enzyme *acetaldehyde dehydrogenase*, and as a result they have a low tolerance for alcoholic beverages. They exhibit symptoms of acetaldehyde poisoning even after consuming small amounts of alcohol. They are also at greater risk of developing liver diseases caused by exposure to acetaldehyde.

Energy is an important, perhaps the most important, component of all chemical reactions. Chemical reactions either absorb or release energy. The metabolism of ethanol releases energy, but ethanol is not a good energy source because it contains no nutrients—just "empty" calories. Indeed, alcohol consumption can contribute to weight gain while also contributing to malnutrition. ●

In chapters 2 and 3 you learned that elements and compounds have characteristic chemical properties—the tendency to react with other compounds or elements in a **chemical reaction**. *In a chemical reaction, bonds are broken and new bonds are formed, creating different substances.* We witness chemical reactions every day, whether it is rust forming on a piece of metal exposed to air, food being cooked, or burning wood in a fire. In this chapter you will study chemical reactions—changes in matter—which are at the foundation of chemistry.

Chemical reactions are carried out routinely in the laboratory. For example, the reaction between solutions of potassium iodide (KI) and lead nitrate ($Pb(NO_3)_2$) react to produce the yellow solid lead iodide (PbI_2), shown in Figure 4-1. A chemical reaction also occurs when hydrogen peroxide is placed on a wound. The hydrogen peroxide (H_2O_2) reacts to form water (H_2O) and oxygen (O_2) in a chemical reaction. The bubbles you observe in this reaction arise from the oxygen gas produced in the reaction.

Chemical reactions also take place in a biological cell; these reactions are known as biochemical reactions. Indeed, a cell is only living so long as it is carrying out chemical reactions. All the chemical reactions that occur in a biological cell are collectively known as metabolic reactions, or **metabolism**. These reactions produce the substances and the energy that make it possible for us to walk, talk, breathe, grow, and reproduce.

In this chapter you will also learn that energy transfer is intimately connected with chemical reactions. When chemical bonds are broken and formed in a chemical reaction, energy is transferred *to* or *from* the surroundings. You will also consider how fast chemical reactions occur, known as **chemical kinetics**. And finally, you will learn about reversible reactions and what it means when a chemical reaction is at equilibrium. To understand the important relationship between energy and chemical reactions, reaction rates, and equilibrium, we must first learn how to quantify matter. Hence, we begin with the mole, an important counting unit in chemistry that gives us information about the *number* of atoms, ions, or molecules in a sample of matter.

Figure 4-1 A chemical reaction occurs when mixing the clear, colorless solutions potassium iodide and lead nitrate. The product is the yellow solid lead iodide. [Richard Megna/Fundamental Photographs]

■ 4.1 The Mole: Counting and Weighing Matter

Atoms, ions, and molecules are incredibly small in size, so how can we ever know how many of these particles are present in any given sample of matter? The answer is that there is a quantitative relationship between the *number* of particles (atoms, ions, or molecules) and their *mass*. Since we know the mass of a single atom—average atomic mass—we can calculate the number of atoms in a macroscopic sample of this element from its total mass. Indeed, your body mass is the sum total of the mass of all the atoms, molecules, and ions in your body. In chemistry it is particularly important to know how many particles are present in a given sample because matter reacts according to the *number* of atoms, ions, or molecules present. Indeed, many fundamental concepts in chemistry were first discovered only after careful measurements of quantities of matter were observed and analyzed.

Formula Mass and Molecular Mass

In chapter 2 you learned that the average atomic mass of an element is given in the periodic table of the elements, in atomic mass units (amu). It follows that the *mass of one molecule, known as a* **molecular mass**, *is the sum of the individual atomic masses of its component atoms (given in the molecular formula). Similarly, the mass of one formula unit of an ionic compound, known as a* **formula mass**, *is calculated from the atomic masses of its component ions (given in the formula unit).*

> Molecular mass = sum of atomic masses of all atoms in the molecular formula (amu).
> Formula mass = sum of atomic masses of all ions in the formula unit (amu).

> Note that the mass of an ion is the same as the mass of the element from which it was derived.

For example, a single molecule of water, H_2O, has a molecular mass of 18.02 amu, calculated as follows:

> To avoid rounding errors, do not round until you have the final answer.

Atom Type	Number of Atoms		Average Atomic Mass		Total
H	2	×	1.008 amu/atom	=	2.016 amu/atom
O	1	×	16.00 amu/atom	=	+16.00 amu/atom
					18.016 amu/molecule
					18.02 molecular mass after rounding

WORKED EXERCISES Calculating Molecular Mass and Formula Mass

4-1 Calculate the molecular mass of caffeine, which has the molecular formula $C_8H_{10}N_4O_2$.

4-2 Calculate the formula mass of calcium chloride, $CaCl_2$, the active ingredient in DampRid, a drying agent that absorbs water from the air (Figure 4-2).

Solutions

4-1 $C_8H_{10}N_4O_2$

Atom Type	Number of Atoms		Atomic Mass (from Periodic Table)		Total
C	8	×	12.01	=	96.08 amu
H	10	×	1.008	=	10.08 amu
N	4	×	14.01	=	56.04 amu
O	2	×	16.00	=	32.00 amu
					194.20 amu/molecule

The mass of one molecule of caffeine is 194.20 amu, known as its molecular mass.

Figure 4-2 DampRid contains the drying agent calcium chloride ($CaCl_2$), which removes humidity from the air by reacting with water.
[Catherine Bausinger]

4-2 $CaCl_2$

Atom Type	Number of Atoms		Atomic Mass		Total
Ca	1	×	40.08	=	40.08 amu
Cl	2	×	35.45	=	+70.90 amu
					110.98 amu/formula unit

The mass of one formula unit of $CaCl_2$ is 110.98 amu, known as its formula mass.

PRACTICE EXERCISES

1   Calculate the molecular mass of methane, CH_4.
2   Calculate the formula mass of aluminum oxide, Al_2O_3.
3   Glucose—blood sugar—has the molecular formula $C_6H_{12}O_6$. Calculate the molecular mass of glucose.

Molar Mass and Counting Particles

Patients are routinely given an intravenous (IV) solution when they are admitted to the hospital. An IV solution must contain the correct number of dissolved ions. A physiological saline solution, for example, contains sodium (Na^+) and chloride (Cl^-) ions. If the number of sodium and chloride ions in the solution is too great or small, the patient could die. There are many times when it is important to know the *number* of atoms, ions, or molecules in a sample of matter. *Since there is a relationship between the number of atoms in a sample and the mass of the sample, we can calculate the number of particles in the sample by weighing it.*

The Mole A sample of matter on the macroscopic scale is composed of an enormous number of atoms, ions, or molecules. When counting large numbers of things, we often use *counting units*, such as a dozen, a case, and so on. Several examples of counting units are shown in Table 4-1. The chemist's counting unit is the **mole**, abbreviated mol, which represents the number 6.02×10^{23}, usually applied to, and known as **Avogadro's number**, much like the number 12 is the number in one *dozen*. Avogadro's number was determined from the number of carbon-12 atoms in exactly 12.00 grams of carbon-12. Thus, we have a relationship between the mole, a counting unit, and the mass (grams) of any substance we want to count. **Figure 4-3** shows what *one* mole of three familiar compounds and elements looks like. Each sample in this photograph contains 6.02×10^{23} atoms, ions, or molecules—one mole.

Figure 4-3 One mole of some common substances: sodium chloride (NaCl), water (H_2O), and copper (Cu) (reading clockwise from the top). [Richard Megna/ Fundamental Photographs]

TABLE 4-1 Examples of Counting Units and the Number They Represent

Name of Counting Unit	Number
Pair	2
Dozen	12
Case	24
Ream	500
Googol	1×10^{100}
Mole	6.02×10^{23}

Since 1 mole represents 6.02×10^{23} (Avogadro's number), we can use it as a conversion between units of moles and number of particles:

$$1 \text{ mol of atoms} = 6.02 \times 10^{23} \text{ atoms}$$
$$1 \text{ mol of molecules} = 6.02 \times 10^{23} \text{ molecules}$$
$$1 \text{ mol of formula units} = 6.02 \times 10^{23} \text{ formula units}$$

To count atoms by weighing, we must know the mass of one mole of these atoms. The mass of one mole of any element or compound is known as its **molar mass**, given in units of grams per mole (g/mol). Since the mass of an element is defined relative to carbon-12, the molar mass of an element is numerically equal to the atomic mass of the element, but the units are different: grams per mole (g/mol) for molar mass versus amu for atomic mass. For example, the atomic mass of calcium, Ca, is 40.08 amu, so the molar mass of calcium is 40.08 grams per mole. In other words, there are 6.02×10^{23} calcium atoms in one mole of calcium, which has a mass of 40.08 g.

The molar mass of a compound has the same numerical value as its formula mass or molecular mass except the units are g/mol rather than amu. Thus, a molecule of water has a molecular mass of 18.011 amu, so one mole of water has a molar mass of 18.011 g/mol. *Molar mass is a conversion factor between the moles of a substance and the mass of a substance.* The molar mass of water is 18.011 g/mole, so the conversion factors are

$$\frac{18.011 \text{ g } H_2O}{1 \text{ mol } H_2O} \quad \text{or} \quad \frac{1 \text{ mol } H_2O}{18.011 \text{ g } H_2O}$$

In the next section you will see how the mole is a unit central to many types of calculations.

WORKED EXERCISES   The Mole and Avogadro's Number

4-3   What is the molar mass of gold? Include units in your answer. What number of gold atoms does this represent?

4-4   How many silver atoms are there in 1 mol of silver? What is this number called?

Solutions

4-3   From the periodic table, we see that the atomic mass of gold, Au, is 197.0 amu. The molar mass of gold has the same numerical value, but different units: 197.0 g/mol. One mole of gold is 6.02×10^{23} gold atoms, which have a mass of 197.0 grams.

4-4   There are 6.02×10^{23} atoms in *one* mole of any element. Thus, there are 6.02×10^{23} silver atoms in one mole of silver, Ag. The number 6.02×10^{23} is called Avogadro's number.

PRACTICE EXERCISES

4   How many marbles are there in 1 mole of marbles?

5   Which has a greater mass, one mole of Ping-Pong balls or one mole of basketballs? Explain.

6   What is the average atomic mass of one mercury atom? What is the mass of 1 mole of mercury atoms? What is the molar mass of mercury?

7   Previously you calculated the molecular mass of caffeine as 194.20 amu. What is the molar mass of caffeine? How many caffeine molecules are there in one mole of caffeine?

Converting between Number of Moles and Number of Particles In a blood test, such as the one shown in Figure 4-4, the number of moles or millimoles is reported for various molecules and ions in 1 L of blood and used for

Figure 4-4 Blood test results for a 51-year-old female. The first column shows measured values for several ions and molecules, some in units of millimoles per liter. The second column shows the normal range for these ions and molecules. [Catherine Bausinger]

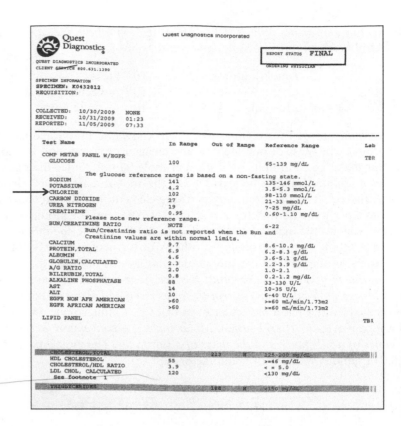

diagnostic purposes. To assess patient health, doctors and nurses evaluate the number of chloride ions reported in a blood test and compare it to the normal range. In the blood test shown, the patient has 102 millimoles of chloride ions (Cl^-) in every liter of her blood. In other words, she has 0.102 mole of chloride ions in every liter of her blood. You can compare this value to the normal range for this ion, in the column on the far right, which is 98 mmol to 110 mmol in every 1 L of blood. The test results show that the patient's chloride levels are within the normal range. If you were interested in knowing how many chloride ions correspond to 0.102 mol, a calculation using Avogadro's number as a conversion factor would be required. You would find that it is a very large number, and for this reason, chemists prefer to refer to the number of atoms in units of moles rather than the actual number of ions, much like a baker refers to *dozens* of eggs rather than the actual number of eggs.

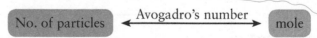

To interconvert between moles and ions, atoms, or molecules, we use the conversion $6.02 \times 10^{23} = 1$ mole. For example, to calculate the number of chloride ions in 0.102 mole of chloride ions, use dimensional analysis by multiplying the supplied unit, 0.102 mol, by the correct form of the conversion factor that allows moles to cancel:

$$0.102 \; \cancel{mol} \times \frac{6.02 \times 10^{23} \; Cl^- \; atoms}{1 \; \cancel{mol}} = 6.14 \times 10^{22} \; Cl^- \; atoms$$

This value should make you appreciate why we prefer to say the patient has 102 mmol of chloride ions rather than 6.14×10^{22} chloride ions in 1 L of her blood. Both represent the same number of chloride ions, but the former is a simpler unit to report.

| WORKED EXERCISES | Interconverting between Number of Moles and Number of Particles |

4-5 How many silver atoms are there in 5.6 moles of silver?

4-6 How many moles is 2.3×10^{21} gold atoms?

Solutions

4-5 Use dimensional analysis to set up the calculation using the correct form of the conversion factor (Avogadro's number) that will allow moles to cancel:

$$5.6 \ \cancel{mol} \times \frac{6.02 \times 10^{23} \ \text{silver atoms}}{1 \ \cancel{mol}} = 3.4 \times 10^{24} \ \text{silver atoms}$$

4-6 Use dimensional analysis to set up the calculation using the correct form of the conversion factor (Avogadro's number) that will allow number of particles to cancel:

$$2.3 \times 10^{21} \ \cancel{\text{gold atoms}} \times \frac{1 \ \text{mol}}{6.02 \times 10^{23} \ \cancel{\text{gold atoms}}} = .0038 \ \text{mol}$$

PRACTICE EXERCISES

8 How many ammonia (NH_3) molecules are there in 0.56 mole of ammonia?

9 Using the blood test results shown in Figure 4-4, how many millimoles of carbon dioxide, CO_2, are there in every liter of this patient's blood? Is the patient's blood CO_2 level normal? How many CO_2 molecules are there in every liter of this patient's blood?

10 How many moles of caffeine does 1.100×10^{27} caffeine molecules represent?

Interconverting between Moles and Mass Interconverting between mass and moles is one of the most routine calculations performed in the chemical laboratory. The key to solving this type calculation is recognizing that the molar mass of an element or compound *is* a conversion factor between the mass and moles of that substance. As with any calculation involving conversions, using the correct conversion factor—described in Chapter 1—provides a reliable and straightforward method for arriving at the solution. Follow the three step guidelines below for converting between grams and moles of an element or compound.

$$\boxed{\text{mole}} \ \overset{\text{molar mass}}{\longleftrightarrow} \ \boxed{\text{grams of substance}}$$

To interconvert between moles and grams of ions, atoms or molecules, we must use the molar mass or the inverted form of molar mass as a conversion factor. Calculating the molecular mass or formula mass of a substance was described on page 143, and the molar mass has the same numerical value. For example, to calculate the number of moles in 0.30 g of caffeine, the amount consumed by the average person in a day, we use the inverted form of the molar mass of caffeine, 194.20 g/mol. We use dimensional analysis by multiplying the supplied unit, 0.30 g, by the correct form of the conversion factor that allows grams to cancel:

$$0.30 \ \cancel{g} \times \frac{1 \ \text{mol}}{194.20 \ \cancel{g}} = 0.0015 \ \text{mol}$$

So, the average person consumes 0.0015 mole of caffeine a day.

Figure 4-5 A can of Red
Bull contains 80 mg of
caffeine. An eight-ounce
cup of coffee contains
between 100 and
200 mg of caffeine.
[Catherine Bausinger]

<u>WORKED EXERCISES</u>   Interconverting between Mass and Moles

4-7   Sulfur reacts with oxygen, producing a blue flame as it forms sulfur dioxide, a
compound with a noxious odor. How many grams of sulfur are there in 5.600 mol
of sulfur?

4-8   In a 250 mL can of Red Bull there are 80. mg of caffeine (Figure 4-5). How many
moles of caffeine, $C_8H_{10}N_4O_2$, are in 80. mg of caffeine?

Solutions

4-7   First determine the molar mass of sulfur because this will be the conversion
between moles and mass. Remember, molar mass has the same numerical value as
atomic mass, molecular mass, and formula mass but has units of g/mol instead of
amu. According to the periodic table, the molar mass of sulfur is 32.07 g/mol.

Next, express the molar mass of S as two possible conversion factors:

$$\frac{32.07 \text{ g S}}{1 \text{ mol S}} \quad \text{or} \quad \frac{1 \text{ mol S}}{32.07 \text{ g S}}$$

Using dimensional analysis, multiply the supplied unit by the appropriate form of
the conversion factor that allows the supplied units to cancel, leaving the requested
unit. Note that the supplied unit here is moles and the requested unit is grams, so
we use the conversion factor shown in red above:

$$5.600 \text{ mol S} \times \frac{32.07 \text{ g S}}{1 \text{ mol S}} = 179.6 \text{ g S} \quad \text{(four significant figures)}$$

Therefore, 5.600 moles of sulfur have a mass of 179.6 g.

4-8   First calculate the molar mass of caffeine, $C_8H_{10}N_4O_2$, from the atomic masses of
its constituent atoms:

$$8 \text{ C} + 10 \text{ H} + 4 \text{ N} + 2 \text{ O} = (8 \times 12.01) + (10 \times 1.008) + (4 \times 14.01) + (2 \times 16.00)$$
$$= 194.20 \text{ g/mol}$$

The molar mass of caffeine and its inverted form are the two possible conversion
factors:

$$\frac{194.20 \text{ g } C_8H_{10}N_4O_2}{1 \text{ mol } C_8H_{10}N_4O_2} \quad \text{or} \quad \frac{1 \text{ mol } C_8H_{10}N_4O_2}{194.20 \text{ g } C_8H_{10}N_4O_2}$$

Note that molar mass is always given in units of *grams* per mole but the
supplied unit here is given in *milli*grams. So the calculation will also require a
metric conversion from milligrams to grams (shown first).

Set up the calculation by multiplying the supplied unit, in milligrams, by the
correct form of the metric conversion factor that allows milligrams to cancel and
multiplying by the correct form of the molar mass conversion factor that allows
grams to cancel. The final answer will be in units of moles, the requested unit.

$$80. \text{ mg caffeine} \times \frac{1 \text{ g caffeine}}{10^3 \text{ mg caffeine}} \times \frac{1 \text{ mol caffeine}}{194.20 \text{ g caffeine}} = 4.1 \times 10^{-4} \text{ mol caffeine.}$$

<u>PRACTICE EXERCISES</u>

11   How many moles of mercury are there in 159 g of mercury?

12   How many moles of water are there in 45 g of water?

13   Calculate the number of moles of ibuprofen in a 200. mg dose of Advil. The molecular formula
for ibuprofen is $C_{13}H_{18}O_2$.

14 About 18 million people in the United States are diabetic. A fasting glucose ($C_6H_{12}O_6$) level of greater than 126 mg of glucose in every deciliter of blood is usually indicative of diabetes. How many moles of glucose are there in 126 mg of glucose?

15 How many grams of mercury are there in 0.666 mol of mercury?

16 What is the mass, in grams, of 1.9 mol of water?

17 How many grams of potassium ions, K^+, are there in 1 liter of the patient's blood shown in the blood test in Figure 4-4?

18 Which contains a greater number of molecules: 1.0 g of water (H_2O) or 10. g of glucose ($C_6H_{12}O_6$)?

Converting between Mass and Number of Particles

You have learned how to interconvert between grams and moles and also how to interconvert between moles and number of particles. Occasionally you may need to know the *number* of atoms, molecules, or ions in a given *mass* of sample, which requires both of these types of calculations. *This type of calculation requires two steps, involving two conversion factors: Avogadro's number, which serves as a conversion between moles and the number of any particle, and molar mass, the conversion between moles and mass of a substance.*

$$\boxed{\text{No. of particles}} \xleftrightarrow[]{\text{Avogadro's number}} \boxed{\text{mole}} \xleftrightarrow[]{\text{molar mass}} \boxed{\text{grams of substance}}$$

Notice that the mole is at the center of this type of calculation. Thus, if you need to determine the number of copper, Cu, atoms in a 5.0 g sample of copper, first convert the mass of copper into moles of copper and then convert the moles of copper into the number of copper atoms. This can be done in one step using dimensional analysis, by multiplying two conversion factors: Avogadro's number and molar mass.

WORKED EXERCISE Interconverting between Mass and Number of Particles

4-9 A person is considered diabetic when he or she has more than 126 mg of glucose ($C_6H_{12}O_6$) in every deciliter of blood, how many glucose molecules are there in 126 mg of glucose?

Solution

4-9 Determine the conversion factors needed. One form of Avogadro's number will be used for one of the conversion factors:

$$\frac{6.02 \times 10^{23} \text{ molecules}}{1 \text{ mol}} \quad \text{or} \quad \frac{1 \text{ mol}}{6.02 \times 10^{23} \text{ molecules}}$$

The other conversion factor involves the molar mass of glucose, which must first be calculated from its constituent atoms:

$$(6 \times 12.01) + (12 \times 1.008) + (6 \times 16.00) = 180.16 \text{ g/mol}$$

The molar mass, like any conversion factor, can be expressed in two ways:

$$\frac{1 \text{ mol glucose}}{180.156 \text{ g glucose}} \quad \text{or} \quad \frac{180.156 \text{ g glucose}}{1 \text{ mol glucose}}$$

Set up the calculation by multiplying the supplied unit, 126 mg of glucose, by the metric conversion between g and mg so that mg cancel. Then multiply by the form of the conversion factor for molar mass that allows grams to cancel. Finally, multiply by the conversion between moles and Avogadro's number that allows moles to cancel. The calculation can be done in three sequential steps or in one step that multiplies the appropriate conversion factors:

$$126 \text{ mg glucose} \times \frac{1 \text{ g}}{10^3 \text{ mg}} \times \frac{1 \text{ mol glucose}}{180.156 \text{ g glucose}} \times \frac{6.02 \times 10^{23} \text{ molecules}}{1 \text{ mol}}$$

$$= 4.21 \times 10^{20} \text{ glucose molecules}$$

Thus, 126 mg of glucose contains 4.21×10^{20} glucose molecules. Note that the first conversion factor is a metric conversion, the second is the molar mass conversion factor used in its inverted form, and the third conversion factor is Avogadro's number.

PRACTICE EXERCISES

19 What is the mass of 6.5×10^{20} copper atoms?

20 A normal dose of ibuprofen is 200. mg. How many ibuprofen molecules are there in a single dose of ibuprofen? The molecular formula of ibuprofen is $C_{13}H_{18}O_2$.

21 How many formula units of table salt are there in 1.0 g of sodium chloride? How many sodium ions is this? How many chloride ions is this?

4.2 The Law of Conservation of Mass and Balancing Chemical Equations

Now that you know how to count atoms, ions, and molecules by weighing them, we can explore what happens when they are combined in a chemical reaction. Chemical reactions are extremely important in chemistry and biochemistry. Each of our cells carries out millions of chemical reactions every second, including the reactions that convert the proteins, fats, and carbohydrates that we eat into energy. Our cells also carry out the chemical reactions that consume energy needed to build proteins, cell membranes, and DNA from smaller biomolecules. These reactions are the biochemical reactions of metabolism. In the remainder of this chapter you will learn how the composition of a substance is changed in a chemical reaction. To do this, we begin with learning how to write and balance a chemical equation.

The Law of Conservation of Mass

On the macroscopic scale, what we typically observe in a chemical reaction is one or more of the following: a change in color, the formation of a precipitate (see Figure 4-1), an increase in temperature, the evolution of a gas (bubbles), or a noticeable smell or aroma.

Viewed on the atomic scale, particles are in constant motion, colliding with the walls of their container and occasionally colliding with one another, the kinetic molecular view of matter, described in Chapter 1. You learned that molecular motion is proportional to the kinetic energy of the molecules. Thus, at higher temperatures, atoms, ions, and molecules—particles—have greater kinetic energy and are, therefore, moving faster.

In most collisions, atoms and/or molecules bounce off one another and no reaction occurs. Occasionally, however, *the kinetic energy of two colliding molecules or atoms is sufficient and occurs in the proper orientation for a chemical reaction to take place.* Consider, for example, the reaction between nitrogen (N_2) and oxygen (O_2). Mixing these two gases at high

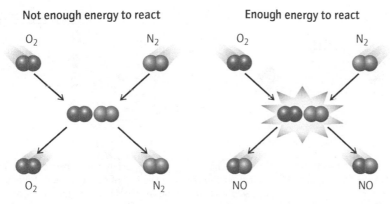

Figure 4-6 The collision between a molecule of oxygen (O_2) and a molecule of nitrogen (N_2) results in a reaction to produce two molecules of nitrogen oxide (NO) only if the collision occurs with enough energy and in the proper orientation.

temperature results in the occasional sufficiently energetic collision between a nitrogen molecule and an oxygen molecule. In the collision, the N—N triple bond breaks and the O—O double bond breaks as new N—O bonds are formed, producing nitrogen oxide, NO, molecules as illustrated in Figure 4-6. In the chemical reaction, nitrogen (N_2) and oxygen (O_2) molecules have been transformed into nitrogen oxide (NO) molecules, a different substance.

The compounds or elements that react are known as the **reactants** and the compounds or elements produced are known as the **products**. In this example, the elements nitrogen (N_2) and oxygen (O_2) are the reactants, and the compound nitrogen oxide (NO) is the product. A change in composition has occurred: the reactants, nitrogen, N_2, and oxygen, O_2, are clearly different from the product, nitrogen oxide, NO.

Notice, however, that every *atom type* in the reactants (N and O) also appears in the products (N and O)—they are just bonded to different atoms. *In a chemical reaction, the number and types of atoms in the product(s) are the same as in the reactant(s).* Hence, the atoms themselves have not been altered. No new atoms have been created, and no atoms have been destroyed! This fundamental principle that applies to all chemical reactions is known as the **law of conservation of mass**: *Matter can be neither created nor destroyed.*

> Only in nuclear reactions do atoms change into other atoms. In chemical reactions, the atoms themselves remain intact throughout the reaction—effectively only the bonds change.

> The *law of conservation of mass*: Matter can be neither created nor destroyed.

Consider another example, the reaction between methane (CH_4) and oxygen (O_2), a type of reaction known as a **combustion reaction**. This reaction produces the familiar blue flame seen on a gas burner. When a gas stove is turned on, methane gas (CH_4) flows out of the gas jets on the burner and reacts with the oxygen (O_2) in the air. A spark from the stove initiates the reaction. Every one molecule of methane reacts with two molecules of oxygen (O_2) to form one molecule of carbon dioxide (CO_2) and two molecules of water (H_2O), as illustrated in Figure 4-7. The law of conservation of matter is observed because the same number of each atom type is seen in both the reactants and the products: carbon (1), hydrogen (4), and oxygen (4).

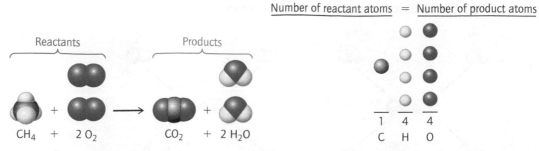

Figure 4-7 The reaction of methane (CH_4) and oxygen (O_2), produces carbon dioxide (CO_2) and water (H_2O), and like all chemical reactions obeys the law of conservation of mass. The number of each atom type in the reactants is the same as in each atom type in the products: carbon (1), hydrogen (4), and oxygen (4).

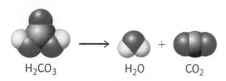

Figure 4-8 The decomposition of carbonic acid (H_2CO_3) to water (H_2O) and carbon dioxide (CO_2) is an example of a reaction with one reactant and two products.

Although the combustion of methane involves two reactants that go on to form two products, there is sometimes only one reactant and/or one product involved. For example, a reaction involved in the transport of carbon dioxide out of the body, in the lungs, is shown in Figure 4-8. One carbonic acid molecule (H_2CO_3) decomposes to produce one water molecule (H_2O) and one carbon dioxide molecule (CO_2). In this example, we have one reactant and two products.

Writing a Chemical Equation

A chemical reaction is described by a **chemical equation**, which conveys the identity of the reactants and the products *and* their relative *proportions*. Other pertinent information about the reaction, such as the physical state of the reactants and products, may also be shown.

In a chemical equation the reactants and products are represented by their chemical formula or chemical structure. A reaction arrow ($\rightarrow$) is used to separate the *reactants* on the left side of the arrow from the *products* on the right side of the arrow. For example, the chemical equation shown below represents the reaction between methane and oxygen to produce carbon dioxide and water, described in Figure 4-7.

$$CH_4(g) \; + \; 2\,O_2(g) \; \longrightarrow \; CO_2(g) \; + \; 2\,H_2O(g)$$

The whole numbers you see before some of the chemical formulas are known as **coefficients** and indicate the molar ratio of the individual reactants and products. When no number is shown, the coefficient is assumed to be 1. Returning to the initial example, in Figure 4-6, we see that *one* mole of nitrogen reacts with *one* mole of oxygen to produce *two* moles of nitrogen oxide, which is indicated in the chemical equation by placing the coefficient 2 in front of NO. A 1 can be assumed to exist in front of N_2 and O_2.

$$N_2 + O_2 \rightarrow 2\,NO$$

In summary, the parts of a chemical equation include the

- **Reactants.** The compounds or elements that are combined in a chemical reaction are shown to the left of the arrow and are referred to as the *reactants*.

- **Products.** The compounds or elements produced in a chemical reaction are shown to the right of the arrow and are referred to as the *products*.
- **A + sign** is used to separate more than one reactant or more than one product in a chemical equation.
- **Reaction arrow.** An arrow pointing from left to right separates the reactants from products. The arrow symbolizes the transformation of reactants into products in the chemical reaction.
- **Coefficients.** The coefficients indicate the lowest whole number ratio of each reactant and each product.
- **Physical state.** The physical state of each compound or element is sometimes included in parentheses following the formula of the compound: solid (s), liquid (l), gas (g), or aqueous solution (aq).

WORKED EXERCISE   Chemical Equations and the Law of Conservation of Mass

4-10 Answer the questions for the chemical equation below:

$$4 \ Fe(s) + 3 \ O_2(g) \rightarrow 2 \ Fe_2O_3(s)$$

a. What are the reactant(s)?
b. What are the product(s)?
c. What are the coefficients in this equation and what do they signify?
d. What is meant by the symbols in parentheses?
e. How many total iron atoms (Fe) are on both sides of the equation? Has any iron been created or destroyed?

Solution

4-10 a. The reactants appear to the left of the reaction arrow: iron (Fe) and oxygen (O_2).
   b. The product appears to the right of the reaction arrow: iron(III) oxide (Fe_2O_3).
   c. The coefficients indicate that *four* iron atoms react with *three* oxygen molecules to produce *two* iron(III) oxide formula units.
   d. The abbreviations in parentheses indicate states of matter: iron and iron(III) oxide are in the solid (s) state and oxygen, O_2, is in the gas (g) state.
   e. There are four iron atoms on the reactant side and four iron atoms on the product side, so no new iron atoms have been created and none have been destroyed. We count the iron atoms on the product side by noting that there are two iron atoms in every one Fe_2O_3 formula unit, and since there are two iron(III) oxide formula units, as indicated by the coefficient 2, we have a total of four (2×2) iron atoms on the product side.

PRACTICE EXERCISE

22 Answer the questions below for the following chemical equations:
   i. $C_3H_8(g) + 5 \ O_2(g) \rightarrow 3 \ CO_2(g) + 4 \ H_2O(g)$
   ii. $C_4H_6(g) + 2 \ H_2(g) \rightarrow C_4H_{10}(g)$
      a. Label the reactant(s).
      b. Label the product(s).
      c. Give the coefficients in each equation and state what they signify.
      d. Indicate the physical state (s, l, or g) of each reactant and product.

Balancing a Chemical Equation

We have seen that the law of conservation of mass requires that the total number of each type of atom on the reactant side of a chemical equation must equal the total number of each type of atom on the product side of the equation. *A chemical equation that contains the proper coefficients such that there are an equal number of each type of atom on both sides of the equation is known as a balanced equation.*

For example, we determine that the chemical equation below is balanced by counting each atom type, on each side of the equation. To count the number of one type of atom on one side of the equation, multiply the coefficient by the subscript for that atom type in the molecular formula or formula unit. Remember, when there is no coefficient, it is assumed to be 1. Then add up the results for all atoms of that type in the reactants and likewise in the products. For example, the number of oxygen (O) atoms on each side of the equation below is 18, calculated as shown beneath the equation. Coefficients are shown in black, and the relevant subscripts are shown in red.

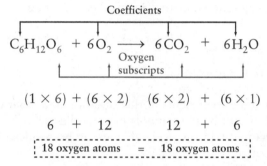

Law of conservation of mass

Coefficients

$$C_6H_{12}O_6 + 6O_2 \longrightarrow 6CO_2 + 6H_2O$$

Oxygen subscripts

$$(1 \times 6) + (6 \times 2) \quad (6 \times 2) + (6 \times 1)$$

$$6 + 12 \quad\quad 12 + 6$$

18 oxygen atoms = 18 oxygen atoms

Similar calculations for the carbon atoms (C) and hydrogen atoms (H) show that there are six carbon atoms and 12 hydrogen atoms on each side of the equation. This chemical equation is therefore balanced.

To balance a chemical equation, insert the appropriate whole-number coefficients before the reactants and/or products as needed. A systematic step-by-step process for balancing a chemical equation is described in the guidelines box. The law of conservation of mass is further demonstrated in Using Molecular Models 4-1: Balancing a Chemical Equation.

Guidelines for Balancing a Chemical Equation

Step 1: Balance the equation one atom type at a time by inserting whole-number coefficients. Systematically place a coefficient in front of a reactant **or** a product as necessary to arrive at an equal number of one atom type on both sides of the equation. Then repeat the process with the next atom type. Note the following important points:

- Insert coefficients to increase the number of atoms, but never change the numerical subscripts in a chemical formula because this changes the identity of the chemical compound.

- Remember that inserting a coefficient alters the number of every atom type in the formula that it precedes. Multiply the coefficient by the subscript for a particular atom type to determine the number of that atom type. *For example, placing a 2 in front of CO_2 changes the number of carbon atoms from one to two (2 × 1) and also changes the number of oxygen atoms from two to four (2 × 2).*

- Although it does not matter in what order you balance each atom type, it is often easier if you balance an atom type that appears in more than one compound or element on one side of the equation last. *For example, since O appears in both products of a combustion reaction: H_2O and CO_2, it is best to balance O last.*

Step 2: Check that the coefficients are the lowest whole-number ratio possible. Convention requires that coefficients be the lowest whole-number ratio; therefore no fractions may appear in the final balanced equation. Furthermore, if every coefficient can be divided by a common divisor, do so. *For example, the coefficients 4, 8, and 12 can all be divided by 4, so divide each coefficient by 4 and use the coefficients 1, 2, and 3 instead.*

Using Molecular Models 4-1 Balancing a Chemical Equation

Construct the Reactants

1. Obtain one black carbon atom, four red oxygen atoms, and four light blue hydrogen atoms. Obtain four bent bonds and four straight bonds.

2. Construct one model of methane, CH_4, and two models of oxygen, O_2. Remember, an oxygen molecule contains an oxygen-oxygen double bond requiring two bent bonds. Imagine that these models are the reactants for a combustion reaction. Answer Inquiry Question 5, below.

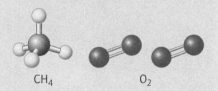

CH₄   O₂

Construct the Products

3. Simulate a chemical reaction by breaking *all* the bonds in your methane and oxygen molecules.

4. Using *only* these atoms and bonds, construct as many carbon dioxide, CO_2, and water, H_2O, molecules as you can. Remember that CO_2 contains two C=O double bonds, requiring two bent bonds per double bond. Answer Inquiry Questions 6 through 10.

Inquiry Questions

5. How many C, O, and H atoms are present in the reactants? Fill in the table in the next column.

Atom Type	Number on Reactant Side
C	
H	
O	

6. How many CO_2 molecules were you able to build? Therefore, what coefficient should be placed before CO_2 in the balanced equation?

7. How many water molecules were you able to build? Therefore, what coefficient should be placed before H_2O in the balanced equation?

8. Were any atoms left over after the exercise? Why or why not? What law does this demonstrate?

9. How many C, O, and H atoms are present in the products? Fill in the table below. How do these numbers compare with those on the reactant side in question 5?

Atom Type	Number on Reactant Side	Number on Product Side
C	1	
H	4	
O	4	

10. Write a balanced equation for the reaction between methane and oxygen simulated here.

WORKED EXERCISES   Balancing Chemical Equations

4-11 The hydrogen fuel cell, used in cars such as the Honda FCX Clarity (Figure 4-9), employs a revolutionary new type of car engine that doesn't use gasoline but instead generates energy according to the following unbalanced equation. Balance the equation.

$$\underline{\quad} H_2(g) + \underline{\quad} O_2(g) \rightarrow \underline{\quad} H_2O(g) + \text{energy}$$

4-12 Balance the following equation showing the combustion of propane on a barbecue grill:

$$\underline{\quad} C_3H_8(g) + \underline{\quad} O_2(g) \rightarrow \underline{\quad} CO_2(g) + \underline{\quad} H_2O(g)$$

Solution

4-11 Step 1: Balance the equation one atom type at a time by inserting whole number coefficients (multipliers). Note that there are already two hydrogen atoms on each side of the equation, so H is already balanced. There are two oxygen atoms on the left side and only one on the right. Since fractions are not allowed, we place a 2 in front of H_2O to balance O. While the O atoms are now balanced, the H atoms are no longer balanced because we have 2 H on the left and 4 H on the right. Place the coefficient 2 in front of H_2 to balance the H atoms:

$$2\,H_2 + O_2 \rightarrow 2\,H_2O$$

H:	$2 \times 2 =$	$2 \times 2 =$
	4	4
O:	$1 \times 2 =$	$2 \times 1 =$
	2	2

Step 2: Check that the coefficients cannot be divided by a common factor (divisor). There is no common divisor.

Figure 4-9 The reaction of hydrogen (H_2) and oxygen (O_2) produces water (H_2O) and energy in cars such as the Honda FCX Clarity, which have a hydrogen fuel cell instead of a gasoline engine. Notice the "gas" station is a "hydrogen" station. [ZUMA Press, Inc./Alamy.]

4-12 Step 1: Balance the equation one atom type at a time by inserting whole number coefficients. Remember that in a combustion reaction it is easiest if you balance O last because it appears in *two* compounds on the product side. Beginning with C, insert the coefficient 3 in front of CO_2:

$$C_3H_8 + O_2 \rightarrow 3\,CO_2 + H_2O \quad \text{(carbon is now balanced with 3 C's}$$
$$\text{on both sides of the equation)}$$

Balance hydrogen by inserting the coefficient 4 in front of water:

$$C_3H_8 + O_2 \rightarrow 3\,CO_2 + 4\,H_2O \quad \text{(hydrogen is now balanced with eight}$$
$$\text{H atoms on both sides of the equation)}$$

Balance oxygen by determining that there are $(3 \times 2) + (4 \times 1) = 10$ oxygen atoms on the product side. Thus, we need to place a 5 as a coefficient in front of O_2 on the reactant side so that there will also be $5 \times 2 = 10$ oxygen atoms on the reactant side:

$$3\ C_3H_8 + 5\ O_2 \rightarrow 3\ CO_2 + 4\ H_2O \quad \text{(oxygen is now balanced with 10 O atoms on both sides of the equation)}$$

Step 2: Check that the coefficients cannot be divided by a common factor (divisor). There is no common divisor for the coefficients: 3, 5, 3, and 4; therefore, you have the lowest set of whole-number coefficients and a balanced equation.

PRACTICE EXERCISE

23 Balance the following chemical equations by inserting the appropriate coefficients into the blanks. If the coefficient is a 1, leave it blank.

a. ___ $N_2O_5(g) \rightarrow$ ___ $NO_2(g) +$ ___ $O_2(g)$

b. ___ $C_6H_{12}O_2(l) +$ ___ $O_2(g) \rightarrow$ ___ $CO_2(g) +$ ___ $H_2O(g)$

c. ___ $Mg(s) +$ ___ $HCl(aq) \rightarrow$ ___ $MgCl_2(aq) +$ ___ $H_2(g)$

d. ___ $NH_3(g) +$ ___ $O_2(g) \rightarrow$ ___ $N_2(g) +$ ___ $H_2O(l)$

Mole-to-Mole Stoichiometry Calculations From a balanced equation, it is possible to calculate the number of moles of product produced if the number of moles of each reactant is known. Moreover, since there is a relationship between mass and moles, we can also determine the mass of products formed if we know the mass of the reactants. This latter calculation, known as a **stoichiometry** calculation, is routinely performed in the laboratory.

Reaction stoichiometry calculations always begin with a balanced chemical equation, because *the coefficients in a chemical equation represent the molar ratio of each reactant and product.* For example, one mole of ethanol, C_2H_6O, reacts with three moles of oxygen to form two moles of carbon dioxide and three moles of water, according to the balanced equation for the combustion of ethanol shown below:

$$C_2H_6O(l) + 3\ O_2(g) \rightarrow 2\ CO_2(g) + 3\ H_2O(l)$$

Ethanol   Oxygen   Carbon dioxide   Water

In this section we describe how to calculate the number of *moles* of products formed if the number of *moles* of reactants is given. In the Extension Topic on page 158 we describe how to determine the *mass* of products formed if the *mass* of reactants is given. Stoichiometry calculations are solved using dimensional analysis.

Suppose that 5.0 moles of ethanol undergoes the reaction described above. How many moles of oxygen are required for the complete combustion of ethanol, and how many moles of carbon dioxide can be produced?

Step 1: Select the appropriate coefficients from the balanced chemical equation to construct a conversion factor between the substances involved in the calculation. Since the first part of the question asks about the molar relationship between ethanol and oxygen, use these coefficients to construct the following conversion factors:

$$\frac{1 \text{ mol ethanol}}{3 \text{ mol oxygen}} \quad \text{or} \quad \frac{3 \text{ mol oxygen}}{1 \text{ mol ethanol}}$$

Step 2: Using dimensional analysis, multiply the supplied moles of ethanol by the conversion factor from step 1 that allows moles of ethanol to cancel, leaving moles of oxygen:

$$5.0 \; \text{mol ethanol} \times \frac{3 \; \text{mol oxygen}}{1 \; \text{mol ethanol}} = 15 \; \text{mol oxygen}$$

Thus, 15 moles of oxygen are required to completely burn 5.0 moles of ethanol.

Use the same process to determine the moles of carbon dioxide formed in a reaction where 5.0 moles of ethanol react with 15 moles of oxygen. This time, construct a conversion factor between ethanol and carbon dioxide (or oxygen and carbon dioxide) using the coefficients from the balanced equation:

$$\frac{1 \; \text{mol ethanol}}{2 \; \text{mol carbon dioxide}} \quad \text{or} \quad \frac{2 \; \text{mol carbon dioxide}}{1 \; \text{mol ethanol}}$$

Then multiply the supplied moles of ethanol by the conversion factor above that allows moles of ethanol to cancel, leaving moles carbon dioxide:

$$5.0 \; \text{mol ethanol} \times \frac{2 \; \text{mol carbon dioxide}}{1 \; \text{mol ethanol}} = 10. \; \text{mol carbon dioxide}$$

Thus, combustion of 5.0 moles of ethanol requires 15 moles of oxygen and can produce 10. moles of carbon dioxide. Note that in practice, we usually obtain less than the calculated amount of product due to side reactions and other mechanical issues related to performing a chemical reaction and isolating the product.

PRACTICE EXERCISES

24 If six moles of glucose were to react with excess oxygen, how many moles of carbon dioxide can be produced? How many moles of water can be produced? The balanced equation is shown below. (Note, when there is excess oxygen the reaction will continue until it runs out of glucose.)

$$C_6H_{12}O_6 + 6 \; O_2 \rightarrow 6 \; CO_2 + 6 \; H_2O$$

25 How many moles of HCl would be required to react completely with 5.60 moles of magnesium? How many moles of hydrogen gas can be produced from such a reaction?

$$Mg(s) + 2 \; HCl(aq) \rightarrow MgCl_2(aq) + H_2(g)$$

Extension Topic 4-1   Stoichiometry Calculations

It is important to recognize that the coefficients in a balanced chemical equation represent mole ratios of reactants and products, not mass ratios. However, since there is a relationship between the moles of a substance and its mass, as described in section 4.1, moles can be converted into mass using the molar mass of the substance as a conversion factor. For example, suppose you wanted to calculate the mass of carbon dioxide (CO_2) produced when 12.0 g of ethanol (C_2H_6O) is burned in excess oxygen (in other words, you won't run out of oxygen). First you would need a balanced equation:

$$C_2H_6O(l) + 3 \; O_2(g) \rightarrow 2 \; CO_2(g) + 3 \; H_2O(l)$$

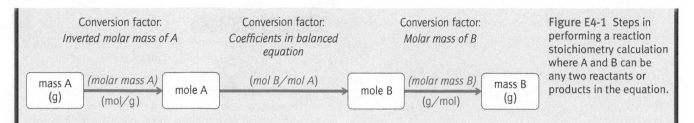

Conversion factor: *Inverted molar mass of A*   Conversion factor: *Coefficients in balanced equation*   Conversion factor: *Molar mass of B*

Figure E4-1 Steps in performing a reaction stoichiometry calculation where A and B can be any two reactants or products in the equation.

Then perform a stoichiometry calculation according to the following three steps, which are summarized in Figure E4-1:

Step 1: Convert the mass of the supplied substance into moles of supplied substance. Here we need to convert grams of ethanol to moles of ethanol. The molar mass of ethanol, C_2H_6O, must first be calculated from its constituent atoms:

$$(2 \times 12.01 \text{ g/mol}) + (6 \times 1.008 \text{ g/mol}) + (1 \times 16.00 \text{ g/mol}) = 46.07 \text{ g/mol}$$

Convert the mass of ethanol into moles of ethanol using the appropriate form of the molar mass of ethanol as a conversion factor so grams of ethanol cancel. This type of calculation was described in section 4.1.

$$12.0 \text{ g } C_2H_6O \times \frac{1 \text{ mol } C_2H_6O}{46.07 \text{ g } C_2H_6O} = 0.260 \text{ mol } C_2H_6O$$

Step 2: Convert the moles of supplied substance to moles of requested substance. Here we need to convert the moles of ethanol (a reactant) calculated in step 1 to moles of carbon dioxide (a product), using their respective *coefficients* from the balanced equation to construct a conversion factor, as described on page 157 (Mole-to-Mole calculation). (Note that coefficients are exact numbers, so they do not affect significant figures.)

$$0.260 \text{ mol } C_2H_6O \times \frac{2 \text{ mol } CO_2}{1 \text{ mol } C_2H_6O} = 0.520 \text{ mol } CO_2$$

Step 3: Convert moles of the requested substance, calculated in step 2, to mass of the requested substance. Here we need to convert moles of carbon dioxide, from step 2, to grams of carbon dioxide. The molar mass of carbon dioxide, CO_2, must first be calculated from its constituent atoms:

$$(1 \times 12.01 \text{ g/mol}) + (2 \times 16.00 \text{ g/mol}) = 44.01 \text{ g/mol}$$

Convert the moles of carbon dioxide from step 2 into grams of carbon dioxide using the appropriate form of the molar mass of carbon dioxide as a conversion

factor, that allows moles of carbon dioxide to cancel. This type of calculation was described in section 4.1.

$$0.520 \text{ mol } CO_2 \times \frac{44.01 \text{ g } CO_2}{1 \text{ mol } CO_2} = \textbf{22.9 g } CO_2$$

As with any dimensional analysis problem involving several conversion factors, the calculation does not need to be done in three sequential steps as described here. Instead, multiply the supplied value by the three conversion factors to arrive at the same final answer. Remember round only at the end to avoid rounding errors.

$$12.0 \text{ g } C_2H_6O \times \frac{1 \text{ mol } C_2H_6O}{46.07 \text{ g } C_2H_6O} \times \frac{2 \text{ mol } CO_2}{1 \text{ mol } C_2H_6O}$$
$$\times \frac{44.01 \text{ g } CO_2}{1 \text{ mol } CO_2} = 22.9 \text{ g } CO_2$$

WORKED EXERCISE E4.1 Reaction Stoichiometry Calculation

Calculate the mass of water produced if 12.1 g of ethanol undergoes the reaction described previously:

$$C_2H_6O(l) + 3 O_2(g) \rightarrow 2 CO_2(g) + 3 H_2O(l)$$

Ethanol   Oxygen   Carbon dioxide   Water

Solution

Step 1: Convert the mass of the supplied substance into moles of supplied substance. Here we need to convert grams of ethanol to moles of ethanol using the molar mass of ethanol as a conversion factor, as shown previously:

$$12.1 \text{ g } C_2H_6O \times \frac{1 \text{ mol } C_2H_6O}{46.07 \text{ g } C_2H_6O} = 0.263 \text{ mol } C_2H_6O$$

Step 2: Convert moles of supplied substance to moles of requested substance. Here we need to convert moles of ethanol calculated in step 1 to moles of water using the coefficients for water and ethanol from the balanced equation to create a conversion factor:

$$0.263 \text{ mol } C_2H_6O \times \frac{3 \text{ mol } H_2O}{1 \text{ mol } C_2H_6O} = 0.789 \text{ mol } H_2O$$

Step 3: Convert moles of the requested substance, calculated in step 2, to mass of the requested substance.

Here we must convert moles of water to mass of water using the molar mass of water as a conversion factor:

$$0.789 \ \text{mol H}_2\text{O} \times \frac{18.01 \ \text{g H}_2\text{O}}{1 \ \text{mol H}_2\text{O}} = 14.2 \ \text{g H}_2\text{O}$$

Alternatively, doing the calculation all in one step:

$$12.1 \ \text{g C}_2\text{H}_6\text{O} \times \frac{1 \ \text{mol C}_2\text{H}_6\text{O}}{46.07 \ \text{g C}_2\text{H}_6\text{O}} \times \frac{3 \ \text{mol H}_2\text{O}}{1 \ \text{mol C}_2\text{H}_6\text{O}}$$

$$\times \frac{18.01 \ \text{g H}_2\text{O}}{1 \ \text{mol H}_2\text{O}} = 14.2 \ \text{g H}_2\text{O}$$

Thus, combustion of 12.1 g of ethanol produces up to 14.2 g of water.

PRACTICE EXERCISES

E4.1 How many grams of oxygen are needed to convert 12.0 g of ethanol into products in the reaction described previously?

E4.2 Balance the chemical equation below and then determine how many grams of oxygen are needed to react completely with 500.0 g of propane from a barbecue grill. How many grams of CO_2 will be produced?

___ $C_3H_8(g)$ + ___ $O_2(g) \rightarrow$ ___ $CO_2(g)$ + ___ $H_2O(g)$ (unbalanced)
  Propane

E4.3 How many grams of carbon dioxide are formed when 10.0 g of glucose, $C_6H_{12}O_6$, undergo combustion to form carbon dioxide and water? Begin by writing the complete balanced equation.

■ 4.3 Energy and Chemical Reactions

When you pick up this book, you have done *work*. To do work requires *energy*. In chapter 1 you learned that **energy is the capacity to do work and work is the act of moving an object over a distance against an opposing force.** The cells in our body do work, and therefore they require energy. Our cells convert the potential chemical energy of food molecules into usable energy through a series of chemical reactions. The study of energy transfer in biological cells is known as **bioenergetics.**

The foods we eat contain molecules with *chemical potential energy*, in the form of covalent bonds. Carbohydrates, proteins, and triglycerides (fats) are large biomolecules that release energy as they undergo a series of chemical reactions that break them down into smaller molecules, ultimately producing carbon dioxide and water. The energy released is used to drive other chemical reactions in the cell that require energy, such as those involved in picking up this book.

Heat Energy

Energy exists in several forms, including chemical, mechanical, and heat energy. A chemical reaction is usually accompanied by a transfer of energy, either from the surroundings to the reaction or from the reaction to the surroundings. Some or all of the energy transferred in a chemical reaction appears in the form of *heat*. Consider, for example, the gas grill used for a summer barbecue, where combustion of charcoal or propane involves the transfer of heat from the reaction to the surroundings. Once you ignite the propane, it reacts with the oxygen in the air, forming carbon dioxide, water, and heat. The reaction can be represented by the following balanced chemical equation:

$$C_3H_8(g) + 5 \ O_2(g) \rightarrow 3 \ CO_2(g) + 4 \ H_2O(g) + \text{heat energy}$$
  Propane

Heat is obviously a key product of this reaction; if it weren't, your hamburgers would never cook. *Chemical reactions transfer heat energy by either absorbing heat from the surroundings or releasing heat to the surroundings.*

It is important to note that in a chemical reaction not *all* of the energy is transferred in the form of heat. For example, in the engine of your car, much of the energy released in the combustion of gasoline is converted into mechanical energy, which propels your car, while some is converted to heat (hence the warm engine). In a biological cell, glucose (from the digestion of carbohydrates) undergoes a series of chemical reactions wherein the energy from combustion is transferred into usable forms of energy. As with a car engine, some of this energy is transferred in the form of heat, but most is not.

Exothermic and Endothermic Reactions We have seen that combustion reactions release heat to the surroundings. Some reactions, however, absorb heat from the surroundings. For example, photosynthesis is a series of reactions that absorb energy overall. Photosynthesis is the biochemical pathway—a sequence of chemical reactions—that occurs in a plant to produce carbohydrates. The energy is supplied by the sun in the form of electromagnetic radiation (see chapter 2), another form of energy.

In chapter 3 you learned that a covalent bond contains chemical potential energy. Two atoms sharing electrons—a covalent bond—are lower in energy than two separate atoms. *Thus, breaking a bond requires an input of energy, whereas making a bond results in an output of energy.* Therefore, whether heat is *released* or *absorbed* in a chemical reaction depends on the difference in potential energy between the products and the reactants. The difference often appears in the form of heat on either the reactant or product side of the reaction. This fundamental observation that energy is conserved in a chemical reaction, is given in the first law of thermodynamics:

The first law of thermodynamics: Energy the can be neither created nor destroyed, only transformed.

The difference in potential energy between the reactants and products of a chemical reaction can be measured and is known as the **change in enthalpy** of the reaction, $\Delta H_{reaction}$, where the Greek letter Δ, pronounced "delta", in science always means "change." When the value for $\Delta H_{reaction}$ is negative (less than 0), it means the reaction releases energy, which is known as an **exothermic reaction**. When the value for $\Delta H_{reaction}$ is positive (greater than 0), it means the reaction absorbs energy, which is known as an **endothermic reaction**.

| Exothermic reaction: | Heat is released; | $\Delta H_{reaction} < 0$ |
| Endothermic reaction: | Heat is absorbed; | $\Delta H_{reaction} > 0$ |

For example, the combustion of propane just described is an exothermic reaction, $\Delta H_{reaction} < 0$, where heat is released to the surroundings because the potential energy of the products is lower than the potential energy of the reactants as illustrated in Figure 4-10b.

Exothermic and endothermic reactions can be illustrated in an energy diagram, as shown in Figure 4-10. In an **energy diagram**, the y-axis represents

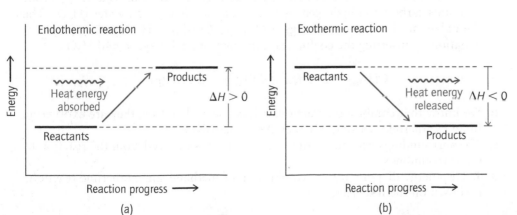

(a)   (b)

Figure 4-10 Energy diagrams: (a) An endothermic reaction: products are higher in energy than reactants, so heat is absorbed. (b) An exothermic reaction: products are lower in energy than reactants, so heat is released.

energy and the *x*-axis represents the progress of the reaction from reactants to products. In an exothermic reaction, products are *lower* in energy than reactants. In an endothermic reaction, products are *higher* in energy than reactants. The difference in energy between the energy of the reactants and the products is ΔH. The change in enthalpy is ac-counted for in the form of heat and covalent bonds, in accordance with the first law of thermodynamics.

An energy diagram helps us see that in an endothermic reaction (Figure 4-10a) the reactants are lower in energy than the products; and therefore, heat is absorbed from the surroundings (the surroundings become cooler). We say the change in enthalpy is positive ($\Delta H > 0$). In an endothermic reaction, energy must be continuously supplied to sustain the reaction. In an exothermic reaction (Figure 4-10b) the reactants are higher in energy than the products; therefore, heat is released to the surroundings (the surroundings become warmer). We say the change in enthalpy is negative ($\Delta H < 0$). It follows that if a reaction is exothermic, the reverse reaction is endothermic, by the same numerical value of ΔH, only the sign of ΔH (+ or −) changes. This follows from the *first law of thermodynamics*.

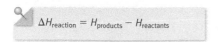
$\Delta H_{reaction} = H_{products} - H_{reactants}$

WORKED EXERCISES | Enthalpy of Reaction

4-13 For the combustion of methane (CH_4), answer the following:
   a. Write a balanced chemical equation.
   b. Does this reaction absorb or release heat? Is $\Delta H < 0$ or is $\Delta H > 0$? Is the reaction exothermic or endothermic?
   c. Do the surroundings become warmer or cooler in this reaction? Explain.
   d. Construct an energy diagram for the reaction.
   e. Label the axes, reactants, and products.
   f. Show how to find the value of ΔH on the energy diagram.

4-14 The reaction that occurs in a hydrogen fuel cell is shown below:

$2 H_2 + O_2 \rightarrow 2 H_2O$. The change in enthalpy for the reaction is $\Delta H = -118$ kcal.

   a. Is this reaction exothermic or endothermic? How can you tell?
   b. The reverse reaction, known as electrolysis, converts water into hydrogen and oxygen. Is electrolysis exothermic or endothermic? What is the value for the change in enthalpy for electrolysis? What law is illustrated in this example?

Solutions

4-13 a. We know that oxygen (O_2) is the other reactant in a combustion reaction and that the products of a combustion reaction where the reactant is a hydrocarbon (contains carbon and hydrogen) is carbon dioxide (CO_2) and water (H_2O). Thus, the unbalanced reaction is $CH_4(g) + O_2(g) \rightarrow CO_2(g) + H_2O(g)$. Balance the equation by inserting the coefficient 2 in front of both oxygen and H_2O:

$$CH_4(g) + 2\ O_2(g) \rightarrow CO_2(g) + 2\ H_2O(g)$$

   b. We know that combustion reactions release heat; therefore, they are exothermic. In an exothermic reaction $\Delta H < 0$ (ΔH is negative).
   c. The surroundings become warmer because heat is released from the reaction to the surroundings.
   d.–f. The energy diagram for the combustion of methane, an exothermic reaction, is shown on the next page:

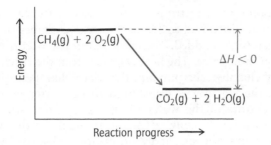

4-14 a. The reaction is exothermic because the change in enthalpy ($\Delta H = -118$ kcal) is negative.

   b. Since electrolysis is the reverse reaction, it is endothermic and $\Delta H = +118$ kcal, according to the first law of thermodynamics.

PRACTICE EXERCISES

26 Indicate whether each of the following reactions is an endothermic or an exothermic reaction.

   a. $8 H_2S(g) \rightarrow 8 H_2(g) + S_8(s)$; heat is absorbed

   b. $2 CO_2 \rightarrow 2 CO + O_2$; $\Delta H = +566$ kJ

27 Provide an energy diagram for the reaction shown in (26b).

   a. Label ΔH.

   b. Label the reactant and the products.

   c. Provide a label for the x- and y-axes.

28 How does an endothermic reaction differ from an exothermic reaction?

Calorimetry

Calorimetry is a technique used to measure enthalpy changes (ΔH) in chemical reactions, using an apparatus known as a **calorimeter**. A *bomb* calorimeter, illustrated in **Figure 4-11**, is designed to measure the heat released in a *combustion* reaction.

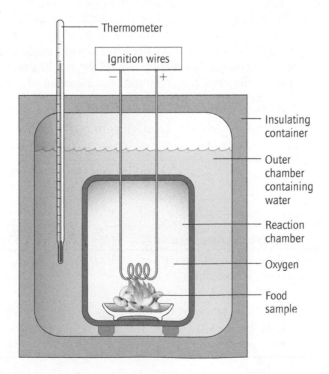

Figure 4-11 A cross section of a bomb calorimeter, used to measure the change in enthalpy of a combustion reaction.

Figure 4-12 The nutritional label on a package of potato chips. The label shows the total amount of heat released in calories when this food is consumed. The mass of each type of food molecule is listed separately as fat, carbohydrate, and protein. [Catherine Bausinger]

Recall from chapter one that 1 Calorie = 1 kcal = 1×10^3 calorie. The capital C distinguishes a nutritional Calorie from the other calorie.

Bomb calorimetry is used to measure the caloric content of food. A food sample is dried and sealed in the inner chamber of the bomb calorimeter and oxygen (O_2) is added. The combustion reaction is initiated with a spark from the ignition wires. The heat released from the reaction warms the water in the outer chamber surrounding the inner chamber. All the heat energy released from the reaction is transferred to the water as kinetic energy, causing the temperature of the water to rise. The temperature change, ΔT, the specific heat of water (1 cal/g · °C), and the mass of the water are used to calculate the amount of heat energy (in calories) absorbed by the water. According to the first law of thermodynamics, the heat absorbed by the water is equal to the heat released in the reaction, $\Delta H_{reaction}$, assuming no heat is lost. Such a calculation provides the caloric content of the food sample to a good approximation. Bomb calorimetry measurements for the combustion of carbohydrates, proteins, and fats are reported as a total on the nutritional labels you see on most packaged foods (**Figure 4-12**).

In accordance with the first law of thermodynamics, the energy released during the metabolism of a particular food is equal to the enthalpy of combustion, ΔH, measured by a calorimeter. The main difference is that combustion in a calorimeter occurs as a single reaction and the energy is released entirely in the form of heat, whereas inside our cells, a series of chemical reactions occurs and the energy released is transferred to and stored in biological molecules as covalent bonds rather than being released entirely as heat. In the Chemistry in Medicine: Calorimetry and the Care of Patients on a Mechanical Ventilator at the end of the chapter, you can read about how calorimetry can be used to determine human energy requirements, especially for individuals on a mechanical ventilator.

Table 4-2 compares the average Calorie content per mass (Cal/g) for the three basic food types: carbohydrates, proteins, and fats. As you can see, both the type and amount of food you eat determines your caloric intake. Notice that fats provide twice as many Calories per gram as carbohydrates and proteins. Indeed, the high caloric content of fat is why fat molecules so effectively serve as the body's long-term supply of chemical potential energy.

If we know both the mass and the type of food—carbohydrate, protein, or fat—we can calculate the total caloric content of a particular food sample using the values from Table 4-2. Calculate the number of calories from each food type by multiplying the mass of each type of food by its caloric content and then add up these values to obtain the total caloric content of the food item, as shown below. Food labels generally round the total calories to the tens place. Table 4-3 shows the caloric content and composition of some foods.

$$\text{grams of fat} \times 9 \text{ Cal/g} = \text{Calories from fat}$$
$$\text{grams of carbohydrate} \times 4 \text{ Cal/g} = \text{Calories from carbohydrate}$$
$$\text{grams of protein} \times 4 \text{ Cal/g} = \text{Calories from protein}$$

Total Calories (rounded to tens place)

TABLE 4-2 Caloric Content of Food

Biomolecule	Representative Foods	Caloric Content (Cal/g)
Carbohydrates	Rice, potatoes, bread, vegetables, fruit, whole milk	4
Proteins	Fish, meat, dairy products, beans, legumes, whole milk	4
Fats	Oils, butter, margarine, animal fats, whole milk	9

TABLE 4-3 Caloric Content and Composition of Some Foods*

Food	Carbohydrate (g)	Protein (g)	Fat (g)	Calories
Potato, baked (~3.6 ounces)	27	2.5	Trace	110
Rice, brown, cooked (1 cup)	46	5	2	220
Pasta, plain, cooked (1 cup)	32	5	1	115
Apple, raw (medium)	21	0	1	90
Beef, ground (3 ounces)	0	15	18	222
Cheese, cheddar (1 ounce)	0	7	9	113
Chicken breast roasted, no skin (medium)	0	54	6	250
Flounder, baked (3.4 ounces)	0	21	5	130
Olive oil (1 tbsp)	0	0	14	119
Avocado (1/2 cup)	9	2	18	185
Mayonnaise (1 tbsp)	0	0	11	100
Sour cream (1 tbsp)	1	0	3	26
Cream cheese (1 tbsp)	0	1	5	51
Butter (1 tsp)	0	0	4	36

*Values from *The Joy of Cooking*, Rombauer et al., Scribner 1997.

WORKED EXERCISES Calculating the Caloric Content of a Sample of Food

4-15 Calculate the caloric content of six ounces of low-fat strawberry yogurt from the nutritional label shown in Figure 4-13.

Solution

4-15 The label shows that 6 oz of strawberry yogurt contains 30 g carbohydrate, 5 g protein, and 1.5 g fat. Use the values provided in Table 4-2 to determine the total number of Calories supplied by each type of biomolecule:

$$\text{Calories provided by fat} = 1.5\text{ g} \times \frac{9\text{ Cal}}{1\text{ g}} = 13.5\text{ Cal}$$

$$\text{Calories provided by carbohydrate} = 30\text{ g} \times \frac{4\text{ Cal}}{1\text{ g}} = 120\text{ Cal}$$

$$\text{Calories provided by protein} = 5\text{ g} \times \frac{4\text{ Cal}}{1\text{ g}} = 20\text{ Cal}$$

Add these three Calorie values, rounding the final answer to the tens place, to obtain the total Calories for the yogurt:

$$13.5\text{ Cal} + 120\text{ Cal} + 20\text{ Cal} = 153.5\text{ Cal, rounded to }150\text{ Cal}$$

This answer agrees with the *Calories* shown at the top of the nutritional label for this yogurt.

Figure 4-13 The nutritional label on a 6-oz container of low-fat strawberry yogurt. [Catherine Bausinger]

PRACTICE EXERCISES

29 A piece of angel food cake contains 33 g carbohydrate, 3 g protein, and 0 g fat. How many Calories does a piece of angel food cake contain?

30 A cup of pasta contains 32 g carbohydrates, 5 g protein, and 1 g fat. How many Calories does a cup of pasta contain?

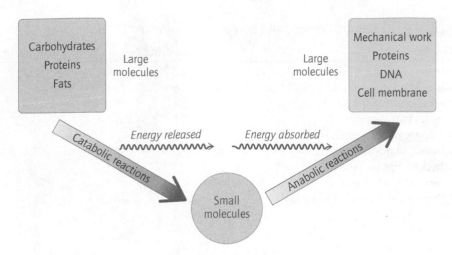

Figure 4-14 An overview of catabolic and anabolic reactions, which together constitute metabolism. The breakdown of large molecules during catabolism releases energy used to build larger molecules from small molecules during anabolism.

An Overview of Energy and Metabolism

The cells in our body are constantly performing reactions—biochemical reactions. A sequence of biochemical reactions leading to a particular product is called a **biochemical pathway**, where the product of one reaction becomes the reactant in the next reaction and so forth. Many biochemical pathways are common to many different organisms. For example, glycolysis is a 10-step biochemical pathway that our cells use to metabolize sugar and produce energy, but the same biochemical pathway is used by many microorganisms to produce energy.

There are two basic types of biochemical pathways: catabolic and anabolic, which together are known as **metabolism. Catabolic pathways** convert large molecules, such as carbohydrates, proteins, and fats from our diet, into smaller molecules (shown on the left side of Figure 4-14). **Anabolic pathways** build larger molecules, such as proteins, lipids, and DNA, from smaller molecules (shown on the right side of Figure 4-14). The most important feature distinguishing catabolic and anabolic pathways is the direction in which energy flows overall in these pathways. *Catabolic pathways release energy, whereas anabolic pathways absorb energy overall.*

Anabolic and catabolic reactions are linked because *the energy required to build molecules via anabolic pathways comes from the energy released in catabolic pathways* (illustrated by the red wavy arrows in Figure 4-14). In biochemical reactions, energy is not transferred in the form of heat; instead it is transferred in the form of chemical energy to and from specialized energy-carrier molecules. These specialized molecules capture the energy released in catabolic reactions in the form of chemical potential energy. The most important of these energy-transfer molecules is *a*denosine *tri*phosphate, ATP. You will learn more about the role of ATP and other aspects of bioenergetics in chapter 15.

PRACTICE EXERCISES

31 What are the two main distinctions between catabolic and anabolic pathways?

32 State whether each of the following processes is catabolic or anabolic. State whether each releases energy or absorbs energy overall:

a building muscle protein from small amino acid molecules
b. a bear burning fat during hibernation
c. glucose molecules reacting to form glycogen, a large biological polymer

4.4 Reaction Kinetics

In this section we consider another important aspect of chemical reactions: how fast they occur, a topic known as **kinetics**. The **reaction rate**—how fast a reaction proceeds—is determined by measuring either the consumption of reactants, or the formation of products, over time. The ability to control reaction rates is especially important in chemistry. In some reactions, we depend on a fast reaction rate and in others, a slow reaction rate. For example, we put food in a refrigerator to slow the rate of reactions that lead to food spoilage. And most of us hope the reactions associated with aging occur slowly. In contrast, we want some reactions to be fast. For example, an air bag deployed in a head-on collision needs to inflate rapidly. An air bag inflates as a result of a chemical reaction that quickly produces nitrogen gas, according to the equation shown below. The air bag is completely inflated within 40 milliseconds after the reactant, sodium azide, is ignited:

$$2\ NaN_3(s) \rightarrow 2\ Na(s) + 3\ N_2(g)$$

If this reaction had a slower reaction rate, the air bag would not be deployed in time to protect the passengers in the car. In living cells it is critical that reactions occur at a rate necessary to maintain physiological processes, and that is why almost all biochemical reactions require enzymes to increase the rate of the reaction. In order to understand enzymes and reaction rates, we must first consider the energy required to initiate a chemical reaction, known as the activation energy, E_A.

Activation Energy, E_A

Earlier in the chapter you learned that when reactant molecules collide with *enough energy* and in the correct spatial orientation, a chemical reaction can occur. The amount of energy that must be attained by the reactants in order for a reaction to occur is called the **activation energy**, abbreviated E_A. Until the reactants acquire the necessary activation energy, they will only bounce off one another without reacting. Using the example of combustion on a gas burner, the activation energy is supplied by the spark from the stove when you turn the dial (accompanied by an audible clicking sound), which initiates the reaction. Once the activation energy is attained, the reaction is self-sustaining because the reaction is exothermic.

A complete energy diagram traces the energy pathway leading from reactants to products. On an energy diagram, the activation energy is represented by a "hill" that reactants must climb in order to become products, as illustrated in **Figure 4-15a** for an exothermic reaction and Figure 4-15b for an

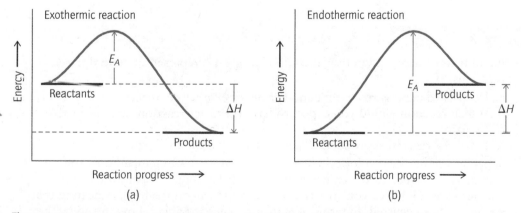

Figure 4-15 Energy diagrams for (a) an exothermic and (b) an endothermic reaction. The height of the "hill" represents the activation energy, E_A, needed for the reactions to occur. In these examples, both reactions have the same change in enthalpy, $\Delta H_{reaction}$.

endothermic reaction. In both exothermic and endothermic reactions, the activation energy, E_A, is the distance between the energy of the reactants and the energy at the top of the "hill" in the energy diagram, measured along the y-axis, as indicated by the red arrow in Figure 4-15. Imagine climbing a snowy hill so that you can slide down the other side. You need a minimum amount of energy to reach the top of the hill (your "activation energy"), but once there, you can easily slide down the other side to get to the bottom, which may be lower (exothermic) or higher (endothermic) than the point where you started, but the top of the hill is still the highest energy point.

The activation energy, E_A, for a reaction is proportional to the rate of the reaction. The *lower* the activation energy, E_A, the *faster* the rate of reaction because a greater number of collisions will lead to products. Thus, an energy diagram with a lower activation energy, E_A—a smaller hill—represents a faster reaction. For example, Figure 4-16 shows energy diagrams for two exothermic reactions with the same change in enthalpy, ΔH; however, Figure 4-16a represents a slower reaction than Figure 4-16b because E_A is greater in that reaction.

Factors that affect the activation energy, E_A, for a reaction have no effect on the potential energy of either the reactants or the products, and therefore they have no effect on the change in enthalpy, ΔH, in that reaction. *The activation energy, E_A, is related to the rate of a reaction, whereas the change in enthalpy, ΔH, is related to the difference in potential energy between the reactants and the products.*

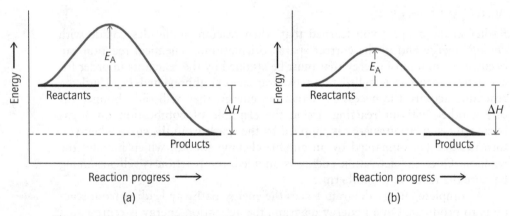

(a)   (b)

Figure 4-16 Energy diagrams for two exothermic reactions with the same change in enthalpy $\Delta H_{reaction}$. The activation energy, E_A, is greater for reaction (a) than (b), so reaction (a) is slower than reaction (b).

WORKED EXERCISES | Energy Diagrams and Activation Energy, E_A

4-16 Examine the energy diagrams shown in Figure 4-17, representing two different reactions.

    a. Do these diagrams represent endothermic or exothermic reactions?

    b. Which reaction would you expect to have the faster reaction rate, (a) or (b)? Explain.

    c. Label the reactants, products, E_A, and ΔH in the energy diagrams.

4-17 Suppose you mix two colorless chemicals together at room temperature and no change is observed. However, when heat is added and the temperature of the reaction rises to 40 °C, your reaction mixture suddenly turns blue, indicating that a reaction has occurred. In terms of activation energy, explain how the added heat caused the reaction to proceed.

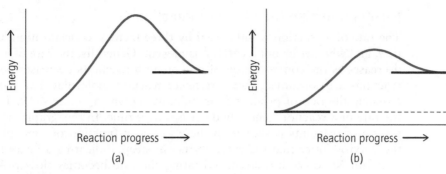

Figure 4-17

Solutions

4-16 a. These diagrams represent endothermic reactions because the energy of the products is greater than the energy of the reactants (higher on the y-axis).

    b. The reaction represented by diagram (b) would be the faster reaction because it has the smaller activation energy, E_A, a lower hill.

    c. Figure 4-18 shows the correctly labeled energy diagrams.

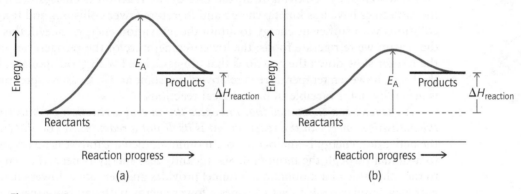

Figure 4-18

4-17 When two reactants are mixed together, there may be insufficient energy available to reach the activation energy required for the reaction to proceed: collisions occur but with insufficient energy. When heat is added, molecules acquire more kinetic energy, which leads to more frequent and more energetic collisions. In this example, the activation energy apparently is reached by heating to 40 °C.

PRACTICE EXERCISES

33 Define the activation energy for a reaction.

34 Draw energy diagrams for two exothermic reactions with different activation energies but the same change in enthalpy, ΔH.

    a. Label each axis.

    b. Indicate which reaction would proceed at the faster rate.

    c. Indicate the activation energy, E_A using an arrow parallel to the y-axis.

    d. Indicate the change in enthalpy of the reaction, ΔH, using an arrow parallel to the y-axis.

    e. Label the horizontal line on the graph that represents the reactants. Do the same for the products.

Factors Affecting the Rate of a Reaction

The rate of a reaction is influenced by three factors: concentration, temperature, and whether or not a catalyst is present. Generally, the rate of a reaction decreases as the concentration of one or both reactants decreases because as reactants are consumed, there are fewer reactant molecules that can collide, reducing the rate of product formation, as shown in Figure 4-19. Therefore, the rate of a reaction generally decreases over time. In this graph the concentration of reactants is shown on the y-axis as a function of time, plotted on the x-axis. Notice that the curve starts out steep, indicating a faster rate, and over time becomes less steep, indicating the rate becomes slower. Although not all reactions follow this type of reaction kinetics, this example illustrates the importance of the relationship between the concentration of reactant(s) and the rate of a reaction.

Recall from the kinetic molecular theory, presented in chapter 1, that as heat is added to a substance, its particles move faster and we observe an increase in temperature. The greater kinetic energy of the particles allows reactant molecules or atoms to collide more frequently and with greater energy, increasing the likelihood that the reactants will attain the activation energy. A common rule of thumb is that for every 10 °C increase in temperature, the rate of a reaction doubles. Conversely, if the temperature of a reaction is lowered, by removing heat, the rate of the reaction decreases because the molecules have less kinetic energy and therefore fewer collisions and fewer collisions with sufficient energy to attain the activation energy. Indeed, this is the reason we refrigerate foods: the lower temperature of the refrigerator and the freezer slow down the reactions that are associated with food spoilage. In the body, however, temperature is relatively constant at 37 °C, so temperature is generally not a variable in biochemical reactions.

A ***catalyst*** *is a substance that increases the rate of a reaction by lowering the activation energy of the reaction but is itself not a reactant.* If you imagine the activation energy is analogous to a mountain, then a catalyst is analogous to a tunnel through the mountain. Significantly less energy is needed to cross to the other side of a mountain if a tunnel provides an alternative, lower-energy pathway. Similarly, a catalyst provides a lower-energy pathway from reactants

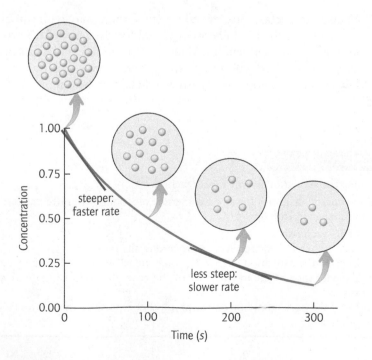

Figure 4-19 The graph shows how the rate of a reaction typically decreases as the concentration of reactants decreases, indicated in the steepness of the curve.

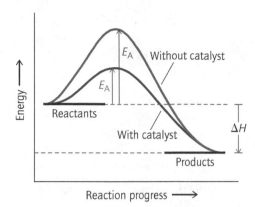

Figure 4-20 Energy diagram showing the effect of a catalyst. The activation energy is lower in the presence of a catalyst, causing the reaction rate to increase.

to products. The energy diagram shown in Figure 4-20 illustrates how a catalyst lowers the activation energy (purple curve) compared to the reaction without a catalyst (red curve). Since a *catalyst reduces the energy of activation, E_A, the rate of the reaction is increased* in the presence of a catalyst.

By definition, a catalyst is not a reactant in the reaction because it does not itself undergo a chemical change. This means that a single catalyst molecule or atom can be reused over and over in a reaction, acting on many reactant molecules. Consequently, catalysts are required in very small amounts relative to the amount of reactants. Catalysts are therefore not shown as a reactant in a chemical equation, but instead the name of the catalyst is often written above the reaction arrow. Coefficients, therefore, do not apply to catalysts.

A catalyst also does not affect the change in enthalpy, ΔH, for a reaction. A catalyst only alters the *pathway* leading from reactants to products but not the potential energy of the reactants and the products, as seen in Figure 4-20. Notice the energy of the reactants and products is the same for both the red and the purple lines.

Catalysts increase the rate of a reaction rate in various ways, but in general they make it easier for reactants to collide with one another and react. Often they provide a surface that brings the reactants together and in closer proximity and/or in the proper orientation for the reaction to occur.

In a biological cell, chemical reactions must occur at normal body temperature, 37 °C, and at a given concentration. Therefore, to increase the rate of a biochemical reaction, a catalyst is necessary. Biological catalysts are molecules called **enzymes**. Most enzymes are proteins, large biomolecules made up of carbon, hydrogen, nitrogen, oxygen, and sulfur atoms. Enzymes are capable of remarkable rate enhancements, ranging from 10^5 to 10^{17} times faster than in the absence of the enzyme. An enzyme, like a receptor, contains a pocket or groove, known as the **active site**, where the reactant molecule(s) bind. The enzyme reduces the freedom of motion available to the reactant(s), thereby forcing reactants into a spatial orientation conducive to reaction and lowering the activation energy for the reaction.

Enzymes are so important to our metabolism that when a person does not produce a particular enzyme or produces a less effective enzyme, there are usually health consequences. Depending on how important the enzyme is, the results range from minor to debilitating. For example, lactose intolerance is a condition found in people who do not produce enough of the enzyme *lactase*, which catalyzes the reaction that converts milk sugar, lactose, into the two simpler sugars, glucose and galactose, illustrated in Figure 4-21.

You can usually recognize a reference to an enzyme because it ends in *-ase*. By convention, many enzyme names are also *italicized*.

$$\text{Lactose} \xrightarrow{\text{lactase}} \text{galactose} + \text{glucose}$$

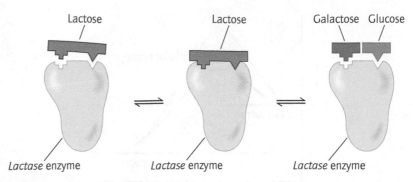

Figure 4-21 This schematic figure shows how the enzyme lactase catalyzes the breakdown of lactose into glucose and galactose.

As a result, individuals with lactose intolerance experience gastrointestinal discomfort when they consume dairy products, foods with a high lactose content. You will learn more about enzymes in chapter 13 when you study proteins.

WORKED EXERCISES | Factors Affecting the Rate of a Reaction

4-18 A variety of catalysts exist, and new ones are being developed all the time. The catalytic converter in a car, for example, uses a platinum catalyst to convert carbon monoxide (CO) into carbon dioxide (CO_2) according to the reaction

$$2\,CO(g) + O_2(g) \xrightarrow[\text{Heat}]{\text{Pt catalyst}} 2\,CO_2(g)$$

Would this reaction be faster or slower if the platinum catalyst were not present?

4-19 For the following reaction, state whether the indicated change in conditions would increase or decrease the rate of the reaction. Explain why.

$2\,H_2O_2(aq) \rightarrow 2\,H_2O(l) + O_2(g)$: More H_2O_2 is added to the reaction.

Solutions

4-18 The reaction would be slower without Pt because it is the catalyst, and a catalyst increases the rate of a reaction.

4-19 The rate of the reaction would increase because the concentration of the reactant has been increased.

PRACTICE EXERCISES

35 For the following reactions, state whether the indicated change in conditions would increase or decrease the rate of the reaction. Explain why.

a. $2\,H_2O_2(aq) \rightarrow 2\,H_2O(l) + O_2(g)$: The temperature is increased.

b. $2\,H_2O_2(aq) \rightarrow 2\,H_2O(l) + O_2(g)$: Sodium iodide, a catalyst, is added to the reaction.

36 Explain why the rate of a reaction usually decreases with time.

37 Enzymes are biological _____, which increase the rate of a biochemical reaction.

38 Draw an energy diagram showing an exothermic reaction with and without a catalyst. Label each curve, and mark the activation energy with a vertical arrow labeled E_A in both curves. Label the change in enthalpy, ΔH. How is E_A different in these two reactions? Is ΔH different in these two reactions? Which reaction occurs at a faster rate?

4.5 Chemical Equilibrium

In this chapter, you have seen many examples of reactions that "go to completion," meaning that all of the reactants are completely converted into products. At the end of the reaction only products are present; no reactants remain. For example, in the combustion of methane, the reactants methane and oxygen are completely consumed in the reaction and only carbon dioxide and water are found at the end of the reaction (unless an excess of one of the reactants is used).

Many reactions are **reversible**, which means that both the forward and the reverse reactions occur simultaneously. In other words, reactants combine to form products—the forward reaction—*and* products combine to form reactants—the reverse reaction. At the end of the reaction, both reactants and products are present. The chemical equation for a reversible reaction is indicated with two half-headed arrows pointing in opposite directions, as shown in the reversible reaction between H_2 and I_2 to form HI:

$$H_2(g) + I_2(g) \underset{\text{reverse}}{\overset{\text{forward}}{\rightleftharpoons}} 2\ HI(g)$$

When hydrogen (H_2) and iodine (I_2) are combined, they react to form hydrogen iodide (HI)—the forward reaction. As reactants are consumed and the concentration of reactants decreases, the rate of the forward reaction decreases. As the concentration of products increases, the rate of the reverse reaction increases: HI decomposes into H_2 and I_2. Eventually, the rate of the forward and the reverse reactions becomes equal, and the concentration of the reactants and products no longer changes—concentrations of reactants and products are constant.

At this point the reaction is said to have reached a state of **equilibrium**. *At equilibrium, the forward and reverse reactions continue to proceed, but since the rate of the forward reaction is the same as the rate of the reverse reaction, the concentration of reactants and products is constant.* Hence, equilibrium is often called a *dynamic* equilibrium. In some reversible reactions, such as the reaction shown above, the product concentrations exceed those of the reactants at equilibrium, and we say the reaction favors products, as illustrated in Figure 4-22a. In other reversible reactions, the concentration of reactants exceeds products at equilibrium, and we say the reaction favors reactants, as illustrated in Figure 4-22b for the reaction of nitrogen and oxygen to form nitrogen oxide. In both types of reactions, both products and reactants are present at equilibrium, a characteristic of a reversible reaction at equilibrium.

In a reversible reaction, a mixture of reactants and products is present at equilibrium regardless of whether the reaction begins with reactants (H_2 and I_2) or products (HI). In other words, if you start with H_2 and I_2, the amount of H_2 and I_2 will decrease and the concentration of HI will increase until the equilibrium concentrations are reached. Similarly, if you start with HI, the concentration of HI will decrease and the concentration of H_2 and I_2 will increase until the same equilibrium concentrations are reached. In the end, the same mixture of HI, H_2, and I_2 will be found at equilibrium, provided the temperature remains constant.

Le Châtelier's Principle

A reversible reaction at equilibrium can be disturbed by adding or removing a reactant or product (a concentration change) or by adding or removing heat from the reaction (a temperature change). *Le Châtelier's principle predicts that a reaction at equilibrium will respond to a disturbance in such a way as*

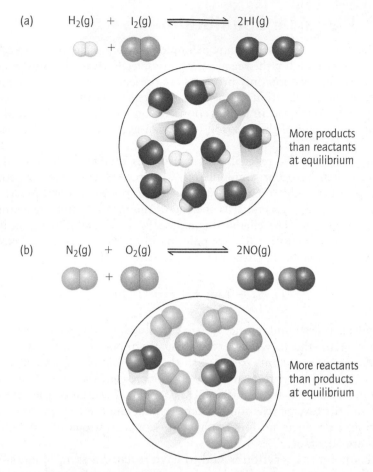

(a) $H_2(g)$ + $I_2(g)$ $\rightleftharpoons$ $2HI(g)$

More products than reactants at equilibrium

(b) $N_2(g)$ + $O_2(g)$ $\rightleftharpoons$ $2NO(g)$

More reactants than products at equilibrium

Figure 4-22 Dynamic states of equilibrium. (a) A reaction where products are favored at equilibrium. (b) A reaction where reactants are favored at equilibrium. Both products and reactants are present at equilibrium, a characteristic of a reversible reaction.

to counteract the disturbance and establish a new equilibrium. For example, if additional H_2, a *reactant*, is *added* to the reaction described above, after equilibrium has been reached, the reaction will respond by counteracting the added H_2 with an increase in the rate of the *forward* reaction so as to consume the added H_2 and produce more products. We say the reaction *shifts to the right* (the forward direction), as shown in part (a) below. If instead the *reactant*, H_2, is *removed* from the reaction, after equilibrium has been reached, the reaction will respond by increasing the rate of the reverse reaction so as to restore some of the H_2 that has been removed. We say the reaction *shifts to the left* (the reverse reaction), as shown in part (b). In each case, a new equilibrium is reached.

(a)
Add more H_2
$\downarrow$
$H_2(g) + I_2(g) \xrightleftharpoons[\text{shift to the right}]{} 2\,HI(g)$

(b)
Remove some H_2
$\uparrow$
$H_2(g) + I_2(g) \xrightleftharpoons[\text{shift to the left}]{} 2\,HI(g)$

The results are the opposite if *product* is added or removed: A reversible reaction at equilibrium will respond in a way that counteracts the change in product concentration, as shown on the next page.

(a)   (b)

Add more HI   Remove some HI

$H_2(g) + I_2(g) \xrightleftharpoons[\text{shift to the left}]{} 2\,HI(g)$ $H_2(g) + I_2(g) \xrightleftharpoons[]{\text{shift to the right}} 2\,HI(g)$

If the temperature of a reaction at equilibrium is changed by adding or removing heat (cooling), the direction of the shift depends on whether the reaction is exothermic or endothermic. In an exothermic reaction, heat is released and appears as a product; therefore, *adding heat* to the reaction will cause it to *shift to the left*, so as to consume the added heat, as shown in part (a) below. If an exothermic reaction is cooled, the reaction will *shift to the right*, so as to restore some of the heat that has been removed, as shown in part (b).

Exothermic reactions

(a)   (b)

Add heat   Remove heat

$H_2(g) + I_2(g) \xrightleftharpoons[\text{shift to the left}]{} 2\,HI(g) + HEAT$ $H_2(g) + I_2(g) \xrightleftharpoons[]{\text{shift to the right}} 2\,HI(g) + HEAT$

The results are the opposite for an endothermic reaction, for the same reasons, as illustrated in the reactions below:

Endothermic reactions

(a)   (b)

Add heat   Remove heat

$HEAT + H_2(g) + I_2(g) \xrightleftharpoons[]{\text{shift to the right}} 2\,HI(g)$ $HEAT + H_2(g) + I_2(g) \xrightleftharpoons[\text{shift to the left}]{} 2\,HI(g)$

Le Châtelier's principle explains the regulation of many biochemical reactions. For example, in chapter 2 you learned that the molecule hemoglobin (Hb) binds oxygen (O_2) and delivers it to cells throughout the body. Why does hemoglobin bind oxygen when passing through the lungs but then release oxygen when hemoglobin arrives at oxygen-starved tissues? The answer is predicted by Le Châtelier's principle. The reaction in which hemoglobin (Hb) reacts with oxygen to form oxyhemoglobin (HbO_2) is a reversible reaction that can be expressed by the following chemical equation:

$$\underset{\text{hemoglobin}}{Hb} + \underset{\text{oxygen}}{O_2} \rightleftharpoons \underset{\text{oxyhemoglobin}}{HbO_2}$$

In the lungs, where the concentration of oxygen is high, the equilibrium shifts to the right to counteract the high concentration of oxygen, producing more oxyhemoglobin (product), as shown in part (a) below. Conversely, in muscle and other tissues where the oxygen concentration is low, the equilibrium shifts to the left to counteract the decreased oxygen concentration, releasing oxygen to the tissues, as shown in part (b) below. The hemoglobin-oxygen/ oxyhemoglobin equilibrium in red blood cells is just one of many important examples of chemical equilibrium in biological cells.

(a)

Lungs: high O_2

$$Hb + O_2 \rightleftharpoons HbO_2$$
Hemoglobin   Oxygen   shift right   Oxyhemoglobin

(b)

Muscles, organs: low O_2

$$Hb + O_2 \rightleftharpoons HbO_2$$
Hemoglobin   Oxygen   shift left   Oxyhemoglobin

WORKED EXERCISES   Understanding Chemical Equilibrium

4-20 What is the relationship between the rate of the forward reaction and the rate of the reverse reaction for a reaction at equilibrium?

4-21 The reaction between nitrogen and oxygen to form nitrogen oxide (NO) is an endothermic and reversible reaction. Predict whether the equilibrium shifts to the left or to the right in response to the following changes.

$$N_2(g) + O_2(g) \rightleftharpoons 2\,NO(g) \quad \Delta H > 0$$

a. Heat is added.
b. The reaction is cooled.
c. Additional nitrogen (N_2) is added.
d. Additional NO is added.
e. Oxygen (O_2) is removed.
f. NO is removed.

4-22 Formic acid, CH_2O_2, is produced by bees as a defense against predators and commercially as a preservative and antibacterial agent in livestock feed. In water, formic acid undergoes the following reversible reaction:

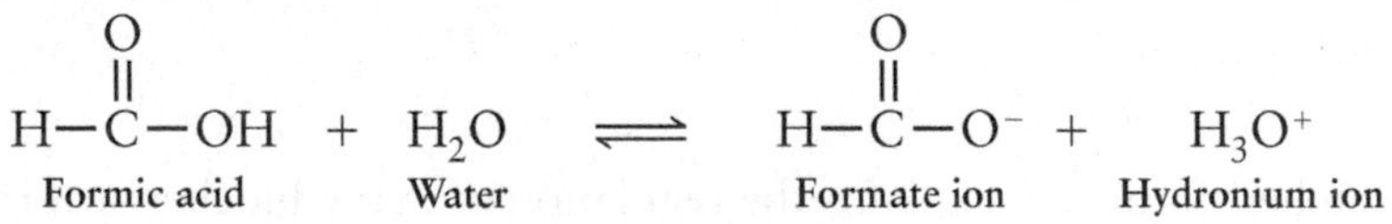

$$\underset{\text{Formic acid}}{H-\overset{\overset{\displaystyle O}{\|}}{C}-OH} + \underset{\text{Water}}{H_2O} \rightleftharpoons \underset{\text{Formate ion}}{H-\overset{\overset{\displaystyle O}{\|}}{C}-O^-} + \underset{\text{Hydronium ion}}{H_3O^+}$$

a. What substances are present at equilibrium?
b. At equilibrium, is the concentration of reactants and products constant or changing?
c. What is the meaning of the two opposing arrows in this equation?
d. How will the equilibrium shift if additional formic acid is added to the reaction?
e. How will the equilibrium shift if formate ion is removed from the reaction?

Solutions

4-20 The rate of the forward reaction is the same as the rate of the reverse reaction at equilibrium.

4-21 a. Since the reaction is endothermic, it helps us predict the direction of the shift if we show heat as a reactant in the equation:

$$\text{Heat} + N_2(g) + O_2(g) \rightleftharpoons 2\,NO(g)$$

     Therefore, added heat shifts the reaction to the right so as to consume the added heat.

    b. If heat is removed, the equilibrium will shift to the left to restore heat.

    c. Nitrogen is a reactant, so adding nitrogen will shift the reaction to the right to consume the added reactant.

    d. NO is a product, so adding more NO will shift the equilibrium to the left to consume the added product.

    e. Oxygen is a reactant, so removing O_2 will shift the equilibrium to the left to replenish the lost reactant.

    f. NO is the product, so removing it will shift the equilibrium to the right to replenish the lost product.

4-22 a. Formic acid, water, formate ion, and hydronium ion are all present at equilibrium.

    b. The concentration of reactants and products is constant at equilibrium.

    c. The two opposing arrows indicate that both the forward and reverse reactions are occurring simultaneously.

    d. Since formic acid, a reactant, is added, the equilibrium will shift to the right.

    e. Since formate ion, a product, is removed, the equilibrium will shift to the right.

PRACTICE EXERCISES

39 Why do the concentrations of reactants and products remain constant at equilibrium? Does this mean the forward and reverse reactions have stopped? What is meant by a *dynamic* equilibrium?

40 For the reaction shown below, what is happening in the reverse reaction? What is happening in the forward reaction? When are the concentrations of Hb, O_2, and HbO_2 constant?

$$Hb + O_2 \rightleftharpoons HbO_2$$

41 Coal is composed of primarily carbon, C. Coal reacts with hydrogen gas (H_2) to produce natural gas, methane (CH_4), in an exothermic reversible reaction. Predict whether the equilibrium shifts to the left or to the right in response to the following changes:

$$C(s) + 2\ H_2(g) \rightleftharpoons CH_4(g) \qquad \Delta H < 0$$

    a. raising the temperature
    b. adding more H_2 to the reaction
    c. adding more methane to the reaction
    d. removing carbon
    e. removing CH_4

42 Consider the reversible reaction shown below, which occurs in the blood:

$$\underset{\text{Carbonic acid}}{H_2CO_3(aq)} \rightleftharpoons \underset{\text{Carbon dioxide}}{CO_2(g)} + \underset{\text{Water}}{H_2O(l)}$$

    a. Carbon dioxide is produced as a waste product in the tissues of the body. Does this additional CO_2 in the blood cause the reaction above to shift *to the left* or *to the right*? Explain.

    b. In the lungs, carbon dioxide is removed as it is exhaled. Does the removal of CO_2 cause the reaction above to shift *to the left* or *to the right*? Explain.

Chemistry in Medicine Calorimetry and the Care of Patients on a Mechanical Respirator

On May 27, 1995, Christopher Reeve, well known for his role as Superman in the 1978 blockbuster movie, was paralyzed from the neck down after being thrown from his horse and landing on his head. For the remainder of his life, Reeve was confined to a wheelchair and unable to breathe without the assistance of a mechanical respirator. Reeve passed away on October 10, 2004. Patients such as Reeve on mechanical respirators cannot eat on their own and must have their caloric and nutritional needs determined through methods based on the principles of calorimetry.

Several conditions can lead to the permanent need for a mechanical respirator, including severe brain injury, spinal cord injury, and some neurological diseases. Patients on mechanical respirators cannot eat on their own, and a medical professional must manage feedings. It is crucial not to underfeed or overfeed these patients. Malnourishment caused by underfeeding is common in patients who require permanent mechanical respirators and, in severe cases, can lead to coma and death. Then again, overfeeding increases oxygen consumption and metabolic rate. The ventilator and lungs must work harder, possibly causing respiratory muscle fatigue or even respiratory failure.

How can a health care professional manage the feeding of these patients? It can be difficult to estimate the number of Calories that a patient on a mechanical respirator

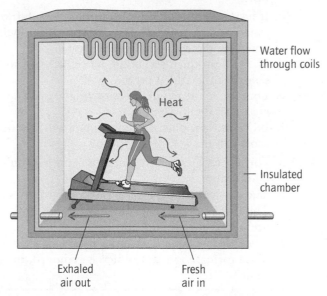

Figure 4-23 A person exercising in a whole-room human calorimeter. Heat released by the person is measured as a rise in the temperature of the water in the coils enclosing the room.

Christopher Reeve (1952–2004) at a conference at the Massachusetts Institute of Technology, March 2, 2003. Reeve sustained a spinal cord injury in 1995 that required him to use a mechanical respirator for the remainder of his life. [Richard Ellis/Getty Images]

requires, especially without knowing how much energy that patient expends. Calorimetry provides a convenient method for determining a patient's energy expenditure, which can then be used to calculate the proper number of Calories that the patient should consume in a day.

Recall that carbohydrates react with oxygen to form carbon dioxide, water, and energy—a combustion reaction:

$$\text{Carbohydrates} + O_2 \rightarrow CO_2 + H_2O + \text{energy}$$

By measuring the heat output (**direct calorimetry**) or the oxygen uptake (**indirect calorimetry**), a health care professional can determine the appropriate number of Calories a person on a mechanical respirator should be consuming.

In the 1890s, professors Atwater and Rosa at Wesleyan University designed and built the first human calorimeter. Figure 4-23 shows the basic layout of a human calorimeter. The energy output (expended energy) of the person inside the calorimeter was determined from the heat radiated from his or her body. The calorimeter consisted of an airtight room large enough for a person to live in. The calorimeter was designed to precisely measure energy expenditure in humans participating in various activities such as resting, sleeping, jogging on a treadmill, and riding a stationary bicycle. Through their experiments, Atwater and Rosa confirmed that the first law of thermodynamics—the law of conservation of energy—did indeed govern the transformation of matter and energy in the human body.

In a typical whole-room calorimeter, the room is insulated in such a way that any heat radiated by the individual is absorbed by a known mass of water flowing through coils surrounding the room. The change in the temperature of the water is attributed to the heat radiated by the individual. An atmosphere with a constant and natural composition of air is maintained by continually adding oxygen and removing moisture and exhaled carbon dioxide from the air.

Direct calorimetry remains the gold standard for the accurate measurement of heat production, but it requires considerable cost, time, and engineering skills. It is impractical to use direct calorimetry to evaluate the dietary needs of patients on mechanical respirators. A more practical and common approach is to use *indirect calorimetry*. This technique relies on portable equipment that is more convenient for routine measurements.

Indirect calorimetry measures the patient's oxygen uptake rather than heat production. Metabolic reactions that produce energy require oxygen, so measuring a patient's *oxygen* uptake is a convenient and accurate *indirect* way of measuring his or her energy output. The amount of oxygen that is inhaled or exhaled is measured using a variable-volume container called a *spirometer*. In a technique known as closed-circuit spirometry, the patient breathes in 100 percent oxygen from a prefilled spirometer (Figure 4-24). The patient continues to rebreathe oxygen from the spirometer (usually for less than a min); hence the term *closed circuit*. Gases exhaled by the patient include carbon dioxide and unused oxygen. The exhaled carbon dioxide is removed by a canister of potassium hydroxide within the breathing circuit. The patient's

Figure 4-24 A patient breathing into a spirometer. [Yoav Levy/ Phototake]

oxygen uptake is then calculated from the decrease in volume of the spirometer, as described in Figure 4-25.

Clearly the first law of thermodynamics governs food Calories consumed and Calories expended. The knowledge gained from research using calorimeters has contributed to the current nutrition recommendations of the USDA (United States Department of Agriculture) as well as many advances in exercise physiology and medical care, including the vital task of managing the Caloric intake of patients on a mechanical respirator.

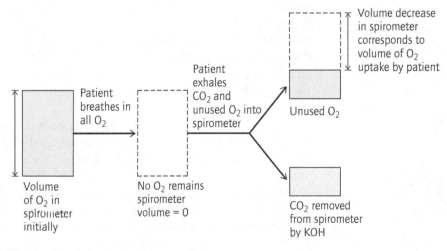

Figure 4-25 How closed-circuit calorimetry works.

Summary

The Mole: Counting and Weighing Matter

- Molecular mass is the mass of a single molecule in units of amu. It is calculated from the sum of the atomic masses of the component atoms of the molecule.
- Formula mass is the mass of the formula unit of an ionic compound in units of amu. It is calculated from the mass of the individual ions that make up the formula unit of the ionic compound.
- The mole is a counting unit that represents the number 6.02×10^{23}, much like a dozen represents the number 12.
- The molar mass of a compound is numerically equivalent to the molecular mass or the formula mass, but in units of grams per mole (g/mol rather than amu).
- Avogadro's number, 6.02×10^{23}, represents the number of atoms, ions, or particles in one mole of any substance. It can be used as a conversion between the number of moles of a substance and the number of atoms, molecules, or ions of the substance.
- Molar mass is a conversion between the mass, in grams, of a substance and the number of moles of the substance.
- To interconvert between moles and mass, use molar mass (or inverted molar mass) as a conversion factor. To interconvert between moles and number of particles, use Avogadro's number as a conversion.
- To interconvert between mass and number of particles, use both molar mass and Avogadro's number as conversion factors.

The Law of Conservation of Mass and Balancing Chemical Equations

- In a chemical reaction bonds break and new bonds are formed, producing different compounds or elements.
- The law of conservation of mass states that matter can be neither created nor destroyed.
- A chemical equation is used to represent a chemical reaction. Reactants, on the left side of the arrow, and products, on the right side of the arrow, are represented by their chemical formulas or structures separated by a reaction arrow. Whole-number coefficients are placed before each chemical formula to indicate the molar ratio of each reactant and product.
- In a balanced chemical equation, the total number of each type of atom on the reactant side must equal the total number of each type of atom on the product side.
- Balancing a chemical equation requires systematically inserting whole-number coefficients in front of the reactants and/or products so that the number of each type of atom is the same on the reactant side as on the product side.
- A mole-to-mole stoichiometry calculation can be performed given a balanced equation. In a mole-to-mole stoichiometry calculation we can determine the number of moles of any reactant or product given the number of moles of any other reactant or product.

Energy and Chemical Reactions

- Energy is the capacity to do work or to transfer heat.
- Chemical bonds have potential energy. Carbohydrates, proteins, and fats are the important food molecules with high potential energy in their covalent bonds.

- The energy transferred in a chemical reaction often appears in the form of heat energy.
- Most chemical reactions absorb or release heat.
- The first law of thermodynamics states that energy can be neither created nor destroyed, only transformed.
- The change in enthalpy, ΔH, is a measure of the heat released or absorbed in a chemical reaction.
- An exothermic reaction releases heat energy into the surroundings; $\Delta H < 0$ (ΔH is negative) because the products are lower in energy than the reactants.
- An endothermic reaction absorbs heat energy from the surroundings; $\Delta H > 0$ (ΔH is positive) because the products are higher in energy than the reactants.
- An energy diagram illustrates the change in energy during a chemical reaction. The x-axis represents the reaction progress and the y-axis represents energy.
- Calorimetry is a technique used to measure enthalpy changes, ΔH, using an apparatus called a calorimeter. A bomb calorimeter measures the change in enthalpy for a combustion reaction.
- The caloric content of foods—the heat released during their combustion— has been determined by calorimetry. Carbohydrates and proteins release approximately 4 Cal/g, whereas fats release 9 Cal/g.
- Metabolism consists of two types of biochemical pathways, catabolic and anabolic. Catabolic pathways convert large molecules into smaller molecules, whereas anabolic pathways convert small molecules into larger ones. Catabolic pathways release energy, whereas anabolic pathways absorb energy overall.
- In biochemical reactions, energy is not transferred as heat; instead, it is transferred in the form of chemical energy to and from specialized energy-carrier molecules, such as adenosine triphosphate, ATP.

Reaction Kinetics
- Chemical kinetics is the study of reaction rates—how fast reactants are converted into products.
- A chemical reaction can proceed only when the reactant molecules collide with enough energy and in the correct spatial orientation.
- The activation energy, E_A, is the minimum amount of energy that must be attained by the reactants for a chemical reaction to occur.
- The activation energy is related to the rate of a reaction; the greater the activation energy, the slower the reaction.
- In an energy diagram, the activation energy appears as a hill connecting reactants to products. The higher the "hill" is, the greater the activation energy and the slower the reaction.
- The three major factors that affect the rate of a reaction are concentration, temperature, and whether or not a catalyst is present.
- The rate of a reaction usually decreases as the concentration of reactants decreases.
- As the temperature of a reaction increases, the rate of a reaction also increases.
- The addition of a catalyst to a reaction lowers the activation energy, thereby increasing the rate of the reaction.
- A catalyst increases the rate of a reaction by lowering the activation energy of the reaction.

- A catalyst is not a reactant because it is chemically unchanged during a reaction. Instead, a catalyst facilitates the reaction, often by bringing reactants together in the proper orientation.
- A catalyst does not affect ΔH but lowers E_A.
- Biological catalysts are called enzymes. Almost all biochemical reactions require an enzyme.

Chemical Equilibrium

- A reversible reaction is a reaction in which the forward and the reverse reaction are occurring simultaneously.
- At equilibrium, the rate of the forward reaction is equal to the rate of the reverse reaction; therefore the concentration of reactants and products is constant at equilibrium.
- Le Châtelier's principle predicts that a reversible reaction at equilibrium will respond to a disturbance in such a way as to counteract the disturbance. Thus, adding a reactant or removing a product will result in a shift to the right. Removing a reactant or adding a product will cause a shift to the left.
- Exothermic reactions will shift to the left with added heat, whereas endothermic reactions will shift to the right. Conversely, exothermic reactions will shift to the right when heat is removed, whereas endothermic reactions will shift to the left.

Key Words

Activation energy, E_A The minimum amount of energy that must be attained by the reactant(s) in order for a chemical reaction to occur.

Active site The location an enzyme where the reactants bind.

Anabolic pathways Biochemical reactions that convert smaller molecules into larger molecules such as proteins and DNA. Anabolic reactions consume energy overall.

Avogadro's number The number 6.02×10^{23}, which represents the number of particles in 1 mole of any substance.

Balanced equation A chemical equation in which the coefficients indicate there are an equal number of each type of atom on both sides of the equation.

Biochemical pathway A sequence of chemical reactions that occurs in a biological cell wherein the product of one reaction becomes the reactant in the next reaction.

Bioenergetics The field of study concerned with the transfer of energy in reactions occurring in living cells.

Calorimeter An apparatus used to measure the change in enthalpy, ΔH, for a chemical reaction. It can be used to measure the Caloric content of foods.

Calorimetry The experimental technique that uses a calorimeter to measure enthalpy changes in chemical reactions.

Catabolic pathways Reactions that convert large molecules, such as carbohydrates, proteins, and fats, into smaller molecules. Catabolic reactions release energy overall.

Catalyst A substance that increases the rate of a reaction by lowering the activation energy for the reaction. A catalyst is not consumed in the reaction but facilitates the reaction.

Change in enthalpy, ΔH The heat energy transferred in a chemical reaction under certain defined conditions.

Chemical equation A symbolic representation of a balanced chemical reaction that includes a reaction arrow, the formulas of the reactants and products, and whole-number coefficients.

Chemical kinetics The study of reaction rates—how fast reactants are converted into products.

Chemical reaction The transformation of one or more substances into one or more products. Chemical bonds are broken and new ones are formed in a chemical reaction. However, atoms are unchanged.

Coefficient The whole numbers placed in front of the element symbol, molecular formula, or formula unit of each reactant and product in a chemical equation. Coefficients represent the molar ratio of each of the reactants and products. The coefficient 1 is assumed and not written in.

Combustion reaction An exothermic reaction in which a substance containing C, H, and O reacts with oxygen (O_2) to produce carbon dioxide (CO_2) and water (H_2O).

Endothermic reaction A reaction that absorbs heat from the surroundings and therefore has a positive change in enthalpy, $\Delta H > 0$, because the products are higher in energy than the reactants.

Energy The ability to do work or transfer heat.

Energy diagram A diagram that depicts the energy changes during the course of a chemical reaction. Energy appears on the y-axis, and the energy pathway from reactants to products is shown along the x-axis.

Enzyme A biological catalyst. An enzyme is usually a large protein containing a pocket or groove where the reactants bind to the enzyme.

Equilibrium When the forward and reverse rates of reaction in a reversible reaction are equal, so the concentration of reactants and products is constant.

Exothermic reaction A reaction that releases energy to the surroundings and has a negative change in enthalpy, $\Delta H < 0$, because the products are lower in energy than the reactants.

Formula mass The mass of one formula unit of an ionic compound in units of amu.

Kinetics The study of reaction rates: how fast reactions occur.

Law of conservation of mass The universal observation that matter can be neither created nor destroyed.

Le Châtelier's principle A principle that governs reversible reactions at equilibrium when a disturbance is introduced. A reversible reaction at equilibrium responds to a disturbance by shifting to the right or the left in such a way as to counteract the disturbance.

Metabolism The biochemical pathways of catabolism and anabolism.

Molar mass The mass of one mole of any element or compound in units of g/mol.

Mole A counting unit that represents the number 6.02×10^{23}, usually applied to atoms, ions, or molecules.

Molecular mass The mass of one molecule in units of amu calculated from the sum of the individual atomic masses of the atoms in the molecular formula.

Products The compounds or elements that are produced in a chemical reaction. They are shown to the right of the arrow in a chemical equation.

Reactants The compounds or elements that are combined in a chemical reaction. They are shown to the left of the arrow in a chemical equation.

Reaction rate The change in the concentration of reactants or products over time.

Reversible reaction A reaction in which reactants are converted into products and products are converted into reactants simultaneously. A reversible reaction is represented in a chemical equation by two opposing half-headed reaction arrows.

Stoichiometry Molar ratios between reactants and/or products in a chemical reaction.

Work The act of moving an object over a distance against an opposing force.

Additional Exercises

How Does the Body Handle Alcoholic Beverages?

43 Which of the following are true for a chemical reaction?
   a. Bonds never break.
   b. Bonds are formed.
   c. The chemical structures of the reactants change.
   d. The chemical structures of the reactants are unchanged.

44 Metabolism refers to the chemical reactions that occur in _____. The chemical reactions produce _____ that enable us to survive.

45 What compound is produced in the first step of the metabolism of ethanol in liver cells?

46 Why does acetaldehyde build up when a person consumes too much alcohol? What causes a hangover?

47 What is the function of *alcohol dehydrogenase* in the metabolism of ethanol?

48 Is energy released or absorbed in the metabolism of ethanol?

49 Why can alcohol consumption lead to weight gain?

The Mole: Counting and Weighing Matter

50 What is the molecular mass of a compound?

51 What is the formula mass of a compound?

52 What is the molecular mass of butane, C_4H_{10}?

53 Budesonide is the active pharmaceutical ingredient in the Pulmicort Turbuhaler, an inhaler used in daily prophylactic treatments for adults and children with asthma. The molecular formula for budesonide is $C_{25}H_{34}O_6$. What is the molecular mass of budesonide?

54 What is the formula mass of sodium chloride, used in saline solutions?

55 What is the formula mass of magnesium sulfate, $MgSO_4$?

56 How many M&M's would there be in one mole of M&M's?

57 What is Avogadro's number?

58 How many pieces of paper are in a ream of paper?

59 How many eggs are in a googol of eggs?

60 What is the mass of one mole of oxygen O_2?

61 What is the mass of one mole of nickel atoms?

62 Calculate the molar mass of the following elements:
   a. aluminum, Al   c. calcium, Ca
   b. hydrogen, H_2 d. nitrogen, N_2

63 Calculate the molar mass of the following compounds:
   a. Benicar, $C_{29}H_{30}N_6O_6$, used to treat hypertension
   b. Prevacid, $C_{16}F_3H_{14}N_3O_2S$, used to treat gastroesophageal reflux disease (GERD)

64 Celebrex ($C_{17}F_3H_{14}N_3O_2S$) is used to treat rheumatoid arthritis. Which is the correct molar mass of Celebrex?
   a. 323 g/mol
   b. 438.4 amu/molecule
   c. 381.38 g/mol
   d. 381.38 amu/molecule

65 $Al(OH)_3$ is often used as an antacid. Which is the correct molar mass of aluminum hydroxide?
   a. 78.00 g/mol
   b. 78.00 amu
   c. 43.99 g/mol
   d. 43.99 amu

66 Calculate the molar mass of the following compounds:
   a. Li_2CO_3, a drug used to treat bipolar disorder
   b. $CaSO_4$, used in plaster of Paris casts
   c. $BaSO_4$, used for CT scans of the esophagus

67 Which has a smaller mass: one mole of feathers or one mole of chemistry books? Which is a larger number: one mole of feathers or one mole of chemistry books?

68 Which has a lighter mass: one mole of platinum or one mole of zirconium?

69 Which has a greater mass: one mole of hydrogen (H_2) or one mole of zinc?

70 How many atoms are there in 0.25 mole of any element?

71 What is the difference between formula mass and molar mass?

72 How many molecules are there in 0.28 mol of Synthroid, a drug used to treat hypothyroidism?

73 How many moles do 8.65×10^{18} molecules of Crestor represent? Crestor is used to reduce the risk of cardiovascular disease.

74 How many molecules are there in 1.38 moles of Lexapro? Lexapro is a selective serotonin reuptake inhibitor (SSRI) used to treat depression.

75 How many moles do 1.45×10^{20} formula units of $MgSO_4$ represent?

76 How many moles are there in 2.00 g of calcium chloride?

77 How many grams of hydrogen peroxide, H_2O_2, are there in

   a. 1.0 mole   **c.** 10. moles

   b. 0.50 mole   **d.** 0.678 mole

78 Blood tests measuring the amount of creatinine ($C_4H_7N_3O$) are used to determine renal health. For women, the normal range of creatinine is between 0.5 and 1.0 mg per deciliter of blood. How many moles of creatinine are there in 1.0 mg of creatinine?

79 If a patient has an abnormal electrocardiogram (EKG), the doctor may order a blood calcium test. The normal range of calcium in the blood is between 8.8 and 10.4 mg per deciliter of blood. How many moles of calcium are there in 8.8 mg of calcium?

80 Physicians monitor the amount of potassium in patients undergoing kidney dialysis. The normal range of potassium in the blood is between 3.5 and 5.2 mmol of potassium per liter of blood. How many grams of potassium are there in 3.5 mmol of potassium?

81 Physicians use blood sodium levels to monitor kidney or adrenal gland diseases. The normal range of sodium in the blood is between 136 and 145 mmol of sodium per liter of blood. How many grams of sodium are there in 145 mmol?

82 How many water molecules are there in 5.6 g of H_2O?

83 A dose of Tylenol (acetaminophen) is 325 mg. How many acetaminophen molecules are in one dose of acetaminophen? The molecular formula for acetaminophen is $C_8H_9NO_2$.

84 A low dose (80 mg) of aspirin is used to help prevent cardiovascular disease. How many aspirin molecules are in an 80-mg dose of aspirin? The molecular formula for aspirin is $C_9H_8O_4$.

85 How many formula units of silver nitrate, $AgNO_3$, are in 0.70 g of silver nitrate?

86 A vitamin D deficiency can lead to rickets in children. A blood test can measure the amount of vitamin D (in the form of a compound called calcifediol or 25-hydroxycholecalciferol, abbreviated 25(OH)D. Normal levels of vitamin D are greater than 20. ng/mL of blood. The molecular formula for 25(OH)D is $C_{27}H_{44}O_2$. How many moles of 25(OH)D are there in 20. ng? How many molecules of 25(OH)D are in 20. ng?

87 In 2010, in the United States, lead poisoning affected 230,000 children under the age of six. They had lead levels between 5 to 9 mcg/dL of blood. How many moles of lead are there in 5.0 mcg? How many atoms of lead are there in 5.0 mcg?

The Law of Conservation of Mass and Balancing Chemical Equations

88 What two criteria are necessary for a chemical reaction to occur?

89 Which of the following are examples of common changes seen on the macroscopic scale in a chemical reaction?

   a. a change in color

   b. a change in temperature

   c. the formation of a precipitate

   d. bubbles forming with the evolution of a gas

90 State the law of conservation of mass.

91 What do the abbreviations (s), (g), and (aq) stand for in a chemical equation?

92 For the following chemical equations, identify

   i. the reactants and products

   ii. the coefficients for each reactant and product

   iii. the physical state of each reactant and each product

   a. $2\ C_4H_{10}(g) + 13\ O_2(g) \rightarrow 8\ CO_2(g) + 10\ H_2O(g)$

   b. $C_6H_{12}O_6(aq) \rightarrow 2\ CO_2(g) + 2\ CH_3CH_2OH(aq)$

   c. $2\ Al(s) + 3\ I_2(s) \rightarrow 2\ AlI_3(s)$

   d. $2\ H_2O_2(l) \rightarrow 2\ H_2O(l) + O_2(g)$

93 During a chemical reaction, why must the total number of atoms of one kind on the reactant side equal the total number of atoms of that same kind on the product side? This is an application of the law of _____ of _____.

94 Balance the following chemical equations:

   a. $N_2H_4(l) + N_2O_4(l) \rightarrow N_2(g) + H_2O(l)$

   b. $Li(s) + O_2(g) \rightarrow Li_2O(s)$

   c. $H_2(g) + N_2(g) \rightarrow NH_3(g)$

   d. $KClO_3(s) \rightarrow KClO_4(s) + KCl(s)$

95 Balance the following combustion equations:

   a. $C_6H_{14}(l) + O_2(g) \rightarrow CO_2(g) + H_2O(l)$

   b. $C_2H_6O\ (g) + O_2(g) \rightarrow CO_2(g) + H_2O(l)$

   c. $C_3H_6(g) + O_2(g) \rightarrow CO_2(g) + H_2O(l)$

96 The following reactions are not balanced properly. Change the coefficients to reflect a properly balanced equation:

   a. $2\ CO(g) + 4\ H_2(g) \rightarrow 2\ CH_4O(g)$

   b. $\frac{1}{2}\ N_2(g) + \frac{1}{2}\ O_2(g) \rightarrow NO(g)$

   c. $Na(s) + Cl_2(g) \rightarrow NaCl(s)$

   d. $Al(s) + HCl(aq) \rightarrow AlCl_3(aq) + H_2(g)$

97 When iron reacts with oxygen, rust forms, as shown by the following chemical equation. If 12 moles of iron react in an excess of oxygen, how many moles of iron(III) oxide are formed?

$4\ Fe(s) + 3\ O_2(g) \rightarrow 2\ Fe_2O_3(s)$

a. 8 moles
b. 12 moles
c. 6 moles
d. 4 moles

98 The following equation shows the combustion of acetylene in an arc welder. If seven moles of acetylene undergo combustion in an excess of oxygen, how many moles of carbon dioxide can be formed?

$$C_2H_4(g) + 3\ O_2(g) \rightarrow 2\ CO_2(g) + 2\ H_2O(l)$$

a. 4 moles
b. 7 moles
c. 14 moles
d. 21 moles

Extension Topic: Stoichiometry Calculations

E4.4 In the reaction shown, determine how many grams of carbon dioxide can be produced when the body metabolizes 15.0 g of glucose. Assume an excess of oxygen.

$$C_6H_{12}O_6(aq) + 6\ O_2(g) \rightarrow 6\ CO_2(g) + 6\ H_2O(l)$$
Glucose

E4.5 Balance the following equation and then determine how many grams of copper are produced when excess aluminum reacts with 10.5 g of copper(II) oxide.

$$Al(s) + CuO(s) \rightarrow Al_2O_3(s) + Cu(s)$$

E4.6 Balance the following equation and then determine how many grams of water are produced when burning 1.05 g of butane with a butane lighter. Assume excess oxygen is present.

$$C_4H_{10}(g) + O_2(g) \rightarrow CO_2(g) + H_2O(l)$$
Butane

Energy and Chemical Reactions

99 Define bioenergetics.

100 What kind of energy is stored in the covalent bonds of food? How do dietary carbohydrates and fats provide energy for cells?

101 Define ΔH.

102 What is the difference between an exothermic and an endothermic reaction?

103 Why is the reverse of an exothermic reaction always endothermic?

104 Where does the heat from burning propane in your barbecue grill come from?

105 State whether the following chemical reactions are endothermic or exothermic:

a. $2\ H_2(g) + O_2(g) \rightarrow 2\ H_2O(l) + heat$
b. $2\ C_2H_2(g) + 5\ O_2(g) \rightarrow 4\ CO_2(g) + 2\ H_2O(l)$
c. $2\ CO_2(g) + heat \rightarrow 2\ CO(g) + O_2(g)$
d. $CH_3OH(l) + heat \rightarrow CO(g) + 2\ H_2(g)$

106 Which processes release energy?
a. breaking chemical bonds
b. making chemical bonds
c. photosynthesis
d. anabolic biochemical pathways
e. catabolic biochemical pathways

107 What is the name of the apparatus used to measure the caloric content of various substances, including food?

108 What purpose does the water in the outer chamber of a calorimeter serve? How is a calorimeter used to measure the caloric content of a food?

109 Use Table 4-2 to calculate the total Calories in the following foods:
a. one cup of almonds containing 27 g protein, 71 g fat, and 28 g carbohydrate
b. a banana containing 1 g protein, 1 g fat, and 27 g carbohydrate
c. a 1-oz serving of cheddar cheese containing 7 g protein, 9 g fat, and 0 g carbohydrate
d. a glazed doughnut containing 2 g protein, 10 g fat, and 23 g of carbohydrate
e. a 3-oz serving of broiled swordfish containing 22 g protein, 4 g fat, and 0 g carbohydrate

110 Explain why drinking milk supplies your body with energy but drinking water does not.

111 Walking one mile burns about 40 Calories of energy. If you live two miles away from work and you walk to and from work each day for five days, how many Calories would you burn each week?

112 An average-size woman who is moderately active needs to eat about 2,000 Calories per day in order to maintain her weight. If her diet consists of 30 percent protein, 40 percent carbohydrate, and 30 percent fat, how many grams of each type of food should she eat each day?

113 If you wanted to lose 2.0 lb of fat, how many Calories would you have to burn, considering that fat is 15 percent water? There are 454 g/lb.

114 What serves as the important energy-carrier molecule in catabolic and anabolic biochemical processes?

115 What are the two types of metabolic reaction pathways?

116 List the three major biomolecules that are obtained through the diet and metabolized into smaller molecules.

117 Which of the following apply to catabolic reactions?
a. release energy
b. absorb energy
c. build larger molecules from smaller molecules
d. break down larger molecules into smaller molecules

118 In the body, the synthesis of proteins requires energy. Is this an anabolic or a catabolic process?

119 The combustion of pentane, C_5H_{12}, produces carbon dioxide and water. Write the balanced equation for this reaction. Be sure to include heat appropriately (as a product or reactant).

120 If heated, zinc carbonate, $ZnCO_3$, will decompose into zinc oxide, ZnO, and carbon dioxide. Write the balanced equation for this reaction. Include "heat" appropriately as a product or reactant.

121 The invention and development of the steam engine showed that heat can be used to perform work, and therefore heat is a form of energy. Explain, using calorimetry and the energy content of foods, why heat is a form of energy.

Reaction Kinetics

122 Define the term *reaction kinetics*.

123 What does *activation energy* refer to in a chemical reaction? What happens if the reactants do not have the required activation energy?

124 Which reaction is faster, the one with the larger or the smaller E_A?

125 Consider the energy diagram shown below.
a. Does it illustrate an exothermic or an endothermic reaction?
b. Label the following parts of the energy diagram:
　i. reactants
　ii. products
　iii. ΔH
　iv. E_A
c. How might the energy diagram differ if a catalyst were added to the reaction? Sketch a curve that shows how the reaction would be different in the presence of a catalyst.

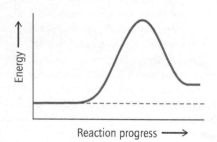

126 What effect would the following have on the rate of a chemical reaction?
a. decreasing the temperature
b. adding a catalyst
c. increasing the concentration of a reactant

127 Does a catalyst affect the value of ΔH? Does a catalyst affect the value of E_A?

128 What are biological catalysts called?

129 How do enzymes catalyze reactions?

130 Indicate whether the following statements are TRUE or FALSE.
＿＿ a. A catalyst is always used in high concentration.
＿＿ b. A catalyst speeds up the rate of a reaction.
＿＿ c. A catalyst lowers the activation energy for a reaction.
＿＿ d. A catalyst is chemically unaltered during a chemical reaction and therefore is reused over and over again in a reaction.
＿＿ e. Biological catalysts are known as enzymes.

131 Why are enzymes required for biochemical reactions? What role do enzymes play in biochemical reactions?

132 What enzyme is missing or ineffective in a person with lactose intolerance?

133 Reaction A is an endothermic reaction that occurs twice as fast if a catalyst is added. Draw side-by-side energy diagrams for reaction A with and without a catalyst. Which has the higher activation energy? Explain.

134 The combustion of fats is a slower process than the combustion of carbohydrates.
a. Are these reactions exothermic or endothermic?
b. Draw side-by-side energy diagrams for each, showing a difference in activation energies.
c. Which reaction has the higher activation energy, the combustion of carbohydrates or fats? Explain.

Chemical Equilibrium

135 What does it mean for a reaction to be reversible?

136 At the end of a reversible reaction, we find
a. only reactants
b. only products
c. both reactants and products

137 At equilibrium, the concentration of reactants and products are
a. constant
b. changing

138 State Le Châtelier's principle.

139 The reversible reaction of acetic acid shown below occurs in the bottle of vinegar sitting on your kitchen shelf.

$$\underset{\text{Acetic acid}}{\text{H}-\overset{\overset{\text{H}}{|}}{\underset{\underset{\text{H}}{|}}{\text{C}}}-\overset{\overset{\text{O}}{||}}{\text{C}}-\text{OH}} + \underset{\text{Water}}{\text{H}_2\text{O}} \rightleftharpoons \underset{\text{Acetate}}{\text{H}-\overset{\overset{\text{H}}{|}}{\underset{\underset{\text{H}}{|}}{\text{C}}}-\overset{\overset{\text{O}}{||}}{\text{C}}-\text{O}^-} + \underset{\text{Hydronium ion}}{\text{H}_3\text{O}^+}$$

a. What substances are present at equilibrium?
b. Is the concentrations of acetic acid and acetate constant or changing at equilibrium?
c. How does the equilibrium change if more acetate is added to the reaction?
d. If more acetic acid is added, does the reaction shift to the left or to the right?

140 A reversible reaction is shown below:

$$\text{NH}_4^+ + \text{H}_2\text{O} \rightleftharpoons \text{NH}_3 + \text{H}_3\text{O}^+$$

a. What substances are present at equilibrium?
b. Is the concentration of ammonia, NH_3, and NH_4^+ constant or changing at equilibrium?
c. If more ammonia is added to the reaction, which way does the reaction shift?
d. If more NH_4^+ is added to the reaction, which way does the reaction shift?

141 The following reaction plays an important role in the blood:

$$\text{H}_2\text{CO}_3 + \text{H}_2\text{O} \rightleftharpoons \text{HCO}_3^- + \text{H}_3\text{O}^+$$

a. Identify two ways to shift the reaction to the right.
b. Identify two ways to shift the reaction to the left.

142 Carbon monoxide binds to hemoglobin 200 times more tightly than oxygen does. This process sets up a competing equilibrium within the body:

$$HbO_2 \rightleftharpoons O_2 + Hb$$
$$Hb + CO \rightleftharpoons HbCO$$

When a patient suffers from carbon monoxide poisoning, there is too much of the HbCO complex in their blood. Using Le Châtelier's principle, explain how treating the patient with oxygen affects both equilibria shown above to produce more HbO_2 and less HbCO.

Chemistry in Medicine: Calorimetry and the Care of Patients on a Mechanical Ventilator

143 Why is it important to know how many of calories per day to feed a person on a mechanical respirator?

144 Are the final products of the combustion of food the same whether the food is burned in a calorimeter or the human body? Explain.

145 What is the difference between direct and indirect calorimetry? Which is the more practical method to use on patients who are on mechanical respirators? Explain.

Answers to Practice Exercises

1 $(1 \times 12.01) + (4 \times 1.008) = 16.04$ amu

2 $(2 \times 26.98) + (3 \times 16.00) = 101.96$ amu

3 $(6 \times 12.01) + (12 \times 1.008) + (6 \times 16.00) = 180.16$ amu

4 There are 6.02×10^{23} marbles in one mole of marbles.

5 One mole of basketballs has a greater mass than one mole of Ping-Pong balls because one basketball weighs more than one Ping-Pong ball and there is the same number of balls in both samples.

6 One Hg atom has an atomic mass of 200.6 amu. The mass of one mole of Hg atoms is 200.6 g. The molar mass of Hg is 200.6 g/mol.

7 The molar mass of caffeine is 194.20 g/mol. There are 6.02×10^{23} caffeine molecules in 1 mole of caffeine.

8 $0.56 \text{ mol} \times \dfrac{6.02 \times 10^{23} \text{ molecules}}{1 \text{ mol}} = 3.4 \times 10^{23}$ molecules

9 Figure 4-4 shows that the patient's blood CO_2 is 27 mmol per liter, which is within the 21–33 mmol per Liter range considered normal. To calculate the number of molecules of CO_2 in 27 mmol, use a metric conversion and Avogadro's number as a conversion factor:

$$27 \text{ mmol} \times \dfrac{1 \text{ mol}}{10^3 \text{ mmol}} \times \dfrac{6.02 \times 10^{23} \text{ molecules}}{1 \text{ mol}}$$
$$= 1.6 \times 10^{22} \text{ molecules}$$

10 $1.100 \times 10^{27} \text{ molecules} \times \dfrac{1 \text{ mol}}{6.02 \times 10^{23} \text{ molecules}}$
$= 1,827$ moles

11 $159 \text{ g} \times \dfrac{1 \text{ mol}}{200.6 \text{ g}} = 0.793$ mol

12 2.5 mol water

13 Molar mass of ibuprofen: $(13 \times 12.01) + (18 \times 1.008) + (2 \times 16.00) = 206.27$ g/mol

$$200. \text{ mg} \times \dfrac{1 \text{ g}}{10^3 \text{ mg}} \times \dfrac{1 \text{ mol}}{206.27 \text{ g}} = 9.70 \times 10^{-4} \text{ mol}$$

14 Molar mass of glucose: 180.16 g/mol (see Exercise 3).

$$126. \text{ mg} \times \dfrac{1 \text{ g}}{10^3 \text{ mg}} \times \dfrac{1 \text{ mol}}{180.16 \text{ g}} = 6.99 \times 10^{-4} \text{ mol}$$

15 $0.666 \text{ mol} \times \dfrac{200.6 \text{ g}}{1 \text{ mol}} = 134$ g

16 34 g

17 According to Figure 4-4, there are 4.2 mmol of potassium ions in every liter of the patient's blood, and the molar mass of potassium is 39.10 g/mol. This corresponds to 0.16 g of potassium ions in every liter of the patient's blood.

18 $1.0 \text{ g} \times \dfrac{1 \text{ mol}}{18.016 \text{ g}} = 0.056$ mol water compared to

$10. \text{ g} \times \dfrac{1 \text{ mol}}{180.16 \text{ g}} = 0.056$ mol glucose. They contain the same number of molecules since the number of moles is the same.

19 $6.5 \times 10^{20} \text{ atoms} \times \dfrac{1 \text{ mol}}{6.02 \times 10^{23} \text{ atoms}} \times \dfrac{63.55 \text{ g}}{1 \text{ mol}}$
$= 0.069$ g of copper

20 $200. \text{ mg} \times \dfrac{1 \text{ g}}{10^3 \text{ mg}} \times \dfrac{1 \text{ mol}}{206.27 \text{ g}} \times \dfrac{6.02 \times 10^{23}}{1 \text{ mol}}$
$= 5.84 \times 10^{20}$ ibuprofen molecules

21 $1.0 \text{ g} \times \dfrac{1 \text{ mol}}{58.44 \text{ g}} \times \dfrac{6.02 \times 10^{23}}{1 \text{ mol}} = 1.0 \times 10^{22}$ formula units of NaCl. Therefore, there are 1.0×10^{22} Na^+ ions and 1.0×10^{22} Cl^- ions.

22 i. a. Reactants: C_3H_8 and O_2.
   b. Products: CO_2 and H_2O.
   c. For every 1 C_3H_8 that reacts with 5 O_2, there are 3 CO_2 and 4 H_2O produced.
   d. All are gases.
ii. a. Reactants: C_4H_6 and H_2.
   b. Product: C_4H_{10}.
   c. For every 1 C_4H_6 that reacts with 2 H_2, there is 1 C_4H_{10} produced.
   d. All are gases.

23 a. $\underline{2} \; N_2O_5(g) \rightarrow \underline{4} \; NO_2(g) + \underline{} \; O_2(g)$
   b. $\underline{} C_6H_{12}O_2(l) + \underline{8} \; O_2(g) \rightarrow \underline{6} \; CO_2(g) + \underline{6} \; H_2O(aq)$
   c. $\underline{} \; Mg(s) + \underline{2} \; HCl(aq) \rightarrow \underline{} \; MgCl_2(aq) + \underline{} \; H_2(g)$
   d. $\underline{4} \; NH_3(g) + \underline{3} \; O_2(g) \rightarrow \underline{2} \; N_2(g) + \underline{6} \; H_2O(l)$

24 There would be 36 moles of water and 36 moles of carbon dioxide produced if six moles of glucose were to react in excess oxygen:

$$6 \text{ mol glucose} \times \frac{6 \text{ mol } CO_2}{1 \text{ mol glucose}} = 36 \text{ mol } CO_2$$

25 $5.60 \text{ mol Mg} \times 2 \text{ mol HCl}/1 \text{ mol Mg} = 11.2 \text{ mol HCl}$
$5.60 \text{ mol Mg} \times 1 \text{ mol } H_2/1 \text{ mol Mg} = 5.60 \text{ mol } H_2$

Extension Topic 4-1: Stoichiometry Calculations

E4.1

$$12.0 \text{ g } C_2H_6O \times \frac{1 \text{ mol } C_2H_6O}{46.07 \text{ g } C_2H_6O} \times \frac{3 \text{ mol } O_2}{1 \text{ mol } C_2H_6O}$$

$$\times \frac{32.00 \text{ g } O_2}{1 \text{ mol } O_2} = 25.0 \text{ g } O_2$$

E4.2

$$C_3H_8(g) + 5 \, O_2(g) \rightarrow 3 \, CO_2(g) + 4 \, H_2O(g)$$

$$500.0 \text{ g } C_3H_8 \times \frac{1 \text{ mol } C_3H_8}{44.094 \text{ g } C_3H_8} \times \frac{5 \text{ mol } O_2}{1 \text{ mol } C_3H_8}$$

$$\times \frac{32.00 \text{ g } O_2}{1 \text{ mol } O_2} = 1{,}814 \text{ g } O_2$$

$$500.0 \text{ g } C_3H_8 \times \frac{1 \text{ mol } C_3H_8}{44.094 \text{ g } C_3H_8} \times \frac{3 \text{ mol } CO_2}{1 \text{ mol } C_3H_8}$$

$$\times \frac{44.01 \text{ g } CO_2}{1 \text{ mol } CO_2} = 1{,}497 \text{ g } CO_2$$

E4.3

$$C_6H_{12}O_6 + 6 \, O_2 \rightarrow 6 \, CO_2 + 6 \, H_2O$$

$$10.0 \text{ g } C_3H_{12}O_6 \times \frac{1 \text{ mol } C_6H_{12}O_6}{180.16 \text{ g } C_6H_{12}O_6}$$

$$\times \frac{6 \text{ mol } CO_2}{1 \text{ mol } C_6H_{12}O_6} \times \frac{44.01 \text{ g } CO_2}{1 \text{ mol } CO_2} = 14.7 \text{ g } CO_2$$

26 a. endothermic   b. endothermic

27

Products
2 CO + O_2

$\Delta H = + 566$ kJ

Reactant
2 CO_2

Energy →

Reaction progress →

28 An endothermic reaction absorbs heat energy ($\Delta H > 0$), whereas an exothermic reaction releases heat energy ($\Delta H < 0$). In an endothermic reaction the products are higher in energy than the reactants. In an exothermic reaction the products are lower in energy than the reactants.

29 144 Cal, or 140 Cal when rounded to the tens place, the common practice with nutritional labels.

30 157 Cal, or 160 Cal when rounded to the tens place.

31 Catabolic reactions break down larger molecules into smaller molecules and release energy. Anabolic reactions build larger molecules from smaller molecules and absorb energy.

32 a. anabolic, absorbs energy
   b. catabolic, releases energy
   c. anabolic, absorbs energy

33 The activation energy is the amount of energy that must be supplied to the reactants for the reaction to occur.

34 (b) The energy diagram with the smaller "hill" is the faster reaction because it has the smaller energy of activation, E_A.

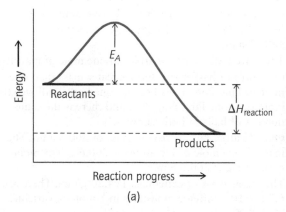

(a)

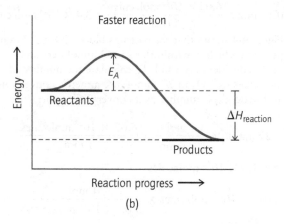

(b)

35 a. An increase in temperature increases the rate of the reaction because the kinetic energy of the reactants increases, so the number of collisions increases, leading to a greater likelihood of reaction.
   b. Adding a catalyst, increases the rate of the reaction because a catalyst facilitates the reaction by lowering the energy of activation, E_A.

36 The rate of a reaction decreases with time because there are fewer reactants as the reaction progresses, and thus fewer collisions.

37 catalysts

38 E_A is smaller in the reaction with the catalyst; ΔH is the same in both reactions. The reaction with the catalyst occurs at a faster rate.

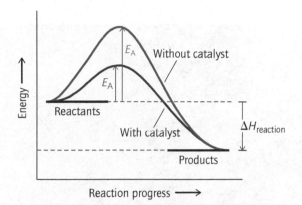

39 The concentration of the reactants and products is constant at equilibrium because the rate of the forward reaction is equal to the rate of the reverse reaction. No, the reactions have not stopped; the concentrations simply have stopped changing. Dynamic means both the forward and the reverse reactions are still occurring.

40 In the reverse reaction, HbO_2 is decomposing into Hb and O_2. In the forward reaction, Hb reacts with O_2 to form HbO_2. The concentration of Hb, O_2, and HbO_2 are constant at equilibrium.

41 a. Since $\Delta H < 0$, this is an exothermic reaction, which means heat is a product. Thus, raising the temperature will shift the equilibrium to the left to consume the added heat.
b. H_2 is a reactant, so adding more will shift the equilibrium to the right to consume the added reactant.
c. Methane is a product, so adding more will shift the equilibrium to the left to consume the added product.
d. Carbon is a reactant, so removing a reactant will shift the equilibrium to the left to restore lost reactant.
e. Methane is a product, so removing methane will cause the equilibrium to shift to the right to restore lost product.

42 a. Since CO_2 is a product, adding product will cause the equilibrium to shift to the left to consume added product.
b. Since CO_2 is a product, removing some product will cause the equilibrium to shift to the right to restore the removed product.

Scuba diving is a fun and safe activity when standard safety measures are followed.
[© J.W. Alker/Imagebroker/Alamy]

5

Changes of State and the Gas Laws

Diver Rescue and "the Bends"

Every year thousands of people go scuba diving into the ocean depths, enticed by coral reefs, exotic fish, and famous shipwrecks. Although most scuba divers are well trained and avoid safety blunders, about five in every 100,000 divers has an accident. Consider the following scenario: On a diving trip, your companion surfaces looking confused and weak, complaining of a headache, itching, and joint pains. The diving instructor immediately radios for help, and a helicopter arrives, transporting your companion to the nearest medical facility for treatment. What happened? The short answer is that your companion probably spent too much time at a particular depth and ascended too quickly, causing a condition known as decompression sickness or "the bends."

The bends occurs when a person experiences a sudden and significant change in pressure. At sea level, you experience pressure from the mass of air above you, known as atmospheric pressure. At sea level, atmospheric pressure is 1 atm, a unit of pressure. For every 30 feet that you descend beneath the surface of the sea, you experience an additional 1 atm of pressure as a result of the mass of air *and* seawater above you. Air is a mixture of primarily nitrogen and oxygen gases, which dissolve in the blood in proportion to pressure. At greater depths, pressure is higher, so more molecules of nitrogen and oxygen are dissolved in the blood.

When a diver ascends back to the surface, the pressure on the diver *decreases*, whereupon the additional oxygen and nitrogen molecules come out of solution. If a diver ascends too quickly, the dissolved nitrogen gas will diffuse out of the blood too quickly, forming bubbles in the bloodstream. Think of how carbon dioxide bubbles emerge from a bottle of a carbonated beverage, a solution at high pressure, when you twist open the cap, exposing the beverage to the lower atmospheric pressure. Similarly, nitrogen bubbles diffuse out of the diver's blood and can clog small blood vessels, disrupt circulation, and cause severe pain when they expand within the closed spaces of joints. These are symptoms of the bends.

An important part of scuba training is learning how long you can stay at a particular depth and how to ascend safely. In the event that a case of the bends does occur, there is a treatment. See Chemistry in Medicine: Hyperbaric Oxygen Therapy (HBOT) at the end of this chapter to learn how the bends is treated. In this chapter, you will learn about the unique properties of gases and gain a more in-depth understanding of not only why scuba diving can affect the body the way it does, but how gases behave in respiration and other medical applications involving gases. ●

In this chapter, you will learn about changes of state, the physical change of going from one state of matter to another state of matter. In the second part of the chapter, you will focus on the unique characteristics of the gas state. You will see why a steam burn is more severe than a burn from scalding water, how gases in an anesthetic affect patient recovery time, and why hyperbaric oxygen is an effective treatment for the bends.

Figure 5-1 Carbon dioxide (CO_2) in the solid phase—known as dry ice—sublimes (s → g) at −78 °C (−109 °F). When dry ice comes into contact with water at room temperature, it appears as a fog as water vapor condenses on the rapidly subliming carbon dioxide vapors. [Catherine Bausinger]

> Although *evaporation* may occur at any temperature, *vaporization* is defined as the change of state that occurs at the boiling point of a liquid.

■ 5.1 Changes of State

In chapter 1 you were introduced to the three physical states of matter: solid, liquid, and gas. In this section we explore **changes of state**—the process of going from one physical state to another physical state. We have all seen water condense on the glass of a cold drink on a warm, humid day or a fog emerge from a block of dry ice in water, as shown in Figure 5-1. These are both examples of changes of state: gas to liquid and solid to gas, respectively. A change of state, or phase change, is a *physical change*, not a *chemical change*, because *covalent bonds* are not formed or broken in the process. For example, the covalent bonds in water molecules remain intact in all three states: solid (ice), liquid, and gas (steam).

Although covalent bonds remain intact during a change of state, intermolecular forces of attraction between molecules do change. Recall from chapter 3 that nonpolar molecules interact through dispersion forces, the weakest of the intermolecular forces of attraction, whereas polar molecules also have dipole-dipole or hydrogen-bonding forces of attraction, where the latter are the strongest of the intermolecular forces of attraction between molecules. Remember that even the strongest intermolecular forces of attraction are significantly weaker than covalent bonding forces.

In the gas phase there are no intermolecular forces of attraction because the molecules or atoms are too far apart to interact as a result of their greater kinetic energy. In the gas phase there is mostly empty space. Intermolecular forces of attraction exist in the liquid phase and are even more abundant in the solid phase.

Energy and Changes of State

The terms we use to describe the different changes of state are summarized in Figure 5-2, many of which are probably already familiar to you. A change of state from the *solid* to the *liquid* phase is known as **melting**; the reverse is known as **freezing**. A change of state from the *liquid* to the *gas* phase is known as **vaporization** or **evaporation**; the reverse is known as **condensation**. A *solid* can also undergo a change of state directly to the *gas* phase without first becoming a liquid, a process known as **sublimation**. For example, dry

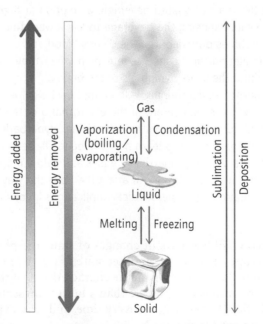

Figure 5-2 Changes of state shown in red require the addition of heat energy. Changes of state shown in blue require the removal of heat energy.

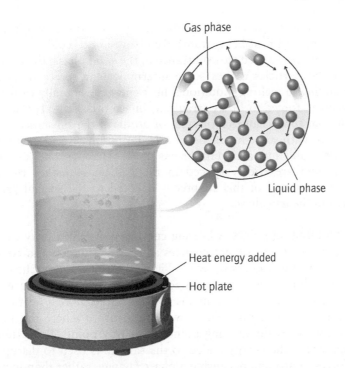

Gas phase

Liquid phase

Heat energy added

Hot plate

Figure 5-3 A hot plate provides the heat energy necessary for molecules in the liquid phase to break away from the intermolecular forces of attraction that hold them together, enabling them to enter the gas phase, where they are far apart from one another.

ice (CO_2) changes directly from the solid phase to the gas phase at room temperature. The reverse process, changing from the *gas* phase to the *solid* phase, is known as **deposition**. An example of deposition is the formation of snowflakes in the upper atmosphere where water vapor becomes a solid in clouds.

Energy is an important factor in a change of state. During melting, vaporization, and sublimation intermolecular forces of attraction are *disrupted* as the kinetic energy of the particles increases (temperature therefore increases too). Conversely, during condensation, freezing, and deposition, intermolecular forces of attraction are *created* as the kinetic energy of the particles decreases (temperature therefore decreases too). Remember, an increase or a decrease in kinetic energy occurs when there is a transfer of energy to or from the substance. Energy must be *added* for *melting*, *vaporization*, and *sublimation* to occur; while energy must be *removed* for *freezing*, *condensation*, and *deposition* to occur (Figure 5-2). For example, to turn liquid water into steam, you must *add* energy to the liquid, perhaps using heat energy from a stove or hot plate (Figure 5-3) or electromagnetic radiation from a microwave oven.

Have you ever wondered why you feel cold when you step out of the shower, even when the room is warm? The reason involves a change of state and energy transfer. Recall from chapter 1 that heat energy always flows from hot to cold, so heat flows *out* of your warm body to evaporate the water on your skin. Evaporation is a change of state from the liquid to the gas phase. You experience this as "drying off" because the water molecules enter the gas phase and expand to fill the room and you feel cold because heat is being transferred out of your body. This phenomenon is known as **evaporative cooling**. It's also why you sweat after vigorous exercise and in hot weather. Heat is removed from the surroundings during evaporation (l → g), and sometimes those surroundings are your body.

Condensation (gas → liquid), the reverse of vaporization and evaporation, requires that heat energy be *removed* from the gas phase. For example, if you place your hand over a pot of boiling water, the steam (water in the gas phase) condenses on your skin (water in the liquid phase) because your hand is at a lower temperature than the steam. You can feel the heat being transferred from the steam to your hand as well as the liquid water droplets forming on

Steam is another term for water in the gas phase. When a substance is a liquid at room temperature (25 °C), it is often referred to as a vapor when it is in the gas phase (at higher temperature).

your hand as water molecules in the gas phase condense on your hand, another example of heat being transferred from hot to cold.

When energy is added to a substance in the solid phase, the greater kinetic energy of the particles causes some intermolecular forces of attraction to break, resulting in melting. Additional heat energy eventually causes all of the intermolecular forces to break, resulting in vaporization. The disruption of intermolecular forces causes molecules or atoms of the substance to be spaced farther apart, with some interaction in the liquid phase and complete separation and negligible interaction in the gas phase. The fact that particles are not interacting and are widely spaced in the gas phase makes the gas phase unique. The last half of this chapter is devoted to the special properties of substances in the gas phase.

Melting and Boiling Points A **heating curve** is a graphical way to show how the temperature of a substance changes as energy is added at a constant rate. A heating curve for water is shown in Figure 5-4. Notice how the temperature increases in a linear fashion as you add heat, except in two places along the curve where the temperature remains constant even though heat continues to be added at the same rate. The lower temperature where the first plateau occurs is known as the **freezing point** or **melting point** of the substance. At this temperature, the energy added to the sample is used to disrupt intermolecular forces of attraction, causing a phase change, rather than increasing the kinetic energy of the particles, and this is why the temperature remains constant. The temperature at which *a substance undergoes a change of state from the solid to the liquid is known as the melting point*. Once the sample has melted completely, the temperature begins to rise again in a linear fashion. The higher temperature where the second plateau occurs is known as the **normal boiling point** of the substance. At this temperature, the energy added to the sample is used to disrupt the remaining intermolecular forces of attraction, causing a phase change, rather than increasing the kinetic energy of the particles, and this is why the temperature remains constant. At this temperature we observe the liquid boiling; hence the term *boiling point*. The liquid undergoes a change of state to the gas phase. Once the sample is completely vaporized, the temperature begins to rise again.

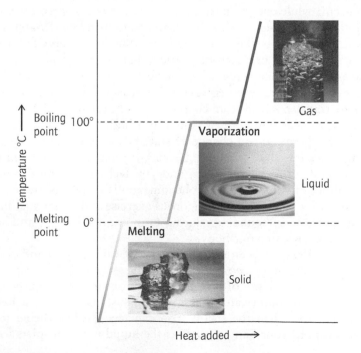

Figure 5-4 A heating curve for water showing the melting point and boiling point of water. [Left to right: © Corbis, © moodboard/Alamy, Martyn F. Chillmaid/ Science Source.]

TABLE 5-1 Melting Points and Boiling Points of Some Common Elements and Compounds

Substance	Melting Point (°C)	Boiling Point (°C)	Type of Substance
Oxygen	−219	−183	Nonpolar element
Nitrogen	−210	−196	Nonpolar element
Propane	−188	−42.0	Nonpolar covalent compound
Methane	−182	−164	Nonpolar covalent compound
Ethanol	−114	78	Polar covalent compound
Chloroform	−64	61	Covalent compound
Water	0	100	Polar covalent compound
Acetic acid	17	118	Polar covalent compound
Phenol	41	182	Covalent compound

The melting point and boiling point of a substance are unique physical properties of a substance, as shown in Table 5-1 for a few representative compounds. For example, the freezing point of water is 0 °C and the boiling point of water is 100 °C. Nonpolar covalent compounds tend to have lower melting and boiling points because they have the weakest intermolecular forces of attraction, requiring the least amount of energy to disrupt. In contrast, compounds with O—H and N—H bonds, which have the ability to hydrogen bond, have the strongest intermolecular force of attraction. These compounds have higher melting and boiling points because it takes more energy to disrupt these stronger intermolecular forces of attraction. Indeed, this is why water is a liquid and methane is a gas at room temperature.

The amount of energy (calories) required to melt a solid at its *melting point* is known as the **heat of fusion** or **enthalpy of fusion**: ΔH_{fus}. Similarly the amount of energy required to vaporize a liquid at its *boiling point* is known as the **heat of vaporization**, or **enthalpy of vaporization**: ΔH_{vap}. Generally, heats of vaporization are greater than heats of fusion because the former requires that all intermolecular forces of attraction be disrupted. For example, the heat of vaporization for water is 540 calories per gram, whereas the heat of fusion for water is 80 calories per gram. Heats of fusion and heats of vaporization, much like melting points and boiling points, represent physical properties of a substance.

A *cooling* curve has the appearance of an inverted heating curve. It begins with the substance in the gas phase, and the temperature decreases as heat is removed at a steady rate. The freezing and boiling points occur at the same temperatures as observed in the heating curve for a given substance and also appear as horizontal plateaus. The amount of heat that must be *removed* for condensation to occur is equal to the heat of vaporization, ΔH_{vap}, for that substance. In other words, 540 calories per gram of water must be *added* to the liquid to achieve vaporization, so 540 calories per gram of water must be *removed* to achieve condensation, and both occur at 100 °C, the boiling point of water. Similarly, 80 calories per gram of water must be *added* to melt the sample, so 80 calories per gram of water must be *removed* to freeze the sample, the enthalpy of fusion, ΔH_{fus}, for water, and both occur at 0 °C, the freezing point of water.

Steam Burns Why are steam (water vapor) burns more severe than burns from scalding (boiling) water? It is not because steam is at a higher temperature but rather because steam undergoes a change of state from the gas to the

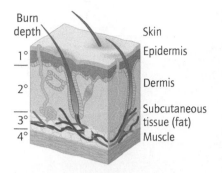

Figure 5-5 The severity of a skin burn is indicated by the depth of tissue damage: epidermis (1°), dermis (2°), fat tissue (3°), or muscle (4°).

liquid phase when it comes in contact with the lower-temperature skin. Remember, during the gas to liquid phase change, 540 cal per gram of water, ΔH_{vap}, is transferred, in this case, to the skin. For example, 10 g of steam will transfer 5,400 calories of heat to the skin, and then an additional 630 calories is transferred to the skin as the liquid cools from 100 °C to 37 °C (body temperature). In contrast, for 10 g of boiling *liquid* water, only 630 calories is transferred to the skin because the water is already in the liquid phase when it comes in contact with the skin (no phase change).

Figure 5-5 shows the degree of severity of a skin burn based on the depth of the skin damage: a first-degree burn affects only the top layer of skin (epidermis), a second-degree burn affects the epidermis and some or all of the dermis, a third-degree burn extends into the fat tissue beneath the dermis, and a fourth-degree burn extends into the muscle.

PRACTICE EXERCISES

1. Identify the term that describes the changes of state listed below:
   a. gas → solid
   b. solid → liquid
   c. liquid → gas

2. Which of the phase changes in the question above involve(s) the transfer of heat from the sample to the surroundings?

3. In which state are intermolecular forces of attraction the greatest? In which state are they negligible? Explain.

4. What phase change occurs when a liquid is heated to its boiling point? What phase change occurs when a gas is cooled to its boiling point?

5. What occurs on the atomic level during vaporization? Why is the boiling point of a small hydrocarbon significantly lower than the boiling point of water?

6. What change of state occurs during the process of sublimation?

7. What change of state is the reverse of vaporization?

8. Explain why a steam burn at 100 °C causes more damage to the skin than a burn from boiling water at 100 °C.

9. A microwave oven transfers energy to water molecules with electromagnetic radiation, another form of energy. Explain what is happening to the water molecules from a kinetic-molecular view.

10. When a child has a high fever, you might be advised to place the infant in a bathtub of tepid water. The temperature of the water in the bathtub need not be cold; in fact, it isn't advisable to immerse the infant in cold water (to avoid hypothermia). Explain how the evaporative cooling effect works here to reduce the infant's body temperature with merely tepid water.

Vapor Pressure

While shopping, a friend passes a bottle of cologne to you to smell. Simply by placing your nose above the open bottle, you are able to smell the fragrance. How is this possible, given that the liquid cologne never actually comes in contact with your nose? The answer is that at the surface of the liquid, some of the cologne molecules gain sufficient kinetic energy to enter the gas phase, a process known as evaporation. We call it evaporation instead of vaporization because the phase change is occurring below the boiling point of the liquid. Simultaneously, some of the cologne molecules in the gas phase directly above the liquid lose kinetic energy and return to the liquid phase, as illustrated in Figure 5-6, creating an equilibrium between the liquid and gas phases:

$$l \rightleftharpoons g$$

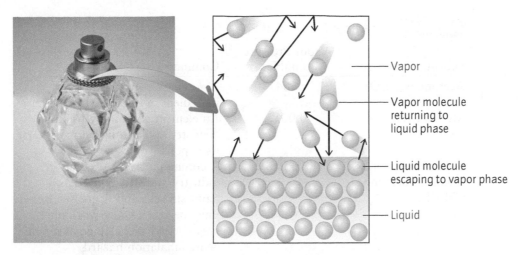

Figure 5-6 Vapor pressure is the pressure exerted by molecules in the liquid-gas equilibrium. Vapor pressure increases with temperature. Volatile liquids such as perfume have a higher vapor pressure than less volatile compounds. [Photo: Catherine Bausinger]

The change of state described above from the liquid to the gas phase is a result of vapor pressure. *Vapor pressure is the pressure exerted by a gas in the liquid-gas equilibrium.* As you know from the kinetic molecular view of matter, atoms or molecules in the gas phase are continuously colliding against the walls of their container, creating pressure. Pressure (P) is defined as the force per unit area exerted by *gas* particles colliding against the walls of the container, as illustrated in Figure 5-7. The greater the kinetic energy of the molecules, the greater the pressure.

Pressure, P, is a measure of force applied over a given area:

$$\text{Pressure } (P) = \frac{\text{force}}{\text{area}}$$

If you have ever had the misfortune of breaking a bone, you know firsthand the pain that excessive *pressure* can cause. When a bone breaks, a force applied to a given area of the bone exceeds the strength of the bone.

Vapor pressure is another physical property of a substance that depends on the chemical structure and on the kinetic energy of the substance as measured by its temperature. A liquid is said to be **volatile** if it has a high vapor pressure; which means its molecules enter the gas phase readily. Volatile compounds have high vapor pressures because they have weak intermolecular forces of attraction, making it easier for the molecules to enter the gas phase. The cologne you smelled is a volatile liquid, which is why you were able to smell its odor so readily. Moreover, since a gas expands to fill its container, once the cologne molecules are in the gas phase, they fill the room, eventually reaching the olfactory receptors in your nose, which you detect as a hopefully pleasant odor. On the other hand, breathing vapors from toxic liquids that have a high vapor pressure can be dangerous depending on the length of exposure. The vapor pressure at 20 °C for some familiar commercial substances is listed in Table 5-2.

The vapor pressure of a substance increases with temperature. At higher temperatures, molecules in the liquid phase have greater kinetic energy and therefore their vapor pressure is higher, allowing them to enter the gas phase more readily. You have experienced this phenomenon if you have ever filled your car with gasoline on a hot day. Gasoline is a volatile liquid, which means

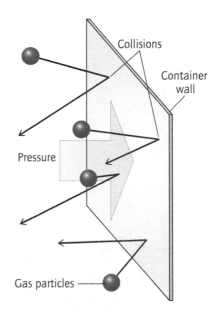

Figure 5-7 The pressure of a gas is a measure of the force per unit area exerted by the gas particles as they collide with the walls of the container.

TABLE 5-2 The Vapor Pressure of Some Substances at 20 °C

Substance	Vapor Pressure (mmHg)	Common Uses/Notes
Acetone (C_3H_6O)	184	A common solvent and an ingredient in nail polish remover
Mercury (Hg)	0.0012	An element commonly used in dentistry and also in thermometers; its vapor is toxic
Methylene chloride (dichloromethane, CH_2Cl_2)	350	A common solvent used in industry and widely used as a paint stripper and degreaser; care must be taken when working with methylene chloride as it is an acute inhalation hazard
Phenol (C_6H_6O)	0.36	Used to make commercial plastics, nylon, and detergents

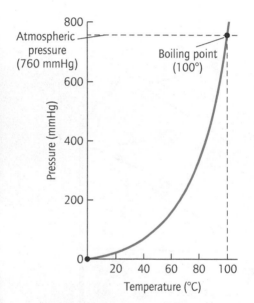

Figure 5-8 Graph showing the vapor pressure of water at different temperatures. Boiling occurs when the vapor pressure equals atmospheric pressure.

gasoline molecules have a high vapor pressure. On a hot day, the gasoline molecules have an even *higher* vapor pressure, allowing more gasoline molecules to enter the gas phase and be more likely to reach your nose, where you can detect them.

Figure 5-8 shows how the vapor pressure of water increases with temperature. Boiling occurs when the vapor pressure of a substance equals atmospheric pressure. *The **normal boiling point** of a liquid is defined as the temperature at which the vapor pressure of a liquid equals 1 atm, the atmospheric pressure at sea level.* The normal boiling point for water, for example, is 100 °C, as shown in Figure 5-8. Substances with a *higher* vapor pressure (more volatile substances) will have similarly shaped curves, but their boiling points will be lower because the curves cross the 760 mmHg (1 atm) line at a lower temperature.

Atmospheric Pressure

Like other physical properties, pressure can be measured and reported in a variety of different units. Some common pressure units are given in Table 5-3, along with their conversion to units of *atmospheres* (atm). The most common unit of pressure in medicine is millimeters of mercury, mmHg.

Molecules of air in the atmosphere create a downward pressure as a result of gravity. This pressure is known as **atmospheric pressure** or **barometric pressure** and it is created by the force of the mass of air at any given place

TABLE 5-3 Common Units of Pressure and Their Conversion to Units of Atmospheres

Pressure Units Unit	Abbreviation	Conversions
Atmospheres	atm	
Torr	torr	1 atm = 760 torr (exact)
Millimeters of mercury	mmHg	1 atm = 760 mmHg (exact)
Pounds per square inch	psi or lb/in.^{2}	1 atm = 14.70 psi
Pascal	Pa	1 atm = 1.013×10^5 Pa

(area) on the earth (Figure 5-9). Atmospheric pressure is commonly measured in units of atmospheres using a device called a **barometer**. At sea level, the atmospheric pressure from the column of air above you is 1 atmosphere (1 atm). As you move to higher altitudes, atmospheric pressure decreases because there is less air above you. For example, at a mile high (5,280 feet), the altitude of Denver, Colorado, the atmospheric pressure is 0.83 atm, as shown in Figure 5-10. Hikers often experience difficulty breathing at higher altitudes because a lower atmospheric pressure means less oxygen is breathed in with every breath. However, the body can compensate over time by producing more red blood cells. This is why hikers who climb Mount Everest (the peak is at 29,000 ft) spend time at incrementally higher base camps to get acclimated. Indeed, many athletes train at high altitudes such as Denver so that their bodies will produce more red blood cells, improving their performance at athletic competitions held at sea level.

Sudden changes in pressure can lead to decompression sickness. Symptoms of decompression sickness include localized deep pain, dizziness, nausea, and shortness of breath. Deep-sea divers experience significant *increases* in pressure, which under certain circumstances can lead to decompression sickness, "the bends," as described in this chapter's opening vignette.

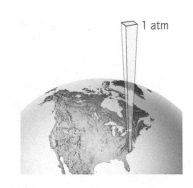

Figure 5-9 At sea level, the column of air above you applies a force per unit area known as atmospheric pressure, which is equal to one atmosphere (1 atm).

At sea level, atmospheric pressure is 760 mmHg and water boils at 100 °C. At higher altitudes, atmospheric pressure is lower, so the boiling point of water and other liquids is lower. For example, in Denver atmospheric pressure is 560 mmHg; therefore, water boils at 92 °C rather than 100 °C.

9,144 m (30,000 ft) — 0.27 atm

Denver, the mile-high city, 1,609 m (5,280 ft) — 0.83 atm

Sea level 0 m (0 ft) — 1 atm

−250 m (−820 ft) — 25 atm

Deep sea −2,000 m (−6,562 ft) — 200 atm

Figure 5-10 Atmospheric pressure at sea level is 1 atm (760 mmHg). At higher altitudes atmospheric pressure decreases because there is a smaller column of air above you. Conversely, pressure increases by 1 atm for every 30 ft you dive beneath sea level as a result of the increased pressure from the water above you.

WORKED EXERCISES   Vapor Pressure

5-1 Consult Table 5-2 to answer the following questions:
   a. Which has a higher vapor pressure, mercury or methylene chloride? Which is the more volatile substance?
   b. Which would you expect to have a higher boiling point, phenol or acetone? Explain.

5-2 Intraocular pressure is the pressure inside the eye. A typical intraocular pressure is 15 mmHg. What is the intraocular pressure in units of atmospheres?

Solutions

5-1 **a.** Methylene chloride has a higher vapor pressure than mercury at 20 °C , therefore methylene chloride is more volatile than mercury.
   b. Phenol has a lower vapor pressure than acetone, so you would expect phenol to have a higher boiling point than acetone.

5-2 **Step 1: Identify the conversion.** Use Table 5-3 to find the conversion between atm and mmHg:

$$1 \text{ atm} = 760 \text{ mmHg}$$

Step 2: Express the conversion as two possible conversion factors:

$$\frac{1 \text{ atm}}{760 \text{ mmHg}} \quad \text{or} \quad \frac{760 \text{ mmHg}}{1 \text{ atm}}$$

Step 3: Multiply the supplied unit by the conversion factor that allows the supplied units to cancel. Use the conversion factor that allows mmHg to cancel in order to obtain an answer in the requested unit, atm:

$$15 \ \cancel{\text{mmHg}} \times \frac{1 \text{ atm}}{760 \ \cancel{\text{mmHg}}} = 0.020 \text{ atm}$$

PRACTICE EXERCISES

11 Define vapor pressure.

12 How is the vapor pressure of a substance related to its boiling point?

13 If you ski in the Rocky Mountains and take a break for lunch, you often find that hot beverages are not as hot as you are accustomed to at sea level. Explain.

14 Define normal boiling point. Will the boiling point of a liquid be higher or lower at high altitude?

15 Skunks produce a compound with a very noticeable odor. Would you expect this substance to be volatile? Would you expect it to have a high or a low boiling point? Would you expect it to have a high vapor pressure or a low vapor pressure?

16 An autoclave uses water vapor at high pressure to sterilize medical and laboratory equipment. Sterilization is a process that destroys bacteria, viruses, and other transmittable agents. One way to sterilize something is by applying high temperatures. The pressure in an autoclave is much greater than atmospheric pressure. Would water boil at a temperature greater than or less than 100 °C in an autoclave? Speculate on why items are sterilized more effectively in an autoclave than in a dishwasher.

17 What does a barometer measure?

18 Is the atmospheric pressure higher or lower at high altitudes? Explain.

19 When a golfer hits a shot off the tee in Denver, it travels farther than it would in Florida. Explain this observation.

20 Tennis balls manufactured for use in communities at high altitude, such as Denver, are designed with less bounce. Provide an explanation for this practice.

■ 5.2 The Gas Laws

Gases are compressible, which gives them unique physical properties. To understand the macroscopic properties of a gas, it is useful to consider the kinetic-molecular view of a gas:

- The particles of a gas are in constant, random motion. Gas particles move at high speeds in straight lines and in random directions, filling the entire volume of the container they occupy.
- The total volume of the gas particles themselves is negligible compared to the volume of the container. In other words, the container consists of mostly empty space. This is why a gas is easily compressed.
- The attractive forces between the particles of a gas are negligible. Gas particles are too far apart for intermolecular forces of attraction to occur.
- The temperature of a gas depends on the average kinetic energy of the gas particles. Recall from section 1.1 that the faster the gas particles move, the greater their kinetic energy and the higher the temperature of the gas: $KE = \frac{1}{2}\ mv^2$.

The *macroscopic* properties of a gas can be described by the following four interrelated variables:

- pressure (P),
- volume (V),
- temperature (T), and
- number of moles (n)

These variables are all interrelated, so changing one affects the others in a predictable way. The relationship between any two variables, while the other two are constant, is expressed in the gas laws: Boyle's law, Charles's law, and Avogadro's law.

Pressure-Volume Relationship of Gases: Boyle's Law

Imagine a gas in a closed volume of space such as the barrel and plunger shown in **Figure 5-11a**. What happens if you depress the plunger, reducing the volume of the barrel, as shown in Figure 5-11b? Keep in mind that the number of moles of gas, *n*, within the barrel remains constant (assuming no leaks) and the temperature, *T*, does not change. The farther you depress the plunger, the more resistance you feel because the pressure of the gas *increases* as its volume *decreases*.

According to kinetic molecular theory, as the volume of the barrel decreases, there is less wall area with which the gas molecules collide. Therefore, collisions are more frequent, resulting in an increase in pressure, *P*. As *the volume of a gas decreases, the pressure of the gas increases, and as the volume of a gas increases, the pressure decreases if temperature, T, and number of moles of gas, n, are constant*. In other words, pressure and volume are *inversely proportional* to each other: as one goes up, the other goes down. The pressure-volume relationship of gases is known as **Boyle's law** and can be written mathematically as

$$V \propto \frac{1}{P}$$

where the symbol $\propto$ means "proportional to."

To see an application of Boyle's law, consider the average pair of human lungs, which can expand to hold about 6.0 L of air, although only a small

> *Pressure* is the most common unit for representing the amount of a gas, much like *mass* is the most common unit used to represent the amount of a solid.

> In science, a "law" is a well-accepted hypothesis that has been thoroughly tested and confirmed over time and has reliable predictive value.

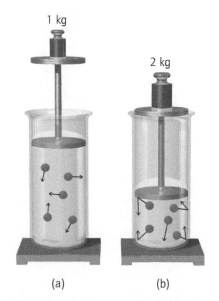

Figure 5-11 Boyle's law: (a) the volume of a gas at a particular pressure; (b) as pressure increases, volume decreases if *T* and *n* are constant: $V \propto 1/P$.

fraction of this capacity is used during normal breathing. Suppose that at a pressure of 1 atm, the air in a pilot's lungs occupies 1 L. If a pilot flying a plane climbs rapidly to an altitude where the pressure is 0.5 atm while holding his breath, his lungs would double in volume to 2 L.

It follows from Boyle's law that when a gas with an initial pressure, P_i, and an initial volume, V_i, undergoes a change to a final pressure, P_f, and final volume, V_f, these variables are related by the following equation:

$$P_i V_i = P_f V_f \,(n \text{ and } T \text{ are constant})$$

> Consult Appendix A for a review of algebra.

We can use this equation to determine any one of the pressure/volume variables if the other three are known, as demonstrated in the second worked exercise below.

The pressure-volume relationship for a gas is demonstrated every time you inhale and exhale (Figure 5-12). The lungs are an elastic organ within an airtight chamber. The diaphragm, also an elastic structure, is located at the base of this chamber. When you inhale, your diaphragm moves downward, causing your lungs to expand (V increases). According to Boyle's law, as the volume of the lungs increases, the pressure within the lungs decreases (inverse relationship). Consequently, the decreased pressure in the lungs allows the surrounding air, which is at atmospheric pressure, to enter the lungs.

When you exhale, your diaphragm moves upward, causing the volume of your lungs to decrease. According to Boyle's law, as the volume of the lungs decreases, the pressure within the lungs increases. The increased pressure in the lungs causes air and carbon dioxide in the lungs to be expelled into the surroundings, where the pressure is lower (1 atm).

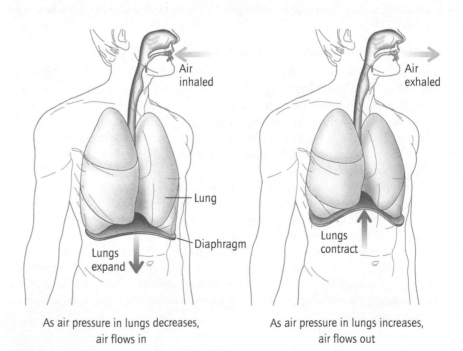

As air pressure in lungs decreases, air flows in

As air pressure in lungs increases, air flows out

Figure 5-12 When you inhale, your diaphragm moves down, allowing your lungs to expand (greater volume) and causing the pressure inside your lungs to decrease. The pressure in the lungs is below atmospheric pressure, so air enters the lungs. When you exhale, your diaphragm moves up, causing your lungs to compress (smaller volume) and the pressure inside your lungs to increase. The pressure in the lungs is above atmospheric pressure, forcing the air out of the lungs.

WORKED EXERCISES | The Pressure-Volume Relationship: Boyle's Law

5-3 An air bubble forms at the bottom of a lake, where the total pressure is 2.52 atm. At this pressure, the bubble has a volume of 3.31 mL. When the bubble rises to the surface, where the pressure is 1.00 atm, will the bubble have a larger volume or a smaller volume? Explain. Assume that the temperature and the moles of gas (n) within the air bubble are constant. No calculation necessary.

5-4 A nitrogen bubble forms in the left knee joint of a deep-sea diver as she ascends. At a depth of 52 ft the pressure is 3.15 atm and the bubble has a volume of 0.015 mL. Assuming constant temperature and constant number of moles of nitrogen in the bubble, what will be the volume of this bubble at the surface of the sea, where the pressure is 1.00 atm?

Solutions

5-3 The volume of the bubble will be greater at the surface because the pressure decreases as the bubble moves from the bottom of the lake to the surface of the lake (the pressure decreases from 2.52 atm to 1.00 atm). Boyle's law states that volume and pressure are inversely related; therefore the volume of the bubble will increase as its pressure decreases.

5-4 Both P and V are changing, whereas n and T are constant, so we use the initial-final pressure-volume equation:

$$P_i V_i = P_f V_f \, (n \text{ and } T \text{ are constant})$$

Use the following steps to solve this problem:

Step 1: Define the supplied variables and identify the requested variable. The problem indicates that $P_i = 3.15$ atm, $V_i = 0.015$ mL, and $P_f = 1.00$ atm. You are asked to find the final volume, V_f.

Step 2: Algebraically isolate the requested variable on one side of the equation. Use algebra to manipulate the equation above to isolate V_f:

$$V_f = \frac{P_i V_i}{P_f} \, (n \text{ and } T \text{ are constant})$$

This is done algebraically by dividing both sides by P_f.

Step 3: Substitute the values from step 1 into the equation in step 2 and solve for the requested variable:

$$V_f = \frac{3.15 \text{ atm} \times .015 \text{ mL}}{1.00 \text{ atm}} = 0.047 \text{ mL}$$

PRACTICE EXERCISES

21 Suppose you were 60 ft beneath the surface of the ocean scuba diving and took a deep breath of air from your tank and held your breath. Why is it not a good idea to start ascending without exhaling?

22 What happens to the volume of your lungs when you inhale? How does it affect the pressure in the lungs? Why does air enter the lungs as a result? What is different when you inhale at high altitude?

23 The pressure gauge on a patient's full 10.7-L oxygen tank reads 8.5 atm. At constant temperature, how many liters of oxygen can the patient's tank deliver at a pressure of 0.92 atm?

24 The average pair of human lungs can expand to hold about 6.0 L of air, although only a small fraction of this capacity is used during normal breathing. Suppose that at a pressure of 1 atm, the air in a pilot's lungs occupies 0.55 L. If a pilot flying an airplane ascends rapidly to a pressure of 0.45 atm while holding his breath, could his lungs rupture?

Volume-Temperature Relationship of Gases: Charles's Law

Imagine a gas contained in a barrel with a plunger as shown in Figure 5-13a. If the plunger is free to move, what happens to the plunger if the gas is heated? Assume the number of moles of gas is unchanged and the pressure is maintained at 1 atm. The volume of the gas in the barrel will expand as the temperature increases, causing the plunger to move up as shown in Figure 5-13b.

According to kinetic molecular theory, as the temperature of a gas increases, the kinetic energy of the particles of gas increase, causing the volume occupied by the gas to increase if pressure is constant and the number of gas molecules is constant (no gas escapes). As this example shows, the volume and temperature of a gas are *directly proportional*: as one goes up, so does the other (Figure 5-13). The volume-temperature relationship of a gas is known as **Charles's law**, written mathematically as

$$V \propto T$$

An application of Charles's law can be demonstrated by heating a balloon and observing its volume increase or cooling the balloon and observing its volume decrease. In general, when a gas with an initial temperature, T_i, and an initial volume, V_i, undergoes a change to a final temperature, T_f, and final volume, V_f, these variables are related by the following equation, which follows from Charles's law:

$$\frac{V_i}{T_i} = \frac{V_f}{T_f} \text{ (} n \text{ and } P \text{ are constant; } T \text{ must be in kelvins)}$$

This equation can be used to determine any one of the temperature or volume variables if the other three are known, provided the temperature is in kelvins and the moles of gas and the pressure of the gas are constant. An example is demonstrated in the second worked exercise below. Recall that the kelvin scale is an absolute scale where K = °C + 273.15.

2 kg

2 kg

(a) T = 200 K   (b) T = 400 K

Figure 5-13 Charles's law: (a) the volume of a gas at a particular temperature; (b) as temperature increases, volume increases if P and n are constant: V ∝ T.

| WORKED EXERCISES | The Volume-Temperature Relationship of Gases: Charles's Law |

5-5 Explain what happens to the volume of a balloon blown up at room temperature if it is placed in the freezer? Assume no gas escapes and the pressure is atmospheric pressure at all times.

5-6 A gas is warmed until its final volume is 14.2 L. Originally the gas occupied a volume of 9.60 L at 68 °C. What is its final temperature, in °C, assuming P and n remain constant?

Solutions

5-5 The volume of the balloon will decrease in the freezer in proportion to the decrease in temperature because according to Charles's law • volume and temperature are directly proportional.

5-6 Since P and n are constant and V and T are the variables, we can use Charles's law to predict the final temperature, T_f:

$$\frac{V_i}{T_i} = \frac{V_f}{T_f} \ (n \text{ and } P \text{ are constant; } T \text{ is in kelvins})$$

Use the following steps to solve this problem:

Step 1: Determine what the supplied variables are and what the requested variable is. The problem indicates that $V_i = 9.60$ L, $T_i = 68$ °C, and $V_f = 14.2$ L and asks that you solve for T_f. To use Charles's law, the initial temperature must first be converted into kelvins, even though your final answer is requested in °C:

$$T_i = 68° + 273.15 = 341 \text{ K}$$

Step 2: Algebraically isolate the requested variable on one side of the equation. Use algebra to manipulate the equation so that the requested variable, T_f, is isolated on one side of the equality. Since the variable you need to solve for is in the *denominator*, cross multiply so that T_f moves to the numerator. Then divide both sides by V_i to isolate T_f:

$$T_f = T_i \times \frac{V_f}{V_i}$$

Step 3: Substitute the values given in step 1 into the equation from step 2 and solve for the requested variable. Substitute the values for T_i, V_i, *and* V_f and solve for T_f:

$$T_f = 341 \text{ K} \times \frac{14.2 \ \cancel{L}}{9.60 \ \cancel{L}} = 504 \text{ K}$$

Then convert the temperature from kelvin to °C because that is the requested unit of temperature:

$$°C = 504 - 273.15 = 231 \text{ °C}$$

PRACTICE EXERCISES

25 Use kinetic molecular theory to explain why the volume of a gas increases when its temperature increases if P and n are constant.

26 If you drew up 2.0 mL of air in a syringe with a freely movable plunger on a day when the clinic was at room temperature, 25 °C, and then placed the syringe in a freezer at −20. °C, would the plunger move up or down? Would the final volume in the syringe be greater than, less than, or equal to the initial 2.0 mL of air in the syringe? Assume the plunger does not leak air and is free to move within the barrel of the syringe. No calculation necessary.

27 Would the density of a gas change if you heated it? Would the density increase or decrease? Assume pressure and the number of moles is constant. Remember, $d = m/v$. No calculation necessary.

28 Neon, a noble gas, is cooled from 76 °C to 38 °C. If the original volume of neon gas was 19.5 L, what is the final volume in liters, assuming P and n are constant?

29 For exercise 26 above, calculate the final volume of the syringe after it has been in the freezer.

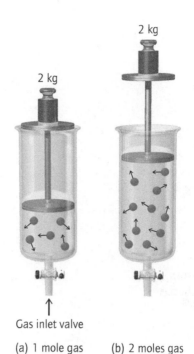

2 kg

2 kg

Gas inlet valve

(a) 1 mole gas   (b) 2 moles gas

Figure 5-14 Avogadro's law:
(a) the volume of 1 mole of a gas
(b) after adding an additional 1 mole
of gas, the volume increases if P and T
are constant $V \propto n$.

Volume-Mole Relationship of Gases: Avogadro's Law

Most of us have filled a balloon by blowing air into it, demonstrating that there is a direct relationship between the number of moles of a gas and its volume if the pressure and temperature are constant. As we add air to a balloon, we are increasing the number of moles of nitrogen and oxygen (air) in the balloon, so the volume increases. Imagine a gas contained in a barrel with a plunger, as shown in Figure 5-14a. If additional gas is added to the plunger through a gas inlet valve, what happens to the plunger if it is free to move? Assume the pressure remains at 1atm and the temperature is constant. The volume of the gas in the barrel will expand as additional gas molecules are added, causing the plunger to move up, as illustrated in Figure 5-14b. The *direct proportionality* between moles of a gas and the volume of the gas is known as **Avogadro's law**. Avogadro's law can be written mathematically as

$$V \propto n \ (P \text{ and } T \text{ are constant})$$

It follows from Avogadro's law that when a gas with an initial number of moles, n_i, at an initial volume, V_i, undergoes a change to a final number of moles, n_f, and final volume, V_f, these variables are related by the equation

$$\frac{V_i}{n_i} = \frac{V_f}{n_f}$$

WORKED EXERCISES   Volume-Mole Relationship: Avogadro's Law

5-7   If you add helium to a balloon, how will the volume of the balloon change if the temperature and pressure remain constant?

5-8   A woman with emphysema has lungs with a volume of 5.0 L that fill with 0.20 mole of air when she inhales. When she exhales, her lung volume decreases to 2.0 L. How many moles of gas remain in her lungs after she exhales? Assume constant temperature and pressure.

Solutions

5-7   The volume of the balloon will increase, inflating the balloon, because volume is directly proportional to the number of moles of a gas and the added helium is an increase in the number of moles of helium.

5-8   Since P and T are constant and V and n are the variables changing, we can use Avogadro's law to calculate the number of moles of air in the woman's lungs after exhaling:

$$\frac{V_i}{n_i} = \frac{V_f}{n_f}$$

Step 1: Define the supplied variables and determine the requested variable. The problem indicates that $V_i = 5.0$ L, $n_i = 0.20$, and $V_f = 2.0$ L. The requested variable you are asked to solve for is n_f.

Step 2: Algebraically isolate the requested variable on one side of the equation. Use algebra to manipulate the equation above so that the requested variable, n_f, is isolated on one side of the equation. Since the variable you need to solve for is in the denominator, you must first cross multiply so that the requested unit, n_f, moves to the numerator. Then divide both sides by V_i to isolate n_f:

$$n_f = n_i \times \frac{V_f}{V_i}$$

Step 3: Substitute the values given in step 1 into the equation in step 2 and solve for the requested variable. Substitute the values for n_i, V_i, and V_f and solve for n_f:

$$n_f = 0.20 \text{ mol} \times \frac{2.0 \text{ L}}{5.0 \text{ L}} = 0.080 \text{ mol}$$

PRACTICE EXERCISES

30 Use kinetic molecular theory to explain why the volume of a gas will decrease when some of the gas, n, is removed if P and T are constant.

31 Imagine you are riding your bike and you run over a nail that punctures one of your tires. What happens to the volume of the tire? Explain in terms of Avogadro's law. After you repair the tire with a patch, you fill the tire with a handheld tire pump. What does the tire pump do? What happens to the volume of the tire? Explain in terms of Avogadro's law.

32 A construction worker arrives at the emergency room with a punctured lung. What has happened to the air in his lung? What variables: P, V, T, or n have changed as a result of this injury?

33 If 1.5 moles of neon occupy a volume of 33.6 L, what is the final volume of neon if an additional 0.5 mole of neon gas is added under constant temperature and pressure?

34 A child with asthma has a lung volume of 3.5 L containing 0.50 mole of air when he inhales. After he exhales, the number of moles of air in his lungs decreases to 0.15 mole. What is the volume of air in his lungs after he exhales, assuming the temperature and pressure in his lungs is constant?

The Ideal Gas Law

We have seen how volume is affected by changing one of the following variables: temperature, pressure, or number of moles. Each of these relationships to volume is one of the gas laws:

$$V \propto \frac{1}{P} \text{ (Boyle's law)}$$

$$V \propto T \text{ (Charles's law)}$$

$$V \propto n \text{ (Avogadro's law)}$$

By combining all three of the above gas laws, we can write a single mathematical relationship, known as the **ideal gas law:**

$$V \propto \frac{nT}{P}$$

This proportionality ($\propto$) can be converted to an equality ($=$) by adding the proportionality constant, R:

$$V = \frac{RnT}{P}$$

where R is known as the **universal gas constant,** a value that is the same for all ideal gases:

$$R = 0.08206 \frac{\text{L} \cdot \text{atm}}{\text{mol} \cdot \text{K}}$$

Figure 5-15 The ideal gas law is a single mathematical relationship that combines Boyle's, Charles's, and Avogadro's laws: $PV = nRT$. The combined constants are the universal gas constant, R.

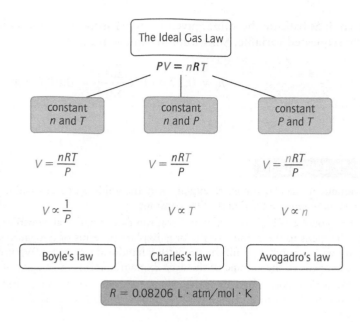

In an ideal gas the particles do not interact at all and the ideal gas law applies. Most real gases at 0 °C and 1 atm behave like an ideal gas.

If we rearrange this equation algebraically by multiplying both sides by P, we get the common form of the ideal gas law:

$$PV = nRT$$

The ideal gas law is a combination of all three gas laws: Boyle's, Charles's, and Avogadro's laws, as summarized in **Figure 5-15**. The ideal gas law can be used to determine one variable (P, V, T, or n) when the other three are known. When using the universal gas constant, R, units for the variables must match those in the universal gas constant: volume in liters, pressure in atm, and temperature in Kelvin.

WORKED EXERCISE   The Ideal Gas Law

5-9   A caisson is a pressurized, watertight structure used in the construction of dams and bridges. If a worker in a caisson inhales 125 mL of pressurized air at 1.8 atm and 27 °C, how many moles of air is he inhaling?

Solution

5-9   Since P, T, and V are given and R is a known constant, we can use the ideal gas law to solve for n:

$$PV = nRT$$

Use the following steps to solve this problem:

Step 1: Define the variables and identify the requested variable. The problem indicates that $P = 1.8$ atm, $V = 125$ mL, and $T = 27$ °C and requests the number of moles, n. You must also convert units into L, K, and atm if other units are given. Here you must first convert 27 °C into kelvin and mL into L.

$$K = 27\ °C + 273.15 = 300.\ K$$

$$125\ \cancel{mL} \times \frac{1\ L}{10^3\ \cancel{mL}} = 0.125\ L$$

Step 2: Isolate the requested unit on one side of the equation by algebraically rearranging the ideal gas law. You need to solve for n, so divide both sides of the ideal gas equation by R and T:

$$n = \frac{PV}{RT}$$

Step 3: Substitute the values for P, V, and T given in Step 1, into the rearranged ideal gas law equation shown in step 2. Then solve for the unknown, n.

$$n = \frac{(1.8 \, \cancel{atm})(0.125 \, \cancel{L})}{\left(0.08206 \, \frac{\cancel{L} \cdot \cancel{atm}}{mol \cdot \cancel{K}}\right)(300. \, \cancel{K})} = 9.1 \times 10^{-3} \, mol$$

In the calculation above, all units cancel except moles. Since moles is in the denominator of the denominator, this is the same as moles in the numerator.

PRACTICE EXERCISES

35 Calculate the volume occupied by 1.5 moles of helium gas at a pressure of 1.0 atm and 85 °C.

36 Your patient is suffering from the bends and needs to undergo hyperbaric oxygen treatment, which requires that she spend some time in a hyperbaric chamber with pure oxygen at 2.0 atm. If the temperature in the hyperbaric chamber is ambient temperature (25 °C) and the hyperbaric chamber has a total volume of 250 L, how many moles of oxygen do you need to add to the chamber initially? Assume the patient occupies 100. liters of the chamber volume. Ignore the amount of oxygen the patient will be consuming through breathing.

37 A piece of dry ice with a mass of 50.0 g is placed inside a balloon and sealed. After all the carbon dioxide has sublimed, what will the volume of the balloon be at 22 °C and 730 mmHg? *Hint:* You will first have to convert the mass of CO_2 into moles of CO_2 using the molar mass of CO_2 as a conversion factor.

38 Gay-Lussac's law is another gas law and shows that when pressure, P, increases, temperature, T, increases if volume, V, and the number of moles, n, is constant.
    a. Write Gay-Lussac's law as a mathematical proportionality.
    b. Show how this relationship is also supported by the ideal gas law.
    c. Using Gay-Lussac's law, explain why it is dangerous to heat an aerosol can.

Molar Volume of a Gas at STP From the ideal gas law we can calculate the **molar volume of a gas** at STP, the volume occupied by *one mole* of a gas under a standard set of conditions. A standard set of reference conditions is used to make comparisons between different gases. *Standard temperature and pressure*, known as **STP**, is 0 °C (273.15 K) and 1 atm. *The molar volume of any gas at STP is 22.4 L.* This is approximately the volume of a cube into which a basketball would fit, as shown in **Figure 5-16**. (Note that the moles of gas inside the basketball are actually greater than one mole because the basketball is pressurized so that it bounces.) *The molar volume of a gas does not depend on the identity or molar mass of the gas.* For example, one mole of helium has the same volume as one mole of oxygen, which is 22.4 L at 0 °C and 1 atm. The molar volume of a gas is a conversion between the number of moles of a gas (n) and the volume (V) of a gas at STP, as shown in the worked exercise below.

Figure 5-16 The molar volume of a gas is 22.4 L at STP (0° C and 1 atm), which is approximately the *volume* of a cube into which a basketball would fit. [© 2001 Richard Megna/Fundamental Photographs, NYC]

Molar volume of a gas: 22.4 L = 1 mol (at STP)

WORKED EXERCISE Calculations with Molar Volume at STP

5-10 What is the volume occupied by 3.2 moles of nitrogen at STP? Would the volume be different if the gas were argon instead?

Solution

5-10 This exercise can be solved using dimensional analysis.

Step 1: Write the two conversion factors for molar volume. Since 22.4 L = 1 mol at STP:

$$\frac{22.4 \text{ L}}{1 \text{ mol}} \qquad \frac{1 \text{ mol}}{22.4 \text{ L}}$$

Step 2: Multiply the supplied unit by the appropriate conversion factor that will allow the supplied unit (mol) to cancel, leaving only the requested unit (L).

$$3.2 \text{ mol} \times \frac{22.4 \text{ L}}{1 \text{ mol}} = 72 \text{ L at STP}$$

Thus, 3.2 moles of nitrogen at STP occupy a volume of 72 L. If the gas were argon, it would also occupy 72 L because all gases have the same molar volume at STP.

PRACTICE EXERCISES

39 How many moles of xenon are there in 34.9 L at STP?

40 How many liters do 4.50 moles of chlorine occupy at STP?

Density of a Gas at STP Since we know that the molar volume of any gas at STP is 22.4 L, we can also determine the density of a gas at STP. Recall that density = mass/volume. By substituting *molar* mass for mass and *molar* volume for volume, we have the following equation for density:

$$\text{Density} = \frac{\text{molar mass}}{\text{molar volume}}$$

From this equation we see that the density of a gas is proportional to its molar mass; in other words, the greater the molar mass, the greater the density of the gas.

For example, the density of helium at STP is

$$d_{\text{He}} = \frac{4.00 \dfrac{\text{g}}{\text{mol}}}{22.4 \dfrac{\text{L}}{\text{mol}}} = 0.179 \text{ g/L}$$

The density of air, a mixture of primarily nitrogen and oxygen, has a molar mass of 28.8 g/mol, so it has a density of 1.29 g/L at STP. A gas with a molar mass less than air will also be less dense than air and so it will float in air. This is why helium and hydrogen balloons float. Helium balloons are more practical, however, since hydrogen is a flammable gas.

PRACTICE EXERCISES

41 Calculate the density of hydrogen gas, H_2, and then explain why H_2 floats in air.

42 In a house fire, carbon dioxide gas is produced during combustion. Will carbon dioxide sink or float in air? The molar mass of carbon dioxide is 44.01 g/mol. What is the density of carbon dioxide at STP? Ignore the fact that the house is not at STP during the fire.

■ 5.3 Gas Mixtures and Partial Pressures

We have learned how moles, pressure, volume, and temperature are interrelated for a *pure* gas—a gas composed of only one element or compound. In this section we will consider the fundamental law that governs gas *mixtures*, a combination of two or more gases. For example, dry air is a mixture of the gases nitrogen and oxygen as well as some argon and carbon dioxide, in the relative percentages shown in the pie chart in **Figure 5-17**.

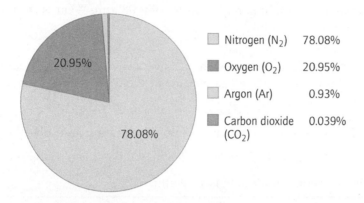

Figure 5-17 Composition of dry air.

■	Nitrogen (N$_2$)	78.08%
■	Oxygen (O$_2$)	20.95%
■	Argon (Ar)	0.93%
■	Carbon dioxide (CO$_2$)	0.039%

Dalton's law of partial pressures states that for a mixture of gases, each gas exerts a pressure independent of the other gases and each gas will behave as if it alone occupied the total volume. The pressure exerted by one gas in a gas mixture is known as the partial pressure of that gas, P_n. The total pressure of a gas mixture, therefore, is equal to the sum of the partial pressures of each gas in the mixture:

$$P_{\text{total}} = P_1 + P_2 + P_3 + \ldots P_n$$

where $P_1, P_2, P_3, \ldots, P_n$ represent the partial pressure of each gas in the mixture.

Consider the scenario depicted in **Figure 5-18**, in which we combine the contents of a cylinder of pure oxygen gas at 1.0 atm with the contents of a cylinder with the same volume of pure nitrogen at 2.0 atm. According to

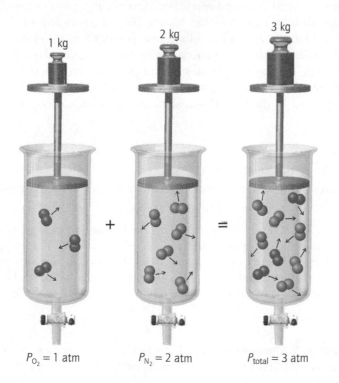

P_{O_2} = 1 atm P_{N_2} = 2 atm P_{total} = 3 atm

Figure 5-18 Dalton's law of partial pressure is illustrated in this figure. When two gases are combined the total pressure is equal to the sum of the partial pressures of each gas.

Dalton's law of partial pressures, the total pressure of the gas mixture will be equal to the sum of the partial pressures of each individual gas:

$$P_{total} = 1.0 \text{ atm} + 2.0 \text{ atm} = 3.0 \text{ atm}$$

WORKED EXERCISE Dalton's law and Partial Pressures of Gases

5-11 The air in an average adult lung contains oxygen at a partial pressure of 0.18 atm, nitrogen at a partial pressure of 0.77 atm, carbon dioxide at a partial pressure of 0.06 atm, and water vapor at a partial pressure of 0.08 atm. What is the total air pressure in an average adult lung?

Solution

5-11 Since the partial pressures are provided, use Dalton's law of partial pressures to calculate the total pressure of the air (a mixture of these gases) in the lungs:

$$P_{air} = P_{oxygen} + P_{nitrogen} + P_{carbon\ dioxide} + P_{water}$$

Substitute the partial pressures for each gas into the equation and solve:

$$P_{air} = 0.18 \text{ atm} + 0.77 \text{ atm} + 0.06 \text{ atm} + 0.08 \text{ atm} = 1.09 \text{ atm}$$

The pressure in an average adult lung is 1.09 atm.

PRACTICE EXERCISES

43 An air tank contains a mixture of nitrogen and oxygen with a total pressure of 3.00 atm. If the partial pressure of oxygen is 0.63 atm, what is the partial pressure of nitrogen inside the tank?

44 Often helium replaces nitrogen in scuba tanks prepared for divers who are diving more than 150 ft below the surface of the sea. A tank is prepared containing only oxygen and helium for a scuba diver who is going to descend 230 ft below the ocean surface. At that depth, the diver breathes a gas mixture that has a total pressure of 8.0 atm. If the partial pressure of oxygen in the tank at that depth is 1.6 atm, what is the partial pressure of helium at that depth?

Henry's Law

Blood is a mixture containing water and many other ions and molecules. Blood also contains several dissolved gases. For example, oxygen, nitrogen, and carbon dioxide are dissolved gases in the blood. The concentration of some of these gases in the blood is shown in a blood test. *Henry's law states that the number of gas molecules dissolved in a liquid is directly proportional to the partial pressure of the gas.* Thus, as the partial pressure of a gas increases, the number of gas molecules dissolved in solution increases. At constant temperature, Henry's law can be expressed by the following mathematical expression:

$$P = kC$$

where

P = the partial pressure of the gas,

k = Henry's constant, a proportionality constant unique for each gas, and

C = the concentration of the dissolved gas

Henry's law is especially useful in the field of anesthesiology, where gaseous anesthetics are inhaled to put a patient to sleep (Figure 5-19). The various anesthetics have different Henry's constants, k, and therefore at a given pressure they achieve different concentrations of dissolved anesthetic in the blood. The *smaller* the Henry's constant, the *higher* the concentration of dissolved anesthetic in the blood:

$$C = \frac{P}{k}$$

For example, the anesthetic diethyl ether has a relatively small Henry's constant, k; thus relatively high concentrations of the anesthetic are dissolved in the

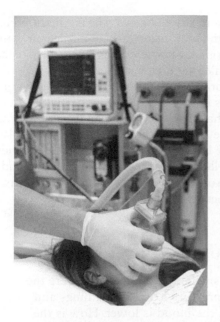 

Figure 5-19 Anesthetics are gases that are used to put a patient to sleep during surgical procedures. [Left: Antonia Reeve/Science Source. Right: Blair Seltz/Science Source.]

blood. Because most of the diethyl ether is dissolved in the blood, it does not reach the brain as quickly as an anesthetic with a larger Henry's constant. For anesthetics with larger Henry's constants, the gas diffuses through tissues directly, allowing it to reach the brain more rapidly. Therefore, a patient anesthetized with diethyl ether takes a longer time to become anesthetized and a longer time to regain consciousness. Desflurane (Suprane), a common anesthetic, has a larger Henry's constant than diethyl ether, so a patient anesthetized with desflurane is anesthetized more rapidly and regains consciousness faster.

WORKED EXERCISE | Henry's Law

5-12 You have learned that at higher altitudes atmospheric pressure is lower. Considering Henry's law, would you expect the concentration of oxygen in your blood to be higher or lower if you were at an altitude of 14,000 ft above sea level?

Solution

5-12 Henry's law states that there is a direct relationship between the pressure and the concentration of a gas in solution:

$$C = \frac{P}{k}$$

Therefore at higher altitude, where atmospheric pressure, P, is lower, you would expect a corresponding decreased concentration, C, of dissolved oxygen in your blood. Indeed, this is why oxygen masks are worn at very high altitudes.

PRACTICE EXERCISES

45 At 30 ft beneath the surface of the ocean, would you expect the concentration of dissolved nitrogen in a person's blood to be higher or lower than the amount of dissolved nitrogen at the surface of the ocean? Explain using Henry's law.

46 Isoflurane (Forane), commonly used for veterinary anesthesia, has a larger Henry's constant (k) than diethyl ether. Which anesthetic would allow an anesthetized animal to regain consciousness more quickly, isoflurane or diethyl ether? Explain.

47 Two patients similar in build and weight were both anesthetized with the same amount of desflurane, but one patient is in a hospital in Boston and the other is in a hospital in Denver. Considering Henry's law, which patient would you expect to become anesthetized in a shorter time? Explain.

48 Where would you expect a freshly poured glass of champagne to have more bubbles, in London or at the top of the Swiss Alps? Explain.

As you know, the air we breathe is a mixture of gases. The pie chart in Figure 5-20 shows the partial pressures of the four major gases in inhaled air compared to exhaled air. As you can see, there is a greater percentage of carbon dioxide and a lower percentage of oxygen in exhaled air as compared to inhaled air.

Notice that the total pressure of both gas mixtures is 760 mmHg (1 atm) at sea level, in accordance with Dalton's law. There is a higher partial pressure of oxygen in inhaled air because we inhale more oxygen than we exhale. Figure 5-20 shows the exchange of these gases in the lungs. Oxygen inhaled through the lungs (red) dissolves in the blood (Henry's law), and some binds to hemoglobin, which transports it to cells that require oxygen. Carbon dioxide gas is a waste product of cells and is also dissolved in the blood (Henry's law) and transported to the lungs (blue), where it is exhaled.

WORKED EXERCISE Applying Gas Laws to the Human Body

5-13 When an asthmatic has an asthma attack, a smaller amount of air enters the lungs, and therefore a lower oxygen partial pressure exists in the lungs and consequently the concentration of oxygen in the blood is lower. How is the concentration of oxygen in the blood affected during an asthma attack? Explain using Henry's law.

Solution

5-13 When an asthmatic has an asthma attack, the partial pressure of oxygen in the lungs is less because of the decreased amount of air inhaled, and therefore, according to Henry's law, the concentration of dissolved oxygen in the blood is less.

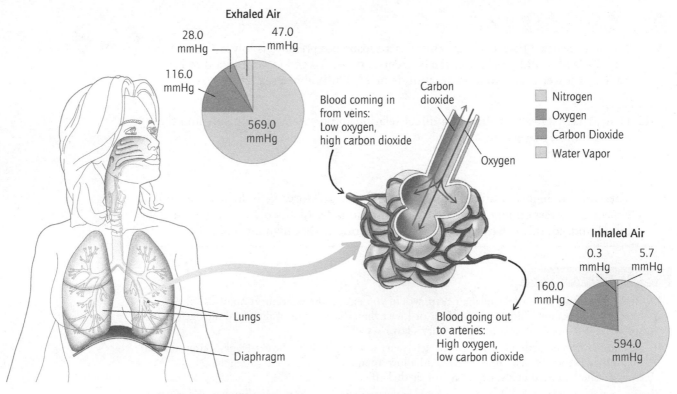

Figure 5-20 Gas exchange in the lungs. The pie charts show that inhaled air contains higher levels of oxygen (red) and lower levels of carbon dioxide (blue) than exhaled air. The oxygen from inhaled air is delivered to oxygen-depleted tissues via the arteries. Carbon dioxide waste produced in cells is delivered to the lungs to be exhaled via the veins.

PRACTICE EXERCISE

49 In the lungs, the normal oxygen partial pressure ranges between 0.1 atm and 1.4 atm.

    a. Which would you expect a person climbing Mount Everest to be more concerned with: the partial pressure of oxygen in her lungs rising above 1.4 atm or falling below 0.1 atm? Explain.

    b. Consider a scuba diver at 150 ft below the surface of the sea breathing compressed air. Which would you expect the diver to be more concerned with: the partial pressure of oxygen in his lungs rising above 1.4 atm or falling below 0.1 atm? Explain.

Chemistry in Medicine Hyperbaric Oxygen Therapy (HBOT)

Hyperbaric oxygen therapy (HBOT) uses high-pressure oxygen to treat medical conditions, such as the bends, described in the opening vignette. Other conditions in which HBOT is used include carbon monoxide poisoning, healing of diabetic wounds, and infection with necrotizing fasciitis (flesh-eating bacteria). These are all conditions in which tissues are deprived of oxygen, known as *hypoxia*. Here we discuss the use of HBOT in the treatment of the bends, carbon monoxide poisoning, and healing of diabetic wounds.

The hyperbaric chamber is a clear plastic tube about 7 ft long, like the one shown in **Figure 5-21**. A patient is placed on a padded table that slides into the tube. The chamber is then gradually filled with pure oxygen to a pressure of 1.5 to 3 atm.

Treating the Bends

HBOT has been used to treat the bends in divers since the 1940s. While in the hyperbaric chamber, the patient breathes oxygen at a higher pressure than atmospheric pressure which causes the nitrogen bubbles in the diver's blood to dissolve, as expected from Henry's law (greater pressure means more gas dissolves). The chamber is then depressurized slowly to allow nitrogen to emerge as bubbles small enough to circulate easily to the lungs, where it is exhaled.

In a case of the bends, oxygen gas does not present the same problem as nitrogen gas because nitrogen is an inert gas, so it does not react chemically with other substances in the blood, as oxygen does. Although some inhaled oxygen dissolves in the blood, most of the oxygen molecules bind to the iron atoms in hemoglobin (Hb). One hemoglobin protein molecule can bind up to four oxygen molecules to form oxyhemoglobin, HbO_2. Because an oxygen molecule must be released from an oxyhemoglobin molecule before it can diffuse out of the blood, oxygen gas is released from the blood more gradually than nitrogen gas. As a diver ascends to the surface or as a hyperbaric chamber returns to atmospheric pressure, the excess oxygen that has dissolved in the blood comes out of solution more slowly than nitrogen. This enables the diver to safely exhale the excess oxygen.

Treating Carbon Monoxide Poisoning

It has been estimated that more than 40,000 people each year in the United States seek medical attention for poisoning from carbon monoxide gas, a colorless, odorless gas. Common sources of carbon monoxide poisoning include wood-burning stoves, motor vehicle exhaust, and home furnaces that are not functioning properly.

Carbon monoxide (CO) also binds to hemoglobin, producing carboxyhemoglobin (COHb). The electrostatic attraction of hemoglobin for carbon monoxide is 240 times stronger than for oxygen. Therefore, when carbon monoxide binds to hemoglobin, it displaces the oxygen in hemoglobin, so the level of oxygen available to tissues drops, creating hypoxia. Because oxygen uptake by the brain is also reduced, the victim may become unconscious, sustain brain damage, or die.

Figure 5-21 A hyperbaric oxygen therapy (HBOT) chamber. [© Jason Cohn/Reuters/Corbis]

Our cells naturally produces some carbon monoxide, but normal COHb levels in the body are less than 5 percent of total hemoglobin. Carbon monoxide is toxic at levels of COHb above 25 percent, and levels above 70 percent are often fatal. Carbon monoxide poisoning is treated either by administering pure oxygen at atmospheric pressure or by applying HBOT. With a 10 percent COHb level, HOBT will reduce the time needed to drive out half the carbon monoxide in the blood from five hours to 30 minutes. Once carbon monoxide is eliminated, oxygen can once again bind to hemoglobin, becoming available to tissues, most importantly the brain.

Treating Diabetic Wounds

Of the 11 million Americans with diabetes, an estimated 25 percent will develop diabetic wounds on their feet. Diabetes is a disease associated with abnormally high blood sugar levels, resulting in a number of complications, including neuropathy, a form of nerve damage that causes lack of sensation in the limbs. A diabetic patient with neuropathy is less likely to feel a wound in his or her extremities, particularly the feet, and therefore is less likely to seek treatment. Moreover, a diabetic wound is more likely to become infected because bacteria feed on the higher glucose levels found in a diabetic patient's blood. Infections restrict blood supply to the wound and surrounding tissue, so tissues become hypoxic. Hypoxia and lack of treatment cause the wounds to heal more slowly. Oxygen is essential to cell function because it is needed for cellular respiration, which provides energy for the cell. By oxygenating tissues near the wound, HBOT stimulates the formation of capillary blood vessels, enabling blood flow to hypoxic tissues and aiding tissue repair.

HOBT is a medical procedure in which higher oxygen pressure is used to treat acute problems, such as the bends and carbon monoxide poisoning, as well as chronic problems, such as diabetic wounds. Understanding how gases behave within the human body was critical in the development of HOBT.

Summary

Summary of Gas Laws

Boyle's law	$V \propto \dfrac{1}{P}$
Charles's law	$V \propto T$
Avogadro's law	$V \propto n$
Ideal gas law	$PV = nRT$
Dalton's law	$P_{total} = P_1 + P_2 + P_3 \ldots + P_n$
Henry's law	$P = kC$ or $C = \dfrac{P}{k}$

Changes of State

- The physical change of going from one state to another state is called a change of state and involves a transfer of energy to or from the surroundings.
- Covalent bonds are not broken or formed during a change of state, only intermolecular forces of attraction between molecules.
- Changes of state include solid to liquid (melting), solid to gas (sublimation), liquid to gas (vaporization and evaporation), liquid to solid (freezing or fusion), gas to liquid (condensation), and gas to solid (deposition).
- The heat of fusion, ΔH_{fus}, is the amount of energy that must be added to melt a solid at its melting point, which is the same amount of energy that must be removed to freeze a liquid.
- The heat of vaporization, ΔH_{vap}, is the amount of energy that must be added to vaporize a liquid at its boiling point, which is the same amount of energy that must be removed to condense a gas.
- A heating curve is a graph that shows how the temperature of a substance increases with the addition of heat energy, showing two temperature plateaus at the melting point and the boiling point of the substance.
- The melting point of a substance is the temperature at which it undergoes a change of state between the liquid and solid phases.

- The boiling point of a substance is the temperature at which it undergoes a change of state between the liquid and gas phases.
- Evaporation of water cools the surroundings because heat from the surroundings supplies the necessary energy for the phase change from the liquid to the gas.
- A steam burn causes more damage to the skin than a burn from scalding water because of the added heat transferred to the skin as a result of the phase change (g → l).
- Vapor pressure is the pressure exerted at a given temperature by molecules in the gas phase that are in equilibrium with molecules in the liquid phase. The vapor pressure of a liquid increases with temperature.
- When the vapor pressure of a liquid equals atmospheric pressure, boiling occurs.
- A liquid with a higher vapor pressure enters the gas phase more easily and is said to be more volatile.
- Pressure is the amount of force applied over a given area: P = force/area.
- Pressure can be measured in a variety of units: atm, torr, millimeters of mercury (mmHg), and pounds per square inch.
- Molecules of air exert a pressure known as atmospheric pressure, which can be measured by a barometer.
- Atmospheric pressure decreases as you go to higher altitudes because there is less air. The atmospheric pressure at sea level is 1 atm.

The Gas Laws

- Gases are unique because they are compressible, which is a result of the negligible volume that gas particles occupy compared to the volume of their container.
- According to the kinetic molecular view, the particles of a gas are in constant, random motion; the attractive forces among the particles of a gas are negligible; the volume of a gas is mostly empty space; and the temperature of a gas depends on the kinetic energy of the gas particles.
- Four interrelated variables describe the physical properties of a gas: pressure (P), volume (V), temperature (T), and the amount of gas (n).
- Boyle's law states that the pressure and the volume of a gas are inversely related when T and n are constant:

$$V \propto \frac{1}{P}$$

- This relationship can be used to calculate initial and final volume and pressure conditions of a gas: $P_1V_1 = P_2V_2$ (n and T are constant).
- Breathing is an application of Boyle's law: as the diaphragm moves up, the volume of the lungs decreases, so the pressure increases, and we exhale. The reverse occurs when we inhale.
- Charles's law states that the volume and the temperature of a gas are directly proportional to each other when P and n are constant:

$$V \propto T$$

- This relationship can be used to calculate initial and final volume and temperature conditions of a gas:

$$\frac{V_i}{T_i} = \frac{V_f}{T_f} \ (n \text{ and } P \text{ are constant; } T \text{ must be in kelvins})$$

- Avogadro's law states that the number of moles and the volume of a gas are directly proportional when T and P are constant:

$$V \propto n$$

- This relationship can be used to calculate initial and final volume and moles for a gas:

$$\frac{V_i}{n_i} = \frac{V_f}{n_f} \ (P \text{ and } T \text{ are constant})$$

- The ideal gas law combines Boyle's, Charles's, and Avogadro's laws into a single mathematical relationship: $PV = nRT$.
- The universal gas constant, R, is equal to 0.08206 L · atm/mol · K.
- Standard temperature and pressure, STP, is 0 °C and 1 atm.
- One mole of any gas occupies a volume of 22.4 L at STP.
- The density of a gas at STP is the molar mass of the gas divided by 22.4 L; hence the density of a gas is proportional to its molar mass.

Gas Mixtures and Partial Pressures

- Dalton's law states that the sum of the partial pressures of each gas in a mixture of gases is equal to the total pressure:

$$P_{\text{total}} = P_1 + P_2 + P_3 + \ldots P_n$$

- Henry's law states that the concentration of a gas dissolved in solution is directly proportional to the partial pressure of the gas: $C = kP$
- Anesthetics with a larger Henry's constant allow the patient to be anesthetized and regain consciousness faster because less of the anesthetic gas is dissolved in the blood.

Key Words

Atmospheric pressure The pressure exerted by the atmosphere (air) at any given place on the earth as a result of gravity.

Avogadro's law The direct relationship between the number of moles and the volume of a gas if the temperature and the pressure are constant.

Barometer A device used to measure the atmospheric pressure.

Barometric pressure See *atmospheric pressure*.

Boyle's law The inverse relationship between the pressure and the volume of a gas if the temperature and number of moles are constant and temperature is given in kelvins.

Change of state The process of going from one state of matter (s, l, g) to another state of matter.

Charles's law The direct relationship between the temperature and the volume of a gas if the pressure and number of moles are constant and temperature is given in kelvins.

Condensation A change of state from the gas phase to the liquid phase. Requires removal of heat energy.

Dalton's law of partial pressures The observation that for a mixture of gases each gas exerts a pressure independent of the other gases and each gas will behave as if it alone occupied the total volume:

$$P_{\text{total}} = P_1 + P_2 + P_3 + \ldots P_n$$

Deposition A change of state from the gas phase directly to the solid state. Requires the removal of heat energy.

Enthalpy of fusion, ΔH_{fus} See heat of fusion.

Enthalpy of vaporization, ΔH_{vap} See heat of vaporization.

Evaporation A change of state from the liquid to the gas phase at any temperature. It occurs when atoms and/or molecules have enough kinetic energy to leave the surface of a liquid and enter the gas phase.

Evaporative cooling The cooling of the surroundings that occurs when a liquid evaporates (liquid to gas), as a result of heat being transferred from the surroundings to the liquid to achieve the change of state.

Freezing A change of state from liquid to solid. Requires the removal of heat energy.

Freezing point See *melting point*.

Heat of fusion, ΔH_{fus} The amount of energy that must be added to melt a solid or removed to freeze a liquid.

Heat of vaporization, ΔH_{vap} The amount of energy that must be added to vaporize a liquid or removed to condense a gas.

Heating curve A graphical way to show how the temperature of a substance changes as energy is added at a constant rate, where a horizontal plateau occurs as energy is used to effect a phase change rather than an increase in kinetic energy.

Henry's law The observation that the number of gas molecules dissolved in a liquid is directly proportional to the partial pressure of the gas: $P = kC$.

Ideal gas law A combination of Boyle's, Charles's, and Avogadro's laws: $PV = nRT$.

Melting A change of state from solid to liquid.

Melting point The temperature at which a substance undergoes a change of state between the liquid and solid phases. Same as the freezing point.

Molar volume of a gas The volume occupied by one mole of any gas at STP, which is 22.4 L.

Normal boiling point The temperature at which a pure liquid changes to the gas phase or a gas changes to the liquid phase at atmospheric pressure.

Partial pressure The pressure exerted by one gas in a mixture of gases.

Pressure A measure of the amount of force applied over a given area: $P = force/area$.

STP (standard temperature and pressure) A standard set of reference conditions for a gas: 0 °C and 1 atm.

Sublimation A change of state directly from solid to gas. Requires the addition of heat energy.

Universal gas constant The proportionality constant, R, in the ideal gas law:

$$R = 0.08206 \frac{L \cdot atm}{mol \cdot K}$$

Vapor pressure The pressure of a gas in equilibrium with the liquid phase at a given temperature.

Vaporization A change of state from liquid to gas that occurs at the boiling point of the liquid.

Volatile A liquid with a high vapor pressure, which therefore enters the gas phase more readily.

Additional Exercises

Diver Rescue and "the Bends"

50 What does atmospheric pressure measure? At sea level, atmospheric pressure is _____.

51 The mass of _____ and _____ contribute to the pressure you experience when diving beneath the surface of the sea.

52 The pressure when diving beneath the sea is (greater than or less than) atmospheric pressure.

53 What are some of the symptoms of "the bends"? Why do nitrogen bubbles form in capillaries when a person gets the bends?

Changes of State

54 Are covalent bonds or intermolecular forces of attraction affected during a change of state?

55 Identify the following as a change of state or a chemical reaction:
a. rust forming on metal
b. steam coming off a cup of coffee
c. molten lava becoming rock
d. the wick of a candle burning

56 For which changes of state does energy need to be added?

57 For which changes of state does energy need to be removed?

58 Identify the following processes as evaporation or vaporization:
a. liquid water on a hot sidewalk in summer that turns into steam
b. liquid water turning into steam in a pot of boiling water

59 Identify the change of state indicated in the following:
a. melting snow
b. dry ice (solid CO_2) becoming a gas
c. formation of icicles
d. formation of steam in the shower
e. formation of water droplets on the outside of a glass containing an iced drink on a hot summer day

f. fumes coming from the gas tank as you fill your car with gasoline
g. formation of frost on a cold windshield overnight

60 Identify the changes of state in the following processes:
a. Your friend's perfume produces an aroma that fills the room.
b. Making ice cream from milk in an ice cream maker.
c. Wax dripping from a candle.
d. Naphthalene, a solid, used in moth balls, producing fumes in the closet to deter moths.
e. Your breath fogs up on your car windshield on a cold winter day.

61 Mercury is the only metal that is a liquid at room temperature. It is a toxic metal that is readily absorbed by the lungs. What phase change is occurring that makes this metal particularly dangerous?

62 Humans sweat or shiver to regulate their body temperature. How does the body cool itself by sweating? How does the body warm up by shivering?

63 At the melting point and the boiling point of a substance, what happens as energy is added?

64 Figure 5-4 shows a heating curve for water. What portion of the curve represents the formation of liquid water from ice? Why does the temperature remain constant even though heat continues to be added? What is this point called?

65 Why is ethanol, CH_3CH_2OH, a liquid at room temperature, while carbon dioxide, CO_2, is a gas at room temperature?

66 Why are heats of fusion, ΔH_{fus}, generally less than heats of vaporization, ΔH_{vap}?

67 Why is steam able to cause a skin burn?

68 Which liquid is volatile—one with a low vapor pressure or one with a high vapor pressure?

69 Define the term *pressure*.

70 Using Table 5-2, predict which substance would have a higher boiling point, acetone or methylene chloride.

71 Using Table 5-2, predict which substance would have a lower boiling point, mercury or phenol.

72 Gasoline is flammable. If the vapor pressure of gasoline is 6.9 psi, what is this pressure in mmHg? Why is it necessary to avoid electrical sparks when filling your car with gasoline, even if the gasoline goes directly from the pump into your gas tank?

73 Do you expect the boiling point of water at a base camp on Mount Everest to be higher or lower than the boiling point of water on the beach in Honolulu? Explain.

74 What is the difference between the unit *atmosphere* and *atmospheric pressure*?

75 Why is the atmospheric pressure at 14,000 ft above sea level 0.69 atm, whereas the pressure at 14,000 ft below sea level is 470 atm?

76 Arterial blood gas measurements are used to determine the amount of oxygen in the blood of critically ill patients. The normal range for the arterial partial pressure of oxygen is between 75 mmHg and 100 mmHg. Convert 75 mmHg to and torr.

77 Arterial blood gas measurements can also determine the amount of carbon dioxide in the blood, an indicator of respiratory function. The normal range of arterial partial pressure for carbon dioxide is between 4.7×10^3 Pa and 6.0×10^3 Pa. Convert 4.7×10^3 Pa to mmHg and atmospheres.

78 If a critically ill patient has an arterial partial pressure of oxygen that is less than 60 mmHg, the patient will need supplemental oxygen. If a patient has an arterial partial pressure of oxygen of 5.42×10^3 Pa, does this patient require supplemental oxygen?

79 A condition known as respiratory alkalosis occurs when a critically ill patient is given too many breaths by a mechanical ventilator. If a patient has respiratory alkalosis, the arterial partial pressure of carbon dioxide is less than 35 mmHg. If a critically ill patient has an arterial partial pressure of carbon dioxide of 5.69×10^3 Pa, does the number of breaths given by the mechanical ventilator need to be adjusted?

The Gas Laws

80 Use kinetic-molecular theory to explain the following:
a. Molecules in the gas phase cannot form hydrogen bonds.
b. On a cold day, the tire pressure in your car decreases.
c. The volume of a container filled with a gas contains mostly empty space.

81 Imagine you are hiking in the Rocky Mountains near the Continental Divide at 12,000-ft elevation. You drink all the water in your 1-L plastic water bottle, put the cap back on the bottle, and carry the bottle back to Denver (5,280 ft in elevation). When you arrive back in Denver, the water bottle has collapsed. Explain why the empty water bottle has collapsed.

82 When you buy bags of potato chips in Denver (5,280-ft elevation), sometimes the bags look like filled balloons; the package has expanded as if it were ready to burst. These bags are packaged at sea level. Explain why

the bag expands in Denver but not in Los Angeles. (Assume that the temperature is the same in the two cities.)

83 A child holds a helium balloon with a volume of 1.1 L at a pressure of 0.91 atm. A little while later the volume of the balloon is 3.1 L. Did the child go up or down in altitude with the balloon?

84 William Trubridge broke the world record in free diving (diving underwater without the use of supplemental oxygen) by diving to a depth of 121 m. Assume that he takes a breath that fills his lungs to 3.6 L at the surface of the water (1 atm). To what volume does that correspond to when he reaches the depth of 121 m underwater (13.2 atm)?

85 On inhalation, the pressure of the air in the lungs _____ as the volume _____. (increases or decreases)

86 On exhalation, the pressure of the air in the lungs _____ as the volume _____. (increases or decreases)

87 Summarize how Boyle's law (the *PV* relationship) describes breathing.

88 Why do the labels on cans of baked beans tell you not to heat the unopened can directly on the stove?

89 When you bake a cake, you add leavening agents that create pockets of carbon dioxide. Why does a cake rise when you put it in the oven to bake?

90 If a balloon filled with air is put into a container of liquid nitrogen ($-196\ °C$), what happens to the size of the balloon? Using Charles's law, explain this result.

91 A girl with cystic fibrosis has lungs with a volume of 2.5 L that fill with 0.10 mole of air when she inhales. When she exhales, her lung volume decreases to 2.0 L. How many moles of gas remain in her lungs after she exhales? Assume constant temperature and pressure.

92 A boy with pneumonia has lungs with a volume of 1.9 L that fill with 0.08 mole of air when he inhales. When he exhales, his lung volume decreases to 1.5 L. How many moles of gas remain in his lungs after he exhales? Assume constant temperature and pressure.

93 A man with asthma has lungs with a volume of 5.5 L that fill with 0.30 mole of air when he inhales. When he exhales, his lung volume decreases to 3.5 L. How many moles of gas remain in his lungs after he exhales? Assume constant temperature and pressure.

94 A healthy male has lungs with a volume of 6.0 L that fill with 0.28 mole of air when he inhales. After he exhales, the number of moles of air in his lungs decreases to 0.20 mole. What is the volume of air in his lungs after he exhales, assuming constant temperature and pressure?

95 Nitrogen gas was heated from 12 °C to 34 °C. The original volume of nitrogen was 8.7 L. Find the new volume in liters assuming *P* and *n* remain the same.

96 Argon gas is cooled from 120. °C to 52 °C. The original volume of argon was 122 L. Find the new volume in liters, assuming *P* and *n* remain the same.

97 A gas occupies a volume of 18 L at a pressure of 4.3 atm and a temperature of 501 K. How many moles of gas are there?

98 A sample of nitrogen gas that contains 0.026 mole is collected at 2.7 atm and 12.0 °C. What is the volume of the gas?

99 A sample of oxygen that contains 7.00×10^{-3} mole has a pressure of 712 mmHg and a volume of 205 mL. What is the temperature of the sample?

100 What is the volume of 1 mole of carbon dioxide at STP?

101 How many moles of gas are present in 11.2 L at STP?

102 What volume does 4.1 mole of helium occupy at STP?

103 If argon occupies 15.3 L at STP, how many moles of argon are present?

104 How many moles of hydrogen occupy 220. L at STP?

105 What volume is occupied by 0.2 mole of neon at STP?

106 Most hot air balloons are flown in the early morning or evening, when outside air temperatures are cooler. Explain why the hot air in a balloon makes it float.

107 A patient undergoes laparoscopic surgery. During laparoscopic surgery, a technique that is minimally invasive, a surgeon uses a small incision in the abdomen to perform the operation. During the procedure, the patient's abdomen is filled with 5.58 L of carbon dioxide at 25 mmHg and 21 °C. How many moles of carbon dioxide have been added to her abdomen?

Gas Mixtures and Partial Pressures

108 State Dalton's law.

109 An air tank contains a mixture of nitrogen and oxygen at a total pressure of 4.30 atm. If the partial pressure of nitrogen is 0.92 atm, what is the partial pressure of oxygen in the tank?

110 A mixture of neon and argon gases has a total pressure of 2.42 atm. The partial pressure of the neon alone is 1.81 atm. What is the partial pressure of the argon?

111 When a patient hyperventilates, he or she is breathing faster than normal and taking in too much oxygen and not eliminating enough carbon dioxide. A treatment for hyperventilation is to have the patient cover his or her nose and mouth with a small paper bag and breathe in and out slowly. This technique increases the amount of carbon dioxide in the bloodstream. Explain what happens to the partial pressure of carbon dioxide inside the bag as the patient breathes into the bag.

112 Consider two people who have the same build and weight. One of them goes scuba diving and the other climbs Mount Hood in Oregon. Which one of them has more dissolved oxygen in the blood?

113 Would you expect a glass of soda to have more bubbles on the beach or in the mountains?

114 Why would a patient anesthetized with desflurane regain consciousness more quickly than one anesthetized with diethyl ether?

115 In normal air, the partial pressure of oxygen is 0.18 atm, the partial pressure of nitrogen is 0.77 atm, and the partial pressure of carbon dioxide is 0.05 atm. Scuba divers often use tanks with an enhanced mixture of only oxygen and nitrogen. If the total pressure in the scuba tank is 3.83 atm and the partial pressure of oxygen is 1.38 atm, what is the partial pressure of nitrogen? Is the percentage of nitrogen in the tank more or less than what is found in normal air?

116 A 50:50 nitrous oxide–oxygen mixture is often used as a pain reliever and a sedative in dentistry. If the partial pressure of oxygen is 0.50 atm, what is the partial pressure of nitrous oxide? What is the total pressure?

Chemistry in Medicine: Hyperbaric Oxygen Therapy (HBOT)

117 When might a patient need hyperbaric oxygen therapy?

118 When a patient is treated for the bends with HBOT, the chamber is pressurized to 1.5–3 atm of pure oxygen. What effect does this increase in pressure have on the nitrogen bubbles in the patient's blood? Using Henry's law, explain why the increase in pressure has this effect.

119 How does the hemoglobin in your blood prevent oxygen gas from causing the bends?

120 Why is carbon monoxide poisonous? What does it do when it enters the bloodstream?

121 What is the advantage of using HBOT over administering oxygen at atmospheric pressure to treat carbon monoxide poisoning?

122 How does HBOT help patients with diabetic wounds?

123 A scuba diver ascends too quickly and develops the bends.
a. A nitrogen bubble forms in the elbow. At a depth of 60 ft, where the pressure is 2.81 atm, the bubble has a volume of 0.021 mL. Assuming a constant temperature and a constant number of moles of nitrogen in the bubble, what volume will the bubble have at the surface of the sea, where the pressure is 1.00 atm?
b. The scuba diver is put into a hyperbaric oxygen chamber where the pressure is 2.25 atm. What is the volume of the nitrogen bubble once the patient is in the hyperbaric chamber?

Answers to Practice Exercises

1   a. deposition
    b. melting
    c. vaporization or evaporation

2   a deposition (gas to solid)

3   Intermolecular forces of attraction are greatest in the solid phase because molecules have less kinetic energy and are therefore closer together. Intermolecular forces of attraction are negligible in the gas phase because molecules are too far apart to interact with one another.

4   Vaporization occurs when a liquid is heated to its boiling point. Condensation occurs when a vapor is cooled to its boiling point.

5   During vaporization intermolecular forces of attraction between molecules in the liquid phase are broken, and molecules move apart from one another. The boiling point of a hydrocarbon is lower than that of water because the intermolecular forces of attraction between hydrocarbon molecules (dispersion forces) are much weaker than the intermolecular forces of attraction between water molecules (hydrogen bonding); therefore less heat must be added to break these forces of attraction and so they enter the gas phase at a much lower temperature.

6   solid → gas

7   gas → liquid (condensation)

8   When steam comes in contact with the skin, the amount of heat transferred to the skin includes both the heat of vaporization (heat removed during the phase change from vapor to liquid) and the heat transferred as the steam is cooled from 100 °C to 37 °C on the body. When boiling water comes in contact with the skin, no heat of vaporization is transferred to the skin, only the amount of heat transferred as liquid water is cooled from 100 °C to the 37 °C of the body.

9   When food is heated in a microwave oven, energy from the electromagnetic radiation in the microwave region of the electromagnetic spectrum is transferred to water molecules in the food, increasing their kinetic energy. This increase in kinetic energy, as with any increase in energy, means the food is at a higher temperature and thus cooked.

10 The heat of vaporization comes from the infant's body to evaporate the water, thereby lowering his body temperature. This is a result of the relatively high heat of vaporization of water and is known as the evaporative cooling effect.

11 Vapor pressure is the pressure exerted by molecules in the liquid-gas equilibrium.

12 A substance with a higher vapor pressure will have a lower boiling point because its molecules will achieve a vapor pressure equal to atmospheric pressure—boiling—at a lower temperature.

13 The boiling point of water is lower at altitudes above sea level because the atmospheric pressure is lower. The vapor pressure of the beverage equals the atmospheric pressure at a lower temperature and therefore boils at a lower temperature.

14 The normal boiling point is the temperature at which a liquid boils when at 1 atm. The boiling point of a liquid is lower at high altitude.

15 Since the skunk odor is so noticeable it is probably volatile. This means it has a low boiling point and therefore a high vapor pressure.

16 In an autoclave, pressure is greater than atmospheric pressure, so water boils at a temperature higher than the normal boiling point, 100 °C. The higher-temperature water kills transmissible agents such as bacteria, viruses, and spores more effectively than a dishwasher, where the pressure is atmospheric pressure and water boils at a lower temperature.

17 A barometer measures atmospheric pressure at a particular location on earth.

18 Atmospheric pressure is lower at higher altitudes because there is less air above you.

19 Denver is one mile above sea level, whereas Florida is at sea level. The atmospheric pressure in Denver is therefore lower than it is in Florida. The lower atmospheric pressure allows the golf ball to have more "lift" and carry farther since there is not as much atmospheric pressure on the ball in Denver as there is in Florida.

20 Special tennis balls are made for communities at higher altitude because regular balls would have too much bounce at the lower atmospheric pressure at higher altitudes.

21 As you ascend, pressure decreases, so the volume of air in your lungs would increase, potentially rupturing them.

22 As you inhale, the diaphragm moves down so the volume of your lungs increases, which decreases the pressure inside the lungs, causing air, which is at a higher pressure (1 atm), to enter the lungs. At high altitudes, atmospheric pressure is lower, so less air enters the lungs with each inhalation. You compensate by taking deeper breaths, which increases the volume of the lungs and decreases the pressure inside the lungs more, so that more air will enter the lungs.

23 99 L

24 No, his lungs will not rupture, because the air in the pilot's lungs will occupy a volume of 1.2 L, which is less than the 6.0 L capacity of his lungs.

25 An increase in temperature means that the kinetic energy of the gas particles is greater. The volume of the gas therefore expands if the pressure is to remain constant.

26 The plunger would move down because the volume of the barrel decreases as the temperature decreases, a direct proportionality. The final volume in the syringe would be less than 2.0 mL.

27 Yes, the density of the gas would change because density is based on volume: $d = m/V$ and volume changes in proportion to temperature. If a gas were heated, its volume increases (Charles's law) but the number of moles, and therefore the mass, does not. An increase in volume means density decreases because volume is in the denominator of the density equation.

28 17 L

29 1.7 mL

30 If gas molecules are removed, fewer molecules collide with the walls of the container, so the volume occupied by the gas will decrease to keep the number of collisions with the walls (P) constant (Avogadro's law).

31 The volume of your bike tire decreases as a result of air molecules leaking out through the puncture hole and because as n decreases, volume decreases, according to Avogadro's law. The pump introduces more air molecules into the tire, which according to Avogadro's law increases the volume of the tire. Once the maximum tire volume has been reached, additional air causes the tire pressure to increase.

32 n has decreased, causing the volume in the construction worker's lung to decrease (collapse).

33 45 L because $n_i = 1.5$ moles and $n_f = 1.5$ moles + 0.5 moles = 2.0 moles.

34 1.1 L

35 44 L

36 12 mol. Volume is 250 L − 100 L = 150 L.

37 29 L. There are 1.14 mol of CO_2. Note that pressure must be converted from mmHg to atm.

38 **a.** $P \propto T$ when n and V are constant.
b. $PV = nRT$, and if V, n, and R are all constants, this equation becomes $P = \text{constant} \times T$ where the constant $= nR/V$.
c. Since the volume of the can and its contents are constant, increasing T will result in an increase in P. If the pressure exceeds the strength of the can's wall, the can will explode.

39 1.56 mol

40 101 L

41 0.0900 g/L. Since this value is less than the density of air, 1.29 g/L, hydrogen will float in air.

42 The density of carbon dioxide at STP is 1.96 g/L, which is more dense than air (1.29 g/L), so it will sink in air if we ignore the fact that the temperature in the room is higher than STP.

43 2.37 atm

44 6.4 atm

45 The concentration of dissolved nitrogen gas in the blood would be greater at 30 ft below sea level because pressure is greater below the ocean surface and Henry's law shows that pressure and concentration are proportional: $P = kC$, so as pressure increases, so does the concentration of dissolved nitrogen.

46 Isoflurane. Since Henry's constant, k, is larger for isoflurane, the concentration, C, of isoflurane in the bloodstream would be smaller because $C = P/k$. A smaller concentration of the anesthetic in the blood means more anesthetic in the tissues and the more rapidly it gets to and from the brain, thus allowing the animal to regain consciousness faster.

47 The patient in Denver. The atmospheric pressure in Denver is less than in Boston. According to Henry's law, at lower pressure, P, there is a lower concentration of anesthetic dissolved in the blood. The less anesthetic there is dissolved in the blood, the better able the anesthetic is to reach the tissues and the faster it gets to and from the brain.

48 In the Swiss Alps, where the atmospheric pressure is lower because the mountains are at a higher altitude. The pressure in the champagne bottle is the same before it is opened. Once the bottle is opened, the lower atmospheric pressure in the Swiss Alps allows more carbon dioxide gas to come out of solution, which we observe as more bubbles.

49 **a.** Falling below 0.1 atm because atmospheric pressure is lower at higher altitudes, so the person would be more concerned with lung pressure being too low than too high.
b. Rising above 1.4 atm because as you dive below sea level, the pressure increases, and so too does the pressure in the lungs. So the person would be more concerned with lung pressure being too high than too low.

A display of foods containing unsaturated fats. [Tracey Kusiewicz/Foodie Photography]

6 Organic Chemistry: Hydrocarbons

Good Fats . . . Bad Fats . . . What Does It All Mean?

"Eat a diet low in fat and cholesterol" used to be the standard recommendation for a heart-healthy diet. However, studies showing a link between the amount of fat in a person's diet and his or her risk of developing heart disease, breast cancer, or colon cancer has changed this recommendation. We now know that the *total* amount of fat in a person's diet—high or low—is not the culprit in heart disease and cancer but the *type* of fat that matters. The landmark Nurses' Health Study, which followed more than 100,000 nurses over several years with surveys and blood tests, showed that women who consumed the most *saturated* and *trans* fats were more likely to develop heart disease and type II diabetes than women who consumed the least amount of these fats—thus their label as "bad fats." In fact, unsaturated fats were actually found to lower the risk of developing heart disease and cancer—thus the label "good fats." The recommendation now is that the key to a healthy diet is not to decrease *total* fat but to substitute *good fats* for *bad fats*.

Saturated fats are found in both animal and vegetable sources. Animal sources high in saturated fats include whole milk, cheese, butter, ice cream, and red meat. The vegetable oils high in saturated fat are coconut, palm, and palm kernel oil. Coconut oil has the highest saturated fat content of any food—even higher than lard (pig fat) or beef (Figure 6-1).

What is the difference in the chemical structure of a good fat and a bad fat? Fats, like most biological compounds, are composed of *organic* molecules—compounds containing *carbon* atoms. Saturated fats contain long carbon chains *without* any carbon-carbon double bonds, as you can see from the structure of a representative fat shown in Figure 6-2, derived from three palmitic acid molecules. The long saturated hydrocarbon chains in saturated fats allow them to pack together closely, creating more intermolecular forces of attraction. As a result, saturated fats have a higher melting point, which gives them a waxy solid consistency at room temperature.

Unsaturated fats are found in most vegetable oils such as canola, olive, and peanut oil—but not in tropical oils such as coconut oil. Nuts, avocados, and cold-water fish, such as tuna and salmon, are high in unsaturated fats. An unsaturated fat contains one or more carbon-carbon double bonds in its long hydrocarbon chains. A fat with one carbon-carbon double bond is known as a *monounsaturated* fat and a fat with more than one carbon-carbon double bond is known as a *polyunsaturated*

Figure 6-1 Saturated fats are found in both animal sources, such as bacon and lard (pig fat), and vegetable sources, such as coconut oil. [Catherine Bausinger]

Figure 6-2 Space-filling model of a simple saturated fat derived from three palmitic acid molecules, shown on top, which contain no carbon-carbon double bonds, allowing for close packing of the fat molecules.

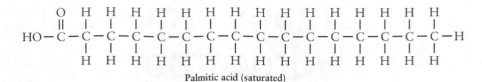

Palmitic acid (saturated)

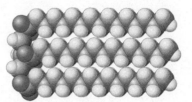

Saturated fat

fat. Unsaturated fats produced in nature have a type of carbon-carbon double bond that creates a bend or kink in the overall shape of the molecule, as seen in the representative unsaturated fat shown in Figure 6-3, derived from three linolenic acid molecules. Therefore, unsaturated fats do not pack as closely as saturated fats. Consequently, there are fewer intermolecular forces of attraction between unsaturated fat molecules, and as a result they have a lower melting point, which gives them an oily *liquid* consistency at room temperature.

Trans fats are also unsaturated fats but they have a different shape around their carbon-carbon double bond(s), which gives them a molecular shape more like a saturated fat than an unsaturated fat. Trans fats are not abundant in nature but they are produced as an undesired by-product of the chemical reaction that converts carbon-carbon double bonds into carbon-carbon single bonds, known as a catalytic hydrogenation reaction. The food industry employs catalytic hydrogenation reactions to prepare "partially hydrogenated vegetable oils," and some margarines, which have a longer shelf life and a consistency that consumers prefer—somewhere between an oily liquid and a waxy solid. Trans fats are even less healthy than saturated fats.

As you can see, an understanding of organic chemistry provides us with the information necessary for understanding current issues in health and nutrition; in this case, the chemistry of "good fats" and "bad fats." ●

Linolenic acid (polyunsaturated)

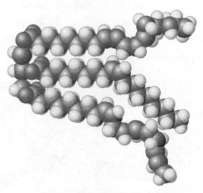

Figure 6-3 Space-filling model of a simple unsaturated fat derived from three linolenic acid molecules, shown on top, which contain several carbon-carbon double bonds (highlighted), preventing close packing of the fat molecules.

Unsaturated fat

Organic chemistry is the branch of chemistry that focuses on carbon-containing compounds and their chemical reactions. Carbon is unique in its ability to form four covalent bonds, especially to other carbon atoms, which leads to tremendous molecular diversity. Organic compounds also contain many C—H bonds. Compounds that contain exclusively carbon and hydrogen atoms are known as **hydrocarbons**. Further molecular diversity is created by the fact that carbon forms covalent bonds to other nonmetal elements, such as oxygen, nitrogen, phosphorous, and sulfur—atoms often referred to as **heteroatoms**. About 95 percent of all known compounds are organic compounds. Most biological compounds are organic, and therefore, to understand the chemistry of biological systems, an understanding of organic chemistry is essential.

Historically, when **organic compounds** were first studied, it was believed that they could *not* be prepared in the laboratory and that only a living plant or animal could produce an organic compound. This is the origin of the term *organic*, which means "from living things." Compounds that do not contain carbon were known as **inorganic compounds**. Today, almost any organic compound can be prepared in the laboratory, even extremely complex organic molecules. Nevertheless, the terms *organic* and *inorganic* remain with us today to distinguish these two basic classes of compounds.

Organic compound:	Contains carbon
Inorganic compound:	Does not contain carbon

Many compounds produced in nature are synthesized in the laboratory. Some are used as **pharmaceuticals**—drugs used for therapeutic purposes. Plants and animals are rich sources of medicinally valuable organic compounds, such as Taxol, the lifesaving anti-cancer drug first isolated from the yew tree (Figure 6-4). Most pharmaceuticals, however, are synthesized in the laboratory. For example, Lipitor (atorvastatin), the best-selling pharmaceutical in the history of medicine, used for the treatment of high cholesterol, is entirely synthetic; it is not produced naturally by any plant or animal (Figure 6-5). You will see examples of pharmaceuticals throughout this chapter and the next as you are introduced to the fundamental principles of organic chemistry.

Figure 6-4 The yew tree, from which the lifesaving anti-cancer drug Taxol was first isolated. [Thomas & Pat Leeson/ Science Source]

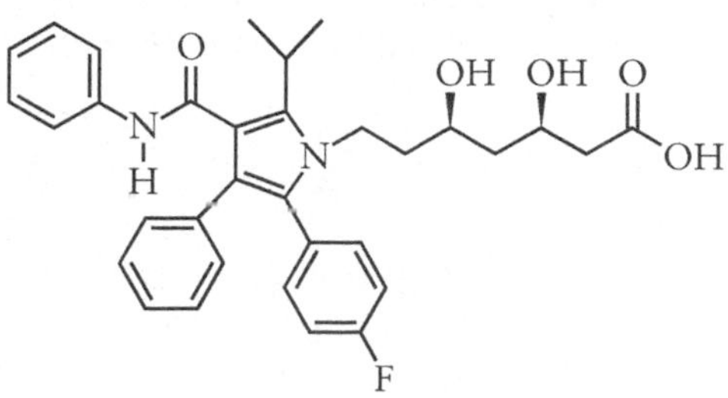

Lipitor

Figure 6-5 Lipitor (atorvastatin), a synthetic drug used for the treatment of high cholesterol.

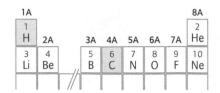

Figure 6-6 Hydrocarbons are organic molecules that contain only carbon and hydrogen atoms.

■ 6.1 Introduction to Hydrocarbons

In this chapter we will focus on a class of organic compounds known as hydrocarbons, compounds composed of exclusively *carbon* and *hydrogen* atoms (Figure 6-6). The octet rule, described in chapter 3, applies to organic compounds because they are composed of covalent bonds. With four valence electrons, carbon always forms four bonds:

- four single bonds,

- a double bond and two single bonds,

- a triple bond and one single bond, or
- two double bonds (not common)

Hydrogen atoms always form one bond (—H).

Types of Hydrocarbons

Hydrocarbons are divided into four classes: alkanes, alkenes, alkynes, and aromatic hydrocarbons, as shown in Figure 6-7. **Alkanes** contain only carbon-carbon *single bonds* (and C—H bonds) and include cycloalkanes, which contain ring structures. **Alkenes** contain one or more carbon-carbon *double bonds*, and **alkynes** contain one or more carbon-carbon *triple bonds*. **Aromatic hydrocarbons** are distinguished by their unique six-carbon ring structure containing three carbon-carbon double bonds. Aromatic hydrocarbons will be described in section 6.6.

A **saturated hydrocarbon** contains the maximum number of hydrogen atoms for a given number of carbon atoms: it is "saturated with hydrogen." For example, the alkane C_2H_6, shown in Figure 6-7, has *six* hydrogen atoms, which is the maximum number of hydrogen atoms that a molecule with *two* carbon atoms can have. The general formula for a saturated hydrocarbon is C_nH_{2n+2}, where *n* is the number of carbon atoms in the molecular formula. Alkanes, with the exception of cycloalkanes, are saturated hydrocarbons.

> Saturated hydrocarbon: C_nH_{2n+2} *n* = number of carbon atoms.

Cycloalkanes (alkanes containing rings), alkenes, alkynes, and aromatic compounds are classified as **unsaturated hydrocarbons** because they contain fewer than the maximum number of hydrogen atoms for a given number of carbon atoms, *n*. Figure 6-7 shows the two-carbon alkene, C_2H_4, has a carbon-carbon double bond and the two-carbon alkyne, C_2H_2, has a carbon-carbon triple bond; therefore they must have fewer than the *six* hydrogen atoms of a saturated hydrocarbon with two carbons.

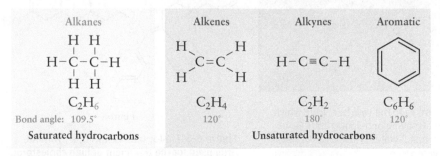

Figure 6-7 Classification of hydrocarbons.

As a result of carbon's unique ability to bond to up to four other carbon atoms, we find both straight chain and branched chain hydrocarbons. **Straight-chain hydrocarbons** have a chemical structure characterized by a chain of carbon atoms in which each carbon has a bond to two other carbon atoms (except the carbon atoms at each end of the chain). For an alkane, this means that each carbon atom has two C—H bonds (except the carbons at each end, which have three C—H bonds), as shown below. A **branched-chain hydrocarbon** has one or more carbon atoms in the chain with bonds to three or four carbon atoms, creating a "branch" in the chain. The branched-chain hydrocarbon shown below has the branching carbons highlighted in red. In this example there are two branches, and the branching carbon atoms have bonds to three carbon atoms:

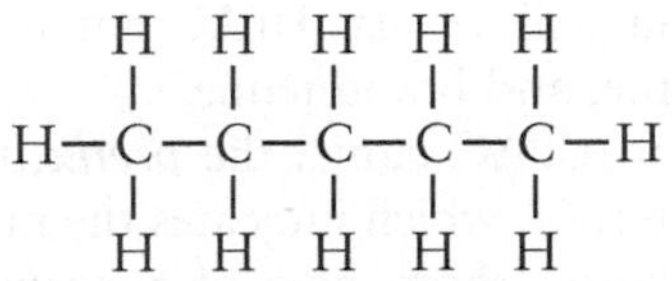

```
                                    H
                                    |
                              H — C — H
                                    |
                              H — C — H
                                    |
                      H   H         |   H   H
                      |   |         |   |   |
                H — C — C — C — C — C — H
                      |   |     |   |   |
                      H   |     H   H   H
                          |
                      H — C — H
                          |
                          H
```

```
          H   H   H   H   H
          |   |   |   |   |
    H — C — C — C — C — C — H
          |   |   |   |   |
          H   H   H   H   H
```

Straight-chain alkane 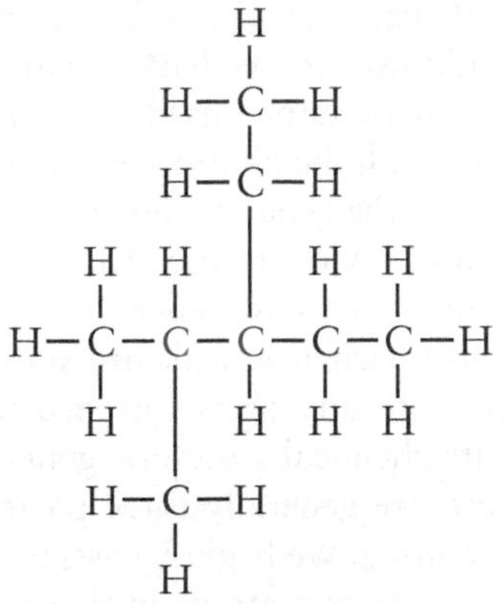   Branched-chain alkane

You are already familiar with the terms *unsaturated* and *saturated* in the context of "good" and "bad" fats, respectively, described in the opening vignette. Saturated fats contain only carbon-carbon single bonds and carbon-hydrogen bonds in the long straight-chain hydrocarbon part of their structure, whereas unsaturated fats contain one or more carbon-carbon double bonds in their long straight-chain hydrocarbon part of their structure. Strictly speaking, fats are not hydrocarbons because they contain heteroatoms (six oxygen atoms); nevertheless, a significant portion of their structure is hydrocarbon-like, causing them to have physical properties resembling those of a hydrocarbon. You will learn more about fats when you study lipids in chapter 12, a broad class of biological compounds that includes fats.

Physical Properties of Hydrocarbons

In chapter 3 you learned that carbon-carbon and carbon-hydrogen bonds are nonpolar covalent bonds. Therefore, hydrocarbons are nonpolar molecules, regardless of their structure. Recall that nonpolar molecules interact through dispersion forces, the weakest of the intermolecular forces of attraction. Thus, hydrocarbons have some of the lowest boiling points compared to other compounds with a comparable molar mass. In chapter 3 you learned that dispersion forces increase with the number of electrons and surface area in a molecule. Thus, hydrocarbons with more carbon atoms will have higher boiling points than hydrocarbons with fewer carbon atoms. For hydrocarbons with the same molecular formula, straight-chain hydrocarbons have higher boiling points than branched-chain hydrocarbons because they have more surface area, resulting in more dispersion forces.

In chapter 8 you will study mixtures and learn why hydrocarbons do not dissolve in water, instead forming two layers, as observed for oil and vinegar. The insolubility of hydrocarbons in water gives hydrocarbons *hydrophobic* properties—from the Latin "water fearing." Lipids, the class of biological molecules that do not dissolve in water, including fats, cholesterol, and steroids, contain a significant hydrocarbon component as part of their chemical structure, which gives them their hydrophobic properties.

Naming Simple Organic Compounds

Many compounds have common names that were given to them when they were first discovered. As the number of organic compounds identified increased, a more systematic naming system had to be developed. Every organic compound has been assigned a unique name based on a systematic set of rules created by the organization known as the IUPAC (International Union of Pure and Applied Chemistry). For example, C_2H_4 has the common name ethylene and the IUPAC name ethene. Similarly, C_2H_2 has the common name acetylene and the IUPAC name ethyne. Some common names are used more often than the IUPAC names, and so it is important to know some common names as well.

In addition to its IUPAC name, an active pharmaceutical ingredient is given a generic name, used to identify the drug, and a brand name, which is associated with the pharmaceutical company that makes the drug. For example, ibuprofen is the generic name for the over-the-counter analgesic sold under the brand names Motrin and Advil. Levothyroxine is the generic name for the thyroid hormone sold under the brand name Synthroid. The IUPAC names for these two pharmaceuticals are shown in Table 6-1. If you read the literature supplied with any prescription drug, you will see its IUPAC name listed as well as its chemical structure, generic name, and brand name.

There are generally three parts to an IUPAC name: the *prefix*, the *root*, and the *ending*. We begin by assigning the root, which indicates the number of contiguous carbon atoms in the main hydrocarbon chain of a straight-chain or branched-chain hydrocarbon. We then change the ending of the root to indicate the type of compound: Alkanes end in *-ane*, alkenes end in *–ene*, and alkynes end in *-yne*. In the next chapter, we will learn additional endings for hydrocarbons that contain heteroatoms. For branched-chain hydrocarbons we add a prefix before the root to indicate the type and location of the branch or branches. The rules for assigning prefixes are described in section 6.5. In this section we describe the basic rules for naming straight-chain alkanes, alkenes, and alkynes according to the IUPAC rules.

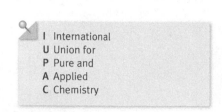

I International
U Union for
P Pure and
A Applied
C Chemistry

TABLE 6-1 Names of Some Common Pharmaceuticals

Skeletal Line Structure	Brand Name	Generic Name	IUPAC Name	Chemical Formula	Type of Drug
	Motrin, Advil	Ibuprofen	2-[4-(2-methylpropyl)] phenylpropanoic acid	$C_{13}H_{18}O_2$	Analgesic
	Synthroid	Levothyroxine	(S)-2-amino-3-[4-(4-hydroxy-3,5-diiodophenoxy)-3,5-diiodophenyl] propanoic acid	$C_{15}H_{11}I_4NO_4$	Synthetic thyroid hormone

Photos by Catherine Bausinger

Guidelines for Naming Straight-Chain Hydrocarbons

Rule 1: Assign the root. Count the number of contiguous carbon atoms in the hydrocarbon chain. Use Table 6-2 to assign the root name based on the number of contiguous carbon atoms. These names come from the corresponding Latin or Greek numbers. Memorize these names because they form the basis for naming most organic compounds.

For example, according to Table 6-2, the root for the straight-chain alkane shown is pentane because it contains five carbon atoms.

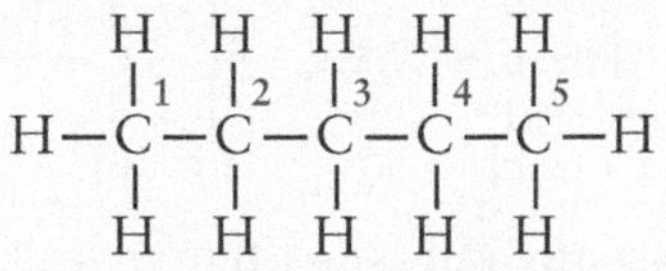

Pentane

Rule 2: Assign the ending. The ending in an IUPAC name specifies the type of compound it is: alkanes end in *-ane*, alkenes end in *-ene*, and alkynes end in *-yne*.

For example, the root for the straight-chain alkene shown below is butane because it contains four contiguous carbon atoms (see Table 6-2). Since it also contains a carbon-carbon double bond in the chain, the ending is changed from -ane to -ene forming the IUPAC name butene.

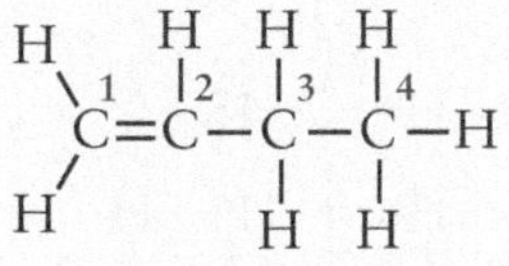

But-1-ene

Rule 3: Assign a locator number indicating the position of the first carbon of the carbon-carbon double or triple bond. When the hydrocarbon chain has more than three carbons, number the chain starting from the end closer to the carbon-carbon double or triple bond. Then place a locator number separated by hyphens (-) between the root and the ending indicating where the first carbon atom of the double or triple bond appears in the chain.

In the example above, we begin numbering the chain from the left end because that end is closer to the double bond. Since the double bond is between C(1) and C(2), we insert the locator "1" between the root and the ending. Thus, the IUPAC name is but-1-ene.

TABLE 6-2 IUPAC Names of Straight-Chain Alkanes with One to 10 Carbons

Number of Carbon Atoms	IUPAC Name
1	Methane
2	Ethane
3	Propane
4	Butane
5	Pentane
6	Hexane
7	Heptane
8	Octane
9	Nonane
10	Decane

WORKED EXERCISES   Saturated and Unsaturated Hydrocarbons

6-1 For each of the structures shown below, identify whether the molecule is an *alkane*, *alkene*, or *alkyne*. Explain why they are classified as *hydrocarbons*. Indicate whether each compound is *saturated* or *unsaturated*. Assign the IUPAC name to each compound.

a.

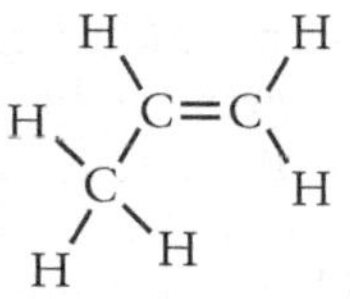

b.

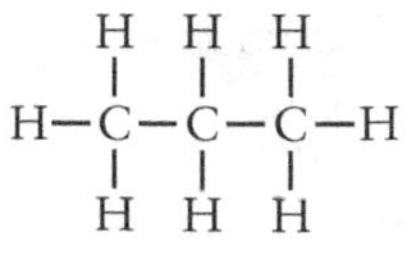

6-2 What is the molecular formula for a saturated hydrocarbon with seven carbon atoms?

6-3 Name the straight-chain alkanes shown below:

a.
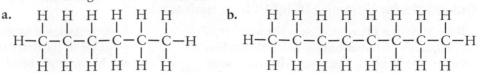

$$H-\overset{\overset{\displaystyle H}{|}}{\underset{\underset{\displaystyle H}{|}}{C}}-\overset{\overset{\displaystyle H}{|}}{\underset{\underset{\displaystyle H}{|}}{C}}-\overset{\overset{\displaystyle H}{|}}{\underset{\underset{\displaystyle H}{|}}{C}}-\overset{\overset{\displaystyle H}{|}}{\underset{\underset{\displaystyle H}{|}}{C}}-\overset{\overset{\displaystyle H}{|}}{\underset{\underset{\displaystyle H}{|}}{C}}-\overset{\overset{\displaystyle H}{|}}{\underset{\underset{\displaystyle H}{|}}{C}}-H$$

b.

$$H-\overset{\overset{\displaystyle H}{|}}{\underset{\underset{\displaystyle H}{|}}{C}}-\overset{\overset{\displaystyle H}{|}}{\underset{\underset{\displaystyle H}{|}}{C}}-\overset{\overset{\displaystyle H}{|}}{\underset{\underset{\displaystyle H}{|}}{C}}-\overset{\overset{\displaystyle H}{|}}{\underset{\underset{\displaystyle H}{|}}{C}}-\overset{\overset{\displaystyle H}{|}}{\underset{\underset{\displaystyle H}{|}}{C}}-\overset{\overset{\displaystyle H}{|}}{\underset{\underset{\displaystyle H}{|}}{C}}-\overset{\overset{\displaystyle H}{|}}{\underset{\underset{\displaystyle H}{|}}{C}}-\overset{\overset{\displaystyle H}{|}}{\underset{\underset{\displaystyle H}{|}}{C}}-H$$

6-4 Name the straight-chain alkenes and alkynes shown below:

a.

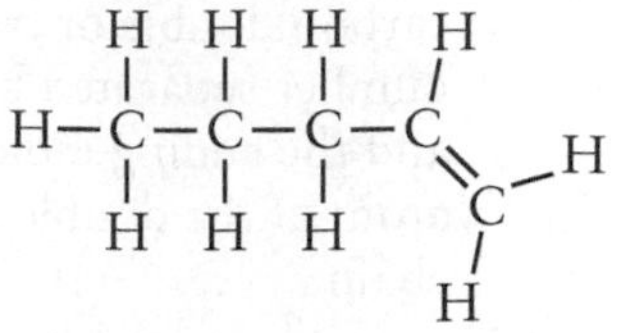

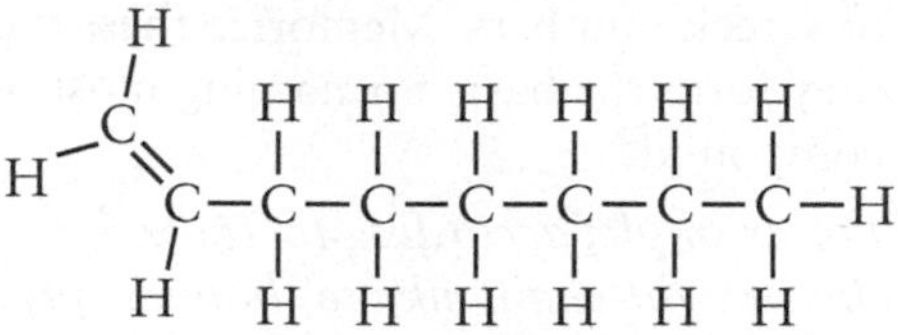

c. 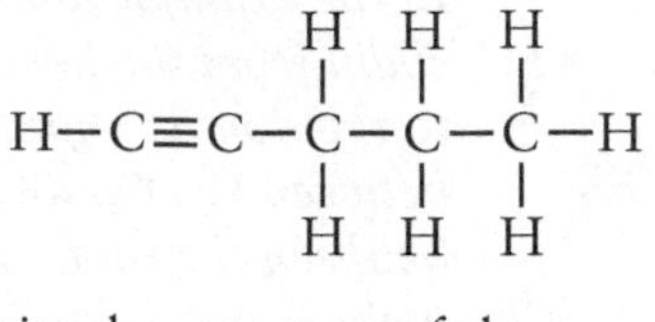 $H-C\equiv C-\overset{\overset{\displaystyle H}{|}}{\underset{\underset{\displaystyle H}{|}}{C}}-\overset{\overset{\displaystyle H}{|}}{\underset{\underset{\displaystyle H}{|}}{C}}-\overset{\overset{\displaystyle H}{|}}{\underset{\underset{\displaystyle H}{|}}{C}}-H$

d. 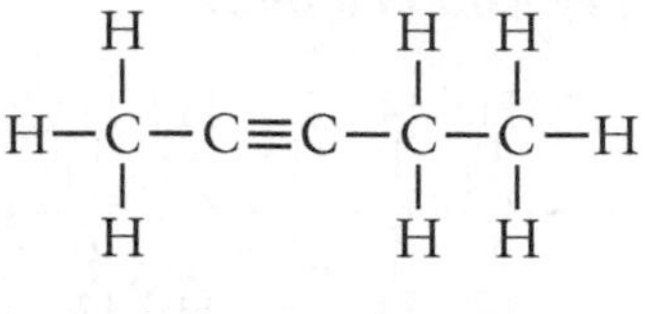 $H-\overset{\overset{\displaystyle H}{|}}{\underset{\underset{\displaystyle H}{|}}{C}}-C\equiv C-\overset{\overset{\displaystyle H}{|}}{\underset{\underset{\displaystyle H}{|}}{C}}-\overset{\overset{\displaystyle H}{|}}{\underset{\underset{\displaystyle H}{|}}{C}}-H$

6-5 Write the structure of the compounds with the following IUPAC names:

a. propane   b. hex-1-ene   c. non-3-yne

6-6 State whether each of the following compounds is a straight-chain alkane or a branched-chain alkane:

a.

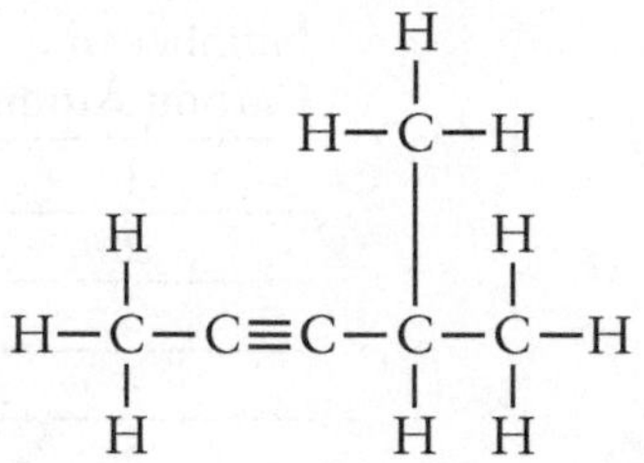

b. 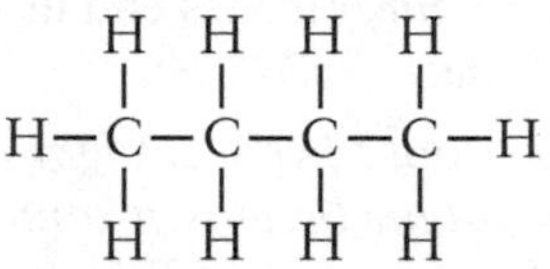 $H-\overset{\overset{\displaystyle H}{|}}{\underset{\underset{\displaystyle H}{|}}{C}}-\overset{\overset{\displaystyle H}{|}}{\underset{\underset{\displaystyle H}{|}}{C}}-\overset{\overset{\displaystyle H}{|}}{\underset{\underset{\displaystyle H}{|}}{C}}-\overset{\overset{\displaystyle H}{|}}{\underset{\underset{\displaystyle H}{|}}{C}}-H$

c.

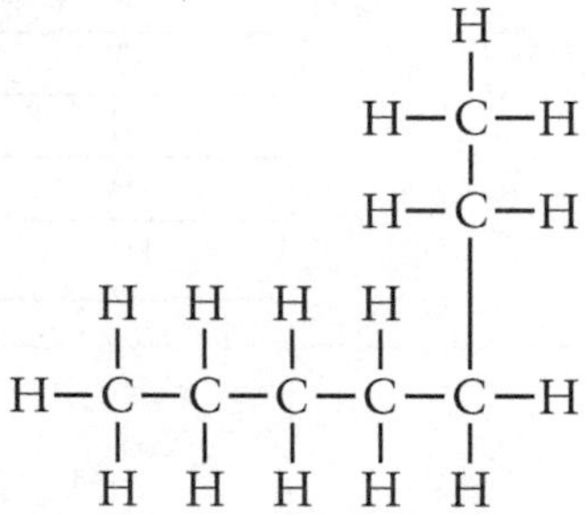

d.

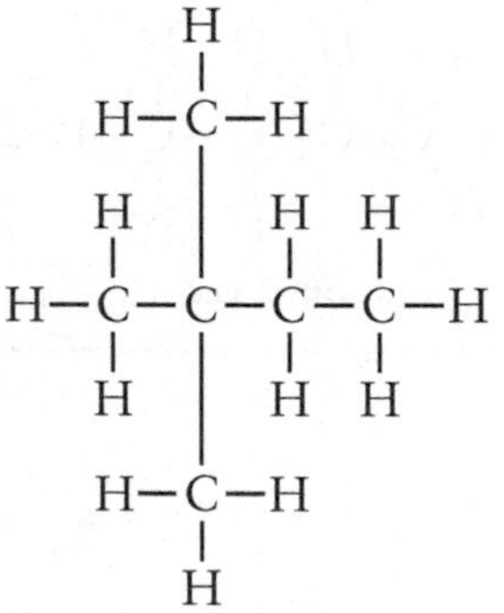

6-7 Based on molecular formula alone, which of the following compounds is a *saturated* hydrocarbon?

a. C_5H_8 b. C_5H_{10} c. C_5H_{12}

Solutions

6-1 **a.** The compound is an alkene because it contains a carbon-carbon double bond. It is a hydrocarbon because it contains only carbon and hydrogen atoms. It is unsaturated because it contains a carbon-carbon double bond and therefore fewer hydrogens (six) than the maximum number of hydrogens (eight) for a compound with three carbons. Its IUPAC name is propene because it contains a chain of three carbons (root = *propane*) with a double bond, so the ending is changed from *-ane* to *-ene*. No locator number is needed because there is only one place the double bond can be.

b. The compound is an alkane because it contains only carbon-carbon single bonds (and C—H bonds). It is a hydrocarbon because it contains only carbon and hydrogen atoms. It is saturated because there are only single bonds, and it has the maximum number of hydrogen atoms (eight) for a hydrocarbon with three carbons. The IUPAC name is propane because it contains a chain of three carbons, which has the root *propane*. The ending remains -*ane* because it is an alkane.

6-2 The formula for a saturated hydrocarbon is C_nH_{2n+2}. Substituting the value $n = 7$ in the formula $2n + 2$ gives the number of hydrogen atoms in the formula: $(2 \times 7) + 2 = 16$. Therefore, the formula for a saturated hydrocarbon with seven carbon atoms is C_7H_{16}.

6-3 **a.** This compound has six contiguous carbon atoms, so according to Table 6-2 the root name is *hexane*. The compound has no double or triple bond and so the ending remains -*ane*.

b. There are eight contiguous carbon atoms, so the root name is *octane* (see Table 6-2). The compound is an alkane, so the ending remains -*ane*.

6-4 **a.** There are five contiguous carbon atoms, so the root is *pentane* (see Table 6-2). Since there is a carbon-carbon double bond in the molecule, the compound is an alkene and the ending is changed from -*ane* to -*ene*, giving *pentene*. We then add a locator number to indicate where the double bond first appears when numbering the chain of carbons starting from the right side because that is the side closer to the double bond. Since the carbon-carbon double bond is between C(1) and C(2), the locator number -1- is inserted between the root and the ending: pent-1-ene.

b. There are eight contiguous carbon atoms, so the root is *octane* (see Table 6-2). Since there is a carbon-carbon double bond in the molecule, the compound is an alkene and the ending is changed from -*ane* to -*ene*, giving *octene*. We then add a locator number to indicate where the double bond first appears when numbering the chain of carbons starting from the left side because that is the side closer to the double bond. Since the carbon-carbon double bond is between C(1) and C(2), the locator number -1- is inserted between the root and the ending: oct-1-ene.

c. There are five contiguous carbon atoms, so the root is *pentane* (see Table 6-2). There is a carbon-carbon triple bond in the molecule, meaning the compound is an alkyne and so the ending is changed from -*ane* to -*yne*, giving *pentyne*. We then add a locator number to indicate where the triple bond begins. Number the chain of carbons starting from the left side because that is the side closer to the triple bond. Since the carbon-carbon triple bond is between C(1) and C(2), we insert the locator number -1- in between the root and the ending: pent-1-yne.

d. There are five contiguous carbon atoms, so according to Table 6-2 the root is *pentane*. Since there is a carbon-carbon triple bond in the molecule, the ending is changed from -*ane* to -*yne*, giving *pentyne*. Next, add a locator number to indicate where the triple bond begins. We number the chain of carbons starting from the left side because that is the side closer to the triple bond. Since the carbon-carbon triple bond is between C(2) and C(3), we insert the locator number -2- in between the root and the ending: pent-2-yne.

6-5 **a.** The name propane indicates a chain of three carbons and that the structure is an alkane (all single bonds). We add hydrogens to every carbon until they each have four bonds.

```
      H   H   H
      |   |   |
  H — C — C — C — H
      |   |   |
      H   H   H
```

b. The root is *hexane*, which indicates a chain of six carbons (see Table 6-2). Since the ending is *-ene* we know there is a carbon-carbon double bond somewhere in the molecule. The locator number -1- indicates that the carbon-carbon double bond is between C(1) and C(2). Thus, write a chain of six carbons with a double bond between the first two carbons and then add hydrogens to every carbon until each carbon has four bonds:

$$
\begin{array}{ccccccc}
H & & H & H & H & H \\
| & & | & | & | & | \\
C & = & C - C - C - C - C & - & H \\
| & & | & | & | & | \\
H & & H & H & H & H
\end{array}
$$

c. The root is *nonane*, which indicates a chain of nine carbons (see Table 6-2). Since the ending is *–yne*, we know there is a carbon-carbon triple bond somewhere in the molecule. The locator number -3- indicates that the triple bond is between C(3) and C(4). Thus, write a chain of nine carbons with a triple bond between the third and fourth carbon. Add C—H bonds so that every carbon atom has a total of four bonds.

$$
\begin{array}{ccccccccc}
& H & H & & & H & H & H & H & H \\
& |_1 & |_2 & _3 & _4 & |_5 & |_6 & |_7 & |_8 & |_9 \\
H - & C - & C - & C \equiv & C - & C - & C - & C - & C - & C - H \\
& | & | & & & | & | & | & | \\
& H & H & & & H & H & H & H
\end{array}
$$

6-6 a. Branched-chain hydrocarbon because one of the carbon atoms in the chain has a bond to more than two carbons. In this example, the branch is located on the second carbon of the longest contiguous chain when counting starting from the right side.
   b. Straight-chain hydrocarbon (chain of four) because every carbon is bonded to no more than two carbons.
   c. Straight-chain hydrocarbon (chain of seven) because every carbon is bonded to no more than two carbons. The chain can be drawn in more than one way, including turning. The key is whether or not there are carbon branches.
   d. Branched-chain hydrocarbon because the second carbon from the left is bonded to four carbons, creating two branches off the second carbon.

6-7 Compound (c) is the only saturated hydrocarbon because it contains five carbon atoms and 12 hydrogen atoms, which fits the formula C_nH_{2n+2}. (a) and (b) have less than the $2n + 2$ hydrogen atoms so they are unsaturated hydrocarbons.

PRACTICE EXERCISES

1 Which of the following are unsaturated hydrocarbons? (More than one correct choice is possible.)
   a. alkanes   **b.** alkenes   **c.** alkynes   **d.** aromatic hydrocarbons

2 Which of the following are saturated hydrocarbons? (More than one correct choice is possible.)
   a. C_7H_{14} **b.** C_6H_{14} **c.** C_3H_8 **d.** C_2H_2

3 By the modern definition, what distinguishes an organic compound from an inorganic compound?

4 How many covalent bonds does carbon always form in a molecule? Explain why.

5 Name the following hydrocarbons:

a.
$$
\begin{array}{cccc}
H & H & H & H \\
| & | & | & | \\
H - C - C - C - C - H \\
| & | & | & | \\
H & H & H & H
\end{array}
$$

b.
$$
\begin{array}{cccc}
H & H & & H \\
| & | & & | \\
H - C - C = C - C - H \\
| & & | & | \\
H & & H & H
\end{array}
$$

c.
$$
\begin{array}{ccccc}
H & & & H & H \\
\backslash & & & | & | \\
& C = C - C - C - H \\
/ & & | & | & | \\
H & & H & H & H
\end{array}
$$

d.
$$
\begin{array}{ccccc}
H & H & H & H \\
| & | & | & | \\
H - C - C - C - C - C \equiv C - H \\
| & | & | & | \\
H & H & H & H
\end{array}
$$

6 For each compound shown below, identify whether it is an alkane, alkene, or alkyne. Does every carbon atom in each molecule have four bonds? Indicate whether each compound is a straight-chain or a branched-chain hydrocarbon.

a.

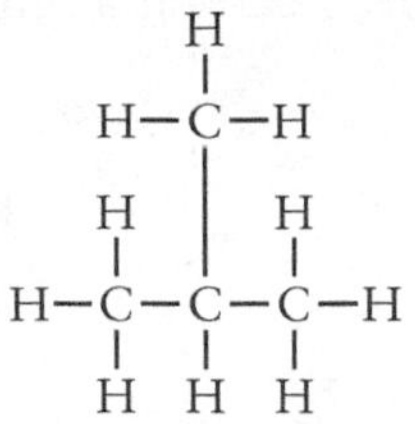

$$\begin{array}{c} H \\ | \\ H-C-H \\ \end{array}$$

$$\begin{array}{ccc} H & H \\ | & | \\ H-C-C-C-H \\ | & | & | \\ H & H & H \end{array}$$

b.

$$\begin{array}{ccc} H & & H \\ | & & | \\ H-C-C\equiv C-C-H \\ | & & | \\ H & & H \end{array}$$

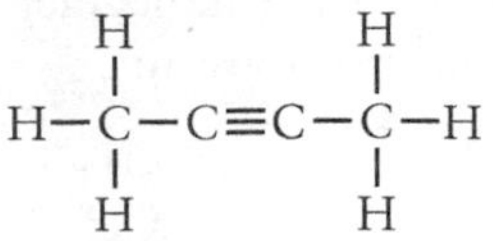

■ 6.2 Writing Alkane and Cycloalkane Structures

Alkanes provide the carbon backbone for more complex organic compounds, where heteroatoms replace some hydrogen and carbon atoms on the carbon backbone. Therefore, understanding the structure of alkanes is the foundation for understanding the structure of all organic compounds. In this section you will learn about the structure of saturated hydrocarbons and the common shorthand methods for writing their structures.

Alkanes contain only carbon-carbon single bonds and carbon-hydrogen bonds. The simplest alkane is methane, CH_4, the natural gas used in gas stoves and most gas heating systems (Figure 6-8). Methane is shown in **Figure 6-9a** as a Lewis structure and as a ball-and-stick model, where the latter provides a three-dimensional representation of the molecule. Remember, Lewis structures do not necessarily illustrate molecular shape. You learned in chapter 3 that methane has a tetrahedral molecular shape with H—C—H bond angles of 109.5°, which is evident in the ball-and-stick model. *Indeed, every carbon atom in an alkane has a tetrahedral shape*, as seen in ethane (C_2H_6) and propane (C_3H_8), shown in Figures 6-9b and 6-9c. The tetrahedral shape of the carbon atoms in an alkane gives these molecules an overall zigzag shape when there are more than two carbon atoms present, as you can see in the ball-and-stick model of propane in Figure 6-9c.

Because every bond and atom is shown, writing Lewis structures for large organic molecules is tedious and time consuming. Two simpler notations in common practice for writing chemical structures are *condensed structures* and *skeletal line structures*.

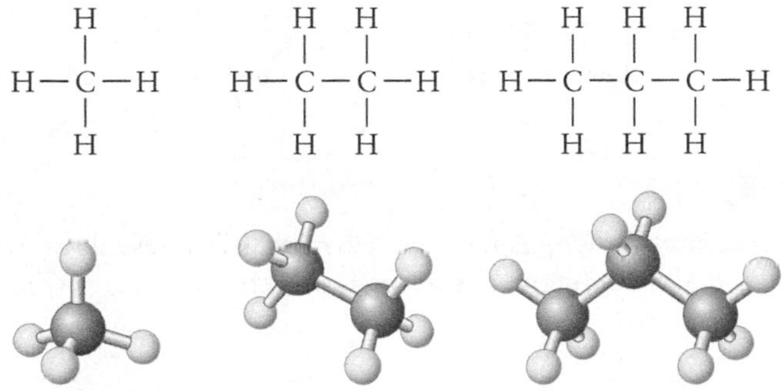

Figure 6-8 Methane, CH_4, is the simplest alkane and is present in natural gas, the fuel used in gas stoves.
[Tetra images/Punchstock]

$$\begin{array}{c} H \\ | \\ H-C-H \\ | \\ H \end{array}$$

$$\begin{array}{cc} H & H \\ | & | \\ H-C-C-H \\ | & | \\ H & H \end{array}$$

$$\begin{array}{ccc} H & H & H \\ | & | & | \\ H-C-C-C-H \\ | & | & | \\ H & H & H \end{array}$$

(a) Methane, CH_4 (b) Ethane, C_2H_6 (c) Propane, C_3H_8

Figure 6-9 Lewis structure and ball-and-stick models of the simplest alkanes: (a) methane, (b) ethane, and (c) propane.

Condensed Structures

To write a **condensed** structure, we begin at one end of the molecule and work our way to the other end of the molecule, writing each carbon atom and its attached hydrogen atom(s) as a group:

$$C$$
$$CH$$
$$CH_2$$
$$CH_3$$

Bonds are omitted except at branch points. For example, the condensed structure for the straight-chain alkane pentane, C_5H_{12}, can be written as

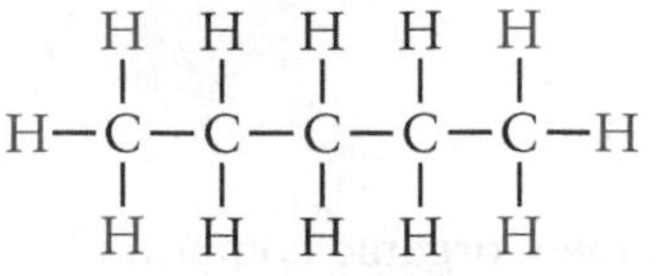

$$= CH_3CH_2CH_2CH_2CH_3 = CH_3(CH_2)_3CH_3$$

Lewis structure   Condensed structure   Condensed structure
(abbreviated)

For long straight-chain alkanes, the repeating CH_2 groups can be abbreviated further by writing a single CH_2 group enclosed in parentheses, followed by a subscript indicating the number of consecutive CH_2 groups. Thus, an alternative condensed structure for pentane is $CH_3(CH_2)_3CH_3$, where the subscript 3 following the parentheses enclosing CH_2 indicates that there are three consecutive CH_2 groups in the chain.

The condensed structure for a branched-chain hydrocarbon is written the same way as a straight-chain alkane except branch points along the chain are shown by including the branching bond(s) as a line pointing up or down from the carbon chain at its branch point. The branch itself is then written as a sequence of carbon atoms and its attached hydrogen atoms. The condensed structure for a branched-chain alkane is shown below. The carbon atom at the branch point is indicated in red. Note that a carbon group with a branch will have fewer than two hydrogen atoms; here it is a CH group.

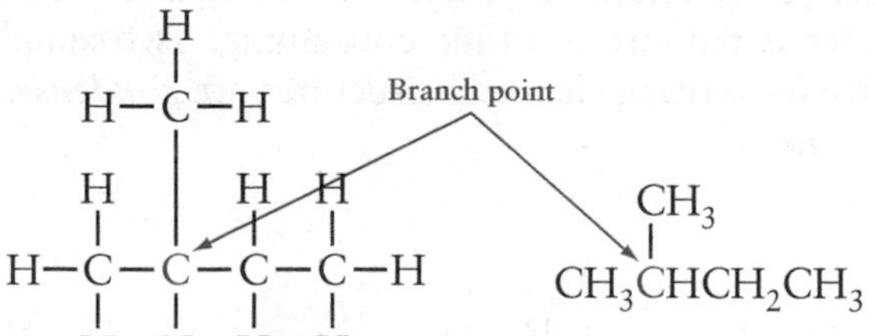

Lewis structure   Condensed structure

WORKED EXERCISES | Writing Condensed Structures

For each condensed structure shown, determine whether it represents a *straight-chain* alkane or a *branched-chain* alkane, then write a Lewis structure:

6-8 $CH_3(CH_2)_5CH_3$

6-9
$$CH_3$$
$$|$$
$$CH_3CCH_2CH_3$$
$$|$$
$$CH_3$$

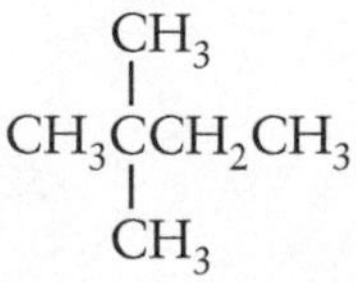

Solutions

6-8 Straight-chain alkane because each carbon in the chain has bonds to no more than two carbons. The subscript 5 following the CH_2 in parentheses indicates that there are five consecutive CH_2 groups. CH_3 groups are always on each end of the chain:

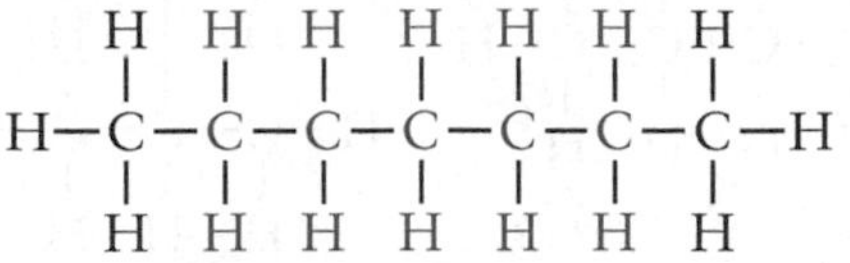

$$
\begin{array}{ccccccc}
H & H & H & H & H & H & H \\
| & | & | & | & | & | & | \\
H-C-&C-&C-&C-&C-&C-&C-H \\
| & | & | & | & | & | & | \\
H & H & H & H & H & H & H
\end{array}
$$

6-9 Branched-chain alkane because there is a carbon with a bond to more than two carbons. There are two CH_3 groups branching from the second carbon atom along the main chain, indicated by the two bonds projecting above and below the second carbon atom, shown in red.

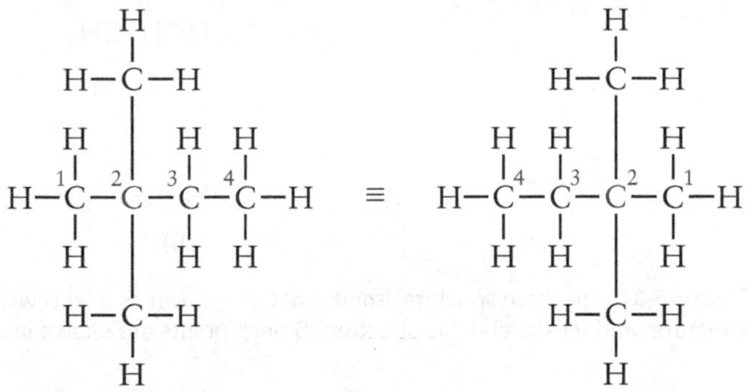

$$
\begin{array}{ccc}
& H & \\
& | & \\
& H-C-H & \\
H & | & H\ H \\
| & | & |\ | \\
H-\overset{1}{C}-\overset{2}{C}-\overset{3}{C}-\overset{4}{C}-H \\
| & | & |\ | \\
H & | & H\ H \\
& H-C-H & \\
& | & \\
& H &
\end{array}
\quad \equiv \quad
\begin{array}{ccc}
& H & \\
& | & \\
& H-C-H & \\
H\ H & | & H \\
|\ | & | & | \\
H-\overset{4}{C}-\overset{3}{C}-\overset{2}{C}-\overset{1}{C}-H \\
|\ | & | & | \\
H\ H & | & H \\
& H-C-H & \\
& | & \\
& H &
\end{array}
$$

PRACTICE EXERCISES

7 For each condensed structure shown below, determine whether it represents a *straight-chain* alkane or a *branched-chain* alkane, then write a Lewis structure:

   a. $CH_3CH_2CH_2CH_2CH_2CH_3$ **b.** 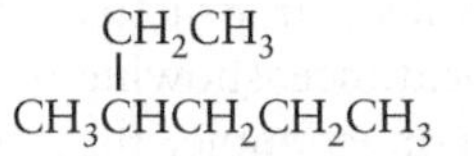 $\begin{array}{c} CH_2CH_3 \\ | \\ CH_3CHCH_2CH_2CH_3 \end{array}$ **c.** $\begin{array}{c} CH_3 \\ | \\ CH_3CHCH_2CHCH_2CH_3 \\ | \\ CH_3 \end{array}$

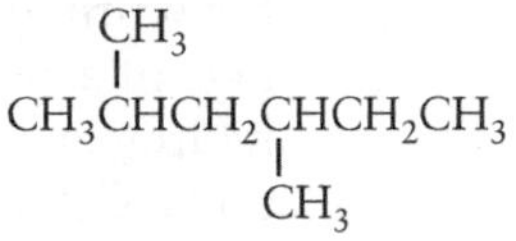

8 Write a condensed structure for the following Lewis structure:

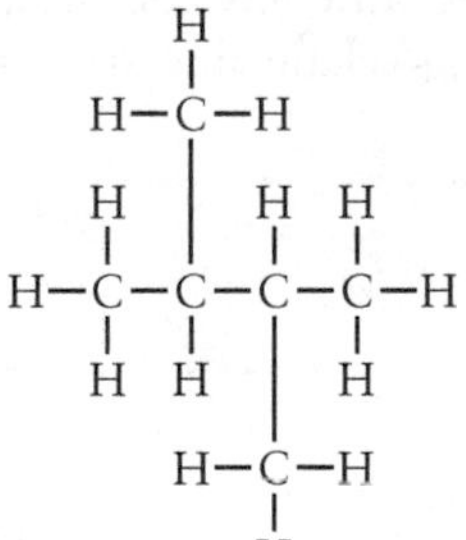

$$
\begin{array}{ccc}
& H & \\
& | & \\
& H-C-H & \\
H & | & H\ H \\
| & | & |\ | \\
H-C-&C-&C-C-H \\
| & | & |\ | \\
H & H & |\ H \\
& H-C-H & \\
& | & \\
& H &
\end{array}
$$

9 Write a condensed structure and a Lewis structure for the molecule shown as a ball-and-stick model below:

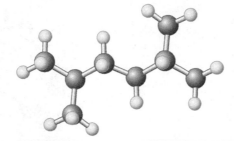

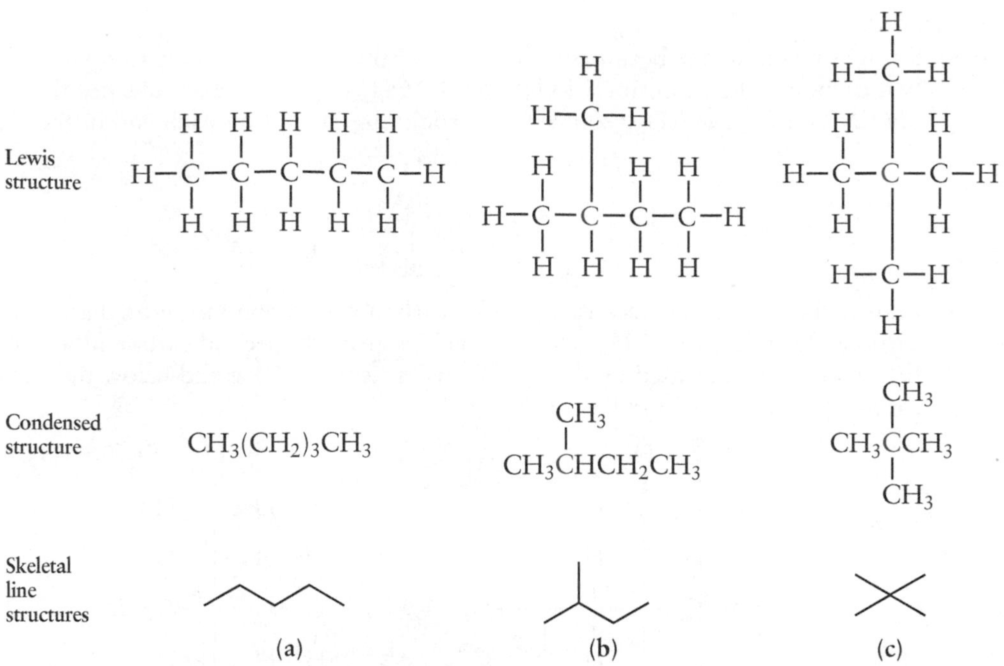

Lewis structure

Condensed structure

$CH_3(CH_2)_3CH_3$

$CH_3CHCH_2CH_3$ with CH_3 branch

CH_3CCH_3 with CH_3 above and CH_3 below

Skeletal line structures

(a)   (b)   (c)

Figure 6-10 The three structural isomers of C_5H_{12} drawn as a (a) Lewis structure, (b) condensed structure, and (c) skeletal line structure. Branch points are shown in red.

Skeletal Line Structures

You have seen how writing condensed structures simplifies the task of writing Lewis structures. **Skeletal line structures** are an even more efficient shorthand for writing large molecular structures. Skeletal line structures have the advantage of a clean appearance, showing only carbon atom linkages uncluttered by hydrogen atoms. For example, three hydrocarbons with the molecular formula C_5H_{12} are shown as a Lewis structure, a condensed structure, and a skeletal line structure in Figure 6-10. As you can see, the skeletal line structure is the easiest to write and has the least cluttered appearance. The general guidelines for writing skeletal line structures are described in the Guidelines box below.

Guidelines for Writing Skeletal Line Structures

1. Carbon-carbon bonds in a contiguous chain of carbon atoms are written in a zigzag fashion, with the symbol for carbon, C, omitted as well as all hydrogen atoms and C—H bonds, even though they are still understood to be present. A carbon atom is implied wherever two lines (representing carbon-carbon bonds) come together at a point *and* at the terminal ends of a line. For example, pentane is represented by the following zigzag arrangement of lines, where every point represents a carbon atom with two or three hydrogen atoms:

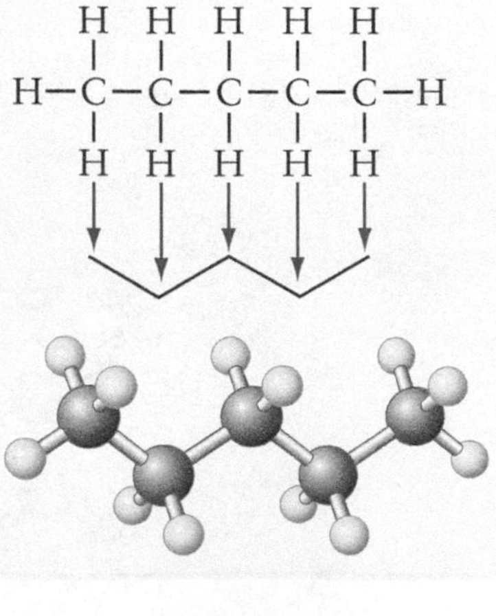

2. Double bonds are written as *two* parallel lines, $=$, and triple bonds are written as *three* parallel lines, $\equiv$.

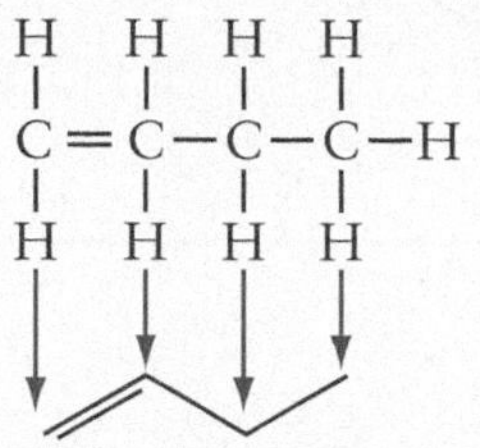

$$H \atop |$$

H   H H H
 | | | |
C = C — C — C — H
 | | | |
H   H H H

3. Carbon branches are drawn above the zigzag chain when the zigzag is up and below the zigzag chain when the zigzag is down, as shown in pink:

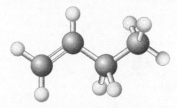

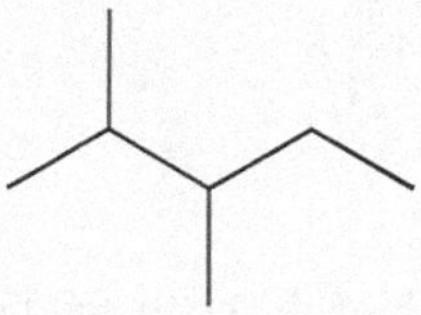

Wherever a heteroatom (an atom other than carbon and hydrogen) appears, it must be written in at a point in order to distinguish it from a carbon atom. All hydrogen atoms bonded to heteroatoms (OH, NH$_2$, etc.) must also be written in.

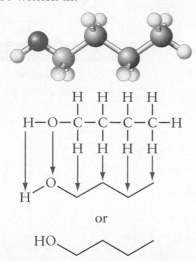

        H   H   H   H
        | | | |
H — O — C — C — C — C — H
        | | | |
        H   H   H   H

or

HO〜〜〜

4. To determine the number of hydrogen atoms on a particular carbon atom from a skeletal line structure, count the number of bonds shown and subtract this value from 4 because the octet rule tells us a carbon atom must always have four bonds.

WORKED EXERCISES   Writing Skeletal Line Structures

6-10 Write a Lewis structure for the skeletal line structure shown below:

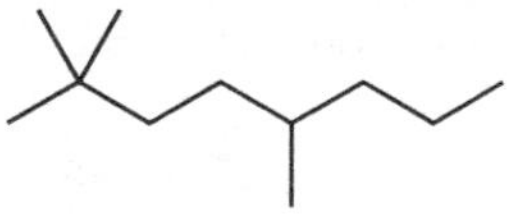

6-11 Indicate how many hydrogen atoms are on each of the carbon atoms in the skeletal line structure shown:

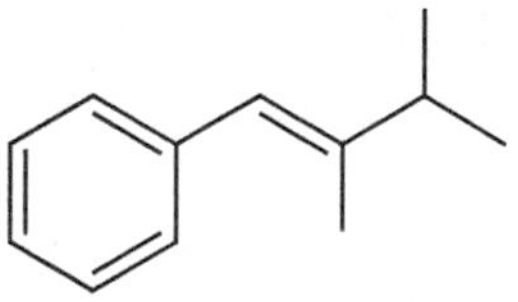

Solutions

6-10 Since each line in a skeletal line structure represents a carbon-carbon bond, place a carbon atom at every intersection of two lines and at the end of a line. Add C—H bonds to each carbon atom until it has a total of four bonds. In a Lewis structure the branch points can be written above or below the linear main chain, whereas in a skeletal line structure they are placed in the same direction as the zigzag of the branching carbon.

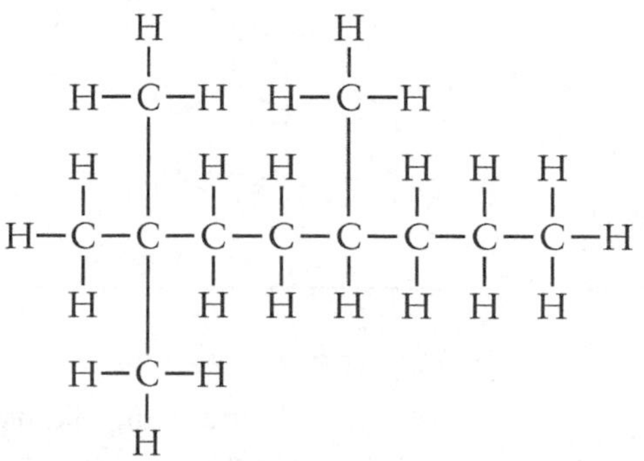

6-11 The number of hydrogen atoms on each carbon is indicated in red.

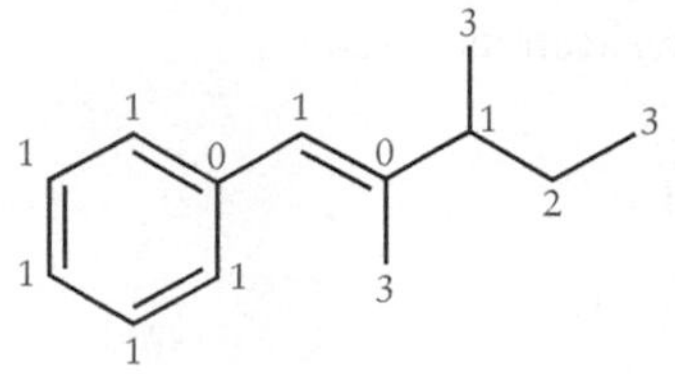

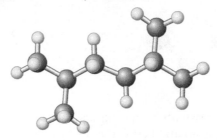

PRACTICE EXERCISES

10 Write the skeletal line structure for the ball-and-stick model shown below. Note, some hydrogen atoms are hidden from view by carbon atoms

11 Write the Lewis structure and the condensed structure for the skeletal line structures shown:

a.   **b.**

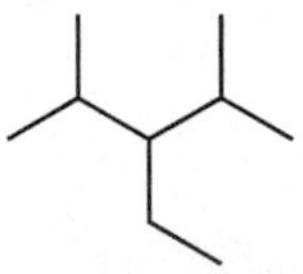

12 Write a skeletal line structure that corresponds to each of the condensed structures shown:

a. $CH_3(CH_2)_8CH_3$ **b.** CH_3
                                          |
                              $CH_3CHCHCH_3$
                                          |
                                        CH_3

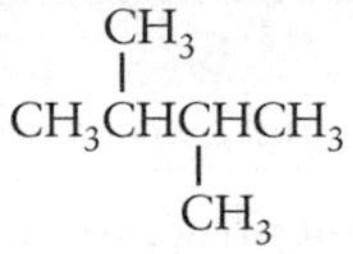

13 Indicate the number of hydrogen atoms on each of the carbon atoms in the skeletal line structures below:

a.   **b.** 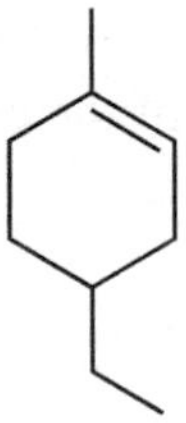   **c.**

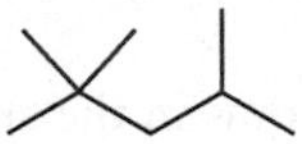

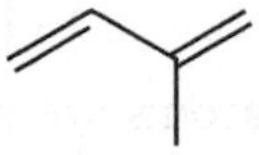

d.

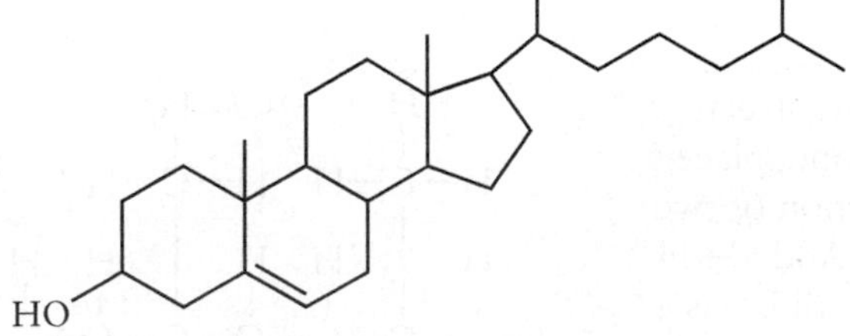

Cholesterol

Cycloalkanes

A **cycloalkane** is an alkane whose chain of carbon atoms is joined in a way that forms a ring. You can make a model of a cycloalkane from the corresponding straight-chain alkane simply by removing one hydrogen atom from

Remove two hydrogen atoms and join two carbon atoms as C—C bond

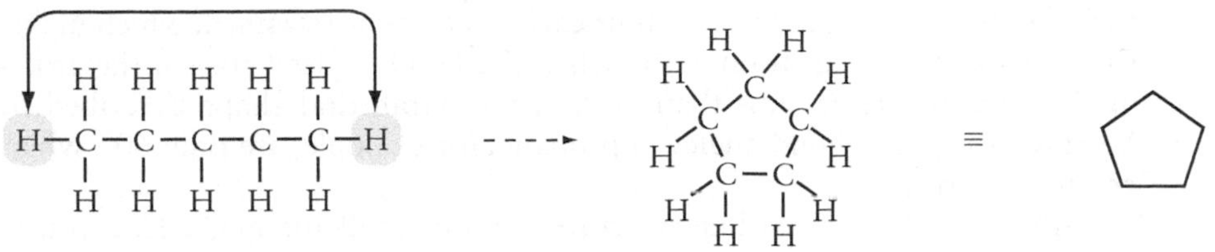

Pentane, C_5H_{12} Cyclopentane, C_5H_{10}

Figure 6-11 To make a cycloalkane with five carbon atoms (cyclopentane) from the corresponding straight-chain alkane with five carbon atoms (pentane): remove a hydrogen atom from both ends of the chain and form a carbon-carbon single bond between the carbon atoms.

the first and the last carbon in the chain and joining these two carbon atoms as a carbon-carbon single bond, as illustrated in Figure 6-11. An unbranched cycloalkane is composed of only CH_2 groups. Cycloalkanes are classified as unsaturated hydrocarbons because they contain fewer than the maximum number of hydrogen atoms for a given number of carbon atoms, n. The structure of a cycloalkane is always written as a skeletal line structure. For example, a cycloalkane with five carbon atoms is written as a pentagon, a five-sided figure.

Guidelines for Naming Cycloalkanes

Name cycloalkane the same way as an alkane by counting the number of carbon atoms in the ring. Insert the term "cyclo" in front of the root without a space or dash. Thus, a six-carbon ring is named *cyclo*hexane, a five-carbon ring is named *cyclo*pentane, and so forth, as shown below. The structure of the cycloalkanes containing three to eight carbon atoms are shown and named in Table 6-3.

TABLE 6-3 Cycloalkane Formulas, Skeletal Structures, and IUPAC Names

Number of Carbon Atoms in Ring	Molecular Formula	Structure	Name
3	C_3H_6		cyclopropane
4	C_4H_8		cyclobutane
5	C_5H_{10}	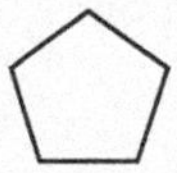	cyclopentane
6	C_6H_{12}	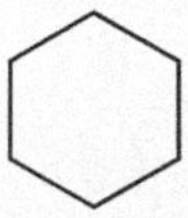	cyclohexane
7	C_7H_{14}	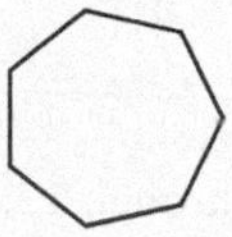	cycloheptane
8	C_8H_{16}	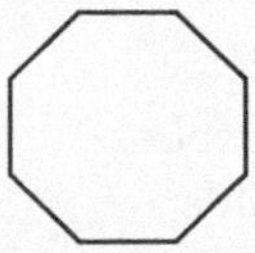	cyclooctane

(a) Ball-and-stick model

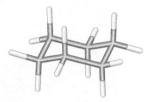

(b) Tube model

Figure 6-12 Three-dimensional models of Cyclohexane.

In nature we find mainly cyclopentane and cyclohexane containing structures because rings with three or four carbon atoms are strained, which makes them less stable. Ring strain exists when the bond angles between the atoms in the cycloalkane are less than 109.5°, the tetrahedral shape described by VSEPR theory. The bond angles in propane, for example, are only 60°, which creates ring strain.

Although the skeletal line structure for a cycloalkane might lead you to believe that cycloalkanes are flat planar structures, with the exception of cyclopropane, they are not. For example, if you look at a model of cyclohexane from the side, as shown in Figure 6-12, you can see that it is not planar. Work through Using Molecular Models 6-1: Hexane and Cyclohexane to gain more familiarity with cycloalkane structures.

Cycloalkanes are common in nature and provide the skeletal backbone for many molecules. Steroids, for example, are defined by their characteristic four cycloalkane rings, as shown in Figure 6-13.

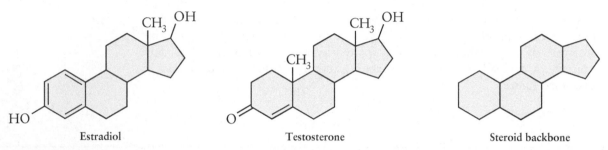

Estradiol   Testosterone   Steroid backbone

Figure 6-13 Steroids, such as estradiol and testosterone, are defined by their characteristic four cycloalkane rings, known as a steroid backbone (shown in blue).

Using Molecular Models 6-1 Hexane and Cyclohexane

Construct Hexane, C_6H_{14}

1. Obtain six black carbon atoms, 14 light blue hydrogen atoms, and 19 straight bonds.

2. Construct a model of hexane, C_6H_{14} and answer Inquiry Questions 5 through 7.

Construct Cyclohexane, C_6H_{12}

3. Remove one hydrogen atom from C(1) and one hydrogen atom from C(6) in hexane.

4. Use one of the two remaining bonds to join C(1) and C(6). Your model should look similar to the tube model shown. It may require some twisting to get it into this orientation. Twist until it resembles a lounge chair. Answer Inquiry Questions 8 through 12.

Inquiry Questions

5. How many C—H bonds are there on carbons 1 and 6 of hexane? How many C—H bonds are there on all the other carbon atoms?

6. Is there free rotation around every C—C bond in hexane?

7. What are the H—C—C bond angles at every carbon in hexane?

8. How many C—H bonds are there on all the carbon atoms in cyclohexane? What makes cyclohexane an unsaturated hydrocarbon and hexane a saturated hydrocarbon?

9. Is there free rotation around every C—C bond in cyclohexane? Is there some flexibility in cyclohexane?

10. What are the H—C—C bond angles in cyclohexane? Does there appear to be any ring strain in the model?

11. Rotate the model so that you are viewing it from the side. Does the molecule appear flat? Can you force it to be flat?

12. Rotate the model so that you have a top view of the ring. Does it look like the skeletal line structure used to represent cyclohexane?

WORKED EXERCISE · Writing Cycloalkane Structures

6-12 Write the skeletal line structure for a cycloalkane containing four carbon atoms in a ring. What is the IUPAC name of this compound? The H—C—C bond angles are close to 90°. Would you expect this molecule to have ring strain?

Solution

6-12

The IUPAC name for this compound is cyclobutane because it is a ring (cyclo) containing four (butane) carbon atoms with only carbon-carbon single bonds (ane). The preferred bond angle for a tetrahedral center is 109.5°, so the 90° bond angle in cyclobutane creates ring strain. Indeed, you will find that making a model of this compound with your model kit is difficult for this reason.

PRACTICE EXERCISES

14 Name the cycloalkanes below. Which of these cycloalkanes is most common in nature? What is the molecular formula for each cycloalkane? Why do the formulas not fit the C_nH_{2n+2} formula for a saturated hydrocarbon?

a.    b. 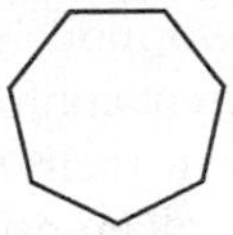   c.

15 Examine the bond angles in the ball-and-stick model of cyclopropane shown below. What is the preferred C—C—C bond angle for an alkane and why? What is the actual bond angle in cyclopropane? Why might you risk breaking pieces of your model kit if you tried to make a model of cyclopropane? Why is this ring size rare in nature?

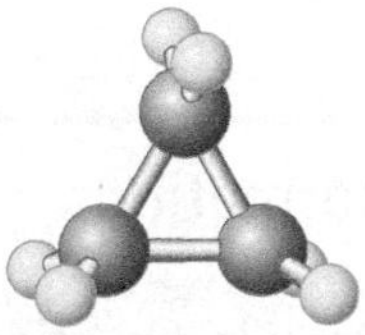

■ 6.3 Alkane Conformations and Structural Isomers

In the introduction to this chapter we saw that carbon has the capacity to form bonds to other carbon atoms and that every carbon atom has four bonds. This capacity for carbon to form many bonds to other carbons means that the molecular formula of an organic compound with more than three carbons provides insufficient information to determine the structure of the molecule. This is why a Lewis structure, a skeletal line structure, a condensed structure, or an IUPAC name are needed to identify an organic compound.

Conformations

In chapter 1 you learned that because molecules have kinetic energy they are in constant motion. *Alkanes are also freely rotating about each of their carbon-carbon single bonds.* An alkane has many different rotational forms, known as **conformations.** One conformation is converted to another simply by rotation about one or more carbon-carbon single bonds. Whenever we write a three-dimensional representation of a molecule or build a model, we are

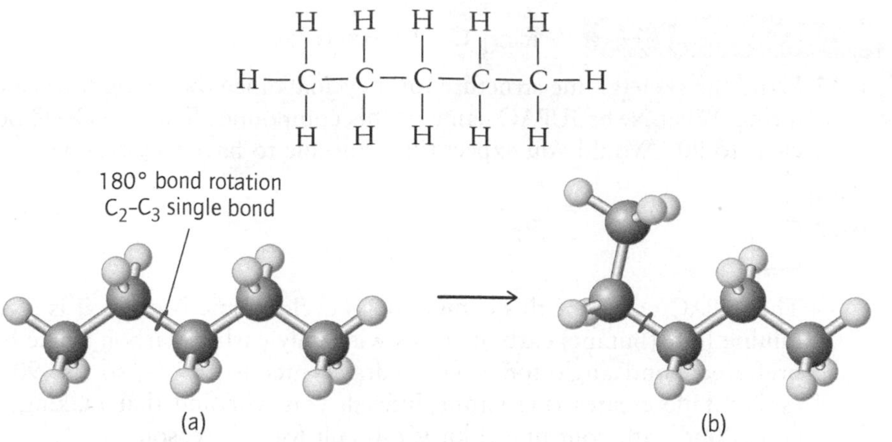

$$H-\underset{\underset{H}{|}}{\overset{\overset{H}{|}}{C}}-\underset{\underset{H}{|}}{\overset{\overset{H}{|}}{C}}-\underset{\underset{H}{|}}{\overset{\overset{H}{|}}{C}}-\underset{\underset{H}{|}}{\overset{\overset{H}{|}}{C}}-\underset{\underset{H}{|}}{\overset{\overset{H}{|}}{C}}-H$$

180° bond rotation
C$_2$-C$_3$ single bond

(a)   (b)

Figure 6-14 Two of many conformations of pentane shown as ball-and-stick models. Rotating 180° around the carbon-carbon single bond labeled with a red ring in (a) results in the conformation shown in (b).

choosing one of many possible **conformations**—a single snapshot in time of a moving object. For example, if you rotate 180° about the C(2)-C(3) bond in the zigzag conformation of pentane shown in Figure 6-14a, the conformation shown in Figure 6-14b results. Although these two models appear different on first inspection, they are not; they are merely different conformations of pentane. The convention is to write hydrocarbon chains in their zigzag conformation; however, this is only one of many conformations, albeit usually the most stable. Work through Using Molecular Models 6-2: Alkane Conformations to gain experience with the free rotation available to the carbon-carbon single bonds of an alkane.

Using Molecular Models 6-2 Alkane Conformations

Construct the Zigzag Conformation of Pentane

1. Obtain five black carbon atoms, 12 light blue hydrogen atoms, and 16 straight bonds.

2. Construct a model of the Lewis structure shown below:

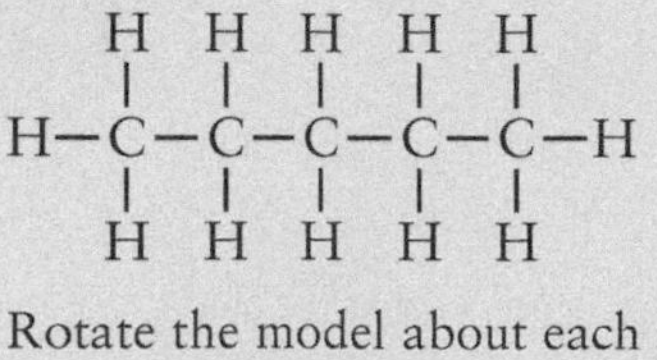

$$H-\underset{\underset{H}{|}}{\overset{\overset{H}{|}}{C}}-\underset{\underset{H}{|}}{\overset{\overset{H}{|}}{C}}-\underset{\underset{H}{|}}{\overset{\overset{H}{|}}{C}}-\underset{\underset{H}{|}}{\overset{\overset{H}{|}}{C}}-\underset{\underset{H}{|}}{\overset{\overset{H}{|}}{C}}-H$$

3. Rotate the model about each of the carbon-carbon single bonds until you arrive at the zigzag conformation shown:

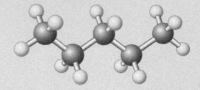

Inquiry Questions

4. What is the molecular shape around each carbon atom? How does the molecular shape around each carbon atom affect the overall shape of the molecule?

5. What are the H—C—H bond angles? Why aren't the bond angles 90° as they appear in the Lewis structure?

6. Did you have to break a C—C or C—H bond to arrive at the zigzag conformation?

7. Do you arrive at different conformations by rotating around C(2)-C(3) or C(3)-C(4)? Do the models appear different? Are they different molecules?

8. Why do molecules rotate about their carbon-carbon single bonds?

Structural Isomers

Although there is only *one* alkane with the molecular formula CH_4, C_2H_6, and C_3H_8, there are *two* alkanes with the molecular formula C_4H_{10}, three alkanes with the molecular formula C_5H_{12}, and five alkanes with the molecular formula C_6H_{14}. Compounds with the *same* molecular formula but a *different* bonding arrangement of atoms are known as **structural isomers**. Structural isomers are different chemical compounds, with different physical and chemical properties and different IUPAC names.

Consider Figure 6-10, which shows the three structural isomers of C_5H_{12}: the three unique ways that five carbon atoms and 12 hydrogen atoms can be assembled while adhering to the octet rule. One of the C_5H_{12} structural isomers has a chain of five carbon atoms with no branching, and therefore it is referred to as the **straight-chain isomer**. It has the IUPAC name pentane. The other two structural isomers have branching along the main chain; hence they are known as **branched-chain isomers**. In a branched-chain alkane at least one hydrogen atom in the chain has been substituted with a carbon atom or chain of carbon atoms. In Figure 6-10b there is a one-carbon branch on the second carbon and in Figure 6-10c there are two one-carbon branches on the second carbon. The rules for naming branched-chain hydrocarbons are described in section 6.5. To give you experience building and recognizing structural isomers, work through Using Molecular Models 6-3: Structural Isomers.

Since structural isomers are different chemical compounds, they have different physical and chemical properties. For example, each of the C_5H_{12} structural isomers has a different boiling point (a physical property). The boiling points are all relatively low as expected because they are all hydrocarbons. However, the straight-chain isomer (Figure 6-10a) has the highest boiling point (35°C), whereas the most highly branched isomer (Figure 6-10c) has the lowest boiling point (9.5°C). The greater surface area of a straight-chain isomer increases dispersion forces and hence the higher boiling point.

> Keep in mind that every structural isomer exists in many different conformations, so there is more than one way to draw any given structural isomer. The definitive test that you are looking at structural isomers is that bonds have to be *broken* and *remade* to convert one structural isomer into another.

Using Molecular Models 6-3 Structural Isomers

Construct the Two Structural Isomers of C_4H_{10}

1. Obtain four black carbon atoms, 10 light blue hydrogen atoms, and 13 straight bonds.
2. Construct a model of butane, the straight-chain alkane with the formula C_4H_{10}.
3. Draw a Lewis structure of this straight-chain structural isomer and answer Inquiry Question 5.
4. Using **only** the atoms from the model you constructed in step 2, construct the branched-chain isomer of C_4H_{10} and then answer Inquiry Questions 6 through 9.

Inquiry Questions

5. Is rotation about the C(2)-C(3) and C(3)-C(4) of the straight-chain isomer possible? Are there many different conformations of this structural isomer? Does one of the conformations have an overall zigzag shape?
6. Did you have to break and remake bonds to build the branched-chain isomer from the straight-chain isomer? How does the fact that you had to break and make bonds indicate you made structural isomers and not different conformations?
7. Draw a Lewis structure of the branched-chain structural isomer.
8. How are the carbon atoms connected differently in the branched-chain isomer than in the straight-chain isomer?
9. Is there any other way to assemble four carbons and 10 hydrogens—structural isomers?

WORKED EXERCISES   Alkane Conformations and Structural Isomers

6-13 Do the following pairs of models represent different conformations of the same molecule or are they structural isomers? If they are structural isomers, explain what makes them different.

a.

b.

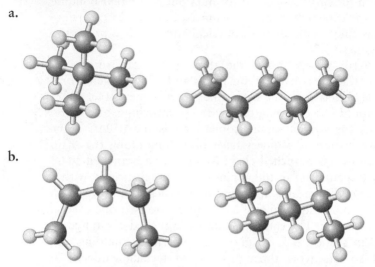

6-14 Which of the following pairs represent structural isomers? Explain.

a.

$$
\begin{array}{c}
H \\
| \\
H-C-H \\
| \\
H\ H\ H\ |\ H \\
|\ \ |\ \ |\ \ |\ \ | \\
H-C-C-C-C-C-H \\
|\ \ |\ \ |\ \ |\ \ | \\
H\ H\ H\ H\ H
\end{array}
\qquad
\begin{array}{c}
H \\
| \\
H-C-H \\
| \\
H\ |\ H\ H\ H \\
|\ \ |\ \ |\ \ |\ \ | \\
H-C-C-C-C-C-H \\
|\ \ |\ \ |\ \ |\ \ | \\
H\ H\ H\ H\ H
\end{array}
$$

b.

$$
\begin{array}{c}
H \\
| \\
H-C-H \\
| \\
H\ |\ H\ H \\
|\ \ |\ \ |\ \ | \\
H-C-C-C-C-H \\
|\ \ |\ \ |\ \ | \\
H\ H\ H\ H
\end{array}
\qquad
\begin{array}{c}
H \\
| \\
H-C-H \\
| \\
H\ |\ H \\
|\ \ |\ \ | \\
H-C-C-C-H \\
|\ \ |\ \ | \\
H\ |\ H \\
| \\
H-C-H \\
| \\
H
\end{array}
$$

Solutions

6-13 a. The two ball-and-stick models represent structural isomers because they do not have the same carbon atom connectivity. Although both models contain five carbon atoms and 12 hydrogen atoms, in the first model one carbon atom is bonded to four other carbon atoms (a branched-chain hydrocarbon), whereas in the second model all five carbon atoms are bonded to at most two carbon atoms (a straight-chain hydrocarbon).

b. The two ball-and-stick models represent two different conformations of pentane. The carbon atoms are connected to each other in exactly the same way: a straight chain of five carbon atoms, each with two hydrogen atoms except for each carbon on the end of the chain, which has three hydrogen atoms. They differ only in rotation about carbon-carbon bond(s), so they are different conformations.

6-14 a. These pairs are two different views of the *same* compound: a chain of five carbon atoms with a branch on the second carbon from one end. One structure is simply the flipped view of the other structure.

> **b.** These pairs represent a pair of structural isomers because they have the same formula, C_5H_{12}, but the carbon and hydrogen atoms are joined differently: the structure on the left has a one-carbon branch on the second carbon, whereas the structure on the right has two one-carbon branches on the second carbon.

PRACTICE EXERCISES

16 Which of the following pairs represent different conformations of the same model? Which of the following pairs represent structural isomers? For any pairs that are structural isomers, explain what makes them different.

a.

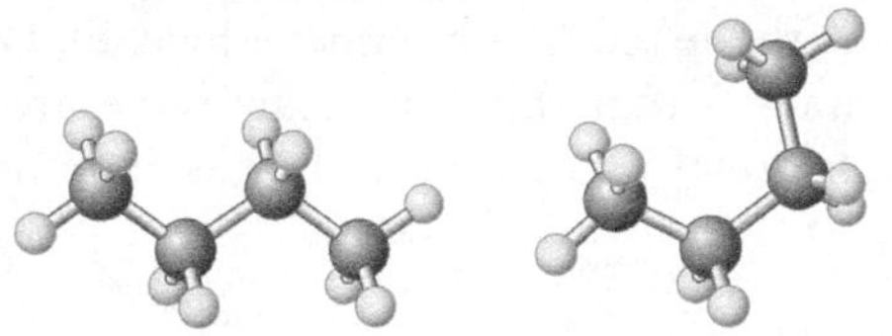

b.

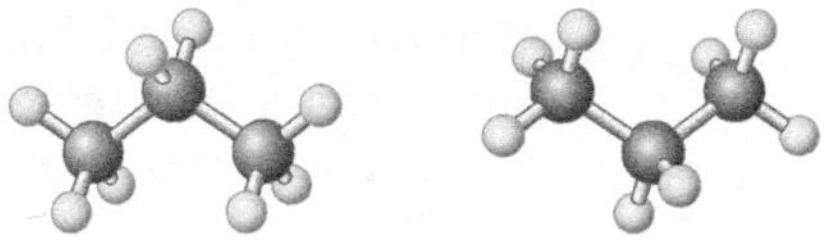

c.

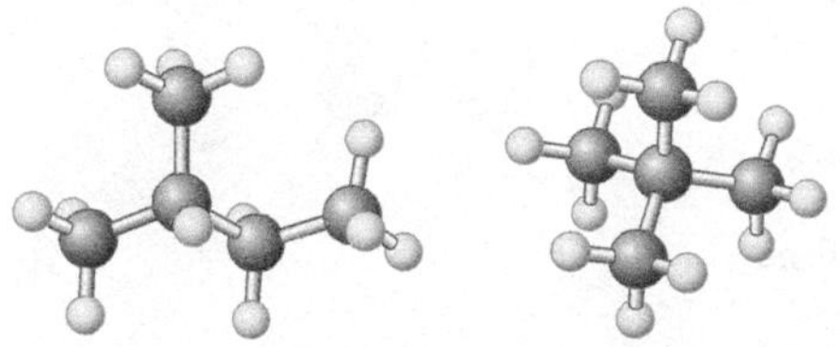

17 Would you expect the number of structural isomers to increase with increasing number of carbon atoms in the molecular formula? Why is there only one structural isomer of C_2H_6?

18 Which of the following represent a pair of structural isomers? Explain.

a.

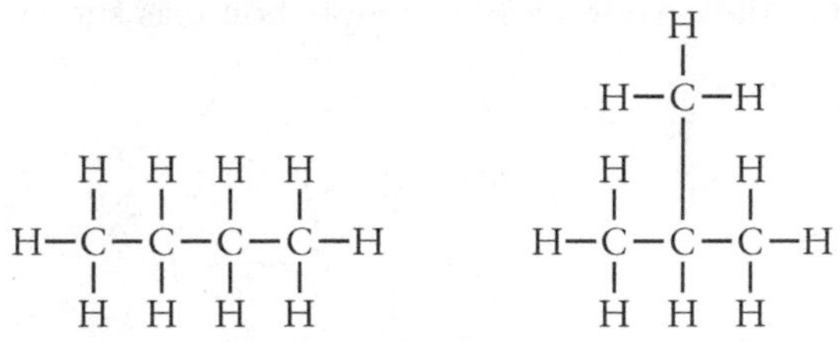

b.

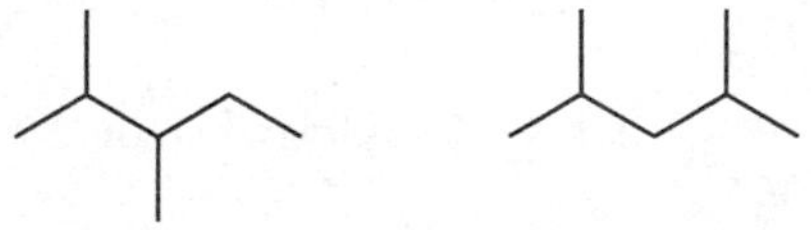

c.

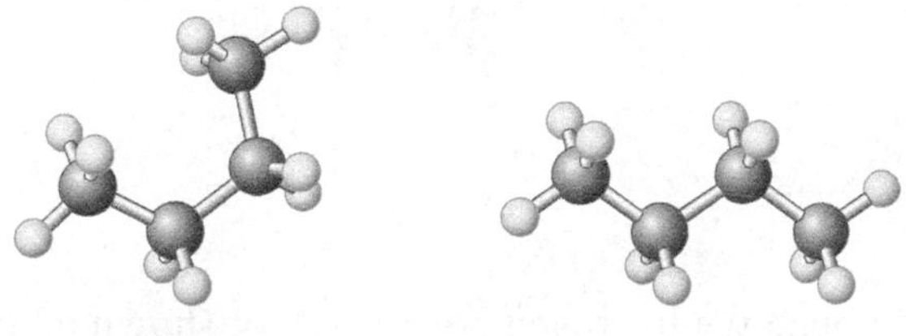

d.

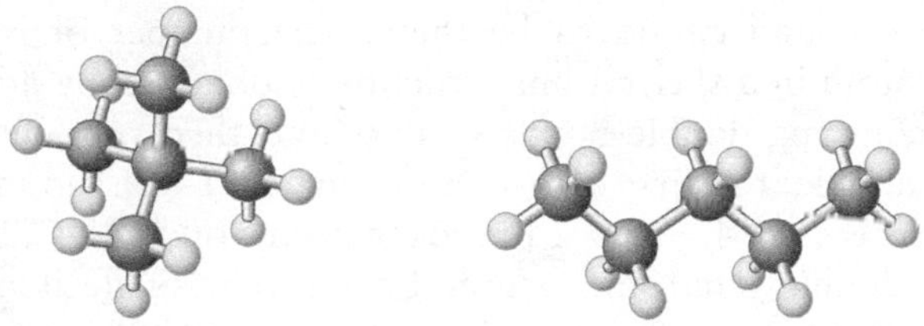

19 Which pairs of compounds in the previous exercise would you expect to have different physical properties?

20 Write the Lewis structure of the five structural isomers of C_6H_{14}. Label and name the straight-chain isomer. Be sure you have not repeated any structures by writing two different views or conformations of the same structural isomer.

■ 6.4 Alkenes and Alkynes

Alkenes, alkynes, and aromatic hydrocarbons are *unsaturated* hydrocarbons because they contain one or more carbon-carbon double or triple bonds, and therefore they have fewer than the maximum number of hydrogen atoms per carbon. In this section we describe the structural characteristics of alkenes and alkynes, and in section 6.6 we will describe the structural characteristics of aromatic hydrocarbons.

Alkenes

An **alkene** is a hydrocarbon that contains one or more C—C double bonds. In chapter 3 you were introduced to ethene (common name: ethylene), C_2H_4, the simplest alkene. Recall from chapter 3 that the molecular shape around the C—C double bond of ethene is trigonal planar, with H—C—H bond angles of 120°.

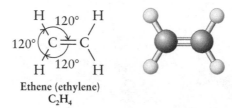

Ethene (ethylene)
C_2H_4

> Ethylene, a gas, is a plant hormone that promotes ripening in fruit.

In the condensed structure for an alkene, the double bond may be drawn in, or inferred, as shown for but-1-ene below. The skeletal line structure is written in the customary zigzag fashion showing four carbon atoms and writing parallel lines to signify the carbon-carbon double bond, as shown below:

IUPAC name	but-1-ene
Lewis structure	$\underset{H}{\overset{H}{\underset{|}{C}}}=\underset{H}{\overset{H}{\underset{|}{C}}}-\underset{H}{\overset{H}{\underset{|}{C}}}-\underset{H}{\overset{H}{\underset{|}{C}}}-H$
Condensed structure	$\overset{1}{CH_2}=\overset{2}{CH}\overset{3}{CH_2}\overset{4}{CH_3}$ or $\underset{1}{CH_2}\underset{2}{CH}\underset{3}{CH_2}\underset{4}{CH_3}$
Skeletal line structure	

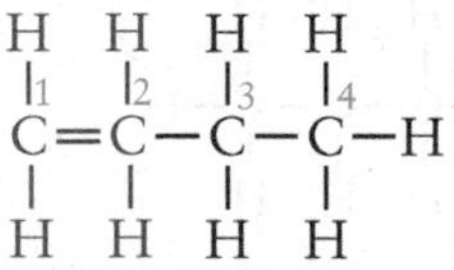

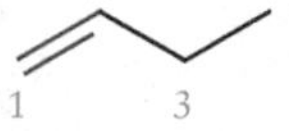

Remember that although the hydrogen atoms are not shown in a skeletal line structure, they are understood to be there. The number of hydrogen atoms on a carbon atom in a skeletal line structure is obtained by adding all the bonds shown, counting double bonds as two, and then subtracting this sum from 4. For example, the first carbon atom in but-1-ene shows only a double bond; hence there are $4 - 2 = 2$ hydrogen atoms on C(1). C(2) shows one carbon-carbon double bond and one carbon-carbon single bond for a total of three; hence there is $4 - 3 = 1$ hydrogen atom on C(2).

An alkene containing two C—C double bonds is called a **diene**, where *di-* means "two." An alkene containing several C—C double bonds is called a **polyene**, where *poly-* means "many"—in this case, many double bonds. Polyenes with alternating double and single bonds are known as **conjugated** polyenes. Polyenes with extensive conjugation have a particularly stable

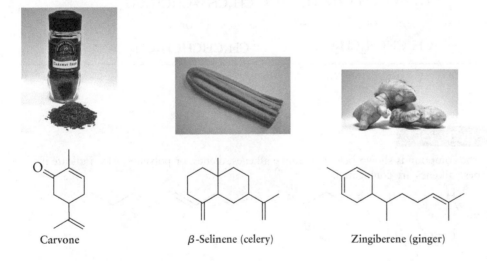

β-Carotene, a source of vitamin A

Figure 6-15 β-Carotene is the conjugated polyene that gives carrots their distinctive orange color. [© Sue Wilson/Alamy]

arrangement of electrons. Conjugation also causes these compounds to absorb light in the visible range, and therefore these compounds tend to be brightly colored. Indeed, it is conjugated polyenes that give carrots, tomatoes, and other brightly colored fruits and vegetables their distinctive colors. The hydrocarbon β-carotene, from which vitamin A is derived and which gives carrots their orange color, is an example of a conjugated polyene (Figure 6-15). Other polyenes without conjugation are also found in fruits, vegetables, and spices as shown in the structure of carvone, found in caraway seeds, β-selinene, found in celery, and zingiberene, found in ginger (Figure 6-16).

Figure 6-16 Some dienes and polyenes found in various fruits, vegetables, and spices. [Catherine Bausinger]

Carvone

β-Selinene (celery)

Zingiberene (ginger)

WORKED EXERCISES   Alkenes

6-15 Accutane is used to treat severe acne. Aside from the heteroatoms, is Accutane a simple alkene, diene, or polyene? Is it a conjugated polyene? How many carbon-carbon double bonds does Accutane contain?

Accutane
(isotretinoin)

6-16 Indicate whether the compound below is a simple alkene, a diene, or a polyene. Is it conjugated? How many hydrogen atoms are there on each carbon in the alkene?

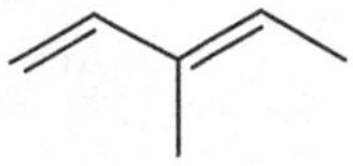

6-17 Provide the IUPAC name for the following simple alkenes:

a.    b. $CH_3CH = CHCH_3$

6-18 Provide a skeletal line structure and a condensed structure for pent-1-ene and pent-2-ene. Are these two compounds structural isomers?

Solutions

6-15 Accutane is a conjugated polyene containing five carbon-carbon double bonds.

6-16 This compound is a conjugated *di*ene because there are *two* carbon-carbon double bonds separated by one carbon-carbon single bond. The number of hydrogen atoms on each carbon is indicated below.

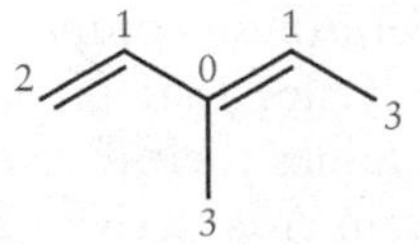

6-17 a. hept-1-ene   **b.** but-2-ene

6-18 These alkenes are structural isomers.

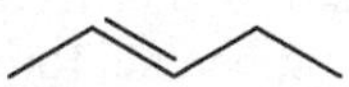

$$CH_2 = CHCH_2CH_2CH_3$$
or
$$CH_2CHCH_2CH_2CH_3$$

$$CH_3CH = CHCH_2CH_3$$
or
$$CH_3CHCHCH_2CH_3$$

PRACTICE EXERCISES

21 Identify the compounds shown below as simple alkenes, dienes, or polyenes. Also indicate if any of these alkenes are conjugated.

a. 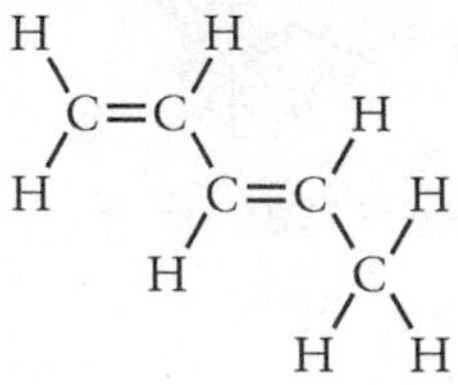   b. 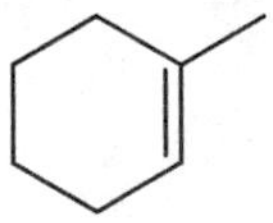   c.

22 What is the bond angle around each of the carbon atoms in compound (a) of exercise 21?

23 The structure of vitamin A is shown below:

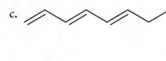

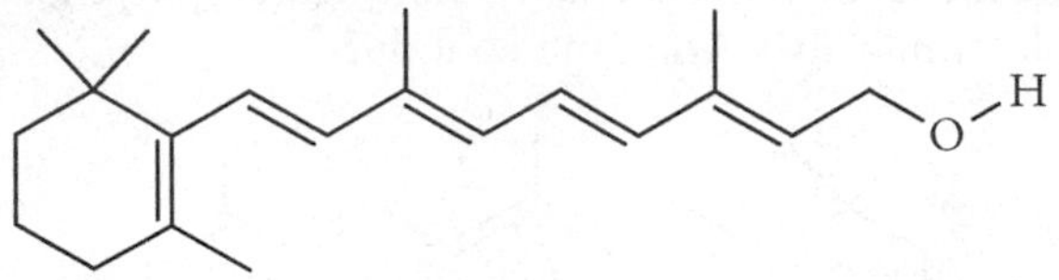

Retinol
(vitamin A)

a. How many carbon-carbon double bonds does vitamin A have?
b. In what way does vitamin A resemble β-carotene?

24 Lycopene is the pigment that gives many vegetables such as tomatoes their rich red color. What structural characteristic of lycopene is responsible for its red color? How many carbon-carbon double bonds does lycopene contain?

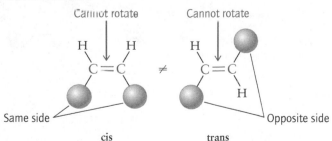

Lycopene

Geometric Isomers

The most significant difference between carbon-carbon single bonds and carbon-carbon double bonds is that double bonds cannot rotate freely about the double bond (Figure 6-17). This means that the four atoms or groups of atoms directly attached to the carbon-carbon double bond are fixed in space relative to one another. Consequently, when there are two *different* atoms or groups of atoms on *both* double-bond carbon atoms, they can be arranged in *two* possible orientations, giving rise to two geometric isomers. *Geometric isomers have the same chemical formula* and *the same connectivity of atoms but a different three-dimensional orientation of atoms as a result of the restricted rotation about the carbon-carbon double bond.*

To identify geometric isomers, compare the two larger groups on each of the double-bond carbon atoms (shown as spheres in Figure 6-17). If they are on the *same side* of the double bond, it is the **cis** geometric isomer. If instead they are on *opposite sides* of the double bond, it is the **trans** geometric isomer. Note that for the cis-trans naming system to apply, there must also be one hydrogen on each double bond carbon.

> *cis:*   The larger groups on both double-bond carbons are on the *same side* of the double bond.

> *trans:*   The larger groups on both double-bond carbons are on the *opposite sides* of the double bond.

The pair of geometric isomers shown in Figure 6-18 whose IUPAC names are *cis*-but-2-ene and *trans*-but-2-ene are shown as ball-and-stick models, space-filling models, and Lewis structures. As you can see, the four carbon atoms in the chain are connected in the same way. However, there are two possible spatial arrangements as a result of the restricted rotation of the double bond: either the larger groups on each double-bond carbon (shown circled in blue in the Lewis structure) are on the *same side* of the double bond or they are on *opposite sides* of the double bond. The difference in the overall molecular

Geometric Isomers:

Cannot rotate   Cannot rotate

cis   trans

Same side   Opposite side

Figure 6-17 Carbon-carbon double bonds cannot rotate freely about the double bond. Geometric isomers exist when two different atoms or groups of atoms are found on both carbon atoms of the carbon-carbon double-bond. The *cis* isomer has the larger groups (shown here as black spheres) on the same side of the double bond, whereas the *trans* isomer has them on the opposite sides of the double bond.

Figure 6-18 The overall shape of a pair of geometric isomers is distinctly different, as seen here in the ball-and-stick and space-filling models of (a) *cis*-but-2-ene and (b) *trans*-but-2-ene. In the Lewis structures the two larger groups on the double-bond carbons are circled.

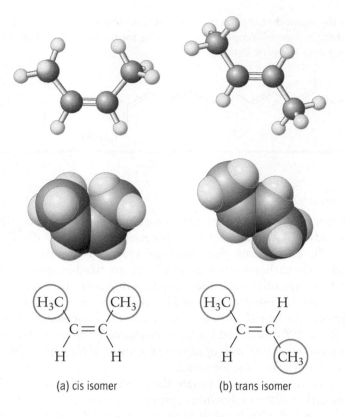

(a) cis isomer   (b) trans isomer

shape of these two geometric isomers is most apparent when you compare the ball-and-stick and space-filling models shown in Figure 6-18. To gain greater familiarity with geometric isomers, work through Using Molecular Models 6-4: Geometric Isomers.

Geometric isomers are named according to the IUPAC rules for naming alkenes with the added prefix *cis-* or *trans-* included in front of the root name.

Using Molecular Models 6-4 Geometric Isomers

Construct a Model of Butane

1. Obtain eight black carbon atoms, 16 light blue hydrogen atoms, 23 straight bonds, and four bent bonds.

2. Construct a model of butane, C_4H_{10} and answer Inquiry Question 5.

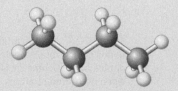

Construct *cis-* and *trans*-but-2-ene

3. Remove one hydrogen atom and its C—H bond from both C(2) and C(3) of the model you

made above. Next, replace the C(2)–C(3) single bond with two bent bonds to represent a double bond.

4. Build a model of the other geometric isomer. Your models should look like the ball-and-stick models shown below. Answer Inquiry Questions 6 through 12.

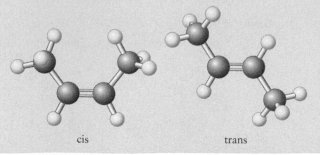

cis   trans

Inquiry Questions

5. Is there free rotation around C(2)–C(3) of butane? Does the molecule look different when rotating around this bond? Does butane have a geometric isomer? Explain.

6. Does each carbon have the same type and number of bonds in both geometric isomers?

7. Which of your models is the cis isomer? How can you tell?

8. Which of your models is the trans isomer? How can you tell?

9. Superimpose the cis and trans models on top of each other. Without breaking bonds but rotating about any single bonds, can you make the two models overlay? Why or why not?

10. Try rotating about the C(2)–C(3) double bond. Were you successful? Why not?

11. What must you do to turn the model of the cis isomer into the trans isomer? Do you have to break bonds to do so?

12. From your models, how can you tell the cis and trans models are geometric isomers and not simply different conformations?

WORKED EXERCISE | Assigning IUPAC Names to Geometric Isomers

6-19 Draw the skeletal line structure for each of the following alkenes:

    a. oct-1-ene   **b.** *trans*-oct-4-ene   **c.** *cis*-oct-2-ene

Solution

6-19 a.

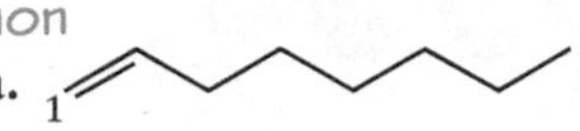

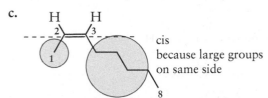

Write a zigzag structure showing eight points and place two parallel lines between the first two (or last two) carbon atoms.

b.

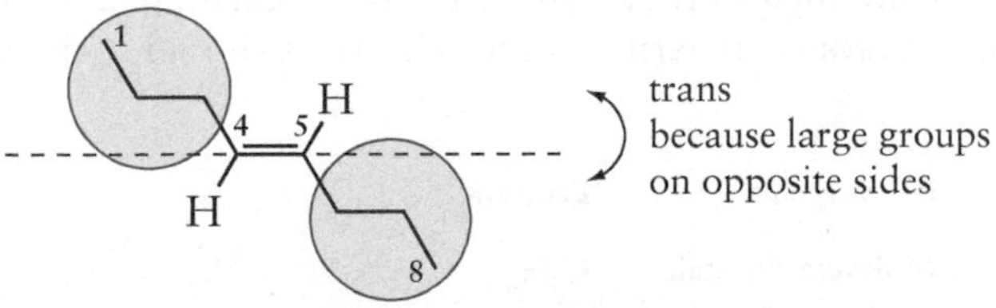

trans
because large groups
on opposite sides

Write a zigzag structure with eight points and place two parallel lines between the fourth and the fifth carbon atoms. The structure is the trans isomer when drawn as a zigzag because the two larger groups (shown in pink) appear on opposite sides of the double bond.

c.

cis
because large groups
on same side

A cis double bond cannot be drawn as a zigzag. Begin by writing two parallel lines to represent the double bond between C(2) and C(3). Then add the two large groups on the same side of the double bond (the two H atoms will therefore also be on the same side of the double bond). You determine that the two large groups are a —CH_3 and a —$CH_2CH_2CH_2CH_2CH_3$ because you need a chain of eight carbons altogether.

PRACTICE EXERCISES

25 For the following pairs of compounds, indicate if they are cis-trans isomers or the same compound. If they are cis-trans isomers, indicate which is the cis isomer and which is the trans isomer and explain why.

a.

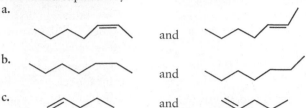

and

b.

and

c.

and

26 Write the structure of *cis*-pent-2-ene and *trans*-pent-2-ene. How are these two compounds different?

Alkynes

Alkynes contain one or more carbon-carbon triple bonds. The simplest alkyne is ethyne, CH≡CH, also known by its common name, acetylene. Alkynes have a linear shape around the carbon atoms of the carbon-carbon triple bond and an H—C—C bond angle of 180°. Therefore, the zigzag convention used for writing skeletal line drawings of alkanes and alkenes is typically not used for triple bonds.

180°

H—C≡C—H

Ethyne (acetylene)
C_2H_2

When writing the skeletal line structure of an alkyne use three parallel lines and form 180° bond angles around the triple bond carbon(s). Sometimes the triple-bond carbon atoms are written in for clarity as shown in the example of propyne below.

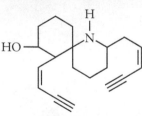

Histrionicotoxin

Figure 6-19 Histrionicotoxin, a compound with two triple bonds (alkyne), is the neurotoxin produced by the poison dart frog. The shape of the molecule is similar to the neurotransmitter acetylcholine, which transmits nerve impulses in the brain. Its structural similarity to acetylcholine accounts for the neurotoxin's paralytic effect. [iStockphoto/Thinkstock]

IUPAC name	Propyne
Molecular formula	C_3H_4
Lewis structure	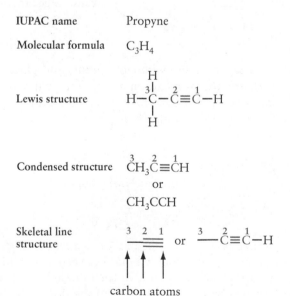
Condensed structure	$\overset{3}{C}H_3\overset{2}{C}\equiv\overset{1}{C}H$ or CH_3CCH
Skeletal line structure	

carbon atoms

Alkynes are not common in nature, although they are found in some compounds. For example, the poison dart frog (Figure 6-19), native to the South American rain forest, produces the toxic alkyne known as histrionicotoxin to ward off predators. This molecule has two triple bonds (and two double bonds), as seen in the triple bond with a linear arrangement of atoms in the skeletal line structure shown in Figure 6-19.

WORKED EXERCISE | Assigning IUPAC Names to Alkynes

6-20 Write the Lewis structure, condensed structure, and skeletal line structure for but-1-yne and but-2-yne.

Solution

6-20 The root for these two alkynes is *butane*, which indicates four carbon atoms in the chain. The *-yne* ending indicates that a triple bond is present. The -1- in but-1-yne indicates that the triple bond is located between C(1) and C(2). The -2- in but-2-yne indicates that the triple bond is between C(2) and C(3). Generally, a bond attached to a triple-bond carbon is shown in a linear arrangement rather than as a zigzag in the skeletal line structure.

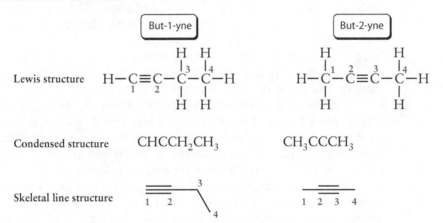

PRACTICE EXERCISES

27 Write the skeletal line structure and the Lewis structure for the following alkynes:
  a. oct-3-yne
  b. pent-1-yne

28 Name the straight-chain hydrocarbons shown, all containing five carbons. What is the molecular formula for each? Which is a saturated hydrocarbon?

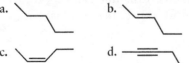

29 Provide the IUPAC name for the following alkynes:
  a. 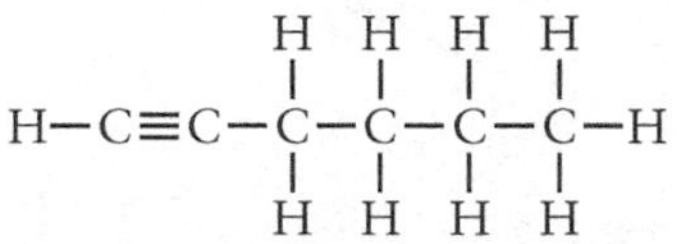
$$H-C\equiv C-\overset{\displaystyle H}{\underset{\displaystyle H}{C}}-\overset{\displaystyle H}{\underset{\displaystyle H}{C}}-\overset{\displaystyle H}{\underset{\displaystyle H}{C}}-\overset{\displaystyle H}{\underset{\displaystyle H}{C}}-H$$
  b.

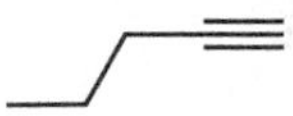

■ 6.5 Naming Branched-Chain Hydrocarbons

A branched-chain hydrocarbon has one or more hydrogen atoms along its main chain *substituted* with a carbon branch containing one or more carbons, known as a **substituent** because it "substitutes" for what would normally be a

TABLE 6-4 IUPAC Root Names for Alkanes with 1 to 10 Contiguous carbon Atoms

Number of Carbon Atoms	IUPAC Root Name	Condensed Structure	Skeletal Line Structure
1	Methane	CH_4	Not applicable
2	Ethane	CH_3CH_3	—
3	Propane	$CH_3CH_2CH_3$	
4	Butane	$CH_3(CH_2)_2CH_3$	
5	Pentane	$CH_3(CH_2)_3CH_3$	
6	Hexane	$CH_3(CH_2)_4CH_3$	
7	Heptane	$CH_3(CH_2)_5CH_3$	
8	Octane	$CH_3(CH_2)_6CH_3$	
9	Nonane	$CH_3(CH_2)_7CH_3$	
10	Decane	$CH_3(CH_2)_8CH_3$	

hydrogen atom in a straight-chain hydrocarbon. The root name is assigned based on the number of carbon atoms in the longest contiguous chain—known as the main chain or parent chain—using Table 6-4. The rules for identifying the main chain in a branched hydrocarbon are described below. A prefix is then inserted in front of the root that indicates where and what substituents are found on the main chain.

An alkyl substituent consists of only carbon and hydrogen atoms, and is named according to the number of carbon atoms, using the same guidelines used to name the main chain, given in Table 6-4, except the ending is changed from -*ane* to -*yl*. A locator number is inserted in front of each substituent name to indicate where on the main chain it is located.

Most of the effort in naming branched-chain hydrocarbons goes into constructing the prefix. Don't let long and complicated names scare you: the rules are systematic and logical. After a little practice, you'll be able to name branched hydrocarbons. A step-by-step process for naming branched hydrocarbons is described in the Guidelines box below.

> Structural isomers always have different IUPAC names.

Guidelines for Naming Branched-Chain Hydrocarbons

Rule 1a: Locate and assign the root for the main chain. Count the number of carbon atoms in the longest contiguous chain of carbon atoms and assign this sequence as the main chain. Assign the root name to this sequence of carbon atoms in accordance with Table 6-4. *Note that the main chain is not always drawn from left to right.*

For example, the compound shown below has a main chain composed of nine carbon atoms (not seven as drawn horizontally from left to right), with a branch off the third and sixth carbon atoms, so the root for this branched alkane is *nonane*.

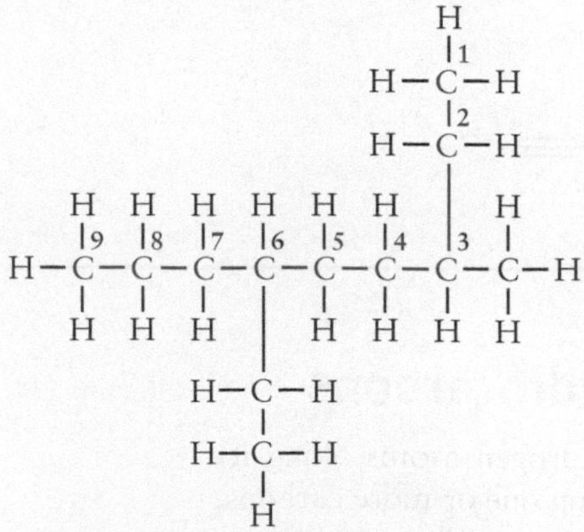

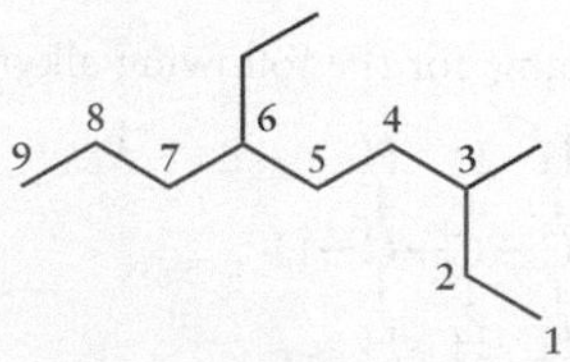

...nonane

Rule 1b: If there is a carbon-carbon double or triple bond in the molecule, select the longest contiguous chain that contains both carbon atoms of the double or triple bond as the main chain even if there is a longer chain in the molecule.

For example, the compound shown below has a main chain composed of four carbon atoms because both carbon atoms of the double bond must be included as part of the main chain even though there is a longer five-carbon chain.

Remember, the ending is changed for an alkene from -ane to -ene to signify the molecule is an alkene. A locator number is also inserted between the root and the ending to indicate where the first carbon of the double bond appears in the main chain. So the example below has a root and ending:
. . .but-1-ene, *with a branch at C(2).*

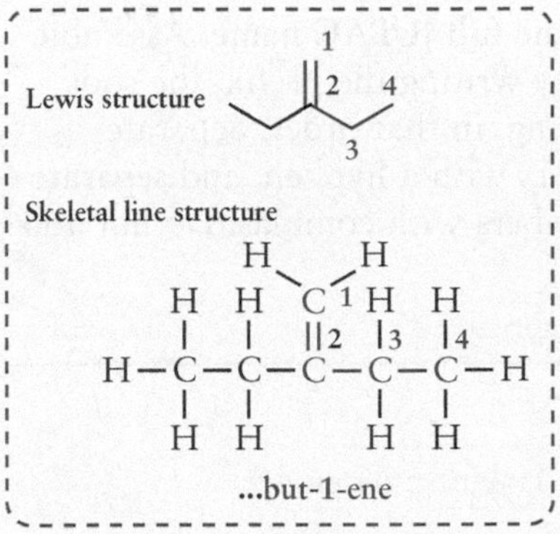

Rule 1c: If a molecule has two possible main chains with the same number of carbon atoms, choose the chain with the greater number of *substituents* as the main chain. *For example, the correct five-carbon main chain in the example below is the one highlighted in red because there are two substituents on the pentane main chain, and not the one highlighted in blue, which has only one substituent on the pentane main chain.*

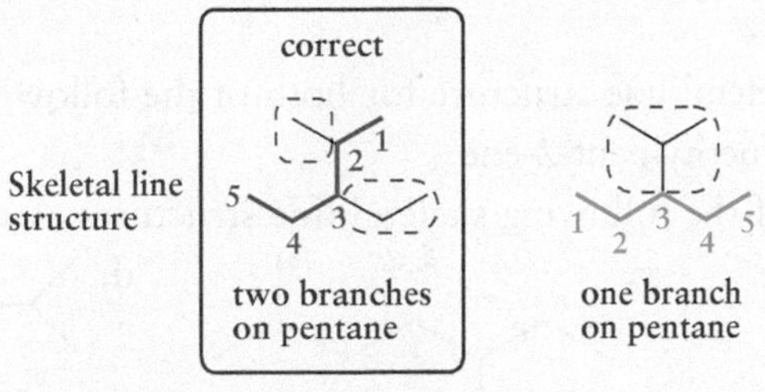

Rule 2: Name each substituent. Assign a name to each substituent based on the number of carbon atoms it contains, using Table 6-4. Change the ending from *-ane* to *-yl* to signify that it is a substituent and not the main chain. A *-yl* ending always signifies a substituent.

For example, if the substituent is a —CH_3, the substituent name is methyl, not methane. If the substituent is —CH_2CH_3, the substituent name is ethyl. Methyl and ethyl are the most common substituents. In the example below there is a one-carbon branch—methyl—and a two-carbon branch—ethyl—on a nonane main chain.

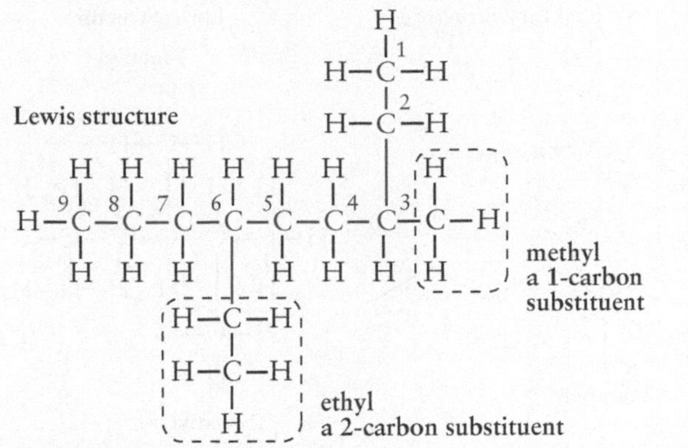

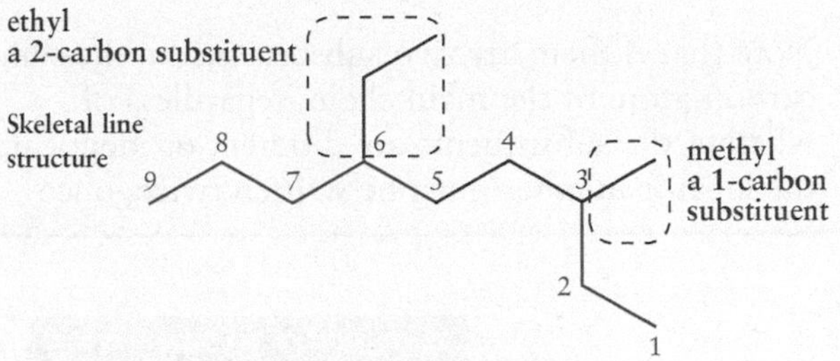

6-ethyl-3-methylnonane

Rule 3: Assign a locator number to each substituent. Number the main chain starting from the end closer to the first substituent (first branch point) unless there is a multiple bond, in which case numbering must begin from the end closer to the multiple bond. Assign a locator number to each substituent based on where along the main chain it is located. If there are *two* substituents on the same carbon atom, cite the locator number twice, once for each substituent. Place each locator number before its associated substituent name separated by a hyphen. *In the example above, there is a 3-methyl substituent and a 6-ethyl substituent.*

Rule 4a: Assemble the prefix. List the substituent names in *alphabetical order* preceded by each of their locator numbers. Place the prefix before the root. Letters should be separated from numbers by a hyphen, and numbers separated from numbers by commas. In the example above, the prefix is *6-ethyl-3-methyl* . . .

Rule 4b: Use a multiplier if a substituent appears more than once. If the same substituent appears more than once along the main chain, insert the multiplier *di-* (for 2), *tri-* (for 3), or *tetra-* (for 4) in front of the repeating substituent name to indicate how many times the substituent appears along the main chain. Place locator numbers in front of the *multiplier*, separated by commas, corresponding to the location of each of the repeated substituents *For example, the main chain below has two methyl* substituents at C(2) and C(3), *so write 2,3-dimethyl . . .*

Skeletal Line Structure | **Lewis Structure**

3-methyl

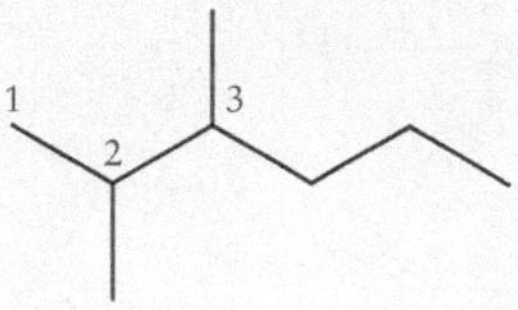

2-methyl

3-methyl

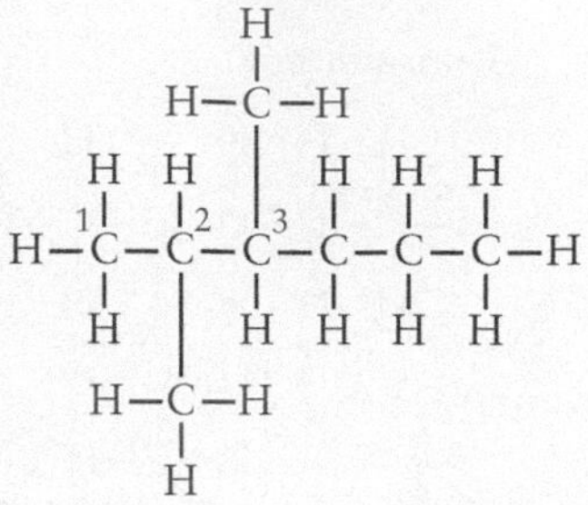

2-methyl

2,3-dimethylhexane

Note that if there are two substituents on the same carbon atom of the main chain, regardless of whether the substituents are different or identical, the locator number must be written twice, once for each substituent. *For example, the prefix 2,2-dimethyl . . . indicates that there are two methyl substituents along the main chain and that they are both located on the second carbon of the chain.*

Skeletal Line Structure | **Lewis Structure**

2-methyl 2-methyl

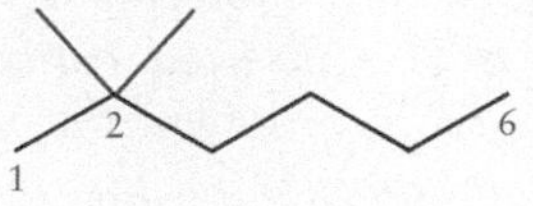

2-methyl

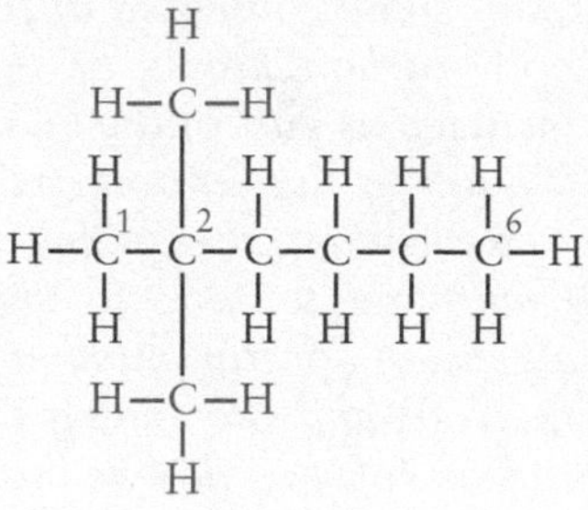

2-methyl

2,2-dimethylhexane

Rule 5: Assemble the full IUPAC name. Assemble the IUPAC name by writing the prefix, the root name, and the ending, in that order. Separate numbers from letters with a hyphen, and separate numbers from numbers with commas. Do not add spaces.

WORKED EXERCISES Naming Branched-Chain Hydrocarbons

6-21 Provide the IUPAC name for the Lewis structure shown:

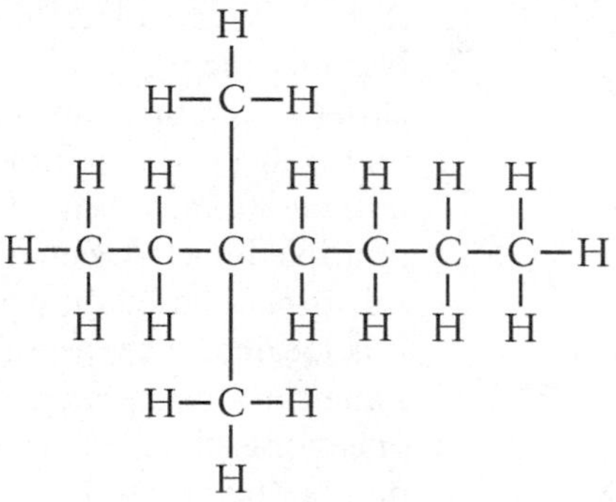

6-22 Provide a Lewis structure and a skeletal line structure for both of the following:

a. 2,2-dimethylnonane **b.** 2-methylpent-2-ene

6-23 Provide an IUPAC name for each of the following skeletal line structures:

a. 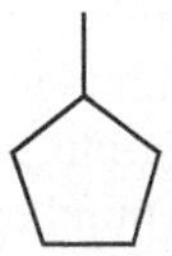 **b.** 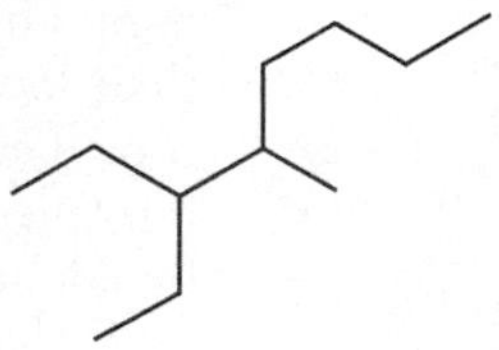 **c.** 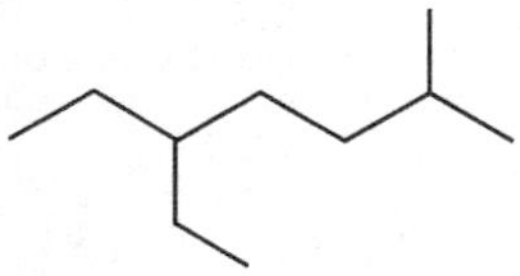 **d.** 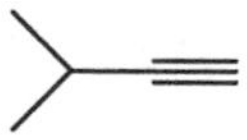

Solutions

6-21 The IUPAC name is 3,3-dimethylheptane. The longest contiguous chain of carbons is seven carbons, so the root for the main chain is heptane. The heptane main chain is numbered from the end closer to the first substituent, so numbering starts from the left as drawn. The prefix is 3,3-dimethyl because there are two —CH_3 groups on carbon 3. Assembling the prefix, main chain, and ending gives 3,3-dimethylheptane. Note that numbers are separated by a comma and numbers and letters are separated by a hyphen.

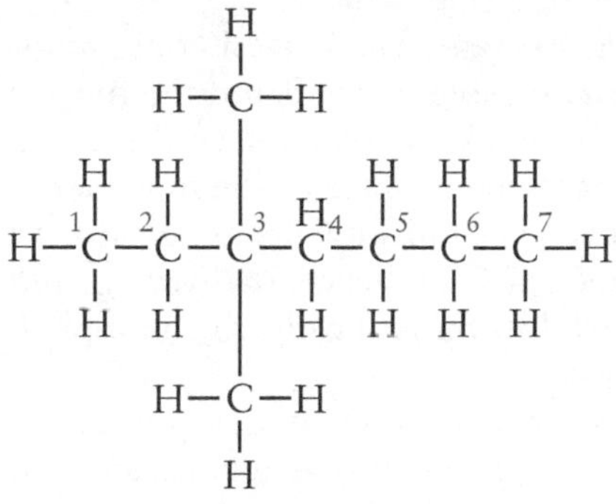

$$
\begin{array}{c}
\text{H} \\
| \\
\text{H}-\text{C}-\text{H} \\
\end{array}
$$

3,3-dimethylheptane

6-22 a. From the name we determine the root is nonane, so write a chain of nine carbons. Number the chain from one end to the other. It does not matter from which end you start numbering. The prefix is 2,2-dimethyl, so replace two hydrogen atoms on carbon 2 of the nonane chain with CH_3 groups (one carbon branches). In a Lewis structure, write one substituent above the chain and one below. In a skeletal line structure, write both substituents in the same direction as the corresponding point in the carbon chain.

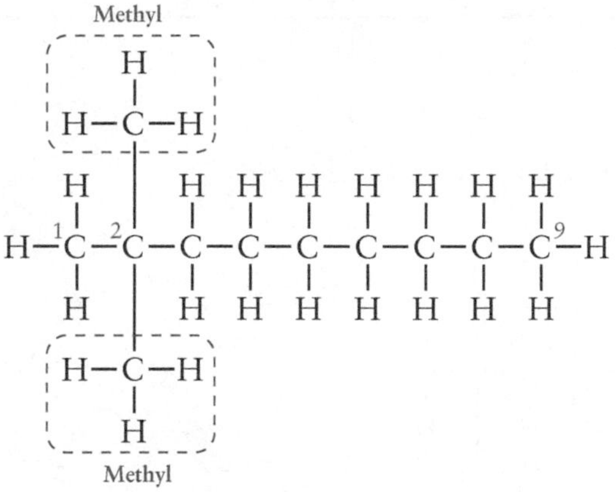

Methyl

Methyl

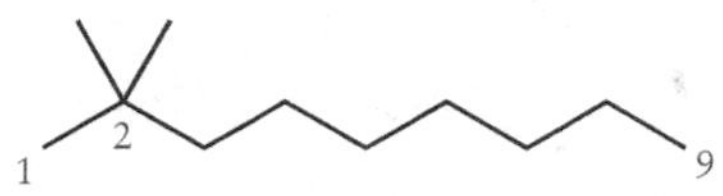

b. From the name we determine the root is pentane and write a chain of five carbons and number starting from either end. We see that the ending is -2-ene, so we know there is a double bond between C(2) and C(3) because the *-ene* ending indicates that there is a double bond and the 2- indicates the double bond is between C(2) and C(3). The prefix *2-methyl-* indicates that a hydrogen is replaced with a methyl group ($-CH_3$) at carbon 2. Since it is an alkene, determine whether there are two possible geometric isomers. If there are, determine if you have to make adjustments to show cis or trans. In this case, since C(2) has two identical groups, it does not have a geometric isomer.

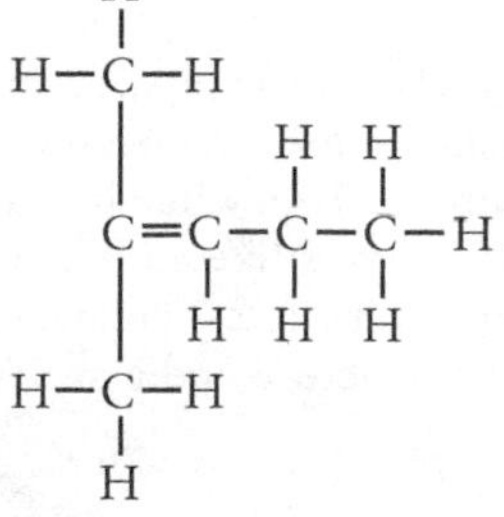

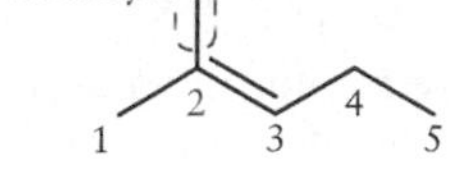

2-methyl

6-23 a. Methylcyclopentane. The main chain is a ring containing five carbons; hence the root cyclopentane. There is a one-carbon branch, so we add the prefix methyl . . . No locator number is needed for a cycloalkane with only one branch because every position on a ring is the same.

   b. 3-Ethyl-4-methyloctane. The longest contiguous chain has eight carbons; hence, the root is *octane*. Note that it is not the five-carbon chain that you see along the horizontal. It is an alkane, so we keep the *-ane* ending. Start numbering from the end of the octane chain closer to the first branch, which is from the left as drawn here. Determine what substituents are on the main chain and where. There is a 2-carbon branch at C(3); hence *3-ethyl* . . . , and a one-carbon branch at C(4); hence *4-methyl*. List these substituents in alphabetical order and place them in front of the root.

   c. 5-ethyl-2-methyl-heptane. The longest contiguous chain has seven carbon atoms; hence the root is *heptane*. It is an alkane, so we keep the *-ane* ending. Start numbering from the end closer to the first substituent, which is from the right as drawn here. There are two substituents along the main chain: a one-carbon branch at C(2); hence *2-methyl* . . . , and a two-carbon branch at C(5); hence *5-ethyl*. List the substituents in alphabetical order and place them in front of the root.

   d. 3-methylbut-1-yne. The longest contiguous chain containing the triple bond has four carbon atoms, so the root is *butane*. The ending is changed to *butyne* because there is a triple bond. Number the main chain from the end closer to the triple bond. Since the triple bond occurs between C(1) and C(2), we insert a -1- between the root and the ending: but-1-yne. There is a one-carbon branch at C(3), so we insert the prefix 3-methyl in front of the root: *3-methylbut-1-yne*.

PRACTICE EXERCISES

30 Write the IUPAC name for the following branched-chain alkanes:

   a.   b.

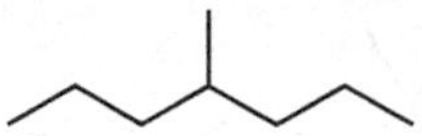

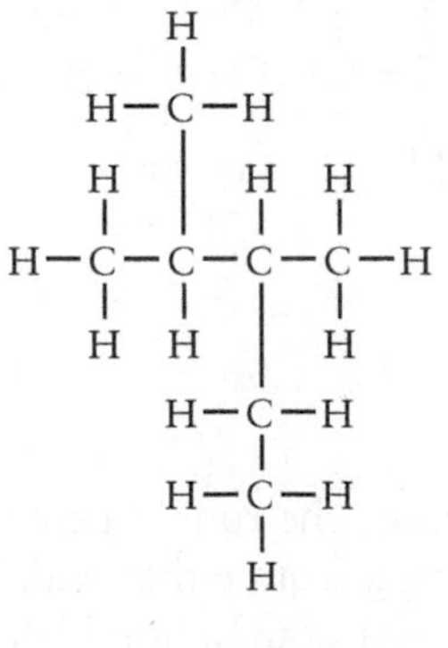

31 Write the Lewis structure and the skeletal line structure for the branched-chain alkanes with the following IUPAC names:

   a. 2-methyloctane   b. 3-ethylhexane
   c. ethylcyclopentane   d. 3,3-diethyl-5-methylheptane

32 Name the following branched-chain alkenes (there are no cis-trans isomers):

   a.   b. 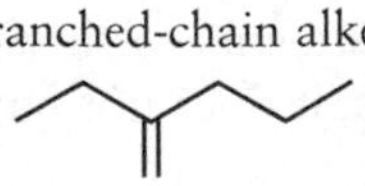   c.

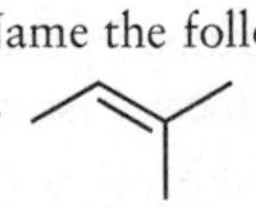

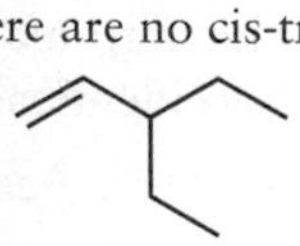

33 Octane ratings on gasoline are an indication of how well the fuel burns compared to a mixture of isooctane (common name) and heptane—hence the term *octane rating*. Isooctane has the IUPAC name 2,2,4-trimethylpentane, which has an octane rating of 100, while heptane has an octane rating of 0. Thus, a fuel with an octane rating of 87 behaves like an 87:13 mixture of 2,2,4-trimethylpentane and heptane. A higher octane rating results in less engine knock (premature ignition of the fuel in the cylinder). Provide a Lewis structure and a skeletal line structure for both isooctane and heptane.

34 Write the structure of cyclohexene.

6.6 Aromatic Hydrocarbons

In the nineteenth-century, chemists isolated a number of hydrocarbons that had strong aromas and therefore came to be known as **aromatic** compounds. Vanillin is one such fragrant aromatic compound (Figure 6-20), albeit not a hydrocarbon. Later, aromatic compounds were found to consist of a unique unsaturated hydrocarbon ring containing six carbon atoms and three conjugated double bonds. Although the term *aromatic* remains with us today, we now define aromatic hydrocarbons by their chemical structure, not their aroma.

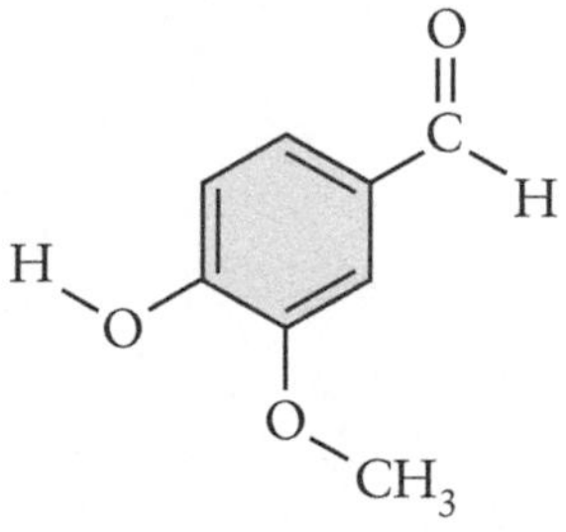

Benzene

The simplest aromatic hydrocarbon is benzene, always represented as a skeletal line structure, as shown in Figure 6-21a. From the skeletal line structure each carbon atom in benzene appears to have one carbon-carbon single bond and one carbon-carbon double bond (and one C—H bond). However, the six electrons from the second pair of electrons in each double bond are actually distributed evenly across all six carbon atoms, which we refer to as *delocalized electrons*. Benzene is an example of a molecule with a structure that is an exception to the otherwise *localized* electrons typically seen in a Lewis structure. All the carbon-carbon bonds in benzene are the same length and strength: longer and weaker than a double bond but shorter and stronger than a single bond. The delocalization of six electrons in the ring is sometimes represented by dashed lines in the skeletal line structure, as shown in Figure 6-21b, or a circle in the center of a hexagon, as shown in Figure 6-21c. The delocalization of electrons in an aromatic ring makes it particularly stable and therefore aromatic compounds are widely distributed in nature. The stability of an aromatic ring makes them much less likely to undergo chemical reactions that involve disrupting the aromatic ring.

Each carbon atom in benzene has one C—H bond, and the H—C—C bond angle around each carbon atom is 120°, characteristic of a trigonal planar center. Thus, benzene is a flat, rigid, two-dimensional molecule, evident in the three views of the ball-and-stick and space-filling models shown in Figure 6-22. The condensed structure used to represent a benzene ring is C_6H_5— or Ph— (phenyl).

Naming Substituted Benzenes

Benzene is the IUPAC name for C_6H_6, and it is also the root name for hydrocarbons containing substituted benzene rings. Three examples of a benzene ring with one substituent are shown below. Notice that the root for each is *benzene*. The substituent is named according to the number of carbon atoms

Figure 6-20 Vanillin is the aromatic compound in vanilla beans that gives them their sweet, pleasant aroma.
[Catherine Bausinger]

Benzene is a carcinogen: a compound that causes or promotes cancer.

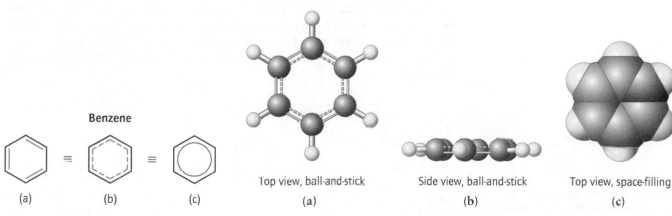

Benzene

(a)   (b)   (c)

Figure 6-21 Three ways to write the structure of benzene.

Top view, ball-and-stick (a)   Side view, ball-and-stick (b)   Top view, space-filling (c)

Figure 6-22 Models showing the overall shape of benzene: (a) ball-and-stick model of top view showing $1\frac{1}{2}$ bonds for each C—C bond; (b) side view of ball-and-stick model showing its overall flat shape; (c) top view of a space-filling model.

that it contains, with the ending changed to *-yl* to indicate that it is a substituent and not the main chain. No locator number is needed for one substituent because every position on a ring is identical.

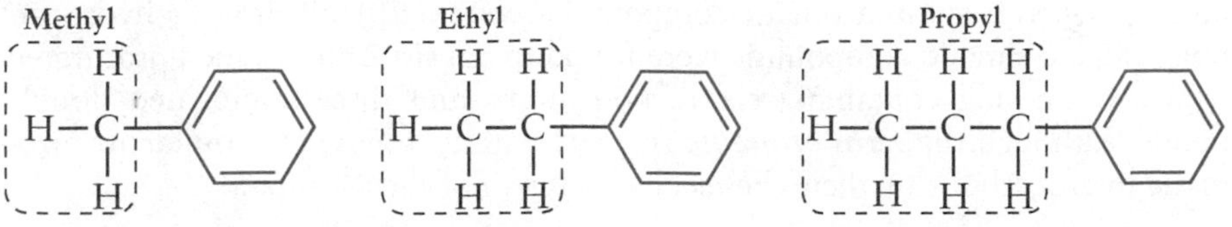

Methylbenzene (toluene) | Ethylbenzene | Propylbenzene

The benzene derivative 1-ethyl-3-methylbenzene contains two substituents: an ethyl substituent ($—CH_2CH_3$) and a methyl substituent ($—CH_3$), located at C(1) and C(3). Locator numbers *are* necessary to indicate the relative position of the substituents when there is more than one substituent. The benzene ring is numbered beginning at one substituent and working clockwise or counterclockwise, whichever gives a lower number to the second substituent. If there are two equal choices, numbering begins on the carbon containing the substituent that appears first alphabetically, as in the example below (*e* before *m*).

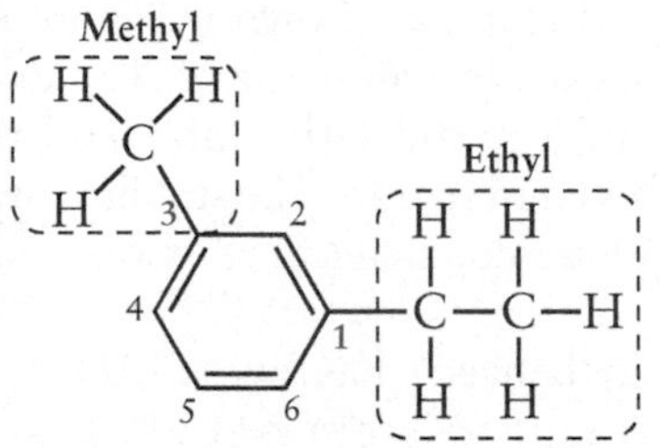

1-Ethyl-3-methylbenzene

Another naming system still in common use today uses the terms *ortho* (*o*) for 1,2; *meta* (*m*) for 1,3; and *para* (*p*) for 1,4 substitution.

When two identical substituents are present, the multiplier *di-* is inserted in front of the substituent name, as in *di*methylbenzene. Furthermore, the designation 1,2- indicates that the two CH_3 groups are on adjacent carbon atoms (carbons 1 and 2); the designation 1,3 indicates that the CH_3 groups are on the first and third carbon atoms; and the designation 1,4 indicates that the CH_3 groups are on the first and fourth carbon atoms (Figure 6-23), three structural isomers.

Several common over-the-counter (OTC) analgesics contain an aromatic ring with two substituents (Figure 6-24), including aspirin, ibuprofen (Motrin, Advil), and acetaminophen (Tylenol). These molecules are not hydrocarbons because they contain heteroatoms; however, they do contain a disubstituted benzene ring. To gain more familiarity with benzene and substituted benzenes, work through Using Molecular Models 6-5: Benzene, Ethylbenzene, and Dimethylbenzene.

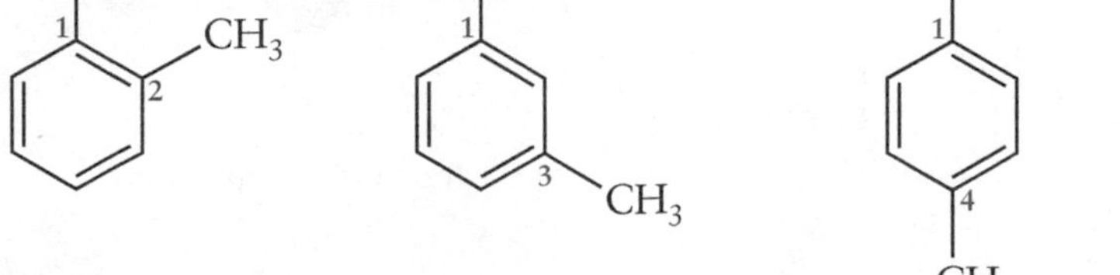

1,2-Dimethylbenzene | 1,3-Dimethylbenzene | 1,4-Dimethylbenzene

Figure 6-23 Structural isomers of dimethylbenzene differ in the relative position of the two methyl substituents on the benzene ring: 1,2- or 1,3-, or 1,4-.

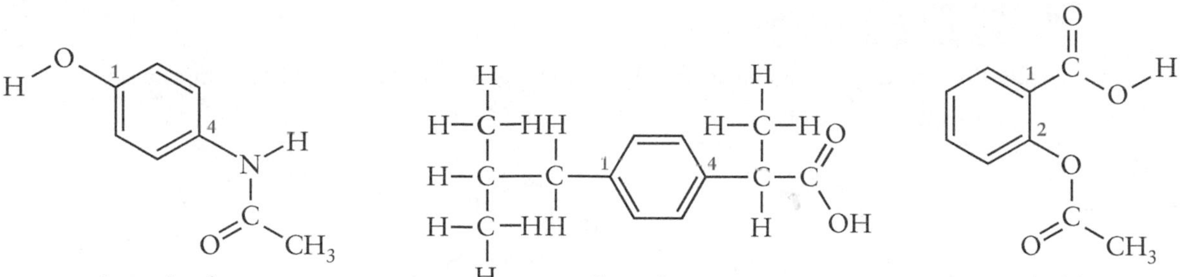

Acetaminophen   Ibuprofen   Aspirin

Figure 6-24 Structures of the common over-the-counter (OTC) analgesics acetaminophen (Tylenol), ibuprofen (Advil and Motrin), and aspirin. They all contain an aromatic ring with two substituents.

Using Molecular Models 6-5 Benzene, Ethylbenzene, and Dimethylbenzene

Construct Benzene

1. Obtain eight black carbon atoms, 10 light blue hydrogen atoms, 18 straight bonds, and six bent bonds.

2. Construct a model of benzene, C_6H_6, by joining six carbon atoms with alternating single and double bonds and adding a C—H bond to each carbon. Use two bent bonds for every double bond. Answer Inquiry Questions 5 through 13.

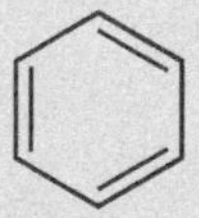

Benzene

Construct Ethylbenzene

3. Construct a model of ethylbenzene by removing one hydrogen atom from your model of benzene and replacing it with a two-carbon chain—an ethyl group. Answer Inquiry Question 14.

Construct Structural Isomers of Dimethylbenzene

4. Remove the ethyl group from the model you constructed in step 3 and remove another H atom from the benzene ring. Replace the ethyl group and the H atom with two methyl groups using the atoms from the ethyl group. Answer Inquiry Questions 15 through 19.

Inquiry Questions

5. Describe the overall shape of your benzene model.

6. Viewed from the side, what does benzene look like?

7. Viewed from above, what does benzene look like?

8. Are the C—H bonds in benzene in the same plane as the C—C bonds? Is the molecule flat?

9. Does benzene have the same shape as cyclohexane? Explain.

10. What are the H—C—C and C—C—C bond angles around each carbon atom in benzene?

11. Is there free rotation around any of the C—C bonds in benzene? Why not?

12. Does benzene have more or less flexibility than cyclohexane? Explain.

13. You constructed your model using alternating single and double bonds. Although this is the only way to make a ball-and-stick model of benzene, why is this not a completely accurate representation of benzene?

14. Why is the model of ethylbenzene the same regardless of which hydrogen atom is replaced with the ethyl group?

15. How many structural isomers of dimethylbenzene are there?

16. How does 1,3-dimethylbenzene differ from 1,2-dimethylbenzene?

17. How does 1,4-dimethylbenzene differ from 1,2-dimethylbenzene and 1,3-dimethylbenzene?

18. Why are there no structural isomers named 1,5- or 1,6-dimethylbenzene?

19. Why does 1,1-dimethylbenzene not exist, whereas 1,1-dimethylcyclohexane does exist?

WORKED EXERCISE   Benzene and Substituted Benzenes

6-24 Provide the IUPAC name for the following substituted aromatic compounds:

a.

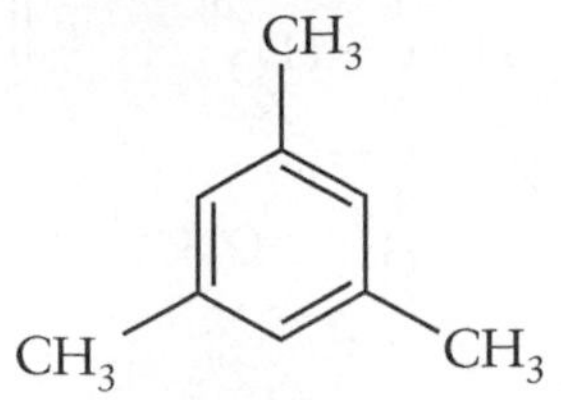

CH₃

CH₃   CH₃

b.

CH₂CH₃

CH₂CH₃

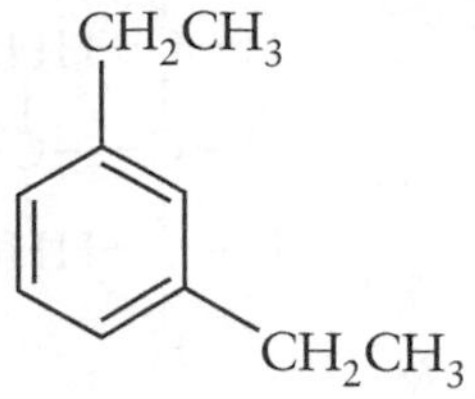

Solution

6-24 a. The IUPAC name is 1,3,5-trimethylbenzene. The root is *benzene* because the main chain is the aromatic ring. There are three identical one-carbon branches—methyl substituents—so we use the multiplier *tri-*. The substituents are at positions 1, 3, and 5 when we number starting at one of the ring carbons containing a methyl group.

b. The IUPAC name is 1,3-diethylbenzene (or *m*-diethylbenzene). The root is *benzene*. There are two ethyl (two carbon) substituents on the ring, so the prefix *diethyl* is placed in front of the root name. The relative positions of the two ethyl substituents are 1,3 when you count along the shortest path between carbons containing substituents.

PRACTICE EXERCISES

35 Provide the IUPAC names for the following substituted aromatic compounds:

a.
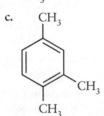

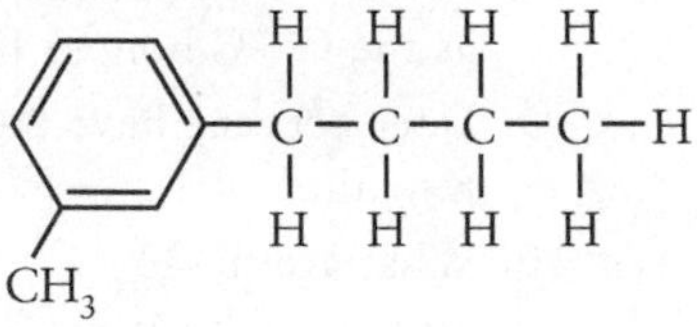

b. $C_6H_5CH_2CH_3$

c.

CH₃

CH₃

CH₃

d.
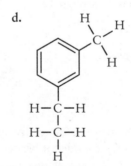

36 Provide a skeletal line structure for the following compounds:
   a. 1,4-diethylbenzene   **b.** 1-ethyl-4-methylbenzene

Chemistry in Medicine   The Chemistry of Vision

In memory of Ingeborg Vogel (1935–2011), my mother, who suffered from macular degeneration in the last decade of her life.

The leading cause of blindness in people over the age of 60 is age-related macular degeneration (ARMD). It is estimated that 11 million people in the United States have some form of ARMD, and that this number is expected to double by the year 2050. Macular degeneration is a condition in which a person loses his or her central vision, preventing them from being able to read, recognize faces, drive, and see detail. **Figure 6-25** shows what a person with ARMD sees compared to someone with normal vision.

We can better understand this disease if we consider the chemistry of vision, a process that involves an important chemical reaction initiated by light.

Figure 6-25 Image of what a person with age-related macular degeneration sees compared to someone with normal vision, showing loss of central vision. [National Eye Institute/National Institutes of Health]

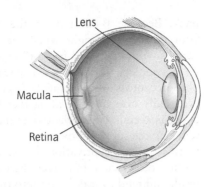

Figure 6-26 Parts of the eye: lens, retina, and macula. The macula is located at the center of the retina right behind the lens.

The retina, located at the back of the eye, contains millions of special photoreceptor cells known as rods and cones. In the center of the retina is a small area known as the macula, which contains a high concentration of cones. Light focused on the macula enables us to see small detail and color. ARMD is caused by a deterioration of the tissue that supports the macula (**Figure 6-26**).

On the surface of the cell membranes of rods and cones is a conjugated polyene known as retinal, which is covalently bonded to the protein opsin, forming a protein complex known as rhodopsin, illustrated in **Figure 6-27**. When light in the visible region of the electromagnetic spectrum (see chapter 2) is absorbed by rhodopsin, a chemical reaction occurs in which a cis double bond in retinal is converted into a trans double bond, as shown in the reaction in Figure 6-27. This reaction is known as an *isomer*ization reaction because the molecule is converted

from one geometric *isomer*—the cis—to the other geometric isomer—the trans. As a result of this cis-to-trans isomerization reaction, the overall shape of the protein-retinal complex changes significantly. The change in shape of the protein-retinal complex initiates a nerve impulse that travels along the optic nerve to the brain. Nerve impulses from the rods and cones are then interpreted in the brain as a visual image—sight.

Perhaps you noticed the similarity between the chemical structure of retinal and β-carotene, shown on page 249. This is because retinal is synthesized from β-carotene in the liver. β-carotene is the molecule that gives carrots, sweet potatoes, squash, and other yellow or orange vegetables their color. It is also an antioxidant. Polyenes such as β-carotene absorb visible light in the green/blue range of the electromagnetic spectrum and so appear as the complementary color—red/yellow—which is reflected

Retinal

Rhodopsin

Figure 6-27 The chemistry of vision: light energy initiates the isomerization of the cis double bond (labeled) in rhodopsin to the trans double bond. This notable change in the overall shape of the protein causes a nerve impulse to be sent to the brain, which is interpreted as part of a visual image.

back. Because β-carotene is the precursor for retinal, your mother was right when she told you eating your carrots would be good for your eyes!

In people with age-related macular degeneration, the photoreceptors of the macula begin to malfunction and eventually stop working altogether. Cellular waste and fat deposits accumulate in the tissues that support the macula, reducing the supply of oxygen to the macula. As a response to this oxygen deprivation, blood vessels begin to grow around the macula. However, these new blood vessels are weak and tend to leak blood into the macula, further preventing the rods and cones from absorbing light—which prevents the isomerization reaction necessary for central vision.

There is no cure for macular degeneration, although treatments exist that can in some cases slow the progression of the disease. The cause of macular degeneration is unknown, but several risk factors have been identified, including age, smoking, exposure to sunlight, a diet deficient in β-carotene and lutein, light-colored eyes, and other hereditary factors. The best thing you can do today to prolong the health of your macula is to regularly wear sunglasses when outdoors, stop smoking if you are a smoker, and eat your vegetables.

Summary

Introduction to Hydrocarbons

- Organic compounds are composed of molecules containing one or more carbon atoms.
- Carbon always forms four bonds.
- Hydrocarbons are organic compounds composed of only carbon and hydrogen atoms.
- The four types of hydrocarbons are alkanes, alkenes, alkynes, and aromatic hydrocarbons.
- Alkanes, but not cycloalkanes, are saturated hydrocarbons, which fit the formula C_nH_{2n+2}, where n is the number of carbon atoms.
- Cycloalkanes, alkenes, alkynes, and aromatic hydrocarbons are unsaturated hydrocarbons.
- Hydrocarbons can be straight chain, each carbon bonded to no more than two carbons, or branched chain, one or more carbon atoms bonded to three or four carbons, creating branches along the chain.
- Hydrocarbons are nonpolar compounds with low boiling points because they interact by dispersion forces, the weakest of the intermolecular forces of attraction.
- Hydrocarbons are not soluble in water, giving them hydrophobic properties.
- Organic compounds have common names and systematic names created by the IUPAC.
- The IUPAC name for an organic compound contains a prefix, a root, and an ending.
- The number of carbon atoms in the chain determines the root of the IUPAC name; the type of compound determines the ending.
- The number of carbons in and the location of branches determines the prefix in a branched hydrocarbon.
- Cycloalkanes are named according to the number of carbon atoms in the ring preceded by "cyclo."

Writing Alkane and Cycloalkane Structures

- Alkanes are saturated hydrocarbons containing only carbon-carbon single bonds and C—H bonds.
- Alkanes have a tetrahedral molecular shape around each of their carbon atoms, giving a chain of three or more carbon atoms a zigzag appearance. Bond angles are 109.5°.

- Condensed structures simplify the process of writing organic structures because each carbon atom and its attached hydrogen atom(s) is listed in sequential order and bonds are omitted except where a branch occurs.
- Skeletal line structures are a shorthand method for writing organic structures in which the carbon and hydrogen atom symbols are omitted. Carbon atoms are represented at the intersection of lines, which represent carbon-carbon bonds, and the ends of lines. Heteroatoms and any hydrogen atoms bonded to them are written in.
- Cycloalkanes are ring structures depicted as skeletal line structures, which are drawn as three- to eight-sided polygons.

Alkane Conformations and Structural Isomers

- Molecules are in constant motion and rotating around each of their carbon-carbon single bonds because of their kinetic energy.
- Conformations of a molecule are different rotational forms of the same compound. Alkanes can rotate freely about their carbon-carbon single bonds and therefore give rise to many different conformations.
- Structural isomers are different compounds that have the same chemical formula but a different connectivity of atoms, as for example straight-chain and branched-chain isomers.
- Structural isomers have different physical and chemical properties. A straight-chain isomer has a higher boiling point than a branched-chain isomer because it has more dispersion forces due to greater surface area.

Alkenes and Alkynes

- Alkenes are unsaturated hydrocarbons containing one or more double bonds.
- The carbon atoms in a double bond have a trigonal planar geometry with bond angles of 120°.
- Alkenes with two double bonds are known as dienes.
- Alkenes with two or more alternating double and single bonds are known as conjugated polyenes.
- Alkenes are assigned a root based on the number of carbon atoms, and the ending is changed to -ene. A locator number indicating where along the chain the first double-bond carbon appears is inserted between the root and the ending.
- Some alkenes have a geometric isomer, a compound with the same chemical formula and the same connectivity of atoms but a different three-dimensional orientation in space as a result of the restricted rotation about the double bond.
- The cis geometric isomer has the two large groups on the double bond carbons on the same side of the double bond; the trans geometric isomer has the large groups on the opposite sides of the double bond.
- Alkynes are unsaturated hydrocarbons containing one or more triple bonds.
- A triple bond has a linear geometry with bond angles of 180°. Triple bonds are not common in nature.
- Alkynes are named like alkenes except the ending is changed to -yne.

Naming Branched-Chain Hydrocarbons

- The prefix for an IUPAC name indicates the type of substituents and where they are along the main chain.
- Substituents are named like the root, based on the number of carbon atoms in the substituent, except the ending is changed to -yl. A locator number is placed in front of each substituent name.

Aromatic Hydrocarbons

- Aromatic hydrocarbons are unsaturated hydrocarbons containing a six-membered ring with three double bonds alternating with three single bonds.
- The simplest aromatic hydrocarbon is benzene, C_6H_6, a flat molecule with unique stability due to delocalization of the double-bond electrons over the entire ring.
- Benzene is a flat, rigid molecule.
- Aromatic compounds do not undergo reactions that disrupt the aromatic ring.
- Aromatic compounds are named using the root name *benzene*.
- Substituents on a benzene ring are named in the same way as substituents in branched hydrocarbons. Locator numbers must be provided when there are two or more substituents, giving the relative position of the substituents.

Key Words

Alkane A hydrocarbon containing only carbon-carbon single bonds and C—H bonds.

Alkene A hydrocarbon containing one or more carbon-carbon double bonds.

Alkyne A hydrocarbon containing one or more carbon-carbon triple bonds.

Aromatic hydrocarbon A six-membered ring written as alternating double and single bonds, although the double bond electrons are actually distributed evenly over all six carbon atoms in the ring.

Benzene The simplest aromatic hydrocarbon, C_6H_6.

Branched-chain isomer A hydrocarbon in which there are one or more carbon atoms bonded to more than two carbons, creating branches in the chain.

Cis An alkene geometric isomer in which the two large groups on the carbon-carbon double bond are on the same side of the double bond.

Condensed structure A shorthand notation for writing the chemical structure of a molecule such that each carbon atom and its attached hydrogen atoms are written as a group: C, CH, CH_2, or CH_3. Bonds are omitted except at branch points.

Conformations Different rotational forms of the same molecule resulting from free rotation about carbon-carbon single bonds.

Conjugated Two or more alternating carbon-carbon double and single bonds.

Cycloalkane An alkane whose chain of carbon atoms forms a ring structure.

Diene An alkene with two carbon-carbon double bonds.

Geometric isomers Compounds with the same chemical formula and the same connectivity of atoms but a different spatial orientation as a result of the restricted rotation about a double bond.

Heteroatom An atom in an organic molecule that is not carbon or hydrogen.

Hydrocarbon An organic compound composed of exclusively carbon and hydrogen atoms.

Inorganic compound A compound that does not contain carbon.

Organic chemistry The branch of chemistry devoted to the study of carbon-containing compounds and their chemical reactions.

Organic compound A compound containing carbon atoms.

Pharmaceutical A drug used for therapeutic purposes. Most are organic compounds.

Polyene A molecule with more than two carbon-carbon double bonds.

Saturated hydrocarbon A hydrocarbon with only carbon-carbon single bonds and C—H bonds that fits the formula C_nH_{2n+2}. Does not include cycloalkanes.

Skeletal line structure A shorthand notation for writing chemical structures in which carbon and hydrogen atom symbols are not written and C—H bonds are omitted. Carbon-carbon bonds are written as lines in a zigzag format.

Straight-chain isomer A hydrocarbon in which every carbon atom is bonded to at most two other carbon atoms, creating a straight chain without branches.

Structural isomers Compounds that have the same chemical formula but differ in the connectivity of the atoms. Structural isomers are different compounds that exhibit different physical and chemical properties and have different IUPAC names.

Substituent A carbon branch along the main chain where a hydrogen atom has been substituted with a chain of one or more carbons.

Trans An alkene geometric isomer in which the two large groups on the carbon-carbon double bond are on the opposite side of the double bond.

Unsaturated hydrocarbon A compound containing one or more carbon-carbon double or triple bonds, or a cycloalkane, or an aromatic hydrocarbon. It has fewer than $2n + 2$ hydrogens, where n is the number of carbon atoms.

Additional Exercises

Good Fats . . . Bad Fats . . . What Does It All Mean?

37 What structural feature distinguishes "good" fats from "bad" fats?

38 List two animal and two vegetable sources of saturated fats. What are some animal and vegetable sources of unsaturated fats?

39 Which kind of fat has more carbon-carbon double bonds—monounsaturated fats or polyunsaturated fats?

40 What structural feature of unsaturated fats contributes to the liquid consistency of these fats at room temperature?

41 What process in the food industry produces trans fats? Why does the food industry use this process?

42 What type of organic compound contains only carbon and hydrogen atoms?

43 What is a heteroatom?

44 How do we currently describe the chemical difference between an organic compound and an inorganic compound? What was the original definition?

45 Indicate whether each of the following statements is true or false.
   a. Organic compounds cannot be synthesized in the laboratory.
   b. Pharmaceuticals can be isolated from plants and animals.
   c. Pharmaceuticals can be prepared in the chemical laboratory.
   d. Only a living organism can produce an organic compound.

Introduction to Hydrocarbons

46 List the four types of hydrocarbons.

47 Determine from the molecular formula which of the following are unsaturated hydrocarbons?
   a. C_3H_8 **b.** C_4H_{10} **c.** C_5H_{10} **d.** C_2H_2

48 Determine from the molecular formula which of the following are saturated hydrocarbons?
   a. C_3H_6 **b.** C_5H_{12} **c.** C_2H_6 **d.** C_4H_8

49 Circle all of the statements below that apply to hydrocarbons. There is more than one correct answer.
   a. They are hydrophobic.
   b. They are capable of hydrogen bonding.
   c. They are insoluble in water.
   d. They generally have low boiling points.

50 What type of intermolecular force of attraction exists between hydrocarbon molecules? Is this force of attraction strong or weak compared to the other types of intermolecular forces?

51 Does water or methane have a higher boiling point? Explain why.

52 The analgesic acetaminophen is the active pharmaceutical ingredient in Tylenol. Which name is the generic name and which is the brand name?

53 What are the three parts of an IUPAC name?

54 For each of the Lewis structures shown below, identify whether it is an alkane, alkene, or alkyne.

Explain why they are classified as hydrocarbons. Indicate whether each compound is saturated or unsaturated. Assign the IUPAC name to each compound.

a.

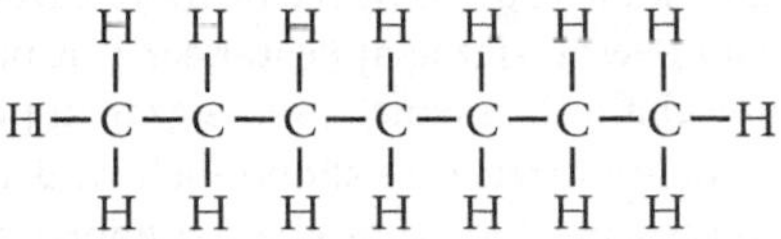

b.

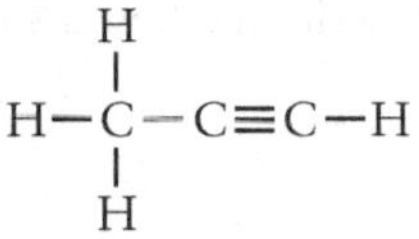

c.

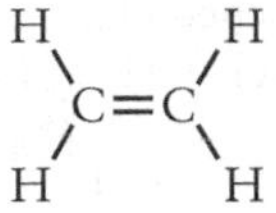

55 What is the correct IUPAC name for the following compound?

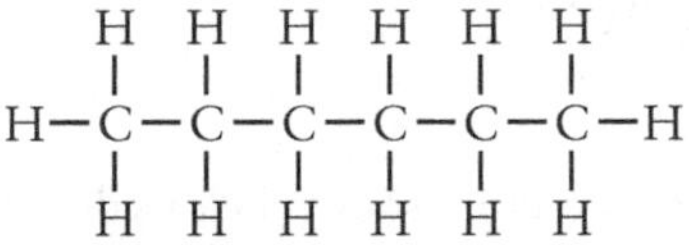

   a. butane   **b.** pentane
   c. hexane   **d.** heptane

56 What is the correct IUPAC name for the following compound, ignoring cis and trans?

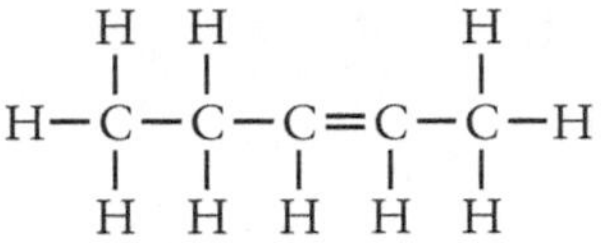

   a. pent-1-ene   **b.** pentane
   c. pent-2-ene   **d.** pent-3-ene

57 Provide the IUPAC name for the following compounds:

a.

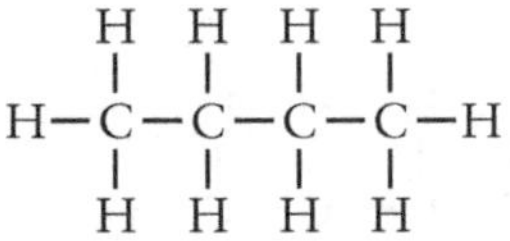

b.

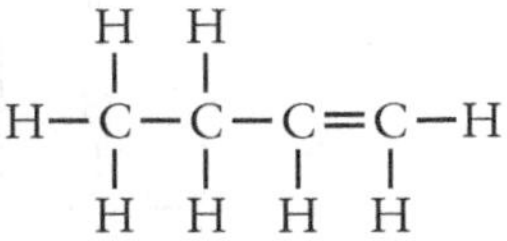

c.

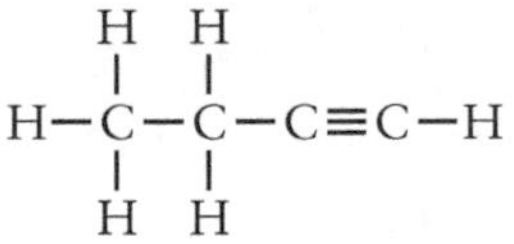

58 Provide the IUPAC name for the following compounds, ignoring cis and trans:

a.

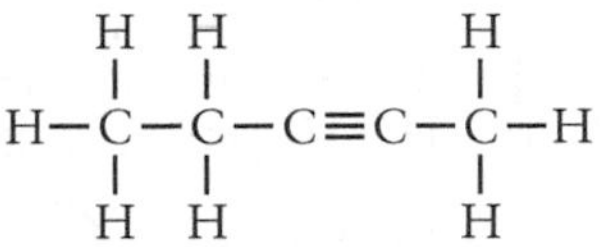

b.

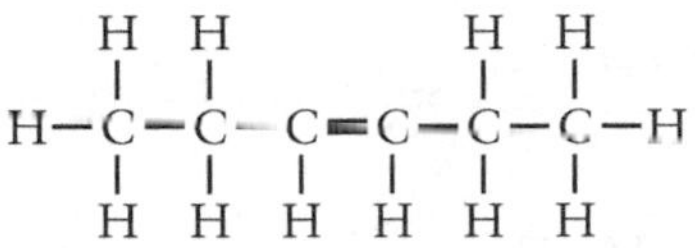

c.

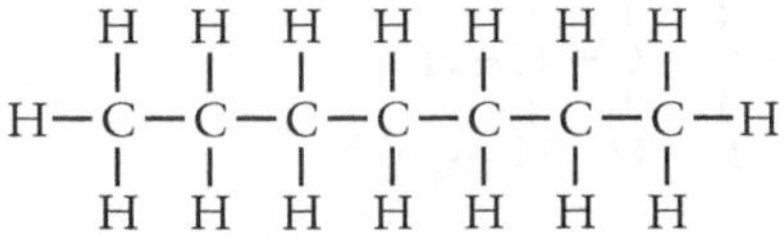

Writing Alkane and Cycloalkane Structures

59 What is an alkane? What is a cycloalkane? Which of the two is a saturated hydrocarbon?

60 What is the molecular shape of a carbon atom in an alkane: tetrahedral, trigonal planar, or linear?

61 Why does an alkane chain of three or more carbon atoms have an overall zigzag appearance? Is it possible for the alkane chain to assume a non-zigzag appearance?

62 For each condensed structure shown below, determine whether it represents a straight-chain alkane or a branched-chain alkane, then write the Lewis structure.

a. $CH_3(CH_2)_6CH_3$

b. 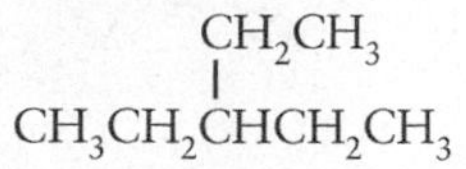
$$CH_2CH_3$$
$$CH_3CH_2CHCH_2CH_3$$

c. 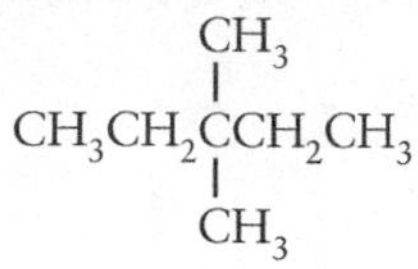
$$CH_3$$
$$CH_3CH_2CCH_2CH_3$$
$$CH_3$$

63 For each condensed structure shown below, determine whether it represents a straight-chain alkane or a branched-chain alkane, then write the Lewis structure.

a. 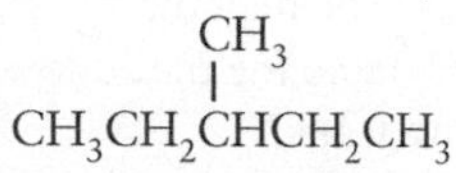
$$CH_3$$
$$CH_3CH_2CHCH_2CH_3$$

b. 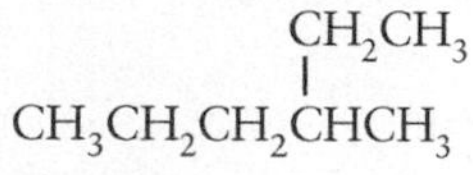
$$CH_2CH_3$$
$$CH_3CH_2CH_2CHCH_3$$

c. $CH_3(CH_2)_3CH_3$

64 Write the condensed structures for the following Lewis structures:

a.

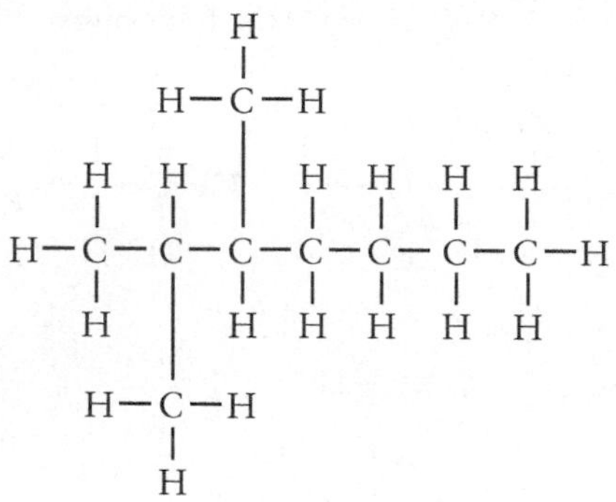

b.

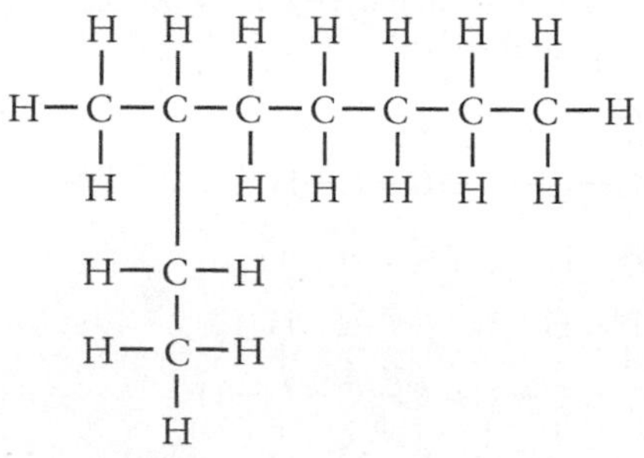

c.

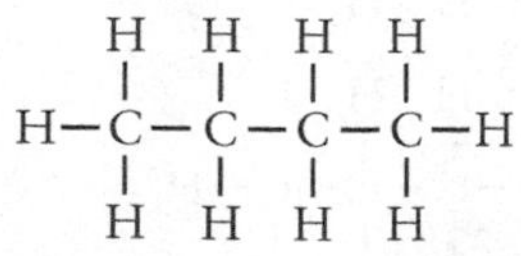

d.

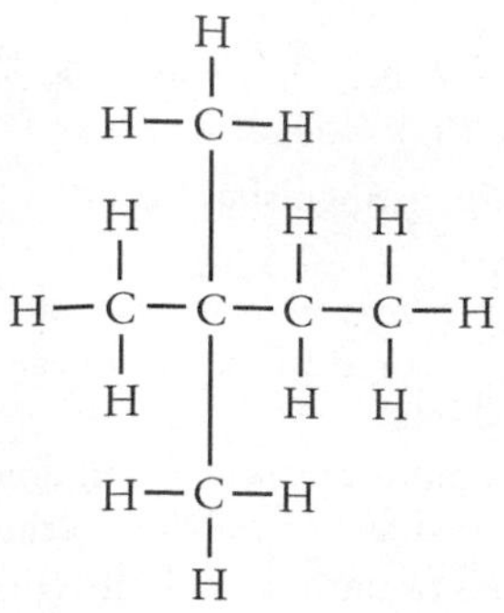

e.

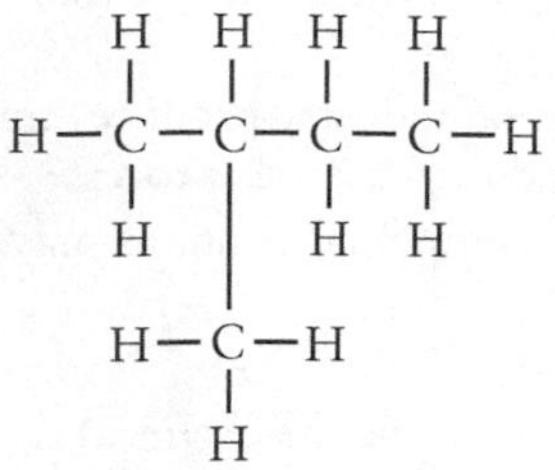

f.

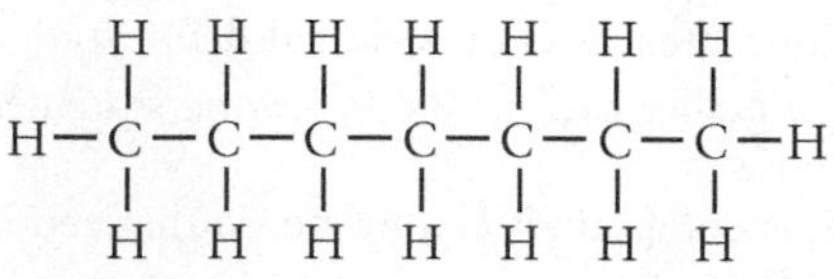

65 Provide skeletal line structures for the following Lewis structures:

a.

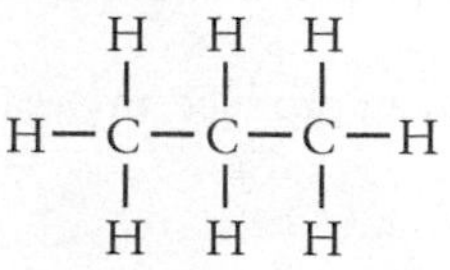

b.

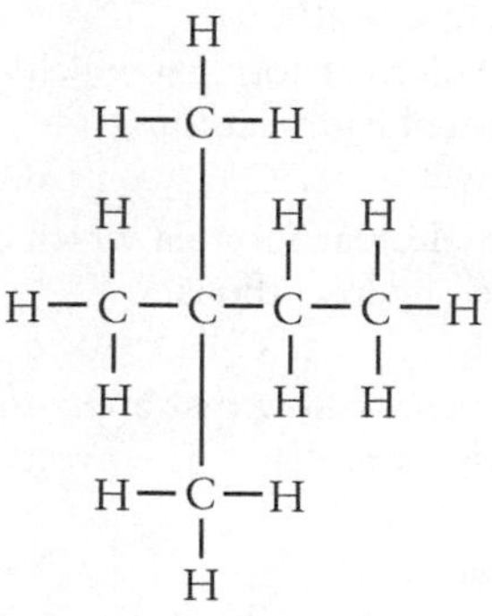

c.

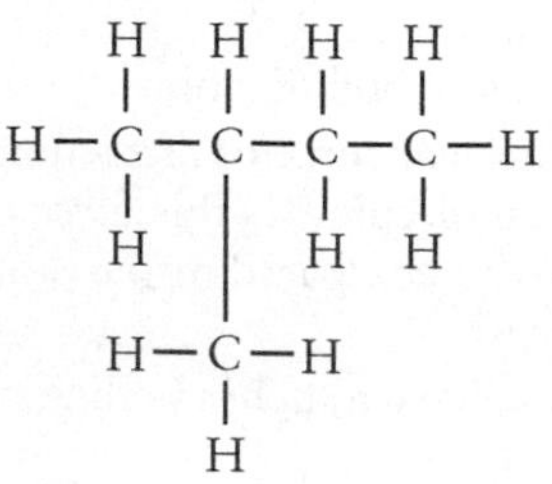

d.

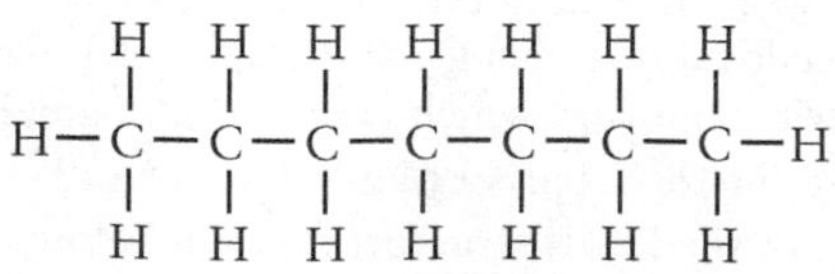

e.

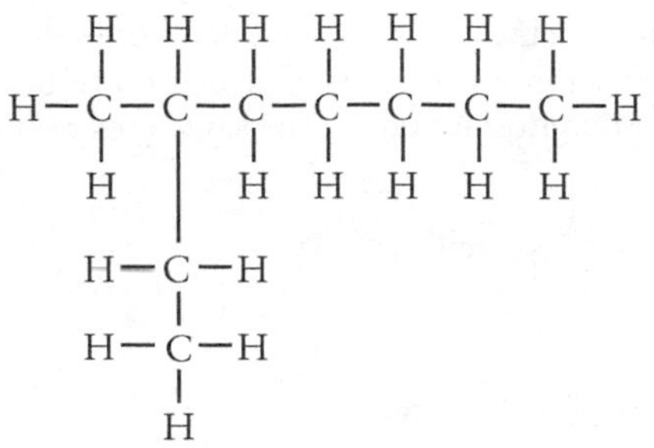

```
        H   H   H   H   H   H   H
        |   |   |   |   |   |   |
   H — C — C — C — C — C — C — C — H
        |   |   |   |   |   |   |
        H   |   H   H   H   H   H
            |
        H — C — H
            |
        H — C — H
            |
            H
```

66 Provide skeletal line structures for the following condensed structures:

a.

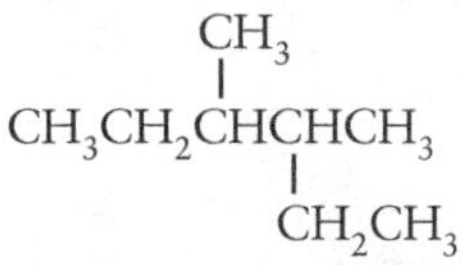

$$CH_3CH_2CHCHCH_3$$

with CH_3 above and CH_2CH_3 below

b.

$$CH_3CHCH_2CH_2CCH_2CH_2$$

with CH_3, CH_3 above and CH_3 below

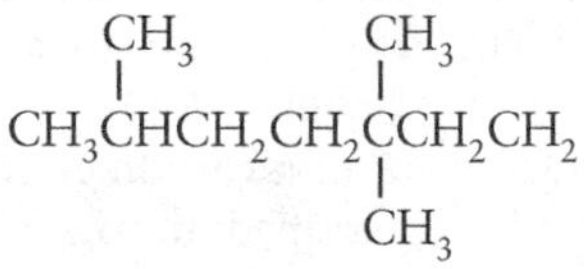

c. $CH_3(CH_2)_{12}CH_3$

67 The structure of menthol, found in peppermint and used to relieve minor throat irritations, is shown below. Indicate the number of hydrogen atoms on each of the carbon atoms in menthol. What heteroatom is found in menthol? Note the wedged and dash notation indicates the three-dimensional orientation of groups on tetrahedral carbon centers.

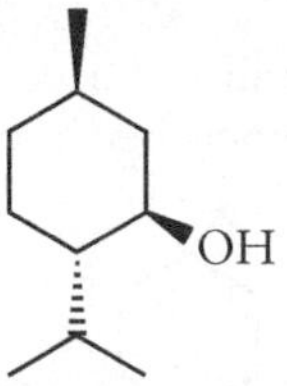

68 Geraniol, shown below, is the main component of rose oil and citronella. It is used in perfumes and in fruit flavorings. Indicate the number of hydrogen atoms on each of the carbon atoms in the skeletal line structure. What heteroatom is found in geraniol?

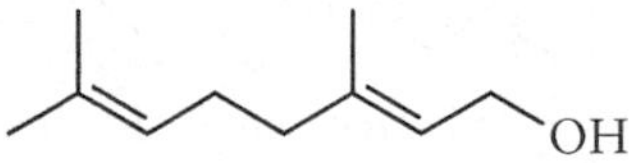

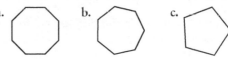

69 Provide a Lewis structure and a condensed structure for the following skeletal line structures:

a. 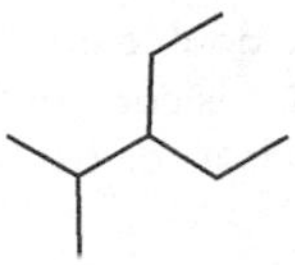   **b.** 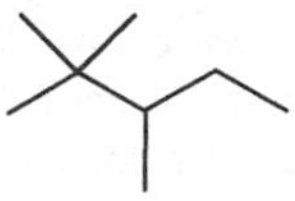   **c.**

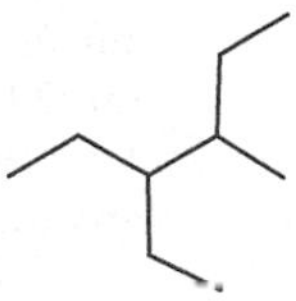

70 Provide the IUPAC name for each of the following compounds:

a. 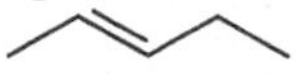   **b.**

71 Write the skeletal line structures and molecular formulas for the pairs of compounds listed below. For each pair,

indicate how their molecular formulas are different. In what way are they the same? How are their structures different?

a. hexane, cyclohexane

b. propane, cyclopropane.

72 Provide skeletal line structures for the following cycloalkanes:

a. cyclopropane

b. cyclohexane

c. cyclobutane

d. methylcyclobutane.

73 Provide the IUPAC names for the following cycloalkanes:

a.   **b.**   **c.**

74 Some large rings are found in nature. How many carbon atoms are in the cycloalkane shown below?

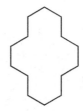

75 Write the skeletal line structure for a cycloalkane containing three carbon atoms in the ring and one containing five carbon atoms in the ring. Which is more common in nature and why?

76 How many carbon atoms are there in the cycloalkane shown below? How many hydrogen atoms does each carbon atom have? How is the number of hydrogen atoms in this cycloalkane different from the corresponding straight-chain hydrocarbon with the same number of carbon atoms?

77 Muscone is a compound used in the perfume industry. Does this compound contain a cycloalkane? How many carbon atoms are in the ring? What is the heteroatom in this molecule?

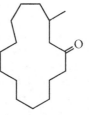

Muscone

Alkane Conformations and Structural Isomers

78 Indicate whether each of the following statements is true or false:

a. Two different conformations of a compound will have a different connectivity of atoms.

b. One conformation of a molecule can be converted into another conformation without having to break any bonds.

c. Carbon-carbon single bonds are not free to rotate.

d. Alkanes are stationary molecules.

79 Draw two different conformations of C_5H_{12}.

80 Provide the IUPAC name for the following compounds and indicate whether the following pairs of compounds are

i. different compounds because they have different chemical formulas

ii. different compounds because they are structural isomers

iii. identical compounds, but different conformations or views

a.

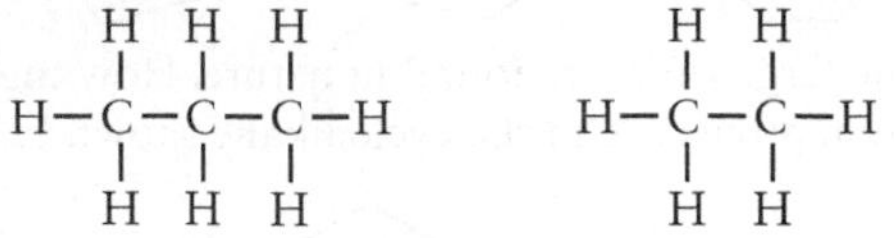

      H   H   H   H   H
      | | | | |
  H—C—C—C—H   H—C—C—H
      | | | | |
      H   H   H   H   H

b.

      H   H   H   H   H   H   H
      | | | | | | |
  H—C—C—C—C—H   H—C—C—C—H
      | | | | | | |
      H   H   H   H   H   H
                                       |
                                   H—C—H
                                       |
                                       H

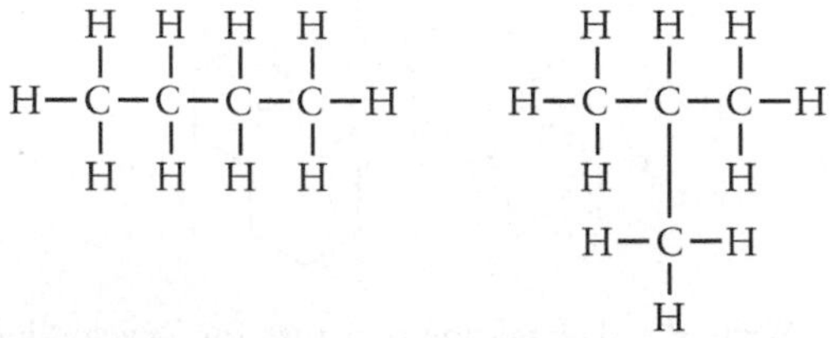

c.

      H   H   H   H   H   H   H   H
      | | | | | | | |
  H—C—C—C—C—H   H—C—C—C—C—H
      | | | | | | | |
      H   H   H   H   H   H   H
                                       |
                                   H—C—H
                                       |
                                       H

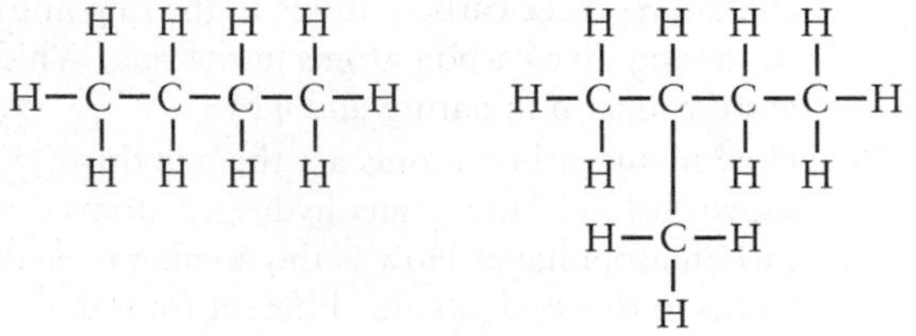

d.

      H   H   H   H   H
      | | | | |
  H—C—C—C—C—H   H—C—H
      | | | | |
      H   H   H   H   H | H
          | | | | |
      H—C—H   H—C—C—C—C—H
          | | | | |
          H   H   H   H   H

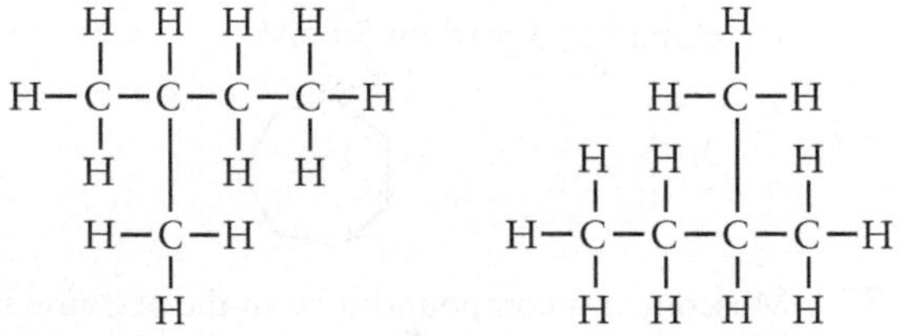

81 Write the Lewis structure for the two structural isomers of C_4H_{10}. Label the branched-chain and straight-chain structural isomer. Name the straight-chain isomer.

82 There are 75 structural isomers for $C_{10}H_{22}$ and only two structural isomers for C_4H_{10}. Why does $C_{10}H_{22}$ have so many more structural isomers than C_4H_{10}?

83 Explain how the structural isomers below are different from one another. What similarity do structural isomers have?

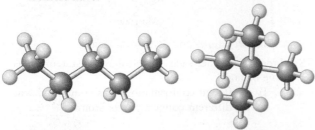

84 Indicate whether the following pairs of skeletal line structures represent a pair of structural isomers or two different conformations of the same compound:

a.

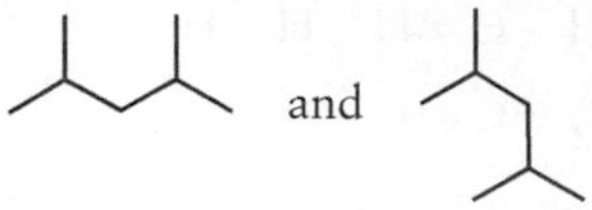

b.

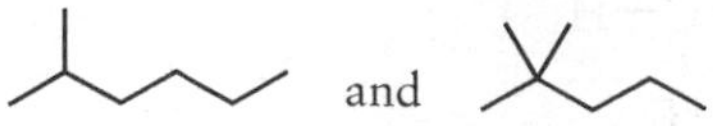

c.

d.

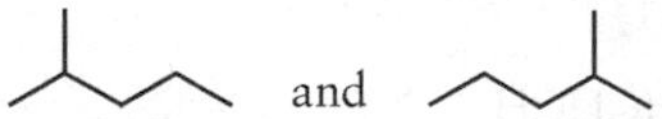

Alkenes and Alkynes

85 What is the molecular shape of a carbon atom in the triple bond of an alkyne: tetrahedral, trigonal planar, or linear? What is the bond angle?

86 What is the molecular shape of a carbon atom in the double bond of an alkene: tetrahedral, trigonal planar, or linear? What is the bond angle?

87 Indicate whether each of the following alkene structures should be classified as a simple alkene, a diene, or a polyene. Identify the conjugated dienes and polyenes.

a.   **b.**

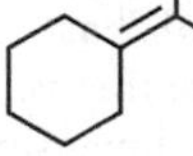

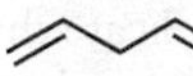

c. 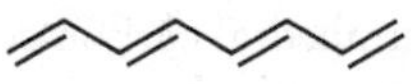   **d.**

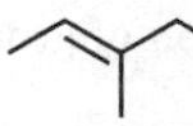

88 Indicate the number of hydrogen atoms bonded to each of the carbon atoms in the structure shown below:

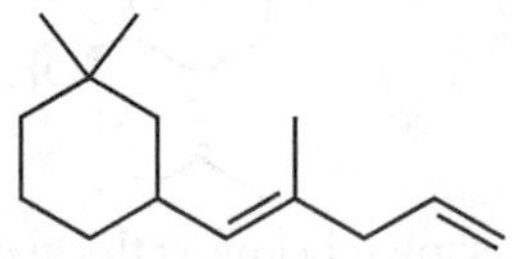

89 Which of the following pairs of alkenes are geometric isomers? Label the cis and the trans.

a.

b.

c.

90 Is the rotational freedom about a carbon-carbon double bond different from that about a carbon-carbon single bond?

91 The structure of linoleic acid, found in plants and animals, is shown below. Are the double bonds cis or trans? How can you tell? Would this compound be classified as a trans fatty acid or is it the natural cis fatty acid?

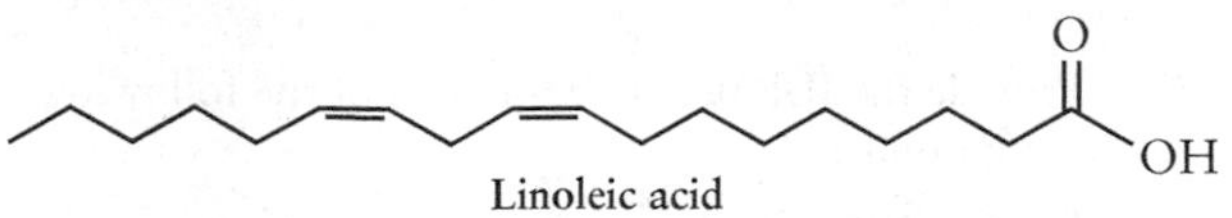

Linoleic acid

92 Provide skeletal line structures for the following alkenes:

a. propene

b. ethene

c. hept-1-ene

d. *trans*-oct-2-ene

e. cyclopentene

93 Provide the skeletal line structure for the following alkenes:

a. *cis*-hex-3-ene

b. *trans*-pent-2-ene

c. *cis*-pent-2-ene

d. *trans*-hept-3-ene

94 Write the skeletal line structure for the following alkynes:

a. pent-2-yne

b. pent-1-yne

c. hept-3-yne

d. oct-4-yne

95 Name the following alkynes:

a.

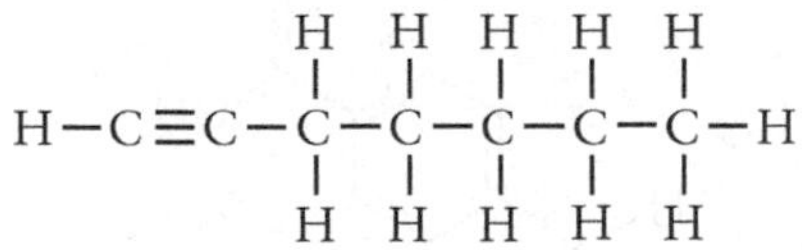

b.

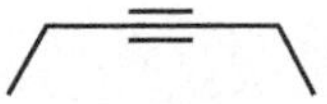

c.

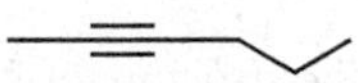

96 Indicate the C—C—C bond angle around the central carbon atom in propyne:

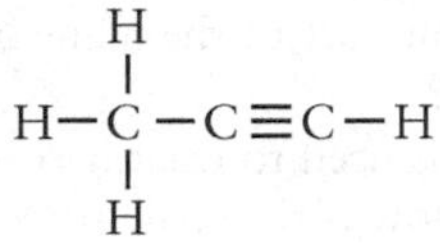

Naming Branched-Chain Hydrocarbons

97 Name the following branched-chain hydrocarbons:

a.

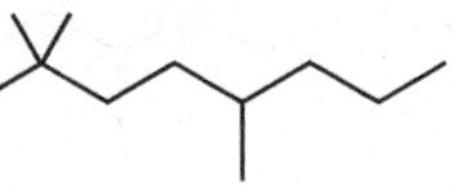

b.

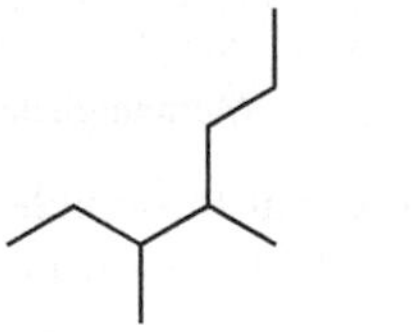

c.

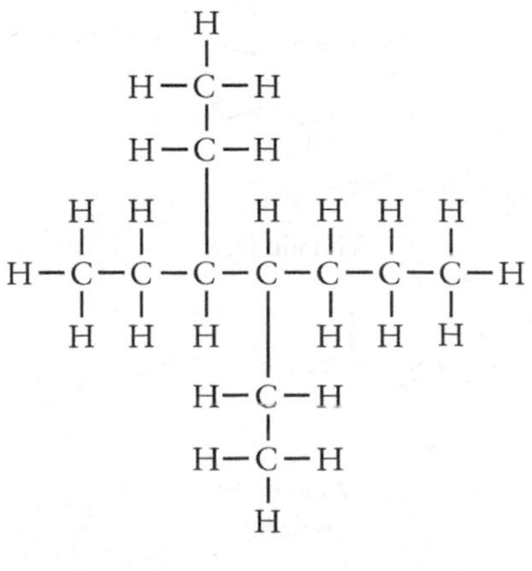

d.

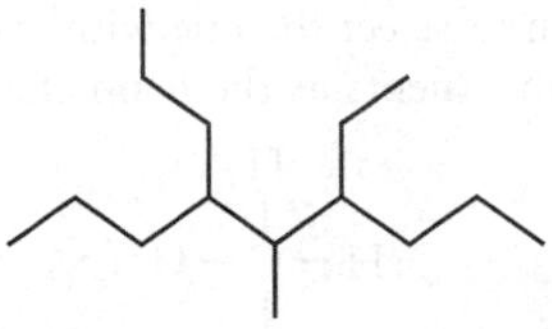

e.

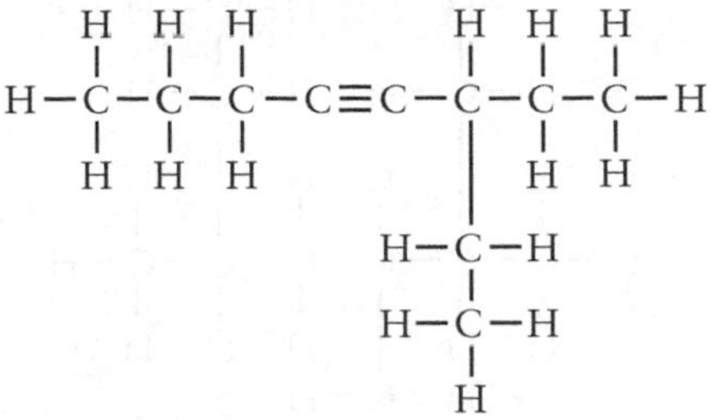

f.

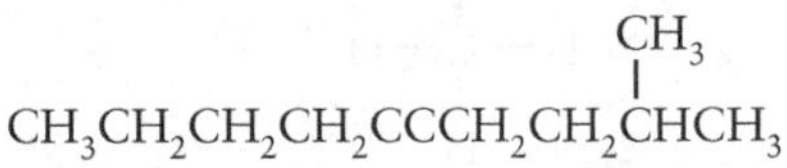

$$CH_3$$

$$CH_3CH_2CH_2CH_2CCCH_2CH_2CHCH_3$$

98 Name the following branched-chain hydrocarbons:

a.

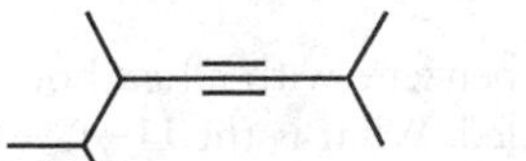

b.

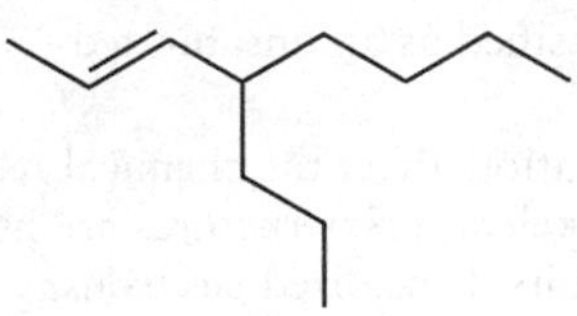

c.

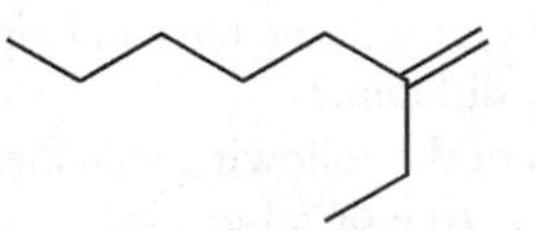

99 Write the skeletal line structure for all the structural isomers of C_6H_{14}. Provide the IUPAC name for each structural isomer. Can two structural isomers have the same IUPAC name?

100 Provide skeletal line structures for the IUPAC names below:

a. *cis*-4-ethylhex-2-ene

b. 2,2-dimethyl-3-ethylnonane

101 Name the following alkenes (there are no cis/trans isomers):

a.

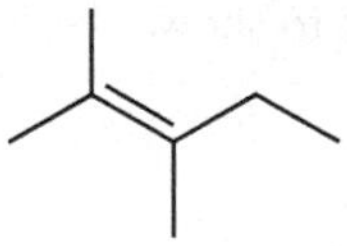

b.

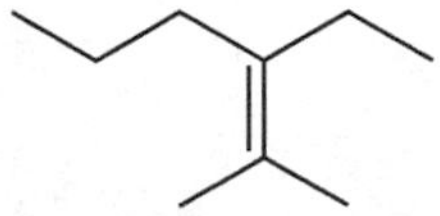

c.

d.

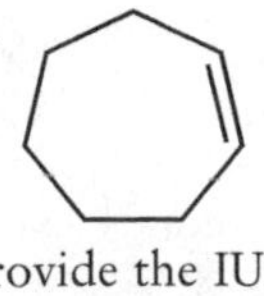

102 Provide the IUPAC name for the following compound. (Remember, when there are two contiguous carbon

chains of equal length to select the one with the greater number of substituents as the main chain.)

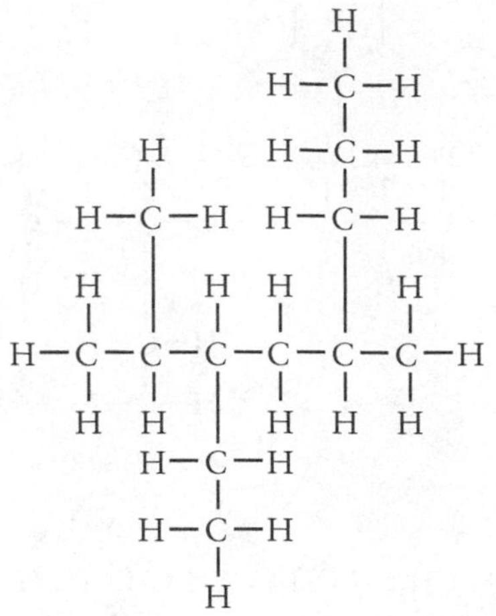

$$
\begin{array}{c}
\text{H} \\
\mid \\
\text{H-C-H} \\
\end{array}
$$

Aromatic Hydrocarbons

103 Write the structure of benzene with all carbon and hydrogen atoms included. What is the H—C—C bond angle at every carbon atom in the molecule?

104 Why is benzene classified as an unsaturated hydrocarbon?

105 How does delocalization affect the chemical reactivity of benzene? What skeletal line structures are used to depict benzene with its delocalized electrons?

106 What do benzene and cyclohexane have in common? In what ways are they different?

107 Indicate whether each of the following statements about benzene, C_6H_6, is true or false.
   a. Aromatic rings are stable and are found in many compounds throughout nature.
   b. Exposure to benzene, C_6H_6, can cause cancer.
   c. Benzene is a flat molecule.
   d. Benzene does not really have alternating single and double bonds; rather, all carbon-carbon bonds are a bond and a half.

108 Write the structure of 1-ethyl-2-methylbenzene.

109 Which of the following are acceptable ways to represent benzene? For those selected as acceptable, explain what they are intended to show.

   a.    **b.**

   c.    **d.**

   e. C_6H_6

110 Name the following aromatic compounds:

   a. 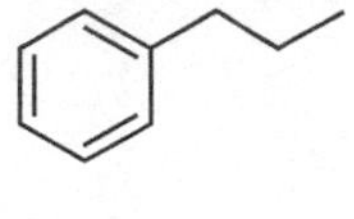   **b.**

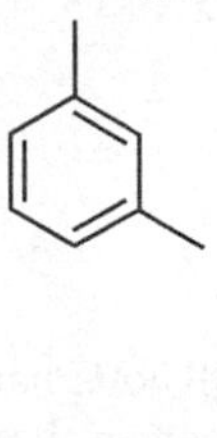

   c.

111 Many insect repellents contain DEET, whose chemical structure is shown below. Circle the aromatic ring and indicate whether the substitution is 1,2-, 1,3-, or 1,4-.

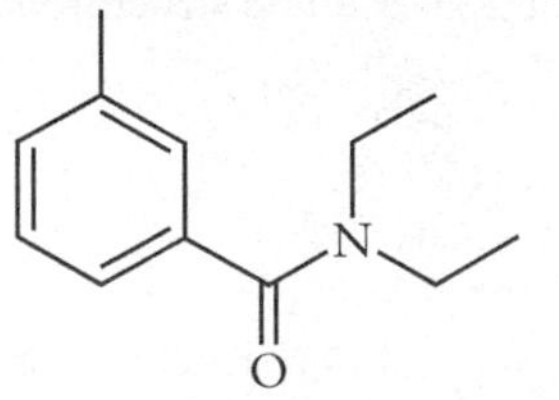

DEET

112 PABA, found in many sunscreens, is an abbreviation for *para-aminobenzoic acid*. The structure of PABA is shown on the next page.

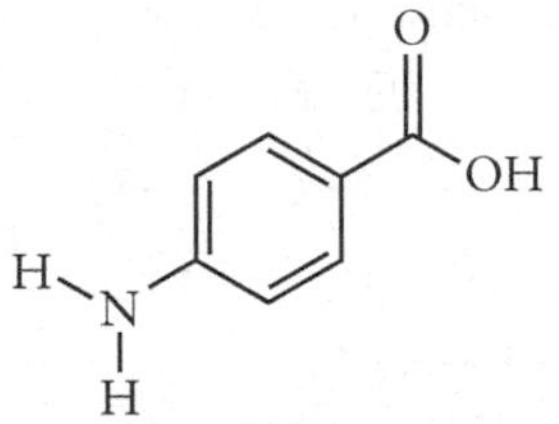

PABA

   a. Circle the aromatic ring in the structure of PABA. What part of the root name *benzene* appears in the IUPAC name of PABA?
   b. Is the substitution on the aromatic ring 1,2-; 1,3-; or 1,4-? What part of the name PABA reflects this substitution?

113 Methadone is used to treat heroin and morphine addiction. How is the structure of methadone similar to methamphetamine? How is it different? Circle the aromatic rings in both molecules.

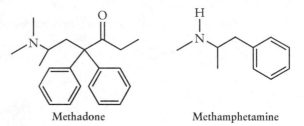

   Methadone   Methamphetamine

114 The chemical structures for the four fat-soluble vitamins (vitamins D, A, E, and K) are shown below:

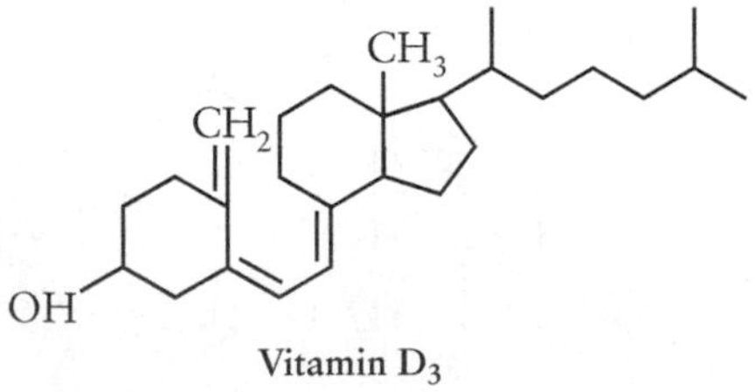

Vitamin D₃

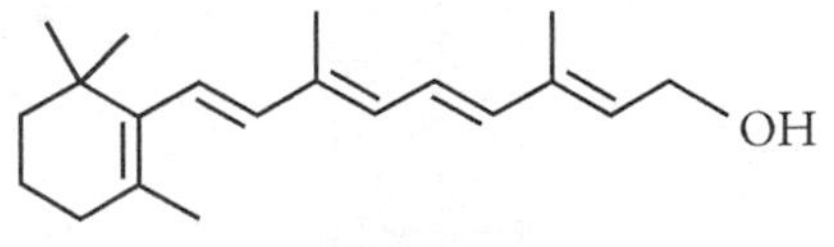

**Vitamin A
(retinol)**

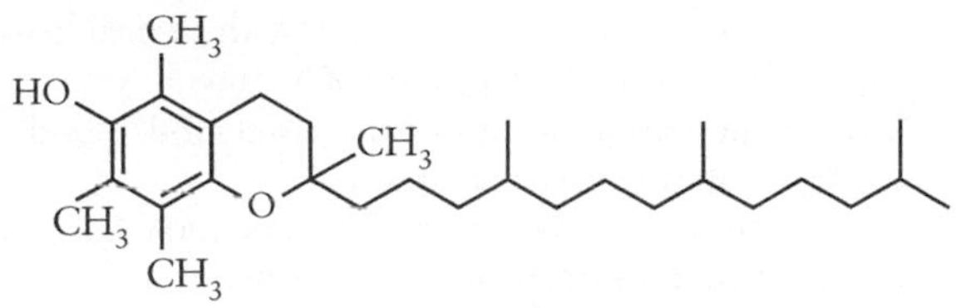

Vitamin E

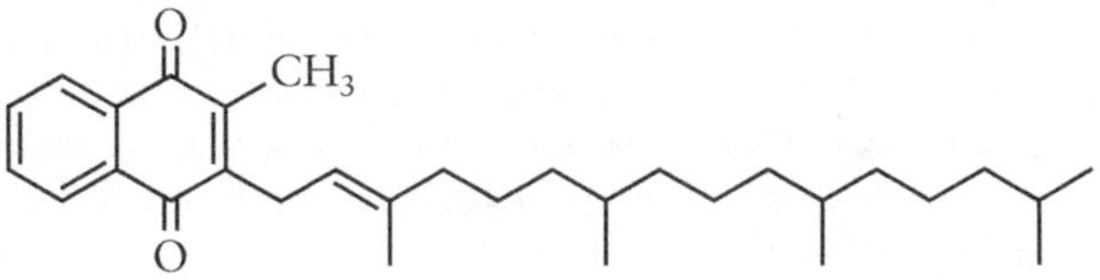

Vitamin K

a. Which vitamins contain aromatic rings?
b. How many carbon atoms are there in the long hydrocarbon chain in vitamin E? Is this a straight-chain hydrocarbon or a branched-chain hydrocarbon?
c. There are two heteroatoms in vitamin E; what are they?
d. Which vitamins are conjugated polyenes?

Chemistry in Medicine: The Chemistry of Vision

115 The structure of rhodopsin is shown below. Answer the questions by referring to the corresponding lettered arrow next to the structure.

a. Is this double bond cis or trans?
b. Is this double bond cis or trans?
c. If this double bond were to undergo an isomerization reaction, would the overall shape of the molecule change? What is the physiological result?
d. Which double bond undergoes an isomerization reaction during the chemical process of vision? Where does the energy for this reaction come from?

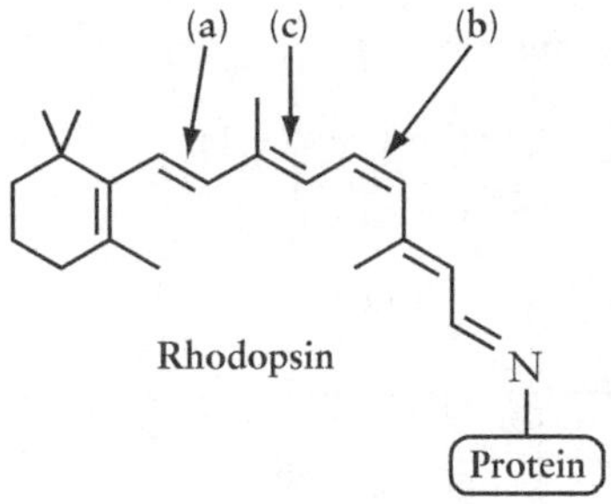

Rhodopsin

116 What is age-related macular degeneration?
117 On the surface of what type of cells is the protein rhodopsin found?
118 What is a cis-trans isomerization reaction?
119 What type of isomers are rhodopsin and bathrhodopsin?
120 How does a diet containing β-carotene support the health of your rods and cones?

Answers to Practice Exercises

1 (b), (c), and (d)
2 (b) and (c)
3 An organic compound contains one or more carbon atoms. An inorganic compound does not contain carbon.
4 Carbon always forms four bonds because it has four valence electrons and needs an additional four electrons to achieve an octet—a stable arrangement of electrons.
5 **a.** butane **b.** but-2-ene **c.** but-1-ene **d.** hex-1-yne
6 **a.** Each carbon has four bonds in this branched-chain alkane.
  b. Each carbon has four bonds in this straight-chain alkyne.
7 **a.** straight chain:

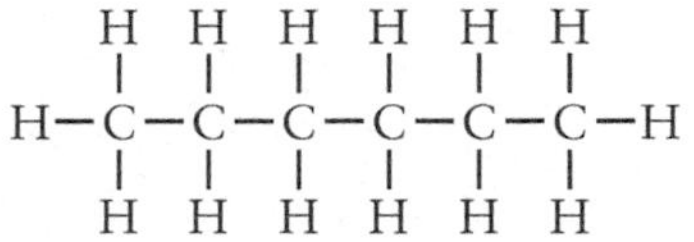

b. branched chain:

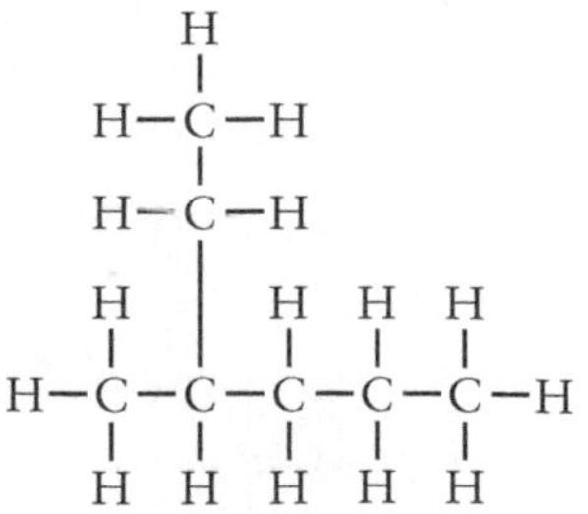

c. branched chain:

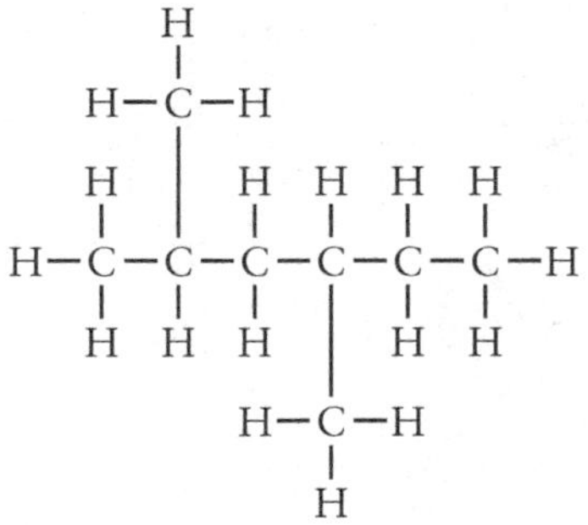

8

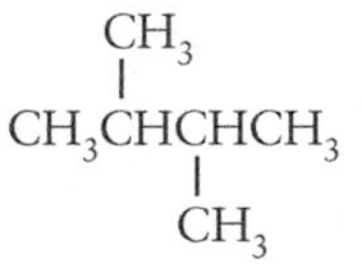

9

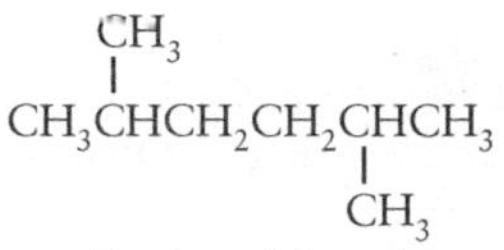

Condensed formula

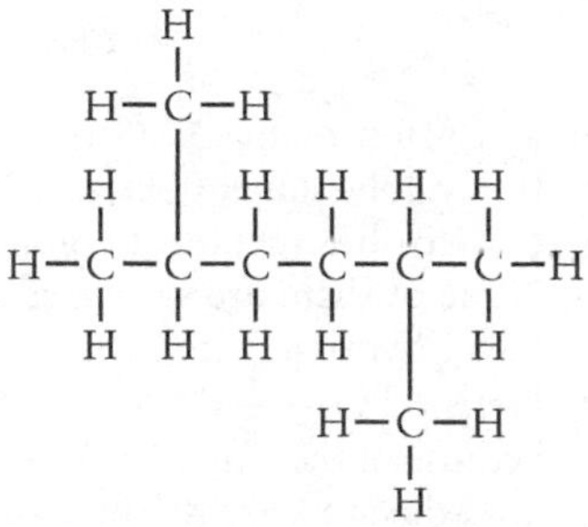

Structural formula

10

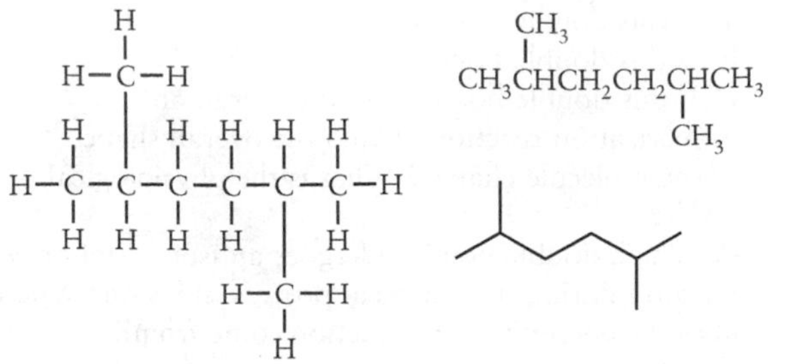

$$CH_3CHCH_2CH_2CHCH_3$$

11 a.

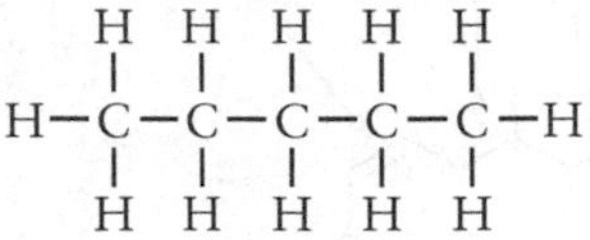

$$CH_3(CH_2)_3CH_3$$

b.

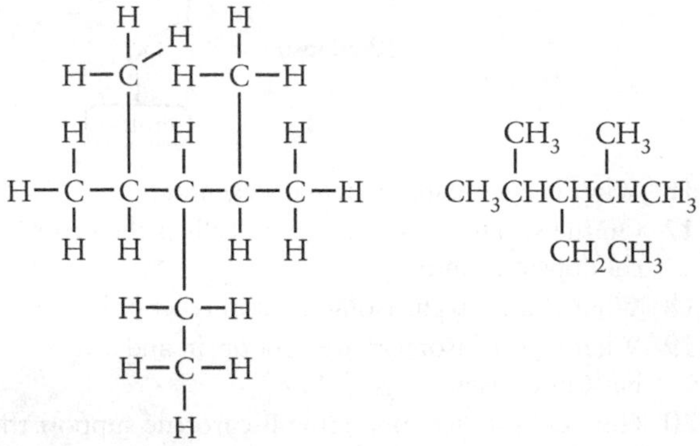

$$CH_3CHCHCHCH_3$$

12 a.    **b.**

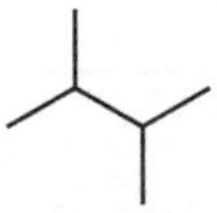

13 a. 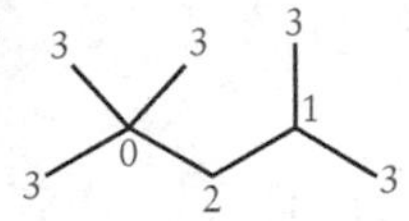   **b.** 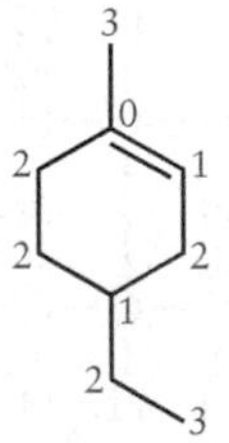   **c.**

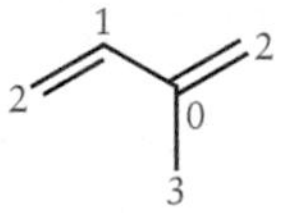

d.

HO

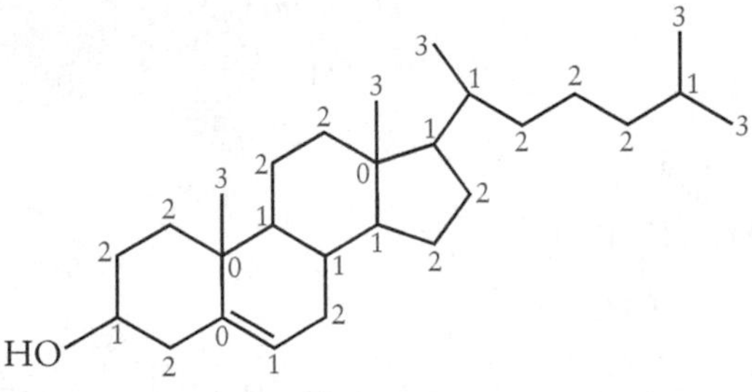

Cholesterol

14 a. Cyclopropane; C_3H_6;
b. Cycloheptane; C_7H_{14};
c. Cyclohexane; most common in nature; C_6H_{12};
None of them are saturated because a ring has two fewer hydrogen atoms than $2n + 2$.

15 Preferred C—C—C bond angle for an alkane is 109.5° because of the requirements of a tetrahedral carbon according to VSEPR. The actual bond angle in an equilateral triangle is 60°. You might break pieces on your model kit as you try to force the bonds into such acute angles. This is

known as ring strain and accounts for a three-membered ring not being a common ring size in nature.

16 a. Different conformations of the same model, both are different rotational forms of pentane.
b. Different conformations of the same molecule, both are different conformations of propane.
c. Different structural isomers because the atoms are connected in a different way but they both have the same molecular formula, C_5H_{12}.

17 The more carbon atoms there are, the more possibilities there are for branching. There is only one structural isomer of C_2H_6 because there is only one way to arrange two carbon atoms and six hydrogen atoms and adhere to the octet rule.

18 a. Structural isomers because they both have the same formula, C_4H_{10}, but the atom connectivity is different. The first isomer is a straight chain isomer and the second is a branched chain isomer.
b. Structural isomers because they both have the same formula, C_7H_{16}, but the connectivity is different: methyl branches on C(2) and C(3) verses methyl branches on C(2) and C(4).
c. These are different conformations of the same compound.
d. These are structural isomers because they both have the same formula, C_5H_{12}, but the atoms are connected differently. One molecule has two branches on the second carbon, and the other molecule is a straight-chain alkane—no branches.

19 (a), (b), and (d) because structural isomers are different compounds.

20 The first one is the straight-chain isomer, hexane.

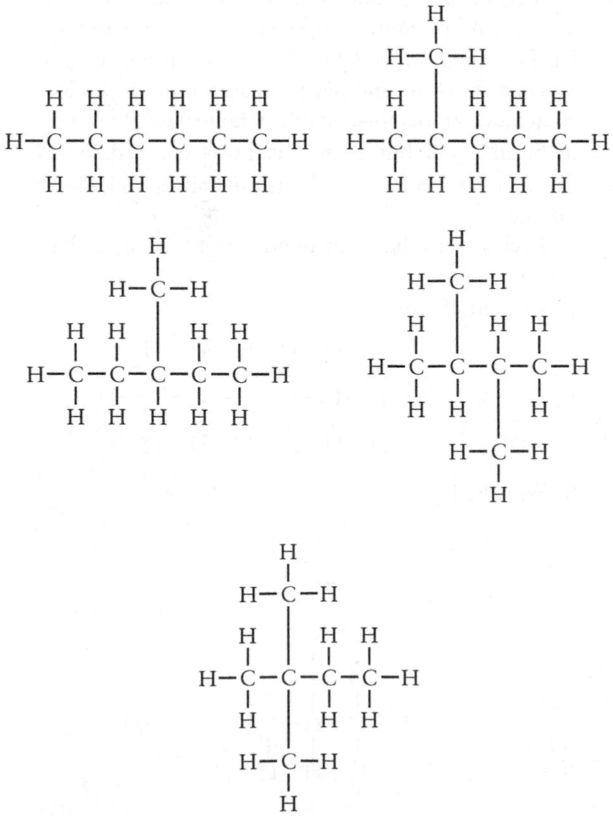

21 **a.** diene; conjugated
  b. simple alkene
  c. polyene; conjugated

22 The bond angles are 120° for the carbon atoms with a double bond and 109.5° for the tetrahedral carbon on the far right, as drawn.

23 **a.** Five.
  b. Vitamin A is essentially half of a β-carotene molecule with an added OH group.

24 The extensive conjugated polyene structure gives lycopene its red color. Lycopene has 13 carbon-carbon double bonds.

25 **a.** Cis/trans isomers. The alkene on the left is the cis isomer because the larger groups on each double-bond carbon are on the same side of the double bond. The alkene on the right is the trans isomer because these same groups are on opposite sides of the double bond.
  b. These are the same alkane drawn as different conformations.
  c. These are different conformations or views of the same alkene. This alkene does not have a geometric isomer because one of the double-bond carbon atoms has two hydrogen atoms (the same group or atom).

26 They are geometric isomers and differ in the spatial orientation of the larger groups on the double-bond carbons:

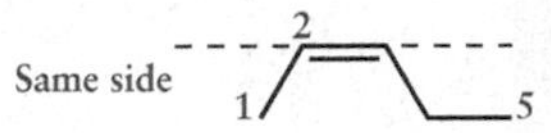

Same side

cis-pent-2-ene

Opposite sides

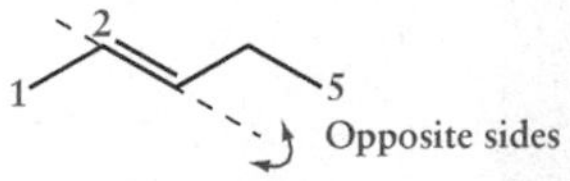

trans-pent-2-ene

27 **a.**

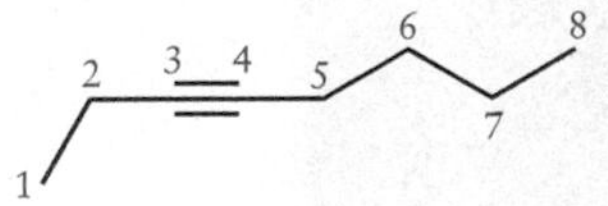

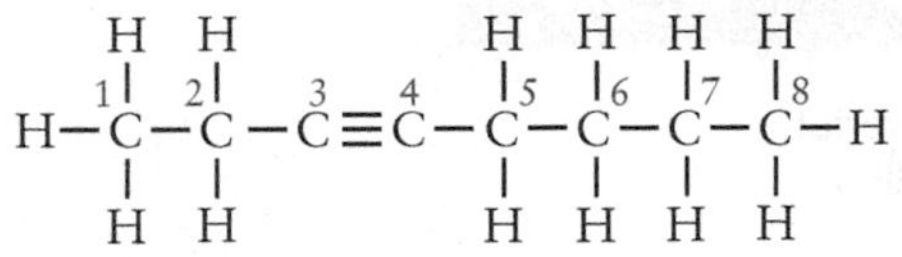

oct-3-yne

  b.

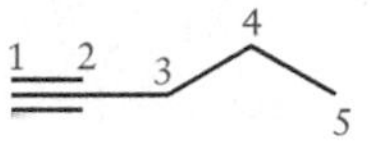

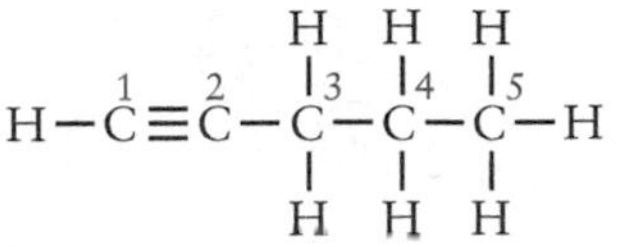

pent-1-yne

28 **a.** pentane, C_5H_{12}, a saturated hydrocarbon
  b. *trans*-pent-2-ene, C_5H_{10}
  c. *cis*-pent-2-ene, C_5H_{10}
  d. pent-2-yne, C_5H_8

29 **a.** hex-1-yne   **b.** pent-1-yne

30 **a.** 4-methylheptane
  b. 2,3-dimethylpentane

31 **a.**

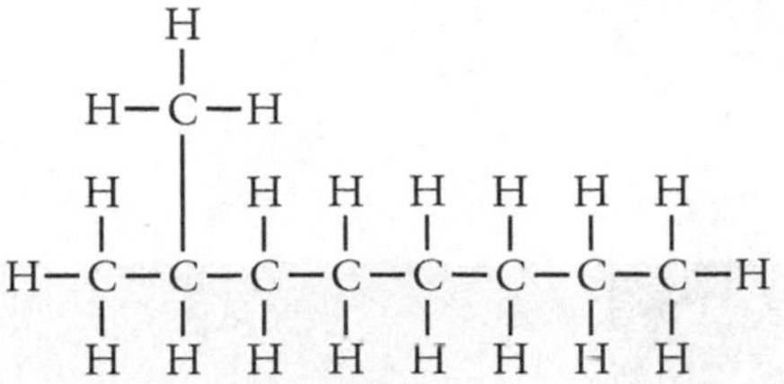

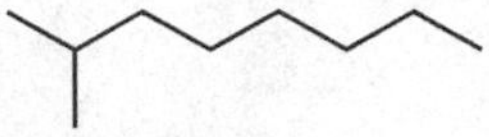

  b.

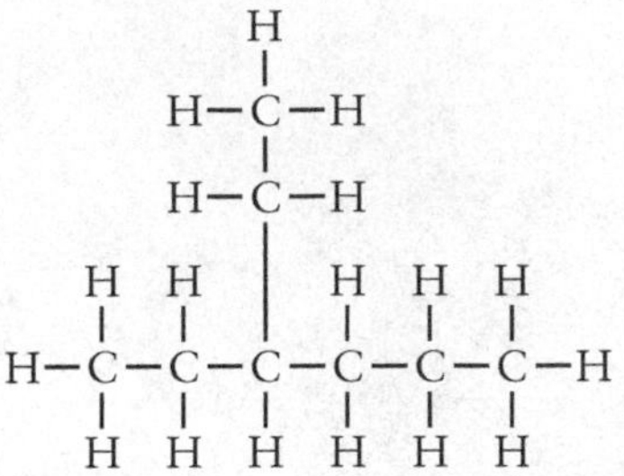

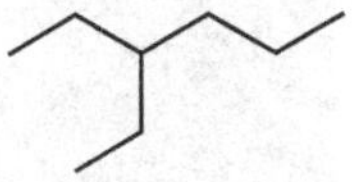

  c.

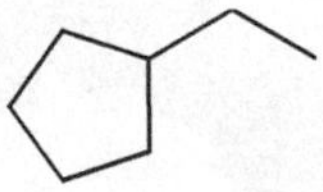

  d.

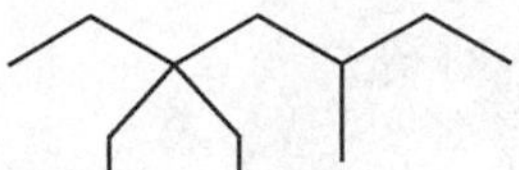

32 **a.** 2-methylbut-2-ene
  b. 2-ethylpent-1-ene
  c. 3-ethylpent-1-ene

33   Isooctane
  (**2,2,4-trimethylpentane**)   Heptane

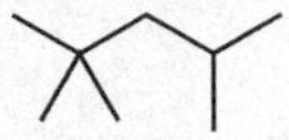

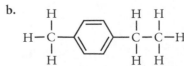

$$CH_3CCH_2CHCH_3$$

$CH_3(CH_2)_5CH_3$

34

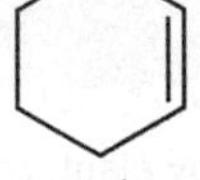

35 **a.** 1-butyl-3-methylbenzene
  b. ethylbenzene
  c. 1,2,4-trimethylbenzene (number in a way that gives the lowest set of locator numbers)
  d. 1-ethyl-3-methylbenzene (If the locator numbers are the same regardless of which substituent you start numbering from, start with the one that appears earlier in the alphabet.)

36 **a.**

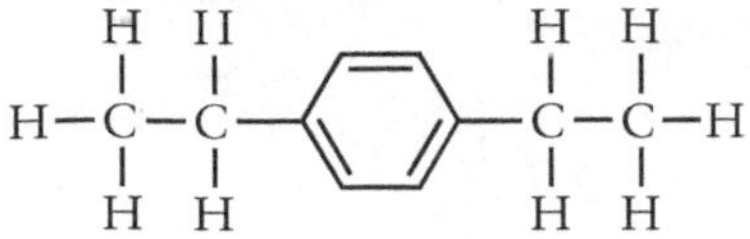

  b.

Papaver somniferum, commonly known as the poppy plant, contains morphine, a powerful analgesic. An amine functional group in the model shown is key to morphine's painkilling effects. [© John Glover/Alamy]

7 Organic Chemistry and Biomolecules

Pain and the Opioid Analgesics

Have you ever suffered from a headache and reached for the medicine cabinet for some aspirin? Perhaps you have spent a night in the hospital, where a nurse administered a narcotic, an even stronger painkiller. You are not alone—most people suffer from pain at some time in their life, and many diseases are associated with pain. Pain medicines are known as analgesics, a combination of the Greek words *an,* meaning "without," and *algesic,* meaning "pain."

Pain medications are organic compounds. They contain heteroatoms such as nitrogen and oxygen covalently bonded to carbon. The common groupings of atoms found in organic molecules, most of which contain one or more heteroatoms, are known as *functional groups.* The functional groups in a molecule determine its chemical and physical properties. Indeed, it is the functional groups in an analgesic that determine how it interacts with pain receptors in the brain to reduce pain.

The strongest analgesics are the opioid analgesics, first isolated from the opium poppy plant, *Papaver somniferum.* The opioid analgesics include morphine, codeine, and hydrocodone (Figure 7-1). As of 2010, hydrocodone, sold under the brand name Vicodin, was the most prescribed pharmaceutical in the United States. All the opioid analgesics contain an amine functional group, highlighted in blue in Figure 7-1, which is essential to their analgesic properties. The opioid analgesics can also induce a state of euphoria. Morphine, heroin, and hydrocodone can also lead to dependence and, with repeated use, addiction; hence they are classified as narcotics and their distribution is carefully monitored.

Modifying one or more of the functional groups in morphine, but retaining the amine functional group, produces compounds known as morphine *analogs.* Even slight changes to the functional groups in morphine can have a profound effect on the analgesic and addictive properties of a morphine analog. For example, codeine (Figure 7-1b) differs structurally from morphine (Figure 7-1a) at only one functional group. Where codeine has an *ether* and an *aromatic* ring, morphine has a *phenol.*

(a) Morphine (b) Codeine (c) Hydrocodone (Vicodin)

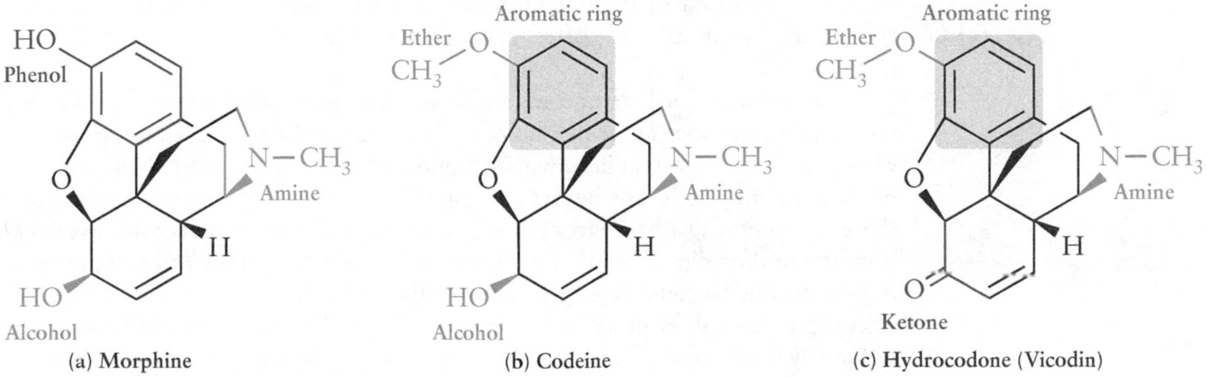

Figure 7-1 The amine functional group in the opioid analgesics morphine, codeine, and hydrocodone (Vicodin) is essential to their analgesic properties. Also, changing one or more functional groups, affects both the analgesic and addictive properties of these drugs.

279

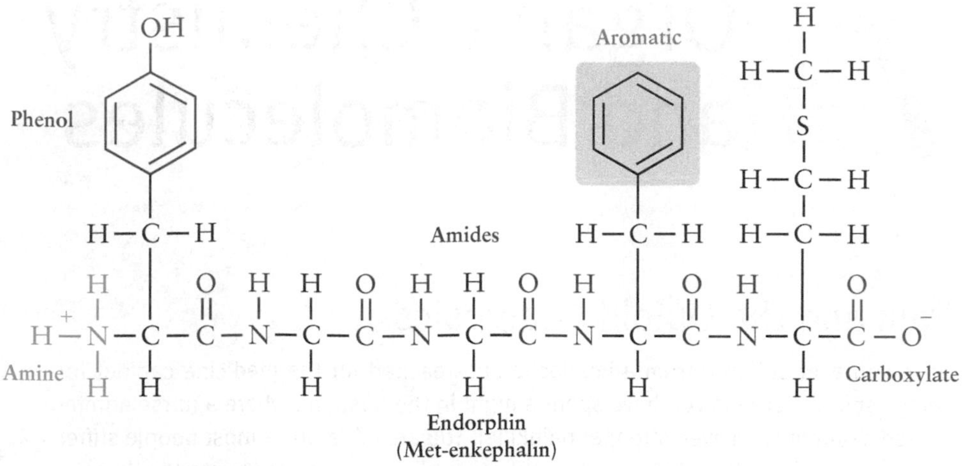

Figure 7-2 Endorphins, such as met-enkephalin, are the body's endogenous analgesics.

Endorphin
(Met-enkephalin)

This structural difference results in codeine having only 15 percent of the potency of morphine as a painkiller, and less addictive properties. Vicodin (Figure 7-1c) has the same structure as codeine except it has a ketone where codeine has an alcohol, which gives it much stronger analgesic properties and also greater addictive properties.

Opioids work by binding to the opioid receptors, large protein molecules located on the surface of brain cells. Morphine and its analogs bind to opioid receptors, initiating a sequence of biological events that, among other things, leads to reduced pain sensation. Why do neurons contain opioid receptors? The sensation of pain is a signal to the brain that it is in danger and needs to act to avoid further injury or to rest and recover from the injury. However, if you were being chased and had to run away to save your life, you could not afford to be hindered by the excruciating pain of your injury. So as humans evolved, we developed endogenous (internal) analgesics with the corresponding receptors on brain cells. The body's endogenous analgesics are organic compounds known as endorphins. The word *endorphin* is a combination of the words *endo*genous (internal) and m*orphin*e. The structure of the endorphin met-enkephalin is shown in Figure 7-2. Both endorphins and opioids bind to the opioid receptors.

Functional groups clearly impact the chemical behavior of an organic compound. In this chapter you will learn the major heteroatom-containing functional groups in organic chemistry and learn to recognize them in pharmaceutical compounds as well as the important biomolecules. ●

In chapter 6 you learned about the structure and properties of hydrocarbons—molecules containing only carbon and hydrogen atoms. In this chapter you will be introduced to organic compounds that contain oxygen (O), nitrogen (N), and phosphorus (P), often referred to as **heteroatoms**—atoms that are not C or H.

The presence of heteroatoms in an organic molecule impacts the physical and chemical properties of the molecule because C—O, C—N, and in particular H—O and H—N, are polar covalent bonds, whereas C—H and C—C are nonpolar covalent bonds. The permanent dipole created by these heteroatoms causes these compounds to have stronger intermolecular forces of attraction: hydrogen bonding and/or dipole-dipole interactions. Consequently, they have higher boiling points and in some cases greater solubility in water. The solubility properties of compounds will be described in the next chapter. Furthermore, molecules with polar covalent bonds are more likely to undergo chemical reactions. Chemical reactions of organic functional groups is described in chapter 10.

To understand the physical and chemical properties of the vast number of organic compounds, we categorize them according to the type of functional groups they contain. **Functional groups** are groups of atoms and covalent bonds that "function" in a predictable way in certain chemical reactions. You have already been introduced to three hydrocarbon functional groups in chapter 6: carbon-carbon double bonds (alkenes), carbon-carbon triple bonds (alkynes), and aromatic rings ($-C_6H_5$). Note that alkanes are not classified as a functional group. Although alkanes do undergo a limited number of reactions, they serve primarily as the carbon backbone of organic molecules.

In this chapter, you will be introduced to the major heteroatom-containing functional groups. You will see these functional groups in the important biomolecules as well as in pharmaceuticals. You will also learn about their relative chemical reactivity and some of their physical properties.

7.1 Alcohols and Ethers

$$H-\overset{..}{\underset{..}{O}}-R \qquad R-\overset{..}{\underset{..}{O}}-R'$$

Alcohol   Ether

In this section we introduce the functional groups derived from water (H_2O), an inorganic compound, formed by replacing *one* of the hydrogen atoms with carbon atoms to make an **alcohol**, R—O—H, as shown in Figure 7-3a, or replacing *both* hydrogen atoms with carbon atoms to create an **ether**, R—O—R, as shown in Figure 7-3b. We use —R to signify a carbon atom or a chain of carbons of any length. When two R groups are part of the definition, we use —R and —R' to indicate that the carbon chains are not necessarily the same.

Since alcohols and ethers are derived from water, they have the same molecular shape as water at the oxygen center: *bent* with a bond angle of approximately 109.5° because the oxygen atom has two bonding and two nonbonding pairs of electrons.

The antidepressant Effexor, shown in Figure 7-4, is an example of a pharmaceutical that contains both an alcohol and an ether functional group. When viewing the structure of Effexor, we recognize an ether functional group because there is an oxygen atom with two single bonds to carbon atoms: a CH_3 group and an aromatic ring carbon. The alcohol functional group in Effexor is distinguished by an oxygen atom with one single bond to a tetrahedral carbon and a bond to hydrogen. The O—H group is known as a **hydroxyl group**.

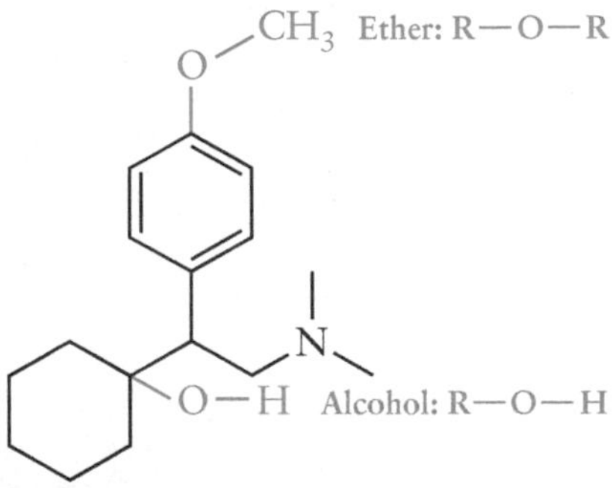

Recall from chapter 6 that when writing skeletal line structures, only hydrogen atoms attached to carbon atoms and the carbon atoms themselves are omitted. Thus, when writing the skeletal line structure for an alcohol or an ether, the oxygen atom must be written in to denote an oxygen atom rather than a carbon atom, as shown below. In addition, hydrogen atoms bonded to heteroatoms must always be shown. Therefore, the hydroxyl group of an alcohol must be written either as O—H or OH. To gain further experience with alcohols and ethers, work through the exercises in Using Molecular Models 7-1: Alcohols and Ethers.

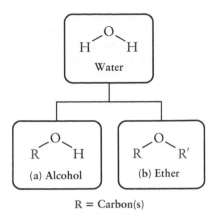

R = Carbon(s)

Figure 7-3 Alcohols and ethers are structurally related to water, H_2O.

The nonbonding electrons on heteroatoms are typically omitted from Lewis structures, especially in complex molecules. Nevertheless, they are understood to be present: typically oxygen has two, and nitrogen has one nonbonding pair of electrons.

Figure 7-4 The antidepressant Effexor contains an alcohol (a hydroxyl group) and an ether functional group.

Alcohol

Ether

or

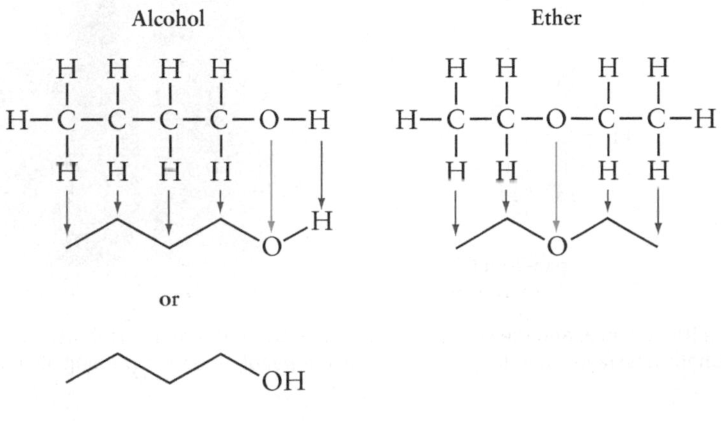

Figure 7-5 Alcohols are classified as primary (1°), secondary (2°), or tertiary (3°) depending on the number of R groups (shaded) on the carbon atom bearing the —OH group.

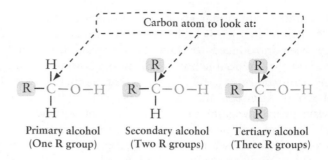

Primary alcohol
(One R group)

Secondary alcohol
(Two R groups)

Tertiary alcohol
(Three R groups)

Alcohols R—OH

Alcohols are further subdivided into three classes depending on the number of R groups attached to the carbon atom bearing the hydroxyl group (—OH), as illustrated in **Figure 7-5**:

- A primary (1°) alcohol has the hydroxyl group on a carbon atom with *one* R group and two hydrogen atoms;
- A secondary (2°) alcohol has the hydroxyl group on a carbon atom with *two* R groups and one hydrogen atom;
- A tertiary (3°) alcohol has the hydroxyl group on a carbon atom with *three* R groups and zero hydrogen atoms

Differentiating the subclasses of alcohols becomes important when predicting the chemical reactivity of alcohols, described in chapter 10.

A ball-and-stick model and a Lewis structure for three simple alcohols are shown in **Figure 7-6**. The simplest alcohol, known as methanol, is found in deicing fluids (Figure 7-6a) used to remove ice from car windshields. Ethanol is found in alcoholic beverages, fuel, and disinfectants (Figure 7-6b). Ethanol is a primary alcohol because it has one R group (a —CH₃ group) and two hydrogen atoms on the carbon atom bearing the hydroxyl group. Propan-2-ol, also known by its common name isopropanol, is found in rubbing alcohol.

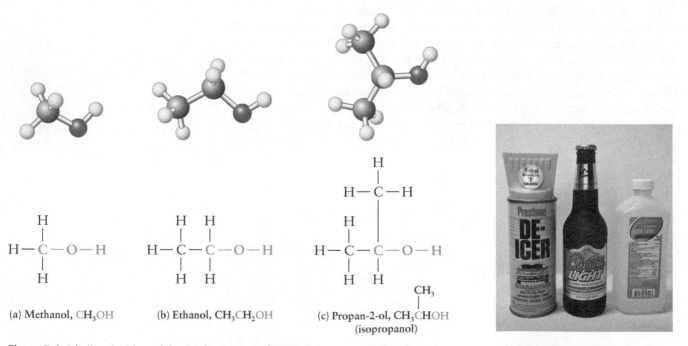

(a) Methanol, CH₃OH

(b) Ethanol, CH₃CH₂OH

(c) Propan-2-ol, CH₃CHOH
(isopropanol)

Figure 7-6 A ball-and-stick model, a Lewis structure, the IUPAC name, and the condensed structure for three simple alcohols, R—O—H: (a) methanol, found in de-icer; (b) ethanol, found in alcoholic beverages; and (c) propan-2-ol (isopropanol), found in rubbing alcohol.
[Catherine Bausinger]

Propan-2-ol (Figure 7-6c) is a secondary alcohol because it has two R groups (both —CH$_3$ groups) and one hydrogen atom on the carbon atom bearing the hydroxyl group. The drug Effexor (Figure 7-4) contains a tertiary alcohol because the carbon atom bearing the hydroxyl group has three R groups and no hydrogen atoms. As you can see from this more complex structure, the R groups are part of the rest of the Effexor molecule. To determine the IUPAC name of an alcohol, follow the steps given in the Guidelines box below.

Guidelines for Naming Alcohols

Rule 1: Assign the root. Select the longest contiguous chain of carbon atoms that also contains the hydroxyl group (—OH) and assign the root name based on the number of *carbon* atoms in this chain, using Table 6-2.

For example, the root name for the alcohol shown below is **hexane** *because there are six contiguous carbon atoms in the chain of carbon atoms containing the hydroxyl group.*

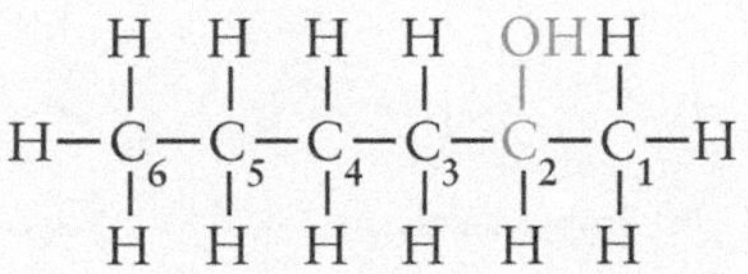

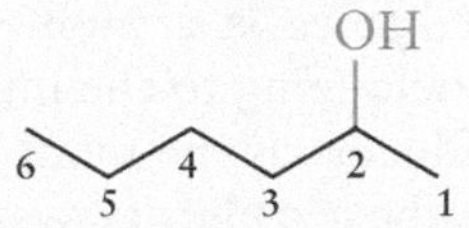

Hexan-2-ol

Rule 2: Assign the ending. Since the chain is an alkane containing a hydroxyl group, we drop the *-e* at the end of the root and add *-ol* to signify that the molecule is an alco*hol* rather than an alk*ane*.

For the example above we change the name from hexane to hexanol.

Rule 3: If the chain has more than two carbons, assign a locator number to indicate which carbon along the chain contains the bond to the hydroxyl group. Number the carbon atoms starting from the end of the chain closer to the hydroxyl group. Then insert a locator number, separated by hyphens, between the root and the ending: *alkan-#-ol.*

For the example above, we insert a -2- into the name: hexan-2-ol, because the hydroxyl group is located on the second carbon in a contiguous chain of six carbons.

Rule 4: Assign a prefix if there are alkyl substituents on the main chain. If the main chain contains alkyl substituents, assign the prefix name according to the rules for naming branched chain hydrocarbons described in section 6.5, Naming Branched Hydrocarbons.

Alcohols in Nature and Biochemistry Alcohols are widely distributed in nature. For example, cholester*ol*, an important steroid introduced in chapter 6, is a secondary alcohol because the carbon bearing the —OH group has *two* R groups and one hydrogen atom (Figure 7-7). Recall that while H atoms are not shown in a skeletal line structure, they are present.

Alcohols containing *two* hydroxyl groups are known as *diols*, those containing *three* hydroxyl groups are known as *triols*, and those containing many hydroxyl groups are known as *polyols*. The most common diols are ethylene

Figure 7-7 The structure of cholesterol, an important steroid produced in the body, contains a secondary alcohol: the hydroxyl group is bonded to a carbon with two R groups and one hydrogen atom.

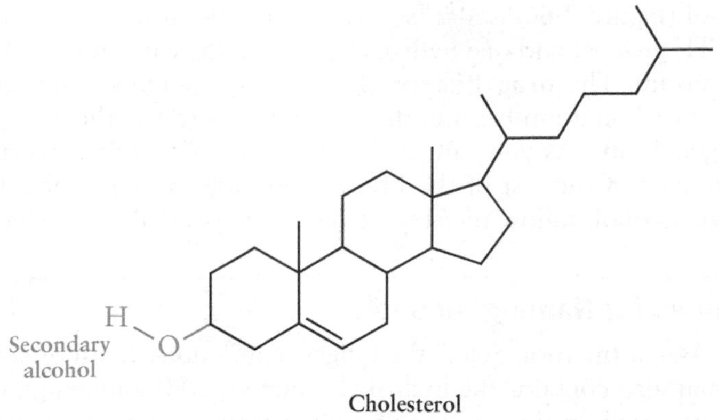

Cholesterol

glycol (a) and propylene glycol (b), both of which are found in antifreeze. Glycerol (c) is an important triol involved in the metabolism of glucose and fats.

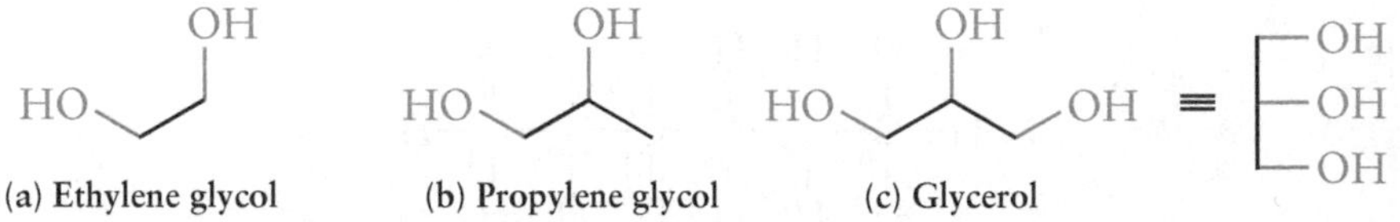

(a) Ethylene glycol   (b) Propylene glycol   (c) Glycerol

Glucose is an example of a polyol; it contains several hydroxyl groups, as shown in **Figure 7-8**. Glucose is a **monosaccharide**, also known as a simple sugar, a compound belonging to the important class of biomolecules known as **carbohydrates**. Glucose is the main source of energy for our cells and is transported through the circulatory system to our tissues, where it is referred to as "blood sugar." You will note that one hydroxyl group in glucose is not highlighted in Figure 7-8 because the carbon atom bearing the hydroxyl group does not contain only H and R groups; instead it has a bond to another oxygen atom, so it is not classified as an alcohol. You will learn about this functional group in chapter 11 when you study carbohydrates. You will also notice that the ring is drawn slightly different from what you saw in chapter 6 for cycloalkanes. This is the convention used for writing carbohydrate structures, where the three-dimensional orientation of the hydroxyl groups is a key component of their structure. Carbohydrates are the subject of chapter 11, where this topic will be further explored.

Lactose, found in milk, is another type of carbohydrate known as a **disaccharide**. (**Figure 7-9**). Lactose is a polyol containing seven hydroxyl groups, as highlighted in the figure.

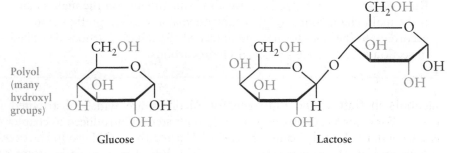

Glucose

Lactose

Figure 7-8 The structure of the simple sugar glucose, a polyol, highlighting the many hydroxyl groups (—OH).

Figure 7-9 The structure of the disaccharide lactose, another polyol, highlighting the many hydroxyl groups. This carbohydrate is found in milk.

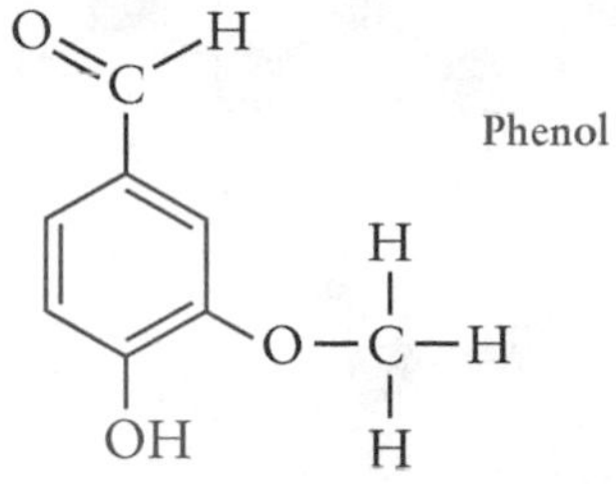

Vanillin
(from vanilla beans)

Phenol

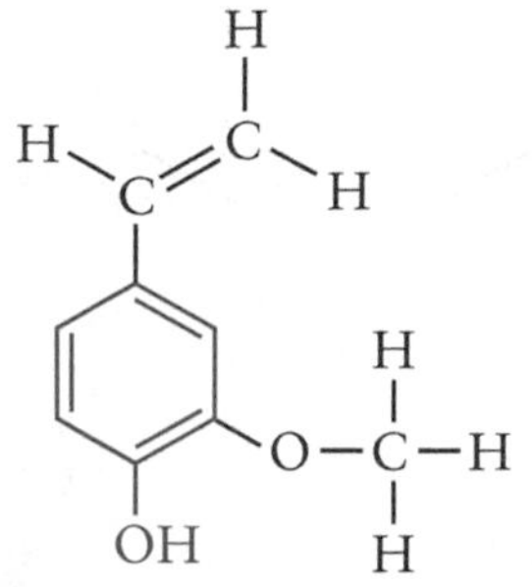

Eugenol
(from cloves)

Figure 7-10 Vanillin, found in vanilla beans, and eugenol, found in cloves, are both fragrant substances containing a phenol functional group. [Catherine Bausinger]

Phenols

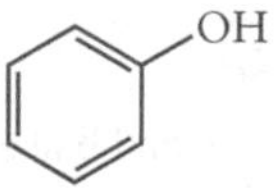

A hydroxyl group directly attached to an aromatic ring carbon is not an alcohol but a related functional group, known as a **phenol**. A phenol is defined as the aromatic ring together with the hydroxyl group and is highlighted in pink in the structures below. The aromatic ring together with the hydroxyl group act as one unit, which causes phenols to have a different chemical reactivity than alcohols. For example, phenols lose the hydrogen ion, H^+, of their hydroxyl group more readily than do alcohols. Phenols are used as disinfectants. Phenols are also found in nature, as for example the fragrant phenols vanillin, from vanilla beans, and eugenol, from cloves, shown in Figure 7-10. You may recall from the opening vignette that morphine contains a phenol (Figure 7-1a).

The phenol functional group is responsible for the antioxidant properties of compounds such as vitamin E (Figure 7-11). A phenol can quench (remove by reacting with) **free radicals**—unstable atoms or molecules with an odd number of nonbonding electrons that cause damaging oxidation reactions.

Phenols are named according to the rules for naming aromatic compounds described in section 6.6 except the root is *phenol* rather than *benzene* and numbering always begins at the carbon atom bearing the hydroxyl group. Number in the direction, counterclockwise or clockwise, that assigns the other substituents the lowest set of locator numbers. For example, the substituted phenol shown in Figure 7-12 has the IUPAC name 3-ethylphenol because there is a two carbon chain on the phenol at C(3).

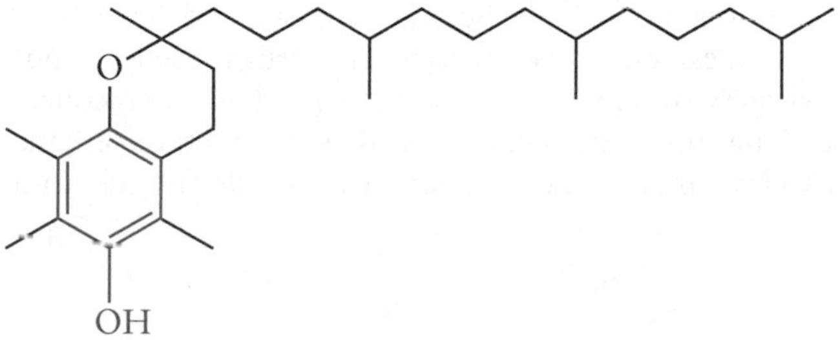

Vitamin E
(antioxidant)

Figure 7-11 The antioxidant properties of vitamin E are a result of the phenol functional group in its structure.

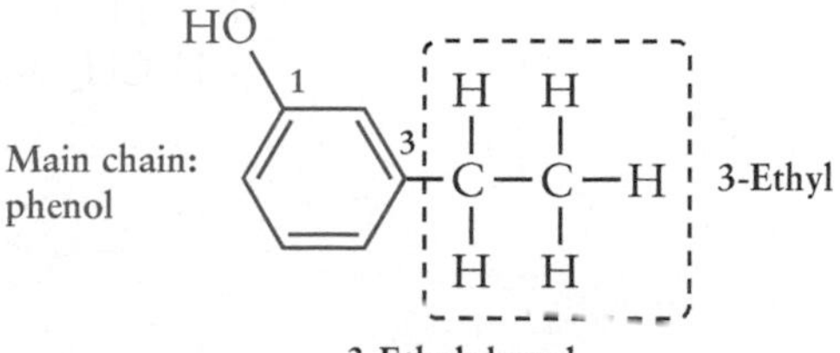

Main chain: phenol

3-Ethyl

3-Ethylphenol

Figure 7-12 Phenols are named according to the rules for naming aromatic compounds but using the root *phenol* instead of *benzene*, as shown here for 3-ethylphenol.

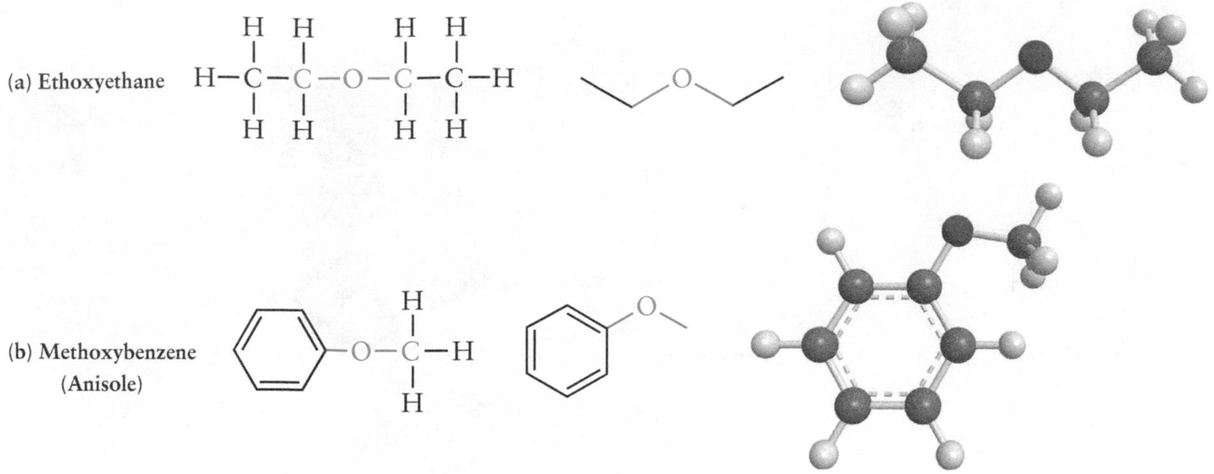

(a) Ethoxyethane

(b) Methoxybenzene
(Anisole)

Figure 7-13 The IUPAC name, Lewis structure, skeletal line structure, and ball-and-stick model for two ethers, R—O—R′.

Ethers

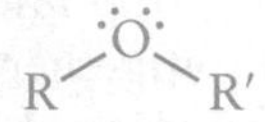

Two simple ethers are shown in **Figure 7-13**. Ethers are functional groups containing an oxygen with two single bonds to carbon atoms: R—O—R′. Ethers may be symmetrical, with identical R groups, as shown in Figure 7-13a, or they may be unsymmetrical, with two different R groups, as shown in Figure 7-13b. In contrast to alcohols, the —R groups in an ether can be aromatic and still be classified as an ether. Ethers are named as substituted alkanes, therefore, the ending in their IUPAC name is -*ane*. The IUPAC rules for naming ethers are described in the Guidelines box on the next page.

Ethers are common functional groups that are found in nature as well many pharmaceutical drugs. The opioid narcotics, described in the opening vignette, all have an ether functional group in one of their rings. Codeine and Vicodin have a second ether functional group where morphine has a phenol, as seen in Figure 7-1. The structure of Toprol, one of the most commonly prescribed pharmaceuticals, used for the treatment of hypertension (high blood pressure), has two ether functional groups and a secondary alcohol in its structure as highlighted in Figure 7-14.

Ethers are relatively unreactive functional groups and for this reason they often serve as solvents in chemical reactions. Solvents are liquids that dissolve compounds or elements without reacting with them.

Low-molecular-mass ethers are volatile liquids or gases and have been used since the 1800s as anesthetics. Anesthetics are drugs used to put patients to sleep during surgery so that they feel no pain and have no recollection of the procedure. The first anesthetic was the ether ethoxyethane $(CH_3CH_2OCH_2CH_3)$. Today, many anesthetics, such as desflurane and

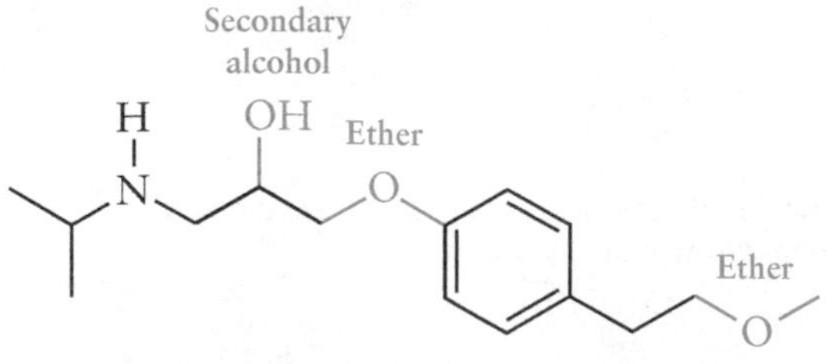

Figure 7-14 The structure of Toprol, a drug used for the treatment of hypertension, highlighting the ether and alcohol functional groups.

Toprol (nonsalt form)

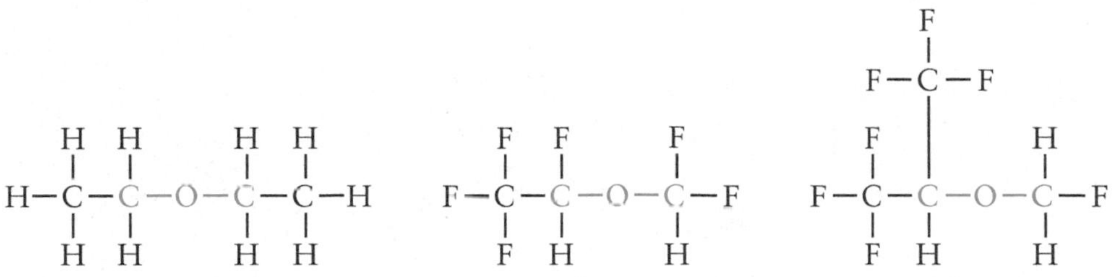

Figure 7-15 The Lewis structure of the anesthetics ethoxyethane, desflurane, and sevoflurane, highlighting the ether functional group.

sevoflurane, are fluorinated compounds in which F atoms replace H atoms (Figure 7-15). Recall from chapter 5 that these anesthetic ethers reach the brain faster allowing the patient to come in and out of consciousness more rapidly. They are also less flammable than ethoxyethane and have fewer side-effects.

Guidelines for Naming Ethers

Rule 1: Assign the root. Since an ether has an oxygen bonded to two R groups, we assign the main chain to the R group with the greater number of contiguous carbons, assigning the root name based on the number of carbon atoms that it contains using Table 6-2.

*For example, in the ether shown below, the oxygen has a bond to a one-carbon R group and a three-carbon R group. The longer, three-carbon R group is assigned as the main chain and therefore the root is **propane**.*

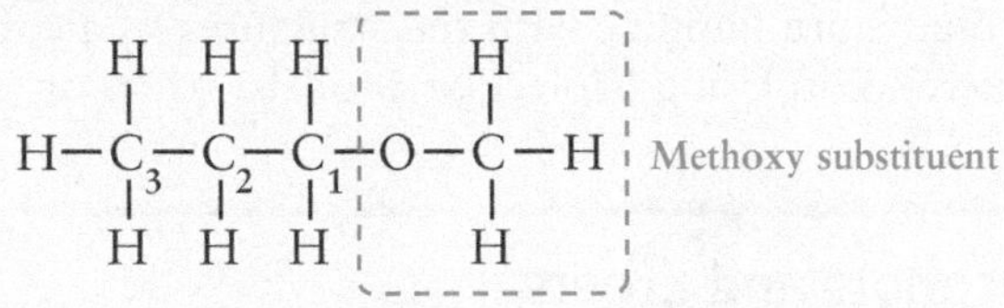

Methoxypropane

Rule 2: Assign the ending. The ending remains *-ane* since an ether is treated as a substituted alkane.

*In the example above, the root and ending is **propane**.*

Rule 3: Assign a prefix. The other R group is named as a substituent and appears in the prefix. Whereas a hydrocarbon substituent, —R, is named *alkyl* (section 6.5), an —OR substituent is named *alkoxy*. Name the —OR group (R with shorter chain) according to the number of carbon atoms in the —OR group, using Table 6-2, and changing the ending from *-ane* to *-oxy*. In this example, the substituent is a *methoxy* substituent.

Rule 4: When the main chain has three or more carbon atoms, assign a locator number indicating the location of the alkoxy substituent on the main chain. A locator number is used to indicate which carbon atom on the main chain contains the alkoxy substituent. Number the main chain starting from the end closer to the —OR group and place the locator number followed by a hyphen in front of the alkoxy substituent name.

*In the example above, the main chain has three carbon atoms (propane) and the methoxy substituent is located on C(1), so the IUPAC name is **1-methoxypropane**.*

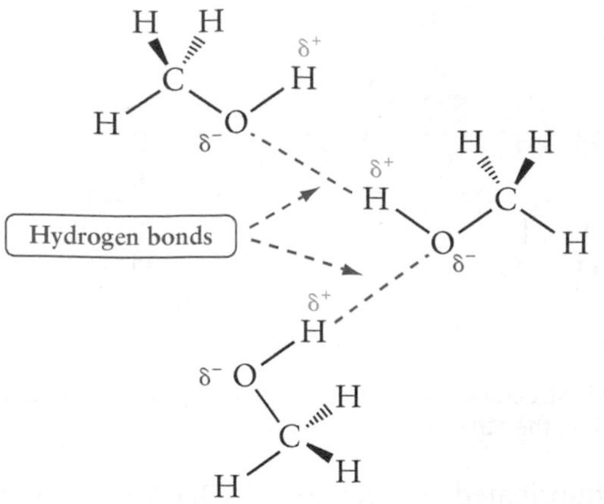

Figure 7-16 Hydrogen bonding between three methanol molecules.

Physical Properties of Alcohols, Phenols, and Ethers

Alcohols and phenols form hydrogen bonds, the strongest of the intermolecular forces of attraction. Figure 7-16 illustrates hydrogen bonding between three methanol molecules. Hydrogen bonding accounts for the higher boiling points of alcohols and phenols compared to that of other molecules of similar size. For example, the alcohols methanol, ethanol, and 2-propanol are all liquids at room temperature, whereas the comparably sized alkanes ethane, propane, and butane are gases at room temperature.

Since ethers do not have an O—H bond, they interact through dipole-dipole intermolecular forces of attraction as a result of the weakly polar C—O bond. Therefore ethers have lower boiling points than alcohols of similar size. Yet ethers have slightly higher boiling points than alkanes of comparable size because they have a permanent dipole and alkanes do not.

To become more familiar with the structures of alcohols and ethers, perform the exercises in Using Molecular Models 7-1.

Using Molecular Models 7-1 Alcohols and Ethers

Construct Water, Methanol, and Methoxymethane

1. Obtain two black carbon atoms, six light blue hydrogen atoms, one oxygen atom, and eight straight bonds.

2. Construct a model of water, H_2O. Answer Inquiry Questions 5 through 7.

3. Using the water molecule you made above, remove one hydrogen atom and replace it with a carbon atom attached to three hydrogen atoms (—CH_3). Answer Inquiry Questions 8 through 11.

4. Using the model you made in (3), replace the other hydrogen atom on oxygen with a carbon atom and three hydrogen atoms (—CH_3). Answer the Inquiry Questions 12 through 13.

Inquiry Questions

5. What is the H—O—H bond angle in water?

6. What is the shape of the water model?

7. Is water organic or inorganic? Explain.

8. What functional group does methanol contain?

9. What is the O—C—H bond angle in the methanol model? What is the molecular shape around the carbon atom?

10. What is the H—O—C bond angle in the methanol model? What is the molecular shape around the oxygen atom?

11. Is methanol organic or inorganic? Explain.

12. What is the C—O—C bond angle in the ether model? What is the molecular shape around the oxygen atom?

13. What is the molecular shape around each carbon atom in the ether model?

WORKED EXERCISES | Alcohols and Ethers

7-1 Write the Lewis structure, condensed structure, and the skeletal line structure for butan-2-ol.

7-2 Which of the following molecules is an ether?
   a. CH_3CH_2OH **b.** $CH_3CH_2CH_3$ **c.** $CH_3CH_2OCH_3$ **d.** $HOCH_2CH_2OH$

7-3 Which of the compounds in the previous problem is a diol?

7-4 Write a skeletal line structure for the alcohol shown below. Indicate whether it is a 1°, 2°, or 3° alcohol. Add the nonbonding electrons to the oxygen atom.

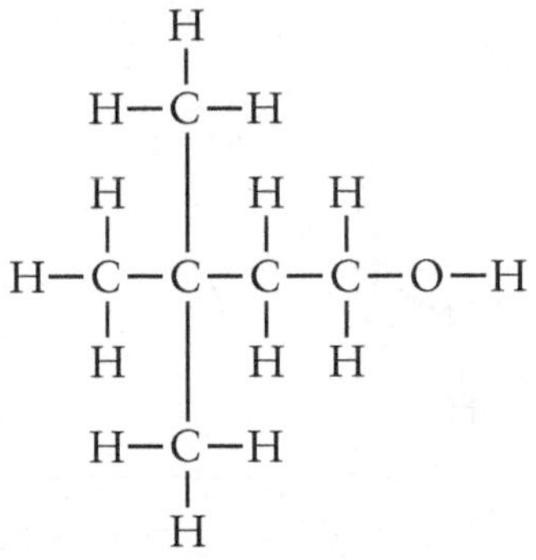

7-5 Name the alcohol shown below. Is this a primary, secondary, or tertiary alcohol?

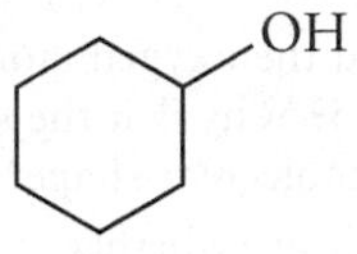

7-6 The antidepressant Prozac is a pharmaceutical with the chemical structure shown below. Highlight and label the ether functional group. Highlight and label the aromatic rings.

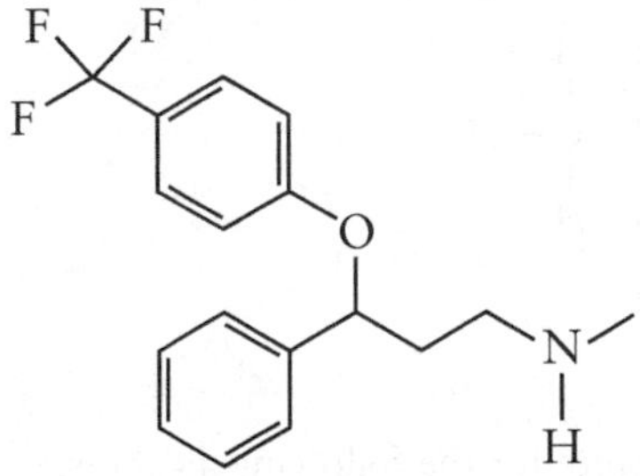

Prozac

Solutions

7-1

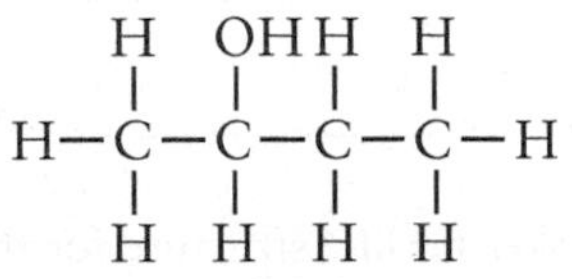

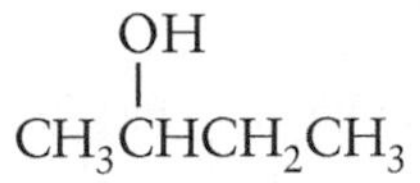

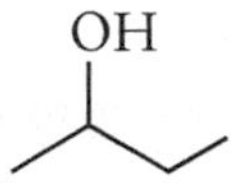

7-2 **c.** $CH_3CH_2OCH_3$ because it contains an oxygen atom with single bonds to two carbon atoms.

7-3 **d.** $HOCH_2CH_2OH$, because it contains two hydroxyl groups in the molecule.

7-4

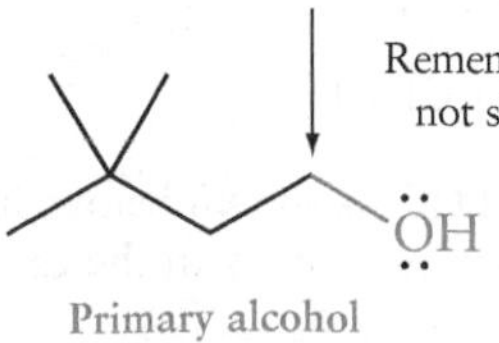

Remember there are two hydrogens on this carbon not shown because it is a skeletal line structure.

Primary alcohol

7-5 It is a secondary alcohol, because the carbon bearing the OH group has bonds to two carbon atoms and one hydrogen atom.

Rule 1: Assign the root. There are six carbon atoms in a ring in the main chain containing the hydroxyl group, so the root name is *cyclohexane*.

Rule 2: Assign the ending. Since there is a hydroxyl group on the main chain, we drop the *-e* and add the ending *-ol*, giving us *cyclohexanol*.

Rule 3: Assign a locator number to indicate the location of the hydroxyl group. A locator number is not necessary when there are no alkyl substituents on the ring, because every position is the same.

7-6

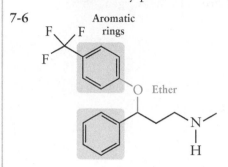

PRACTICE EXERCISES

1   What is the molecular shape around the oxygen atom in an alcohol and an ether? What is the bond angle around the oxygen atom? Why is it the same for an alcohol and an ether? What inorganic compound has the same molecular shape?

2   For each of the alcohols shown, indicate whether it is a 1°, 2°, or 3° alcohol.

a.

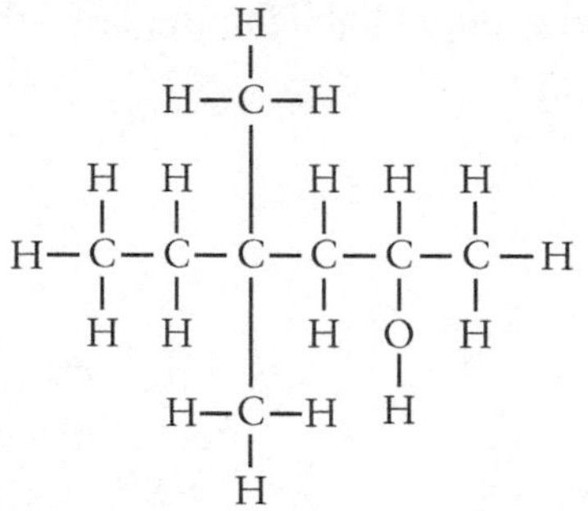

```
                    H
                    |
                H—C—H
                    |
    H  H           H  H  H
    |  |           |  |  |
H—C—C—C—C—C—C—H
    |  |           |  |  |
    H  H           H  O  H
                    |    |
                H—C—H   H
                    |
                    H
```

b. ⟍⟋⟍OH   c. OH on isopropyl   d. OH on tert-butyl

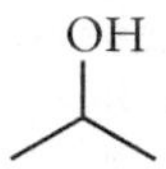

3   Provide a skeletal line structure for the following alcohols.
    a. octan-3-ol   **b.** heptan-1-ol   **c.** heptan-2-ol

4   Show how one ethanol molecule might hydrogen bond to two other ethanol molecules, using dashed lines to represent the hydrogen bonds. How does hydrogen bonding affect the boiling point of an alcohol?

5   Monosaccharides belong to a class of biomolecules known as _____, which have many _____ groups.

6   Is the structure shown below a diol, triol, or polyol? Write a skeletal line structure for this molecule and highlight the alcohol functional groups. Indicate whether they are primary, secondary, or tertiary. Would you expect this compound to have a higher or a lower boiling point than propan-1-ol? Explain.

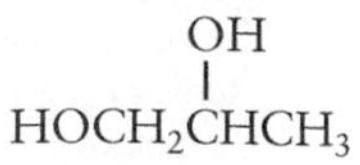

$$\underset{\displaystyle HOCH_2CHCH_3}{\overset{\displaystyle OH}{\vert}}$$

7   The structure of Δ^9-tetrahydrocannabinol, Δ^9-THC, is shown below. It is the psychoactive substance found in *cannabis* (marijuana), which interacts with the cannabinoid receptors in the brain.

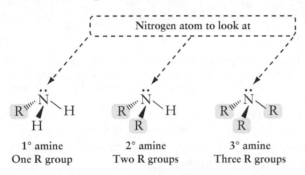

Δ^9-THC

a. Locate and identify the ether and phenol functional groups in this molecule.

b. What part of the chemical name for this compound suggests that there is a phenol present in the molecule?

■ 7.2 Amines

An **amine** is a functional group derived from the inorganic compound ammonia (NH_3), by replacing *one*, *two*, or all *three* of the hydrogen atoms with an —R group. The —R groups can be either alkyl or aryl (aromatic) groups. Amines are further subdivided into three classes depending on the number of —R groups attached to the nitrogen atom, as shown below:

Nitrogen atom to look at

1° amine
One R group

2° amine
Two R groups

3° amine
Three R groups

- A primary (1°) amine has *one* R group and two hydrogen atoms bonded to the nitrogen atom, RNH_2.
- A secondary (2°) amine has *two* R groups and one hydrogen atom bonded to the nitrogen atom, R_2NH.
- A tertiary (3°) amine *three* R groups and zero hydrogen atoms bonded to the nitrogen atom, R_3N.

Note that the classification system for identifying primary, secondary, and tertiary amines is different from the system for identifying alcohols: in an alcohol we look at the carbon atom bearing the —OH group, whereas in an amine we look at the nitrogen atom itself.

Figure 7-17 shows ball-and-stick models and the corresponding Lewis structures and condensed structures for ammonia (7-17a), and some simple amines (Figure 7-17b-d), and also an aromatic amine (Figure 7-17e). Since amines are derived from ammonia, they have the same molecular shape as ammonia at the nitrogen center: *trigonal pyramidal* with bond angles of approximately 109.5°, because nitrogen has three bonds and one nonbonding pair of electrons. The IUPAC system for naming amines is described in the Guidelines box on the next page.

As with alcohols, differentiating the subclasses of amines is important when predicting the chemical reactivity and physical properties of amines. For example, a tertiary amine does not undergo many of the reactions that primary and secondary amines undergo. Furthermore, primary and secondary

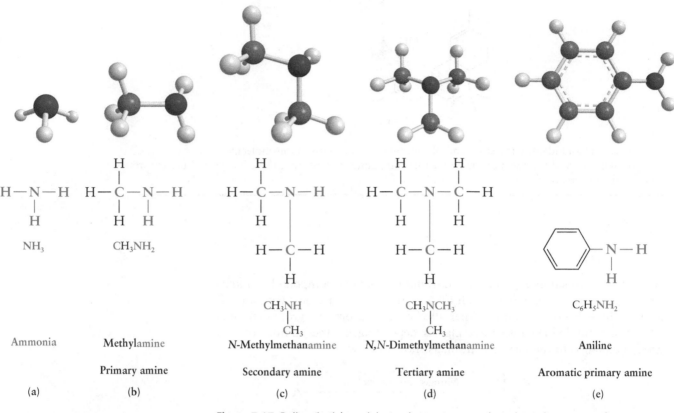

Ammonia   Methylamine   N-Methylmethanamine   N,N-Dimethylmethanamine   Aniline

                 Primary amine   Secondary amine   Tertiary amine   Aromatic primary amine

   (a)   (b)   (c)   (d)   (e)

Figure 7-17 Ball-and-stick models, Lewis structures, and condensed structures, for (a) ammonia, (b) methylamine, a primary amine (c) *N*-methylmethanamine, a secondary amine (d) *N,N*-dimethylmethanamine, a tertiary amine, and (e) aniline, an aromatic amine.

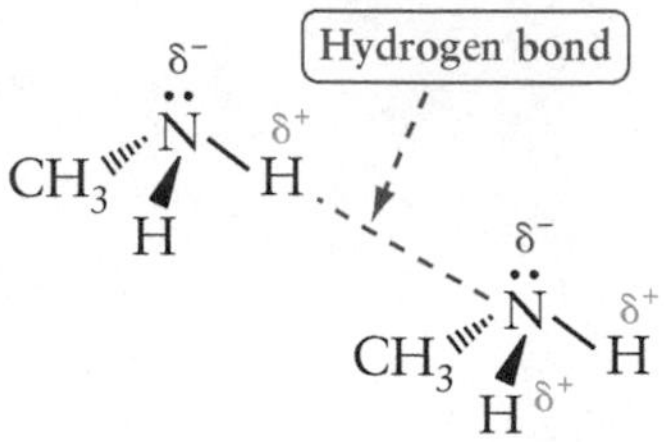

Figure 7-18 Hydrogen bonding between two primary amines. Tertiary amines cannot form hydrogen bonds.

amines, but not tertiary amines, are capable of hydrogen bonding because they contain the very polar N—H bond, as illustrated in **Figure 7-18**. Consequently, tertiary amines have lower boiling points than primary or secondary amines of similar molecular mass.

In the opening vignette for this chapter you learned that the amine functional group is essential to the pharmacological effects of the opioid narcotics. Indeed, amines are some of the most important functional groups in pharmaceuticals. For example, the antidepressant Prozac, shown in **Figure 7-19**, gets much of its pharmacological activity from its secondary amine functional group.

Guidelines for Naming Amines

Rule 1: Assign the root. Since an amine has up to three R groups, assign the —R group with the most contiguous carbons as the main chain and assign the root name according to Table 6-2

For example, the tertiary amine below has three —R groups on nitrogen. We choose the R group with the most contiguous carbons, three carbons, as the main chain, and therefore the root name is **propane**.

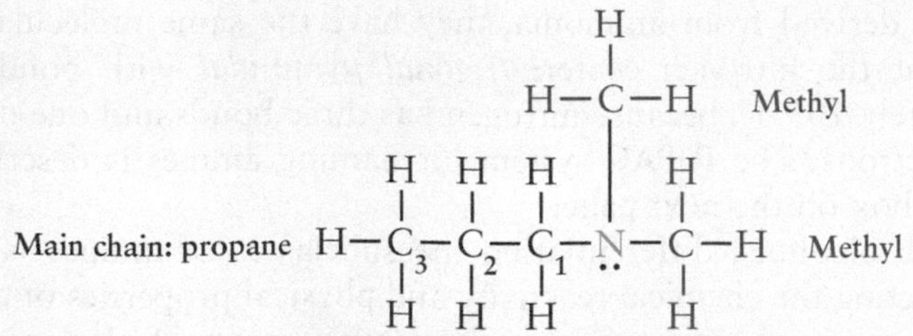

N,N-Dimethylpropan-1-amine

Rule 2: Assign the ending. Since the main chain is an alkane containing an amine, we drop the *-e* from the root name and add the ending *-amine* to signify the molecule is an amine.

In the example above, we change the root name from propane to propanamine to signify the molecule is an amine.

Rule 3: Assign a locator number to indicate which carbon along the main chain contains the bond to the nitrogen atom. If the main chain has more than two carbon atoms, number the main chain starting from the end closer to the nitrogen atom. Then insert a locator number, separated by hyphens, between the root and the ending: *alkan-#-amine.*

In the example amine above, we insert a -1- into the name, giving us propan-1-amine, because the nitrogen atom is bonded to the first carbon of the propane chain.

Rule 4: Assign the prefix. For secondary and tertiary amines, we name the other R group(s) attached to the nitrogen atom as though they were substituents; however, instead of a locator number, we use the prefix *N-* to signify that the substituent is bonded to the nitrogen atom and not to a carbon atom in the main chain. For a tertiary amine, list these R groups in alphabetical order. If the R groups are identical, use the multiplier *di-*. If there are also alkyl substituents along the main chain, name those as described in section 6.5 and place them after the *N-*substituents in the prefix name.

In the example above, we see that there are no alkyl branches on the main chain, but nitrogen has two identical one-carbon R groups. Hence, we insert the prefix N,N-dimethyl. The complete IUPAC name is N,N-dimethylpropan-1-amine.

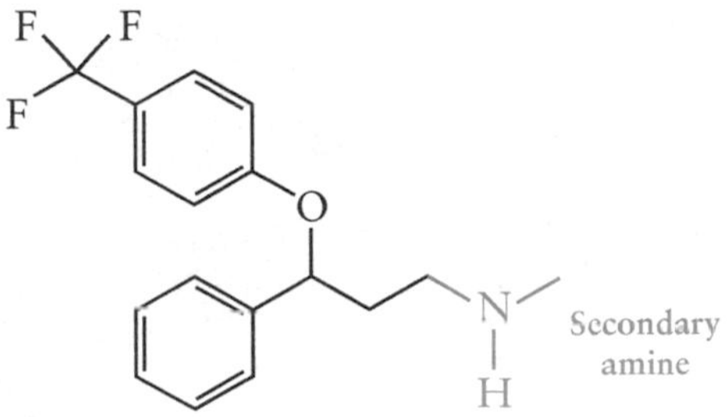

Prozac

Figure 7-19 Prozac, an antidepressant that derives its pharmacological activity from the secondary amine functional group.

The amine functional group is present in many endogenous brain chemicals, including the neurotransmitters dopamine and epinephrine (adrenaline). The Chemistry in Medicine section at the end of this chapter describes how dopamine is involved in schizophrenia and Parkinson's disease. The number of physiologically active amines is extensive and includes amines found in nature, known as **alkaloids**, such as the alkaloids shown in Figure 7-20.

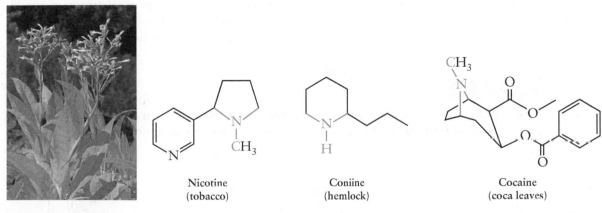

Nicotine
(tobacco)

Coniine
(hemlock)

Cocaine
(coca leaves)

Figure 7-20 Alkaloids are amines derived from plants, such as nicotine, found in the tobacco plant (shown in photo); coniine, a poison found in hemlock leaves; and cocaine, found in the leaves of the coca plant. [© WILDLIFE GmbH/Alamy]

Ionic and Neutral Forms of an Amine

One of the most important characteristics of amines is their chemical behavior as organic bases. A base is a compound that accepts a hydrogen ion, H^+, from water. Amines become polyatomic cations when they accept a hydrogen ion, to form an N—H bond. This reversible reaction is illustrated below for a primary amine:

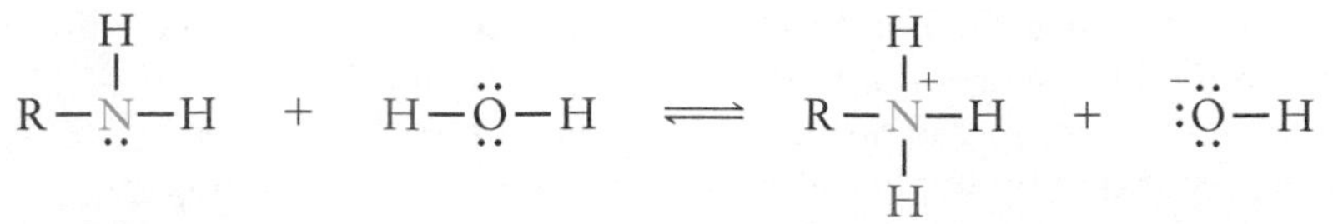

Amine (neutral)   Polyatomic cation (positively charged ion)   Hydroxide ion

Hence, the formation of an N—H bond converts an amine, a neutral compound, into its ionic form. You will learn more about this important class of reactions in chapter 9, Acids and Bases. Whether the equilibrium favors the neutral amine, the reactant, or the ionic form, the product, depends on the environment. In our cells, amines are usually in their ionic form, but you should be able to recognize both forms of an amine. Work through Using Molecular Models 7-2: Amines to gain familiarity with this characteristic reaction of amines.

Using Molecular Models 7-2 Amines

Construct Ammonia and Methanamine

1. Obtain one black carbon atom, ten hydrogen atoms, two nitrogen atoms, one oxygen atom, and twelve straight bonds.

2. Construct a model of ammonia, NH_3. Answer Inquiry Questions 5 through 6.

3. Construct a model of water, H_2O. Answer Inquiry Question 7.

4. Using your original model from step 2, replace one of the hydrogen atoms with a —CH₃ group to build methanamine. Answer Inquiry Questions 8 through 12.

Inquiry Questions

5. What is the H—N—H bond angle in ammonia? What is the shape of the model?

6. Is ammonia an organic or an inorganic compound? Explain.

7. Show what happens when ammonia acts as a base and reacts with water by breaking a hydrogen atom off water and attaching it to nitrogen. You will need an extra straight bond to do this. What two new ions have you made? What bond did you have to break? What new bond did you make?

8. What functional group does the model of methanamine contain?

9. What is the bond angle around the nitrogen atom in methanamine? What is the molecular shape around the nitrogen atom?

10. Is methanamine capable of hydrogen bonding?

11. Is methanamine an organic or an inorganic compound? Explain.

EXERCISES   Amines

7-7   Brain cells communicate through chemical messengers known as neurotransmitters. Neurotransmitters are organic amines such as dopamine, shown below:

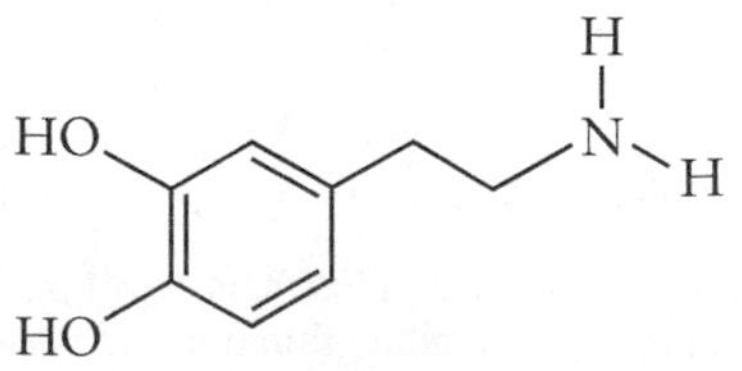

Dopamine

a. Highlight the amine functional group.

b. What is the molecular shape around the amine nitrogen?

c. Write the ionic form of dopamine.

7-8 Identify the following amines as primary, secondary, or tertiary amines. Which of these amines can form hydrogen bonds and why?

a.   **b.**

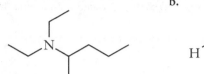

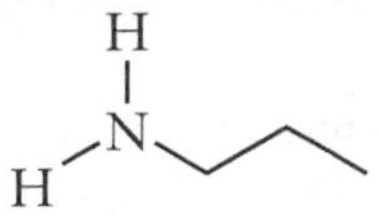

7-9 Name the amines in the previous problem (7-8).

Solutions

7-7 a.

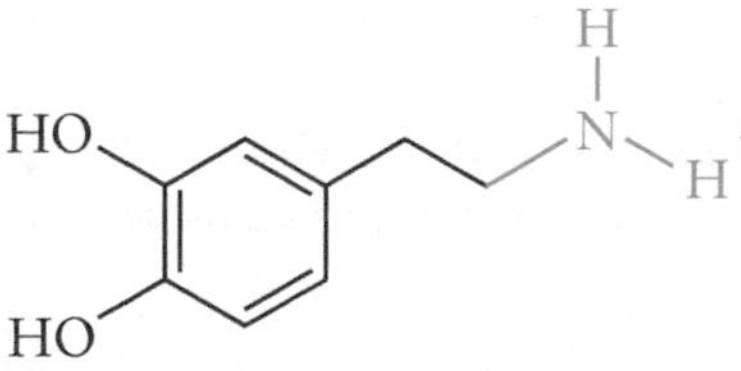

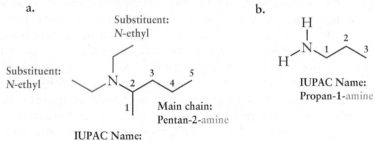

Dopamine

b. The molecular shape around the nitrogen atom is trigonal pyramidal.

c.

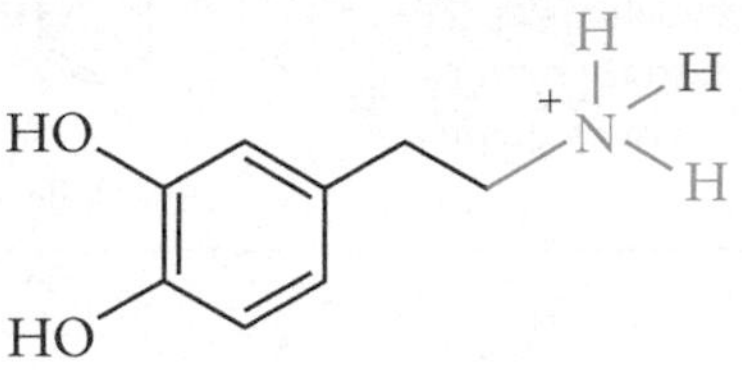

HO

HO

$\overset{+}{N}$ H H H Amine (ionic form)

Dopamine

7-8 a. tertiary; cannot form hydrogen bonds because it does not have any N—H bonds

     b. primary; can form hydrogen bonds because it has two N—H bonds

7-9 a.   **b.**

Substituent:
N-ethyl

Substituent:
N-ethyl

Main chain:
Pentan-2-amine

IUPAC Name:
N,N-Diethylpentan-2-amine

IUPAC Name:
Propan-1-amine

PRACTICE EXERCISES

8 Epinephrine (adrenaline) is a hormone secreted when the body is under stress. Illicit drugs that are structurally similar to epinephrine are highly addictive stimulants. For example, methamphetamine, an illegal drug, is similar in structure to epinephrine. However, pseudoephedrine, which is structurally similar to epinephrine, is a decongestant and sold over the counter at pharmacies.

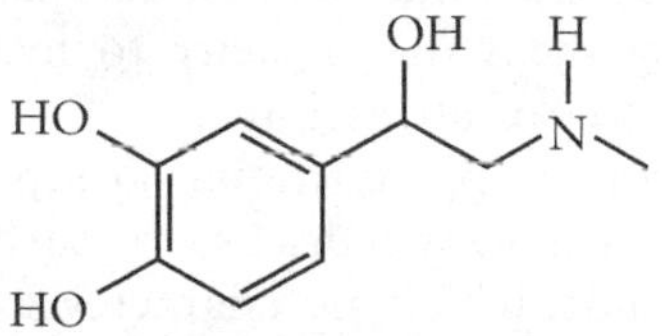

Epinephrine
(adrenaline)

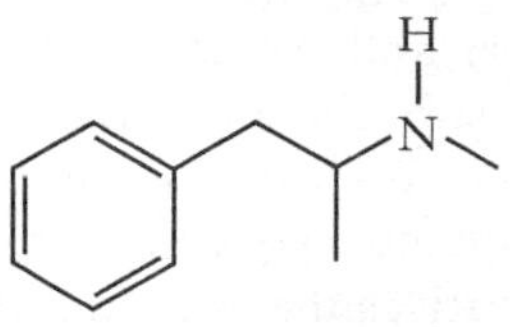

Methamphetamine

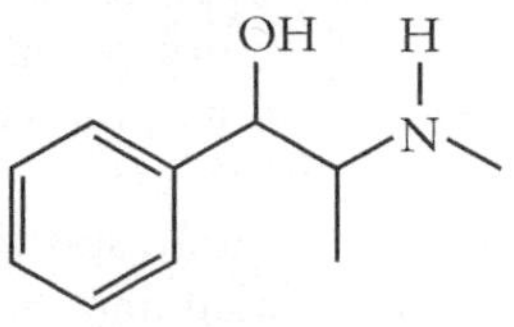

Pseudoephedrine

   a. Highlight and label all the amine functional groups in these molecules. Indicate whether they are primary, secondary, or tertiary amines.

   b. In what way is methamphetamine structurally similar to pseudoephedrine? How is it different? Explain by comparing the functional groups in each molecule.

   c. What is the molecular shape around the nitrogen atom in these molecules?

   d. Which of these compounds also contains a phenol functional group?

   e. What part of the name of these compounds suggests that they contain an amine functional group?

9 Indicate whether the following amines are primary, secondary, or tertiary amines.

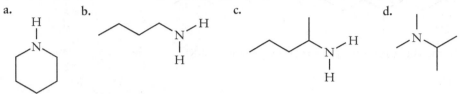

10 Name the amines b through d in the previous problem.

11 What is the molecular formula for each of the amine isomers shown below? What type of isomers are they? Which amine would you expect to have the lowest boiling point and why?

   a. $(CH_3)_3N$ (*N,N*-dimethylmethanamine)

   b. $CH_3CH_2NHCH_3$ (*N*-methylethanamine)

   c. $CH_3CH_2CH_2NH_2$ (propan-1-amine)

12 Psilocybin, shown in the margin, is an alkaloid produced by various species of mushrooms that are known to have psychedelic properties like LSD: hallucinations and euphoria. Identify the tertiary amine functional group in this molecule. What two characteristics qualify this compound as an alkaloid?

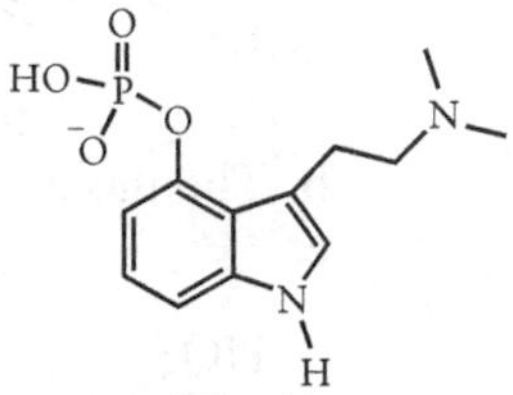

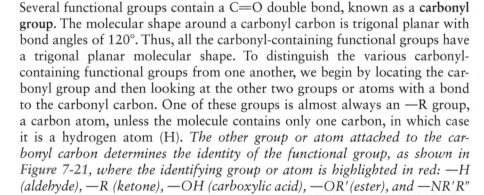

Psilocybin

■ 7.3 Carbonyl-Containing Functional Groups

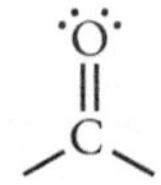

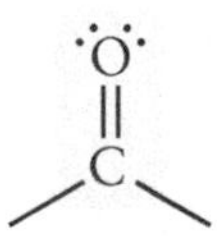

A carbonyl group

Several functional groups contain a C=O double bond, known as a **carbonyl group**. The molecular shape around a carbonyl carbon is trigonal planar with bond angles of 120°. Thus, all the carbonyl-containing functional groups have a trigonal planar molecular shape. To distinguish the various carbonyl-containing functional groups from one another, we begin by locating the carbonyl group and then looking at the other two groups or atoms with a bond to the carbonyl carbon. One of these groups is almost always an —R group, a carbon atom, unless the molecule contains only one carbon, in which case it is a hydrogen atom (H). *The other group or atom attached to the carbonyl carbon determines the identity of the functional group, as shown in Figure 7-21, where the identifying group or atom is highlighted in red: —H (aldehyde), —R (ketone), —OH (carboxylic acid), —OR′(ester), and —NR′R″ (amide).*

Since oxygen is more electronegative than carbon, a carbonyl group has a polar covalent bond. Hence, all functional groups containing a carbonyl group have a permanent dipole and dipole-dipole intermolecular forces of attraction. Consequently, compounds with these functional groups will have higher boiling points compared to those of alkanes of similar size. Two of these functional groups, the carboxylic acid and some amides, also have the more polar O—H and N—H bond, giving them the capacity to hydrogen bond, the strongest of the intermolecular forces of attraction.

Let us consider each of the carbonyl-containing functional groups separately, starting with aldehydes and ketones. Then we will discuss the carboxylic acid and its derivatives the ester and the amide, which are characterized by a heteroatom bonded to the carbonyl group.

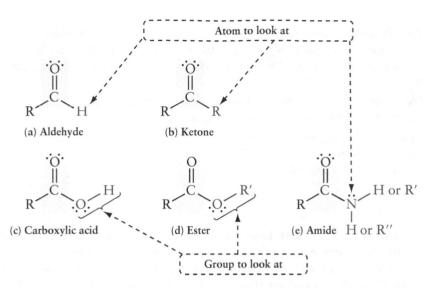

Figure 7-21 Common functional groups that contain a carbonyl group. The functional groups are identified by the atom or group of atoms (shown in pink) attached to the carbon atom of the carbonyl group (highlighted in yellow).

Aldehydes

The **aldehyde** functional group is defined by a carbonyl group with a bond to a hydrogen atom (H) and an R group, except in the case of methanal (formaldehyde, H_2CO), which has the carbonyl group bonded to two H atoms. The condensed notation for an aldehyde is RCHO. When writing the skeletal line structure for an aldehyde, the carbon atom of the carbonyl group is not shown, but the double bond and the oxygen must be written in. Sometimes the hydrogen atom on the carbonyl group is also written in to emphasize the H atom, which is characteristic of an aldehyde. For example, the skeletal line structure for cinnamaldehyde, $C_6H_5CH\!=\!CHCHO$, the compound responsible for the fragrance and taste of cinnamon, is shown in Figure 7-22.

Aldehydes can often be recognized by the *-al* ending in their name. The IUPAC rules for naming aldehydes are described in the Guidelines box below.

Figure 7-22 Cinnamaldehyde is responsible for the fragrance and taste of cinnamon. [© vario images GmbH & Co.KG/Alamy]

Cinnamaldehyde

$C_6H_5CH\!=\!CHCHO$

Guidelines for Naming Aldehydes

Rule 1: Assign the root. Select the longest contiguous chain of carbon atoms that also contains the carbonyl carbon and assign the root name based on the number of carbon atoms using Table 6-2. When counting, begin with the carbonyl carbon.

*For example, the root name for the aldehyde shown below is **propane** because there are three contiguous carbon atoms in the chain, including the carbonyl carbon.*

Propanal

> **Rule 2: Assign the ending.** Change the ending on the root by dropping the *-e* and adding *-al* to signify that the molecule is an *al*dehyde. No locator number is needed for an aldehyde because it is understood to be at C(1).
>
> *For example, to assign the root and ending for the aldehyde above, we change the name from propane to propanal.*
>
> **Rule 3: Assign the prefix.** Name any alkyl substituents along the main chain as described in section 6.5 by numbering the main chain beginning with the carbonyl carbon.

Common names are frequently used for carbonyl compounds with one or two carbon atoms. Carbonyl compounds containing one carbon have the root *form-* and those containing two carbon atoms have the root *acet-* (see **Table 7-1**).

Formaldehyde is a carcinogen. Funeral embalmers use formaldehyde as a preservative for human tissue.

A ball-and-stick model, Lewis structure, condensed structure, and IUPAC name for three simple aldehydes are shown in **Figure 7-23**. The simplest aldehyde, methan*al* has two hydrogen atoms (H) because it contains only one carbon, the carbonyl carbon (Figure 7-23a). Methanal is also known by its common name, *formaldehyde*. The condensed structural formula for formaldehyde is HCHO. Ethan*al*, shown in Figure 7-23b, has two carbon atoms: the carbonyl carbon and a —CH_3 group, the R group. The condensed structural formula for ethanal is CH_3CHO. Ethanal is also known by its common name, *acetaldehyde*. An aldehyde with a three-carbon chain is known by the IUPAC name propan*al*, whose condensed structure is CH_3CH_2CHO.

Although aldehydes are a relatively reactive functional group, they are found in nature. For example, 11-*cis*-retinal, which binds to the protein rhodopsin on the surface of the rods and cones of our eyes, is an aldehyde (**Figure 7-24**). Glyceraldehyde (**Figure 7-25**) is an important aldehyde intermediate in glycolysis, a 10-step biochemical pathway that converts glucose into energy in many organisms as well as our cells. Note that glyceraldehyde is also a diol and the simplest carbohydrate.

Earlier in this chapter you learned that monosaccharides with five or six carbon atoms are polyols. In their open-chain form, they also contain either an aldehyde or a ketone functional group. For example, glucose, the most common monosaccharide, contains an aldehyde at C(1) in its open-chain form, shown in **Figure 7-26**.

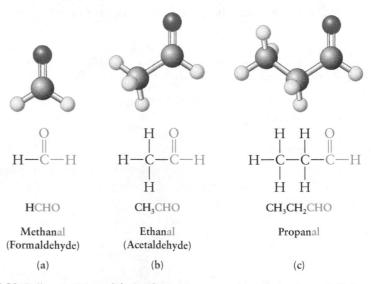

Figure 7-23 Ball-and-stick models, Lewis structures, condensed structures, IUPAC names, and common names for the three simplest aldehydes.

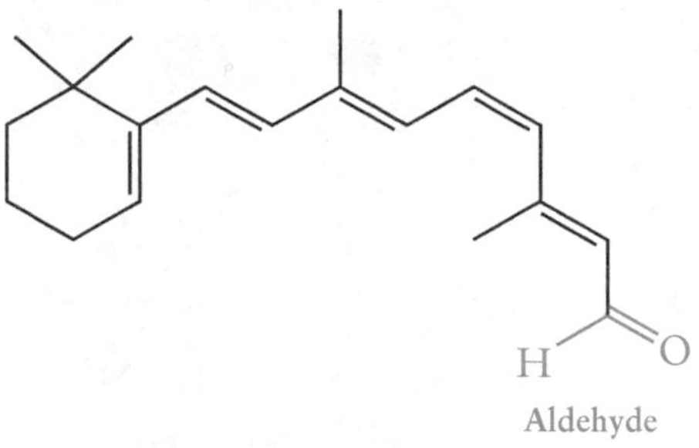

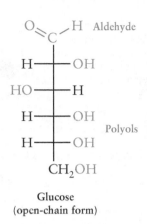

Figure 7-24 The aldehyde 11-*cis*-retinal binds to the protein rhodopsin on the surface of the rods and cones of our eyes.

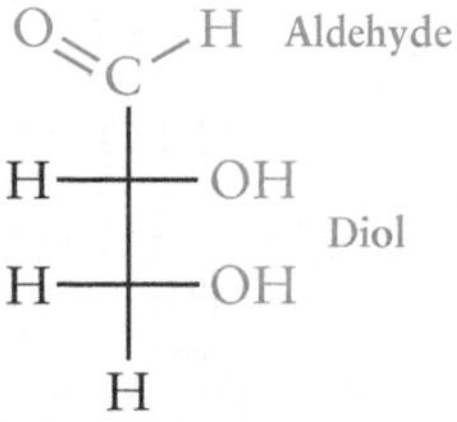

11-*cis*-Retinal

Glyceraldehyde

Figure 7-25 Glyceraldehyde, an important intermediate of metabolism, is an aldehyde.

Glucose (open-chain form)

Figure 7-26 The open-chain form of glucose, the most common monosaccharide, contains an aldehyde at C(1).

Ketones

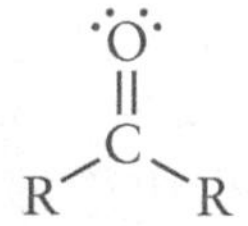

The **ketone** functional group is similar to the aldehyde except that the carbonyl carbon has bonds to *two* R groups (no hydrogen atoms). The condensed notation for a ketone is RCOR. When you write a skeletal line structure for a ketone, the carbonyl carbon does not need to be written in, but the oxygen atom and the double bond do. Figure 7-27 shows a ball-and-stick model, Lewis structure, condensed notation, and the IUPAC name for three simple ketones. The simplest ketone, propan-2-one, is also known by its common name *acetone*. You may have seen acetone being used around the laboratory as a solvent for cleaning glassware. Many organic compounds are soluble

TABLE 7-1 Common Names of Carbonyl Compounds with One and Two Carbon Atoms

Functional Group	One-carbon chain:		Two-carbon chain:	
Aldehyde	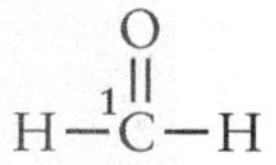 $H-\overset{O}{\underset{1}{C}}-H$	formaldehyde	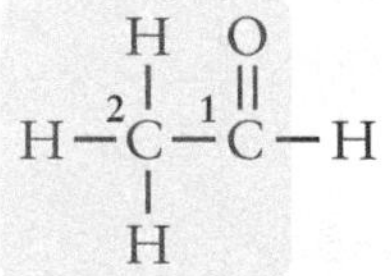 $H-\overset{H}{\underset{H}{\overset{2}{C}}}-\overset{O}{\overset{1}{C}}-H$	acetaldehyde
Carboxylic acid	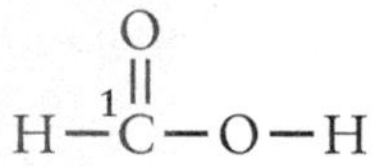 $H-\overset{O}{\underset{1}{C}}-O-H$	formic acid	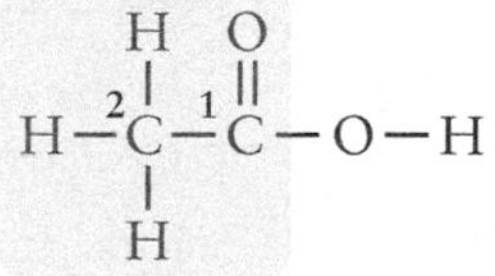 $H-\overset{H}{\underset{H}{\overset{2}{C}}}-\overset{O}{\overset{1}{C}}-O-H$	acetic acid
Ester	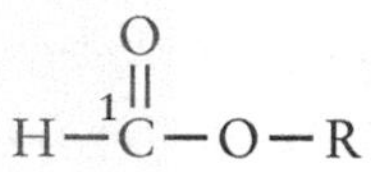 $H-\overset{O}{\underset{1}{C}}-O-R$	formate	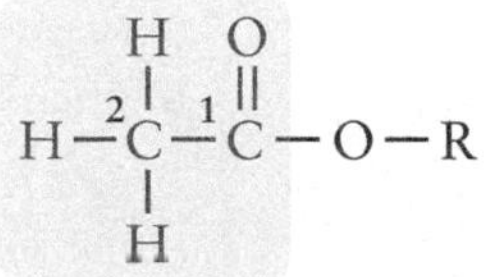 $H-\overset{H}{\underset{H}{\overset{2}{C}}}-\overset{O}{\overset{1}{C}}-O-R$	acetate
Amide	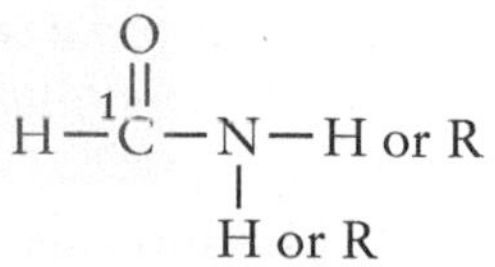 $H-\overset{O}{\underset{1}{C}}-N-H$ or R, H or R	formamide	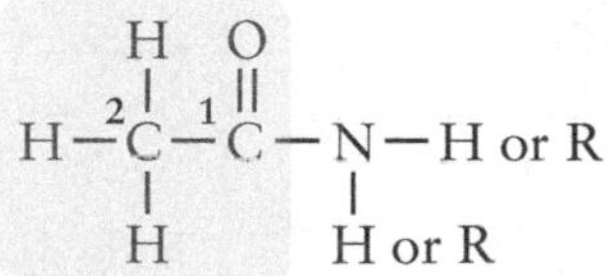 $H-\overset{H}{\underset{H}{\overset{2}{C}}}-\overset{O}{\overset{1}{C}}-N-H$ or R, H or R	acetamide

Figure 7-27 Ball-and-stick models, Lewis structures, condensed structures, and IUPAC names for three simple ketones.

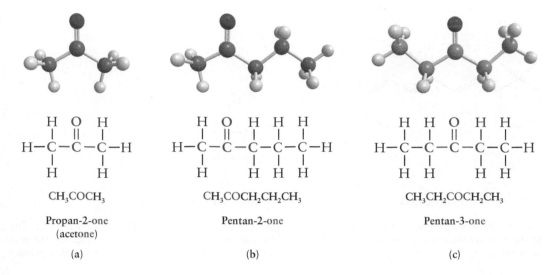

CH₃COCH₃

Propan-2-one
(acetone)

(a)

CH₃COCH₂CH₃

Pentan-2-one

(b)

CH₃CH₂COCH₂CH₃

Pentan-3-one

(c)

Figure 7-28 Acetone, the common name for propan-2-one, is a common solvent for cleaning glassware in the laboratory. It is also the solvent used in many nail polish removers. [Catherine Bausinger]

in acetone, and so it is a good cleaning agent. Acetone is the solvent in many fingernail polish removers because it dissolves nail polish (Figure 7-28). A compound can often be identified as a ketone by the *-one* ending in its name. For example, hydrocod*one* (Vicodin), testoster*one*, and acet*one* can all be recognized as ketones from their name. The IUPAC rules for naming ketones are described in the Guidelines box below.

Guidelines for Naming Ketones

Rule 1: Assign the root. Select the longest contiguous chain of carbon atoms that also contains the carbonyl carbon and assign the root name based on the number of carbon atoms it contains using Table 6-2.

For example, the root name for both ketones shown is **pentane** *because there are five contiguous carbon atoms in the chain, including the carbonyl carbon.*

Pentan-2-one   Pentan-3-one

Rule 2: Assign the ending. Change the ending on the root by dropping the *-e* and adding *-one* to signify that the molecule is a ket*one*.

*In the example above, change the name from pentane to pentan**one** to signify the molecules are ketones.*

Rule 3: Assign a locator number to indicate which carbon along the main chain is the carbonyl carbon. Number the main chain from the end closer to the ketone. Then insert a locator number, separated by hyphens, between the root and the ending: *alkan-#-one*.

*In the first example above, we see that the carbonyl carbon is at C(2) when numbering from the end closer to the carbonyl group, and in the second example it is at C(3) when numbering from either end. We then insert the locator numbers between the root and the ending such that the first ketone has the IUPAC name pentan-**2**-one and the second ketone has the IUPAC name pentan-**3**-one.*

Rule 4: Assign a prefix. Name any alkyl substituents along the main chain as described in section 6.5.

Figure 7-29 The structure of the sex hormone testosterone, highlighting the ketone functional group. [© K. L. Howard/Alamy]

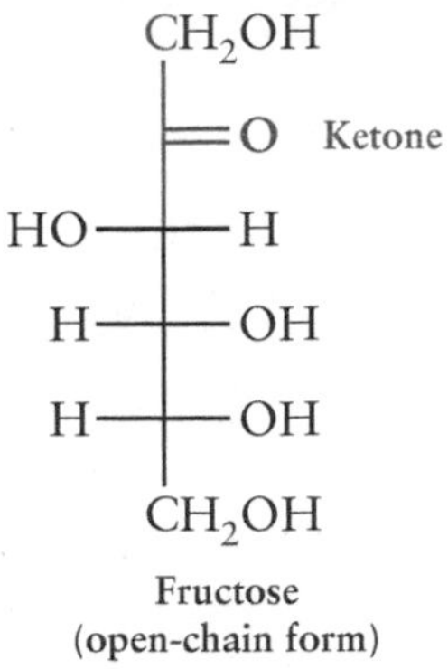

Ketone

Testosterone

CH₂OH

O Ketone

HO——H

H——OH

H——OH

CH₂OH

Fructose
(open-chain form)

Figure 7-30 The structure of fructose, the most common ketone-containing monosaccharide, highlighting the ketone functional group at C(2).

Ketones are less reactive than aldehydes and widely distributed throughout nature. For example, a ketone functional group is part of the structure of the sex hormone testosterone (Figure 7-29). Previously you learned that a monosaccharide in its open-chain form contains either an aldehyde or a ketone. Fructose is the most common ketone-containing monosaccharide, as highlighted in Figure 7-30. Ketones are functional groups found in many pharmaceuticals. For example, the structure of Vicodin, shown in Figure 7-1c, differs from that of the other opioids in that it has a ketone functional group where the others have an alcohol. Prednisone (Figure 7-31) is a synthetic steroid used as an immunosuppressant, a substance that reduces the activity of the immune system. Prednisone is used to treat conditions that improve when the immune system is suppressed, such as asthma and organ rejection following a transplant. Prednisone has three ketones as part of its structure.

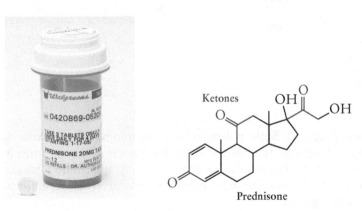

Ketones

Prednisone

Figure 7-31 The structure of prednisone, a synthetic steroid used as an immunosuppressant, with its three ketones highlighted in brown. [© A. T. Willett/Alamy]

WORKED EXERCISES   Identifying and Naming Aldehydes and Ketones

7-10 Identify the following compounds as aldehydes or ketones. Write their IUPAC names. Write a condensed structural formula for (a).

a.

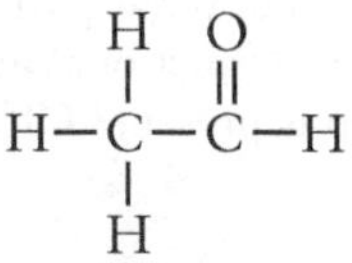

b.

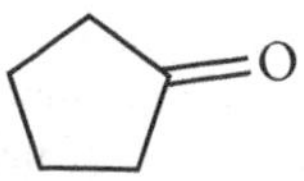

7-11 Zofran (ondansetron) is used to treat nausea and vomiting in cancer patients and also morning sickness in pregnant women. Highlight and identify the ketone functional group in this molecule.

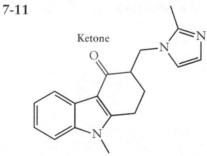

7-12 Name the two compounds shown and predict which would have a higher boiling point. Explain why.

or

Solutions

7-10 a. An aldehyde. The IUPAC name is *ethanal* because there are two carbon atoms in the chain and the ending is changed to *-al* to reflect that it is an aldehyde. The condensed structural formula is CH_3CHO. Note —CHO always stands for an aldehyde, and in a condensed structure it will be at one end of the structure.

   b. A ketone. The IUPAC name is *cyclopentanone* because the main chain contains five carbon atoms in a ring, cyclopentane, and the ending is changed to *-one* to reflect that it is a ketone. No locator number is required because the ring contains no substituents, so every position is the same.

7-11

Ketone

7-12 Propan-1-ol and propanal. Both molecules have approximately the same molecular mass, but propan-1-ol is an alcohol and therefore has the ability to form hydrogen bonds, whereas propanal has dipole-dipole interactions, which are a weaker intermolecular force of attraction than hydrogen bonding. Therefore, the alcohol has the higher boiling point.

Hex-3-enal

Figure 7-32 Hex-3-enal is found in both coriander and strawberries. [Catherine Bausinger]

PRACTICE EXERCISES

13 What is the H—C—H bond angle in formaldehyde? What is the molecular shape of formaldehyde?

14 It was recently discovered that one of the volatile organic compounds present in both coriander and strawberries is hex-3-enal (**Figure 7-32**). In cooking chemistry this means that these ingredients should be good pairings in recipes, a combination that might otherwise not have been considered. Highlight the aldehyde functional group in this molecule. What part of the name suggests it contains an aldehyde? This compound contains another functional group; what is it? What part of the name suggests this other functional group is present? Offer a possible explanation for why there is a 3 in the IUPAC name.

15 Rank the compounds below from lowest to highest boiling point and explain your ranking.

   a. butanal
   b. butan-2-ol
   c. pentane

16 Name the following aldehydes and ketones. Provide the common name for (e).

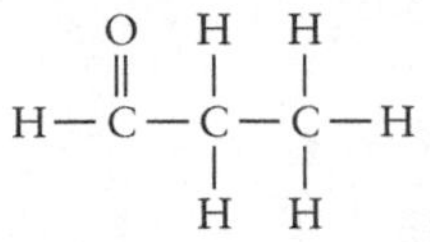

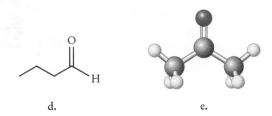

a.

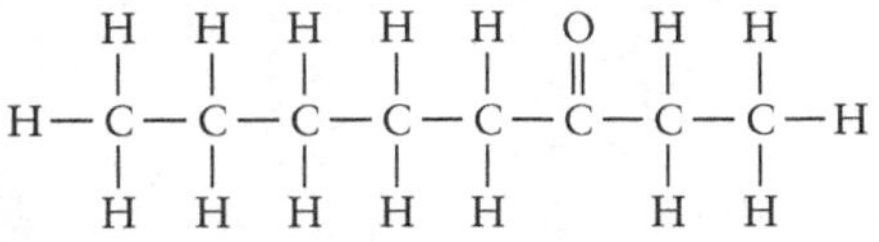

b.

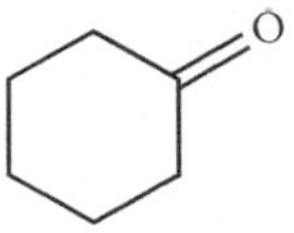

c.

d.

e.

17 The open-chain form of galactose, a monosaccharide, is shown below. Highlight and label all the functional groups in this molecule.

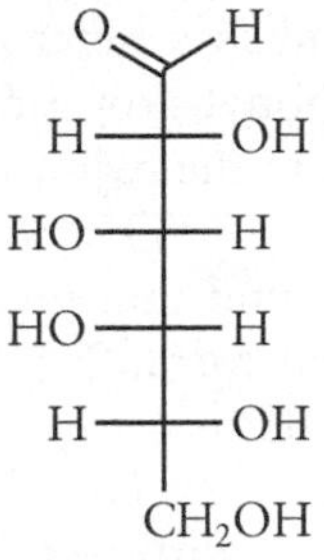

Galactose

18 The structure of estrone, a female sex hormone, is shown below. Highlight and identify all of the functional groups.

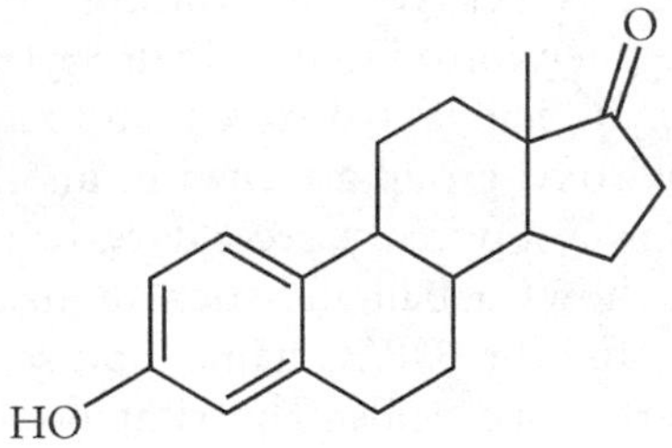

Estrone

19 Dihydroxyacetone (DHA), also known as glycerone, is the active ingredient in sunless tanning products. DHA has the same molecular formula as glyceraldehyde, an important intermediate in the metabolism of glucose. Highlight and label all the functional groups in DHA and glyceraldehyde. What is the molecular formula for DHA and glyceraldehyde? What type of isomers are DHA and glyceraldehyde?

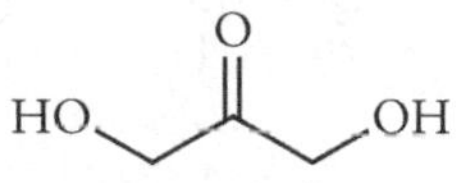

DHA

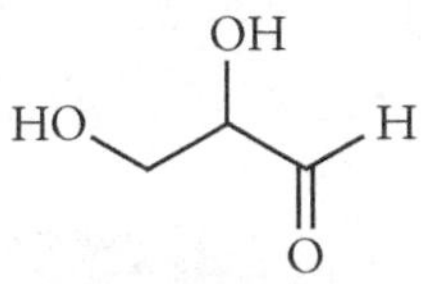

Glyceraldehyde

Figure 7-33 Ball-and-stick model, Lewis structure, condensed structure, IUPAC name, and common name for the most common carboxylic acid. Acetic acid gives vinegar (the bottom layer in the photo of oil and vinegar) its distinctive taste. [George Mattei/ Science Source]

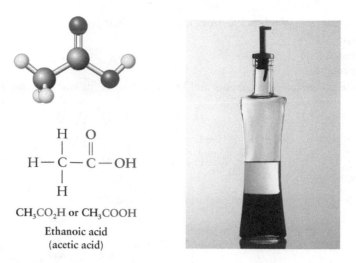

$$H-\underset{\underset{H}{|}}{\overset{\overset{H}{|}}{C}}-\overset{\overset{O}{||}}{C}-OH$$

CH$_3$CO$_2$H or CH$_3$COOH
Ethanoic acid
(acetic acid)

Carboxylic Acids

In the next three sections we will describe functional groups containing a carbonyl group with a bond to a heteroatom. A **carboxylic acid** is a functional group containing a carbonyl group directly bonded to a hydroxyl group, as shown for ethanoic acid in the ball-and-stick model, Lewis structure, and condensed structure in Figure 7-33. Ethanoic acid, more commonly known as acetic acid, is the compound that gives vinegar its distinctive taste and odor. The condensed structural formula for a carboxylic acid can be written two ways: RCO$_2$H or RCOOH.

Despite the presence of an O—H group, the chemical reactivity of a carboxylic acid is significantly different from that of an alcohol. Hence, alcohols (ROH) and carboxylic acids (RCOOH) are classified as entirely different functional groups. *The carbonyl group together with the O—H group act as one unit in the carboxylic acid functional group.* Figure 7-34 summarizes the structural differences between the various hydroxyl-containing functional groups that have been introduced: a hydroxyl group attached to a carbonyl group is a carboxylic acid; a hydroxyl group attached to an aromatic ring is a phenol; and a hydroxyl group attached to an alkyl carbon with one, two, or three carbon atoms is a primary, secondary, or tertiary alcohol, respectively.

Figure 7-35 shows a ball-and-stick model, a Lewis structure, the condensed structure, and the IUPAC name (and some common names) for some representative carboxylic acids. The simplest carboxylic acid is methanoic

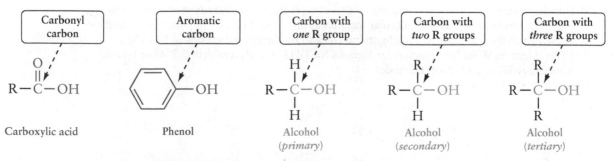

Figure 7-34 A summary of the structural differences between the various hydroxyl (—OH) containing functional groups.

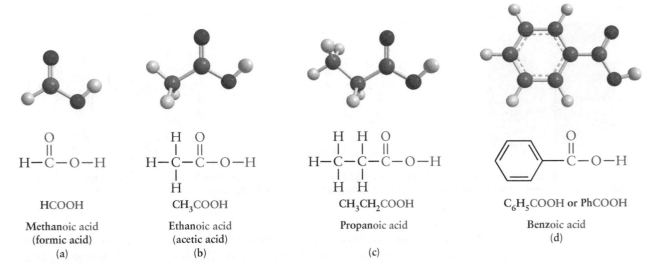

Figure 7-35 Ball-and-stick model, Lewis structure, condensed structure, IUPAC name, and common name for some representative carboxylic acids.

acid, whose common name is formic acid. It is unique because it contains only one carbon atom and therefore has no R group (HCOOH). Ethanoic acid, also known by the common name acetic acid, has two carbon atoms, and propanoic acid has three carbon atoms. The simplest aromatic carboxylic acid is benzoic acid, a compound where the R group is an aromatic ring. Benzoic acid is used as a food preservative because it inhibits the growth of mold, yeast, and some bacteria. Carboxylic acids can often be recognized by the *-oic acid* ending in their name. The rules for naming carboxylic acids are described in the Guidelines box below.

Guidelines for Naming Carboxylic Acids

Rule 1: Assign the root. Select the longest contiguous chain of carbon atoms that also contains the carbonyl carbon and assign the root name based on the number of carbon atoms it contains using Table 6-2.

*For example, the root name for the carboxylic acid shown below is **butane** because there are four contiguous carbon atoms in the main chain, including the carbonyl carbon.*

$$
\begin{array}{ccccc}
& H & H & H & O \\
& | & | & | & || \\
H- & \overset{4}{C} - & \overset{3}{C} - & \overset{2}{C} - & \overset{1}{C} -O-H \\
& | & | & | & \\
& H & H & H &
\end{array}
$$

Butanoic acid

Rule 2: Assign the ending. Change the ending by dropping the *-e* in the root and adding *-oic acid* to signify the molecule is a carboxylic acid. No locator number is needed for a carboxylic acid because it is always at C(1).

*In the example above, we change the name from butane to butan**oic acid** to signify the molecule is a carboxylic acid.*

Rule 3: Assign a prefix. Name any substituents on the main chain as described in section 6.5 by numbering the chain starting at the carbonyl carbon.

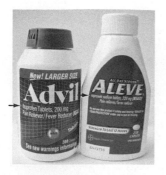

Figure 7-36 Ibuprofen, the active ingredient in Motrin and Advil, is a carboxylic acid produced in its neutral form; whereas naproxen, the active ingredient in Aleve, is produced in its ionic form. [Catherine Bausinger]

Ionic and Neutral Forms of a Carboxylic Acid In section 7.2 you learned that amines are characterized by their chemical behavior as organic bases: capable of *accepting* a hydrogen ion H^+ from water. Carboxylic acids are characterized by their chemical behavior as organic acids: capable of *donating* a hydrogen ion (H^+) to water, as shown in the equilibrium reaction below. An acid is defined as a substance that donates a hydrogen ion (H^+) to water to form H_3O^+. Upon donating a hydrogen ion, the carboxylic acid becomes a carboxylate ion, $RCOO^-$, a polyatomic anion. You will learn about acids and bases in chapter 9. Other common reactions of carboxylic acids are described in chapter 10.

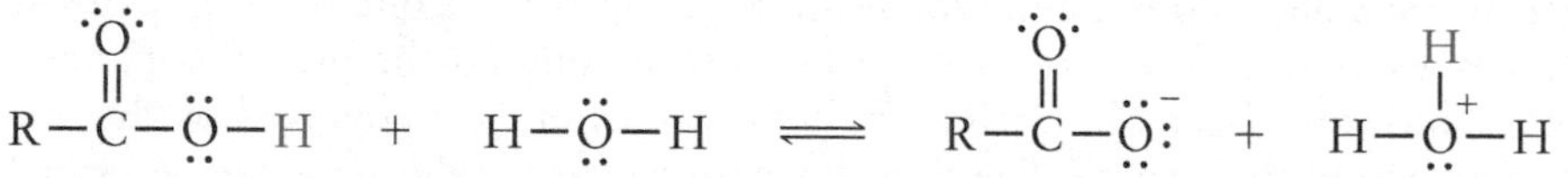

The reaction of a carboxylic acid in water, shown above, is reversible, and at equilibrium favors either reactants or products, depending on the environment. In a biological cell, equilibrium favors the products so you will see carboxylic acids written in their ionized form in most biochemical applications. You should become accustomed to recognizing both the neutral and the ionic form of a carboxylic acid.

For example, the analgesic ibuprofen, the active ingredient in Motrin and Advil, is a carboxylic acid produced in its neutral form, whereas the analgesic naproxen, the active ingredient in Aleve, is produced in its ionic form (Figure 7-36).

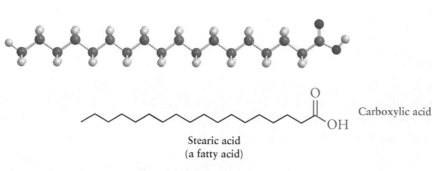

Stearic acid
(a fatty acid)

Figure 7-37 Skeletal line structure and ball-and-stick model of a fatty acid (stearic acid), which contains a long hydrocarbon chain and a carboxylic acid functional group on one end.

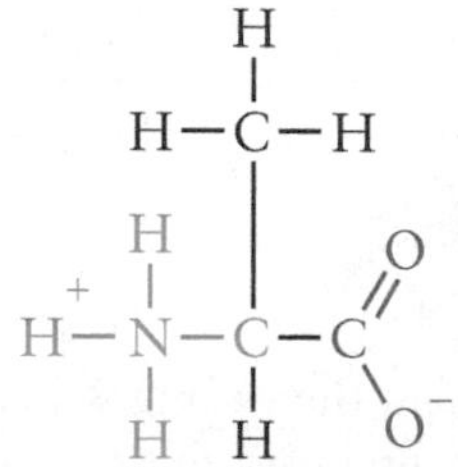

Figure 7-38 Amino acids, the building blocks for making proteins, contain a carboxylic acid and an amine functional group, as shown here in the structure of alanine. Both the amine and the carboxylic acid are in their ionized forms in the cell.

The structure drawings:

Citric acid (neutral form) and Citrate (ionized form)

Figure 7-39 The structure of citric acid, a tricarboxylic acid, shown in both its neutral and ionized forms. [Catherine Bausinger]

Since ionic compounds must be electrically neutral, we find them accompanied by an oppositely charged ion. In the case of naproxen, the cation is sodium, Na^+, shown as part of its chemical structure in Figure 7-36. Note that often only the carboxylate ion is shown.

Carboxylic acids are widely distributed in nature, where they are found in several important biomolecules. **Fatty acids**, involved in the metabolism of fats, are carboxylic acids containing an —R group that is a long unbranched hydrocarbon chain. For example, stearic acid is a carboxylic acid with a chain of 18 carbons, where C(1) is the carbonyl carbon of the carboxylic acid functional group, as shown in the ball-and stick model and the skeletal line structure in Figure 7-37. In the cell, this carboxylic acid would exist in its ionized form.

Amino acids, the building blocks for making **proteins**, contain a carboxylic acid *and* an amine functional group, as shown in Figure 7-38 for the amino acid alanine. In the cell, the carboxylic acid and the amine functional groups of an amino acid are both in their ionized forms, as shown in Figure 7-38.

Some compounds have two or more carboxylic acid functional groups in their structure. For example, citric acid is a *tri*carboxylic acid, a compound that contains *three* carboxylic acid functional groups. This molecule is found in citrus fruits. Citric acid is also a key metabolic intermediate in the citric acid cycle, an important biochemical pathway involved in energy production in the cell. Lewis structures of the neutral form of citric acid and its fully ionized form, citrate, are shown in Figure 7-39.

Hydrogen Bonding in Carboxylic Acids Carboxylic acids are capable of hydrogen-bonding intermolecular forces of attraction because they contain the strong O—H bond dipole and also a partial negative charge on both oxygen atoms of the carboxylic acid, as shown in Figure 7-40. Like alcohols, carboxylic acids have higher boiling points than other compounds with a comparable number of atoms and therefore, are often in liquid or solid form.

The ionized form of a carboxylic acid, known as a carboxylate ion, has very different physical properties than its neutral form because of the negative charge. Indeed, a carboxylic acid is a molecule, whereas a carboxylate ion is an ion. The ion is soluble in water whereas carboxylic acids with five or more carbon atoms do not dissolve in water. You will learn more about solubility in chapter 8.

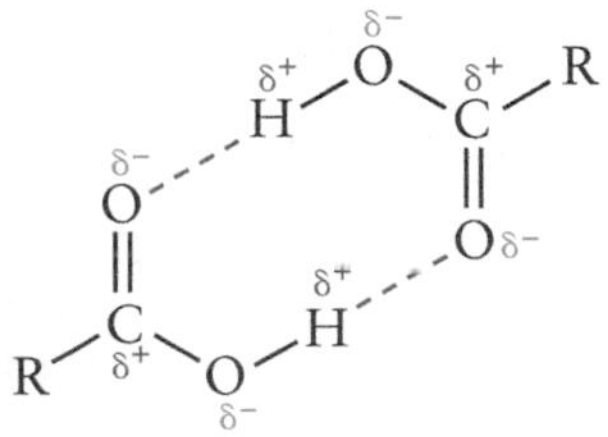

Figure 7-40 Hydrogen bonding between two carboxylic acids.

WORKED EXERCISES | Carboxylic Acids |

7-13 The structure of Singulair, a drug used to prevent asthma attacks and treat seasonal nasal allergies, is shown below. Highlight and label the carboxylic acid and the alcohol functional groups. Is the carboxylic acid in its neutral or ionic form? How were you able to distinguish the carboxylic acid from the alcohol?

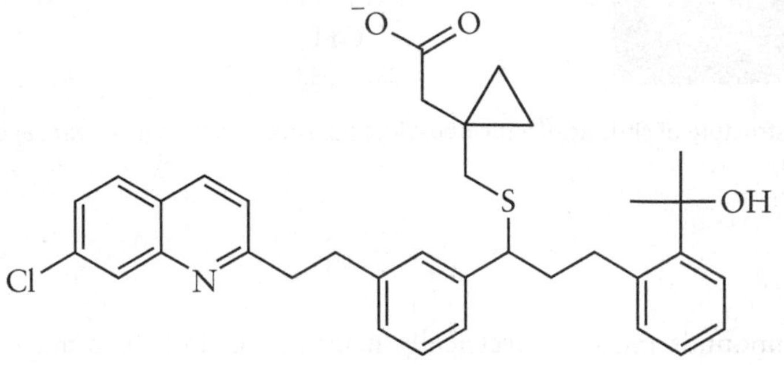

Singulair

7-14 Provide a skeletal line structure for hexanoic acid.

7-15 Provide the IUPAC name for the following carboxylic acids. What is the common name for the carboxylic acid in (a)?

a.

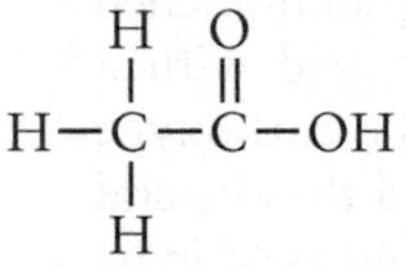

b. $CH_3(CH_2)_5CO_2H$

c.

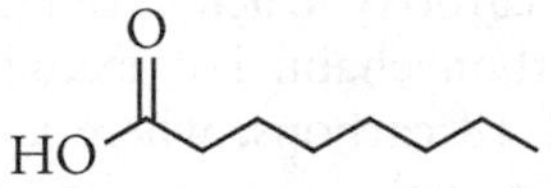

7-16 What type of biological molecule is palmitic acid (common name)? What part of its name suggests it is a carboxylic acid? Explain.

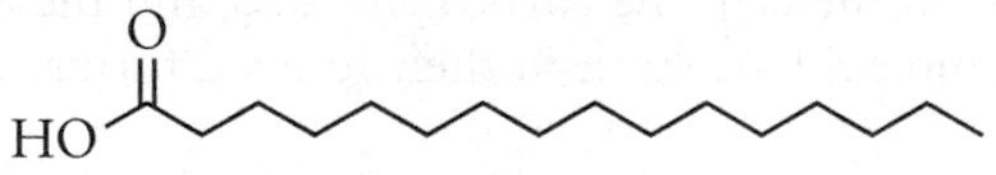

Palmitic acid

7-17 Highlight and identify the functional groups in valine. Would you expect valine to be used to build carbohydrates, proteins, or fats?

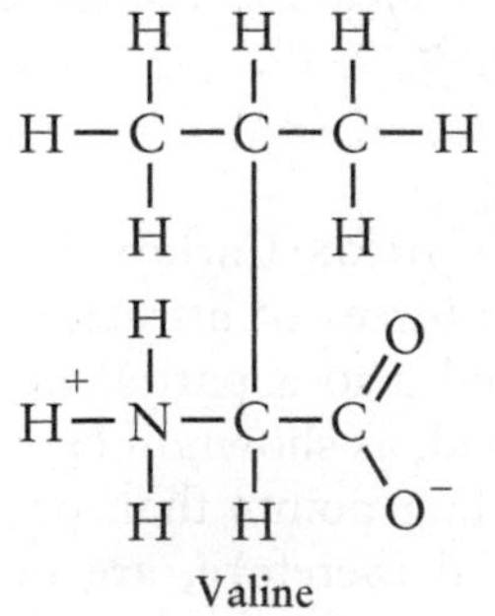

Valine

Solutions

7-13 The carboxylic acid is in its ionized form: $RCOO^-$. A carboxylic acid is different from an alcohol in that it contains a carbonyl group that is directly bonded to a hydroxyl group, O—H group, whereas an alcohol has a hydroxyl group bonded to an alkyl carbon.

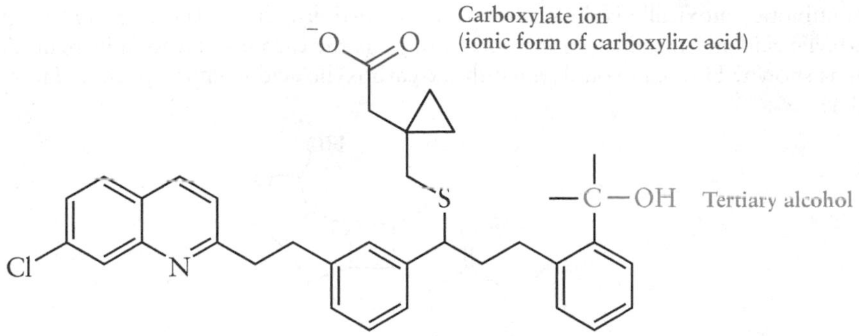

Carboxylate ion
(ionic form of carboxylizc acid)

—C—OH Tertiary alcohol

Cl

N

Singulair

7-14

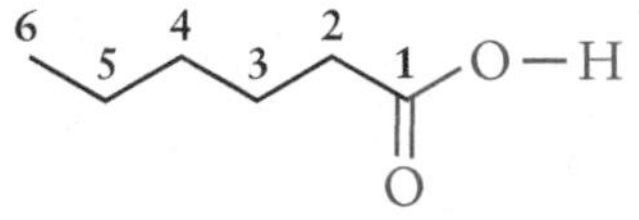

Hexanoic acid

7-15 a. IUPAC name: Ethanoic acid; common name: acetic acid. The chain contains two carbon atoms, so the root is *ethane*. The ending is changed by dropping the *-e* and adding *-oic acid*. The common name is acetic acid because it contains two carbon atoms and is a carboxylic acid.

b. Heptanoic acid. The chain contains seven carbon atoms, so the root is *heptane*. The ending is changed by dropping the *-e* and adding *-oic acid*.

c. Octanoic acid. The chain contains eight carbon atoms, so the root is *octane*. The ending is changed by dropping the *-e* and adding *-oic acid*.

7-16 Palmitic acid is a fatty acid because it contains a carboxylic acid at the end of a long straight-chain hydrocarbon. The ending *-ic acid* in palmi*tic acid* indicates it is a carboxy*lic acid*.

7-17 Amino acids, such as valine, are used to build proteins.

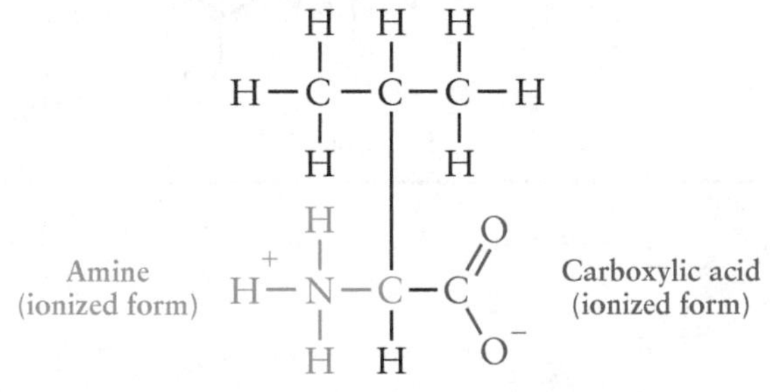

Amine
(ionized form)

Carboxylic acid
(ionized form)

Valine

PRACTICE EXERCISES

20 What is the difference between an alcohol, a phenol, and a carboxylic acid? How can you readily distinguish these three functional groups?

21 Supply the IUPAC names for the following carboxylic acids:

a.

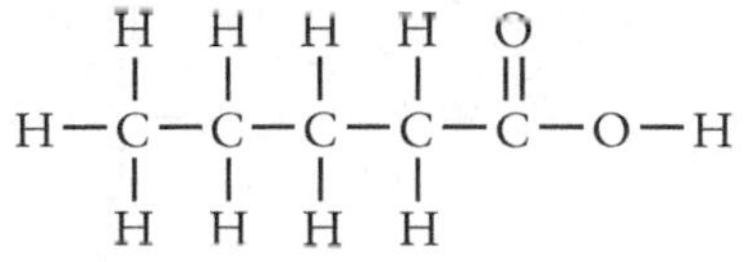

b.

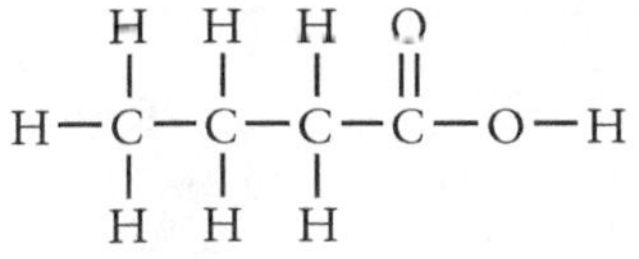

22 Citric acid contains three carboxylic acid functional groups (Figure 7-39). What is the other functional group in citric acid? How can you distinguish it from a carboxylic acid?

23 The antibiotic amoxicillin is the sixth-most prescribed drug in the United States. Identify the carboxylic acid and the phenol functional groups. Is the carboxylic acid in its neutral or ionic form as shown? How can you distinguish the carboxylic acid from the phenol? How are they similar?

Amoxicillin

24 The structure of Synthroid, used to treat hypothyroidism (low-functioning thyroid), is shown below. Highlight the phenol, the ether, the amine, and the carboxylic acid. Is the carboxylic acid in its neutral or ionic form? Is the amine in its neutral or ionic form?

Synthroid

Esters   R—C—O—R'

The general structure of an **ester** is shown in Figure 7-41a. An ester is similar to a carboxylic acid (Figure 7-41c) except that in an ester the carbonyl carbon has a bond to an alkoxy group (—OR') rather than a hydroxyl group

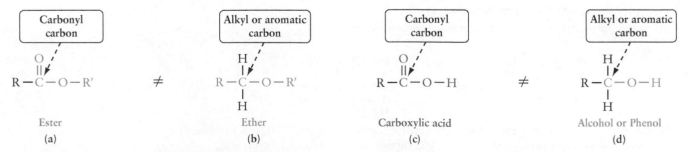

Figure 7-41 Comparison of the structural differences between (a) Esters, (b) Ethers, (c) Carboxylic acids, and (d) Alcohols and Phenols.

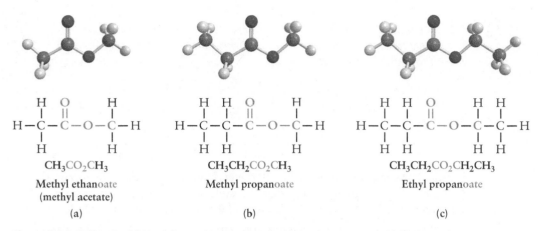

$CH_3CO_2CH_3$

Methyl ethanoate
(methyl acetate)

(a)

$CH_3CH_2CO_2CH_3$

Methyl propanoate

(b)

$CH_3CH_2CO_2CH_2CH_3$

Ethyl propanoate

(c)

Figure 7-42 Ball-and-stick models, Lewis structures, condensed structure, and IUPAC names of three representative esters.

(—OH). Indeed, esters can be prepared from carboxylic acids in a chemical reaction that you will learn about in chapter 10. Hence, esters are classified as **carboxylic acid derivatives**. Note that an ester is also different from an ether (Figure 7-41b) because it contains a carbonyl group directly attached to the —O—R′ group. The carbonyl group and the alkoxy group act together as one unit to define an ester functional group. This comparison is analogous to the difference between a carboxylic acid (Figure 7-41c) and an alcohol (Figure 7-41d).

Figure 7-42 shows a ball-and-stick model, a Lewis structure, the condensed structure, and the IUPAC name (and common names where applicable) of three representative esters. Note that an ester has two R groups: one is attached to the carbonyl carbon (unless it has only one carbon HCO_2R) and the other is attached to the oxygen, so the definition of an ester is written in condensed notation as $RCO_2R′$, or $RCOOR′$, where R′ means that the two R groups are not necessarily the same. The IUPAC rules for naming an ester are described in the Guidelines box below.

Note that all molecules, including esters, can be written from left to right or right to left. For example, ethyl methanoate can be written with the —OR group on the left or on the right, so it is important to differentiate the two R groups in an ester when naming an ester or predicting the outcome of a chemical reaction involving an ester:

Main chain   Main chain

Methyl ethanoate

Guidelines for Naming Esters

Rule 1: Assign the root. There are two R groups in an ester: the one containing the carbonyl carbon is the main chain and the other is in the —OR′ group, which is treated as a substituent. Assign the root name based on the number of contiguous carbon atoms in the main chain using Table 6-2.

For example, the root name for the ester shown below is **heptane** *because there are* seven *contiguous carbon atoms in the carbon chain containing the carbonyl group..*

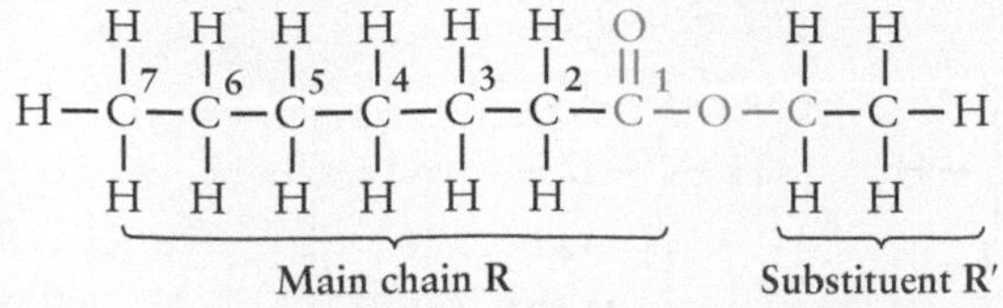

$$
\underset{\text{Main chain R}}{\underbrace{\overset{\displaystyle H}{\underset{\displaystyle H}{H-\overset{7}{C}}}-\overset{\displaystyle H}{\underset{\displaystyle H}{\overset{6}{C}}}-\overset{\displaystyle H}{\underset{\displaystyle H}{\overset{5}{C}}}-\overset{\displaystyle H}{\underset{\displaystyle H}{\overset{4}{C}}}-\overset{\displaystyle H}{\underset{\displaystyle H}{\overset{3}{C}}}-\overset{\displaystyle H}{\underset{\displaystyle H}{\overset{2}{C}}}-\overset{\displaystyle O}{\overset{\|}{\overset{1}{C}}}}}-O-\underset{\text{Substituent R}'}{\underbrace{\overset{\displaystyle H}{\underset{\displaystyle H}{C}}-\overset{\displaystyle H}{\underset{\displaystyle H}{C}}-H}}
$$

Rule 2: Assign the ending. Change the ending on the root by dropping the -*e* and adding -*oate* to signify that the molecule is an ester. No locator number is needed for an ester because the carbonyl carbon is C(1).

In the ester above, we change the root name from heptane to heptanoate.

Rule 3: Assign the prefix. Name the R′ group attached to the oxygen atom as you would a substituent by counting the number of carbon atoms it contains and changing the ending to -*yl*. Insert the name of the R′ group, as a prefix, before the root name followed by a space. If there are alkyl substituents on the main chain, name them as described in section 6.5, and place them in front of the root name as a prefix.

In the example above, the R′ group contains two carbon atoms. Thus, ethane is changed to ethyl to indicate it is the substituent R′ group attached to the oxygen atom. The full IUPAC name is therefore **ethyl** *heptanoate.*

Esters contain a permanent dipole, but they do not have the more polar N—H or O—H bond. Hence, they have dipole-dipole intermolecular forces of attraction, which are weaker than the hydrogen bonding seen in alcohols and carboxylic acids but stronger than the dispersion forces of hydrocarbons. Esters also have stronger dipoles than ethers because of the presence of the carbonyl group. For example, the alcohol, ester, and alkane shown in Figure 7-43 have a comparable molecular mass (72–74 g/mol), but the alcohol has the highest boiling point and the alkane the lowest because of the stronger intermolecular forces of attraction in an alcohol over an ester, and an ester over an alkane, as shown.

Esters in Nature Esters are widely distributed in nature, and the reactions of esters are common in biochemistry. The flavors and fragrances of many fruits and flowers are due to compounds that contain an ester functional group. For example, methyl jasmonate is the ester that gives the jasmine flower its distinctive fragrance (Figure 7-44). The fragrance of an ester depends on the shape of the ester, which is determined by both R groups of the ester. You will

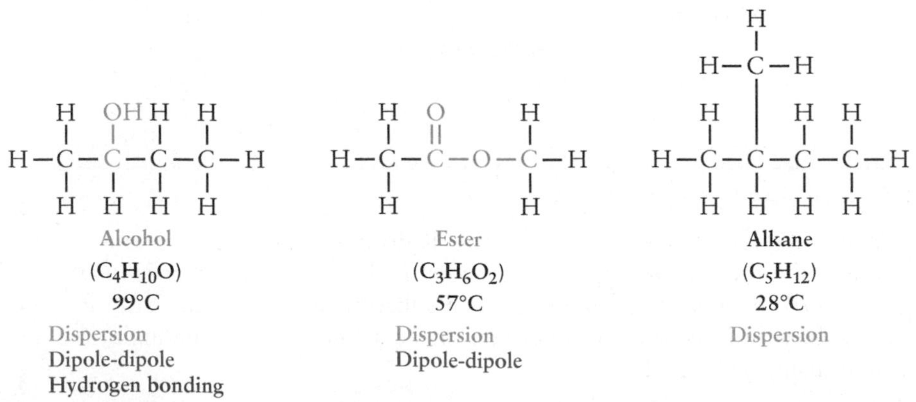

Figure 7-43 Comparison of intermolecular forces of attraction (pink) and boiling points of an alcohol, an ester, and an alkane of similar size.

	Alcohol	Ester	Alkane
Formula	$(C_4H_{10}O)$	$(C_3H_6O_2)$	(C_5H_{12})
Boiling point	99°C	57°C	28°C
Forces	Dispersion, Dipole-dipole, Hydrogen bonding	Dispersion, Dipole-dipole	Dispersion

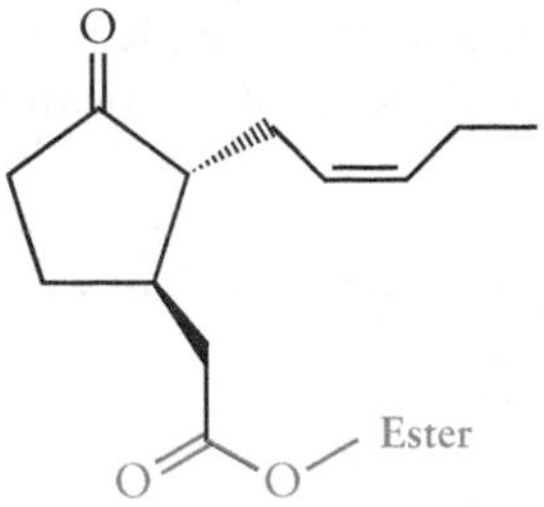

Methyl jasmonate

Figure 7-44 Methyl jasmonate is the ester responsible for the distinctive fragrance of jasmine. [© Nic Miller/Organics image library/ Alamy]

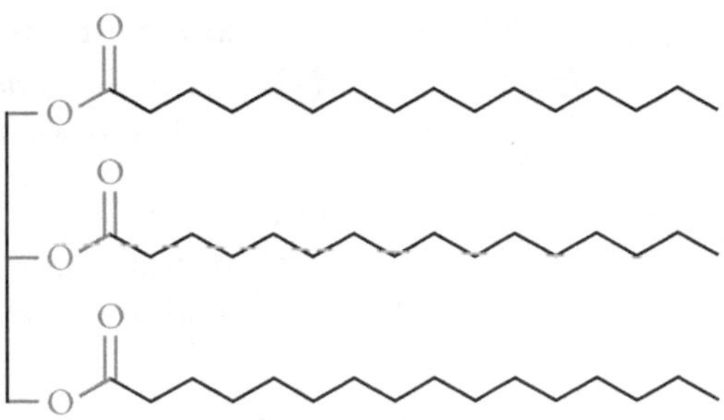

Esters

Triglyceride (fat)

Figure 7-45 Trigylcerides, known more commonly as fats have a structure characterized by three ester functional groups and hydrocarbon chains of varying length and degree of unsaturation.

learn how to prepare esters with different characteristic aromas in chapter 10, by varying the two R groups.

Esters are also found in many of biomolecules. Triglycerides, known more commonly as **fats,** are characterized by three ester functional groups. An example of a triglyceride is shown in **Figure 7-45.** Triglycerides differ in the length and degree of unsaturation in their hydrocarbon chains. Excess triglycerides are stored in adipose cells (fat cells), serving as the body's long-term source of energy.

Many pharmaceuticals contain an ester functional group in their structure. Aspirin (acetyl salicylic acid), for example, contains an ester and a carboxylic acid functional group in addition to an aromatic ring (**Figure 7-46**).

The topical anesthetics procaine and benzocaine contain an ester functional group in their structure (**Figure 7-47**).

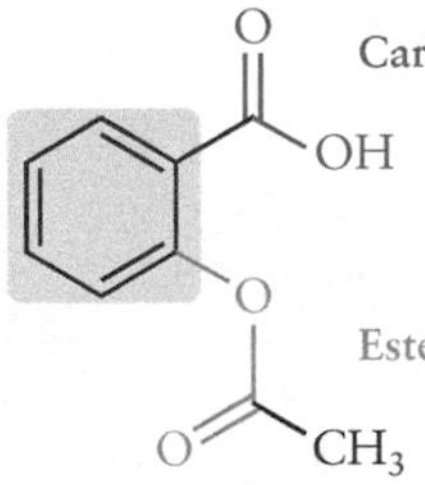

Aspirin

Figure 7-46 The chemical structure of aspirin, highlighting the ester and carboxylic acid functional groups.

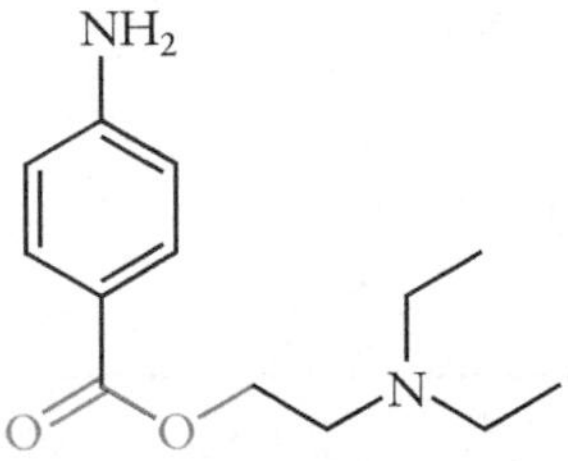

Procaine

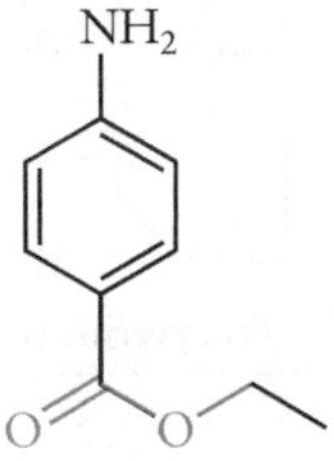

Benzocaine

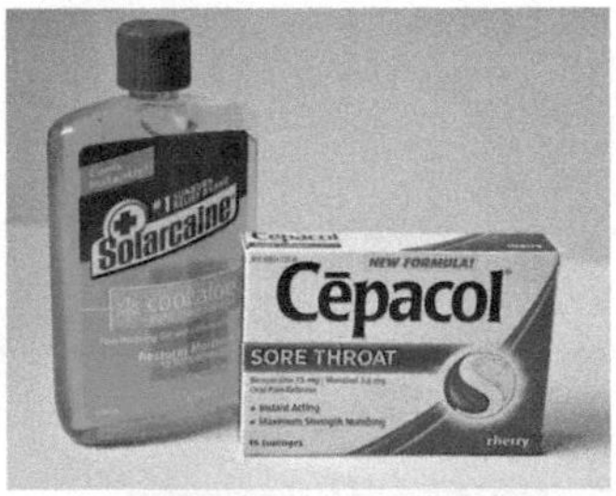

Figure 7-47 The structure of the topical anesthetics procaine and benzocaine, highlighting the ester functional group in each. [Catherine Bausinger]

WORKED EXERCISES Esters

7-18 Highlight the two ester functional groups in the structure of cocaine, shown below. What other functional groups are present in cocaine?

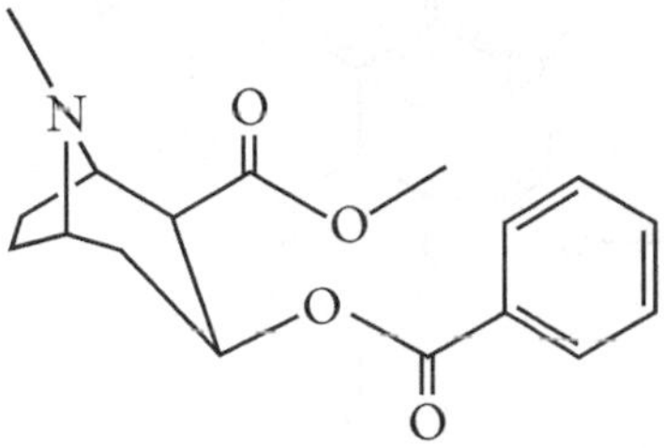

Cocaine

7-19 Write a skeletal line structure for the following esters:
  a. ethyl propanoate   **b.** propyl ethanoate   **c.** ethyl acetate

7-20 Write the general structure of a fat using R, R′, and R″ to signify the long hydrocarbon chains. What do R, R′, and R″ mean?

Solutions

7-18 In addition to the two ester groups, cocaine has an amine functional group and an aromatic ring:

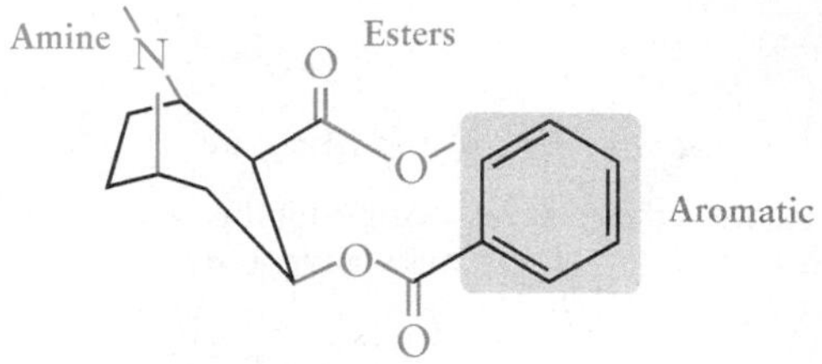

Cocaine

7-19 a.

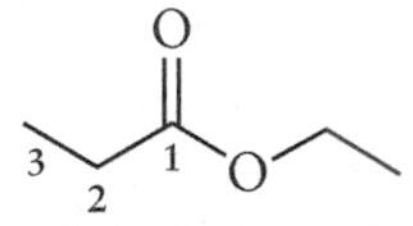

Ethyl propanoate

b.

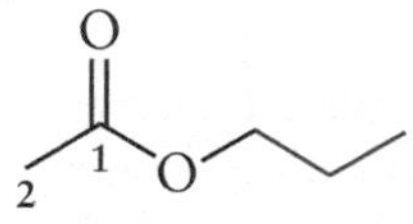

Propyl ethanoate
(propyl acetate)

c.

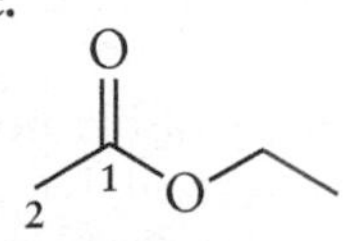

Ethyl ethanoate
(ethyl acetate)

7-20 R means a carbon chain, R′ means a carbon chain not necessarily the same as R, and R″ means a carbon chain not necessarily the same as R or R′. In a fat these R groups are long straight hydrocarbon chains with or without double bonds.

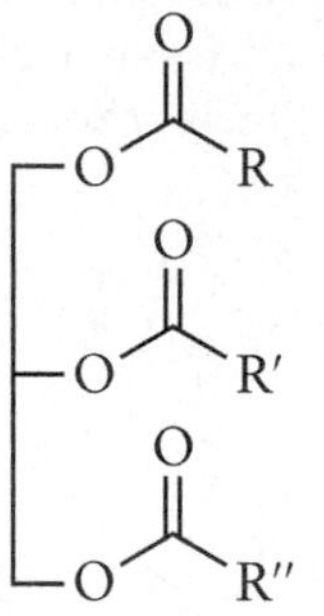

Triglyceride (fat)

PRACTICE EXERCISES

25 The structure of Norvasc, the tenth-most widely prescribed prescription drug in the United States, is shown below. Norvasc is used to treat high blood pressure (hypertension) and chest pain (angina). Highlight the two ester functional groups. Highlight the ether. How can you distinguish an ether from an ester?

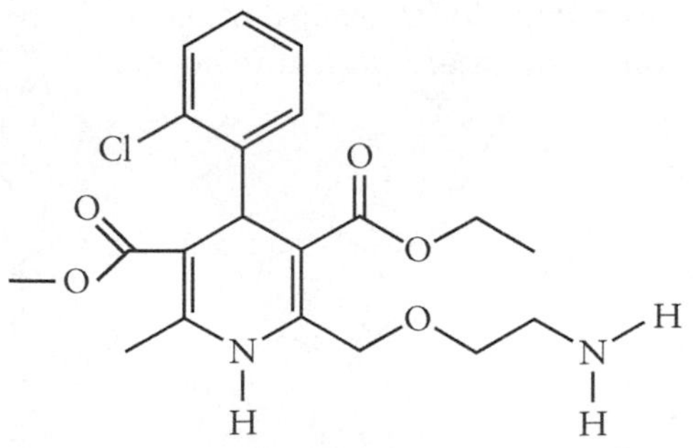

Norvasc

26 To what class of biomolecules does the compound below belong? Highlight the ester functional groups.

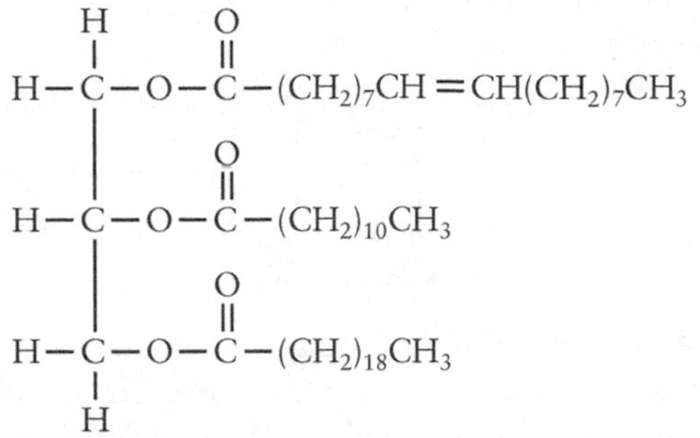

$$H-\overset{\overset{\displaystyle H}{|}}{\underset{|}{C}}-O-\overset{\overset{\displaystyle O}{||}}{C}-(CH_2)_7CH=CH(CH_2)_7CH_3$$

$$H-\overset{|}{\underset{|}{C}}-O-\overset{\overset{\displaystyle O}{||}}{C}-(CH_2)_{10}CH_3$$

$$H-\overset{|}{\underset{\underset{\displaystyle H}{|}}{C}}-O-\overset{\overset{\displaystyle O}{||}}{C}-(CH_2)_{18}CH_3$$

27 Write the IUPAC name for the esters shown. Include the common name for the ester in (c).

a.

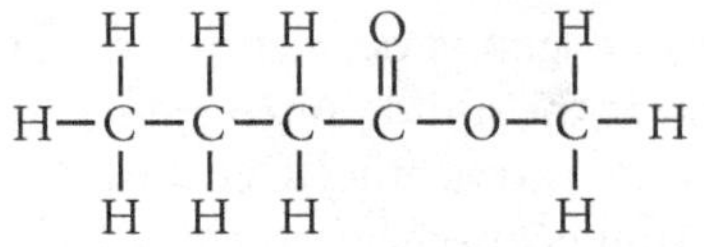

$$H-\overset{\overset{\displaystyle H}{|}}{\underset{\underset{\displaystyle H}{|}}{C}}-\overset{\overset{\displaystyle H}{|}}{\underset{\underset{\displaystyle H}{|}}{C}}-\overset{\overset{\displaystyle H}{|}}{\underset{\underset{\displaystyle H}{|}}{C}}-\overset{\overset{\displaystyle O}{||}}{C}-O-\overset{\overset{\displaystyle H}{|}}{\underset{\underset{\displaystyle H}{|}}{C}}-H$$

b.

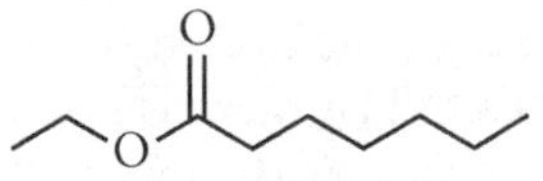

c.

$HCO_2CH_2CH_3$

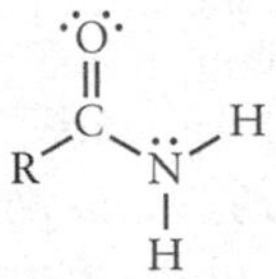

Amides

The general structure of an **amide** is shown in **Figure 7-48**. Since nitrogen forms three bonds, the nitrogen atom in an amide will have one bond to the carbonyl carbon and two bonds to either H atoms or R groups. Like esters, amides are carboxylic acid derivatives because they can be prepared from carboxylic acids in a chemical reaction (chapter 10). Note that an am*ide* differs from an am*ine* in having a carbonyl group directly bonded to the nitrogen atom, as shown in Figure 7-48. The carbonyl group together with the nitrogen atom act as one unit to define an amide.

Amides and amines have distinctly different chemical properties, and so they are classified as different functional groups. Amides are generally less reactive than amines. Amides are often recognized by the *-ide* ending in their name. The IUPAC rules for naming amides are described in the Guidelines box below. Work through Using Molecular Models 7-3: Carboxylic Acids and Their Derivatives to gain more familiarity with amides and other carboxylic acid derivatives.

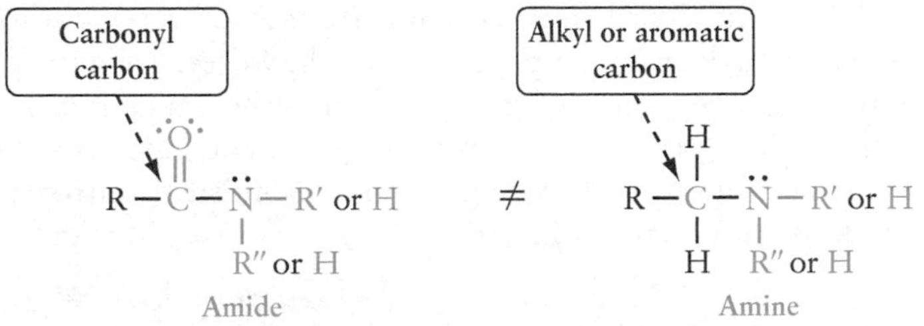

Figure 7-48 The structure of an amide (left) compared to an amine (right).

Guidelines for Naming Amides

Rule 1: Assign the root. Select the longest contiguous chain of carbon atoms that also contains the carbonyl carbon and assign the root name based on the number of carbon atoms using Table 6-2.

*For example, the root name for the amide shown below is **hexane** because there are six contiguous carbon atoms in the R group containing the carbonyl group.*

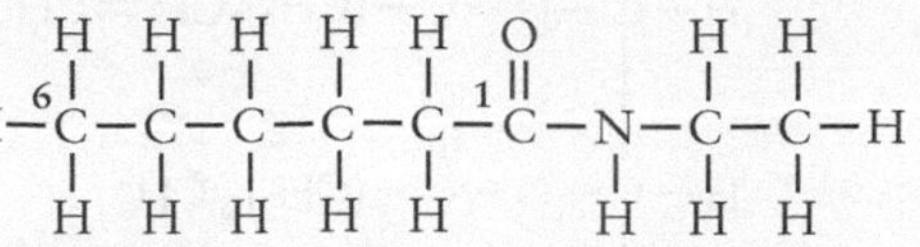

$$
\underset{\text{N-Ethylhexanamide}}{
\overset{\displaystyle H \underset{6}{\overset{H}{\underset{|}{\overset{|}{C}}}} - \overset{H}{\underset{|}{\overset{|}{C}}} - \overset{H}{\underset{|}{\overset{|}{C}}} - \overset{H}{\underset{|}{\overset{|}{C}}} - \overset{H}{\underset{|}{\overset{|}{C}}} - \overset{O}{\underset{1}{\overset{\|}{C}}} - N - \overset{H}{\underset{|}{\overset{|}{C}}} - \overset{H}{\underset{|}{\overset{|}{C}}} - H}{}
}
$$

Rule 2: Assign the ending. Change the ending on the root by dropping the *-e* and adding *-amide* to signify that the molecule is an amide. No locator number is needed for an amide because the carbonyl carbon is C(1).

In the example above, we change the root name from hexane to hexanamide.

Rule 3: Assign the prefix. If the amide nitrogen is bonded to two H atoms: $RCONH_2$, there is no *N-* prefix. If the amide nitrogen has one or two R' groups: $RCONHR'$ or $RCONR'_2$, treat the R groups as substituents, but instead of a locator number, use the prefix *N-* to indicate that the substituent is bonded to the nitrogen atom and not a carbon atom in the main chain. If there are two R groups, list them in alphabetical order. If the R groups are identical, use the multiplier *di-*. If there are also alkyl substituents along the main chain, name those as described in section 6.5 and list them after the *N*-substituents.

*In the example above, there are no alkyl groups on the main chain, but nitrogen has one substituent with two carbon atoms in its chain. Hence we add the prefix N-ethyl. The complete IUPAC name is thus **N-ethylhexanamide**.*

The condensed structural formula for an amide is $RCONH_2$, $RCONHR'$, or $RCONR'_2$, depending on whether the nitrogen atom has zero, one, or two R groups attached. A ball-and-stick model, a Lewis structure, the condensed structure, and the IUPAC name (and common names where applicable) for four representative amides are shown in **Figure 7-49**. Methanamide (Figure 7-49a), also known by its common name formamide, is the simplest amide, containing only one carbon atom, the carbonyl carbon, so it has an H rather than an R group attached to the carbonyl carbon. The nitrogen in methanamide has bonds to two hydrogen atoms. Ethanamide (Figure 7-49b), known also by its common name, acetamide, has two carbon atoms in the main chain containing the carbonyl carbon and two hydrogen atoms on the nitrogen. The amides in Figures 7-49c and 7-49d both contain a chain of three carbon atoms, including the carbonyl carbon, a propanamide, however, the amide in (c) has one R group on nitrogen, an ethyl group, whereas the amide in (d) has two R groups, both methyl groups. When writing a skeletal line structure for an amide, the nitrogen atom and hydrogen atoms bonded to nitrogen must be written in because nitrogen is a heteroatom.

Amides in Nature The amide functional group is an important and common functional group in biochemistry. It is the repeating functional group in peptides and proteins, which are formed in the ribosomes of cells, where anywhere from two to thousands of amino acids are joined via amide functional groups.

Small peptides, such as the endorphins—the body's natural painkillers, described in the opening vignette—are also formed from amino acids.

"Runner's high" is believed to be caused by the release of endorphins during intense exercise. In the opening vignette you learned that endorphins bind to the brain's opioid receptors.

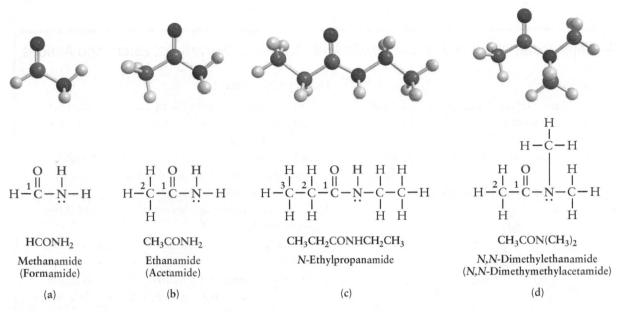

Figure 7-49 Ball-and-stick models, Lewis structures, condensed structures, IUPAC names, and common names for four representative amides.

Metenkephalin, for example, is derived from five amino acids to form four amide functional groups, as shown in Figure 7-50.

Some pharmaceuticals contain amide functional groups. For example, the penicillin antibiotics have two amide functional groups. The characteristic feature of the penicillin class of antibiotics is that they contain a ring of four atoms containing an amide functional group (known as a β-lactam), as in the broad-spectrum antibiotic amoxicillin shown in Figure 7-51. The other nitrogen-containing functional group in amoxicillin is an amine. Since rings with only four atoms are strained (chapter 6), the ring amide of penicillin antibiotics reacts with a key enzyme that bacteria need for building their cell walls. This reaction inactivates the enzyme, thereby destroying the bacteria. Since our cells do not have cell walls, penicillin does not interfere with any important enzymes required by our cells.

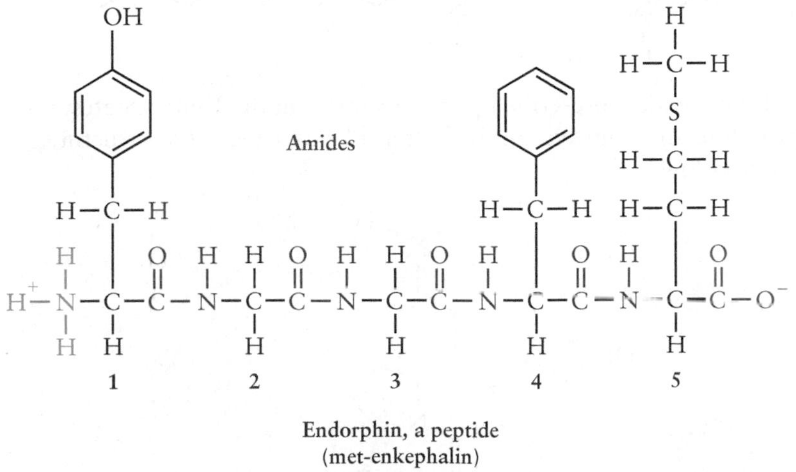

Endorphin, a peptide
(met-enkephalin)

Figure 7-50 The small peptide met-enkephalin is derived from five amino.

Amoxicillin

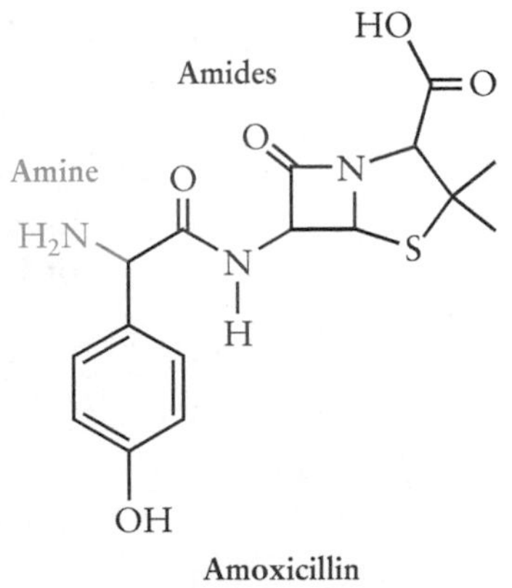

Figure 7-51 The structure of Amoxicillin, an antibiotic from the penicillin class of antibiotics, is characterized by a ring of four atoms containing an amide functional group.

Using Molecular Models 7-3 Carboxylic Acids and Their Derivatives, Esters and Amides

Construct Acetic Acid

1. Obtain three black carbon atoms, seven hydrogen atoms, two oxygen atoms, one nitrogen atom, ten straight bonds, and two bent bonds.

2. Make a model of ethanoic acid (acetic acid) as shown below.

3. Answer Inquiry Questions 7 through 11.

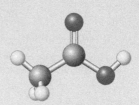

Acetic acid

Construct Methyl Acetate

4. Remove the —OH group on the carbonyl carbon and replace with an —OCH$_3$ group.

5. Answer Inquiry Questions 12 and 13.

Construct N-methylacetamide

6. Remove the —OCH$_3$ group from your model of methyl acetate and replace with an —NHCH$_3$ group. Answer Inquiry Questions 14 and 15.

Inquiry Questions

7. What functional group does acetic acid contain?

8. Why did you have to use bent bonds to make a carbonyl group?

9. What is the bond angle around the carbonyl carbon in acetic acid? What is the bond angle around the other carbon atom?

10. How is the functional group in acetic acid different from an alcohol?

11. Which atom does acetic acid lose to become the acetate ion?

12. When you replace the hydroxyl group on acetic acid with an —OCH$_3$ group, what new functional group have you made?

13. How is this new functional group different from a carboxylic acid? How is it different from an ether?

14. What new functional group have you made in your model when you replaced —OCH$_3$ with —NHCH$_3$?

15. How is this functional group different from an ester? How is it different from an amine?

WORKED EXERCISES   Amides

7-21 Which of the following nitrogen-containing compounds are *not* amides? Explain why.

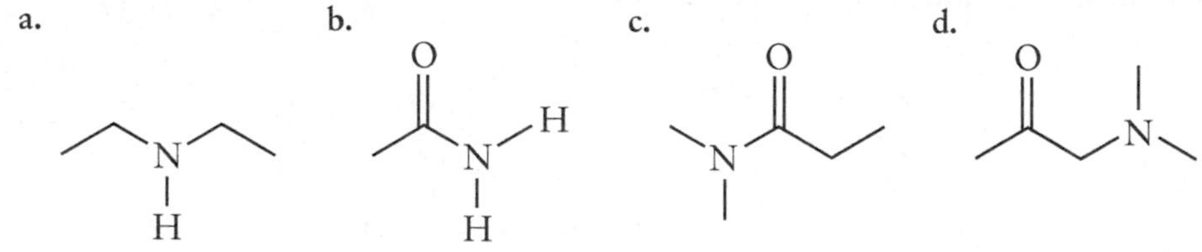

a.   b.   c.   d.

7-22 Lisinopril, the third-most widely prescribed pharmaceutical in the United States, is used to treat heart failure and improve survival after a heart attack. The structure of Lisinopril is shown below.

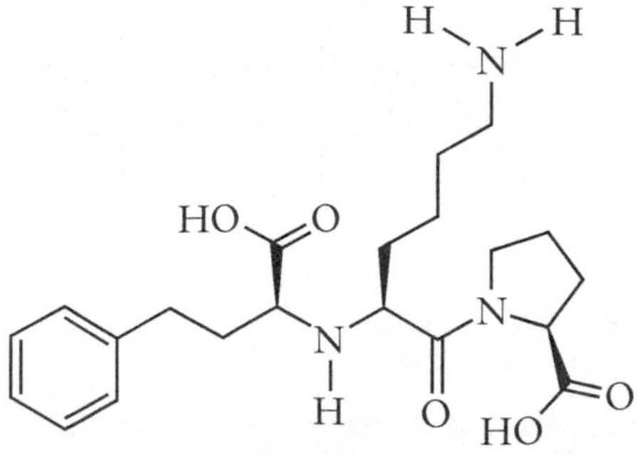

Lisinopril

    a. Highlight and label the amide functional group.

    b. Highlight and label the two amine functional groups.

    c. How can you distinguish the amines from the amide in this molecule?

    d. Highlight and label the two carboxylic acid functional groups.

7-23 Write the Lewis structure and the skeletal line structure for *N*-ethylbutanamide.

Solution

7-21 Structures (a) and (d) are not amides because they do not contain a carbonyl carbon directly attached to a nitrogen atom. (a) is an amine and (d) contains an amine and a ketone.

7-22 Look for a carbonyl group, and then determine what is attached directly to the carbonyl carbon. If it is a nitrogen atom, then it is an amide, of it is an OR′, then it is an ester, if it is an OH then it is a carboxylic acid. If there is no carbonyl group, but there is a nitrogen atom, then it is an amine.

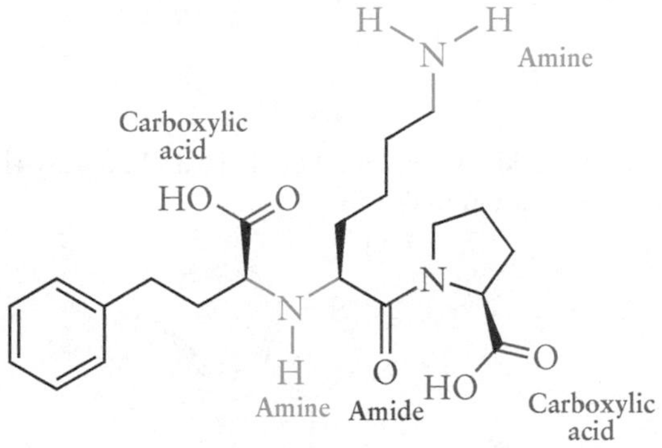

Lisinopril

7-23 The root name is *butane*, which means there are four carbon atoms in the main chain, including the carbonyl carbon. The *-amide* ending means C(1) is part of an amide functional group. Therefore, write a chain of four carbon atoms, where the first carbon atom is a carbonyl group with a bond to nitrogen. To determine what is attached to the nitrogen atom, look at the prefix. The prefix is *N-ethyl*, which means there is one two-carbon chain and a hydrogen atom attached to the nitrogen atom, shown in pink. Remember the structure can be written from left to right or from right to left. The N and the H on N must be written in the skeletal line structure.

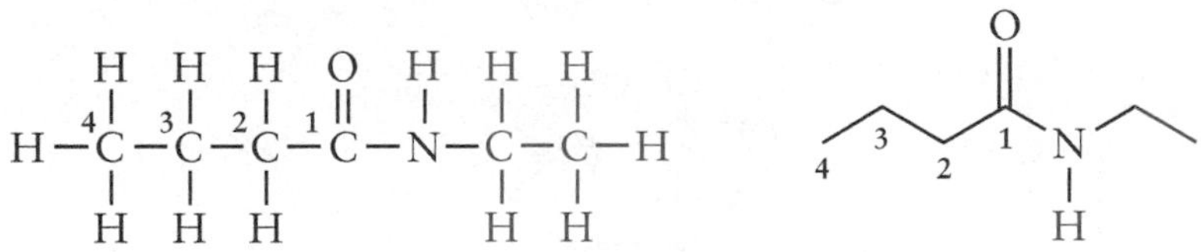

N-Ethylbutanamide

PRACTICE EXERCISES

28 Write the Lewis structure and the skeletal line structure for the condensed structures shown below. Which compound is *not* an amide?

    a. $CH_3CH_2CONHCH_3$ **b.** CH_3CONCH_3 **c.** $CH_3CH_2NH_2$ **d.** $CH_3CH_2CH_2CH_2CONH_2$
                                     |
                                     CH_3

29 Identify the functional groups represented below:

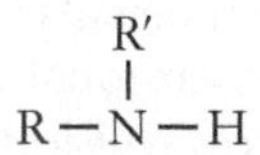

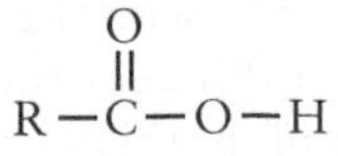

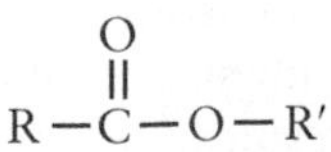

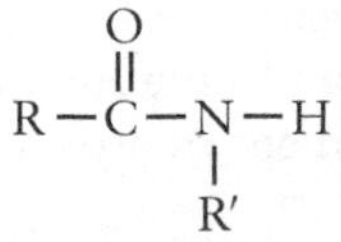

30 The structure of vitamin B$_5$, pantothenic acid, is shown below. Highlight and identify all the functional groups.

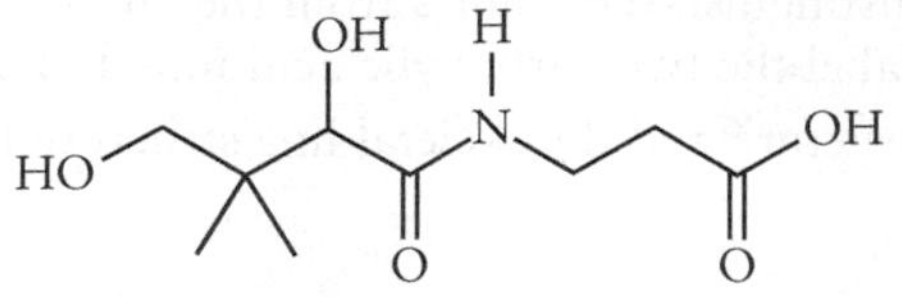

31 The structure of atenolol, a drug used to treat angina, is shown below. Highlight and label the amide, amine, ether, and secondary alcohol functional groups.

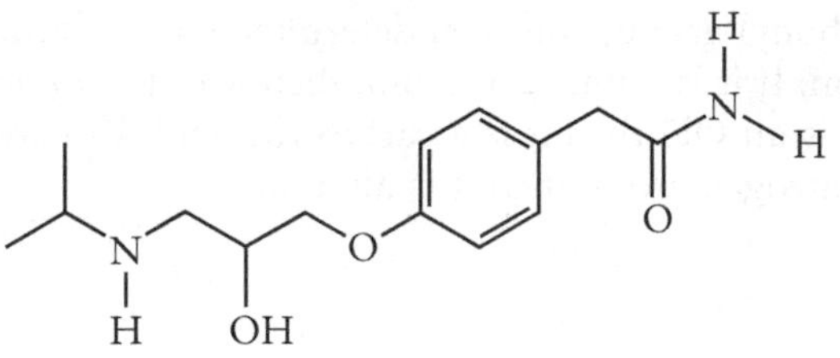

Atenolol

32 Turkey contains the amino acid tryptophan. Highlight and identify the functional groups that characterize this molecule as an amino acid.

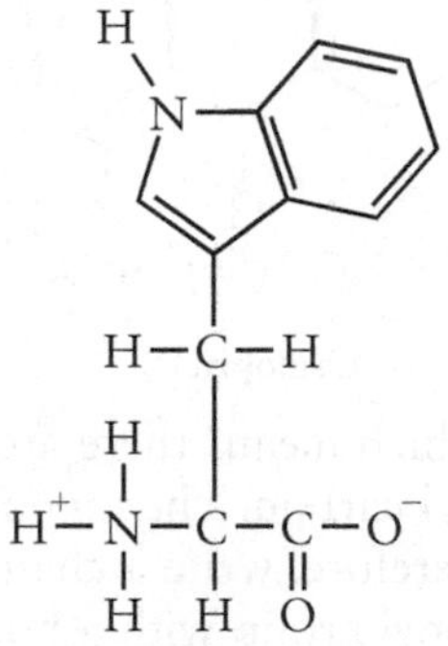

Tryptophan

33 Identify the amino acid, the carbohydrate, and the fatty acid among the three biomolecules shown.

a.

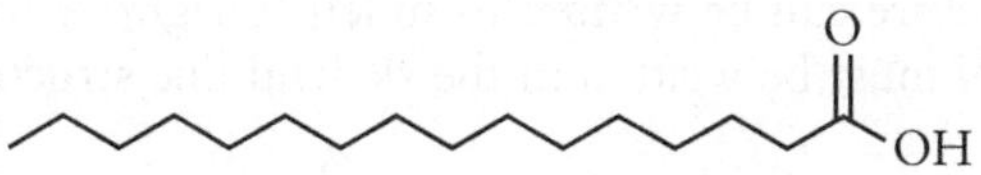

b.

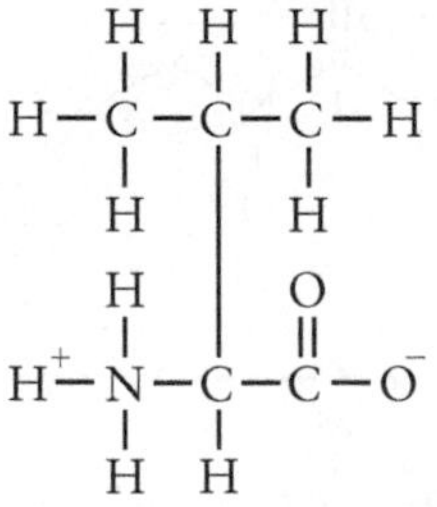

c.

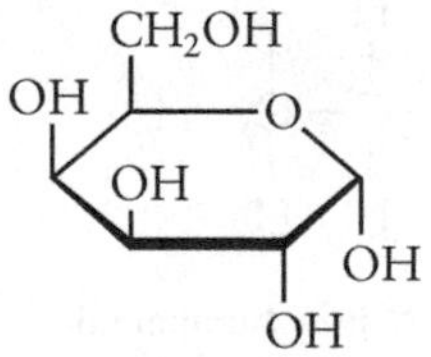

34 Write the Lewis structure and skeletal line structure for *N*-ethylhexanamide.

■ 7.4 Stereoisomers

In chapter 6 you learned about structural isomers: compounds with the same molecular formula but a different bonding arrangement of the atoms. You also learned about geometric isomers: compounds with the same molecular formula and bonding arrangement of atoms but a different three-dimensional arrangement of atoms as a result of a double bond. In this section you will learn about stereoisomers: compounds with the same molecular formula and the same bonding arrangement of atoms but a different three-dimensional arrangement of the atoms as a result of chirality. Chirality is a symmetry property of some objects and molecules. Stereoisomers, the subject of this section, are important in nature and also in pharmaceutical compounds. To understand stereoisomers, we must first learn what it means for an object or a molecule to be chiral.

> *Stereoisomers* are molecules with the same chemical formula and *the same bonding arrangement of atoms but a different three-dimensional spatial arrangement of atoms as a result of chirality.*

Figure 7-52 Chirality is a property exhibited by an object or a molecule that is nonsuperimposable on its mirror image—like the left hand and the right hand. [Catherine Bausinger]

Chirality

For an example of chirality, we need look no further than our hands. What is the difference between your left hand and your right hand? Clearly, you have the same digits on both hands (like a molecular formula), and the digits appear in the same order—thumb, forefinger, middle, ring, and pinky finger (analogous to the same bonding arrangement of atoms). Yet, your left hand is not identical to your right hand. You realize this when you try to *superimpose* your left hand on top of your right hand (try this): either the digits overlay or the palm and the top of hand overlay, but not both. We say the left hand is *non-superimposable* on the right hand. Yet, the left hand is clearly related to the right hand: the left hand is the *mirror image* of the right hand (Figure 7-52). The **mirror image** of an object (or a molecule) is the reflection you see when it is held up to a mirror. *An object or molecule is chiral only if it is nonsuperimposable on its mirror image.* Indeed, the word *chirality* comes from the Greek word for "handedness." In contrast, an object or a molecule is **achiral** if it is *superimposable* on its mirror image because it is identical to its mirror image.

> A **chiral** object or molecule is nonsuperimposable on its mirror image.
>
> An **achiral** object or molecule is superimposable on its mirror image— because it is identical to its mirror image.

In our everyday life we encounter both chiral and achiral objects. In addition to our hands, other examples of familiar chiral objects include gloves, shoes, and screws. In contrast, a table, a bed, and a ball are examples of *a*chiral objects.

Enantiomers

A pair of nonsuperimposable mirror-image stereoisomers is known as a pair of **enantiomers**. The amino acid L-alanine is chiral because it is nonsuperimposable on its mirror image, D-alanine, illustrated in Figure 7-53. Hence,

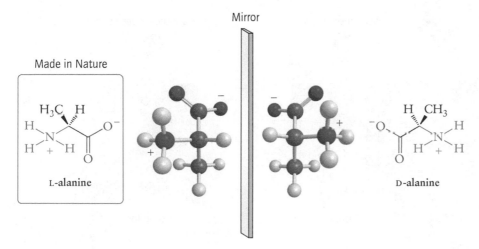

Figure 7-53 L-alanine and D-alanine are enantiomers: nonsuperimposable mirror-image stereoisomers. L-Alanine is found in nature, whereas D-alanine is not.

HO

Safe

HO

L-Dopa

(a)

HO

Unsafe

HO

D-Dopa

(b)

Figure 7-54 (a) L-dopa (b) D-dopa. Enantiomers depicted using the wedge-and-dash notation used to show the difference in the three dimensional arrangement of atoms.

L-alanine and D-alanine are a pair of enantiomers. To distinguish enantiomers by name, we include a one-letter prefix as part of the IUPAC name. For amino acids and monosaccharides we use the one letter abbreviations D- and L-. Often only one of a pair of enantiomers is produced in nature and the other is not. For example, L-alanine is a natural amino acid but D-alanine is not. D-glucose is found in nature, but L-glucose is not. In fact, the 19 chiral amino acids produced in nature are all L-amino acids.

A pair of enantiomers differ only in the three-dimensional orientation of the atoms bonded to one or more tetrahedral carbon atoms in the molecule. Indeed, it is the symmetry properties of the tetrahedral shape that usually causes chirality to exist in molecules. To illustrate the difference in the three-dimensional orientation of the atoms in a pair of enantiomers, we must use the wedge-and-dash notation introduced in chapter 3 for drawing tetrahedral carbon centers. For example, in the structure of L-dopa shown in Figure 7-54a, the amine is projecting toward the viewer (wedged bond) and the hydrogen atom is projecting away from the viewer (dashed bond). In D-dopa, the wedged and dashed atoms/groups are reversed (Figure 7-54b). You will see wedged and dashed bonds used throughout this text to indicate a particular stereoisomer. Work through Using Molecular Models 7-4: Enantiomers to gain hands-on experience with chiral and achiral molecules.

A pair of enantiomers have identical physical and chemical properties, such as boiling point, melting point, and solubility, which makes them difficult to separate in the laboratory. *However, when placed in a chiral environment, such as the body, enantiomers exhibit different physical and chemical properties.* You have experienced this phenomenon if you have ever placed your left hand into a right-handed glove. Similarly, when a chiral drug molecule interacts with the chiral receptors and enzymes in your body, one enantiomer will typically exhibit different physiological effects than its enantiomer because they are different in their three-dimensional arrangement of atoms, as illustrated in Figure 7-55.

Some drugs, such as Darvon and Novrad, are enantiomers used for different indications because they have different pharmacological effects: Darvon is an analgesic and Novrad is as an antitussive (cough suppressant). A profound and devastating example of the different physiological effects of enantiomers was seen with the drug thalidomide. One enantiomer of thalidomide was effective in treating morning sickness, whereas the other enantiomer was later discovered to be a teratogen—a substance that causes birth defects. When thalidomide was first introduced in the 1950's, patients were given a mixture of both

L-enantiomer fits

D-enantiomer does not fit

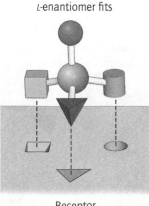
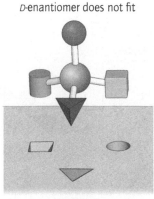

Figure 7-55 Enantiomers display different properties in a chiral environment such as the human body. Showing how a chiral receptor might distinguish one drug enantiomer from the other.

Receptor   Receptor

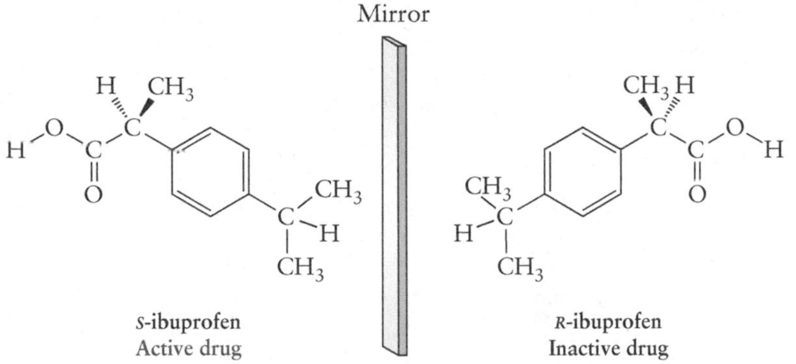

Figure 7-56 Ibuprofen contains s-ibuprofen, the active analgesic, and r-ibuprofen, the inactive enantiomer, known as a racemic mixture.

enantiomers, and the teratogenic effects of the one stereoisomer were revealed. Thalidomide was eventually taken off the market, but not before thousands of women who had already taken the drug had given birth to severely deformed babies with flipper like arms and legs. Today, the Food and Drug Administration (FDA) requires that all chiral drugs be separated into their enantiomers and each be tested separately for safety before a drug comes to market.

Often, particularly with synthetic drugs, both enantiomers are produced as a 50:50 mixture of enantiomers, known as a **racemic mixture**. Ibuprofen, the active ingredient in Advil and Motrin, for example, is sold as a racemic mixture. Only one enantiomer, s-ibuprofen, is the active analgesic; r-ibuprofen, the other enantiomer, is inactive (Figure 7-56). You have seen the letters D- and L- used to distinguish enantiomers. The IUPAC prefixes R- and S- are also used to distinguish enantiomers, especially in synthetic compounds.

When only one enantiomer has medicinal value and the other enantiomer has adverse physiological effects, a racemic mixture cannot be used as a pharmaceutical. The enantiomers must be separated or produced as a single enantiomer in specialized chemical reactions. For example, the drug L-dopa, used to treat Parkinson's disease, is administered as a single enantiomer drug because D-dopa has the negative side effect of lowering a person's white blood cell count, placing the patient at risk of infection.

Enantiomers and Centers of Chirality

You have learned that a molecule is chiral if it is nonsuperimposable on its mirror image. What makes some molecules chiral and others achiral? Chirality in a molecule exists when there are one or more centers of chirality in the molecule. *A center of chirality is a tetrahedral carbon atom with bonds to four* different *atoms or groups of atoms*. A molecule containing one center of chirality is by definition chiral. A molecule with two or more centers of chirality is usually chiral, unless it is superimposable on its mirror image and therefore achiral.

Consider the enantiomers L- and D-alanine. They are chiral because they have one center of chirality (highlighted), four bonds to the different groups/atoms: H, NH_3^+, CO_2^-, and CH_3. In order to draw enantiomers, you must use the wedge-and-dash notation for the bonds to the atoms bonded to the center of chirality. Indeed, the only difference between a pair of enantiomers is that two atoms or groups of atoms on the center of chirality are switched. In the case of D- and L-alanine, shown on the next page, we see that the CH_3 group projects toward you in the L-enantiomer and projects away from you in the D-enantiomer and vice versa for the H atom all on the center of chirality. In fact, to convert one enantiomer into the other, we need only to exchange *any* two bonds at the center of chirality. In this example, the CH_3 and H have been exchanged.

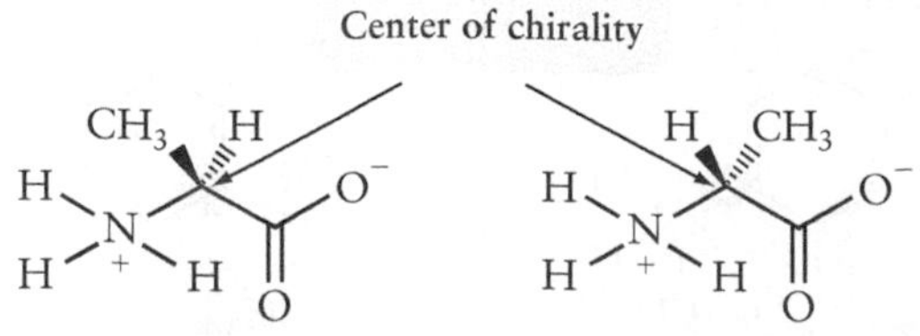

Center of chirality

L-alanine   *D*-alanine

Using Molecular Models 7-4 Enantiomers

Construct an Achiral Molecule: Propan-2-ol

1. Obtain six black carbon atoms, two red oxygen atoms, 16 hydrogen atoms, and 22 straight bonds.
2. Construct two models of propan-2-ol.

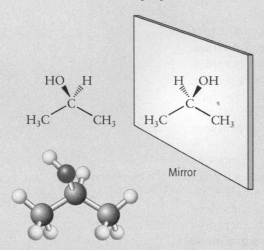

Mirror

3. Imagine that there is a mirror next to one model and orient the other model so that they appear as mirror images. Rotate the models around bonds as needed.
4. Now try to superimpose the two models.

Inquiry Questions

8. Is your model superimposable on its mirror image? You may rotate as many bonds as necessary (any conformation) but do not break or make any bonds.
9. Based on your answer to (8), is propan-2-ol chiral?
10. Based on your answers to (8) and (9), does propan-2-ol have an enantiomer?

11. What is the relationship between your two models: identical or enantiomers?

Construct *D*-Alanine and *L*-Alanine

5. Using Figure 7-53 as a guide, make models of *L*-alanine and *D*-alanine using two blue nitrogen atoms, six black carbon atoms, 14 light blue hydrogen atoms, four red oxygen atoms, 22 straight bonds, and four bent bonds. Be sure to place the atoms in the correct three-dimensional arrangement, as shown in Figure 7-53.
6. Imagine that there is a mirror next to one model and orient the other model so that they appear as mirror images. If you can't get a mirror image, double-check that you didn't accidentally make two D- or two L-alanine models.
7. See if the two mirror image models you constructed are superimposable. You may rotate as many bonds as necessary, but do not break or make any bonds.

Inquiry Questions

12. Are your models superimposable?
13. Based on your answer to (12), are your models chiral?
14. What is the relationship between your two models: identical or enantiomers?
15. Do both molecules have the same molecular formula? Do both molecules have the same structure? Do both models have the same spatial orientation?

WORKED EXERCISES   Chiral Molecules

7-24 Indicate whether the following objects are chiral or achiral:
   a. a grand piano   **b.** a baseball bat   **c.** a spoon   **d.** a corkscrew
7-25 A 50:50 mixture of enantiomers is known as a _____ mixture.

7-26 Phil Mickelson is a left-handed golfer and uses different golf clubs from most other golfers. How are his clubs different from other golf clubs? Would they feel awkward to a right-handed golfer?

7-27 S-Naproxen is the active ingredient in the analgesic Aleve. It is sold as a single-enantiomer drug. What does this mean?

Solutions

7-24 a. A grand piano is chiral because it is nonsuperimposable on its mirror image.
   b. A baseball bat is achiral because it is superimposable on its mirror image—it's the same as its mirror image.
   c. A spoon is achiral because it is superimposable on its mirror image—it's the same as its mirror image.
   d. A corkscrew is chiral—it is either left-handed or right-handed.

7-25 A 50:50 mixture of enantiomers is known as a *racemic* mixture.

7-26 Phil Mickelson's clubs are the mirror image of most golf clubs because he is a left-handed player. If you are right-handed, his clubs would feel awkward because a chiral object (the left-handed golf club) is interacting with a chiral environment (your hand).

7-27 It means the drug contains only s-Naproxen; no r-Naproxen, its enantiomer. This is usually the case when one enantiomer has adverse physiological effects.

PRACTICE EXERCISES

35 Indicate whether the following objects are chiral or achiral:
   a. a sock　　**b.** a glove　　**c.** a basketball

36 Name two types of biomolecules that are chiral.

37 Indicate whether the following statements about a chiral molecule are true or false:
   a. is superimposable on its mirror image
   b. has an enantiomer
   c. may exhibit different chemical properties from its enantiomer in the body
   d. typically exhibits different chemical properties from its enantiomer in an achiral environment

■ 7.5 Phosphate Ester Functional Groups

The final functional groups we will consider in this chapter are those derived from the inorganic substance phosphoric acid (H_3PO_4). These functional groups contain a phosphorus-oxygen double bond, $P{=}O$, as well as three $P{-}O$ single bonds. The functional groups derived from phosphoric acid are found primarily in the important biomolecules 2-deoxyribonucleic acid (DNA), ribonucleic acid (RNA), adenosine triphosphate (ATP), and coenzyme A.

Phosphoric acid (Figure 7-57) has an expanded octet: it contains a central phosphorus atom surrounded by 10 bonding electrons. Recall from chapter 3 that phosphorus can have an expanded octet because it is an element in the third period of the periodic table.

Phosphoric acid, like all acids, tends to donate a hydrogen ion, H^+, to water. Moreover, phosphoric acid can donate one, two, or all three of its hydrogen ions. The number of hydrogen ions lost depends on the environment. In a biological cell, the most abundant form of phosphoric acid is the monohydrogen phosphate ion, HPO_4^{2-}, often called **inorganic phosphate** and abbreviated P_i.

If the hydrogen atom in monohydrogen phosphate is substituted with an R group, the result is an organic molecule with a functional group known as a **phosphate ester**.

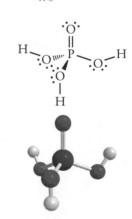

Figure 7-57 Lewis structure and ball-and-stick model of the inorganic compound phosphoric acid, which contains a P atom with an expanded octet: a central phosphorus atom surrounded by 10 bonding electrons.

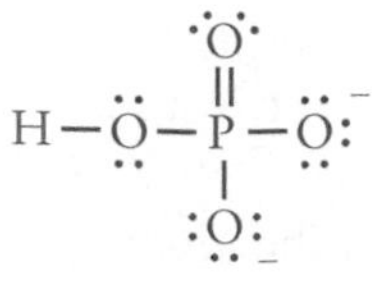

$$
H-\overset{..}{\underset{..}{O}}-\overset{\overset{\overset{..}{O}}{\|}}{\underset{\underset{:\overset{..}{O}:^-}{|}}{P}}-\overset{..}{\underset{..}{O}}:^- \qquad \text{monohydrogen phosphate, } P_i
$$

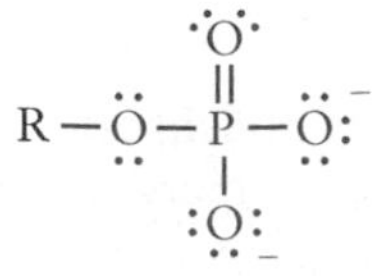

$$
R-\overset{..}{\underset{..}{O}}-\overset{\overset{\overset{..}{O}}{\|}}{\underset{\underset{:\overset{..}{O}:^-}{|}}{P}}-\overset{..}{\underset{..}{O}}:^- \qquad \text{Phosphate ester}
$$

When two hydrogen atoms in hydrogen phosphate are substituted by R groups, the result is a phosphate *diester*—a molecule with *two* phosphate ester functional groups:

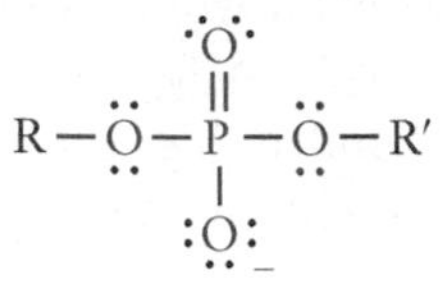

$$
R-\overset{..}{\underset{..}{O}}-\overset{\overset{\overset{..}{O}}{\|}}{\underset{\underset{:\overset{..}{O}:^-}{|}}{P}}-\overset{..}{\underset{..}{O}}-R' \qquad \text{Phosphate diester}
$$

One of the unique characteristics of phosphate esters is that the *phosphate* group ($-PO_3^{2-}$) can form a bond to a second and even a third phosphate group, P_i, to form a diphosphate ester and a triphosphate ester, respectively, as shown in Figure 7-58.

Phosphoanhydride bonds join phosphate groups, which are very important in biochemistry because they are high energy bonds as a result of the unstable arrangement created when two negatively charged oxygen atoms are in close proximity to one another.

The cell stores energy by forming phosphoanhydride bonds. For example, *a*denosine *tri*phosphate (ATP), the main energy storage molecule in the cell, is a triphosphate (Figure 7-59). The —R group is known as adenosine; hence the name *adenosine* triphosphate. We will consider the important energy implications of these molecules in chapter 15.

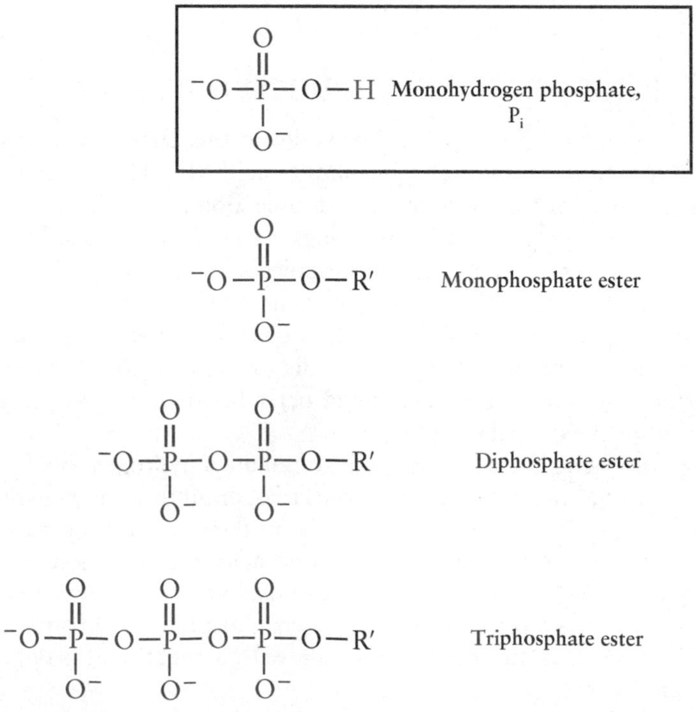

Phosphoanhydride
high energy bonds

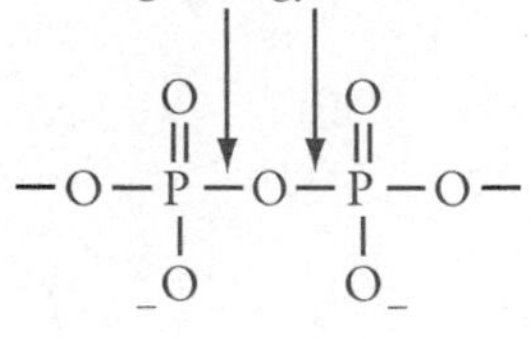

$$
\underset{\underset{O^-}{|}}{\overset{\overset{O}{\|}}{-O-P}}-O-\underset{\underset{O^-}{|}}{\overset{\overset{O}{\|}}{P}}-O-
$$

$$
\underset{\underset{O^-}{|}}{\overset{\overset{O}{\|}}{^-O-P}}-O-H \quad \text{Monohydrogen phosphate, } P_i
$$

$$
\underset{\underset{O^-}{|}}{\overset{\overset{O}{\|}}{^-O-P}}-O-R' \quad \text{Monophosphate ester}
$$

$$
\underset{\underset{O^-}{|}}{\overset{\overset{O}{\|}}{^-O-P}}-O-\underset{\underset{O^-}{|}}{\overset{\overset{O}{\|}}{P}}-O-R' \quad \text{Diphosphate ester}
$$

Figure 7-58 Comparision of the structure of a mono, di, and triphosphate ester to the hydrogen phosphate ion, P_i, from which they are derived.

$$
\underset{\underset{O^-}{|}}{\overset{\overset{O}{\|}}{^-O-P}}-O-\underset{\underset{O^-}{|}}{\overset{\overset{O}{\|}}{P}}-O-\underset{\underset{O^-}{|}}{\overset{\overset{O}{\|}}{P}}-O-R' \quad \text{Triphosphate ester}
$$

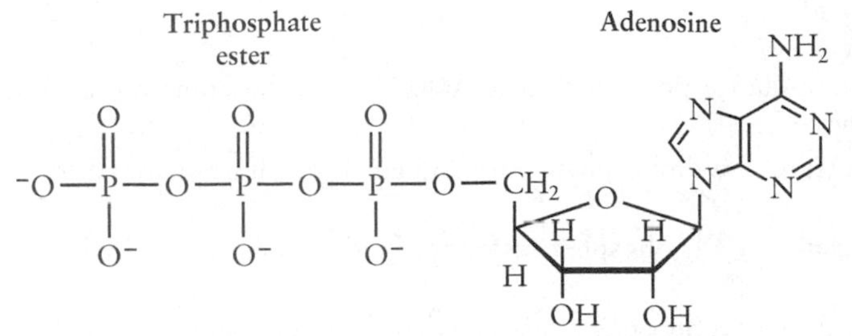

ATP
Adenosine triphosphate

Figure 7-59 The structure of adenosine triphosphate, ATP, a triphosphate ester and the main energy storage molecule in the cell.

WORKED EXERCISE Phosphate Esters

7-28 The structure of the important biomolecule known as coenzyme A, derived from the vitamin pantothenic acid, is shown below:

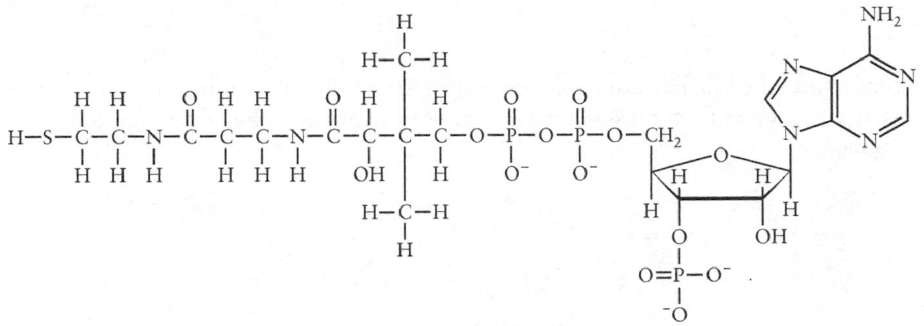

Coenzyme A

a. Highlight the diphosphate diester and the phosphate ester in this molecule.
b. Highlight and label the two amide functional groups.
c. Highlight and label the alcohol functional groups. Indicate if they primary, secondary, or tertiary alcohols.
d. Label the phosphoanydride bonds.
e. Coenzyme A contains adenosine, hence the "A" in the name. Highlight the adenosine component of coenzyme A.

Solution

7-28 a.–d.

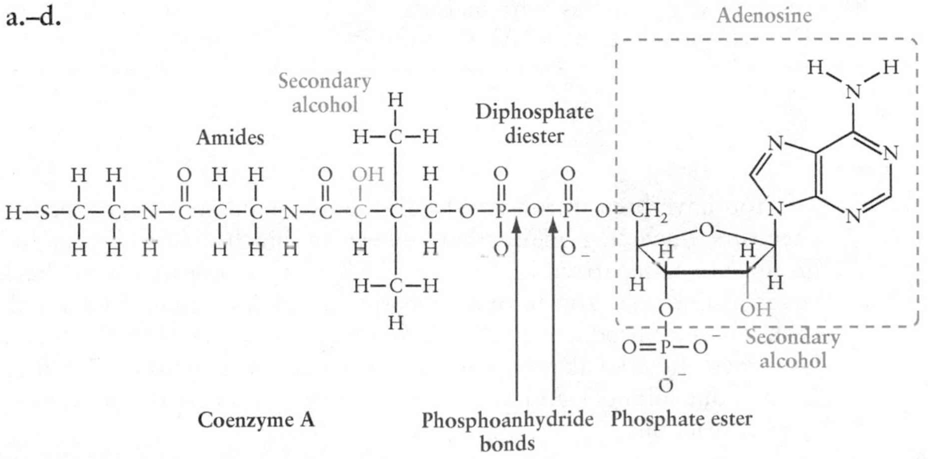

PRACTICE EXERCISES

38 Write the molecular formula for phosphoric acid. What is the most common form of phosphoric acid in the cell?

39 Write the general structure of a diphosphate ester and label the phosphoanhydride bonds.

40 Write the general structure of a triphosphate ester and label the phosphoanhydride bonds.

41 Write the general structure of a phosphate diester. Does it contain phosphoanhydride bonds?

42 In section 7.2 you learned about the amine functional group in psilocybin, an alkaloid produced by various species of mushrooms. Highlight and identify the phosphorus-containing functional group in this molecule.

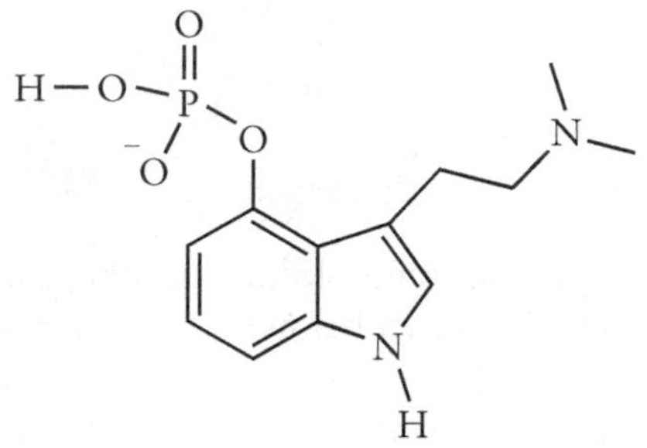

43 The structure of *cyclic* **a**denosine *mono***p**hosphate, cAMP, is shown below. It is an important messenger molecule that transmits the effects of epinephrine (adrenaline) and other hormones.

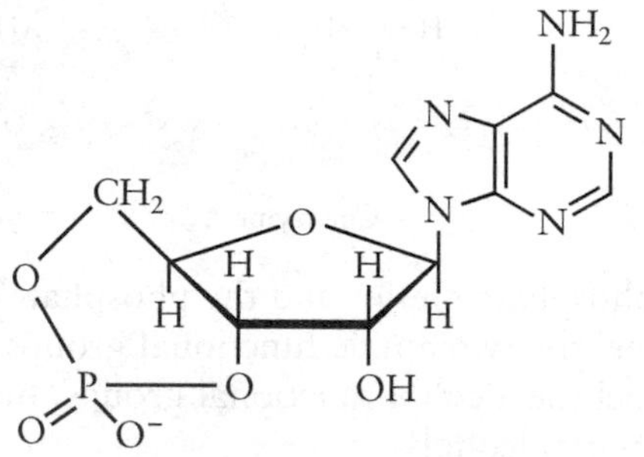

Cyclic adenosine monophosphate,
cAMP

a. This molecule is a monophosphate *di*ester. Explain
b. Why is this molecule called a monophosphate and not a di- or triphosphate?
c. What is the total charge on this polyatomic ion?
d. Why is the molecule called "cyclic" AMP? Define *cyclic*, *A*, *M*, and *P*.

You have been introduced to all the major functional groups in organic chemistry, including many that appear in biochemistry. These functional groups are summarized in Table 7-2, which also appears on the back inside cover of this text. You have also learned to identify these functional groups within natural products and in pharmaceuticals. In chapter 10 you will learn how these functional groups behave in chemical reactions. But first, you'll learn about solutions and acid-base reactions, the simplest and most common type of reactions.

TABLE 7-2 Summary of the Common Functional Groups

Functional Group Name	Structure	Functional Group Name	Structure
Alkene	$\diagdown C=C \diagup$	1° amine	$R-\ddot{N}-H$ with H
Alkyne	$-C\equiv C-$	2° amine	$R-\dot{N}-H$ with R
Aromatic hydrocarbon	(benzene ring)—R	3° amine	$R-\ddot{N}-R$ with R
1° Alcohol	$R-\underset{H}{\overset{H}{C}}-O-H$	Aldehyde	$-\overset{O}{\underset{}{C}}-H$
2° Alcohol	$R-\underset{H}{\overset{R}{C}}-O-H$	Ketone	$R-\overset{O}{\underset{}{C}}-R$
3° Alcohol	$R-\underset{R}{\overset{R}{C}}-O-H$	Carboxylic acid	$-\overset{O}{\underset{}{C}}-O-H$
Phenol	(benzene ring)—OH	Ester	$-\overset{O}{\underset{}{C}}-O-R$
Ether	$R-O-R$		
Phosphate ester	$^{-}O-\underset{O^-}{\overset{O}{P}}-O-R$	Amide	$-\overset{O}{\underset{}{C}}-\ddot{N}-$

R = carbon atom or chain of carbon atoms

Chemistry in Medicine The Role of Dopamine in Parkinson's Disease and Schizophrenia

Schizophrenia is a debilitating mental disorder that affects 1 percent of the population worldwide. Its symptoms include hallucinations, delusions, disordered thinking, movement disorders, and social withdrawal. The disease usually strikes during late adolescence to early adulthood. The cause of schizophrenia is still unknown, although an *excess* of the neurotransmitter dopamine appears to be involved.

Parkinson's disease is somewhat better understood and is believed to arise from *decreased* dopamine activity in another part of the brain. Parkinson's disease is a chronic neurodegenerative disease characterized by a loss of coordination and movement. The disease tends to strike after the age of 60. Symptoms of Parkinson's disease include tremor (shaking), bradykinesia (slow movement), rigidity (stiffness of limbs), and impaired balance. About 60,000

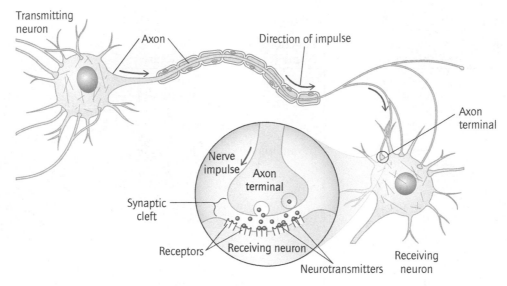

Figure 7-60 Communication between neurons.

new cases of Parkinson's disease are diagnosed in the United States every year.

Schizophrenia and Parkinson's disease are two very different diseases, but both involve an imbalance of dopamine in the brain. Dopamine, a neurotransmitter essential for normal brain function, is involved in the brain's reward and pleasure centers. Dopamine is also involved in regulating movement and emotional responses. Dopamine is an amine, as suggested by the ending in its name.

Dopamine

Although there is no cure for schizophrenia or Parkinson's disease, prescription drugs are available to treat some of the more debilitating symptoms. In order to understand the biochemistry of schizophrenia and Parkinson's disease, we must first understand how nerve cells communicate with one another on a molecular level.

The human brain has about 100 billion (1×10^{11}) nerve cells, known as neurons. Most neurons communicate through chemical messengers known as neurotransmitters: organic molecules characterized by an amine functional group. Communication between neurons begins with an electrical impulse that travels along the axon, the long shaft of a nerve cell. When the electrical impulse reaches the end of the axon, neurotransmitters are released from the neuron into the synaptic cleft, the gap between two neurons, shown in the inset in **Figure 7-60**. Neurotransmitters diffuse to the next neuron, where they bind to specific receptors on the cell membrane of the receiving neuron. In this way electrical impulses travel from one neuron to the

next, creating what is known as a neuronal pathway. The transmission of electrical impulses through a neuronal pathway ultimately leads to a physiological response. The two neuronal pathways in the brain that use dopamine as a neurotransmitter are highlighted in **Figure 7-61**.

The impaired movement associated with Parkinson's disease is believed to result from the loss of dopamine-producing neurons along one of the two neuronal pathways in the brain that rely on dopamine, highlighted in red in Figure 7-61. Neurons differ from most other cells in the body in that they do not regenerate themselves, so the loss of these cells is permanent. Patients with Parkinson's disease show 80 percent less dopamine activity in this neuronal pathway.

Effective prescription drugs are available to treat many of the symptoms of Parkinson's disease. Dopamine itself

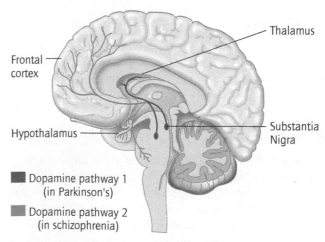

Figure 7-61 Neuronal pathways that involve dopamine. The pathway shown in red controls movement and is involved in Parkinson's disease. The pathway shown in green affects desire, memory, and motivation and is involved in schizophrenia.

cannot be administered as a drug because it can't cross the blood-brain barrier. Instead, the drug L-dopa is administered. The only difference between the chemical structure of L-dopa and dopamine is the carboxylic acid functional group in L-dopa. The carboxylic acid is removed in a biochemical reaction in the neuron that converts L-dopa into dopamine:

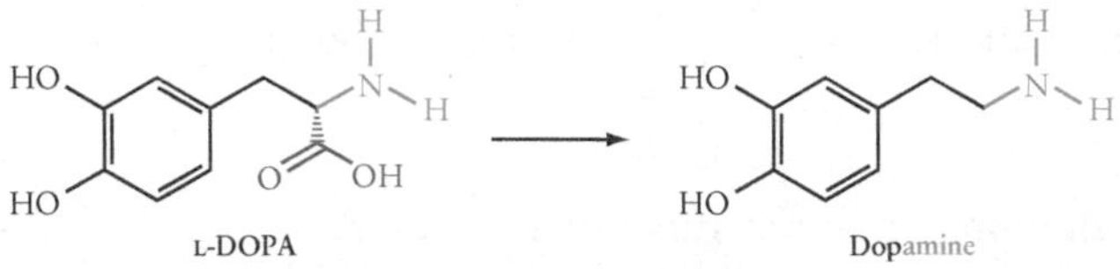

The prolonged use of L-dopa eventually reduces its ability to activate the dopamine receptor. At that point other drugs are usually prescribed. Unfortunately, dopamine receptors eventually become less sensitive to these drugs too and symptoms of the disease return.

The thought disorders associated with schizophrenia are believed to result from the opposite effect: an *excess* of dopamine activity in another neuronal pathway in the brain, highlighted in green in Figure 7-61. Stimulation of this particular neuronal pathway is associated with feelings of reward and desire, memory, and motivation. By binding to the dopamine receptor in this neuronal pathway, antipsychotic medications prevent dopamine from binding to its receptor. The first generation of antipsychotic drugs bound to all dopamine receptors in the brain, including those in the neuronal pathway that control movement (shown in red in Figure 7-61). Consequently, some of the side effects of these drugs were tremors and other symptoms like those from Parkinson's disease. The newest class of antipsychotic medications, known as *atypical antipsychotics*, block dopamine receptors but specifically target a type of dopamine receptor that is found only in the neuronal pathway associated with schizophrenia (shown in green). Therefore, atypical antipsychotic drugs have fewer movement-related side effects.

There have been significant advances in the treatment of these two diseases as scientists have unraveled the chemistry behind each of these diseases. Further research should eventually lead to even better treatment options for these patients in the future.

Summary

Alcohols and Ethers

- An alcohol has the general structure R—O—H, where the —OH group is referred to as a hydroxyl group. The carbon atom bearing the hydroxyl group is an alkyl carbon.

- An ether has the general structure: R—O—R'. The carbon atoms bonded to the oxygen atom can be either alkyl or aromatic carbons.

- An R group is a carbon atom or chain of carbons of undefined length and composition.

- Alcohols and ethers, like water, have a bent molecular shape around the oxygen with approximately 109.5° bond angles.

- Alcohols are further classified as 1°, 2°, and 3° alcohols, depending on the number of R groups attached to the carbon atom bonded to the —OH group: 1° has one R group, 2° has two R groups, and 3° has three R groups.

- Compounds with two hydroxyl groups are known as diols; three hydroxyl groups, triols; and many hydroxyl groups, polyols.

- Monosaccharides and disaccharides are carbohydrates, a class of biomolecules with many hydroxyl groups—polyols.

- A phenol functional group contains an —OH group bonded to an aromatic ring.

- Phenols are present in compounds that have antioxidant properties.

- Anesthetics are low-molecular-weight ethers, many of which have hydrogen atoms replaced with F.

- Alcohols and phenols can hydrogen bond and so they have higher boiling points.

- Ethers have dipole-dipole intermolecular forces of attraction, tend to have low boiling points, and are often used as solvents.

Amines

- Amines are derived from ammonia, NH_3, in which one, two, or all three hydrogen atoms are replaced with R groups: primary amine RNH_2, secondary amine R_2NH, tertiary amine R_3N.
- Amines have a trigonal pyramidal shape around the nitrogen atom with approximately 109.5° bond angles.
- Amines with one or two N—H bonds are capable of hydrogen bonding.
- An amine acts as a base and reacts with water to form a polyatomic cation: RNH_3^+.
- Plant-derived amines are known as alkaloids.
- Many compounds containing an amine functional group are pharmacologically active, such as morphine, dopamine, and epinephrine.
- Amines are part of the chemical structure of amino acids and neurotransmitters.

Carbonyl-Containing Functional Groups

- A carbonyl group is a C=O group in which the carbon atom is also bonded to an R group and one other atom or group, which determines its identity: aldehyde, —H; ketone, —R; carboxylic acid, —OH; ester, —OR′; amide, —NR_2.
- A carbonyl group has a dipole, so functional groups containing a carbonyl group have dipole-dipole intermolecular forces of attraction. Carboxylic acids and some amides can also form hydrogen-bonds.
- Carbonyl-containing functional groups have a trigonal planar molecular shape around the carbonyl carbon and bond angles of 120°.
- Monosaccharides are polyols containing an aldehyde or a ketone functional group.
- Carboxylic acids contain a carbonyl group directly bonded to an OH group, RCOOH or RCO_2H, and behave as one functional group, distinct from an alcohol.
- A carboxylic acid acts as an acid and reacts with water to form a polyatomic anion: $RCOO^-$ or RCO_2^-.
- Fatty acids are important biomolecules involved in the metabolism of triglycerides. They are carboxylic acids with a long hydrocarbon chain for an R group.
- Amino acids, the important building blocks for proteins, contain a carboxylic acid and an amine as part of their structure.
- An ester contains a carbonyl group directly bonded to an —OR′ group: $RCO_2R′$, which is a functional group distinctly different from an ether because it contains a carbonyl group.
- An ester has two R groups, distinguished as R and R′, that can be the same or different.
- Triglycerides are a type of biomolecule that is characterized by its three ester functional groups and three long hydrocarbon R groups.
- Esters and amides are classified as carboxylic acid derivatives because they can be prepared from carboxylic acids.
- Esters have a permanent dipole, so they have dipole-dipole intermolecular forces of attraction.
- Many esters have fragrant aromas.
- An amide functional group contains a carbonyl group directly bonded to nitrogen: $RCONH_2$, RCONHR, or $RCONR_2$, making an amide distinctly different from an amine, which does not contain a carbonyl group.
- Amides are formed between amino acids when peptides and proteins are formed.

Stereoisomers

- Stereoisomers are compounds with the same chemical formula and the same atom connectivity but a different three-dimensional arrangement of the atoms as a result of chirality.

- Chirality is a symmetry property of certain objects and molecules characterized by being nonsuperimposable on their mirror image—like your left hand and your right hand.

- A pair of nonsuperimposable mirror-image stereoisomers are known as enantiomers and are distinguished by the prefixes D- and L- or R- and S-.

- An achiral molecule or object is superimposable on its mirror image; that is, it is the same as its mirror image—such as a ball or pencil.

- Enantiomers have identical physical properties in an achiral environment such as the laboratory. Their differences become apparent in a chiral environment such as the body.

- A 50:50 mixture of enantiomers is known as a racemic mixture.

- Chiral molecules contain one or more centers of chirality. A center of chirality is a tetrahedral carbon bonded to four different atoms or groups of atoms.

- To write the structure of stereoisomers, wedges and dashes are drawn at the center(s) of chirality. Switching any two groups or atoms on the center of chirality forms the other enantiomer.

Phosphate Ester Functional Groups

- Phosphoric acid, H_3PO_4, is an inorganic acid with an expanded octet around P that can lose one, two, or three hydrogen atoms. In the cell it is found primarily as HPO_4^{2-}, known as inorganic phosphate, abbreviated P_i.

- A phosphate ester functional group is derived from monohydrogen phosphate, HPO_4^{2-}, by replacing the hydrogen with an R group.

- The phosphate group of a phosphate ester can be connected to additional phosphate groups, P_i's, via phosphoanhydride bonds to form diphosphate esters and triphosphate esters.

- Phosphate esters are found in some important biological molecules such as adenosine triphosphate (ATP), an important energy carrier molecule in the cell that contains a triphosphate ester.

Key Words

Achiral An object or a molecule that is not chiral. It is superimposable on its mirror image and so identical to its mirror image.

Alcohol A functional group derived from water where one of the H atoms has been replaced with an —R group: R—O—H. The carbon atom bearing the hydroxyl group is an alkyl carbon.

Aldehyde A carbonyl containing functional group characterized by a hydrogen atom bonded to the carbonyl carbon. Also includes formaldehyde which contains two —H atoms bonded to the carbonyl carbon.

Alkaloid A compound containing an amine that is found in nature.

Amide A carbonyl compound with a nitrogen atom bonded to the carbonyl carbon. The nitrogen atom has two additional bonds to either —H or —R.

Amine A functional group derived from ammonia in which one, two, or all three of the hydrogen atoms have been replaced by R groups: RNH_2, R_2NH, or R_3N. There is no carbonyl group bonded to the nitrogen.

Amino acid Biological compounds used to build proteins, characterized by an amine and a carboxylic acid functional group bonded to the same carbon atom.

Analgesic A substance that reduces pain.

Carbohydrate A type of biomolecule that includes monosaccharides—simple sugars—and disaccharides. Carbohydrates are a source of energy for cells.

Carbonyl group A carbon-oxygen double bond, $C=O$.

Carboxylic acid A carbonyl containing functional group that contains an —OH group bonded directly to the carbonyl carbon.

Carboxylic acid derivative Compounds that can be prepared from carboxylic acids, such as esters and amides.

Center of Chirality A tetrahedral carbon atom with bonds to four different groups or atoms. A molecule with one center of chirality is chiral. A molecule with two or more centers of chirality is chiral if it is nonsuperimposable on its mirror image.

Chiral An object or a molecule that is nonsuperimposable on its mirror image.

Common Names Names that have been used traditionally in addition to the IUPAC name, such as the names of carbonyl compounds containing one or two carbon atoms.

Diol A compound containing two hydroxyl groups. A type of alcohol.

Disaccharide A type of carbohydrate derived from two monosaccharides.

Enantiomers A pair of nonsuperimposable mirror-image stereoisomers.

Endorphin An endogenous substance that binds to the opioid receptors in the brain, reducing the sensation of pain.

Ether A functional group derived from water where both of the —H atoms have been replaced with —R groups: R—O—R′. The carbon atom bonded to the oxygen can be alkyl or aromatic.

Ester A carbonyl containing functional group with an OR′ group bonded to the carbonyl carbon.

Fats Triglycerides. Triesters containing three long hydrocarbon chains.

Fatty acid A long straight-chain hydrocarbon with a carboxylic acid at one end.

Functional group A group of atoms and bonds that react in a characteristic way and give a molecule its physical properties. Most functional groups contain a heteroatom such as O, N, or P. Hydrocarbon functional groups include alkenes, alkynes, and aromatic hydrocarbons. Alkanes are not functional groups.

Heteroatom An atom that is not carbon or hydrogen.

Hydroxyl group The —OH group in alcohols, phenols, and carboxylic acids.

Inorganic phosphate Monohydrogen phosphate ion, HPO_4^{2-}, abbreviated P_i. The most abundant form of phosphoric acid in the cell.

Ketone A carbonyl containing functional group in which two R groups (carbon atoms) are bonded to the carbonyl carbon.

Mirror image The image seen when a molecule or object is held up to a mirror.

Monosaccharide The simplest type of carbohydrate, composed of many hydroxyl groups and either an aldehyde or a ketone. Glucose and galactose are common monosaccharides.

Nonsuperimposable When two objects or molecules are overlain and do not match up at every point.

Opioids Morphine and its derivatives that have analgesic properties.

Phenol A functional group composed of an aromatic ring bonded to a hydroxyl group.

Phosphate ester A functional group derived from hydrogen phosphate in which the hydrogen has been replaced by an —R group: $R-O-PO_3^{2-}$.

Phosphoanhydride bond The P—O—P bonds found in, diphosphate, and triphosphate esters. They are high-energy bonds because of the proximity of the negative charges in the phosphate groups.

Polyols Compounds containing many hydroxyl groups.

Protein Large biomolecules formed from amino acids, containing many amide functional groups.

Racemic mixture A 50:50 mixture of enantiomers.

Stereoisomers Compounds with the same chemical formula and the same chemical structure but a different three-dimensional orientation in space. Most stereoisomers are chiral.

Triols Compounds containing three hydroxyl groups—a type of alcohol.

Additional Exercises

Pain and the Opioid Analgesics

44 What is an analgesic?

45 What role do functional groups play in molecules?

46 Name a few opioid analgesics. What functional group do they have in common that is essential to their opioid properties?

47 Why is the distribution of morphine, heroin, and hydrocodone strictly monitored?

48 How does changing the phenol group in morphine to an ether group and an aromatic ring in codeine affect the analgesic properties of codeine compared to those of morphine? What other pharmacological property is affected by these functional group changes?

49 What happens when an opioid binds to an opiate receptor in the brain?

50 Why did humans develop endogenous (internal) analgesics and their corresponding receptors in the brain?

51 Name two types of molecules that bind to opioid receptors in the brain.

52 What impact does a heteroatom, such as O, N, or P, have on the physical properties of a molecule?

Alcohols and Ethers

53 What is the structural difference between an ether and an alcohol?

54 Are 1°, 2°, or 3° alcohols. Rewrite alcohols (a) through (c) as condensed structures.

55 Write a skeletal line structure for the following alcohols and indicate whether they are 1°, 2°, or 3° alcohols.
  a. nonan-1-ol
  b. butan-2-ol
  c. 3-methylheptan-3-ol

56 Which of the following molecules are alcohols? Indicate whether the alcohols are 1°, 2°, or 3°. Which molecule is a diol? Which molecule is a polyol?

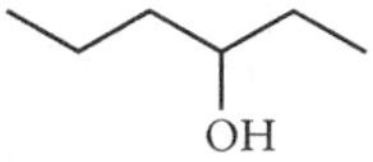

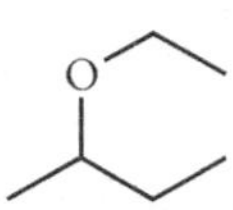

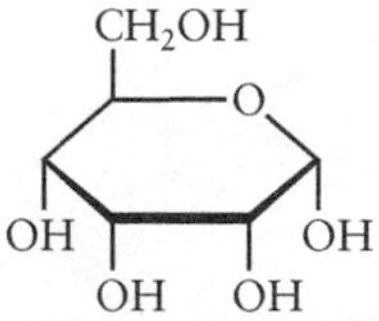

57 Draw the skeletal line structure for butan-1-ol. What is the molecular shape around the oxygen atom? What is the C—O—H bond angle?

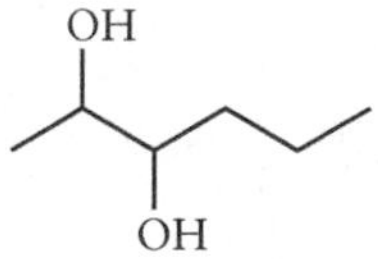

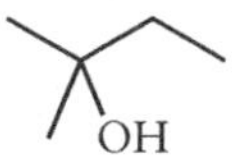

58 Which of the following structures are ethers?

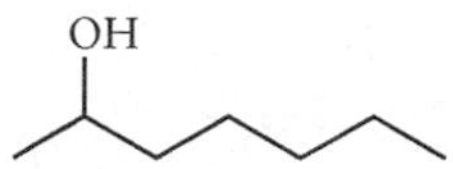

59 Estradiol is a sex hormone and betamethasone is a drug sometimes given to pregnant women to help mature the lungs of a fetus in cases where the child is expected to be born prematurely, as in the case of twins and triplets. The structures for both hormones are shown below. Highlight the alcohol and phenol functional groups. How is a phenol different from an alcohol? What common features do these two compounds have?

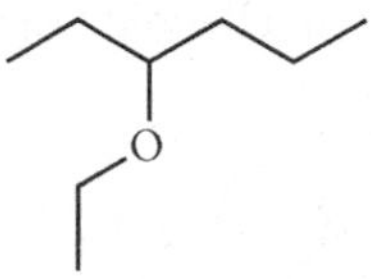

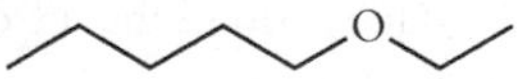

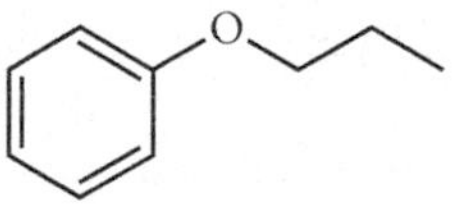

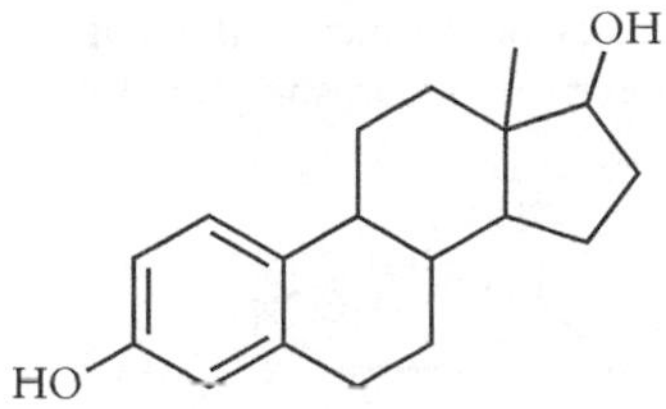

Estradiol

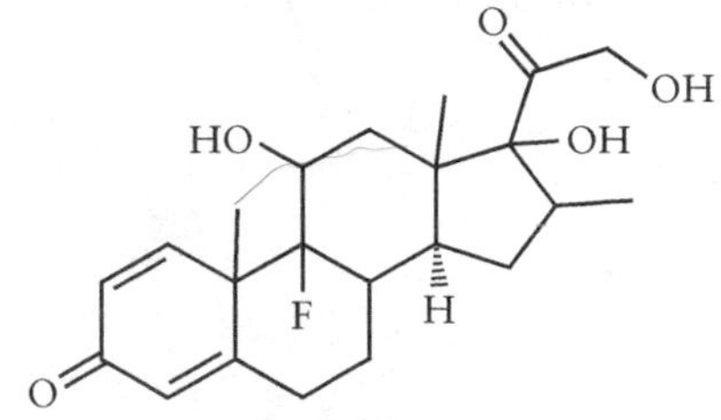

Betamethasone

60 Which of the following compounds do you expect to have the higher boiling point? Explain your answer.

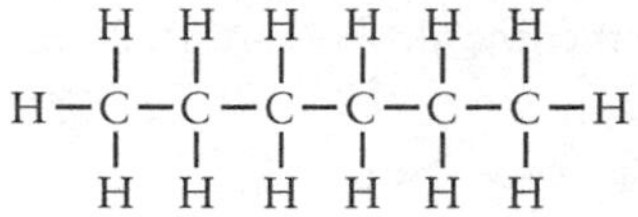

61 Arrange the following compounds in order of increasing boiling point.

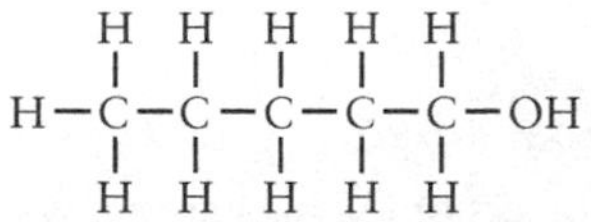

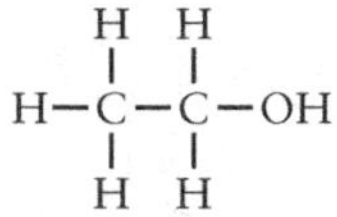

62 Ferulic acid, shown below, is an antioxidant found in coffee, rice, wheat, and oats. What functional group in ferulic acid is responsible for its antioxidant properties? What is an antioxidant?

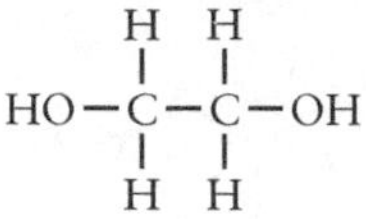

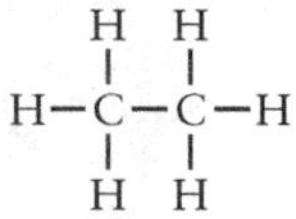

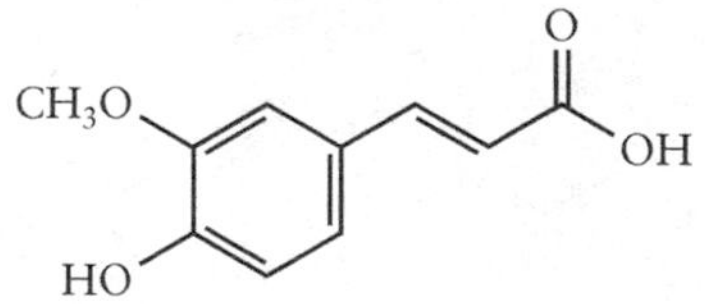

63 Imdur (isosorbide mononitrate), shown below, is used to treat angina. It dilates the blood vessels to reduce blood pressure.

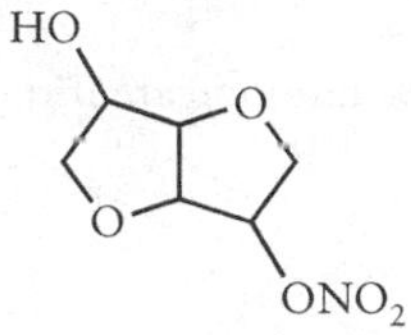

  a. Highlight and identify the ether functional groups.
  b. Highlight and identify the alcohol functional group.

64 Circle and identify the alcohol functional groups in sucrose, also known as table sugar, shown below.

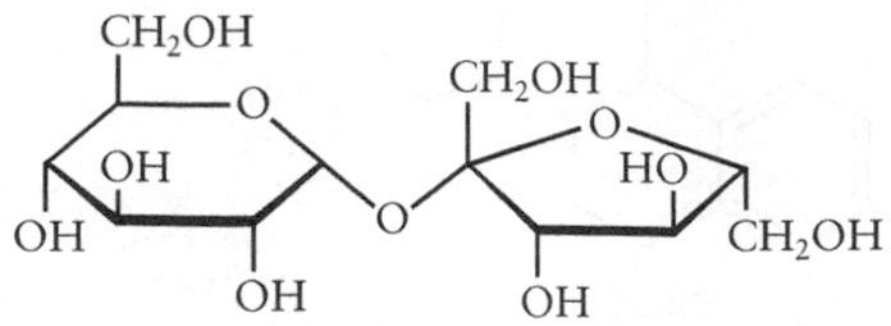

65 Provide the IUPAC name for the following ethers.

a.

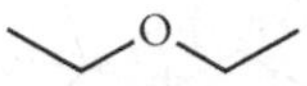

b.

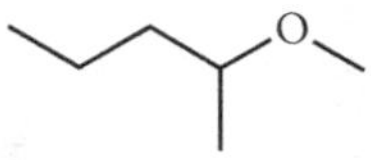

c.

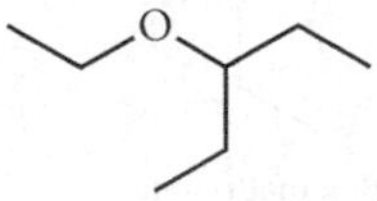

66 Write the skeletal line structures for the following ethers, alcohols, and phenols:
a. 1-methoxypropane
b. 2-ethoxybutane
c. 3-ethylphenol
d. 2-methylpropan-2-ol

67 All seven structural isomers of $C_4H_{10}O$ are shown below. Identify which molecules are alcohols and which are ethers. Indicate whether the alcohols are 1°, 2°, or 3°.

a.   b.

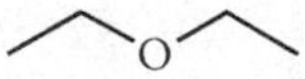

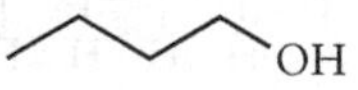

c.    d.

e. 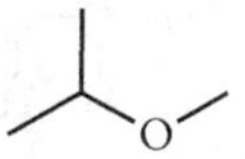   f. 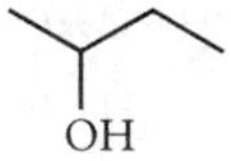   g.

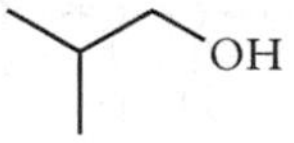

Amines

68 Indicate whether the following amines are 1°, 2°, or 3°. Name each of the amines.

a.   b.

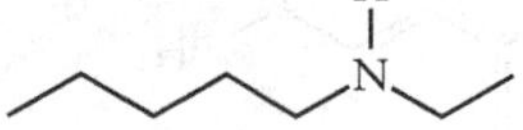

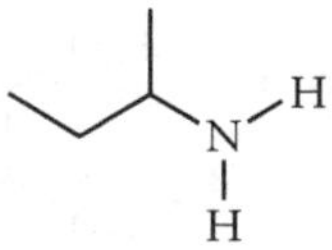

c. 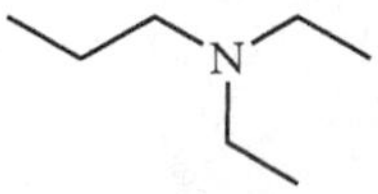   d.

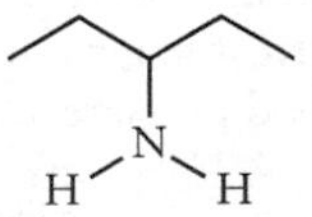

69 Name the amines and indicate if they are 1°, 2°, or 3°.

a.   b.

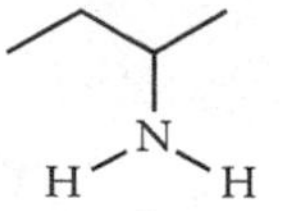

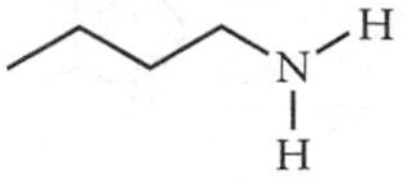

c.

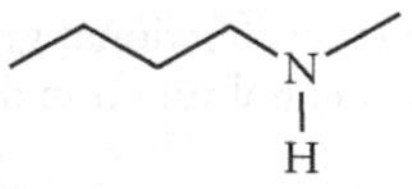

70 Which of the following is a 2° amine? Which amine(s) has/have hydrogen-bonding intermolecular forces of attraction?

a.   b.

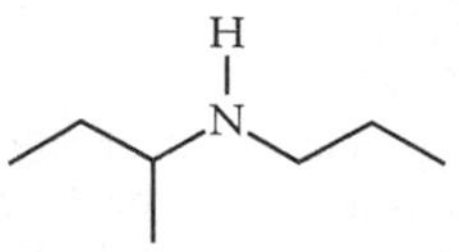

c.

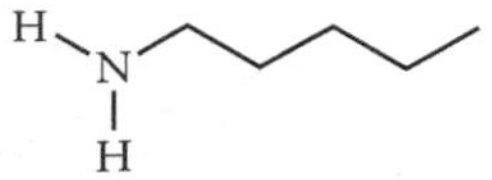

71 Which of the following is a 1° amine? Which amine(s) has/have hydrogen-bonding intermolecular forces of attraction?

a.   b.

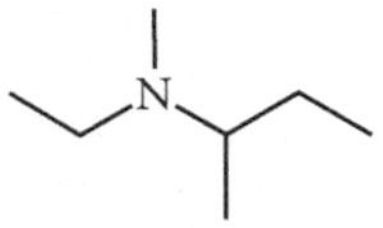

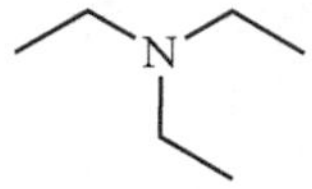

c.

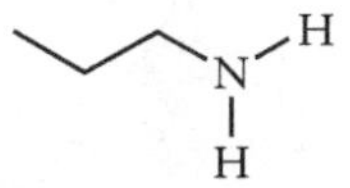

72 Write the polyatomic cation form of the following amines:

a.   b.

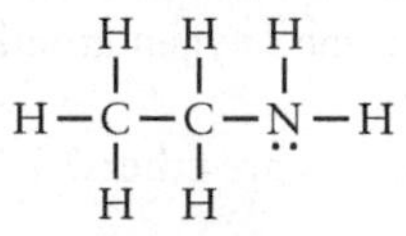

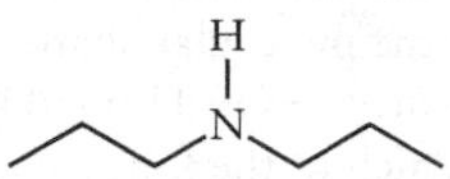

73 Cuscohygrine, shown below, is an alkaloid found in the coca plant.

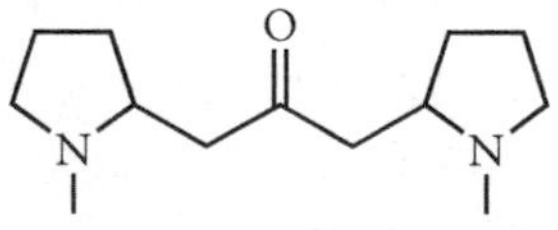

a. Highlight and identify the amine functional groups.
b. Are the amines 1°, 2°, or 3°?
c. What is the molecular shape around the nitrogen atoms in the amines?
d. What is the C—N—C bond angle in the amine?
e. Give the two reasons this compound classified as an alkaloid.

74 Benadryl, shown below, is an over-the-counter antihistamine.

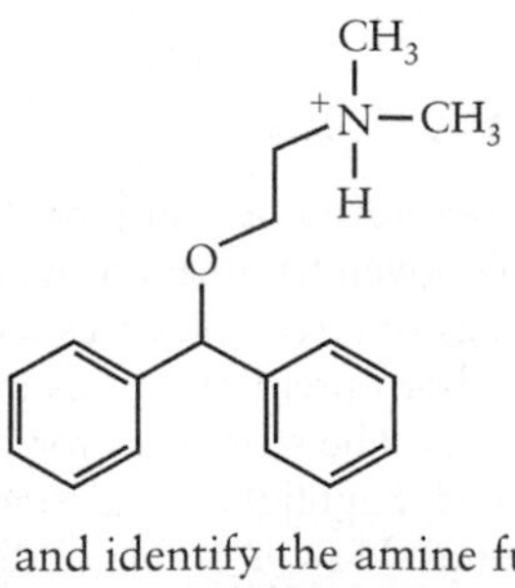

a. Highlight and identify the amine functional group.
b. Is the amine in Benadryl in its neutral or ionic form?
c. What other functional groups are present in Benadryl?

75 Zoloft, shown below, is an antidepressant:

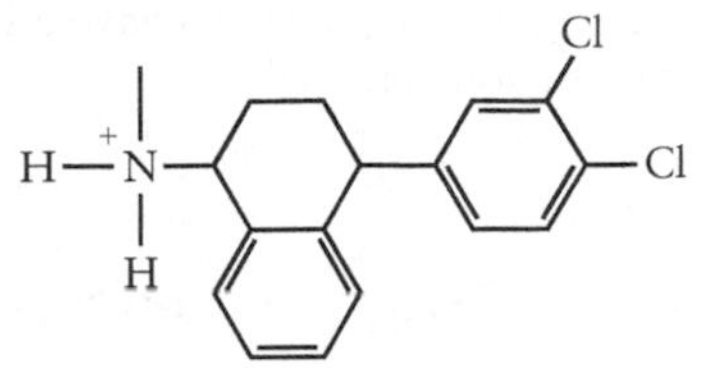

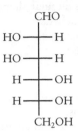

  a. Highlight and identify the amine functional group.
  b. Is the amine in Zoloft in its neutral or ionic form?
  c. Is the amine in Zoloft primary (1°), secondary (2°), or tertiary (3°)?

76 Tamoxifen, shown below, is a drug used to treat breast cancer.

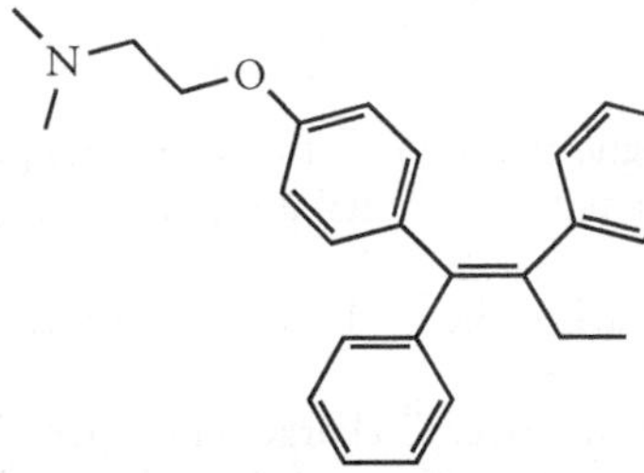

  a. Highlight and identify the amine functional group.
  b. How many —R groups are attached to the nitrogen atom? Is it a primary (1°), secondary (2°), or tertiary(3°) amine?
  c. Is the amine in its neutral or ionic form?
  d. What other functional groups does Tamoxifen contain?

Carbonyl-Containing Functional Groups

77 Which of the following compounds are aldehydes? Which of the following compounds are ketones? Provide the IUPAC name for each compound.

  a.

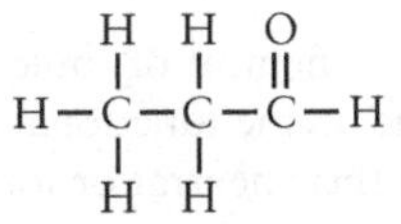

  b.

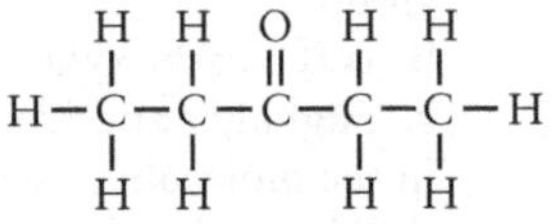

  c.

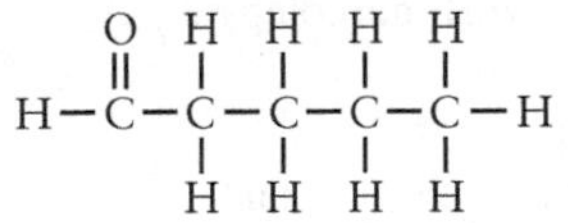

  d.

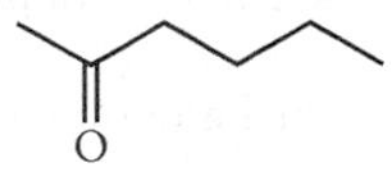

  e.

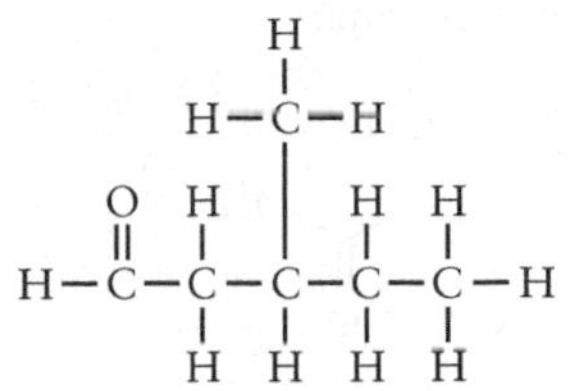

78 Write the condensed structural formulas for the compounds in exercise 77.

79 Mannose is a monosaccharide and is shown in its open-chain form below:

  a. Highlight and identify the carbonyl-containing functional group. Is it an aldehyde or a ketone?
  b. What is the molecular shape around the carbonyl carbon?
  c. What other functional groups are present in mannose?
  d. To which of the following classes of biomolecules does mannose belong: protein, carbohydrate, or fat?

80 Provide the common name for each of the compounds shown below. Which one is a ketone?

  a.   **b.**   **c.**

81 Provide the IUPAC name for each of the molecules shown below. Which has a higher boiling point? What types of intermolecular forces of attraction are present in each molecule?

82 TriCor (fenofibrate), shown below, is used to lower cholesterol levels in patients at high risk of developing cardiovascular disease.

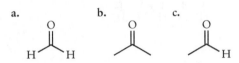

  a. Highlight and identify the ketone functional group. How are you able to distinguish it from an aldehyde?
  b. What is the molecular shape around the carbonyl carbon?
  c. Highlight and identify the ether and the aromatic ring functional groups.

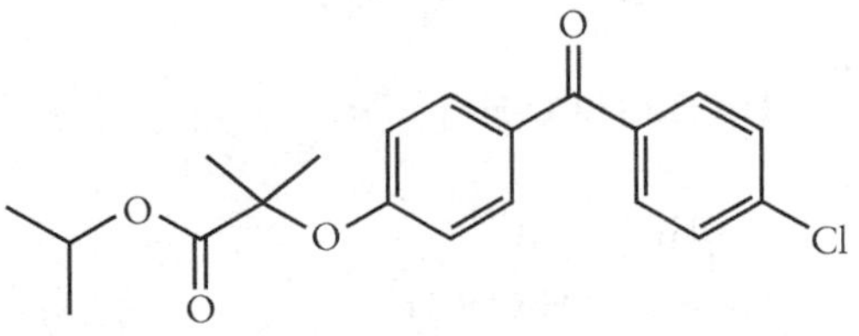

83 Cinnamaldehyde, shown below, occurs naturally in the bark of cinnamon trees and gives cinnamon its flavor:

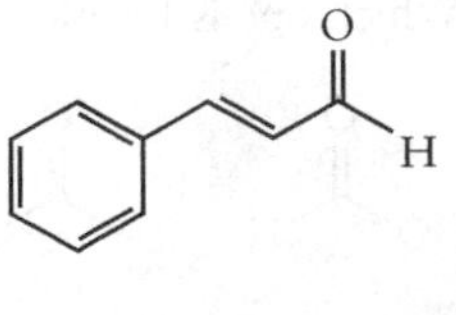

a. Highlight and identify the aldehyde functional group.
b. What is the molecular shape around the carbonyl carbon of the aldehyde?
c. Circle and identify the double bond. Is the double bond cis or trans?

Carboxylic Acids

84 Describe the difference between the following functional groups:
a. a carboxylic acid and an ester
b. a carboxylic acid and an amide
c. an amide and an amine
d. a carboxylic acid and an alcohol

85 Provide the IUPAC name for the following carboxylic acids:

a.
$CH_3CH_2CH_2CH_2CO_2H$

b.

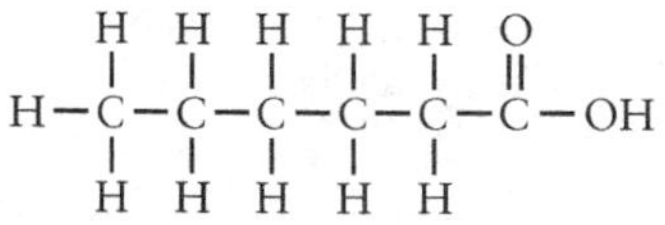

c.

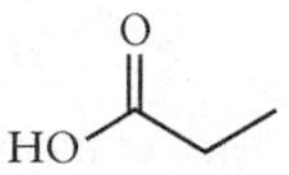

d.

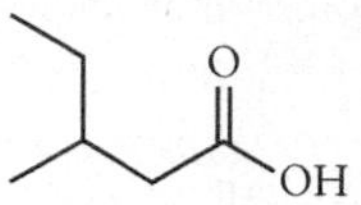

86 Draw the skeletal line structure for the following carboxylic acids:

a.
$CH_3CH_2CH_2CH_2CH_2CO_2H$

b.
$CH_3CH_2CH_2CO_2H$

c.
CH_3
$CH_3CHCH_2CH_2COOH$

87 Draw the skeletal line structure for the following carboxylic acids:
a. pentanoic acid
b. 2-methylpropanoic acid
c. benzoic acid

88 What ions are formed when a carboxylic acid reacts with water? What characteristic of a carboxylic acid makes it an "acid"?

89 Two carboxylic acids are shown below. Which one is acetic acid? Which one is formic acid?

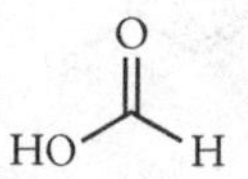

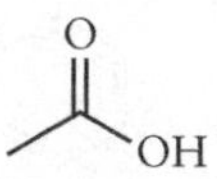

90 Provide an IUPAC name for the following molecules. Which one would you expect to have a higher boiling point? Explain your answer.

a. **b.**

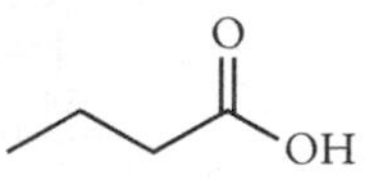

91 The amino acid serine is shown below.

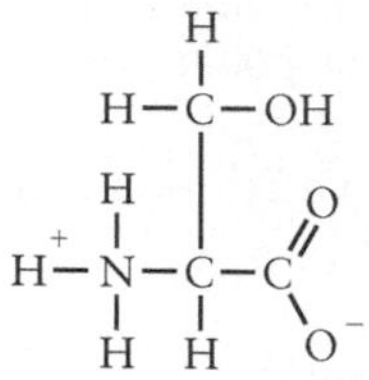

a. Highlight and identify the functional groups.
b. Is the amine functional group in its neutral or ionic form?
c. Is the carboxylic acid functional group in its neutral or ionic form?
d. What is the overall charge on serine?

92 Zyrtec (cetirizine), shown below, is an antihistamine used as an allergy medicine.

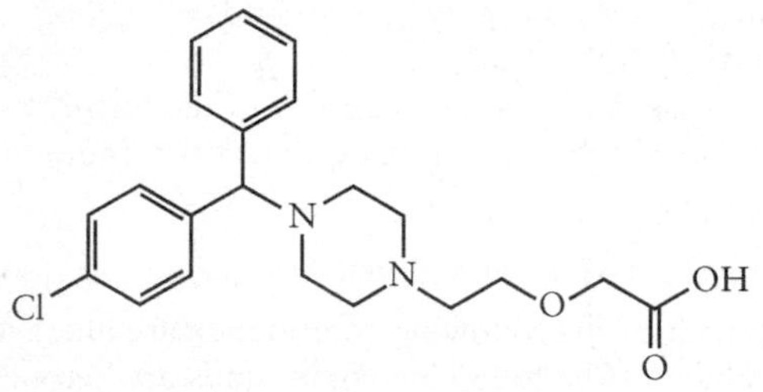

a. Highlight and identify the carboxylic acid functional group.
b. Is the carboxylic acid in its neutral or ionic form?
c. Highlight and identify the amine functional groups in the molecule. Are they in their neutral or ionic form?
d. What other functional groups are present in Zyrtec?

93 Xyrem, shown below, is used to treat extreme daytime sleepiness for people with narcolepsy.

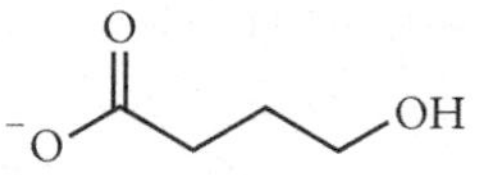

a. Highlight and identify the functional groups present in Xyrem.
b. Is the carboxylic acid functional group in its ionic or neutral form?

94 Tartaric acid, shown below, is one of the main carboxylic acids found in wine.

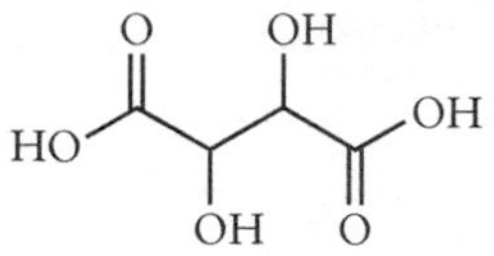

a. Highlight and identify the carboxylic acid functional groups in tartaric acid.
b. What other functional group is present in tartaric acid?

95 The structure of Lyrica, used to treat fibromyalgia, is shown below.

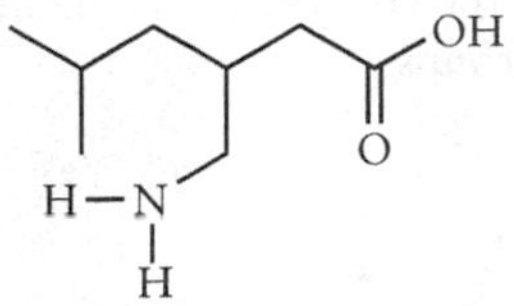

a. Highlight and identify the carboxylic acid functional group in Lyrica.
b. What other functional group is present in Lyrica?
c. Is the carboxylic acid in its neutral or ionic form?
d. Is the other functional group in its neutral or ionic form?

Esters

96 Provide the IUPAC name for the following esters:

a.
$$CH_3CH_2CO_2CH_2CH_3$$

b.
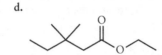

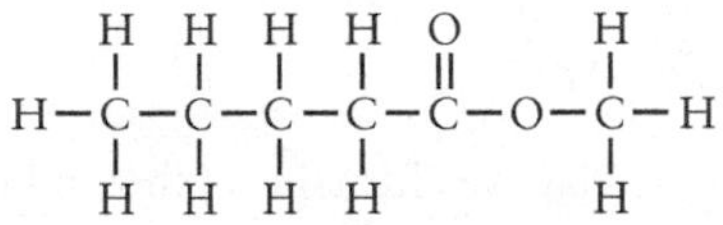

c. 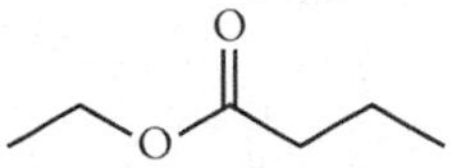　　　　　　　　**d.**

97 Write the skeletal line structure of the following esters:
a. propyl propanoate
b. methyl butanoate
c. ethyl 2,3-dimethylhexanoate

98 Octyl ethanoate (octyl acetate) has the fragrance of oranges. Draw the skeletal line structure of this compound.

99 Provide the IUPAC name of the ester shown below.

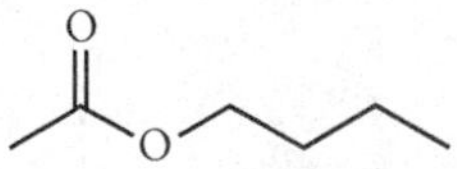

100 Provide the IUPAC name for the two molecules below. Which one has the lower boiling point? What intermolecular forces of attraction are present in each molecule?

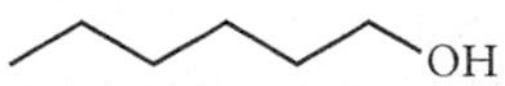

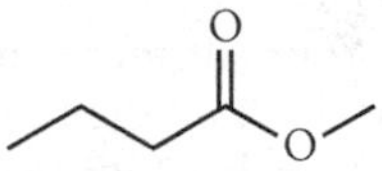

101 A triglyceride is shown below. Highlight and identify the functional groups that are present in all triglycerides. What is a common term for a triglyceride?

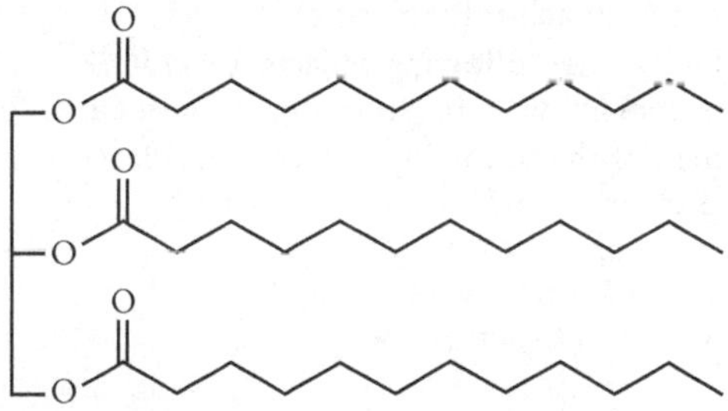

102 Vioxx, a nonsteroidal anti-inflammatory drug (NSAID), was taken off the market by the pharmaceutical company Merck because of concerns that it increases the risk of heart attacks and strokes. Circle and identify the ester in Vioxx.

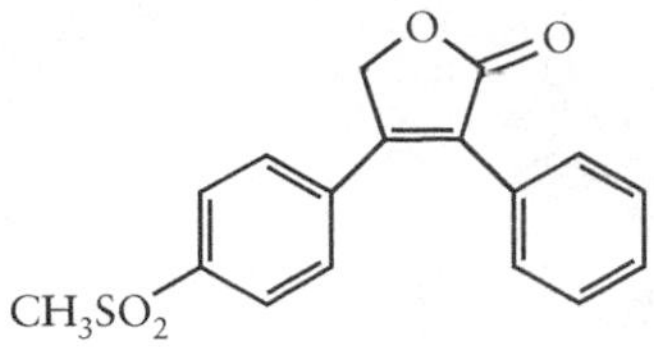

$$CH_3SO_2$$

103 Novocain, shown below, is used as a local anesthetic, particularly in dentistry.

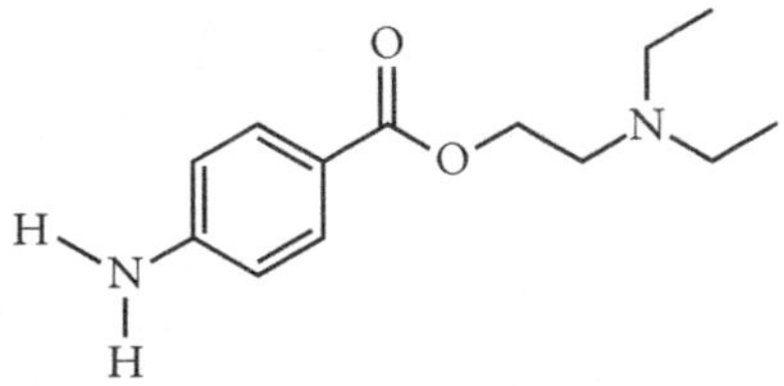

Highlight and identify all the functional groups in Novocain.

104 Concerta, shown below, is a central nervous system stimulant used to treat attention deficit hyperactivity disorder in children over the age of six, adolescents, and adults.

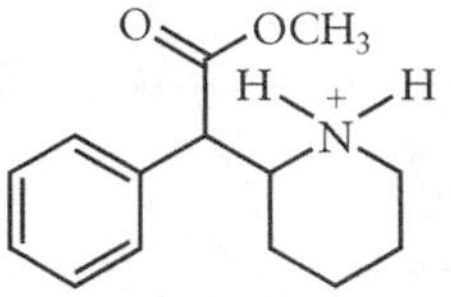

a. Highlight and identify all the functional groups in Concerta.
b. Identify whether the amine is in its neutral or ionic form.

105 Plavix, shown below, is used to treat acute coronary syndrome. Patients with a recent history of myocardial infarction or stroke are also treated with Plavix. Highlight and identify the following functional groups: the amine, the ester, and the aromatic ring. Is the amine in its neutral or ionic form?

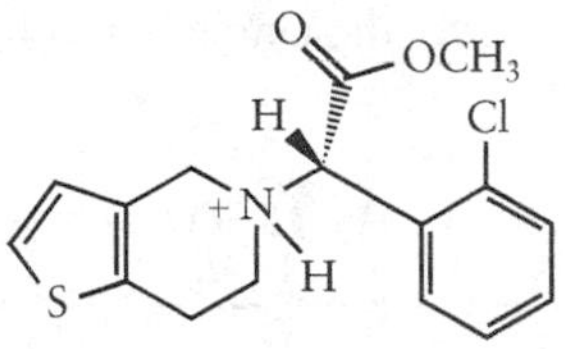

Amides

106 Write the condensed structural formulas for the following amides:

a.

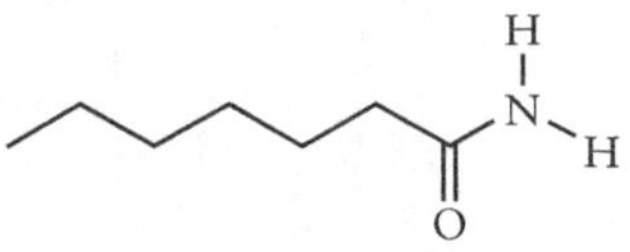

b.

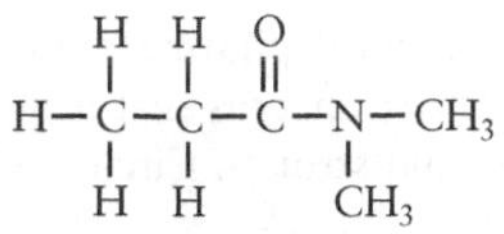

c.

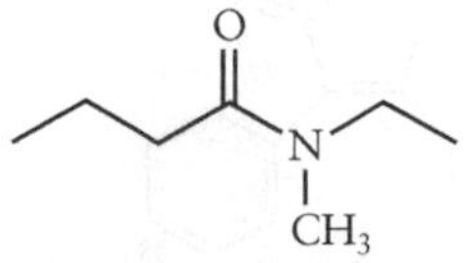

107 Which of the following compounds are amides?

a.

$CH_3CH_2CH_2NHCH_2CH_3$

b.

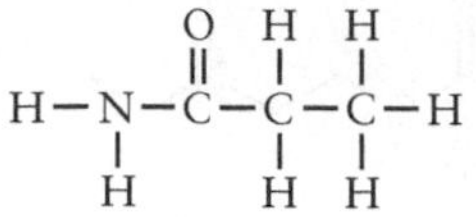

c.

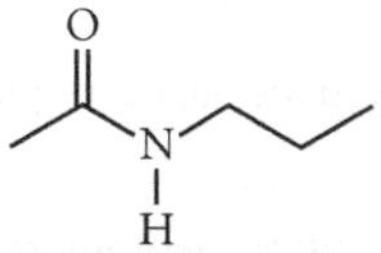

108 Provide the IUPAC name for the following amides:

a.

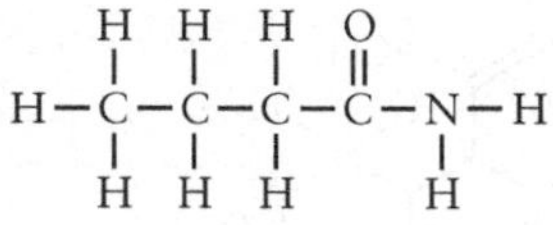

b.

$CH_3CH_2CH_2CH_2CH_2CONH_2$

c.

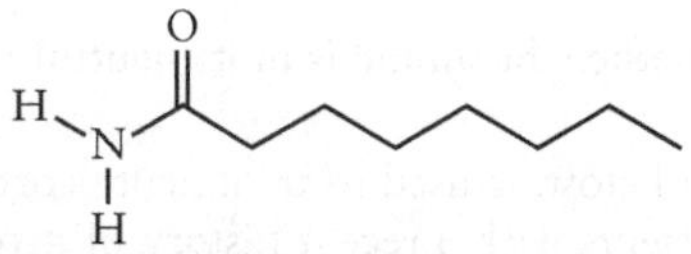

109 The body uses the peptide GHK to promote wound healing. GHK is also used in the cosmetics industry as an anti-aging ingredient. GHK is prepared from the amino acids glycine, histidine, and lysine. Circle and identify the amide bonds in GHK, shown below:

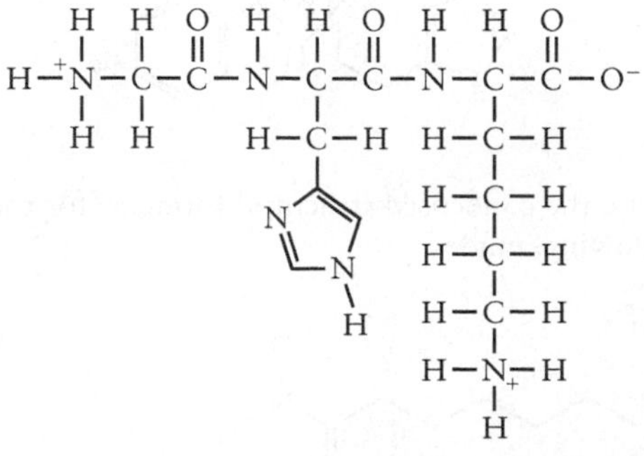

110 Capsaicin, one of the molecules that contributes to the hotness of chili peppers, is shown below. Highlight and identify the ether, the phenol, and the amide functional groups.

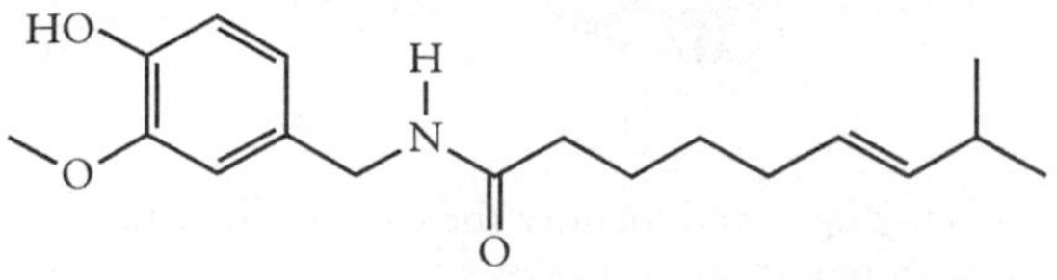

111 Ambien (zolpidem) is used as a sleeping pill. It is fast acting (15 minutes) and eliminated from the body quickly. The structure is shown below. Highlight and identify the amide functional group.

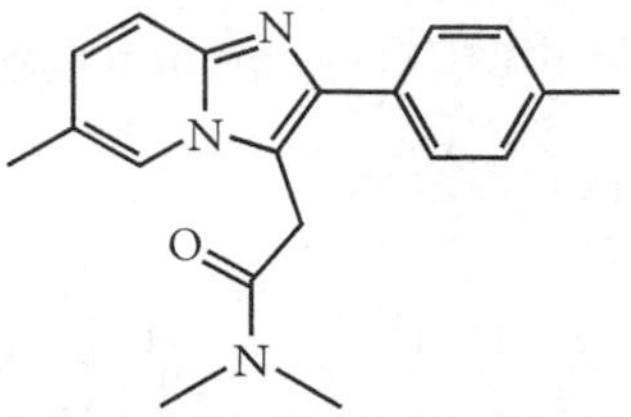

112 The structure of Viagra (sildenafil) is shown below. Highlight and identify the following functional groups: the amide, the ether, and the amine.

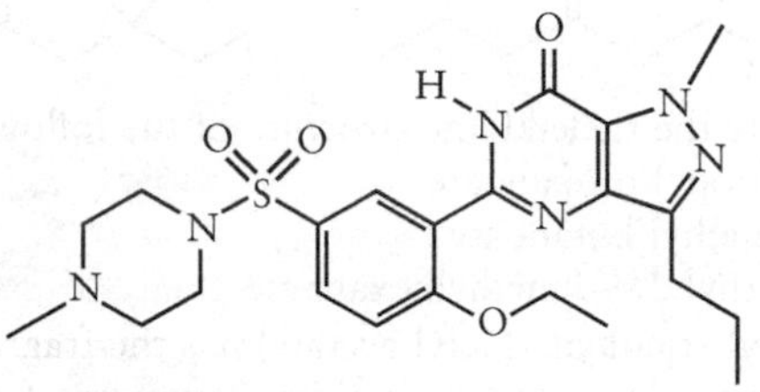

113 What functional groups can be prepared from carboxylic acids?

Stereoisomers

114 Indicate whether the following statements are true or false:
 a. Stereoisomers have different chemical formulas.
 b. A chiral object is superimposable on its mirror image.
 c. Your left foot is nonsuperimposable on your right foot because the foot is chiral.
 d. Your left foot is the mirror image of your right foot.
 e. Enantiomers are a pair of nonsuperimposable mirror-image stereoisomers.

115 What is an achiral molecule?

116 Which of the following objects are chiral?
 a. a corkscrew   **b.** an orange   **c.** a car   **d.** a nail

117 Which of the following objects are chiral?
 a. a chair
 b. a spiral staircase
 c. a conch shell
 d. your right foot

118 Indicate whether the following pairs of molecules are enantiomers:

a.

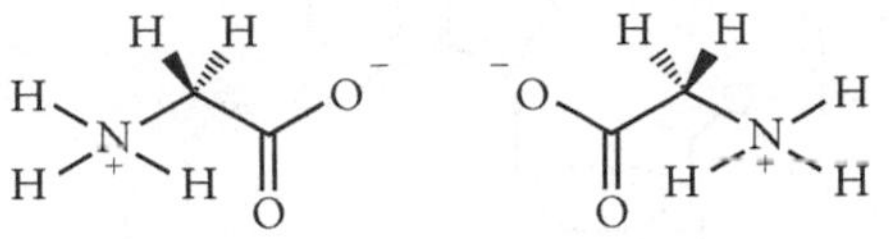

b.

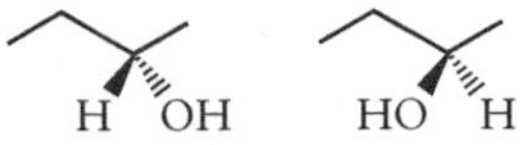

119 Indicate whether the following pairs of molecules are enantiomers:

a.

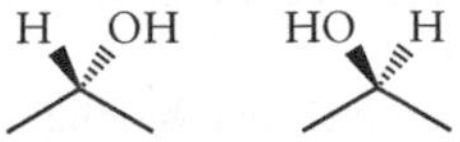

b.

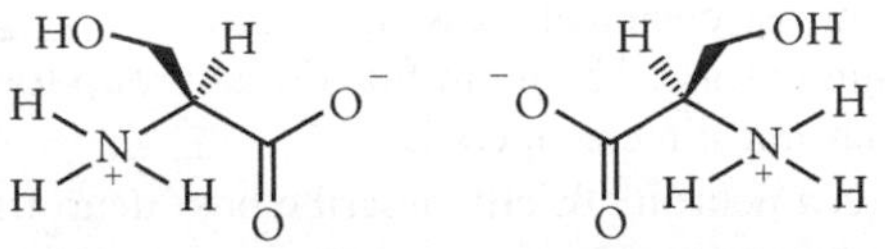

120 Floxin, shown below, is a broad-spectrum antimicrobial agent sold as a racemic mixture. What does this mean? Why are two structures shown for Floxin? What type of stereoisomers are these two compounds?

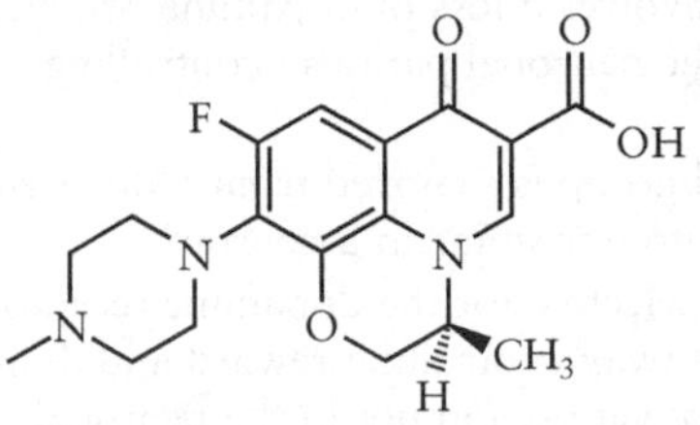

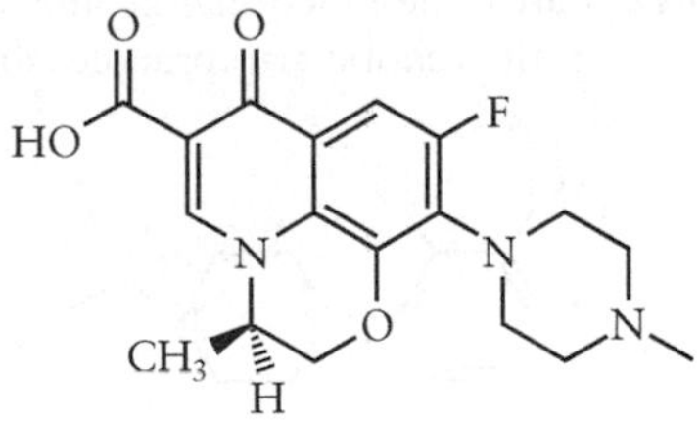

121 D-ethambutol is used to treat tuberculosis. L-ethambutol causes blindness. Should ethambutol be sold as a racemic mixture? Explain why or why not.

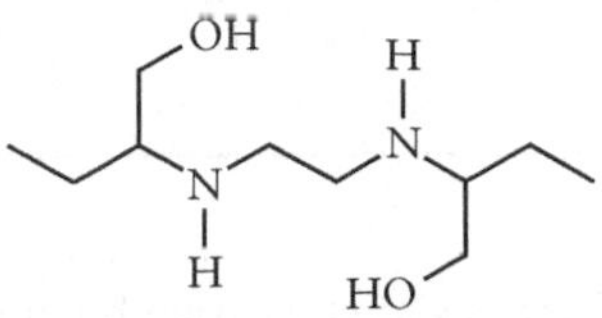

Ethambutol

122 Lexapro contains only the S-enantiomer of citalopram, while Celexa is a racemic mixture of citalopram. Both Celexa and Lexapro are used to treat depression. How is Celexa different from Lexapro?

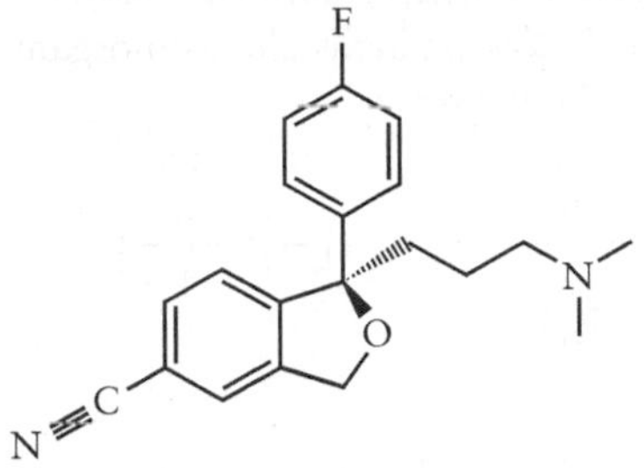

Lexapro

123 Identify the center of chirality in Lexapro. What is the molecular shape and what are the bond angles around a center of chirality?

124 How can you identify a center of chirality in a molecule?

125 There are two enantiomers of carvone. R-carvone smells like spearmint, whereas S-carvone smells like caraway seeds. The structures for both enantiomers are shown below. Identify the center of chirality in both structures. How are these enantiomers different structurally?

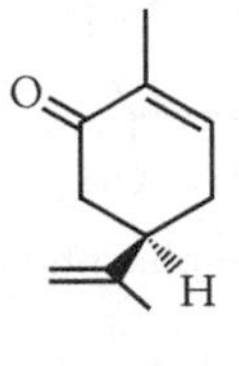

R-carvone 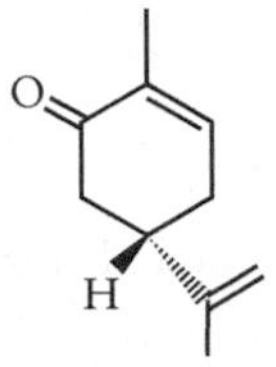   **S-carvone**

126 The structure of L-isoleucine, an amino acid, is shown below. Draw D-isoleucine.

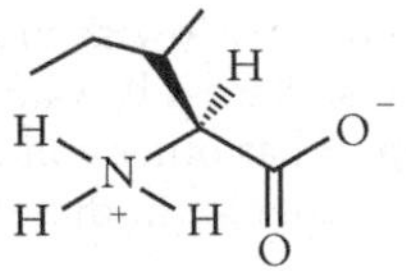

L-isoleucine

127 R-Limonene smells like oranges, whereas its enantiomer smells like lemons. The structure of R-limonene is shown below. Draw S-limonene. Why are most people able to distinguish these two compounds by smell?

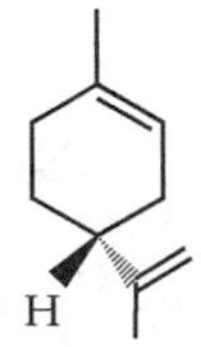

R-limonene

Phosphate Ester Functional Groups

128 Write a Lewis dot structure for the most common form of phosphoric acid in the body and answer the following questions about its structure.

a. What is this ion called? What is its abbreviation?

b. How many electrons surround the central phosphorus atom?

c. What is the molecular shape around the phosphorus atom? Explain your answer.

129 Label the following molecules as inorganic or organic. Explain your answer.

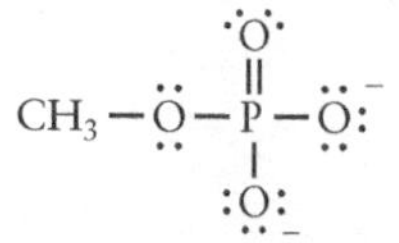

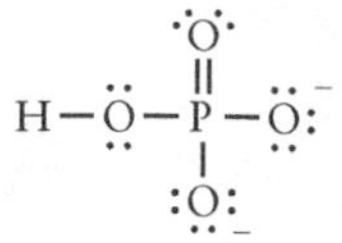

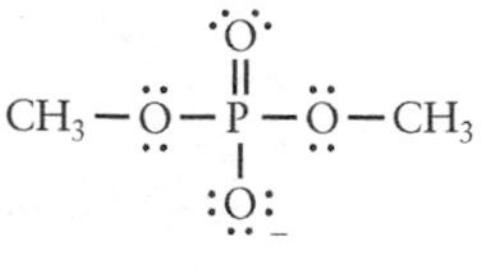

130 Which compound in the previous problem is a phosphate ester? Which is a phosphate diester?

131 The structure of adenosine diphosphate, ADP, is shown below:

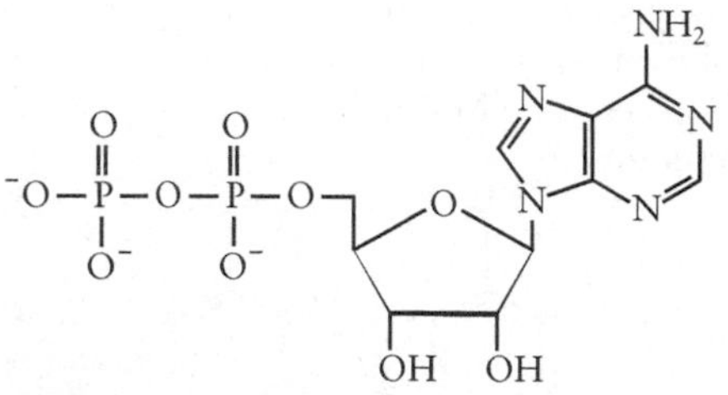

a. Highlight and identify the phosphoanhydride bonds.

b. Why are phosphoanhydride bonds so important in biochemistry?

c. Is this compound a monophosphate ester, a diphosphate ester, or a triphosphate ester? Explain.

132 Identify the structures below as mono-, di-, or triphosphate esters. Place a box around each phosphate group. What other common functional group(s) do you recognize in these molecules?

a.

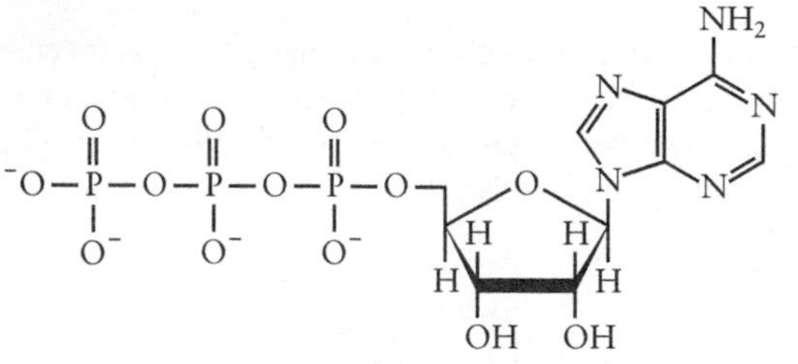

ATP

b.

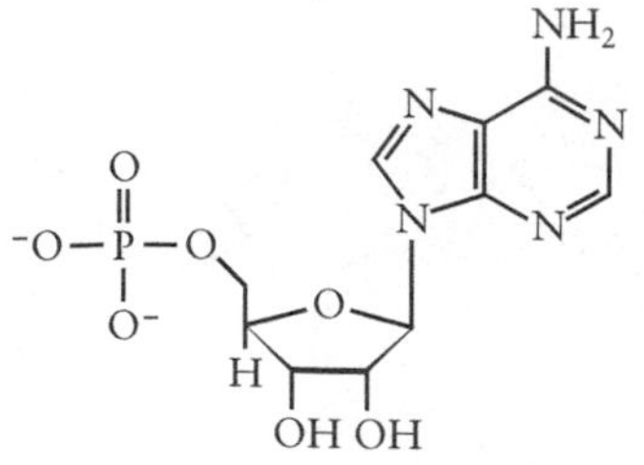

AMP

Chemistry in Medicine: The Role of Dopamine in Parkinson's Disease and Schizophrenia

133 How is dopamine activity different in the brain of someone who has schizophrenia compared to someone who has Parkinson's disease?

134 Write the structure of dopamine, then highlight and label all its functional groups.

135 Neurons use chemical messengers called _____ to communicate. The main functional group found in these chemical messengers is _____.

136 What is a neuron? Briefly describe how neurons communicate with other neurons.

137 Explain why dopamine cannot be administered as an oral drug.

138 How are antipsychotic drugs able to reduce the symptoms of schizophrenia (i.e., what do they do chemically)?

139 Which disease involves a loss of dopamine-producing neurons along the neuronal pathway controlling movement?

140 What functional group is removed from L-dopa when it is converted into dopamine in a neuron?

141 What drugs are selective for the dopamine receptors in the neuronal pathway controlling reward and desire, memory, and motivation and not in the neuronal pathway controlling movement?

142 Circle the functional groups in the atypical antipsychotic aripiprazole, shown below:

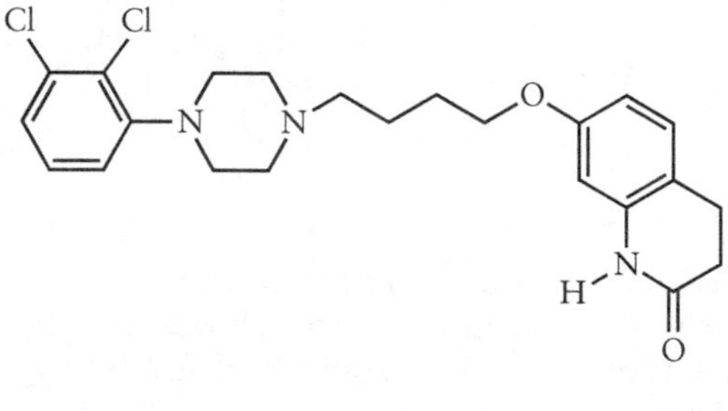

Answers to Practice Exercises

1 The molecular shape is *bent* around the oxygen atom of an alcohol and an ether. The bond angles are approximately 109.5° because they both have a tetrahedral electron geometry (two bonding groups and two nonbonding groups). Alcohols and ethers have the same molecular shape around oxygen as water.

2 **a.** secondary alcohol
   b. primary alcohol
   c. secondary alcohol
   d. tertiary alcohol

3 **a.**

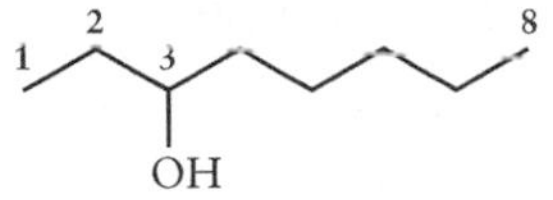

   b.

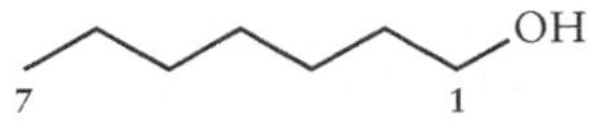

   c.

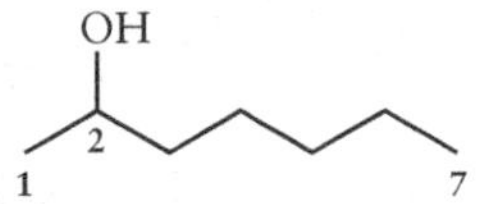

4 Hydrogen bonding increases the boiling point of an alcohol compared to another compound of similar molecular mass but weaker intermolecular forces.

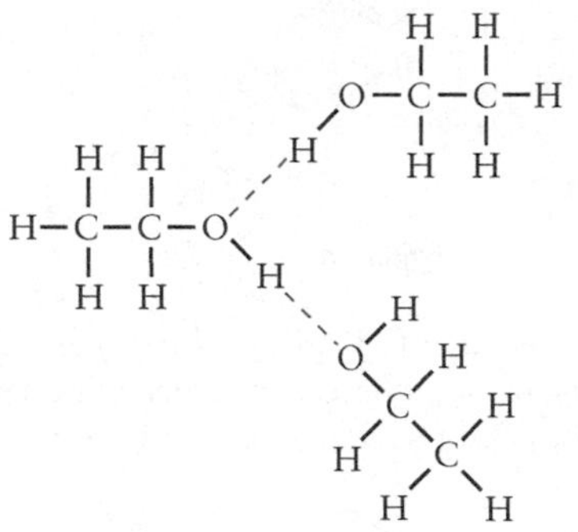

5 Monosaccharides belong to a class of biomolecules known as <u>carbohydrates</u>, which have many <u>alcohol (or hydroxyl)</u> functional groups.

6 The structure shown is a diol because it contains two hydroxyl groups. This compound has a higher boiling point than propan-1-ol because it has *two* hydroxyl groups, whereas propan-1-ol has only one, so there is more hydrogen bonding intermolecular forces of attraction.

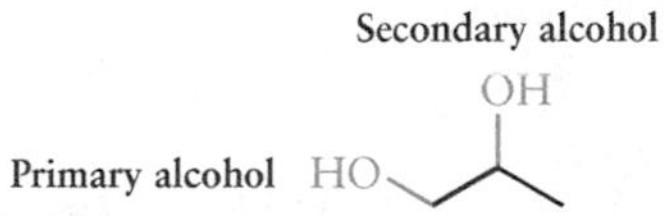

7 **b.** The ending *-ol* in tetrahydrocannabin*ol* indicates an alcohol or a phenol is present.

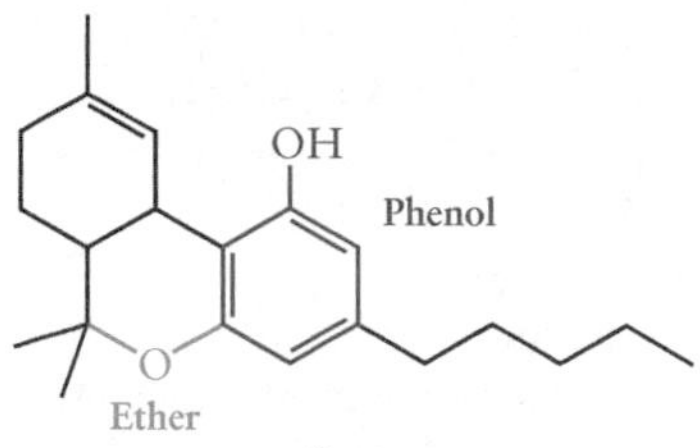

Δ^9-THC

8 **a.**

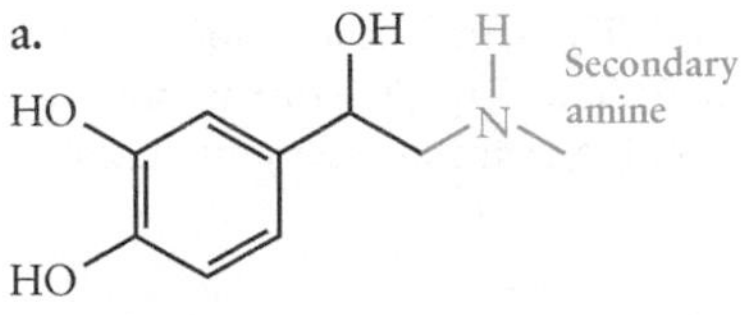

Epinephrine (adrenaline)

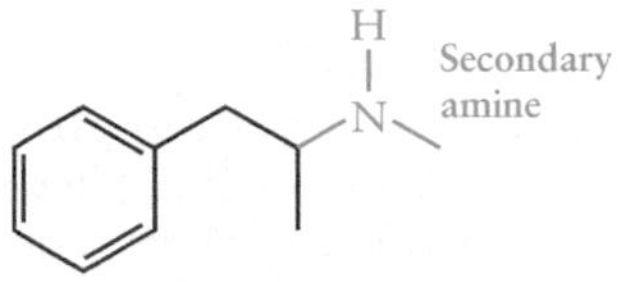

Methamphetamine

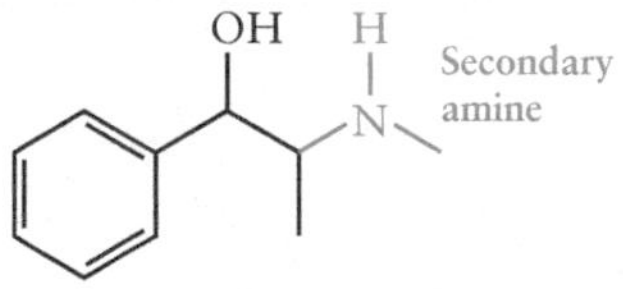

Pseudoephedrine

b. Methamphetamine is similar to pseudoephedrine in having an aromatic ring and a carbon chain with a secondary amine and a methyl group on the carbon chain. It is different in that pseudoephedrine also has a hydroxyl group on the carbon chain but methamphetamine does not.
c. The molecular shape around an amine is trigonal pyramidal.
d. Epinephrine contains a phenol.
e. The ending *-amine* or *-ine* indicates they are amines.

9 **a.** secondary   **b.** primary   **c.** primary   **d.** tertiary

10 **b.** butan-1-amine
   c. pentan-2-amine
   d. N,N-dimethylpropan-2-amine

11 Each amine has the formula C_3H_9N, but the atoms are connected in different arrangements, so they are structural isomers. The tertiary amine (a) would have the lowest boiling point because it does not have an N—H bond and therefore cannot hydrogen bond like the other two amines, so it requires less energy to enter the gas phase.

12 This compound is an alkaloid because it contains an amine and is produced by a plant.

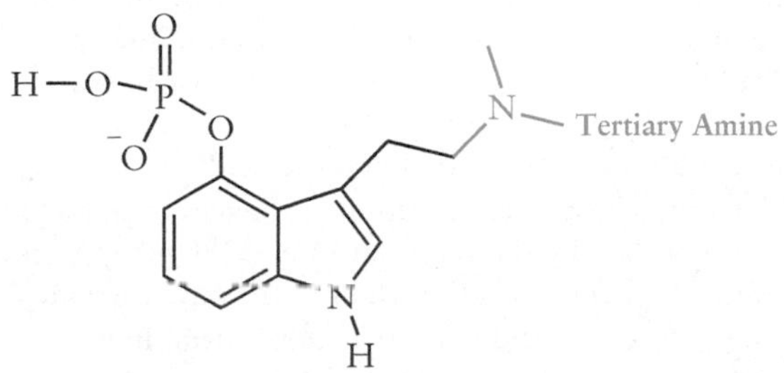

13 The H—C—H bond angle in formaldehyde is 120°. The molecular shape is trigonal planar.

14 The *-al* ending in hex-3-en*al* suggests it is an aldehyde. The other functional group is an alkene (a double bond).

It is a cis double bond. The -ene in the name hex-3-enal suggests it is an alkene. The 3 is a locator number giving the position of the double bond in the main chain.

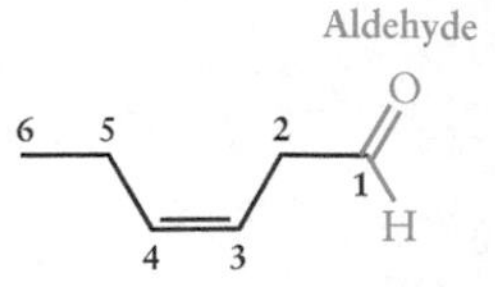

hex-3-enal

15 Pentane, a hydrocarbon (dispersion forces only), has a lower boiling point than butanal, an aldehyde (dipole-dipole forces), which has a lower boiling point than butan-1-ol, an alcohol (hydrogen bonding).

16 a. propanal　　b. octan-3-one　　c. cyclohexanone
　　d. butanal　　e. propan-2-one (common name: acetone)

17

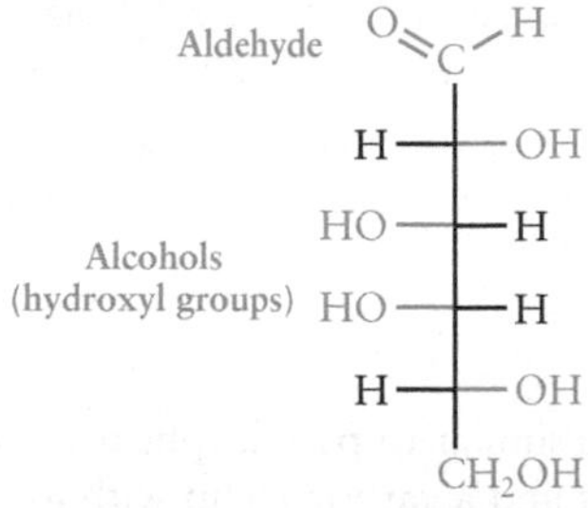

Galactose

18

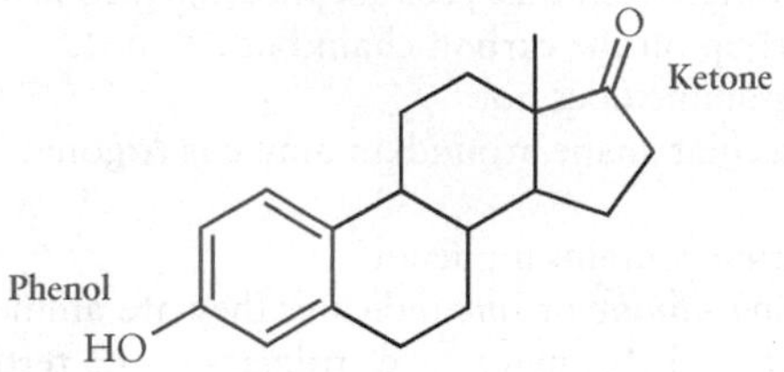

Estrone

19 $C_3H_6O_3$. Dihydroxyacetone and glyceraldehyde are structural isomers.

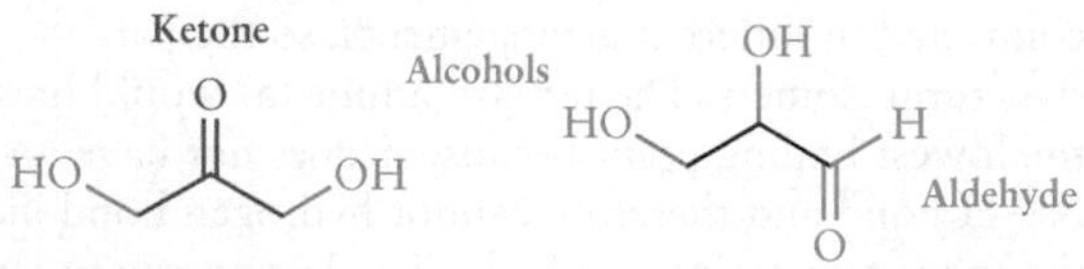

20 They all contain a hydroxyl group, —OH, but in a carboxylic acid it is attached to a carbonyl carbon, in a phenol it is attached to an aromatic ring, and in an alcohol it is attached to an alkyl carbon. They are readily distinguished by looking at the carbon atom bearing the —OH group.

21 a. pentanoic acid　　b. butanoic acid

22 The other functional group in citric acid is an alcohol, distinguished by the fact that the —OH group is bonded to an alkyl carbon rather than a carbonyl carbon.

23 The carboxylic acid is in its neutral form. In a carboxylic acid the —OH is directly bonded to a carbonyl carbon, whereas in a phenol it is directly bonded to an aromatic ring. They are similar only in that they both have an —OH group.

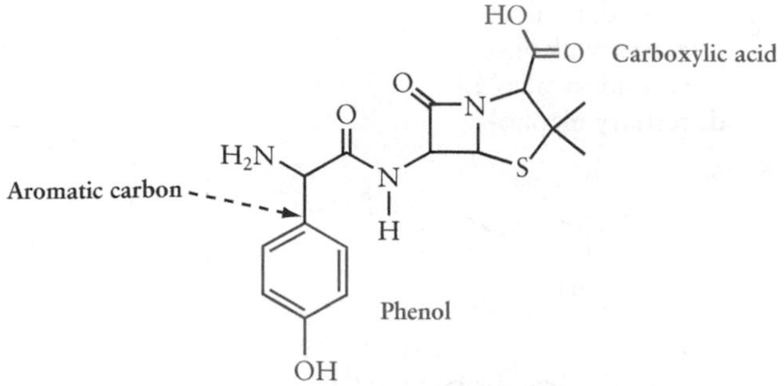

Amoxicillin

24 The carboxylic acid is in its neutral form:

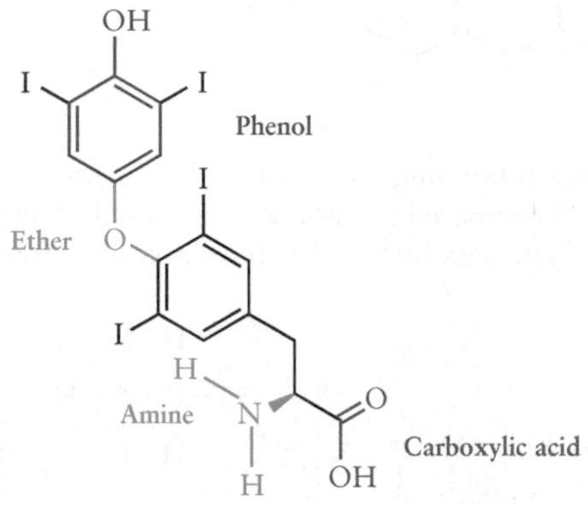

Synthroid

25 An ester has an —OR′ group directly bonded to a carbonyl carbon, whereas an ether has an —OR′ directly bonded to an aromatic ring or an alkyl carbon.

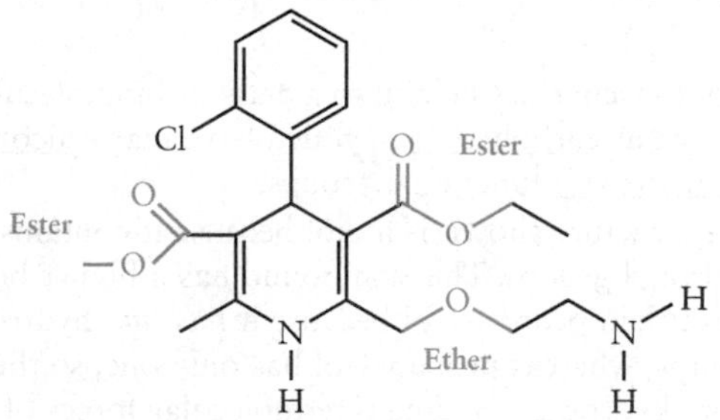

Norvasc

26 This is a triglyceride:

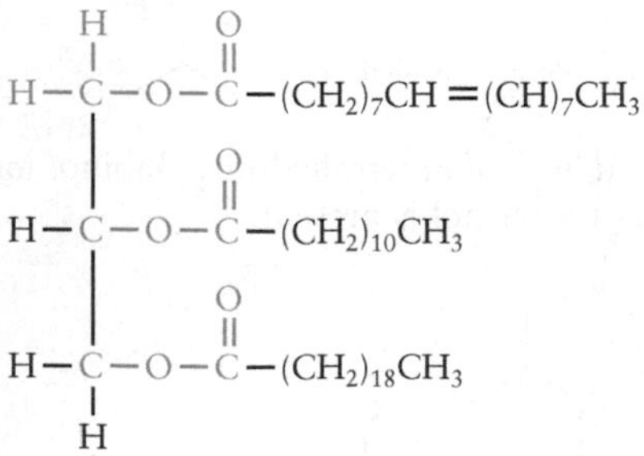

27 a. methyl butanoate
　　b. ethyl heptanoate
　　c. ethyl methanoate (common name: ethyl formate)

28 a.

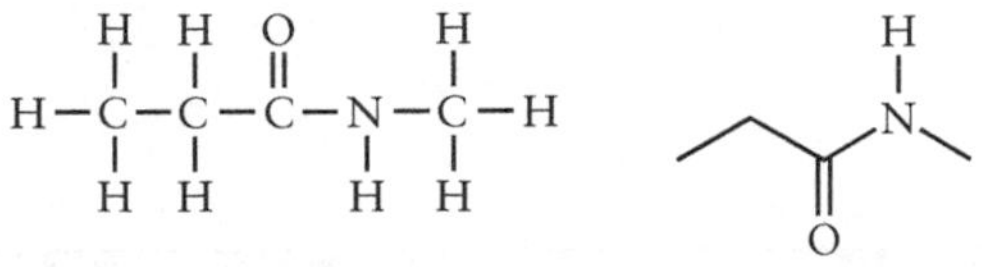

b.

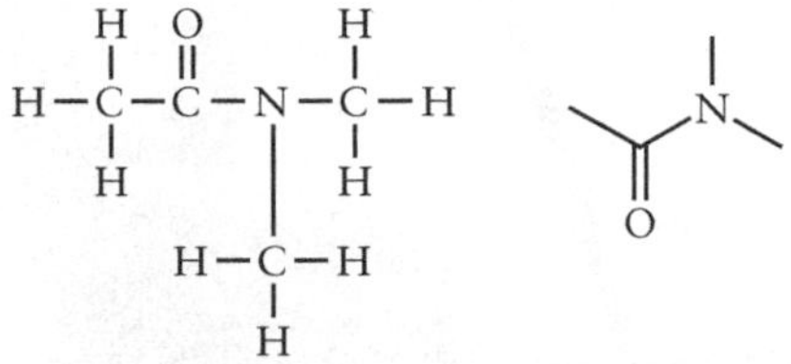

c. This compound is not an amide; it is an amine:

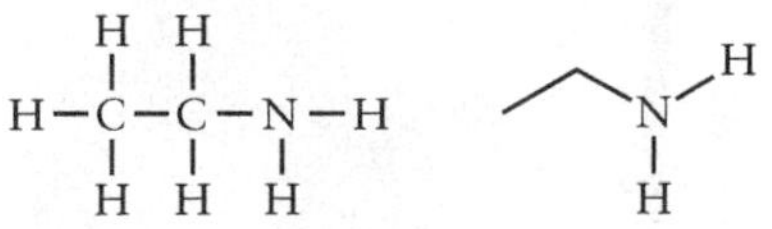

d.

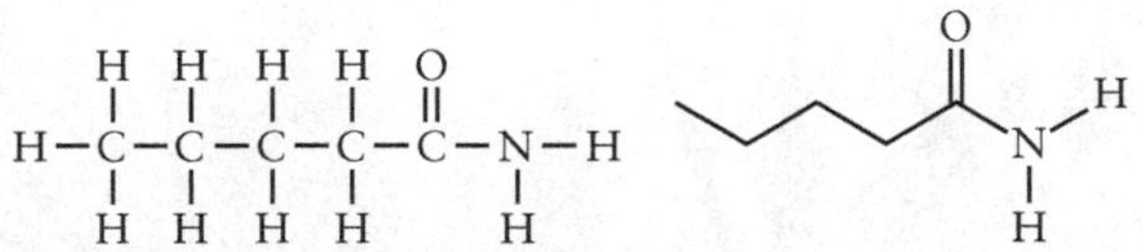

29 a. secondary amine   **b.** carboxylic acid
    c. ester   **d.** amide

30

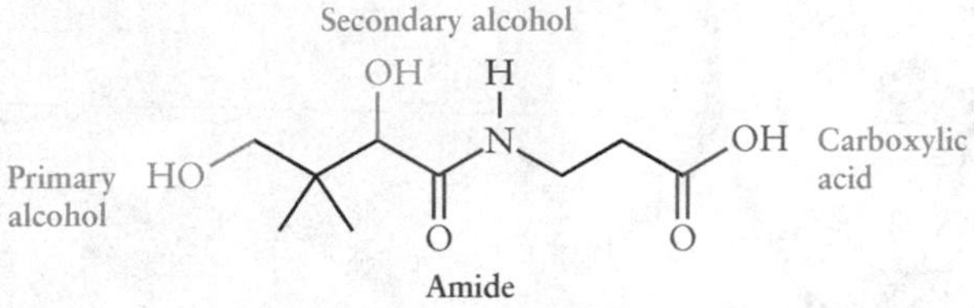

Secondary alcohol
Primary alcohol
Carboxylic acid
Amide

31

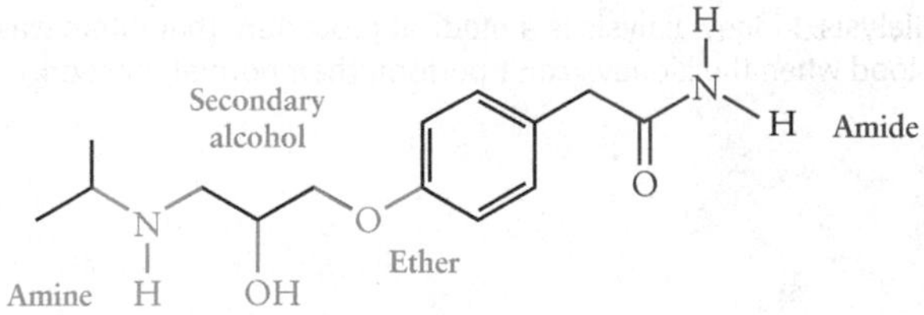

Secondary alcohol
Amine
Ether
Amide
Atenolol

32

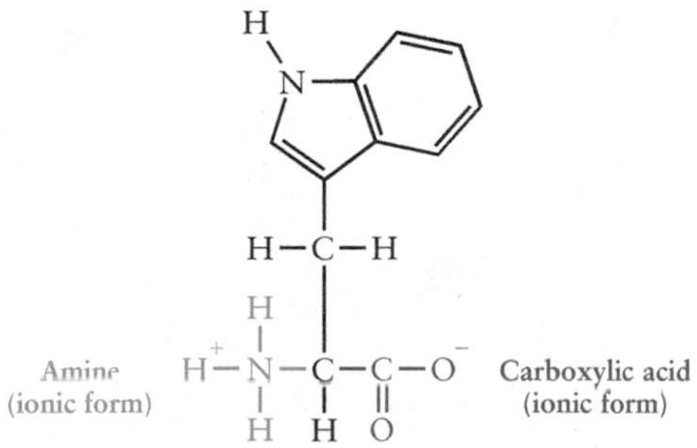

Amine (ionic form)
Carboxylic acid (ionic form)
Tryptophan

33 a. fatty acid   **b.** amino acid
    c. carbohydrate (a monosaccharide)

34

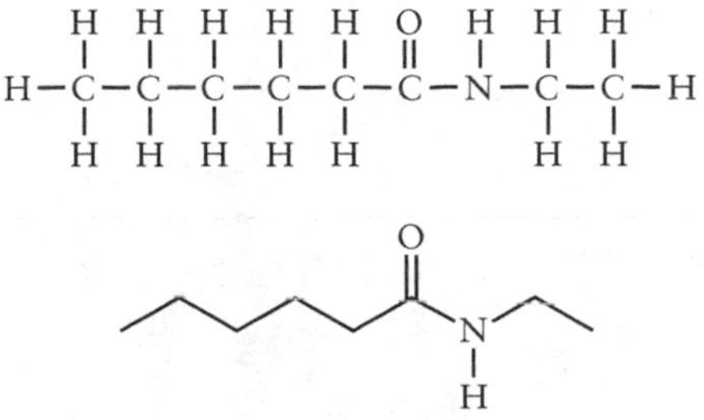

35 a. achiral   **b.** chiral   **c.** achiral
36 amino acids and monosaccharides
37 a. false   **b.** true   **c.** true   **d.** false
38

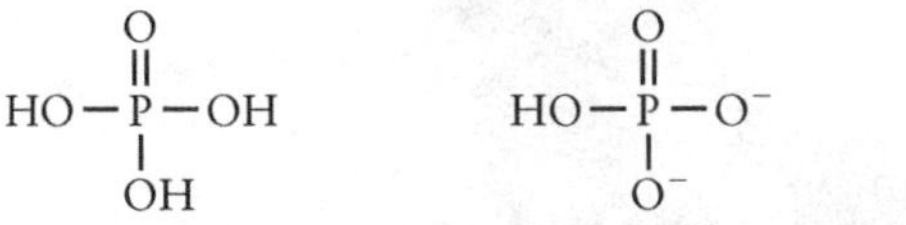

Phosphoric acid   Monohydrogen phosphate, P_i

39 Phosphoanhydride bonds

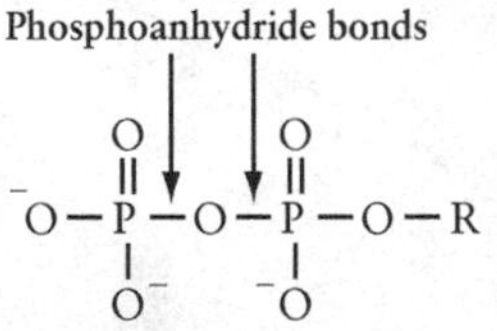

40 Phosphoanhydride bonds

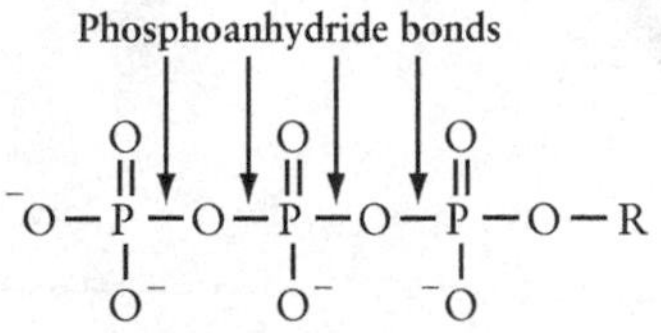

41 Does not contain phosphoanhydride bonds because there is only one P atom.

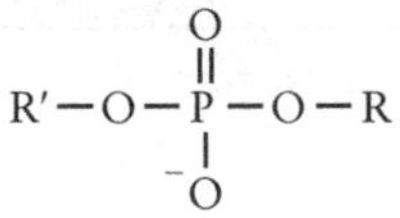

42 Phosphate ester

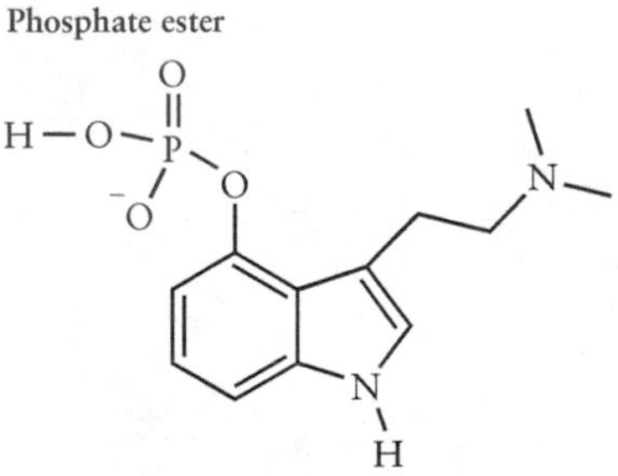

43 a. There are two phosphate esters, because there are two —OR groups (di) on phosphorus.
b. Because there is only one phosphate as noted by the single phosphorus atom.
c. The total charge is −1.
d. Is called cyclic because the phosphate group is part of a ring. *A* indicates adenosine is part of the structure of molecule, and *M* and *P* stand for "mono" and "phosphate," indicating there is only one phosphate group in the molecule.

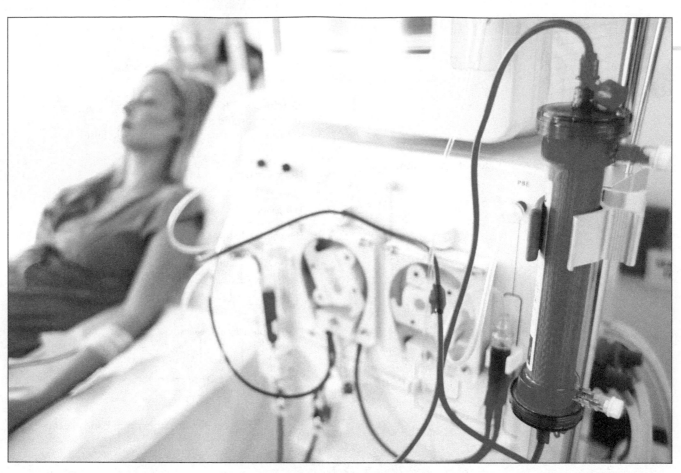

A patient undergoing kidney dialysis. Kidney dialysis is a medical procedure that filters waste and removes water from the blood when the kidneys can't perform their normal function.
[© Science Photo Library/Alamy]

8 Mixtures, Solution Concentrations, and Diffusion

Kidney Disease

In the United States alone, kidney disease affects more than 7.5 million people. There are approximately 73,000 people actively waiting for a donor kidney according to the United Network of Organ Sharing (UNOS). People who have less than 15 percent kidney function must undergo dialysis until a kidney is found. Kidney dialysis is a life-support treatment that filters the blood, performing the function that the damaged kidneys are no longer able to do.

Your kidneys filter over 200 L of blood every day, removing waste products (ions and small molecules) and about 2 L of water. Blood enters the kidneys through the renal artery, where it is filtered by selectively permeable membranes in the millions of tubules that make up the kidney. Waste and surplus water exits the kidneys through the ureters to the bladder, where it is stored until it is eliminated in the urine (Figure 8-1).

Kidney failure, known as renal failure, may occur suddenly—acute renal failure— or it may be of gradual onset—chronic renal failure. Acute renal failure can be caused by severe shock, dehydration, a heart attack, or a severe kidney infection. Chronic renal failure can result from diabetes, high blood pressure, or certain hereditary factors.

When the kidneys stop working altogether, the body retains water and waste products, causing swelling, which is especially noticeable in the hands and feet. Dialysis is the only way to remove the excess water and waste products and must be done on average three times a week. Although dialysis allows many people with kidney failure to live, it is uncomfortable and inconvenient—and often must be done in a hospital. Sometimes kidney disease cannot be cured, and kidney transplants become the only hope for long-term survival in these patients.

In this chapter you will learn about the chemical properties of various types of mixtures and then consider diffusion through selectively permeable membranes, such as the kidneys, cell membrane, and other membranes. ●

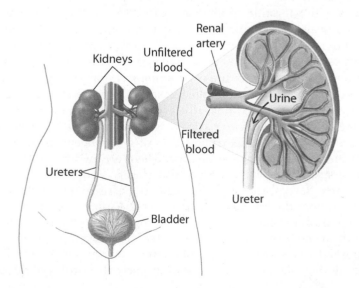

Figure 8-1 Blood enters the kidneys, through the renal artery, where it is filtered through membranes in the kidneys. Waste and excess water exit the kidney through the ureters and are stored in the bladder as urine until eliminated.

In the previous chapters you studied mainly *pure* substances. A pure substance is composed of only one element or one compound. Mixtures are composed of two or more compounds and/or elements. We can prepare mixtures by combining one or more elements or compounds. Mixtures abound in nature. In the body we find primarily aqueous mixtures, those containing water as the main component. Blood, for example, is an aqueous mixture: water is the main component along with many other important atoms, ions, and molecules found in lesser amounts.

In this chapter you will learn how to calculate the concentration of substances in a mixture, the way scientists and medical professionals *quantify* how much of a particular compound or element is present in a mixture. For example, a blood test measures the amount of certain important molecules, ions, and gases in a given volume of the blood, which are indicative of the health of the patient. You will learn about different types of mixtures, including solutions, colloidal dispersions, and suspensions, and which components of a mixture can diffuse through membranes, something your kidneys do as they filter wastes while retaining nutrients in the blood.

■ 8.1 Mixtures

A combination of two or more elements and/or compounds, in any proportion, is known as a **mixture**. Mixtures can be separated into their pure components through *physical* separation techniques. Since each component in a mixture has different physical properties, we can exploit these differences to separate the mixture into its pure components. For example, a mixture of sugar and sand can be separated by first adding water to the mixture. Since sugar dissolves in water and sand does not—a difference in physical properties— the sand can be filtered from the mixture to obtain pure sand. After the water is evaporated from the remaining sugar-water mixture, pure sugar is isolated. Many techniques exist for separating mixtures and new ones are continually being developed.

In a blood test, the different components of a small sample of blood are separated into their components by a sophisticated instrument that also measures the amount of each component of interest in the mixture. These specialized instruments are able to detect illegal substances in the blood, such as narcotic drugs and steroids. This type of separation and analysis is routinely performed at most major athletic competitions such as the Olympics and professional sports.

Mixtures with the same components can differ in the relative amount of each component. For example, gasoline is a familiar mixture composed of several different hydrocarbons such as octane, heptane, and isooctane. The relative amount of each of these hydrocarbons in the mixture varies depending on whether the mixture is *premium* gasoline or *regular* gasoline. Nevertheless, the mixture is still called gasoline because the same components are present in both mixtures, albeit in different amounts.

Fundamentally, there are two basic types of mixtures: *heterogeneous* and *homogeneous* (Figure 8-2). The components of a **homogeneous mixture** are *evenly* distributed throughout the mixture. There are two types of homogeneous mixtures: solutions and colloidal dispersions. A cup of coffee containing water, coffee flavors, caffeine, and sugar is classified as a homogeneous mixture or solution: the amount of caffeine at the top is the same as at the bottom. In contrast, the components of a **heterogeneous mixture** are *unevenly* distributed throughout the mixture. Granite rock is an example of a heterogeneous mixture. If you examine a piece of granite, you will see several different components, distinguished by color and texture, unevenly distributed throughout

(a) Homogeneous mixture

(b) Heterogeneous mixture

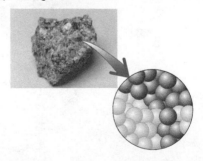

Figure 8-2 (a) Coffee, a homogeneous mixture, with its components evenly distributed throughout the mixture. (b) Granite, a heterogeneous mixture, with its components unevenly distributed throughout the mixture.
[(a) © D. Hurst/Alamy; (b) © sciencephotos/ Alamy]

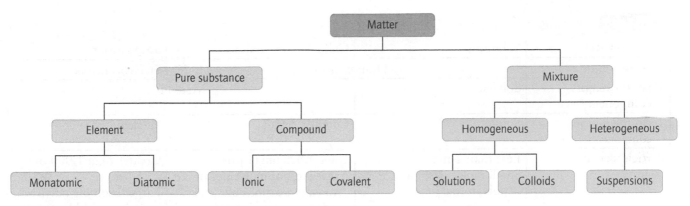

Figure 8-3 Classification of matter.

the rock. A summary of the different classifications of matter are shown in Figure 8-3. Solutions are described in detail in section 8.3, and colloidal dispersions and suspensions are described below.

Solutions, colloidal dispersions, and suspensions differ at the atomic level in the size of their particles:

- **Solutions** contain the smallest particles, known as the solute, that measure less than 1 nm in diameter.
- **Colloidal dispersions** contain particles intermediate in size, known as colloids, which range from 1 nm to 1 μm in size.
- **Suspensions** contain the largest particles, greater than 1 μm.

Colloidal Dispersions

Have you ever spread mayonnaise on bread? Would you describe mayonnaise as a solution? Probably not, since you cannot see through it, yet it is uniform throughout. Chemically, mayonnaise is a colloidal dispersion. In a **colloidal dispersion** the major component is called the **medium**. The minor component(s) are called **colloids** or colloidal particles. Table 8-1 summarizes and compares some of the different properties of solutions, colloids, and suspensions.

Colloids range in size from 1 nm to 1 μm. Colloidal particles do not *dissolve* in their medium because they are significantly larger than the molecules of the medium. Proteins and polysaccharides like starch are large organic molecules that behave as colloids in an aqueous medium. Proteins, for example, are much larger molecules than the water molecules that surround them. Colloidal dispersions have an opaque appearance because the colloidal particles are large and visible yet uniformly distributed throughout the medium. Milk is another example of a colloidal dispersion. Colloids can be large molecules or many small molecules clumped together, known as **aggregates**.

Colloidal dispersions can be found in all three states of matter: *aerosols* such as smoke and fog are solid and liquid colloids, respectively, dispersed in a gas medium; *sols* such as blood and paint are solid colloids dispersed in a liquid medium; *emulsions* such as mayonnaise are liquid colloids in a liquid medium; and *gels* such as gelatin, ointment, and hair gel are liquid colloids in a solid medium.

Suspensions

The particles in a **suspension** are larger than 1 micrometer in diameter and can often be seen with the naked eye or under a light microscope. In the absence of constant stirring, particles in a suspension eventually settle due to gravity and

TABLE 8-1 Comparison of Solutions, Colloidal Dispersions, and Suspensions

Characteristic	Solutions	Colloidal Dispersions	Suspensions
Type of mixture	Homogeneous		Heterogeneous
Term for major component	Solvent	Medium	
Term for minor components	Solute	Colloid	Particle
Particle size and identity	Less than 1 nm Small molecules, ions, or polyatomic ions	1 nm–1,000 nm (1 μm) Aggregates of molecules or ions; large molecules, such as proteins and starch	Greater than 1,000 nm (1 μm) Large insoluble particles, such as red blood cells
Physical appearance	Transparent	Cloudy; individual particles not visible	Cloudy; individual particles visible with visible particles
Particles settle	No		Yes
Separation with filter or centrifuge?	No	Requires special techniques	Yes
Examples	Laboratory solutions [© Ocean/Corbis]	Mayonnaise, blood serum, smoke, hand lotion [Glowimages/Getty Images]	Milk [Catherine Bausinger]

the large size and mass of their particles. Consequently, particles in a suspension are readily separated from the medium by a filter or a centrifuge.

A **centrifuge** is a standard piece of laboratory equipment used to separate the particles in a suspension from the medium. A centrifuge contains receptacles for several test tubes, which are spun at high speed (5,000 rpm), causing the suspended particles to collect at the bottom of the test tube in what is known as a **pellet**, with the remainder of the mixture, known as the **supernatant**, on top. The supernatant is then easily separated from the pellet.

When a test tube containing a sample of whole blood is placed in a centrifuge and spun for a few minutes, the suspended blood cells collect at the bottom of the tube, leaving **blood plasma** as the supernatant, as illustrated in Figure 8-4. Blood plasma contains proteins (fibrinogen, albumin, and globulin), which are colloids, electrolytes (Na^+, K^+, Cl^-), and small molecules (glucose, creatinine), which are solutes. The protein fibrinogen and other clotting factors can be removed from a blood sample to prevent clotting, leaving what is known as **blood serum** (blood plasma without the fibrinogen and other clotting factors).

Some oral medications are prepared as suspensions because they are insoluble in all acceptable media. Due to their fluid properties, suspensions are an ideal medium for patients such as children, who have difficulty swallowing tablets or capsules. However, it is important that oral suspensions be shaken and mixed before administering because the active ingredient settles on standing.

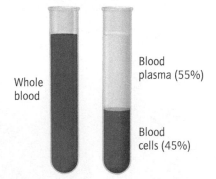

Whole blood

Blood plasma (55%)

Blood cells (45%)

Figure 8-4 Centrifugation of whole blood separates it into blood plasma on top (supernatant) and blood cells on the bottom (pellet).

In the opening vignette you learned that membranes in the kidneys filter your blood. They do so by retaining the suspended blood cells and colloidal proteins while eliminating solute wastes and some water (the solvent). In section 8.4 you will learn about membranes and how water and select solutes can cross a selectively permeable membrane by diffusion.

WORKED EXERCISES Distinguishing Solutions, Colloids, and Suspensions

8-1 Classify each mixture below as *homogeneous* or *heterogeneous*:
a. air
b. coffee grounds in water
c. an alcoholic beverage
d. a fruit smoothie
e. sweat

8-2 Indicate whether the following examples represent a solution, a colloid, or a suspension. Explain your reasoning.
a. a glass of iced tea with very little added sugar
b. clean dry air
c. chalk dust in water
d. hair gel

Solutions

8-1 a. homogeneous because a mixture of gases is uniform throughout
b. heterogeneous because the grounds will settle
c. homogeneous because ethanol and water are evenly distributed throughout the mixture
d. heterogeneous because some components will settle with time
e. a homogeneous mixture of salt uniformly distributed in water

8-2 a. A solution, because sugar dissolves in water.
b. Air is a solution because it is a mixture of gases and gases are soluble in one another. An exception is polluted air, in which case the air is a suspension or a colloid, depending on the size of the pollutant particles.
c. Chalk dust in water is an example of a suspension—after a while the larger chalk particles will settle to the bottom and the smaller ones will float.
d. Hair gel is a liquid colloid in a solid medium.

PRACTICE EXERCISES

1 Explain the difference between a homogeneous and a heterogeneous mixture.

2 Classify each mixture below as *homogeneous* or *heterogeneous*:
a. a chocolate chip cookie
b. calamine lotion
c. a bucket full of rocks and sand
d. a carbonated beverage
e. the contents of an intravenous (IV) bag

3 How does the particle size differ between a solution, a colloidal dispersion, and a suspension.

4 Indicate whether the following examples represent solutions, colloidal dispersions, or suspensions?
a. smoke
b. milk
c. 1 g of sugar in 35 mL of water
d. Pepto-Bismol liquid for the relief of upset stomach—must be shaken before use

5 Explain how blood has characteristics of a solution, a colloidal dispersion, and a suspension.

6 Name two ways in which you could separate the particles in a suspension from the rest of the mixture.

(a)

(b)

Figure 8-5 (a) A teaspoon of sugar, the solute, added to a beaker of water, the solvent, forms an aqueous solution. (b) Water, the solute, dissolved in sugar, the solvent, forms taffy, an extremely sweet candy. [(a) © 2006 Richard Megna/ Fundamental Photographers; (b) © Mitch Wojnarowicz/Amsterdam Recorder/ The Image]

■ 8.2 Solutions

A *solution is a homogeneous mixture containing small molecules, atoms, or ions with a diameter of less than one nm.* To understand solutions we must first introduce a few additional terms that are used to describe a solution. The component present in the greatest amount is called the **solvent** and the component(s) present in lesser amounts are known as the **solute(s)**.

Solution = solvent + solute(s)

A solution may contain one solute or many different solutes. One of the most common solvents is water, and a solution where water as the solvent is known as an **aqueous solution** and abbreviated **aq**.

If you prepare a solution by dissolving one teaspoon of sugar ($C_{12}H_{22}O_{11}$) in an 8-oz glass of water, sugar is the solute and water is the solvent because there is more water than sugar in the solution (Figure 8-5a). Moreover, this is an example of an aqueous solution because water is the solvent. If more sugar were added so that it exceeded the amount of water, then sugar would be the solvent and water would be the solute. This is the case for the mixture known as *taffy*, an extremely sweet candy (Figure 8-5b).

Although we typically think of a solution as a solid solute dissolved in a liquid solvent, solutions in which the solute and/or the solvent are in the other states of matter are also common. For example, air is a solution of nitrogen, oxygen, water vapor, carbon dioxide, and argon. Since nitrogen constitutes 79 percent of the mixture, it is the solvent. The other gases are the solutes. *Mixtures of gases are always homogeneous solutions because each gas is uniformly distributed throughout the mixture.*

Liquid-solvent/liquid-solute solutions are also quite common, such as ethanol and water, the homogeneous mixture that makes up alcoholic beverages. Gas solutes in aqueous solution were described in chapter 5, where you learned that pressure is a key factor in the amount of gas dissolved in the liquid solvent. Thus, a carbonated beverage is packaged under high pressure so that more carbon dioxide molecules, a gas solute, is dissolved in the aqueous beverage—a solution.

Solid-solvent/solid-solute solutions are seen in metal alloys, such as bronze, a homogeneous solution of tin (Sn) and copper (Cu). Since copper is the major component, copper is the solvent. Dental amalgam is a solution of mercury (Hg) in the liquid phase mixed with silver in the solid phase. Table 8-2 lists several familiar solutions. As you can see from the table, examples of all three states of matter—solid (s), liquid (l), and gas (g)—are found for both the solute and the solvent.

The most common types of solutions you will encounter in medicine are aqueous solutions. In an aqueous solution, the solute is uniformly distributed throughout water, the solvent, and we say the solute is *dissolved in* the solvent. The term dissolved means that each of the solute atoms, ions, or molecules is completely surrounded by solvent molecules and evenly distributed throughout the solvent. **Dissolution**, the process of dissolving, is a physical change, not a chemical change. On the macroscopic level, you have undoubtedly witnessed that sugar or salt crystals gradually seem to disappear when added to water as they "dissolve." In the next section we examine what occurs at the atomic level when a solute dissolves in a liquid solvent. We will also consider the factors that determine whether or not a solute will dissolve in a particular solvent. You will see that the answers to these questions, not surprisingly, depend on the chemical structure of the solute and the solvent.

TABLE 8-2 Solute-Solvent Combinations for Some Common Solutions in Various States of Matter

	Solvent	Solute	Solution
Solid Solvent	Cu (s)	Sn (s) and sometimes P (s)	Bronze, a metal alloy
	Ag (s)	Hg (l)	Amalgam used in dentistry
Liquid Solvent	Ethanol (l)	I_2 (s)	Tincture of iodine
	H_2O (l)	Ethanol (l)	Alcoholic beverages
	H_2O (l)	CO_2 (g)	Carbonated beverages
Gas Solvent	N_2 (g)	Naphthalene (g)	The vapor near mothballs
	N_2 (g)	Water droplets (l)	Mist
	N_2 (g)	O_2 (g)	Air

WORKED EXERCISES | Solute and Solvent

8-3 Identify the solute and the solvent in each solution:
   a. an IV bag of 0.9% physiological saline (NaCl)
   b. a chlorinated pool
   c. a tincture of iodine
8-4 Is dissolution a physical or a chemical change?

Solutions
8-3 **a.** solute: NaCl; solvent: water because it is present in the greater amount
   b. solute: chlorine; solvent: water because it is present in the greater amount
   c. solute: iodine; solvent: ethanol because it is present in the greater amount (see Table 8-2)
8-4 A physical change

PRACTICE EXERCISES

7 Identify the solute(s) and solvent in each solution:
   a. a mixture of 5 mL of ethanol and 25 mL of water
   b. 200 g of water containing 6 g of NaCl and 2 g of sugar
   c. a gaseous mixture of 0.005 L of CO_2 and 2 L of O_2

Polarity of the Solute and Solvent: "Like Dissolves Like"

We know that sugar, a covalent compound, dissolves in water. We know that table salt, an ionic compound, also dissolves in water. Both are examples of solid solutes. We know that ethanol, a liquid, is a covalent compound that dissolves in water. We have also seen mixtures where the components do not dissolve, such as a mixture of oil and water or a mixture of barium sulfate in water. Why do some compounds dissolve in a solvent while others do not, forming a heterogeneous mixture instead? *The solubility of a solute in a solvent depends mainly on the polarity of the solute and the solvent.*

In chapter 3 you learned that polarity ranges from the very polar ionic compounds, with full charges (+ and −) on ions, to polar covalent compounds that have strong hydrogen bonding or dipole-dipole intermolecular forces of attraction between separated partial charges ($\delta+$ and $\delta-$), to nonpolar compounds that have no charge separation and exhibit weak dispersion forces. *The general rule of thumb for predicting the solubility of a solute in a solvent is that polar compounds dissolve in polar solvents and nonpolar compounds dissolve in nonpolar solvents, expressed in the well-known saying "like dissolves like."* Thus, sugar dissolves in water because sugar is "like" water—they are both polar substances. Sodium chloride dissolves in water because sodium, Na^+, and chloride, Cl^-, ions are polar, as are most monatomic and polyatomic ions. Similarly, oil dissolves in hydrocarbon solvents because oil is "like" a hydrocarbon and both are nonpolar compounds.

In contrast, a polar and a nonpolar substance do not form a homogeneous mixture. Consider what occurs on the atomic scale when oil and water are combined. When nonpolar molecules encounter polar molecules such as water, they cannot participate in the hydrogen bonding network between the water molecules. Instead, they form a separate layer that "avoids" the water molecules. We say the nonpolar molecules are **hydrophobic**, a Latin word meaning "water fearing." This is known as the **hydrophobic effect**, and it is driven by the natural tendency for all systems, including solutions, to attain the lowest energy. In order for nonpolar and polar molecules to mix, some water molecules would have to sacrifice hydrogen bonds with other water molecules in order to organize themselves around nonpolar oil molecules with which they cannot form hydrogen bonds, hence the mixing of oil and water is energetically unfavorable, and therefore does not occur.

Common polar solvents include water; small alcohols such as methanol, ethanol, and propan-2-ol; and small carboxylic acids such as acetic acid (CH_3CO_2H). These are all liquids that have strong hydrogen-bonding forces of attraction. Think of hydrogen bonding as a kind of "stickiness" that holds polar molecules together.

Common nonpolar solvents include liquid hydrocarbons (C_xH_y)—those with five or more carbons, carbon dioxide (CO_2), elements (for example, N_2, He, and Ar), and carbon tetrachloride (CCl_4). Nonpolar compounds interact through dispersion forces. Think of dispersion forces as a much "less sticky" interaction between "greasy" nonpolar molecules.

Compounds with both a polar functional group and some hydrocarbon character will exhibit intermediate polarity. For example, 1-octanol dissolves in nonpolar solvents, but not water, because the nonpolar hydrocarbon chain of eight carbons dominates the single polar hydroxyl group. However, 1-butanol is intermediate in polarity and is partially soluble in water, because it has four carbon atoms and one hydroxyl group.

One of the most polar solvents is water, and it is the solvent encountered most in biological organisms and in human cells. Blood serum is a solution in which water is the solvent and many of the dissolved solutes and ions are nutrients for the cell. For example, glucose (blood sugar) in the blood provides energy for cells throughout the body. Other solutes in the blood include amino acids and important ions such as Na^+, K^+, and Cl^-. In the next section we will examine what happens at the atomic level when a polar solute dissolves in water to form an aqueous solution.

Dissolution of Ionic Compounds in Aqueous Solution

In chapter 3 you learned about two fundamentally different types of polar solutes: ionic compounds and polar covalent compounds. In this section we consider what happens to the lattice structure of an ionic compound containing monoatomic or polyatomic ions when it is dissolved in water. Recall that

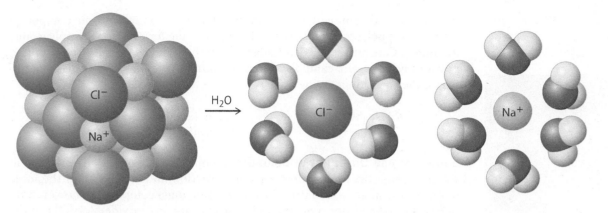

Figure 8-6 The solid sodium chloride, NaCl, an ionic lattice, dissolves in water to produce separated Na^+ and Cl^- ions, each surrounded by many water molecules, creating ion-dipole electrostatic interactions between ions and water.

ions are held together in a lattice structure by the strong electrostatic attraction between opposite charges, known as an ionic bond, to give an overall neutral ionic compound.

When an ionic compound dissolves in water, the electrostatic attractions between ions—the ionic bonds—are broken and substituted with electrostatic interactions between each ion and several water molecules. Every cation is completely surrounded by the partial negatively charged pole of several water molecules, and every anion is surrounded by the partial positively charged pole of several water molecules, as illustrated in **Figure 8-6**. The electrostatic attraction between an *ion* and several *polar* solvent molecules is known as an **ion-dipole** interaction. Cations and anions are thus separately surrounded and dissolved by several solvent molecules such that they are uniformly distributed throughout the solution. Ionic compounds that dissolve in water do so because the solution is lower in energy than the undissolved lattice in water.

Consider what happens on the atomic level when sodium chloride (NaCl) is dissolved in water. The ionic bonds between the sodium cations, Na^+, and the chloride anions, Cl^-, of the lattice break and are replaced by ion-dipole interactions between each of the ions and several water molecules, as illustrated in Figure 8-6.

Although dissolution is a physical process, we can use reaction equations to represent the dissolution process. Write the formula unit for the ionic compound and its physical state—solid—on the left side of the arrow, NaCl (s), and write the separate ions, including their charges, followed by the abbreviation *aq*, enclosed in parentheses, to signify the ions are surrounded by water molecules, on the right side of the arrow:

$$\underset{\text{(1 formula unit)}}{NaCl(s)} \longrightarrow \underset{\text{(1 sodium ion)}}{Na^+(aq)} + \underset{\text{(1 chloride ion)}}{Cl^-(aq)}$$

The subscripts in the formula unit of an ionic compound appear as coefficients on the right side of the equation. The formula unit NaCl indicates sodium and chloride ions are in a 1:1 ratio, so no coefficients are needed (see equation above). However, when calcium chloride, $CaCl_2$, for example, is dissolved in water, *two* chloride ions and one calcium ion are released into solution for every one $CaCl_2$ formula unit:

$$\underset{\text{(1 formula unit)}}{CaCl_2(s)} \longrightarrow \underset{\text{(1 calcium ion)}}{Ca^{2+}(aq)} + \underset{\text{(2 chloride ions)}}{2\ Cl^-(aq)}$$

When writing an equation to show the dissolution of an ionic compound containing a polyatomic ion, remember to keep the formula for the polyatomic ion intact. Consult Table 3-1 for a list of formulas and charges for the common polyatomic ions. For example, when sodium sulfate, Na_2SO_4, is dissolved in water, the equation is written as:

$$\underset{\text{(1 formula unit)}}{Na_2SO_4(s)} \longrightarrow \underset{\text{(2 sodium ions)}}{2\,Na^+(aq)} + \underset{\text{(1 sulfate ion)}}{SO_4^{2-}(aq)}$$

Saturated and Unsaturated Solutions Soluble ionic compounds dissolve in water because the ion-dipole forces of attraction are stronger than the ionic bonds of the intact solid lattice in water. Some ionic compounds, however, do not dissolve in water. For example, magnesium hydroxide, $Mg(OH)_2$, the active ingredient in milk of magnesia, used to relieve the symptoms of heartburn, is insoluble in water. Radioactive barium sulfate, $BaSO_4$, a suspension taken orally to image the gastrointestinal tract or as a "barium enema," used to screen for colon cancer (Figure 8-7), is another example of an insoluble ionic compound. Insoluble ionic compounds have stronger electrostatic interactions between the cation and the anion in the solid ionic lattice than as dissolved ions. There are many reasons for this, including the magnitude of the charge (+2 and −2 rather than +1 and −1) and the size of the ion. Insoluble ionic compounds form suspensions, a heterogeneous mixture consisting of large aggregates of ionic lattice suspended in water.

Even for ionic compounds that are soluble in water, there is a limit to the amount of solute that will dissolve in a given volume of solvent. You may have noticed this when adding "too much" sugar to a glass of iced tea. Initially the sugar dissolves in the iced tea—we say the solution is **unsaturated**. As you continue to add solute (the sugar), the solution eventually becomes cloudy, as undissolved solute appears suspended in solution, or forms a precipitate at the bottom of the glass (Figure 8-8). *A solution in which the solute no longer dissolves in the solvent is known as a **saturated solution**, a heterogeneous mixture.* A solution becomes saturated when there are no more solvent molecules left to surround—dissolve—the solute. You may see a separate phase, or even a precipitate, such as solid sugar crystals at the bottom of a glass of iced tea. Kidney stones and gout are diseases caused by saturated solutions of calcium salts such as calcium phosphate, $Ca_3(PO_4)_2$, and calcium oxalate, CaC_2O_4, forming in the kidneys and joints. When these calcium salts precipitate, they form kidney "stones" or crystals in the joints, as in the case of gout.

Electrolytes Dissolved ions in aqueous solution are referred to as **electrolytes**, especially in biological applications. Solutions containing electrolytes conduct electricity, which is the origin of the term *electrolyte*. Some physiologically important electrolytes are listed in Table 8-3. Our cells, which are composed mainly of water, contain many electrolytes that are critical to cellular function.

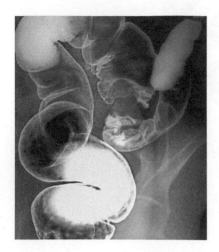

Figure 8-7 X-ray image of a patient with colon cancer, after receiving a barium enema—a suspension of radioactive barium sulfate, $BaSO_4$, in water. [ZEPHYR/Getty Images]

TABLE 8-3 Physiologically Important Electrolytes

Sodium (Na^+)
Potassium (K^+)
Calcium (Ca^{2+})
Magnesium (Mg^{2+})
Chloride (Cl^-)
Hydrogen phosphate (HPO_4^{2-})
Hydrogen carbonate (HCO_3^-)

Figure 8-8 Two solutions with different amounts of the same solute and solvent: Left: An unsaturated solution (homogeneous) showing all the solute dissolved in the solvent. Right: a saturated solution (heterogeneous) showing the solute precipitated from solution. [Catherine Bausinger]

The most important of these electrolytes are Na^+, K^+, Ca^{2+}, Mg^{2+}, and Cl^-. For example, sodium ion, Na^+, and potassium ion, K^+, are involved in regulating blood pressure and are important in the transmission of electrical signals in nerve cells. The most important electrolytes are the macronutrients and micronutrients shown in Figure 2-15 and described in chapter 2.

Electrolytes are added to many popular sports drinks such as Gatorade, Propel, and Vitamin water. Electrolytes are also added to drinks for infants and young children such as Pedialyte (Figure 8-9). These beverages help replenish essential electrolytes lost while perspiring during physical exertion or from diarrhea and vomiting.

Proper electrolyte balance is regulated by hormones in the body, with the kidneys playing an important role in eliminating excess electrolytes. A routine blood test measures the concentration of each of these electrolytes. When they are outside the normal range, it is often an indicator of organ malfunction. See Chemistry in Medicine: Blood Chemistry and the Diagnosis of Disease at the end of this chapter to learn more about electrolytes and their significance.

Ionic compounds that do not dissolve in water are nonelectrolytes. In the next section you will learn that dissolved molecules are also nonelectrolytes, and in the next chapter you will be introduced to some weak electrolytes.

Figure 8-9 Pedialyte, a solution containing dissolved ions, is often recommended for diarrhea or vomiting to replenish essential electrolytes in infants and small children. [Catherine Bausinger]

WORKED EXERCISES   Dissolution of Ionic Compounds

8-5 What ions are dissolved in solution when sodium phosphate, Na_3PO_4, is added to water? Show the dissolution process as an equation.

8-6 Student A adds 50 g of sodium nitrate, $NaNO_3$, to 100 mL of water and sees that she has a clear colorless homogeneous solution. Student B adds 100 g of sodium nitrate to the same volume of water, and observes an opaque heterogeneous mixture. In this example, student A has a(n) _____ solution and student B has a(n) _____ solution (choose *saturated* or *unsaturated*). Why were the results different even though the solute and solvent were the same in both cases? Which mixture contains a precipitate and why?

8-7 When a solid sample of potassium iodide, KI, dissolves in water, it looks like the solid is "disappearing." What is happening on the atomic scale? Provide an illustration showing ion-dipole interactions. Is this solution an electrolyte? Explain.

8-8 Which of the following solvents are polar?
  a. carbon dioxide
  b. hexane
  c. water
  d. methanol
  e. ethylene glycol ($HOCH_2CH_2OH$)
  f. CCl_4

Solutions

8-5 Sodium ions, Na^+, and phosphate ions, PO_4^{3-}, a polyatomic ion, are dissolved in solution. The subscript 3 after Na in the formula unit of the ionic compound indicates that there are three sodium ions for every one phosphate ion dissolving in solution; therefore, the coefficient 3 is placed before Na^+ on the right side of the equation:

$$Na_3PO_4(s) \longrightarrow 3\,Na^+(aq) + PO_4^{3-}(aq)$$

8-6 In this example, student A has an *unsaturated* solution and student B has a *saturated* solution. Student B has the same solute but twice as much, creating a heterogeneous mixture because it exceeds the amount of water available to dissolve all the solute. Eventually the solid suspended in the mixture prepared by student B will settle at the bottom of the flask, where it is referred to as a precipitate.

8-7 The ionic bonds in the lattice are broken and replaced by ion-dipole interactions between each ion and several water molecules. This solution is a strong electrolyte because it contains charged ions dissolved in solution.

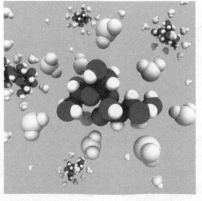

8-8 (c), (d), and (e) are polar because they have —OH groups and very little hydrocarbon.

PRACTICE EXERCISES

8 Explain the meaning of the saying "like dissolves like." What "like" properties is this statement referring to? How is an ion "like" water?

9 For the following ionic compounds write an equation showing what occurs when the solid dissolves in water. Include the appropriate coefficients.
a. $MgCl_2$ **b.** K_3PO_4 **c.** $NaHCO_3$

10 Based on the solubility characteristics of solutions, why do kidney stones form?

11 On the macroscopic scale how does a saturated solution look different from an unsaturated solution? Explain the difference on the atomic level.

Dissolution of Molecular Compounds in Solution

When a covalent compound or a diatomic element dissolves in a solvent, the covalent bonds remain intact, and only the intermolecular forces of attraction between solute molecules are disrupted as new solute-solvent intermolecular forces of attraction are formed. Indeed, this is why dissolving is classified as a physical change and not a chemical change because the chemical composition of the molecules remains the same after dissolution.

Polar Solvents and Polar Solute Molecules Polar molecules dissolve in polar solvents ("like dissolves like"). When a polar solute molecule dissolves in a polar solvent, the intermolecular forces of attraction between the solute molecules are disrupted and new intermolecular forces of attraction, usually hydrogen bonding forces, are formed between the solute molecules and the solvent molecules. This is illustrated in Figure 8-10, which shows the intermolecular forces of attraction between sucrose molecules (table sugar), a polar solute,

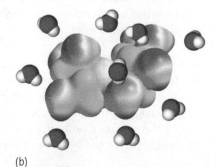

Figure 8-10 Computer-generated (a) space-filling models of sucrose (colored) dissolved in water (white). (b) electron density diagram showing how several water molecules surround one sucrose molecule.

(a)   (b)

and water, a polar solvent. Molecules of solid sucrose in the crystalline lattice exhibit strong hydrogen-bonding intermolecular forces of attraction as a result of the many hydroxyl groups in a sucrose molecule. When sucrose is dissolved in water, these hydrogen bonds are broken and replaced by new hydrogen bonds between an individual sugar molecule and several water molecules. In Figure 8-9 you can see that each sucrose molecule (colored) is surrounded by many water molecules (white). Polar solute *molecules* dissolve in water because the hydrogen-bonding interactions between solute and solvent are stronger than those in the lattice surrounded by water. Hence, the dissolved solution is lower in energy than the undissolved heterogeneous mixture. In contrast to *ions* dissolved in solution, dissolved molecules in aqueous solution do not conduct electricity because there are *no* electrically charged species in solution. Thus, solutions containing dissolved molecules are classified as **nonelectrolytes**.

Nonpolar Solvents and Nonpolar Solute Molecules Nonpolar solutes dissolve in nonpolar solvents ("like dissolves like"). Nonpolar solutes interact with nonpolar solvents through weak dipole-dipole interactions or dispersion forces. When a nonpolar solute is dissolved in a nonpolar solvent, the solute is distributed uniformly throughout the mixture and dispersion forces of attraction form between solute and solvent. Hydrocarbons with five or more carbon atoms, such as pentane and hexane, are common nonpolar solvents because they are liquids at room temperature. Dry-cleaning solvents, for example, are nonpolar and clean clothing by dissolving dirt with a nonpolar solvent rather than water; hence the term "dry" cleaning. The most common dry-cleaning solvent is tetrachloroethylene, which has similar solubility properties to those of hydrocarbons but the added advantage of being nonflammable.

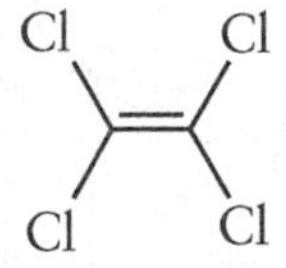

Tetrachloroethylene

You may have worked with a nonpolar solvent when cleaning up after using an oil-based paint, which is a nonpolar substance. Oil-based paints do not dissolve in water, so nonpolar solvents such as paint thinner are used instead. Turpentine was formerly used to dissolve oil-based paints but has been replaced with safer, less toxic solvents.

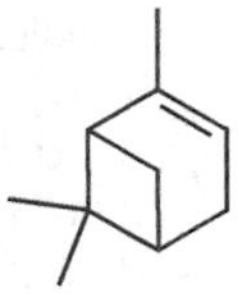

Turpentine

WORKED EXERCISES | Solubility of Solute in Solvent

8-9 What is the difference between a covalent compound and an ionic compound when dissolved in water?

8-10 What characteristics allow methanol to dissolve in water? Provide an illustration at the atomic level of a molecule of methanol dissolved in water.

Solutions

8-9 Covalent bonds do not break when a covalent compound is dissolved in water, only intermolecular forces of attraction are broken. Ionic bonds do break when an ionic compound is dissolved in water, releasing the charged ions of the lattice to form an electrolyte—individual ions surrounded by water molecules.

8-10 Methanol dissolves in water because it is polar, like water. Methanol is polar as a result of its polar O—H bond and very little hydrocarbon structure (only one carbon). The illustration shows one methanol molecule hydrogen bonding to two water molecules:

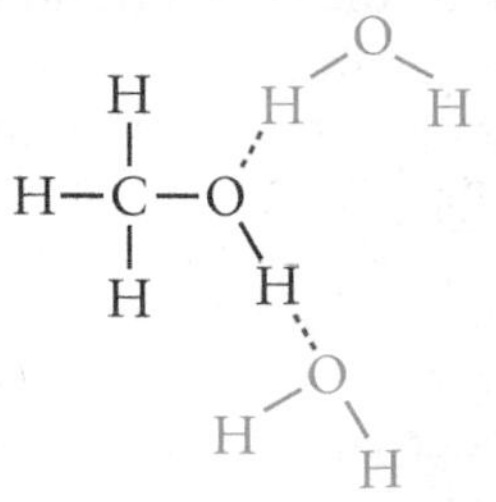

PRACTICE EXERCISES

12 Urine is an aqueous solution containing urea, and various other polar molecules and ions. Illustrate how a molecule of urea dissolves in water by showing hydrogen bonding to several water molecules.

$$H-N-\overset{\displaystyle \overset{O}{\|}}{C}-N-H$$

Urine

13 Which of the following solutes would you expect to dissolve in water and why?

   a. NaCl
   b. CCl_4
   c. gasoline (C_xH_y)
   d. lactose (a monosaccharide; Figure 7-9)
   e. CH_3OH
   f. acetic acid, CH_3CO_2H
   g. Na_3PO_4
   h. β-carotene (a polyene; Figure 6-15)
   i. cholesterol (a steroid; Figure 7-7)

14 Which of the following solutes would you expect to dissolve in hexane and why?

   a. NaCl
   b. CCl_4
   c. gasoline
   d. lactose
   e. CH_3OH
   f. acetic acid, CH_3CO_2H
   g. Na_3PO_4
   h. β-carotene
   i. cholesterol

15 For the following solutes in aqueous solution, indicate whether the solute is an electrolyte or a nonelectrolyte and why.

   a. Glucose, $C_6H_{12}O_6$
   b. KCl
   c. CH_3CH_2OH
   d. $CaCl_2$

■ 8.3 Concentration

We have seen that solutions can contain varying amounts of a solute. The amount of solute in a given volume of solution is known as its **concentration**. A range of solution concentrations is illustrated in Figure 8-11 for five solutions containing increasing amounts of the same blue-colored solute in the same volume of solution. The solution concentrations increase from left to right. The lighter-colored solutions contain less solute and are said to be more **dilute**; the darker solutions contain more solute and are said to be more concentrated.

Several methods are used to measure the concentration of a solution. One method, known as a *colorimetric technique*, measures the intensity of the color of the solution to determine its concentration. For example, the Bradford protein assay measures the amount of protein in solution by adding a special colored dye that binds to the protein, and then the intensity of the colored dye is measured to determine the concentration of protein in solution.

Quantitative measurements of concentration are routine in science and medicine. For example, a blood test is a quantitative measurement of the concentration of important solutes in a sample of blood serum, a useful indicator of a patient's health. The ability to calculate and interpret solution concentrations is an important part of a health care professional's responsibilities.

The concentration of a solution is a quantitative measure of the amount a particular solute in a given volume of solution. Thus, concentration is always

Figure 8-11 Solutions ranging from dilute to concentrated as a result of increasing amounts of a blue solute in the same volume of solution. [© 2008 Richard Megna/Fundamental Photographs]

a ratio, where the numerator represents the *amount* of a solute and the denominator represents the *volume* of total solution (solutes + solvent):

$$\text{Solution concentration} = \frac{\text{amount of solute}}{\text{volume of solution}}$$

Hence, units of concentration are also a ratio: a unit representing the amount of solute in the numerator and a unit of volume for the solution in the denominator. Various units are used to represent the amount of solute (the numerator), but most fall into one of two categories:

- *mass* of solute, and
- *moles* of solute

The units for the volume of solution—the denominator—are typically metric units of volume: L, dL, mL. Thus, the various units of *concentration* differ primarily in how the units of the solute are given: *mass* or *moles*. The units of concentration most commonly encountered in the medical field are

- mass/volume (m/v)
- % mass/volume (% m/v)
- moles/volume (mol/L, *M*)
- equivalents/volume (eq/L)

Preparing a Solution with a Specified Concentration

It is important to recognize that *all* concentration units are defined in terms of the volume of total *solution*, not the volume of the *solvent*. For example, to prepare a 5 g/L aqueous solution of sodium chloride, you cannot simply add 5 g of sodium chloride to 1 L of water, because then you would have a solution volume greater than 1 L and therefore a concentration less than 5 g/L.

Figure 8-12 shows the steps for preparing a solution with a specified concentration using a volumetric flask. First calculate and then weigh out the amount of solute needed (Figure 8-12a). Next, transfer the solute to an appropriately sized volumetric flask (Figure 8-12b). Volumetric flasks come in a variety of sizes, each with a mark on the neck of the flask representing one precise volume. Finally, add solvent to the volumetric flask up to the mark on the neck of the flask and mix it (Figure 8-12c). For example, to prepare the solution described above, add 5 g of sodium chloride to a 1-L volumetric flask and then add distilled water up to the mark on the neck of the flask while swirling. This ensures that the total volume of solution, sodium chloride plus water, is equal to 1 L.

(a)

(b)

(c)

Figure 8-12 Steps in the preparation of a solution using a volumetric flask: (a) Obtain the amount of solute (solid or liquid) needed and a volumetric flask (a 50-mL volumetric flask is shown), (b) transfer the solute to the volumetric flask, and (c) add solvent to the mark on the flask while mixing to dissolve. [Catherine Bausinger]

Mass/Volume Concentration

The concentration of many important solutes in a blood test are reported as *mass per volume* (*m/v*), a common unit of concentration in medicine. The mass per volume of a solution gives the ratio of the mass of the solute as a metric unit of mass, per volume of solution, in a metric unit of volume. In a blood test, the solute is a molecule or ion of interest, the solvent is water, and the solution is known as blood serum. Some medications are also administered in solution form, and their concentration is often also given as a *mass per volume* concentration. The general equation for calculating the concentration of a solution as a mass per volume is

$$\frac{mass}{volume} = \frac{mass\ of\ solute}{volume\ of\ solution}$$

For example, a patient whose iron concentration is 125 μg/dL has 125 micrograms of iron in every one deciliter of her blood. Note that when there is no numerical value in front of a unit, it is assumed to be 1. The / in the concentration units means "per," "divided by," or "for every." Since concentration is a ratio, it can be used as a conversion factor in calculations involving the amount of solute or the volume of solution. The next set of Worked Exercises shows how to perform this type of calculation.

Blood tests normally report glucose, cholesterol, and creatinine in units of mg/dL and protein in units of g/dL. These are all units of mass/volume, but since they differ in the amount of solute, different metric prefixes are preferred, as shown in Figure 8-13. This patient has cholesterol and triglyceride concentrations above normal, which is why these two solutes are highlighted in the blood test.

In medicine the dL is such a frequently used unit of volume that it will often save you one conversion step if you memorize the conversion between a dL and a mL:

100 mL = 1 dL

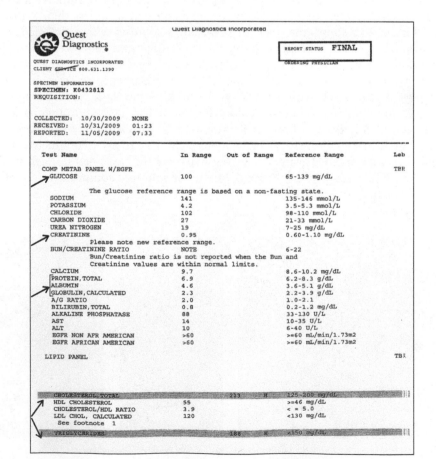

Figure 8-13 Blood test results from a 51-year-old female. The first column lists solutes of interest, the second column shows the concentration in units of m/v, mg/dL, and g/dL, and the third column shows the normal concentration range for each solute.

[Catherine Bausinger]

WORKED EXERCISES Mass/Volume Concentration

8-11 The concentration of an intravenous (IV) L-dopa solution, used to treat Parkinson's disease, contains 300. mg L-dopa in 250. mL of total solution. What is the m/v concentration of this solution in milligrams per milliliter (mg/mL)?

8-12 A patient is prescribed a 50 mg dose of a medicine that is only available as a 10 mg/mL aqueous solution. What volume, in milliliters, of this solution should you administer to the patient?

8-13 What is the mass, in milligrams, of L-dopa in 50 mL of the solution described in exercise 8-11?

8-14 For a solution with a chloride ion concentration of 137 μg/dL, what is the m/v concentration in units of g/mL?

Solutions

8-11 You are asked to calculate the concentration, given the mass of the solute and the total volume of solution. Use the equation for calculating a mass/volume concentration and substitute the supplied mass and supplied volume into the equation and solve the fraction:

$$\frac{m}{v} = \frac{300.\text{ mg}}{250.\text{ mL}} = 1.20\text{ mg/mL}$$

The final answer has units of mg/mL as requested, which is a typical m/v concentration unit.

8-12 You are asked to determine the volume of solution needed given the mass of solute (the medicine) prescribed and the concentration of the solution supplied. Remember, concentration is a conversion factor that can be used as is or in its inverted form. Using dimensional analysis, multiply the mass of the solute supplied by the correct form of the concentration conversion factor that allows mg to cancel leaving volume. This requires that we use the inverted form of concentration:

$$50\text{ mg} \times \frac{1\text{ mL}}{10\text{ mg}} = 5\text{ mL of solution}$$

Thus, you would give the patient 5 mL of the available 10 mg/mL solution of medicine.

8-13 You are asked to calculate the mass of L-dopa, the solute, in a given volume of a solution of L-dopa whose concentration was calculated previously to be 1.20 mg/mL. Using dimensional analysis, multiply the total volume of solution given by the correct form of the concentration conversion factor that allows volume to cancel, leaving only units of mass:

$$50\text{ mL} \times \frac{1.20\text{ mg}}{1\text{ mL}} = 60\text{ mg}$$

8-14 In this exercise you are asked to convert the concentration into other metric units, thus two metric conversions are required: one for the numerator (μg to g) and one for the denominator (dL to mL). Thus, you need to multiply the supplied concentration by two metric conversions, in any order:

$$\frac{137\ \mu\text{g}}{1\text{ dL}} \times \frac{1\text{ dL}}{100\text{ mL}} \times \frac{1\text{ g}}{10^6\ \mu\text{g}} = 1.37 \times 10^{-6}\text{ g/mL}$$

The first ratio represents the supplied concentration, the second is the metric dL-to-mL conversion factor, and the third is the metric g-to-μ conversion factor. Always use the form of the conversion factor that allows the supplied unit to cancel, thereby leaving the requested unit.

16 Isuprel, used to treat asthma, comes in IV form as 4 mg in every 500 mL. What is the concentration of this solution in mg/mL?

17 Lidocaine, a common local anesthetic, comes in IV form in a concentration of 4 g/L. What is the concentration of this solution in milligrams per deciliter?

18 For a solution with a concentration of 9.9 mg/dL of glucose, how many milligrams of glucose are in 3.5 dL of solution?

19 For a solution with a concentration of 122 mg/dL of iron, how many deciliters of solution contain 28 mg of iron?

20 If a doctor orders a 25.0 mg dose of lidocaine be given to a patient and the drug is available in IV form in a concentration of 4.00 mg/mL, how many milliliters of lidocaine solution should you give the patient?

% mass/volume concentration

Solutions used in intravenous (IV) therapy often have concentrations given in units of percent mass per volume, **% m/v**. For example, the most commonly administered IV solution, physiological saline (NaCl), has a concentration of 0.9% m/v (Figure 8-14). *To calculate the concentration of a solution in % m/v, the units of the solute must be in grams (g) and the units of solution must be in milliliters (mL).* The resulting ratio is then solved and multiplied by 100 to obtain a percentage:

$$\% \frac{m}{v} = \frac{g\ solute}{mL\ solution} \times 100$$

A % symbol always stands for "per 100," which means "divided by 100." For example, a 12% m/v solution means 12 g/100 mL, or solving the fraction, 0.12 g/mL.

For example, to calculate the % m/v concentration for a solution with 25 mg of glucose dissolved in 50. mL of total solution:

(1) Convert the mass of the solute from mg to g, a metric conversion:

$$25\ \cancel{mg} \times \frac{1\ g}{10^3\ \cancel{mg}} = 0.025\ g\ of\ glucose$$

(2) Substitute the mass, in grams, and the volume, in mL, into the concentration equation displayed above:

$$\frac{0.025\ g}{50.\ mL} \times 100 = 0.050\%\ m/v$$

For calculations where you need to use % m/v as a conversion factor, you must turn the % value into the corresponding g/mL ratio. Since % means "per 100 mL," a 2% solution is written as 2 g/100 mL:

$$2\% = \frac{2\ g}{100\ mL}$$

Or you can turn the fraction into its decimal value:

$$2\% = \frac{2\ g}{100\ mL} = 0.02\ g/mL$$

Otherwise, calculations employing % m/v as a conversion factor are set-up the same way as those shown for m/v concentrations in the previous section.

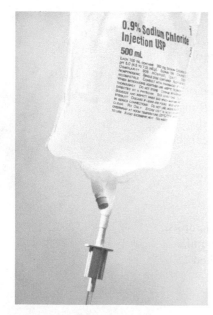

Figure 8-14 The most common intravenous (IV) solution is 0.9% sodium chloride, also known as physiological normal saline. The *0.9% m/v* is a concentration unit that indicates 0.9 gram of NaCl in every 100 of solution. [dtimiraos/Getty Images]

TABLE 8-4 Some Common IV Solutions and Their Concentration	
Solution Concentration	**Common Name**
0.90% NaCl	Normal saline (NS)
0.45% NaCl	Half-normal saline (1/2NS)
5% dextrose	D5W
5% dextrose, 0.9% NaCl	D5NS
0.60% NaCl, 0.31% sodium lactate, 0.03% KCl, 0.02% $CaCl_2$	Lactated Ringer's (LR)

For example, to calculate the mass of sodium chloride in 500 mL of a physiological saline solution with a concentration of 0.90% m/v, use dimensional analysis to multiply the supplied unit, 500. mL, by the appropriate form of the concentration conversion factor:

$$500. \text{ mL} \times \frac{0.90 \text{ g}}{100 \text{ mL}} = 4.5 \text{ g of NaCl}$$

Table 8-4 lists some common IV solutions and their concentrations in % mass/volume. These solutions are used to treat dehydration and electrolyte imbalances. For example, the solute in a 5% dextrose solution is the sugar D-glucose. This solution is often given to patients at risk of having low blood sugar.

The terms **ppm** (parts per million) and **ppb** (parts per billion) are analogous to % mass/volume except instead of "per 100" they stand for "per million" and "per billion," respectively. As with % m/v, the mass must be in *grams* and the volume must be in *milliliters*. Units of ppm and ppb are used instead of % when the concentration of the solute is very low, to avoid decimal values with many zeros as for example in environmental applications, such as the allowable amount of lead or radioisotopes dissolved in water.

$$\% = \frac{\text{g of solute}}{\text{mL of solution}} \times 100$$

$$\text{ppm} = \frac{\text{g of solute}}{\text{mL of solution}} \times 1{,}000{,}000$$

$$\text{ppb} = \frac{\text{g of solute}}{\text{mL of solution}} \times 1{,}000{,}000{,}000$$

WORKED EXERCISES Calculations Involving % m/v, ppm, and ppb

8-15 A solution is prepared by dissolving 0.15 g of NaCl in enough water to produce a total solution volume of 275 mL. What is the concentration of this solution in % m/v?

8-16 What is the mass, in grams, of solute in 3.0 L of a 5.0% aqueous dextrose solution, a common IV solution?

8-17 What volume of a 10.% solution of sugar contains 3.5 g of sugar?

8-18 The maximum acceptable level of lead in drinking water in Illinois is 50 ppb. What is the maximum mass of lead, in grams, permitted in every liter of water coming from a sink in Illinois?

Solutions

8-15 In this question you are being asked to calculate a concentration. Thus, use the equation that defines % m/v concentration (note mass must be in grams and solution volume in mL; otherwise perform a metric conversion first):

$$\% \text{ m/v} = \frac{\text{grams solute}}{\text{mL solution}} \times 100$$

Substitute the mass of solute (it must be in grams) and volume of solution (it must be in mL) into the equation and solve:

$$\frac{0.15 \text{ g}}{275 \text{ mL}} \times 100 = 0.055\%$$

8-16 You are supplied with the concentration and the volume of solution and asked to calculate the mass of solute. Using dimensional analysis, first convert the supplied volume unit, 3.0 L, into mL. Then multiply by the correct form of the concentration conversion factor that allows mL to cancel, leaving only grams:

$$3.0 \text{ L} \times \frac{10^3 \text{ mL}}{\text{L}} \times \frac{5.0 \text{ g}}{100 \text{ mL}} = 150 \text{ g}$$

Note that the concentration, 5.0%, has been written as the ratio 5.0 g/100 mL because % means "per 100."

8-17 You are supplied with the concentration and the mass of solute and asked to solve for the volume of solution. Using dimensional analysis, multiply the supplied mass of solute, 3.5 g, by the correct form of the concentration conversion factor that allows grams to cancel, leaving only mL of solution:

$$3.5 \text{ g} \times \frac{100 \text{ mL}}{10. \text{ g}} = 35 \text{ mL}$$

The concentration, 10.%, is written as the ratio 10. g/100 mL because that is what the % symbol means. In this exercise, we use the inverted form of the concentration conversion factor because grams must be in the denominator in order to cancel with the supplied unit of grams.

8-18 In this problem we must first determine the numerical value in m/v that 50 ppb represents. Note that *ppb* stands for "per billion," as % stands for "per 100." Remember, a billion is 1×10^9.

$$50 \text{ ppb} = \frac{50 \text{ g}}{1 \times 10^9 \text{ mL}}$$

Using dimensional analysis, first convert the supplied unit, 1 L, into mL by performing a metric conversion, and then multiply by the concentration conversion factor that allows mL to cancel, leaving grams of lead:

$$1 \text{ L} \times \frac{10^3 \text{ mL}}{1 \text{ L}} \times \frac{50 \text{ g}}{1 \times 10^9 \text{ mL}} = 0.00005 \text{ g} = 5 \times 10^{-5} \text{ g}$$

Thus, no more than 0.00005 g of lead is allowed in every one liter of drinking water in Illinois. As you can see, using ppb as a concentration unit avoids all those zeros when there is a trace amount of solute.

PRACTICE EXERCISES

21 Calculate the concentration in % m/v of a carbonated beverage that contains 28 g of sucrose in 315 mL of beverage.

22 What volume, in milliliters, of a physiological saline solution contains 2.50 g of NaCl? Assume a physiological saline solution has a concentration of 0.900% m/v.

23 When a scientist prepares a 100. mL solution with a given concentration, she must mix the correct amount of solute and solvent. The solute is weighed, but why can she not simply add 100 mL of solvent to the solute? Provide the steps you would take to properly prepare 100 mL of a 2% aqueous glucose solution using a 100. mL volumetric flask.

24 What is the mass of solute in 1.0 L of a 0.45% NaCl (% m/v) solution used for IV therapy? (Note: A metric conversion is involved in this calculation.)

You have been introduced to the most common units of concentration encountered in medicine based on the *mass of solute*: m/v and % m/v. In the next section, you will be introduced to the most common units of concentration based on *moles of solute*: molarity and equivalents/liter.

Molar Concentration

The unit of concentration most often used in chemistry is **molar concentration**, also known as **molarity** and abbreviated **M**. The molar concentration of a solution is defined as the *moles* of solute per liter of solution as shown below. Note that the unit of volume in the denominator must be in liters.

$$M = \frac{\text{moles of solute}}{\text{liters of solution}}$$

In chapter 4 you learned that the mole is a counting unit that allows us to keep track of the number of atoms, ions, or molecules in a sample of matter. We can also keep track of molecules, atoms, or ions in a solution by calculating the molar concentration of the solution because molarity represents the *number* of solute particles rather than the *mass* of the solute particles in one liter of solution. The molarity of a solution is a more useful concentration unit than a mass-based unit when you are doing chemical reactions, where it is important to combine the correct *number* of reacting molecules, atoms, or ions. For example, a 1 *M* aqueous sodium chloride solution will have the same number of Na^+ ions in a given volume as a 1 *M* solution of glucose will have glucose molecules.

Blood tests typically report concentrations of calcium, sodium, potassium, and chloride ions and also carbon dioxide in units of *milli*moles per liter (mmol/L = m*M*). When the prefix *milli-* is applied to the base unit of the mole, we have the following standard metric conversion:

$$10^3 \text{ mmol} = 1 \text{ mol}$$

The units millimole and micromole are used when the concentration of solute is very small to avoid many zeros in the numerical value.

In calculations involving molarity as a conversion factor, you should always replace the abbreviation M with moles/L so that units can cancel. Do not mix up M and mol, as they mean something entirely different. Some sample calculations using molarity are shown in the Worked Exercises below.

> To calculate the molar concentration of a solution when the mass rather than the moles of solute is provided, remember that you can convert the mass of any substance into moles using its molar mass as a conversion factor, as described in Chapter 4.

WORKED EXERCISES Molarity Calculations

8-19 A solution with a total volume of 1.5 L contains 0.018 mole of carbon dioxide (CO_2). What is the molar concentration (*M*) of this solution?

8-20 Calculate the number of moles of glucose in 500. mL of a 0.50 *M* glucose solution.

8-21 How many milliliters of a 0.5 *M* glucose solution contains 0.2 mole of glucose?

Solutions

8-19 In this exercise you are asked to calculate a molar concentration—moles per liter, given the moles of solute and the liters of solution. Substitute these values into the molarity equation and solve:

$$M = \frac{\text{moles of solute}}{\text{liters of solution}}$$

$$\frac{0.018 \text{ mol}}{1.5 \text{ L}} = 0.012 \text{ mol/L} = 0.012 \; M$$

8-20 The volume of the solution supplied (500. mL) must first be converted into liters since molarity always has units of liters for the volume of solution. Using dimensional analysis, perform a metric conversion on the supplied volume of solution, 500. mL, and then multiply by the correct form of the supplied concentration, a conversion factor, that allows volume to cancel:

$$500. \; \cancel{mL} \times \frac{1 \; \cancel{L}}{10^3 \; \cancel{mL}} \times \frac{0.50 \text{ mol}}{1 \; \cancel{L}} = 0.25 \text{ mol glucose}$$

Note that we must turn the abbreviation M into mol/L when using molar concentration in a calculation so that units can cancel.

8-21 Multiply the moles of solute supplied, 0.2 mol, by the inverted form of molar concentration as a conversion factor so that moles of solute cancels, leaving liters of solution. Then perform a metric conversion from liters to milliliters as requested:

$$0.2 \; \cancel{mol} \times \frac{1 \; \cancel{L}}{0.5 \; \cancel{mol}} \times \frac{10^3 \text{ mL}}{1 \; \cancel{L}} = 400 \text{ mL}$$

PRACTICE EXERCISES

25 A mixture with a volume of 3.0 L contains 0.023 mole of oxygen (O_2). What is the molar concentration (molarity) of oxygen in this solution?

26 How many moles of lithium ions, Li^+, are there in 2.5 L of 1.2 M aqueous lithium chloride?

27 How many liters of a 1.25 M aqueous glucose solution contain 0.5 mole of glucose?

Equivalents per Liter

For an electrolyte, the *moles of charge per liter* are reported as **equivalents per liter, eq/L**. An equivalent, abbreviated eq, is the moles of charge, which depends on the magnitude of charge on the ion and the number of moles of ion. Equivalents per liter (eq/L) are simply calculated by multiplying the molar concentration of the solution, M, by the numerical value of the charge on the ion:

$$\frac{\text{eq}}{\text{L}} = \frac{\text{mol}}{\text{L}} \times |\text{charge}| = M \times |\text{charge}|$$

where the symbol | | stands for "absolute value", which means we remove the sign. Thus, for ions with a +1 or −1 charge, the equivalent per liter is equal to the moles per liter, M. For ions with a +2 or −2 charge, the eq/L is two times the molarity. For example, a 1 M solution of calcium ions (Ca^{2+}) is equal to

$$1 \; M \times 2 = 2 \text{ eq/L}$$

TABLE 8-5	Common Ions and Equivalents/L	
Charge	Common Ions	Equivalents/L
+1	$Na^+, K^+, H^+, Li^+, NH_4^+$ (ammonium)	1
+2	Ca^{2+}, Mg^{2+}	2
+3	$Fe^{3+}, Cr^{3+}, Al^{3+}$	3
−1	$Cl^-, F^-, C_2H_3O_2^-$ (acetate)	1
−2	$O^{2-}, S^{2-}, CO_3^{2-}$ (carbonate)	2
−3	PO_4^{3-} (phosphate), $C_6H_5O_7^{3-}$ (citrate)	3

Table 8-5 lists some common ions, their charge, and the concentration in eq/L for a 1 M solution. Blood tests report electrolyte concentrations in meq per liter (meq/L) because the solute concentration is small. The electrolyte concentrations reported in a blood test include calcium (Ca^{2+}), sodium (Na^+), potassium (K^+), and chloride (Cl^-) ions. Equivalents per liter (eq/L) is a useful concentration unit when we want to assess electrolyte balance since it measures the amount of charge, regardless of the mass of or type of ions.

WORKED EXERCISES · Converting Molar Concentration to eq/L or meq/L

8-22 Consider a blood test that shows a calcium ion concentration of 9 mmol/L. Convert this concentration into units of meq/L of Ca^{2+}. Note that meq/L is equal to mmol/L × charge, analogous to eq/L being equal to mol/L × charge.

8-23 What is the concentration in eq/L for a 2.5 M solution of Fe^{3+} ions?

8-24 How many equivalents per liter of sodium ions are there in a 0.50 M solution of Na_2SO_4? Hint: you will need to consider the dissolution equation and the number of moles of Na^+ produced for every mole of Na_2SO_4 in solution.

Solutions

8-22 Substitute the millimolar concentration into the equation for meq/L. The charge on a calcium ion is +2, so we use the value without the sign, 2, in the equation:

$$\frac{meq}{L} = mM \times charge$$

$$\frac{meq}{L} = 9 \ mM \times 2 = 18 \ meq/L$$

8-23 The molar concentration is 2.5 M, and the charge on the ion is +3:

$$2.5 \ M \times 3 = 7.5 \ eq/L$$

8-24 In this problem we must calculate the number of moles of Na^+ in solution for every 1 mole of Na_2SO_4:

$$Na_2SO_4(s) \rightarrow 2 \ Na^+(aq) + SO_4^{2-}(aq)$$

The dissolution equation shows that two moles of Na^+ enter solution for every 1 formula unit of Na_2SO_4:

$$0.50 \ M \ Na_2SO_4 \times \frac{2 \ M \ Na^+}{1 \ M \ Na_2SO_4} = 1.0 \ M \ Na^+$$

Next, multiply the molar concentration of Na^+ by the numerical value of the charge on a sodium ion, 1:

$$\frac{eq}{L} = M \times charge = 1.0\ M \times 1 = 1.0\ eq/L$$

PRACTICE EXERCISES

28 How many equivalents per liter of phosphate ion (PO_4^{3-}) are there in a 0.25 M solution of phosphate ions?

29 How many equivalents per liter of magnesium ion are there in a 0.10 M solution of magnesium hydroxide?

30 How many equivalents per liter of chloride ion are there in a 0.50 M solution of $FeCl_3$?

Solution Dilution

You have learned how to prepare solutions with a given concentration by weighing the pure solute and adding the appropriate amount of solvent—such as water—using a volumetric flask. A solution with a given concentration can also be prepared by diluting a more concentrated solution. *Dilution is a method for preparing a less concentrated solution from a more concentrated solution.* Knowing how to prepare a solution by dilution is important because many solutes are not readily available in their pure form and are instead only available as **stock solutions**—preprepared solutions of known concentration. Many medicines and household products are available as stock solutions. For example, hydrogen peroxide is sold in pharmacies as a 3% aqueous solution and commercial bleach is available as a 5.5% aqueous solution of NaOCl. Sometimes stock solutions are used because the solution is not readily prepared from the solute. For example, the preparation of an aqueous solution of hydrochloric acid, HCl, from the pure solute is difficult because HCl is a gas. However, aqueous solutions of 12 M HCl, known as concentrated HCl, are commercially available.

To prepare a dilute solution with a volume V_2 and concentration C_2 from a more concentrated solution with a concentration C_1 and a volume V_1, use the dilution equation:

$$C_1 \times V_1 = C_2 \times V_2$$

where C_1 = concentration of the more concentrated solution (stock solution),
     V_1 = volume of the more concentrated solution (stock solution),
     C_2 = concentration of the dilute solution,
     V_2 = volume of the dilute solution

This equation can be applied to any concentration units: m/v, %m/v, molarity, M, provided C_1 and C_2 have the same units and V_1 and V_2 have the same units. For example, if molarity, M, is the concentration unit used for C_1 and C_2, the equation can be rewritten

$$M_1 \times V_1 = M_2 \times V_2$$

Imagine we were to prepare 100 mL of a 1% m/v saline solution from a 10% saline stock solution. We would first need to calculate how much of the stock solution to add to a 100 mL volumetric flask. The steps for solving a dilution calculation, illustrated with this example, are described in the Guidelines box on the next page.

Guidelines for Solving a Dilution Calculation

Step 1. Begin by determining the three supplied variables and the one variable that needs to be calculated. Remember from algebra, that if you have one equation, you can solve for at most one unknown. Applied to a dilution calculation, any one of these variables can be determined if the other three are known. In this example:

$C_1 = 10\%$ concentration of stock solution

$C_2 = 1\%$ desired concentration of dilute solution

$V_1 = ?$ the variable being solved: how much of the stock solution we need to dilute

$V_2 = 100$ mL   the requested final volume of the dilute solution

Step 2. Rearrange the dilution equation so that the variable you are solving for—the unknown—is isolated on one side of the equality, by dividing both sides of the equation by the appropriate variable(s). In this example, we want to solve for V_1, so we isolate it on one side of the equality by dividing both sides by C_1:

$$V_1 = \frac{C_2 \times V_2}{C_1}$$

Step 3. Substitute the supplied variables listed in step 1 into the algebraically rearranged equation from step 2 and solve for the unknown variable. In this example:

$$V_1 = \frac{1\% \times 100 \text{ mL}}{10\%} = 10 \text{ mL}$$

Step 4. Prepare the dilute solution by transferring the calculated volume of stock solution, from step 3, to a volumetric flask with volume V_2. Then add water to the mark, with mixing. In this example, you would transfer 10 mL of the 10% stock saline solution to a 100. mL volumetric flask and then add water, the solvent, to the mark on the 100. mL volumetric flask, along with mixing. You now have a 1% saline solution.

Note that the amount of solute removed from the concentrated solution is the same amount of solute that is in present in the dilute solution since only solvent was added to the solution.

WORKED EXERCISES   Dilution Calculations

8-25 A student needs 250. mL of a 1.0 M HCl solution, and all he can find is concentrated HCl, a stock solution with a concentration of 12 M. How should he prepare 250. mL of 1.0 M HCl using the HCl stock solution?

8-26 A student transferred 50. mL of a 2.5 M NaOH stock solution to a 500. mL volumetric flask and added water to the mark. What is the molar concentration, M, of the solution she prepared?

8-27 How many moles of NaOH are in 50. mL of the stock solution in the previous question? How many moles of NaOH are in the 500. mL of the dilute solution in the previous question?

Solutions

8-25 **Step 1.** Begin by writing the three supplied variables and determine the variable that needs to be calculated:

$$V_1 = ?$$
$$C_1 = 12 \ M$$
$$V_2 = 250. \text{ mL}$$
$$C_2 = 1.0 \ M$$

Step 2. Algebraically rearrange the dilution equation so that the variable you are solving for is isolated on one side of the equality, by dividing both sides of the equation by the appropriate variable(s). Here we want to isolate V_1 on one side of the equation so we divide both sides by C_1:

$$V_1 = \frac{C_2 \times V_2}{C_1}$$

Step 3. Substitute the supplied variables listed in step 1 into the algebraically rearranged equation in step 2 and solve for the unknown variable determined in step 1. In this example:

$$V_1 = \frac{C_2 \times V_2}{C_1} = \frac{1.0\ M \times 250.\ \text{mL}}{12\ M} = 21\ \text{mL}$$

Step 4. Transfer 21 mL of the stock solution to a 250 mL volumetric flask and fill the volumetric flask to the mark with water while mixing.

8-26 Step 1. Begin by writing the three supplied variables and determine the variable that needs to be calculated:

$$V_1 = 50.\ \text{mL}$$
$$C_1 = 2.5\ M$$
$$V_2 = 500.\ \text{mL}$$
$$C_2 = ?$$

Step 2. Rearrange the dilution equation so that the variable you are solving for is isolated on one side of the equality, by dividing both sides of the equation by the appropriate variable(s). Begin with the dilution equation:

$$C_1 \times V_1 = C_2 \times V_2$$

Rearrange the equation so that C_2, the unknown variable, is isolated on one side of the equality. This is done by dividing both sides by V_2:

$$C_2 = \frac{C_1 \times V_1}{V_2}$$

Step 3. Substitute the supplied variables determined in step 1 into the rearranged equation in step 2 and solve for the unknown variable determined in step 1. In this example:

$$C_2 = \frac{2.5\ M \times 50.\ \text{mL}}{500.\ \text{mL}} = 0.25\ M$$

Therefore, the diluted solution prepared as described above has a concentration of 0.25 M.

8-27 To determine the number of moles of solute in 50. mL of 2.5 M NaOH, we use molar concentration as a conversion factor, remembering to use mol/L for M, but we must also convert mL to L so that liters will cancel in the denominator of the conversion factor:

$$50.\ \text{mL} \times \frac{1\ \text{L}}{10^3\ \text{mL}} \times \frac{2.5\ \text{mol}}{1\ \text{L}} = 0.13\ \text{mol solute}$$

To determine the concentration of the diluted solution, we can perform the same type of calculation using the total volume of the dilute solution and the molar concentration calculated in 8-26:

$$500.\ \text{mL} \times \frac{1\ \text{L}}{10^3\ \text{mL}} \times \frac{0.25\ \text{mol}}{1\ \text{L}} = 0.13\ \text{mol solute}$$

As expected, the number of moles of solute does not change in a dilution.

31 What volume of a 2.0 *M* glucose stock solution would you need in order to prepare 50. mL of a 1.0 *M* glucose solution?

32 How would you prepare 100. mL of a 0.90% physiological saline solution from a stock solution of 10.% m/v saline solution?

33 What is the molar concentration of a lidocaine solution prepared by diluting 10. mL of a 2.0 *M* lidocaine stock solution to 25 mL in a 25 mL volumetric flask?

Oral Medications

Oral medications are substances given by mouth and administered, especially to children, in solution form. For example, NyQuil Cold & Flu is an over-the-counter aqueous solution with a concentration of 21.7 mg/mL of acetaminophen, 1 mg/mL of dextromethorphan HBr, 0.42 mg/mL doxylamine succinate, and 0.1 mL/mL ethanol (Figure 8-15). The recommended *dose* of NyQuil is 15.0 mL every 6 hours. Thus, a patient taking 15.0 mL of NyQuil is receiving 326 mg of acetaminophen, calculated as follows:

$$15.0 \; \cancel{mL} \times \frac{21.7 \; mg}{1 \; \cancel{mL}} = 326 \; mg \; acetaminophen$$

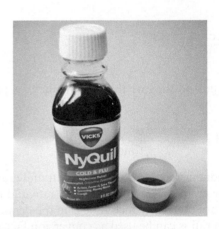

Figure 8-15 NyQuil, an over-the-counter oral medication taken by mouth in liquid form, is an aqueous solution containing acetaminophen, an analgesic, and several other solutes.
[Catherine Bausinger]

WORKED EXERCISE Calculating How Much Oral Drug Suspension to Administer

8-28 An order is given for 500. mg of amoxicillin (Figure 8-16) to be administered to a patient every 6 hours. For a suspension with a concentration of 200. mg/5.00 mL, what volume should be administered to the patient every 6 hours?

Solution

8-28 The supplied unit is the doctor's order of 500. mg of amoxicillin (the solute). The concentration of the oral medication is also supplied (200. mg/5.00 mL). Calculate the volume of suspension required using dimensional analysis by multiplying the supplied mass of solute by the correct form of concentration as a conversion factor:

$$500. \; \cancel{mg} \times \frac{5.00 \; mL}{200. \; \cancel{mg}} = 12.5 \; mL \; solution$$

Thus, the patient should receive 12.5 mL of the oral medication every 6 hours.

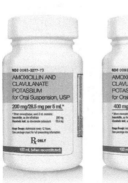

Figure 8-16 Amoxicillin, an antibiotic, and Clavulanate potassium are prepared as an oral suspension with an amoxicillin concentration of 200. mg/5.00 mL or 400. mg/5.0 mL and a Clavulanate concentration of 28.5 mg/5 mL or 57 mg/5 mL.
[rvlsoft/Shutterstock]

34 An order is given for 60. mg of Tylenol to be administered to a patient every 4 hr. The pharmacy supplies a 40. mg/mL solution of Tylenol. How much of the Tylenol solution should you administer to the patient every 4 hr?

35 Tegretol is commonly used as an anticonvulsant and mood-stabilizing drug. The order is given to administer 250. mg as needed. The medicine is only available as a solution with a concentration of 20.0 mg/mL. How many milliliters of the Tegretol solution should be administered to the patient as needed?

36 An order is given to administer 20 mg of Atarax, an antihistamine, every 4 hr. The solution has a concentration of 2 mg/mL. How many milliliters should you administer to the patient every 4 hr?

IV Solutions

Intravenous (IV) administration directly into a vein is often used to deliver a pharmaceutical gradually over a period of time because the body metabolizes the drug rapidly, the drug is not stable when given orally, or the patient is unconscious or unable to take medications orally. For IV drug administration, a solution with a known concentration of the medicine is infused at a specified volume (mL) of solution per unit of time (minutes or hours), known as a **flow rate**.

$$\text{flow rate} = \frac{\text{solution volume}}{\text{time}}$$

The flow rate selected on the IV depends on the rate that the medicine is prescribed by the physician:

$$\text{rate medicine prescribed} = \frac{\text{solute mass (medicine)}}{\text{time}}$$

The rate at which the drug should be administered is usually given in milligrams per minute, micrograms per minute, or grams per hour. Both flow rate and the rate at which the medicine is delivered are ratios in which time is in the denominator, as is the case for any "rate." Both ratios can be used as conversion factors. For example, multiplying concentration by flow rate, yields the rate the medicine is prescribed by the physician because solution volume cancels.

$$\frac{\text{solute mass}}{\text{solution volume}} \times \frac{\text{solution volume}}{\text{time}} = \frac{\text{solute mass}}{\text{time}}$$

$$\text{concentration} \times \text{flow rate} \qquad = \text{rate medicine delivered}$$

Similarly, multiplying the inverted form of concentration by the rate prescribed yields the flow rate because solute mass cancels:

$$\text{flow rate} = \frac{\text{solution volume}}{\text{solute mass}} \times \frac{\text{solute mass}}{\text{time}}$$

WORKED EXERCISES   IV Solution Calculations with Flow Rate and Concentration

8-29 An order is given to infuse 50. units per hour of heparin, an anticoagulant. The IV bag supplied has a concentration of 250 units/mL (Figure 8-17). At what flow rate in milliliters per hour should the IV solution be infused into the patient? Note that proteins are sometimes reported in international units, IU, or simply "units." Treat international units the same as any other solute unit.

8-30 Isuprel is used to treat asthma, chronic bronchitis, and emphysema. A patient is receiving 30. mL/hr of a solution with a concentration of 8.0 µg/mL of Isuprel in D5W. How many micrograms per hour is the patient receiving?

Figure 8-17 Two IV solutions of heparin: Left 500 mL of 25,000 units (50 units/mL). Right: 500 mL of 1000 units (2 units/mL). [Scott Linnett/ ZUMA Press/Newscom]

Solutions

8-29 In this exercise you are asked to calculate the flow rate (mL/hr) and supplied with the concentration (250 units/mL) and the rate the medicine should be administered (50. units/hr). It is important to recognize that the supplied terms are conversion factors that can be used in either of two forms:

$$\text{Rate medicine delivered: } \frac{50. \text{ units}}{1 \text{ hr}} \quad \text{or} \quad \frac{1 \text{ hr}}{50. \text{ units}}$$

$$\text{Concentration: } \frac{250 \text{ units}}{1 \text{ mL}} \quad \text{or} \quad \frac{1 \text{ mL}}{250 \text{ units}}$$

The flow rate will also be a ratio:

$$\text{flow rate} = \frac{\text{solution volume}}{\text{time}} = \frac{\text{mL}}{\text{hr}}$$

In this type of problem, look at the units required in the answer, flow rate (mL/hr), to determine how to multiply the conversion factors so that units cancel, leaving only mL in the numerator and hr in the denominator. To do this, select the correct form of the concentration conversion factor that has mL in the numerator and multiply by the correct form of prescribed rate that has hr in the denominator:

$$\frac{1 \text{ mL}}{250 \text{ units}} \times \frac{50. \text{ units}}{1 \text{ hr}} = 0.20 \frac{\text{mL}}{\text{hr}}$$

inverted concentration × prescribed rate = flow rate

8-30 In this exercise you are asked to calculate the rate the medicine is administered, in micrograms of solute (the medicine) every hour (µg/hr) given the flow rate (30. mL/hr) and the concentration (8.0 µg/mL). The two conversion factors supplied can be used in either of two forms:

$$\text{Flow rate: } \frac{30. \text{ mL}}{1 \text{ hr}} \quad \text{or} \quad \frac{1 \text{ hr}}{30. \text{ mL}}$$

$$\text{Concentration: } \frac{8.0 \text{ µg}}{1 \text{ mL}} \quad \text{or} \quad \frac{1 \text{ mL}}{8.0 \text{ µg}}$$

The rate the medicine is administered, micrograms per hour, is also a ratio:

$$\text{Mass Isuprel per hour} = \frac{\text{µg}}{\text{hr}}$$

To set up the calculation, multiply the conversion factor that has micrograms in the numerator by the conversion factor that has hours in the denominator so that mL cancel:

$$\frac{8.0 \text{ µg}}{1 \text{ mL}} \times \frac{30. \text{ mL}}{1 \text{ hr}} = 240 \frac{\text{µg}}{\text{hr}}$$

concentration × flow rate = rate Isoprel delivered

Thus, the patient is receiving 240 micrograms of Isuprel per hour. Isn't dimensional analysis great!

37 Digoxin is often used to treat a variety of heart conditions, for example, heart failure, atrial flutter, or atrial fibrillation. An order is given to administer 250 μg digoxin by IV every 5 min. The digoxin solution concentration is 2.5 μg/mL. What should you set the flow rate to in milliliters per minute?

38 Droperidol is often used to treat postoperative nausea. An order is given to administer 0.275 mg droperidol by IV over 3.0 hr. The IV bag contains 2.5 mg/mL droperidol. What should you set the flow rate to in milliliters per hour?

39 A diabetic patient suffering from hyperglycemia is receiving an insulin solution by IV at a rate of 25 mL/hr. The concentration of the insulin solution is 0.4 unit/mL. How many units of insulin per hour is the patient receiving?

■ 8.4 Osmosis and Dialysis

Our bodies are composed mainly of water, and this abundant substance serves as the solvent or medium for many different aqueous mixtures throughout the body. Most of these aqueous mixtures are enclosed within a membrane that separates the mixture on the inside of the membrane from the mixture on the outside of the membrane. The most important membranes in our body are **cell membranes,** which function to separate and control what enters and leaves the cell, a property of selectively permeable membranes (Figure 8-18). A **selectively permeable membrane** allows only certain substances to pass through the membrane at certain times. Colloids and suspended particles are too large to pass through a selectively permeable membrane; thus, proteins and cellular organelles remain inside the cell. Whether or not a particular solute can pass through a membrane depends on the nature of the membrane and the charge, polarity, and size of the solute. The chemical composition of cell membranes will be described in Chapter 12, Lipids.

Permeability of the Cell Membrane

Cell membranes vary in their permeability, which depends on several characteristics of the membrane itself. However, the most important factors in determining whether a solute can pass through a cell membrane are the size and polarity of the solute. The interior of a cell membrane is nonpolar, so the most permeable substances are small nonpolar molecules such as oxygen (O_2) and carbon dioxide (CO_2), followed by small polar neutral molecules such as water, followed by larger polar molecules such as glucose. The least permeable solutes

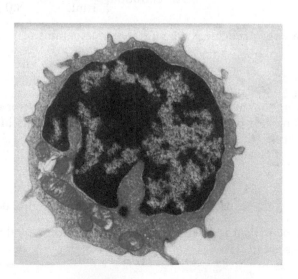

Figure 8-18 Micrograph of a normal cell showing the cell membrane. [Biophoto Associates/ Science Source]

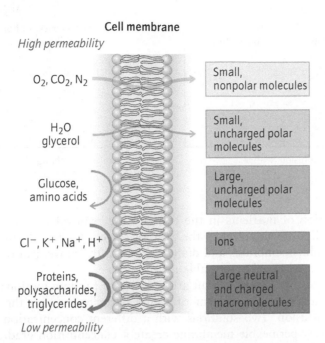

Cell membrane

High permeability

O_2, CO_2, N_2 → Small, nonpolar molecules

H_2O glycerol → Small, uncharged polar molecules

Glucose, amino acids → Large, uncharged polar molecules

Cl^-, K^+, Na^+, H^+ → Ions

Proteins, polysaccharides, triglycerides → Large neutral and charged macromolecules

Low permeability

Figure 8-19 The membrane permeability of various substances depends on their size and polarity.

are the highly polar charged ions such as chloride (Cl^-), sodium (Na^+), and potassium (K^+). The relative permeability of the cell membrane toward various solutes is summarized in Figure 8-19. Colloids and suspensions are unable to cross the cell membrane at all because of their large size. Ions require assistance to cross the membrane, a process that often requires energy and specialized proteins. Small nonpolar solutes and water can pass freely through the cell membrane by diffusion.

It is the selective permeability of the cell membrane that allows the concentration of ions to be significantly different on the inside of the cell compared to the outside of the cell, as shown in Table 8-6. The different concentrations of ions on either side of the cell membrane is essential for the transmission of nerve impulses and other cellular signaling processes. The following terms describe the *relative* concentration of two solutions separated by a selectively semipermeable membrane:

- **hypertonic**—the solution with the higher solute concentration (lower water concentration),
- **hypotonic**—the solution with the lower solute concentration (higher water concentration),
- **isotonic**—solutions with equal solute concentrations

Dialysis

Osmosis and dialysis describe the movement of solvent and/or solutes across a selectively permeable membrane. These are important processes that control

TABLE 8-6 Concentration of Electrolytes Inside and Outside a Blood Cell

Electrolyte	Concentration Inside the Cell (mM)	Concentration Outside the Cell (mM)
Na^+	10	140
Cl^-	4	100
K^+	140	4
Ca^{2+}	1×10^{-4}	2.5

Figure 8-20 Dialysis: the spontaneous movement of solute particles across a selectively permeable membrane from (a) a region of higher concentration (left) to a region of lower concentration (right) until (b) both solutions have the same concentration.

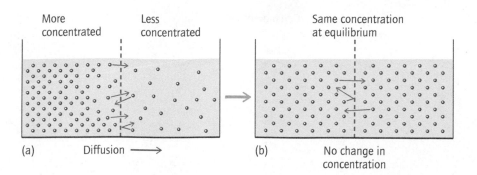

the distribution of nutrients in the cell and the removal of waste products from the cell. **Osmosis** refers to the flow of *solvent* (water) through a selectively permeable membrane, and **dialysis** refers to the movement of small *solutes* through a selectively permeable membrane.

The solutes in a liquid solvent are in constant random motion as a result of kinetic energy. The movement of solute particles through the solvent is known as **diffusion**. Two solutions with a different concentration separated by a selectively permeable membrane create a **concentration gradient**. Diffusion of the solute always occurs from the hypertonic solution to the hypotonic solution (in the direction of the concentration gradient) until both solutions have an equal concentration. This movement of solute particles across a selectively permeable membrane is known as **dialysis**.

Figure 8-20a shows two solutions with different concentrations separated by a selectively permeable membrane. The solution on the left has a higher solute concentration (hypertonic solution) than the solution on the right (hypotonic solution). Dialysis occurs until both solutions have the same concentration (isotonic solutions) (Figure 8-20b). At equilibrium (Figure 8-20b), both solutions have the same concentration. Note that dialysis involves the diffusion of only certain small solute particles. Colloids (proteins) and suspended particles do not cross the cell membrane.

WORKED EXERCISE   Dialysis

8-31 A selectively permeable dialysis membrane separates an aqueous mixture A containing urea, a small molecule, and albumin, a large protein, from pure water (Figure 8-21). After dialysis, what do the two compartments look like? Does urea cross the membrane? Does albumin cross the membrane? Explain.

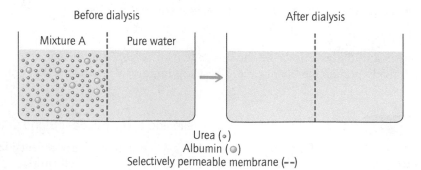

Figure 8-21

Urea (○)
Albumin (◉)
Selectively permeable membrane (--)

Solution

8-31 Since urea is a small solute molecule, it will diffuse (dialysis) across the selectively permeable membrane from solution A to the pure water compartment: from higher solute concentration (hypertonic solution) to lower solute concentration (hypotonic solution) until the concentrations are equal (isotonic). Albumin is a colloid, so it is unable to cross the membrane because of its large size (Figure 8-22).

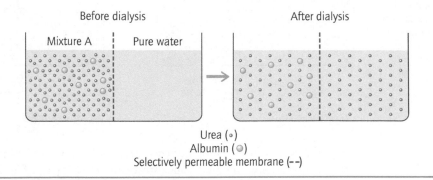

Figure 8-22

PRACTICE EXERCISES

40 Consider the following solution separated from pure water by a selectively permeable membrane (Figure 8-23). One compartment contains a solution of sucrose (a small molecule) and glycogen (a polysaccharide), whereas the other compartment contains pure water. Describe how you could separate sucrose from glycogen based on the principles of dialysis.

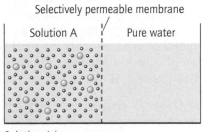

Figure 8-23 Solution A is an aqueous mixture of sucrose (∘) and glycogen (⊙).

41 Consider two solutions that are separated by a selectively permeable membrane that is designed for dialysis. One side of the membrane contains 0.45% saline solution (1/2 NS) and the other contains a physiological saline solution (0.90% m/v NaCl). Which solution is hypertonic? Which solution is hypotonic? In which direction will ions diffuse why? How will the concentrations of the two solutions change after dialysis.

42 What does the term *selectively permeable* mean?

Osmosis

Water, the solvent or medium in most biological mixtures, can cross a selectively permeable membrane when solutes cannot. The diffusion of only water across a selectively permeable membrane is referred to as **osmosis**. Osmosis occurs when there is a concentration gradient between two solutions separated by a selectively permeable membrane. Osmosis, the diffusion of water across a membrane, always occurs from the solution with the lower solute concentration (higher water concentration), the hypotonic solution, to the solution with the higher solute concentration (lower water concentration), the hypertonic solution, until both solutions have the same concentration (isotonic). So instead of the solute diffusing (dialysis) from the more concentrated to the less concentrated solution, water diffuses (osmosis) from the less

Figure 8-24 Osmosis (a) Before osmosis: two aqueous solutions with different concentrations are separated by a selectively permeable membrane. (b) After osmosis: water has diffused across the membrane from the solution with the lower solute concentration to the solution with the higher solute concentration, creating solutions with equal concentrations. The volume of the hypertonic solution increases as a result of the influx of water.

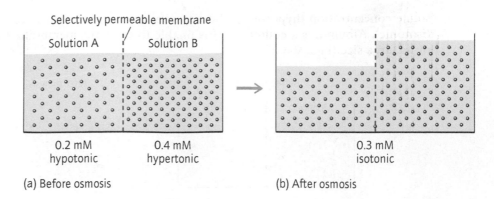

Selectively permeable membrane

Solution A | Solution B

0.2 mM
hypotonic

0.4 mM
hypertonic

0.3 mM
isotonic

(a) Before osmosis   (b) After osmosis

concentrated to the more concentrated solution, achieving the same end result: solutions with equal concentrations, as shown in Figure 8-24.

As shown in Figure 8-24, *the hypertonic solution also increases in volume as a result of osmosis.* Figure 8-24a shows two solutions of equal volume separated by a selectively permeable membrane. Water (the solvent) will diffuse in the direction of the more concentrated solution. Thus, if solution A has a concentration of 0.2 m*M* and solution B has a concentration of 0.4 m*M*, water will diffuse from solution A to solution B. Osmosis will continue until the solute concentration on both sides of the membrane is the same, in this case, 0.3 m*M* (Figure 8-24b). At that point, the net flow of water will stop. After osmosis, the solution that was initially hypertonic solution, will have a greater volume, as a result of the added water molecules that have diffused into this solution.

Osmosis is governed by the total *number* of solute particles, not their mass, size, or identity. For example, a solution may contain 2 mmol of glucose and 3 mmol of urea, but only the total concentration, 5 mmol, determines the direction of osmosis. Therefore, molar concentration (molarity), a mole-based concentration unit, is used to report concentrations for these applications.

Reverse Osmosis Osmosis can be stopped or prevented by applying external pressure on the hypertonic solution. The minimum amount of pressure that must be applied to stop osmosis is called the **osmotic pressure**. If a pressure *greater* than the osmotic pressure is applied to the hypertonic solution, **reverse osmosis** occurs, and in this way water can be forced to diffuse against a concentration gradient. In areas of the world where fresh water is scarce, reverse osmosis is used to produce fresh water from seawater. Reverse osmosis is also used in hemodialysis, a life-support treatment for individuals with kidney disease, where it is used to force excess water out of the blood.

Osmosis in Red Blood Cells Red blood cells placed in an isotonic solution maintain their healthy biconcave shape, as shown in Figure 8-25a. A person receiving IV fluids must always be given a solution isotonic with red blood cells, such as a physiological saline solution: 0.9% m/v saline or a 5% m/v glucose solution. Consider what happens to a red blood cell if it is immersed in a hypertonic or a hypotonic solution.

When a red blood cell is immersed in a *hypotonic* solution, Figure 8-25b, water diffuses through the cell membrane by osmosis from the solution to the inside of the cell, where the solute concentration is greater. Consequently, the volume of the red blood cell increases. Red blood cells immersed in a hypotonic solution will swell until they eventually burst—an event known as **hemolysis**.

When a red blood cell is immersed in a *hypertonic* solution, Figure 8-25c, water diffuses through the cell membrane by osmosis in the opposite direction: from the inside of the cell to the solution, where the solute concentration is greater. Consequently, the volume of the red blood cell shrinks and the cell assumes a shrunken appearance, a process known as **crenation**.

Osmolarity is a term that defines, in moles per liter, the number of particles in a solution that contribute to its osmotic pressure.

Red blood cell in:

(a) Isotonic solution   (b) Hypotonic solution   (c) Hypertonic solution

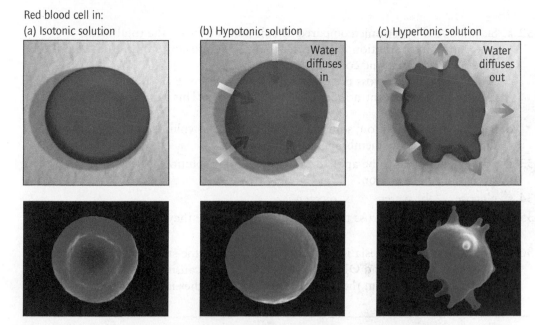

Figure 8-25 Red blood cells immersed in (a) an isotonic solution maintain their healthy biconcave shape because there is no net flow of water into or out of the cell. (b) a hypotonic solution swell as a result of water entering the cell by osmosis, (c) a hypertonic solution shrink as a result of water diffusing out of the cell by osmosis. [David M. Phillips/Science Source]

WORKED EXERCISES   Osmosis

8-32 Consider the two solutions A and B shown in **Figure 8-26**, which are separated by a selectively permeable membrane.

  a. Which is the hypertonic solution?

  b. Which is the hypotonic solution?

  c. If the solutes cannot cross the membrane, which direction will water diffuse: from solution A to B or from B to A? What is this process called?

  d. After osmosis which solution will have the greater volume: A or B?

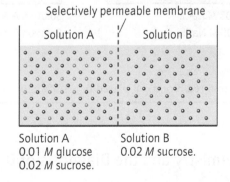

Selectively permeable membrane

Solution A   Solution B

Solution A
0.01 M glucose
0.02 M sucrose.

Solution B
0.02 M sucrose.

Figure 8-26

8-33 To which solution in the example above would you need to apply pressure in order to prevent osmosis?

8-34 Which solute in each pair is more likely to cross the cell membrane and why?

  a. Cl^- or glucose

  b. glucose or water

  c. oxygen or K^+

8-35 Why are proteins unable to cross the cell membrane?

8-36 Milk of magnesia works as a laxative by drawing water out of tissues through osmosis and increasing the water content of the stool. Explain what is occurring in terms of concentration and osmosis.

Solutions

8-32 a. Solution A is hypertonic compared to solution B because the total solute concentration of solution A is greater than the solute concentration of solution B.

b. Solution B is hypotonic compared to solution A.

c. Water will diffuse across the membrane from solution B to solution A in order to equalize the concentrations of the two solutions. This process is known as osmosis.

d. The hypertonic solution, solution A, will increase in volume as a result of water diffusing across the membrane into this solution.

8-33 Pressure would need to be applied to the hypertonic solution A because water is diffusing into this solution.

8-34 a. glucose   **b.** water   **c.** oxygen

8-35 Proteins are unable to cross the cell membrane because they are colloids—too large to cross the membrane.

8-36 Ingesting Milk of Magnesia introduces $Mg(OH)_2$ into the stool, which increases the number of Mg^{2+} ions and OH^- ions in solution, thus causing water to cross the membrane of the colon in the direction toward the higher ion concentration, thus acting as a laxative.

PRACTICE EXERCISES

43 Consider the solutions shown in Figure 8-27. Indicate the direction in which water flows through the membrane by osmosis. If no net flow occurs, state so.

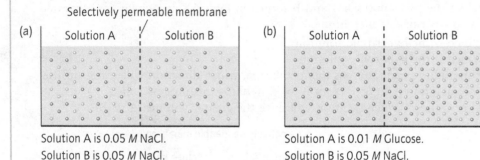

Solution A is 0.05 M NaCl.
Solution B is 0.05 M NaCl.

Solution A is 0.01 M Glucose.
Solution B is 0.05 M NaCl.

Figure 8-27

44 What happens to a red blood cell placed in a 1.5% m/v solution of sodium chloride?

Chemistry in Medicine Blood Chemistry and the Diagnosis of Disease

Almost everyone has had a blood test taken at some point in his or her life. It is a routine medical test, readily performed, and provides a wealth of information. A blood test is often called a blood chemistry panel because it measures the concentration of important atoms, ions, and molecules in the blood.

Blood is a complex mixture that contains nutrients, gases, hormones, and other molecules and ions that are transported to tissues throughout the body while also transporting waste products from cells to the kidneys, where they are filtered and eliminated from the body.

Abnormal levels or the presence of certain ions and compounds in the blood are often the first sign of organ malfunction. Problems with the kidneys, liver, adrenal glands, heart, and other organs can often be diagnosed with a simple blood test. To understand a blood chemistry panel, let us consider the composition of blood.

The Composition of Blood

Earlier in the chapter you learned that centrifugation of whole blood can be used to separate blood cells from the supernatant, blood plasma. Blood plasma that has had fibrinogen and other clotting factors removed is known as serum. Let us consider the components of blood serum that are measured in a typical blood test.

Blood = plasma + blood cells

Plasma = serum + fibrinogen and other clotting factors

Blood serum is a light-yellow-colored liquid composed of 90 percent water. Serum is not red because it does not contain red blood cells. The chemical substances dissolved in blood serum, grouped according to type of substance, include

- small molecules such as glucose, amino acids, creatinine, and urea. These small molecules are all neutral organic compounds.
- large biomolecules such as proteins; for example, albumin and other enzymes, and lipids such as triglycerides and cholesterol. These are all large organic molecules.
- electrolytes (Na^+, K^+, Ca^{2+}, Cl^-, HCO_3^-, etc.). These ions include most of the macronutrients and also important polyatomic ions.
- dissolved gases (CO_2, N_2, O_2). Although most of the oxygen in the blood is bound to hemoglobin, some is dissolved in the serum. These are nonpolar elements and compounds.

In the second column of the blood test shown in Figure 8-12, you will see the normal range of concentrations for the substances measured in this blood test. You will notice the units of concentration used in a blood test include mole-based units such as the mmol/L (and meq/L) and mass-based units such as mg/dL and g/dL. The numerator gives the mass or moles of solute or colloid, and the denominator gives the volume of the total solution—serum.

Substances of Interest in Blood Serum

Consider some of the important substances measured in a blood panel. Urea, referred to as blood urea nitrogen or BUN, has the molecular formula CH_4N_2O and the structure shown below. It is the final product of amino acid metabolism and is eliminated in the urine. High serum BUN levels are usually an indication of kidney problems.

Normal concentration: 7–20 mg/dL

$$\begin{array}{c} O \\ \| \\ H-N-C-N-H \\ | \quad\quad | \\ H \quad\quad H \end{array}$$

Urea, CH_4N_2O

Glucose, commonly referred to as blood sugar, is measured to determine if a patient is *hypoglycemic* (low blood sugar), or *hyperglycemic* (high blood sugar). Since a person's glucose concentration fluctuates throughout the day, depending on when and what they last ate, this test is

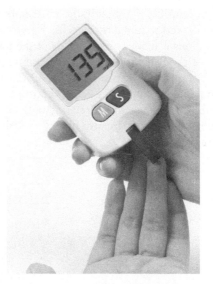

Figure 8-28 Diabetics monitor their blood glucose concentration throughout the day by testing a drop of their blood with a glucometer, an easy-to-use handheld device. [iStockphoto/ Thinkstock]

often given after the person has been fasting for 12 hours. Fasting glucose levels above 126 mg/dL are usually an indicator of diabetes. Diabetics must monitor their glucose levels throughout the day, typically by taking a drop of their blood and measuring glucose concentration using a glucometer, as shown in **Figure 8-28**.

Albumin, whose large organic structure is illustrated in **Figure 8-29**, is the most abundant protein found in the blood. The concentration of albumin regulates the amount of water in tissues and blood through osmosis.

Since many proteins are produced by the liver, assessing the concentration of several key liver proteins in the serum is a common method for evaluating liver function. When liver enzymes are higher than normal in the blood, it often is a sign of liver disease, shock, or dehydration, whereas low levels may suggest hemorrhage (bleeding), diarrhea, infection, or other liver diseases.

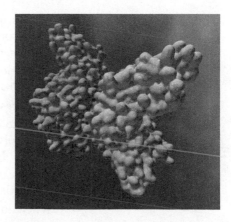

Figure 8-29 A space-filling model of the protein albumin. [© Bodwell Communications, Inc./Phototake]

Potassium ion, K^+, is arguably the most important cation in the body and required by all cells. The concentration of potassium ions is higher on the inside of the cell than on the outside of the cell. It is the opposite for sodium ions, Na^+. This concentration gradient is important for proper signaling in neurons and muscle cells. Abnormal potassium levels are observed when the adrenal glands malfunction. Increased potassium levels are also observed during cardiac arrest (heart attack).

As you can see, modern medicine is all about chemistry since our bodies are made up of chemical substances whose concentrations are carefully regulated. Too high or too low a concentration of any one substance is often associated with disease. Here you have seen that these substances are measured by a simple blood test, which provides a convenient and noninvasive window into the health of a patient.

Summary

Mixtures

- A mixture is a combination of two or more compounds and/or elements in any proportion.
- A mixture can be separated into its pure components through physical separation techniques.
- A homogeneous mixture has its components uniformly distributed throughout the mixture. Includes solutions and colloidal dispersions.
- A heterogeneous mixture has its components unevenly distributed throughout the mixture. Suspensions are heterogenous mixtures.
- Solutions, colloidal dispersions, and suspensions differ in the size of their particles.
- Colloidal dispersions are mixtures that contain particles, known as colloids, ranging in size from 1 nm to 1,000 nm. They are larger than solutes but smaller than particles of a suspension. Colloids are evenly distributed throughout the medium in a colloidal dispersion.
- Colloids are very large molecules, such as proteins or polysaccharides, or many small molecules clumped together, known as an aggregate.
- A suspension is a mixture in which the particles are larger than 1 micrometer. The particles in a suspension eventually settle.
- The particles of a suspension can be separated from the mixture by filtration or centrifugation, whereupon they appear as a pellet at the bottom of the centrifuge tube.
- A centrifuge can separate suspended particles from the rest of the mixture, known as the supernatant.

Solutions

- A solution is composed of a solvent and one or more solutes. The solvent is the component present in the greatest amount and the solute(s), the components in the lesser amount(s).
- A solution in which water is the solvent is known as an aqueous (aq) solution.
- Mixtures of gases are always homogeneous solutions.
- Polar solvents tend to dissolve polar solutes, and nonpolar solvents tend to dissolve nonpolar solutes. This general observation is described by the saying "like dissolves like."
- Hydrocarbons placed in water display hydrophobic—water-fearing—properties by separating into a separate layer.
- When an ionic compound dissolves in solution, the ionic bonds of the lattice break and the individual ions become separately surrounded by several water molecules, forming ion-dipole interactions.

- A solution of dissolved ions is a strong electrolyte because ions have a full charge and therefore conduct electricity.
- A precipitate is an insoluble solid that forms at the bottom of some heterogenous mixtures and saturated solutions.
- A saturated aqueous solution exceeds the amount of water available to dissolve the solute, resulting in a precipitate.
- Certain ionic compounds are insoluble in water and form a precipitate.
- When a molecular solute dissolves in solution, the covalent bonds of the molecules remain intact and only intermolecular forces of attraction between molecules are broken and replaced by new intermolecular forces of attraction with the solvent.
- Molecules (with the exception of acids) dissolved in water are nonelectrolytes because they are uncharged solutes.
- Nonpolar solutes dissolve in nonpolar solvents forming dispersion intermolecular forces of attraction.

Concentration

- Concentration is a quantitative measure of the amount of a solute per total volume of solution (solute + solvent).
- The solution with the higher concentration is said to be more concentrated and the solution with the lower concentration is said to be more dilute.
- The concentration of a solution is a ratio of the amount of solute in the numerator and the total volume of solution in the denominator.
- The most commonly encountered units of concentration in the medical field are mass/volume, % mass/volume, moles/L (M), and eq/L.
- A solution is prepared by adding a measured mass of solute to a volumetric flask of the appropriate size and adding solvent to the mark on the neck of the flask with mixing.
- m/v concentrations are represented by a variety of metric mass units for the solute and metric volume units for the solution; however, % m/v is always expressed as *grams* of solute per 100 *milliliters* of total solution.
- A percentage, %, means per 100, so where that appears in a concentration, the denominator is 100 mL.
- A ppm means per 1,000,000 so where it appears in a concentration, the denominator is 10^6. It is used for dilute solutions.
- A ppb means per 1,000,000,000 so where it appears in a concentration, the denominator is 10^9. It is used for very dilute solutions.
- Molar concentration, also known as molarity and abbreviated M is a unit the means moles of solute per liter of solution.
- Equivalents per liter, eq/L, is calculated by multiplying the molarity times the numerical value of the charge on the ion, to give moles of charge per liter.
- Dilute solutions can be prepared from more concentrated stock solution—solutions with a known concentration—by dilution.
- Dilution calculations use the equation $C_1V_1 = C_2V_2$.
- Flow rate is the volume of solution administered over a given unit of time, as with IV infusion.
- The rate at which a certain mass of drug is to be given per unit of time is a ratio used to calculate flow rates.
- Flow rate $\times$ concentration = rate the drug is administered over time.

Osmosis and Dialysis

- Cell membranes are selectively permeable and function to control what enters and leaves the cell.

- A concentration gradient is created when two solutions with a different concentration are separated by a selectively permeable membrane.
- Colloids and suspended particles are too large to cross a selectively permeable membrane.
- Some solutes can cross a selectively permeable membrane, a process known as dialysis. Solutes cross a selectively permeable membrane from the more concentrated solution to the less concentrated solution by diffusion.
- Water diffuses across a selectively permeable membrane by a process known as osmosis, from the less concentrated solution to the more concentrated solution.
- The size and polarity of a solute are the most important factors in determining whether a solute can cross a cell membrane.
- Osmosis and dialysis regulate the distribution of nutrients in the cell and the removal of waste products from the cell.
- Simple diffusion is the spontaneous movement of a molecule or an ion from a region of higher concentration to a region of lower concentration.
- The relative concentrations of solutions on either side of a selectively permeable membrane can be described as hypertonic (having the higher concentration), hypotonic (having the lower concentration), or isotonic (having the same concentration).
- Osmosis can be stopped by applying an external pressure, known as osmotic pressure, on the hypertonic solution. Applying a pressure greater than the osmotic pressure results in reverse osmosis: water diffuses against the concentration gradient from the hypertonic to the hypotonic solution.
- Red blood cells placed in a hypertonic solution will shrivel up (crenation) as a result of water leaving the cell through osmosis.
- Red blood cells placed in a hypotonic solution will burst (hemolysis) as a result of water entering the cell.
- Red blood cells placed in an isotonic solution will have no net flow of water into or out of the cell and thereby maintain their natural biconcave shape.

Key Words

Aggregates Many small molecules or ions clumped together. They constitute the particles of a suspension.

Aqueous solution A homogeneous solution in which water is the solvent

Blood plasma Whole blood that has had the blood cells removed through centrifugation.

Blood serum Blood plasma that has had the fibrinogen and other clotting factors removed.

Cell membrane The selectively permeable barrier that separates the aqueous mixture inside the cell from the aqueous mixture outside the cell and controls what enters and leaves the cell.

Centrifuge A standard piece of laboratory equipment used to separate the particles in a suspension from the medium. A centrifuge spins test tubes at high speed, 5,000 rpm, causing the suspended particles to collect at the bottom—the pellet.

Colloid A particle with a diameter between 1 nm and 1,000 nm dispersed in a medium. Colloids include large molecules such as proteins and polysaccharides.

Colloidal dispersion A homogeneous mixture containing colloids—large molecules or aggregates—with a diameter between 1 nm and 1,000 nm, and evenly distributed throughout the medium. Colloidal dispersions have an opaque appearance.

Concentration A quantitative measure of the amount of solute per total solution (solute + solvent). Some common concentration units are mg/dL, % m/v, and moles/L.

Concentration gradient A region with a higher concentration and a region with a lower concentration often separated by a membrane.

Crenation The shrinkage of red blood cells when immersed in a hypertonic solution as water exits the cell, across the cell membrane, by osmosis.

Dialysis Diffusion of certain solutes from a region of higher solute concentration to a region of lower solute concentration across a selectively permeable membrane. Also, a life-support treatment that performs the functions the kidneys can no longer perform.

Diffusion The movement of particles through a solvent as a result of kinetic energy.

Dilute A solution with less solute dissolved in the solvent.

Dilution The process of preparing a more dilute solution from a more concentrated solution.

Dissolution The process of dissolving a solute in a solvent—a physical process. On the macroscopic scale, the solute appears to disappear in the solvent. On the atomic level, the solvent surrounds each of the solute particles: individual ions in the case of an ionic compound, and individual molecules in the case of a covalent compound.

Electrolyte An aqueous solution containing dissolved ions. The solution can conduct electricity because of the charged ions in solution.

Equivalents per liter (eq/L) A unit representing the moles of charge in every liter. Calculated by multiplying the molar concentration of an ion by the numerical value of the charge on the ion.

Flow rate The volume of solution per unit time in which an IV solution is administered.

Hemolysis The swelling and bursting of red blood cells when immersed in a hypotonic solution as a result of water entering the cells through the cell membrane by osmosis.

Heterogeneous mixture A mixture whose components are unevenly distributed throughout the mixture, such as a suspension or a saturated solution.

Homogeneous mixture A mixture whose components are evenly distributed throughout the mixture, such as a solution or a colloidal dispersion.

Hydrophilic A molecule or part of a molecule that is attracted to water because it is polar.

Hydrophobic A molecule or part of a molecule that is nonpolar and therefore does not dissolve in water.

Hydrophobic effect The tendency for nonpolar molecules to separate from water so as to avoid mixing with polar water molecules, an energetically favorable arrangement.

Hypertonic solution The solution with the higher solute concentration.

Hypotonic solution The solution with the lower solute concentration.

Intravenous (IV) The administration of a solution through a vein.

Ion-dipole The electrostatic interaction between an ion and the opposite partial charge on a polar molecule, such as water.

Isotonic solutions Solutions that have equal solute concentrations.

Like-dissolves-like A general rule of thumb that predicts that polar solutes will dissolve in polar solvents and nonpolar solutes will dissolve in nonpolar solvents.

Medium The component of a colloidal dispersion that is present in the greater quantity. Analogous to the solvent in a solution.

Membrane A barrier between two environments such as the cell membrane.

Mixture A combination of two or more elements and/or compounds in any proportion.

Molar concentration A unit of concentration represented by the moles of solute per liter of solution: mol/L, abbreviated *M*. Also referred to as molarity.

Molarity See *molar concentration*.

% m/v A unit of concentration that represents the grams of solute per 100 mL of solution. Thus a 1% solution contains 1 g of solute per 100 mL of solution.

Nonelectrolyte A solution composed of neutral solute molecules or an insoluble ionic solid, which therefore does not conduct electricity.

Oral medication Medications given by mouth in solution or suspension form.

Osmosis Simple diffusion of water from a region of lower solute concentration (higher water concentration) to a region of higher solute concentration (lower water concentration) across a selectively permeable membrane.

Osmotic pressure The minimum pressure that must be applied to the hypertonic solution to prevent osmosis.

Pellet The precipitated particles of a suspension after centrifugation.

Percentage Parts per 100, as for example grams of solute in 100 mL of solution.

Permeability The ability of substances to pass through a barrier, such as a membrane.

ppb Parts per billion: typically refers to grams of a solute in 10^9 mL of solution.

ppm Parts per million: typically refers to grams of a solute in 10^6 mL of solution.

Precipitate The insoluble solid that eventually settles at the bottom of the container of a heterogeneous mixture or saturated solution.

Reverse osmosis The application of pressure greater than the osmotic pressure to a hypertonic solution, so water crosses a membrane against the concentration gradient.

Saturated solution A solution with a solute concentration that exceeds the capacity of the solvent to dissolve the solute, resulting in a precipitate.

Selectively permeable membrane A membrane that allows for the passage of certain molecules and ions across the membrane while preventing the passage of others.

Serum Blood plasma that has had the protein fibrinogen removed, thus preventing the blood from clotting.

Simple diffusion The spontaneous movement of molecules or ions from a region of higher concentration to a region of lower concentration.

Solute The component(s) in a solution that is(are) present in the lesser amount. The solute is dissolved in the solvent. The solute has a size less than 1 nm in diameter.

Solution A homogeneous mixture that contains one or more solutes, dissolved in the solvent.

Solvent The component in a solution that is present in the greatest amount and dissolves the solute(s). Water is the solvent in an aqueous solution.

Stock solution A pre-prepared solution with a known concentration from which less concentrated (dilute) solutions can be prepared.

Supernatant The part of a suspension that remains after the suspended particles have been removed by centrifugation.

Suspension A heterogeneous mixture containing particles larger than 1,000 nm (1 micrometer), which are unevenly distributed throughout the medium, and will eventually settle.

Unsaturated solution A homogeneous solution in which all the solute particles are dissolved in the solvent (no precipitate).

Additional Exercises

Kidney Disease

45 What is the function of the kidneys?

46 How do the kidneys filter blood?

47 What substances do the kidneys filter?

48 What organ has failed for a person to need dialysis?

49 Briefly describe how the kidneys act as a selectively permeable membrane.

50 What are some symptoms of kidney disease?

51 What is renal failure?

52 What can cause acute renal failure?

53 What can cause chronic renal failure?

54 What role do the bladder and the ureters serve in the body?

55 When the kidneys have failed, what alternative is there for the patient other than dialysis?

Mixtures

56 What is the difference between a mixture and a pure substance?

57 Indicate whether the following statements are true or false.
a. Mixtures cannot be separated into their components through physical means.
b. Mixtures contain only one component.
c. Mixtures can differ in the relative amounts of each component.
d. A solution is a type of mixture.

58 What is the difference between a heterogeneous and a homogeneous mixture?

59 Classify the following mixtures as heterogeneous or homogeneous.
a. tomato juice
b. a cup of tea with honey
c. a bowl of chicken noodle soup
d. urine
e. gasoline
f. blood

60 What are the two types of homogeneous mixtures?

61 For a colloidal dispersion, what are the particles called? What is the component found in the greatest amount called?

62 Which of the following mixtures have particles with the largest size: solutions, colloidal dispersions, or suspensions?

63 Which of the following mixtures have particles in size ranging from 1 nm to 1 μm: solutions, colloidal dispersions, or suspensions?

64 An IV solution contains sodium chloride. Approximately how big are the particles in solution?

65 What are aggregates in a colloidal dispersion?

66 Why are mixtures of proteins or polysaccharides in water classified as colloidal dispersions?

67 When the kidneys filter blood, do they remove proteins? Explain.

68 When the kidneys filter blood, do they remove blood cells? Explain.

69 When the kidneys filter blood, do they remove urea (NH_2CONH_2)? Explain.

70 Identify the following as either a colloidal dispersion or a solution:
a. a glass of milk
b. a glass of lemonade
c. a carbonated soda
d. marshmallows
e. a glass of wine
f. a spoonful of mayonnaise
g. whipped cream

71 Identify the following as either a solution or a colloid:
a. 5% dextrose
b. Jell-O
c. blood
d. smoggy air
e. mercury amalgam used in dentistry
f. butter cookies

72 In what states of matter are the colloids and the mediums in the following colloidal dispersions?
a. mayonnaise
b. smoke
c. blood
d. gelatin

73 What piece of laboratory equipment can be used to separate particles out of a suspension?

74 What are the two components of a suspension called after it has been centrifuged.

75 Why must some medications be prepared as suspensions?

76 List at least one example each of a solute, a colloidal particle, and a suspended particle in whole blood.

Solutions

77 What are the two components of a solution called?

78 Which of the two components of a solution is present in the greater amount?

79 Identify the solute and the solvent in each solution below. State whether the solute is a gas, solid, or liquid.
a. an IV solution of aqueous glucose
b. carbonated water
c. a tincture of iodine (a small amount of I_2 in ethanol)
d. an IV solution of aqueous $MgSO_4$, used to prevent seizures in pregnant women with preeclampsia

80 Identify one solute and the solvent in each solution below. For (a)–(c), consult Table 8-2.
a. bronze
b. lemon-lime soda
c. beer
d. 3 mL isopropyl alcohol dissolved in 20 mL of aqueous solution
e. dry air

81 Explain the statement "like dissolves like."

82 Which of the following household substances would dissolve in water, based on your everyday experience?
a. bleach
b. butter
c. waterproof mascara
d. regular mascara

83 Are hydrocarbons soluble in water? Explain your answer in terms of the hydrophobic effect.

84 What happens when a hydrocarbon is mixed with water, and why?

85 Which of the following compounds would dissolve in water? Explain your answer.
a. $CH_3CH_2CH_2CH_2CH_2CH_3$
b.
c. CH_3OH
d.

HO——H
H——OH
H——OH
H——OH
CH_2OH

e.

86 Which of the following compounds would dissolve in pentane? Explain your answer.
a. $CH_3CH_2CH_2CH_2CH_2CH_3$
b.

c.

Turpentine

d. CH_3OH
e.

f. cholesterol
g. tetrachloroethene ($CCl_2=CCl_2$)

87 Which of the following solvents are polar? Which are nonpolar?
a. CCl_4　　　　d. carbon dioxide
b. propan-2-ol　　e. octane
c. acetic acid

88 Write equations to describe what happens when the following compounds dissolve in water. Do these equations represent a chemical change or a physical change?
a. $MgSO_4$　　　c. KCl
b. Na_3PO_4　　d. $CaCO_3$

89 Write the equations to describe what happens when the following compounds dissolve in water:
a. $CaCl_2$　　　c. K_3PO_4
b. NH_4Cl　　　d. $NaCl$

90 Ethanol and KCl both dissolve in water. What happens to the ionic bond when KCl dissolves? What types of interactions allow KCl to dissolve? What happens to the covalent bonds in ethanol when it dissolves? What type of interactions allow ethanol to dissolve?

91 Children's acetaminophen is sold as an oral suspension. If it sits in the medicine cabinet for a while, what will form in the bottom of the bottle? Why do you need to shake the bottle before dispensing the medicine?

92 You can make rock candy by dissolving sugar in warm water until no more sugar will dissolve. At this point in the process, do you have an unsaturated solution or a saturated solution?

93 What causes gout?

94 What causes kidney stones?

95 Which of the following compounds dissolved in water will form a strong electrolyte?
a. glucose　　　c. K_3PO_4
b. NaCl　　　　d. ethanol

96 Which of the following compounds dissolved in water will form a strong electrolyte?
a. Li_3PO_4　　　c. dextrose
b. $MgCl_2$　　　d. sodium hydroxide

97 Indicate whether the following statements are true or false.
a. When a molecule dissolves in water, the covalent bonds of the molecule break.
b. Hydrogen bonding aids in dissolving polar molecules in water.
c. Solutions containing electrolytes conduct electricity.
d. Dissolving an ionic compound in water is a chemical change.

Concentration

98 In your own words explain what information about a solution *concentration* indicates.

99 How does the colorimetric technique determine the concentration of a solution?

100 A patient's blood test shows that his cholesterol level is 161 mg/dL. What is this concentration in grams per liter? Is cholesterol ($C_{27}H_{46}O$) a molecule or an electrolyte?

101 How many grams of glucose would you need to prepare 5.0 mL of a glucose solution that had a concentration of 78 mg/dL?

102 A patient's blood test shows that her hemoglobin level is 14 g/dL. How many milligrams of hemoglobin would there be in 2.5 mL of the patient's blood?

103 Ampicillin is an antibiotic that can be administered intravenously. It is sold in vials that contain 2.0 grams of ampicillin. Sterile water is added to the vial to make a solution that has a concentration of 250. mg/mL? What is the total volume of a solution containing 2.0 g of ampicillin?

104 The maximum acceptable level of arsenic in water is 0.05 ppm. At that level, how many grams of arsenic would there be in 1 L of water?

105 In New Zealand, the legal blood alcohol limit is 400 μg of alcohol per 1 L of breath. What is this alcohol concentration in ppb?

106 You are asked to prepare 1.0 L of a 3.3% dextrose (m/v) solution for IV therapy. How much dextrose do you need to weigh out to prepare this solution?

107 How many grams of sodium chloride would you need to prepare 500. mL of a 0.90% normal saline (m/v) solution for IV therapy?

108 How much sodium chloride is in 250 mL of a 0.90% saline (m/v) solution used for IV therapy?

109 How many mL of total solution can be prepared from 36.0 g of dextrose if you prepare a 3.3% dextrose (m/v) solution for IV therapy?

110 How many liters of a 0.90% normal saline (m/v) solution for IV therapy can be prepared from 45 grams of sodium chloride?

111 How many mL of a 0.90% saline (m/v) solution for IV therapy can be prepared from 27.5 grams of sodium chloride?

112 A patient's blood test shows that he has a bilirubin level of 0.012 mol/L. How many moles of bilirubin are there in 5.0 mL of her blood?

113 A patient's blood test shows that she has elevated levels of LDL (low-density lipoproteins) at 4.14 mM. How many moles of LDL are there in 8.00 mL of blood?

114 What is the molarity of a solution that has 0.36 mole of dextrose in 250 mL of water?

115 What is the molarity of a solution that has 0.24 mole of glucose in 185 mL of water?

116 What is the volume of a 2.2 M solution containing 0.036 mole of NaCl?

117 What is the volume of a 0.56 M solution containing 2.5 moles of glucose?

118 How many moles of calcium ions (Ca^{2+}) are there in 0.52 L of 0.45 M $CaCl_2$?

119 A patient's blood test reports a magnesium level of 0.9×10^{-3} M. Convert this concentration to equivalents per liter of Mg^{2+}.

120 A patient's blood test reports a potassium level of 4.0×10^{-3} M. Convert this concentration to equivalents per liter of K^+.

121 How many equivalents per liter of sulfate ion (SO_4^{2-}) are there in a 0.15 M solution of $MgSO_4$? How many equivalents per liter of Mg^{2+} are there in this solution?

122 Sometimes KCl will be added to IV solutions. It will be added in concentrations of 0.020 eq/L. How many moles of K^+ will there be in 1 L of solution?

123 A solution of calcium ions contains 0.010 eq/L of calcium ion. How many moles of calcium ion does this solution contain? If this were a calcium phosphate solution, how many moles of phosphate ion would be in the solution?

124 A student needs to prepare 100. mL of a 0.50 M NaOH solution from a 2.0 M NaOH stock solution. How should she prepare this solution?

125 A student transferred 30. mL of a 4.5 M HCl stock solution to a 250. mL volumetric flask and then added water to the mark. What is the molar concentration, M, of the dilute solution?

126 How would you prepare 100. mL of a 0.45% half-normal saline solution from a stock solution of 0.90 % m/v normal saline?

127 What is the concentration of a dextrose solution prepared by diluting 15 mL of a 2.0 M dextrose solution to 25 mL using a 25 mL volumetric flask?

128 An order is given for 50. mg of Vistaril, an antihistamine, to be administered every 3 hr. The suspension contains 25 mg/5.0 mL. What volume of the suspension, in milliliters, should be administered to the patient every 3 hr?

129 An order is given to administer 187.5 mg of Betapen-VK, an oral formulation of penicillin. The suspension provided contains 125 mg of Betapen-VK in 5.0 mL. What dose of the oral formulation, in milliliters, should be administered to the patient?

130 An order is given to administer 15 g of lactulose, often used for treating complications of liver disease. The syrup contains 10. g/15 mL. What volume of syrup, in milliliters, should be given to the patient?

131 An order is given to administer methylprednisolone, an anti-inflammatory drug, by IV at a rate of 250. mg every 6 hr. The IV bag contains 125 mg methylprednisolone in every 0.02 mL. What should the flow rate be in milliliters per hour?

132 Morphine is a potent painkiller that acts directly on the central nervous system to relieve pain. An order is given to administer 2.0 mg of morphine to a patient as needed for breakthrough pain. The solution supplied contains 2.0 mg/mL of morphine. If 2.0 mg of morphine should be administered over a period of 5 min, what should the flow rate be in milliliters per minute?

133 Nipride is often ordered for patients who have experienced severe heart failure. A patient is receiving a Nipride solution by IV at a rate of 60. mL/hr. The concentration of the Nipride solution is 50. mg/250 mL D5W. How many milligrams of Nipride per hour is the patient receiving?

134 Aminophylline is used in the treatment of bronchial asthma. A patient is receiving an aminophylline solution by IV at a rate of 22 mL/hr. The concentration of the solution is 2.0 g/1.0 L D5W. How many grams of aminophylline per hour is the patient receiving?

135 Nitroglycerine, used to treat patients with angina, is prescribed for a patient by IV at a rate of 225 μg/min. The IV bag contains 50. mg nitroglycerine/250 mL D5W. How many milliliters per hour should the patient receive?

Osmosis and Dialysis

136 What does *permeability* in a selectively permeable membrane mean?

137 What factors determine whether a solute can pass through a cell membrane?

138 What is the primary function of the cell membrane?

139 What is diffusion?

140 What is a concentration gradient?

141 What is dialysis?

142 What is the difference between osmosis and dialysis?

143 What is the difference between hypertonic, hypotonic, and isotonic solutions?

144 What is osmotic pressure?

145 In reverse osmosis, is a pressure applied to the hypertonic solution that is greater than or less than the osmotic pressure? Explain why.

146 In reverse osmosis, which way does water flow relative to the concentration gradient?

147 Consider the solutions shown in Figure 8-30 separated by a selectively permeable membrane. Indicate the direction of osmosis between solution A and solution B or state if no net flow occurs. In which solution will the volume increase?

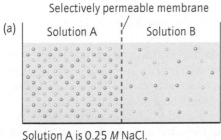

Selectively permeable membrane

(a) Solution A | Solution B

Solution A is 0.25 *M* NaCl.

Solution B is 0.05 *M* NaCl.

(b) Solution A | Solution B

Solution A is 0.02 *M* sucrose.

Solution B is 0.02 *M* glucose and 0.01 *M* sucrose.

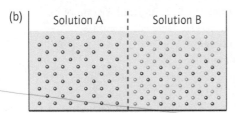

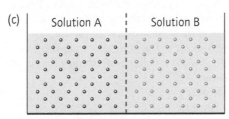

(c) Solution A | Solution B

Solution A is 0.02 *M* sucrose.

Solution B is 0.02 *M* glucose.

Figure 8-30

148 Preeclampsia is a condition that only occurs in pregnancy. The main symptoms are high blood pressure and protein in the urine; however, if not treated, it can cause seizures, a stroke, or liver rupture in the mother. The only treatment for this condition is to deliver the baby. Pregnant women who have preeclampsia are at increased risk of hemolysis of red blood cells. What happens to the red blood cells?

149 In which direction does water flow in osmosis between a hypertonic and a hypotonic solution separated by a selectively permeable membrane?

150 When dialysis is performed, why do colloidal particles not cross the selectively permeable membrane but solute particles do?

151 If two solutions on either side of a selectively permeable membrane are isotonic, do you expect dialysis to occur? Explain your answer.

152 Consider the dialysis shown in Figure 8-31: in one compartment is a solution of creatine (a small molecule) and globulin (a type of large protein molecule), in the other compartment is pure water. The two compartments are separated by a selectively permeable dialysis membrane. Describe how you could separate creatine from globulin based on the principles of dialysis.

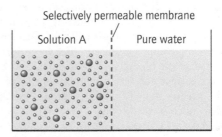

Selectively permeable membrane

Solution A | Pure water

Solution A is an aqueous mixture of creatine (∘) and globulin (●).

Solution B is pure water.

Figure 8-31

153 What happens if you apply pressure to seawater separated from fresh water by a selectively permeable membrane? What is this process called? If pressure is not applied, what are the results after osmosis?

Chemistry in Medicine: Blood Chemistry and the Diagnosis of Disease

154 What does a blood chemistry panel measure?

155 What are the major components of blood?

156 How is whole blood different from blood plasma?

157 How is blood serum different from blood plasma?

158 How is whole blood different from blood serum?

159 Give some examples of small molecules that are found in blood. What are some electrolytes found in blood?

160 A patient's blood test showed his BUN level was 250 mg/L. The normal range for BUN levels is between 7 and 18 mg/dL. Is this patient's BUN level in the normal range? If it isn't within the normal range, what organ might be affected?

161 A patient's fasting glucose level was 0.085 g/dL. The normal range for fasting glucose levels is between 70 and 110 mg/dL. Is this patient's glucose level in the normal range? Does this patient have diabetes?

162 Why is potassium ion considered one of the most important ions in the body?

Answers to Practice Exercises

1 A heterogeneous mixture has the components unevenly distributed throughout the mixture, whereas a homogeneous mixture has the components evenly distributed throughout the mixture.

2 a. heterogeneous   d. homogeneous
   b. heterogeneous   e. homogeneous
   c. heterogeneous

3 A solution has solute particles smaller than 1 nm. A colloidal dispersion has colloids ranging in size from 1 nm to 1,000 nm. A suspension has particles larger than 1,000 nm.

4 a. colloidal dispersion   c. solution
   b. colloidal dispersion   d. suspension

5 Blood has the characteristics of a solution because it contains solute particles smaller than 1 nm, dissolved in water, such as sodium, and potassium ions and small molecules, such as creatinine. It is a colloidal dispersion because it contains colloids, particles ranging from 1 nm to 1,000 nm in size, uniformly distributed in the blood, such as proteins and fats, and it is a suspension because it contains large particles greater than 1,000 nm, such as blood cells suspended in the blood.

6 The particles of a suspension can be separated using a centrifuge or by filtration.

7 a. solute: ethanol; solvent: water
   b. solutes: NaCl and sugar; solvent: water
   c. solute: CO_2; solvent: O_2

8 "Like dissolves like" refers to the general observation that polar solvents dissolve polar solutes, and nonpolar solvents dissolve nonpolar solutes. The "like" property is polarity. An ion has a full charge and water has partial charges, so they are both polar.

9 a. $MgCl_2$ (s) → Mg^{2+} (aq) + 2 Cl^- (aq)
   b. K_3PO_4 (s) → 3 K^+ (aq) + PO_4^{3-} (aq)
   c. $NaHCO_3$ (s) → Na^+ (aq) + HCO_3^- (aq)

10 Kidney stones form when saturated solutions of calcium salts form in the kidneys.

11 A saturated solution contains a precipitate, or suspended particles, whereas an unsaturated solution is clear. The amount of solute in a saturated solution has exceeded the number of solvent molecules available to dissolve the solute.

12

13 a. Soluble, because NaCl is an ionic compound that forms ion-dipole interactions with water
    b. Insoluble, because it is a nonpolar compound (C—Cl bond dipoles cancel)

c. Insoluble, because it is a hydrocarbon and hydrocarbons are nonpolar
d. Soluble, because lactose is a monosaccharide, and monosaccharides are polyols, which are polar functional groups
e. Soluble, because methanol has one polar hydroxyl group and only one carbon
f. Soluble, because a carboxylic acid is a polar functional group and there are few carbon and hydrogen atoms in the molecule
g. Soluble, because it is an ionic compound that forms ion-dipoles in water
h. Insoluble, because it has mainly hydrocarbon character
i. Insoluble, because it contains a lot of carbon and hydrogen atoms and only one hydroxyl group

14 b, c, h, and i because hexane is a nonpolar solvent and these compounds are nonpolar and form dispersion forces. The other compounds are polar and won't dissolve in a nonpolar solvent because they form hydrogen bonds and dipole-dipole interactions.

15 a. nonelectrolyte; molecular compound
    b. electrolyte; ionic compound
    c. nonelectrolyte; molecular compound
    d. electrolyte; ionic compound

16 0.008 mg/mL

17 400 mg/dL

18 35 mg of glucose

19 0.23 dL

20 6.25 mL of lidocaine solution

21 8.9% (m/v) sucrose

22 278 mL of 0.900% saline solution

23 Mixing 100 mL of solvent with the calculated amount of solute would result in a total solution volume greater than 100 mL. To properly prepare a 100 mL volume of a 2% aqueous glucose solution, begin by weighing 2 g of glucose and transferring it to a 100. mL volumetric flask. Add distilled water, with mixing, up to the mark on the neck of the volumetric flask. You have prepared a solution with a volume of 100 mL that contains 2 g of glucose.

24 4.5 g of NaCl

25 7.7×10^{-3} M

26 3.0 moles of Li^+ ions

27 0.4 L of solution

28 0.75 eq/L

29 0.20 eq/L

30 1.5 eq/L since there are three chloride ions for every $FeCl_3$ in solution.

31 25 mL of the stock solution

32 Transfer 9.0 mL of the 10% stock solution to a 100-mL volumetric flask. Add water to the 1 volumetric flask to the mark on the neck of the flask while mixing.

33 0.80 M

34 Administer 1.5 mL of the Tylenol solution every four hours.

35 Administer 12.5 mL of Tegretol solution as needed

36 Administer 10 mL of Atarax solution to the patient every 4 hrs.

37 20 mL/min

38 0.037 mL/hr

39 10 units/hr

40 Sucrose will dialyze to the fresh water side until both solutions have the same concentration. This could be repeated with fresh water, to cause more sucrose to dialyze.

41 The 0.45% solution is hypotonic, and the 0.90% solution is hypertonic. Ions will diffuse from the 0.90% solution to the 0.45% solution until the concentration on both sides of the membrane is the same. Thus, after dialysis the concentration of the hypertonic solution has decreased and the concentration of the hypotonic solution has increased.

42 *Selectively permeable* refers to a membrane that allows only certain particles to pass through.

43 **a.** The solutions are isotonic—the same concentration, so there is no net flow of water in either direction.

b. Water flows from solution B (hypertonic solution—0.05 M) to solution A (hypotonic solution—0.01 *M*).

44 Since 1.5% m/v saline is hypertonic relative to a red blood cell, water will exit the red blood cell through osmosis, causing it to shrink—crenation.

Every year pets die from accidental poisoning from ethylene glycol, found in antifreeze. Two tablespoons could be lethal to a 10-pound dog, and one teaspoon is lethal to an 8-pound cat. The LD_{50} depends on the species of pet. [StockLite/Shutterstock]

9 Acids and Bases

Ethylene Glycol Poisoning in Pets

Several hundred pets die each year from accidental ethylene glycol poisoning. Attracted to its sweet taste, they ingest antifreeze or deicing fluid that has leaked or spilled onto a garage floor. An animal's sensitivity to ethylene glycol depends on the species: 22 mL could be lethal to a 10-pound dog, while 5 mL is lethal for an 8-pound cat. Ethylene glycol is also toxic to humans. Ethylene glycol, $HOCH_2CH_2OH$, is a colorless, odorless diol found in antifreeze and deicing fluids. When ingested, ethylene glycol is metabolized into the toxic organic acids oxalic acid and glyoxylic acid. Oxalic acid is also found in high concentration in the leaves of rhubarb and that's why even rabbits don't eat the leaves of this plant. Once oxalic acid reaches the kidneys, it forms insoluble calcium oxalate salts that cause permanent damage to the kidneys eventually leading to death. Death from ethylene glycol poisoning is confirmed by the presence of calcium oxalate crystals in the kidneys during an autopsy.

In this chapter you will learn that the body regulates the acidity and alkalinity of the various bodily fluids. The acidity and alkalinity of blood is regulated within a very narrow range so that cells can function properly. Outside the normal range breathing becomes difficult, and cells are irreversibly damaged.

The symptoms of ethylene glycol poisoning in pets appears 30–120 minutes after ingesting it and include trouble walking, lethargy, and vomiting. If a veterinarian suspects ethylene glycol poisoning and not too much time has already elapsed, the pet can be treated with drugs that counteract the effects of the toxic acids. For example, other alcohols could be administered to the pet to slow the metabolism of ethylene glycol. In humans hemodialysis can be performed to eliminate the unmetabolized ethylene glycol and toxic organic acids from the blood.

Propylene glycol is used today in many commercial products that formerly contained ethylene glycol because the metabolism of propylene glycol does not produce toxic substances. ●

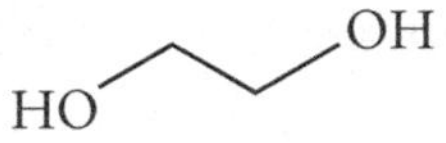

Ethylene glycol

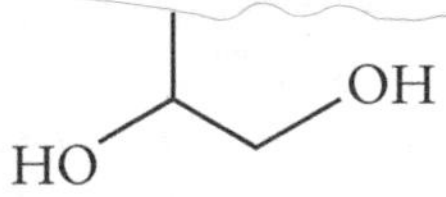

Propylene glycol

Acids and bases play a critical role in health and medicine. For example, the gastric juices in our stomach contain hydrochloric acid (HCl), an acid that together with enzymes allows us to digest our food. Many drugs, such as morphine and other amines, are organic bases. In this chapter you will learn about the chemical reactions of acids and bases, as well as how pH is used as a measure of the acidity and alkalinity of solutions. At the end of the chapter, you will learn how buffers in the blood maintain the narrow range of pH that is so critical to health.

(a)

(b)

Figure 9-1 Litmus paper. (a) Blue litmus paper turns red in the presence of acid. (b) Red litmus paper turns blue in the presence of base. [Catherine Bausinger]

9.1 Acids and Bases

In chapter 4 you learned the basic principles governing chemical reactions. In this chapter we apply these principles to the simplest, fastest, and most common of all reactions: the reactions of acids and bases in aqueous solution. Acids are known for their sour taste, their ability to neutralize bases, and their ability to turn blue litmus paper red (Figure 9-1a). Acids are found all around us, from the vinegar in our salad dressing to the carbonic acid in our blood. Bases are known for their bitter taste, slippery feel, ability to neutralize acids, and their ability to turn red litmus paper blue (Figure 9-1b). Bases are less common in foods because of their bitter taste, but they are found in cayenne peppers and chocolate, for example. The base sodium hydroxide, NaOH, also known as lye, is the main ingredient in Drano, a product used to unclog drains. Hydrogen carbonate ion, HCO_3^-, is the base found in baking soda, some over-the-counter antacids, and in the blood.

Acids and bases are defined by their chemical behavior in water. Two definitions, focusing on different aspects of the chemical reaction of an acid or a base in water, are used to define these important substances—the Arrhenius definition and the Brønsted-Lowry definition.

The Arrhenius Definition of Acids and Bases: Hydronium Ion and Hydroxide Ion

Acids *According to the Arrhenius definition, an* **acid** *is a substance that produces hydrogen ions (H⁺) in aqueous solution.* For example, hydrochloric acid (HCl) is an acid because it undergoes the following reaction in aqueous solution to produce hydrogen ions (H^+):

$$H-Cl(aq) \longrightarrow H^+(aq) + Cl^-(aq)$$

Acid   Hydrogen ion

A hydrogen ion is a hydrogen atom without its electron; thus, it is simply a proton (H^+). A hydrogen ion always bonds to a water molecule in aqueous solution, to form a **hydronium ion**, H_3O^+ (Figure 9-2):

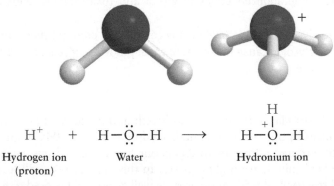

$$H^+ \quad + \quad H-\overset{..}{\underset{..}{O}}-H \quad \longrightarrow \quad H-\overset{\overset{H}{\overset{+|}{|}}}{\underset{..}{O}}-H$$

Hydrogen ion   Water   Hydronium ion
(proton)

Figure 9-2 Protons (H⁺) released in aqueous solution bond to water molecules forming hydronium ions, H_3O^+.

Chemists use the terms *proton* and *hydrogen ion* interchangeably to represent H^+, but know that in aqueous solution, it is the polyatomic hydronium ion (H_3O^+) that is the actual species dissolved in solution. Thus, acids produce hydronium ions (H_3O^+) in solution.

Bases *According to the Arrhenius definition, a* **base** *is a substance that produces hydroxide ions (OH⁻) in aqueous solution.* For example, the ionic compound sodium hydroxide (NaOH) is classified as a base because when it is dissolved in water it releases hydroxide ions, OH^-, into solution according to the dissolution equation:

$$NaOH(s) \longrightarrow Na^+(aq) + OH^-(aq)$$

Base	(dissolved cation)	Hydroxide ion
(ionic compound)		(dissolved anion)

In chapter 3, you learned that the hydroxide ion is a polyatomic anion found in various ionic compounds (salts). For instance, the hydroxide ion is the anion component in ionic compounds like calcium hydroxide, $Ca(OH)_2$, sodium hydroxide, NaOH, and magnesium hydroxide, $Mg(OH)_2$. In chapter 8 you learned that when an ionic compound dissolves in water, the ionic lattice breaks apart into its respective cations and anions, each separated and surrounded by water molecules. *Hydroxide ion–containing salts are therefore bases because in aqueous solution they release hydroxide ions.*

When a proton from an acid, H^+, is combined with a hydroxide ion, OH^-, from a base, they react to form a molecule of water, H_2O. Because water is a neutral substance, the reaction is called a *neutralization* reaction:

$$H^+(aq) + OH^-(aq) \longrightarrow H_2O(l)$$

Hydrogen ion	Hydroxide ion	Water

This is the type of reaction that occurs when you take antacids to relieve acid indigestion. We examine neutralization reactions in greater detail in section 9.3, after we consider in more detail the behavior of acids and bases in water.

Brønsted-Lowry Definition of Acids and Bases: Proton Transfer

Another more widely applicable definition of an acid and a base is taken from the perspective of the proton (H^+) being transferred. *According to the Brønsted-Lowry definition, an acid is a proton donor, and a base is a proton acceptor.*

Acid: A proton (H^+) donor

Base: A proton (H^+) acceptor

Acids Consider hydrochloric acid (HCl) in water described previously. According to the Brønsted-Lowry definition, HCl is an *acid* because it *donates* a proton (H^+) to water, and water is a *base* because it *accepts* a proton from HCl, the acid. In other words a proton is transferred from the acid (HCl) to the base (water). The products in this reaction are the hydronium ion (H_3O^+) and chloride ion (Cl^-):

Hydrochloric acid	Water	Hydronium ion	Chloride ion
(acid)	(base)		

Remember that when a proton is transferred from one species to another, the charge on both species changes because a hydrogen ion, H^+, has a positive charge, which is lost from the acid and gained by the base.

Consider the organic acid acetic acid found in vinegar. When acetic acid is placed in water, it donates a proton to water. The products are the hydronium ion, consistent with the Arrhenius definition of an acid, and the acetate ion, a polyatomic ion that has the same structure as the acid but with one less proton.

Acetic acid	Water	Hydronium ion	Acetate ion
(acid)	(base)		

Consider another example. The ammonium ion, NH_4^+, a polyatomic ion, behaves as a Brønsted-Lowry acid because it transfers a proton to water, and produces the hydronium ion and ammonia, NH_3.

$$
\underset{\substack{\text{Ammonium ion} \\ \text{(acid)}}}{H-\overset{\overset{\displaystyle H}{|}{}^+}{N}-H} \quad + \quad \underset{\substack{\text{Water} \\ \text{(base)}}}{H-\overset{..}{\underset{..}{O}}-H} \quad \rightleftharpoons \quad \underset{\substack{\text{Hydronium ion}}}{H-\overset{\overset{\displaystyle H}{|}{}^+}{\underset{..}{O}}-H} \quad + \quad \underset{\substack{\text{Ammonia}}}{H-\overset{..}{N}-H}
$$

Bases One advantage of the Brønsted-Lowry definition of a base (a proton acceptor) is that it allows us to see how molecular compounds can act as bases. For example, ammonia (NH_3) is a base because it accepts a proton (H^+) from water, and water acts as an acid because it donates a proton to ammonia, the base:

$$
\underset{\substack{\text{Ammonia} \\ \text{(base)}}}{H-\overset{..}{\underset{\displaystyle H}{N}}-H} \quad + \quad \underset{\substack{\text{Water} \\ \text{(acid)}}}{H-\overset{..}{\underset{..}{O}}-H} \quad \rightleftharpoons \quad \underset{\substack{\text{Hydroxide ion}}}{{}^-\overset{..}{\underset{..}{:O}}-H} \quad + \quad \underset{\substack{\text{Ammonium ion}}}{H-\overset{\overset{\displaystyle H}{|}{}^+}{N}-H}
$$

Notice that hydroxide ion is one of the products in this reaction, consistent with the Arrhenius definition of a base. As you can see, acids and bases exhibit opposite behaviors in aqueous solution.

Conjugate Acid-Base Pairs According to the Brønsted-Lowry definition, an acid transfers a proton to a base, and the products formed are related to the acid and the base by a proton; hence, the products themselves are an acid and a base. After transferring a proton the acid becomes what is known as the **conjugate base** of the acid. Similarly, after receiving a proton the base becomes what is known as the **conjugate acid** of the base. An acid and its conjugate base and a base and its conjugate acid are known as a **conjugate acid-base pair**. By this definition, water and hydroxide ion (OH^-) are a conjugate acid-base pair, and water and hydronium ion (H_3O^+) are a conjugate acid-base pair. Therefore, in the reaction of an acid or a base with water, two conjugate acid-base pairs can be identified. For example, in the reaction of ammonia and water described above, NH_3 and NH_4^+ are the main conjugate acid-base pair and H_2O and OH^- are also a conjugate acid-base pair:

conjugate acid-base pair

$$
\underset{\substack{\text{Ammonia} \\ \text{(base)}}}{H-\overset{..}{\underset{\displaystyle H}{N}}-H} \quad + \quad \underset{\substack{\text{Water} \\ \text{(acid)}}}{H-\overset{..}{\underset{..}{O}}-H} \quad \rightleftharpoons \quad \underset{\substack{\text{Hydroxide ion} \\ \text{(conjugate base)}}}{{}^-\overset{..}{\underset{..}{:O}}-H} \quad + \quad \underset{\substack{\text{Ammonium ion} \\ \text{(conjugate acid)}}}{H-\overset{\overset{\displaystyle H}{|}{}^+}{N}-H}
$$

conjugate acid-base pair

Generally, bases are neutral molecules or negatively charged polyatomic anions. When the base is neutral, the conjugate acid is a polyatomic cation, as in this example. When the base is negatively charged, the conjugate acid is a neutral molecule.

Similarly, for acetic acid in water the conjugate acid-base pairs are:

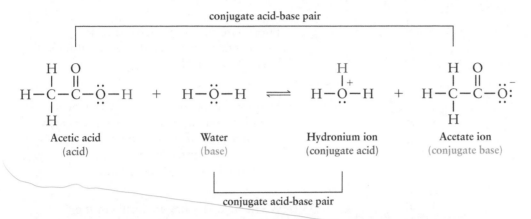

Generally, acids are neutral molecules or positively charged polyatomic cations. When the acid is neutral, the conjugate base is an anion, as seen in this example. When the acid is positively charged, the conjugate base is neutral.

In aqueous solution, water acts as a base in the presence of an acid, whereas it acts as an acid in the presence of a base. A molecule that can act as both an acid or a base is known as an **amphoteric** molecule, an important property of water that will be explored further in section 9.2.

Commonly Encountered Acids and Bases Gastric juice produced in the stomach contains hydrochloric acid, HCl. The high concentration of hydronium ions in the stomach facilitates the digestion of food. Stomach acid is produced in the mucosa, the inner lining of the stomach, which can withstand the high concentration of hydronium ions that would otherwise damage most tissues (Figure 9-3). All cells and biological systems are sensitive to the concentration of hydronium ions. For example, a gunshot wound to the stomach is often fatal, not because of the damage it causes to the stomach, but because the acidic contents of the stomach spill into the abdominal cavity, damaging tissue that cannot withstand the high concentration of hydronium ions in the stomach. Tissue damage from an acid or a base is known as a chemical burn.

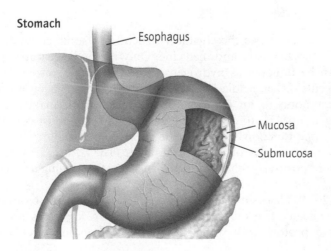

Figure 9-3 HCl (gastric juice) is produced in the mucosal lining of the stomach.

TABLE 9-1 Common Acids in Biochemistry and Medicine

Molecule	Lewis Structure*	Biological Role
Hydrochloric acid (HCl)	H—Cl	HCl is the main component of gastric juice, found in the stomach.
Phosphoric acid (H_3PO_4)	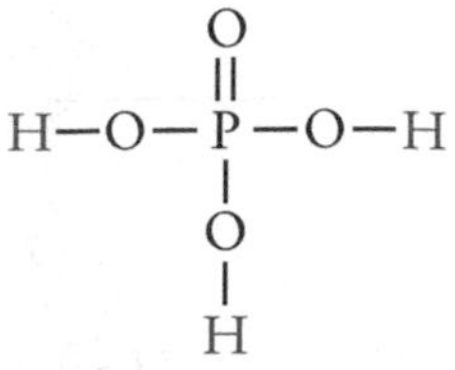 (see structure) $H-O-P(=O)-O-H$ with $O-H$ below	Phosphate esters are derived from phosphoric acid. They are found in ATP, DNA, and RNA.
Carbonic acid (H_2CO_3)	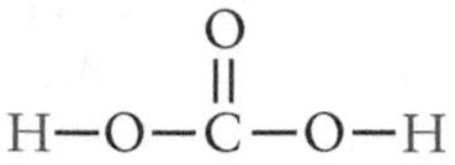 $H-O-C(=O)-O-H$	Dissolved carbon dioxide (CO_2) reacts with water to produce carbonic acid, a component of the carbonic acid–hydrogen carbonate buffer in the blood and essential for controlling blood pH.
Acetic acid ($C_2H_4O_2$)	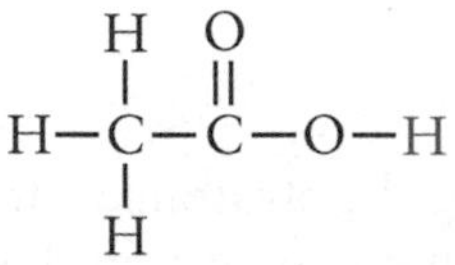 $H_3C-C(=O)-O-H$	Acetic acid is a key molecule in the metabolism of carbohydrates and fats.
Citric acid ($C_6H_8O_7$)	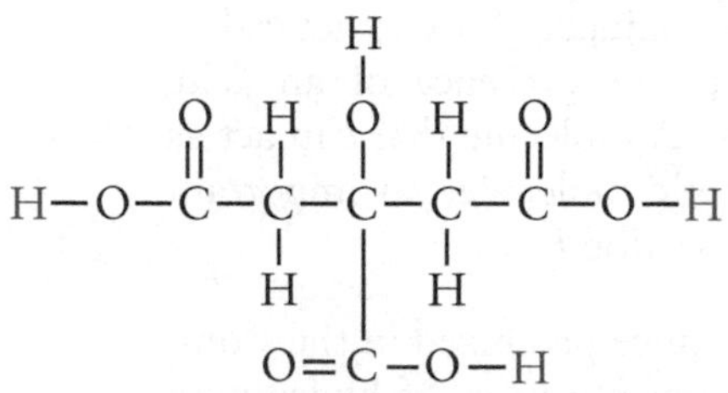 (see structure)	Citric acid is formed in the citric acid cycle, a biochemical pathway involved in energy production. Also a common component of fruits and vegetables.
Lactic acid ($C_3H_6O_3$)	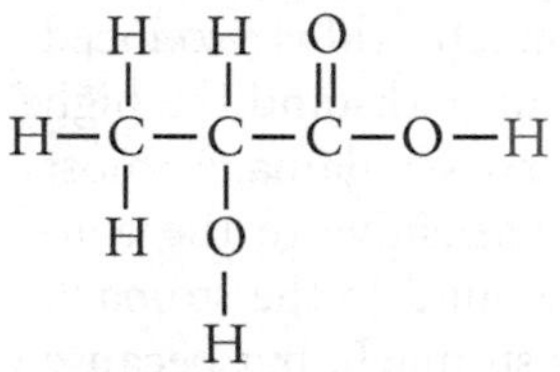 (see structure)	Lactic acid is produced in anaerobic biochemical processes, such as during intense exercise. Microorganisms also produce lactic acid as a waste product of metabolism.
Pyruvic acid ($C_3H_4O_3$)	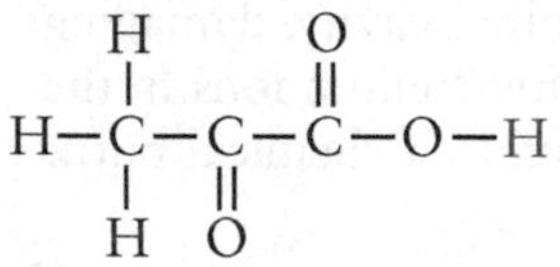 (see structure)	Pyruvic acid is produced in the metabolism of glucose.

*The hydrogen atoms highlighted in red can be released as protons, H^+, in aqueous solution, known as the acidic hydrogens.

Table 9.1 shows the chemical structure of some common acids encountered in biochemistry and medicine. Only the hydrogen atoms highlighted in red can be donated as hydrogen ions, H^+, often referred to as the "acidic hydrogen(s)." The acidic hydrogen is usually a hydrogen atom with a polar covalent bond to an electronegative atom, which makes it easier for the hydrogen ion, H^+, to dissociate from the acid. The most common acidic neutral *organic* functional group is the carboxylic acid, introduced in chapter 7. Organic derivatives of phosphoric acid, H_3PO_4, are also acidic. The most common positively charged acids include the conjugate acids of amines, RNH_3^+.

Table 9.2 lists some commonly encountered organic bases in biochemistry and medicine. The most common neutral *organic* bases are amines, which accept a proton from water using the nonbonding electrons on nitrogen.

TABLE 9-2 Ammonia and Common Amine Bases in Biochemistry and Medicine

Molecule	Structure*	Biological Role
Ammonia (NH_3)	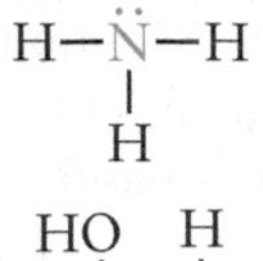	Ammonia is one of the end products of amino acid metabolism.
Epinephrine (Adrenaline)	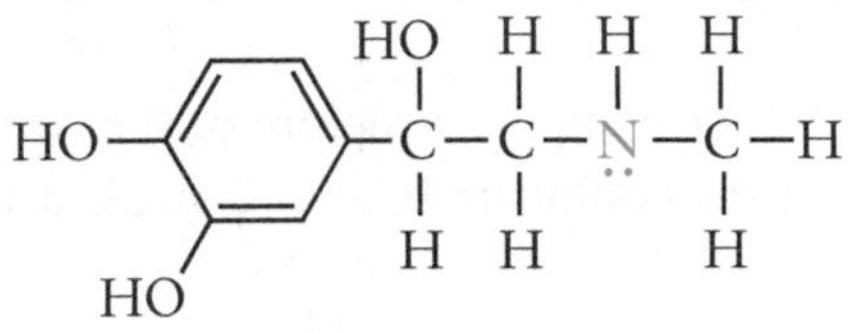	Epinephrine is a hormone and a neurotransmitter.
Dopamine	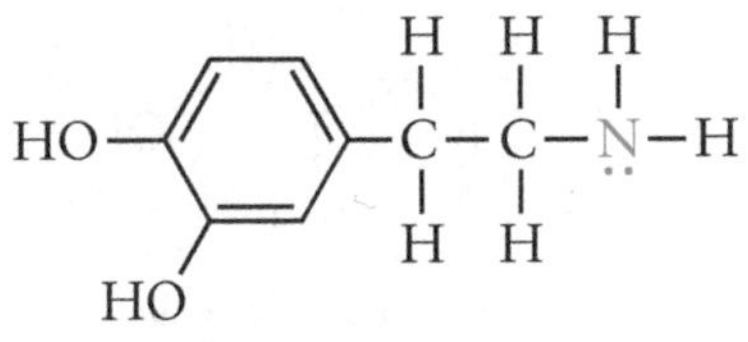	Dopamine is a hormone and a neurotransmitter.
Adenine	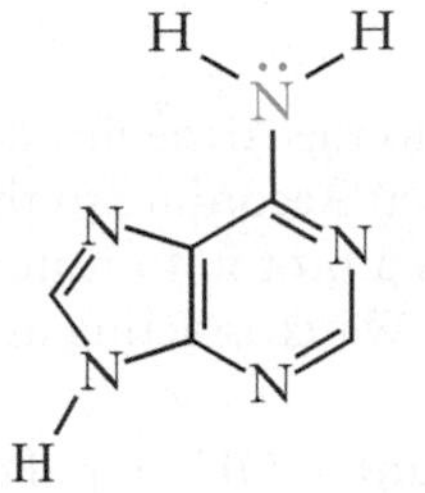	Adenine is part of the structure of ATP and the structure of DNA and RNA.

*The amine nitrogen and its nonbonding pair of electrons is shown in blue.

The amine nitrogen and its nonbonding pair of electrons are highlighted in blue in the structures shown in Table 9.2. Common negatively charged organic bases include the carboxylate ion, $RCOO^-$, the conjugate base of a carboxylic acid, dihydrogen phosphate, $H_2PO_4^-$, and hydrogen phosphate, HPO_4^{2-}, the conjugate base of phosphoric acid and dihydrogen phosphate, respectively.

WORKED EXERCISES — Writing Reaction Equations for Acids and Bases in Water

9-1 Based on the reaction shown of HF in water, is HF an acid or a base? Explain in terms of the Arrhenius definition of an acid or base. Explain in terms of the Brønsted-Lowry definition of an acid or base. What are the conjugate acid-base pairs in this reaction? Is water acting as an acid or a base in this reaction?

$$H-F \ + \ H-\overset{..}{\underset{..}{O}}-H \ \rightleftharpoons \ H-\overset{\overset{H}{|}}{\underset{..}{O}}{}^+-H \ + \ F^-$$

9-2 Write the complete chemical equation that describes what happens when potassium hydroxide, KOH, is added to water. Is KOH an acid or a base and explain why in terms of the Arrhenius definition of an acid or a base? Is potassium hydroxide a molecule or an ionic compound?

9-3 Lactic acid is a carboxylic acid that is produced in oxygen depleted muscle tissue during intense exercise. The buildup of lactic acid causes the sensation of sore muscles. Using Table 9-1, write the complete chemical equation for the reaction that occurs when lactic acid is placed in water. Label the conjugate acid-base pairs. Is water acting as an acid or a base? Explain why lactic acid is considered an acid in terms of both the Arrhenius and Brønsted-Lowry definitions.

9-4   Write the complete chemical equation for the reaction between methanamine, CH_3NH_2, and water. Label the conjugate acid-base pairs. Is water acting as an acid or a base? Explain why methanamine is considered a base in terms of both the Arrhenius and Brønsted-Lowry definitions.

9-5   Write the complete chemical equation for the reaction between acetate CH_3COO^-, and water. Is acetate an acid or a base? How do you know? Label the conjugate acid-base pairs.

9-6   Referring to Table 9-2, provide the structure of the conjugate acid of dopamine.

9-7   Write the structure of the conjugate base of benzoic acid.

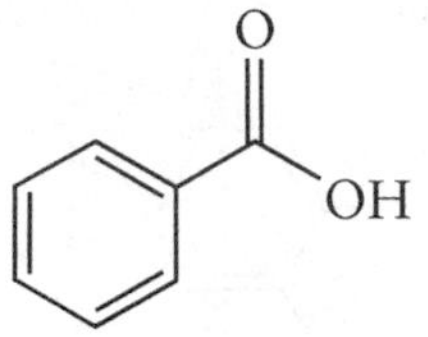

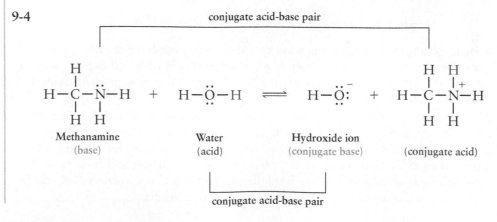

Benzoic acid

Solutions

9-1   HF is an acid. According to the Arrhenius definition, it is an acid because hydronium ion is a product. According to the Brønsted-Lowry definition, HF is an acid because it donates a proton to water. HF/F^- and H_2O/H_3O^+ are the two conjugate acid-base pairs. Water is acting as a base because it accepts a proton from HF.

9-2 $KOH(s) + H_2O(l) \rightarrow K^+(aq) + OH^-(aq)$. Potassium hydroxide is a base because it produces hydroxide ions, OH^-, in solution. KOH is an ionic compound, a salt.

9-3

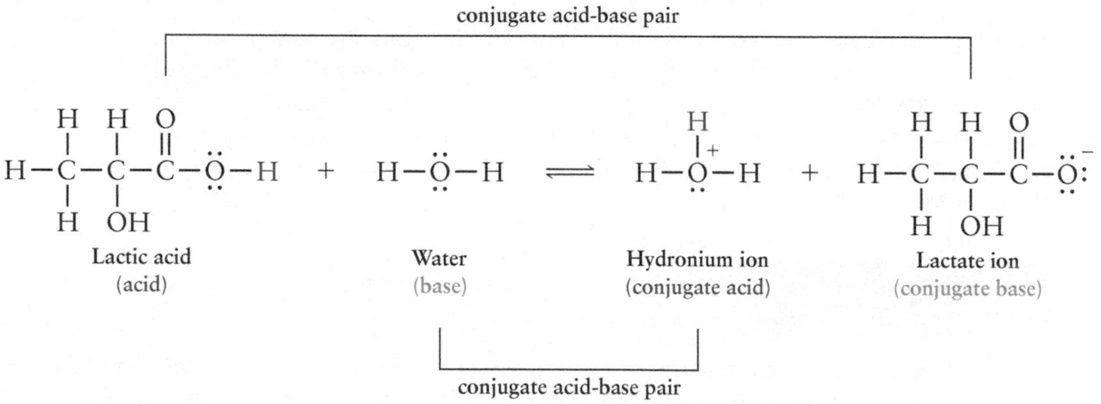

conjugate acid-base pair

H—C—C—C—O—H + H—O—H ⇌ H—O—H + H—C—C—C—O:

| Lactic acid | Water | Hydronium ion | Lactate ion |
| (acid) | (base) | (conjugate acid) | (conjugate base) |

conjugate acid-base pair

Water is acting as a base. Lactic acid is an acid according to the Arrhenius definition because it produces hydronium ions in solution. Lactic acid is a Brønsted-Lowry acid because it donates a proton to water in aqueous solution.

9-4

conjugate acid-base pair

H—C—N—H + H—O—H ⇌ H—O: + H—C—N—H

| Methanamine | Water | Hydroxide ion | |
| (base) | (acid) | (conjugate base) | (conjugate acid) |

conjugate acid-base pair

Water is acting as an acid. Methanamine is a base according to the Arrhenius definition because it produces hydroxide ion in aqueous solution. Methanamine is a Brønsted-Lowry base because it accepts a proton.

9-5 $CH_3COO^-(aq)$ + H_2O ⇌ $CH_3COOH(aq)$ + $OH^-(aq)$

     Base   Acid   Conjugate acid   Conjugate base

Acetate is a base because according to the Arrhenius definition it produces hydroxide ions in solution and, according to the Brønsted-Lowry definition, because it accepts a proton from water. Moreover, acetate is negatively charged ion, and bases are usually negatively charged or neutral, whereas acids are positively charged or neutral. CH_3COO^-/CH_3COOH are a conjugate acid-base pair. Water and hydroxide ion are also a conjugate acid-base pair.

9-6

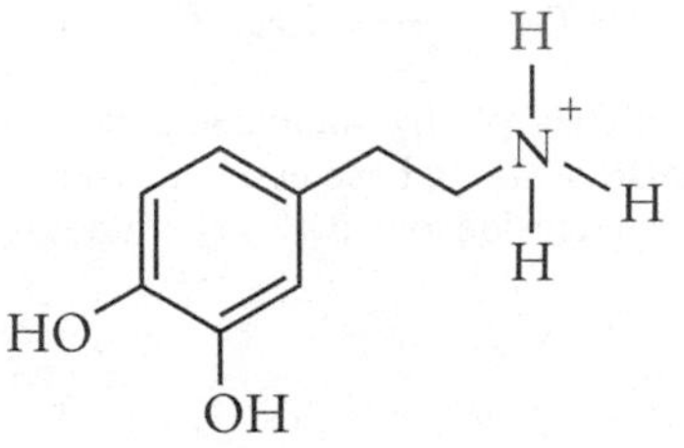

9-7

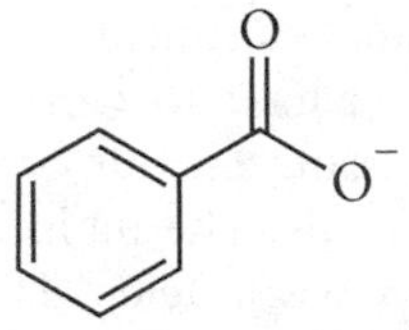

Benzoate ion

PRACTICE EXERCISES

1 According to the Arrhenius definition, a base produces _____ in solution. According to the Brønsted-Lowry definition, a base _____ a proton from/to water.

2 Which of the compounds below are bases? Which compound is amphoteric, and what does that mean?

  a. KOH
  b. $CH_3CH_2NHCH_3$
  c. HBr
  d. H_2O
  e. $Ca(OH)_2$
  f. $CH_3CH_2CH_2COOH$

3 For each of the following chemical equations, label the conjugate acid-base pairs. Indicate whether water acts as an acid or a base in the reaction.

  a. $NH_4^+ + H_2O \rightarrow H_3O^+ + NH_3$
  b. $HCO_3^- + H_2O \rightarrow OH^- + H_2CO_3$
  c. $HI + H_2O \rightarrow I^- + H_3O^+$

4 Write the chemical equation that represents the reaction of pyruvic acid ($CH_3COCOOH$) with water. Label the conjugate acid-base pairs. Are hydronium or hydroxide ions formed? What functional groups does pyruvic acid contain?

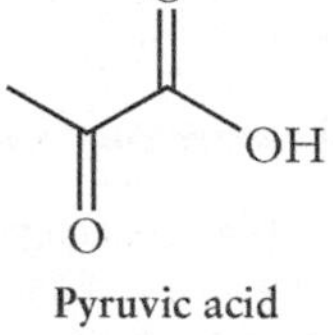

5 Write the chemical equation that represents the reaction of N-methylmethanamine, $(CH_3)_2NH$, in water. Is N-methylmethanamine an acid or a base? Label the conjugate acid-base pairs. Is hydronium or hydroxide ion formed? What functional group does N-methylmethanamine contain?

6 Write the chemical equation that shows what happens when $Sr(OH)_2$ is dissolved in water. Is this compound an ionic or a covalent compound? Is it an acid or a base? Explain.

Pyruvic acid

(a)   (b)

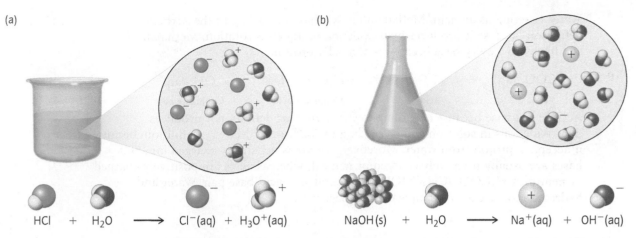

$HCl + H_2O \longrightarrow Cl^-(aq) + H_3O^+(aq)$ $NaOH(s) + H_2O \longrightarrow Na^+(aq) + OH^-(aq)$

Figure 9-4 (a) HCl, a strong acid, dissociates completely into hydronium ions (H_3O^+) and chloride ions (Cl^-) when dissolved in water. (b) NaOH, a strong base and ionic compound, dissociates completely into sodium ions (Na^+) and hydroxide ions (OH^-) in aqueous solution.

Strengths of Acids and Bases

Spilling battery acid on bare skin can cause a serious chemical burn that requires medical attention; on the other hand, we don't have to worry about the consequences of spilling oil and vinegar salad dressing on our bare skin. Clearly, differences exist among the various types of acids. The main distinction between acids is whether they are strong acids or weak acids. The same distinction exists for bases: there are strong bases, and there are weak bases. *Acids and bases are classified as either strong or weak depending on the extent to which they dissociate in water—in other words, the extent to which the forward reaction proceeds.* Battery acid is a strong acid, whereas acetic acid, the acid in vinegar, is a weak acid. Both strong acids and strong bases can be highly corrosive. As dilute solutions, however, most strong acids and bases are generally safe to handle.

Strong Acids and Strong Bases *A strong acid, HA, dissociates completely into its conjugate base (A^-) and hydronium ions (H_3O^+) in aqueous solution.* Little if any undissociated acid, HA, exists in solution. We say "the reaction has gone to completion," and we use a single headed forward arrow when writing the chemical equation. For example, **Figure 9-4a** shows the complete dissociation of hydrochloric acid, HCl, a strong acid, into hydronium ions, H_3O^+, and chloride ions, Cl^-. Virtually no HCl molecules exist in solution. Strong acids are therefore also strong electrolytes because they form ions, that is, charged species, in aqueous solution. Table 9-3 lists the common strong acids, which are all inorganic compounds. In the laboratory, we find strong acids such as nitric acid, HNO_3, and sulfuric acid, H_2SO_4. Exposure to any of these acids in concentrated form can cause severe chemical burns.

Hydrochloric acid, HCl, is the only strong acid found in the body, where it is often called "gastric juice," because it is produced in the stomach and aids with digestion. In the stomach, the concentration of HCl ranges from 0.01 M to 0.1 M. Since HCl is a strong acid, all of the HCl in the stomach is ionized, so it is actually the hydronium ion (and its conjugate base) that has a concentration of 0.01 M to 0.1 M. While this concentration of hydronium ions does not damage the stomach lining, it is damaging to the more delicate lining of the esophagus (**Figure 9-5**). This is why chronic acid reflux—the upward movement of stomach acid into the esophagus—can damage the esophagus.

Similarly, a **strong base** is defined as a base that is completely ionized in aqueous solution (Figure 9-4b). Most strong bases are salts containing the

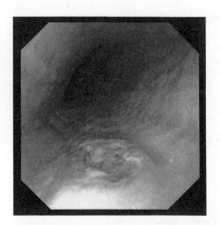

Figure 9-5 High concentrations of hydronium ions can damage the delicate lining of the esophagus as shown here in this esophageal ulcer. [David M. Martin, M.D./Science Source]

Chronic acid reflux disease is formally known as **G**astro **E**sophageal **R**eflux **D**isorder (GERD).

TABLE 9-3 Strong Acids		
Formula	Name	Conjugate Base
HNO_3	Nitric acid	NO_3^-
H_2SO_4	Sulfuric acid	HSO_4^-
$HClO_4$	Perchloric acid	ClO_4^-
HCl	Hydrochloric acid	Cl^-
HBr	Hydrobromic acid	Br^-
HI	Hydroiodic acid	I^-

TABLE 9-4 Strong Bases	
Formula	Name
$Ba(OH)_2$	Barium hydroxide
$Ca(OH)_2$	Calcium hydroxide
LiOH	Lithium hydroxide
KOH	Potassium hydroxide
NaOH	Sodium hydroxide (lye)
$Sr(OH)_2$	Strontium hydroxide

hydroxide ion and a Group 1A or Group 2A metal cation. We have seen that when a salt containing hydroxide ion dissolves in water it releases hydroxide ions. Strong bases are also strong electrolytes because they form ions in solution. Table 9-4 lists some strong bases, which are all ionic compounds containing the hydroxide ion.

WORKED EXERCISES | Strong Acids and Bases

9-8 Nitric acid, HNO_3, is a strong inorganic acid used in the laboratory.

   a. Write the chemical equation that shows the reaction of nitric acid in water.
   b. Label the conjugate acid-base pairs.
   c. Is aqueous nitric acid an electrolyte? Explain.
   d. Are there any molecules of nitric acid present in solution?
   e. To what extent is nitric acid dissociated?

9-9 Lithium hydroxide is an ionic compound.

   a. Write the equation for the dissolution of lithium hydroxide in water.
   b. What about this dissolution process indicates lithium hydroxide is a base, according to the Arrhenius definition?
   c. What about this dissolution process indicates LiOH is a strong base?
   d. Is LiOH (aq) a strong electrolyte?

Solutions

9-8 a. $HNO_3(aq) + H_2O(l) \rightarrow H_3O^+(aq) + NO_3^-(aq)$
   b. HNO_3 and NO_3^- are a conjugate acid-base pair and so are H_2O and H_3O^+.
   c. Yes, because it produces many ions in solution.
   d. No, all the nitric acid, HNO_3, is completely dissociated into H_3O^+ and NO_3^- in aqueous solution.
   e. Nitric acid is completely dissociated.

9-9 a. $LiOH(s) + H_2O(l) \rightarrow Li^+(aq) + OH^-(aq)$
   b. It is a base because hydroxide ions, OH^-, are produced in solution, the Arrhenius definition of a base.
   c. We know it is a strong base because soluble salts dissociate completely in solution.
   d. Yes, because LiOH dissociates completely into ions in aqueous solution: $Li^+(aq)$ and $OH^-(aq)$.

PRACTICE EXERCISES

7 HBr is a strong acid. Write the chemical equation that shows the reaction of HBr in water. Label the conjugate acid-base pairs. What does it mean that HBr is a strong acid? Explain by listing all species in solution. Is this solution a strong, weak, or nonelectrolyte?

8 Do all strong acid molecules transfer a proton to solution or only some acid molecules?

9 Does a strong base dissociate completely or partially in aqueous solution?

10 Write the equation for the dissolution of KOH in water. Is the solution a strong, weak, or nonelectrolyte? Is KOH a strong or a weak base? How do we know?

11 Perchloric acid, $HClO_4$, is a strong acid. Write the equation for the reaction that occurs when perchloric acid is added to water. Is the concentration of molecular $HClO_4$ high or low? What is the formula for the conjugate base? Is there a high concentration or a low concentration of the conjugate base in solution relative to the acid?

Weak Acids and Weak Bases *Weak acids and bases differ from strong acids and bases in the extent to which they dissociate in water.* Weak acids and weak bases undergo a reversible reaction in water that favors the reactants at equilibrium. We use the equilibrium arrows (two opposing half-headed arrows) described in chapter 4 to represent the equation for the reaction of a weak acid or a weak base with water (Figure 9-6). When a weak acid or a weak base dissolves in water, a mixture of both products and reactants is present at equilibrium. Moreover, at equilibrium, the concentration of the reactants is greater than the concentration of products: we say the equilibrium lies to the left.

For example, consider the reversible reaction between acetic acid (CH_3COOH), a weak acid, and water, which produces hydronium ions (H_3O^+) and acetate ions (CH_3COO^-), the conjugate base of acetic acid. The majority of acetic acid molecules do not dissociate, and so at equilibrium the solution contains more acetic acid (CH_3COOH) molecules than acetate ions (CH_3COO^-) and hydronium ions (H_3O^+). Because the extent of dissociation is small, acetic acid is classified as a weak acid. Indeed, this is why you don't need to worry about spilling oil and vinegar (acetic acid) salad dressing on your bare skin.

$$
\begin{array}{c}
\underset{\substack{\text{Acetic acid}\\\text{(weak acid)}}}{\text{H}-\overset{\overset{\text{H}}{|}}{\underset{\underset{\text{H}}{|}}{\text{C}}}-\overset{\overset{\text{O}}{\|}}{\text{C}}-\text{O}-\text{H}}
\;+\;
\underset{}{\text{H}-\overset{\cdot\cdot}{\underset{\cdot\cdot}{\text{O}}}-\text{H}}
\;\rightleftharpoons\;
\underset{\text{Hydronium ion}}{\text{H}-\overset{\overset{\text{H}}{|}}{\underset{\underset{\cdot\cdot}{|}}{\text{O}}}{}^{+}-\text{H}}
\;+\;
\underset{\substack{\text{Acetate ion}\\\text{(conjugate base)}}}{\text{H}-\overset{\overset{\text{H}}{|}}{\underset{\underset{\text{H}}{|}}{\text{C}}}-\overset{\overset{\text{O}}{\|}}{\text{C}}-\text{O}^{-}}
\end{array}
$$

A weak acid is a weak electrolyte because the concentration of ions in solution is small (see Figure 9-6).

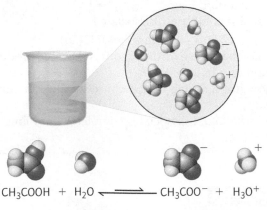

$$CH_3COOH + H_2O \rightleftharpoons CH_3COO^- + H_3O^+$$

Incomplete dissociation of a weak acid, CH_3OOH

Figure 9-6 Acetic acid is a weak acid, dissociating to a small extent; mainly the molecular neutral form is present in solution at equilibrium. Since few product ions are in solution, it is also a weak electrolyte.

Organic compounds containing carboxylic acid functional groups ($RCOOH$) are weak acids. Carbonic acid (H_2CO_3), phosphoric acid (H_3PO_4), and organic derivatives of phosphoric acid are also weak acids. The conjugate acids of amines, RNH_3^+, are weak acids. Water is also a weak acid.

Weak bases, like weak acids, react with water in a reversible reaction that produces an equilibrium mixture of reactants and products. The equilibrium lies to the left so the concentration of hydroxide ions is low. At equilibrium an aqueous solution containing a weak base consists of mainly the un-ionized base. Weak bases are weak electrolytes because they form few ions in solution. For example, for every 100 molecules of ammonia, NH_3, a weak base, only four hydroxide ions are formed in aqueous solution:

$$NH_3(aq) \quad + \quad H_2O \quad \rightleftharpoons \quad OH^-(aq) \quad + \quad NH_4^+(aq)$$

Weak base   Hydroxide ion   Conjugate acid

Perform *Using Molecular Models 9-1: Equilibrium of a Weak Base* to gain further insight into the reaction of a weak base with water.

Using Molecular Models 9-1   Simulating the Reaction of a Weak Base in Water

The Forward Reaction

1. Obtain 1 red oxygen atom, 5 light-blue hydrogen atoms, and 1 blue nitrogen atom. Obtain 6 straight bonds.

2. Construct a model of ammonia, NH_3, and a model of water, H_2O.

3. Using only the models of ammonia and water created in step 2, simulate the *forward* reaction of ammonia reacting with water by forming a model of the ammonium ion and a model of the hydroxide ion. Answer inquiry questions 5–7.

$$NH_3 + H_2O \rightarrow NH_4^+ + OH^-$$

The Reverse Reaction

4. Using only the ammonium ion and hydroxide ion created in step 3, simulate the *reverse* reaction of the ammonium ion reacting with hydroxide ion, by forming ammonia and water. Answer inquiry questions 8–10.

$$NH_4^+ + OH^- \rightarrow NH_3 + H_2O$$

Inquiry Questions

5. Did you transfer a hydrogen ion? Which model donated the hydrogen ion? Which model accepted the hydrogen ion?

6. Which substance is the acid, NH_3 or H_2O? How do you know?

7. What are the conjugate acid-base pairs? How does a conjugate acid differ from the base?

8. Did you transfer a hydrogen ion? What model donated a hydrogen ion? Which model accepted a hydrogen ion? What is another name for a hydrogen ion?

9. Which new bond did you *make* in order to form ammonia and water?

10. Which acted as the acid in the reverse reaction, NH_4^+ or OH^-?

The weak bases we see most often in biochemistry are ammonia, NH_3, organic amines, RNH_2, R_2NH, and R_3N, hydrogen carbonate ion, HCO_3^-, and the conjugate bases of carboxylic acids such as acetate ion, CH_3COO^-, the conjugate base of acetic acid. Weak bases have a number of common applications and biological functions. Dilute solutions of ammonia are often used as cleaning fluids, although exposure to high levels of concentrated gaseous ammonia can cause lung problems and even death. The amine functional group is found in a number of pharmaceuticals, as well as in the important neurotransmitters of the brain.

There are many more weak acids than there are strong acids. Among the many weak acids, is an enormous range of acid strengths. For example, both acetic acid (CH_3COOH) and formic acid ($HCOOH$) are classified as weak acids, but formic acid is the stronger acid of the two. This means that formic acid produces more hydronium ions in aqueous solution than acetic acid: specifically, 10 times more hydronium ions. The extent to which a weak acid dissociates in solution can be measured, as described in Extension Topic 9-1.

Similarly, there is a range of base strengths among the weak bases. For example, ammonia, NH_3, is a weak base, but acetate ion, CH_3COO^-, is a weaker base, which means it is even less dissociated than ammonia in water.

The underlying reason why some acids and bases are weaker than others is a result of their chemical structure. An acid that produces a lower-energy, more stable, conjugate base dissociates to a greater extent than an acid with a less stable conjugate base.

Extension Topic 9-1 The Acid Ionization Constant, K_a, and the Strength of a Weak Acid

We have seen that the reaction of a weak acid in water is a reversible reaction, and in chapter 4 we learned that the concentration of reactants and products is constant for a reversible reaction at equilibrium. Therefore, the concentration of the products divided by the concentration of reactants is a constant, known as the **acid dissociation constant, K_a.** *The value of K_a is a quantitative measure of the extent to which an acid dissociates.* For the reversible reaction of a weak acid HA in water, the chemical equation is:

$$HA(aq) + H_2O(l) \rightleftharpoons H_3O^+(aq) + A^-(aq)$$

for which the acid dissociation constant, K_a, a unitless constant, is defined as:

$$K_a = \frac{[H_3O^+] \times [A^-]}{[HA]}$$

where brackets [ ] signify the concentration is given in units of moles per liter, M. This equation shows that an

acid with a larger K_a is stronger than an acid with a smaller K_a. The range of K_a values extends from 1×10^{-2} to 1×10^{-50}.

To avoid the use of scientific notation, scientists calculate and report pK_a values. The pK_a of an acid is calculated by performing the following mathematical operations:

$$pK_a = -\log K_a$$

Thus, the range of pK_a values is 2 to 50. The pK_a values for hundreds of weak acids are tabulated in pK_a reference tables, making it possible to compare the acid strengths of various weak acids. Since most biological acids are weak acids, quantifying their relative strengths can be useful.

Note that since we are multiplying the log by −1, a "negative log," the larger the value of K_a, the smaller the value of pK_a, *an inverse relationship.* For example, three weak acids are listed in the accompanying table, with their K_a and corresponding pK_a values.

Weak Acid	K_a	pK_a	Relative Acid Strength
Formic Acid, $HCOOH$	1.8×10^{-4} (larger number)	3.74 (smaller number)	Stronger weak acid
Acetic Acid, CH_3COOH	1.8×10^{-5}	4.74	
Phenol, C_6H_5OH	1.3×10^{-10} (smaller number)	9.89 (larger number)	Weaker weak acid

WORKED EXTENSION EXERCISES

E9.1 What does K_a measure in terms of the dissociation of an acid? Why is K_a a constant?

E9.2 What is the pK_a of ascorbic acid, which has a K_a of 7.94×10^{-5}?

E9.3 What is the pK_a of lactic acid, which has a K_a of 1.38×10^{-4}?

E9.4 Which is the stronger acid in the previous two questions?

E9.5 The pK_a of phosphoric acid is 2.1. Is phosphoric acid a stronger or a weaker acid than formic acid?

Solutions

E9.1 K_a measures the extent of dissociation by dividing the concentration of the product by the concentration of the reactants—the acid—at equilibrium. It is a constant because at equilibrium the concentration of reactants and products is no longer changing.

E9.2 $pK_a = -\log[K_a] = -\log(7.94 \times 10^{-5}) = 4.100$

E9.3 $pK_a = -\log[K_a] = -\log(1.38 \times 10^{-4}) = 3.860$

E9.4 Lactic acid is a stronger acid than ascorbic acid because it has a lower pK_a (larger K_a).

E9.5 Phosphoric acid has a $pK_a = 2.1$ while formic acid has a $pK_a = 3.7$. A lower pK_a means a larger K_a, which means a more dissociated acid, and hence, phosphoric acid is a stronger acid than formic acid.

E9.1 Hydrogen cyanide, HCN, has a $K_a = 4 \times 10^{-10}$. What is the pK_a of HCN?

E9.2 What is in the numerator in the mathematical expression that defines K_a? What is in the denominator? When comparing two acids, will the weaker acid have a larger or a smaller denominator? Does a larger denominator result in a larger or a smaller K_a?

E9.3 Carbonic acid has a $K_a = 4.2 \times 10^{-7}$. What is the pK_a of carbonic acid? Is carbonic acid a stronger or a weaker acid than acetic acid?

WORKED EXERCISES | Weak Acids and Bases

9-10 Carbonic acid is a weak acid. Explain what a weak acid is in terms of the relative concentration of reactants and products. Is this solution a strong or a weak electrolyte? What is the conjugate base of carbonic acid?

$$H_2CO_3 \;+\; H_2O \;\rightleftharpoons\; H_3O^+ \;+\; HCO_3^-$$

9-11 Phosphoric acid is a weak acid. Our genetic material, DNA and RNA, is a derivative of phosphoric acid. In aqueous solution, phosphoric acid undergoes the following reversible reaction:

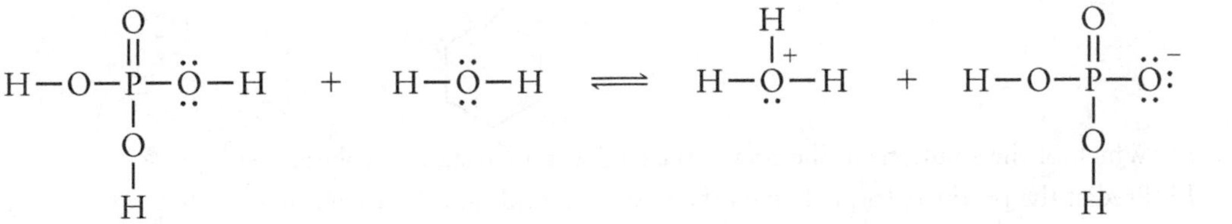

Phosphoric acid   Dihydrogen phosphate

    a. At equilibrium, is this solution a strong or a weak electrolyte? Explain.
    b. What substances are present at equilibrium?
    c. At equilibrium is the concentration of dihydrogen phosphate ion or phosphoric acid greater? Does the equilibrium lie to the left or to the right? Explain.
    d. Is the concentration of hydronium ions high or low? Explain.

9-12 Write the chemical equation showing the reaction of ethanamine in water. Label the conjugate acid-base pairs. At equilibrium, is the concentration of ethanamine or its conjugate acid greater? Is this solution a strong, weak, or nonelectrolyte? Explain.

9-13 In exercise 9-11 water acts as a base, but in exercise 9-12 it acts as an acid. Explain this seeming contradiction.

Solutions

9-10 A weak acid produces few hydronium ions and its conjugate base at equilibrium, so there is mainly carbonic acid, H_2CO_3, in solution, and very little hydronium ion, H_3O^+, and hydrogen carbonate ions, HCO_3^-. As a weak acid, it produces few ions, and therefore, the solution is a weak electrolyte. The conjugate base of carbonic acid is hydrogen carbonate ion, HCO_3^-.

9-11 a. As a weak acid, phosphoric acid produces few ions in solution, and therefore, the solution is a weak electrolyte.

b. Phosphoric acid, H_3PO_4, dihydrogen phosphate, $H_2PO_4^-$, and hydronium ions, H_3O^+, are all present in solution at equilibrium.

c. At equilibrium the concentration of phosphoric acid, the reactant, is much greater than dihydrogen phosphate, the conjugate base, because weak acids are only dissociated to a small extent in solution. At equilibrium, therefore, there is a greater concentration of the reactants than the products; the equilibrium lies to the left.

d. The concentration of hydronium ions, a product, is low because phosphoric acid is a weak acid, and weak acids are weakly dissociated in solution and the equilibrium lies to the left.

9-12 $CH_3CH_2NH_2(aq)$ + H_2O $\rightleftharpoons$ $CH_3CH_2NH_3^+(aq)$ + $OH^-(aq)$
      Weak base   Conjugate acid

The conjugate acid-base pairs are $CH_3CH_2NH_2$ and $CH_3CH_2NH_3^+$ and also H_2O and OH^-. At equilibrium the concentration of ethanamine is greater than its conjugate acid because an amine is a weak base. A solution of ethanamine in water is a weak electrolyte because it produces few ions in solution.

9-13 Water is an amphoteric molecule, which means that it can act as either an acid (donating a proton) or a base (accepting a proton). In exercise 9-11 water is in the presence of an acid, therefore water acts as a base. In exercise 9-12 water is in the presence of a base, therefore water acts as an acid.

PRACTICE EXERCISES

12 Indicate whether each of the following compounds is a *strong acid*, a *weak acid*, a *strong base*, or a *weak base*.

    a. LiOH   **b.** HNO_3 **c.** CH_3COO^- **d.** [piperidine structure]   **e.** CH_3CH_2COOH

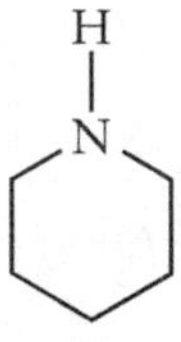

13 Which of the solutions in the previous exercise is a strong electrolyte? Explain.

14 Predict the products formed when the following acids and bases are dissolved in aqueous solution. Indicate whether the acid or base is strong or weak. Indicate whether the acid or base is a strong or a weak electrolyte.

    a. HNO_3 **b.**   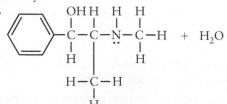   **c.** NaOH

15 Adenine, shown below, is a part of the structure of DNA and RNA. Would you classify adenine as an acid or a base? Explain.

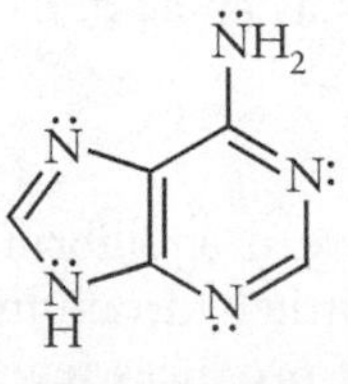

Le Châtelier's Principle In chapter 4 you learned that a reaction at equilibrium can be disturbed by changing the temperature or changing the concentration of a reactant or product. Le Châtelier's principle predicts that when a reaction at equilibrium is disturbed, the reaction responds by shifting in the direction that restores equilibrium: either the forward direction (a shift to the right, $\rightleftharpoons$) or the reverse direction (a shift to the left, $\rightleftharpoons$). Le Châtelier's principle, therefore, can be applied to reactions of weak acids and bases whose equilibrium has been disturbed.

Consider, for example, the reaction of pyruvic acid, a weak acid, in aqueous solution. At equilibrium, the forward and the reverse reaction rates are the same, so the concentration of products and reactants is constant. If more of one of the products, such as the hydronium ion, is *added*, Le Châtelier's principle predicts that the reaction will respond by *shifting to the left*, to consume the added H_3O^+ until a new equilibrium is established:

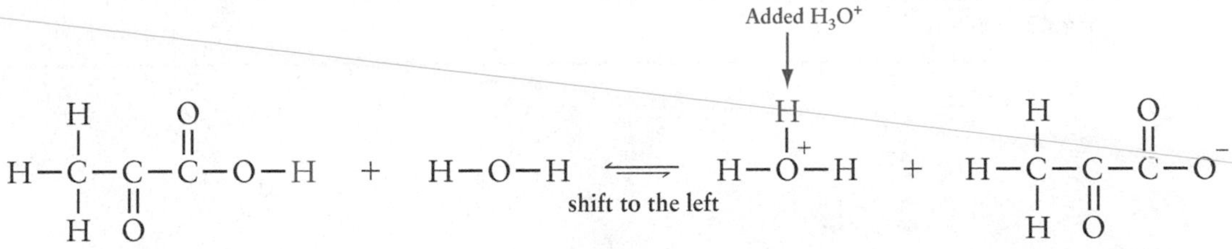

Pyruvic acid
(weak acid)

Pyruvate
(weak conjugate base)

Figure 4-23 summarizes the direction a reaction shifts in order to re-establish equilibrium when reactants or products are added or removed. In biochemical systems equilibrium is constantly adjusting to disturbances, according to Le Châtelier's principle, as the surrounding conditions warrant. We will see further applications of Le Châtelier's principle in section 9.4 when we study blood buffers.

WORKED EXERCISE | Acid–Base Equilibrium and Le Châtelier's Principle

9-14 Consider the reversible reaction, which occurs in our blood, shown below for the weak acid, carbonic acid:

$$H_2CO_3 + H_2O \rightleftharpoons H_3O^+ + HCO_3^-$$

Carbonic acid

Hydrogen
carbonate ion

    a. Does an increase in carbonic acid cause a *shift to the left* or *to the right*? Explain. How does this change affect the concentration of hydronium ion?

    b. Does a decrease in carbonic acid cause a *shift to the left* or *to the right*? Explain. How does this change affect the concentration of hydronium ion?

Solution

9-14 **a.** An increase in the concentration of carbonic acid causes a *shift to the right*, in order to consume the added carbonic acid, a reactant. This shift increases the concentration of products, so the concentration of hydronium ions increases.

    b. The removal of carbonic acid causes a *shift to the left*, in order to produce more carbonic acid, a reactant. This shift decreases the concentration of products, so the concentration of hydronium ions decreases.

PRACTICE EXERCISES

16 Lactic acid is produced in muscle cells during intense exercise when oxygen in the cell has been depleted. Write the equilibrium reaction for this weak acid in aqueous solution.

$$
\begin{array}{c}
\text{H} \quad \text{O} \\
| \quad \; \parallel \\
\text{H}-\text{C}-\text{C}-\text{O}-\text{H} \\
| \\
\text{OH}
\end{array}
$$

Lactic acid

a. What substances are present at equilibrium?

b. Is the concentration of lactic acid and lactate ion constant or changing at equilibrium?

c. If additional hydronium ions are introduced, in which direction will the equilibrium shift?

d. If the person starts to exercise more intensely, in which direction will the equilibrium shift?

17 Write the equilibrium reaction for ammonia, NH_3, in aqueous solution.

a. If additional hydroxide ions, OH^-, are added, will the reaction shift to the left or to the right? Explain.

b. If some ammonia is removed from the solution, will the reaction shift to the left or to the right? Explain.

9.2 pH

Have you ever tested the pH of a fish tank, a pool, or a solution in a high school experiment? If so, you were measuring the concentration of hydronium ions (H_3O^+) in solution. *The pH of a solution is a quantitative measure of the concentration of hydronium ions, H_3O^+, in aqueous solution.* Aqueous solutions range in pH from 0 to 14. A pH less than 7 indicates an *acidic* solution, a pH greater than 7 indicates a *basic* solution, also known as an **alkaline** solution, and a pH equal to 7 indicates a *neutral* solution. Special dye coated papers, known as **pH indicator paper** can be used to approximate the pH of a solution. pH indicator paper changes color according to the concentration of hydronium ions in solution (**Figure 9-7**).

The pH inside our cells (intracellular), extracellular fluids (outside the cell), our blood, and most solutions in our body are carefully regulated. Table 9-5 lists the normal pH range of some important body fluids. Blood and intracellular fluids have a pH close to neutral: 7.35–7.45. When the pH of blood rises or falls out of the normal range, biochemical reactions cannot proceed normally, and serious medical conditions result. A blood pH that drops below 7.35 causes a condition known as **acidosis**, and a blood pH above 7.45 causes a condition known as **alkalosis**. Acidosis and alkalosis are life-threatening conditions. In this section you will learn how to calculate pH, interpret pH values, and learn how pH affects the environment of the cell.

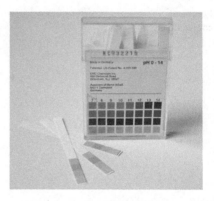

Figure 9-7 pH indicator strips are coated with special dyes that turn a particular color depending on the pH. The pH indicator strip is simply compared to the color key (shown) to determine the pH, to a close approximation. [Catherine Bausinger]

TABLE 9-5 pH of Common Body Fluids

Body Fluid	Normal pH Range
Gastric fluid	0.5–2
Urine	5–8
Saliva	6.5–7.5
Muscle cells	6.7–6.8
Blood	7.35–7.45

Water

The chemical properties of water are key to understanding how acids and bases behave in aqueous solution. In section 9.1 you learned that water can act as either an acid or a base, a property of amphoteric molecules. Indeed, even in pure water, a few water molecules react with another water molecule in a reaction known as the **autoionization of water**:

$$
\text{H}-\ddot{\text{O}}-\text{H} \;+\; \text{H}-\ddot{\text{O}}-\text{H} \;\rightleftharpoons\; \text{H}-\ddot{\ddot{\text{O}}}\!:^{-} \;+\; \text{H}-\overset{\text{H}}{\underset{..}{\text{O}}}\!-\text{H}^{+}
$$

(acid)   (base)   Hydroxide ion   Hydronium ion

The autoionization of water produces an equal concentration of hydronium ions (H_3O^+) and hydroxide ions (OH^-). In pure water the concentration of these ions is constant at $1 \times 10^{-7}\, M$ (at 25 °C):

$$[H_3O^+] = 1.0 \times 10^{-7}\, M$$
$$[OH^-] = 1.0 \times 10^{-7}\, M$$

Brackets [ ] indicate that the units of concentration are given in moles per liter, *M*, molarity. If we multiply the concentration of these two constants, we obtain the equilibrium constant known as the **ion-product constant of water**, K_w:

$$K_w = [H_3O^+] \times [OH^-] = (1.0 \times 10^{-7}\, M) \times (1.0 \times 10^{-7}\, M) = 1 \times 10^{-14}$$

If we add acid to pure water, we disturb the equilibrium for the autoionization of water by increasing the amount of hydronium ions (a product) in solution. Le Châtelier's principle predicts a shift to the left will occur as a result of the added hydronium ions, which causes the hydroxide ion concentration to decrease proportionately, because by definition the ion-product constant remains constant at 1×10^{-14}:

$$K_w = [H_3O^+] \times [OH^-] = 1 \times 10^{-14}$$

For example, when the hydronium ion concentration is increased 10-fold to $1 \times 10^{-6}\, M$, the hydroxide ion concentration decreases 10-fold to $1 \times 10^{-8}\, M$, in order to maintain the ion-product constant at 1×10^{-14}:

$$[H_3O^+] \times [OH^-] = 1 \times 10^{-14}$$
$$(1 \times 10^{-6}) \times (1 \times 10^{-8}) = 1 \times 10^{-14}$$

Similarly, when a base is added to pure water increasing the hydroxide ion concentration, the equilibrium for the autoionization of water shifts to the left to consume the added hydroxide ion causing the hydronium ion concentration to decrease proportionately in order to maintain the ion-product constant. *Thus, the quantitative relationship between the hydronium ion concentration and the hydroxide ion concentration are related through the ion product constant in aqueous solution, as illustrated in* Table 9-6.

Using the equation $[H_3O^+] \times [OH^-] = 1 \times 10^{-14}$, we can calculate the hydroxide ion concentration if we know the hydronium ion concentration and vice versa because they are mathematically related through the ion-product constant. For example, if $[OH^-] = 2.6 \times 10^{-12}$, we calculate the hydronium ion concentration as follows:

$$[H_3O^+] \times [OH^-] = 1 \times 10^{-14}$$
$$[H_3O^+] = \frac{1 \times 10^{-14}}{[OH^-]} = \frac{1 \times 10^{-14}}{2.6 \times 10^{-12}} = 3.8 \times 10^{-3}\, M$$

Calculating pH

In Table 9-6 you can see that the hydronium ion concentration ranges from $1\, M$ to $1 \times 10^{-14}\, M$, a range that spans 14-orders of magnitude. An order of magnitude is a factor of 10. We can avoid scientific notation by calculating the pH from the hydronium ion concentration. *The pH of an aqueous solution is defined as the negative logarithm of the hydronium ion concentration.* In other words, the *logarithm* ($\log_{10}x$) of the hydronium ion concentration multiplied by -1, which is written as:

$$pH = -\log_{10}[H_3O^+]$$

TABLE 9-6 The Relationship between pH and the Concentration of H_3O^+ and OH^- in Solution

	pH	$[H_3O^+]$	$[OH^-]$
Battery acid — HCl, 0.1 M —	0	1×10^0	1×10^{-14}
Gastric juice —	1	1×10^{-1}	1×10^{-13}
	2	1×10^{-2}	1×10^{-12}
Cola —	3	1×10^{-3}	1×10^{-11}
	4	1×10^{-4}	1×10^{-10}
Rain —	5	1×10^{-5}	1×10^{-9}
	6	1×10^{-6}	1×10^{-8}
Saliva —	7	1×10^{-7}	1×10^{-7}
Human blood, tears —	8	1×10^{-8}	1×10^{-6}
Baking soda —	9	1×10^{-9}	1×10^{-5}
Milk of magnesia —	10	1×10^{-10}	1×10^{-4}
	11	1×10^{-11}	1×10^{-3}
Household bleach, Household ammonia	12	1×10^{-12}	1×10^{-2}
NaOH, 0.1 M —	13	1×10^{-13}	1×10^{-1}
	14	1×10^{-14}	1×10^0

More acidic ↑ / More basic ↓

For example, if the hydronium ion concentration is 1×10^{-7}, the logarithm is the exponent, -7, which when multiplied by -1 changes the value to $+7$:

$$pH = -\log_{10}(1 \times 10^{-7}) = 7.0$$

Note that while calculating the log of 1×10^n does not require a calculator, you will need a calculator to compute the logarithm of a number other than "1×10^n", and the pH will not be a whole number. For example, the pH of a solution with a hydronium ion concentration of 1×10^{-3} has a pH = 3.0 while a solution with a hydronium ion concentration of 3.8×10^{-3} M has a pH = 2.42.

Table 9-6 shows the pH for various solutions with a 10-fold decrease in hydronium ion concentration and a corresponding 10-fold increase in hydroxide ion concentrations as we move down each row. Since the hydroxide ion concentration is inversely related to the hydronium ion concentration through the ion-product constant, we can calculate the pH of a solution if we know the hydroxide ion concentration, by first calculating the hydronium ion concentration from the hydroxide ion concentration, and then substituting this value into the pH equation, as demonstrated in the worked exercises below.

It is important to understand that the pH scale is a logarithmic scale, which means that 1 pH unit represents a 10-fold change in hydronium ion and hydroxide ion concentration—one order of magnitude. Thus, a solution with a pH of 2 is 10 times more acidic than a solution with a pH of 3 and 100 times more acidic than a solution with a pH of 4. Consider that if the pH of a lake changes by just 1 pH unit, most fish eggs will not hatch, and many adult fish will die. If the pH of your blood rises or falls by more than one tenth of a pH unit, you might find yourself in an ambulance headed to the hospital.

🔍 The reverse process, calculating the hydronium ion concentration, $[H_3O^+]$, from the pH, is done by taking the inverse log, 10^x, of the negative pH value:

$$[H_3O^+] = 10^{-pH}$$

For example, a solution with a pH = 5.0, has $[H_3O^+] = 1 \times 10^{-5}$. Most scientific calculators have the $\log_{10}x$ (log) and 10^x (inverse log) functions on the same key. See the *Mathematical Review: Logarithms* or the math appendix for a review of the $\log_{10}x$ function.

Mathematical Review of Logarithms

We have seen that the pH scale is a logarithmic scale. The **logarithm** ($\log_{10}$) of a number is the power to which 10 must be raised to equal the number. For example, the logarithm of the number 100 (10^2) is 2.

Number	Logarithm
1	$\log_{10}(10^0) = 0$
10	$\log_{10}(10^1) = 1$
100	$\log_{10}(10^2) = 2$
1000	$\log_{10}(10^3) = 3$
and so on . . .	
0.1	$\log_{10}(10^{-1}) = -1$
0.01	$\log_{10}(10^{-2}) = -2$
And so on . . .	

Most scientific calculators have a key labeled "log" or "log x" that calculates the logarithm of a number.

WORKED EXERCISE

Find the logarithm of the following numbers:
   a. 6.3×10^4 b. 3.4×10^{-6} c. 5.2×10^{-10}
   d. 100,000   e. 0.000234

Solution
   a. 4.80   b. -5.47 c. -9.28 d. 5.0   e. -3.631

WORKED EXERCISES — pH and Calculations Involving the Ion-Product Constant

9-15 What is the hydroxide ion concentration for an aqueous solution with a hydronium ion concentration of $5.2 \times 10^{-6}\,M$? Show your work.

9-16 A blood sample has a hydronium ion concentration of $4.0 \times 10^{-8}\,M$. What is the pH of this blood sample? Is this blood sample within the normal pH range for blood?

9-17 A commercial liquid antacid has a hydroxide ion concentration of $4.8 \times 10^{-3}\,M$. What is the pH of this solution?

9-18 What is the pH of a solution having a hydronium ion concentration of 0.02 M?

Solutions

9-15 You are supplied with the hydronium ion concentration: $[H_3O^+] = 5.2 \times 10^{-6}$, and asked to calculate $[OH^-]$, so use the equation for the ion-product constant to calculate $[OH^-]$:

$$[H_3O^+] \times [OH^-] = 1 \times 10^{-14}$$

First, rearrange the equation algebraically so that the variable you are solving for is isolated on one side of the equality. In this exercise we rearrange algebraically to solve for $[OH^-]$ by dividing both sides by $[H_3O^+]$:

$$[OH^-] = \frac{1 \times 10^{-14}}{[H_3O^+]}$$

Then, substitute the supplied value for $[H_3O^+]$ into the equation and solve for $[OH^-]$:

$$[OH^-] = \frac{1 \times 10^{-14}}{5.2 \times 10^{-6}} = 1.9 \times 10^{-9} \, M$$

9-16 Substitute $4.0 \times 10^{-8} \, M$ for $[H_3O^+]$ in the formula for pH and solve. Use the LOG function on your calculator to compute the logarithm:

$$pH = -\log[4.0 \times 10^{-8}] = 7.40$$

This value is within the normal range for the pH of blood: 7.35–7.45.

9-17 You are given a hydroxide ion concentration ($4.8 \times 10^{-3} \, M$) but need the hydronium ion concentration to calculate pH. So, this calculation will require two steps: first, calculate the hydronium ion concentration from the hydroxide ion concentration using the ion-product constant, and then calculate the pH using the hydronium ion concentration calculated in the first step. To calculate $[H_3O^+]$ from $[OH^-]$:

$$[H_3O^+] \times [OH^-] = 1 \times 10^{-14}$$

$$[H_3O^+] = \frac{1 \times 10^{-14}}{[OH^-]} = \frac{1 \times 10^{-14}}{4.8 \times 10^{-3}} = 2.1 \times 10^{-12} \, M$$

Next, calculate the pH by substituting the value calculated above for $[H_3O^+]$ into the pH equation:

$$pH = -\log[H_3O^+] = -\log(2.1 \times 10^{-12}) = 11.68$$

9-18 Calculate the pH by substituting the value given for $[H_3O^+]$ into the equation for pH:

$$pH = -\log[H_3O^+] = -\log(2 \times 10^{-2}) = 1.7$$

PRACTICE EXERCISES

18 Fill in the empty cells in the table representing three solutions whose hydronium or hydroxide ion concentration is provided. In the third column, indicate whether each solution is acidic, basic, or neutral.

	$[H_3O^+]$	$[OH^-]$	Acidic, basic, or neutral
Solution 1	1.0×10^{-11}		
Solution 2		9.1×10^{-9}	
Solution 3	3.3×10^{-6}		

19 What is the pH of a urine sample that has a hydronium ion concentration of $5.6 \times 10^{-7} \, M$?

20 What is the pH of a saliva sample that has a hydroxide ion concentration of $3.1 \times 10^{-8} \, M$?

Physiological pH

The pH of arterial blood ranges between 7.35 and 7.45 and is referred to as **physiological pH**. Many medical conditions, including kidney disease, stroke, and depression of the respiratory center by drugs, narcotics, and poisons, can raise or lower the pH of the blood.

Changes in intracellular pH disrupt cell function by causing functional groups such as carboxylic acids and amines in biomolecules to change between their neutral and ionized forms. At physiological pH, carboxylic acids, amines, and phosphate esters are in their ionized forms shown in Table 9-7.

TABLE 9-7 Neutral and Ionized Forms of Acidic and Basic Functional Groups

Functional Group	Neutral Form	Ionized Form at Physiological pH
Carboxylic acid	$$\begin{matrix} & O \\ & \| \\ R-&C-OH \end{matrix}$$	$$\begin{matrix} & O \\ & \| \\ R-&C-O^- \end{matrix}$$
Phosphate ester	$$\begin{matrix} & O \\ & \| \\ RO-&P-OH \\ & \| \\ & OH \end{matrix}$$	$$\begin{matrix} & O \\ & \| \\ RO-&P-O^- \\ & \| \\ & O^- \end{matrix}$$
Amine	$$\begin{matrix} & \ddot{N} \\ R-&N-H \\ & \| \\ & H \end{matrix}$$	$$\begin{matrix} & H \\ & \| \\ R-&\overset{+}{N}-H \\ & \| \\ & H \end{matrix}$$

Why is it important that certain functional groups in a molecule are ionized? In addition to enabling the molecule to bind to enzymes and chemically react, the charge on these molecules ensures that they remain within the cell. In the last chapter you learned that charged substances cannot freely pass through the cell membrane, thus charged species are trapped within the cell, allowing a biochemical pathway that occurs within the cell to proceed. Even a small change in pH lowers the concentration of the ionized forms. As a consequence essential biochemical reactions within the cell do not take place. In section 9.4 you will learn how your body maintains pH, which is so critical to health.

When a patient presents with symptoms of an acid-base disorder, such as heavy breathing, fatigue, and confusion, the first step is to measure the pH of their arterial blood. Arterial blood pH is measured with a blood gas analyzer such as the one illustrated in Figure 9-8. Often, the pH of the patient's urine is tested, too.

Specialized papers incorporating pH-sensitive dyes have been designed to indicate by color small differences in pH. The pH of urine is often tested using dipsticks containing dyes like pH indicator paper, made specifically for testing urine. In a chemistry laboratory setting, pH probes are often used because they are a more accurate and quantitative way to measure the pH of a solution. The pH probe is connected to an electronic meter or a computer. Once calibrated the probe is placed in the sample, and the pH meter or computer displays the pH of the solution.

Figure 9-8 A blood gas analyzer is an instrument used to test the pH of a sample of arterial blood. [© BSIP/Phototake]

9.3 Acid-Base Neutralization Reactions

You learned that acids donate a proton to water, and bases accept a proton from water in aqueous solution. Let us consider the reaction that occurs when an acid is combined with a base, known as a **neutralization reaction**. Neutralization reactions always go to completion—the forward reaction—regardless of the strength of the acid or the base. We describe three common types of neutralization reactions that differ slightly in products, depending on the type of base used:

- an ionic compound containing hydroxide ion, OH^-
- an ionic compound containing hydrogen carbonate, HCO_3^-, or carbonate ion, CO_3^{2-}
- ammonia, NH_3, or an organic amine (RNH_2, R_2NH, R_3N)

When Hydroxide Ion Is the Base

When an acid (H^+) and hydroxide ion (OH^-) are combined in an equal molar ratio, a solution of H_2O and a neutral salt (an ionic compound) is formed in the reaction; hence the term neutralization. Consider the neutralization reaction that occurs between an aqueous solution of NaOH, the base, containing dissolved hydroxide ions, $OH^-(aq)$ and sodium ions, $Na^+(aq)$, and an aqueous solution of HCl, the acid, containing dissolved hydrogen ions, H^+, and chloride ions, Cl^-. The hydroxide ions, $OH^-(aq)$ react with the hydrogen ions, $H^+(aq)$, in a 1:1 ratio to form water, H_2O, a neutral product, while sodium, Na^+, and chloride ions, Cl^-, remain dissolved in solution. Because the Na^+ and Cl^- ions remain dissolved throughout the reaction and do not play a role in the reaction, they are referred to as **spectator ions**. In the *complete balanced* equation, the spectator ions are written as a new ionic compound, NaCl (aq), where "aq" following the formula unit indicates they are dissolved ions in solution:

$$NaOH(aq) \quad + \quad HCl(aq) \quad \longrightarrow \quad H_2O \quad + \quad NaCl(aq)$$

| Base | Acid | Water (neutral) | Salt (neutral) |

Therefore, the **net reaction** in a neutralization reaction between a hydroxide ion–containing salt and *any* acid involves only the hydroxide ion and the hydrogen ion:

$$OH^-(aq) \quad + \quad H^+(aq) \quad \longrightarrow \quad H_2O$$

| **Hydroxide ion** (base) | **Hydrogen ion** (acid) | **Water** (neutral) |

For acids that can donate more than one proton or salts with more than one hydroxide ion in the formula unit, coefficients are needed to balance the complete balanced equation. In such cases, follow the principles for balancing an equation described in chapter 4. Examples are shown in the worked exercises.

WORKED EXERCISES Neutralization Reactions with Hydroxide Ion Salts

9-19 For a neutralization reaction between an aqueous solution of acetic acid and an aqueous solution of calcium hydroxide, answer the following questions.
  a. Which of these solutions is acidic? What ions are dissolved in the acidic solution?
  b. Which solution is basic? What ions are dissolved in the basic solution? What is the molar ratio of cations and anions in this solution?
  c. What is the net reaction that occurs when these two solutions are combined in a neutralization reaction?
  d. Write the complete balanced equation for the neutralization reaction.
  e. What are the spectator ions in this neutralization reaction?
  f. Why is this reaction classified as a neutralization reaction?

9-20 Write the complete balanced equation and the net reaction for the reaction between potassium hydroxide, KOH, and sulfuric acid, H_2SO_4. Note that sulfuric acid loses two protons for every one molecule of H_2SO_4 in a neutralization reaction. What are the spectator ions?

Solutions

9-19 a. An aqueous solution of acetic acid, CH_3COOH, is acidic because it contains a carboxylic acid, which donates a proton to water and produces hydronium ions in solution. The solution contains dissolved H_3O^+, and dissolved acetate ions, CH_3COO^-, the conjugate base of acetic acid.

b. The calcium hydroxide, $Ca(OH)_2$, solution is basic because it releases some dissolved OH^- ions. The other ion in solution is dissolved Ca^{2+}, although to a limited extent because $Ca(OH)_2$ is not very soluble in water. The ratio of hydroxide ions to calcium ions is 2:1, as indicated in the formula unit.

c. The net reaction is $H^+(aq) + OH^-(aq) \rightarrow H_2O(l)$.

d. The complete balanced equation is:

$$Ca(OH)_2(aq) + 2\ CH_3COOH(aq) \rightarrow 2\ H_2O(l) + Ca(CH_3COO)_2(aq).$$

e. The spectator ions are calcium ion, Ca^{2+}, and acetate ion, CH_3COO^-.

f. It is classified as a neutralization reaction because neutral water and a neutral salt are produced from a mixture of an acid and a base.

9-20 Complete balanced equation: $2\ KOH(aq) + H_2SO_4(aq) \rightarrow 2\ H_2O(l) + K_2SO_4(aq)$
Net reaction: $OH^-(aq) + H^+(aq) \rightarrow H_2O(l)$
Spectator ions: K^+ and SO_4^{2-} (sulfate ion)

PRACTICE EXERCISES

21 For a neutralization reaction between an aqueous solution of hydrobromic acid, HBr, and an aqueous solution of potassium hydroxide, KOH, answer the following questions.

a. Which solution is acidic? What ions are dissolved in the acidic solution?
b. Which solution is basic? What ions are dissolved in the basic solution?
c. What is the net reaction when the two solutions are combined in a neutralization reaction?
d. Write the complete balanced equation for the neutralization reaction.
e. What are the spectator ions in this reaction?

22 Complete and balance the neutralization reactions shown.

a. $HF(aq) + LiOH(aq) \rightarrow$
b. $HF(aq) + Al(OH)_3(aq) \rightarrow$
c. $HCOOH(aq) + Ba(OH)_2(aq) \rightarrow$

23 Use Le Châtelier's principle to explain how hydroxide salts with low solubility in water, such as aluminum hydroxide, $Al(OH)_3$, and magnesium hydroxide, $Mg(OH)_2$, are still effective at neutralizing stomach acid, HCl, even though the concentration of dissolved hydroxide ion at any time is small.

Bases that Contain Carbonate or Hydrogen Carbonate Ion

When an acid is combined with an ionic compound containing the carbonate ion, CO_3^{2-}, or the hydrogen carbonate ion, HCO_3^-, a neutral solution containing H_2O, a salt, *and* carbon dioxide, CO_2, gas are formed. The hydrogen carbonate ion and the carbonate ion react as bases by accepting one or two protons, respectively, from the acid to produce carbonic acid H_2CO_3. Carbonic acid (H_2CO_3) is an unstable compound that immediately decomposes to carbon dioxide, CO_2, and water, H_2O:

$$\underset{\text{Carbonic acid}}{H_2CO_3(aq)} \rightleftharpoons \underset{\text{Water}}{H_2O(l)} + \underset{\text{Carbon dioxide}}{CO_2(g)}$$

Therefore, the products of a neutralization reaction between an acid and an ionic compound containing the carbonate or hydrogen carbonate ion are water, a neutral salt, and carbon dioxide, CO_2. Since carbon dioxide is a gas, we observe bubbling from this type of neutralization reaction as the carbon dioxide gas comes out of solution (Figure 9-9). For example, the reaction

Figure 9-9 A neutralization reaction between hydrochloric acid (HCl) and calcium carbonate ($CaCO_3$) showing the carbon dioxide, CO_2, gas bubbling out of solution. [© 1987 Paul Silverman/ Fundamental Photographs]

TABLE 9-8 Active Ingredients in Common Over-the-Counter Antacids

Brand Name	Active Ingredient(s)
Amphogel	Aluminum hydroxide, $Al(OH)_3$
Phillips' Milk of Magnesia	Magnesium hydroxide, $Mg(OH)_2$
Maalox, Mylanta	$Al(OH)_3$ and $Mg(OH)_2$
Tums, Rolaids	Calcium carbonate, $CaCO_3$

between sodium bicarbonate ($NaHCO_3$), found in baking soda, and hydrochloric acid (HCl) produces water, sodium chloride (NaCl), and carbon dioxide (CO_2):

$$HCl(aq) \quad + \quad NaHCO_3(aq) \quad \longrightarrow \quad H_2O \quad + \quad NaCl(aq) \quad + \quad CO_2(g)$$

Acid   Base   Water   Salt   Carbon dioxide
(neutral)   (neutral)   (neutral)

Sodium ions from the base and chloride ions from the acid are spectator ions, shown in the complete balanced chemical equation as the salt NaCl (aq). The net reaction is:

$$H^+(aq) \quad + \quad HCO_3^-(aq) \quad \longrightarrow \quad H_2O \quad + \quad CO_2(g)$$

Water   Carbon dioxide

Antacid tablets, taken to relieve heartburn and acid indigestion, work by supplying the base for a neutralization reaction with stomach acid, HCl. Some common over-the-counter antacids and their active ingredients are listed in Table 9-8. Tums and Rolaids contain the carbonate ion, CO_3^{2-}, a weak base similar to the hydrogen carbonate ion, HCO_3^-. The neutralization reaction involves a 1:2 ratio between the active ingredient in Tums ($CaCO_3$) and stomach acid, HCl:

$$2\ HCl(aq) \quad + \quad CaCO_3(aq) \quad \longrightarrow \quad H_2O(l) \quad + \quad CaCl_2(aq) \quad + \quad CO_2(g)$$

Acid   Base   Water   Salt   Carbon dioxide

Calcium ions, Ca^{2+}, and chloride ions, Cl^-, are the spectator ions that are combine to form the neutral salt $CaCl_2$ in the complete balanced equation. Thus, the net reaction is:

$$2\ H^+(aq) + CO_3^{2-}(aq) + \longrightarrow H_2O(l) + CO_2(g)$$

WORKED EXERCISE	Neutralization Reactions Using the Hydrogen Carbonate Ion as the Base

9-21 Answer the questions below regarding the neutralization reaction that occurs when vinegar is combined with an aqueous solution of baking soda. You can try this experiment at home, but stand back when you mix the two solutions.

a. What is the acid? How can you tell this is the acid?

b. Is baking soda, $NaHCO_3$, an ionic or a molecular compound? What ions are dissolved in an aqueous solution of baking soda? Is baking soda an electrolyte? Is $NaHCO_3$ the acid or the base in this neutralization reaction? How can you tell?

c. Write the complete balanced equation for this neutralization reaction.

d. Write the *net* reaction that occurs in this neutralization reaction?

e. What are the spectator ions in this neutralization reaction?

f. Why are bubbles observed in this reaction? What does it mean when bubbles are no longer seen?

Solution

9-21 a. Acetic acid is the acid. It is an acid because it has a carboxylic acid functional group, and carboxylic acids donate a proton to water in aqueous solution.

b. Sodium hydrogen carbonate is an ionic compound because it is composed of a cation, Na^+, and an anion, HCO_3^-. These ions are dissolved in aqueous solution. $NaHCO_3$ is an electrolyte, because it forms dissolved ions in solution. Sodium hydrogen carbonate is a base because it is a negatively charged ion and accepts a proton from water in aqueous solution.

c. $CH_3COOH(aq) + NaHCO_3(aq) \rightarrow H_2O(l) + CH_3COONa(aq) + CO_2(g)$

d. $H^+(aq) + HCO_3^-(aq) \rightarrow H_2O(l) + CO_2(g)$

e. The spectator ions are sodium ion, Na^+, and acetate ion, CH_3COO^-.

f. Bubbling is observed because carbon dioxide (CO_2), which is a gas, is produced in the reaction. The bubbles are CO_2 molecules coming out of solution. When the bubbling stops, it means the neutralization reaction is over.

PRACTICE EXERCISES

24 Write the complete balanced neutralization reaction that occurs between HI and potassium hydrogen carbonate, $KHCO_3$? What is the net reaction? What are the spectator ions?

25 If there is a nitric acid, HNO_3, spill in the laboratory, will pouring calcium carbonate, $CaCO_3$, on the spill neutralize it? Write the balanced neutralization reaction that occurs between calcium carbonate, $CaCO_3$, and nitric acid, HNO_3. What is the net reaction? What are the spectator ions?

Ammonia and Organic Amines as Bases

We have seen that ammonia and amines are common weak bases seen in biochemistry. These nitrogen containing compounds with a lone pair of electrons on nitrogen react with strong or weak acids in a neutralization reaction to form ammonium salts, $RNH_3^+X^-$. The conjugate acid, RNH_3^+, is a weak acid, which together with the X^- ion forms an ionic compound known as an ammonium salt:

$$RNH_2(s) \quad + \quad HCl(aq) \quad \longrightarrow \quad RNH_3^+Cl^-(aq)$$

Amine base   Acid   Ammonium salt
                                     (weak acid)

This type of neutralization reaction is often used to convert a water insoluble amine into a water soluble salt, especially for pharmaceutical applications. Larger amines are generally insoluble in water due to the hydrocarbon character of the rest of the molecule. However, reacting an amine with an acid forms the conjugate acid of the amine, RNH_3^+, a polyatomic cation, which because of its full charge *is* soluble in water. The conjugate base formed in the reaction serves as the anion of the product salt. When the acid used in the neutralization reaction is HCl, the product is known as the hydrochloride salt of the amine: $RNH_3^+Cl^-$. If you look at the drug label accompanying a prescription drug, you will see that if it contains an amine it is probably a salt form of the amine. For example, ephedrine, a common decongestant, is prepared as the hydrochloride salt of ephedrine by reacting ephedrine with HCl in the last step of production:

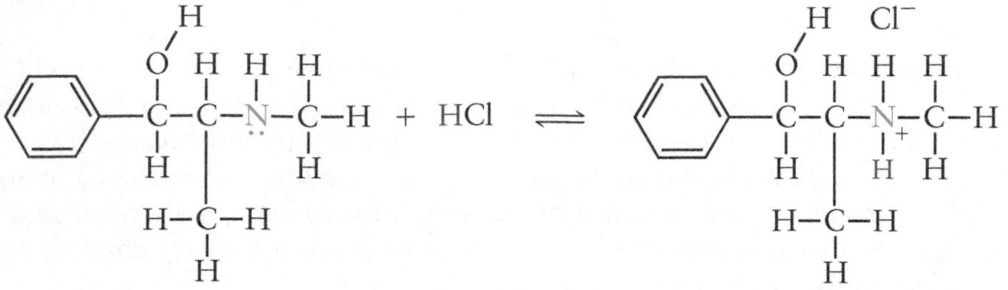

Base (ephedrine)   Acid   Hydrochloride salt of ephedrine

WORKED EXERCISE Acid–Base Neutralization Reactions with Amines as the Base

9-22 Using Table 9-2, write the neutralization equation for the reaction between epinephrine and acetic acid.

Solution
9-22

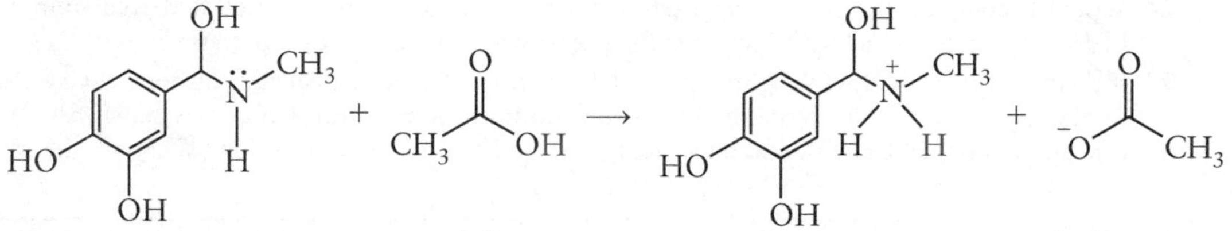

Base (Epinephrine)   Acid (Acetic acid)   Conjugate acid (Cation)   Conjugate base (Anion)

Ionic compound

PRACTICE EXERCISE

26 Zoloft, a drug used to treat depression in adults, is sold as the hydrochloride salt of the amine shown. Write the reaction of this molecule with HCl, showing the formation of the hydrochloride salt. Why is the hydrochloride salt more soluble than the amine?

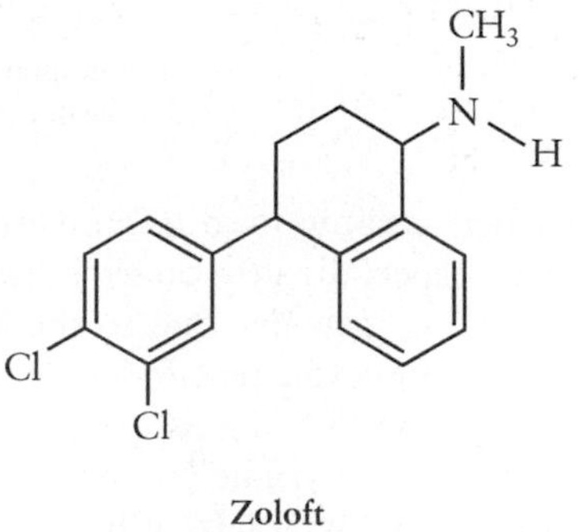

Zoloft
Neutral organic compound
(insoluble in water)

■ 9.4 Buffers

If the pH of a person's blood falls below 7.35 or rises above 7.45, cells cannot function properly and a severe medical condition known as acidosis or alkalosis, respectively, arises. Yet, the blood is constantly absorbing acids and bases from the foods we eat and the metabolic reactions that produce H^+. How then are we able to maintain the pH of blood within such a narrow range? The answer is that blood contains buffers. *A **buffer** is a solution that resists changes in pH by neutralizing any added acid (H^+) or base (OH^-).*

Let us consider how a buffer works, and then consider some applications of this in our blood. We have seen that when even a small amount of acid or base is added to water, the pH can change significantly. However when an acid or a base is added to a buffer solution, the pH changes only slightly. *Chemically, a buffer is a mixture of a weak acid, HA, and its conjugate base, A^-; or a weak base, B, and its conjugate acid, BH^+.* When an acid, H^+, is added to a buffer, such as a weak acid and its conjugate base, HA/A^-, the weak base component of the buffer, A^-, neutralizes the added acid to form HA, producing more of the buffer. Thus, the concentration of H^+ and OH^- is not affected, as shown below. Remember, it is the concentration of H^+ and OH^- that determines pH.

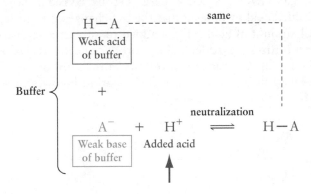

If instead of a base, OH^-, is added to a buffer, the weak acid component of the buffer, HA, neutralizes the added base, OH^-, to form water and A^-, producing more of the buffer but not more OH^-, as shown below.

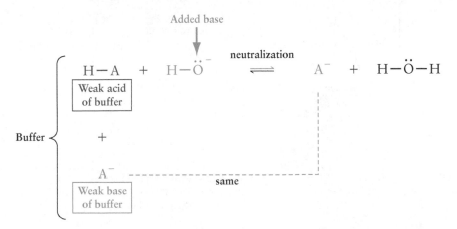

There are many different types of buffers and each is suitable for maintaining a certain pH range. Table 9-9 lists some common buffers and some important physiological buffers found inside cells (intracellular) and in the blood (extracellular). Consider, for example, an acetic acid–acetate buffer

TABLE 9-9 Buffer Systems and Their Components

Buffer System	Weak Acid	Weak Base	Function
Acetic acid–acetate	CH_3COOH	CH_3COO^-	Common laboratory buffer
Carbonic acid–hydrogen carbonate	H_2CO_3	HCO_3^-	Extracellular fluid buffer
Phosphate	H_3PO_4	$H_2PO_4^{2-}$	Intracellular fluid buffer; and lab applications
Hemoglobin	HHb (hemoglobin with bound H^+)	$Hb(O_2)$ (hemoglobin with bound oxygen)	Red blood cell buffer

used in the laboratory to maintain a pH between 4 and 5. Acetic acid (CH_3COOH) is a weak acid, and its conjugate base, the acetate ion (CH_3COO^-), is a weak base. To make an acetic acid–acetate buffer that maintains a pH of 4.7, we mix approximately equal molar amounts of acetic acid and sodium acetate in aqueous solution. Recall sodium acetate is an ionic compound that completely dissociates in solution into sodium ions, Na^+, and acetate ions, CH_3COO^-, a weak base. Let us consider how this buffer responds to added acid or base so that the pH changes very little.

If a small amount of acid (H^+) is added to an acetic acid–acetate buffer, acetate ion neutralizes the added acid to produce acetic acid, the other component of the buffer. Consequently, the pH, a measure of H_3O^+ concentration, changes very little, as shown in the equation below and schematically in Figure 9-10.

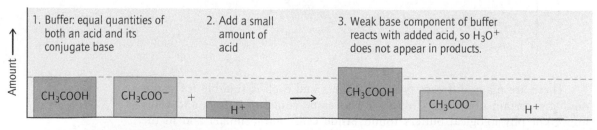

Figure 9-10 How the acetic acid–acetate buffer equilibrium shifts upon addition of a small amount of acid (H^+).

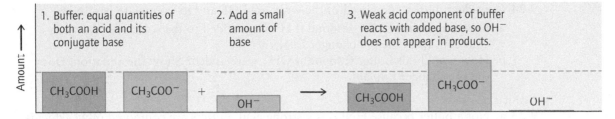

Figure 9-11 How the acetic acid–acetate buffer equilibrium shifts upon addition of a small amount of base (OH⁻).

Similarly, if a small amount of base is added to the acetic acid–acetate buffer, acetic acid neutralizes the added base to produce acetate ion, the other component of the buffer. Consequently, the pH, a measure of H_3O^+ concentration, which is related to the OH^- concentration through the ion product constant, does not change, as shown in the equation below and schematically in Figure 9-11.

The additional acetic acid or acetate ion formed when an acid or a base reacts with a component of the buffer does not appreciably affect the pH since acetic acid is a weak acid, and acetate is a weak base—only ionized to a small extent.

The amount of added acid or base that a buffer can neutralize is known as the **buffer capacity**. The greater the concentration is of the weak acid and the conjugate base of the buffer system, the greater is its buffer capacity. The addition of protons or hydroxide ions in excess of the buffer capacity will cause the pH to change.

WORKED EXERCISES   Understanding Buffers

9-23 Which of the following aqueous mixtures is a buffer? Explain why or why not. You can refer to Tables 9-3 and 9-4.

a. HNO_3 (nitric acid) and $NaNO_3$ (sodium nitrate)
b. NaCl and KCl
c. Citric acid and sodium citrate
d. CH_3CH_2COOH and CH_3CH_2COONa
e. KOH and H_2O
f. H_3PO_4 and $H_2PO_4^-$

9-24 Consider a buffer composed of the weak acid HF and its conjugate base NaF.

    a. How would this buffer respond if H^+ were added to the solution? Show the equation. Does the pH change? Explain.

    b. How would this buffer respond if OH^- were added? Show the equation. Does the pH change? Explain.

Solutions

9-23 a. Not a buffer because HNO_3 is a strong acid. Buffers are composed of weak acids.

    b. Not a buffer because the ions Na^+, Cl^-, and K^+ are not acids or bases.

    c. Yes, citric acid is an organic compound with three carboxylic acids, hence a weak acid, and citrate is the conjugate base of citric acid.

    d. Yes, propanoic acid is a weak acid and sodium propanoate is its conjugate base.

    e. No, because KOH is a strong base and water is the solvent, not the buffer

    f. Yes, this is the phosphate buffer. H_3PO_4, phosphoric acid, is a weak acid, and $H_2PO_4^-$ is the conjugate base of phosphoric acid, a weak base.

9-24 a.

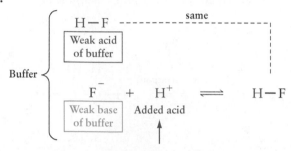

The weak base component of the buffer, F^-, neutralizes the added hydrogen ions, H^+, to form HF, the other component of the buffer. The concentration of hydronium and hydroxide ions remains relatively unchanged, so the pH does not change.

    b.

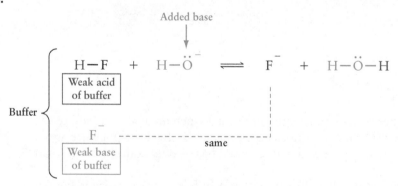

The weak acid component of the buffer, HF, neutralizes the added hydroxide ion, OH^-, to form F^-, the other component of the buffer. The concentration of hydronium and hydroxide ions remains relatively unchanged, so the pH does not change.

PRACTICE EXERCISE

27 A buffer can also be made from a weak base and its conjugate acid. For example, ammonia, NH_3, mixed with an equal molar amount of ammonium chloride, NH_4Cl, forms the NH_3/NH_4^+ buffer.

    a. What role does NH_4Cl play in this buffer?

    b. How would this buffer react if H^+ were added? Show the equation.

    c. How would this buffer react if OH^- were added? Show the equation.

    d. Why does the pH not change significantly when hydronium or hydroxide ions are added to the buffer?

    e. If an excess of hydronium ions or hydroxide ions were added, exceeding the buffer capacity, what would happen?

Acid-Base Homeostasis

Maintenance of proper pH in the body is known as **acid-base homeostasis,** where the term **homeostasis** generally refers to a balanced system. Acid-base homeostasis is maintained by buffers and by regulation of breathing rate (ventilation), as well as secretions by the kidneys. The primary buffers in intracellular fluids are the phosphate buffer, the hemoglobin buffer system in red blood cells, and protein buffers. The primary buffer in the blood is the carbonic acid (H_2CO_3)-hydrogen carbonate ion (HCO_3^-) buffer.

Consider the carbonic acid/hydrogen carbonate buffer in the blood. Our cells produce carbon dioxide (CO_2), a molecule that is able to freely pass through cell membranes because it is small and nonpolar, as an end-product of metabolism. Most of the carbon dioxide in solution reacts with water to form carbonic acid (H_2CO_3) according to the following important reversible reaction:

$$CO_2(g) + H_2O \rightleftharpoons H_2CO_3(aq) \qquad \text{Equation 9-1}$$

Note that this reaction is not an acid-base reaction. The kidneys release hydrogen carbonate ion (HCO_3^-) into the blood and, together with dissolved carbonic acid from Equation 9-1, form the carbonic acid–hydrogen carbonate buffer system.

The body responds immediately to acid-base imbalances with the carbonic acid–hydrogen carbonate buffer and by regulating breathing (ventilation) rate. Figure 9-12, shows how increased respiration removes carbon dioxide via the lungs, causing a shift to the *left* in Equation 9-1, according to Le Châtelier's principle. This shift in turn consumes carbonic acid, causing the buffer reaction shown to shift to the left, thereby consuming H^+ and preventing

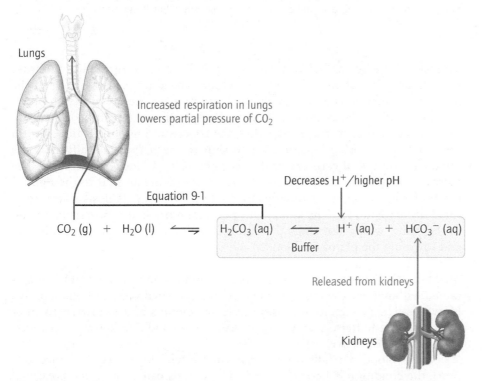

Figure 9-12 The body responds to a decrease in pH in the blood with the carbonic acid–hydrogen carbonate buffer and by regulating breathing (ventilation) rate. Increased respiration rate helps remove protons, because it shifts Equation 9-1 to the left by eliminating CO_2 through the lungs when exhaling. This in turn causes a shift to the left in the buffer equation, which consumes protons and raises the pH. Added HCO_3^- from the kidneys also shifts the equilibrium to the left.

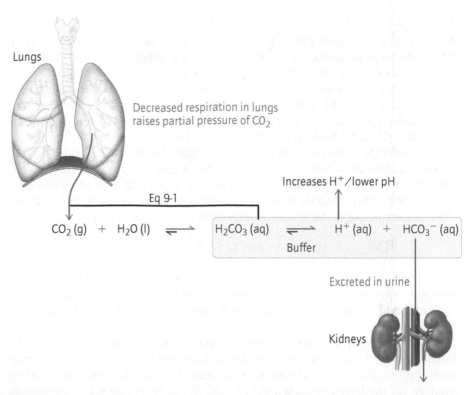

Figure 9-13 Our body responds to an increase in pH in the blood with the carbonic acid–hydrogen carbonate buffer and by regulating breathing (ventilation) rate. Decreased respiration produces more CO_2 which causes a shift to the right in Equation 9-1, producing more carbonic acid. The buffer responds to the increase in carbonic acid by shifting to the right consuming carbonic acid and producing more protons, thus lowering the pH.

acidosis (decrease in OH^-/increase in H^+). Eventually the kidneys release more hydrogen carbonate ion into the blood which also causes a shift to the *left* in the buffer equation, consuming H^+ and raising the pH of the blood.

Figure 9-13 shows how a decreased respiration rate causes the partial pressure of CO_2 to increase, increasing the concentration of CO_2 dissolved in the blood and causing Equation 9-1 to shift to the *right*. This shift increases the concentration of carbonic acid, according to Le Châtelier's principle. The increase in carbonic acid, H_2CO_3, causes a corresponding shift to the right in the buffer equation, generating protons and preventing alkalosis (increase in OH^-/decrease in H^+). The kidneys also excrete more hydrogen carbonate ion in the urine, which also shifts the buffer equation to the right, producing H^+ and lowering the pH of the blood.

Acidosis We have seen that under normal circumstances, blood pH is maintained within a narrow range by the carbonic acid–hydrogen carbonate buffer. However, when a condition causes excessive amounts of acid or base to enter the blood, the buffer capacity is exceeded, and a condition known as *acidosis* or *alkalosis*, respectively results.

When blood pH falls below its normal range (higher H_3O^+ concentration), the condition is known as **acidosis**. Acidosis can occur when breathing is too weak or too slow, preventing carbon dioxide from being removed from the lungs rapidly enough. The excess carbon dioxide that builds up in the blood causes equation 9-1 to shift to the right, which in turn results in a shift to the right in the buffer equation shown in Figure 9-13, causing an increase in H^+ concentration. Although hydrogen carbonate ion neutralizes some of the excess H^+ produced, if it exceeds the buffer capacity, the pH drops. This condition is known as **respiratory acidosis** and is often seen with a severe head injury, emphysema or asthma, narcotic use, and sometimes while under anesthesia.

Metabolic acidosis is another form of acidosis, caused by a metabolic disorder rather than weak respiration. Metabolic acidosis can result from kidney failure, uncontrolled diabetes, or starvation. Any of these conditions may lead to an excess of H_3O^+ entering the bloodstream, exceeding the buffer capacity, and causing blood pH to drop. Treatment of any form of acidosis usually includes IV infusion of hydrogen carbonate ion.

Alkalosis When blood pH rises *above* its normal range, the condition is known as **alkalosis**. A rise in blood pH can occur when breathing becomes too fast, as when someone hyperventilates, causing carbon dioxide to be removed from the lungs too fast. The rapid removal of carbon dioxide causes Equation 9-1 to shift to the left, reducing the amount of carbonic acid dissolved in the blood. The decreased concentration of carbonic acid causes the buffer equation Figure 9-12 to shift to the left, consuming H^+, and thereby increasing the pH. This condition is known as **respiratory alkalosis**. Hyperventilation can result from anxiety, hysteria, altitude sickness, or intense exercise. Breathing into a paper bag often corrects the problem because the patient is forced to inhale exhaled air that has a higher concentration of carbon dioxide than normal air does. The carbon dioxide inhaled restores dissolved CO_2 levels to the blood and shifts the equilibrium in Equation 9-1 to the right. Increased carbonic acid levels in turn cause the buffer equation in Figure 9-12 to shift to the right, increasing H^+/lowering pH.

Excessive vomiting, ingestion of excessive amounts of antacids, and some adrenal gland diseases can cause **metabolic alkalosis**. Metabolic alkalosis may be treated by intravenous administration of a dilute solution of hydrochloric acid, HCl, or by hemodialysis.

As you can see, buffer systems in the body are critical to maintaining acid-base homeostasis. The reactions of acids and bases are just one type, out of many types of chemical reactions. In the next chapter, you will study some of the important organic reactions in biochemistry.

WORKED EXERCISE   Understanding Buffers

9-25 Answer the following questions about a patient entering the emergency room presenting with weak breathing and respiratory acidosis.
   a. Is the patient's blood pH above or below the normal range?
   b. Is the concentration of H^+ too high or too low in this patient?
   c. As a result of the patient's weak breathing, in which direction has the equilibrium in Equation 9-1 shifted? Explain why the shift occurs and why it is in this direction.
   d. How will the buffer in the carbonic acid–hydrogen carbonate equation shift as a result of the equilibrium shift in (c)? How does this affect the concentration of H^+ in the blood?
   e. You are asked to give the patient an IV infusion of _____. What chemical reaction occurs in the blood when you add this IV solution?

Solution
9-25 **a.** The blood pH is below the normal range in a patient with acidosis.
   b. The concentration of H^+ ion is too high in this patient.
   c. Weak breathing prevents carbon dioxide from being eliminated from the lungs rapidly enough, causing an increase in the partial pressure of CO_2 and

therefore an increase in the concentration of CO_2 in the blood. Added CO_2 causes the equilibrium shown in Equation 9-1 to shift to the right, according to Le Châtelier's principle, thereby increasing the concentration of carbonic acid (H_2CO_3).

    d. The increased concentration of carbonic acid causes the equilibrium in the carbonic acid–hydrogen carbonate buffer equation to shift to the right to produce more H^+ lowering the pH, and causing acidosis.

    e. IV infusion of hydrogen carbonate ion, HCO_3^-, is recommended because adding this component of the buffer will shift the buffer equation to the left, consuming protons in the process and thereby raising the pH of the blood.

PRACTICE EXERCISES

28 Answer the following questions for the carbonic acid–hydrogen carbonate buffer in the blood.
    a. Carbonic acid is an unstable compound that is formed when _____ and dissolved _____ react.
    b. Write the reaction that shows how this buffer will respond to the addition of a small amount of acid (H^+).
    c. Write the reaction that shows how this buffer will respond to the removal of a small amount of acid (H^+).
    d. If someone presents with alkalosis, why are they asked to breath into a bag?

29 What is the difference between respiratory alkalosis and metabolic alkalosis? What is common to both of these conditions?

30 How is breathing important to acid-base homeostasis?

31 What role do the kidneys play in acid-base homeostasis?

Chemistry in Medicine pH and Oral Administration of Drugs

Imagine as a child being told that you can no longer eat sweets, and you must make regular injections into your body several times a day. This is a common scenario for children who have been diagnosed with type I diabetes. According to the National Diabetes Education Program (NDEP), over 180,000 Americans under the age of 20 have diabetes. In addition, many adults are diagnosed with type II diabetes each year and may eventually have to give themselves injections or rely on an insulin pump.

Diabetics must receive insulin injections at regular intervals. Insulin is a hormone synthesized by the pancreas, which signals cells to transport blood sugar across the cell membrane and into the cell, thereby reducing the concentration of blood sugar circulating in the blood. In many children diagnosed with type I diabetes, the pancreas has either partially or completely stopped producing insulin, and blood glucose levels rise steeply after they eat a meal because sugar is not able to enter cells. High blood sugar levels cause a host of medical problems ranging in severity from frequent urination to coma. In adult-onset diabetes, type II diabetes, insulin is produced by the pancreas, but too few or ineffective insulin receptors prevent glucose from entering cells.

Insulin is a small protein and protein drugs cannot be administered orally; therefore, they are injected under the skin. The main reason protein drugs cannot be taken orally is that they degrade in the acidic environment of the gastrointestinal tract. Figure 9-14 shows an overview of the human digestive system. Digestion is a complex process. One critical factor in the normal functioning of the digestive tract is its pH. The pH of the mouth and esophagus is weakly acidic, about pH = 6.8. The pH of the stomach is very acidic, in the range of pH = 1–3. The mucosal tissue of the small intestine is alkaline, pH = 8.5. In the small intestine, food is mixed with bile, pancreatic juices, and enzymes. As the pH changes in the small intestine, enzymes are activated that break down proteins, carbohydrates, and fats, so that the reaction products can be absorbed into the circulatory system.

It takes only seconds for food to pass through the esophagus, and by the time a protein reaches your stomach, the gastric glands in the stomach lining have begun to secrete gastric juice—a strongly acidic solution (pH = 1–3) containing enzymes, hydrochloric acid, and mucous. Under these acidic conditions, the shape of insulin is altered, so that it no longer can carry out its normal cellular functions.

Figure 9-15 The enteric coating, on a pill shown here in orange.
[© 2008 Richard Megna/Fundamental Photographs]

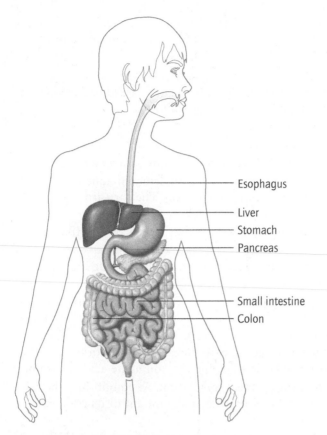

Figure 9-14 The digestive system. Digestion is a complex process in which pH plays an important role.

Esophagus
Liver
Stomach
Pancreas
Small intestine
Colon

As a first step in overcoming the pH issues of orally administered protein drugs like insulin, scientists have developed enteric coatings—barriers applied to oral medications that control where in the digestive tract the medication is absorbed (Figure 9-15). The enteric coating creates a surface on the tablet that is stable in the acidic environment of the stomach, protecting the protein from exposure to the low-pH of the surroundings.

Large molecules that contain carboxylic acid functional groups are widely used as enteric coatings. In an acidic environment, the carboxylic acid functional groups are in their neutral carboxylic acid form and the coating remains intact. In a basic environment, the carboxylic acid functional groups of the enteric coating react with hydroxide ions to form carboxylate ions (the conjugate base of a carboxylic acid), which is soluble in water and causes the enteric coating to dissolve and thus release the active drug. These coatings enable a protein to pass through the acidic environment of the stomach and into the basic environment of the small intestine. The enteric coating then dissolves, depositing the protein in its unaltered state and enabling it to perform its function.

Once the intact protein is deposited in the small intestine, the challenge remains to get the intact protein across the membrane barrier of the small intestine, without damaging intestinal cells. These are chemistry problems of another type, still under investigation. With active research continuing in this area, there is hope that one day diabetics will be able to simply swallow a pill, rather than inject themselves several times a day every day.

Summary

Acids and Bases

- According to the Arrhenius definition, an acid is a substance that produces hydrogen ions (H^+) in solution, and a base is a substance that produces hydroxide (OH^-) ions in solution.
- Hydrogen ions, H^+, also known as protons, form a bond to water molecules in aqueous solution, creating hydronium ions, H_3O^+.
- The Brønsted-Lowry definition of an acid and a base is broader and defines an acid as a proton donor and a base as a proton acceptor.
- The product formed when an acid loses a proton is known as the conjugate base. The product formed when a base accepts a proton is known as the conjugate acid.

- For an acid or a base dissolved in water, two conjugate acid-base pairs can be identified. One pair is always water and hydroxide ion or water and hydronium ion.
- A conjugate acid-base pair differs by a hydrogen ion.
- Ionic compounds containing the hydroxide ion are bases because they release hydroxide ion into solution upon dissolution.
- Amines are common organic bases.
- Carboxylic acids are common organic acids.
- Water is an amphoteric molecule because it can act as both an acid and a base.
- A strong acid and a strong base dissociate completely in aqueous solution; no un-ionized acid or base is present in solution. The forward reaction goes to completion.
- The strong acids are inorganic compounds that include HCl, HBr, HI, HNO_3, H_2SO_4, and $HClO_4$.
- The strong bases are ionic compounds containing hydroxide ion and a group 1A or 2A metal cation.
- A weak acid dissociates to a small extent in solution. At equilibrium there is a mixture of mainly the un-ionized acid, a small amount of its conjugate base, and hydronium ions.
- Weak acids include carboxylic acids, carbonic acid, phosphoric acid, organic derivatives of phosphoric acid, and the conjugate acids of amines.
- A weak base ionizes to a small extent in solution. At equilibrium there is a mixture of mainly the un-ionized base, a small amount of its conjugate acid, and hydroxide ions.
- Weak bases include ammonia, amines, hydrogen carbonate ion (HCO_3^-), carbonate ion (CO_3^{2-}), and the conjugate base of carboxylic acids ($RCOO^-$).
- There is a wide range of acid strengths among the weak acids, which is related to the stability of the conjugate base.
- Le Châtelier's principle allows one to predict how the equilibrium reaction for a weak acid or a weak base will shift if a reactant or product is added or removed. The equilibrium shifts in the direction that restores equilibrium.

pH

- pH is a quantitative measure of the hydronium ion concentration in solution.
- A pH less than 7 indicates an acidic solution, a pH greater than 7 indicates a basic solution, and a pH equal to 7 indicates a neutral solution.
- The autoionization of water is a reversible reaction that occurs in pure water between two water molecules to produce a hydroxide ion and a hydronium ion.
- At 25 °C, the concentration of hydronium and hydroxide ions in pure water is 1×10^{-7} M, due to the autoionization of water.
- At equilibrium the product of the hydronium ion concentration and the hydroxide ion concentration is a constant, known as the ion-product constant for water K_w, with a value of 1.0×10^{-14} at room temperature. It is a unitless constant.
- The concentration of hydroxide ions and hydronium ions in aqueous solution are related by the ion product constant according to the following equation: $[OH^-] \times [H_3O^+] = 1 \times 10^{-14}$. Brackets indicate concentration is given in units of moles per liter, M.
- The pH of a solution can be calculated from the hydronium ion concentration using the pH equation: $pH = -\log[H_3O^+]$.

- Physiological pH is the pH of arterial blood: pH = 7.35–7.45.
- pH can be approximated using pH paper or more accurately using a pH meter.

Acid-Base Neutralization Reactions

- The reaction between and acid and a base is known as a neutralization reaction because a neutral compound, like water, is formed.
- The products of a neutralization reaction involving hydroxide ion are water and a salt. The net reaction is $H^+ + OH^- \rightarrow H_2O$.
- The products of a neutralization reaction involving carbonate or hydrogen carbonate ion are water, a salt, and CO_2 gas. The net reaction is either $H^+ + HCO_3^- \rightarrow H_2O + CO_2$ or $2\,H^+ + CO_3^{2-} \rightarrow H_2O + CO_2$.
- Antacids are hydroxide or carbonate ion bases that react with stomach acid, HCl, in a neutralization reaction that produces water and a salt or water, a salt, and CO_2, thereby relieving acid indigestion.
- An amine—a weak base—reacts with an acid, to form a salt composed of the conjugate acid of the amine and the conjugate base of the acid: $RNH_2 + HX \rightarrow RNH_3^+X^-$.
- Amine salts are soluble in water because they have a full charge, whereas amines are generally insoluble in water because they are neutral.

Buffers

- A buffer is a solution that resists changes in pH upon addition of a small amount of acid or base.
- A buffer contains a weak acid and its conjugate base, or a weak base and its conjugate acid.
- When acid, H^+, is added to a buffer, the weak base component of the buffer reacts with the added H^+ to form more of the acid component of the buffer; thus no H^+ is formed, and the pH changes only slightly.
- When base, OH^-, is added to a buffer, the weak acid component of the buffer reacts with the added OH^- to form more of the base component of the buffer; thus no OH^- is formed, and the pH changes only slightly.
- The primary buffering system of the blood consists of carbonic acid and hydrogen carbonate ion.
- Acid-base homeostasis is the maintenance of the proper pH in blood and other bodily fluids.
- Respiration rate controls the amount of CO_2 dissolved in blood, which affects the reversible reaction of $CO_2 + H_2O$ to form carbonic acid, which in turn affects the buffer equation and the amount of protons in solution.
- Buffer capacity is the amount of acid or base that can be added to a buffer and still maintain the pH. Exceeding the buffer capacity causes an increase or decrease in the pH of solution.
- A disturbance in the buffer system of blood can cause acidosis (pH below 7.35) or alkalosis (pH above 7.45), both serious medical conditions.

Key Words

Acid According to the Arrhenius definition, a substance that produces hydrogen ions, H^+, in solution. According to the Brønsted-Lowry definition, a proton donor.

Acid-base homeostasis Maintenance of the proper pH balance in the blood and other bodily fluids.

Acid Dissociation Constant, K_a The product of the concentration of products divided by the concentration of reactants for the reaction of an acid with water at equilibrium:

$$K_a = \frac{[H_3O^+] \times [A^-]}{[HA]}$$

Acidosis A serious medical condition characterized by an arterial blood pH below 7.35.

Alkaline A solution with a pH greater than 7; a basic solution.

Alkalosis A serious medical condition characterized by an arterial blood pH above 7.45.

Amphoteric molecule A molecule, such as water, that can act as either an acid or a base.

Autoionization of water The reaction that occurs in pure water between two water molecules to produce an equal and small concentration of hydronium ions (H_3O^+) and hydroxide ions (OH^-).

Base According to the Arrhenius definition, a substance that produces hydroxide ions, OH^-, in solution. According to the Brønsted-Lowry definition, a proton acceptor.

Buffer A solution that resists changes in pH upon addition of a small amount of acid or base. A buffer is composed of a weak acid and its conjugate base, or a weak base and its conjugate acid.

Buffer capacity The amount of acid or base that can be added to a buffer before the pH will begin to change. Buffer capacity is directly related to the amount of buffer components in solution.

Conjugate acid A base after it has accepted a hydrogen ion, H^+. It is neutral or positively charged.

Conjugate acid-base pair Molecules and ions that are related by the transfer of a proton: HA/A^- or B/BH^+.

Conjugate base An acid after it has donated a hydrogen ion, H^+. It is neutral or negatively charged.

Homeostasis A balanced system.

Hydronium ion A water molecule with a bond to a proton, H_3O^+, produced when an acid is added to water.

Ion-product constant, K_w The constant 1.0×10^{-14}, which represents the product of the hydronium ion concentration and the hydroxide ion concentration in an aqueous solution at 25 °C.

Metabolic acidosis A medical condition resulting when blood pH falls below normal as a result of a metabolic disorder.

Metabolic alkalosis A medical condition resulting when blood pH rises above normal as a result of a metabolic disorder.

Net reaction Represents the actual species reacting in a neutralization reaction. For example: $H^+ + OH^- \rightarrow H_2O$.

Neutralization reaction The chemical reaction between an acid and a base to form a neutral compound like water and a salt, or water, a salt, and CO_2.

pH $pH = -\log[H_3O^+]$. A way to express the concentration of hydronium ion that avoids using scientific notation.

pH indicator paper Special dye-coated papers that can be used to determine the approximate pH of a solution by a visible color change.

Physiological pH The normal pH of arterial blood, which is 7.35–7.45.

Respiratory acidosis A condition in which blood pH falls below normal as a result of weak breathing and ineffective removal of CO_2 from the blood.

Respiratory alkalosis A condition in which blood pH rises above normal because CO_2 is eliminated too quickly as a result of hyperventilating or breathing too fast.

Spectator ions An ion dissolved in solution that is not involved in the chemical reaction.

Strong acid An acid that is fully dissociated in water.

Strong base A base that is fully dissociated in water such as hydroxide ion—containing salts of group 1A or 2A metal cations.

Weak acid An acid that dissociates to a small extent in aqueous solution. At equilibrium the acid, and a small amount of hydronium ions, and the conjugate base are present.

Weak base A base that dissociates to a small extent in aqueous solution. At equilibrium, the base, and a small amount of hydroxide ions, and the conjugate acid are present.

Additional Exercises

Ethylene Glycol Poisoning in Pets

32 What functional groups are present in ethylene glycol, $HOCH_2CH_2OH$?

33 What toxic substance is ethylene glycol metabolized into when it is ingested?

34 Why is ingesting ethylene glycol fatal? Why are pets attracted to ethylene glycol?

35 Write the chemical equation for the reaction of oxalic acid, $HOOCCH_2CH_2COOH$, with water?

36 What are the symptoms of ethylene glycol poisoning?

37 How do veterinarians treat pets with ethylene glycol poisoning? How do doctors treat people who have ethylene glycol poisoning?

38 What compound has replaced ethylene glycol in commercial products? How is its chemical structure different from ethylene glycol?

Acids and Bases

39 What are some general characteristics of acids? What are some general characteristics of bases?

40 According to the Arrhenius definition, an acid is a substance that produces _____ when dissolved in water, and a base is a substance that produces _____ when dissolved in water.

41 What polyatomic ion is formed when a proton is released in aqueous solution? Write the Lewis dot structure for this ion.

42 What is the product formed when H^+ from an acid reacts with OH^- from a base? What is this reaction called and why?

43 According to the Brønsted-Lowry definition, an acid is a substance that _____ a proton, and a base is a substance that _____ a proton.

44 Write the equation that represents the reaction of lactic acid in water. (You will find the structure of lactic acid in Table 9.1.) Label the conjugate acid-base pairs. Does water act as an acid or a base in this reaction?

45 Write the equation that represents the reaction of carbonic acid in water. (You will find the structure of carbonic acid in Table 9.1.) Label the conjugate acid-base pairs. Does water act as an acid or a base in this reaction?

46 Write the equation that represents the reaction of ephedrine in water. (You will find the structure of ephedrine in Table 9.2.) Label the conjugate acid-base pairs. Does water act as an acid or a base in this reaction?

47 Write the equation that represents the reaction of dopamine in water. (You will find the structure of dopamine in Table 9.2.) Label the conjugate acid-base pairs. Does water act as an acid or a base in this reaction?

48 Write the equation that represents the dissolution of potassium hydroxide in water. Why is potassium hydroxide classified as a base?

49 Write the equation that represents the dissolution of calcium hydroxide in water. Why is calcium hydroxide classified as a base?

50 Which of the following compounds are acids?
a. CH_3CH_2COOH
b. KOH
c. H_3O^+
d. $CH_3CH_2CH_2NH_2$
e. $CH_3CH_2NH_3^+$

51 Which of the following compounds are bases?
a. $CH_3CH_2CH_2CH_2COOH$
b. $CH_3CH_2CH_2COO^-$
c. HI
d. $CH_3CH_2CH_2NHCH_3$
e. $Mg(OH)_2$

52 Write the structure of the conjugate base of the following acids.
a. $CH_3CH_2CH_2COOH$
b. HCl
c. H_3O^+
d. CH_3CH_2COOH
e. H_2O

53 Write the structure of the conjugate acid of the following bases.
a. H_2O
b. OH^-
c. $CH_3CH_2NHCH_3$
d. $CH_3CH_2CH_2CH_2COO^-$
e. HCO_3^-

54 Why is hydrochloric acid, HCl, classified as a strong acid?

55 What happens if you spill a concentrated solution of a strong acid on your skin? What strong acid is present in the human body?

56 Hydrobromic acid, HBr, is a strong acid. Write the equation for the reaction of hydrobromic acid and water. Are there any molecules of hydrobromic acid present in solution? What species are present in solution?

57 Write the equation for the reaction of hydroiodic acid, HI, and water. Identify the conjugate acid-base pair. Why is HI classified as a strong acid?

58 Which group of metals form hydroxide salts that are strong bases?

59 Barium hydroxide, $Ba(OH)_2$, is a strong base. Write the equation for the dissolution of barium hydroxide in water. Why is this considered a dissolution rather than a reaction?

60 Strontium hydroxide is a strong base. Write the equation for the dissolution of strontium hydroxide in water. Why is strontium hydroxide classified as a strong base?

61 How do weak acids differ from strong acids in water?

62 How do weak bases differ from strong bases in water?

63 For each of the following chemical equations, indicate whether it represents the dissociation of a strong acid, a strong base, a weak acid, or a weak base.
a. $Ca(OH)_2(s) + H_2O \rightarrow Ca^{2+}(aq) + 2\ OH^-(aq)$
b. $HI + H_2O \rightarrow H_3O^+(aq) + I^-(aq)$
c. $HCOO^- + H_2O \rightleftharpoons HCOOH + OH^-(aq)$

64 For each of the following chemical reactions, indicate whether the equation represents the dissociation of a strong acid, a strong base, a weak acid, or a weak base.
a. $H_3PO_4 + H_2O \rightleftharpoons H_3O^+(aq) + H_2PO_4^-(aq)$
b. $NH_3 + H_2O \rightleftharpoons NH_4^+(aq) + OH^-(aq)$
c. $HClO_4 + H_2O \rightarrow H_3O^+(aq) + ClO_4^-(aq)$

65 Write the equation for the reaction of the following acids in water. State whether the acid is a strong acid or a weak acid. Is the reaction reversible? Identify the acid, base, conjugate acid, and conjugate base.
a. HCl b. HNO_3 c.

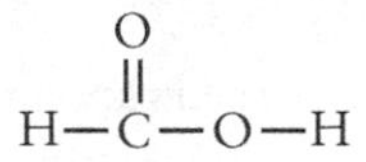

$$\begin{matrix} & O & \\ & \| & \\ H-&C&-O-H \end{matrix}$$

d. H_2CO_3

66 Sulfuric acid, H_2SO_4, is a strong acid. In two sequential steps it can lose both protons when it reacts with water. Write the equation for when H_2SO_4 loses one proton. Once the first proton is lost, the weak acid HSO_4^- is formed. Write the equilibrium reaction for this weak acid in water.

67 The structure of malonic acid is shown below.

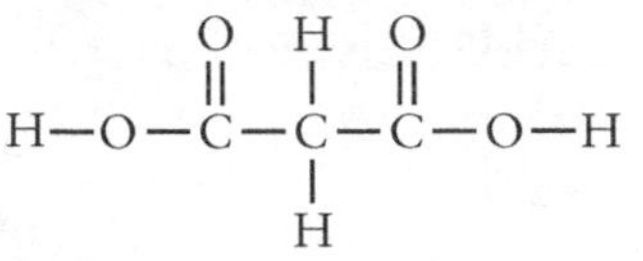

$$\begin{matrix} & O & H & O & \\ & \| & | & \| & \\ H-O-&C&-C&-C&-O-H \\ & & | & & \\ & & H & & \end{matrix}$$

a. Highlight the acidic protons in malonic acid that can dissociate.
b. Write the dissociation reaction for one of the acidic protons in malonic acid.

68 Butanoic acid has an acrid taste and an unpleasant odor. It is found in rancid butter and Parmesan cheese. In aqueous solution butanoic acid undergoes the following reversible reaction:

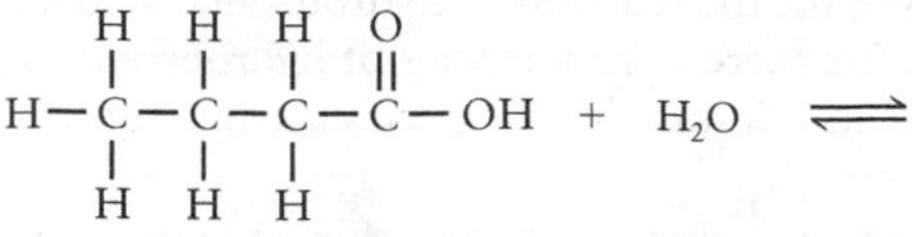

Butanoic acid   Water

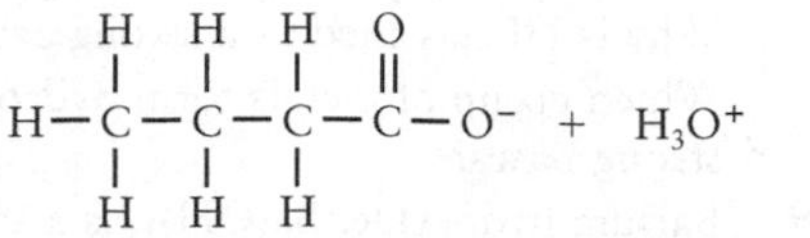

Butanoate   Hydronium ion

a. What substances are present at equilibrium?
b. At equilibrium, are reactants or products favored, and why?
c. Is the concentration of butanoic acid and butanoate ion constant or changing at equilibrium?
d. What is the significance of the two opposing arrows in this equation?
e. Label the conjugate acid-base pairs.

69 Propanoic acid prevents the growth of mold and some bacteria. It is used as a preservative in animal feed and in some consumer foods. In aqueous solution propanoic acid undergoes the following reversible reaction:

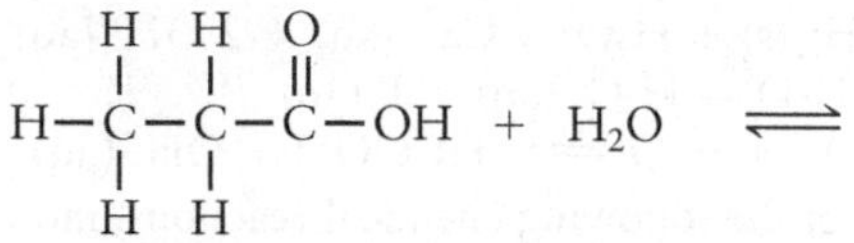

Propanoic acid   Water

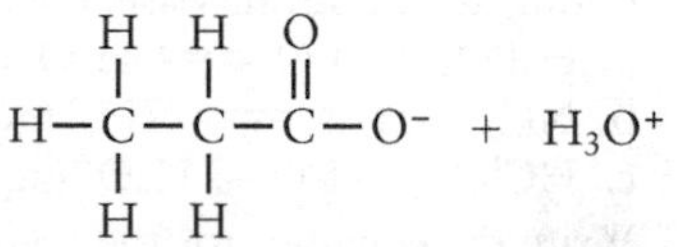

Propanoate   Hydronium ion

a. What substances are present at equilibrium?
b. At equilibrium are reactants or products favored, and why?
c. Is the concentration of propanoic acid and propanoate ion constant or changing at equilibrium?
d. What is the significance of the two opposing arrows in the equation?
e. Label the conjugate acid-base pairs.

70 The reversible reaction of acetic acid shown below occurs in the bottle of vinegar sitting on your kitchen shelf.

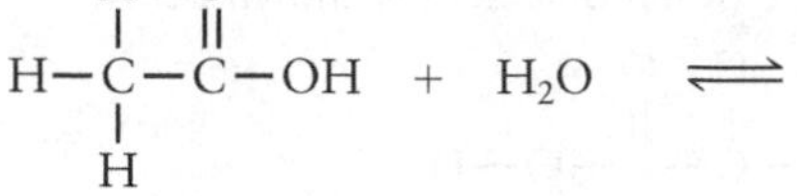

Acetic acid   Water

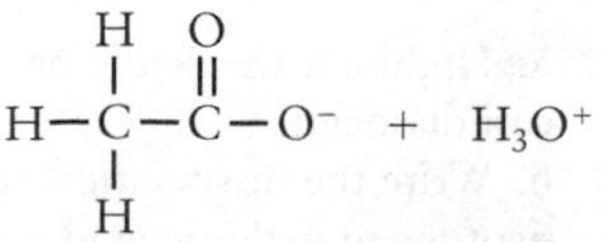

Acetate   Hydronium ion

a. What substances are present at equilibrium?
b. Are the concentrations of acetic acid and acetate constant or changing at equilibrium?
c. At equilibrium are reactants or products favored, and why?
d. Label the conjugate acid-base pairs.

71 Trichloroacetic acid is used to treat genital warts. Write the equilibrium reaction for this weak acid in aqueous solution.

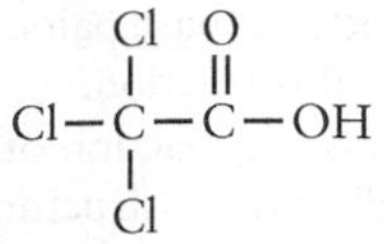

Trichloroacetic acid

a. What substances are present at equilibrium?
b. Are the concentrations of trichloroacetic acid and trichloroacetate constant or changing at equilibrium?
c. What happens to the equilibrium if more trichloroacetate is added to the reaction?
d. If more hydronium ions are added to the reaction, does the reaction shift to the left or to the right?

72 The conjugate base of benzoic acid is used as a preservative. Write the equilibrium reaction for this weak acid in aqueous solution.

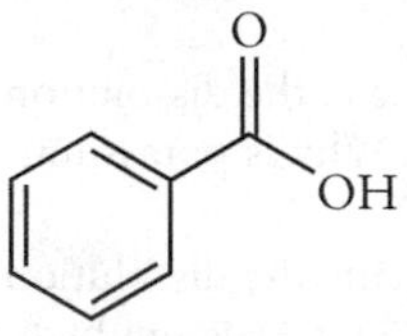

Benzoic acid

a. What substances are present at equilibrium?
b. At equilibrium, are reactants or products favored, and why?
c. Are the concentrations of benzoic acid and benzoate constant or changing at equilibrium?
d. What happens to the equilibrium if more hydronium ions are added to the reaction?
e. Label the conjugate acid-base pairs.

73 N,N-dimethylmethanamine, $(CH_3)_3N$, is produced when plants and animals decompose, and it has a very unpleasant smell. Write the equilibrium reaction for this amine in aqueous solution.

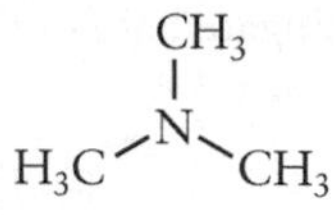

N,N-dimethylmethanamine

a. If OH^- is added, will the reaction shift to the left or to the right? Explain.
b. If some N,N-dimethylmethanamine is removed, will the reaction shift to the left or to the right? Explain.

pH

74 pH indicates the concentration of _____.
75 What is pH indicator paper? What is red litmus paper?
76 Indicate whether the following solutions are acidic, basic, or neutral based on the pH listed.
   a. pH = 8.5   b. pH = 6.3   c. pH = 7.0
   d. pH = 1.1   e. pH = 10.8

77 Indicate whether the following solutions are acidic, basic, or neutral based on the pH listed.
    a. pH = 7.4   **b.** pH = 1.9   **c.** pH = 14.0
    d. pH = 3.5   **e.** pH = 10.7

78 Using Table 9-5, indicate whether the following body fluids are acidic, basic, or neutral.
    a. saliva   **b.** gastric fluid   **c.** blood   **d.** urine

79 What is the difference between acidosis and alkalosis?

80 Write the chemical equation for the autoionization of water? In this reaction how do the concentrations of hydronium and hydroxide ion compare to each other?

81 Fill in the empty cells in the following table. Assume the temperature is 25 °C.

$[H_3O^+]$	$[OH^-]$	Is the solution acidic, neutral, or basic?
1.0×10^{-3}		
	1.0×10^{-2}	
1.0×10^{-7}		
1.0×10^{-5}		
	1.0×10^{-5}	

82 Fill in the empty cells in the following table. Assume the temperature is 25 °C.

$[H_3O^+]$	$[OH^-]$	Is the solution acidic, neutral, or basic?
	1.0×10^{-10}	
	1.0×10^{-1}	
1.0×10^{-6}		
1.0×10^{-11}		
	1.0×10^{-4}	

83 What is the pH of apple juice that has an $[H_3O^+] = 3.2 \times 10^{-4}$? Calculate the $[OH^-]$. Is apple juice acidic, basic, or neutral?

84 What is the pH of a bleach solution that has an $[OH^-] = 3.16 \times 10^{-2}$? Is bleach acidic, basic, or neutral?

85 What is the pH of milk that has an $[H_3O^+] = 3.2 \times 10^{-7}$? Calculate the $[OH^-]$. Is milk acidic, basic, or neutral?

86 What is the pH of a blood sample that has an $[OH^-] = 2.2 \times 10^{-7}$? Is the pH of this blood sample within the normal range?

87 Define physiological pH. How does it compare to the pH of water?

88 Draw the structure for the following compounds at physiological pH.

89 At physiological pH, are amines in their neutral or ionized form? What are the advantages of the form present at physiological pH?

90 At physiological pH, are carboxylic acids in their neutral or ionized form? Explain.

91 How is the arterial blood pH of a patient measured? How is the pH of a patient's urine measured?

Acid-Base Neutralization Reactions

92 What type of reactants define a neutralization reaction? What characteristic(s) of the product gives the reaction its name?

93 Write the balanced equations for the neutralization reaction of HBr with each of the following bases:
    a. NaOH   **b.** $Al(OH)_3$ **c.** $Sr(OH)_2$

94 Barium hydroxide, $Ba(OH)_2$, is used as a homeopathic drug. Write the balanced equation for the neutralization of barium hydroxide with hydrochloric acid. Identify the spectator ions in this reaction.

95 Write the balanced equation for the neutralization reaction of HNO_3 with the following compounds:
    a. KOH   **b.** $Mg(OH)_2$ **c.** $Al(OH)_3$

96 Baking soda is used as a leavening agent in cakes. Explain why cakes rise when baking soda is mixed with an acidic ingredient such as butter.

97 How do antacid tablets relieve heartburn and indigestion?

98 Maalox and Mylanta, both antacids, contain aluminum hydroxide as their active ingredient. Write the balanced equation for the neutralization of hydrochloric acid with aluminum hydroxide, $Al(OH)_3$. Identify the spectator ions in this reaction.

99 Benadryl is sold as the hydrochloride salt of the amine shown. Write the reaction of this molecule with HCl, showing the formation of the hydrochloride salt. Why is this product more soluble than the amine?

100 Razadyne, used to treat mild to moderate dementia in patients with Alzheimer's disease, is sold as the hydrobromide salt of the amine shown. Write the reaction of this molecule with HBr, showing the formation of the hydrobromide salt. Why is this product more soluble than the amine?

Buffers

101 What is the definition of a buffer?

102 What happens to the pH of solution when an acid is added to a buffer?

103 What happens to the pH of solution when a base is added to a buffer?

104 Which of the following represents a buffer? Explain.
   a. HCl and HNO_3
   b. CH_3COOH and CH_3COONa
   c. NaOH and KCl
   d. LiCl and H_2O

105 What does "buffer capacity" refer to?

106 Whenever you eat, a small amount of acid enters your bloodstream. Explain how the buffers in your blood react with the small amount of added acid to maintain a constant pH.

107 Explain why a small amount of base added to your bloodstream does not affect the pH of your blood.

108 Which component of the carbonic acid (H_2CO_3)-hydrogen carbonate (HCO_3^-) blood buffer would react with each of the following to maintain a constant pH?
   a. NaOH   b. HCl   c. NH_3 d. CH_3COO^-

109 Hyperventilation affects the concentration of CO_2 in the blood.
   a. Does the concentration of CO_2 dissolved in the blood increase or decrease in a patient who is hyperventilating?
   b. How does the concentration of CO_2 of a patient who is hyperventilating affect the pH of their blood?

110 An emergency medical team evaluates a marathon runner and determines that he has alkalosis. What should be given to the marathoner to decrease the pH of his blood?

111 A patient has severe diarrhea, which results in an excessive loss of sodium hydrogen carbonate from the body. How is the pH of the patient's blood affected by the loss of this ion? What component of the hydrogen carbonate–carbonic acid buffer should be given to the patient to bring his blood pH back in the normal range?

Chemistry in Medicine: pH and Oral Administration of Drugs

112 In which organ of the body is insulin produced? What type of molecule is insulin? What biological function does insulin have?

113 Why do people with type I diabetes need to take insulin?

114 How is type I diabetes different from type II diabetes?

115 Why do drugs that are proteins need to be injected rather than administered orally?

116 Which is a more acidic environment, the mouth and esophagus or the stomach?

117 In the stomach, does a protein keep its original shape and cellular function?

118 Why does the stomach secrete more gastric juice when proteins are consumed?

119 What is the function of the enteric coating of a pill?

120 Where in the digestive system is the enteric coating of a pill degraded and why?

Extension Topic: The Acid Ionization Constant, K_a, and the Strength of a Weak Acid

E9.6 Pyruvic acid has a K_a of 3.16×10^{-3}. What is its pK_a?

E9.7 Phosphoric acid has a K_a of 7.59×10^{-3}. What is its pK_a?

E9.8 Citric acid has a K_a of 8.13×10^{-4}. What is its pK_a?

E9.9 Which is the strongest acid in the previous three questions? Which is the weakest?

E9.10 Some weak acids and their pK_a's are listed. Which compound is the strongest acid? Which compound is the weakest acid?

Benzoic acid, C_6H_5COOH	pK_a 4.2
Phosphoric acid, H_3PO_4	pK_a 2.1
Carbonic acid, H_2CO_3	pK_a 6.4
Tartaric acid	pK_a 3.0

E9.11 Some weak acids and their pK_a's are listed below. Which compound is the strongest acid? Which compound is the weakest acid?

Malic acid, found in apple juice	pK_a 3.4
Oxalic acid, found in kidney stones	pK_a 1.2
Caffeic acid, found in coffee	pK_a 4.6
Gallic acid, found in mango peels	pK_a 4.4

Answers to Practice Exercises

1 According to the Arrhenius definition, a base produces <u>hydroxide ions</u> in solution. According to the Brønsted-Lowry definition, a base <u>accepts</u> a proton from water.

2 a, b, and e are bases. Water is amphoteric, which means it can act as either an acid or a base.

3 a. Water acts as a base

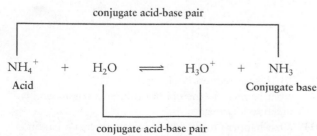

b. Water acts as an acid

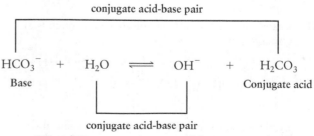

c. HI and I^- are a conjugate acid-base pair and so are H_2O and H_3O^+. Water is acting as a base.

4

conjugate acid-base pair

Carboxylic acid → Ketone →
Pyruvic acid + H_2O ⇌ H_3O^+ + Pyruvate ion
Hydronium ion

conjugate acid-base pair

5

R$_2$NH is an amine functional group

conjugate acid-base pair

CH_3—N—H + H—O—H ⇌ CH_3—N—H + $:O$—H
Base
Conjugate acid   Hydroxide ion
Conjugate base

conjugate acid-base pair

6 $Sr(OH)_2(s) + H_2O \rightarrow 2\ OH^-(aq) + Sr^{2+}(aq)$

$Sr(OH)_2$ is an ionic compound and a base because it produces hydroxide ions in solution (the Arrhenius definition of a base).

7 $HBr + H_2O \rightarrow H_3O^+ + Br^-$

HBr and Br^- are a conjugate acid-base pair and so are H_2O and H_3O^+.

HBr is a strong acid, which means the forward reaction goes to completion, that only H_3O^+ and Br^- are in solution, and no molecular HBr is in solution. A strong acid is also a strong electrolyte because it is completely dissociated, so there is a high concentration of dissolved ions in solution.

8 A strong acid transfers all or almost all of its protons to water.

9 A strong base produces many hydroxide ions in aqueous solution, hence the term "strong."

10 $KOH(s) + H_2O \rightarrow K^+(aq) + OH^-(aq)$

KOH is a strong electrolyte because it forms many ions in solution. KOH is a strong base. We know KOH is a strong base because it is an ionic compound composed of hydroxide ions, which are released from the lattice when placed in water.

11 $HClO_4 + H_2O \rightarrow H_3O^+ + ClO_4^-$

The concentration of $HClO_4$ is low because perchloric acid is a strong acid and therefore completely dissociated. The conjugate base of perchloric acid ($HClO_4$) is ClO_4^-. There is a high concentration of the conjugate base because a strong acid is completely dissociated in solution.

12 a. strong base **b.** strong acid **c.** weak base
d. weak base **e.** weak acid

13 (a) because it is an ionic compound that dissolves in water producing ions and (b) because it is a strong acid, which dissociates completely into ions, H_3O^+ and NO_3^-.

14 a. The products are H_3O^+ and NO_3^-. HNO_3 is a strong acid and a strong electrolyte.

b. The reactant is a weak base, and therefore also a weak electrolyte.

c. Na^+ and OH^-. NaOH is a strong base. It is a strong electrolyte.

15 Due to the amine and other nitrogen atoms, it is classified as a base, that is, a proton acceptor.

16

Lactic acid

Lactate ion

a. At equilibrium, lactic acid, lactate ion (the conjugate base of lactic acid), and hydronium ion are all present.

b. At equilibrium the concentration of these substances is constant.

c. Since hydronium ion is a product, adding more will shift the equilibrium to the left, toward the reactant side, in order to consume the added hydronium ions, in accordance with Le Châtelier's principle.

d. As the person exercises more intensely, more lactic acid, the reactant, is produced, causing the equilibrium to shift to the right, toward the product side, in order to consume the added lactic acid, in accordance with Le Châtelier's principle.

17 $NH_3 + H_2O \rightleftharpoons OH^- + NH_4^+$

a. If more OH^-, one of the products, is added, the equilibrium will shift to the left to consume the added product, according to Le Châtelier's principle.

b. If ammonia, NH_3, a reactant, is removed, the equilibrium will shift to the left to restore the lost ammonia, in accordance with Le Châtelier's principle.

18

$[H_3O^+]$	$[OH^-]$	
1.0×10^{-11}	1.0×10^{-3}	basic
1.1×10^{-6}	9.1×10^{-9}	acidic
3.3×10^{-6}	3.0×10^{-9}	acidic

19 pH = 6.25

20 pH = 6.49. Remember you must first convert the hydroxide ion concentration to a hydronium ion concentration using the ion product constant.

21 a. HBr is acidic; it contains H_3O^+ and Br^- ions.
b. KOH is basic; it contains OH^- and K^+ ions.
c. Net reaction: $H^+ + OH^- \rightarrow H_2O$
d. Complete balanced equation: $HBr(aq) + KOH(aq) \rightarrow H_2O + KBr(aq)$
e. The spectator ions are K^+ and Br^-.

22 a. $HF(aq) + LiOH(aq) \rightarrow H_2O + LiF(aq)$
b. $3\ HF(aq) + Al(OH)_3(s) \rightarrow 3\ H_2O + AlF_3(aq)$
c. $2\ HCOOH(aq) + Ba(OH)_2(s) \rightarrow 2\ H_2O + Ba(HCOO)_2$

23 Even though little $Al(OH)_3$ dissolves, and therefore dissociates to produce OH^- ions in solution, the small amount of hydroxide ion that is released into solution reacts immediately with the stomach acid, HCl, present in solution in a neutralization reaction. The neutralization reaction consumes all the hydroxide ion, which according to Le Châtelier's principle, causes more hydroxide ions to be released into solution, until eventually all of the $Al(OH)_3$ has reacted with HCl.

24 Complete balanced equation: $HI(aq) + KHCO_3(aq) \rightarrow H_2O(l) + CO_2(g) + KI(aq)$
Net reaction: $H^+ + HCO_3^- \rightarrow H_2O + CO_2$
Spectator ions: K^+ and I^-

25 Yes, pouring calcium carbonate on nitric acid will neutralize it.
Balanced equation: $2\ HNO_3(aq) + CaCO_3(s) \rightarrow H_2O + CO_2(g) + Ca(NO_3)_2(aq)$
Net reaction: $2\ H^+ + CO_3^{2-} \rightarrow H_2O + CO_2$
Spectator ions: calcium, Ca^{2+}, and nitrate ions, NO_3^-

26

Zoloft
(insoluble in water)
(neutral organic compound)

$+\ \ H\!-\!Cl\ \longrightarrow$

Zoloft hydrochloride salt
(soluble in water)
(polyatomic ion)

27 a. NH_4^+ is the acid component of the buffer that reacts with added hydroxide ions that enter solution.
b. The base component of the buffer, NH_3, would react with the added protons:

$$H^+ + NH_3 \rightleftharpoons NH_4^+$$

c. The acid component of the buffer, NH_4^+, would react with the added hydroxide ions:

$$OH^- + NH_4^+ \rightleftharpoons NH_3 + H_2O$$

d. The ions that affect pH, OH^- and H_3O^+, are neutralized by one of the buffers. Furthermore, the buffer itself is a weak base, so it is not very dissociated in solution, and hence, it produces few hydroxide ions in solution.
e. The pH of the solution would change significantly because not enough NH_3 or NH_4^+, the buffer components, is available to neutralize all the added ions.

28 a. Carbonic acid is an unstable compound that is formed when $\underline{H_2O}$ and dissolved $\underline{CO_2}$ react.
b. How this buffer will respond to the addition of a small amount of acid (H^+):

$$H^+\ \text{added}$$
$$\downarrow$$
$$H_2CO_3(aq) \rightleftharpoons H^+(aq) + HCO_3^-(aq)$$
equilibrium shifts
to the left

c. How this buffer will respond to the removal of a small amount of acid (H^+):

equilibrium shifts
to the right
$$H_2CO_3(aq) \rightleftharpoons H^+(aq) + HCO_3^-(aq)$$
$$\downarrow$$
$$H^+\ \text{removed}$$

d. Someone with alkalosis has too high a pH (too much OH^-/too little H^+). Breathing into a bag causes the patient to breath in exhaled air which contains more CO_2, which causes a shift to the right in Equation 9-1, consuming the added CO_2, and producing more carbonic acid, H_2CO_3. By increasing the concentration of carbonic acid, the equation below shifts to the right, increasing the concentration of H^+ and thereby reducing the pH of the patient's blood.

$$H_2CO_3(aq) \rightleftharpoons H^+(aq) + HCO_3^-$$

29 Respiratory alkalosis occurs when breathing becomes too fast, such as from hyperventilation, causing carbon dioxide to be removed by the lungs at too rapid a rate. Metabolic alkalosis can result from vomiting and other metabolic, nonbreathing related, issues. Both conditions involve an increase in hydroxide ion/decrease in hydronium ion raising the pH above 7.45.

30 Breathing regulates the amount of carbon dioxide in the blood, and carbon dioxide reacts with water to form carbonic acid, which is the acid component of the carbonic acid-hydrogen carbonate buffer. Changes in carbon dioxide shift the carbonic acid–hydrogen carbonate buffer, thereby regulating pH.

31 The kidneys secrete HCO_3^- into the blood, which is one component of the carbonic acid–hydrogen carbonate buffer that responds to changes in blood pH.

Extension Practice Exercises

E1 $pK_a = 9.4$

E2 The product concentrations are in the numerator: $[H_3O^+][A^-]$: $[HA]$: the acid concentration is in the denominator. A weaker acid will have a smaller numerator and a larger denominator because it is less dissociated. A larger denominator means a smaller value, and, therefore, a smaller K_a.

E3 Carbonic acid $pK_a = 6.38$. Since carbonic acid has a higher pK_a/lower K_a than acetic acid (4.74), carbonic acid is a weaker acid than acetic acid.

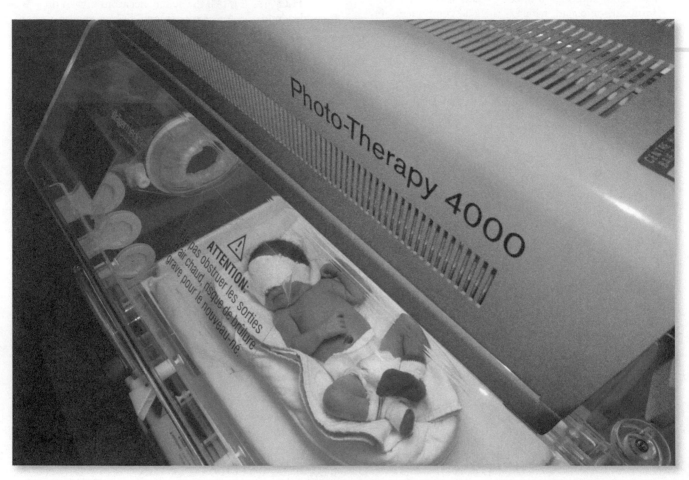

The ultraviolet light used in phototherapy degrades the vitamin riboflavin. Vitamins are organic compounds that are required by our cells in order to perform the metabolic reactions necessary for living. [© CBW/Alamy]

The Reactions of Organic Functional Groups in Biochemistry

Are You Getting Your Vitamins?

Are you eating green leafy vegetables? Do you take a daily multivitamin? Why are vitamins important? Vitamins, which allow cells to perform the metabolic reactions necessary for living, are organic compounds required daily in trace quantities. While our cells can synthesize many molecules, they cannot synthesize most vitamins, so they must be supplied on a regular basis through the diet, much like the macronutrients and micronutrients. Insufficient vitamin intake or poor vitamin absorption can lead to vitamin deficiency diseases.

For example, riboflavin, also known as vitamin B_2, is a vitamin found in liver, leafy green vegetables, almonds, eggs, milk, cheese, meat, and fortified cereals (Figure 10-1). Men need about 1.7 mg of riboflavin a day, and women require about 1.3 mg a day. Since riboflavin is a water-soluble vitamin, it does not build up in the body, so it is difficult to get too much riboflavin. Riboflavin is essential in the metabolism of carbohydrates, proteins, and fats. Insufficient riboflavin intake has symptoms that include inflammation of the tongue (glossitis) and skin (dermatitis), lesions in the mouth, and cataracts in the eyes. Riboflavin deficiencies often occur in newborn babies undergoing phototherapy (light therapy), a common treatment for newborns with jaundice, a condition characterized by a yellow coloring of the skin and eyes. The ultraviolet light used in phototherapy degrades riboflavin, depleting the newborn of this important vitamin.

Historically, vitamins have been divided into two categories based on their solubility characteristics: *water-soluble* and *fat-soluble vitamins*, as shown in Table 10-1. All the B-complex vitamins and vitamin C are water-soluble. Water-soluble vitamins are readily eliminated through the urine, and for this reason must be replenished on a daily basis. Vitamins A, D, E, and K are the fat-soluble vitamins. The fat-soluble vitamins are stored in tissues where they are soluble, and they are eliminated less

Outline

Riboflavin

Figure 10-1 Green leafy vegetables like spinach are rich in riboflavin, also known as vitamin B_2. [Catherine Bausinger]

TABLE 10-1 Water-Soluble and Fat-Soluble Vitamins

Water-Soluble Vitamins		Fat-Soluble Vitamins
Thiamin	B_1	A
Riboflavin	B_2	D
Niacin	B_3	E
Pantothenic acid	B_5	K
Pyridoxamine, pyridoxal, pyridoxine	B_6	
Biotin	B_7	
Folic acid	B_9	
Cobalamin	B_{12}	
Ascorbic acid	C	

readily in the urine because they are not soluble in water. Indeed, this is why fat-soluble vitamins generally should not be taken in excess.

While most natural foods are rich in vitamins, most processed foods are not. Some processed foods, like white flour, are *enriched* in niacin, riboflavin, and folic acid, which means vitamins are added to the processed flour to replace the vitamins lost during processing. Other foods like milk are *fortified* with vitamins A and D, which means the food has been supplemented with these vitamins.

Vitamin supplements are available in many different forms, the most popular being multivitamins available in pill form. Recently, vitamin supplements have even become available as gummies, an alternative for children and adults who have difficulty swallowing pills. While a healthy nutritious diet can provide all the vitamins a person needs, some people benefit from supplementation. For example, pregnant women are generally advised by their doctors to take prenatal vitamins, which are high in folic acid (vitamin B_9), a vitamin that has been shown to prevent neural tube defects in a developing fetus if it is taken regularly within the first 28 days of pregnancy. In addition to many micronutrients, prenatal vitamins also include other B vitamins such as riboflavin, thiamin, niacin, and vitamin B_{12}; vitamin C; and the fat-soluble vitamins D and E. In this chapter you will see that vitamins are important because they play a critical role in biochemical reactions that take place in our cells. In this chapter you will learn some of the most important chemical reactions in biochemistry. ●

In chapter 4 you were introduced to the basic principles governing chemical reactions and in the previous chapter you saw how those principles were applied to acid-base reactions, the simplest and most common of chemical reactions. In this chapter we add the reactions of organic functional groups, especially as they apply to biochemical reactions.

To understand organic reactions it is important that you be able to readily identify the common functional groups introduced in chapters 6 and 7, summarized here in Table 10-2. *Functional groups are the reactive sites in a molecule.* Generally, it is only one functional group in a molecule that undergoes a change in a particular chemical reaction; the other functional groups in the molecule are not involved, and therefore, are chemically unchanged. With the exception of combustion reactions, the alkane portion of a molecule is rarely involved in chemical reactions, as discussed in chapter 6. *The benefit of learning the major functional groups and their corresponding chemical properties is that you can then predict the outcome of a chemical reaction for any molecule based on the functional groups that it contains.*

TABLE 10-2 Common Organic Functional Groups*

Name	Structure	Name	Structure
Alkene	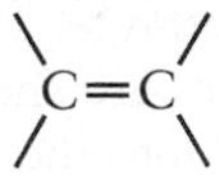 $\overset{\diagdown}{\underset{\diagup}{C}}=\overset{\diagup}{\underset{\diagdown}{C}}$	Amine	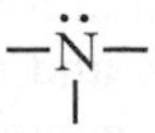 $-\overset{..}{\underset{\mid}{N}}-$
Alkyne	$-C\equiv C-$	Aldehyde	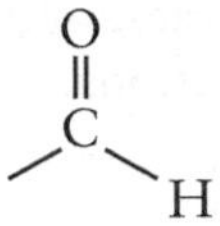 $\overset{O}{\overset{\parallel}{\underset{\diagdown H}{C}}}$
Aromatic ring	 (benzene ring)	Ketone	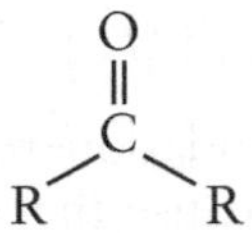 $\overset{O}{\underset{R\diagup\diagdown R}{\overset{\parallel}{C}}}$
1° Alcohol	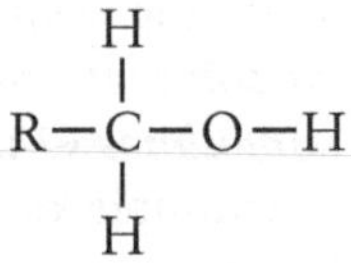 $R-\overset{H}{\underset{H}{\overset{\mid}{\underset{\mid}{C}}}}-O-H$	Carboxylic acid	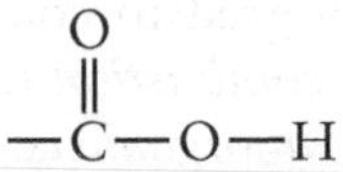 $-\overset{O}{\overset{\parallel}{C}}-O-H$
2° Alcohol	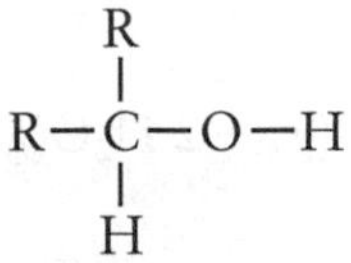 $R-\overset{R}{\underset{H}{\overset{\mid}{\underset{\mid}{C}}}}-O-H$	Ester	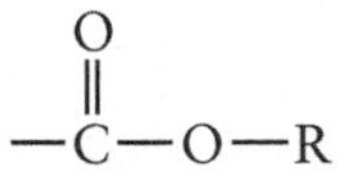 $-\overset{O}{\overset{\parallel}{C}}-O-R$
3° Alcohol	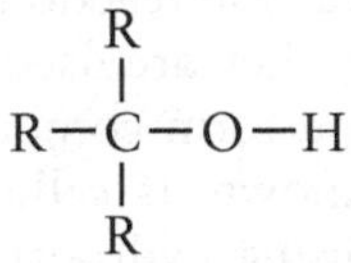 $R-\overset{R}{\underset{R}{\overset{\mid}{\underset{\mid}{C}}}}-O-H$	Amide	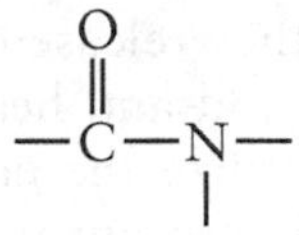 $-\overset{O}{\overset{\parallel}{C}}-\overset{\mid}{N}-$
Phenol	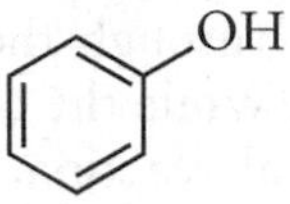 (phenol ring with OH)	Thioester	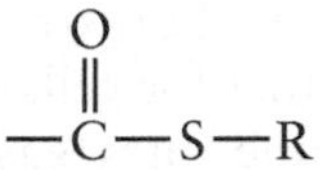 $-\overset{O}{\overset{\parallel}{C}}-S-R$
Ether	$R-O-R$		

*R is one or more carbon atoms.
"—" represents a bond to hydrogen or carbon.

Almost all biochemical reactions are catalyzed by a specific enzyme whose primary function is to increase the rate of the reaction. Enzymes are remarkable catalysts because they not only significantly increase the rate of the reaction, converting a particular reactant—known as the **substrate**—to products, but they also ensure that the correct stereoisomer is produced when a chiral product is formed.

While hundreds of thousands of reactions occur in a biological cell, they fall into six basic reaction types, listed in Table 10-3. If you learn these basic

TABLE 10-3 Common Biochemical Reactions and the Associated Major Class of Enzymes

Type of Reaction	Class of Enzyme
Oxidation–reduction reactions	Oxidoreductases
Group transfer reactions	Transferases
Hydrolysis reactions	Hydrolases
Hydration and dehydration reactions	Lyases
Isomerization reactions	Isomerases
Condensation reactions and ATP/ADP hydrolysis	Ligases

reaction types, you can understand almost all of the reactions encountered in biochemistry. Associated with each of the six basic reaction types is a major class of enzymes that catalyze this type of reaction. Indeed, enzymes are classified according to the type of reaction they catalyze. The six major reaction types and the associated major class of enzymes are shown in Table 10-3. In this chapter we will consider reactions that fall into the first four of these categories, and in chapter 15 you will see examples of the other two classes of reactions.

10.1 Oxidation–Reduction Reactions

Oxidation–reduction reactions are some of the most important reactions because they are involved with energy transfer. In chapter 4 you were introduced to combustion reactions, a type of oxidation–reduction reaction. *All combustion reactions require oxygen, O_2, and complete combustion converts an organic compound containing C, H, and sometimes O into carbon dioxide (CO_2) and water (H_2O),* as illustrated in the example of the combustion of methane with a Bunsen burner (Figure 10-2):

$$CH_4(g) + 2\ O_2(g) \rightarrow CO_2(g) + 2\ H_2O(g)$$

In chapter 4 you also learned that combustion reactions are exothermic—they release heat energy. Indeed, this is why they are used every day to cook food and heat our homes. Our cells also perform combustion reactions to utilize the energy released, in a process known as **cellular respiration,** an important part of metabolism. Instead of using hydrocarbons as a fuel, the cell burns carbohydrates and fats taken in through the diet. The primary fuel for cells is glucose, that is, blood sugar. While the combustion of methane on a gas burner occurs in one chemical reaction, cellular respiration occurs over a sequence of many biochemical reactions, some of which are oxidation–reduction reactions, ultimately leading to the same end products: carbon dioxide, water, and energy. As with all combustion reactions, cellular respiration requires oxygen (O_2), and indeed this is the reason we breathe. Like a car engine, which converts some of the energy from the combustion of gasoline into mechanical energy (moving the car) rather than heat, the cell also converts energy from the combustion of glucose into other usable forms of energy. You will see applications of oxidation–reduction reactions when you study metabolism and energy transfer in greater detail in chapter 15.

Chemical Definitions of Oxidation and Reduction

An oxidation–reduction reaction (often abbreviated **redox**) *is characterized by the transfer of electrons from one reactant to another reactant.* The reactant losing electrons is said to undergo **oxidation,** and the reactant gaining electrons is said to undergo **reduction.** *By definition, where there is oxidation, there must also be reduction.*

Oxidation: loss of electrons

Reduction: gain of electrons e^-

The transfer of electrons that occurs in an oxidation–reduction reaction is more obvious in redox reactions involving inorganic compounds, so we begin with an inorganic example. Consider the reaction seen in the beaker

Figure 10-2 The flame from a Bunsen burner is produced by reacting methane with oxygen—a combustion reaction. Combustion reactions are a type of oxidation–reduction reaction.

[Martyn F. Chillmaid/Science Source]

To Help You Remember:
LEO the lion goes **GER**
Loss of Electrons is Oxidation (**LEO**)
Gain of Electrons is Reduction (**GER**)

shown in Figure 10-3, where a strip of magnesium metal, Mg, has been placed into an aqueous solution of hydrochloric acid, HCl. The reaction between Mg and HCl produces dissolved magnesium chloride, $MgCl_2$ (aq), and hydrogen gas, H_2 (g), which is seen bubbling out of solution in the photo. The overall chemical equation, illustrated with Lewis dot symbols so that you can follow the transfer of electrons (pink), is shown below:

Oxidation–Reduction: $Mg: + 2\,H-\ddot{\underset{\cdot\cdot}{Cl}}: \longrightarrow H-H + Mg^{2+} + 2\,:\ddot{\underset{\cdot\cdot}{Cl}}:^-$

In this reaction, the two valence electrons from a magnesium atom, Mg, are transferred to the two hydrogen ions, H^+, from two dissociated hydrochloric acid molecules, HCl, forming the covalent bond in the hydrogen molecule, H_2. Since magnesium loses electrons (two electrons) to become a magnesium cation, Mg^{2+}, magnesium has undergone oxidation. The two hydrogen ions (H^+) have been reduced because they gain two electrons to form a molecule of hydrogen, H—H. The chloride ion, Cl^-, does not undergo a change, it is a spectator ion. You can think of the reaction in two parts, the oxidation half and the reduction half, which occur simultaneously:

Oxidation: $Mg: \longrightarrow Mg^{2+} + 2\,e^-$

Reduction: $2\,e^- + 2\,H^+ \longrightarrow H-H$

Figure 10-3 Magnesium metal, Mg, reacts with aqueous hydrochloric acid, HCl, in an oxidation–reduction reaction that produces dissolved magnesium ions, Mg^{2+}, chloride ions, Cl^-, and hydrogen gas, H_2, which is seen here emerging from solution as bubbles. [Andrew Lambert Photography/Science Source]

Reduction Reactions of Organic Compounds

Redox reactions involving *organic compounds* also involve the transfer of electrons. However, since organic compounds are more complex and have many more electrons, it is easier to simply follow the change in the number of hydrogen atoms and/or oxygen atoms that occurs in going from reactants to products. To determine whether an organic substrate is undergoing oxidation or reduction, focus on the functional group undergoing a change. *If the functional group change involves an increase in the number of oxygen atoms and/or a decrease in the number of hydrogen atoms, then it has lost electrons, and therefore has been oxidized. The opposite is true for a functional group that has been reduced.*

$\uparrow$ O
$\downarrow$ H   **Oxidation:** *Loss of electrons. Observe decrease in hydrogen atoms and/or increase in oxygen atoms.*

$\downarrow$ O
$\uparrow$ H   **Reduction:** *Gain of electrons. Observe increase in hydrogen atoms and/or decrease in oxygen atoms.*

Let us consider an oxidation–reduction reaction in detail, knowing that typically you will simply apply the definitions above to determine whether the substrate has undergone oxidation or reduction. The reaction of 2-propanol (isopropanol) with bleach forms 2-propanone (acetone), water, and chloride ion:

Oxidation   Reduction

$$
\underset{\substack{\text{2-Propanol}\\ C_3H_8O\\ \text{(26 electrons)}}}{\text{H}_3\text{C}-\overset{\text{OH}}{\underset{\text{H}}{\text{C}}}-\text{CH}_3} + \underset{\substack{\text{Bleach}\\ \text{(14 electrons)}}}{\text{Na}^+\ :\overset{..}{\underset{..}{\text{O}}}-\overset{..}{\underset{..}{\text{Cl}}}:} \longrightarrow \underset{\substack{\text{2-Propanone}\\ C_3H_6O\\ \text{(24 electrons)}}}{\text{H}_3\text{C}-\overset{..\ \text{O}\ ..}{\text{C}}-\text{CH}_3} + \underset{\substack{\text{Chloride}\\ \text{ion}\\ \text{(8 electrons)}}}{\text{Na}^+\ :\overset{..}{\underset{..}{\text{Cl}}}:{}^-} + \underset{\substack{\text{Water}\\ \text{(8 electrons)}}}{\overset{..}{\underset{..}{\text{O}}}\big\langle\substack{\text{H}\\ \text{H}}}
$$

(16 electrons)

In this reaction we see that a secondary alcohol is converted to a ketone. At the same time, bleach (ClO^-) is converted into chloride ions, Cl^-, and water. What makes this reaction an oxidation–reduction reaction?

(1) Focusing on the functional group transformation of the substrate from an alcohol to a ketone, we see a decrease in the number of hydrogen atoms in the molecule, hence, an oxidation. If we look closely and count valence electrons we see that the alcohol has 26 electrons and the ketone has 24. This is a loss of electrons and the fundamental definition of an oxidation.

(2) Bleach is an aqueous solution containing hypochlorite ions, ClO^-. In this reaction it gains two electrons and splits apart into two stable species, water and a chloride ion. Counting valence electrons, we see that bleach, ClO^-, has 14 electrons while the sum of electrons in the chloride ion, Cl^-, and water is 16 electrons. This is a gain of electrons and the fundamental definition of a reduction.

(3) The sodium ion, Na^+, is a spectator ion, dissolved in solution and unchanged throughout the reaction.

> Typically all we need to do to recognize that the substrate is being oxidized or reduced is to note a decrease or increase in the hydrogen (or oxygen) atoms. It is not important that we count electrons, but here we do so to demonstrate that electrons have indeed been transferred from the alcohol to bleach.

In an oxidation–reduction reaction, the reactant that causes the substrate to be oxidized is referred to as the **oxidizing agent**, which is itself being reduced in the process. In the example above, bleach is the oxidizing agent. Bleach is a well-known *oxidizing agent,* which you may have added to the washer to remove stains from your clothes to make them look "white." Table 10-4 lists some other common oxidizing agents. Since there are a variety of oxidizing agents, each capable of oxidizing the substrate and yielding the same end result, the identity of the oxidizing agent is often not specified in a chemical reaction; instead, the general abbreviation [O] is placed above the reaction arrow to indicate an oxidizing agent is present as a reactant. For example, the oxidation of ethanol can be written as:

Oxidation

$$
\underset{\substack{\text{Ethanol}\\ C_2H_6O}}{\text{H}_3\text{C}-\text{CH}_2-\overset{..}{\underset{..}{\text{O}}}-\text{H}} \xrightarrow{[O]} \underset{\substack{\text{Ethanal}\\ C_2H_4O}}{\text{H}_3\text{C}-\overset{..\ \text{O}\ ..}{\text{C}}-\text{H}} + \text{H}^+ + \text{H}:^-
$$

Similarly, the reactant that causes a substrate to be reduced is referred to as **reducing agent**, which is itself being oxidized in the process. Hydrogen, H_2, is a common reducing agent, and others are shown in Table 10-4. Since there are a variety of reducing agents that reduce a substrate with the same end result, the identity of the reducing agent is often not specified; instead, the

TABLE 10-4 Common Oxidizing and Reducing Agents*

Oxidizing Agents

Name	Formula
Bleach	NaOCl
Oxygen	O_2
Hydrogen peroxide	H_2O_2
Chromium(VI)	Cr^{6+}
Nicotinamide Adenine Dinucleotide	NAD^+
Flavin Adenine Dinucleotide	FAD

Reducing Agents

Name	Formula
Hydrogen gas	H_2
Sodium borohydride	$NaBH_4$
Nicotinamide Adenine Dinucleotide	NADH
Flavin Adenine Dinucleotide	$FADH_2$
Vitamin C	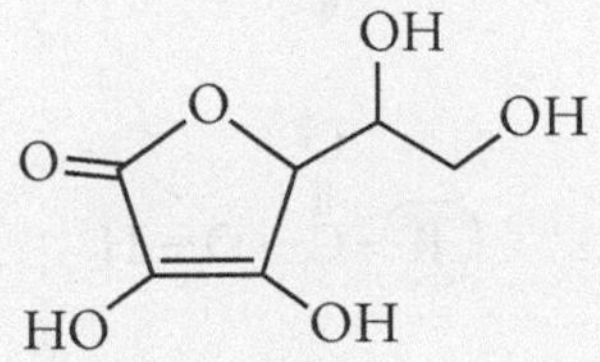

*Shaded items are biological agents.

general abbreviation [H] is placed above the reaction arrow to indicate a reducing agent is present as a reactant. For example, the reduction of 2-propanone to 2-propanol, the reverse of the reaction shown earlier, can be accomplished with a reducing agent, as shown in the equation below:

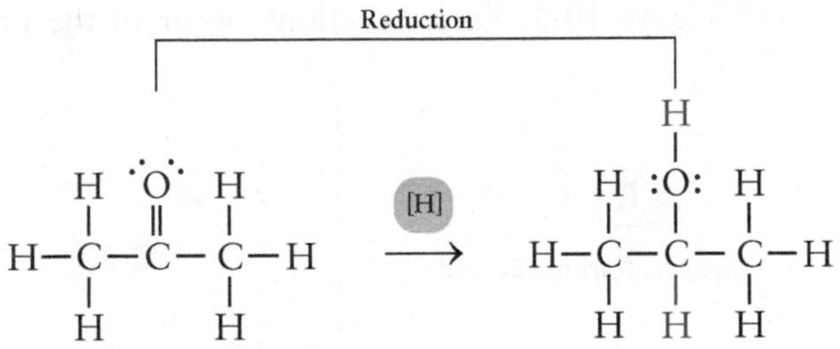

Common Oxidation and Reduction Reactions of Functional Groups

In chapter 7 you were introduced to the common functional groups and learned that each functional group has a characteristic chemical behavior that is generally independent of the structure of the rest of the molecule. This is particularly true for oxidation–reduction reactions. Thus, by learning how a given functional group behaves in the presence of an oxidizing or reducing agent, we can predict the outcome for an oxidation or reduction reaction of any molecule that contains this functional group.

Beginning with functional groups containing carbon-oxygen bonds, Figure 10-4 shows the outcome of primary and secondary alcohols, and aldehydes reacting with an oxidizing agent. The first column shows that primary alcohols are oxidized to aldehydes and that aldehydes can be oxidized to carboxylic acids. The second column shows that secondary alcohols are oxidized to ketones. In these

Figure 10-4 Oxidation of functional groups containing a carbon-oxygen bond.

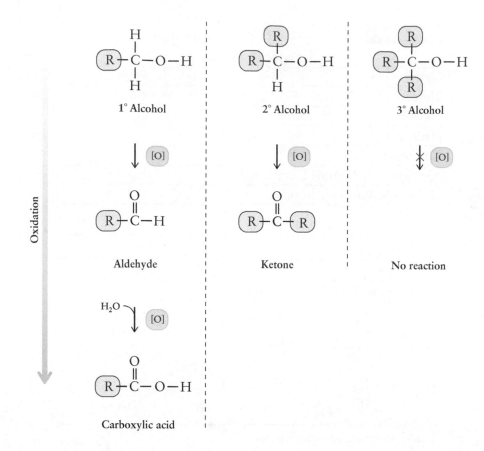

reactions, note that the number of hydrogen atoms decreases and/or the number of oxygen atoms increases (oxidation). The third column shows that tertiary alcohols do not react with oxidizing agents. As you can see, being able to recognize functional groups greatly simplifies the task of learning organic reactions.

Figure 10-5 shows the corresponding reductions, that is, the reverse of the reactions shown in Figure 10-4. These reactions occur in the presence of a

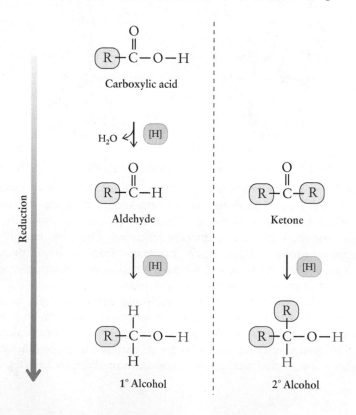

Figure 10-5 Reduction of functional groups containing a carbon-oxygen bond.

reducing agent. A carboxylic acid can be reduced to an aldehyde, and an aldehyde can be reduced to a primary alcohol. Similarly, ketones can be reduced to secondary alcohols.

Next, consider the oxidation–reduction reactions of hydrocarbon functional groups (Figure 10-6). An oxidizing agent is required to convert an alkane into an alkene and an alkene into an alkyne. The reverse of each of these reactions is a reduction. Thus, alkynes can be reduced to alkenes, and alkenes can be reduced to alkanes.

Catalytic Hydrogenation The reduction of an alkene to an alkane is a common reduction reaction of hydrocarbons. In the laboratory, this reaction is typically carried out using hydrogen gas (H_2) as the reducing agent in the presence of a metal catalyst (Pd, Pt, Ni, etc.), in a reaction known as a **catalytic hydrogenation**:

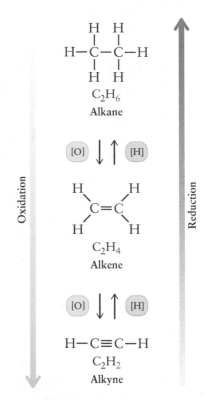

Figure 10-6 Hydrocarbon oxidation and reaction reactions.

Catalytic hydrogenation reactions are common in the commercial food industry where they are used to prepare "partially hydrogenated" fats. Recall from chapter 6 that most vegetable oils contain unsaturated fats comprised of *cis* double bonds and that these fats are healthier than saturated fats derived from animal sources. However, unsaturated fats have a shorter shelf-life and an oily consistency at room temperature. Catalytic hydrogenation of unsaturated fats reduces some of the double bonds to single bonds, according to the reaction above, creating a "partially hydrogenated fat" with a longer shelf-life and a more solid, spreadable consistency. It is required that these foods be labeled as "hydrogenated" or "partially hydrogenated" to indicate they have been chemically modified. Unfortunately, during the catalytic hydrogenation process, some double bonds are not reduced, but are instead isomerized from *cis* to *trans* double bonds, as shown below. Whereas fats containing *cis* double bonds are healthy, those containing *trans* double bonds are less healthy than even saturated fats, and these are the infamous "*trans* fats." Indeed, *trans* fats have been banned from most restaurants. The current trend in the food industry is to produce processed foods without *trans* fats, which now carry the label "no trans fats."

> Note that while alkanes are not considered a functional group, under the right conditions they can be oxidized to an alkene.

WORKED EXERCISES Oxidation and Reduction Reactions

10-1 Place an [H] or an [O] above each reaction arrow below to indicate whether an oxidizing agent or a reducing agent is required for the reaction shown. Identify the functional group(s) in the reactant and the product.

a.

b.

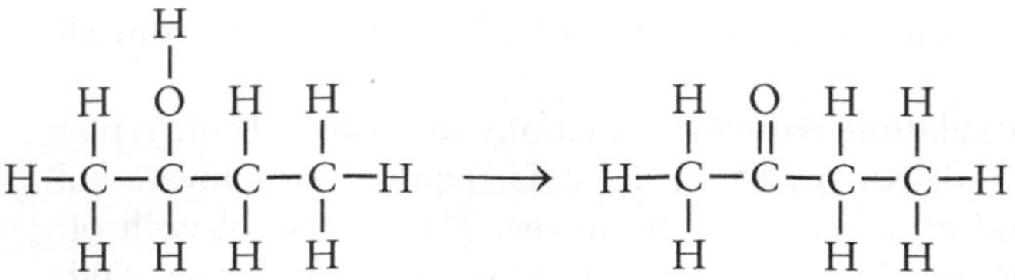

10-2 Predict the product formed in the following reaction. How has the functional group changed?

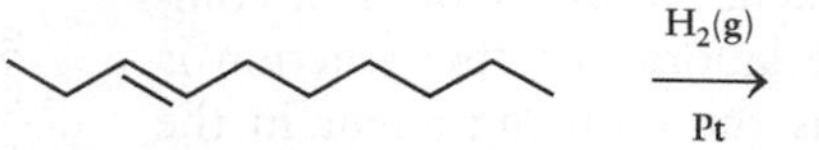

10-3 Predict the structure of the product formed in each of the following reactions. Write the name of the functional groups involved. What is meant by [O] and [H] above the arrows?

a.

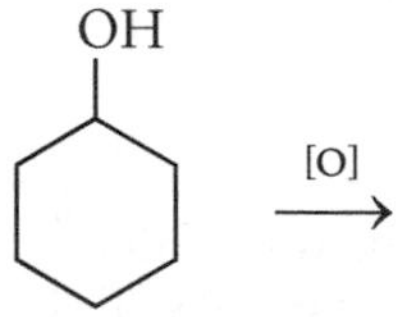

b.

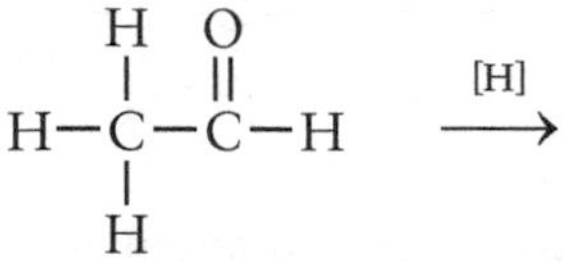

c.

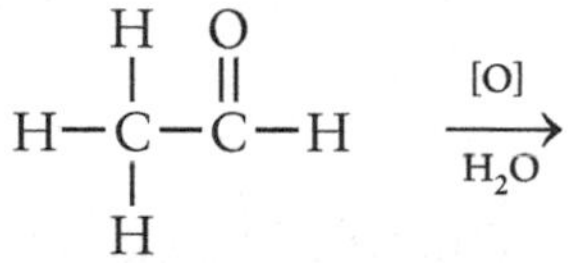

Solutions

10-1

a.

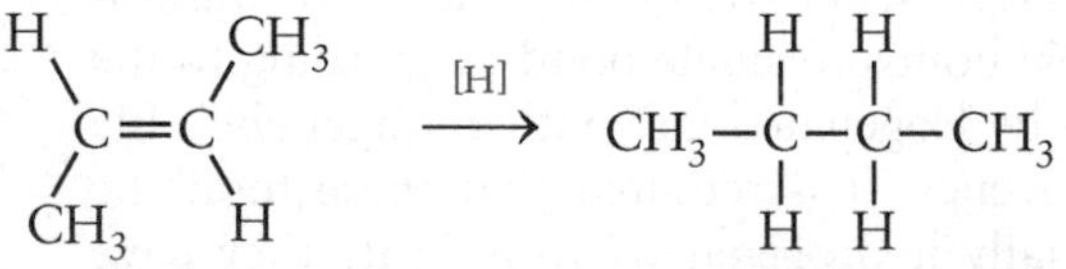

Alkene   Alkane

The reactant is an alkene, and the product is an alkane. The product has more hydrogen atoms than the reactant. Since there is an increase in the number of hydrogen atoms, the transformation is a reduction, and therefore, a reducing agent, [H], is required.

b.

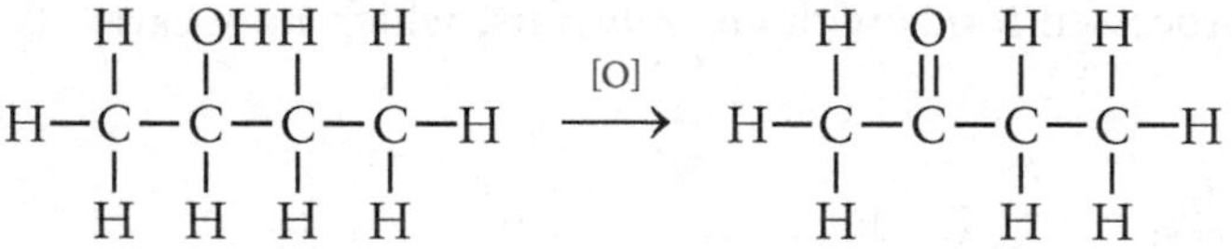

Secondary alcohol   Ketone

The reactant is a secondary alcohol, and the product is a ketone. The product has fewer hydrogen atoms than the reactant. Since there is a decrease in hydrogen atoms, the reaction is an oxidation, and an oxidizing agent, [O], is required.

10-2 3-Decene is subjected to the conditions of a catalytic hydrogenation, which means it is reduced. Reduction of an alkene produces an alkane; hence, 3-decene is reduced to decane.

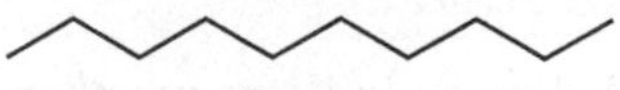

Decane

10-3 a.

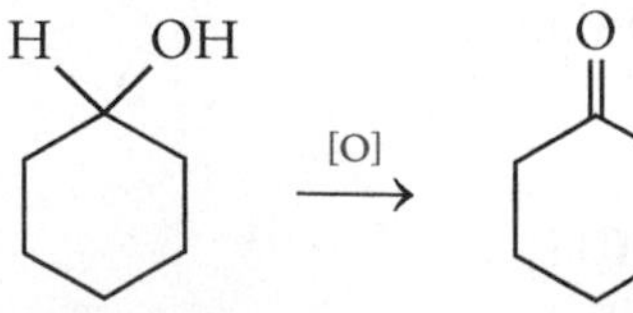

Secondary alcohol   Ketone

The [O] above the arrow indicates that an oxidizing agent is present; hence, the alcohol undergoes oxidation. A secondary alcohol is oxidized to a ketone (Figure 10-4). Write the structure of the product by changing only the functional group, leaving the rest of the molecule looking exactly the same. Therefore, cyclohexanol is oxidized to cyclohexanone. A word of caution when interpreting skeletal line structures: Remember the C—H bonds are not shown in skeletal line structures, so you need to remember that they are there when counting H atoms.

b.

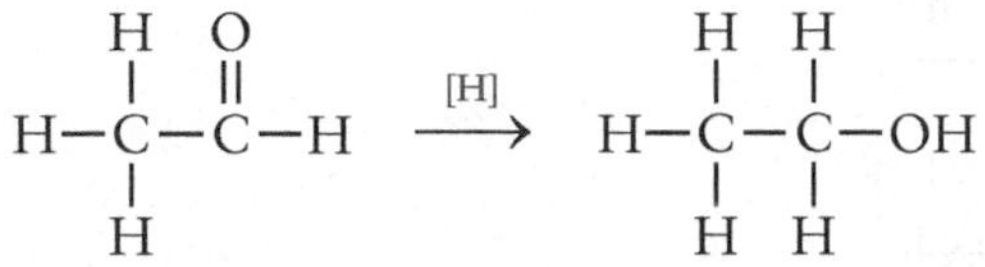

Aldehyde → Primary alcohol

The [H] above the arrow signifies a reduction, which means the aldehyde is reduced. An aldehyde is reduced to a primary alcohol (Figure 10-5). Write the product by making the change described to only the functional group (aldehyde) involved. In other words, the —CH₃ group is unchanged.

c.

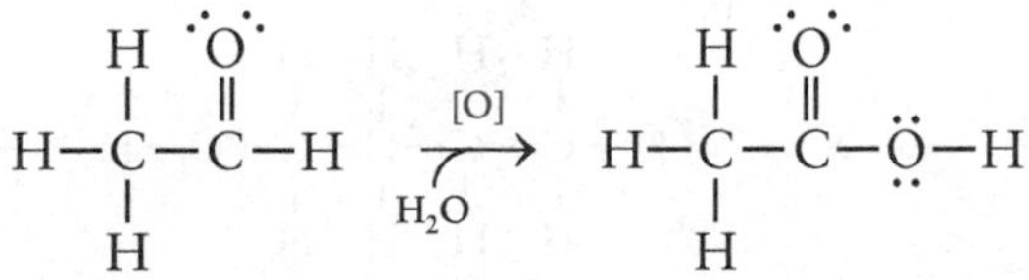

Aldehyde → Carboxylic acid

The reactant is an aldehyde. The [O] above the arrow signifies that it undergoes oxidation, which in the case of an aldehyde requires water and results in the formation of a carboxylic acid (Figure 10-4). Note that aldehydes can be reduced to a primary alcohol or oxidized to a carboxylic acid, so take a moment to think about which reaction it is.

PRACTICE EXERCISES

1   Determine whether the substrate in the following reactions is undergoing an oxidation or a reduction. What is another name for the type of reaction shown in (c)?

a.

b.

c.

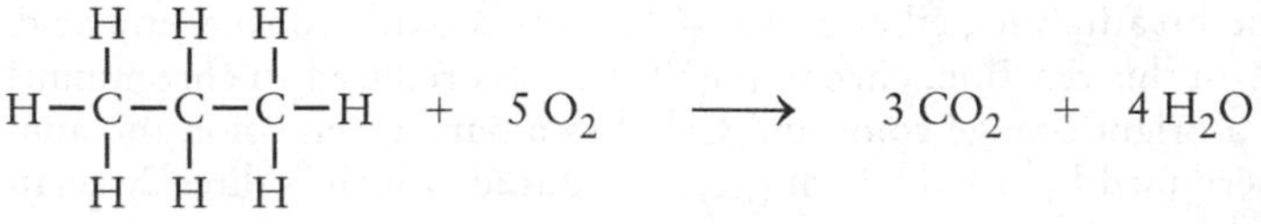

d.

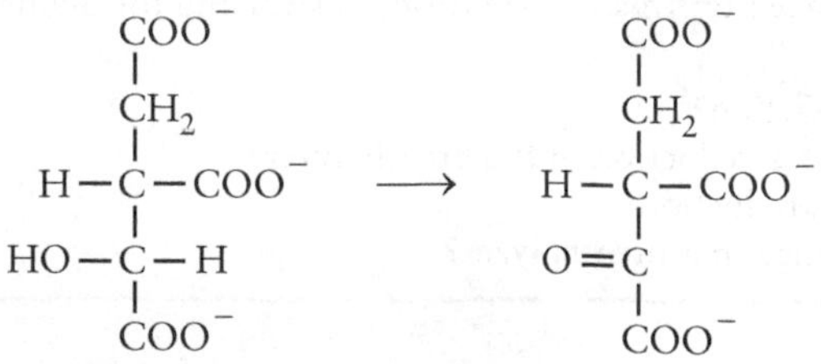

Figure 10-7 Silver plating is observed when copper is placed in a solution of silver chloride, as a result of an oxidation–reduction reaction. [© 1986 Peticolas/Megna/Fundamental Photographs]

2 If copper, Cu, is placed in a solution of silver chloride (AgCl), silver plating is seen on the strip of copper. The net reaction equation is shown below and illustrated in the photograph in Figure 10-7. Which reactant is oxidized? Which reactant is reduced? Write the two half reactions.

$$2\,Ag^+(aq) + Cu(s) \rightarrow Cu^{2+}(aq) + 2\,Ag(s)$$

3 Predict the structure of the product formed in the following reactions:

a.

$$\xrightarrow{[H]}$$

b. *Hint:* The carboxylic acid is unchanged.

$$\xrightarrow[Pt]{H_2(g)}$$

4 Predict the structure of the product formed in the following reactions. If no reaction occurs, state why.

a.

$$\xrightarrow{[H]}$$

b.

$$\xrightarrow{[O]}$$

c.

$$\xrightarrow{[O]}$$

5 Predict the structure of the product formed in the following reactions:

a.

$$\xrightarrow[H_2O]{[O]}$$

b.

$$\xrightarrow{[H]}$$

c.

$$\xrightarrow{[H]}$$

Figure 10-8 A breathalyzer determines blood alcohol (ethanol) levels by oxidizing ethanol with chromium(VI) in an oxidation–reduction reaction. [Yellow Dog Productions/The Image Bank/Getty Images]

6 Law enforcement officials use the *breathalyzer* to measure blood alcohol levels (Figure 10-8). This device employs an oxidation–reduction reaction to determine the amount of ethanol in a person's breath. In the breathalyzer, ethanol (CH_3CH_2OH) is oxidized to acetic acid (CH_3CO_2H) by Cr^{6+}. In this reaction, chromium(VI) Cr^{6+}, is reduced to chromium(III), Cr^{3+}. Since Cr^{6+} has a bright orange color and Cr^{3+} has a dark green color, the amount of Cr^{3+} produced is determined by the shade of green produced, which is directly related to the amount of alcohol in a person's breath (remember from chapter 8 that color intensity is one way to measure concentration). Based on this information, answer the following questions.

a. What is the oxidizing agent in this reaction?
b. Is ethanol undergoing an oxidation or a reduction in the breathalyzer?
c. Is Cr^{6+} undergoing an oxidation or a reduction?
d. What is responsible for the color change in a breathalyzer?

Coenzymes in Biochemical Oxidation–Reduction Reactions. We have seen hydrogen gas and bleach used as reducing and oxidizing agents, respectively; however, these are obviously not the reagents our cells use to perform redox reactions! Instead, our cells use **coenzymes** to transfer electrons in oxidation–reduction reactions. In chapter 4 you learned that enzymes are biological catalysts. The enzymes that catalyze oxidation–reduction reactions—*oxidoreductases*—require a coenzyme to achieve their catalytic activity. While the enzyme is chemically unchanged in the reaction, the coenzyme *does* undergo a chemical change in the reaction. Whereas enzymes are proteins, coenzymes are organic molecules, derived from vitamins. The coenzymes for *oxidoreductases* essentially shuttle electrons between molecules in a biological cell. Thus, a molecule in one part of the cell can undergo oxidation (loss of electrons) by transferring electrons to the coenzyme (reducing the coenzyme). The reduced coenzyme then transfers the electrons to another molecule, often in another part of the cell, which gains the electrons from the coenzyme and is reduced (the coenzyme is oxidized). The coenzyme then becomes available again. The shuttling of electrons from one molecule or atom to another is central to cellular respiration, the biochemical pathway that harvests energy from glucose.

One of the coenzymes seen in oxidation–reduction reactions involving carbon-oxygen bonds is *n*icotinamide *a*denine *d*inucleotide, $NAD^+/NADH$, a substance produced by the cell from niacin, vitamin B_3, with the chemical structure shown in **Figure 10-9**. NAD^+ is the oxidized form of the coenzyme and NADH is the reduced form of the coenzyme. In its NADH form, the coenzyme has two more electrons than in its NAD^+ form.

The convention when writing biochemical pathways is to show only the structure of the substrate and the product on either side of the main reaction arrow (the straight arrow). The abbreviations for the coenzymes are written on the reactant and product sides of a curved arrow that intersects the main reaction arrow. This convention makes it easier to focus on the functional group changes in a sequence of reactions in a biochemical pathway. At the

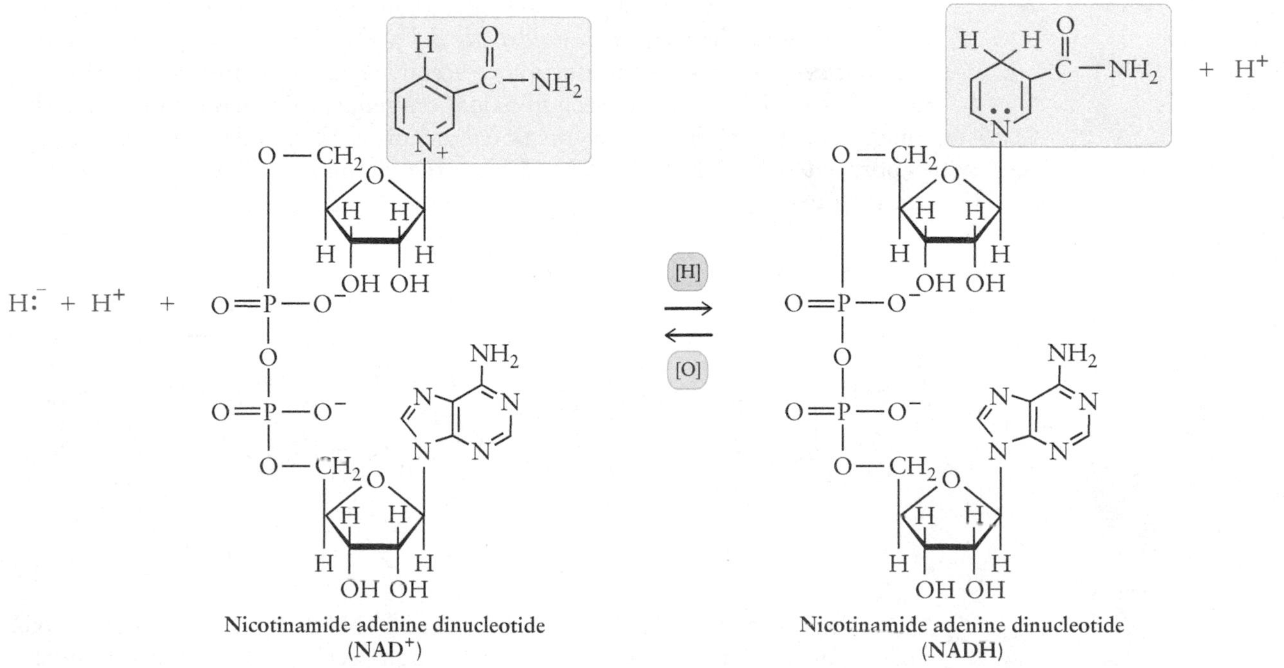

Foods high in the vitamin niacin include rice, wheat bran, peanuts, and some fish, such as anchovies and tuna.

Figure 10-9 The structure of nicotinamide adenine dinucleotide in its oxidized form, NAD^+, gains H^- and H^+ in going to its reduced form $NADH + H^+$. The reverse reaction is an oxidation.

same time it makes it possible to keep track of the other reactants and coenzymes involved in the biochemical sequence.

Consider as an example what happens in the body when you consume alcohol. The first two steps in the metabolism of ethanol require NAD^+ as the oxidizing agent, which is reduced to NADH and a proton, as indicated on either side of the curved arrows. The oxidation of ethanol (a primary alcohol) to acetaldehyde (an aldehyde) is shown in the first reaction, and the oxidation of acetaldehyde to acetic acid (a carboxylic acid) is shown in the second reaction. The second reaction also requires water as a reactant, as indicated by the curved arrow merging into the main reaction arrow. The reactions are catalyzed by *alcohol dehydrogenase* and *acetaldehyde dehydrogenase*, respectively, enzymes that require NAD^+/NADH as a coenzyme. Acetic acid is ultimately further oxidized to CO_2 and H_2O in subsequent catabolic reactions.

Initial Oxidation Steps in Ethanol Catabolism

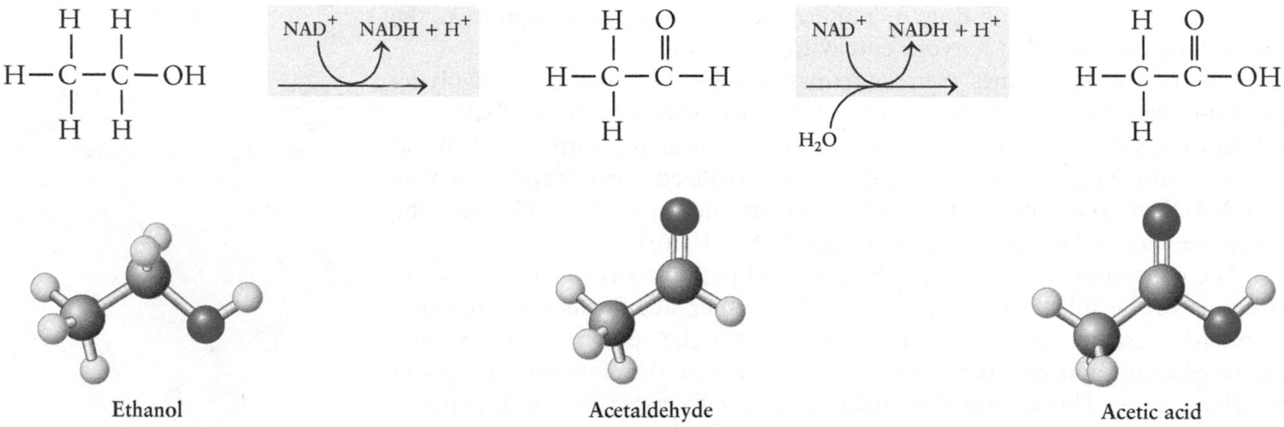

Ethanol   Acetaldehyde   Acetic acid

Recall from chapter 4 that it is the build-up of acetaldehyde, the middle compound in sequence above, that is responsible for the hangover that often accompanies the consumption of too much ethanol in too short a period of time. This occurs because the enzyme *acetaldehyde dehydrogenase* in the second reaction cannot keep up with the production of acetaldehyde in the first reaction.

One of the coenzymes seen in oxidation–reduction reactions involving the metabolism of hydrocarbon functional groups is *f*lavin *a*denine *d*inucleotide, abbreviated **FAD/FADH₂**, derived from the vitamin riboflavin and described in the opening story.

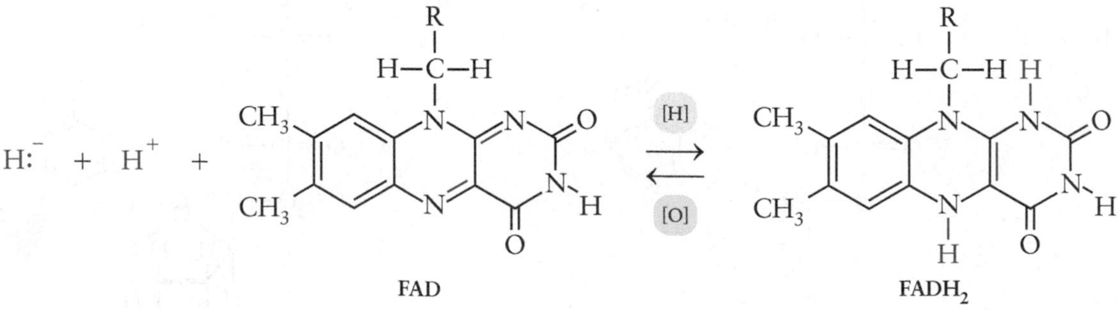

FAD   FADH₂

Like the coenzyme NAD^+, which is reduced to NADH + H^+, FAD is an oxidizing agent that is reduced to FADH₂ upon receiving two electrons in the form of a hydride ion, H^-, and a proton, H^+, as shown in the forward reaction above. The reverse reaction also occurs.

WORKED EXERCISES Coenzymes in Biochemical Oxidation–Reduction Reactions

10-4 The following reaction is one of the 10-steps of glycolysis, the first biochemical pathway in the catabolism of glucose:

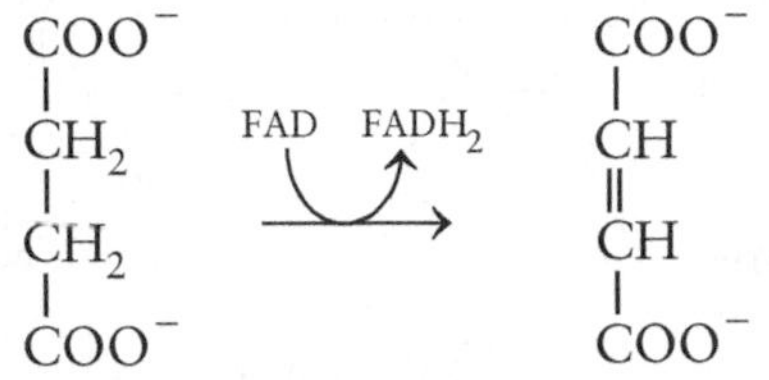

$$
\begin{array}{ccc}
\text{COO}^- & & \text{COO}^- \\
| & & | \\
\text{CH}_2 & \text{FAD} \quad \text{FADH}_2 & \text{CH} \\
| & \curvearrowright & \parallel \\
\text{CH}_2 & \longrightarrow & \text{CH} \\
| & & | \\
\text{COO}^- & & \text{COO}^- \\
\text{Succinate} & & \text{Fumarate}
\end{array}
$$

a. Is succinate oxidized or reduced in this reaction? How can you tell?
b. What is the coenzyme and is it undergoing an oxidation or a reduction?
c. Is FAD an oxidizing agent or a reducing agent?
d. What functional group change occurs in the reaction? Are the carboxylic acid functional groups affected in this reaction?

10-5 When muscle cells are depleted of oxygen, they form lactic acid, which occurs when pyruvic acid is reduced. This reaction replenishes the cell's supply of NAD$^+$. Lactic acid is responsible for the feeling of sore muscles following intense exercise. Predict the structure of lactic acid based on the reaction shown:

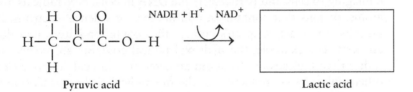

Pyruvic acid   Lactic acid

Solutions

10-4 a. Succinate is oxidized because the product, fumarate, has fewer hydrogen atoms than the substrate, hence an oxidation.
b. The coenzyme FAD is reduced.
c. FAD is an oxidizing agent because it causes the substrate to get oxidized (it is itself reduced).
d. An alkane is converted into an alkene. The carboxylic acid functional groups are unaffected in the reaction.

10-5 In this reaction we see that NADH is oxidized to NAD$^+$, so we know that the substrate is undergoing a corresponding reduction. The substrate has two functional groups, a ketone and a carboxylic acid. Although both ketones and carboxylic acids can be reduced (Figure 10-5), we know the product is an acid, so we can assume it is the ketone that is reduced, not the carboxylic acid. We know that ketones are reduced to secondary alcohols. Thus, we change only the ketone functional group, leaving the rest of the molecule unchanged, when writing the structure of the product alcohol.

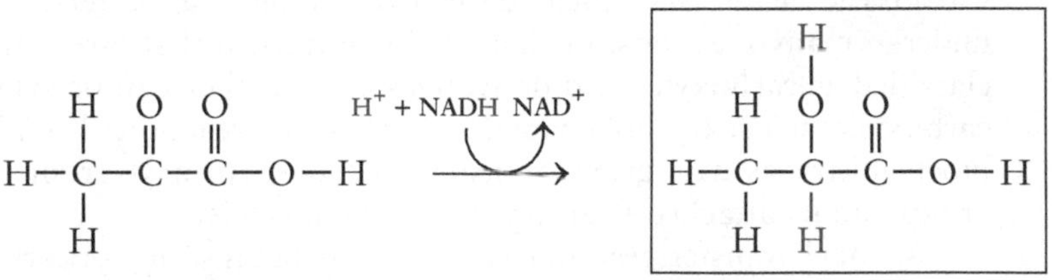

$$
\begin{array}{ccc}
\text{H} \quad \text{O} \quad \text{O} & & \quad\quad \text{H} \\
| \quad \parallel \quad \parallel & \text{H}^+ + \text{NADH} \quad \text{NAD}^+ & \quad\quad | \\
\text{H}-\text{C}-\text{C}-\text{C}-\text{O}-\text{H} & \curvearrowright\!\!\longrightarrow & \text{H} \quad \text{O} \quad \text{O} \\
| & & | \quad | \quad \parallel \\
\text{H} & & \text{H}-\text{C}-\text{C}-\text{C}-\text{O}-\text{H} \\
& & | \quad | \\
& & \text{H} \quad \text{H}
\end{array}
$$

Pyruvic acid   Lactic acid

PRACTICE EXERCISES

7 For the oxidation–reduction reactions shown below, determine whether the substrate has undergone an oxidation or a reduction. Indicate whether the coenzyme shown has undergone an oxidation or a reduction.

a.

$$H_2O \quad + \quad \underset{\text{Acetaldehyde}}{H-\overset{\overset{\displaystyle H}{|}}{\underset{\underset{\displaystyle H}{|}}{C}}-\overset{\overset{\displaystyle O}{\|}}{C}-H} \quad \overset{NAD^+ \quad NADH + H^+}{\underset{\Large\curvearrowright}{\longrightarrow}} \quad \underset{\text{Acetic acid}}{H-\overset{\overset{\displaystyle H}{|}}{\underset{\underset{\displaystyle H}{|}}{C}}-\overset{\overset{\displaystyle O}{\|}}{C}-OH}$$

b.

$$\underset{\quad O}{\diagdown\!\diagup\!\diagdown\!\diagup\!\diagdown\!\diagup\!\diagdown\!\diagup\!\overset{\displaystyle SCoA}{\overset{\|}{C}}} \quad \overset{FAD \quad FADH_2}{\underset{\Large\curvearrowright}{\longrightarrow}} \quad \underset{\qquad O}{\diagdown\!\diagup\!\diagdown\!\diagup\!\diagdown\!\diagup\!\diagdown\!\overset{\displaystyle SCoA}{\overset{\|}{C}}}$$

8 Which coenzyme is most likely involved in a biochemical oxidation of an alcohol?
    a. FAD   **b.** $FADH_2$ **c.** NAD^+ **d.** NADH

9 Which coenzyme is most likely involved in a biochemical reduction of an alkene?
    a. FAD   **b.** $FADH_2$ **c.** NAD^+ **d.** NADH

$$H-\ddot{\underset{\displaystyle ..}{O}}\cdot$$
Hydroxyl radical

Figure 10-10 Fruits and berries contain compounds known as antioxidants. Antioxidants are reducing agents that prevent the damaging effects of free radical oxidation reactions in the cell. [Catherine Bausinger]

Antioxidants Some oxidation reactions produce free radicals, which can be damaging to cells and particularly the DNA in cells. **Free radicals** are molecules, atoms, or ions that contain an odd number of valence electrons. Radicals are unstable and reactive species because they contain an atom that does not have an octet. For example, the hydroxyl radical contains an oxygen atom with 7 rather than 8 electrons. To attain an octet, free radicals remove electrons from other molecules—a reduction of the free radical and an oxidation of the other molecule—in a type of chain reaction that can ultimately damage the cell.

Antioxidants are substances that are believed to prevent the destructive oxidation of substances in the cell. Antioxidants work by stopping the chain reaction initiated by free radicals or preventing the formation of the radical species in the first place. Antioxidants work because they are reducing agents, so they are themselves oxidized in the process, thereby preventing the unwanted oxidations. Antioxidants have received a great deal of attention in recent years because of the role they are thought to play in preventing cancer and cardiovascular disease. Antioxidants are also thought to slow the outward signs of aging. Fruits, particularly berries, vegetables, and tea are high in antioxidants such as vitamin C, vitamin E, and glutathione (Figure 10-10). Our cells also contain enzymes whose primary function is to scavenge free radicals before they can do any oxidative damage.

■ 10.2 Group Transfer Reactions: Esterification and Amidation Reactions

Carboxylic acids and their derivatives—esters, thioesters, and amides—undergo many reactions. In chapter 7 you learned that esters and amides are classified as carboxylic acid derivatives because they can be synthesized from carboxylic acids. In this chapter we introduce reactions broadly defined in biochemistry as **acyl group transfer reactions**. An **acyl group** is a carbonyl group and its attached R group, as shown in yellow.

An acyl transfer reaction is so named because it appears like the acyl group on a carboxylic acid moves from the —OH of the carboxylic acid to the heteroatom of an alcohol (HOR), thiol (HSR), or amine (HNR_2) to

$$\underset{\text{Acyl group}}{R-\overset{\overset{\displaystyle O}{\|}}{C}-}$$

$$\underset{\text{Acetyl group}}{H-\overset{\overset{\displaystyle H}{|}}{\underset{\underset{\displaystyle H}{|}}{C}}-\overset{\overset{\displaystyle O}{\|}}{C}-}$$

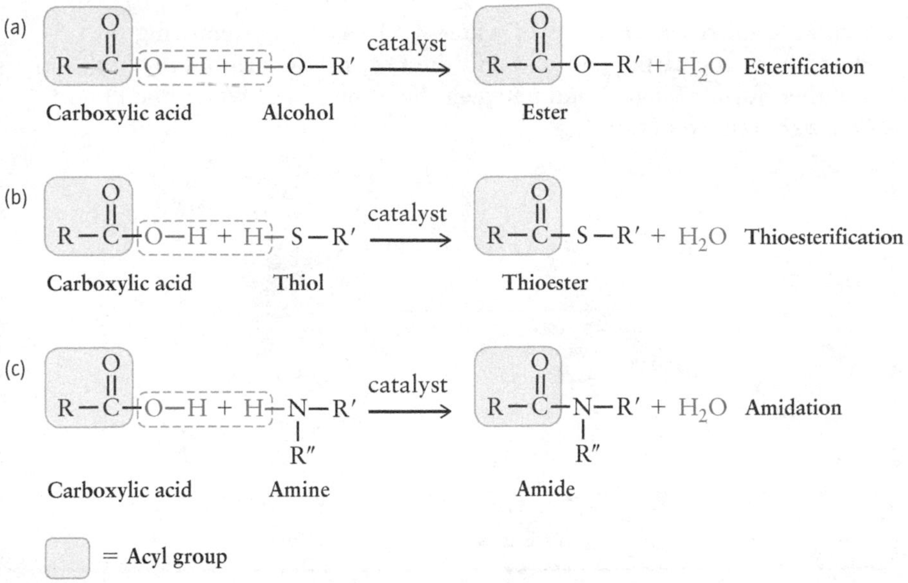

Figure 10-11 Acyl group transfer reactions occur between a carboxylic acid and an alcohol, thiol, or amine to form the corresponding carboxylic acid derivatives: esters, thioesters, and amides, respectively. Water is also produced.

produce an ester, thioester, or amide, respectively, as shown in Figure 10-11. A molecule of water is produced in these reactions, too. *Transferase* enzymes are the major class of enzymes that catalyze acyl transfer reactions in biochemistry.

When the —R group bonded to the C=O group is —CH$_3$, the acyl group is called an **acetyl group**.

Esters and Thioesters

Figure 10-11 shows that an ester and water are formed in the reaction of an alcohol with a carboxylic acid, in the presence of a catalyst. In an analogous manner, thioesters are formed by the reaction of a thiol with a carboxylic acid. A **thiol** is a functional group analogous to an alcohol, with sulfur instead of oxygen: R—S—H. A **thioester** is a functional group analogous to an ester, with sulfur instead of oxygen attached to the carbonyl group as shown here.

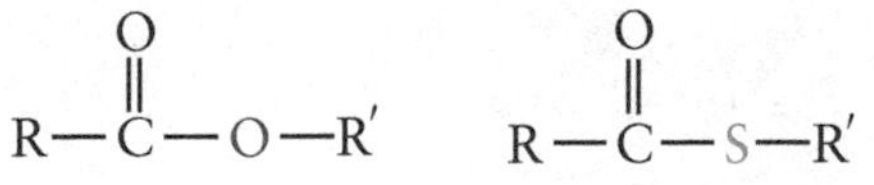

$$R-\overset{\overset{\displaystyle O}{\|}}{C}-O-R' \qquad R-\overset{\overset{\displaystyle O}{\|}}{C}-S-R'$$

Ester   Thioester

To predict the chemical structure of the ester or thioester formed in the reaction between a carboxylic acid and an alcohol or a thiol, follow the guidelines given below. These reactions are called **esterification reactions**, a type of acyl group transfer reaction.

Guidelines: Predicting the Products Formed in an Esterification Reaction

Step 1: Write the carboxylic acid with its OH group *pointing to the right,* and place the alcohol or thiol next to it on the right with its H—O or H—S group *pointing to the left* as shown.

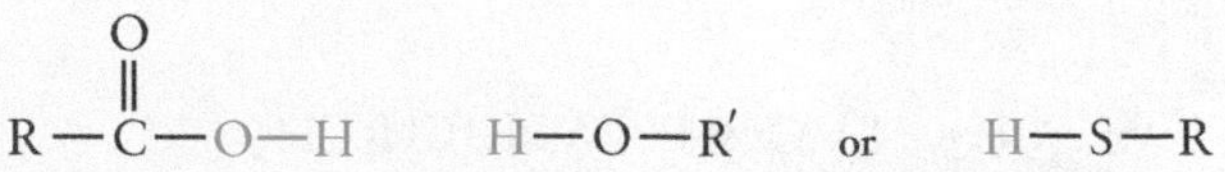

$$R-\overset{\overset{\displaystyle O}{\|}}{C}-O-H \qquad H-O-R' \qquad \text{or} \qquad H-S-R$$

Carboxylic acid   Alcohol   Thiol

Step 2: Remove the elements of water, OH and H, by removing the H—O from the carboxylic acid and the H from the alcohol or thiol, and then form a single bond between the atoms from which the H and OH were removed.

Remove OH and H and combine to make H₂O

$$R-\overset{\overset{\displaystyle O}{\|}}{C}-\underset{}{[O-H} \quad H]-O-R'$$

Form a single bond between these two atoms to make an ester

$$R-\overset{\overset{\displaystyle O}{\|}}{C}-O-R' \;+\; H-O-H$$

Ester   Water

Do not interpret these guidelines as the actual order in which bond-breaking and bond-making occurs; these are merely guidelines to help you determine the structure of the product.

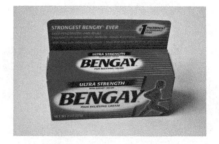

Figure 10-12 Esters, such as oil of wintergreen, are found in muscle relaxants like Bengay, which gives it its characteristic scent. [Catherine Bausinger]

Esters in Fragrances In chapter 7 you learned that the characteristic fragrance of fruits and perfumes is often due to the presence of esters. Some over the counter muscle relaxers such as Bengay contain the ester found in oil of wintergreen (Figure 10-12). These and other esters are readily prepared in the laboratory in esterification reactions by heating together a carboxylic acid and an alcohol in the presence of an acid catalyst (H^+). Different scents can be created by varying the structure of the carboxylic acid and the alcohol. The resulting esters have different shapes that can be detected by the olfactory receptors in our nose, which our brain interprets as different scents. For example, acetic acid reacts with 1-octanol to form 1-octyl acetate (IUPAC name 1-octyl ethanoate), an ester with the fragrance of oranges; whereas the reaction of acetic acid with 3-methyl-1-butanol forms an ester with the fragrance of bananas. Table 10-5 shows the alcohol and carboxylic acid that react to form some simple esters and their associated fragrance.

TABLE 10-5 Esters, the Carboxylic Acids and Alcohols from Which They Were Formed, and Their Fragrance

Ester	Carboxylic Acid	Alcohol	Fragrance
	CH_3COOH	$HOCH_2(CH_2)_6CH_3$	Orange
	CH_3COOH	$HOCH_2CH_2\overset{\overset{\displaystyle CH_3}{\|}}{C}HCH_3$	Banana
	$CH_3CH_2CH_2COOH$	$HOCH_3$	Apple
	$CH_3CH_2CH_2COOH$	$HOCH_2CH_3$	Pineapple

Biological Esterification Reactions In chapter 7 you learned that fats are esters derived from three fatty acids and glycerol. Our cells store fat, our long term supply of fuel, in the form of glycerol esters, known as **triacylglycerols**, also known as **triglycerides**, or more simply, as fats. The esterification of three fatty acids (the same or different) and one glycerol molecule produces a fat, according to the following esterification reaction:

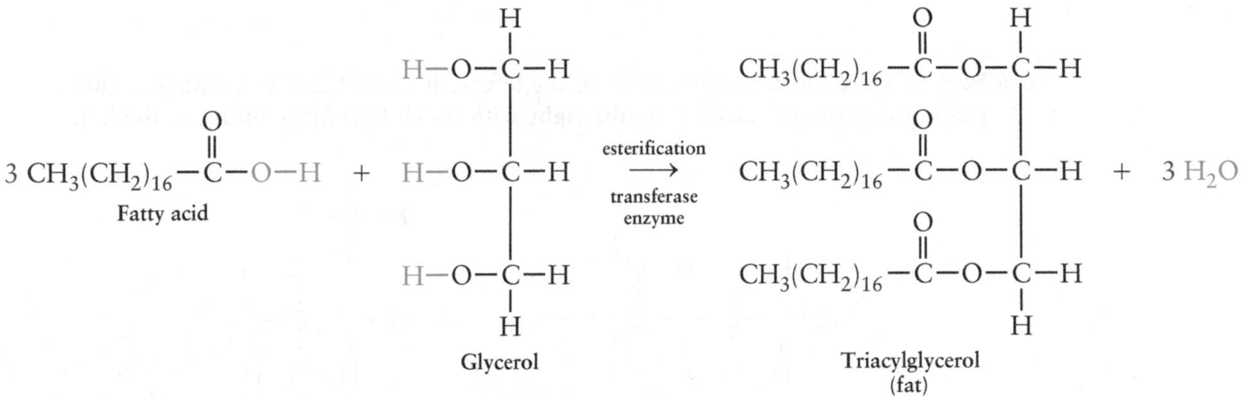

Our cells transport cholesterol in the form of cholesterol esters, formed in esterification reactions between cholesterol and a fatty acid, as shown:

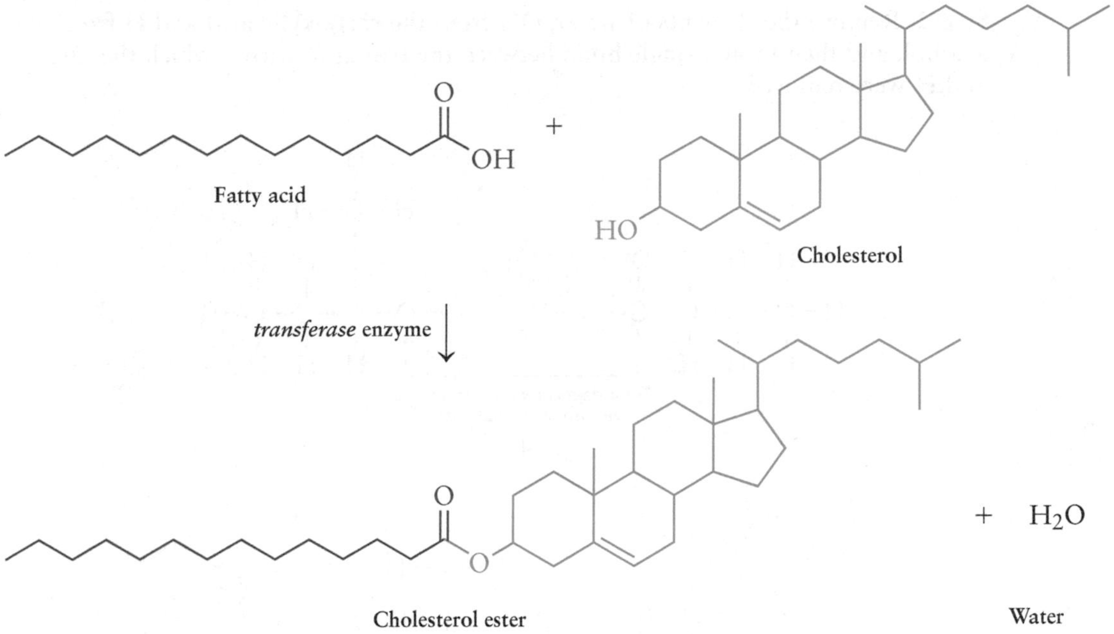

WORKED EXERCISES Predicting the Products of Esterification Reactions

10-6 Write the structure of the product formed in the following esterification reaction.

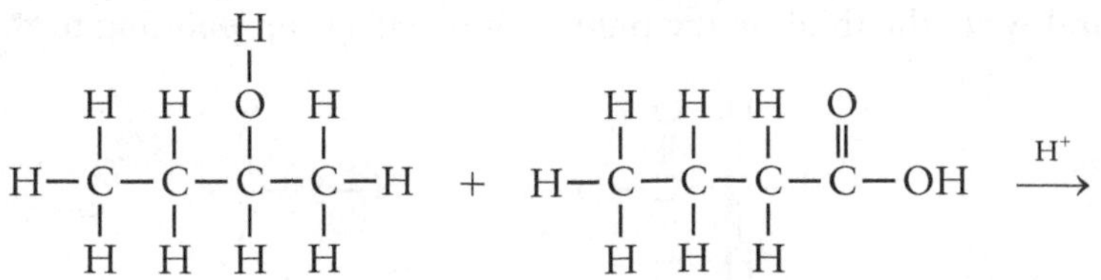

10-7 Write the structure of the product formed in the following esterification reaction involving a thiol.

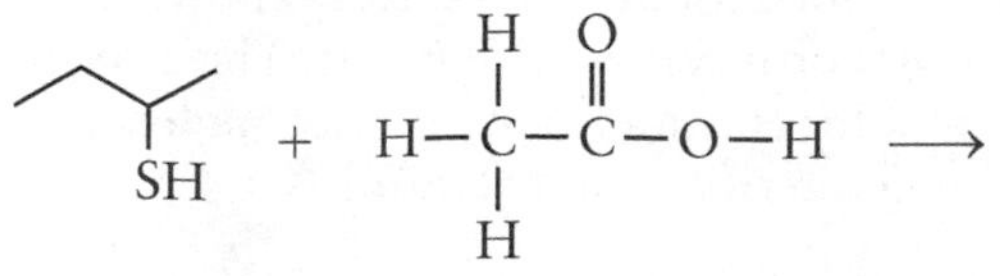

$$
\underset{\text{SH}}{\diagup\diagup\diagup} \quad + \quad \underset{\underset{\text{H}}{|}}{\overset{\overset{\text{H}\quad\text{O}}{|\quad\|}}{\text{H}-\text{C}-\text{C}-\text{O}-\text{H}}} \quad \longrightarrow
$$

Solutions

10-6 Step 1: Write the carboxylic acid on the left with its OH group pointing to the right, and write the alcohol on the right with its HO group pointing to the left.

$$
\underset{\underset{\text{H}\;\text{H}\;\text{H}}{|\;\;|\;\;|}}{\overset{\overset{\text{H}\;\text{H}\;\text{H}\;\text{O}}{|\;\;|\;\;|\;\;\|}}{\text{H}-\text{C}-\text{C}-\text{C}-\text{C}-\text{O}-\text{H}}}
$$

Carboxylic acid

$$
\underset{\underset{\text{H}\;\text{H}\;\text{H}}{|\;\;|\;\;|}}{\overset{\overset{\text{H}-\text{C}-\text{H}}{|}}{\text{H}-\text{O}-\text{C}-\text{C}-\text{C}-\text{H}}}
$$

Secondary alcohol

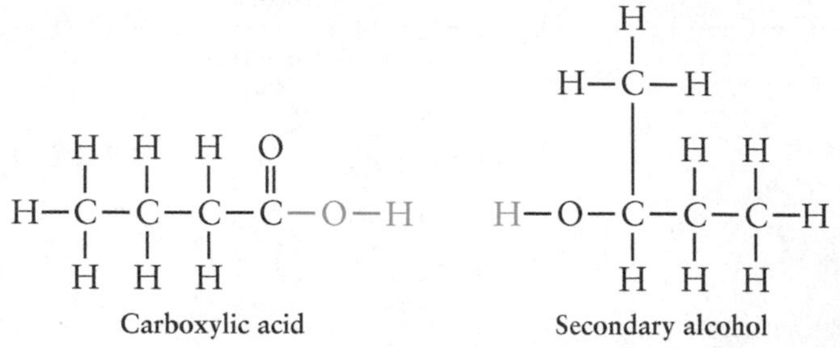

Remember, molecules are moving around and freely rotating about their single bonds, so feel free to flip them and rotate their bonds to get them into the conformation you need to help you predict the structure of the product. *However, never change the atom connectivity because then you have changed the identity of the molecule.*

Step 2: Remove the elements of water, OH from the carboxylic acid and H from the alcohol, and then form a single bond between the two atoms from which the OH and H were removed.

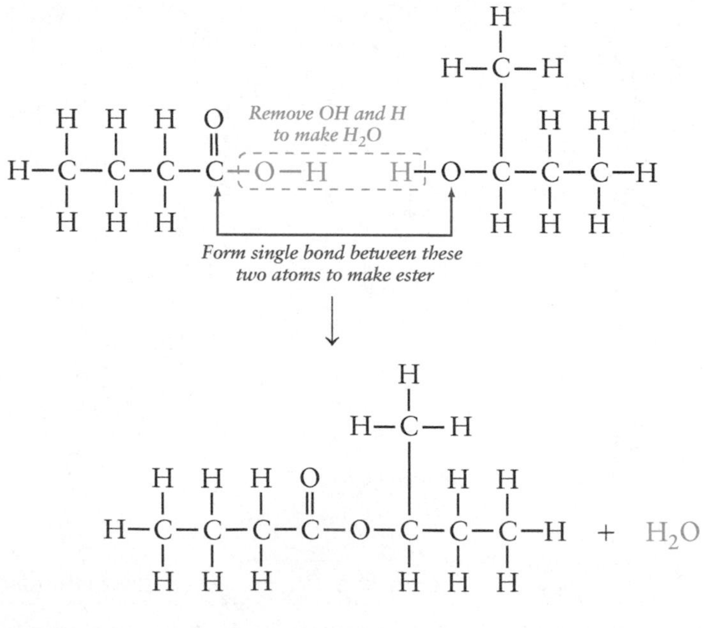

Remove OH and H to make H₂O

$$
\text{H}-\text{C}-\text{C}-\text{C}-\text{C}\!\!+\!\!\text{O}-\text{H} \quad \text{H}\!+\!\text{O}-\text{C}-\text{C}-\text{C}-\text{H}
$$

Form single bond between these two atoms to make ester

↓

$$
\underset{\underset{\text{H}\;\text{H}\;\text{H}}{|\;\;|\;\;|}}{\overset{\overset{\text{H}\;\text{H}\;\text{H}\;\text{O}}{|\;\;|\;\;|\;\;\|}}{\text{H}-\text{C}-\text{C}-\text{C}-\text{C}-\text{O}-\text{C}-\text{C}-\text{C}-\text{H}}} \quad + \quad \text{H}_2\text{O}
$$

10-7 Step 1: Write the carboxylic acid on the left with its OH group pointing to the right, and write the thiol on the right with its HS group pointing to the left.

$$
\underset{\underset{\text{H}}{|}}{\overset{\overset{\text{H}\quad\text{O}}{|\quad\|}}{\text{H}-\text{C}-\text{C}-\text{O}-\text{H}}} \qquad \text{H}-\text{S}-\diagdown
$$

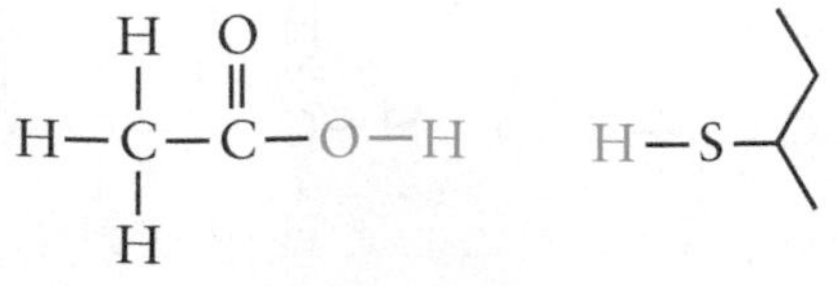

Step 2: Remove the elements of water, OH and H, from the carboxylic acid and the thiol, and then form a single bond between the two atoms from which the OH and H were removed.

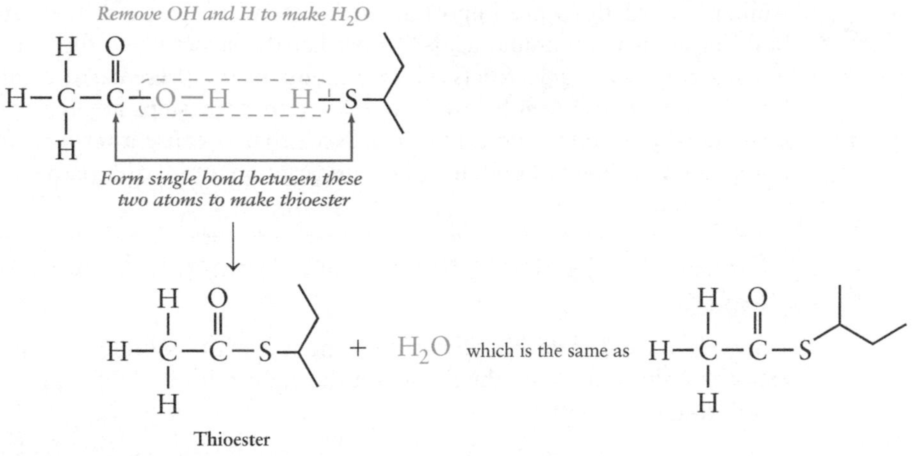

Thioester

PRACTICE EXERCISES

10 Write the structures of the products formed in the following reactions.

a.

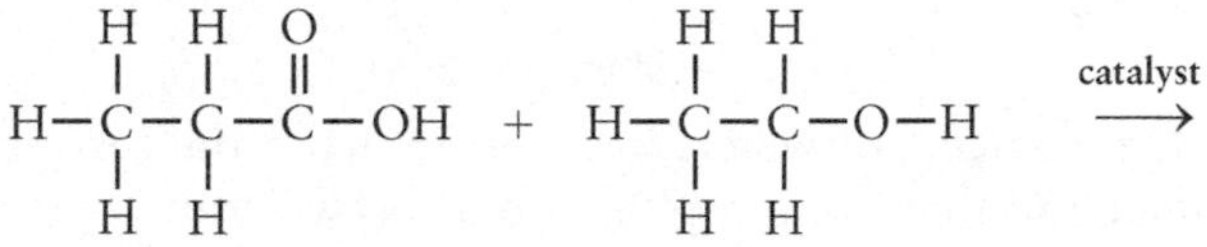

b.

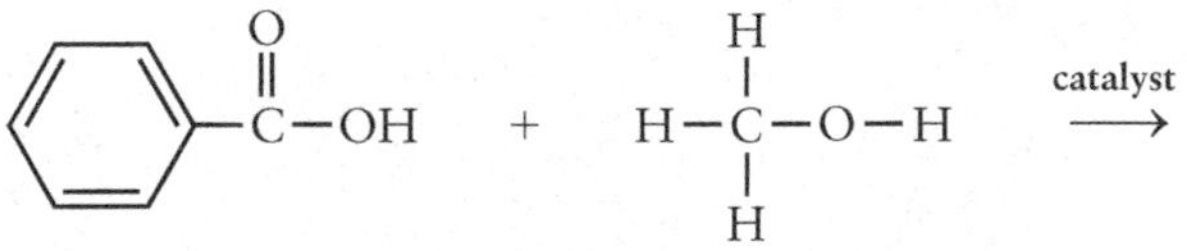

c.

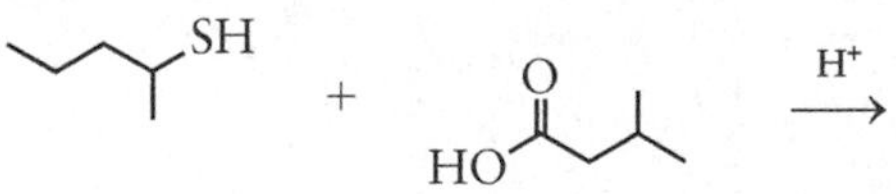

11 What is the structure of the ester formed when butanoic acid and ethanol undergo an esterification reaction? What familiar smell does this ester have?

12 What are the structures of the carboxylic acid and the alcohol that produce the following ester in an esterification reaction?

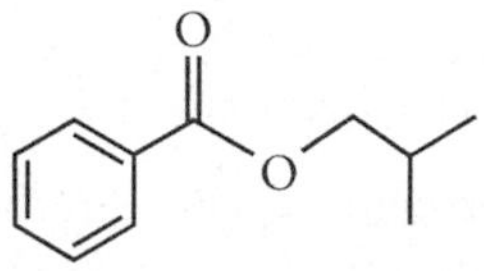

13 What is the structure of the carboxylic acid and the alcohol that upon heating would produce the ester shown below?

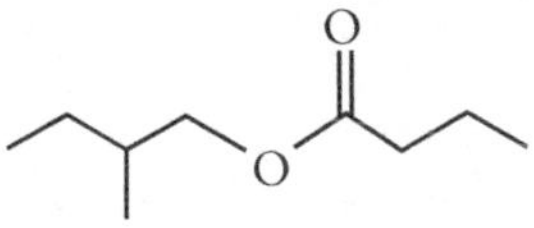

14 What is the structure of the thioester formed in the esterification reaction between $CH_3CH_2CH_2COOH$ and CH_3CH_2SH?

Amidation Reactions

Figure 10-11 shows an **amidation** reaction: the formation of an amide from a carboxylic acid and a primary or secondary amine, in the presence of a catalyst. Amidation reactions are important in biochemistry because they are used to build proteins from amino acids. To predict the structure of the amide formed when a carboxylic acid reacts with an amine in the presence of a catalyst, follow the guidelines given below. You will see that the steps are analogous to the formation of an ester and a thioester. Note that because a tertiary amine does not contain an N—H bond, it cannot undergo an amidation reaction.

Guidelines: Predicting the Products Formed in an Amidation Reaction

Step 1: Write the carboxylic acid on the left with its OH group pointing to the right, and write the amine on the right with its NH group pointing to the left.

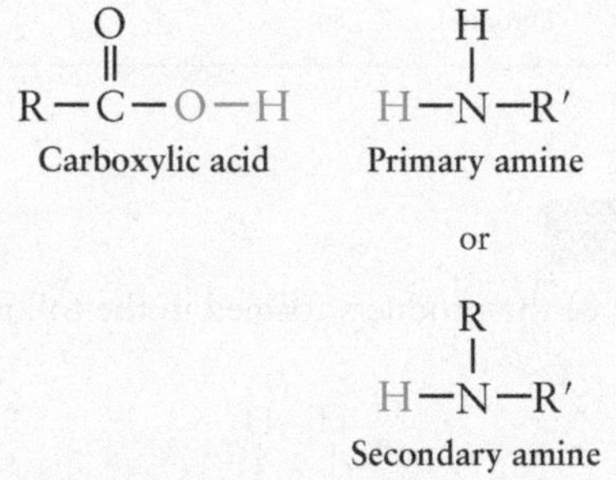

$$
\underset{\text{Carboxylic acid}}{\overset{\displaystyle\overset{O}{\parallel}}{R-C-O-H}} \qquad \underset{\text{Primary amine}}{\overset{\displaystyle\overset{H}{\mid}}{H-N-R'}}
$$

or

$$
\underset{\text{Secondary amine}}{\overset{\displaystyle\overset{R}{\mid}}{H-N-R'}}
$$

Step 2: Remove the elements of water, OH and H, from the carboxylic acid and the amine, and then form a single bond between the two atoms from which the OH and H were removed.

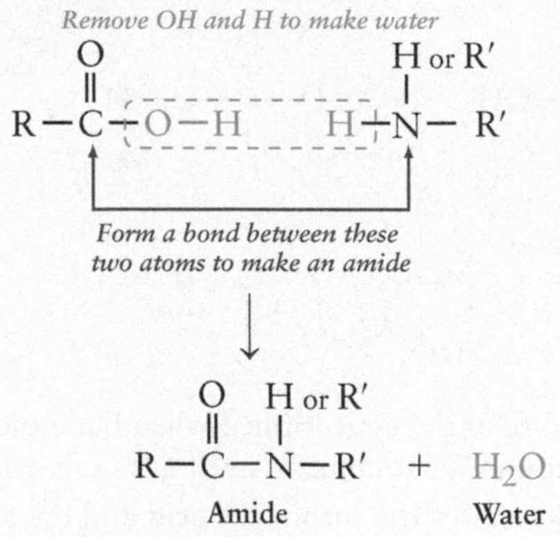

Remove OH and H to make water

$$
\underset{\substack{\text{Form a bond between these}\\\text{two atoms to make an amide}}}{R-\overset{\overset{O}{\parallel}}{C}-O-H \quad H-\overset{\overset{H \text{ or } R'}{\mid}}{N}-R'}
$$

$$\downarrow$$

$$
\underset{\text{Amide}}{R-\overset{\overset{O}{\parallel}}{C}-\overset{\overset{H \text{ or } R'}{\mid}}{N}-R'} \quad + \quad \underset{\text{Water}}{H_2O}
$$

WORKED EXERCISE   Amidation Reactions

10-8 Write the structure of the products formed in the following reaction between a carboxylic acid and an amine.

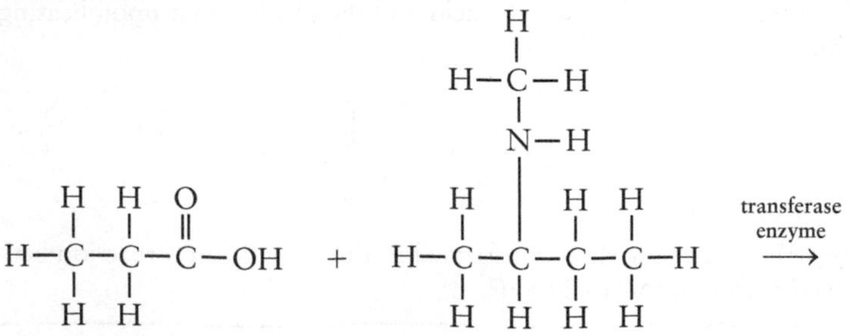

$$
\underset{}{\overset{\overset{\displaystyle H \;\; H \;\; O}{\mid \;\; \mid \;\; \parallel}}{H-C-C-C-OH}}\;\; + \;\; \underset{}{H-\overset{\overset{H}{\mid}}{C}-\overset{\overset{H-\overset{\overset{H}{\mid}}{C}-H}{\mid}}{C}-\overset{\overset{H}{\mid}}{C}-\overset{\overset{H}{\mid}}{C}-H} \quad \overset{\text{transferase enzyme}}{\longrightarrow}
$$

Solution

Step 1: Write the carboxylic acid on the left with its OH group pointing to the right, and write the amine on the right with its HN group pointing to the left.

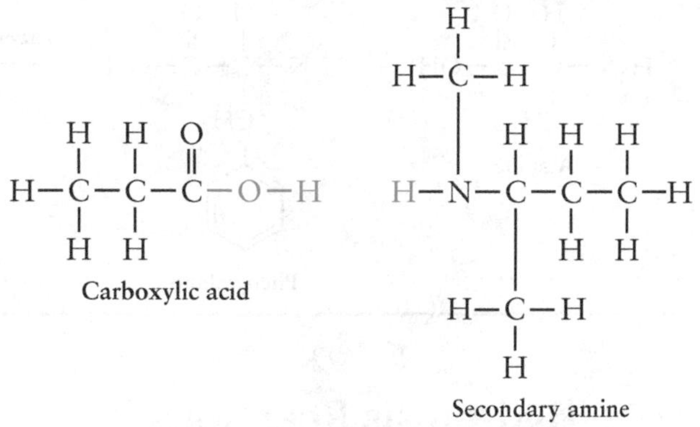

$$
\begin{array}{ccc}
& & \text{H} \\
& & | \\
& & \text{H—C—H} \\
& & | \\
\text{H H O} & & \text{H H H} \\
|\ \ |\ \ || & & |\ \ |\ \ | \\
\text{H—C—C—C—O—H} & \ \ \ \ & \text{H—N—C—C—C—H} \\
|\ \ | & & |\ \ \ \ |\ \ | \\
\text{H H} & & \text{H H} \\
\text{Carboxylic acid} & & \text{H—C—H} \\
& & | \\
& & \text{H} \\
& & \text{Secondary amine}
\end{array}
$$

Step 2: Remove the elements of water, OH and H, from the carboxylic acid and the amine, and then form a bond between the two atoms from which the H and OH were removed.

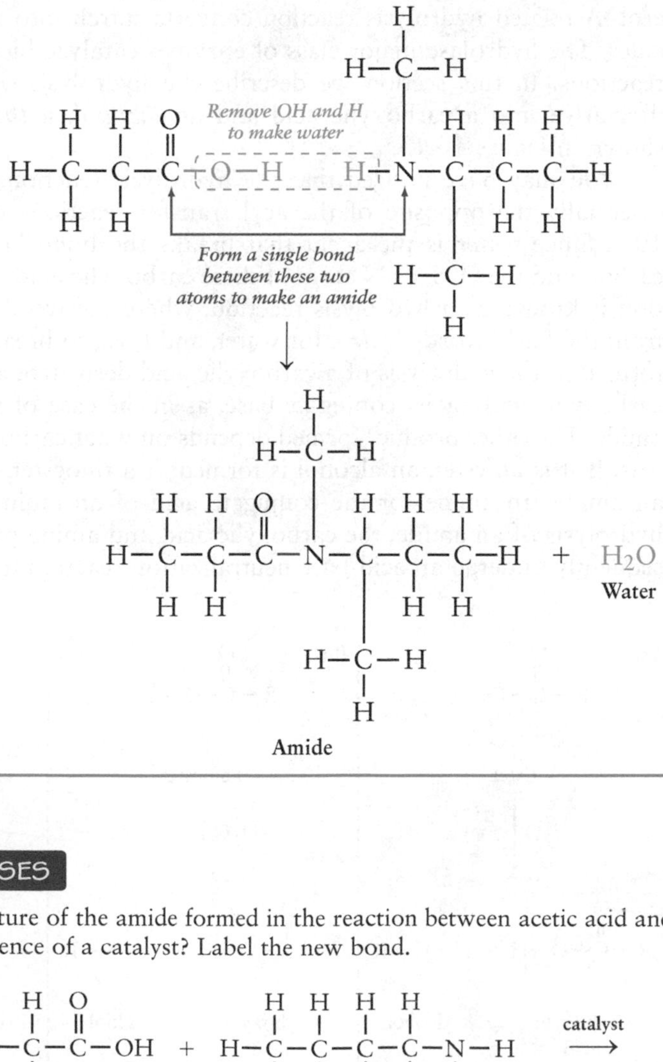

Remove OH and H to make water

$$
\text{H—C—C—C} {\overset{\text{O}}{}} \text{—O—H} \quad \text{H—N—C—C—C—H}
$$

Form a single bond between these two atoms to make an amide

$$
\begin{array}{c}
\text{H} \\
| \\
\text{H—C—H} \\
| \\
\text{H H O | H H H} \\
\text{H—C—C—C—N—C—C—C—H} \quad + \quad \text{H}_2\text{O} \\
\text{H H | H H} \quad\quad \text{Water} \\
\text{H—C—H} \\
| \\
\text{H} \\
\text{Amide}
\end{array}
$$

PRACTICE EXERCISES

15 What is the structure of the amide formed in the reaction between acetic acid and butan-1-amine in the presence of a catalyst? Label the new bond.

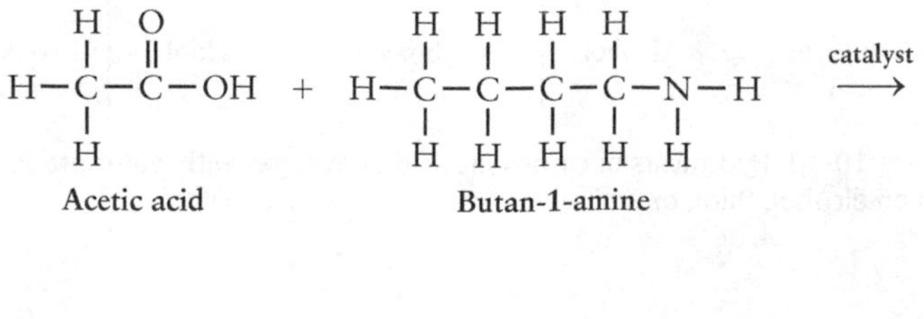

$$
\begin{array}{ccc}
\text{H O} & & \text{H H H H} \\
|\ \ || & & |\ \ |\ \ |\ \ | \\
\text{H—C—C—OH} & + & \text{H—C—C—C—C—N—H} & \xrightarrow{\text{catalyst}} \\
| & & |\ \ |\ \ |\ \ |\ \ | \\
\text{H} & & \text{H H H H H} \\
\text{Acetic acid} & & \text{Butan-1-amine}
\end{array}
$$

16 Two different products can be formed in the amidation reaction between two amino acids. The products are known as dipeptides. Write the structure of the two dipeptides that can be formed when alanine reacts with phenylalanine in the presence of a catalyst. What type isomers are the two dipeptides?

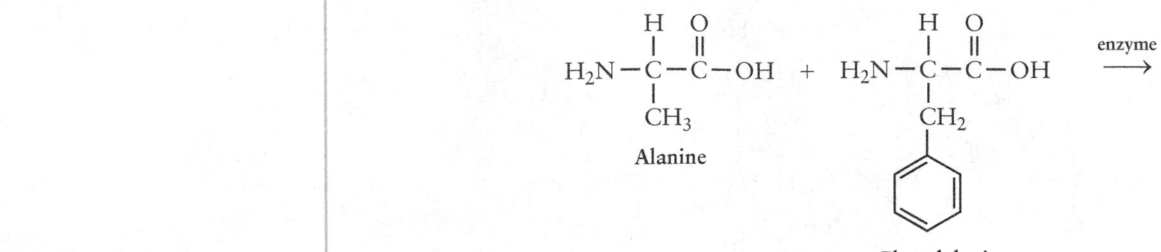

$$H_2N-\overset{\overset{\displaystyle H}{|}}{\underset{\underset{\displaystyle CH_3}{|}}{C}}-\overset{\overset{\displaystyle O}{||}}{C}-OH \quad + \quad H_2N-\overset{\overset{\displaystyle H}{|}}{\underset{\underset{\displaystyle CH_2}{|}}{C}}-\overset{\overset{\displaystyle O}{||}}{C}-OH \quad \overset{enzyme}{\longrightarrow}$$

Alanine   Phenylalanine

■ 10.3 Hydrolysis Reactions

Hydrolysis reactions are some of the most common reactions in both the laboratory and in biochemistry. It is the first step, known as digestion, in the metabolism of proteins into amino acids and fats into fatty acids and glycerol. A related hydrolysis reaction converts starch into many glucose molecules. The hydrolase major class of enzymes catalyze biochemical hydrolysis reactions. In this section we describe the hydrolysis of a carboxylic acid derivative into a carboxylic acid and an *alcohol*, a *thiol*, or an *amine* as shown in Figure 10-13.

You may have noticed that the hydrolysis reactions described here are essentially the opposite of the acyl transfer reactions described in section 10.2. Since *water* is the agent that breaks the bond between the carbonyl carbon and the O, S, or N atom of these carboxylic acid derivatives, the reaction is known as a **hydrolysis reaction**, where the word "hydrolysis" comes from the Latin words *hydro*, for water, and *lysis*, to break. Two products are formed in the hydrolysis of a carboxylic acid derivative and one of these is a carboxylic acid, or its conjugate base, as in the case of the hydrolysis of an amide. The other product formed depends on what carboxylic acid derivative it is. If it is an ester, an alcohol is formed; if a thioester, a thiol is formed; if an amide, an amine (or the conjugate acid of an amine) is formed. In the hydrolysis of an amide, the carboxylic acid and amine products formed subsequently undergo an acid-base neutralization reaction to form the conjugate

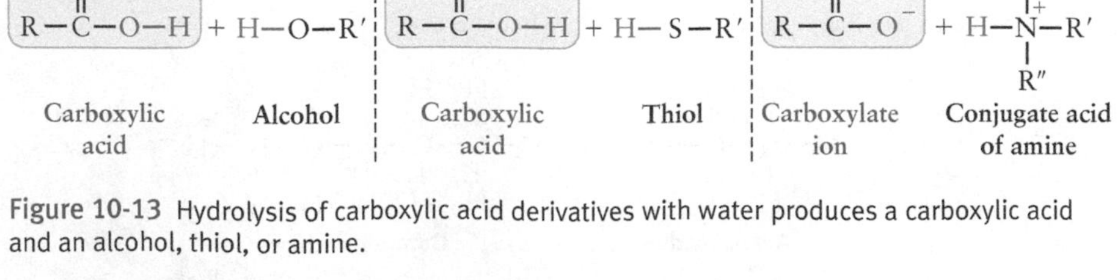

Figure 10-13 Hydrolysis of carboxylic acid derivatives with water produces a carboxylic acid and an alcohol, thiol, or amine.

base of the acid and the conjugate acid of the amine (Figure 10-13c). To help you predict the structure of the products formed in the hydrolysis of a carboxylic acid derivative, follow the guidelines below. To gain more familiarity with hydrolysis reactions, perform Molecular Modeling Exercise 10-1: Hydrolysis of an Ester on page 478.

Guidelines: Predicting the Hydrolysis Products of an Ester, Thioester, or Amide

Step 1: *Break* the single bond between the carbonyl carbon and the heteroatom (O, S, or N) of an ester, thioester, or amide, to produce two partial structures: an acyl group fragment (shown in dashed box) and a heteroatom—R fragment (shown in pink).

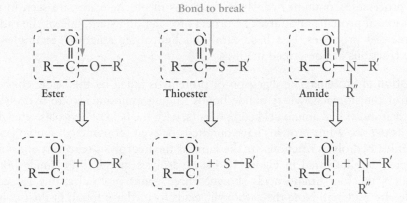

Step 2: Add an —OH group to the carbonyl carbon of the acyl group fragment, forming a carboxylic acid. The new bond is a C—O bond. Water supplies the OH group.

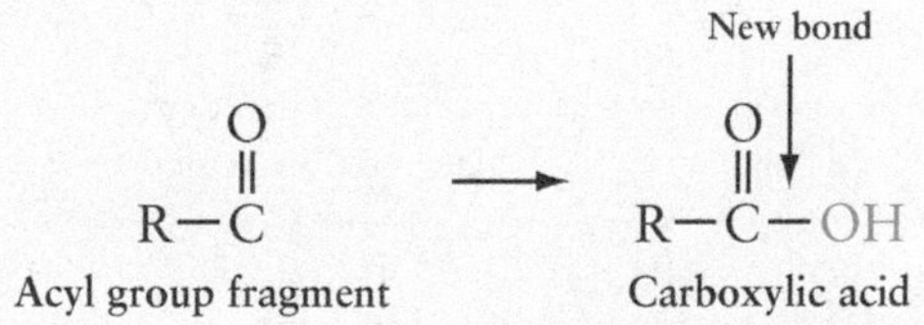

New bond

O   O
‖   ‖
R—C   ⟶   R—C—OH

Acyl group fragment   Carboxylic acid

Step 3: Add a hydrogen atom, H—, to the heteroatom—R′ fragment to form an alcohol, thiol, or amine. The new bond is a heteroatom—H bond. Water supplies the H atom.

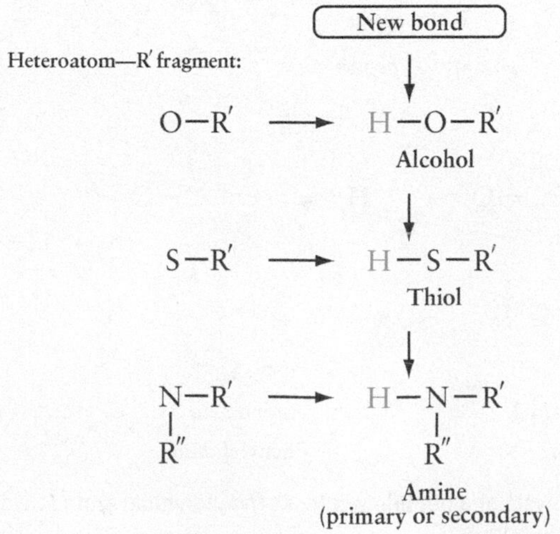

Heteroatom—R′ fragment: New bond

O—R′   ⟶   H—O—R′
Alcohol

S—R′   ⟶   H—S—R′
Thiol

N—R′   ⟶   H—N—R′
| |
R″   R″

Amine
(primary or secondary)

Step 4: In the case of an amide, the carboxylic acid and amine produced react in an acid-base neutralization reaction to form the conjugate base of the carboxylic acid and the conjugate acid of the amine.

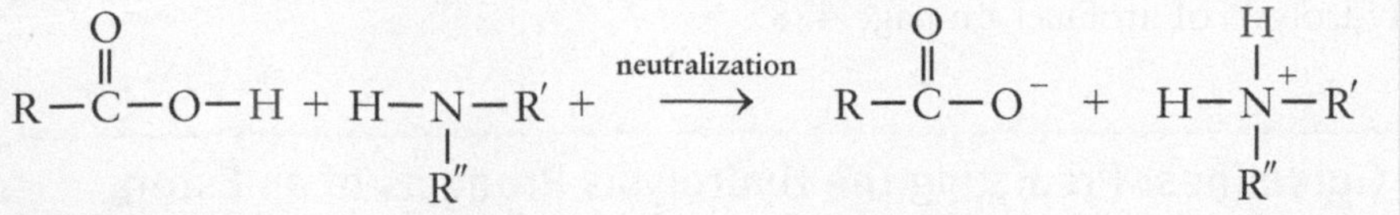

$$\underset{\text{Carboxylic acid}}{R-\overset{\displaystyle O}{\overset{\|}{C}}-O-H} + \underset{\text{Amine}}{H-\overset{\displaystyle }{\underset{\underset{R''}{|}}{\overset{|}{N}}}-R'} + \xrightarrow{\text{neutralization}} \underset{\text{Carboxylate ion}}{R-\overset{\displaystyle O}{\overset{\|}{C}}-O^-} + \underset{\text{Ammonium ion}}{H-\overset{\displaystyle H}{\underset{\underset{R''}{|}}{\overset{|}{\overset{+}{N}}}}-R'}$$

Applications of Hydrolysis Reactions

Hydrolysis reactions are common in biochemistry as well as in many industrial processes. Common hydrolysis reactions in biochemistry are seen in the digestion of proteins and fats. The process for making soap, one of the oldest commercial processes, is a base catalyzed hydrolysis reaction. Examples of these reactions are described below.

Digestion of Proteins The digestion of proteins is aided by the acidic environment of the stomach, where amide bonds joining amino acids are hydrolyzed into their individual amino acid components, with the help of hydrolase enzymes such as *pepsin*. While proteins have hundreds or even thousands of amino acids, the amide hydrolysis reactions are the same as the hydrolysis reactions of a small tetrapeptide, illustrated in Figure 10-14. Hydrolysis at the three amide bonds produces the four amino acids shown. Three water molecules, one for each amide, are required. Note the carboxylic acids formed are found in their conjugate base form ($RCOO^-$), and the amines, in their conjugate acid form (RNH_3^+).

Formation of Soaps Soap has been prepared for centuries by the hydrolysis of vegetable oils. Vegetable oils are triglycerides, which upon heating in the presence

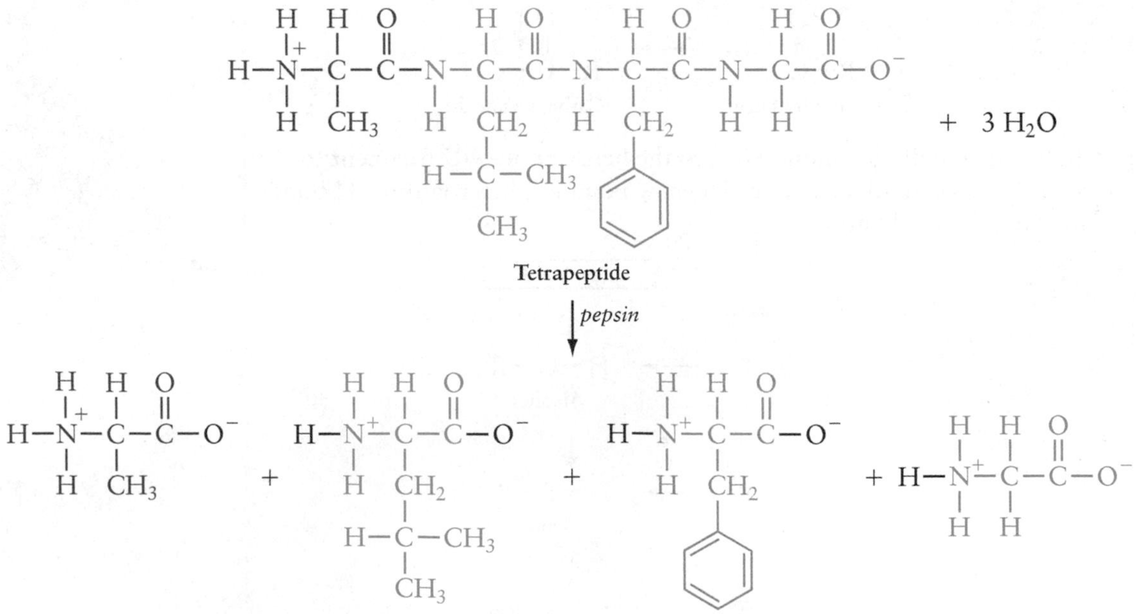

Figure 10-14 Hydrolysis of a peptide produces the individual amino acid components by breaking all the amide bonds.

of a strong base (HO⁻), produce glycerol and three fatty acids, in their conjugate base form, as shown:

Saponification

H₂C—O—$\overset{\overset{\textstyle O}{\|}}{C}$—(CH₂)₁₄CH₃

HC—O—$\overset{\overset{\textstyle O}{\|}}{C}$—(CH₂)₁₄CH₃ + 3 NaOH $\xrightarrow[\text{H}_2\text{O}]{}$ HC—OH + 3 Na⁺ ⁻O—$\overset{\overset{\textstyle O}{\|}}{C}$—(CH₂)₁₄CH₃

H₂C—O—$\overset{\overset{\textstyle O}{\|}}{C}$—(CH₂)₁₄CH₃

Triglyceride (Fat)   Glycerol   Fatty acid carboxylate salts
(soap)

Consequently, the hydrolysis of an ester in aqueous base came to be known as a *sapo*nification reaction, where "sapo" comes from the Greek word for soap. In the soap-making process, glycerol is removed at the end of the reaction, and the remaining fatty acid carboxylate salts are what we call "soap."

Digestion of Fats Hydrolysis is the first step in the digestion of fats, which occurs in the small intestine. The reaction is similar to a saponification reaction, except it occurs in a mildly basic environment and with the help of lipase enzymes. Lipases belong to the hydrolase major class of enzymes and catalyze the hydrolysis of fats into glycerol and fatty acids. The fatty acids produced are then transported to either muscle cells or fat cells (Figure 10-15). In muscle cells, fatty acids are metabolized to produce energy. In fat cells, fatty acids and glycerol are reesterified as described in section 10.2, to form triglycerides, which are stored in fat cells.

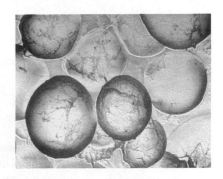

Figure 10-15 Triglycerides are stored in fat cells (adipocytes). [© ISM/Phototake]

WORKED EXERCISE | Hydrolysis of an Ester

10-9 Write the structure of the products formed in the following hydrolysis reaction.

$$\underset{\substack{|\\H}}{\overset{\substack{H\\|}}{H}-\overset{\substack{H\\|}}{\underset{\substack{|\\H}}{C}}-\overset{\substack{H\\|}}{\underset{\substack{|\\H}}{C}}-\overset{\substack{O\\\|}}{C}-O-\overset{\substack{H\\|}}{\underset{\substack{|\\H}}{C}}-H} \quad + \quad H_2O \quad \xrightarrow{\text{acid catalyst}}$$

Solution

The products are butanoic acid an methanol.

Step 1: *Break* the single bond between the carbonyl carbon and the O atom of the ester, to produce two partial structures: an acyl group and an O-R fragment.

Break this bond

$$H-\overset{\substack{H\\|}}{\underset{\substack{|\\H}}{C}}-\overset{\substack{H\\|}}{\underset{\substack{|\\H}}{C}}-\overset{\substack{H\\|}}{\underset{\substack{|\\H}}{C}}-\overset{\substack{O\\\|}}{C}-O-\overset{\substack{H\\|}}{\underset{\substack{|\\H}}{C}}-H$$

ester

↓

Partial structures: $H-\overset{\substack{H\\|}}{\underset{\substack{|\\H}}{C}}-\overset{\substack{H\\|}}{\underset{\substack{|\\H}}{C}}-\overset{\substack{H\\|}}{\underset{\substack{|\\H}}{C}}-\overset{\substack{O\\\|}}{C}$ + $O-\overset{\substack{H\\|}}{\underset{\substack{|\\H}}{C}}-H$

Acyl group fragment   Heteroatom—R′ fragment

Step 2: *Add* an —OH group to the carbonyl carbon of the acyl group fragment forming a carboxylic acid. The new bond is a C—O bond. Water supplies the OH group.

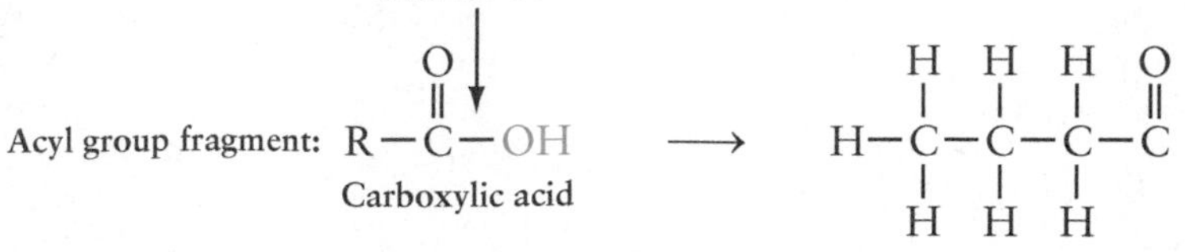

Step 3: *Add* a hydrogen atom, H—, to the oxygen of the heteroatom—R′ fragment to form an alcohol. The new bond is an O—H bond. Water supplies the H atom.

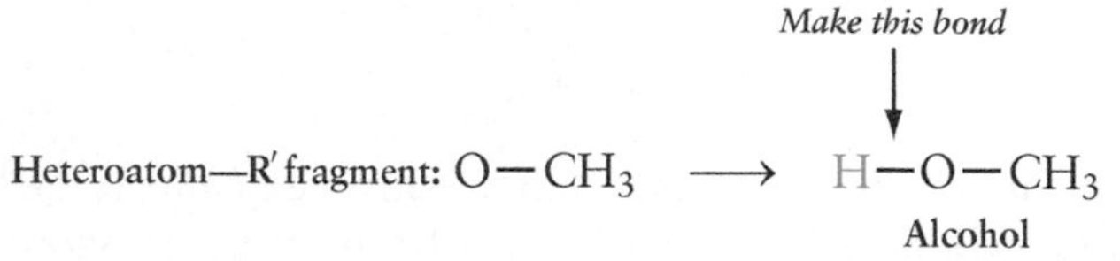

The products are butanoic acid and methanol.

WORKED EXERCISE   Hydrolysis of an Amide

10-10 Write the structure of the products formed in the following hydrolysis reaction.

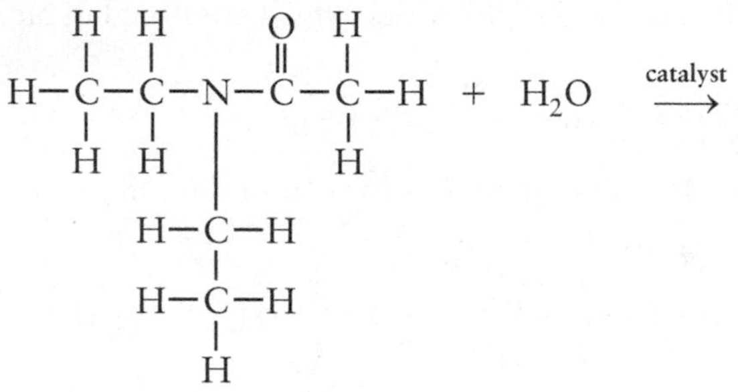

Solution

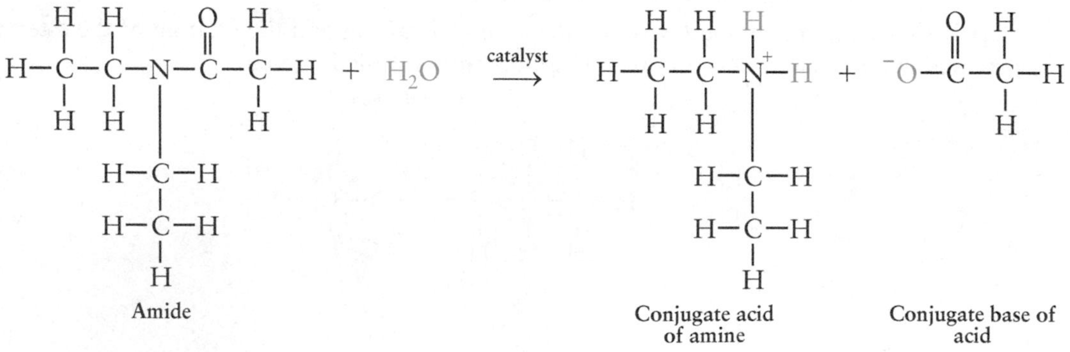

Remember that hydrolysis of the amide initially produces an amine and a carboxylic acid. These products then undergo a subsequent neutralization reaction to form the conjugate acid of the amine and the conjugate base of the carboxylic acid (a carboxylate ion) as shown—also known as an amine salt.

WORKED EXERCISE | Hydrolysis of a Thioester

10-11 Write the structures of the products formed in the hydrolysis of the thioester shown.

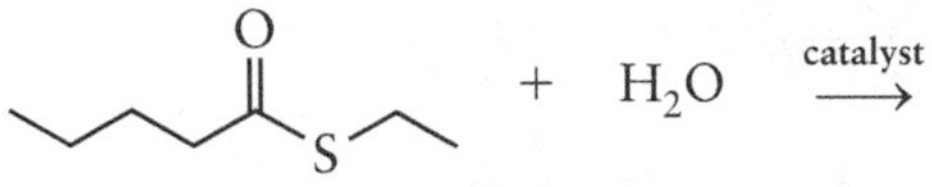

Solution

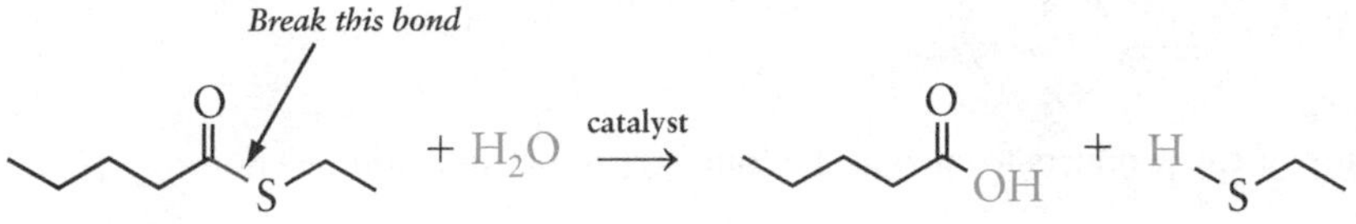

Break this bond

Thioester	Carboxylic acid	Thiol

PRACTICE EXERCISES

17 Write the structure of the products formed in the hydrolysis of the amide shown below, at physiological pH.

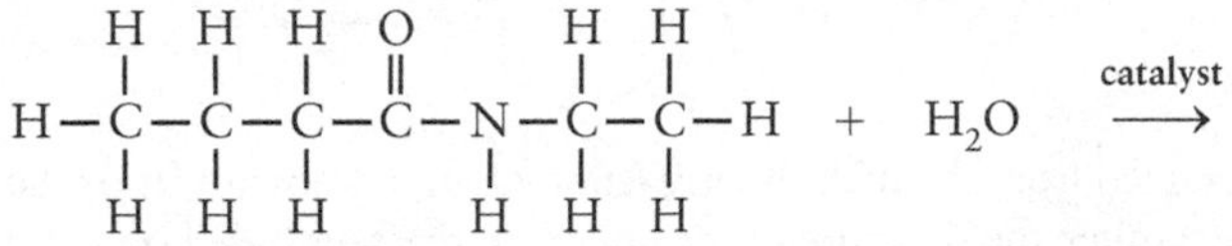

18 Identify the bond that is broken in a hydrolysis reaction of each carboxylic acid derivative. Write the structures of the products formed.

a.

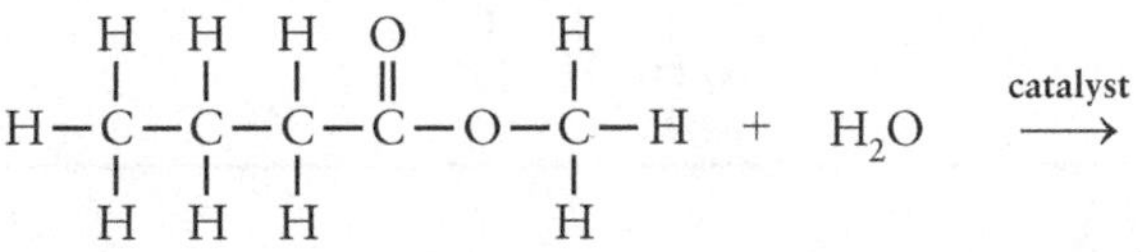

b.

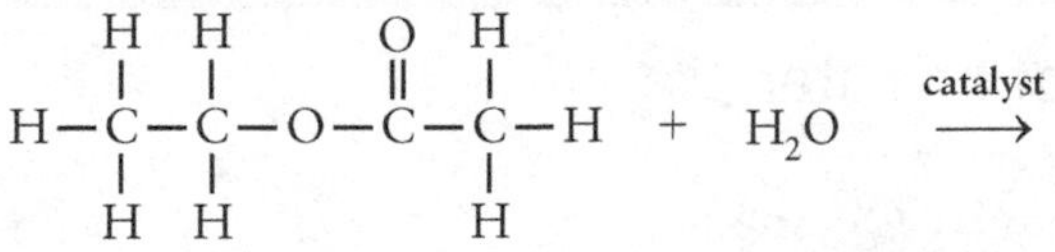

c.

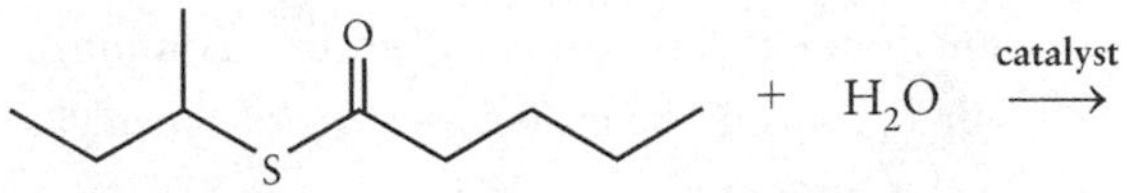

19 Hydrolysis of both ester functional groups in heroin produces morphine. What is the name of the other product formed, which is detected by drug-sniffing dogs, who are trained to detect its strong odor. *Hint:* It is found in most kitchens. Indicate which bonds are broken in these hydrolysis reactions. Why are two water molecules required to convert heroin to morphine?

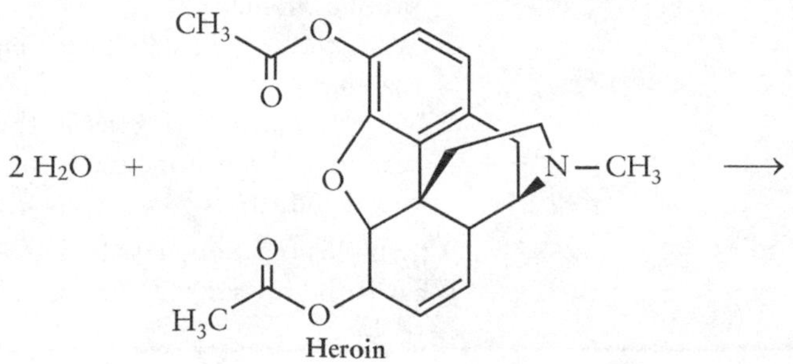

Heroin

20 How do the products differ in structure for the following two hydrolysis reactions?

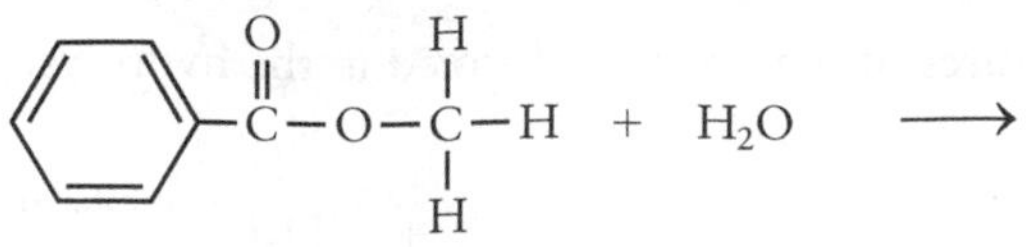

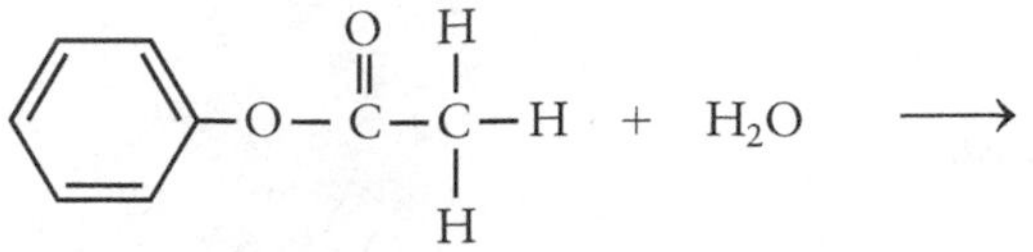

$$\text{C}_6\text{H}_5-\overset{\overset{\displaystyle O}{\|}}{\text{C}}-\text{O}-\overset{\overset{\displaystyle H}{|}}{\underset{\underset{\displaystyle H}{|}}{\text{C}}}-\text{H} \;+\; \text{H}_2\text{O} \;\longrightarrow$$

$$\text{C}_6\text{H}_5-\text{O}-\overset{\overset{\displaystyle O}{\|}}{\text{C}}-\overset{\overset{\displaystyle H}{|}}{\underset{\underset{\displaystyle H}{|}}{\text{C}}}-\text{H} \;+\; \text{H}_2\text{O} \;\longrightarrow$$

21 Write the structure of the products formed in the following saponification reaction:

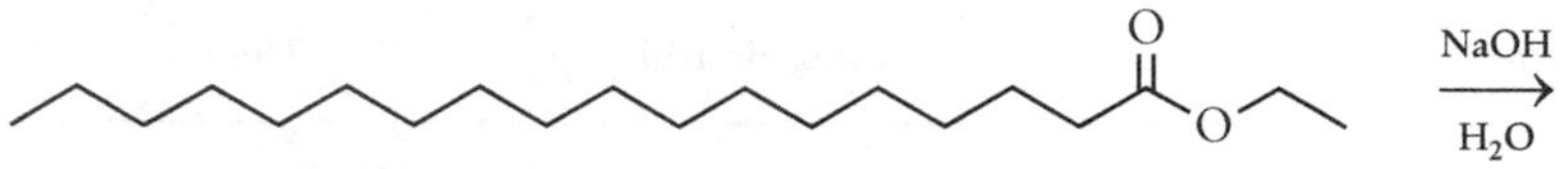

$$\xrightarrow[\text{H}_2\text{O}]{\text{NaOH}}$$

22 Cyclic esters are known as lactones. Write the structure of the product formed in the hydrolysis of the lactone shown. *Hint:* Only one product is formed, and it does not contain a ring.

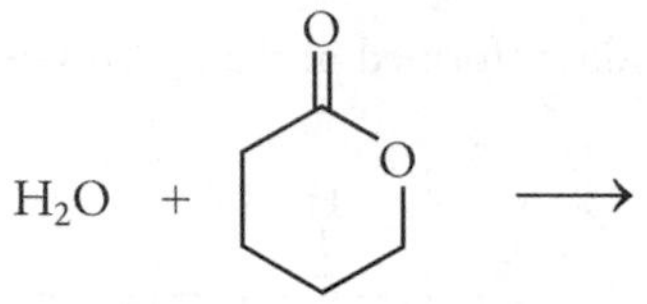

$$\text{H}_2\text{O} \;+\; \longrightarrow$$

23 The products obtained from the hydrolysis reaction of an unknown carboxylic acid derivative are shown below. Determine the structure of the unknown reactant based on the structure of the products.

$$+ \;\text{H}_2\text{O} \;\longrightarrow\; \text{H}-\overset{\overset{\displaystyle H}{|}}{\underset{\underset{\displaystyle H}{|}}{\text{C}}}-\overset{\overset{\displaystyle H}{|}}{\underset{\underset{\displaystyle H}{|}}{\text{C}}}-\overset{\overset{\displaystyle H}{|}}{\underset{\underset{\displaystyle H}{|}}{\text{C}}}-\overset{\overset{\displaystyle O}{\|}}{\text{C}}-\text{O}^{-} \;+\; \text{CH}_3\text{CH}_2\overset{+}{\text{N}}\text{H}_3$$

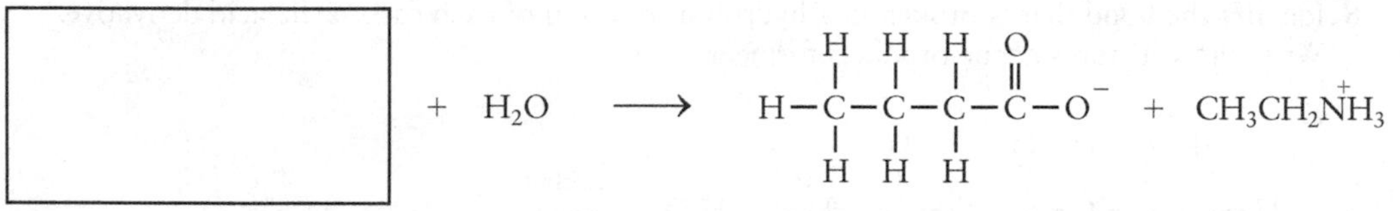

Using Molecular Models 10-1 Hydrolysis of an Ester

1. Obtain 4 black carbon atoms, 10 light-blue hydrogen atoms, 3 red oxygen atoms, 14 straight bonds, and 2 bent bonds.
2. Make a model of ethyl acetate, $\text{CH}_3\text{CO}_2\text{CH}_2\text{CH}_3$. Answer Inquiry Questions 5–6.

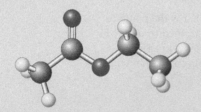

3. Make a model of a water molecule.
4. Simulate a *hydrolysis* reaction between ethyl acetate and water using only the atoms from the two models you made. Answer Inquiry Questions 7–8.

Inquiry Questions

5. Locate the carbonyl group in ethyl acetate. What functional group does ethyl acetate contain?
6. What atoms in the model are part of the acyl group?
7. In the simulated hydrolysis reaction:
   a. What bond did you break in ethyl acetate? What bond did you break in water?
   b. What two new bonds did you make?
   c. What functional group is present in each of the product models?
   d. Write the complete balanced equation for the reaction.
   e. Were any atoms left over at the end of the reaction? What principle from chapter 4 does this demonstrate?
8. Why is this reaction classified as a hydrolysis? What is the reverse reaction called?

10.4 Hydration and Dehydration Reactions

We have seen water used as a reactant in hydrolysis reactions where it splits a carboxylic acid derivative into a carboxylic acid and an alcohol, a thiol, or an amine. Hydration reactions are another class of reactions that use water as a reactant. *In a hydration reaction the elements of water, H and OH, add to an alkene to produce an alcohol.*

The reverse of a hydration reaction, elimination of the elements of water, H and OH, from an alcohol to form an alkene is known as a **dehydration reaction**. Hydration and dehydration reactions are catalyzed by the major class of enzymes known as lyases.

Hydration

In a **hydration reaction** water reacts with an alkene to produce an alcohol. The H and OH atoms from a molecule of water form bonds to each of the carbon atoms of the alkene and the carbon-carbon double bond becomes a carbon-carbon single bond:

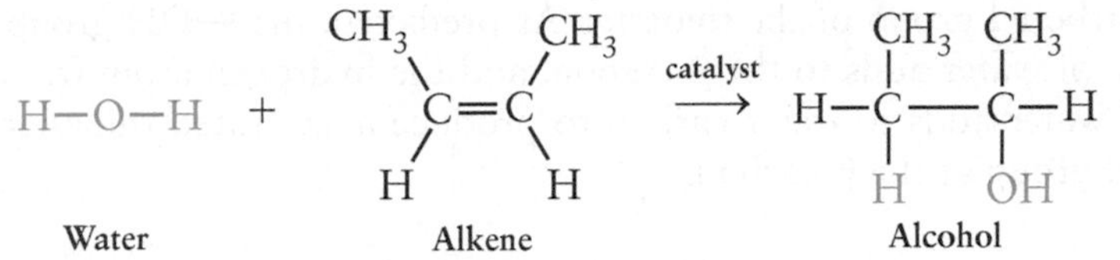

As you can see, water is incorporated into the molecular structure of the product resulting in the formation of an alcohol (ROH).

In the case of an unsymmetrical alkene, the OH from a molecule of water adds to the double-bond carbon with fewer hydrogen atoms, and the H atom from a molecule of water adds to the double-bond carbon with more hydrogen atoms. For example, hydration of propene forms propan-2-ol, not propan-1-ol, as shown below:

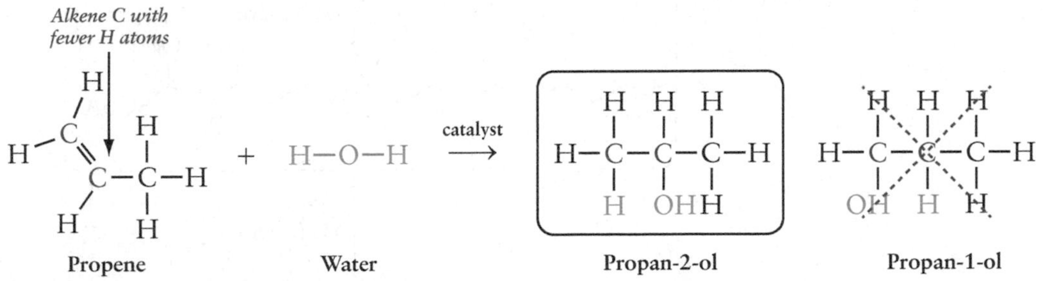

When a reaction forms one structural isomer (propan-2-ol) preferentially over another structural isomer (propan-1-ol) as in the reaction above, it is known as a **regioselective reaction**. Hydration of an alkene is a regioselective reaction, known as **Markovnikov's rule**.

Biochemical Hydration Reactions In biochemical hydration reactions, a carbonyl group is usually adjacent to the carbon-carbon double bond, which directs the outcome of the reaction. The —H atom from water adds to the double-bond carbon *closer to the carbonyl group*, known as the α carbon; and the —OH group from water adds to the double-bond carbon *farther* *from the carbonyl group*, known as the β carbon. The carbonyl group itself is unchanged in the biochemical hydration reaction.

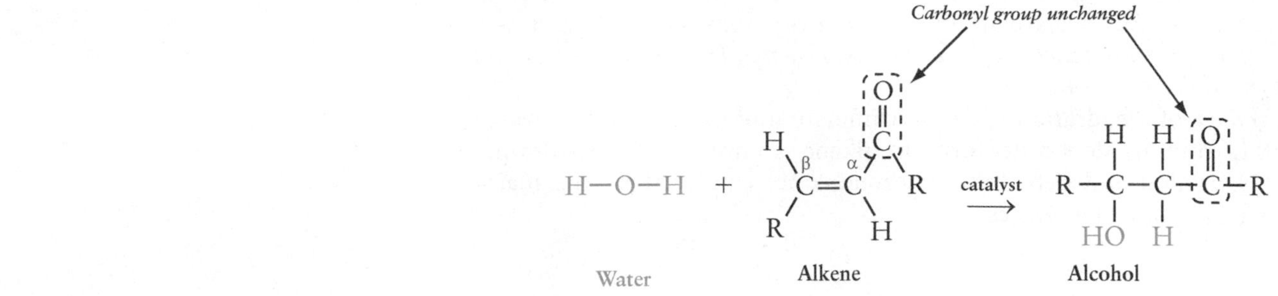

A biochemical hydration reaction is seen in the biochemical pathway that occurs in muscle and liver cells when fatty acids are metabolized for energy. This metabolic pathway supplies hibernating animals with all the energy they need during the long winter months of hibernation. A lyase enzyme catalyzes the hydration reaction that occurs at the carbon-carbon double bond adjacent to the carbonyl group of the thioester. As predicted, the —OH group from a molecule of water adds to the β carbon, and the hydrogen atom from a molecule of water adds to the α carbon to produce a saturated thioester with a hydroxyl group at the β carbon:

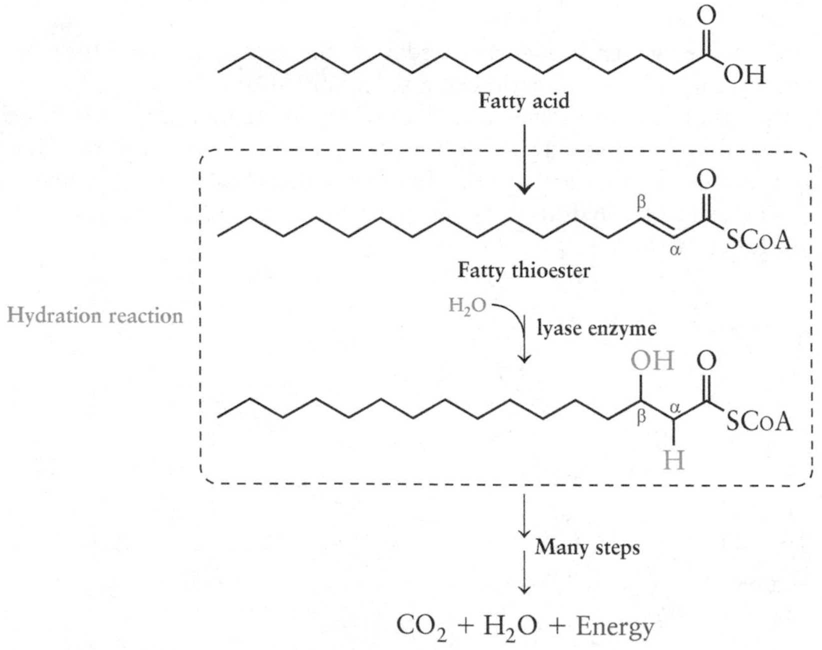

Figure 10-16 provides a summary of reactions where water is a reactant, described in this chapter.

Dehydration

The loss of a water molecule from an alcohol to produce an alkene is known as a **dehydration reaction**, essentially the reverse of a hydration reaction. In a dehydration reaction —OH and —H are *eliminated* from the substrate to form a molecule of water, H_2O, and a carbon-carbon double bond is formed

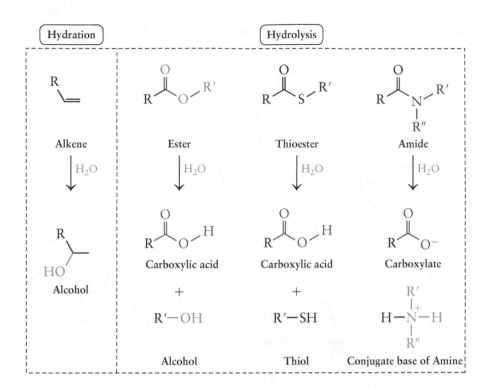

Figure 10-16 Summary of hydrolysis and hydration reactions described in this chapter.

between the carbon atoms that formerly had the —OH group and —H atom, as illustrated in the dehydration of propan-2-ol to form propene:

$$H-C-C-C-H \xrightarrow{\text{catalyst}} \text{Propene} + H-O-H$$

Propan-2-ol   **Propene**   **Water**

If the substrate in a dehydration reaction can lead to two possible alkene products (structural isomers), the alkene with fewer hydrogen atoms on the double bond is formed preferentially over the alkene with more hydrogen atoms, an observation known as **Zaitsev's rule**. Zaitsev's rule is illustrated in the dehydration of butan-2-ol, which produces mainly *trans*-but-2-ene and not but-1-ene:

Butan-2-ol   →(acid catalyst)   **Zaitsev product** / *trans*-But-2-ene (two H's on double bond) + But-1-ene (three H's on double bond) + H_2O

Biochemical Dehydration In biochemical dehydration reactions, the hydroxyl group is usually located β to a carbonyl group. Dehydration then occurs with elimination of an α hydrogen and the β hydroxyl group, forming water and an α-β unsaturated carbonyl compound. The carbonyl group itself is unchanged, but necessary for the biochemical dehydration, as shown in the example below.

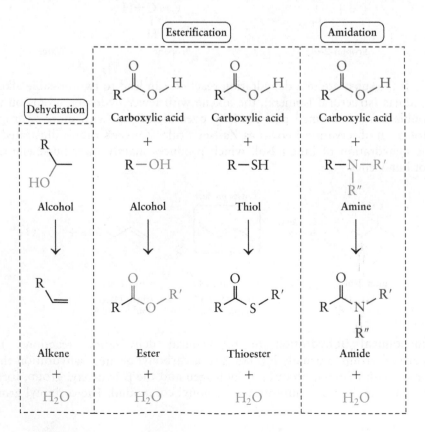

β-Hydroxy ketone   α–β-Unsaturated ketone

In many organisms, and our cells, glucose is metabolized to a molecule known as pyruvate, in a 10-step biochemical sequence known as glycolysis. The ninth step of this biochemical pathway is a dehydration reaction catalyzed by the enzyme *enolase*.

Note that the hydroxyl group at the β carbon and a hydrogen atom on the α carbon are eliminated to form the α-β unsaturated carboxylic acid. The elimination of water is shown next to a curved arrow, using the common biochemical convention. Notice also that the rest of the molecule is unchanged. You can see once again how important it is to focus on the functional group involved in the reaction.

Figure 10-17 provides a summary of reactions where water is eliminated (a product), described in this chapter.

Figure 10-17 Summary of reactions described in this chapter in which water is a product; water is eliminated.

WORKED EXERCISES | Predicting the Product of a Hydration or Dehydration Reaction

10-12 Predict the main product formed in the following reaction.

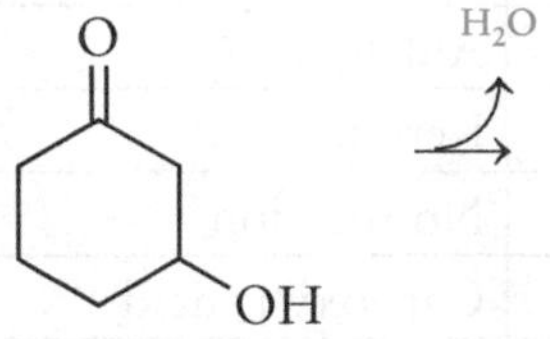

10-13 Why is a hydration reaction not an oxidation or reduction?

Solutions

10-12

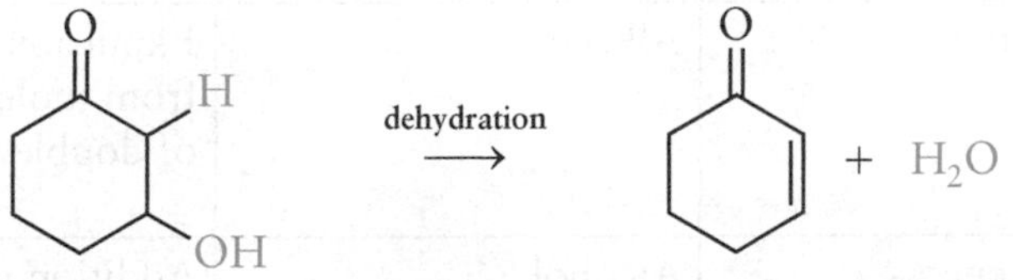

Remember to remove the β OH group and an α hydrogen when deciding where to place the double bond in the product.

10-13 In a hydration reaction the number of oxygen atoms *and* the number of hydrogen atoms increases, therefore it is neither an oxidation nor a reduction.

PRACTICE EXERCISE

24 In a step of an important biochemical pathway known as the citric acid cycle, the reaction shown below occurs.

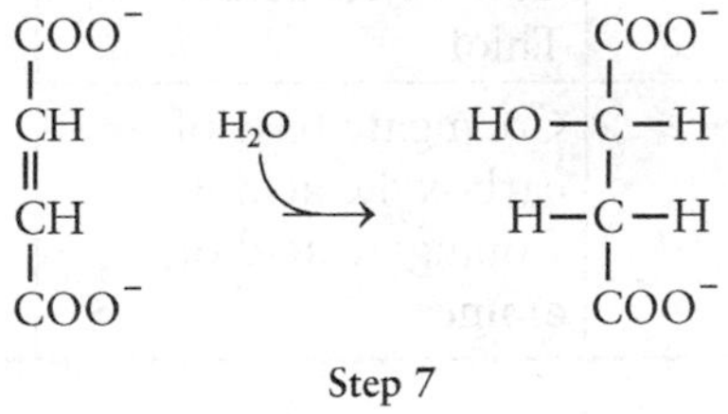

Step 7

a. Is this a hydration or a dehydration reaction?
b. Indicate the OH group in the product structure that came from a molecule of water.
c. What functional groups are unchanged in the reaction?
d. How were you able to determine whether this was a hydration or dehydration reaction?

In this chapter you have seen that biochemistry at its core *is* organic chemistry. The types of reactions described in this chapter will be seen throughout the remaining chapters of the text as you learn about the role of each of the biomolecules in metabolism. Table 10-6 summarizes the types of organic reactions seen in biochemistry that were described in this chapter.

TABLE 10-6 Review of Reactions Discussed in this Chapter

Class of Reaction	Type of Reaction	Reactant Functional Group	Product Functional Group	Change Observed
Oxidation–Reduction (Redox)	Oxidation [O] ⟶	Alkane	Alkene	Decrease in hydrogen or increase in oxygen; loss of electrons.
		1° Alcohol	Aldehyde	
		2° Alcohol	Ketone	
		3° Alcohol	No reaction	
		Aldehyde	Carboxylic acid	
	Reduction [H] ⟶	Alkene	Alkane	Increase in hydrogen or decrease in oxygen; gain of electrons.
		Aldehyde	1° Alcohol	
		Ketone	2° Alcohol	
		Carboxylic acid	Aldehyde	
Dehydration–Hydration	Dehydration H_2O	Alcohol	Alkene	Expulsion of H and OH from molecule; formation of double bond.
	Hydration H_2O	Alkene	Alcohol	Addition of H and OH to molecule; loss of double bond.
Acyl Group Transfer	Amidation	Carboxylic acid + Amine	Amide + Water	Acyl group migrates from O to N: water released.
	Esterification	Carboxylic acid + Alcohol	Ester + Water	Acyl group migrates from O to O or O to S; H_2O released.
		Carboxylic acid + Thiol	Thioester + Water	
Hydrolysis		Ester	Carboxylic acid + Alcohol	Water is a reactant.
		Thioester	Carboxylic acid + Thiol	
		Amide	Conjugate base of carboxylic acid + Conjugate acid of amine	

Chemistry in Medicine Phenylketonuria—A Metabolic Disorder

You've seen the warning label on many artificially sweetened products: "PHENYLKETONURICS: CONTAINS PHENYLALANINE" (Figure 10-18). What does this mean, and what is a phenylketonuric? Phenylketonuria (PKU) is a rare metabolic disorder that affects about 1 in 10,000 infants born every year. PKU is an autosomal recessive trait, meaning that two parents that are carriers of the trait but do not themselves have the disease have a 25 percent chance of having a child with phenylketonuria. Two percent of the population are carriers of the gene. In

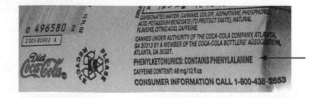

Figure 10-18 Diet soft drinks containing aspartame carry a warning for phenylketonurics, who are unable to metabolize the phenylalanine produced when aspartame is hydrolyzed in the stomach. [Bianca Moscatelli/W.H. Freeman]

the United States, babies are screened for PKU with a simple blood spot card within three days of birth. If left untreated, PKU leads to mental retardation and a much shortened life expectancy. It is estimated that 1 percent of patients in mental institutions have phenylketonuria.

PKU is caused by a deficiency in the enzyme *phenylalanine hydroxylase (PAH)*, an *oxidoreductase* enzyme that catalyzes the first step in the metabolism of the amino acid phenylalanine. Without *PAH*, phenylalanine accumulates, and it is eventually converted to phenylpyruvate by what is normally a seldom used alternative metabolic pathway. While the biochemical basis for mental retardation is still not clear, newer evidence suggests that it is the build-up of phenylalanine itself that is neurotoxic.

Catabolism of Phenylalanine

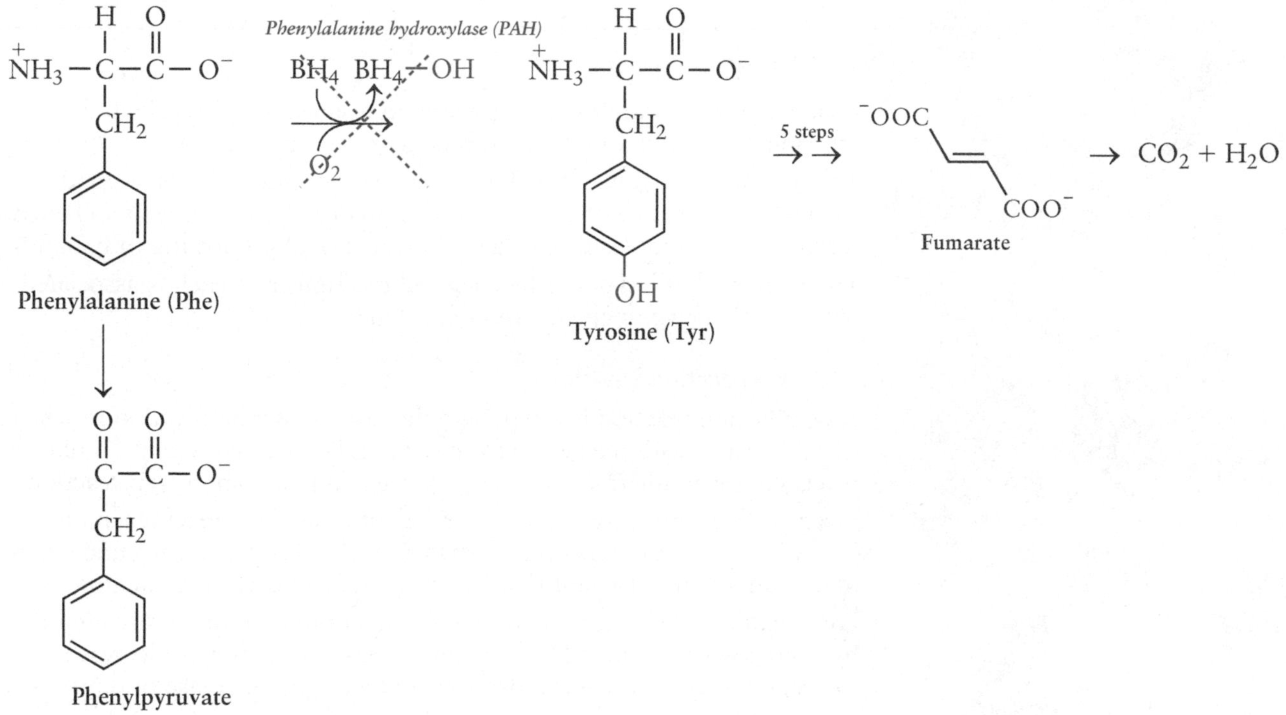

PAH is an enzyme that uses BH_4 (tetrahydrobiopterin) as a coenzyme. Think of this coenzyme as similar in function to FAD and NAD^+. *PAH* and its coenzyme BH_4 catalyze the oxidation of phenylalanine (Phe) to tyrosine (Tyr), another amino acid. Oxygen is also a necessary reactant, as indicated by the other curved arrow. Oxygen oxidizes the substrate and the coenzyme. When a person is deficient in *PAH*, they cannot synthesize tyrosine from phenylalanine, allowing toxic levels of phenylalanine to accumulate in the blood. High levels of phenylalanine are an indicator of PKU. A normal blood phenylalanine level is about 1 mg/dL. Individuals with PKU have phenylalanine levels ranging from 6 mg/dL to 80 mg/dL.

The label on a can of diet soda warns that it contains aspartame (Nutrasweet), and should not be consumed by phenylketonurics. Aspartame is hydrolyzed in the stomach, to the amino acids aspartic acid and phenylalanine. It is the phenylalanine produced in this hydrolysis reaction that cannot be tolerated by phenylketonurics (**Figure 10-19**).

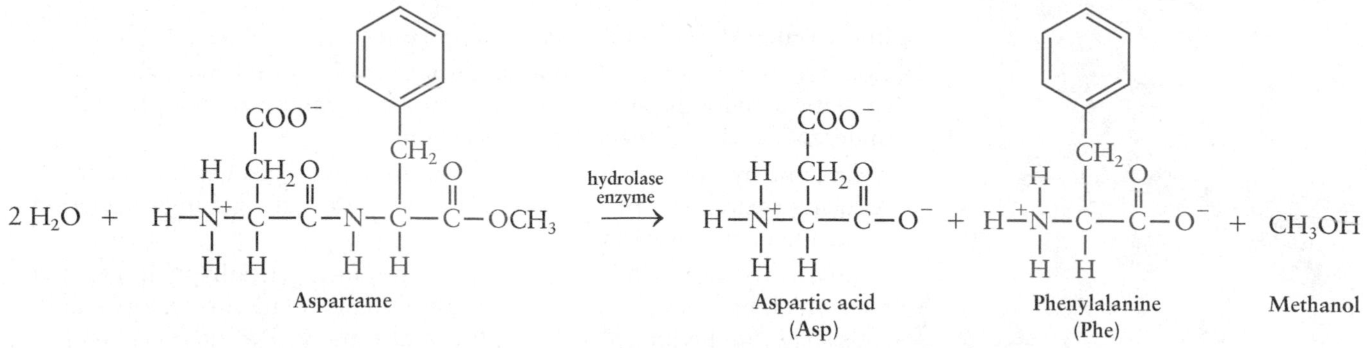

Figure 10-19 The hydrolysis of aspartame produces the amino acids aspartic acid and phenylalanine, and methanol.

Until the 1930s, before the biochemistry of PKU was understood, people with this metabolic disorder became mentally incapacitated and remained so for the rest of their lives. Today, if diagnosed at birth, PKU can be managed through a strict diet, which must be followed for life. The goal is to maintain blood phenylalanine levels between 2 and 10 mg/dL. This is accomplished by severely restricting, or altogether eliminating, foods that contain phenylalanine. Such foods include beef, chicken, fish, and dairy products. In addition, any diet foods that contain aspartame must be avoided. People with PKU must also take tyrosine supplements because they cannot synthesize this amino acid.

Summary

- Functional groups are the reactive sites in a molecule. Typically only one functional group in a molecule reacts in any given reaction.
- The benefit of learning a few functional groups and their corresponding chemical properties is that you can then predict the outcome of a chemical reaction for any molecule based on the functional groups that it contains.
- There are six basic types of biochemical reactions and each is associated with one of the six major classes of enzymes.

Oxidation–Reduction Reactions

- A combustion reaction is a type of oxidation–reduction reaction in which $C_xH_yO_z$ reacts with oxygen to form carbon dioxide and water. Combustion reactions are exothermic, so energy is released in a combustion reaction.
- Electrons are transferred from one reactant to another reactant in an oxidation–reduction reaction. The reactant that loses electrons undergoes oxidation and the reactant that gains electrons undergoes reduction.
- For organic molecules, oxidation and reduction are more easily identified by noting whether there has been an increase in hydrogens/decrease in oxygens—a reduction—or a decrease in hydrogens/increase in oxygens— an oxidation.
- The reactant that is reduced is known as the oxidizing agent, generically specified as [O].
- The reactant that is oxidized is known as the reducing agent, generically specified as [H].
- In a biological cell, oxidation–reduction reactions require the participation of a coenzyme like NAD^+/NADH or FAD/$FADH_2$.
- In the cell, NAD^+ oxidizes primary alcohols to aldehydes, aldehydes to carboxylic acids, and secondary alcohols to ketones.
- In the cell, NADH reduces carboxylic acids to aldehydes, aldehydes to primary alcohols, and ketones to secondary alcohols.
- In the cell, FAD oxidizes alkanes to alkenes.
- In the cell, $FADH_2$ reduces alkenes to alkanes.
- Coenzymes are required by oxidoreductase enzymes to achieve their catalytic activity. Coenzymes are non-protein organic molecules that undergo a chemical change in the reaction.
- A catalytic hydrogenation is a reduction using hydrogen gas (H_2) as the reducing agent and a metal catalyst like Pd or Pt. Catalytic hydrogenation of an alkene produces and alkane.
- Catalytic hydrogenation of vegetable oils produces partially hydrogenated fats, which have fewer double bonds, resulting in a fat with a more solid consistency and a longer shelf life. It can also produce some *trans* fats as a result of some *cis* double bonds isomerizing to *trans* double bonds.

- Antioxidants prevent the harmful oxidation of substances in the cells by reducing free radicals or preventing their formation. Antioxidants are reducing agents.

Group Transfer Reactions: Esterification and Amidation Reactions

- Acyl group transfer reactions convert carboxylic acids and alcohols, thiols, or amines into their derivatives—esters, thioesters, and amides, respectively.
- Acyl transfer reactions are also known as esterification and amidation reactions.
- An acyl group is a carbonyl group and its attached R group.
- An ester can be formed from a carboxylic acid and an alcohol, in the presence of a catalyst, in an esterification reaction. Water is also a product.
- Many esters are fragrant and found in fruits, flowers, and perfumes.
- Fats, also known as triglycerides, are triesters formed when glycerol reacts with three fatty acid molecules.
- Cholesterol reacts with a fatty acid to form a cholesterol ester.
- A thioester is formed when a carboxylic acid and a thiol react in a thioesterification reaction. Water is also a product.
- An amide is formed when a carboxylic acid and an amine react in an amidation reaction. Water is also a product. Amidation reactions are important in building proteins from amino acids.

Hydrolysis Reactions

- Carboxylic acid derivatives (esters, thioesters, and amides) react with water in the presence of a catalyst in a hydrolysis reaction to produce a carboxylic acid and either an alcohol, thiol, or amine.
- Hydrolysis of an ester produces a carboxylic acid and an alcohol.
- Hydrolysis of a thioester produces a carboxylic acid and a thiol.
- Hydrolysis of an amide initially produces a carboxylic acid and an amine, which go on to react in a neutralization reaction to form the conjugate base of the carboxylic acid and the conjugate acid of the amine, a salt.
- Digestion of proteins occurs in the acidic environment of the stomach, where amide bonds are hydrolyzed into their individual amino acid components, with the aid of hydrolase enzymes like pepsin.
- When an ester is heated in the presence of hydroxide ion (OH^-), a strong base, the hydrolysis products are an alcohol and the conjugate base of a carboxylic acid. This type of hydrolysis reaction is known as a saponification reaction.
- Soap is prepared by the saponification of fats, producing glycerol and fatty acid salts, soap.
- Hydrolysis of fats is the first step in the metabolism of fats, known as digestion, and catalyzed by lipase enzymes. The fatty acids produced are either metabolized to produce energy in liver and muscle cells or they are re-esterified in fat cells where they are stored.

Hydration and Dehydration Reactions

- Water is a reactant in a hydration reaction and a product in a dehydration reaction.
- In a hydration reaction the elements of water, OH and H, add to an alkene to form an alcohol. The OH and H replace the double bond.
- In an unsymmetrical alkene Markovnikov's rule predicts that the H atom from water adds to the double-bond carbon with more hydrogen atoms.

- A reaction which produces one structural isomer in preference over another structural isomer is known as a regioselective reaction.
- In a biochemical hydration reaction, a carbonyl group is usually adjacent to the double bond, which directs the OH group to the β carbon and the H atom to the α carbon.
- In a dehydration reaction the elements of water are lost from an alcohol to produce an alkene. It is essentially the reverse of a hydration reaction.
- Dehydration reactions produce the alkene with fewer hydrogen atoms on the double-bond carbons, according to Zaitsev's rule.
- In a biochemical dehydration reaction, the β OH group and an α H atom are eliminated as water, producing an α-β unsaturated carbonyl compound.

Key Words

Acyl group A carbonyl group and its attached R group.

Acetyl group A carbonyl group and its attached CH_3 group.

Acyl group transfer reactions A type of reaction that transfers the acyl group from a carboxylic acid to the heteroatom of an alcohol, thiol, or amine to produce a carboxylic acid derivative: ester, thioester, or amide, respectively. Water is also a product.

Amidation reaction The reaction of a carboxylic acid and an amine, in the presence of a catalyst, that produces an amide. Amidation reactions join amino acids to form proteins.

Antioxidants Substances found in fruits, vegetables, and tea that prevent the damaging effects of oxidation by reducing harmful oxidizing agents in the cell, or preventing their formation.

Catalytic hydrogenation A reduction carried out using hydrogen gas (H_2) in the presence of a metal catalyst (Pd, Pt). Catalytic hydrogenation is used to convert alkenes to alkanes.

Cellular respiration The combustion reactions cells perform to convert food into energy.

Coenzyme A nonprotein organic molecule derived from a vitamin required by some enzymes to achieve their catalytic activity. All the oxidoreductase enzymes require a coenzyme.

Dehydration reaction The elimination of the elements of water (H and OH) from an alcohol to produce an alkene. It is the reverse of a hydration reaction. Produces preferentially the Zaitsev product.

Esterification reaction The reaction of a carboxylic acid and an alcohol, in the presence of a catalyst, that produces an ester. Esterification reactions join fatty acids and glycerol to make fats.

FAD/FADH$_2$ Flavin adenine dinucleotide. A coenzyme used in oxidation–reduction reactions of carbon-carbon bonds. Derived from the B vitamin riboflavin.

Free radical A molecule, atom, or ion that contains an odd number of valence electrons. Free radicals are unstable and reactive species because there is an atom without an octet. In biochemistry, some free radicals are responsible for cell damage.

Hydration reaction The addition of the elements of water (H and OH) to an alkene to form an alcohol and turning the double bond into a single bond. It is the reverse of a dehydration reaction. Follows Markovnikov's rule when the alkene is unsymmetrical.

Hydrolysis reaction A reaction that uses water to split a carboxylic acid derivative (ester, thioester, or amide) into a carboxylic acid and an alcohol, thiol, or amine.

Markovnikov's rule The observation that the major product formed in a hydration reaction of an unsymmetrical alkene has the OH group adding to the double-bond carbon with fewer hydrogens and the H atom adding to the double-bond carbon with more hydrogens.

NAD$^+$/NADH Nicotinamide adenine dinucleotide. A coenzyme seen in oxidation–reduction reactions of carbon-oxygen bonds. Derived from the B vitamin niacin.

Oxidation The reactant that loses electrons in an oxidation–reduction reaction. Observed in organic compounds as a product with fewer hydrogen atoms and or more oxygen atoms than the substrate.

Oxidation–reduction reaction A reaction characterized by the transfer of electrons from one reactant to another.

Oxidizing agent The reactant in an oxidation–reduction reaction that gets reduced; so called because it oxidizes the substrate.

Reducing agent The substance in an oxidation–reduction reaction that gets oxidized; so called because it reduces the substrate.

Redox An abbreviation for oxidation–reduction.

Regioselective reaction A reaction that produces one structural isomer preferentially over another structural isomer in a chemical reaction, as seen with Markovnikov's rule and Zaitsev's rule.

Reduction The reactant that gains electrons in an oxidation–reduction reaction. Observed in organic compounds as a product with more hydrogen atoms and or fewer oxygen atoms than the substrate.

Saponification reaction The hydrolysis of an ester in the presence of hydroxide ion to produce an alcohol and the conjugate base of a carboxylic acid. If the reactant is a fat, the conjugate base of the carboxylic acids produced are soap.

Substrate The organic reactant that binds to the enzyme and undergoes a chemical change in a biochemical reaction.

Thioester A functional group similar to an ester, but where sulfur replaces the oxygen bonded to the carbonyl group: RCOSR'. Thioesters are formed in esterification reactions between a carboxylic acid and a thiol. Water is also a product.

Thiol A functional group similar to an alcohol, but sulfur replaces oxygen: R—S—H.

Triacylglycerol A chemical term for a fat, also known as a triglyceride.

Triglyceride A chemical term for a fat, also known as a triacylglycerol.

Zaitsev's rule The observation that in the dehydration of an alcohol elimination produces the alkene with fewer hydrogen atoms on the double-bond carbon atoms.

Additional Exercises

Are You Getting Your Vitamins?

25 Why are vitamins necessary for living?

26 Why do we need to take in vitamins through our diet? What are some sources of vitamins?

27 What role does riboflavin play in the body?

28 What are the symptoms of a riboflavin deficiency?

29 Why do some newborns need light therapy? What happens to riboflavin when a baby undergoes light therapy?

30 Name the four fat-soluble vitamins, and explain what is meant by "fat soluble."

31 Why does the body need to replenish the water-soluble vitamins frequently? How is this done?

32 What is the difference between "enriching" and "fortifying" processed foods with vitamins?

33 What is the "function" of a functional group?

34 Write the general structure of the following functional groups:
a. aldehyde **c.** amide
b. carboxylic acid **d.** alcohol

35 Write the general structure of the following functional groups:
a. ketone **c.** thioester
b. ester **d.** amine

36 What role do enzymes play in biochemical reactions?

37 Define the term substrate?

Oxidation–Reduction Reactions

38 Why are oxidation–reduction reactions some of the most important reactions in nature?

39 What products are formed when an organic substance containing carbon, hydrogen, and oxygen undergoes a complete combustion reaction?

40 How do our cells perform combustion reactions? What is the biochemical process called?

41 The reactant that undergoes loss of electrons in an oxidation–reduction reaction is said to undergo _____. [oxidation or reduction]

42 The reactant that gains electrons in an oxidation–reduction reaction is said to undergo _____. [oxidation or reduction]

43 For the following inorganic reactions, write the oxidation half reaction separate from the reduction half reaction, and show electrons as Lewis dots. Label the oxidation and the reduction. If there is a spectator ion, identify it.

a. $Zn(s) + 2 HCl(aq) \rightarrow H_2(g) + ZnCl_2(aq)$
b. $2 Na(s) + Cl_2(g) \rightarrow 2 NaCl(s)$

44 If an organic reactant gains hydrogen atoms, has it undergone oxidation or reduction?

45 If an organic reactant loses hydrogen atoms, has it undergone oxidation or reduction?

46 If an organic reactant gains oxygen atoms, has it undergone oxidation or reduction?

47 If an organic reactant loses oxygen atoms, has it undergone oxidation or reduction?

48 What is a reducing agent? What is an oxidizing agent?

49 For the oxidation–reduction reactions shown below, determine whether the organic reactant shown has undergone an oxidation or a reduction by placing either an H or an O in the brackets above the arrow where indicated. Explain how you determined whether the substrate was oxidized or reduced.

a.

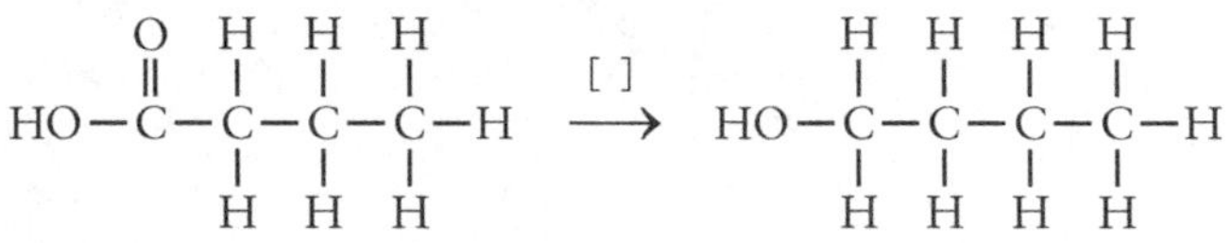

b.

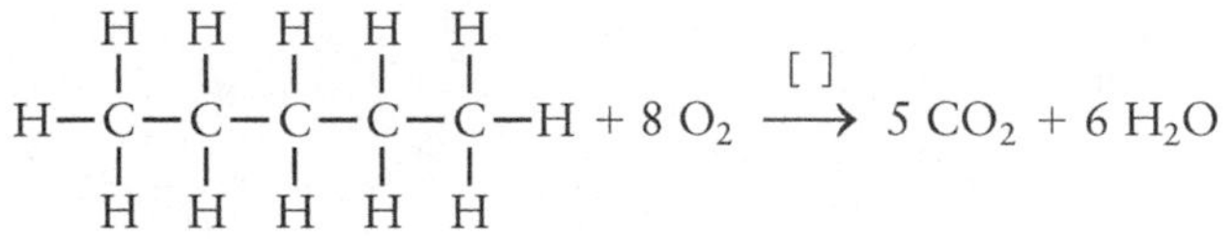

c.

d.

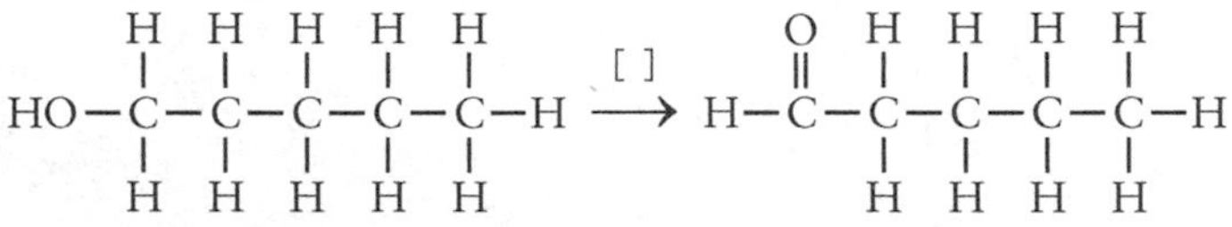

50 Write the structure of the product formed in the following reaction. What type of reaction is this?

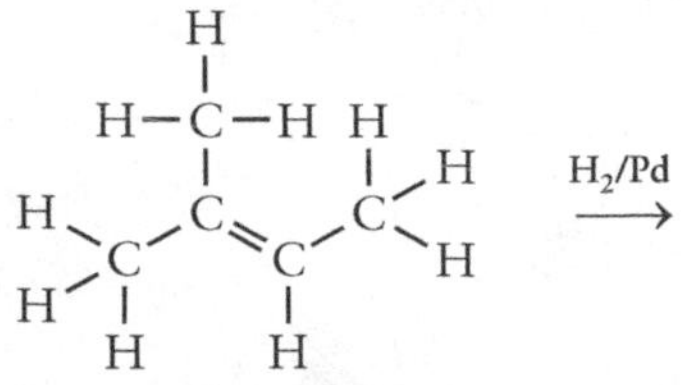

51 Write the structure of the product formed in the following reaction. What type of reaction is this?

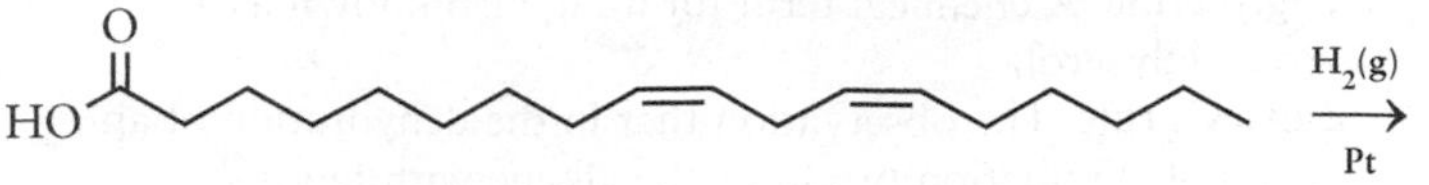

$$\xrightarrow[\text{Pt}]{\text{H}_2(g)}$$

52 Why does the food industry prepare "partially hydrogenated" fats? What does it mean chemically to partially hydrogenate a fat?

53 Which is more likely to be a liquid?
 a. a saturated fat
 b. an unsaturated fat

54 Write the structure of the product formed in the following oxidation–reduction reactions. Label the functional group in the reactant that changes, and identify the new functional group in the product. If the functional group is an alcohol, identify the type of alcohol. If there is no reaction, state so.

 a.

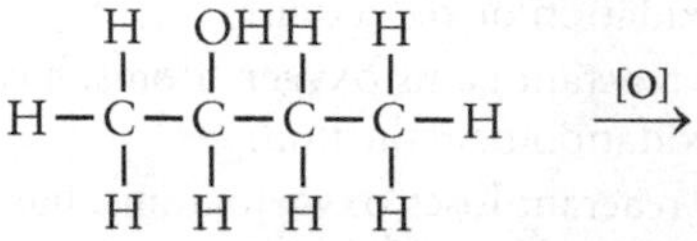

$$\xrightarrow{[O]}$$

 b.

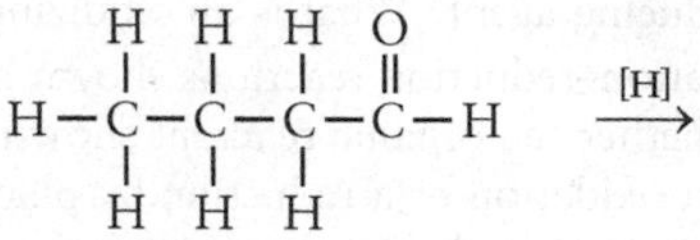

$$\xrightarrow{[H]}$$

 c.

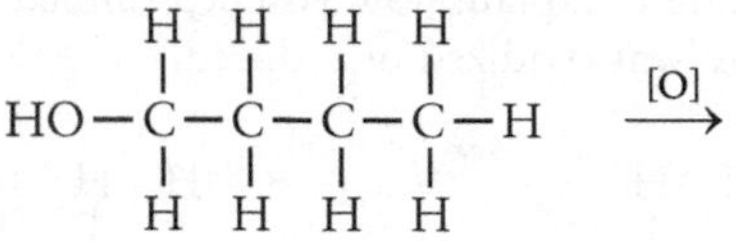

$$\xrightarrow{[O]}$$

 d.

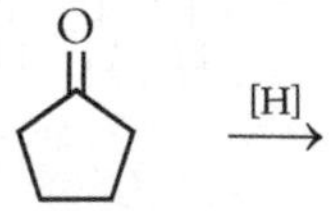

$$\xrightarrow{[H]}$$

 e.

$$\xrightarrow{[O]}$$

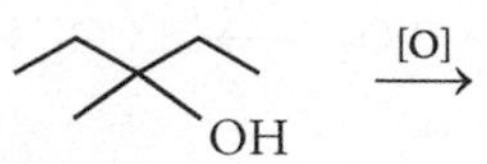

55 Write the structure of the organic product formed in the following reactions:

 a.

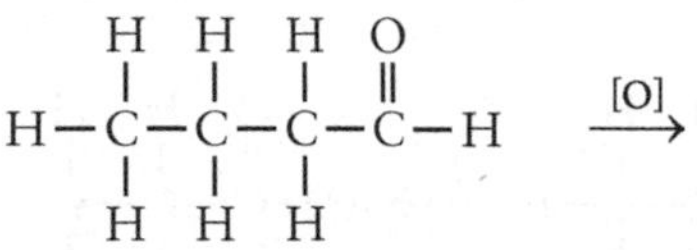

$$\xrightarrow{[O]}$$

 b.

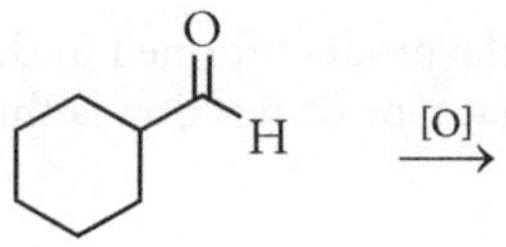

$$\xrightarrow{[O]}$$

 c.

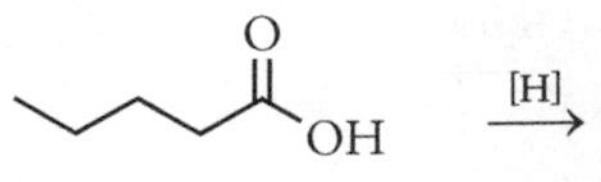

$$\xrightarrow{[H]}$$

 d.

$$\xrightarrow{[H]}$$

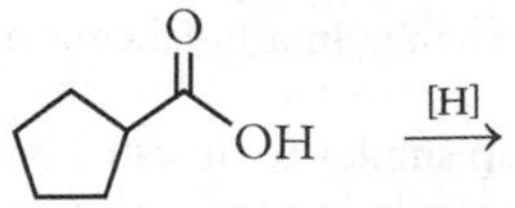

56 What do our cells use as oxidizing and reducing agents?

57 What is the difference between an enzyme and a coenzyme?

58 Referring to Figure 10-9, what is the structural difference between NADH and NAD^+.

59 Write the structure of the product(s) formed in each of the reactions.
 a. $C_3H_8 + 5\ O_2 \rightarrow 3$ _____ $+ 4$ _____

 b.

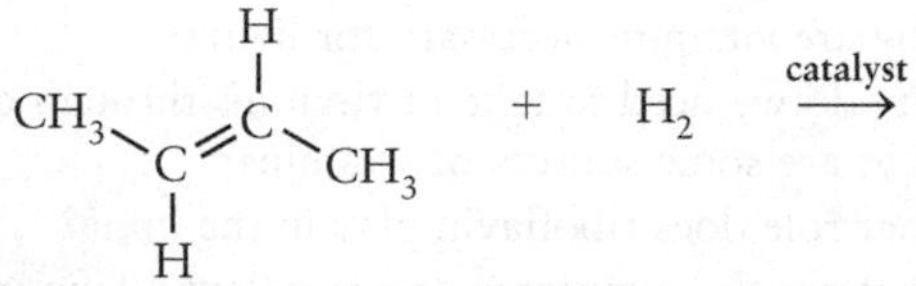

$$+ \quad H_2 \quad \xrightarrow{\text{catalyst}}$$

 c.

$$\overset{NAD^+\ NADH + H^+}{\underset{\curvearrowright}{\longrightarrow}}$$

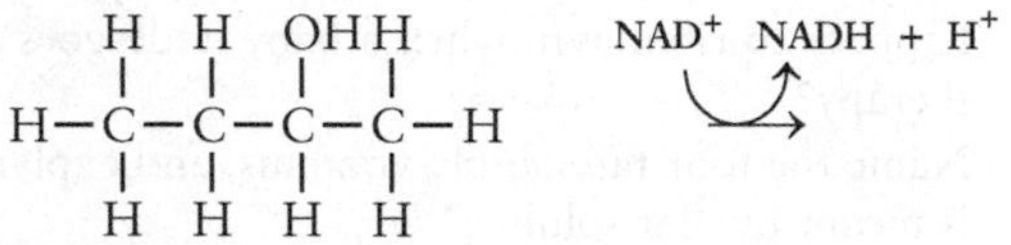

 d. $FADH_2 + NAD^+ \rightarrow$

60 Write the structure of the *reactant* (substrate) that upon reduction would yield the following products.

 a.

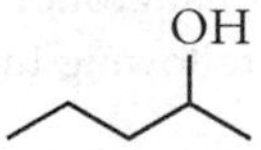

 b.

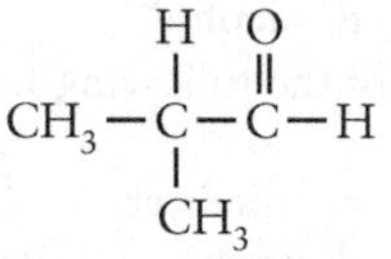

 c. butan-1-ol
 d. $NADH + H^+$

61 Predict the structure of the *reactant* (substrate) that upon oxidation would yield the following products.

 a.

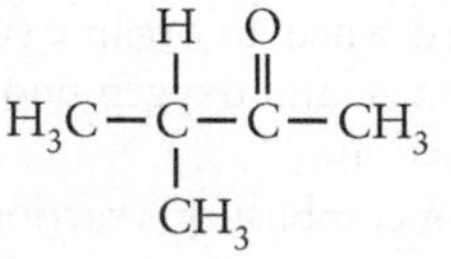

 b.

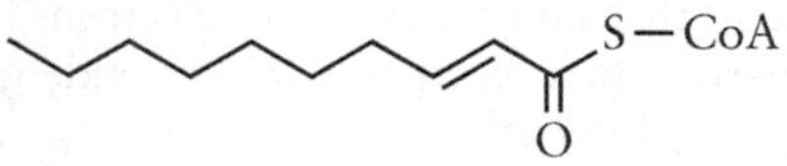

 c. FAD
 d.

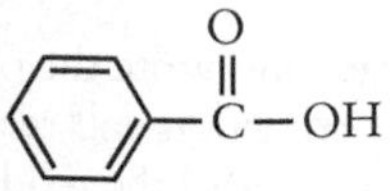

62 Which of the following is not true about an antioxidant?
 a. It is a reducing agent.
 b. It undergoes oxidation.
 c. It prevents the oxidation of other substances in the cell.
 d. It is an oxidizing agent.
 e. They are believed to slow the aging process.

63 Which of the following is not true about a free radical?
 a. It is an unstable species.
 b. It contains an odd number of electrons.
 c. Antioxidants prevent the formation of free radicals.
 d. It contains an octet.

Group Transfer Reactions: Esterification and Amidation Reactions

64 What is an acyl group transfer reaction?

65 What kind of enzyme catalyzes acyl transfer reactions in biochemistry?

66 Write the structure of the products formed in the following reactions. What type of reactions are these?

 a.

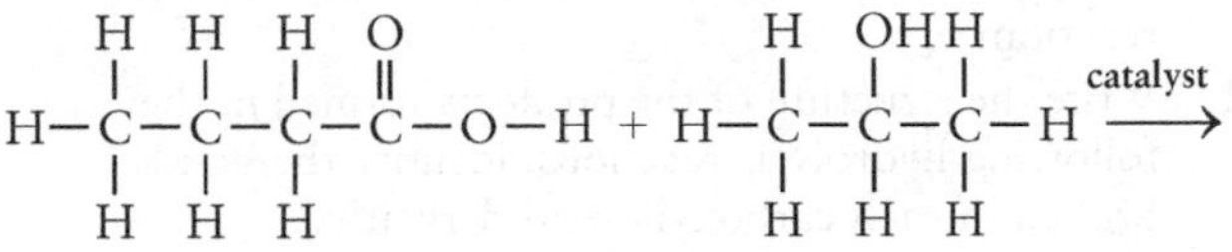

 b.

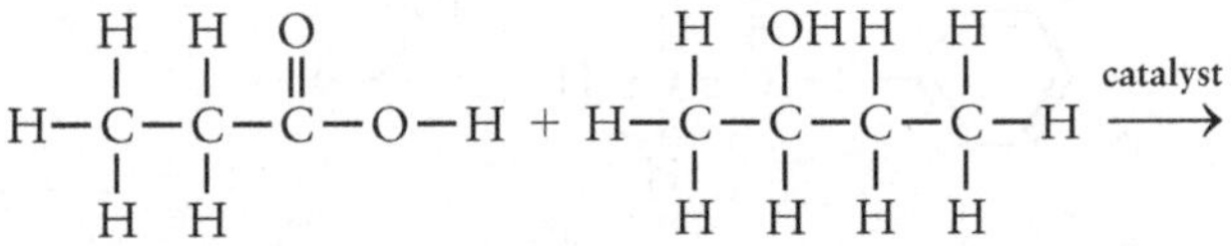

 c.

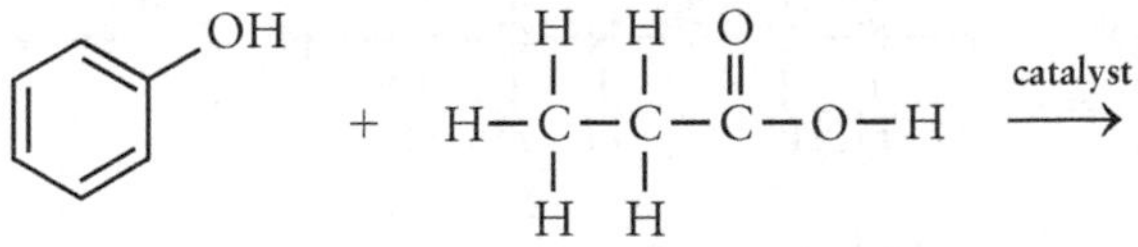

 d.

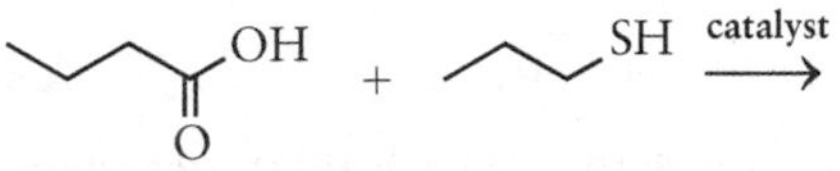

67 Write the structure of the products formed in the following esterification reactions:

 a.

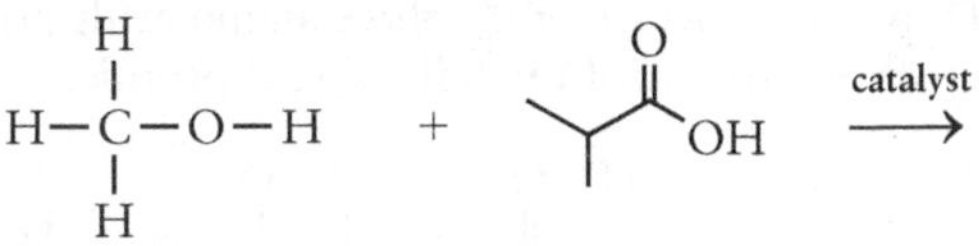

 b.

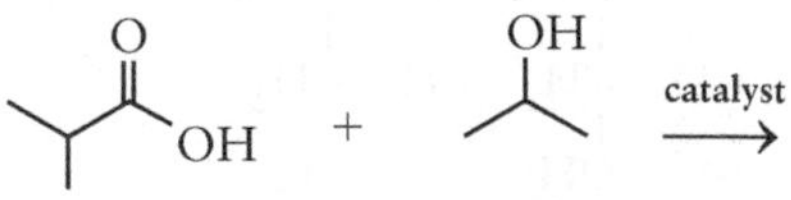

 c.

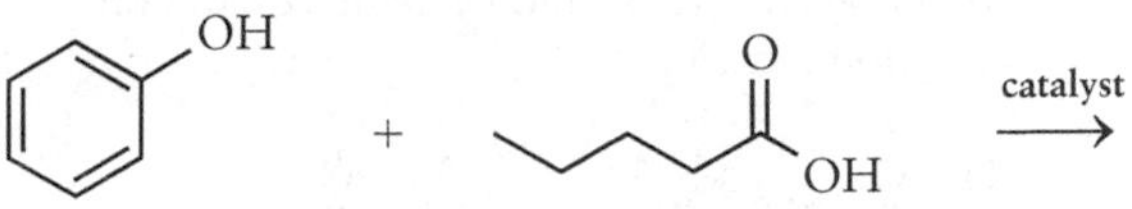

 d.

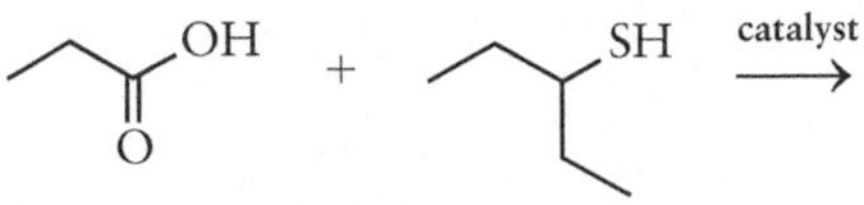

68 What is the structure of the ester that is produced from the reaction of butanoic acid with methanol? What familiar scent does this ester have?

69 What is the structure of the ester that is produced from the reaction of acetic acid and propan-1-ol? What familiar scent does it have?

70 What is the general structure of triacylglycerols or triglycerides? What function do they serve?

71 How many fatty acids react with one glycerol molecule to produce a fat molecule? What type of reaction produces a fat from fatty acids and glycerol?

72 Why are amidation reactions important in biochemistry?

73 Predict the products in the following amidation reactions:

 a.

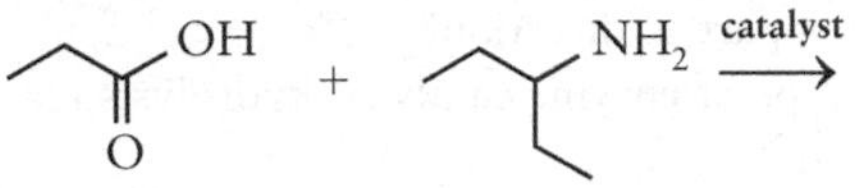

 b.

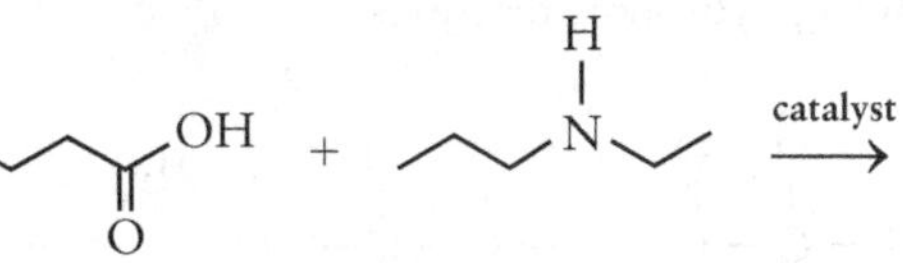

 c.

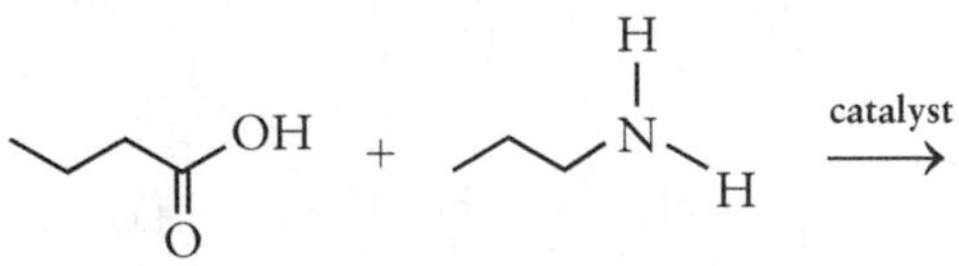

74 Predict the product in the following amidation reactions:

 a.

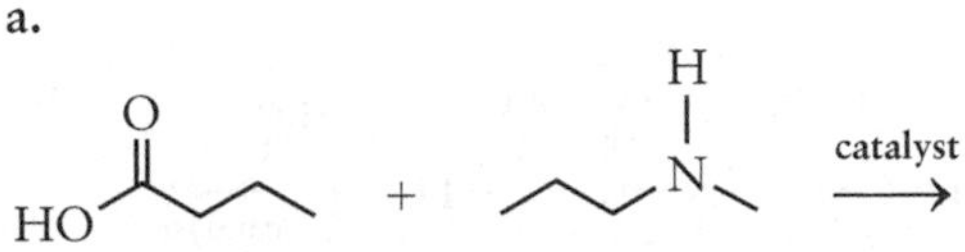

 b.

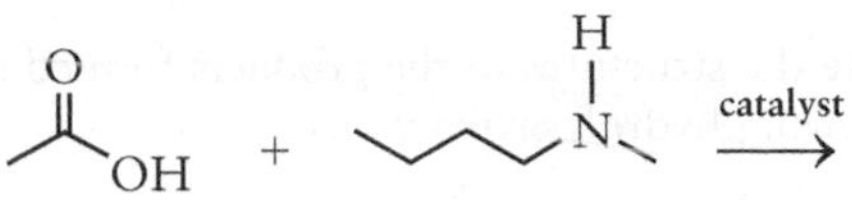

 c.

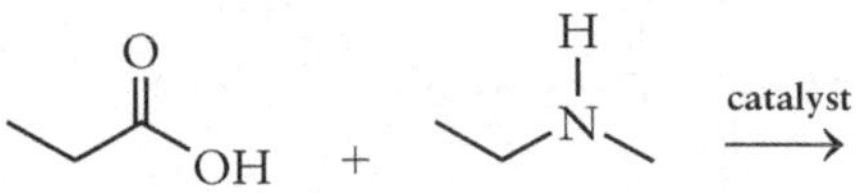

75 Write the structure of the two possible dipeptides formed when glycine and cysteine react in an amidation reaction.

$$
\underset{\text{Glycine}}{\overset{+}{H_3N}-\overset{\overset{\displaystyle H}{|}}{\underset{\underset{\displaystyle H}{|}}{C}}-\overset{\overset{\displaystyle O}{\|}}{C}-O^-}
\;+\;
\underset{\text{Cysteine}}{\overset{+}{H_3N}-\overset{\overset{\displaystyle H}{|}}{\underset{\underset{\displaystyle CH_2}{\underset{\underset{\displaystyle SH}{|}}{|}}}{C}}-\overset{\overset{\displaystyle O}{\|}}{C}-O^-}
\;\xrightarrow{\text{enzyme}}
$$

76 Write the structure of the two possible dipeptides formed when leucine and serine react in an amidation reaction.

$$
\underset{\text{Leucine}}{\overset{+}{H_3N}-C-C-O^-}
\;+\;
\underset{\text{Serine}}{\overset{+}{H_3N}-C-C-O^-}
\;\xrightarrow{\text{enzyme}}
$$

Leucine: CH₂ / H–C–CH₃ / CH₃ side chain. Serine: CH₂ / OH side chain.

Hydrolysis Reactions

77 Provide two examples of hydrolysis reactions that occur as part of digestion?

78 What type of enzyme catalyzes hydrolysis reactions in the body?

79 Write the structures of the products formed in the following hydrolysis reactions.

a.

$$
H-\overset{H}{\underset{H}{C}}-\overset{O}{\overset{\|}{C}}-O-\overset{H}{\underset{H}{C}}-\overset{H}{\underset{H}{C}}-H \;\xrightarrow[\text{catalyst}]{H_2O}
$$

b.

$$
H-\overset{H}{\underset{H}{C}}-\overset{CH_3}{\underset{H}{C}}-S-\overset{O}{\overset{\|}{C}}-\overset{H}{\underset{H}{C}}-\overset{H}{\underset{H}{C}}-H \;\xrightarrow[\text{catalyst}]{H_2O}
$$

c.

$$
H-\overset{H}{\underset{H}{C}}-\overset{H}{\underset{H}{C}}-\overset{H}{\underset{}{N}}-\overset{O}{\overset{\|}{C}}-\overset{H}{\underset{H}{C}}-H \;\xrightarrow[\text{catalyst}]{H_2O}
$$

80 Write the structures of the products formed in the following hydrolysis reactions.

a.

$$
H-\overset{H}{\underset{H}{C}}-\overset{O}{\overset{\|}{C}}-S-CoA \;\xrightarrow{H_2O}
$$

b.

$$
\overset{+}{H_3N}-\overset{\overset{\displaystyle H}{|}}{\underset{\underset{\underset{\displaystyle OH}{|}}{\underset{\displaystyle H-C-H}{|}}}{C}}-\overset{\overset{\displaystyle O}{\|}}{C}-\overset{\overset{\displaystyle H}{|}}{\underset{\underset{\displaystyle H}{|}}{N}}-\overset{\overset{\displaystyle H}{|}}{C}-\overset{\overset{\displaystyle O}{\|}}{C}-O^- \;\xrightarrow{H_2O}
$$

81 The reaction shown below occurs in the citric acid cycle.

$$
\begin{array}{c}
COO^- \\
| \\
H-C-H \\
| \\
HO-C-COO^- \\
| \\
H-C-H \\
| \\
O=C-S-CoA
\end{array}
\;\xrightarrow{H_2O}\;
\underset{\text{Citrate}}{\begin{array}{c}
COO^- \\
| \\
H-C-H \\
| \\
HO-C-COO^- \\
| \\
H-C-H \\
| \\
O=C-O^-
\end{array}}
\;+\; H-SCoA
$$

a. How many carbon atoms does the acyl group being transferred contain?

b. What bond is broken in the substrate?

c. Between what two atoms is the acyl group being transferred?

d. Why is this reaction classified as a hydrolysis reaction?

82 Write the structure of the products formed in the following hydrolysis reactions. Identify the bond broken in each carboxylic acid derivative.

a.

$$
\text{C}_6\text{H}_5-O-\overset{O}{\overset{\|}{C}}-\overset{\overset{\displaystyle H}{|}}{\underset{\underset{\displaystyle H}{|}}{C}}-H \;\xrightarrow{H_2O}
$$

b.

$$
H-\overset{H}{\underset{H}{C}}-\overset{H}{\underset{H}{C}}-\overset{O}{\overset{\|}{C}}-\overset{H}{\underset{H}{N}}-\overset{H}{\underset{H}{C}}-\overset{CH_3}{\underset{H}{C}}-CH_3 \;\xrightarrow{H_2O}
$$

c.

$$
H-\overset{H}{\underset{H}{C}}-\overset{O}{\overset{\|}{C}}-S-\overset{H}{\underset{H}{C}}-\overset{H}{\underset{H}{C}}-H \;\xrightarrow{H_2O}
$$

83 What substances in the stomach aid the digestion of proteins? What chemical reaction is occurring during the digestion of proteins? What type of molecules are produced in the reaction?

84 Draw the structure of the three amino acids produced in the hydrolysis of the following tripeptide:

$$
H-\overset{+}{\underset{H}{N}}-\overset{\overset{\displaystyle H}{|}}{\underset{\underset{\displaystyle CH_3}{|}}{C}}-\overset{\overset{\displaystyle O}{\|}}{C}-\overset{\overset{\displaystyle H}{|}}{\underset{\underset{\displaystyle H}{|}}{N}}-\overset{\overset{\displaystyle O}{\|}}{\underset{\underset{\underset{\displaystyle OH}{|}}{\underset{\displaystyle CH_2}{|}}}{C}}-\overset{\overset{\displaystyle H}{|}}{\underset{\underset{\displaystyle CH_2}{\underset{\underset{\displaystyle SH}{|}}{|}}}{N}}-\overset{\overset{\displaystyle O}{\|}}{C}-O^- \;\xrightarrow{\text{pepsin}}
$$

85 Draw the structure of the three amino acids produced in the hydrolysis of the following tripeptide:

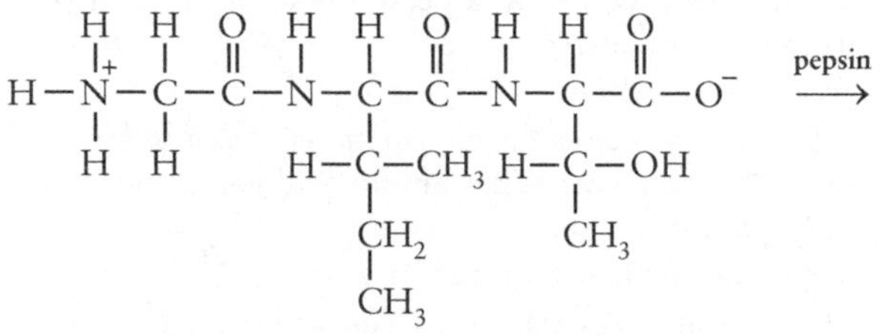

86 What type of reaction produces soap?

87 Where does the digestion of fats occur in the body? What type of enzymes help the digestion of fats?

88 Write the structure of the products formed in the following saponification reaction:

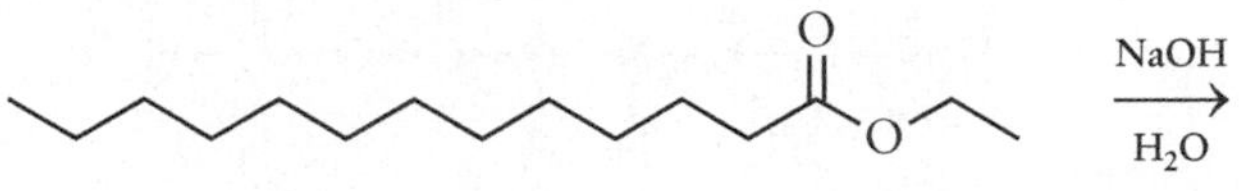

Hydration and Dehydration Reactions

89 What type of enzyme catalyzes hydration and dehydration reactions in the body?

90 For the reactions listed below, indicate whether they are hydration or dehydration reactions, and add a curved arrow showing the role of water in the reaction.

a.

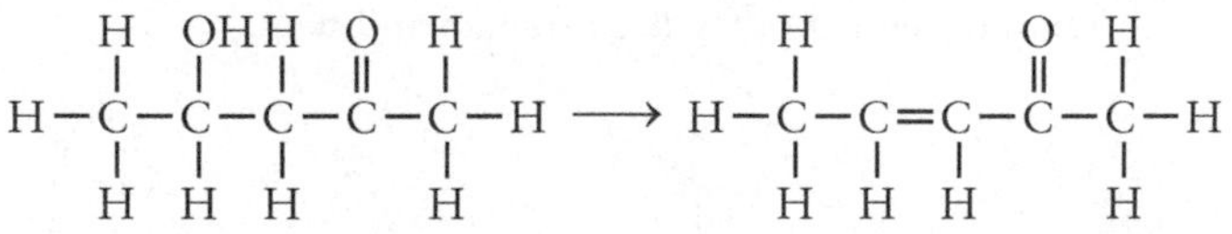

b.

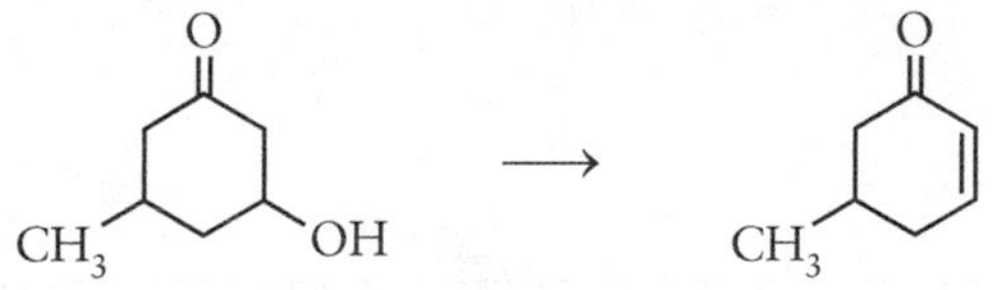

91 For the reactions listed below, indicate whether they represent hydration or dehydration reactions, and show water next to a curved arrow in the equation.

a.

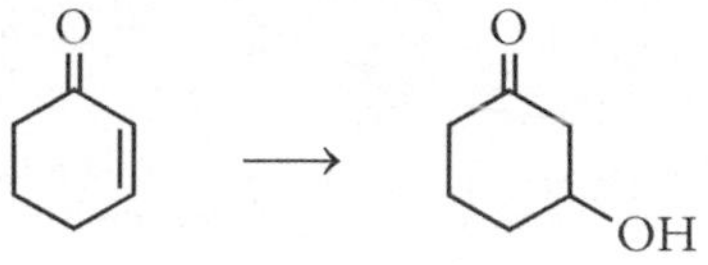

b.

c.

92 Write the product(s) formed in the following reactions.

a.

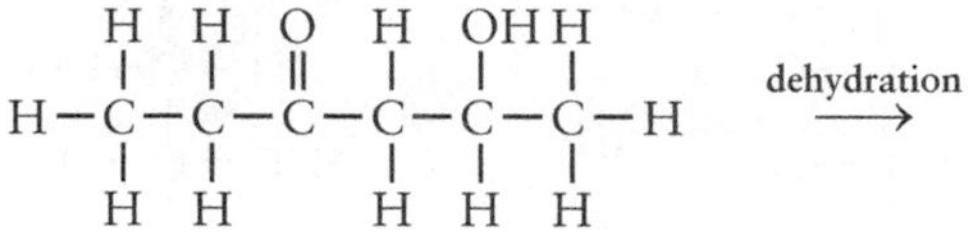

b.

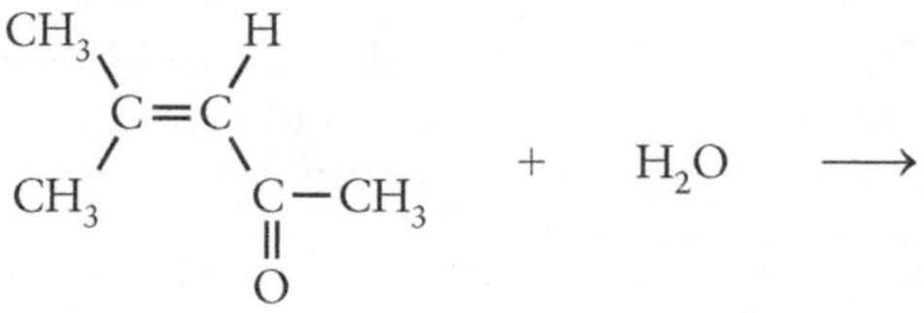

93 Write the product formed in the following reactions.

a.

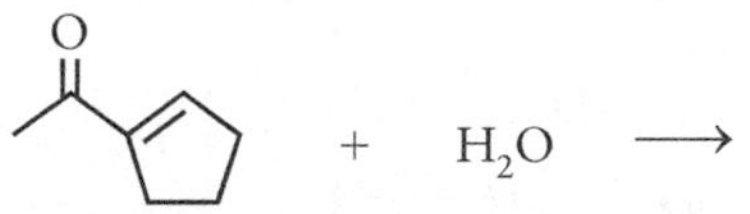

b.

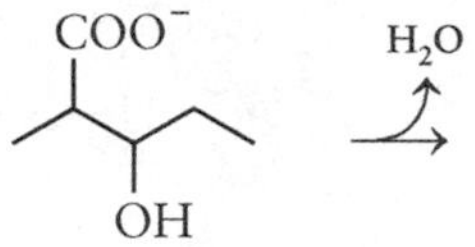

94 Identify the following reactions as oxidation–reduction reactions, hydration or dehydration reactions, hydrolysis reactions, esterification reactions, or amidation reactions.

a.

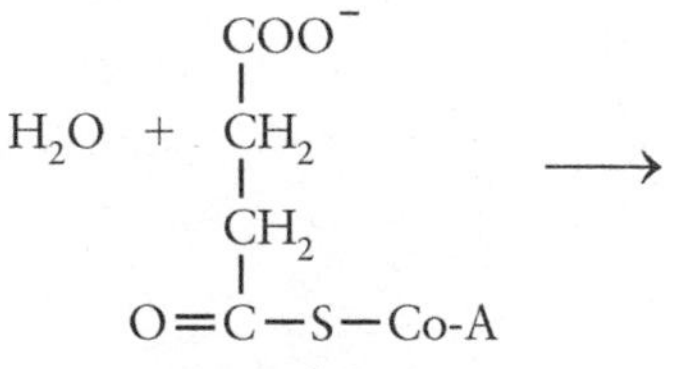

Succinyl-CoA

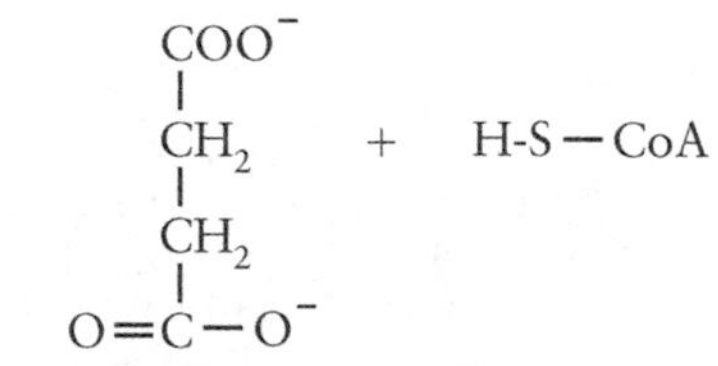

Succinate

b.

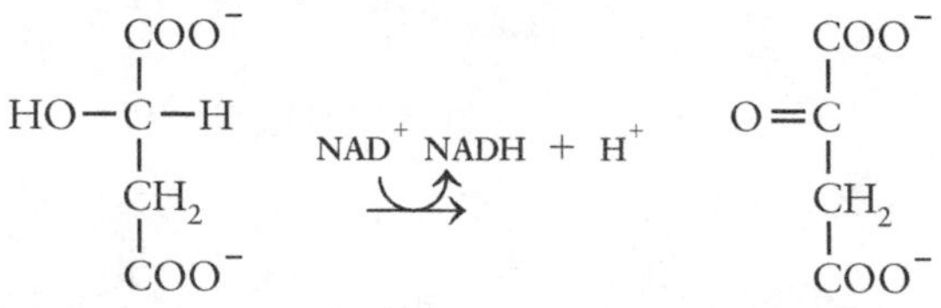

Malate

Oxaloacetate

c.

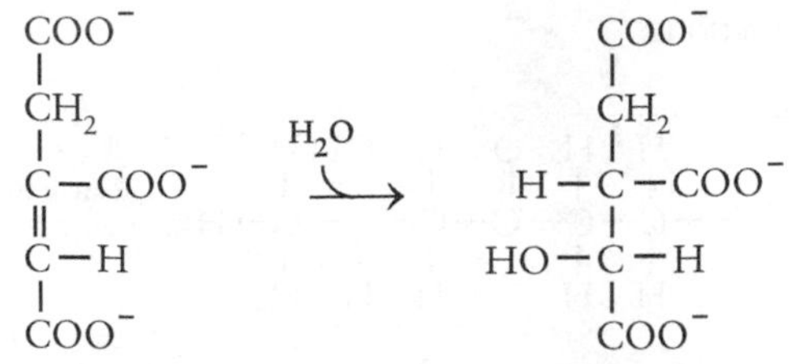

Aconitate   Isocitrate

95   Identify the following reactions as oxidation–reduction reactions, hydration or dehydration reactions, hydrolysis reactions, esterification reactions, or amidation reactions.

a.

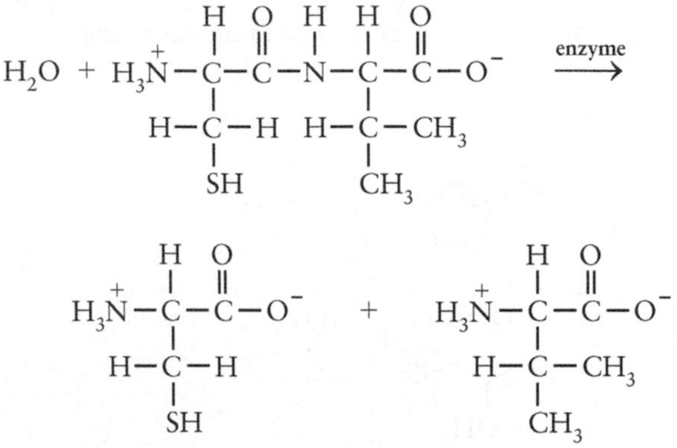

b.

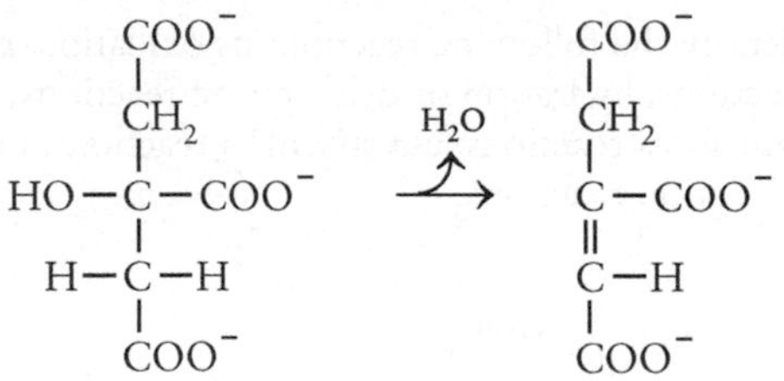

Citrate   Aconitate

Chemistry in Medicine Phenylketonuria—A Metabolic Disorder

96   What would happen to a person who has PKU if it were left untreated?

97   What causes PKU?

98   What reaction does the enzyme *phenylalanine hydrolase* catalyze? What amino acid is produced in this reaction?

99   What coenzyme is required by *PAH*?

100   The structure of aspartame (Nutrasweet) is shown below.

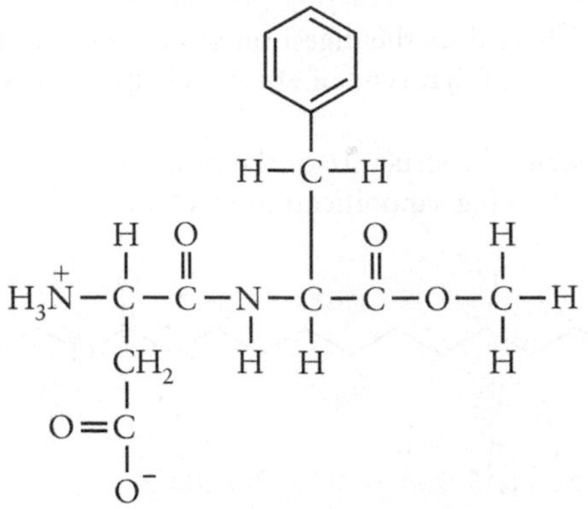

a. Circle the amide functional group in aspartame.
b. Write the structure of the two compounds formed when the amide in aspartame is hydrolyzed in the stomach. Do not hydrolyze the ester.
c. Why should phenylketonurics avoid aspartame?

101   What is the treatment for people with PKU? What supplements do they need to take and why?

Answers to Practice Exercises

1   **a.**

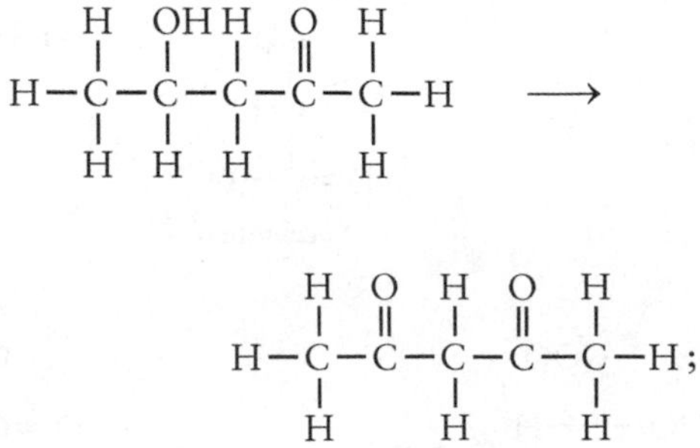

b.

c.

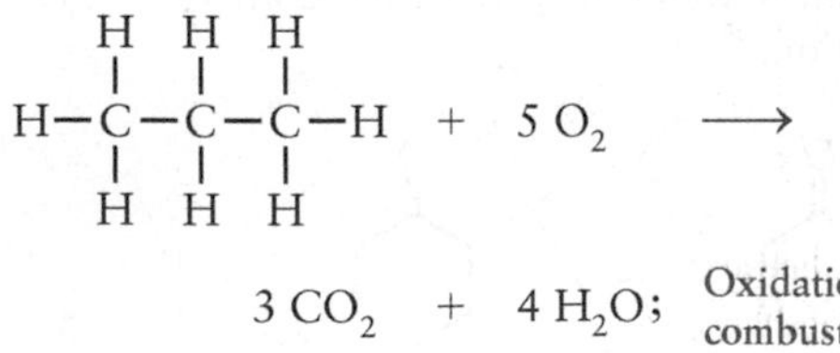

$$3\,CO_2 \;+\; 4\,H_2O;$$ Oxidation; combustion reaction

d.

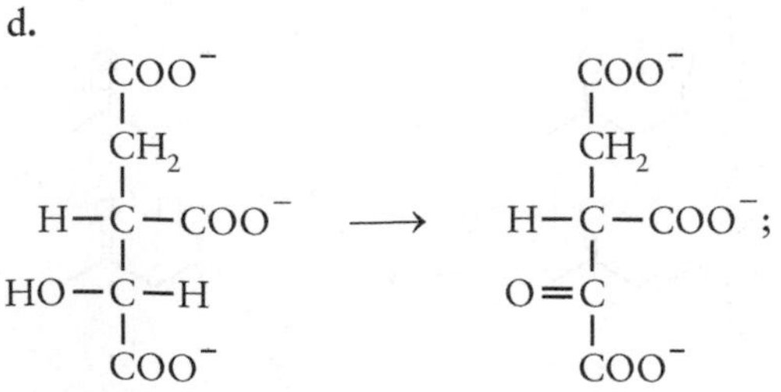

2 Reduction: $2\,Ag^+(aq) + 2e^- \rightarrow 2\,Ag(s)$; each silver ion, Ag^+, gains an electron to become silver metal, Ag. Oxidation: $Cu(s) \rightarrow Cu^{2+}(aq) + 2e^-$; copper metal, Cu, loses two electrons to become copper(II) ion, Cu^{2+}.

3 a.

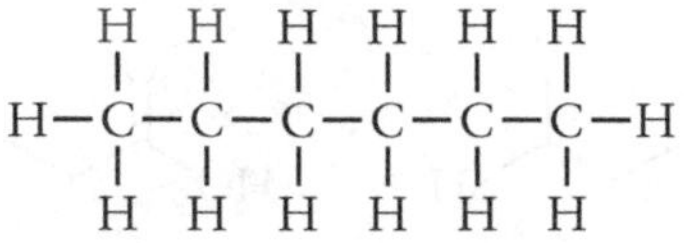

H H H H H H
H—C—C—C—C—C—C—H
H H H H H H

b.

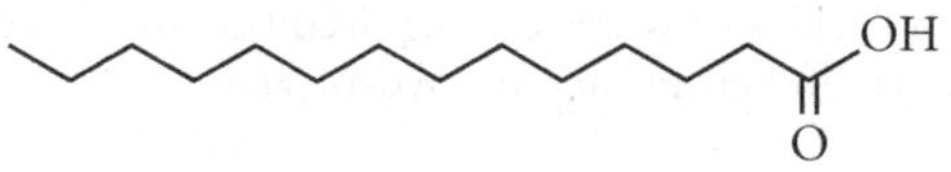

(structure: long-chain carboxylic acid) OH / O

4 a.

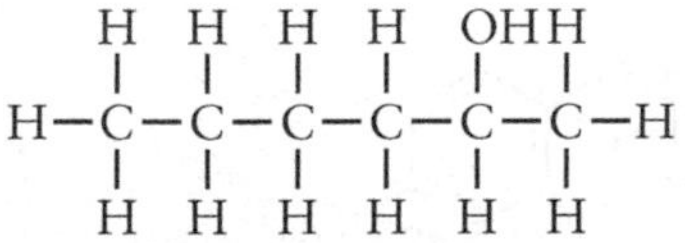

H H H H OHH
H—C—C—C—C—C—C—H
H H H H H H

b.

H H H
H—C—C—C—C
H H H (=O)

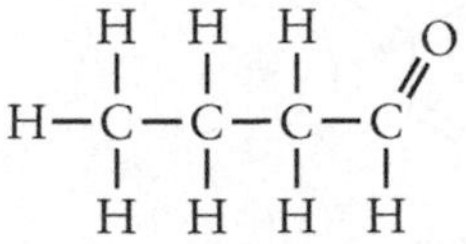

c. No oxidation occurs. The reactant is a tertiary alcohol, which does not undergo oxidation, because there is no hydrogen atom on the carbon bearing the hydroxyl group.

5 a.

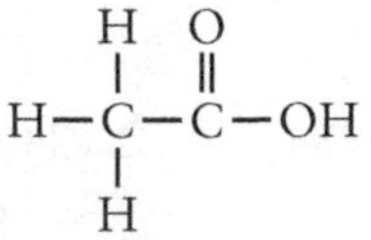

H O
H—C—C—OH
H

b.

H
H—C—H
H | H H
H—C—C—C—C—OH
H H H H

c.

(benzoic acid) OH $\xrightarrow{[H]}$ (benzaldehyde) H

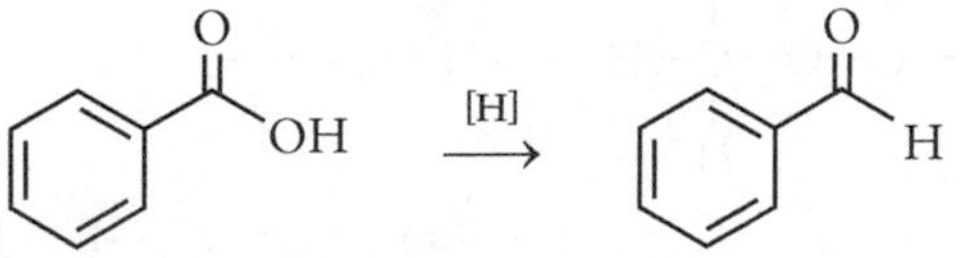

6 a. The oxidizing agent is chromium VI (Cr^{6+}).
b. Ethanol is undergoing an oxidation to acetic acid.
c. Cr^{6+} is undergoing a reduction to Cr^{3+}.
d. Cr^{6+} is orange whereas Cr^{3+} is green, so an orange to green change—a reduction—means that ethanol is oxidized.

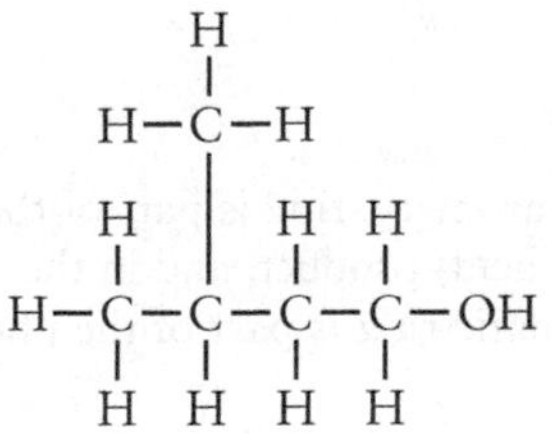

7 a. Acetaldehyde has undergone oxidation (aldehyde to carboxylic acid), and the coenzyme NAD^+ has been reduced to $NADH + H^+$.
b. The substrate has undergone oxidation (alkane to alkene), and the coenzyme FAD has been reduced to $FADH_2$.

8 c. NAD^+

9 b. $FADH_2$

10 a.

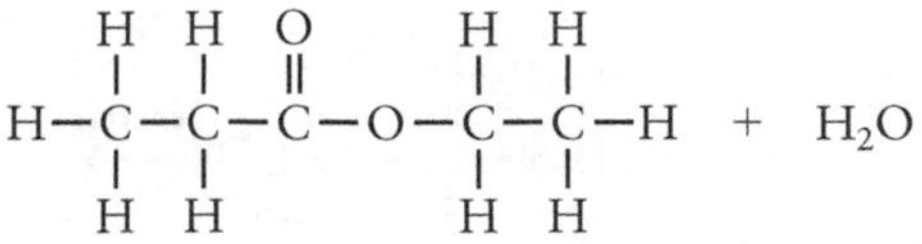

H H O H H
H—C—C—C—O—C—C—H + H_2O
H H H H

b.

(benzene ring)—C(=O)—O—C—H + H_2O
H

c.

(structure: thioester) S

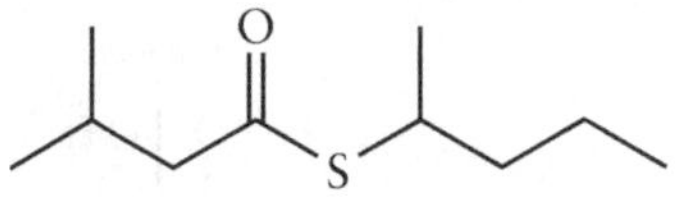

11

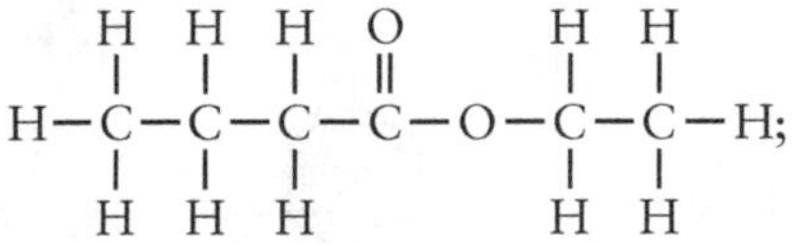

H H H O H H
H—C—C—C—C—O—C—C—H; smells like pineapple
H H H H H

12

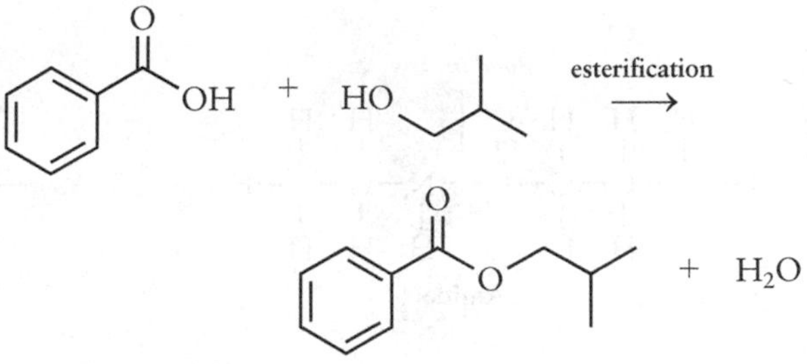

(benzoic acid) OH + HO—(isobutyl) $\xrightarrow{\text{esterification}}$

(benzoate ester) O + H_2O

13

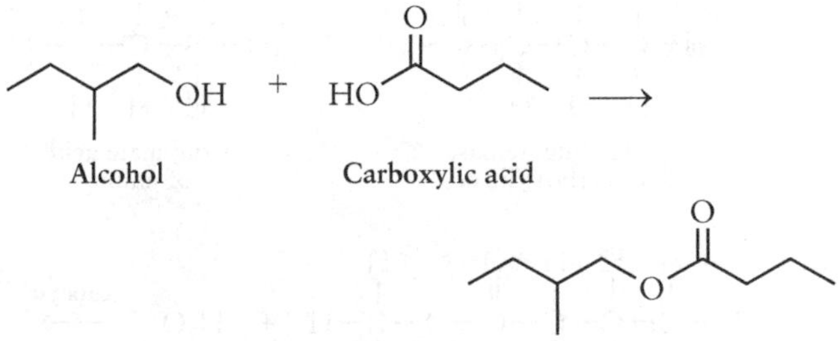

(structure) OH + HO—(butanoic acid) $\longrightarrow$
Alcohol   Carboxylic acid

(structure) O
Ester

14

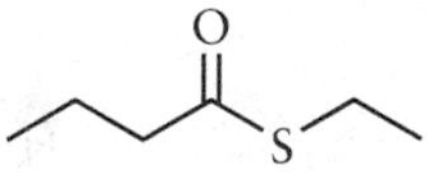

(structure: thioester) O / S

15

H O   H H H H
H—C—C—OH + H—C—C—C—C—N—H $\longrightarrow$
H   H H H H H
Acetic acid   Butan-1-amine
(carboxylic acid)   (primary amine)

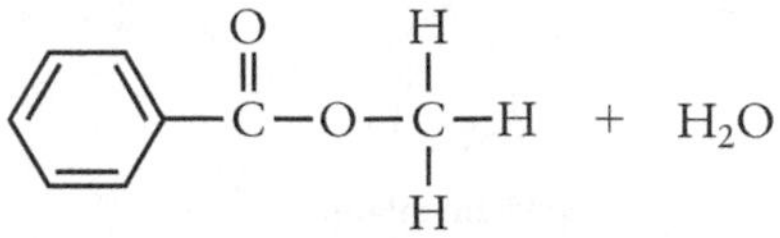

New bond

H O↓ H H H H
H—C—C—N—C—C—C—C—H
H   H H H H H

Amide

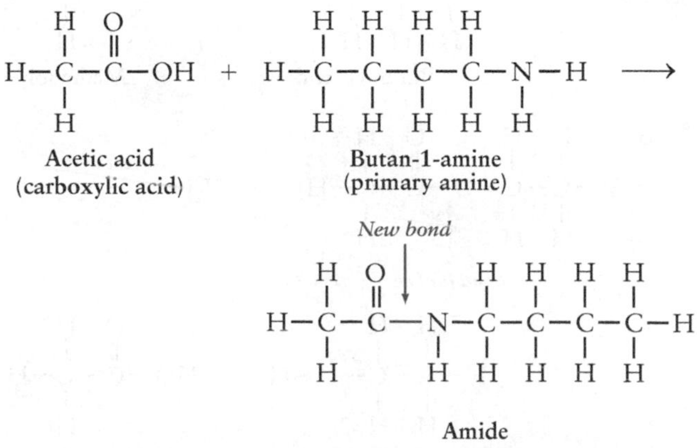

16 Ala-Phe and Phe-Ala are structural isomers.

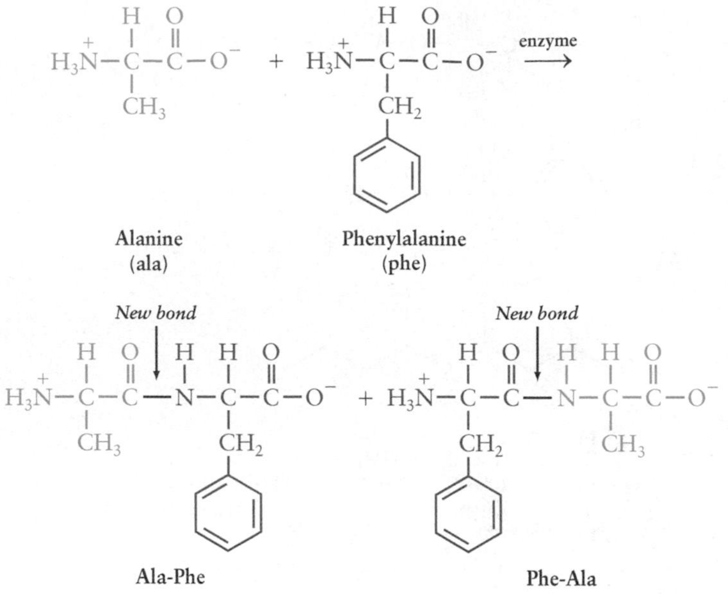

Alanine (ala) + Phenylalanine (phe) → Ala-Phe + Phe-Ala

New bond

17

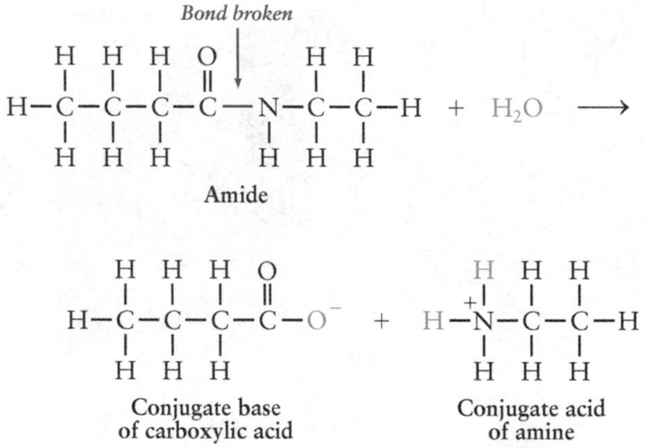

Bond broken

Amide + H₂O →

Conjugate base of carboxylic acid + Conjugate acid of amine

18 a.

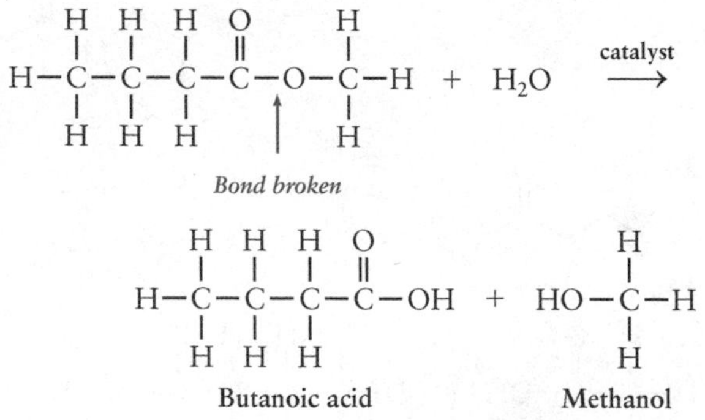

Bond broken

→ Butanoic acid + Methanol

b.

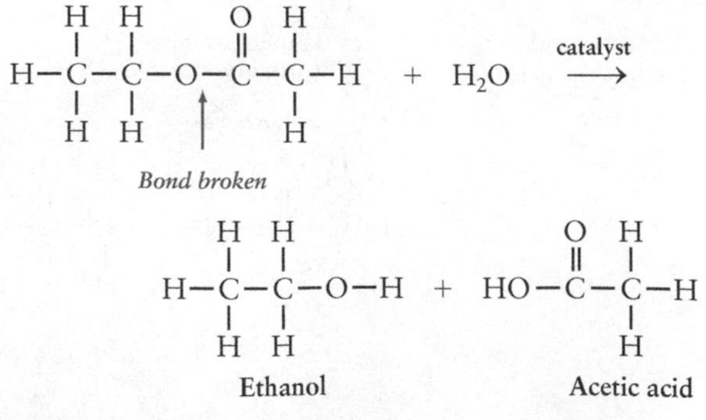

Bond broken

→ Ethanol + Acetic acid

c.

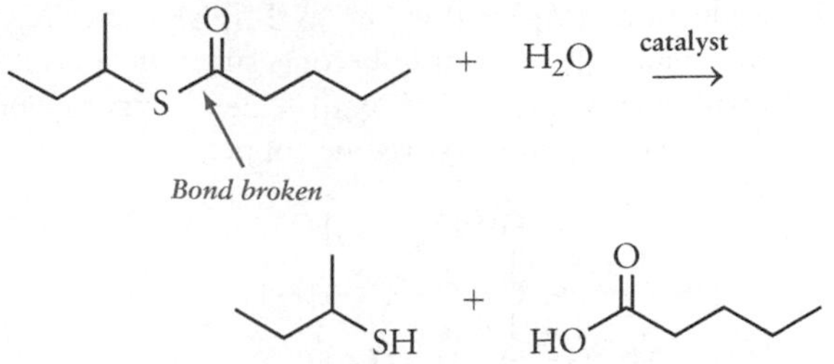

Bond broken

→

19 Two molecules of acetic acid, CH₃COOH, are produced. Two molecules of water are required because there are two esters in heroin that are hydrolyzed.

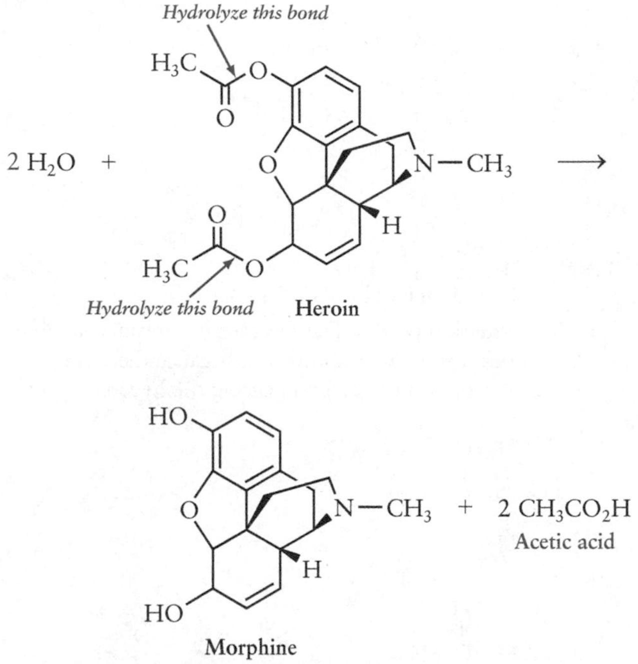

Hydrolyze this bond

2 H₂O + Heroin →

Hydrolyze this bond

Morphine + 2 CH₃CO₂H (Acetic acid)

20 In the first reaction the aromatic ring is part of the carboxylic acid (benzoic acid) product, and in the second reaction the aromatic ring is part of the phenol product.

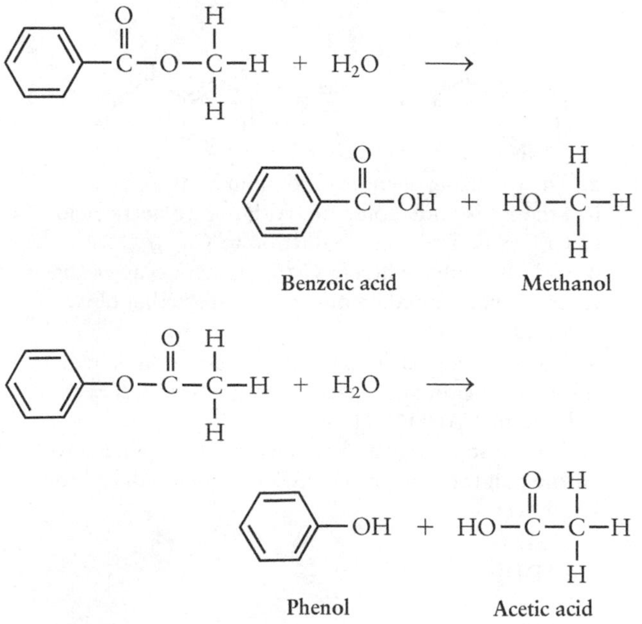

→ Benzoic acid + Methanol

→ Phenol + Acetic acid

21

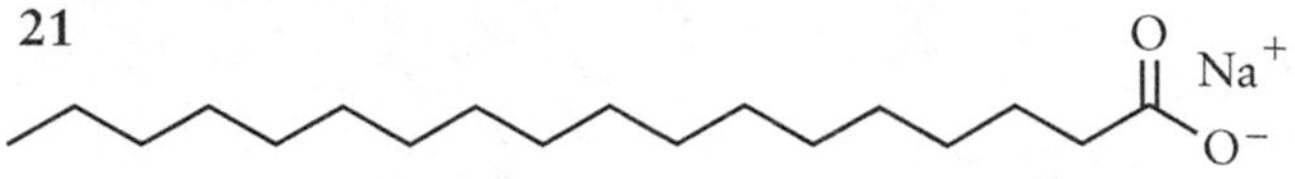

$+ CH_3CH_2OH$

22

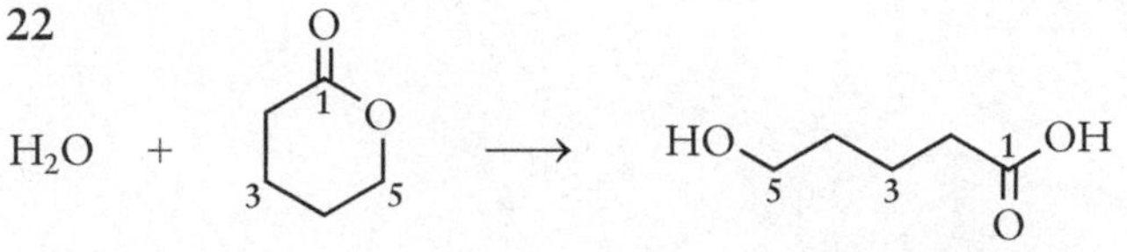

23

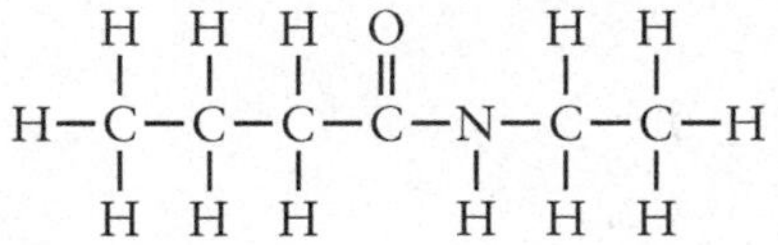

24 a. This is a hydration reaction.
b. Shown in blue.

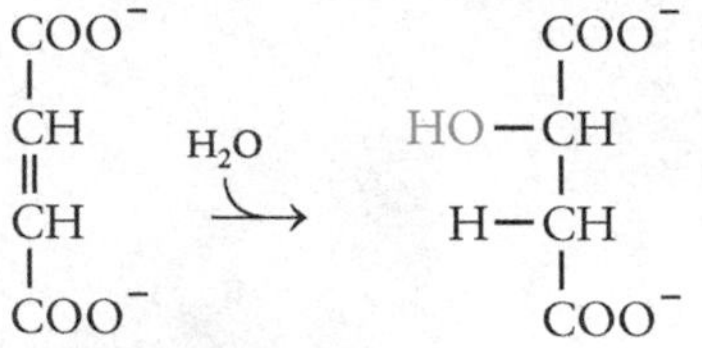

Step 7

c. The carboxylic acids (in their conjugate base form) are unchanged.
d. Water was a reactant, and H and OH were added to the substrate in the reaction.

A patient visiting the dentist. Regular cleanings by a dental hygienist and check-ups with the dentist can prevent tooth decay. [© CandyBox/Alamy]

11 Carbohydrates: Structure and Function

Dental Caries and the Sweet Tooth

Have you seen the dentist lately? Are you brushing your teeth at least twice a day? Tooth decay, known as *dental caries*, is the most common disease in the world. Why do teeth need to be cleaned regularly to avoid tooth decay? Once again, the answer can be explained by chemistry.

In chapter 3 you learned that tooth enamel, the outer part of the tooth above the gums, is composed of calcium hydroxyapatite, $Ca_5(PO_4)_3OH$, an ionic compound composed of calcium ions, Ca^{2+}, phosphate ions, PO_4^{3-}, and hydroxide ion, OH^-. While calcium hydroxyapatite is relatively insoluble in water, it does dissolve when the aqueous environment in the mouth drops below pH = 5, such as when lactic acid is present.

Saliva contains the enzyme *amylase*, which catalyzes the breakdown of complex carbohydrates into glucose. Glucose is the primary fuel for all our cells. Unfortunately, glucose is also a fuel for the anaerobic bacteria that live in our mouths, and the culprit in dental caries. Evidence suggesting that carbohydrates are involved in dental caries is a study showing that Eskimos who ate a traditional diet of meat and fish, but no carbohydrates, had few or no dental caries; whereas Eskimos who had switched to a modern diet had the same amount of dental caries as the non-Eskimo population.

Anaerobic bacteria are organisms that do not require oxygen to produce energy from glucose. Anaerobic bacteria metabolize glucose via an anaerobic biochemical process known as *lactic acid fermentation*, which produces lactic acid, the root cause of dental caries. During lactic acid fermentation, glucose is converted into lactic acid and energy. You may have heard of lactic acid fermentation in another context: it is used in the food industry to produce sourdough bread, yogurt, sauerkraut, pickles, kimchi, and other foods. Indeed, the sour taste of these foods is from the lactic acid produced by microorganisms during lactic acid fermentation.

Lactic acid is a carboxylic acid and, like all acids, donates a hydrogen ion, H^+, to water, causing an increase in the concentration of hydronium ions, H_3O^+, thus a decrease in pH to 4–5. Unfortunately, at this more acidic pH, tooth enamel literally dissolves. Although tooth enamel is constantly being regenerated, at this acidic pH the decay process occurs much faster than the regeneration process. Eventually a cavity is formed, and the only way to prevent further loss of tooth is to have a dentist remove the decayed area and "fill the tooth" with dental amalgam, an inert inorganic mixture that replaces the lost enamel.

Since anaerobic bacteria are always present in the mouth, one way to prevent the progression of dental caries is to minimize the availability of glucose, on which these bacteria thrive. The purpose of brushing your teeth is to remove the colonies of bacteria which form within a hardened biofilm around the teeth, known as dental plaque or tartar, that bathe the teeth in lactic acid. Think of that the next time you brush your teeth! ●

$$\begin{array}{c} COOH \\ | \\ H-C-OH \\ | \\ CH_3 \end{array}$$
Lactic acid

493

In the next four chapters we will consider the structure and function of the important biological molecules, referred to as biomolecules: carbohydrates, lipids, proteins, and nucleic acids. Understanding the structure and function of these biomolecules is the foundation of biochemistry. Then, in chapter 15 we will study the catabolism of carbohydrates, fats (a type of lipid) and proteins with an emphasis on the transfer of energy in these processes. In this chapter we consider the structure and function of the most abundant of the biomolecules: carbohydrates, also known as sugars.

■ 11.1 An Overview of Carbohydrates and Their Function

Have you ever eaten a plate of spaghetti the night before a big race, telling your friends you are "carbo-loading?" What you are doing is consuming carbohydrates the night before the race so that you can burn them on race day when you will need the energy. Indeed, this is the primary function of carbohydrates in our diet: to supply our cells with energy over the short term, making it possible for our cells to perform the many functions required for living. It is the chemical potential energy in the covalent bonds of carbo-hydrates that is converted into usable energy when carbohydrates are metabolized.

Carbohydrates are essential for life. Half of all the earth's carbon exists in the form of carbohydrates. Plants produce carbohydrates and oxygen, O_2, from carbon dioxide, CO_2, and water in an anabolic process that requires energy. The energy is supplied by sunlight, in a famous biochemical pathway known as **photosynthesis**, illustrated in the top half of Figure 11-1.

When we consume starch, a complex carbohydrate, it is metabolized into many glucose molecules, a process that begins in the mouth, with the help of

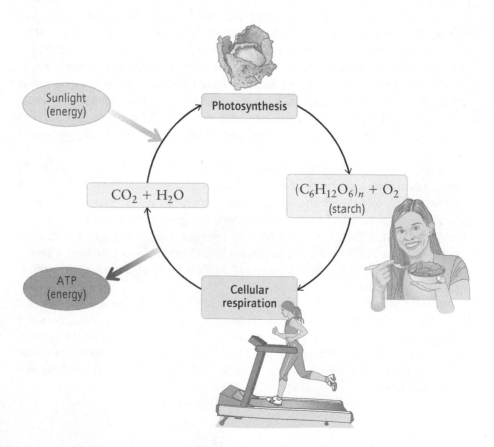

Figure 11-1 Photosynthesis and cellular respiration are the key parts of the carbon cycle: plants produce oxygen and carbohydrates from carbon dioxide and water, using the energy from sunlight (photosynthesis). Mammalian cells convert carbohydrates and oxygen to carbon dioxide and water, generating energy in the form of ATP (cellular respiration).

amylase, an enzyme present in saliva. Digestion continues in the small intestine, where *amylase* and other enzymes are secreted by the pancreas. Glucose then enters the bloodstream where it is commonly referred to as "blood sugar." Given the signal from the hormone insulin, glucose is transported through the cell membrane and into the cell, where it can be metabolized further into carbon dioxide and water, a biochemical process that produces energy for the cell known as **cellular respiration**, shown in the bottom half of Figure 11-1. In essence, the energy from sunlight is stored by plants in the form of starch through photosynthesis, which is then converted into energy, in the form of ATP, by our cells through cellular respiration. Photosynthesis and cellular respiration are part of the carbon cycle illustrated in Figure 11-1.

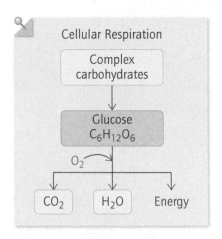

While the primary function of carbohydrates is as a fuel for cells, they also serve many other important functions. Carbohydrates provide the molecular building blocks for the synthesis of more complex molecules such as RNA and DNA, which contain the simple sugars ribose and 2-deoxyribose, respectively, as part of their structural backbone. Carbohydrates are also important in cellular recognition, where they act as molecular markers protruding from the outside of the cell membrane. Immune cells recognize these molecular markers when distinguishing your cells from foreign cells invading the body. A person with type AB blood, for example, cannot donate blood to a person with type O blood because the immune cells of the recipient will mount an attack against these foreign blood cells, because they carry a carbohydrate marker that is different from the recipient's own blood cells. This type of immune response is life threatening, and one of the reasons why blood is always tested and typed before a blood transfusion. In section 11.4 you will learn more about blood types, and why someone with type O blood can donate blood to anyone but cannot receive blood from anyone except another type O blood donor.

Every year, 100 billion (1×10^{11}) metric tons of carbon dioxide and water are converted into carbohydrates through photosynthesis in plants. 1 metric ton = 1,000 kg.

When the structure of carbohydrates were first determined, scientists noted that their composition fit the molecular formula $(CH_2O)_n$, where n is an integer 1, 2, 3. . . . So it was first hypothesized that carbohydrates might be *hydrates* (H_2O) of carbon (C), and they were given the name carbo*hydrate*s. Studies later revealed that carbohydrates are *not* actually hydrates of carbon, but are instead polyols (—OH) with a related formula. Glucose, for example, has the formula $C_6H_{12}O_6$. Although the original hypothesis was modified, as is often the case in science, the term *carbohydrate* remains with us today to describe this important class of biomolecules. Note that the names of most carbohydrates end in –*ose*, as for example gluc*ose*, lact*ose*, and amyl*ose*.

Carbohydrates are subdivided into four categories depending on the number of monosaccharide units produced when the carbohydrate is hydrolyzed. Recall from chapter 10 that a hydrolysis reaction is one that uses water— "hydro"—as a reactant to break bonds—"lysis." Monosaccharides (meaning "one sugar"), cannot be hydrolyzed into simpler carbohydrates; hence, they are also known as "simple sugars." *Di*saccharides (meaning "two sugars") are hydrolyzed into *two* monosaccharides. Oligosaccharides (meaning "a few sugars") are hydrolyzed into *three or more* monosaccharides, and *poly*saccharides (meaning "many sugars") are hydrolyzed into *many* monosaccharides. Polysaccharides are biological polymers of monosaccharides. A polymer, in general, is a large molecule composed of the same or similar repeating smaller units—the monomer. Monosaccharides are the monomer units of polysaccharides. Polysaccharides such as starch and cellulose, for example, are hydrolyzed into thousands of glucose molecules—the monomer unit.

The word *saccharide* comes from the Greek word *sakcharon*, meaning sugar.

Since monosaccharides are the monomer units for all carbohydrates, we begin our study of carbohydrates with the chemical structure of the common monosaccharides.

■ 11.2 The Structure of the Common Monosaccharides

Monosaccharides are defined as carbohydrates that cannot be hydrolyzed into simpler carbohydrates; thus, also known as "simple sugars." Monosaccharides are generally crystalline colorless solids with a sweet taste. A table comparing the relative sweetness of some common monosaccharides and disaccharides is shown in Table 11-1a. Table 11-1b shows the relative sweetness of some sugar substitutes used to sweeten diet drinks and processed foods. You may have heard of some of these sugar substitutes, which are not carbohydrates, and therefore not converted into energy—calories—yet do have an extremely sweet taste. Sugar substitutes typically far exceed the sweetness of sucrose (table sugar), one of the sweetest carbohydrates.

The Chemical Structure of Monosaccharides

The structural features that identify a monosaccharide are:

- the number of carbon atoms in the chain;
- the type of carbonyl-containing functional group: aldehyde (RCHO) or ketone (RCOR);
- the stereochemical configuration of all the centers of chirality.

The common monosaccharides are composed of a straight chain of three to six carbon atoms, with one carbonyl group that is either an aldehyde at C(1) or a ketone at C(2), and a hydroxyl group on all the other carbon atoms. You may recall from chapter 7 that we described monosaccharides as polyhydroxy aldehydes or ketones.

Most monosaccharides are chiral because they contain one or more centers of chirality. Recall from chapter 7 that a center of chirality is a tetrahedral carbon with four *different* atoms or groups of atoms bonded to it. In a monosaccharide the centers of chirality each have one hydroxyl group (—OH) and one hydrogen atom (—H). Since the distinguishing feature of a monosaccharide is often the configuration of its centers of chirality, it is common practice to draw monosaccharides as Fischer projections. In a **Fischer projection** the carbon chain of the monosaccharide is written vertically (from top to bottom) with the carbonyl group at the top. Each hydroxyl group and hydrogen atom on a center of chirality appears as a horizontal bond, either to the left or to the right, intersecting the vertical carbon chain. *In a Fischer projection, a center of chirality occurs wherever a horizontal line intersects a vertical line.* Thus, in the Fischer projection of a monosaccharide, the two possible configurations at centers of chirality are OH group left /H atom right, or vice versa, as shown for D-glyceraldehyde and L-glyceraldehyde below.

> Remember that in condensed notation an aldehyde is written —CHO.

Center of chirality

$$
\begin{array}{ccc}
 & \text{CHO} & & \text{CHO} \\
\boxed{\text{Left OH}} \quad \text{HO} & \!\!-\!\!\!\!+\!\!\!\!-\!\! & \text{H} \qquad\qquad \text{H} & \!\!-\!\!\!\!+\!\!\!\!-\!\! & \text{OH} \quad \boxed{\text{Right OH}} \\
 & \text{CH}_2\text{OH} & & \text{CH}_2\text{OH}
\end{array}
$$

L-glyceraldehyde   D-glyceraldehyde

Enantiomers

Carbon atoms that are not centers of chirality, such as the carbon atom farthest from the carbonyl group, are written in their condensed notation: CH_2OH. Note that the last carbon atom in the chain is not a center of chirality because it contains two identical atoms—two hydrogen atoms (CH_2OH). Although a Fischer projection is drawn as a flat structure, it is understood that carbon atoms with four bonds are still tetrahedral in shape: horizontal bonds

TABLE 11-1 The Relative Sweetness of Some Sugars and Sugar Substitutes Compared to Sucrose

(a)

Sugar	Type of Carbohydrate	Sweetness Relative to Sucrose (100)
Fructose	Monosaccharide	120
Sucrose	Disaccharide	100
Galactose	Monosaccharide	70
Glucose	Monosaccharide	65
Maltose	Disaccharide	50
Lactose	Disaccharide	25

(b)

Sugar Substitute	Sweetness Relative to Sucrose
Saccharin (now banned)	300
Aspartame (NutraSweet and Equal)	180–220
Sucralose (Splenda)	600
Alitame (not approved in the United States)	2,000

are understood to be projecting toward the viewer (wedges), and vertical bonds are understood to be projecting away from the viewer (dashes), indeed, this is why it is called a "projection."

Recall from chapter 7 that a molecule with one center of chirality is chiral because it is nonsuperimposable on its mirror image. Nonsuperimposable mirror image stereoisomers are known as enantiomers, and the centers of chirality have the opposite configuration. D- and L-glyceraldehyde are enantiomers. As you can see in the Fischer projections, D-glyceraldehyde has the OH group at the center of chirality on the right, and L-glyceraldehyde has the OH group on the left. Nature, however, only produces D-glyceraldehyde.

Most monosaccharides have multiple centers of chirality. From the number of centers of chirality, we can calculate the maximum number of stereoisomers possible using the formula 2^n, where n is the number of centers of chirality. Compounds with two centers of chirality (n = 2), for example, can have up to $2^2 = 2 \times 2 = 4$ stereoisomers, as shown in the Fischer projections below for the four stereoisomers of monosaccharides with four carbon atoms and an aldehyde.

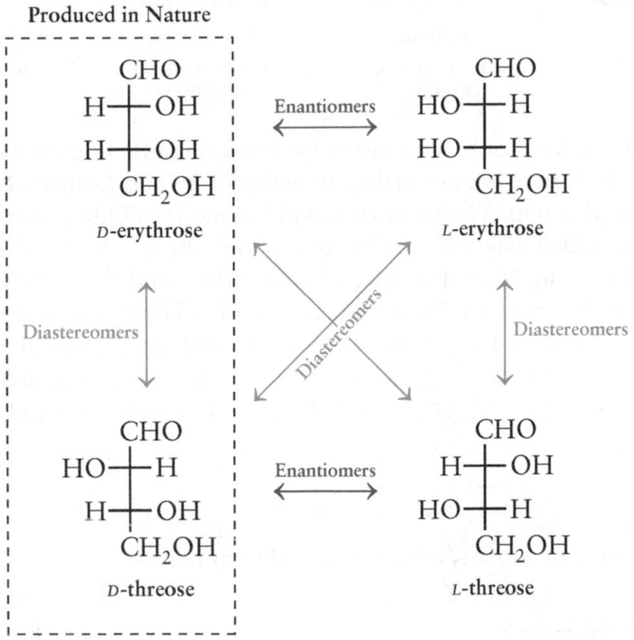

The two carbon atoms in the middle of the molecules are both centers of chirality. Each of these stereoisomers has a different configuration at one or both centers of chirality, easily interpreted by noting the orientation of the OH groups/H atoms at the centers of chirality (intersections of vertical and horizontal lines). Among these four stereoisomers we have two pairs of enantiomers (indicated by the red arrows): stereoisomers with the opposite configuration at every center of chirality. We also have four pairs of **diasteromers**

(indicated by the blue arrows): nonsuperimposable, non-mirror image stereo-isomers. *Diastereomers have a different configuration at one or more centers of chirality but not at every center of chirality, whereas enantiomers have a different configuration at every center of chirality.* Monosaccharides that are enantiomers are identified by the prefixes D- and L- or (+) and (−), while diastereomers are given different names altogether, as seen in the examples of D- and L- erythrose and D- and L-threose.

Since nature produces only D-sugars, it is useful to know how to look at a Fischer projection and determine whether or not it is a D-sugar. The —OH group on the center of chirality farthest from the carbonyl group is on the right in a D-sugar.

> D-**sugar:** The —OH group at the intersection farthest from the carbonyl group is on the *right*.
>
> L-**sugar:** The —OH group at the intersection farthest from the carbonyl group is on the *left*.

Consider the sugars D-threose, D-ribose, and D-glucose. In all four Fischer projections, the last center of chirality has the —OH group on the right, as highlighted in yellow in the structures shown below.

You can assume that a monosaccharide is the D-sugar when it is listed without its D or L designation.

Examples of D-sugars

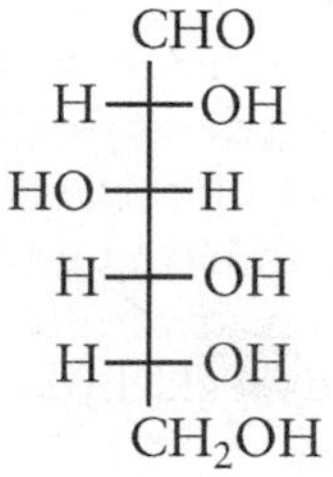

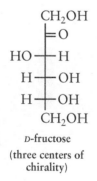

D-glucose
(four centers of
chirality)

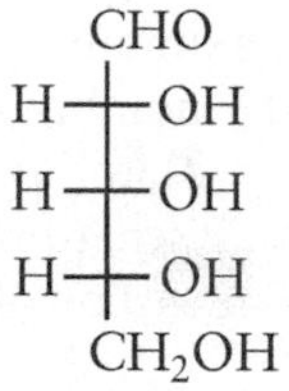

D-ribose
(three centers of
chirality)

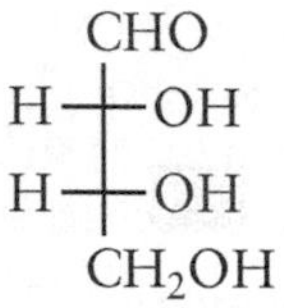

D-erythrose
(two centers of
chirality)

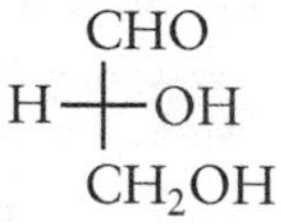

D-glyceraldehyde
(one center of
chirality)

Thus far we have only seen monosaccharides containing an aldehyde carbonyl group, but there are an analogous series of D- and L-sugars containing a ketone carbonyl group. The most common ketone containing monosaccharide is D-fructose, which has six carbon atoms and three centers of chirality, as shown in the margin. Note that the carbonyl carbon and the two carbon atoms at each end of the chain are not centers of chirality. The carbonyl group is not a center of chirality because it is not tetrahedral, and the carbon atoms on each end are not centers of chirality because they contain two hydrogen atoms (identical atoms). By definition, there is only one configuration for a CH_2OH group.

D-fructose
(three centers of
chirality)

WORKED EXERCISE Monosaccharide Structure

11-1 Fischer projections of four monosaccharides are shown below:

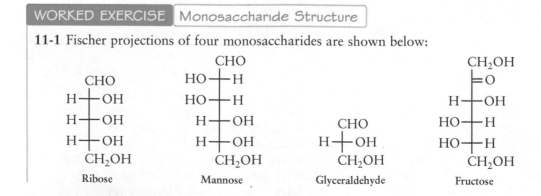

Ribose   Mannose   Glyceraldehyde   Fructose

a. For each monosaccharide, indicate how many carbon atoms it contains.

b. For each monosaccharide, indicate the number of hydroxyl groups; the number of centers of chirality; and whether the carbonyl group is an aldehyde or a ketone.

c. For each monosaccharide, indicate whether it is a D-sugar or an L-sugar. Provide the complete name of the monosaccharide by including the appropriate D- or L-prefix.

d. Why is the hydroxyl group on the carbon atom farthest from the carbonyl group written CH$_2$OH and not as an intersection of a horizontal and vertical line?

e. How is D-mannose different from D-glucose? Is D-mannose an enantiomer or a diastereomer of D-glucose? How can you tell?

f. Are D-ribose and D-mannose stereoisomers? Explain.

g. What makes all of these sugars chiral?

h. Based on the structure shown for D-ribose, write a Fischer projection for L-ribose.

Solution

11-1 a. Ribose, five carbon atoms
Mannose, six carbon atoms
Glyceraldehyde, three carbon atoms
Fructose, six carbon atoms

b. Ribose: four hydroxyl groups; three centers of chirality; aldehyde
Mannose: five hydroxyl groups; four centers of chirality; aldehyde
Glyceraldehyde: two hydroxyl groups; one center of chirality; aldehyde
Fructose: five hydroxyl groups; three centers of chirality; ketone

c. Ribose: D-ribose
Mannose: D-mannose
Glyceraldehyde: D-glyceraldehyde
Fructose: L-fructose

d. The carbon atom farthest from the carbonyl group is not a center of chirality (it has two identical atoms) so the convention for a Fischer projection is to write it in condensed notation.

e. D-mannose and D-glucose differ in only the configuration at C(2), which in the Fischer projection has the hydroxyl group on the left in D-mannose, and on the right in D-glucose. Thus, D-mannose is a diastereomer of D-glucose because it differs at one or more centers of chirality but not at all centers of chirality.

f. D-ribose and D-mannose are not stereoisomers because they do not have the same molecular formula: ribose has five carbons while mannose has six.

g. They all have one or more centers of chirality. In a Fischer projection, the presence of intersecting vertical and horizontal lines indicates centers of chirality.

h. In the Fischer projection of D-ribose all three centers of chirality have the OH group on the right. So, to write the Fischer projection of L-ribose, its enantiomer, we write all three OH groups in the opposite direction, that is to the left:

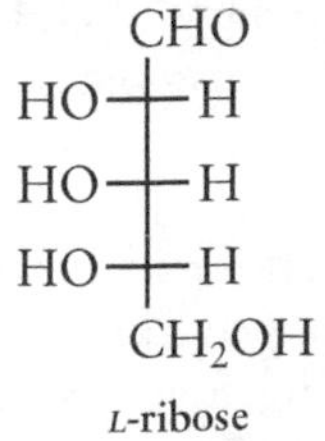

CHO
HO──H
HO──H
HO──H
CH$_2$OH
L-ribose

PRACTICE EXERCISES

1. Dihydroxyacetone is an achiral monosaccharide.

   $$\begin{array}{c} CH_2OH \\ | \\ O \\ | \\ CH_2OH \end{array}$$

   Dihydroxyacetone

   a. What aspect of the Fischer projection at right shows that there are no centers of chirality?

   b. How many carbon atoms does this molecule contain? How many hydroxyl groups?

   c. How is this compound different from D-glyceraldehyde?

   d. What type of isomers are D-glyceraldehyde and dihydroxyacetone?

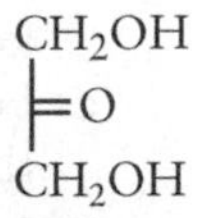

2. Nature produces the following five carbon, aldehyde-containing, $C_5H_{10}O_5$ monosaccharides.

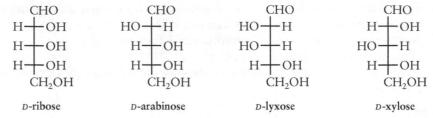

   | D-ribose | D-arabinose | D-lyxose | D-xylose |

   a. How is D-ribose different from D-arabinose? Are they enantiomers or diastereomers?

   b. How is D-ribose different from D-lyxose? Are they enantiomers or diastereomers?

   c. How is D-arabinose different from D-lyxose? Are they enantiomers or diastereomers?

   d. Are these monosaccharides D-sugars or L-sugars, and how can you tell?

   e. How many centers of chirality does each of these sugars contain? Using the formula 2^n, what is the maximum number of stereoisomers? Which stereoisomers are not shown?

   f. Write the Fischer projection of L-arabinose. Is it the enantiomer or a diastereomer of D-arabinose? Is L-arabinose the enantiomer or a diastereomer of D-ribose? Is L-arabinose produced in nature?

Ring Forms of Monosaccharides

Monosaccharides with five or more carbons spontaneously undergo a ring forming reaction in aqueous solution, as shown below for D-glucose. A new carbon-oxygen single bond (shown in green) forms between the oxygen atom (shown in red) on the last center of chirality and the carbonyl carbon. This new bond creates a five or six membered ring with a new hydroxyl group on what was formerly the carbonyl carbon. In the case of D-glucose, shown below, a six membered ring containing five carbons and one oxygen—known as a **pyranose ring**—is formed. The other common ring form is a five membered ring containing one oxygen atom, known as a **furanose ring**.

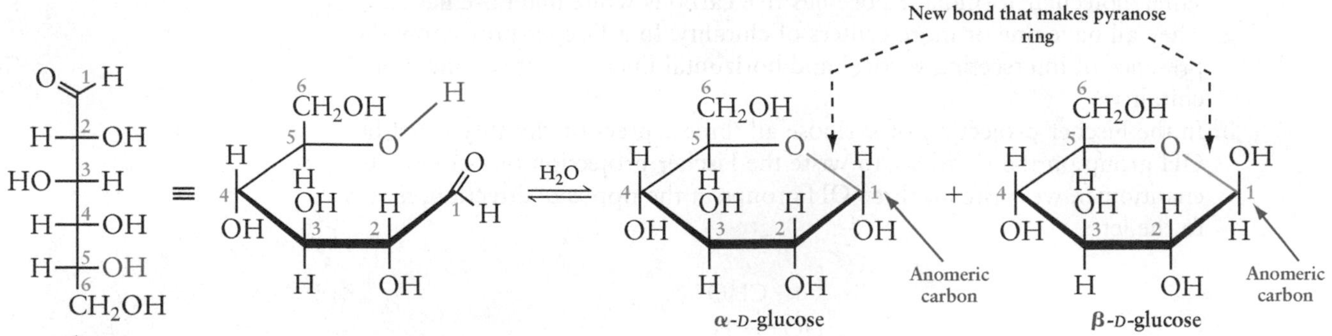

In forming a pyranose or furanose ring, a new center of chirality is created at what was formerly the carbonyl carbon. This new center of chirality, only present in the ring form, is known as the **anomeric center** or anomeric carbon. Since there are always two possible configurations for a center of chirality, two diastereomeric rings are formed, differing in configuration at the anomeric center. The ring form with the OH group at the anomeric center below the ring is known as the **α-anomer**, and the ring form with the OH group above the ring is known as the **β-anomer**.

In aqueous solution, the two ring forms of a monosaccharide are rapidly interconverting via the open-chain form of the monosaccharide, creating an equilibrium mixture of all three forms of the monosaccharide: the open-chain form, the α-anomer, and the β-anomer. The open-chain form is usually depicted as a Fischer projection, and the ring forms are depicted as Haworth projections. At equilibrium, the open-chain form is present in only trace amounts.

You may have noticed that monosaccharide rings are drawn somewhat differently from the ring structures you have encountered in previous chapters, a representation known as a **Haworth projection**. Haworth projections are generally reserved for carbohydrates where stereochemical configuration is often the distinguishing feature. Moreover, Fischer projections are readily translated into Haworth projections in a few simple steps. Convert a Fischer projection into a Haworth projection by turning the Fischer projection 90° in the clockwise direction: *groups that appear on the right side in a Fischer projection appear below the ring in a Haworth projection, and groups that appear on the left side in a Fischer projection appear above the ring in a Haworth projection.* To further simplify the interpretation of Haworth projections, the following additional conventions are observed when writing Haworth projections:

(1) Always write a furanose ring with the ring oxygen at the top, and a pyranose ring with the ring oxygen at the top right, as shown in the furanose and pyranose rings in the margin.

(2) The last center of chirality (farthest from the carbonyl) is the carbon atom with a bond to the ring oxygen and to the CH$_2$OH group, which always appears *above* the ring and at the top left in a D-sugar.

(3) The ring is written so that the anomeric center is positioned at the far right with the —OH group above or below the ring depending on whether it is the β- or α-anomer, respectively.

The most common furanoses are α-D-fructose and β-D-fructose, which form when the —OH group at the last center of chirality in the open-chain form reacts with the ketone, as shown below. The reaction occurs in the same way as described previously for D-glucose, except a ketone rather than an aldehyde is involved, so the anomeric center has a bond to a second R group rather than a hydrogen atom. The furanose with the hydroxyl group below the anomeric center is the α-anomer, α-D-fructose, and the furanose with the hydroxyl group above the anomeric center is the β-anomer, β-D-fructose. The last center of chirality in D-fructose is at C(5), so the CH$_2$OH group attached to C(5) appears above the ring, since this is a D-sugar. The configuration at C(3) and C(4) is unique to D-fructose, readily translated from the Fischer projection of D-fructose, where we see the C(3) hydroxyl is left, and the C(4) hydroxyl is right and, hence, above and below the ring, respectively. To gain more familiarity with the ring forms of monosaccharides, perform *Using Molecular Models 11-1: Converting Between α-D-Glucose and β-D-Glucose.*

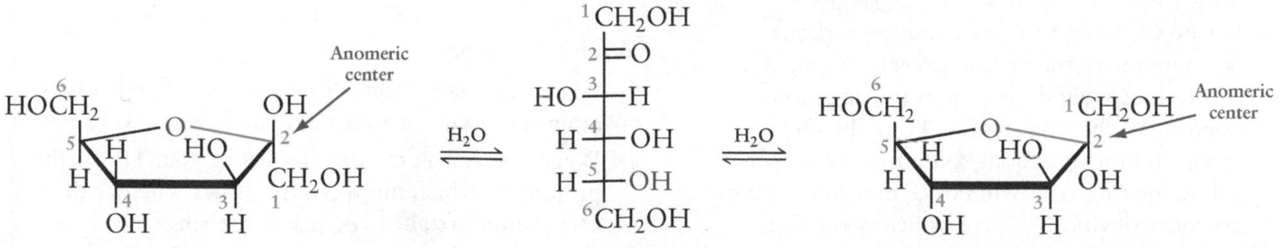

> **α-anomer:** —OH at anomeric center is below the ring
>
> **β-anomer:** —OH at anomeric center is above the ring

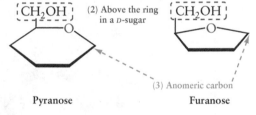

(2) Above the ring in a D-sugar

(3) Anomeric carbon

Pyranose   Furanose

β-D-fructose   ⇌(H$_2$O)   D-fructose   ⇌(H$_2$O)   α-D-fructose

Using Molecular Models 11-1 Converting Between α-D-Glucose and β-D-Glucose

Construct β-D-Glucose

1. Obtain 6 black carbon atoms, 6 red oxygen atoms, 12 light-blue hydrogen atoms, and 24 straight bonds.

2. Construct a model of β-D-glucose:

    a. Form a six-membered ring using five carbon atoms and one oxygen atom as shown in the models below. Let the ring oxygen serve as a reference point by placing it in the position shown in the ball-and-stick model and tube model below (top right according to the convention used for writing a Haworth projection).

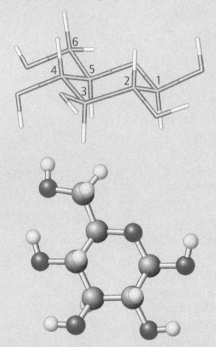

   b. To build the D-sugar, attach the C(6) CH₂OH group *above* the ring on the top left carbon atom, C(5), located next to the ring oxygen.

   c. The carbon atom on the far right of the ring and bonded to two oxygen atoms is the anomeric center. This is C(1), the first carbon in the ring. To create the β-anomer, place the OH group on the anomeric carbon *above* the ring. This OH group and the CH₂OH group from part (c) should be on the same side of the ring, which is the strict definition of a β-anomer, and drawn by convention, above the ring.

   d. Locate C(2) and C(4), by counting carbons beginning with the anomeric center, C(1), and proceeding around the ring in the direction opposite of the ring oxygen. Place the OH groups on these carbons so that they are both below the ring. You will notice that some groups are more obviously above or below the ring,

while other groups appear only slightly above or below the ring. This is normal in rings, but it is ignored when writing a Haworth projection.

   e. At C(3), place the OH group so that it is above the ring.

   f. Add hydrogen atoms to all remaining positions (holes in the model atoms). Answer Inquiry Questions 3–5.

Inquiry Questions

3. Compare your model to the Haworth projection shown below.

    a. Is your molecule flat like this drawing? Explain. From what angle do you need to view your model in order to obtain the vantage point of the Haworth projection shown? What advantages does a Haworth projection have?

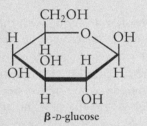

β-D-glucose

   b. Locate the anomeric center in your model. How is it different from all the other carbon atoms in the model?

Conversion to α-D-glucose

4. Convert your model of β-D-glucose into a model of α-D-glucose. Your model should look like the tube model shown below.

5. Write a Haworth projection of α-D-glucose. Answer Inquiry Questions 6–7.

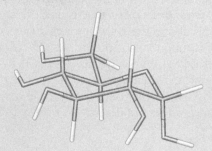

Inquiry Questions

6. To convert the β-anomer to the α-anomer, what two atoms or groups of atoms did you have to exchange?

7. When comparing the tube model to your Haworth projection, which more clearly shows whether the OH groups are above or below the ring?

WORKED EXERCISES | Furanose and Pyranose Forms of a Monosaccharide

11-2 In aqueous solution, D-ribose, the sugar found in RNA, exists in three forms at equilibrium:

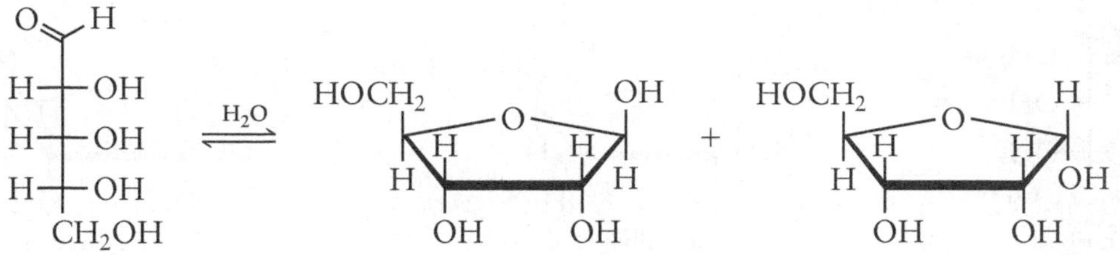

a. How many carbon atoms does ribose contain?

b. Are the ring forms of ribose pyranose or furanose rings?

c. Does ribose contain a ketone or an aldehyde? How can you tell by looking at the Fischer projection? How has this carbonyl group changed in going to the ring forms? Label this carbon in the ring structures. What is this carbon called in the ring form?

d. Identify α-D-ribose and β-D-ribose by filling in the blanks above.

e. Are α-D-ribose and β-D-ribose *diastereomers* or *enantiomers*? Explain.

f. Why are double-headed arrows used in this equation?

g. What part of the ring structures indicates it is a D-sugar? What part of the Fischer projection indicates it is a D-sugar?

11-3 Haworth projections of two monosaccharides are shown below:

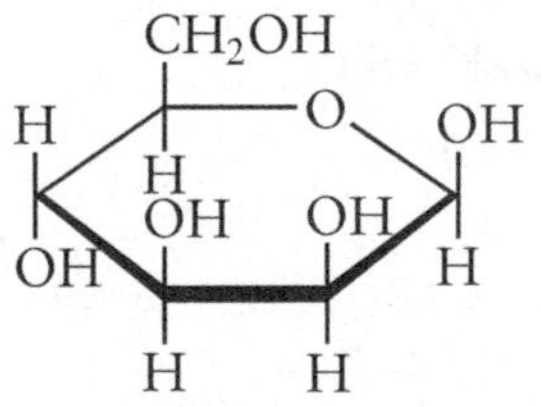

a. How many carbon atoms does each of these monosaccharides contain? What type of rings are they, furanose or pyranose? In the open-chain form are these sugars aldehydes or ketones?

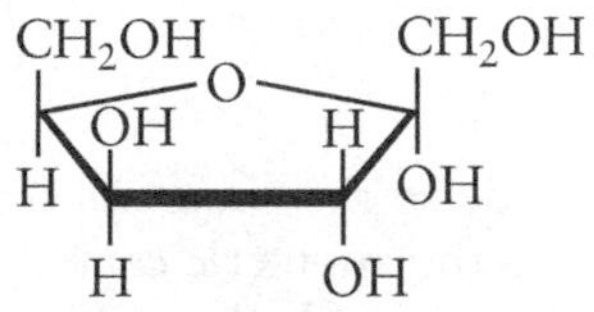

b. Identify the anomeric center in each sugar and then fill in the blanks with the correct label, α- or β-.

c. Why does the anomeric center appear at C(2) in D-sorbose but at C(1) in D-mannose?

d. Are these monosaccharides D-sugars? How can you tell?

e. Label the centers of chirality in each sugar with an *. Which center of chirality disappears when the ring opens to the open-chain form? Explain.

f. Write the open-chain form of each sugar.

g. What is the structural difference between D-mannose and D-glucose? Are D-mannose and D-glucose stereoisomers? If so, are they diastereomers or enantiomers?

h. Structurally, what makes D-sorbose a diastereomer of D-fructose? Why are they not classified as enantiomers?

Solutions

11-2 a. Ribose contains five carbon atoms.

  b. The ring forms of ribose are furanose rings: a five membered ring containing one oxygen atom in the ring.

   c. Ribose contains an aldehyde, as indicated by —CHO in the Fischer projection. In going to the ring forms, the carbonyl group (C=O) becomes the anomeric carbon, the ring carbon bonded to an —OH group and the ring —OR group. See labels below.

   d.

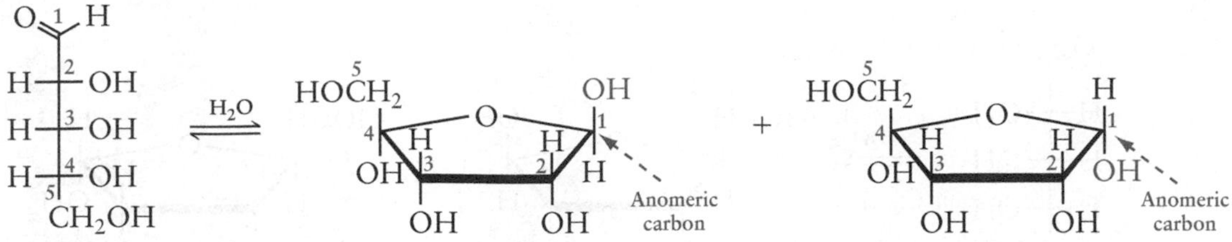

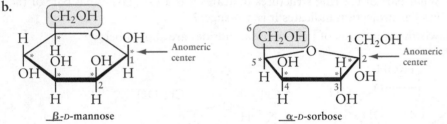

D-ribose   β-D-ribose   α-D-ribose

   e. α-D-ribose and β-D-ribose are diastereomers because they differ in the configuration at one center of chirality, the anomeric center, and not at all centers of chirality.

   f. Double-headed arrows are used in this equation because the reaction is reversible; the rings reopen to the open-chain form of D-ribose.

   g. The last carbon in the chain (CH$_2$OH), bonded to the last center of chirality, is positioned above the ring, in a Haworth projection. In a Fischer projection, a D-sugar has the hydroxyl group on the last center of chirality positioned horizontally on the right.

11-3 a. Both of these monosaccharides have six carbon atoms. D-mannose forms a pyranose ring, whereas D-sorbose forms a furanose ring. D-mannose has an aldehyde at C(1) and D-sorbose has a ketone at C(2) in the open-chain form.

   b.

β-D-mannose   α-D-sorbose

   c. D-mannose is an aldehyde so the anomeric center appears at C(1), whereas D-sorbose is a ketone so the anomeric center appears at C(2).

   d. Yes, they are D-sugars, because the last CH$_2$OH (farthest from the anomeric center) is above the ring in a Haworth projection, as highlighted in the red box.

   e. See asterisks (*) above. The anomeric center disappears as a center of chirality in going to the open-chain form, where it becomes a carbonyl carbon, which is not a center of chirality.

   f.

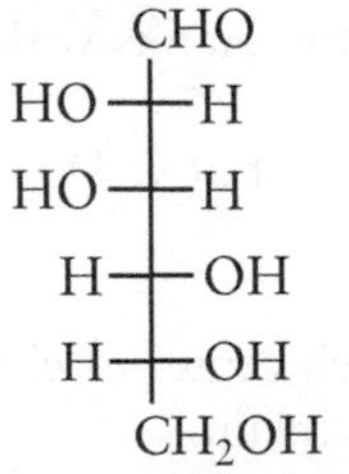

D-mannose 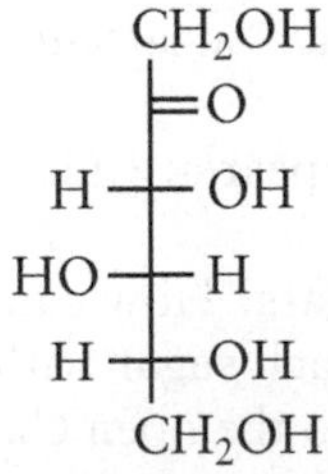   D-sorbose

   g. The only difference between D-mannose and D-glucose is the configuration at C(2) where D-mannose has its OH group on the left and D-glucose has it on the right. Since only one center of chirality is different (and not all), these two sugars are diastereomers.

   h. The fact that D-sorbose and D-fructose have the same chemical formula, and the same connectivity of atoms, but a different configuration at one or more, but not all, centers of chirality makes them diastereomers: C(3) and C(4) are different, but C(5) is the same. They are not enantiomers because they do not have a different configuration at every center of chirality (and they do not have the same name).

3   The two ring forms of D-galactose are shown below.

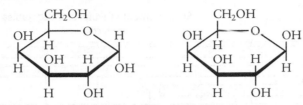

a. Identify which is the α-anomer and which is the β-anomer. Provide the complete names of these two monosaccharides.
b. Number each carbon atom in the ring. Which carbon atom is the anomeric center?
c. Are these furanose or pyranose rings? By convention, where should the ring oxygen be placed?
d. Using the Haworth projections, write the open-chain form of D-galactose.
e. How can you tell these sugars are D-sugars when looking at the Haworth projections? How can you tell they are D-sugars by looking at the Fischer projection?
f. Is D-galactose an enantiomer or a diastereomer of D-glucose? Explain.
g. In aqueous solution, what three structures are present at equilibrium? What are they called? Which structure is present in the lowest concentration at equilibrium?

4   Answer the questions below for the two ring forms of the monosaccharide shown.

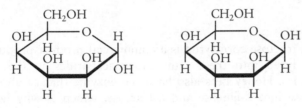

a. Label the anomeric centers with a *.
b. Which structure is the α-anomer and which is the β-anomer?
c. How is the α-anomer of this sugar different from the α-anomer of D-galactose shown in the previous problem?
d. Write the open-chain form of this monosaccharide.

5   For the Haworth projection of D-gulose shown in the margin:

a. List three conventions that indicate that this is a properly drawn Haworth projection.
b. Identify the anomeric center and fill in the blank with the correct α- or β- label.
c. How can you tell that this is a D-sugar?
d. Write the open-chain form of D-gulose. Does the open-chain form contain a ketone or an aldehyde? How can you tell from looking at the Haworth projection?
e. Label the centers of chirality in the Haworth projection with an *. Which center of chirality disappears in going to the open-chain form? Show this by labeling the centers of chirality in the open-chain form in (d).
f. What is the difference between D-gulose and D-glucose? Are D-gulose and D-glucose stereoisomers? If so, are they diastereomers or enantiomers?
g. Is D-gulose produced in nature? Explain.

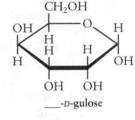

___-D-gulose

11.3 The Structure of Complex Carbohydrates

Now that you have an understanding of the basic structure of a monosaccharide, we can consider the structure of disaccharides and polysaccharides, complex carbohydrates derived from two or many monosaccharides, respectively. Sucrose, the most common disaccharide, is found in table sugar and also in many fruits, as shown in Table 11-2, which also shows the relative amount of glucose and fructose, in various fruits. When we digest sucrose it is hydrolyzed into its two monosaccharide components D-glucose and D-fructose.

TABLE 11-2 Relative Amount of Glucose, Fructose, and Sucrose in Some Fruits and Vegetables

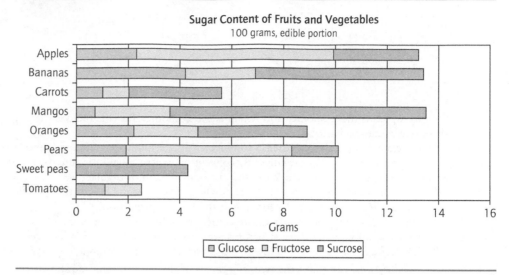

High fructose corn syrup is a commercial product produced when sucrose is hydrolyzed into these same monosaccharides in the presence of certain enzymes. Honey bees also have the enzyme that catalyzes the hydrolysis of sucrose into D-glucose and D-fructose, which is why honey is also high in D-fructose.

Disaccharides

Disaccharides are defined as carbohydrates that upon hydrolysis (reaction with water) yield two monosaccharides: either identical or different monosaccharides. For example, hydrolysis of maltose yields D-glucose, whereas hydrolysis of sucrose yields a 50:50 mixture of D-glucose and D-fructose. Hence, a disaccharide is in part characterized by the identity of its two monosaccharide components.

The two monosaccharide components of a disaccharide are joined by a covalent bond between the anomeric center of one monosaccharide and the oxygen atom of one of the hydroxyl groups on the other monosaccharide. Thus, the other two features that characterize a disaccharide are the configuration of the anomeric center(s); and which hydroxyl group on the second monosaccharide is involved in the covalent bond joining the two monosaccharides. The C—O bond joining the two monosaccharide components is known as a **glycosidic bond** or **glycosidic linkage**. The glycosidic bond is always formed between the anomeric center of one monosaccharide (drawn on the left by convention) and one of the hydroxyl (—OH) groups on the second monosaccharide (drawn on the right or below by convention). *Thus, instead of an —OH group at the anomeric center, a disaccharide has an —OR group at the anomeric center of the first monosaccharide, where "R" is the second monosaccharide.*

In contrast to the anomeric center of a monosaccharide, the anomeric center at a glycosidic linkage is a stable functional group, so it does not open and close in aqueous solution, and therefore, cannot change configuration from the α-anomer to the β-anomer. A catalyst is required to hydrolyze a glycosidic bond. In the laboratory, an acid serves as the catalyst, and in biochemical reactions, a specific enzyme is required to hydrolyze a glycosidic linkage.

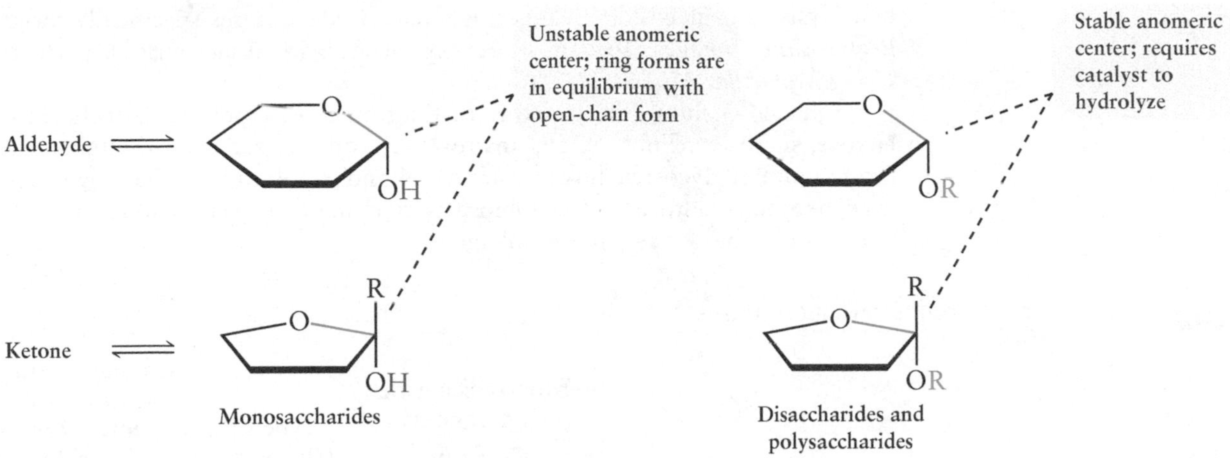

Aldehyde ⇌   Unstable anomeric
center; ring forms are
in equilibrium with
open-chain form

OH

Stable anomeric
center; requires
catalyst to
hydrolyze

OR

Ketone ⇌

R

OH

Monosaccharides

R

OR

**Disaccharides and
polysaccharides**

Consider for example the structure of lactose, the disaccharide found in milk and sometimes referred to as "milk sugar." It is hydrolyzed to D-galactose and D-glucose in the presence of acid or the enzyme *lactase*. The two mono-saccharide components of lactose are joined by a $\beta(1 \rightarrow 4)$ glycosidic linkage between D-galactose and D-glucose, as shown below. The designation $\beta(1 \rightarrow 4)$ signifies that the configuration of the anomeric carbon in D-galactose (sugar on left) is the β-anomer, and that the anomeric center, C(1) of D-galactose, is linked to the hydroxyl group at C(4) of D-glucose (sugar on right). The chemical formula for a disaccharide is $C_{12}H_{22}O_{11}$, which reflects the loss of a molecule of water in its formation (the H from the C(4) —OH, and the —OH from the anomeric center).

Lactose
$C_{12}H_{22}O_{11}$

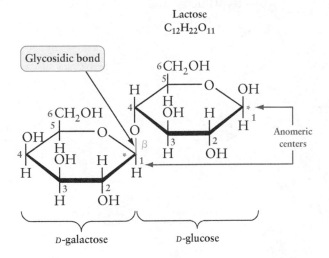

Glycosidic bond

Anomeric
centers

D-galactose   D-glucose

Disaccharides that are connected at both anomeric centers are known as non-reducing sugars.

Note that there are two anomeric centers in a disaccharide. The other anomeric center in lactose (far right) contains a hydroxyl group, and therefore the second ring in lactose opens and closes in aqueous solution to form a mixture of both the α- and β-anomer just as it does in a monosaccharide. Due to the complexity of these structures, however, usually only one anomer is shown, and the other anomer is understood to be present.

Sucrose is unique because the glycosidic linkage occurs between the anomeric centers of both sugars: D-glucose (top) and D-fructose (bottom). Therefore, both rings are stable; neither ring is opening and closing. Hence, both anomeric centers are fixed as the α-glucose anomer and the β-fructose anomer.

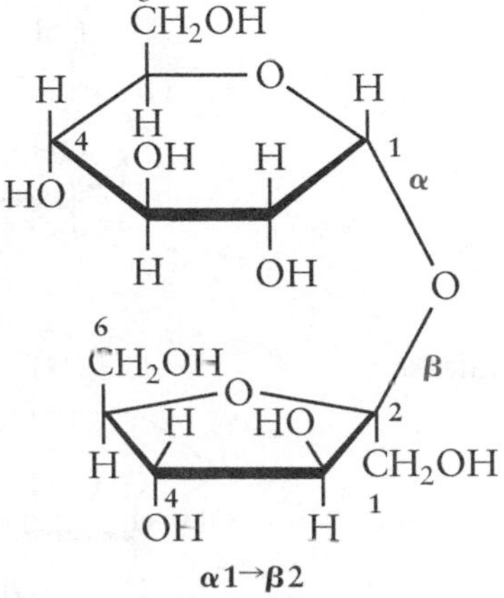

Glycosidases are enzymes that catalyze the hydrolysis of various types of glycosidic linkages. For example, α-glycosidases like *amylase* catalyze the

$\alpha 1 \rightarrow \beta 2$

Sucrose

hydrolysis of α-glycosidic linkages, whereas *β-glycosidases* specifically target β-glycosidic linkages. To gain more experience with disaccharides, perform *Using Molecular Models 11-2: Maltose.*

Table 11-3 shows the structure of the four most common disaccharides: lactose, sucrose, cellobiose, and maltose. As you can see from the table, the most common glycosidic linkages are α-1,4 and β-1,4. Indeed, the only difference between maltose and cellobiose is that maltose has an α-1,4 linkage whereas cellobiose has a β-1,4 linkage.

TABLE 11-3 Structure of the Common Disaccharides

Disaccharide Name	Haworth Projection	Monosaccharide Components (listed from left to right)	Type of Glycosidic Bond	Configuration of Anomeric Center(s) in Glycosidic Bond
Lactose	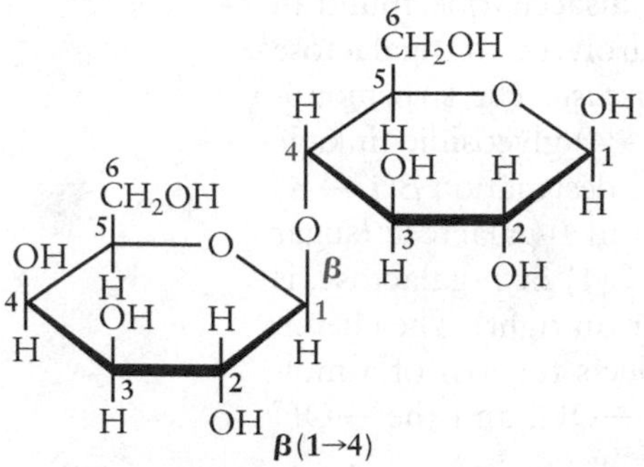	Galactose and Glucose	1, 4	β-
Sucrose	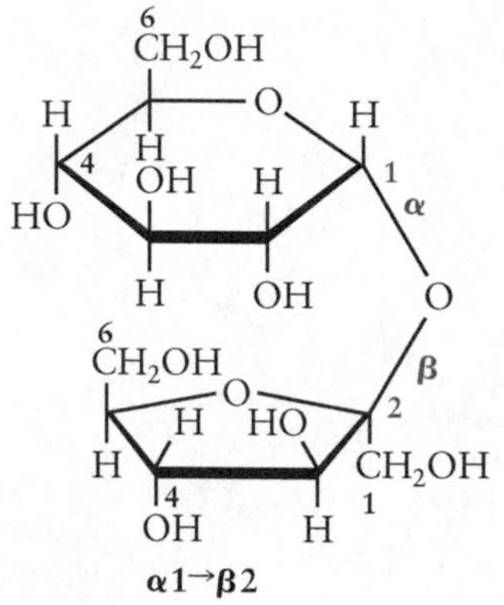	Glucose and Fructose	1, 2	$\alpha 1 \rightarrow \beta 2$
Cellobiose	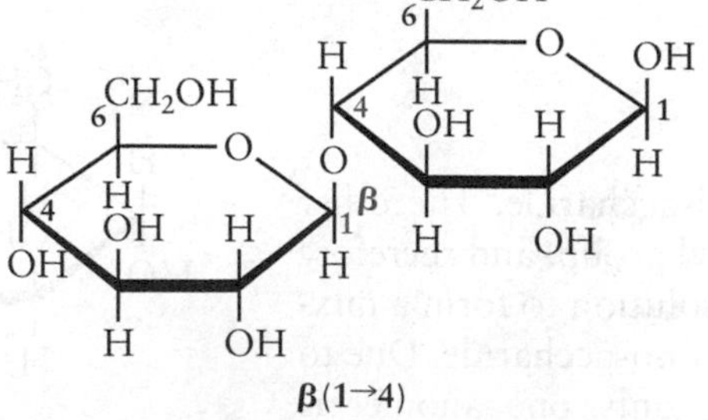	Glucose	1, 4	β-
Maltose	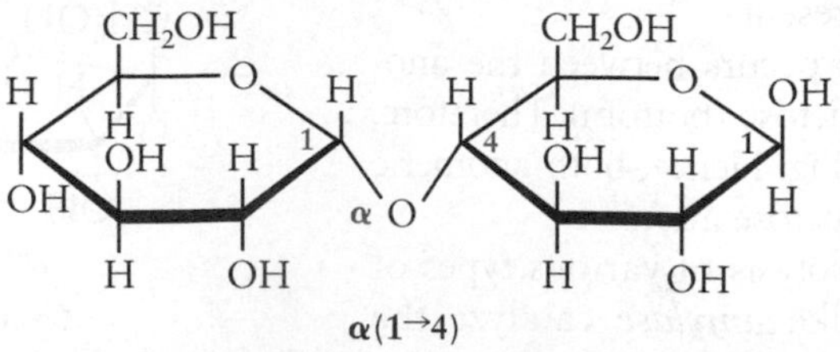	Glucose	1, 4	α-

Using Molecular Models 11-2 Maltose

This modeling exercise requires that you work with a partner and/or use two model kits.

Construct Maltose

1. Build a model of α-D-glucose as described in step 4 of *Using Molecular Models 11-1: Converting Between α-D-Glucose and β-D-Glucose.* Answer Inquiry Question 4.

2. Find a partner who has also built a model of either α- or β-D-glucose.

3. To build the disaccharide maltose, you will need to link your two sugar models together. To help you do this, examine the structure of maltose in Table 11-3, and use α-D-glucose from step 1 as the sugar on the left. Then, determine *where* to connect the two sugars. *Hint:* You will lose a molecule of H_2O in the process of making maltose from two glucose molecules. Answer Inquiry Questions 5-8.

Inquiry Questions

4. Which bond is made and broken as the ring is opened and closed?

5. Did you produce H_2O (H and OH) as a by-product when you joined the two monosaccharides? Where did the H atom come from? Where did the OH group come from?

6. Why is your model classified as a disaccharide? Which bond in your model is the glycosidic bond?

7. Did the configuration of the C(4) hydroxyl group in the second sugar change when the model of maltose was made?

WORKED EXERCISES   Disaccharide Structure

11-4 Answer the questions below for maltose:

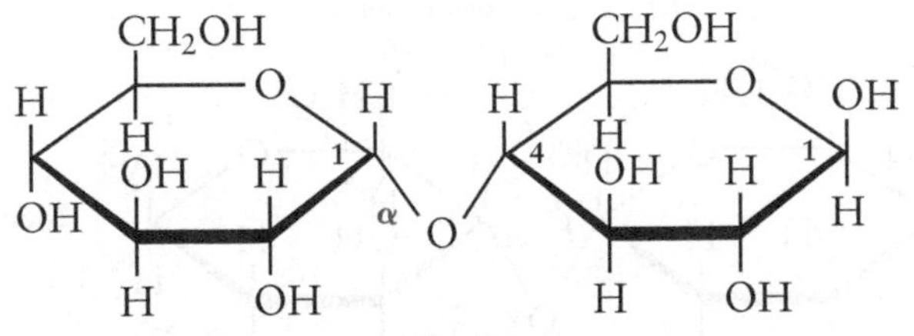

CH₂OH ... CH₂OH

α(1→4)

a. Is maltose a monosaccharide, a disaccharide, or a polysaccharide? Explain.
b. Are the monosaccharide components of this disaccharide the same or different? Name them.
c. Identify the anomeric centers in this carbohydrate with an asterisk. Identify the glycosidic bond.
d. Is the glycosidic linkage α- or β-? Is this linkage stable?
e. Is the anomeric center on the second sugar stable? What impact does the stability of the anomeric center have on the structure of the sugar?
f. What type of linkage does maltose have? $\alpha(1 \rightarrow 2)$, $\alpha(1 \rightarrow 3)$, $\alpha(1 \rightarrow 4)$, $\alpha(1 \rightarrow 6)$, $\beta(1 \rightarrow 2)$, $\beta(1 \rightarrow 3)$, $\beta(1 \rightarrow 4)$, or $\beta(1 \rightarrow 6)$? What does this mean?

11-5 Melibiose is a disaccharide similar to lactose except it has an $\alpha(1 \rightarrow 4)$ linkage. Draw the structure of melibiose.

Solutions

11-4 a. Maltose is a disaccharide because when it is hydrolyzed, *two* monosaccharides are formed.

b. They are the same monosaccharide: D-glucose.

c. See below.

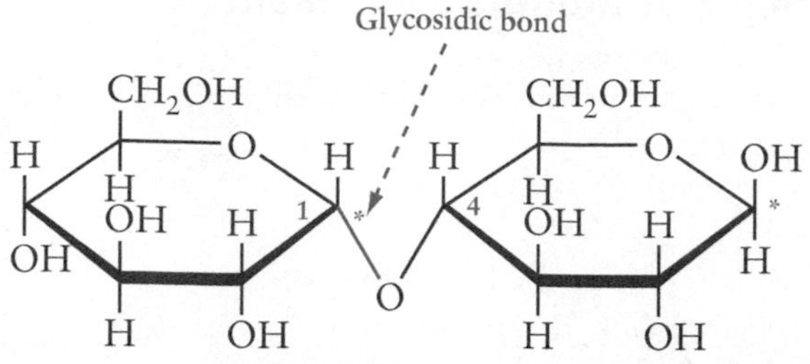

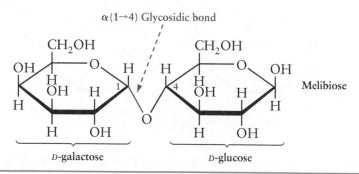

Maltose

d. The glycosidic linkage is α because the —OR group on the anomeric center of the sugar on the left is below the ring. A glycosidic linkage is stable because it contains an anomeric center with an —OR group rather than an —OH group so it cannot open up to its open-chain form.

e. The anomeric center on the second sugar is not stable: it opens and closes via its open-chain form because it contains an anomeric center with an OH group. This means that maltose exists in two forms: the second ring is in both the α- and β-forms, although only the β-form is shown here.

f. Maltose has an $\alpha(1 \rightarrow 4)$ linkage because the anomeric center in the first sugar is at C(1), and the alcohol in the second sugar involved in the glycosidic linkage is at C(4). It is an α-glycosidic linkage because the —OR group on the anomeric center is below the ring.

11-5 We write the structure of D-galactose on the left, and then write the structure of D-glucose on the right. Be sure you have placed each hydroxyl group at C(2)-C(4) correctly because this is the identifying feature of the monosaccharide. Place the oxygen atom on the anomeric center of the first sugar below the ring, α-, and connect it to the oxygen on C(4) of the second sugar, which is below the ring in glucose.

$\alpha(1\rightarrow4)$ Glycosidic bond

Melibiose

D-galactose   D-glucose

PRACTICE EXERCISES

6 Use Table 11-3 to answer the following questions about sucrose.
  a. What simple sugars are formed when sucrose is hydrolyzed?
  b. Sucrose is unusual in that the linkage occurs between the anomeric centers of both monosaccharides: the ___-anomer of D-glucose and the ___-anomer of D-fructose.
  c. Where is sucrose found in nature?

7 Using Table 11-3, answer the following questions for cellobiose.
  a. Why is cellobiose classified as a disaccharide?
  b. What monosaccharides are produced when cellobiose is hydrolyzed?
  c. What type of glycosidic linkage exists between the monosaccharides in cellobiose? How were you able to determine this?
  d. How is cellobiose different from maltose?
  e. Would you expect an α-*glycosidase* or a β-*glycosidase* to catalyze the hydrolysis of cellobiose?

8 Using Table 11-3, answer the following questions for lactose.
   a. Explain why lactose is classified as a disaccharide.
   b. What monosaccharides are produced when lactose is hydrolyzed?
   c. What type of glycosidic linkage exists between the monosaccharides in lactose?
   d. In what two ways does lactose differ from maltose?
   e. Where in your kitchen can you find lactose?

9 Bees hydrolyze sucrose, obtained from the nectar of plants, to make honey. Based on this information, what monosaccharides does honey contain?

10 The blood sugar found in insects is not glucose but trehalose, a disaccharide that can withstand temperature extremes better than glucose. Hydrolysis of trehalose produces only glucose. Trehalose is a sugar with two α glycosidic linkages. What type of linkage does trehalose have between the two monosaccharides? Name two ways that this linkage differs from that in sucrose?

11 Gentiobiose is a disaccharide that is similar to maltose except that it has an $\alpha(1 \rightarrow 6)$ linkage. Draw the structure of gentiobiose.

Polysaccharides

Polysaccharides are biological polymers, where a **polymer** is any large molecule with the same or similar repeating structural units. The repeating unit is known as the **monomer**. The most abundant polysaccharides in plants and animals are polymers of the monomer D-glucose, which differ primarily in whether their glycosidic linkages are $\alpha(1 \rightarrow 4)$ or $\beta(1 \rightarrow 4)$. The important polysaccharides include:

• amylose,
• amylopectin,
• cellulose, and
• glycogen.

Plants synthesize starch and cellulose from glucose. Plants store energy in the form of **starch**, which is a mixture of 20 percent amylose and 80 percent amylopectin. When we consume starch, we hydrolyze it into glucose during digestion. Glucose—blood sugar—is metabolized further into carbon dioxide, water, and energy inside cells, a biochemical process described in chapter 15. Alternatively, in the presence of excess blood sugar, glucose is converted into **glycogen**, the polymeric storage form of glucose in mammals. So, between meals we hydrolyze glucose molecules from stored glycogen as needed. Glycogen provides us with a short term supply of energy that can last approximately a day, assuming normal activity levels. Glycogen is depleted more quickly with vigorous activity. Athletes that follow a practice of carbo-loading the night before a race do so to maximize their glycogen reserves in order to meet the high energy demands of the race.

The three polysaccharides glycogen, amylose, and amylopectin are polymers of glucose joined as primarily $\alpha(1 \rightarrow 4)$ glycosidic linkages. This is the same linkage present in maltose, but there are 250–4,000 glycosidic linkages rather than just one!

Amylose, the minor constituent of starch, is an unbranched polysaccharide: each glucose monomer is connected to the next glucose monomer in an $\alpha(1 \rightarrow 4)$ glycosidic linkage, which gives the polysaccharide an overall helical shape, as illustrated in **Figure 11-2a**. Extensive hydrogen bonding between hydroxyl groups within the polymer supports the helical shape. Its large molecular size makes amylose a colloidal dispersion. Amylose binds to iodine, I_2, to form a starch-iodine complex with a distinctive blue-purple color, a common test for the presence of starch (Figure 11-3).

Amylopectin, the major constituent of starch, is similar to amylose except that it is a branched polymer. Branching occurs approximately every 30 glucose units, wherever an $\alpha(1 \rightarrow 6)$ glycosidic linkage occurs. Branching is indicated by the red hexagons in Figure 11-2b. The overall shape of amylopectin is spherical.

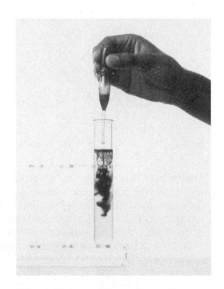

Figure 11-3 Amylose reacts with iodine, I_2, to form a distinctive blue-purple color, a positive test for the presence of starch. [Martyn F. Chillmaid/ Science Source]

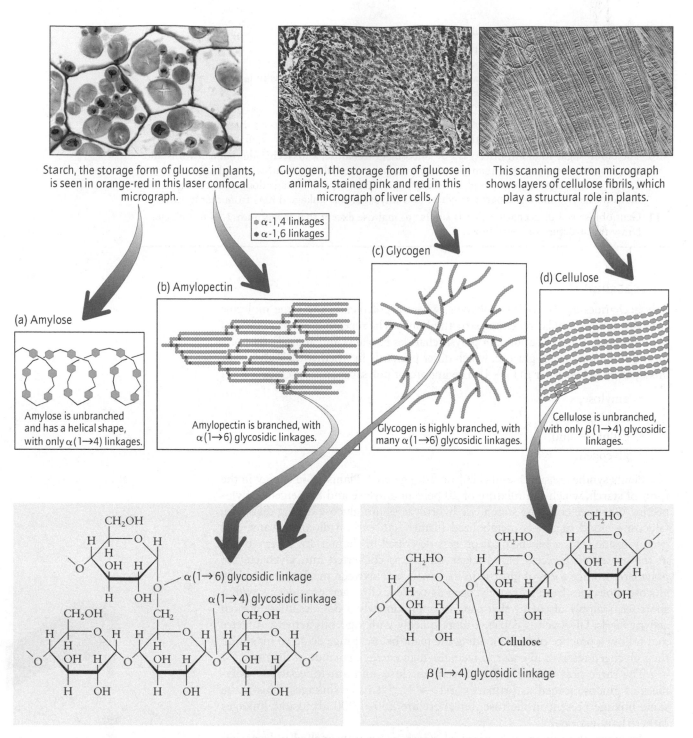

Starch, the storage form of glucose in plants, is seen in orange-red in this laser confocal micrograph.

Glycogen, the storage form of glucose in animals, stained pink and red in this micrograph of liver cells.

This scanning electron micrograph shows layers of cellulose fibrils, which play a structural role in plants.

● α-1,4 linkages
● α-1,6 linkages

(c) Glycogen

(b) Amylopectin

(a) Amylose

(d) Cellulose

Amylose is unbranched and has a helical shape, with only α(1→4) linkages.

Amylopectin is branched, with α(1→6) glycosidic linkages.

Glycogen is highly branched, with many α(1→6) glycosidic linkages.

Cellulose is unbranched, with only β(1→4) glycosidic linkages.

α(1→6) glycosidic linkage

α(1→4) glycosidic linkage

Cellulose

β(1→4) glycosidic linkage

Figure 11-2 The complex carbohydrates starch, glycogen, and cellulose are polymers of D-glucose. They differ in the degree of branching and in the type of glycosidic linkages, α or β. (a) Amylose, a component of starch, is an unbranched polysaccharide composed of α(1 → 4) linkages. (b) Amylopectin, a component of starch, is a branched polysaccharide composed of α(1 → 4) linkages and some α(1 → 6) linkages. (c) Glycogen is a branched polysaccharide composed of α(1 → 4) linkages and some α (1 → 6) linkages. (d) Cellulose is an unbranched polysaccharide composed of β(1 → 4) linkages. [Upper left photo: Dr. Keith Wheeler/Science Source. Middle photo: © Educational Images Ltd./Custom Medical Stock Photo—All rights reserved. Upper right photo: Biophoto Associates/Science Source.]

Glycogen, the polymeric form in which glucose is stored in humans and animals, in liver and muscle cells, serves the same function in animals that starch serves in plants: energy storage. Structurally, glycogen is similar to amylopectin, except that it is even more highly branched: α(1 → 6) glycosidic linkages approximately every 10 glucose units (Figure 11-2c).

Cellulose, the main component of wood, paper, and cotton, is another polysaccharide composed of glucose monomers. However, cellulose contains unbranched $\beta(1 \rightarrow 4)$ glycosidic linkages, which gives cellulose a strong, layered, shape, as illustrated in Figure 11-2d. Cellulose serves a structural role in plants, providing both rigidity and strength. Much of the structural rigidity of cellulose arises from the extensive hydrogen bonding that occurs between sheets of the polysaccharide. Cellulose together with lignin, a non-carbohydrate polymer, provide the bulk of plant matter.

Cellulose, starch, and glycogen can all be hydrolyzed into D-glucose in the presence of an acid. In our body, starch is hydrolyzed in the mouth and small intestine by enzymes that specifically hydrolyze $\alpha(1 \rightarrow 4)$ linkages. These enzymes are known as *α-glycosidases.* Since humans do not have enzymes that hydrolyze $\beta(1 \rightarrow 4)$ linkages, cellulose cannot be digested, and therefore, it is not a source of energy for us. Instead, cellulose plays an important role as dietary fiber, which adds bulk to the contents of the colon and facilitates the absorption of nutrients and water, and also plays a role in regularity.

Some animals and certain insects are able to digest cellulose because bacteria that live in their digestive tracts contain *β-glycosidases.* These animals include cows, giraffes, and other ruminants (animals with a second stomach—a rumin). This is why grass *is* a source of energy for these animals. Termites also have *β-glycosidase* enzymes, and this is why they can eat through a house built of wood.

WORKED EXERCISE | Polysaccharides

11-6 For each statement below, indicate the polysaccharide(s) to which the statement applies. Choose among the following (more than one may apply).

 i. Amylose   ii. Amylopectin   iii. Glycogen   iv. Cellulose

 a. _____ contains $\alpha(1 \rightarrow 4)$ glycosidic linkages.
 b. _____ is a polymer composed of glucose.
 c. _____ provides structural rigidity to plants.
 d. _____ is a branched polysaccharide.
 e. _____ has a helical overall shape.
 f. _____ is a component of dietary fiber.
 g. _____ forms a blue-colored complex with iodine.

Solution
11-6 a. Amylose, amylopectin, glycogen
 b. Amylose, amylopectin, glycogen, cellulose
 c. Cellulose
 d. Amylopectin, glycogen
 e. Amylose
 f. Cellulose
 g. Amylose

PRACTICE EXERCISES

12 For each of the statements below, indicate the polysaccharide(s) to which the statement applies. (More than one may apply.)

 i. Amylose   ii. Amylopectin   iii. Glycogen   iv. Cellulose

 a. _____ provides energy for plants.
 b. _____ is the storage form of glucose in mammalian cells
 c. _____ cannot be digested by humans.
 d. _____ is also known as starch

e. _____ is produced by plants (through photosynthesis).

f. _____ has a helical overall shape.

g. _____ is hydrolyzed by α-*glycosidases*.

13 Dextrins are carbohydrates composed of 3–10 glucose units in $\alpha(1 \rightarrow 4)$ linkages. How are dextrins different from maltose? How are they different from amylose?

14 The rumin is a second compartment in the stomach of certain animals such as giraffes, cattle, and camels. What type of enzyme is present in the bacteria living in the rumin: α-*glycosidases* or β-*glycosidases*? What reaction do these enzymes catalyze?

15 On a molecular level, how are cellulose and cellobiose different? In what way are they similar?

16 On a molecular level, what is the fundamental difference between starch and cellulose?

17 Raffinose is a trisaccharide found in beans, cabbage, and other vegetables. Since people do not have the α-*galactosidase* enzymes needed to hydrolyze the α-glycosidic linkage between galactose and glucose in this sugar, it goes undigested through the stomach. It is then fermented by gas-producing bacteria in the intestines causing gas and bloating. Beano (Figure 11-4) is a product available in stores that can be taken with these foods to prevent gas.

D-galactose

D-glucose

D-fructose

a. How many monosaccharides are produced when raffinose is completely hydrolyzed? Are the same or different sugars produced?

b. Label each of the anomeric centers in raffinose with an asterisk.

c. Label each glycosidic linkage as α- or β-.

d. Which glycosidic linkage requires an α-*galactosidase* as a catalyst for hydrolysis?

e. Chemically, what occurs when a person takes Beano (Figure 11-4)? Which bond is hydrolyzed in the presence of Beano?

Figure 11-4 [Catherine Bausinger]

■ 11.4 The Role of Oligosaccharides as Cell Markers

Organ transplants such as heart transplants, kidney transplants, and liver transplants are quite common today. But did you know that people receiving organ transplants must take immunosuppressant drugs for the rest of their

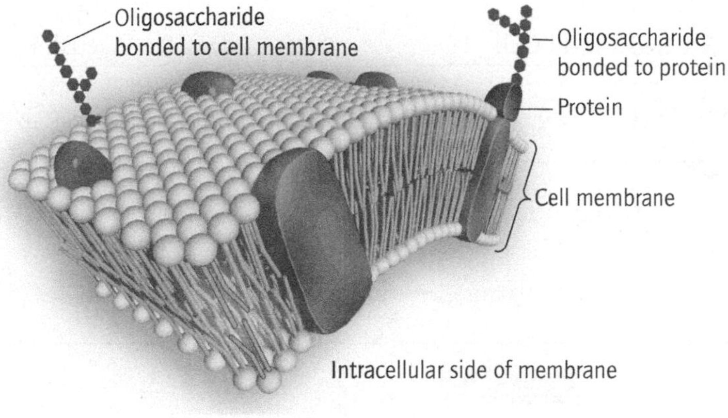

Figure 11-5 The cell membrane, showing oligosaccharide markers (purple hexagons) on the surface of the cell, which allow cells to recognize other cells.

lives so that they won't reject the transplanted organ? Organ rejection after a transplant occurs because the body's immune system recognizes the cells in the transplanted organ as "foreign" and mounts an immune response, attacking and ultimately destroying the transplanted organ. How are our cells able to distinguish our cells from the cells in the transplanted organ? The answer is that cells contain chemical markers on their surface that allow them to distinguish host cells from foreign cells.

Most cell markers are oligosaccharides, that is, carbohydrates composed of a few monosaccharides that identify a cell, much like a fingerprint identifies a person. Covalently bonded to proteins and lipids in the cell membrane, oligosaccharide markers project out from the cell membrane into the extracellular fluid, as represented by the purple hexagons shown in **Figure 11-5**.

One example of cell markers is seen in the four basic blood types: A, B, AB, and O, which differ in the oligosaccharide marker on red blood cells, as shown in Table 11-4. Type O blood cells have a trisaccharide (three monosaccharides) marker. Type A blood cells have a tetrasaccharide (four monosaccharides) marker containing the same three monosaccharide components as Type O but with an additional monosaccharide not present in any of the other blood types. Similarly, type B blood cells have a tetrasaccharide marker different from type A but also containing the same three monosaccharide components as Type O. Type AB blood cells contain a combination of both type A and type B markers.

People with type O blood are referred to as **universal donors** because they can donate blood to recipients with any of the other blood types. The chemical structure of the oligosaccharide marker on Type O blood is common to all the other blood types, so it is recognized as self by people with any blood type. However, people with type O blood will reject all other blood types except O, because the fourth monosaccharide isn't present in their own blood cells so their immune system will recognize them as foreign. The resulting immune response leads to clotting, a life-threatening condition. Therefore, blood compatibility is always determined before any medical procedure that may require a blood transfusion. Conversely, individuals with type AB blood can receive type A, B, or O blood but cannot donate their blood to anyone except others with type AB blood. People with type AB blood are therefore referred to as **universal recipients**.

TABLE 11-4 Blood Type and Oligosaccharide Marker

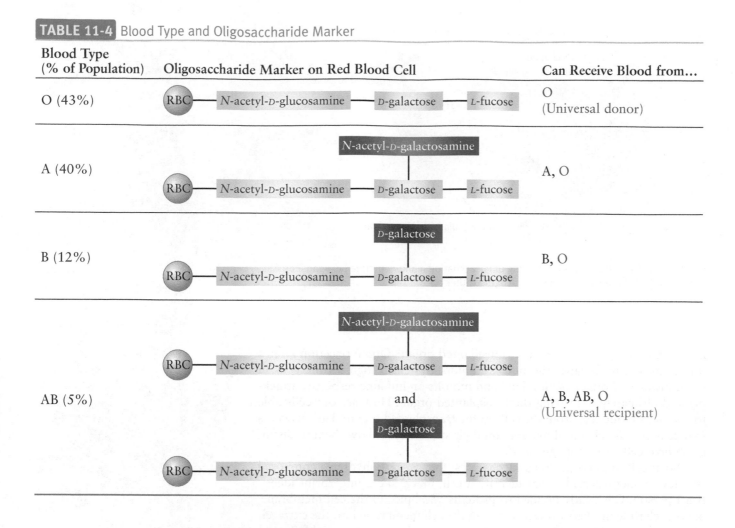

Blood Type (% of Population)	Oligosaccharide Marker on Red Blood Cell	Can Receive Blood from...
O (43%)	RBC — N-acetyl-D-glucosamine — D-galactose — L-fucose	O (Universal donor)
A (40%)	N-acetyl-D-galactosamine / RBC — N-acetyl-D-glucosamine — D-galactose — L-fucose	A, O
B (12%)	D-galactose / RBC — N-acetyl-D-glucosamine — D-galactose — L-fucose	B, O
AB (5%)	N-acetyl-D-galactosamine / RBC — N-acetyl-D-glucosamine — D-galactose — L-fucose and D-galactose / RBC — N-acetyl-D-glucosamine — D-galactose — L-fucose	A, B, AB, O (Universal recipient)

PRACTICE EXERCISES

18 At the molecular level, how is type A blood different from type O blood?

19 Why are people with type AB blood able to receive blood from any donor? They are known as universal _____.

20 How is type B blood different from type O? Can people with type B blood accept a blood donation from someone with type O blood? Can people with type O blood accept a blood donation from someone with type B blood? Explain.

Chemistry in Medicine When Blood Sugar Levels Are too High or too Low

Marcia woke up one morning and walked into the kitchen. As she began assembling the ingredients for her breakfast, her speech suddenly became slurred, and she appeared disoriented. As others watched with concern, her husband quickly responded by pouring her some orange juice and insisted that she drink it. Within seconds, she perked up as though nothing unusual had happened, asking "What's going on? Why are you all standing over me?"

What Marcia experienced is not an uncommon event; in fact, it is something paramedics see quite often. What

happened? Marcia is a diabetic, a condition that prevents her cells from utilizing glucose properly. That morning her blood glucose had dropped to a dangerously low level, making her cells unable to perform some of their basic functions. Without that immediate influx of glucose—the orange juice—Marcia could have lapsed into a coma and even died. How is it possible that a few sips of orange juice was able to bring Marcia back to her normal self and within just a matter of seconds?

Every cell in the body, especially brain cells, need a constant supply of glucose, to produce energy. Marcia's

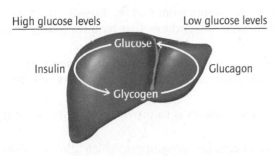

Figure 11-6 Glycogen is stored in liver (shown) and muscle cells. The hormone insulin directs the liver to synthesize glycogen from glucose when blood sugar levels are high. The hormone glucagon directs the liver to degrade glycogen to glucose when blood sugar levels are low.

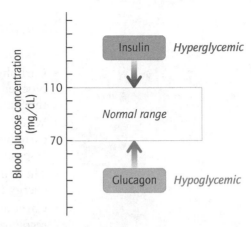

Figure 11-7 Blood sugar levels are carefully regulated in the body so that they are neither too high (hyperglycemic) nor too low (hypoglycemic). When blood glucose levels rise insulin is released. When blood glucose levels drop, glucagon is released.

incident occurred because her blood glucose levels were too low. On the other hand, glucose levels that are consistently too high also pose health risks. In a healthy individual the concentration of glucose in the blood is maintained within a narrow range. Normal fasting glucose levels are 70–110 mg per deciliter (dL), as determined by a Fasting Plasma Glucose (FPG) test, a blood test used to measure blood glucose concentration after a period of fasting. When the concentration of glucose is *below* the normal range, the person is said to be *hypoglycemic*; when it is *above* the normal range, the person is said to be *hyperglycemic*.

The body regulates glucose levels using two important hormones: *insulin* and *glucagon* (Figure 11-6). Special cells in the pancreas release insulin when blood sugar levels rise, such as after a meal. Insulin is a hormone that mobilizes the "glucose transporters" to the cell surface. Glucose transporters are proteins that enable the cell to take up glucose, circulating in the bloodstream, by assisting its passage through the cell membrane. Insulin also directs liver cells to store some of the circulating glucose by converting it into glycogen, the polysaccharide storage form of glucose.

When glucose levels drop, as for example between meals and when we exercise, the pancreas stops producing insulin and releases the hormone glucagon, instead. Glucagon signals liver cells to hydrolyze glucose from stored glycogen—releasing glucose into the blood, where it circulates to cells that need it for energy. Insulin and glucagon work together to maintain blood sugar levels within the normal range, as shown in Figure 11-7.

When the production or utilization of these hormones is compromised, metabolic disorders can result. The most common metabolic disorder is diabetes, characterized by insufficient insulin production or an inability to utilize insulin properly, known as insulin resistance. It is estimated that 7 percent, 21 million children and adults, of the population in the United States have diabetes.

There are two types of diabetes: type I and type II. Type I diabetes, also known as insulin-dependent diabetes mellitus, usually begins in childhood. It is an autoimmune disease in which the body destroys its own insulin-producing cells. These individuals cannot produce enough insulin, and glucagon levels build up. Consequently, glucose cannot enter liver cells where it is stored in the form of glycogen. In the absence of insulin, liver cells start producing glucose through an alternative biochemical pathway—the same one activated during starvation. Glucose is circulating in the blood, but unable to be taken up by cells that need it, because there is insufficient insulin present to effectively mobilize the glucose transporters. People with type I diabetes must manage their diets carefully and, in particular, control the amount of carbohydrates that they consume. They must also receive regular injections of insulin, either by injecting themselves or wearing an insulin pump that automatically injects the insulin on a regular basis. Since insulin is a protein that would be readily digested in the stomach, it cannot be taken orally.

Type II diabetes, or insulin-resistant diabetes, generally appears in adulthood, and even though insulin is produced, cells do not respond to it. For reasons not fully understood, people with type II diabetes are unable to mobilize the glucose transporters. Cells are thus starved of the fuel they need. Here again, we see that understanding biochemical processes at the molecular level provides insight into diseases with symptoms that we observe on a macroscopic level.

Summary

An Overview of Carbohydrates and Their Function

- Carbohydrates are produced by plants from carbon dioxide and water in a biochemical process known as photosynthesis that requires energy from the Sun.
- When we consume carbohydrates, the chemical potential energy in carbohydrates is transferred to the cell in a process known as cellular respiration.
- The primary function of carbohydrates is to provide a short term supply of energy for the cell.
- Carbohydrates serve a role in cellular recognition, which allows cells to recognize other cells.
- Carbohydrates are classified according to the number of monosaccharides that are formed when they are hydrolyzed.
- Monosaccharides, also known as simple sugars, cannot be hydrolyzed into simpler carbohydrates. They are the monomer unit in more complex carbohydrates.
- Disaccharides are hydrolyzed into two monosaccharides, which may be the same or different.
- Polysaccharides are hydrolyzed into many monosaccharides. Cellulose and starch are polysaccharides that are hydrolyzed into thousands of glucose molecules.

The Structure of the Common Monosaccharides

- Monosaccharides are carbohydrates that cannot be hydrolyzed into simpler sugars.
- Monosaccharides are generally crystalline colorless solids with a sweet taste.
- Common monosaccharides include D-glucose, D-fructose, D-ribose, and D-galactose.
- The common monosaccharides contain a chain of three or more carbon atoms, an aldehyde or ketone functional group, and a hydroxyl group on every carbon except the carbonyl carbon.
- Monosaccharides usually have one or more centers of chirality, which each contain a hydroxyl group and a hydrogen atom.
- Monosaccharides with five or more carbon atoms undergo a reaction in aqueous solution that forms a ring, producing either a furanose or a pyranose ring, which is in equilibrium with the open-chain form of the monosaccharide.
- A Fischer projection is a convenient way to convey the configuration of the centers of chirality in a monosaccharide, indicated by a horizontal —OH group on the left and a horizontal hydrogen atom on the right, or vice versa.
- Most monosaccharides are chiral. Enantiomers have the opposite configuration at every center of chirality, bear the same name, but have different prefixes D- and L-. In a D-sugar the —OH group at the last center of chirality, farthest from the carbonyl, is on the right.
- Only D-sugars are produced in nature.
- Diastereomers have the opposite configuration at one or more centers of chirality, but not at all centers of chirality. Diasteromeric monosaccharides have different names, like glucose and galactose.

- The ring form of monosaccharides are typically drawn as Haworth projections, which show the hydroxyl groups above or below the ring. The orientation of the hydroxyl groups above or below the ring indicates the configuration of these centers of chirality and determines the identity of the monosaccharide.
- In the ring form of a monosaccharide, the anomeric center is the carbon atom that is bonded to both the ring oxygen and a hydroxyl group. The hydroxyl group lies above the ring in the β-anomer and below the ring in the α-anomer when drawn as a Haworth projections.
- The α- and β- forms of a monosaccharide are in equilibrium with the open-chain form of the monosaccharide, favoring the ring forms.
- Diastereomers are nonsuperimposable, non-mirror image stereoisomers.

The Structure of Complex Carbohydrates

- Disaccharides are hydrolyzed to two monosaccharides in the presence of a catalyst.
- The common disaccharides are maltose, sucrose, lactose, and cellobiose.
- Disaccharides are joined between the anomeric center of one monosaccharide and the oxygen atom of a hydroxyl group on the second monosaccharide, in a glycosidic bond.
- The anomeric center in a glycosidic linkage is stable in aqueous solution and does not open and close. A catalyst is required to hydrolyze a glycosidic bond.
- Disaccharides are defined by three structural characteristics: (1) the identity of the two monosaccharide components, (2) the carbon atoms involved in the glycosidic bond (3) and whether the glycosidic bond is α- or β-.
- Starch and glycogen are polysaccharides with over 1,000 repeating glucose monomers connected by mainly $\alpha(1 \rightarrow 4)$ linkages. Starch is stored in plants and used for energy by humans and animals. Glycogen is the storage form of glucose in animals.
- Cellulose is a polysaccharide similar to starch but with $\beta(1 \rightarrow 4)$ linkages, giving it a different overall shape. Cellulose provides structure to a plant.

The Role of Oligosaccharides as Cell Markers

- Oligosaccharides projecting from cell membranes into the extracellular fluid serve as cell markers that allow other cells to distinguish the host cell from foreign cells.
- The ABO blood groups are defined by the monosaccharide components of the oligosaccharide markers on red blood cells.

Key Words

Amylopectin The major constituent of starch; composed of primarily $\alpha(1 \rightarrow 4)$ linkages. It is a branched polysaccharide arising from some $\alpha(1 \rightarrow 6)$ linkages.

Amylose The minor constituent of starch; it is an unbranched polysaccharide with $\alpha(1 \rightarrow 4)$ linkages. It has an overall helical shape.

α-anomer The ring form of a monosaccharide wherein the —OH group on the anomeric center is below the ring in a Haworth projection.

Anomeric center The carbon atom in the ring form of a monosaccharide that is bonded to two oxygen atoms: the ring oxygen atom and a hydroxyl group (—OH). It is derived from the carbonyl carbon of the open-chain form of the monosaccharide.

β-anomer The ring form of a monosaccharide wherein the —OH group on the anomeric center is above the ring in a Haworth projection.

Cellulose A polysaccharide that provides structure for a plant, formed of glucose monomer units in $\beta(1 \rightarrow 4)$ linkages. Cannot be digested by humans but can be digested by ruminants. Serves as dietary fiber.

Diastereomers Nonsuperimposable non-mirror image stereoisomers. Monosaccharide stereoisomers that differ at one or more, but not all, centers of chirality, and therefore have different names.

Disaccharides Carbohydrates that, when hydrolyzed, yield two monosaccharides.

Fischer projection A way to represent monosaccharides and other molecules with multiple centers of chirality. Points where horizontal and vertical lines intersect represent centers of chirality.

Furanose ring The five-membered ring form of a monosaccharide containing one ring oxygen, shown with the oxygen at the top in a Haworth projection, by convention.

D-Sugar The enantiomeric form of a monosaccharide produced in nature and distinguished in a Fischer projection by the last center of chirality having the —OH group on the right. In a Haworth projection, a D-sugar has the last CH_2OH group, at the far left, above the ring.

Glycosidic bond The covalent bond that joins two monosaccharides. It always occurs between the anomeric carbon of the first sugar and one of the hydroxyl group oxygen atoms of the second sugar.

Glycosidic linkage See glycosidic bond.

Haworth projection The conventional way to represent a furanose and a pyranose ring that emphasizes the configuration of the hydroxyl groups, by showing them either above or below the ring.

L-Sugar The enantiomeric form of a monosaccharide not produced in nature and distinguished in a Fischer projection by the last center of chirality having the —OH group on the left.

Monomer The repeating structural unit in a polymer.

Monosaccharide A carbohydrate that cannot be hydrolyzed into a simpler carbohydrate. The monomer unit of disaccharides and polysaccharides. Also referred to as a simple sugar.

Photosynthesis The biochemical pathway used by plants to synthesize carbohydrates from carbon dioxide and water, using sunlight as the source of energy.

Polymer A large molecule composed of repeating monomer units.

Polysaccharide A carbohydrate that is hydrolyzed into many monosaccharides. Glycogen, starch, and cellulose are polysaccharides that are hydrolyzed into thousands of glucose molecules.

Pyranose ring The six-membered ring form of a monosaccharide containing one ring oxygen, shown with the oxygen at the top right in a Haworth projection, by convention.

Starch A carbohydrate composed of 20 percent amylose and 80 percent amylopectin. It serves as stored energy for a plant and provides humans and animals with their primary source of energy. The structure is composed of primarily $\alpha(1 \rightarrow 4)$ linkages between glucose monomers.

Universal donor A person with blood type O, who can donate blood to an individual with any other blood type.

Universal recipient A person with blood type AB, who can receive blood from a donor with any other blood type.

Additional Exercises

Dental Caries and the Sweet Tooth

21 How is lactic acid produced from glucose in your mouth?

22 What does it mean for an organism to be anaerobic?

23 What is the root cause of dental caries?

24 What happens to glucose during lactic acid fermentation?

25 How does lactic acid cause tooth decay?

26 How do you prevent the progression of dental caries?

An Overview of Carbohydrates and Their Function

27 What are the four important biomolecules that provide the foundation for biochemistry?

28 What is the primary function of carbohydrates in mammals?

29 How do we obtain energy when eating carbohydrates?

30 What biochemical pathway do plants use to produce carbohydrates? Where do they get the energy for this anabolic process?

31 Where in the body are carbohydrates hydrolyzed to glucose?

32 How does glucose get from the small intestine to cells throughout the body?

33 What happens to glucose once it enters a cell?

34 How does the energy from the Sun become useful energy for a cell?

35 Besides supplying energy, what other functions do carbohydrates serve in our cells?

36 How are carbohydrates involved in cellular recognition?

37 Why did scientists give carbohydrates a name that ended with the word hydrates?

38 What is a polymer? What is a monomer?

39 What is the fundamental difference between monosaccharides, disaccharides, and polysaccharides?

40 What suffix is in the name of most carbohydrates?

The Structure of the Common Monosaccharides

41 Why are monosaccharides called simple sugars?

42 What are some physical properties of monosaccharides?

43 Give an example of a carbohydrate that is sweeter than sucrose?

44 Why do sugar substitutes have no calories? How much sweeter is NutraSweet than sucrose?

45 Are most monosaccharides chiral? What structural feature makes a molecule chiral?

46 What two functional groups are found in all monosaccharides?

47 How many carbon atoms do the common monosaccharides contain?

48 From a Fischer projection, how can you tell the difference between a D-sugar and an L-sugar?

49 Fischer projections of D-arabinose and D-glucose are shown below.

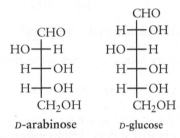

D-arabinose   D-glucose

a. How many carbon atoms does each of these monosaccharides contain?
b. Do these sugars contain an aldehyde or a ketone as part of their structure?
c. What makes these D-sugars?
d. Write the Fischer projection of L-arabinose.

50 The structure of one enantiomer of sorbose is shown below. Is the Fischer projection shown a D-sugar or an L-sugar? Explain.

$$CH_2OH$$
$$=O$$
$$HO-H$$
$$H-OH$$
$$HO-H$$
$$CH_2OH$$

Sorbose

51 What are diastereomers? How are diastereomers different from enantiomers?

52 Why do scientists use Fischer projections and Haworth projections to represent the structure of monosaccharides?

53 How can you identify the anomeric center in a Haworth projection?

54 What is the structural difference between an α- and a β-anomer?

55 Why are both the α- and the β-anomers present in an aqueous solution of a monosaccharide with five or six carbons? What three forms of a monosaccharide are found in solution and why?

56 Indicate whether each structure below is a furanose or a pyranose. Identify the anomeric center in each structure. Indicate whether the α- or β-anomer is shown.

a.

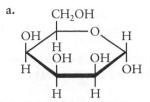

b.

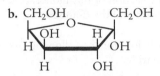

c.

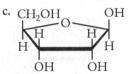

57 Indicate whether each structure below is a furanose or a pyranose. Identify the anomeric center in each structure. Indicate whether the α- or β-anomer is shown.

a.

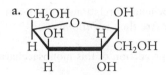

b.

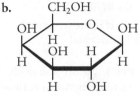

58 The Haworth projection of D-talose is shown below.

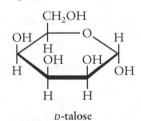

D-talose

a. Number the carbon atoms from 1 to 6.
b. What part of the structure determines that it is a D-sugar?
c. What is the stereochemical relationship between D- and L-talose?
d. Draw the Fischer projection of the open-chain form of D-talose.

59 Answer the questions below for the following two ring forms of a monosaccharide.

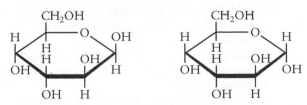

a. Label the anomeric centers with an *.
b. Which structure is the α-anomer and which is the β-anomer?
c. What is the stereochemical relationship between these two structures? Are they *enantiomers* or *diastereomers*?
d. Draw the Fischer projection of this monosaccharide.
e. How do the α-anomer and the β-anomer interconvert?

60   Answer the questions below for the following two ring forms of a monosaccharide.

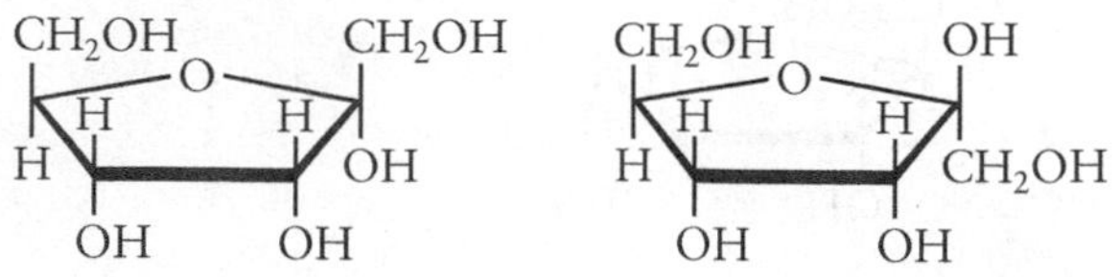

a. Label the anomeric centers with an *.
b. Which structure is the α-anomer, and which is the β-anomer?
c. What is the stereochemical relationship between these two structures? Are they *enantiomers* or *diastereomers*?
d. Draw the Fischer projection of this monosaccharide.

The Structure of the Complex Carbohydrates

61   What is a disaccharide? What is the structural difference between disaccharides and monosaccharides?

62   What three structural features identify a disaccharide?

63   What is the structural difference between maltose and cellobiose? How are these two molecules similar?

64   Isomaltose is similar to maltose except that it contains an α(1 → 6) linkage. Write the structure of isomaltose.

65   The structure of maltulose, found in honey and beer, is shown below.

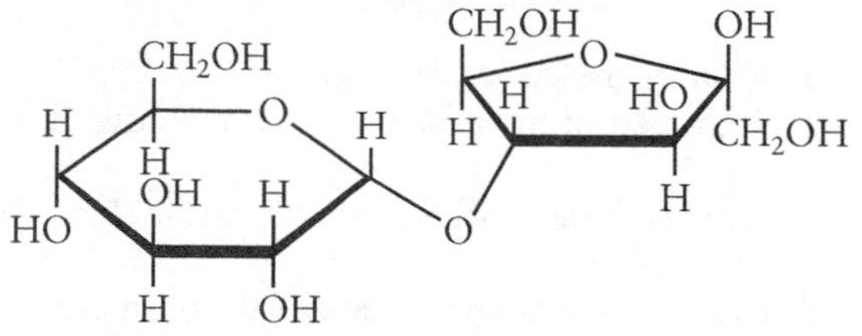

a. Is maltulose a mono-, di-, or polysaccharide? Explain.
b. Label all the anomeric centers in maltulose with an *.
c. Label and identify the glycosidic bond in maltulose.
d. What type of glycosidic linkage does maltulose contain: α or β? Fill in the blanks indicating the type of linkage __(__ → __).
e. How many sugars are formed when maltulose is hydrolyzed? Are they the same or different?

66   Melibiose, shown below, is found in cacao beans.

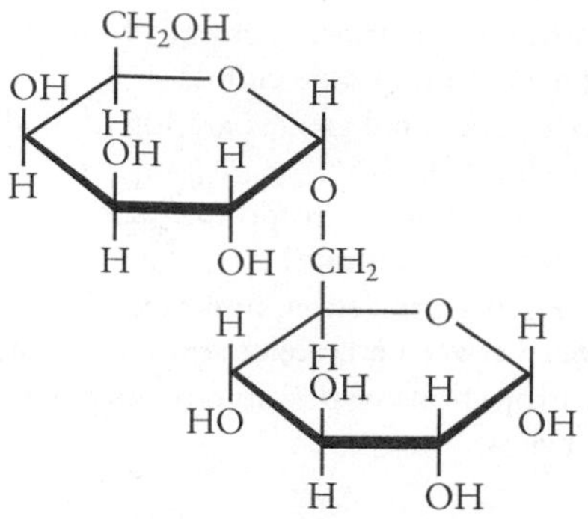

a. Is melibiose a mono-, di-, tri-, or polysaccharide? Explain.
b. Label all the anomeric centers in melibiose with an *.
c. Label and identify the glycosidic bond in melibiose.
d. What type of glycosidic linkage does melibiose contain: α or β? Fill in the blanks indicating the type of linkage __(__ → __).
e. How many sugars are formed when melibiose is hydrolyzed? Are they the same or different?

67   What function does starch have in a plant?

68   What function does cellulose have in a plant?

69   What function does starch have in our body?

70   What function does glycogen have in mammals?

71   What dietary function does cellulose have in our body? Do we derive energy from cellulose? Explain.

72   What are the structural differences between amylose and amylopectin?

73   Which of the following polysaccharides have the same type of glycosidic linkage: amylose, amylopectin, cellulose, or glycogen?

74   Explain why giraffes can use cellulose as a source of energy, while humans can't.

The Role of Oligosaccharides as Cell Markers

75   What are cell markers? How do they help the immune system identify what belongs to the body and what is "foreign"?

76   It has recently been shown that the H1N1 flu virus avoids attack by our immune cells by mutating in a way that changes the location of the oligosaccharide markers on the virus. Explain how this might affect whether or not our immune cells recognize the virus as a virus it has seen before.

77   At the molecular level how is type A blood different from type B blood?

78   At the molecular level how is type B blood different from type O blood?

79   Why can people with type O blood only receive blood from other type O donors? Why are people with type O blood called universal donors?

80   Why are individuals with type AB blood called universal recipients? What blood types can they receive?

81 In 2003, a 17-year-old girl died from complications of a heart-lung transplant. She was mistakenly given organs from a person with type A blood, while she had type O blood. Explain what the difference in blood type was and why this resulted in such a devastating outcome.

Chemistry in Medicine When Blood Sugar Levels Are too High or too Low

82 When an individual's glucose levels fall:
a. Is he or she hyperglycemic or hypoglycemic?
b. What biochemical pathway is activated to increase glucose levels?

c. In a healthy individual, when do glucose levels typically fall?
d. What hormone is secreted when glucose levels fall?
83 What is the difference between type I and type II diabetes?
84 What are normal fasting blood glucose levels?
85 Which cells produce insulin? In which organ are these cells found?
86 Why do individuals with diabetes have to carefully watch their diet? What food must they limit in their diet and why?
87 Why can insulin not be administered orally?
88 What hormone has a complementary role to insulin?

Answers to Practice Exercises

1   a. There are no intersections of vertical and horizontal lines.
b. Three carbon atoms and two hydroxyl groups
c. Dihydroxyacetone has the carbonyl group at C(2) rather than at C(1), and it is a ketone instead of an aldehyde. Also, dihydroxyacetone is achiral (no center of chirality), whereas D-glyceraldehyde is chiral (one center of chirality).
d. They are structural isomers of $C_3H_6O_3$.

2   a. D-ribose has the C(2) hydroxyl group on the right, whereas D-arabinose has it on the left. They are diastereomers because one, and not all centers of chirality are different.
b. D-ribose has the C(2) and C(3) hydroxyl groups on the right, whereas D-lyxose has them on the left. They are diastereomers because two and not all centers of chirality are different.
c. D-arabinose has the C(3) hydroxyl group on the right whereas D-lyxose has it on the left. They are diastereomers because one, and not all centers of chirality are different.
d. They are all D-sugars because the last center of chirality (the last intersection of horizontal and vertical lines) farthest from the carbonyl group has the hydroxyl group on the right.
e. They all have three centers of chirality (three intersections of horizontal and vertical lines). The maximum number of stereoisomers is $2^3 = 2 \times 2 \times 2 = 8$ stereoisomers. The L-sugars are not shown: L-ribose, L-arabinose, L-lyxose, and L-xylose.
f. L-arabinose is the enantiomer of D-arabinose (all centers of chirality are the opposite). L-arabinose is a diastereomer of D-ribose (two of the three centers are different but not all). L-arabinose is not produced in nature.

CHO
H—OH
HO—H
HO—H
CH₂OH

L-arabinose

3   a. α-D-galactose is on the left, and β-D-galactose is on the right.
b. The anomeric center is C(1) at the far right.

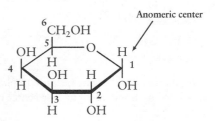

Anomeric center

c. These are pyranose rings because they contain six atoms in the ring: five carbons and one oxygen atom. By convention, the oxygen in a pyranose ring should be placed at the top right.

d.   CHO
H—OH
HO—H
HO—H
H—OH
CH₂OH

D-galactose

e. In a D-sugar, the last carbon (CH₂OH) is above the ring in a Haworth projection. In a D-sugar the last center of chirality farthest from the carbonyl group has the hydroxyl group on the right in a Fischer projection.
f. D-galactose is a diastereomer of D-glucose because it differs in the configuration at only C(4) and not at all centers of chirality.
g. The open-chain form of D-galactose, β-D-galactose, and α-D-galactose. The open-chain form is present in the lowest concentration.

4   a.–b.

CH₂OH   CH₂OH

α-anomer   β-anomer

c. They differ in the configuration of the hydroxyl group at C(2): it is above the ring in this sugar and below the ring in D-galactose. They both have their anomeric hydroxyl group below the ring, α.

d.

CHO
HO—H
HO—H
HO—H
H—OH
CH$_2$OH

5 a. (1) The ring oxygen of the pyranose ring is positioned at top right.
(2) The last carbon (CH$_2$OH) at the far left is above the ring, indicative of a D-sugar.
(3) The anomeric center is at the far right.

b.

CH$_2$OH

← — Anomeric center

α-D-gulose

c. You can tell this is a D-sugar because the last carbon, the CH$_2$OH group, bonded to the last center of chirality, is above the ring.

d. The open-chain form contains an aldehyde. You can tell this from looking at the anomeric center and noting that there is a bond to a hydrogen (—H) atom rather than an —R group, above or below the ring.

CHO
H—OH
H—OH
HO—H
H—OH
CH$_2$OH

D-gulose

e.

CH$_2$OH

← This center of chirality disappears in going to the open-chain form

α-D-gulose

CHO
H—OH
H—OH
HO—H
H—OH
CH$_2$OH

D-gulose

f. D-gulose differs from D-glucose only at the configuration of C(3) and C(4), therefore they are stereoisomers. They are diastereomers because they differ at some but not all centers of chirality.

g. D-sugars are produced in nature, so D-gulose is produced in nature.

6 a. D-glucose and D-fructose
b. the <u>α</u>-anomer of D-glucose and the <u>β</u>-anomer of D-fructose
c. Sucrose is found in fruit and table sugar.

7 a. Cellobiose is a disaccharide because when hydrolyzed, it produces *two* monosaccharides.
b. D-glucose
c. A β(1 → 4) linkage. The sugar on the left is the β-anomer, C(1), and joined to the oxygen at C(4) in the second sugar.
d. Cellobiose has a β(1 → 4) linkage while maltose has an α(1 → 4) linkage.
e. A *β-glycosidase* would catalyze the hydrolysis of cellobiose because cellobiose has a β-glycosidic linkage.

8 a. Lactose is a disaccharide because when hydrolyzed, it produces two monosaccharides.
b. D-galactose and D-glucose
c. a β(1 → 4) linkage
d. Lactose has a β-linkage whereas maltose has an α-linkage, and lactose is composed of different monosaccharides, galactose and glucose, whereas maltose is composed of only glucose.
e. Milk and other dairy products contain lactose.

9 D-glucose and D-fructose

10 The monosaccharides are linked at both their anomeric centers, and since both sugars are glucose, the linkage is α1 → α1. The glycosidic linkage is different from sucrose in that the second sugar has a β-linkage in sucrose. They also differ in that sucrose is composed of D-glucose and D-fructose, whereas trehalose is composed of only D-glucose.

11

CH$_2$OH

Gentiobiose

12 a. i, ii   **b.** iii   **c.** iv   **d.** i, ii   **e.** i, ii, and iv
f. i   **g.** i, ii, and iii

13 Dextrins contain more glucose monomer units than maltose, which has only two; dextrins contain much fewer glucose monomer units than amylose, which is a polymer composed of thousands of glucose monomer units.

14 β-*Glycosidases*; they catalyze the hydrolysis of β-glycosidic linkages.

15 Cellulose is a polysaccharide containing hundreds of glucose units, while cellobiose is a disaccharide containing only two glucose units. They are similar in that they both contain an α(1 → 4) glycosidic linkages between D-glucose monomers.

16 The fundamental difference in structure is the glycosidic linkage, which is $\alpha(1 \rightarrow 4)$ in starch and $\beta(1 \rightarrow 4)$ in cellulose.

17 a. Three monosaccharides are produced upon hydrolysis, and they are three different monosaccharides.

b.–d.

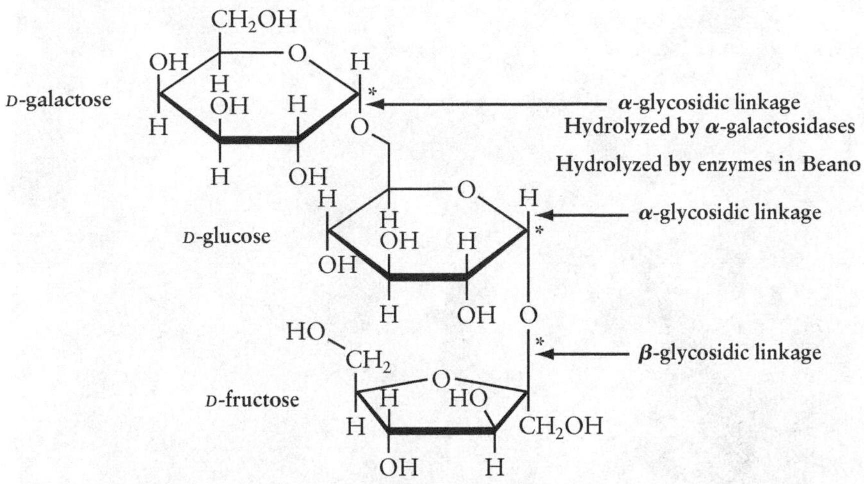

e. The glycosidic linkage between galactose and glucose is hydrolyzed by enzymes in Beano, thereby preventing gas from the incomplete digestion of raffinose.

18 Type A blood has an extra monosaccharide (N-acetyl-D-galactosamine) attached to the core trisaccharide, whereas type O blood only has the trisaccharide.

19 Because they contain the carbohydrate markers present in all other blood types. They are known as universal <u>recipients</u>.

20 Type B blood has an additional monosaccharide (D-galactose) attached to the core trisaccharide whereas type O blood only has the core trisaccharide. Yes, people with Type B blood can accept type O blood because it contains the same carbohydrate markers that are already present in their blood. No, people with Type O blood cannot accept blood from people with Type B blood because type B blood has an additional carbohydrate marker different from what the people with Type O have, and therefore their immune system would mount a potentially fatal immune response against the foreign blood.

Testosterone

Lance Armstrong was stripped of his seven Tour de France titles after testing positive for anabolic steroids. He later admitted to using steroids throughout his professional cycling career. [Pierre Andrieu/AFP/Getty Images]

12

Lipids: Structure and Function

The Effects of Anabolic Steroids on the Body

In 2013 Lance Armstrong was stripped of his seven Tour de France titles after finally admitting to the use of steroids throughout his professional cycling career. Most professional organizations test their athletes regularly for the presence of these substances in their blood, and all professional athletic organizations prohibit the use of testosterone and anabolic steroids by their athletes. While possession of anabolic steroids without a prescription is a federal crime, performance enhancing drugs continue to be widely used by athletes who want to build muscle and increase physical strength. Steroid use not only creates an unfair competitive advantage, but it creates serious health risks, sometimes lethal, to those who use them.

Chemically, a steroid is an organic compound containing the characteristic four fused rings of the steroid backbone, shown in **Figure 12-1** (see also Figure 6-13). Steroids belong to the lipid class of biomolecules, which are defined by their insolubility in water and solubility in organic solvents. Steroids, introduced in chapter 6, have a broad range of physiological functions, including the female and male sex hormones known as estrogens and androgens, respectively. Testosterone is the sex hormone responsible for the development of male sexual characteristics, the *androgenic* effect, during puberty. When testosterone binds to the androgen receptor, it initiates a cascade of biochemical events resulting in protein production, which increases muscle growth, the *anabolic* effect of steroids.

Anabolic steroids are synthetic compounds that mimic the effects of testosterone and related hormones, which bind to the androgen receptor as well as or better than testosterone, and initiate the same cascade of events, as illustrated in **Figure 12-2**. The chemical structure of an anabolic steroid resembles the structure of testosterone, as seen in the anabolic steroid stanozolol shown in Figure 12-1.

Synthetic anabolic steroids were designed to maximize the anabolic effects of testosterone while minimizing its androgenic effects. For instance, anabolic steroids disrupt the normal production of hormones, so in males their use can lead to infertility, gynecomastia (the development of breasts), testicular atrophy (shrinkage of the testes), and premature baldness. Prolonged use of steroids can lead to liver damage, prostate cancer, and cardiovascular disease. In adolescents, the use of steroids can stunt growth because elevated steroid concentrations signal the body that puberty is

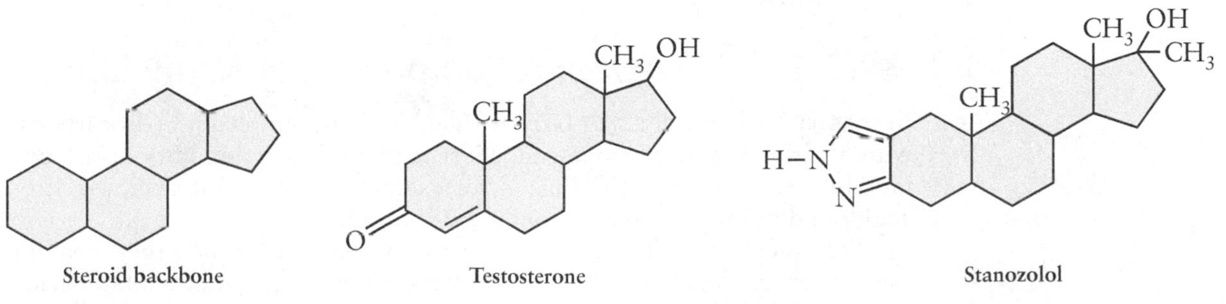

Figure 12-1 The characteristic four fused ring structure of a steroid can be seen in the structure of testosterone, a sex hormone, and Stanozolol, a common anabolic steroid.

Steroid backbone

Testosterone

Stanozolol

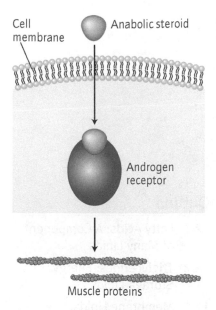

Figure 12-2 Anabolic steroids bind to the androgen receptor, much like testosterone, initiating a sequence of biochemical events that leads to protein synthesis and muscle growth.

over. As a result, bones stop growing. The use of anabolic steroids among high school adolescents has been on the rise and is a serious health concern.

Given that anabolic steroid use is illegal and has such serious negative health effects, why are anabolic steroids even available? Under the supervision of a doctor, anabolic steroids are an effective treatment for some serious medical conditions. For example, anabolic steroids effectively treat muscle wasting diseases, such as those seen in advanced-stage HIV infection (AIDS). The tissue-building properties of anabolic steroids also aid in the recovery time from severe burns. Steroids are also an effective treatment for hypogonadism, a condition in which an adult male produces insufficient testosterone. In all of these applications, the steroid dosage is low and usage is closely monitored by a doctor. In this chapter you will learn about the class of biomolecules known as lipids, to which steroids belong. ●

In chapter 11 you learned about the structure and function of carbohydrates, one of the four major classes of biomolecules. In this chapter we focus on the structure and function of another major class of biomolecules known as lipids. In contrast to the other three major classes of biomolecules, which are defined by their *chemical structure*, lipids are defined by their *physical properties*. *Lipids are biological compounds that do not dissolve in water but are soluble in organic solvents.* It is the extensive hydrocarbon component of their structure that gives lipids their hydrophobic physical properties. Lipids also differ from carbohydrates and the other major classes of biomolecules in that they are not polymers.

Biological organisms contain a variety of lipids that can be subdivided according to their structure and function. In this chapter we will discuss the structure and function of the following types of lipids:

- triglycerides, molecules that provide the body with its long-term supply of energy.
- phospholipids and glycolipids, the main components of mammalian cell membranes.
- steroids, compounds derived from cholesterol, such as the sex hormones, vitamin D, and bile salts.
- eicosanoids, signaling molecules that initiate physiological events, such as the inflammatory response to trauma and infection.

Other lipids include the fat-soluble vitamins which play an important role in cellular processes. In plants, certain lipids also act as pigments that capture light, while others form waterproof coatings such as the waxes on leaves and fruit.

In order to understand the chemical structure of triglycerides (section 12.2), phospholipids and glycolipids (section 12.3), and eicosanoids (Chemistry in Medicine), we must first review and examine in more detail the chemical structure and physical properties of fatty acids from which most lipids, except steroids, are derived.

■ 12.1 Fatty Acids: A Component of Many Lipids

In chapter 7 you learned that **fatty acids** are long straight-chain hydrocarbons, with a carboxylic acid at one end. Most naturally occurring fatty acids contain an even number of carbons, as for example octacosanoic acid, the fatty acid found in beeswax, which contains 28 carbons.

Fatty acids are further characterized by whether or not they contain carbon-carbon double bonds in their hydrocarbon chain: saturated fatty acids contain no carbon-carbon double bonds, whereas unsaturated fatty acids contain one or more double bonds. Fatty acids with one double bond are known

TABLE 12-1 Common Fatty Acids

Fatty Acid (Common Name)	Number of Carbon Atoms: Number of Double Bonds	Condensed Structural Formula	Common Sources	Melting Point (°C)
Myristic	14:0	$CH_3(CH_2)_{12}COOH$	Nutmeg	53
Palmitic	16:0	$CH_3(CH_2)_{14}COOH$	Palm	63
Stearic	18:0	$CH_3(CH_2)_{16}COOH$	Lard (animal fat)	70
Oleic	18:1	$CH_3(CH_2)_7CH{=}CH(CH_2)_7COOH$	Olive oil, corn oil	13
*Linoleic	18:2	$CH_3(CH_2)_4(CH{=}CHCH_2)_2(CH_2)_6COOH$	Grains, eggs, plant oils	−5
*Linolenic	18:3	$CH_3CH_2(CH{=}CHCH_2)_3(CH_2)_6COOH$	Tuna, salmon, herring	−11
Arachidonic	20:4	$CH_3(CH_2)_4(CH{=}CHCH_2)_4(CH_2)_2COOH$	—	−50

*Essential fatty acids.

as **monounsaturated** fatty acids. Oleic acid, the most abundant monounsaturated fatty acid, contains 18 carbon atoms and one *cis* double bond at C(9)-C(10), as shown in Table 12-1. **Polyunsaturated** fatty acids contain more than one carbon-carbon double bond. Examples include linoleic acid and linolenic acid, two fatty acids with 18 carbons and two and three *cis* carbon-carbon double bonds, respectively (Table 12-1).

The overall shape of a fatty acid depends on whether it is saturated or unsaturated, and in the case of an unsaturated fatty acid, whether it contains a *cis* or a *trans* carbon-carbon double bond. Most naturally occurring unsaturated fatty acids contain *cis* double bonds, with the exception of the fatty acids from the milk and meat of ruminants, which contain 2–5% *trans* unsaturated fats. The geometry of a *cis* double bond creates a bend or kink in the otherwise linear zigzag chain of a fatty acid. You can see this difference in shape by comparing the space-filling models of the saturated fatty acid stearic acid (Figure 12-3a) and the monounsaturated fatty acid oleic acid (Figure 12-3b). Interestingly, unsaturated *trans* fatty acids such as elaidic acid (Figure 12-3c) adopt a linear zigzag shape more like a saturated fatty acid than a *cis* unsaturated fatty acid.

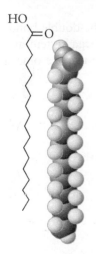

(a) Stearic acid
Melting point 70 °C
(solid at room temperature)

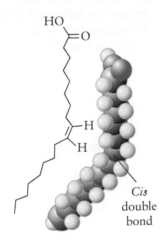

Cis double bond

(b) Oleic acid
Melting point 13 °C
(liquid at room temperature)

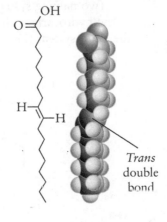

Trans double bond

(c) Elaidic acid
Melting point 46.5 °C
(solid at room temperature)

Figure 12-3 Space-filling models of three 18-carbon fatty acids showing the effect of a carbon-carbon double bond on the overall shape and melting point of the fatty acid: (a) Stearic acid, a saturated fatty acid. (b) Oleic acid, a monounsaturated fatty acid with one *cis* carbon-carbon double bond. (c) Elaidic acid, a monounsaturated fatty acid with one *trans* carbon-carbon double bond.

We have seen that the shape of a molecule greatly influences its physical properties. This is most evident in the physical properties of fatty acids, in particular, their melting points (Table 12-1). Recall from chapter 5 that melting, like boiling, requires that energy be added, because intermolecular forces must be broken in order to go from the solid to the liquid phase. Thus, a higher melting point indicates stronger intermolecular forces of attraction. Fatty acids with more carbon atoms will have higher melting points because they have more dispersion forces due to their greater surface area. This accounts for the differences in melting points observed for a series of saturated fatty acids with 14, 16, and 18 carbon atoms, listed in the first three entries in Table 12-1. Myristic acid (C_{14}) has a melting point of 53 °C; palmitic acid (C_{16}) has a melting point of 63 °C; and stearic acid (C_{18}) has a melting point of 70 °C.

The melting point of a cis unsaturated fatty acid is significantly lower than that of a saturated fatty acid, and it further decreases among a series of fatty acids with an increasing number of *cis* double bonds. You can see this trend in Table 12-1. Stearic acid (18:0), a saturated fatty acid with 18 carbons has a melting point of 70 °C; oleic acid (18:1) a fatty acid with 18 carbons, and one double bond, has a melting point of 13 °C; linoleic acid (18:2), a fatty acid with 18 carbons and two double bonds has a melting point of −5 °C; and linolenic acid (18:3), a fatty acid with 18 carbons and three *cis* double bonds (18:3) has a melting point of −11 °C. Since every *cis* double bond creates an additional "bend" in the overall shape of the molecule, it reduces the number of points of contact between fatty acid molecules, thereby creating fewer dispersion forces, and hence a lower melting point. Consequently, most *cis* unsaturated fatty acids are liquids at room temperature whereas *trans* unsaturated fatty acids and saturated fatty acids are solids at room temperature.

Some unsaturated fatty acids such as linoleic acid and linolenic acid are **essential fatty acids** because they cannot be synthesized in the body from other metabolic intermediates. They must therefore be supplied through the diet. Linoleic and linolenic acids are the well-known omega-6 (ω-6) and omega-3 (ω-3) fatty acids, respectively. Flaxseed, walnuts, salmon, and canola oil are some good sources of these fatty acids. Arachidonic acid, shown in the last row of Table 12-1, is an omega-6 fatty acid with 20 carbons and four *cis* carbon-carbon double bonds which is synthesized in the cell from linoleic acid. Arachidonic acid is the fatty acid from which prostaglandins, important compounds in the immune system, are produced.

Two naming systems are used to indicate the location of the double bonds in the hydrocarbon chain of an unsaturated fatty acid: the **omega** system and the **delta** system (Figure 12-4). Nutritional scientists typically use the omega

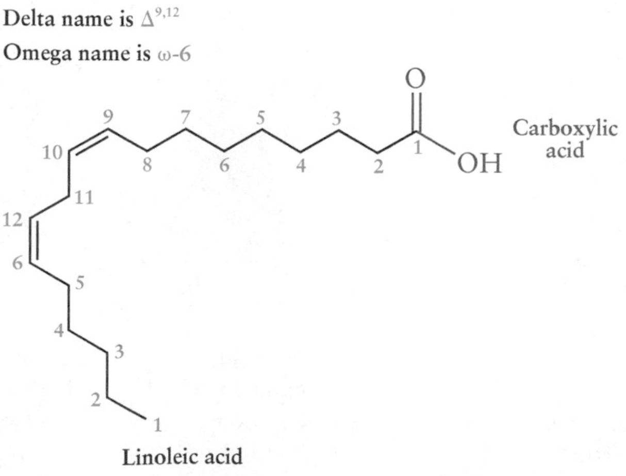

Delta name is $\Delta^{9,12}$

Omega name is ω-6

Linoleic acid

Figure 12-4 A comparison of the delta and omega numbering systems applied to linoleic acid as an example.

system, whereas chemists use the delta system. In the delta system, numbering begins with the carbonyl carbon, C(1), and the location of the first carbon of every double bond in the chain is indicated by a set of superscripts following the Greek capital letter delta (Δ). For example, linoleic acid is a $\Delta^{9,12}$ fatty acid in the delta system because it has two double bonds: one between C(9) and C(10) and another between C(12) and C(13), when starting the numbering from the carbonyl carbon.

In the omega system, numbering begins at the end of the hydrocarbon chain farthest from the carbonyl group and only the location of the first carbon atom containing a double bond is given. This locator number is placed after the Greek lower case letter omega (ω). For example, linoleic acid is an omega-6 (ω-6) fatty acid in the omega system, because the first double bond appears between C(6) and C(7) when starting numbering from the carbon atom farthest from the carbonyl group.

WORKED EXERCISE Fatty Acids

12-1 Both palmit*oleic* acid and palmit*ic* acid are fatty acids containing 16 carbons, but palmitic acid is a saturated fatty acid and palmitoleic acid is a Δ^9 fatty acid.

    a. Write the structure of palmitoleic acid.
    b. Is palmitoleic acid a saturated, monounsaturated, or polyunsaturated fatty acid?
    c. Would you expect palmitoleic acid to have a higher or a lower melting point than palmitic acid? Explain.
    d. What is the name of palmitoleic acid using the omega naming system?

Solution
12-1

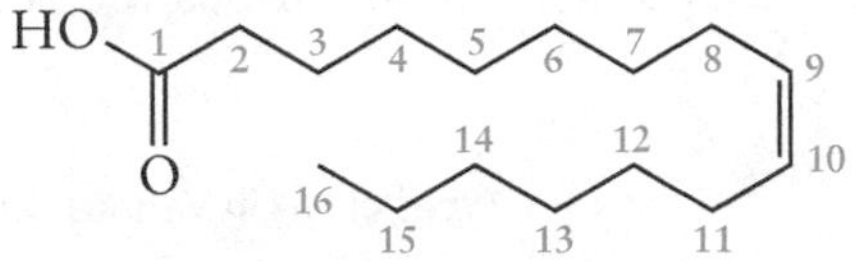

    a. Draw palmitoleic acid by writing a zigzag chain of nine carbon atoms, then draw a *cis* double bond at C(9)-C(10), and then continue with a zigzag chain of six more carbon atoms for a total of 16. Turn C(1) into a carboxylic acid.
    b. Palmitoleic acid is a monounsaturated fatty acid because it contains only one double bond.
    c. You would expect palmitoleic acid to have a lower melting point than palmitic acid because they have the same number of carbon atoms, but palmitoleic acid has a double bond, causing it to have fewer points of contact, thus fewer dispersion forces, and thus entering the liquid phase at a lower temperature (less heat has to be supplied).
    d. To apply the omega system, begin numbering from the end of the chain farthest from the carbonyl group. Then indicate by number where the first carbon is that contains the double bond: omega-7.

PRACTICE EXERCISES

  1 Explain the following physical properties based on structural differences between the indicated fatty acids.
    a. Palmitic acid has a higher melting point than myristic acid.
    b. Oleic acid has a lower melting point than stearic acid.
  2 Why does arachidonic acid have a lower melting point than linolenic acid, even though arachidonic acid has two more carbon atoms?

3 Using Table 12-1, draw the following fatty acids and indicate if it is an omega-6 or an omega-3 fatty acid. Are either of these fatty acids essential fatty acids?

   a. Linolenic acid
   b. Oleic acid

4 Fill in the blanks. In the delta numbering system, arachidonic acid is a $\Delta^{\overline{}\,\overline{}\,\overline{}\,\overline{}}$ fatty acid. In the omega system, it is an omega-_____ fatty acid.

5 Which is more likely to be a liquid at room temperature: linolenic acid or stearic acid? Explain?

6 In what way is the shape of a *cis* unsaturated fatty acid different from the shape of a saturated fatty acid? How does this difference impact its physical properties? Does a *trans* unsaturated fatty acid bear a greater resemblance to a *cis* unsaturated fatty acid or a saturated fatty acid? Explain.

▪ 12.2 Triglycerides: Energy Storage Lipids

You are probably familiar with triglycerides, although you may know them as fats and oils. **Triglycerides**, or simply "**fats**," are the most abundant type of lipid, obtained through our diet from both animal and vegetable sources. Triglycerides are stored in fat cells, also known as **adipocytes** (Figure 10-16), and metabolized in muscle and liver cells. Triglycerides supply the body with energy over the long term, complementing glycogen which is a source of energy over the short term. For example, it is triglycerides that supply hibernating animals with energy during their long winter slumber.

Diets high in saturated fats and *trans* fats have been linked to cardiovascular disease, diabetes, and cancer. Hence, the American Heart Association recommends a diet low in saturated and *trans* fats. The relative amount of saturated and unsaturated fats in various foods is compared in Figure 12-5.

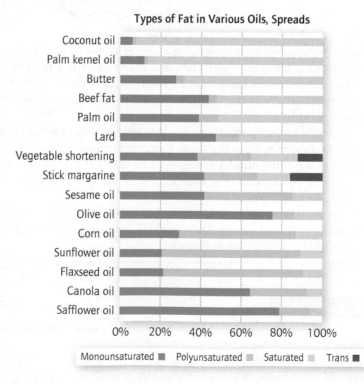

Figure 12-5 A bar graph showing the fat content of some common fats and oils published by Harvard Medical School.

As you can see from the graph, vegetable oils generally have the lowest saturated fat content, while animal fats have the highest. The exceptions are palm oil and coconut oil, which have the highest saturated fat content even though they are derived from plant sources.

The Chemical Structure of Triglycerides

The basic structure of a triglyceride is shown in **Figure 12-6**, where the "tri" in triglyceride refers to the *three* acyl groups. In chapter 10 you learned that triglycerides are derived from one *glycerol* molecule and three *fatty acids* by esterification reactions at each of the three hydroxyl groups of glycerol (blue) and the carboxylic acid of three fatty acids (green). In the example given in Figure 12-6, the triglyceride is derived from three palmitic acid molecules and glycerol. The variety of triglycerides comes from the many different fatty acids available.

Although the ester functional group has a permanent dipole, the extensive hydrocarbon structure of triglycerides makes them nonpolar and therefore insoluble in water.

The physical state (solid or liquid) of a triglyceride at room temperature depends on whether it is derived from saturated or unsaturated fatty acids. Triglycerides derived from *saturated* fatty acids tend to be *solids* at room temperature (25 °C) and, thus, they are commonly referred to as **fats**. These triglycerides are found in meat, whole milk, stick butter, and cheese—animal sources. Tropical oils from plant sources such as coconut and palm oil are also solids because they too are derived from saturated fatty acids. Most plant sources of triglycerides, however, are derived from *unsaturated* fatty acids and so tend to be viscous *liquids* at room temperature, and are referred to as **oils**. For example, canola, sesame, olive, and fish oils contain triglycerides derived from unsaturated fatty acids. Some foods high in unsaturated fats, in particular the essential omega-3 and omega-6 fatty acids are shown in Table 12-2.

Triglycerides have physical properties that resemble the fatty acids from which they are derived. In chapter 6 you learned that saturated fats tend to be solids at room temperature because they are more uniform in shape and therefore can pack more tightly (Figure 6-2). Tighter packing creates more

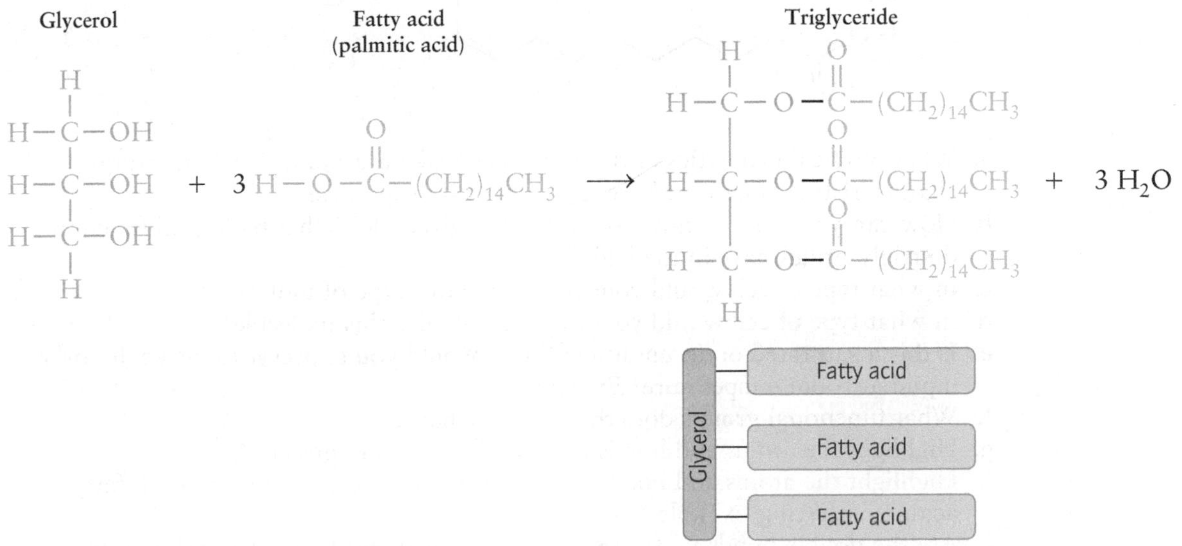

> Triglyceride is a short-hand term for triacylglycerol, a molecule with three acyl groups attached to a glycerol molecule in the form of three (tri) ester functional groups.

Figure 12-6 The esterification reaction that forms a triglyceride from one glycerol molecule and three fatty acid molecules.

TABLE 12-2 Foods High in Unsaturated Fats

Monounsaturated	Polyunsaturated Omega-6	Polyunsaturated Omega-3
Nuts	Soybean oil	Salmon
Canola oil	Corn oil	Trout
Olive oil	Safflower oil	Herring
Sunflower oil		Soybean oil
Safflower oil		Canola oil
		Walnuts
		Flaxseed

intermolecular forces of attraction, resulting in a higher melting point and a waxy solid (think butter) at room temperature. In contrast, unsaturated fats tend to be liquids (oils) at room temperature because of their irregular shape, as a result of *cis* carbon-carbon double bonds, and therefore looser packing (Figure 6-3). Looser packing allows fewer intermolecular forces of attraction, thus resulting in a lower melting point and a liquid (think cooking oil) at room temperature.

WORKED EXERCISE Triglyceride Structure

12-2 Answer the questions below for the following triglyceride.

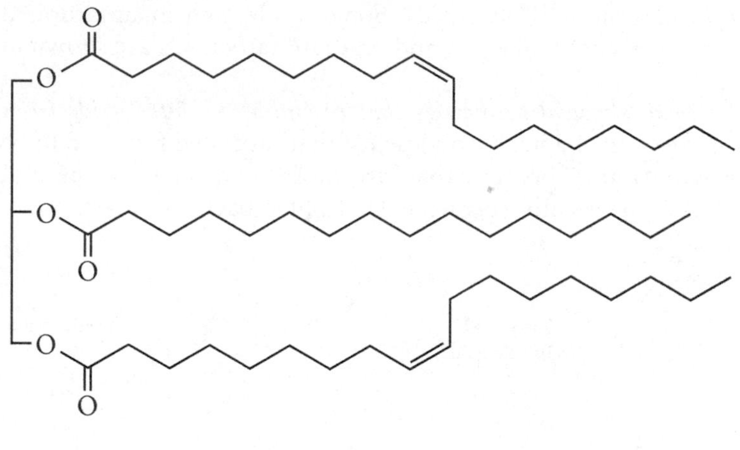

a. What physical properties of this substance make it a lipid? What structural characteristics cause it to have these physical properties?
b. How can you tell that this molecule is a triglyceride? What biological function do triglycerides have in the body?
c. In what type of cell would your body store this type of molecule?
d. In what type of cell would your body metabolize this molecule?
e. Is this a saturated or an unsaturated fat? Would you expect it to be a solid or a liquid at room temperature? Explain.
f. What functional groups does this molecule have?
g. Highlight the atoms and bonds that are derived from glycerol.
h. Highlight the atoms and bonds derived from fatty acids and name each fatty acid, by referring to Table 12-1.
i. Are the double bonds in this triglyceride *cis* or *trans*? Is this a healthy or an unhealthy fat?
j. Is this triglyceride more likely to be found in meat or corn?

Solution

12-2 **a.** It is a lipid because it is found in plants or animals and it is insoluble in water but soluble in organic solvents. The three long nonpolar hydrocarbon chains with few permanent dipoles makes this molecule hydrophobic, and therefore insoluble in water.

b. The three ester functional groups with long hydrocarbon chains on three adjacent carbon atoms is a characteristic of a triglyceride. Triglycerides provide a source of long-term energy for the body.

c. Triglycerides are stored in fat cells, also known as adipocytes.

d. Triglycerides are metabolized in liver and muscle cells.

e. This is an unsaturated triglyceride because it contains one or more carbon-carbon double bonds. Most likely this will be a viscous liquid (oil) at room temperature. The *cis* double bonds prevent close packing of triglyceride molecules, reducing the number of dispersion forces, and hence lowering the melting point.

f. This molecule has three ester functional groups and two carbon-carbon double bonds.

g.–h.

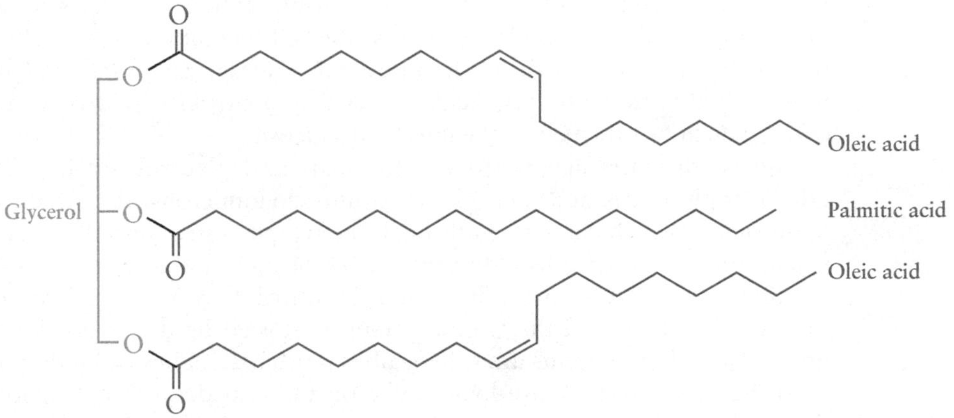

i. The double bonds are *cis* because the R groups on the carbon-carbon double bond are on the same side of the double bond. Unsaturated fats containing *cis* double bonds are healthier than those containing *trans* double bonds or saturated fats.

j. It is more likely found in a vegetable source like corn (but not tropical oils).

PRACTICE EXERCISES

7 Write the structure of a triglyceride composed of linoleic acid, linolenic acid, and oleic acid.
  a. Would this triglyceride be considered a saturated or an unsaturated fat?
  b. Would you expect this triglyceride to be a solid or a liquid at room temperature?
  c. Would you expect this compound to be soluble in water? Explain.

8 Which vegetable oils are high in saturated fat and should, therefore, be included in limited amounts in the diet?

9 Which product is higher in saturated fats: peanut oil or corn oil?

10 Which product is healthier: coconut oil or canola oil? Explain.

11 Are unsaturated fats primarily found in vegetable or animal sources?

12 Explain why saturated fats tend to be solids at room temperature while unsaturated fats tend to be liquids at room temperature.

Digestion of Dietary Fats and Their Transport to Muscle and Fat Cells

Triglycerides stored in fat cells provide the body with energy over the long term while glycogen provides energy over the short term. An average adult man, for example, stores 300 g of glycogen but about 15,000 g of triglycerides.

The glycogen lasts about a day, depending on his activity level, while the triglycerides can last several weeks.

Fatty acid catabolism is the central energy producing pathway for many cells. For example, it provides 80% of the energy needs for heart and liver cells. The highly reduced (saturated or close to saturated) form of the hydrocarbon chain of a fatty acid makes it an excellent fuel—much like the hydrocarbons in gasoline are a good fuel for an automobile engine. Indeed, in chapter 4 you learned that triglycerides provide more than double the energy of carbohydrates and proteins of comparable mass (9 Cal/g rather than 4 Cal/g).

The digestion of dietary fat begins in the small intestine, where triglycerides from the diet enter as large globules of water-insoluble fats. Before degradation reactions can begin, these globules of fat must be turned into a colloid of finely dispersed microscopic fat droplets suspended in water, a process known as **emulsification**. Emulsification brings water-insoluble triglycerides in contact with water-soluble lipase enzymes that catalyze the hydrolysis of triglycerides into glycerol and fatty acids. Emulsification requires **bile salts**, which are released into the small intestine from the gall bladder after a fatty meal. Triglycerides are too large to cross the cell membrane of the cells lining the small intestine known as the **intestinal mucosa**, therefore, triglycerides must be hydrolyzed into fatty acids, glycerol, and partially hydrolyzed triglycerides, which can then enter the intestinal mucosa.

Inside the intestinal mucosa, fatty acids and glycerol are esterified to reform triglycerides and then packaged into **chylomicrons**, the largest of the lipoproteins, which carry them through the lymph system and the circulatory system to their target cells (adipocytes, liver, or muscle cells).

Since triglycerides are not soluble in water, they must be transported through the blood and the lymph system by special lipid carriers, known as **lipoproteins**. Lipoproteins are spherically shaped assemblies of lipids and proteins that, like micelles, are hydrophilic on the outside and hydrophobic on the inside (see section 12.4).

Upon arrival at the capillaries of their target tissues, triglycerides must again be hydrolyzed into fatty acids and glycerol, smaller molecules that can pass through the cell membrane of adipose and muscle cells. Inside an adipocyte, fatty acids and glycerol are converted back into triglycerides and stored until needed. In liver and muscle cells, fatty acids are degraded further, ultimately producing energy in the form of ATP in a biochemical process described in chapter 15.

PRACTICE EXERCISES

13 Catabolism of which biomolecule produces the most energy per gram: proteins, carbohydrates, or triglycerides?

14 The egg in mayonnaise is an emulsifying agent that allows the main ingredients, oil and vinegar, to mix. What is an emulsifying agent?

15 What role does bile serve in the digestion of lipids?

16 Where are intestinal mucosa cells found, and what reactions pertaining to triglycerides occur in these cells?

17 What products are formed when lipase enzymes catalyze the hydrolysis of triglycerides in the small intestine? Why must triglycerides be hydrolyzed before they can enter the intestinal mucosa?

18 What is the function of the chylomicrons?

19 Extracellular (outside the cell) enzymes hydrolyze triglycerides before they enter fat and muscle cells. What happens to these hydrolysis products after they enter adipocytes? What happens to these hydrolysis products after they enter muscle cells?

20 In which organ in the digestive system is bile stored? Why is bile released after consumption of a high-fat meal?

21 Why are carbohydrates readily hydrolyzed, but triglycerides need to be emulsified first?

■ 12.3 Membrane Lipids: Phospholipids and Glycolipids

In chapter 8 you learned that the cell membrane, a part of all living cells, is a selectively permeable membrane. As such, it separates the inside of the cell from the outside of the cell and controls the flow of ions and molecules in and out of the cell. While proteins are attached to and interspersed throughout the cell membrane, and carbohydrates protrude from it, membrane lipids form the bilayer that is the cell membrane. In this section, we describe the structure and function of the cell membrane.

Our cell membranes are composed primarily of two types of lipids: *phospholipids* and *glycolipids*. Cholesterol, a third type of lipid structurally different from phospholipids and glycolipids, is also present in the lipid bilayer, adding structural rigidity to the cell membrane. Membrane lipids and cholesterol have the unique characteristic of being **amphipathic** (pronounced *am-fuh-path-ik*) molecules: they are molecules that contain both a polar (hydrophilic) and a nonpolar (hydrophobic) region.

The Chemical Structure of Phospholipids and Glycolipids

In this section we consider the structural variation and characteristics of the molecules that make up the cell membrane. Then, in the next section we will see how their chemical structure enables them to assemble into the lipid bilayer that forms a cell membrane. Refer to Figure 12-7 to compare the basic structural components of the membrane lipids found in human cells. For comparison, we also show the general structure of a triglyceride (section 12.2), with which you are already familiar.

Whereas triglycerides, or energy storage lipids, contain a glycerol backbone, membrane lipids contain either a glycerol or a sphingosine (pronounced *sfing-guh-seen*) backbone. Figure 12-8 shows the structure of glycerol and sphingosine next to the color-coded schematics used to represent them in Figure 12-7. **Sphingosine** is a straight chain of 18 carbon atoms with a *trans* carbon-carbon double bond at C(4)-C(5), two hydroxyl groups (—OH) at C(1) and C(3), and an amino group (—NH$_2$) at C(2). As you can see, a large portion of the molecule is a hydrocarbon chain; hence, sphingosine itself supplies *one* of the two hydrocarbon "tails" in a sphingosine derived membrane lipid. The other "tail" is derived from a fatty acid, as described on the next page. In contrast, the two hydrocarbon "tails" in a glycerol derived membrane lipid are both derived from fatty acids, much like triglycerides are derived from glycerol and *three* fatty acids.

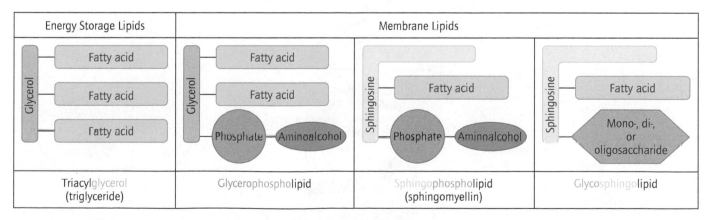

Figure 12-7 A reference chart comparing the basic components of the different types of membrane lipids found in human cells. A comparison to a triglyceride is included. All are derived from one or more fatty acids.

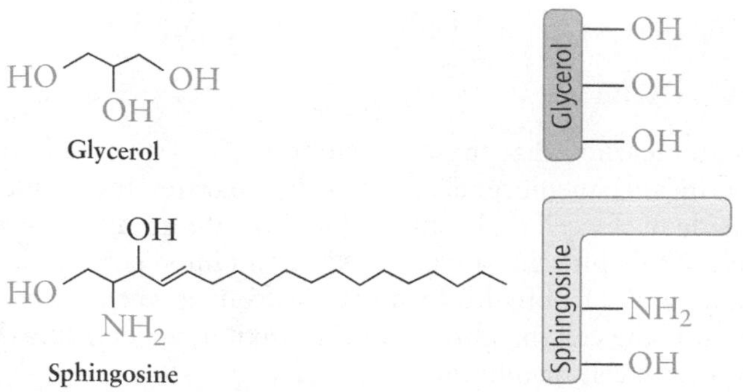

HO⌐⌐OH
      OH
   Glycerol

Glycerol	— OH
	— OH
	— OH

       OH
HO⌐⌐⌐⌐⌐⌐⌐⌐⌐⌐⌐⌐⌐⌐⌐⌐
      NH₂
   Sphingosine

Sphingosine	— OH
	— NH₂
	— OH

Figure 12-8 The chemical structures of *glycerol*, which provides the backbone for *glycero*phospholipids, and *sphingosine*, which provides the backbone for *sphingo*lipids.

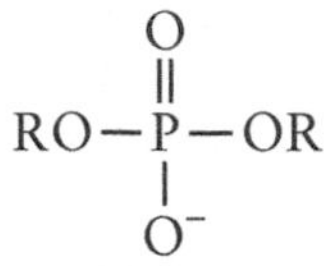

       O
       ‖
RO — P — OR
       |
       O⁻
Phosphate group

Phospholipids Phospholipids (pronounced *fos-foh-lip-ids*) are the most common type of membrane lipids, characterized by a polar phosphate group. The phosphate group is linked to the terminal hydroxyl group (last —OH group) of either sphingosine or glycerol as a phosphate ester (shown in purple in Figure 12-7), like the phosphate esters described in chapter 7. A phospholipid derived from glycerol is known as a ***glycerophospholipid***, and a phospholipid derived from sphingosine is known as a ***sphingophospholipid***. Sphingophospholipids, also known as **sphingomyelins** (pronounced *sfing-goh-mahy-uh-lins*), insulate nerve fibers, creating what is known as the **myelin sheath** (Figure 12-9). Multiple sclerosis (MS) is a serious degenerative disease in which the myelin sheath deteriorates leading to debilitating neurological disorders.

The two nonpolar tails in a glycerophospholipid are derived from two fatty acids that form ester linkages to the other hydroxyl groups in a glycerol molecule. In a sphingophospholipid one nonpolar tail is derived from a fatty acid that forms an amide linkage to the amine group in sphingosine. Recall from chapter 10 that an amine reacts with a carboxylic acid to form an amide analogous to the way an alcohol reacts with a carboxylic acid to form an ester.

The phosphate group in all phospholipids forms two phosphate ester linkages: one to the terminal hydroxyl group of either glycerol or sphingosine and the second to a small amino alcohol, a molecule with an amine (—NH₂) and a hydroxyl (—OH) group, represented as a blue oval in Figure 12-7. *The nitrogen atom in an amino alcohol has four bonds, and thus, a positive charge, which together with the negative charge on the phosphate group creates*

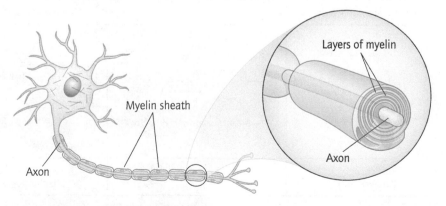

Figure 12-9 A nerve cell showing the myelin sheath that insulates nerve fibers, which is composed of sphingomyelins.

the polar head of a phospholipid. The three amino alcohols from which most phospholipids are derived are ethanolamine, choline, and serine:

Ethanolamine   Choline   Serine
(glycerophospholipids only)

For example, the phospholipid shown in Figure 12-10 contains a glycerol backbone with two ester linkages to two fatty acids (stearic acid), and a phosphate diester linked to the terminal hydroxyl group of glycerol and the hydroxyl group of choline.

Glycolipids All **glycolipids** in human cells are derived from sphingosine, and thus they are **glycosphingolipids**. The "glyco" in *glyco*sphingolipids refers to the carbohydrate component of these lipids, which furnishes the polar region of these amphipathic molecules. Recall that carbohydrates are polyols (many hydroxyl groups), which make them hydrophilic. So instead of a phosphate group, glycolipids contain a carbohydrate, joined between the anomeric carbon and the terminal hydroxyl group of sphingosine. The carbohydrate component of a glycosphingolipid can be either a mono-, di-, or oligosaccharide (3 to 10 monosaccharides). Like sphingophospholipids, one of the nonpolar tails in a glycolipid comes from the amide linkage formed between the amine of sphingosine and a fatty acid, while the other nonpolar tail comes from the long hydrocarbon chain on sphingosine itself.

Glycosphingolipids containing a monosaccharide are known as **cerebrosides** (pronounced *suh-ree-bruh-sahyd*). Cerebrosides derived from D-galactose are found primarily in cells of the central nervous system and the brain, while those containing D-glucose are found in most other tissues. For example, the cerebroside shown in Figure 12-11 is derived from D-galactose, which makes up the polar head, and the fatty acid stearic acid.

> Do not mix up *glycero* and *glycol*. The former refers to a lipid with a glycerol backbone, whereas the latter refers to a lipid with a carbohydrate component.

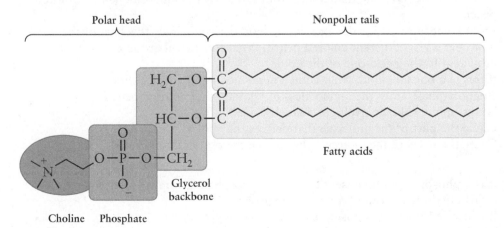

Figure 12-10 The chemical structure of a glycerophospholipid showing the glycerol backbone, two fatty acids, choline, and the phosphate diester. The charged phosphate and nitrogen atom makeup the "polar head," and the two hydrocarbon chains from the fatty acids are the "tails."

Figure 12-11 The chemical structure of a cerebroside showing the sphingosine backbone, one fatty acid, and the monosaccharide D-galactose.

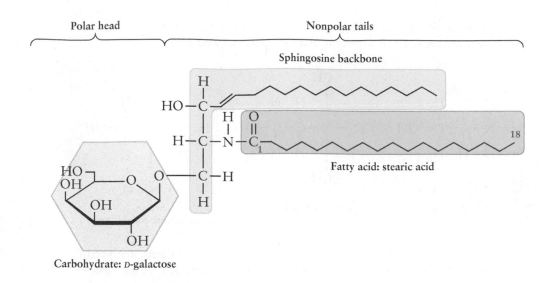

WORKED EXERCISE   Membrane Lipids

12-3 Refer to the membrane lipid shown below to answer the following questions.

a. Is this an example of a phospholipid or a glycolipid? What is the difference?
b. Is this membrane lipid derived from glycerol or sphingosine? Trace the portion of the molecule that comes from glycerol or sphingosine with a colored pencil.
c. From what fatty acid is this membrane lipid derived? Trace this part of the molecule with a different colored pencil. What functional group connects the fatty acid to the backbone of the lipid?
d. If this is a phospholipid, what is the amino alcohol: serine, choline, or ethanolamine? Trace the amino alcohol with another colored pencil.
e. Is this lipid a sphingomyelin or a cerebroside?
f. Circle the part of the molecule that defines the "polar head."
g. Circle the parts of the molecule that define the two nonpolar tails.

Solution

12-3 a. It is a phospholipid because it contains a phosphate group. Glycolipids contain a carbohydrate instead of a phosphate group.
b. It is derived from sphingosine, shown below in green.
c. The fatty acid is palmitic acid, a 16-carbon saturated fatty acid, shown in red. An amide functional group joins the amine group of sphingosine with the carboxylic acid of the fatty acid.
d. The amino alcohol is ethanolamine, $HOCH_2CH_2NH_3^+$, shown in blue.

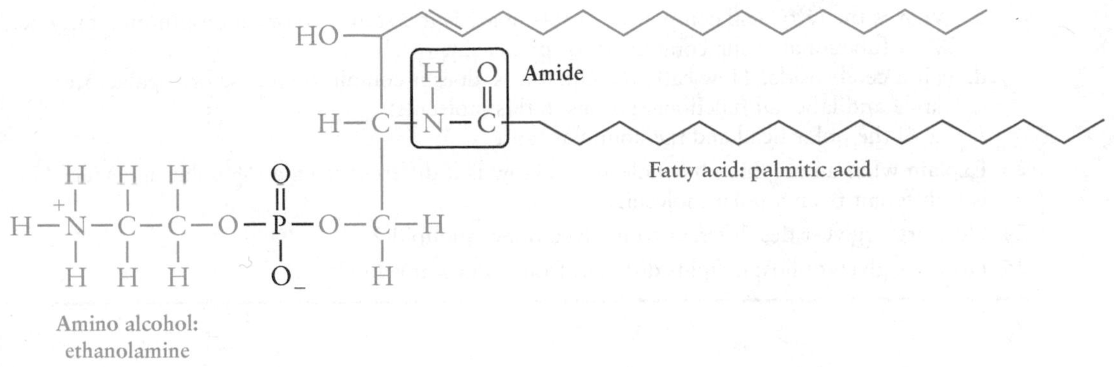

Amide

Fatty acid: palmitic acid

Amino alcohol:
ethanolamine

e. This is a sphingomyelin because it is derived from sphingosine and contains a phosphate group.

f. and g. See below.

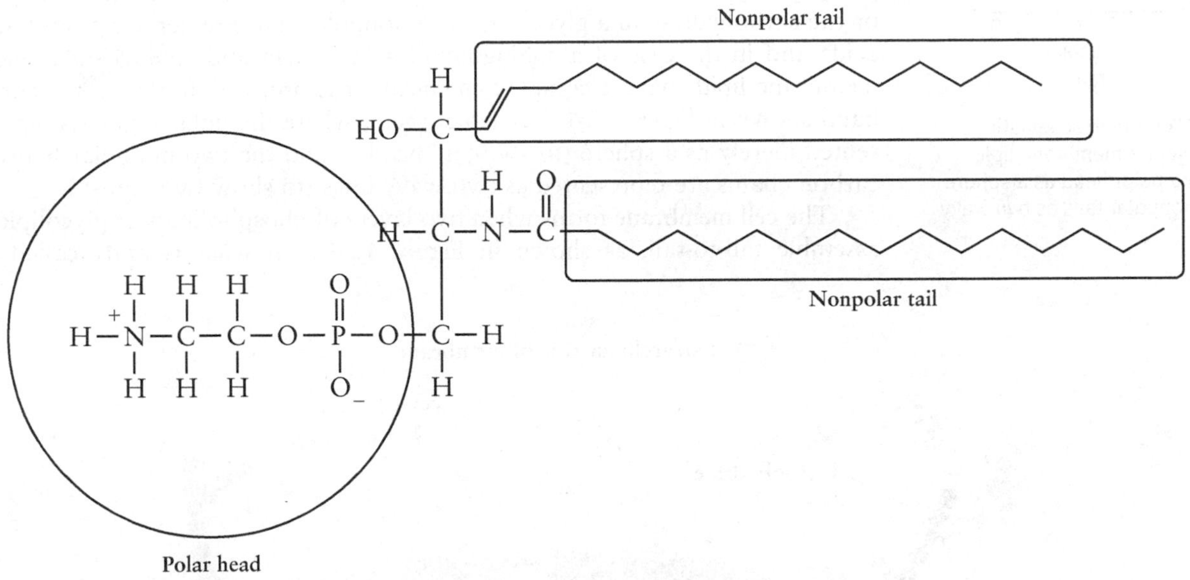

Nonpolar tail

Nonpolar tail

Polar head

PRACTICE EXERCISES

22 For the membrane lipid shown, answer the following questions.

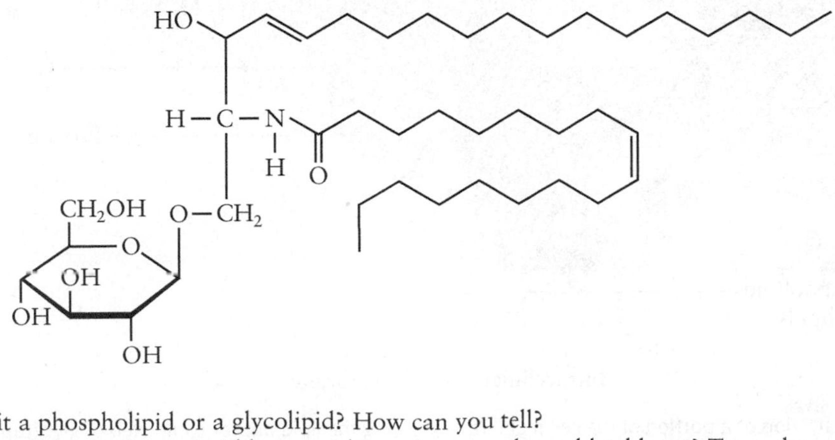

a. Is it a phospholipid or a glycolipid? How can you tell?

b. Does this membrane lipid have a sphingosine or a glycerol backbone? Trace the part of the molecule derived from sphingosine or glycerol.

     c. What is the fatty acid component of this lipid? Is it a saturated or an unsaturated fatty acid? What functional group connects it to sphingosine or glycerol?

     d. Is it a cerebroside? How can you tell? If it is, does it contain D-glucose or D-galactose?

     e. Circle and label all functional groups in this molecule.

     f. Label the polar head and the nonpolar tails.

23 Explain what an amphipathic molecule is. How is it different from a nonpolar molecule? How is it different from a polar molecule?

24 How are triglycerides different from glycerophospholipids?

25 How are glycerophospholipids different from sphingomyelins?

The Cell Membrane

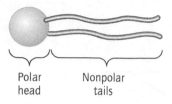

Figure 12-12 Common schematic used to represent a membrane lipid by showing the polar head as a sphere and the two nonpolar tails as two wavy lines.

We have seen that membrane lipids are amphipathic molecules with two non-polar hydrocarbon "tails" and a polar "head." The polar head is the phosphate group together with the positively charged nitrogen in a phospholipid or the carbohydrate in a glycolipid. The nonpolar tails are derived from fatty acids and in the case of a sphingolipid, a fatty acid and sphingosine. Since membrane lipids have a complex chemical structure, we often use the shorthand shown in Figure 12-12 to depict them, where the polar region is represented merely as a sphere (to show a "head"), and the two nonpolar hydrocarbon chains are represented as two wavy lines (to show two "tails").

    The **cell membrane** forms when two layers of phospholipids or glycolipids assemble tail-to-tail, as shown in Figure 12-13, in what is aptly called a

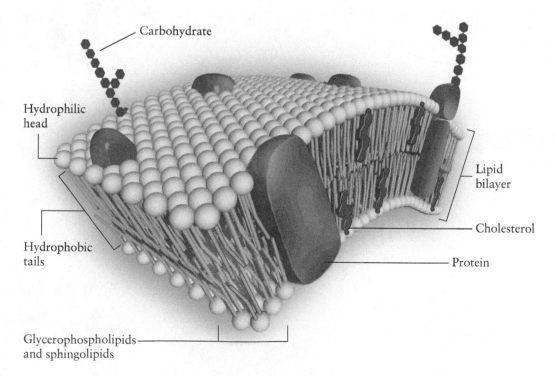

Figure 12-13 An illustration of a portion of the cell membrane showing the lipid bilayer composed of glycerophospholipids, sphingophospholipids, and glycosphingolipids (beige), interspersed with cholesterol (red). Proteins (brown) are also interspersed throughout the membrane, and carbohydrates (dark blue) protrude from the membrane.

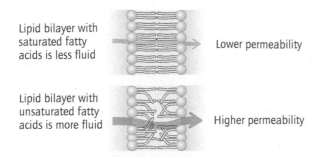

Lipid bilayer with saturated fatty acids is less fluid

Lower permeability

Lipid bilayer with unsaturated fatty acids is more fluid

Higher permeability

Figure 12-14 Fatty acid structure affects the permeability of membranes. Cell membrane permeability increases with membrane lipids derived from unsaturated fatty acids.

bilayer. The lipids that form the bilayer assemble such that their hydrophilic polar heads are in contact with the aqueous environment on either the inside or the outside of the cell, while the nonpolar tails lie in the hydrophobic interior of the bilayer. The width of the cell membrane, the lipid bilayer, is approximately 7 nm wide.

The lipid bilayer has often been described as a "fluid mosaic," because the phospholipid and glycolipid molecules are not covalently bonded to one another, but instead, interact through the weaker noncovalent forces of attraction, creating a flexible fluid-like assembly. Embedded within the cell membrane are molecules of cholesterol (shown in red in Figure 12-13), also an amphipathic molecule. The polar hydroxyl group of cholesterol lies among the polar heads while the hydrocarbon portion of the molecule is situated among the nonpolar tails. Various proteins (shown in brown) can also be found throughout the cell membrane. Some carbohydrate markers are seen protruding from the surface of the cell as described in chapter 11.

Membrane Permeability In chapter 8 you learned that small nonpolar molecules, such as oxygen and carbon dioxide, pass freely through the cell membrane by simple diffusion, while water passes slowly by the process of osmosis. Since the interior of the cell membrane is nonpolar, you can now better understand why ions and polar organic molecules do not diffuse through the cell membrane.

You have also seen that an important function of the cell membrane is to ensure that different concentrations of ions are maintained inside and outside of the cell. Furthermore, the ability to transport ions across the cell membrane is crucial to the survival of the cell. Sometimes special transport proteins facilitate the transfer of ions from one side of the membrane to the other. Specific hormones regulate which ions move in and out of the cell, working with other chemical messengers that interact with proteins on or within the cell membrane.

Cell membranes vary in their permeability depending on characteristics of the membrane lipids, in particular, the extent of unsaturation in the hydrocarbon tails. Cell membranes with more *saturated* phospholipids exhibit less permeability, and less fluidity as shown in Figure 12-14, whereas those with more *unsaturated* phospholipids have greater permeability. This is another example of how saturated lipids pack more closely than unsaturated lipids, in this case affecting the physical property of membrane permeability.

■ 12.4 Cholesterol, Steroids, and Lipoproteins

Steroids are molecules that contain the characteristic steroid ring system: four fused rings labeled A–D in the figure on the next page. Steroids are classified as lipids because they are biomolecules insoluble in water and soluble in organic solvents. However, they are structurally quite different from triglycerides, phospholipids, and glycolipids because they are not derived from fatty

Figure 12-15 The structure of cholesterol, which is the starting material for bile salts, vitamin D, and various hormones.

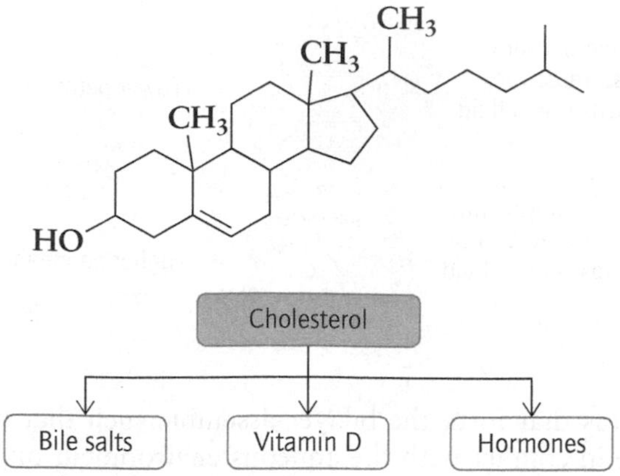

acids. Their hydrophobic physical characteristics arise instead from the fact that the steroid ring system is a hydrocarbon.

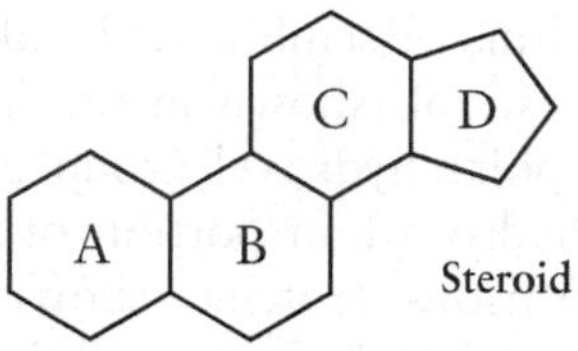

Steroid core

Cholesterol, the most important steroid, not only helps to maintain the integrity of the cell membrane, but serves as the biological precursor for all the other important steroids in the cell, including the steroid hormones, vitamin D, and bile salts (Figure 12-15).

Bile salts, such as cholate, are amphipathic compounds that act as emulsifying agents. The ionized carboxylic acid is the polar end, while the steroid ring system is the nonpolar end of this amphipathic molecule. Cholate and other bile salts (cation component of salt not shown) break up globules of dietary fat in the first stage of digestion, exposing triglycerides to water-soluble enzymes required for hydrolysis. Bile is produced in the liver and stored in the gall bladder, whereupon it is released into the small intestine in response to a high-fat meal.

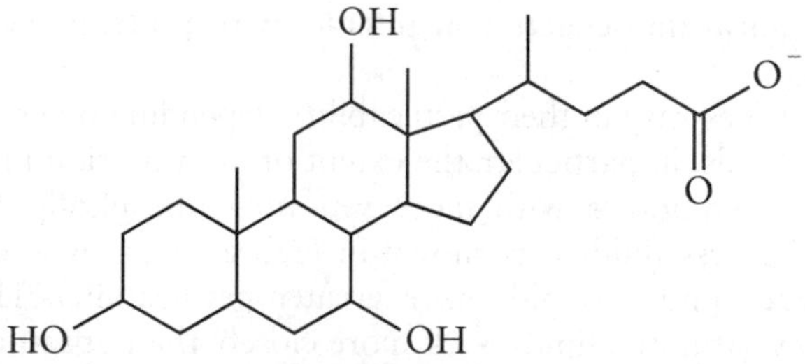

Cholate
(a bile salt)

Steroid Hormones

The important steroid hormones synthesized from cholesterol are summarized in Figure 12-16. The **glucocorticoids** (*gluco*se + adrenal *cort*ex + ster*oid*) are a class of *steroids* produced in the outer part of the adrenal gland, a gland that lies right above the kidneys, known as the **adrenal cortex**. These hormones regulate

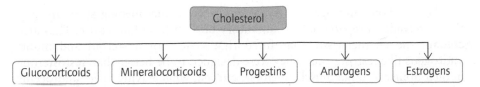

Figure 12-16 A schematic showing the five classes of steroid hormones produced from cholesterol.

glucose metabolism, and have anti-inflammatory and **immunosuppressant** activity. Consequently, they can be used to treat autoimmune conditions as well as allergies, asthma, and other diseases resulting from an overactive immune system.

Cortisol (also known as hydrocortisone) is an important glucocorticoid. Cortisol is released in response to stress, which causes an increase in blood glucose levels, and facilitates fat and protein metabolism. Cortisol also suppresses the immune system. You may be familiar with an over-the-counter ointment of 1% hydrocortisone used to treat dermatitis and other skin irritations (**Figure 12-17**). Prednisone is a related synthetic glucocorticoid that is used as an immunosuppressant for the treatment of asthma and arthritis.

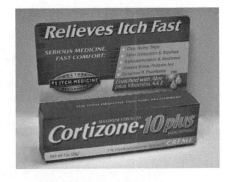

Figure 12-17 A 1% ointment of Hydrocortisone is available over-the-counter for the treatment of dermatitis and other skin irritations. [Catherine Bausinger]

Cortisol
(hydrocortisone)

Prednisone

The **mineralocorticoids** (*mineral* + adrenal *cortex* + ster*oid*) are also produced by the adrenal cortex. These hormones regulate electrolyte (Na^+, K^+, Cl^-) balance in tissues. Aldosterone, the primary mineralocorticoid, acts on the kidneys to increase the reabsorption of sodium ions (Na^+) and the release of potassium ions (K^+), as well as the retention of water, which has the effect of increasing blood pressure and blood volume. Indeed, drugs that inhibit the release of aldosterone are used to treat hypertension (high blood pressure).

Progestins, estrogens, and androgens represent the primary sex hormones. Progestins, such as progesterone, regulate the menstrual cycle and pregnancy. Androgens (testosterone) and estrogens (estradiol) control the development of sexual characteristics in the developing fetus and secondary sexual characteristics during puberty.

Aldosterone
(mineralocorticoid)

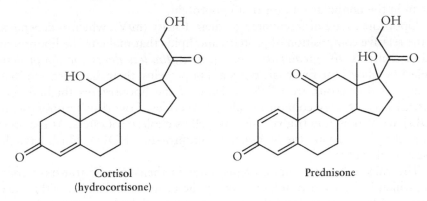

Progesterone
(progestin)

Testosterone
(androgen)

Estradiol
(estrogen)

The anabolic androgenic steroids described in the opening story are synthetic steroids—prepared in the laboratory but not found in nature. They also contain the characteristic four fused ring system of a steroid and mimic androgens like testosterone. They were designed to be better than testosterone in their ability to build muscle (anabolic effects), while having fewer androgenic effects.

Lipoproteins and Cholesterol

We have seen that cholesterol is an important steroid that serves as a starting material for a number of important hormones. Cholesterol is also an important component of cell membranes. However, since cholesterol is insoluble in aqueous solution, it cannot be transported to cells through the circulatory system like blood sugar and other nutrients. Instead, cholesterol and triglycerides are transported through the blood and lymph system by lipoproteins. **Lipoproteins** (pronounced *lip-uh-proh-teen*) are assemblies of both phospholipids and proteins; hence the term lipoprotein (*lip*id + *protein*). However, instead of assembling as a bilayer (two-layers) like the cell membrane, lipoproteins assemble as a spherically shaped monolayer (one layer) of phospholipids with a polar exterior and a nonpolar interior, as shown in Figure 12-18. Triglycerides and cholesterol esters are transported through the circulatory system in the nonpolar interior of a lipoprotein.

Lipoproteins are differentiated by their density (m/V), which is dependent on the relative composition of proteins and lipids that makeup the lipoprotein and its contents: *the greater the lipid content, the less dense the lipoprotein.* Table 12-3 shows the five basic types of lipoproteins, their diameter, and their density. Chylomicrons are the least dense of the lipoproteins (highest lipid content) because they contain mainly triglycerides. Low-*density lipoproteins* (LDLs) are also high in triglycerides, as well as cholesterol esters. At the other end of the spectrum are the *high-density lipoproteins* (HDLs), which have the lowest lipid content.

The various types of lipoproteins differ in their specific transport function, which is determined in large part by the proteins embedded within their structure. Chylomicrons have the largest diameter of the lipoproteins, 1 μm in

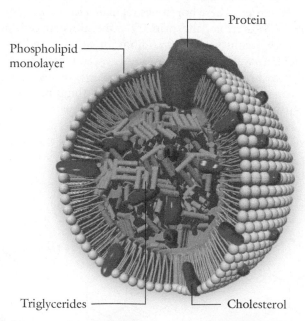

Figure 12-18 A cross section of a lipoprotein: a spherical monolayer of phospholipids with a polar exterior and a nonpolar interior, embedded with proteins (brown) and carrying triglycerides (grey) and cholesterol (red).

TABLE 12-3 Physical Properties of Lipoproteins

Type of Lipoprotein	Chylomicron	VLDL *Very low density lipoprotein*	IDL *Intermediate density lipoprotein*	LDL *Low density lipoprotein*	HDL *High density lipoprotein*
Diameter (nm)	1,000 (1 μm)	70	40	20	10
Density (g/cm^3)	0.95	0.98	1.01	1.04	1.13
Lipid/Protein Ratio	66	11	8	3.8	1.2
Popular Term	"Bad cholesterol"				"Good cholesterol"

diameter (also known as a micron). They function to transport triglycerides from the intestinal mucosa to muscle and fat cells via the lymph system and circulatory system.

Once chylomicrons deliver their contents to cells, they return to the liver as *chylomicron remnants* (leftovers), where they are either degraded, or when there is an excess of fatty acids, packaged into *very low density lipoproteins* (VLDLs). VLDLs also travel through the bloodstream to deliver triglycerides to adipose and muscle cells. Eventually, as they unload cholesterol and triglycerides to cells, they become *intermediate density lipoproteins* (IDLs) and then *low density lipoproteins* (LDLs). In contrast, the role of *high density lipoproteins* (HDLs) is to scavenge cholesterol from the blood and return it to the liver.

You may have heard the terms "bad cholesterol" and "good cholesterol," but in reality, there is only one cholesterol molecule. More accurately, these terms refer to the lipoproteins that carry cholesterol through the bloodstream. High density lipoproteins, HDLs, are considered the "good cholesterol" because they scavenge excess cholesterol from the bloodstream and transport it back to the liver, where it is converted into bile salts and excreted. A higher level of HDLs circulating in the bloodstream is desirable because they remove cholesterol from the blood, reducing the likelihood of plaque formation. Studies have shown that a healthy diet together with exercise increases a person's HDL levels. "Bad cholesterol" refers to low density lipoproteins (LDLs) which are the primary carriers of excess triglycerides to adipose tissue. Hence, high levels of LDLs together with low levels of HDLs are associated with cardiovascular disease.

PRACTICE EXERCISES

26 What structural feature characterizes a steroid?

27 Prednisone reduces itching, swelling, redness, and allergic reactions. Naproxen reduces pain, fever, and swelling. Based on the chemical structure of these two drugs, indicate which one is a steroid.

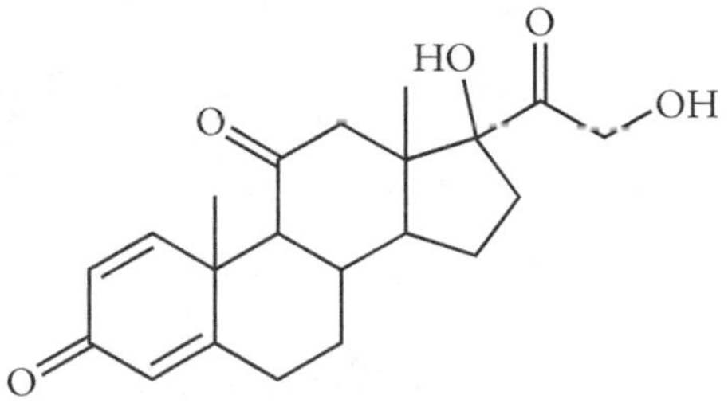

Prednisone

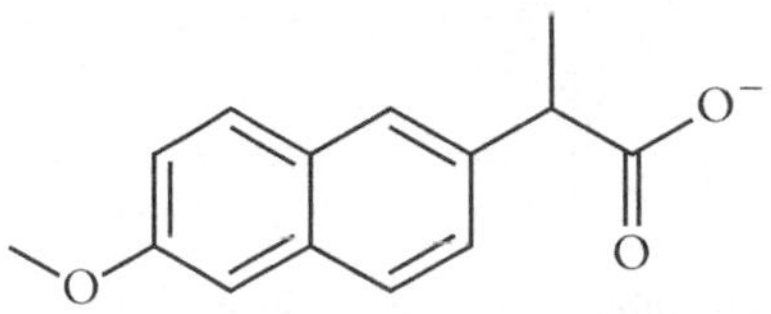

Naproxen
(Aleve)

28 List five compounds synthesized from cholesterol.

29 Sketch a cross section of the lipid portion of a lipoprotein. Label the hydrophobic and hydrophilic portions of the lipoprotein.

30 How are phospholipids in a cell membrane arranged differently from the phospholipids in a lipoprotein?

31 How are LDLs and HDLs different in structure and function?

32 What is the function of a chylomicron? Do chylomicrons have a high or low density? What does this mean?

33 Why do triglycerides have to be transported through the bloodstream, whereas glucose does not?

34 What structural characteristic of bile makes it a good emulsifying agent? *Hint:* Soap is also an emulsifying agent.

Chemistry in Medicine   Inflammation and the Eicosanoids

What do a broken leg, a bee sting, and acute bronchitis have in common? They all elicit the body's inflammatory response. Inflammation is part of the immune response to trauma (a broken leg), invasion by a foreign substance (venom from a bee stinger), or infection (*Streptococcus pneumoniae* infection). Inflammation is part of the body's mechanism for repairing an injury or removing a foreign substance. The typical symptoms of inflammation are redness, swelling, pain, and warmth. The **eicosanoids**, another class of lipids, are signaling molecules that initiate the inflammatory response that produces these symptoms. Thus, drugs used to reduce inflammation generally target specific reactions in the biochemical pathways that produce eicosanoids.

Eicosanoids are produced from arachidonic acid, a fatty acid containing 20 (eicosa) carbon atoms and four *cis* double bonds. Arachidonic acid is an omega-6 fatty

acid, or a $\Delta^{5,8,11,14}$ fatty acid. Eicosanoids have a number of important signaling functions in the body. Here we consider their role in inflammation. Following trauma or an infection, arachidonic acid is hydrolyzed from glycerophospholipids in cell membranes, according to the reaction shown in **Figure 12-19**, catalyzed by the enzyme *phospholipase A₂*. Arachidonic acid is then converted by several biochemical pathways, known as the arachidonic acid cascade, into the various eicosanoids (**Figure 12-20**):

- prostaglandins,
- thromboxanes,
- leukotrienes.

Consider how the arachidonic acid cascade works in the case of a bee sting. Someone stung by a bee experiences several of the classic symptoms of inflammation. In most cases the inflammatory process assists immune cells

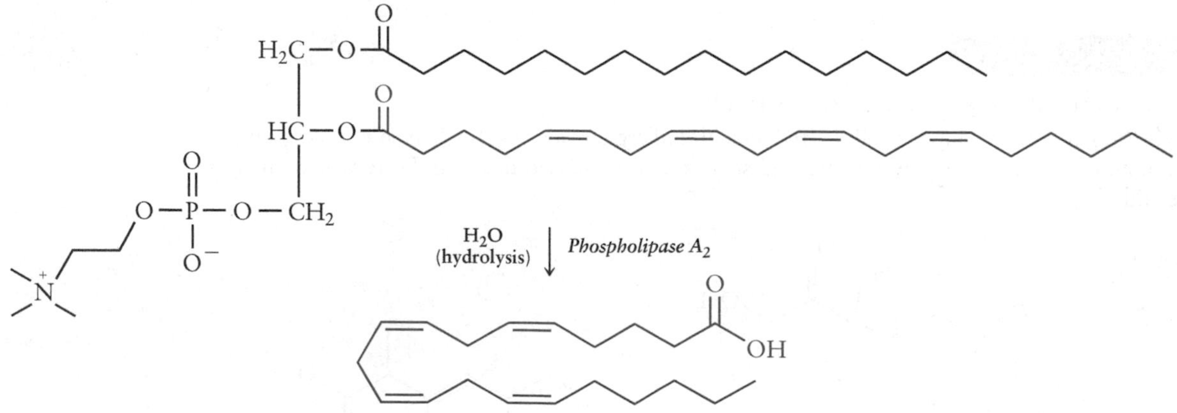

Figure 12-19 Hydrolysis of a membrane lipid to produce arachidonic acid. Note, the other reaction products are not shown.

in their task of destroying the venom from the bee stinger.

Redness—The site of the bee sting is initially pale, and then it eventually turns red. Thromboxanes are short-acting vasoconstrictors (constrict blood vessels) that are released quickly after the bee sting occurs, causing the site to turn pale due to decreased blood flow. Thromboxanes in turn signal nearby cells to release prostaglandins and leukotrienes, which are vasodilators (enlarge blood vessels), causing the area to turn red. The increased blood flow as a result of vasodilation brings immune cells to the damaged tissue.

Swelling—Leukotrienes also make blood vessels more permeable. As a result, plasma leaks into connective tissues, causing them to swell. Increased permeability of blood vessels allows blood, containing immune cells, to enter the damaged tissues more readily.

Pain—Elevated levels of prostaglandins sensitize pain neurons. These neurons signal the brain to get out of harm's way. In other words, run from those bees!

Warmth—The release of prostaglandins also causes fever, or in the case of a localized invasion such as a bee sting, warmth at the site of the bee sting. Elevated temperatures decrease the ability of some pathogens to replicate.

Drugs that inhibit the two main enzymes in the arachidonic acid cascade have been developed to treat inflammatory diseases. *Lipoxygenase* inhibitors are a class of drugs that reduce the production of leukotrienes. They include the anti-asthma medications Singulair and Accolate. Asthma is a complex condition but in large part an inflammatory disease. These medications can also be used to treat inflammation of the bronchial tubes in cases of acute bronchitis.

Nonsteroidal *anti-inflammatory drugs* (NSAIDs) work as inhibitors of *cyclooxygenase,* the key enzyme involved in the formation of prostaglandin H_2, as shown in Figure 12-20. NSAIDs include the well-known over-the-counter medications aspirin, ibuprofen (Motrin, Advil), and naproxen (Aleve), as well as the prescription drug Celebrex used for the treatment of arthritis. By inhibiting *cyclooxygenase,* NSAIDs prevent the formation of prostaglandin H_2, and thereby prevent the formation of prostaglandins and thromboxanes, inflammatory mediators produced later in the arachidonic acid cascade—thus reducing pain (analgesics), fever (antipyretics), and inflammation, all a welcome relief in the case of a broken leg or bee sting.

Other drugs relieve inflammation by inhibiting other key reactions in the arachidonic acid cascade. Some of these drugs are used to treat conditions such as glaucoma, asthma, and pulmonary and ocular hypertension. As you can see, lipids also function as important signaling molecules. Understanding the role of eicosanoids in inflammation at the molecular level has made it possible to develop effective drugs for the treatment of inflammation and many other inflammatory diseases.

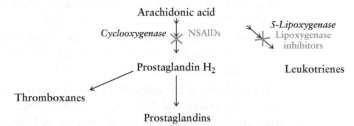

Figure 12-20 The arachadonic acid cascade. NSAIDs are enzyme inhibitors of *cyclooxygenase, and lipoxygenase* inhibitors are enzyme inhibitors of *5-lipoxygenase.* Enzyme inhibitors prevent the formation of all intermediates downstream from the reaction with the inhibitor bound enzyme.

Summary

Fatty Acids: A Component of Many Lipids

- Lipids are a class of biological compounds that are insoluble in water and soluble in organic solvents.
- Fatty acids are long straight-chain hydrocarbons with a carboxylic acid at one end and possibly one or more carbon-carbon double bonds in the hydrocarbon chain.

- Fatty acids are either saturated (have no carbon-carbon double bonds) or unsaturated (have one or more carbon-carbon double bonds).
- Most natural unsaturated fatty acids have a *cis* carbon-carbon double bond, which creates a bend in an otherwise overall linear shaped molecule. Bends in the chain cause the fatty acid to have fewer intermolecular forces of attraction, and therefore a lower melting point. Consequently, unsaturated fatty acids are usually liquids at room temperature.
- The melting point of a fatty acid increases as the number of carbon atoms in the structure increases, due to increased dispersion forces. The melting point of a fatty acid drops significantly with the addition of each double bond.
- Essential fatty acids cannot be synthesized in the body from other metabolic intermediates and so must be supplied through the diet.
- The location of double bonds in an unsaturated fatty acid are identified using the omega or delta system.

Triglycerides: Energy Storage Lipids

- The most abundant lipids in the body are triglycerides, which are stored in fat cells as a long-term source of fuel.
- Triglycerides are derived from one glycerol molecule and three fatty acid molecules, joined by three ester linkages.
- Triglycerides are nonpolar molecules that exhibit hydrophobic properties.
- Triglycerides derived from saturated fatty acids tend to be solids at room temperature (high melting point) and are commonly referred to as fats. Animal sources and tropical oils are high in saturated fats.
- Triglycerides derived from unsaturated fatty acids tend to be liquids at room temperature (low melting points) and are commonly referred to as oils. Vegetable oils, with the exception of tropical oils, are high in unsaturated fats.
- Digestion of triglycerides begins in the small intestine where bile salts emulsify fats so that lipase enzymes can catalyze the hydrolysis of triglycerides into glycerol and fatty acids.
- Fatty acids cross the intestinal mucosa where they are re-esterified into triglycerides and packaged into chylomicrons, which transport triglycerides through the lymph system and the bloodstream to their target cells.
- Fatty acids are metabolized in liver and muscle cells or stored in the form of triglycerides in adipose cells.

Membrane Lipids: Phospholipids and Glycolipids

- The cell membrane is a selectively permeable membrane that separates the inside of the cell from the outside of the cell and controls the flow of ions and molecules in and out of the cell.
- Membrane lipids include glycerophospholipids and sphingolipids, amphipathic molecules with a polar and a nonpolar region within the same molecule.
- A phospholipid derived from glycerol is known as a glycerophospholipid, and one derived from sphingosine is known as a sphingophospholipid. Both contain a phosphate group in a phosphate ester linkage to the terminal hydroxyl group of glycerol or sphingosine.
- The phosphate group in a phospholipid is also joined to one of three amino alcohols via another phosphate ester linkage.

- A phospholipid with a sphingosine backbone is called a sphingomyelin, while a phospholipid with a glycerol backbone is called a glycero-phospholipid.
- Glycolipids found in human cells contain a sphingosine backbone and are called glycosphingolipids.
- Membrane lipids have two long hydrocarbon chains as part of their chemical composition. These long chains are the nonpolar "tail" of the molecule.
- Membrane lipids have a polar "head." In phospholipids the polar head is a phosphate group attached to an amino alcohol. In glycolipids the polar head is a carbohydrate.
- The cell membrane is assembled from two layers, a bilayer, of phospholipids and glycolipids, aligned in a way that brings the polar heads in contact with water on the inside and the outside of the cell and arranges the nonpolar tails in a hydrophobic tail-to-tail arrangement.
- Due to the noncovalent interactions between membrane lipids, the cell membrane is fluid and flexible, which allows certain substances to pass through the membrane.
- Membrane permeability is greater in membranes with more unsaturated lipids.

Cholesterol, Steroids, and Lipoproteins

- Cholesterol is an important steroid that adds rigidity to cell membranes.
- Steroids are characterized by their four fused ring system.
- Cholesterol is an amphipathic molecule with a polar head (—OH) and the nonpolar steroid ring system.
- Cholesterol is the biological precursor (starting material) for bile acids, vitamin D_3, and many important hormones, including the major sex hormones: progestins, androgens, and estrogens.
- The glucocorticoids are a class of steroids produced in the adrenal cortex that regulate glucose metabolism and have anti-inflammatory and immunosuppressant activity.
- The mineralocorticoids are hormones that regulate electrolyte balance in tissues. Aldosterone is one of the primary mineralocorticoids. They are also involved in regulating blood pressure.
- The primary sex hormones are: progestins, androgens, and estrogens.
- Lipoproteins transport cholesterol and triglycerides through the blood and lymph systems.
- Lipoproteins are composed of lipids and proteins and are characterized by their density, which decreases with increasing lipid content.
- Chylomicrons have the largest diameter and the lowest density of the lipoproteins. They are involved in the transport of triglycerides from the intestinal mucosa to their target cells.

Key Words

Adipocytes Fat cells.

Amphipathic Molecules that have both a polar and a nonpolar region within the same molecule. They include membrane lipids, cholesterol, and lipoproteins.

Bile salts Amphipathic molecules derived from cholesterol that act as emulsifying agents during the digestion of lipids.

Cell membrane A selectively permeable barrier that separates the environment of the inside of the cell from the environment of the outside of the cell. Also known as a plasma membrane. Consists of a lipid bilayer composed of phospholipids and glycolipids.

Cerebroside A glycolipid containing a monosaccharide.

Chylomicrons The least dense and largest of the lipoproteins, one micrometer in diameter, that carry triglycerides from the cells lining the small intestine (intestinal mucosa) to their destination through the lymph system and bloodstream.

Delta system A naming system that indicates the location of each carbon-carbon double bond in an unsaturated fatty acid. Numbering starts with the carboxylic acid, and gives the location of the first carbon in each double bond as superscripts, separated by commas, following the Greek letter delta.

Eicosanoids Lipids that function as signaling molecules that initiate physiological events, such as the inflammatory response to trauma and infection.

Emulsification The process whereby fat globules are turned into a colloid of finely dispersed microscopic fat droplets suspended in water.

Essential fatty acids Fatty acids the body cannot synthesize and therefore must be included in the diet. Examples are linoleic acid and linolenic acid.

Fats Triglycerides (triacylglycerols) derived from three fatty acids and one molecule of glycerol. They provide a long-term supply of energy for the body.

Fatty acid A long unbranched hydrocarbon chain containing an even number of carbons with a carboxylic acid (RCO_2H) at one end. Unsaturated fatty acids also contain one or more carbon-carbon double bonds.

Glucocorticoids A class of steroid hormones produced from cholesterol that regulate the metabolism of carbohydrates, proteins, and lipids.

Glycerophospholipid A lipid derived from glycerol with a phosphate diester linkage to glycerol, and one of three amino alcohols.

Glycolipid A membrane lipid containing a carbohydrate as the polar head. Glycolipids in mammals have a sphingosine backbone.

Glycosphingolipid A membrane lipid with a sphingosine backbone and a carbohydrate as a polar head.

Immunosuppressant A substance that reduces the activity of the immune system.

Intestinal mucosa Cells lining the small intestine.

Lipids Biological compounds insoluble in water and soluble in organic solvents. They contain a significant amount of hydrocarbon in their structure.

Lipoproteins Spherical monolayer of phospholipids and proteins assembled so that their polar heads are on the exterior and their hydrophobic tails are on the interior. They function to transport triglycerides and cholesterol through the lymph system and bloodstream.

Mineralocorticoids A class of steroid hormones derived from cholesterol, which regulate ion balance in tissues.

Monounsaturated fatty acid A fatty acid with one carbon-carbon double bond.

Myelin sheath The sphingophospholipids (sphingomyelins) that insulate nerve fibers.

Oils Triglycerides derived from unsaturated fatty acids. They tend to be liquids at room temperature (25 °C) because they have low melting points, due to the *cis* double bonds in their hydrocarbon chains. They are derived from vegetable sources, with the exception of tropical oils.

Omega system A naming system that indicates the location of carbon-carbon double bonds in an unsaturated fatty acid. Numbering starts at the end of the chain opposite the carboxylic acid, and gives the location of the first carbon with a double-bond carbon following the Greek letter omega.

Phospholipid The most common type of membrane lipid. They contain a phosphate group at the terminal end of either a sphingosine or a glycerol backbone.

Polyunsaturated fatty acid A fatty acid with more than one carbon-carbon double bond.

Sphingophospholipid A membrane lipid with a sphingosine backbone and a phosphate group on the terminal hydroxyl group of sphingosine. Also known as a sphingomyelin.

Sphingomyelin A membrane lipid with a sphingosine backbone and a phosphate group at the terminal hydroxyl group of sphingosine. Found in nerve fibers.

Sphingosine A molecule with a long hydrocarbon chain, two hydroxyl groups, and an amine that serves as the backbone of sphingolipids.

Steroid A lipid containing the characteristic four fused ring system of a steroid. Includes molecules like cholesterol, glucocorticosteroids, mineralocorticoids, and the sex hormones.

Triglycerides The most abundant type of lipid in the body, also known as triacylglycerols, or more simply, as fats. They serve as the body's long-term energy supply.

Additional Exercises

The Effects of Anabolic Steroids on the Body

35 To what class of biomolecules do steroids belong?

36 What are some androgenic effects of testosterone?

37 What is the anabolic effect of anabolic steroids? How is this effect initiated in a cell?

38 What are synthetic anabolic steroids designed to do?

39 Identify some side effects of anabolic steroid use. What are some symptoms of prolonged anabolic steroid use?

40 What happens when anabolic steroids are used by adolescents?

41 Give some examples of when the use of anabolic steroids is warranted.

42 Name two physical properties of lipids.

43 What structural feature makes lipids insoluble in water?

44 How do lipids differ from carbohydrates and the other major classes of biomolecules?

Fatty Acids: A Component of Many Lipids

Consult Table 12-1 as needed.

45 Describe the chemical structure of a fatty acid. What functional group do they all contain?

46 What is the structural difference between a saturated fatty acid and an unsaturated fatty acid?

47 What is the structural difference between a monounsaturated fatty acid and a polyunsaturated fatty acid?

48 Classify the following fatty acids as saturated or unsaturated fatty acids:
a. myristic acid
b. arachidonic acid
c. stearic acid
d. oleic acid

49 Classify the following fatty acids as saturated or unsaturated fatty acids:
a. linolenic acid
b. palmitic acid
c. linoleic acid

50 What determines the overall shape of a fatty acid?

51 Explain why stearic acid has a higher melting point than linoleic acid.

52 Arachidic acid and arachidonic acid, shown below, are fatty acids that contain 20 carbons. Explain why arachidic acid is a solid at room temperature, while arachidonic acid is a liquid at room temperature.

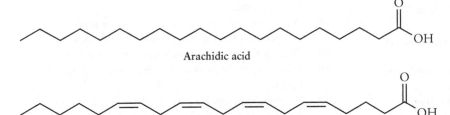

Arachidic acid

Arachidonic acid

53 How does the presence of a *cis* carbon-carbon double bond in a fatty acid affect its melting point?

54 Which fatty acids have stronger intermolecular forces of attraction: saturated or unsaturated fatty acids? Which type of fatty acid would have a higher melting point? Explain.

55 Which fatty acids have stronger intermolecular forces of attraction: unsaturated *trans* fatty acids or unsaturated *cis* fatty acids? Which type of fatty acid would have a higher melting point? Explain.

56 Why are *cis* unsaturated fatty acids generally liquids are room temperature?

57 Why are saturated fatty acids generally solids at room temperature?

58 What is an essential fatty acid? What are some examples of essential fatty acids?

59 What foods are good sources of essential fatty acids?

60 Lauric acid, shown below, is found in coconut oil and palm kernel oil.

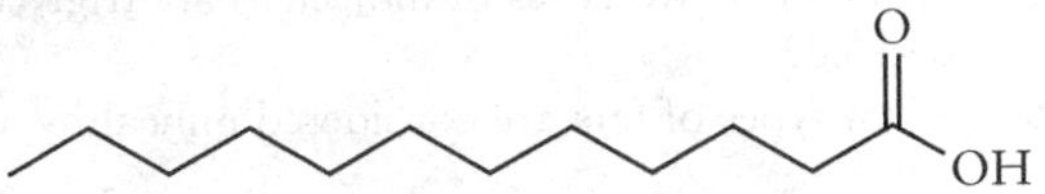

Lauric acid

a. Is lauric acid a saturated or an unsaturated fatty acid?
b. Do you expect lauric acid to be a liquid or solid at room temperature? Explain.

61 Using the delta system and the omega system, what type of fatty acid is oleic acid?

62 Using the delta system and the omega system, what type of fatty acid is linoleic acid?

63 Flaxseed oil is rich in linolenic acid. Linolenic acid is an omega-_____ fatty acid.

64 Docosahexanoic acid, DHA, is a fatty acid found in fish oil and is sold as a dietary supplement.

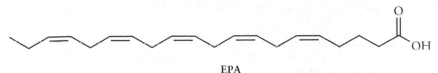

DHA

a. What type of fatty acid is DHA? Use the delta system.
b. What type of fatty acid is DHA? Use the omega system.
c. Would you expect DHA to be a liquid or solid at room temperature? Explain.
d. Is DHA a saturated, monounsaturated, or polyunsaturated fatty acid?

65 Eicosapentaenoic acid, EPA, is a fatty acid found in oily fish and fish oils such as herring, mackerel, or salmon.

EPA

a. What type of fatty acid is EPA? Use the delta system.

b. What type of fatty acid is EPA? Use the omega system.

c. Do you expect EPA to be a liquid or a solid at room temperature? Explain.

Triglycerides: Energy Storage Lipids

66   Describe the chemical structure of a triglyceride.

67   In what type of cells are triglycerides stored? In what type of cells are triglycerides metabolized?

68   From what two types of molecules are triglycerides derived?

69   What types of fats are considered unhealthy and why?

70   What functional group is present in all triglycerides?

71   What physical property of a triglyceride causes it to be classified as a lipid? What part of its chemical structure gives it these physical properties?

72   Would you expect a triglyceride to be soluble in water? Why or why not?

73   What types of triglycerides are classified as fats? Why are they generally solids at room temperature? What are some sources of fats?

74   What types of triglycerides are classified as oils? Why are they generally liquids at room temperature? What are some sources of oils?

75   Coconut oil contains 92 percent saturated fat. Do you expect it to be a solid or a liquid at room temperature? Explain.

76   Canola oil contains 6 percent saturated fat. Do you expect it to be a solid or a liquid at room temperature? Explain.

77   Why is olive oil considered a healthy fat?

78   Which product is higher in saturated fats, lard or coconut oil?

79   Which product is lower in saturated fats, safflower oil or cottonseed oil?

80   Which is a healthier food, soybean oil or sunflower oil?

81   Which is a healthier food, olive oil or butter?

82   Answer the questions below for the following triglyceride.

a. Is this a saturated or unsaturated fat?

b. Would you expect it to be a solid or a liquid at room temperature?

c. What functional groups does this molecule have?

d. Highlight the atoms and bonds that are derived from glycerol.

e. Highlight the atoms and bonds derived from fatty acids and name each fatty acid, by referring to Table 12-1.

f. Is this a healthy or an unhealthy fat?

83   Answer the questions below for the following triglyceride.

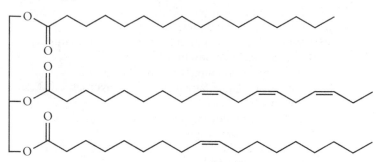

a. Is this a saturated or unsaturated fat?

b. Would you expect it to be a solid or a liquid at room temperature?

c. What functional groups does this molecule have?

d. Highlight the atoms and bonds that are derived from glycerol.

e. Highlight the atoms and bonds derived from fatty acids and name each fatty acid, by referring to Table 12-1.

f. Are the double bonds in this triglyceride *cis* or *trans*? Is this a healthy or an unhealthy fat?

84   Olive oil contains oleic acid. Draw a triglyceride that is derived from only oleic acid.

85   Draw a triglyceride derived from one molecule of oleic acid, one molecule of stearic acid, and one molecule of palmitic acid.

86   Why does the average adult male have more triglycerides than glycogen stored in his body?

87   What is the primary function of triglycerides?

88   Where in the body does the digestion of dietary fat begin?

89   During digestion, how are triglycerides from a meal made more susceptible to hydrolysis by the water-soluble lipases?

90   What substance is released by the gall bladder that aids the digestion of triglycerides?

91   Why are triglycerides unable to cross the cell membrane of the intestinal mucosa? How do triglycerides enter the bloodstream?

92   Are triglycerides soluble in blood? How are triglycerides transported through the lymph system and the bloodstream?

93   Is the interior of a lipoprotein hydrophobic or hydrophilic?

94   What happens when triglycerides arrive at the capillaries of their target tissues? What happens inside an adipocyte?

95   Are triglycerides found on the interior or exterior of a chylomicron? Explain.

96   What is the function of a chylomicron?

Membrane Lipids: Phospholipids and Glycolipids

97 What type of lipids makeup the cell membrane?

98 What lipid molecule provides structural rigidity to the cell membrane?

99 The following molecule is found in soap. Is it polar, nonpolar, or amphipathic? Explain.

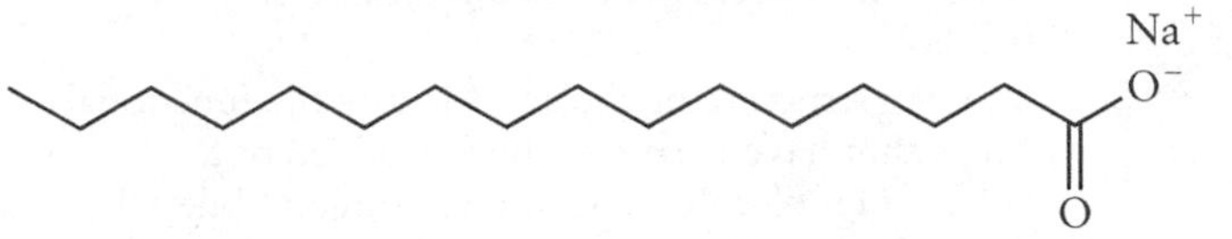

100 How is the structure of glycerol different from sphingosine?

101 How many fatty acids are needed to form a triglyceride? How many fatty acids are needed to form a glycero-phospholipid? How many fatty acids are needed to form a sphingolipid?

102 Which of the following membrane lipids contain a phosphate group?
a. glycolipids
b. sphingomyelins
c. glycerophospholipids

103 Which membrane lipids in human cells are derived from sphingosine?

104 In a phospholipid, what forms the polar head of the molecule?

105 How does the structure of a glycolipid differ from the structure of a phospholipid? Is this difference primarily in the polar head or the nonpolar tails?

106 For the membrane lipid shown below, answer the following questions.

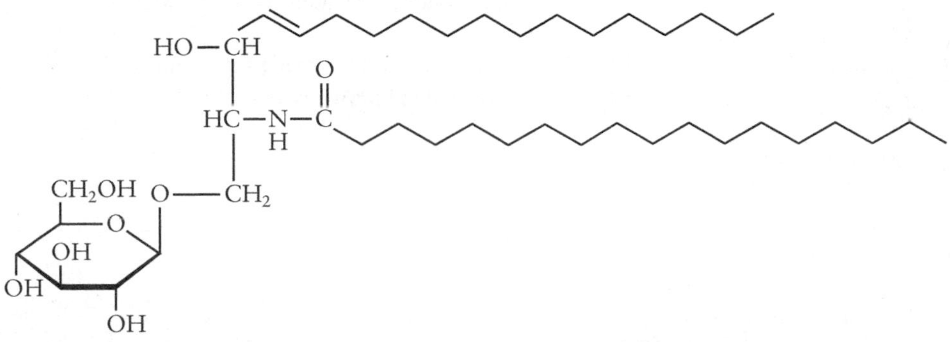

a. Is it a phospholipid or glycolipid?
b. Is it derived from glycerol or sphingosine? Outline the backbone of this lipid.
c. What is the fatty acid from which this lipid is derived? Is the fatty acid component saturated or unsaturated? What functional group connects the fatty acid to the glycerol or sphingosine backbone?
d. If it is a glycolipid, is it derived from glucose or galactose?
e. Circle and label all functional groups.

107 For the membrane lipid shown below, answer the following questions.

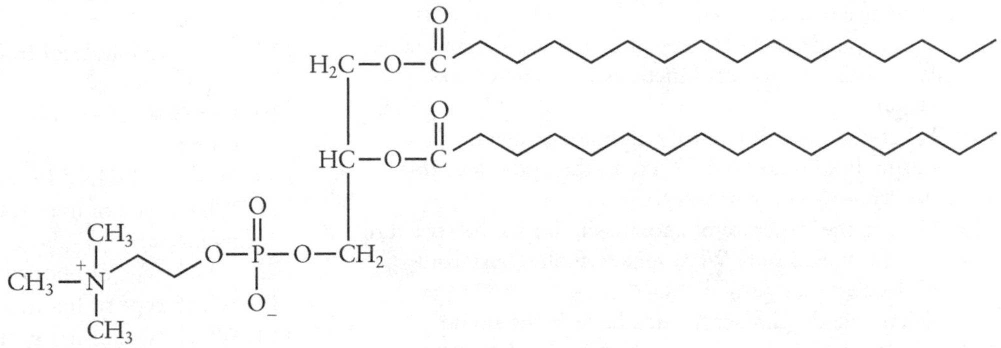

a. Is it a phospholipid or a glycolipid?
b. Is it a sphingomyelin or a glycerophospholipid? Outline the backbone.
c. What is (are) the fatty acid(s) from which this lipid is derived? Are the fatty acid components saturated or unsaturated? What functional group connects the fatty acids to the backbone?
d. Does the lipid contain a phosphate or a carbohydrate?
e. What part of the lipid constitutes the polar head?
f. Circle and label all functional groups.

108 Explain why membrane lipids assemble so that their tails are in the interior of the cell membrane.

109 What type of intermolecular forces of attraction are responsible for the way membrane lipids organize to form a cell membrane? How does this allow the cell membrane to be "fluid?"

110 Draw 12 membrane lipids arranged as a *bilayer*, like a section of cell membrane.

111 What type of molecules can pass freely through the cell membrane?

112 What part of the cell controls the flow of ions and molecules in and out of the cell?

113 Why do cell membranes with more saturated phospholipids have less permeability than membranes with more unsaturated phospholipids?

Cholesterol, Steroids, and Lipoproteins

114 Why are steroids classified as lipids?

115 How are steroids structurally different from triglycerides, phospholipids, and glycolipids? Draw the structural feature of a steroid that characterizes it as a steroid.

116 How is the structure of cholesterol similar to the structure of a fatty acid?

117 What role does cholesterol play in the cell membrane? What other biological functions does cholesterol serve?

118 What compounds emulsify globules of dietary fat during lipid digestion? Where in the body does this occur? Why is it necessary?

119 How is the structure of cholate similar to the structure of a fatty acid salt? What makes cholate a steroid molecule?

120 What role do glucocorticoids have in the body?

121 The structure of prednisone is shown below. What type of hormone is this and where is it produced? What structural characteristic of this molecule makes it a steroid? Circle and identify the functional groups in this molecule.

Prednisone

122 The structure of aldosterone is shown below. What type of hormone is this, and where is it produced? What characteristic of this molecule makes it a steroid? Circle and identify the functional groups in this molecule.

Aldosterone

123 When is cortisol released in the body? What happens when it is released?

124 What role do mineralocorticoids have in the body?

125 Draw the structure of estradiol. Identify the functional groups that have been modified or added to a cholesterol molecule to form a molecule of estradiol.

126 Draw the structure of cholate. Identify the functional groups that have been modified or added to a cholesterol molecule to form a molecule of bile salt (cholate).

127 In what way are progesterone, testosterone, and estradiol similar in structure? How are they different? From what compound are all three of these hormones produced?

128 How is cholesterol transported through the blood and lymph system?

129 How are cell membranes different from lipoproteins?

130 How are LDLs different from HDLs?

131 Which type of lipoprotein has the largest diameter?

132 Which type of lipoprotein is the least dense?

133 Which type of lipoprotein is the most dense?

134 Which type of lipoprotein transports triglycerides?

135 Which type of lipoprotein is considered "good cholesterol?" Explain why.

Chemistry in Medicine: Inflammation and the Eicosanoids

136 What is an eicosanoid?

137 What are the three major classes of eicosanoids?

138 What are the symptoms of inflammation resulting from the production of eicosanoids?

139 Is arachidonic acid a saturated or an unsaturated fatty acid? Are the double bonds *cis*, *trans*, or a combination of both?

140 What important eicosanoids(s) will not be produced under the following conditions?
   a. A *lipoxygenase* inhibitor is present.
   b. A *cyclooxygenase* inhibitor is present.

141 What eicosanoid is formed by the action of *cyclooxygenase* on arachidonic acid?

142 On a molecular level, how does aspirin work?

143 What is the name of the enzyme that catalyzes the release of arachidonic acid from glycerophospholipids? What type of reaction is this?

144 Define an NSAID and provide two examples.

145 What biochemical pathway does an NSAID inhibit?

Answers to Practice Exercises

1 **a.** Both fatty acids are saturated, but myristic acid contains 14 carbons while palmitic acid contains 16. The more carbon atoms, the more dispersion forces and, therefore, palmitic acid has a higher melting point than myristic acid.

b. Both fatty acids contain 18 carbon atoms, but oleic acid is unsaturated, whereas stearic acid is saturated. The *cis* double bond creates a kink in the chain limiting intermolecular forces of attraction between oleic acid molecules, thereby reducing the number of dispersion forces, making it easier to effect a phase change from solid to liquid. Thus, oleic acid has a lower melting point than stearic acid.

2 Arachidonic acid (-50 °C) has four *cis* carbon-carbon double bonds, whereas linolenic acid (-11 °C) has three *cis* carbon-carbon double bonds, and from this data it appears that an additional double bond impacts the melting point more than the addition of two carbon atoms.

3 **a.** Linolenic acid is an omega-3 fatty acid and an essential fatty acid.

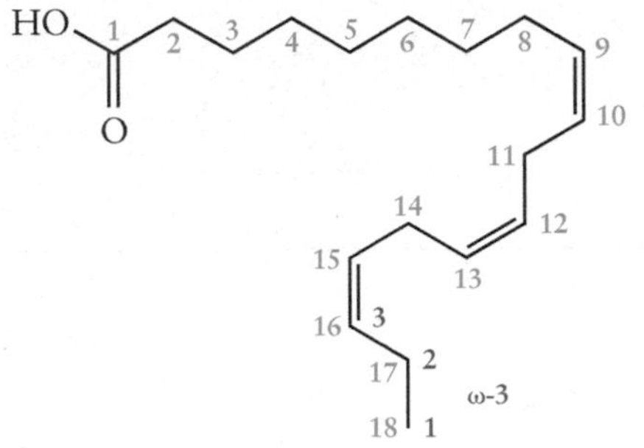

b. Oleic acid is not an omega-3 or omega-6 fatty acid because the first double bond appears at carbon 9 (when counting from the end farthest from the carbonyl carbon). It is not an essential fatty acid.

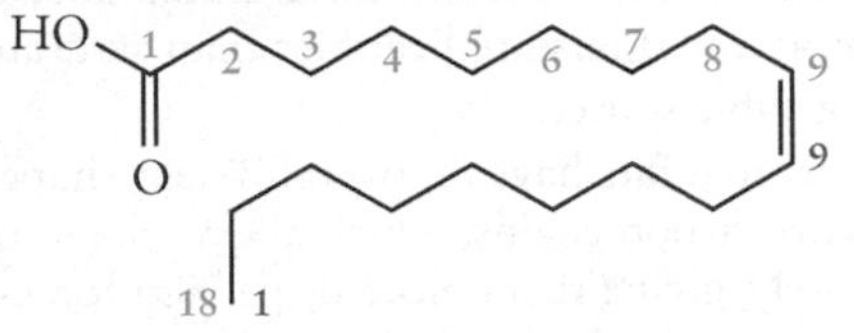

4 $\Delta^{5,8,11,14}$ fatty acid; omega-6 fatty acid

5 Both of these fatty acids contain 18 carbon atoms, but linolenic is more likely to be a liquid since it is polyunsaturated with three *cis* carbon-carbon double bonds, whereas stearic acid is a saturated fatty acid.

6 A saturated fatty acid has a linear overall shape, whereas an unsaturated fatty acid has a bend or kink wherever a *cis* carbon-carbon double bond appears in the molecule. Thus, *cis* unsaturated fatty acids have fewer dispersion forces because there are fewer points of contact between fatty acid molecules. Fewer dispersion forces results in a lower melting and boiling point. A *trans* unsaturated fatty acid has an overall shape more closely resembling a saturated fatty acid, so it has physical properties more similar to saturated fatty acids than *cis* unsaturated fatty acids even though it is unsaturated.

7

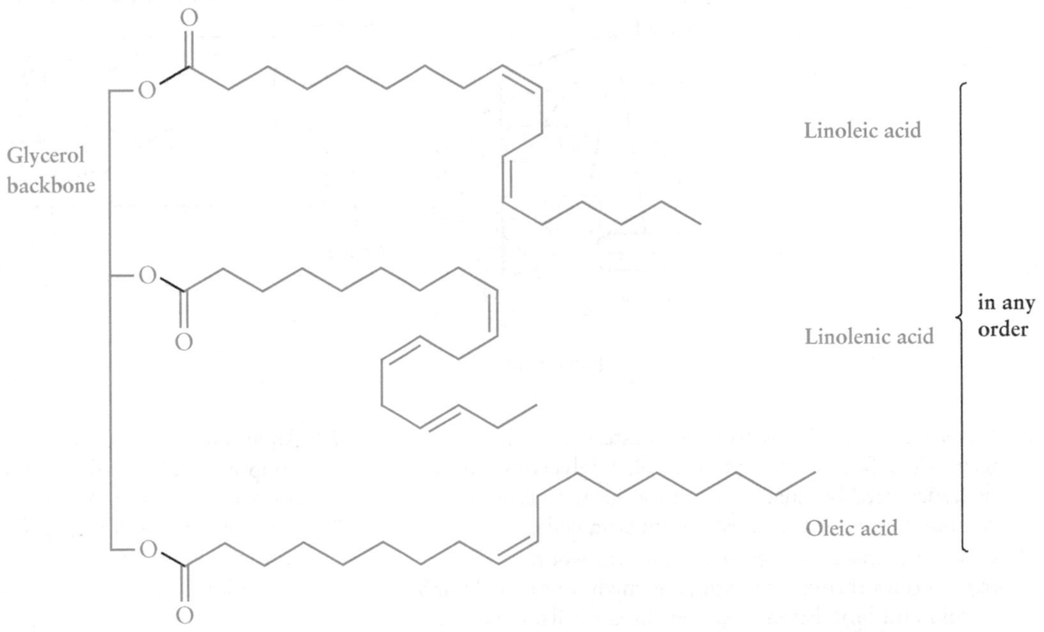

a. This is an example of an unsaturated fat.

b. This triglyceride would be expected to be a liquid (oil) at room temperature because it is an unsaturated fat.

c. This compound would not be soluble in water because of its extensive nonpolar hydrocarbon structure.

8   Palm and coconut oils are vegetable oils high in saturated fats.

9   Peanut oil (18%) is higher in saturated fats than corn oil (13%).

10 Canola oil is healthier than coconut oil because it contains 95 percent unsaturated fats while coconut oil contains only 5 percent unsaturated fats, and unsaturated fats are healthier than saturated fats.

11 vegetable sources

12 Saturated fats have an overall linear shape to their hydrocarbon chains, which allows them to pack more closely, giving them more dispersion forces, and consequently higher melting points, making them solids at room temperature.

13 triglycerides

14 An emulsifying agent is an amphipathic molecule that facilitates the mixing of nonpolar (oil) and polar (vinegar) substances. Proteins in egg serve as an emulsifying agent in mayonnaise.

15 Bile is an amphipathic molecule that serves as an emulsifying agent for globules of dietary fat (triglycerides) in water.

16 Intestinal mucosa cells line the small intestine. Triglycerides are formed in these cells by esterification of glycerol and fatty acids that have crossed the cell membrane from the intestine.

20 Bile is stored in the gall bladder. It is released after consumption of a high-fat meal because it is an emulsifying agent that converts large globules of dietary fat into a colloid of finely dispersed microscopic fat droplets suspended in water, allowing lipase enzymes to catalyze the hydrolysis of triglycerides into glycerol and fatty acids.

21 because carbohydrates are soluble in an aqueous medium and triglycerides are not.

22 a. It is a glycolipid because it contains a carbohydrate rather than a phosphate group.

    b. It has a sphingosine backbone, as shown below in green.

    c. It is derived from oleic acid, an unsaturated fatty acid. An amide functional group connects the amine functional group in sphingosine to the carboxylic acid functional group of the fatty acid.

    d. It is a cerebroside because it contains a monosaccharide rather than an oligosaccharide (many monosaccharides). It contains D-glucose.

    e. See labels below.

    f. See labels below.

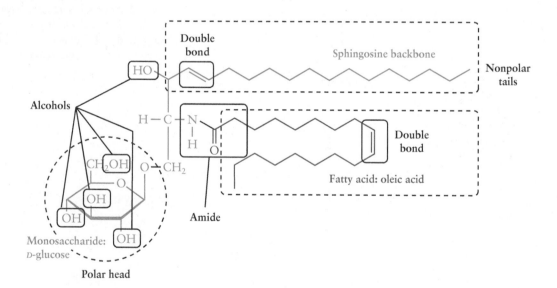

17 Lipases catalyze the hydrolysis of esters in triglycerides to produce fatty acids and glycerol. Triglycerides must be hydrolyzed because they are too large to cross the cell membrane of the intestinal mucosa cells.

18 Chylomicrons are lipoproteins that transport triglycerides through the aqueous medium of the lymph system and bloodstream to their target cells, muscle and fat cells.

19 Upon entering an adipocyte, fatty acids and glycerol undergo esterification reactions to re-form triglycerides, the form in which they are stored. In muscle cells, fatty acids undergo degradation by biochemical pathways that convert fatty acids into carbon dioxide and energy.

23 An amphipathic molecule has a polar region and a nonpolar region within the same molecule, so it has properties of both. A nonpolar substance is hydrophobic and not soluble in water, but it is soluble in nonpolar solvents. In contrast, a polar substance is hydrophilic and soluble in water, but it is not soluble in nonpolar solvents.

24 Triglycerides are nonpolar molecules derived from glycerol and three fatty acids, whereas a glycerophospholipid is an amphipathic molecule that is derived from glycerol, two fatty acids, a phosphate group, and an amino alcohol. In other words, the terminal hydroxyl group is different in these two molecules.

25 Glycerophospholipids have a glycerol backbone and are derived from two fatty acids, whereas sphingomyelins have a sphingosine backbone and are derived from one fatty acid.

26 A steroid is characterized by four fused hydrocarbon rings: three six membered rings and one five-membered ring.

27 Prednisone is a steroid because it contains the characteristic four fused rings of a steroid. Naproxen is not a steroid because it lacks the steroid ring system.

28 Sex hormones like progesterone, testosterone, and estradiol; mineralocorticoids like aldosterone; glucocorticoids like cortisol; bile salts like cholate; and vitamin D.

29

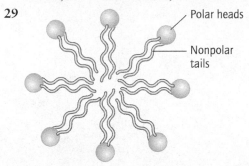

Polar heads

Nonpolar tails

30 In the cell membrane they form a bilayer, whereas in a lipoprotein they form a monolayer.

31 HDLs have a greater protein/lower lipid content than LDLs. HDLs function to scavenge excess cholesterol from the blood, whereas LDLs carry triglycerides and cholesterol to their target cells.

32 Chylomicrons transport triglycerides to muscle and fat cells via the lymph and blood systems. Chylomicrons have the lowest density of all the lipoproteins, which means they have the highest lipid content/lowest protein content.

33 Glucose is soluble in the aqueous medium of the blood, whereas triglycerides are not.

34 The amphipathic properties of bile make it a good emulsifying agent, allowing it to form micelles which dissolve fats in their interior while dissolving in water as a result of the polar heads on their exterior.

Sickle-cell anemia is a condition characterized by crescent-shaped red blood cells (left) caused by a variant form of hemoglobin, an important protein that transports oxygen to tissues throughout the body. [Jackie Lewin, Royal Free Hospital/Science Source]

13 Proteins: Structure and Function

Sickle-cell Anemia: One Wrong Amino Acid

Sickle-cell anemia is the most common inherited blood disorder in the United States. About 70,000 Americans are estimated to have the disease, and 2,000,000 are believed to be carriers of the disease. The incidence is highest in African Americans, where one in 500 has the disease. Sickle-cell anemia gets its name from the crescent (sickle) shape of the abnormal red blood cells in people who have the disease. Normal red blood cells have a biconcave shape that increases the surface area of a red blood cell, optimizing its ability to transport oxygen (see book cover).

Red blood cells contain hemoglobin, a protein that binds oxygen and transports it to tissues throughout the body. Individuals with sickle-cell anemia carry a variant of this protein that causes red blood cells to become misshapen in oxygen-depleted tissues. These sickled red blood cells tend to stick together and aggregate, clogging delicate blood vessels and restricting blood flow, ultimately leading to tissue damage, severe pain, and damage to internal organs. Sickled red blood cells also have a shorter life span than a normal red blood cell—15 days instead of 120 days. Hence, people with sickle-cell anemia have a lower red blood cell count, which causes anemia, a condition characterized by a lack of oxygen-carrying capacity with symptoms of fatigue and lethargy.

Hemoglobin, like all proteins, is a biological polymer constructed from amino acids. Some proteins, like hemoglobin, are composed of more than one chain of amino acids. Hemoglobin is a protein composed of two identical α-chains, each containing 141 amino acids, and two identical β-chains, each containing 146 amino acids, as illustrated in Figure 13-1a.

The only difference between the amino acid sequence in normal hemoglobin and the hemoglobin of individuals with sickle-cell anemia is a single amino acid in

(a)

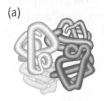 

Normal hemoglobin   Sickle-cell hemoglobin

(b)

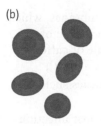 

Normal red blood cells   Sickled red blood cells

Figure 13-1 Normal hemoglobin (Hb) and sickle-cell hemoglobin (HbSc) (a) at the molecular level and (b) its effect on the shape of red blood cells.

both β-chains! The hemoglobin of people with sickle-cell anemia has the amino acid *valine* as the sixth amino acid in both β-chains, where normally the amino acid *glutamic acid* appears. The structure of these two amino acids is shown below:

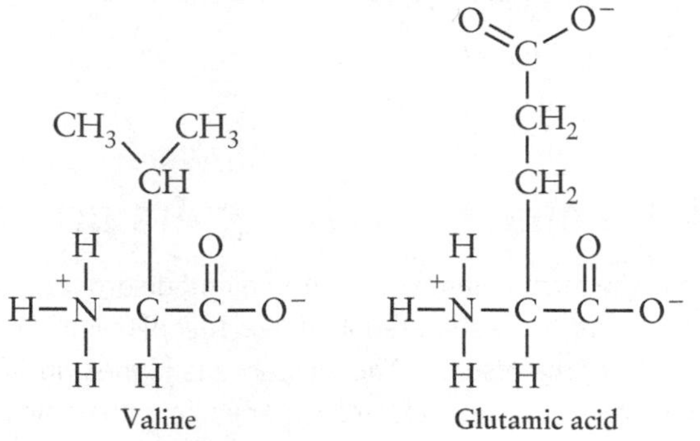

Valine   Glutamic acid

This seemingly small change at the molecular level causes the protein to assume an entirely different conformation that ultimately affects the shape of the red blood cell, as shown in Figure 13-1b. While there is no cure for sickle-cell anemia, there are treatments for its adverse symptoms including:

- Antibiotics taken in the first five years of life prevent life-threatening pneumonia infections.
- Periodic blood transfusions reduce anemia and pain by introducing normal red blood cells into circulation.
- Treatment with hydroxyurea is an effective drug for reducing the number of recurrent episodes of pain and tissue damage, known as "sickle-cell crises."

Sickle-cell anemia is only one of many inherited diseases that are caused by an abnormally shaped protein, which is caused by an error in the amino acid sequence of the protein. In this chapter you will learn about amino acids as well as the structure and function of proteins, arguably the most important of the biomolecules. ●

In the last two chapters you learned about two important classes of biomolecules: carbohydrates and lipids. In this chapter we examine proteins, another class of biomolecules. We will examine the structure of proteins, consider their diverse functions, and then conclude with an examination of enzymes, one of the most important types of proteins.

Proteins have the most diverse functions of all the biomolecules. They serve as:

- enzymes, the biological catalysts required by the reactions of metabolism;
- receptors located within and on the surface of cell membranes, which bind specific ligands, creating a signal that leads to a cascade of biological events (recall from chapter 3 how a molecule of caffeine binds to the adenosine receptor, a protein, to keep you awake);
- structural proteins that provide physical support to tissues such as skin, muscle, blood vessels, hair, and tendons;
- immunoglobulins, antibodies that defend the body against infectious agents;
- transport proteins, such as hemoglobin and myoglobin, which deliver oxygen to cells;

- dietary proteins that supply amino acids for the purpose of building nitrogen-containing biomolecules;
- protein motors that move molecules and ions around within the cell as well as moving body parts such as muscle fibers.

Proteins are biological polymers of amino acids, much like carbohydrates are biological polymers of monosaccharides. Therefore, to understand proteins we begin with the structure of amino acids.

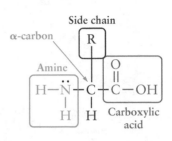

■ 13.1 Amino Acids

Whether a protein is found in a simple bacterium, a frog, or a human being, all proteins are constructed from the same 20 amino acids. As the name implies, all **amino acids** contain an *amine* ($-NH_2$) and a *carboxylic acid* ($-COOH$) functional group as part of their chemical structure. Both functional groups are covalently bonded to the **α-carbon**, as shown in the margin.

pH and Amino Acid Equilibria in Aqueous Solution

You have learned that amines are organic bases and that carboxylic acids are organic acids. Thus, in aqueous solution these functional groups exist in either of two forms: neutral or ionized, depending on the pH. *At physiological pH (pH = 7.3), the amine and the carboxylic acid functional groups of an amino acid are both in their ionized forms: the amine is in its conjugate acid form ($-NH_3^+$), and the carboxylic acid is in its conjugate base form ($-COO^-$), as expected for a base and an acid at neutral pH.* Since both a positive (+) and a negative (−) charge are present *within* the same molecule, the net charge is neutral, and this form of an amino acid is known as a **zwitterion**.

There are two other pH-dependent forms of an amino acid that are in equilibrium with the zwitterion form, as illustrated in Figure 13-2. When the pH increases (more OH^-/less H_3O^+) the equilibrium shifts to the right, and when the pH decreases (less OH^-/more H_3O^+) the equilibrium shifts to the left, in accordance with Le Châtelier's principle. At high pH the conjugate acid of the amine loses a proton and becomes the neutral amine ($-NH_2$), while the carboxylic acid remains in its ionized form ($-COO^-$). Conversely, at low pH, the carboxylate ion accepts a proton to become the neutral carboxylic acid ($-COOH$), while the amine remains in its conjugate acid form ($-NH_3^+$).

> Recall from chapter 9 that when the pH of a solution is less than 7, the concentration of H_3O^+ exceeds the concentration of OH^-. When the pH is greater than 7, the reverse is true. At pH = 7 the concentration of OH^- is equal to the concentration of H_3O^+.

net charge: + 1   0   − 1

pH < 7   pH = 7   pH > 7

Figure 13-2 The three pH-dependent forms of an amino acid. At physiological pH, the zwitterion form predominates. At high pH the equilibrium shifts to the right, and at low pH the equilibrium shifts to the left, creating a charge on the amino acid.

WORKED EXERCISE Zwitterions and Other Forms of an Amino Acid

13-1 The zwitterion form of the amino acid alanine is shown below. At what pH does the zwitterion form predominate? Write the structure of the two forms of alanine that are in equilibrium with the zwitterion form. Under what conditions does the equilibrium shift to these alternate forms of an amino acid? Indicate the net charge on each of these alternate forms of alanine.

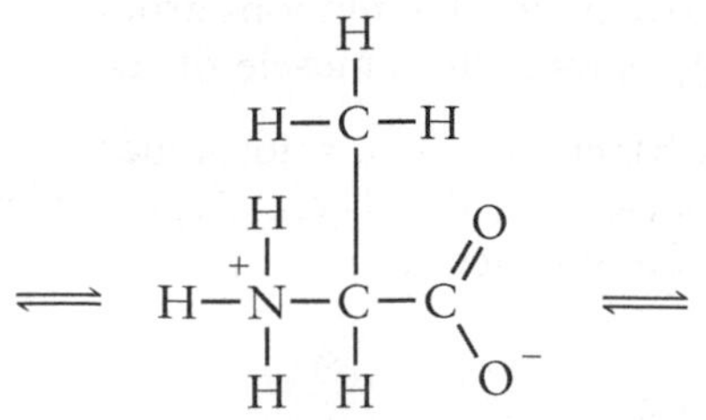

$$
\begin{array}{c}
H \\
| \\
H-C-H \\
\end{array}
$$

$$
\rightleftharpoons \quad
\begin{array}{c}
H \\
+| \quad O \\
H-N-C-C \\
| \quad | \quad \diagdown \\
H \quad H \quad O^-
\end{array}
\quad \rightleftharpoons
$$

Alanine

Solution

13-1 The zwitterion form of an amino acid exists at physiological pH (pH = 7.3–7.4). At high pH, the structure at right predominates (all dissociable protons have dissociated). At low pH, the structure at left predominates (all dissociable protons are attached—shown in red).

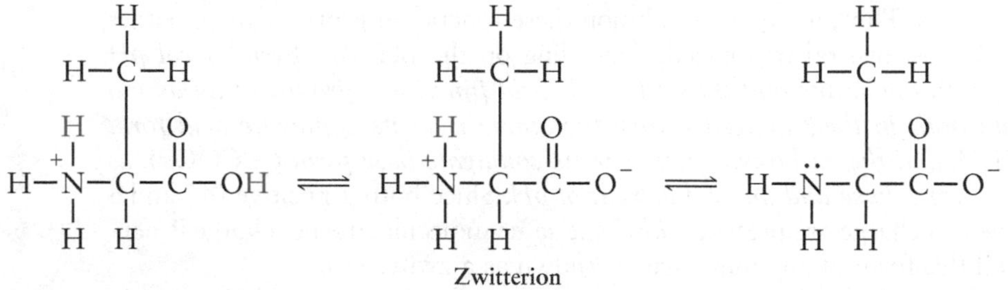

Low pH	Neutral pH	High pH

$$
\begin{array}{c}
H \\
| \\
H-C-H \\
H \quad O \\
+| \quad \| \\
H-N-C-C-OH \\
| \quad | \\
H \quad H
\end{array}
\rightleftharpoons
\begin{array}{c}
H \\
| \\
H-C-H \\
H \quad O \\
+| \quad \| \\
H-N-C-C-O^- \\
| \quad | \\
H \quad H
\end{array}
\rightleftharpoons
\begin{array}{c}
H \\
| \\
H-C-H \\
\quad O \\
\cdot\cdot| \quad \| \\
H-N-C-C-O^- \\
| \quad | \\
H \quad H
\end{array}
$$

Zwitterion

+1   0   −1

PRACTICE EXERCISES

1   The simplest amino acid, glycine, is shown below.

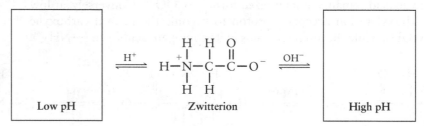

Low pH	Zwitterion	High pH

a. Write the structure of glycine at low pH and at high pH. Explain why the equilibrium shifts to these alternate forms of the amino acid as the pH changes.

b. What is the net charge on glycine at high pH? What is the net charge on glycine at low pH? What is the net charge on the zwitterion?

Amino Acid Side Chains

We have seen that amino acids have an amine and a carboxylic acid functional group both bonded to the α-carbon, a tetrahedral carbon with four bonds. The other two bonds to the α-carbon are a hydrogen atom and an R group. The R group, often referred to as the amino acid **side chain**, is a carbon chain of varying length and branching that in some amino acids also contains a functional group (except glycine which has R = H). Since amino acids differ only in the structure of their side chain, this is the distinguishing feature of an amino acid. **Figure 13-3** shows the structure of the 20 natural amino acids,

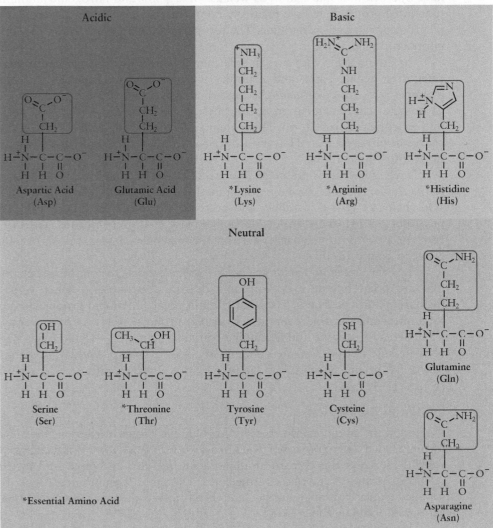

Figure 13-3 The chemical structure of the 20 natural amino acids, and their three letter abbreviation, organized by polarity. Side chains are shown in pink boxes.

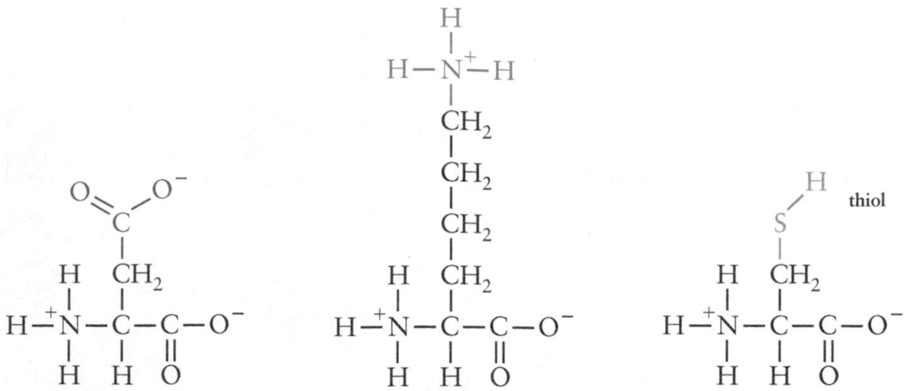

Figure 13-4 Examples of the three types of polar amino acids at physiological pH: (a) aspartic acid has an acidic side chain, which is ionized at physiological pH, (b) lysine has a basic side, which is ionized at physiological pH, and (c) cysteine has a neutral side chain, which is not ionized at any pH.

Aspartic acid
(Asp)
(a)

Lysine
(Lys)
(b)

Cysteine
(Cys)
(c)

with their side chains highlighted in pink boxes, along with their names and three letter abbreviations. All proteins produced in nature are constructed from these 20 amino acids. The structure of the side chain determines whether an amino acid is classified as nonpolar or polar. The nonpolar amino acids have hydrocarbon side chains, as well as some sulfur and nitrogen containing functional groups that have relatively nonpolar covalent bonds, as for example, tryptophan and methionine. The top part of Figure 13-3 shows the nine nonpolar amino acids.

The amino acids with a polar side chain are subdivided further into three categories: acidic, basic, and neutral. Two amino acids, aspartic acid and glutamic acid, are polar *acidic* because their side chains (R groups) contain a carboxylic acid functional group. At physiological pH, carboxylic acids exist in their conjugate base form. These amino acids therefore have a net negative charge at physiological pH even in their zwitterion form, as shown for aspartic acid in Figure 13-4a.

Three amino acids are polar *basic*: lysine, arginine, and histidine. Lysine, for example, contains an amine functional group in its side chain. The basic side chains of lysine, histidine, and arginine are shown in Figure 13-3. At physiological pH, most basic side chains exist in their conjugate acid form creating a net positive charge, even in their zwitterion form, as shown for lysine in Figure 13-4b.

The other six amino acids are polar *neutral* because they have side chains that are polar but have no ionizable functional groups; in other words, they are neither acids nor bases but contain polar functional groups. Serine, threonine, and tyrosine contain a hydroxyl group (—OH) as part of their side chain. Asparagine and glutamine contain an amide (—$CONH_2$) as part of their side chain. Cysteine contains an —S—H group, known as a **thiol** functional group, which is similar to a hydroxyl group, although less polar (Figure 13-4c).

Essential Amino Acids

We obtain the amino acids necessary for building proteins from our diet. The recommended daily allowance (RDA) of protein is 0.36 g per pound of body weight. Eleven of the 20 natural amino acids can also be synthesized by the cell from various metabolic intermediates. The other 9 amino acids must be supplied regularly through the diet, and therefore, are known as **essential amino acids,** and listed in Table 13-1.

The essential amino acids are found in both animal and vegetable proteins. Sources of animal protein include meat, fish, milk, and eggs. Animal sources

TABLE 13-1 The Essential Amino Acids

Amino Acid Name	Three Letter Abbreviation
*Arginine	Arg
Histidine	His
Isoleucine	Ile
Leucine	Leu
Lysine	Lys
Methionine	Met
Phenylalanine	Phe
Threonine	Thr
Tryptophan	Trp
Valine	Val

*Required in children.

(a)

(b)

Figure 13-5 (a) Vegetarian pairings of a grain and a legume can provide a "complete protein" with all of the essential amino acids. (b) Quinoa is a single vegetarian ingredient that contains all of the essential amino acids. [Catherine Bausinger]

are often said to provide "complete protein" because they contain all 9 of the essential amino acids. Single vegetable sources of amino acids are usually missing one or more essential amino acids, quinoa being a notable exception (Figure 13-5b). Rice, for example, is low in the amino acid lysine whereas legumes (beans, soybeans, alfalfa, peas, lentils, and peanuts) are low in the amino acid methionine. A vegetarian diet, therefore, should contain a complementary mixture of plant proteins. Typically a healthy vegetarian meal combines a grain like rice with a legume such as beans. The grain is high in methionine but low in lysine, while the legume is low in methionine but high in lysine. Figure 13-5a illustrates some grain and legume pairings that provide all the essential amino acids.

WORKED EXERCISE The Structure of an Amino Acid

13-2 One of the 20 natural amino acids is shown below.

$$
\begin{array}{c}
\text{O} \\
\parallel \\
\text{C}-\text{NH}_2 \\
| \\
\text{H}-\text{C}-\text{H} \\
| \\
\text{H}-\text{C}-\text{H} \\
\text{H} \quad | \\
| \\
\text{H}-\overset{+}{\text{N}}-\text{C}-\text{C}-\text{O}^- \\
| \quad | \quad \parallel \\
\text{H} \quad \text{H} \quad \text{O}
\end{array}
$$

a. Label the amine functional group. Is it shown in its ionized or neutral form?
b. Label the carboxylic acid functional group. Is it shown in its ionized or neutral form?
c. Write the structure of this amino acid at pH = 1. What is the net charge on the amino acid at this pH?
d. Label the α-carbon.
e. Circle the side chain. Based on the structure of the side chain, what is the name of this amino acid? What is its three letter abbreviation?

    f. Does the side chain contain a functional group? If so, what is the functional
       group?
    g. Is this amino acid *polar* or *nonpolar*? If it is polar, is it *acidic, basic,* or *neutral*?
    h. Using Table 13-1, determine if this amino acid is an essential amino acid. What
       does the term essential amino acid mean?

Solution

13-2 a. The amine functional group shown is in its ionized form, $-NH_3^+$.
    b. The carboxylic acid functional group shown is in its ionized form, $-COO^-$.
    c. The structure of this amino acid at pH = 1 is shown below. At low pH the
       amine is ionized with a +1 charge and the carboxylic acid is unionized with
       no charge. Therefore, the amino acid has a net +1 charge. The side chain is a
       neutral polar side chain so it does not contribute to the overall charge.

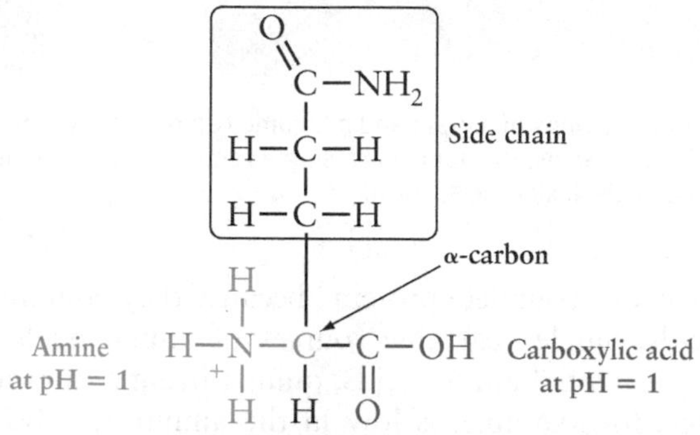

Glutamine
(Gln)

    d. See labels above.
    e. See above. This amino acid is glutamine which has the three letter abbreviation Gln.
    f. The side chain contains an amide functional group ($-CONH_2$).
    g. Glutamine is a polar neutral amino acid.
    h. Glutamine is not an essential amino acid. This means the body can synthesize it from
       other metabolic intermediates and it does not need to be supplied through the diet.

PRACTICE EXERCISES

2   One of the 20 natural amino acids is shown below.

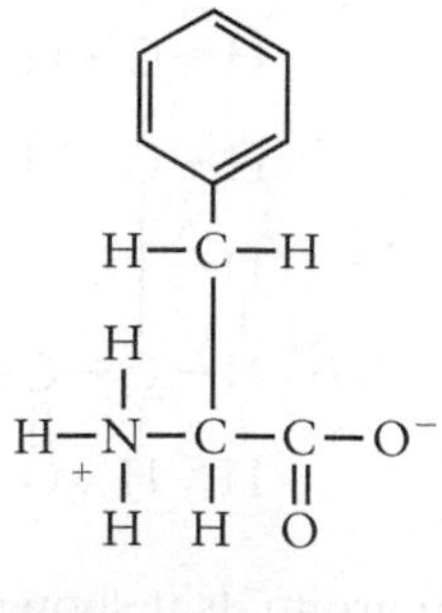

    a. Label the amine functional group. What form is shown here: neutral or ionized?
    b. Label the carboxylic acid functional group. What form is shown here: neutral or ionized?
    c. Identify the α-carbon.
    d. Circle and label the side chain. What is the name of this amino acid? What is its three letter
      abbreviation?
    e. Is this amino acid nonpolar or polar? Explain.
    f. Write the structure of this amino acid at pH = 12. What is the net charge on the amino acid
      at this pH?

3 For each of the following amino acids place an "X" in the boxes that apply.

Amino Acid	Nonpolar	Polar	Acidic Side Chain	Basic Side Chain	Neutral Side Chain	Essential Amino Acid
Cysteine						
Lysine						
Threonine						
Methionine						
Aspartic acid						

4 Which amino acids contain sulfur? Which of these amino acids contains a thiol functional group?

5 One of the 20 natural amino acids is unusual in that its side chain contains a ring that incorporates the amine functional group of the amino acid. What is the name of this amino acid? What is its three letter abbreviation? Write the chemical structure of this amino acid at physiological pH.

6 In the opening story you learned that sickle-cell anemia is caused by the amino acid valine substituting for aspartic acid as the sixth amino acid on the two β-chains of hemoglobin. Write the structure of these two amino acids at physiological pH. Are they polar or nonpolar? If polar, indicate whether it is acidic, basic, or neutral.

7 What single food type contains all the essential amino acids?

Chirality of the Amino Acids

Nineteen of the 20 natural amino acids are chiral because the α-carbon is a center of chirality: a tetrahedral carbon with bonds to four different groups or atoms (chapter 7). The four atoms/groups bonded to the α-carbon in an amino acid are the amine, the carboxylic acid, the R group, and a hydrogen atom. Glycine is the only natural amino acid that is achiral, because it has two hydrogen atoms on the α-carbon (the R group is H). The 19 natural chiral amino acids are L-amino acids. *In a Fischer projection, an L-amino acid has the amine positioned horizontally at the left and the H atom positioned horizontally at the right, while a D-amino acid has the reverse arrangement. By convention, the carboxylic acid is always drawn on the vertical bond at the top, and the side chain is drawn vertically at the bottom. The side chain is drawn in condensed notation, even if there is another center of chirality in the side chain.*

Consider for example, the enantiomers D-alanine and L-alanine, shown below (also see Figure 7-52). The only difference between D-alanine and L-alanine is the configuration of the α-carbon. This is readily apparent in the Fischer projection, where the —NH_3^+ group is on the left in L-alanine and on the right in a D-alanine. Note that in nature, only L-alanine is produced.

> In chapter 11 you learned that the majority of sugars found in biological systems are D-sugars.

$$
\begin{array}{cc}
\text{COO}^- & \text{COO}^- \\
\text{H}_3\overset{+}{\text{N}}\!-\!\!\!+\!\!\!-\text{H} & \text{H}\!-\!\!\!+\!\!\!-\overset{+}{\text{N}}\text{H}_3 \\
\text{CH}_3 & \text{CH}_3 \\
\textit{L-}\text{Alanine} & \textit{D-}\text{Alanine}
\end{array}
$$

Proteins are chiral because they are constructed from L-amino acids. Proteins provide a chiral environment which accounts for many of the different physiological effects observed in the enantiomeric drug molecules described in chapter 7. A chiral drug interacts differently with receptors and enzymes than its enantiomer, because receptors and enzymes are proteins, which are themselves chiral.

WORKED EXERCISES | Chiral Amino Acids

13-3 The Fischer projection of an amino acid is shown below.

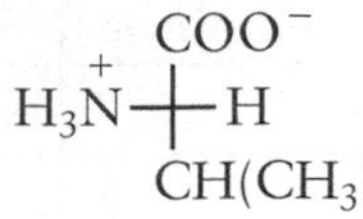

$$
\begin{array}{c}
\text{COO}^- \\
\overset{+}{\text{H}_3\text{N}}\!-\!\!\!\!-\!\!\text{H} \\
\text{CH(CH}_3)_2
\end{array}
$$

a. Label the α-carbon. Is this a center of chirality? How is a center of chirality always represented in a Fischer projection? Is this amino acid chiral?

b. Is this a D- or an L-amino acid? How can you tell? What is the name of this amino acid, including its D- or L-prefix?

c. Is this amino acid produced in nature?

d. Write the Fischer projection of the enantiomer of this amino acid, and name this amino acid.

13-4 Which of the 20 natural amino acids is achiral and why? Write the structure of this amino acid, and label the α-carbon.

Solution

13-3 a. See below. The α-carbon is a center of chirality because it has bonds to four different atoms or groups of atoms. In a Fischer projection, a center of chirality is represented as an intersection of vertical and horizontal bonds. The presence of one center of chirality makes this amino acid chiral.

b. This amino acid is L-because it has the amine group on the left. The name of this amino acid is L-valine because it has a —CH(CH$_3$)$_2$ side chain.

c. Yes, L-amino acids are produced in nature.

d. α-carbon

$$
\begin{array}{c}
\text{COO}^- \\
\text{H}\!-\!\!\!\!-\!\!\overset{+}{\text{NH}_3} \\
\text{CH(CH}_3)_2
\end{array}
$$

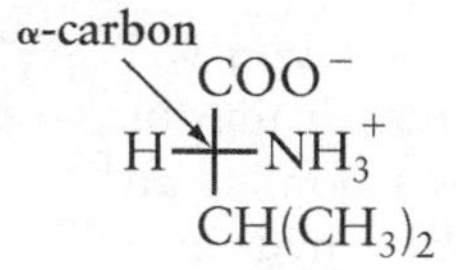

D-Valine

13-4 Glycine is achiral because it has no centers of chirality. The α-carbon is not a center of chirality in glycine because there are two hydrogen atoms bonded to the α-carbon (the R group is a hydrogen atom). A chiral compound must have at least one tetrahedral carbon with bonds to four different atoms or groups.

$$
\begin{array}{c}
\quad\text{H}\quad\text{H} \\
\quad\mid\quad\mid\text{α-carbon} \\
\text{H}\!-\!\text{N}\!-\!\text{C}\!-\!\text{C}\!-\!\text{O}^- \\
\overset{+}{\mid}\quad\mid\quad\parallel \\
\quad\text{H}\quad\text{H}\quad\text{O}
\end{array}
$$

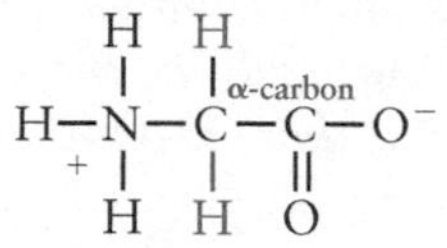

Glycine

PRACTICE EXERCISES

8   Write the Fischer projection of L-cysteine at physiological pH and answer the following questions:

a. Is this amino acid produced in nature? Explain.

b. Identify the α-carbon in this Fischer projection.

c. In a Fisher projection, how can you tell whether an amino acid is D or L?

d. Write the structure of D-cysteine. How is it different from L-cysteine? How is it similar to D-cysteine? What type of stereoisomers are L- and D-cysteine?

e. What functional group does cysteine have in its side chain?

9   Answer the following questions for isoleucine:

a. What four atoms or groups of atoms are bonded to the α-carbon?

b. Which of these atoms or groups of atoms are present in all amino acids?

c. Which of these atoms or groups of atoms are unique to isoleucine?

d. There are only two possible configurations for the atoms or groups of atoms around the α-carbon of any amino acid (or any center of chirality). Show this by writing the Fischer projections for D- and L-isoleucine at physiological pH. Which stereoisomer is produced in nature?

13.2 Peptides

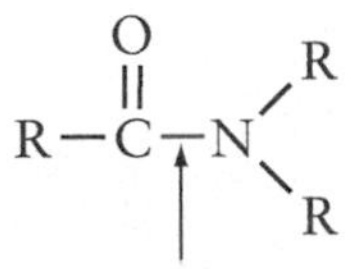

Amino acids are the building blocks for all proteins. In this section we consider how amino acids are joined to form proteins. A **dipeptide** is formed when two amino acids undergo a reaction between the *amine* of one amino acid and the *carboxylic acid* of another amino acid to form an amide bond joining the two amino acids. **Polypeptides**—or **peptides**—contain many amino acids joined as amides. Thus, polypeptides are polymers of amino acids. **Proteins** are large peptides—polypeptides—or several peptides that together serve a particular biological function. Recall from the opening story that hemoglobin is a protein composed of four polypeptide chains, two identical α-chains containing 141 amino acids, and two identical β-chains composed of 146 amino acids.

Amide bond

Peptide Bonds

In chapter 10 you learned that the reaction between a carboxylic acid and an amine to form an amide is known as an **amidation** reaction. Amidation reactions form the carbon-nitrogen bond of an amide functional group. Similarly, when two *amino acids* undergo an amidation reaction, the *carboxylic acid* of one amino acid reacts with the *amine* of the other amino acid to form an amide functional group, and the product is known as a dipeptide, as shown in Figure 13-6. The carbon-nitrogen bond of an amide formed between two amino acids is known as a **peptide bond.** A molecule of water is also expelled in the reaction. Note that the reverse of an amidation reaction is a hydrolysis reaction. Dietary proteins are hydrolyzed in the stomach into their amino acid constituents with the help of enzymes and the low pH of the stomach during digestion. Indeed, this is why eating proteins supplies the body with the amino acid building blocks necessary for building our muscles and other proteins.

A tripeptide, shown in Figure 13-7, is formed when a third amino acid reacts with either the carboxylic acid or the amine of a dipeptide forming another peptide bond. As you can see, an amino acid is capable of forming two peptide bonds, one with its amine functional group and one with its carboxylic acid functional group. Thus, in a peptide, only the first and last amino acids have a free amine and a free carboxylate group, respectively. When

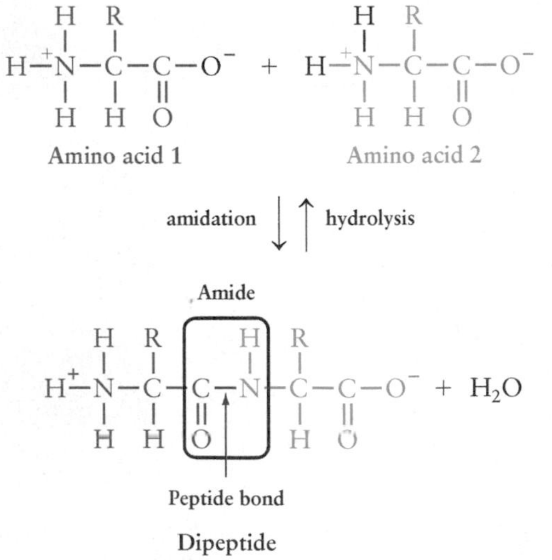

Figure 13-6 An amidation reaction between two amino acids occurs between the carboxylic acid of one amino acid and the amine of the other amino acid, to form a dipeptide. The black box shows the amide functional group formed and the new peptide bond. In the reverse reaction, a hydrolysis reaction, the peptide bond breaks.

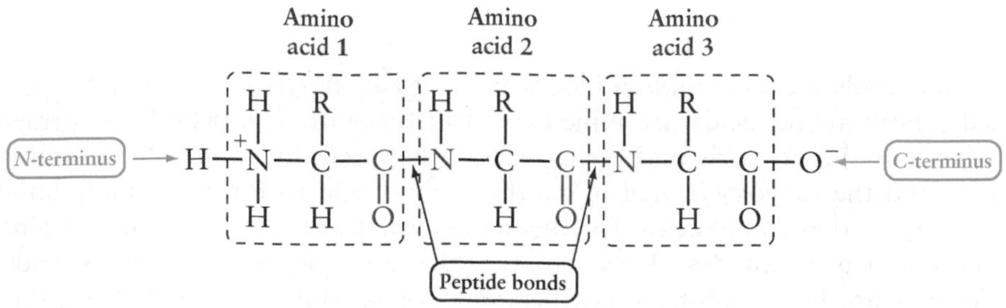

Amino acid 1   Amino acid 2   Amino acid 3

Figure 13-7 A tripeptide, derived from three amino acids, showing the two peptide bonds. The terminal end with a free amine is known as the *N*-terminus, and the terminal end with a free carboxylate is known as the *C*-terminus.

many, often thousands, of amino acids react in this head-to-tail manner, a polypeptide is formed, linking amino acids sequentially with peptide bonds. The terminal end of the polypeptide containing the free amine is known as the **N-terminus,** and the terminal end containing the free carboxylate is known as the **C-terminus** (Figure 13-7). The convention for writing the amino acid sequence for a peptide is to list the amino acids in order from the *N*-terminus to the *C*-terminus, using the amino acid abbreviations. For example, the amino acid sequence Ala-Gly-Val specifies a tripeptide in the order alanine, glycine, valine, where alanine is the *N*-terminus and valine is the *C*-terminus, as shown in Figure 13-8. Note that the amino acids of the polypeptide no longer have an amine and a carboxylic acid functional group, but instead exist as amides, except at the *N*-terminus and the *C*-terminus. To gain more experience with peptides perform *Using Molecular Models 13-1: Building the Dipeptide Gly-Ala.*

Some hormones such as oxytocin, the hormone that signals the start of labor and promotes lactation, are small peptides, and as such, are not classified as proteins. Oxytocin is a *nona*peptide—derived from *nine* amino acids (Figure 13-9). Endorphins, which are released after an injury or intense physical activity and serve as the body's natural analgesics (pain killers), are *penta*peptides—derived from *five* amino acids.

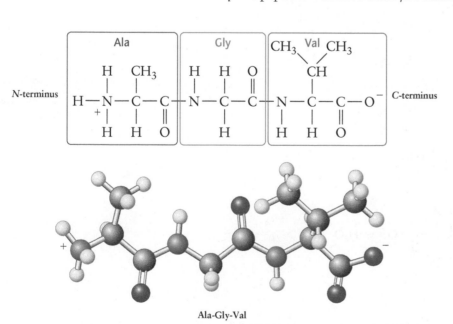

Figure 13-8 The tripeptide Ala-Gly-Val shown as a Lewis structure and a ball-and-stick model.

Figure 13-9 Nursing stimulates the release of the hormone oxytocin, a nonapeptide with the amino acid sequence Cys-Tyr-Ile-Gln-Asn-Cys-Pro-Leu-Gly. [Blend/Punchstock]

Using Molecular Models 13-1 Building the Dipeptide Gly-Ala

Construct Glycine

1. Obtain 2 black carbon atoms, 5 light-blue hydrogen atoms, 2 red oxygen atoms, 1 blue nitrogen atom, 8 straight bonds, and 2 bent bonds.

2. Construct a model of the amino acid glycine at physiological pH. Use Figures 13-2 and 13-3 as a guide.

3. What form should the amine be in at physiological pH? Note that your model kit won't allow you to make a nitrogen atom with four bonds. What form should the carboxylic acid be in at physiological pH?

Construct L-Alanine

4. Obtain 3 black carbon atoms, 7 light-blue hydrogen atoms, 1 blue nitrogen atom, 2 red oxygen atoms, 11 straight bonds, and 2 bent bonds.

5. Construct a model of L-alanine using Figure 13-3 as a reference. Double check that you have constructed L-Ala and not D-Ala (they are mirror images).

Construct the Dipeptide Gly-Ala

6. Join the two amino acids you built above to make the dipeptide Gly-Ala. You should combine the amine of one amino acid with the carboxylic acid of the other amino acid to make an amide. Think carefully about which amine and which carboxylic acid functional groups to join to form the amide, since there are two possibilities. Glycine must be the N-terminus and Ala the C-terminus.

Inquiry Questions

7. In forming the peptide bond in step 6, did you also make water (2 hydrogen atoms and 1 oxygen atom left over)?

8. Write the overall chemical reaction represented in the construction of Gly-Ala from Gly and Ala.

9. Write the chemical structure of the dipeptide Gly-Ala.

10. Label the N-terminus of your dipeptide.

11. Label the C-terminus of your dipeptide.

12. Label the peptide bond in your dipeptide.

13. Is this dipeptide chiral? Explain.

WORKED EXERCISE Peptides

13-5 The endorphin met-enkephalin, described in chapter 7, is a pentapeptide with the amino acid sequence: Tyr-Gly-Gly-Phe-Met.

   a. Why is this peptide classified as a pentapeptide? How many peptide bonds are in a pentapeptide?

   b. Name the amino acid at the N-terminus. How can you identify the N-terminus?

   c. Name the amino acid at the C-terminus. How can you recognize the C-terminus?

   d. What are the side chains in this peptide?

   e. Which are the nonpolar amino acids in this peptide?

   f. Which amino acids in this peptide are chiral? Is the peptide chiral?

   g. Write the chemical structure for this pentapeptide. Label the N-terminus and the C-terminus. Label each amino acid with its three letter abbreviation.

Solution

13-5 a. It is a classified as a pentapeptide because it is derived from five (penta) amino acids. There are four peptide bonds in a pentapeptide (between amino acids 1-2, 2-3, 3-4, and 4-5).

   b. The amino acid at the N-terminus is tyrosine (Tyr). The N-terminus is always the first amino acid in the sequence, by convention, and it has a free amino group ($-NH_3^+$).

   c. The amino acid at the C-terminus is methionine (Met). The C-terminus is always the last amino acid in the sequence, by convention, and it has a free carboxylate group ($-COO^-$).

    d. The side chains from the *N*- to the *C*-terminus are —$CH_2C_6H_4OH$, —H, —H, —$CH_2C_6H_5$, and —$CH_2CH_2SCH_3$.

    e. The nonpolar amino acids are glycine, phenylalanine, and methionine.

    f. The chiral amino acids are all but glycine: tyrosine, phenylalanine, and methionine. The peptide is chiral because at least one of the amino acids is chiral.

    g.

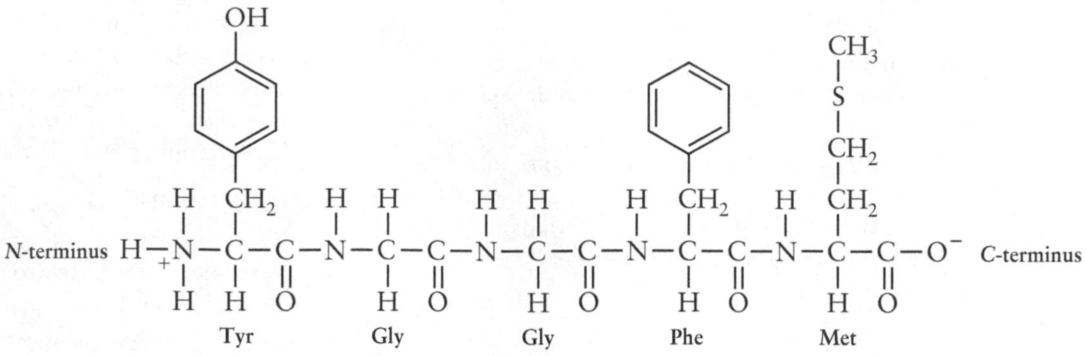

PRACTICE EXERCISES

10 The structural formula for leu-enkephalin, another endorphin, is shown below:

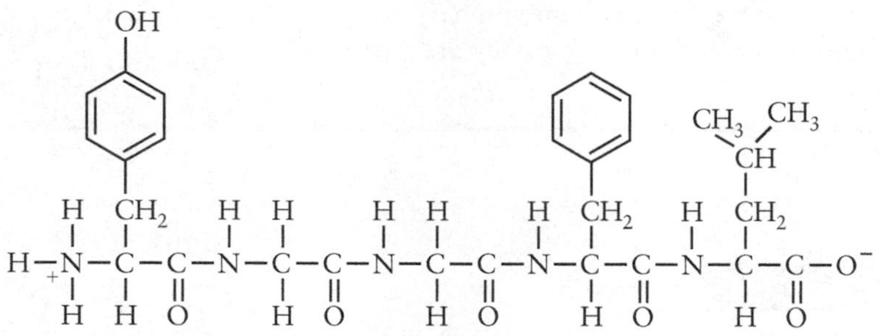

    a. Circle each amino acid in the peptide, and identify it by its three letter abbreviation.

    b. Label the *N*-terminus and the *C*-terminus.

    c. Write the amino acid sequence for this peptide using the three letter abbreviations for the amino acids? Which end of the amino acid should you start from and why?

    d. What are the side chains in this peptide?

    e. Identify each peptide bond.

    f. Is this peptide chiral?

    g. What role do endorphins have in the body? How does the structure of leu-enkephalin differ from that of met-enkephalin, shown in Worked Exercise 13-5?

11 Are the dipeptides Gly-Phe and Phe-Gly different compounds? Are they structural isomers or stereoisomers?

12 Write the structure of the tripeptide Cys-Gly-Val. Label the *N*-terminus and the *C*-terminus.

13 The amino acid sequence for the pregnancy hormone, oxytocin, is given below:

Cys-Tyr-Ile-Gln-Asn-Cys-Pro-Leu-Gly

    a. What is the *N*-terminal amino acid?

    b. What is the *C*-terminal amino acid?

    c. From how many amino acids is this peptide derived?

    d. How many peptide bonds does this peptide contain?

    e. Which amino acid(s) contain(s) a thiol functional group in its side chain?

■ 13.3 Types of Proteins and Protein Architecture

Proteins are composed of anywhere from 50 to thousands of amino acids. The largest protein, titin, found in muscle tissue such as the heart, contains 26,926 amino acids! Most proteins, however, contain less than 2,000 amino acids. Some proteins, such a hemoglobin (see opening story), contain more than one polypeptide chain.

All proteins are built from the 20 L-amino acids. Scientists estimate that there are about one million different proteins in the human body. *The information with the amino acid sequence for every protein the cell needs is encoded in our DNA.* The connection between DNA and proteins explains why diseases involving proteins, such as sickle-cell anemia, are genetic disorders. Since enzymes are proteins, they are also constructed from the information encoded in your DNA. You will learn more about DNA and how it serves as a blueprint for building proteins in Chapter 14 Nucleic Acids: RNA and DNA.

Types of Proteins

There are three general classes of proteins based on their shape and solubility:

- fibrous proteins,
- globular proteins, and
- membrane proteins.

Fibrous proteins tend to be strong and insoluble in water, therefore, they generally serve a structural role in nature. Three basic types of fibrous proteins include keratins, elastins, and collagens. **Keratins** provide structure to hair and fingernails (**Figure 13-10a**). Keratins are also found in feathers, horns, claws, and animal hooves. **Elastins** are fibrous proteins with elastic fibers that provide flexibility to skin, blood vessels, the heart, lungs, intestines, tendons, and ligaments. **Collagens** are the main structural proteins in the body (Figure 13-10b). Collagens provide rigidity to the connective tissue found in skin, tendons, bones, cartilage, and ligaments. Connective tissue is made up of a combination of elastins and collagens. Elastins provide the elasticity, and collagens provide the rigidity, seen in connective tissue. The visible effects of aging—loss of skin flexibility—can be attributed in large part to the declining production of elastins that occurs with age.

(a)

(b)

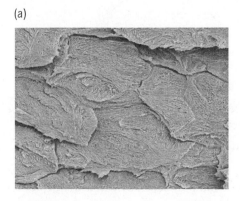
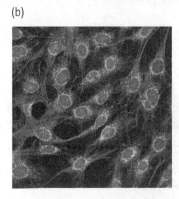

Figure 13-10 Colored scanning electron micrograph (SEM) of two types of fibrous proteins: (a) Layers of keratin in a fingernail. [Susumu Nishinaga/Science Source] (b) Collagens: Fibroblast cells produce collagen, the main structural protein in the body. [Dr. Gopal Murti/Science Source]

Globular proteins are soluble polypeptides folded into complex and spherical shapes. Hemoglobin and insulin are examples of globular proteins. Enzymes, the subject of the next section, are globular proteins.

Membrane proteins are found in part or entirely embedded within the cell membrane. In **transmembrane proteins** the polypeptide crosses the cell membrane several times. The membrane receptors described throughout this text are membrane proteins that convey signals, or they serve as ion channels that shuttle ions in and out of the cell.

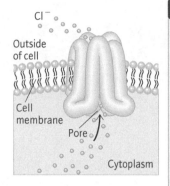

Figure 13-11

14 Answer the following questions about the schematic of a protein shown in Figure **13-11.**
   a. Is this a *globular, fibrous,* or *membrane* protein?
   b. Would you expect amino acids located on the outside of this protein, but within the membrane, to be nonpolar or polar?
   c. Would you expect amino acids in contact with the aqueous environment on the outside of the cell to be polar or nonpolar?
15 State two differences between globular and fibrous proteins.
16 Keratin, collagen, and elastin all belong to what class of proteins?
17 Enzymes are typically _____ proteins.
18 What do collagens and elastins have in common? What is a functional difference between collagens and elastins?

The Three-Dimensional Shapes of Proteins

We have seen that a polypeptide is described in part by its sequence of amino acids from the *N*-terminus to the *C*-terminus. Although the sequence of amino acids for a protein may seem analogous to beads on a necklace, as shown in Figure 13-12, a protein is not a long linear shaped molecule. Instead, the polypeptide(s) fold(s) into a unique three-dimensional shape, known as its **native conformation**, which is necessary for the protein to be able to perform its specific biological function. Imagine the amino acid sequence for a protein to be more like beads on a necklace that fold up on each other (Figure 13-12). To accomplish their diverse array of functions, proteins come in a variety of different shapes and sizes. By analyzing the three-dimensional shape and structure of a protein, scientists can learn how a particular protein works, and eventually learn how diseases like sickle-cell anemia work.

Four hierarchical levels of protein architecture describe the complex three-dimensional shape of a protein:

1. primary structure,
2. secondary structure,
3. tertiary structure, and
4. quaternary structure.

We begin with the primary structure of proteins and then describe these higher levels of protein architecture below.

The Primary Structure of Proteins

The **primary structure** of a protein is defined as its amino acid sequence, or the sequence of amino acids for each of the polypeptides when the protein is composed of more than one polypeptide. Errors in the primary structure of a protein can have significant effects on its overall three-dimensional shape, as you saw for hemoglobin in the case of sickle-cell anemia, described in the opening story.

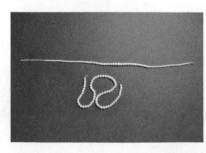

Figure 13-12 The primary structure of a polypeptide is like the beads on a necklace laid out in a linear fashion, whereas the tertiary structure is like the folded necklace (native conformation of protein). [Catherine Bausinger]

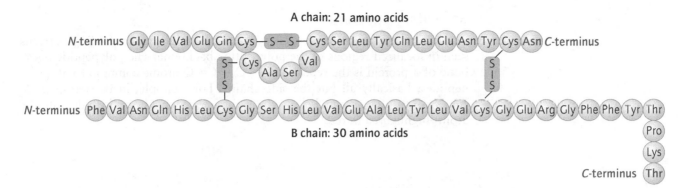

Figure 13-13 The primary structure of human insulin showing the amino acid sequence of its two polypeptide chains (A chain and B chain), with each amino acid represented as a sphere.

Consider the primary structure of human insulin, a relatively small protein. Insulin is composed of two polypeptides: the A chain contains 21 amino acids, and the B chain contains 30 amino acids. The sequence of amino acids from the N-terminus to the C-terminus is given for each polypeptide in Figure 13-13, where each amino acid is represented by a sphere containing its three letter abbreviation.

WORKED EXERCISE | Protein Primary Structure

13-6 Refer to Figure 13-13 to answer the questions below about the primary structure of insulin:

    a. What amino acids are at the N-termini? Why are there two N-termini in insulin?

    b. What is the tenth amino acid in the A chain of insulin?

    c. How many C-termini are there in insulin? What amino acids are at the C-termini?

Solution

13-6 a. The N-terminus on the A chain is glycine (Gly), and on the B chain it is phenylalanine (Phe). There are two N-termini in insulin because insulin is composed of two polypeptides, the A chain and the B chain.

    b. The tenth amino acid on the A chain is valine.

    c. There are two C-termini in insulin, one for each polypeptide. The C-terminus in the A chain is asparagine (Asn), and in the B chain it is threonine (Thr).

PRACTICE EXERCISES

19 If hemoglobin is composed of four polypeptide chains, how many N-termini does hemoglobin have?

20 What are the four levels of protein architecture called?

21 Define the term "native conformation."

22 Cystic fibrosis (CF) is a disease caused by a missing phenylalanine-508 in a protein containing 1,480 amino acids. The protein serves as a chloride ion channel that allows chloride ions to cross the cell membrane. People with cystic fibrosis produce an abnormal form of this protein which does not fold properly as a result of this missing amino acid. Consequently, people with this condition suffer from an accumulation of mucus in the lungs, which causes respiratory infections and difficulty breathing. Many individuals with CF require a lifetime of specialized care and usually die by the age of 30.

    a. Write the structure of the missing amino acid in the protein that causes cystic fibrosis. Assume physiological pH.

    b. Where in the primary structure of the protein is phenylalanine missing?

    c. Would you expect cystic fibrosis to be a genetic disease or an infectious disease? Explain.

    d. Based on the information given in the question, do you think the protein responsible for CF is a fibrous, globular, or membrane protein?

Secondary Structure

The **secondary structure** of a protein describes the regular folding patterns seen in localized regions of the polypeptide backbone. The **polypeptide backbone** of a protein is the repeating N—C$_\alpha$—C=O atoms common to all polypeptides; basically all but the side chains. For example, in the tetrapeptide Ala-Asp-Lys-Met shown below, the polypeptide backbone includes all the atoms highlighted in red.

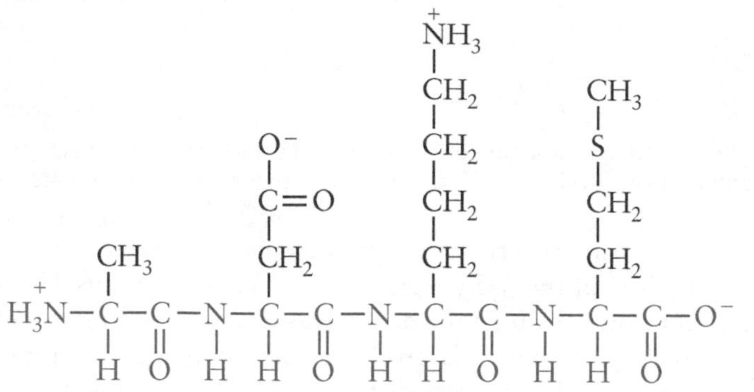

Polypeptide backbone

Since proteins are such large molecules, the polypeptide backbone is often depicted using ribbon diagrams, where the polypeptide backbone is represented as a flat colored ribbon, as seen in Figure 13-14 and other illustrations of proteins throughout the remainder of the chapter.

Secondary protein structure arises from hydrogen bonding between amides in the polypeptide backbone. Hydrogen bonds are formed between the N—H of one amide and the carbonyl group (C=O) of another amide along the polypeptide backbone as shown below:

The two most common types of secondary structure are the α-helix and the β-pleated sheet. An **α-helix** is a segment of the polypeptide that coils into a helix, as seen in Figure 13-14b, where a yellow ribbon traces the polypeptide backbone. Hydrogen bonding (dashed red lines) in an α-helix occurs between the oxygen atom of the amide carbonyl group of one amino acid and the hydrogen atom of an N—H bond on the amide of an amino acid three and a half amino acids farther down the polypeptide chain. The side chains, represented as green spheres labeled "R" in Figure 13-14b, extend out from the α-helix and are not part of the hydrogen bonding network that creates and stabilizes the α-helix.

In a ribbon diagram, an α-helix appears as a helical, colored, flat ribbon. For example, a ribbon drawing of myoglobin, the oxygen-carrying protein in muscle tissue, is shown in Figure 13-14a, which depicts all eight α-helices as red and pink ribbons of varying lengths. Figure 13-14b zooms in on a segment of one of these helices. Indeed, the term "globin" refers to a class of globular proteins that contain eight α-helices, including hemoglobin and myoglobin.

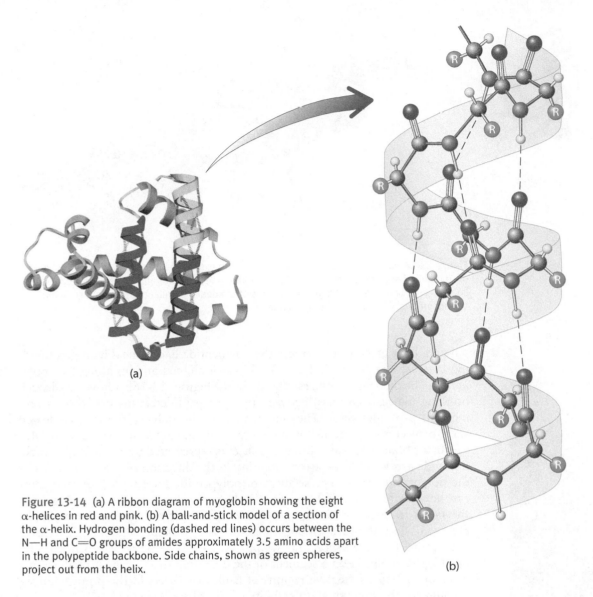

Figure 13-14 (a) A ribbon diagram of myoglobin showing the eight α-helices in red and pink. (b) A ball-and-stick model of a section of the α-helix. Hydrogen bonding (dashed red lines) occurs between the N—H and C=O groups of amides approximately 3.5 amino acids apart in the polypeptide backbone. Side chains, shown as green spheres, project out from the helix.

One common type of membrane protein, characterized by seven α-helices spanning the cell membrane, is illustrated in **Figure 13-15**. Membrane proteins tend to have hydrophobic amino acids on their exterior, since the cell membrane is composed of lipids.

A **β-pleated sheet,** also known as a **β-sheet,** is another type of secondary structure formed when two or more parallel or antiparallel sections of a polypeptide fold in a pattern that looks like a pleated skirt, as seen in **Figure 13-16b,**

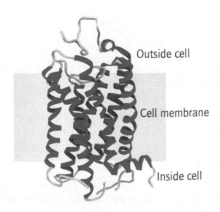

Outside cell

Cell membrane

Inside cell

Figure 13-15 A common type of membrane protein contains seven α-helices (red) that span the width of the cell membrane.

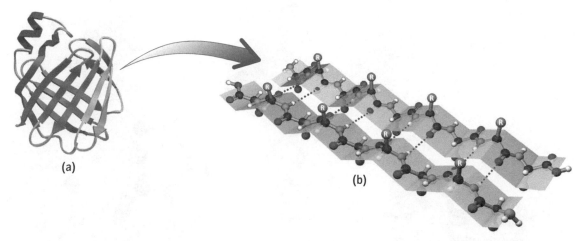

(a)

(b)

Figure 13-16 (a) A ribbon diagram of a protein with extensive antiparallel β-pleated sheets. (b) A ball-and-stick model of a section of β-pleated sheet. Hydrogen bonding (dashed red lines) occurs between N—H and C=O groups of amides in parallel or antiparallel aligned strands of two sections of the polypeptide.

Figure 13-17 Silk fibers contain fibroin, a protein composed of primarily β-sheets which account for the strength and flexibility of silk fibers. [© Massimo Listri/Corbis]

where a light green ribbon traces the polypeptide backbone. Hydrogen bonding between the N—H and C=O groups of aligned amides along two polypeptide backbones stabilizes the β-sheet. Figure 13-16b shows hydrogen bonding (dashed red lines) between two polypeptide chains that form a segment of β-pleated sheet. The side chains on the polypeptide in a β-pleated sheet project above and below, as indicated by the green spheres labeled "R."

In a ribbon diagram, a β-pleated sheet is represented by a straight, colored, flat wide arrow, with the arrow pointing in the direction of the C-terminus of the polypeptide chain. The segment of polypeptide forming a β-pleated sheet can have its arrows either pointing in the same direction—*parallel* β-pleated sheets, or in opposite directions—*antiparallel* β-pleated sheets. For example, Figure 13-16a shows a ribbon diagram of a protein containing extensive antiparallel β-pleated sheets (as well as two small sections of α-helices). Figure 13-16b zooms in on a segment of the β-pleated sheet in the protein shown in Figure 13-16a. Extensive regions of β-pleated sheets in the protein fibroin account for the strength and flexibility of silk fibers (Figure 13-17).

Proteins vary in the extent and type of secondary structure they contain. One protein might have only α-helices throughout its structure, while another may contain some sections of α-helix and some sections of β-pleated sheets. Enzymes typically contain both α-helices and β-pleated sheets. Sections of the protein where there is no secondary structure, the polypeptide backbone folds in no regular discernible pattern and appears as a colored string in a ribbon diagram.

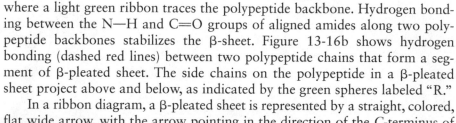

WORKED EXERCISE   Protein Secondary Structure

13-7 Mad cow disease is believed to be caused by an infectious abnormal protein, known as a **prion**, which stands for proteinaceous infectious agent. Prions cause normal membrane proteins in the brain to change from their native conformation, thus preventing them from performing their specific biological function. The ribbon diagrams shown in Figure 13-18 show the difference in secondary structure for the normal protein compared to the prion. Answer the following questions based on these ribbon diagrams.

a. Is the primary structure of the prion different from the normal protein?

b. What type of secondary structure appears in red in these two structures?

Figure 13-18   Normal protein   Prion

c. What type of secondary structure appears in blue in these two structures?
d. Are the β-pleated sheets in the prion parallel or antiparallel? How can you tell?
e. Describe how the secondary structure of the prion is different from the secondary structure of the normal protein.
f. What stabilizes an α-helix? Does an α-helix represent primary or secondary protein structure?

Solution

13-7 a. No, the primary structure—the sequence of amino acids—is the same for both the normal protein and the prion, because the question refers only to a change in conformation (bond rotations). The normal protein and the prion differ only in their three-dimensional shape.
b. α-helices are shown in red.
c. β-pleated sheets are shown in blue.
d. The β-pleated sheets in the prion are antiparallel, as indicated by the opposing directions of the blue arrows. The arrow always points to the C-terminus of the polypeptide.
e. The normal protein has three sections of α-helix and two very small β-pleated sheets, while the prion has two sections of α-helix and an extensive array of antiparallel β-pleated sheets. The conformation of the two proteins is noticeably different in both secondary structure and overall three-dimensional shape.
f. An α-helix is stabilized by hydrogen bonding between amide groups in the polypeptide backbone. An α-helix represents secondary structure.

PRACTICE EXERCISES

23 A ribbon diagram of insulin is shown in Figure 13-19. What type of secondary structure does insulin have? What level of protein architecture does the red ribbon represent: *primary, secondary, tertiary,* or *quaternary*?

24 Add dashes to the partial structures below to show where hydrogen bonding between the two amide functional groups in the polypeptide backbone occurs. Label the partial positive and partial negative charges on the atoms involved in hydrogen bonding. What creates the partial charges on these atoms?

Figure 13-19

25 With respect to the following secondary structures, where are the side chains projecting relative to the polypeptide backbone?

    a. The β-pleated sheet   **b.** The α-helix

26 What is a ribbon diagram?

27 Are side chains involved in the hydrogen bonding that stabilizes an α-helix or a β-pleated sheet?

28 In 2008 the Nobel Prize in chemistry was awarded for the discovery and the characterization of green fluorescent protein (GFP), shown in Figure 13-20, isolated from the bioluminescent jellyfish *A. victoria*. What type of secondary structure is represented by the wide arrows in this ribbon diagram of GFP? What are the arrows pointing toward? Are the segments of β-pleated sheets parallel or antiparallel, and how can you tell?

Figure 13-20

Tertiary Structure

*The **tertiary structure** of a protein describes the complex three-dimensional shape that results from the elaborate folding of a single polypeptide over and above its secondary structure.* Tertiary structure arises from electrostatic interactions between the side chains of a single polypeptide as well as hydrophilic and hydrophobic interactions with its surrounding environment.

For some proteins, tertiary structure also includes **prosthetic groups**: essential non-peptide containing organic molecules or metal ions that form ionic or covalent bonds to the polypeptide. For example, each of the four polypeptides in hemoglobin contains heme as a prosthetic group (**Figure 13-21**), which is in part where *hem*oglobin gets its name. One heme molecule is covalently bonded to each polypeptide chain. At the center of each heme is an iron (Fe) atom that binds oxygen. In proteins that function as enzymes, the prosthetic group is called a **cofactor** or a **coenzyme**.

The fundamental driving force behind tertiary structure—protein folding— is lowering the potential energy of the protein and its surrounding environment. A folded protein and its surrounding environment are lower in potential

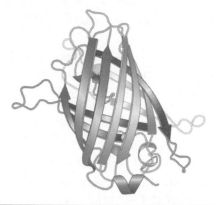

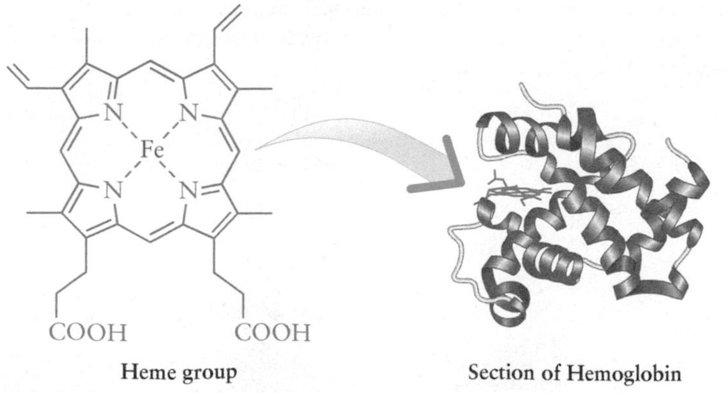

Figure 13-21 Heme, shown in green, is a prosthetic group, a non-peptide organic structure that is an essential part of each of the polypeptide chains in hemoglobin. An iron atom, Fe, lies at the center of each heme.

Heme group   Section of Hemoglobin

energy than the unfolded protein and its surrounding environment. Hence, proteins fold *spontaneously* into their native conformation under physiological conditions. Moreover, the way a protein folds is unique and ideally suited to its biological function.

Folding occurs as a result of favorable electrostatic interactions between side chains, often located far apart in the primary structure of a protein, as well as interactions of side chains with the surrounding environment. For example, enzymes that exist in an aqueous medium will typically fold in a manner that places their hydrophilic polar side chains on the exterior of the protein where they can interact with water, while orienting their hydrophobic side chains on the interior of the protein, where they avoid water, and can interact with other side chains through dispersion forces.

The following electrostatic and covalent interactions within a single polypeptide chain contribute to the tertiary structure of a protein:

- disulfide bridges (covalent bond between two cysteines),
- salt bridges (ionic bonds, *hydrophilic*),
- hydrogen bonding (electrostatic attraction, *hydrophilic*), and
- dispersion forces (electrostatic attraction, *hydrophobic*).

Disulfide Bridges *Disulfide bridges, also known as disulfide bonds, are covalent bonds formed between the side chains on two cysteines located on either the same polypeptide or on different polypeptide chains of the protein (Figure 13-22).* Recall from section 13.1 that cysteine is an amino acid with a thiol functional group (—S—H) on its side chain. Since a disulfide bridge is a covalent bond, it can only be formed or broken in a chemical reaction. The thiol groups on the side chains of two cysteines can undergo oxidation to form a disulfide bond (—S—S—), the strongest type of interaction between side chains. For example, the A chain and the B chain in insulin are joined by two disulfide bonds, shown in orange in Figure 13-13. The A chain also has a disulfide bond between the two cysteines at amino acids 6 and 11.

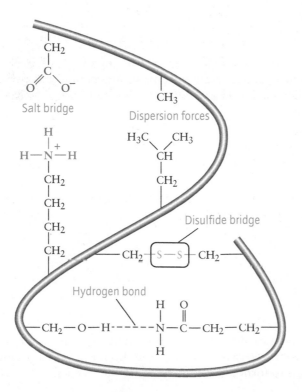

Figure 13-22 Electrostatic interactions (salt bridge, hydrogen bonding, and dispersion forces) and disulfide bonds between side chains create the tertiary structure of a protein.

Salt Bridges Ionic bonds between the side chains of certain amino acids, also known as **salt bridges**, are formed between a positively charged polar, basic side chain and a negatively charged polar, acidic side chain, as illustrated in Figure 13-22. The side chains may be located far apart in the primary structure of the polypeptide(s). Recall that the electrostatic interaction between oppositely charged ions is quite strong because full charges are involved. This ionic interaction is similar to what was described for ionic compounds in chapter 3, except there is no lattice, and the charges are located on a large organic molecule (a very large polyatomic ion).

Hydrogen Bonding We have seen how hydrogen bonding between amides in the polypeptide backbone creates secondary structure. Hydrogen bonding also occurs between side chains of the polypeptide containing O—H or N—H bonds and carbonyl groups, as shown in Figure 13-22. Although hydrogen bonds are weaker than disulfide and ionic bridges, there are usually so many of them that they contribute significantly to the tertiary structure of the protein.

Tertiary structure is also influenced by hydrogen bonding interactions that can occur between polar side chains and the aqueous environment surrounding most proteins. This is why proteins fold in a manner that places polar side chains on the surface of the protein where they can form hydrogen bonds with water molecules, known as *hydrophilic* interactions.

Dispersion Forces Nonpolar side chains in a protein interact with other nonpolar side chains through dispersion forces. (Figure 13-22). Nonpolar side chains are hydrophobic and avoid an aqueous medium. Thus, in an aqueous environment, proteins tend to fold so that sections of the protein with nonpolar side chains arrange themselves so that they are on the interior of the protein, away from the surrounding aqueous environment. In contrast, nonpolar side chains are found on the surface of membrane proteins because their external environment—the cell membrane—is nonpolar.

WORKED EXERCISES | Tertiary Protein Structure

13-8 Consider the side chains on the following pairs of amino acids within a single polypeptide of a protein, and indicate the type of interaction that could occur. Select from among the following choices: *disulfide bridge, salt bridge, hydrogen bonding, and dispersion forces.*

   a. Cysteine and cysteine
   b. Phenylalanine and alanine
   c. Aspartic acid and lysine
   d. Threonine and glutamine

13-9 Which of the following amino acid side chains would be found on the outside of a protein in an aqueous environment?

   a. Methionine
   b. Glutamic acid
   c. Arginine
   d. Leucine
   e. Threonine

Solutions

13-8 a. Disulfide bridge, because cysteine has a thiol functional group on its side chain that makes it possible for two cysteines, even distant in primary structure, to undergo oxidation to form a disulfide bond (R—S—S—R).

b. These are both nonpolar amino acids so their side chains interact by dispersion forces.

c. Aspartic acid is a polar acidic amino acid and lysine is a polar basic amino acid, so at physiological pH they are in their ionized forms. They can form a salt bridge, between the ionized carboxylic acid ($RCOO^-$) and the ionized amine (RNH_3^+) within the polypeptide.

d. Threonine and glutamine can form a hydrogen bond because threonine has an O—H group and glutamine has a carbonyl group and an N—H group.

13-9 Glutamic acid, arginine, and threonine because their side chains contain polar functional groups.

PRACTICE EXERCISES

29 Write the products formed in the following reduction. *Hint:* The reaction is the reverse of the oxidation of two thiols.

$$H-\overset{\overset{\displaystyle H}{|}}{\underset{\underset{\displaystyle H}{|}}{C}}-S-S-\overset{\overset{\displaystyle H}{|}}{\underset{\underset{\displaystyle H}{|}}{C}}-\overset{\overset{\displaystyle H}{|}}{\underset{\underset{\displaystyle H}{|}}{C}}-\overset{\overset{\displaystyle H}{|}}{\underset{\underset{\displaystyle H}{|}}{C}}-H \quad \xrightarrow{[H]}$$

30 In Figure 13-23, define the electrostatic and covalent interactions labeled (a)–(d), as either a *disulfide bridge*, a *salt bridge*, *hydrogen bonding*, or *dispersion forces*. Also indicate the functional groups involved in each of the interactions.

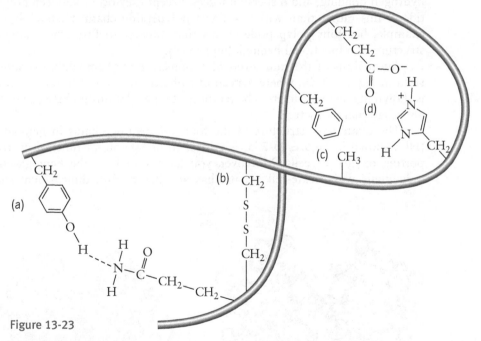

Figure 13-23

31 Using Figure 13-23, and assuming that the interactions are between amino acid side chains on a single polypeptide chain, what level of protein architecture (*primary, secondary, tertiary,* or *quaternary*) do these interactions represent? How can you tell?

Quaternary Structure

Many proteins are comprised of two or more polypeptide chains, giving rise to quaternary protein structure. *The quaternary structure of a protein describes the interactions of two or more polypeptide chains that create the overall three-dimensional shape of the functional protein.* The electrostatic

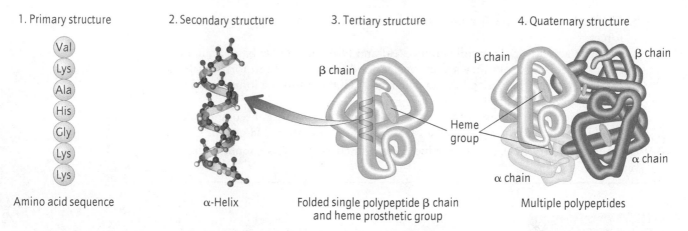

1. Primary structure	2. Secondary structure	3. Tertiary structure	4. Quaternary structure
Amino acid sequence	α-Helix	Folded single polypeptide β chain and heme prosthetic group	Multiple polypeptides

Figure 13-24 The four levels of protein architecture illustrated for hemoglobin. (1) The primary structure refers to the sequence of amino acids in each polypeptide (the two α-chains and the two β-chains). (2) The secondary structure refers to localized regular folding patterns of the polypeptide backbone: α-helices and β-pleated sheets, created from hydrogen bonding between amides in the polypeptide backbone (the eight α-helices). (3) Tertiary structure refers to the overall three-dimensional shape of a single polypeptide chain of a protein created from electrostatic interactions between often distant side chains. Tertiary structure also includes prosthetic groups (the heme group). (4) Quaternary structure refers to any electrostatic and covalent interactions between polypeptides in a protein composed of more than one polypeptide chain (the electrostatic and covalent interactions between the α-chains and β-chains).

and covalent interactions between polypeptide chains are of the same kind that stabilize the tertiary structure of a protein: disulfide bridges, salt bridges, hydrogen bonding, and dispersion forces, except they occur between polypeptide chains rather than within a single polypeptide chain. Hemoglobin, for example, has four polypeptide chains that interact to form the quaternary structure of a functional hemoglobin protein.

An analysis of the four levels of architecture for hemoglobin is summarized in Figure 13-24, where instead of ribbons, the overall tertiary structure is shown in schematic form to better illustrate the relationship between tertiary and quaternary structure.

The quaternary structure of the four polypeptide chains in hemoglobin (Hb), shown in Figure 13-25a, is critical to hemoglobin's function in transporting oxygen to cells. Moreover, you have seen how the hemoglobin in individuals with sickle-cell anemia has an altered three-dimensional shape.

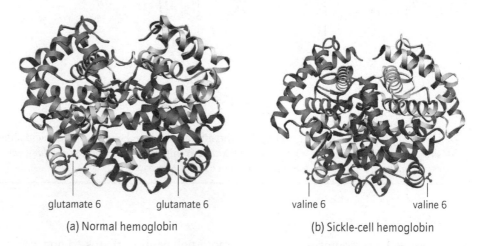

glutamate 6 glutamate 6	valine 6 valine 6
(a) Normal hemoglobin	(b) Sickle-cell hemoglobin

Figure 13-25 Ribbon diagrams showing the different overall shape of (a) normal hemoglobin (Hb), showing the glutamic acid side chains, and (b) sickle-cell hemoglobin (HbSc), showing the valine side chains. The α-chains are shown in orange and yellow, and the β-chains are shown in blue and light blue.

This change in quaternary structure results from a single amino acid substitution, valine, which has a nonpolar side chain, for glutamic acid, an amino acid with a polar acidic side chain, at the sixth amino acid of both β-chains. This substitution of two amino acids in a protein that contains 574 amino acids alters the way in which the protein folds, giving this type of hemoglobin (HbSc) a slightly different tertiary and quaternary structure, as illustrated in Figure 13-25b. Under conditions of low oxygen, blood cells with this variant form of hemoglobin (HbSc) take on a crescent shape rather than the biconcave shape of a normal red blood cell. Consequently, the cells aggregate and lose elasticity, clogging narrow capillaries.

PRACTICE EXERCISES

32 For the following pairs of amino acids, indicate how their side chains might interact to contribute to the quaternary structure of a protein. Choose from among the following choices: *disulfide bridge, salt bridge, hydrogen bonding,* or *dispersion forces.*
   a. Aspartic acid and histidine
   b. Serine and lysine
   c. Leucine and valine
   d. Two cysteines
   e. Tryptophan and isoleucine

33 Which level of protein architecture describes exclusively covalent bonds: *primary, secondary, tertiary,* or *quaternary*?

34 Why is there no quaternary protein structure in a protein composed of a single polypeptide?

Denaturation of a Protein

***Denaturation** of a protein occurs when there is a disruption of the secondary, tertiary, or quaternary structure of a protein so that the protein can no longer perform its function.* Denaturation of a protein does not alter its primary structure, and in some instances can even be reversed. For example, the protein egg albumin found in egg whites can be denatured by heating, as illustrated in **Figure 13-26**. The change from colorless to white observed when an

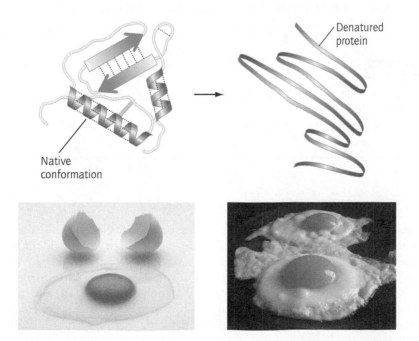

Native
conformation

Denatured
protein

Figure 13-26 Heating an egg denatures the protein egg albumin, irreversibly altering its secondary, tertiary, and quaternary structure. The covalent bonds of its primary structure remain intact. [Photos: left, DAJ/Punchstock; right, Vladimir Glazkov/iStockphoto/Thinkstock]

egg is cooked is the result of the protein being denatured. Cooking an egg is an example of an irreversible denaturation.

Proteins can be denatured in a number of ways, including:

- addition of heat,
- pH changes,
- mechanical agitation,
- addition of detergents, and
- addition of certain metals.

A denatured protein loses its shape and, hence, its biological activity as a result of the disruption in electrostatic interactions. Consider for example the effect of a change in pH on a salt bridge. We have seen that a salt bridge is an ionic interaction between the negatively charged carboxylate ion on the side chain of an acidic polar amino acid (Asp or Glu) and the positively charged conjugate acid of basic polar amino acid (Lys, Arg, or His). If the pH falls, as in the case of acidosis, the equilibrium shifts so that some of the carboxylate ions on the side chain become neutral carboxylic acids, which are then no longer able to form a salt bridge, as shown below left. If instead the pH rises, as in the case of alkalosis, the equilibrium shifts so that the conjugate acid of some of the amines lose a proton and become a neutral amine, which is also unable to form a salt bridge, as shown below right.

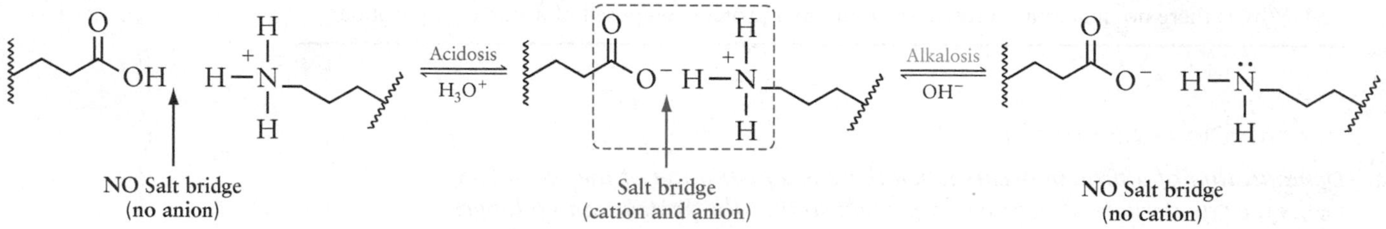

NO Salt bridge
(no anion)

Salt bridge
(cation and anion)

NO Salt bridge
(no cation)

Hydrogen bonding interactions are also disrupted by pH changes because while an amine is a good hydrogen bond acceptor, for example, the conjugate acid of an amine is not. Similarly, a carboxylic acid is a good hydrogen bond donor but a carboxylate ion (the conjugate base of a carboxylic acid) is not.

PRACTICE EXERCISES

35 List three methods for denaturing a protein. What change occurs to a protein when it is denatured? Is denaturation reversible?

36 Which level of protein architecture is unaffected by denaturation: *primary, secondary, tertiary,* or *quaternary structure*?

■ 13.4 Enzymes

Enzymes are essential for life because, in their absence, biochemical reactions are too slow. The importance of specific enzymes is particularly evident in the diseases and conditions caused by a missing or ineffective enzyme, which range from being benign (such as in lactose intolerance) to life threatening, such as Pompe's disease. Babies born with Pompe's disease usually die in their first year of life. Pompe's disease is caused by a deficiency in the enzyme *acid maltase* that results in severe muscle degeneration, affecting motor skills as well as heart and lung function.

How Do Enzymes Work?

In chapter 4 you learned that one way to increase the rate of a reaction is to add a *catalyst*. Nature's catalysts are specialized globular proteins known as enzymes, which increase the rate of biochemical reactions as much as 10^{16} times faster than in the absence of the enzyme. Generally, one enzyme catalyzes one reaction, acting specifically on one particular reactant, referred to as the **substrate**. Enzymes also ensure that the product of the enzyme catalyzed reaction is the correct stereoisomer, when a chiral product is formed.

Like all catalysts, the enzyme itself is chemically unaltered in a biochemical reaction; thus, the same enzyme can be used over and over again to catalyze the transformation of many substrate molecules into product molecules. In a chemical equation, enzymes are therefore not shown as a reactant. Instead, the name of the enzyme is sometimes written above the reaction arrow.

Enzymes are generally named after the substrate or the type of reaction they catalyze, with a change in the ending to "ase." Furthermore, enzyme names are usually italicized. For example, the enzyme *lactase* catalyzes the hydrolysis of lactose into glucose and galactose. Some enzymes have a common name that originates from where it was first isolated. For example, *papain* is an enzyme that was first isolated from the papaya fruit (Figure 13-27). In chapter 10 you learned that biochemists classify enzymes according to which of six general reaction types they catalyze, as summarized in Table 10-3. For example, *oxidoreductases* catalyze oxidation–reduction reactions.

The Enzyme-Substrate Complex Reaction rates for enzyme catalyzed reactions range anywhere from one substrate molecule converted into product every 2 seconds to as high as 10 million substrate molecules every second! To understand how enzymes can have such a profound impact on the rate of a reaction, let us consider how an enzyme and its substrate interact at the molecular level.

An enzyme catalyzed reaction begins with the reversible binding of the substrate (**S**) and the enzyme (**E**) to form an **enzyme-substrate complex** (**ES**), whereupon the reaction occurs, releasing the product, **P**, and the free enzyme, **E**:

$$\text{E} + \text{S} \rightleftharpoons \text{ES} \longrightarrow \text{P} + \text{E}$$

E = Enzyme
S = Substrate
ES = Enzyme-substrate complex
P = Product

In accordance with Le Châtelier's Principle, a greater substrate concentration shifts the equilibrium in the first step to the right. The substrate binds to the enzyme, which is usually much larger in size than the substrate, at a pocket or groove somewhere on or within the protein, known as the **active site** or **binding site**, as illustrated in Figure 13-28. The three-dimensional

Figure 13-27 The enzyme *papain* was first isolated from the papaya fruit. [Catherine Bausinger]

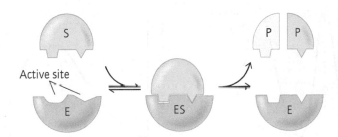

Figure 13-28 The induced-fit model of enzyme substrate binding. In the first step the substrate, S, and enzyme, E, bind reversibly to form the enzyme-substrate complex, ES. Upon binding substrate, the enzyme undergoes conformational changes to attain a better fit between the enzyme and the substrate. After reaction, the enzyme releases the product, P, and regenerating the free enzyme, which is ready to repeat the process with yet another substrate molecule.

> The vast majority of enzymes are proteins, but a few are RNA molecules—known as **ribozymes**.

shape of the active site is roughly complementary to the three-dimensional shape of the substrate, which is the reason we observe such selectivity of the enzyme for the substrate. Substrate selectivity is extremely important in biochemistry because there are so many other organic molecules present in solution at the same time. Within the active site of the enzyme, side chains bind the substrate through electrostatic interactions. Earlier scientific models of enzyme-substrate binding likened the complementary shape of the substrate and the enzyme to that of a lock and a key, aptly known as the lock-and-key model. A more refined model of how an enzyme binds to its substrate describes a less complementary shape, but instead describes a conformational change in the enzyme following binding that improves the "fit" between the substrate and the enzyme, known as the **induced-fit model**. Thus, the enzyme appears to mold itself—induced fit—to the shape of the substrate, as illustrated in the ES complex in Figure 13-28.

While it is part of the enzyme-substrate complex (ES), the substrate and other reactants assume a position and orientation that places the reacting functional group(s) near one another, thereby facilitating the chemical reaction. *It is the placement of the substrate in an optimal geometry for reaction that lowers the energy of activation (E_A) for the reaction, and thus increases the rate of the reaction.* For example, in a hydrolysis reaction, the enzyme brings the substrate and a molecule of water in close proximity and in the proper orientation for bond breaking and making to occur. When the enzyme catalyzed reaction is over, the enzyme releases the product, **P**, and the unbound enzyme, **E**, is free again to bind another substrate molecule, repeating the process. The first of these steps is reversible while the second step, the conversion of the ES complex to product, P, and free enzyme, E, is irreversible.

pH and Temperature Dependence of Enzymes

In order for an enzyme to perform its function, it must be in its native conformation (see Figure 13-26), which is dependent on both temperature and pH. Changes in pH alter the charges on side chains in acidic and basic amino acids of the enzyme, which in turn, affect its shape. For most enzymes, physiological pH (pH = 7.3) is optimal, although some enzymes, such as those found in the stomach, like *pepsin*, function best at low pH (pH = 1 − 2.5) as shown in Figure 13-29a.

Acidosis, described in chapter 9, is a serious medical condition caused by a drop in blood pH that, among other physiological effects, causes enzymes to denature. Some poisons also inflict their damage through pH changes that denature enzymes. Because a denatured enzyme can no longer catalyze reactions, important biochemical pathways are therefore blocked.

Figure 13-29 (a) Enzyme activity as a function of pH for two representative enzymes. *Pepsin* works best at low pH while *salivary amylase*, like most enzymes, works best at physiological pH. (b) Enzyme activity as a function of temperature. The optimal temperature is normal body temperature 37 °C. Denaturation begins to occur at temperatures above 45–50 °C.

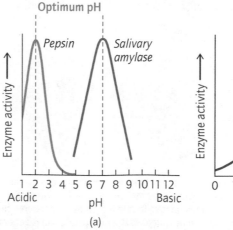
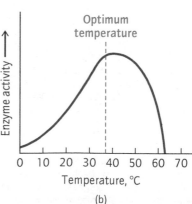

Most enzymes also function best around normal body temperature, 37 °C. Elevated body temperatures, as occurs with a high fever, can denature enzymes, causing reaction rates to decrease (Figure 13-29b). This temperature effect is the opposite of what is observed in reactions carried out in the laboratory, which increase in rate as the temperature is increased (chapter 4). Lower body temperatures also cause enzyme function to decrease and reaction rates to decrease. Consider the case of someone who has fallen into extremely cold waters. The lower body temperature of the victim brought on by the cold water causes all metabolic enzymes to slow. One beneficial effect is that it decreases the person's requirement for oxygen (needed for cellular respiration), thus preserving brain function and allowing the individual to survive longer without brain damage.

WORKED EXERCISE | Enzymes |

13-10 Which of the interactions listed below are involved in binding an enzyme and its substrate?

  a. Disulfide bonds
  b. Hydrogen bonding
  c. Dispersion forces
  d. Salt bridges

Solution

13-10 The interactions involved in binding a substrate and an enzyme include (b) hydrogen bonding, (c) dispersion forces, and (d) salt bridges.

PRACTICE EXERCISES

37 Explain the *induced-fit* model for an enzyme, E, binding to its substrate, S. What does "induced fit" mean?

38 What type of protein is an enzyme? Why are enzymes required for practically all biochemical reactions?

39 What is the *active site* of an enzyme?

40 What is the ES complex?

41 To achieve its maximum catalytic power, what pH is needed for a reaction catalyzed by *pepsin*, an enzyme found in the stomach? Use the graph shown in Figure 13-29a.

42 How is the rate of an enzyme catalyzed reaction affected by the following changes?

  a. An increase in pH
  b. A decrease in pH
  c. A decrease in temperature
  d. An increase in temperature

Enzyme Inhibitors

Enzyme inhibitors are compounds that bind to an enzyme and prevent the enzyme from performing its function. Many poisons and some pharmaceutical drugs work as enzyme inhibitors. *Specific* enzyme inhibitors target and inhibit a specific enzyme without affecting any other enzymes. There are two basic types of specific enzyme inhibitors: competitive inhibitors and noncompetitive inhibitors, illustrated in Figure 13-30.

Competitive Inhibitors *A competitive inhibitor is a substance that competes with the substrate for the active site of the enzyme, because it too has a structure that fits into the active site and can bind to the enzyme* (Figure 13-30b).

Antibiotics and heavy metals work as irreversible enzyme inhibitors. They bind to the enzyme irreversibly, preventing the release of the free enzyme, making the enzyme unavailable to catalyze the reaction, so the substrate cannot be converted into product.

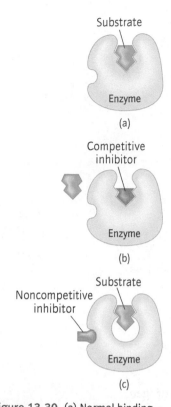

Substrate

Enzyme

(a)

Competitive
inhibitor

Enzyme

(b)

Noncompetitive
inhibitor

Substrate

Enzyme

(c)

Figure 13-30 (a) Normal binding of enzyme and substrate leading to product formation. (b) A competitive inhibitor prevents the substrate from binding to the enzyme by occupying the enzyme binding site, thus preventing product formation. (c) A noncompetitive inhibitor binds at a site on the enzyme other than the active site, causing a change in the conformation of the active site, thereby preventing the substrate from being converted to product, even though substrate may bind to the enzyme.

By binding to the active site, a competitive inhibitor blocks the substrate from binding to the enzyme. As a result, the ES complex cannot form, and no reaction occurs. Enzyme inhibition of one step in a biochemical pathway blocks all subsequent reactions because no product is produced to serve as the substrate for the next reaction in the biochemical pathway. Enzyme inhibition can have beneficial effects, and indeed, this is how many drugs work.

The effectiveness of a competitive inhibitor depends on the relative concentration of substrate and inhibitor in solution. Like the game "musical chairs," both substrate and inhibitor compete for the same active site (the chair) but there is only room for one (someone is left standing in musical chairs). Thus, the higher the concentration of inhibitor, the less likely the substrate will find a free enzyme, E, to bind to. Conversely, the higher the concentration of substrate, S, the more likely the ES complex will form, and therefore product, P, is produced. This is why drugs that function as enzyme inhibitors are given at regular intervals so as to maintain the concentration of inhibitor in solution. To see how a specific enzyme inhibitor is used to treat an enzyme-based disease, read Chemistry in Medicine: ACE Inhibitors as a Treatment for Hypertension.

Noncompetitive Inhibitors *Noncompetitive inhibitors bind to the enzyme at a location other than the active site (Figure 13-30c).* Binding of a noncompetitive inhibitor affects the conformation of the enzyme, preventing it from properly binding the substrate (although the substrate can still bind) and thus preventing the formation of product. In marked contrast to competitive inhibitors, increasing the concentration of substrate will not restore enzyme activity when a noncompetitive inhibitor is bound to the enzyme. Enzyme activity is restored only when the concentration of noncompetitive inhibitor decreases.

Consider the multistep biochemical pathway for cholesterol synthesis shown below. Cholesterol is an endogenous (naturally found in the body) noncompetitive inhibitor of the enzyme *HMG-CoA Reductase* required for its own synthesis. *HMG-CoA Reductase* catalyzes the second step in the biochemical pathway that produces cholesterol.

HMG-CoA $\xrightarrow[\text{reductase}]{\text{HMG-CoA}}$ Mevalonate $\rightarrow\rightarrow\rightarrow\rightarrow\rightarrow$ Cholesterol (noncompetitive inhibitor)

Feedback inhibition

Thus, cholesterol inhibits its own synthesis when its concentration is high, but enzyme activity is restored when cholesterol concentration levels decrease. In this way the cell controls its production of cholesterol—a process known as enzyme **feedback inhibition**. As you know, too much cholesterol can have negative health effects, but so can too little. This is true of many biomolecules. Biochemical pathways are often regulated by noncompetitive inhibition of a key step in the biosynthesis of a particular biomolecule. In this way enzymes serve an important regulatory role in addition to their function as reaction catalysts.

PRACTICE EXERCISES

43 Which type of enzyme inhibitor binds to the active site of an enzyme: a *competitive inhibitor* or a *noncompetitive inhibitor*?

44 For what type of inhibition does an increase in the concentration of substrate not affect enzyme activity: *competitive* or *noncompetitive* inhibition?

Chemistry in Medicine ACE Inhibitors as a Treatment for Hypertension

Enzyme inhibitors have been used as pharmaceuticals in the treatment of a wide range of diseases including cancer, AIDS, diabetes, and heart disease. Some of the earliest enzyme inhibitors were used before their mechanism of action was even understood. Aspirin, for example, was used to treat pain and inflammation long before it was discovered in the 1970s to be an inhibitor of *cyclooxygenase*, a key enzyme involved in several biochemical pathways leading to inflammation.

In this section we consider a class of enzyme inhibitors used to treat chronically elevated blood pressure, a condition known as **hypertension.** Hypertension is a major risk factor in stroke, heart attack, heart failure, and kidney failure. More than half of Americans over age 65 suffer from hypertension.

How the Body Raises Blood Pressure

The body carefully controls blood pressure so that it neither falls too low nor rises too high. Regulation of blood pressure is a complex process involving several biological factors. The concentration of sodium ions (Na^+) in the blood plays an important role in blood pressure, because sodium ions cause water to be retained in the circulatory system through osmosis, which increases blood pressure. Indeed, people with hypertension are advised to eat a low sodium diet. When the kidneys detect a drop in sodium ion, Na^+, concentration they release an enzyme called *renin*. *Renin* acts on a single substrate: angiotensinogen, a polypeptide with 452 amino acids, produced by the liver. *Renin* catalyzes the hydrolysis of the peptide bond between leucine and valine in angiotensinogen to produce the hormone angiotensin I, a polypeptide with 10 amino acids (see **Figure 13-31**).

The next step in this biochemical pathway is a reaction that converts angiotensin I into angiotensin II, catalyzed by *Angiotensin Converting Enzyme* (*ACE*), an enzyme produced in the lungs. *ACE* catalyzes the hydrolysis of the peptide bond between phenylalanine and histidine in

angiotensin I, to produce the hormone angiotensin II, an octapeptide, and the dipeptide His-Leu.

Angiotensin II signals a number of physiological events that raise blood pressure: (1) It causes the muscles surrounding blood vessels to constrict—a process known as **vasoconstriction,** and (2) it triggers the release of the steroid hormone aldosterone from the adrenal cortex, which signals the kidneys to reabsorb sodium ions rather than excreting them through the urine. The renin-angiotensin-aldosterone system, summarized in **Figure 13-32**, shows the central role of the peptide hormone angiotension II in the physiological events that lead to an increase in blood pressure.

Developing a Treatment for Hypertension

Once scientists discovered how the renin-angiotensin-aldosterone pathway is involved in blood pressure, it was hypothesized that a competitive inhibitor of *ACE*—a key enzyme in the second step of the pathway shown in Figure 13-32—might offer a viable treatment for high blood pressure. An *ACE* inhibitor is a molecule that binds to *ACE* and prevents angiotensin I, the natural substrate, from binding to *ACE*, thus blocking the formation of angiotensin II. In the absence of angiotensin II, blood pressure decreases.

One of the first substances shown to be a competitive inhibitor of *ACE* was a polypeptide isolated from the venom of the poisonous pit viper (**Figure 13-33**). Snake venom is well known for containing polypeptides that interfere with blood coagulation and other blood regulation mechanisms. However, since enzymes in the stomach hydrolyze polypeptides, peptide analogs of snake venom are not effective as orally administered drugs.

Instead, compounds that are structurally similar to snake venom but that are not peptides were synthesized and tested. The first effective synthetic *ACE* inhibitor to be used as a drug was captopril, whose structure is shown in Figure 13-33. The sulfur atom in captopril binds to the

Asp-Arg-Val-Tyr-IIe-His-Pro-Phe-His-Leu-Val-Tyr-Ser-Protein
(Angiotensinogen)

Renin

Asp-Arg-Val-Tyr-IIe-His-Pro-Phe-His-Leu + Val-Tyr-Ser-Protein
(Angiotensin I)

Angiotensin Converting Enzyme
(ACE)

Asp-Arg-Val-Tyr-IIe-His-Pro-Phe + His-Leu
(Angiotensin II)

Figure 13-31 The key reactions and enzymes in the renin-angiotensin biochemical pathway that controls blood pressure.

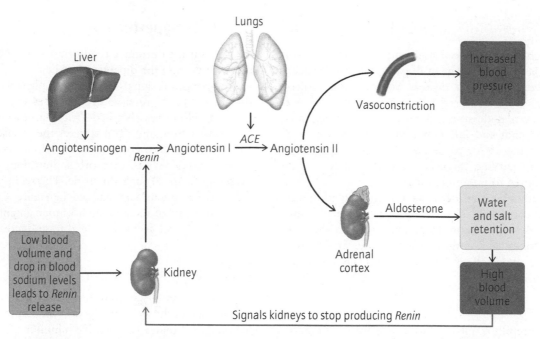

Figure 13-32 The reactions, enzymes, and organs involved in the regulation of blood pressure by the renin-angiotensin-aldosterone system. Blood pressure depends on water and salt retention as well as blood volume.

zinc atom, Zn^{2+}, a cofactor at the active site of *ACE*, to form a strong electrostatic interaction.

Since the introduction of captopril, many new and improved *ACE* inhibitors have been developed for the treatment of hypertension. These *ACE* inhibitors have fewer side effects than captopril today *ACE* inhibitors are some of the most widely prescribed drugs on the market.

Once again we see how understanding a biochemical process has led to the development of treatments for life-threatening conditions. As scientists continue to gain a better understanding of the biochemical reactions that occur in our cells, you will continue to see advances in the field of medicine, and yet another reason for you to learn your chemistry basics!

Figure 13-33 The structure of Captopril, the first effective *ACE* inhibitor produced in the laboratory for the treatment of hypertension, was designed to resemble the structure of the peptides found in the venom of the poisonous Brazilian pit viper. [© blickwinkel/Alamy]

Summary

Amino Acids

- Proteins are the most abundant and have the most diverse functions of all the biomolecules.

- Proteins are constructed from the 20 natural *L*-amino acids.

- An amino acid contains an amine and a carboxylic acid covalently bonded to a tetrahedral carbon atom known as the α-carbon. The α-carbon also contains a hydrogen atom and the side chain, R, which determines the identity of the amino acid.

- The charge on the amine and the carboxylic acid functional groups in an amino acid depend on the pH of the solution. At physiological pH an amino acid exists as a zwitterion, a —COO$^-$ and a —NH$_3$$^+$ together in the same molecule creating a zero net charge.

- The zwitterion is in equilibrium with two other pH dependent forms of the amino acid. At high pH, the free amine and carboxylate exist to create a negatively charged ion. At low pH, the carboxylic acid and the conjugate acid of the amine exist to create a positively charged ion.

- The 20 natural amino acids are classified as either nonpolar or polar depending on their side chains. The polar amino acids can be further subdivided into acidic, basic, and neutral side chains.

- Eleven of the 20 amino acids can be synthesized from various intermediates of metabolism. The other 9 amino acids are known as essential amino acids and need to be supplied through the diet.

- Nineteen of the amino acids are chiral. Glycine is achiral.

- The α-carbon of an amino acid is a center of chirality because it has four bonds to four different atoms or groups of atoms.

- The chiral amino acids found in nature have the L-configuration: the amine is on the left and the H is on the right, when represented as a Fischer projection.

Peptides

- Peptides are molecules derived from two or more amino acids. The amino acids are joined by peptide bonds, an amide functional group formed when the amine on one amino acid reacts with the carboxylic acid of another amino acid in an amidation reaction. A molecule of water is also produced in the reaction.

- An amino acid is capable of forming two peptide bonds, one with its amine functional group and one with its carboxylic acid functional group.

- Peptides contain repeating amide functional groups, except the amino acid at the beginning of the polypeptide which contains a free amine, known as the N-terminus, and the amino acid at the other end of the polypeptide where there is a free carboxylate, known as the C-terminus.

- When writing the sequence of amino acids for a polypeptide, the convention is to begin with the N-terminus and end with the C-terminus. Typically the three letter abbreviations for the amino acids are used.

- Proteins are large polypeptides or several polypeptides folded into their functional three-dimensional shape.

- Some small polypeptides are hormones such as the endorphins and oxytocin.

Types of Proteins and Protein Architecture

- Proteins are composed of anywhere from about 50 to 1,000's of amino acids.

- All proteins are built from the 20 L-amino acids.

- In order for a protein to perform its unique function, the polypeptide(s) must fold into its specific three-dimensional shape.

- There are three general classes of proteins: fibrous proteins (keratins, elastins, and collagens), globular proteins, and membrane proteins.

- The four levels of protein architecture are defined as primary structure, secondary structure, tertiary structure, and quaternary structure.
- The primary structure of a protein is the amino acid sequence of its polypeptide chain(s).
- Secondary protein structure describes regular folding patterns in local regions of the polypeptide backbone, stabilized by hydrogen bonds. The α-helix and the β-pleated sheet are the two most common types of secondary structure.
- Secondary protein structure arises from hydrogen bonding between N—H and C=O groups in the polypeptide backbone.
- The tertiary structure of a protein describes the complex three-dimensional shape that results from the folding of the polypeptide over and above its secondary structure.
- Tertiary structure arises from electrostatic interactions between the side chains in the polypeptide as well as hydrophilic and hydrophobic interactions with the surrounding environment.
- The electrostatic interactions and covalent bonds between amino acid side chains responsible for the tertiary structure of a protein are disulfide bridges, salt bridges, hydrogen bonding, and dispersion forces.
- For some proteins, tertiary structure also includes prosthetic groups: essential non-peptide containing organic molecules or metal ions that form ionic or covalent bonds to the polypeptide.
- The quaternary structure of a protein describes interactions of two or more polypeptide chains that create the overall three-dimensional shape of the functional protein.
- Denaturation of a protein is the disruption of the secondary, tertiary, and quaternary structure of a protein so that it can no longer perform its function.
- Proteins can be denatured with heat, pH changes, mechanical agitation, detergents, and certain metals.
- When a protein is denatured, it loses its shape because electrostatic interactions are broken, such as hydrogen bonding, salt-bridges, and hydrophobic interactions.

Enzymes

- Enzymes are globular proteins that catalyze biochemical reactions. Enzymes are specific for one particular substrate.
- Enzymes work by lowering the energy of activation, E_A, for a reaction.
- An enzyme catalyzed reaction begins with the reversible binding of the substrate (S) to the enzyme (E) to form the enzyme-substrate complex (ES), whereupon the reaction occurs and the free enzyme, E, and the product, P, are released.
- The substrate binds to the enzyme at a pocket or cleft known as the active site or binding site. The shape of the active site is roughly complementary to the shape of the substrate.
- Upon binding substrate, the enzyme makes conformational changes that create a better fit between the substrate and enzyme, known as the "induced-fit model."
- The enzyme-substrate complex places the substrate in an optimal geometry for reaction, which lowers the energy of activation, and thus increases the rate of the reaction.
- Changes in pH or an increase in temperature can alter the shape of an enzyme so that it can no longer perform its function. The pH of solution

affects the charge on acidic and basic side chains so that salt bridges and hydrogen bonding are disrupted.

- Enzyme inhibitors are compounds that bind to the enzyme and prevent the substrate from being converted into product.

- There are two basic types of specific enzyme inhibitors: competitive and noncompetitive inhibitors.

- A competitive inhibitor competes with the substrate for the active site of the enzyme because it also has a structure that is complementary to the active site. The relative concentration of inhibitor and substrate affects enzyme activity.

- Noncompetitive inhibitors bind at a location on the enzyme other than the active site and cause a conformational change in the enzyme that prevents the substrate from being converted to product. Substrate concentration does not affect enzyme activity; only the concentration of the noncompetitive inhibitor affects enzyme activity.

Key Words

α-Carbon The tetrahedral carbon in an amino acid to which the amine, the carboxylic acid, a hydrogen atom, and the R group are attached. It is also a center of chirality in all the natural amino acids except glycine.

α-Helix One of the two major types of protein secondary structure, characterized by a helical shape formed as a result of hydrogen bonding between amides in the polypeptide backbone 3.5 amino acids apart.

Active site The pocket or cleft within an enzyme where the substrate binds and a biochemical reaction takes place. Also known as a binding site.

Amino acid The molecules from which all proteins are formed. They contain both an amine functional group and a carboxylic acid functional group bonded to the same tetrahedral carbon, the α-carbon. Each also has a unique R group that defines it.

β-pleated sheet A type of protein secondary structure in which two or more sections of parallel or antiparallel polypeptides fold in a pattern that looks like a pleated skirt. Hydrogen bonding between amides on adjacent strands stabilize this type of secondary structure.

Binding site See active site.

Coenzyme Essential non-peptide organic molecule that forms ionic or covalent bonds to an enzyme. A type of prosthetic group in the tertiary structure of a protein.

Cofactor A metal ion or organic molecule essential for certain enzymes to achieve their catalytic effect. A type of prosthetic group in the tertiary structure of a protein.

Collagens A type of fibrous protein that serves as the main structural proteins in the body, providing rigidity to connective tissue.

Competitive inhibitor A substance that competes with the substrate for the active site of an enzyme and prevents the substrate from being converted into product.

C-terminus The end of a polypeptide that contains the free carboxylate group.

Denaturation The disruption of secondary, tertiary, and quaternary structure in a protein so that it can no longer perform its function. Denaturation is caused by heat, pH changes, mechanical agitation, detergents, and certain metals.

Dipeptide A peptide derived from two amino acids in an amidation reaction between the carboxylic acid of one amino acid and the amine of the other amino acid.

Disulfide bridge The disulfide bond, R—S—S—R, a covalent bond formed upon oxidation of two thiol functional groups, such as the side chains in two cysteines.

Elastins A type of fibrous protein with elastic fibers that provide flexibility to skin, ligaments, and other tissues that require elasticity.

Enzyme Globular proteins that catalyze biochemical reactions.

Enzyme-substrate complex, ES The complex formed when the substrate is bound to its enzyme.

Essential amino acids Amino acids that must be supplied through the diet because they cannot be synthesized in the body. They include histidine, isoleucine, leucine, lysine, methionine, phenylalanine, threonine, tryptophan, and valine.

Feedback inhibition A type of noncompetitive inhibition where the product of one step in a biochemical pathway is a competitive inhibitor of a step earlier in the biochemical pathway, thereby regulating its own production.

Fibrous protein Strong and water insoluble proteins that generally have a structural role in nature. Includes keratins, elastins, and collagens.

Globular proteins Water-soluble polypeptides that fold into an overall spherical shape. Includes most enzymes.

Induced-fit model The current model of how an enzyme and its substrate bind that involves a conformational change in the enzyme upon binding substrate, which creates a more complementary fit between the enzyme and substrate.

Keratins Fibrous proteins that provide structure to hair, feathers, and other tissues.

Membrane proteins Proteins that are located in part of or entirely within the cell membrane, a hydrophobic environment.

Native conformation The three-dimensional shape of a functional protein.

N-terminus The one end of a polypeptide that contains the free amine.

Noncompetitive inhibitor A substance that binds to an enzyme at a location other than the active site preventing the substrate from being converted to product.

Peptide A molecule composed of two or more amino acids joined by peptide bonds.

Peptide bond The carbon-nitrogen bond of an amide, formed between the carboxylic acid of one amino acid and the amine of another amino acid.

Polypeptide A peptide derived from many amino acids joined by peptide bonds. Polypeptides with more than 50 amino acids are classified as proteins. Many proteins contain more than one polypeptide chain.

Polypeptide backbone The repeating N—C_α—C=O atoms common to all polypeptides. A reference to everything in the polypeptide except the side chains.

Primary structure The amino acid sequence of the polypeptide(s) that makeup a protein.

Prion A proteinaceous infection agent; an infectious abnormal protein. It is believed to be the causative agent in mad cow disease.

Prosthetic groups Non-peptide organic molecules and metal ions that are strongly bound to a protein and essential to its function.

Protein One of the four major biomolecules. A polymer of amino acids. It can be either a large polypeptide or multiple polypeptides, which fold into a unique three-dimensional shape.

Quaternary structure Created by electrostatic interactions between multiple polypeptides in a protein containing more than one polypeptide.

Ribbon drawings A diagram of a protein wherein the polypeptide backbone is depicted as a flat colored ribbon for the purpose of showing secondary structure and the three-dimensional shape of the protein.

Salt bridge Ionic bonds formed between charged side chains in a polypeptide or between side chains on different polypeptides.

Secondary structure The sections of α-helix and β-pleated sheet within a protein formed by hydrogen bonding between parts of the amide functional groups in the polypeptide backbone.

Specific enzyme inhibitors Molecules that bind to a specific enzyme and thereby prevent the enzyme catalyzed reaction from occurring. Includes competitive and noncompetitive inhibitors.

Side chain The unique R group bonded to the α-carbon of an amino acid.

Substrate The main reactant in an enzyme-catalyzed biochemical reaction.

Tertiary structure The complex folding of a polypeptide beyond its secondary structure, showing the overall three-dimensional shape of a single polypeptide.

Thiol A functional group that contains an —S—H group, as seen in the amino acid cysteine.

Tripeptide A peptide derived from three amino acids joined by two peptide bonds.

Additional Exercises

Sickle-cell Anemia: One Wrong Amino Acid

45 What is sickle-cell anemia?

46 How does the sickle shape of red blood cells in sickle-cell anemia cause anemia?

47 What are symptoms of sickle-cell anemia?

48 How many polypeptide chains comprise hemoglobin? What is the total number of amino acids in hemoglobin?

49 What is the difference between the amino acid sequence in the polypeptides that makeup normal hemoglobin compared to the hemoglobin of individuals with sickle-cell anemia?

50 What is the structural difference between valine and glutamic acid? What type of compounds are valine and glutamic acid?

51 What are common treatments for the symptoms of sickle-cell anemia?

Amino Acids

52 What two functional groups are present in every amino acid?

53 Write the structure for each the following amino acids at physiological pH and provide its three letter abbreviation.
 a. tyrosine   b. cysteine

54 Write the structure for each of the following amino acids at physiological pH and provide the three letter abbreviation.
 a. threonine   b. phenylalanine

55 Write the structure of leucine at the following pH values:

 a. pH = 1
 b. physiological pH
 c. pH = 10

56 Write the structure of valine at the following pH values:
 a. pH = 1
 b. physiological pH
 c. pH = 10

57 What is the charge on phenylalanine at the following pH values?
 a. pH = 1
 b. physiological pH
 c. pH = 10

58 What is the charge on methionine at the following pH values?
 a. pH = 1
 b. physiological pH
 c. pH = 10

59 Leucine undergoes the following reversible reaction at pH = 2:

a. What substances are present at equilibrium?
b. If more acid is added, in which direction does the equilibrium shift?

60 Serine undergoes the following reversible reaction at pH = 10:

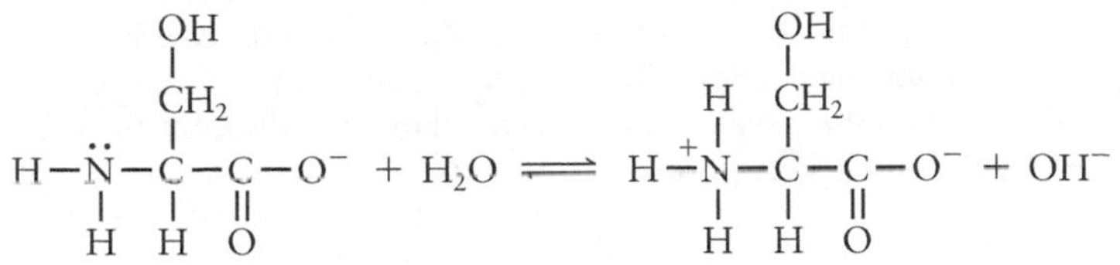

$$
\begin{array}{c}
\text{OH} \\
|\\
\text{CH}_2\\
|\\
\text{H}-\ddot{\text{N}}-\text{C}-\text{C}-\text{O}^- + \text{H}_2\text{O} \rightleftharpoons \\
|\quad|\quad||\\
\text{H}\;\;\text{H}\;\;\text{O}
\end{array}
\quad
\begin{array}{c}
\text{OH}\\
|\\
\text{H}\;\;\text{CH}_2\\
|\quad|\\
\text{H}-\overset{+}{\text{N}}-\text{C}-\text{C}-\text{O}^- + \text{OH}^-\\
|\quad|\quad||\\
\text{H}\;\;\text{H}\;\;\text{O}
\end{array}
$$

a. What substances are present at equilibrium?
b. If more hydroxide ion is added, in which direction does the equilibrium shift?

61 Provide an example of two amino acids that have each of the following characteristics:
a. nonpolar
b. polar basic
c. polar acidic
d. polar neutral

62 Glutamic acid is a polar acidic amino acid. Write the structure of glutamic acid at the following pH values:
a. pH = 1
b. physiological pH

63 Lysine is a polar basic amino acid. Write the structure of lysine at the following pH values:
a. pH = 10
b. physiological pH

64 Identify the following amino acids as *nonpolar, polar basic, polar acidic,* or *polar neutral.*
a. aspartic acid
b. lysine
c. cysteine
d. proline
e. serine
f. tryptophan

65 What does it mean if an amino acid is an essential amino acid?

66 List all the amino acids that have the following functional groups in their side chain:
a. an alcohol   **b.** an amine

67 List all the amino acids that have the following functional groups in their side chain:
a. a carboxylic acid   **b.** an amide

68 Which amino acid contains a thiol group? What is a thiol functional group? What functional group does a thiol most closely resemble?

69 Write the structure of cysteine. What functional group does cysteine contain in its side chain?

70 One of the 20 natural amino acids is shown below.

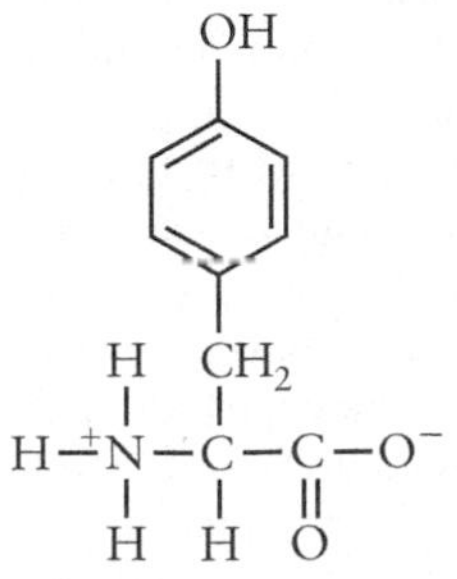

$$
\begin{array}{c}
\text{OH}\\
|\\
\bigcirc\\
|\\
\text{H}\;\;\text{CH}_2\\
|\quad|\\
\text{H}-\overset{+}{\text{N}}-\text{C}-\text{C}-\text{O}^-\\
|\quad|\quad||\\
\text{H}\;\;\text{H}\;\;\text{O}
\end{array}
$$

a. Circle and label the amine functional group. Is it in its neutral or ionized form?

b. Circle and label the carboxylate functional group. How does this group differ from a carboxylic acid?
c. Circle and label the side chain. What is the name of this amino acid?
d. Is this amino acid nonpolar or polar? Explain.
e. Write the structure of this amino acid at pH = 12. What is the net charge on the amino acid at this pH?

71 One of the 20 natural amino acids is shown below.

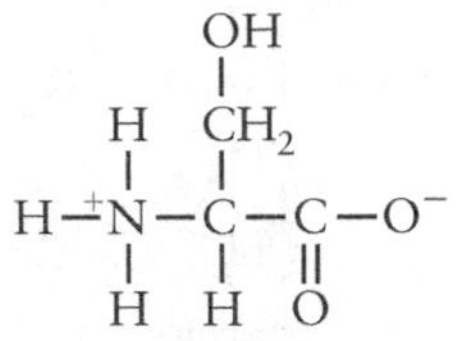

$$
\begin{array}{c}
\text{OH}\\
|\\
\text{H}\;\;\text{CH}_2\\
|\quad|\\
\text{H}-\overset{+}{\text{N}}-\text{C}-\text{C}-\text{O}^-\\
|\quad|\quad||\\
\text{H}\;\;\text{H}\;\;\text{O}
\end{array}
$$

a. Circle and label the amine functional group. Is it in its neutral or ionized form?
b. Circle and label the carboxylate functional group. How does this group differ from a carboxylic acid?
c. Circle and label the side chain. What is the name of this amino acid?
d. Is this amino acid nonpolar or polar? Explain.
e. Write the structure of this amino acid at pH = 12. What is the net charge on the amino acid at this pH?

72 For each of the following amino acids, place an X in the boxes that apply.

Amino Acid	Nonpolar	Polar	Acidic	Basic	Neutral	Essential Amino Acid
Threonine						
Histidine						
Isoleucine						

73 For each of the following amino acids, place an X in the boxes that apply.

Amino Acid	Nonpolar	Polar	Acidic	Basic	Neutral	Essential Amino Acid
Phe						
Asn						
Met						

74 Does plain rice contain all the essential amino acids?

75 How do vegetarians ensure that their diets contain "complete protein?"

76 Does chicken contain all the essential amino acids?

77 The Fischer projection of one of the natural amino acids is shown below.

$$
\begin{array}{c}
\text{COO}^-\\
|\\
\text{H}_3\overset{+}{\text{N}}-\!\!\!\!-\text{H}\\
|\\
\text{CH}_2\\
|\\
\text{CH}_2\\
|\\
\text{S}\\
|\\
\text{CH}_3
\end{array}
$$

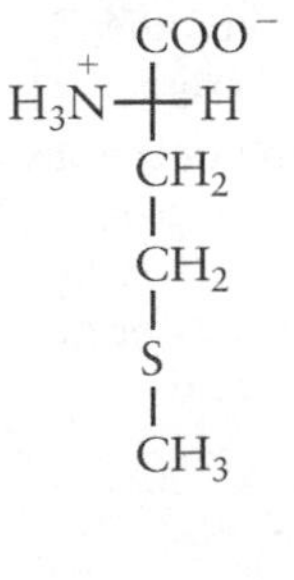

a. Which amino acid is this?

b. Is this a *D*- or *L*-amino acid? Explain.

c. Place an arrow pointing to the α-carbon.

d. Is this molecule chiral? Explain.

e. Write the Fischer projection of the enantiomer of this amino acid. How would it be named?

78 The Fischer projection of an amino acid is shown below.

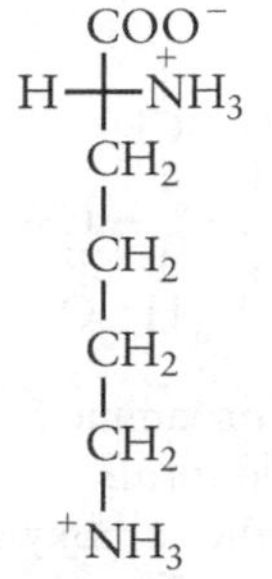

$$
\begin{array}{c}
COO^- \\
| \\
H\!-\!\overset{+}{N}H_3 \\
| \\
CH_2 \\
| \\
CH_2 \\
| \\
CH_2 \\
| \\
CH_2 \\
| \\
{}^+NH_3
\end{array}
$$

a. What amino acid is this?

b. Is this a *D*- or *L*- amino acid? Explain.

c. Place an arrow pointing to the α-carbon.

d. Is this molecule chiral? Explain.

e. Write the Fischer projection of the enantiomer of this amino acid. How would it be named?

79 Write the Fischer projection of *L*-threonine at physiological pH and answer the following questions.

a. Is this amino acid found in nature?

b. Identify the α-carbon in this Fischer projection.

c. How can you tell whether an amino acid is *D* or *L*?

d. Write the structure of *D*-threonine. How is it different from *L*-threonine? What does it have in common with *L*-threonine?

80 Write the Fischer projection for *L*-aspartic acid at physiological pH and answer the following questions.

a. Is this amino acid found in nature?

b. Identify the α-carbon in this Fischer projection.

c. How can you tell whether an amino acid is *D* or *L*?

d. Write the structure for *D*-aspartic acid. How is it different from *L*-aspartic acid? What does it have in common with *L*-aspartic acid? What type of stereoisomers are *L*- and *D*-aspartic acid?

81 Answer the following questions for glutamic acid.

a. What four atoms or groups of atoms are bonded to the α-carbon?

b. Which of these atoms or groups of atoms are present in all amino acids?

c. Which of these atoms or groups of atoms are unique to glutamic acid?

d. There are only two possible configurations for the atoms or groups of atoms bonded to the α-carbon of any amino acid. Show this by writing the Fischer projections for *D*- and *L*-glutamic acid. Are they enantiomers or diastereomers? Indicate which amino acid is produced in nature.

82 Answer the following questions for phenylalanine.

a. What four atoms or groups of atoms are bonded to the α-carbon?

b. Which of these atoms or groups of atoms are present in all amino acids?

c. Which of these atoms or groups of atoms are unique to lysine?

d. There are only two possible configurations for the atoms or groups of atoms around the α-carbon of any amino acid. Show this by writing the Fischer projections for *D*- and *L*-phenylalanine. Indicate which amino acid is produced in nature.

Peptides

83 What is the difference between a dipeptide, a tripeptide, and a polypeptide?

84 Write the structure of the following dipeptides:

a. Ala-Met   b. Asp-Lys

c. Thr-Cys   d. Ser-Gly

85 Write the structure of the following dipeptides:

a. His-Leu   b. Val-Pro

c. Glu-Arg   d. Asn-Ile

86 Write the structure of the tripeptide Leu-Glu-His. Label the *N*- and the *C*-terminus.

87 Write the structure of the tripeptide Tyr-Ala-Cys. Label the *N*- and the *C*-terminus.

88 Palmitoyl-pentapeptide-3 is a component of antiaging creams. It contains the pentapeptide portion shown below in an amide bond to palmitic acid, not shown.

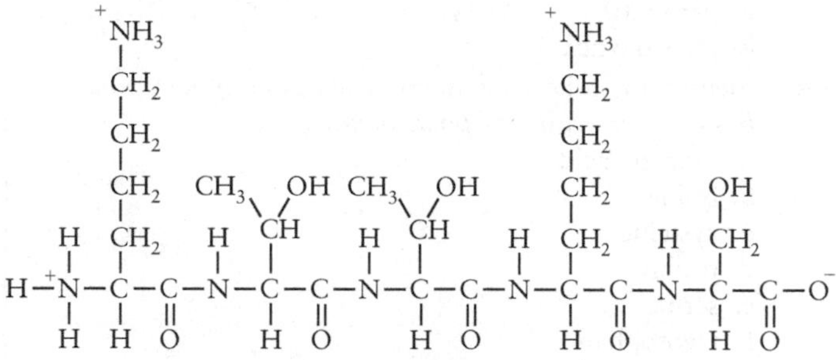

a. What is the amino acid sequence for the peptide portion of palmitoyl-pentapeptide-3?

b. Which end is the *N*-terminus? What is the *N*-terminal amino acid?

c. Which end is the *C*-terminus? What is the *C*-terminal amino acid?

d. Circle all the side chains.

89 The pentapeptide Gln-Tyr-Asn-Ala-Asp is present in human cerebrospinal fluid, and its concentration increases in demyelinating diseases. In these diseases, the myelin sheath of nerves is damaged and impairs the conduction of nerve signals.

a. What is the *N*-terminal amino acid?

b. What is the *C*-terminal amino acid?

c. How many peptide bonds does this pentapeptide contain? From how many amino acids is this pentapeptide derived?

d. Write the structure of this pentapeptide.

e. Circle the side chains.

90 Vasopressin, also known as *antidiuretic hormone* (ADH), controls resorption of water by the kidneys. This hormone is a nonapeptide with a similar structure to oxytocin (Gly-Leu-Pro-Cys-Asn-Gln-Ile-Tyr-Cys), except that it has Phe in place of Ile and Arg in place of Leu. Based on this information, which hormone would you expect to be more polar, oxytocin or vasopressin?

91 The skeletal line structure of the hormone bradykinin is shown below. This hormone signals the contraction of smooth muscle.

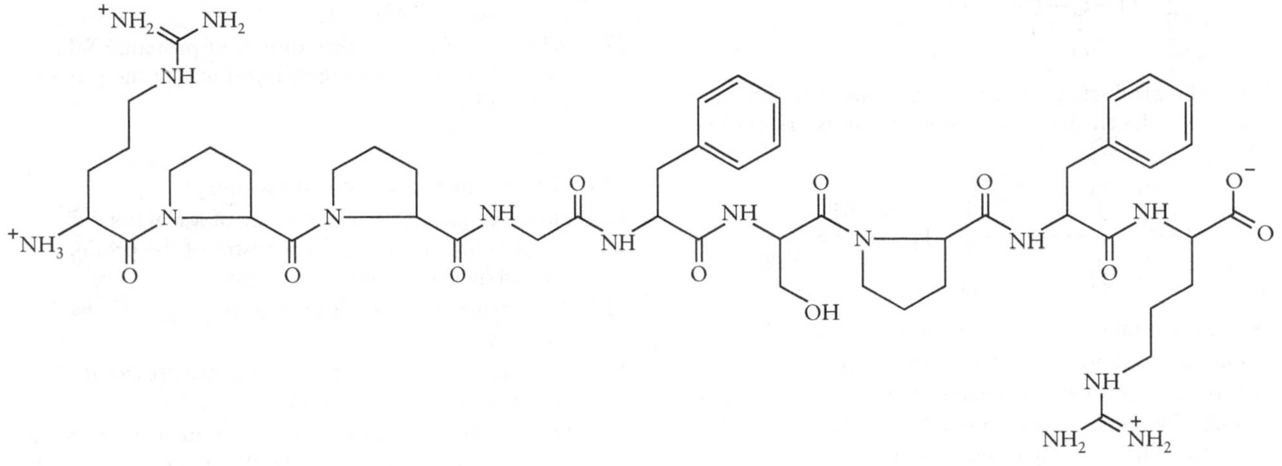

a. How many amino acids does this peptide contain?
b. Referring to Figure 13-3, write the amino acid sequence for bradykinin using the three letter abbreviations for the amino acids.

Types of Proteins and Protein Architecture

92 Where does a cell get the information containing the sequence of amino acids for each polypeptide in a protein?

93 What does the term "native conformation" refer to in a protein? What types of interactions stabilize the protein in its native conformation?

94 What are the three general classes of proteins, distinguished by shape and solubility?

95 What tissues in our bodies are rich in keratins? In animals, what tissues are rich in keratins?

96 What role do elastins play in the body?

97 In what tissues are collagens found in the body? What biological function do collagens play?

98 What shape is a globular protein? What role do globular proteins typically have in nature?

99 Where are membrane proteins found?

100 Since the cell membrane is a hydrophobic environment, what types of amino acids will be found on the exterior of a membrane protein? How is this different from proteins that exist in an aqueous environment?

101 What does the primary structure of a protein refer to?

102 What are the potential consequences of errors in the primary structure of a protein? Provide one example.

103 What are the two most common types of secondary protein structure? What stabilizes secondary protein structure? What part of the polypeptide is involved in secondary structure? What part of the polypeptide is not involved in secondary structure?

104 The structure of one of the polypeptide chains in hemoglobin is shown below.

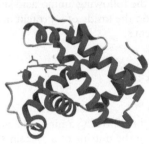

a. There are ___ α-helices and ___ β-pleated sheets in this polypeptide.
b. Heme, shown in green, is part of this polypeptide chain. Is heme a peptide? A non-peptide organic molecule bound to a protein is known as a
_____ _____.

c. What atom lies at the center of the heme group in hemoglobin?

105 What is a prosthetic group? Is it part of the primary, secondary, tertiary, or quaternary structure of a protein?

106 What types of electrostatic interactions are responsible for the tertiary structure of a protein? Provide an example of each type of electrostatic interaction.

107 Why do proteins fold rather than exist as a long linear chain of amino acids?

108 What amino acids need to be present in a protein in order for a disulfide bridge to form? How many amino acids are involved in one disulfide bridge? Would a disulfide bridge between two parts of the same polypeptide be part of the primary, secondary, tertiary, or quaternary structure of the protein? Would a disulfide bridge between two different polypeptide chains be part of the primary, secondary, tertiary, or quaternary structure of the protein?

109 Write the product formed when the thiol shown below is oxidized.

$$2 \quad H-\overset{\overset{\displaystyle H}{|}}{\underset{\underset{\displaystyle H}{|}}{C}}-\overset{\overset{\displaystyle H}{|}}{\underset{\underset{\displaystyle H}{|}}{C}}-S-H \quad \overset{[O]}{\longrightarrow}$$

110 Write the products formed when the disulfide shown below is reduced. What functional group is present in both products?

$$H-\overset{\overset{\displaystyle H}{|}}{\underset{\underset{\displaystyle H}{|}}{C}}-\overset{\overset{\displaystyle H}{|}}{\underset{\underset{\displaystyle H}{|}}{C}}-S-S-\overset{\overset{\displaystyle H}{|}}{\underset{\underset{\displaystyle H}{|}}{C}}-H \quad \overset{[H]}{\longrightarrow}$$

111 Name the amino acid that can form salt bridges.

112 Name all the amino acids that can form hydrogen bonds.

113 Where are nonpolar side chains likely to be found in a protein that exists in aqueous solution? Where are the polar side chains likely to be found?

114 What type of electrostatic interactions do nonpolar side chains have with other nonpolar side chains? Which amino acids have these types of electrostatic interactions?

115 Which of the following amino acid side chains would be found on the inside of a protein in an aqueous environment?
 a. arginine
 b. isoleucine
 c. aspartic acid
 d. valine
 e. proline

116 Which of the following amino acid side chains would be found on the outside of a protein in an aqueous environment?
 a. histidine
 b. leucine
 c. glutamic acid
 d. cysteine
 e. serine

117 What type of interactions stabilize the quaternary structure of a protein?

118 What must be true about the number of polypeptide chains for quaternary structure to exist in a protein?

119 For the following pairs of amino acids within a protein, indicate how their side chains might interact to contribute to the tertiary or quaternary structure of a protein. Choose from among the following: *disulfide bridge, salt bridge, hydrogen bonding,* or *dispersion forces*.
 a. glutamic acid and lysine
 b. tyrosine and lysine
 c. Ile and Ala
 d. two cysteines

120 In the opening story, you learned that valine replaces glutamic acid in the hemoglobin of people who have sickle-cell anemia. Is this a change in primary, secondary, tertiary, or quaternary structure? Why does this small change affect the shape of the protein?

121 What agents can denature a protein? What happens to the structure of a protein when it is denatured? Does it affect the primary structure of a protein?

122 State four ways to denature a protein.

123 Proteins are denatured in your stomach as a result of the low pH. Write the structure of the tripeptide Asp-Trp-Lys with all the functional groups in their correct form at pH = 1.

124 Why does a low pH denature most proteins? What kinds of interactions are changed within the protein as a result?

Enzymes

125 Why are enzymes essential for life?

126 How do enzymes affect the rate of a reaction? Can enzymes affect the stereochemistry of the product formed in a biochemical reaction?

127 An enzyme works by lowering the _____ for a reaction.

128 What types of electrostatic forces are present in the enzyme-substrate complex?

129 Describe how the conformation of an enzyme changes upon binding to its substrate. What is this model of enzyme-substrate binding called?

130 How does an enzyme lower the activation energy for a reaction?

131 Fill in the blank and then indicate what each letter stands for:

$$E \ + \ S \ \rightleftharpoons \ \underline{\hspace{2cm}} \ \longrightarrow \ E \ + \ P$$

132 In the reaction above, which step is reversible? If the concentration of substrate increases how does the equilibrium shift in accordance with Le Châtelier's principle?

133 What is another name for the substrate?

134 Provide two examples of cofactors.

135 Is an *enzyme* chemically changed during an enzyme-catalyzed reaction? Is the substrate chemically altered during an enzyme-catalyzed reaction?

136 What is a prosthetic group? Are cofactors and coenzymes prosthetic groups?

137 What is the optimal pH for most enzymes? What is the optimal temperature for most enzymes?

138 Is the activation energy, E_A, higher or lower for an enzyme-catalyzed reaction compared to an uncatalyzed reaction?

139 What do enzyme inhibitors do to prevent an enzyme from catalyzing a reaction?

140 How does a competitive inhibitor work? Are the molecular shapes of competitive inhibitors similar to or different from the molecular shape of the substrate? Explain.

141 Where on the enzyme do noncompetitive inhibitors bind? How do noncompetitive inhibitors affect the shape of the active site?

142 How does increasing the concentration of the substrate affect enzyme activity in the presence of a competitive inhibitor? How does increasing the concentration of the substrate affect enzyme activity in the presence of a noncompetitive inhibitor?

143 How does cholesterol inhibit its own synthesis within the body? What is this type of enzyme inhibition called?

Chemistry in Medicine: ACE Inhibitors as a Treatment for Hypertension

144 Give some examples of diseases that are treated by drugs that work as enzyme inhibitors.

145 What is hypertension?

146 What are some health risks associated with hypertension?

147 How does the concentration of sodium ions, Na^+, increase blood pressure?

148 Write the structure of the dipeptide Leu-Val. Indicate which bond in angiotensinogen is hydrolyzed by *Renin*.

149 Write the structure for the dipeptide Phe-His. Indicate the bond broken in angiotensin I by *angiotensin converting enzyme (ACE)*.

150 Why does blocking the production of angiotensin II prevent hypertension?

151 Where does Captopril bind to *ACE*? What type of enzyme inhibitor is Captopril?

Answers to Practice Exercises

1

Low pH (+1) ⇌ (H^+) Zwitterion (0) ⇌ (OH^-) High pH (−1)

a. See above. At low pH the equilibrium shifts to the left because there are more hydronium ions in solution, which convert the carboxylate ion into its neutral carboxylic acid form. At high pH the equilibrium shifts to the right because there are more hydroxide ions in solution, which removes a proton from the conjugate acid of the amine and converts it into its neutral amine form.

b. The net charge on glycine at high pH is −1 because the carboxylic acid is in its ionized form (−1) and the amine is in its neutral form (0). The net charge on glycine at low pH is +1 because the amine is in its ionized form (+1) and the carboxylic acid is in its neutral form (0). The zwitterion has no net charge because the carboxylic acid is in its ionized form (−1) and the amine is in its ionized form (+1) so the sum of the two charges adds up to zero.

2

d. Side chain for phenylalanine, Phe

c. α-carbon

a. Amine (ionized form)

b. Carboxylic acid (ionized form)

e. Phenylalanine is a nonpolar amino acid because it has a hydrocarbon side chain.

f.

net charge: −1

3

Amino Acid	Nonpolar	Polar	Acidic Side Chain	Basic Side Chain	Neutral Side Chain	Essential Amino Acid
Cysteine		x			x	
Lysine		x		x		x
Glutamine		x			x	
Methionine	x					x
Aspartic Acid		x	x			

4 Methionine and cysteine contain sulfur. Cysteine contains a thiol (—S—H) functional group.

5 Proline, Pro:

Proline (Pro)

6

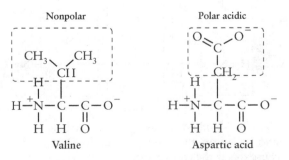

Nonpolar

Valine

Polar acidic

Aspartic acid

7 Meat and fish contain all the essential amino acids.

8

α-carbon

L-Cysteine structure: α-carbon, COO$^-$, H$_3$N$^+$—C—H, CH$_2$SH

L-Cysteine

a. Yes, because this is an L-amino acid (—NH$_3^+$ is on the left in a Fischer projection), and the natural L-amino acids are produced in nature.

b. See above.

c. In a Fischer projection of an L-amino acid, the amine group appears horizontally on the left, whereas in a D-amino acid it appears horizontally on the right.

d. See below. D-cysteine is different from L-cysteine in the configuration at the center of chirality (the α-carbon): the amine is left in an L-amino acid and right in a D-amino acid. They are similar in that they contain the same chemical formula and the same atom and bond connectivity. D- and L-cysteine are enantiomers: mirror image stereoisomers.

D-Cysteine structure: COO$^-$, H—C—NH$_3^+$, CH$_2$SH

D-Cysteine

e. Cysteine has a thiol functional group (—SH) in its side chain.

9 a. The amine: —NH$_3^+$ (or NH$_2$); the carboxylic acid: —CO$_2^-$ (or COOH); an —H atom; and the side chain: —CH(CH$_3$)CH$_2$CH$_3$

b. The amine, carboxylic acid, and H atom are present in all amino acids.

c. The side chain: —CH(CH$_3$)CH$_2$CH$_3$

d.

Produced in nature

L-Cysteine structure: COO$^-$, H$_3$N$^+$—C—H, CH(CH)$_3$CH$_2$CH$_3$

L-Cysteine

D-Cysteine structure: COO$^-$, H—C—NH$_3^+$, CH(CH)$_3$CH$_2$CH$_3$

D-Cysteine

10 a, b.

Peptide structure (N-terminus to C-terminus): Tyr—Gly—Gly—Phe—Leu

c. Tyr-Gly-Gly-Phe-Leu, always write a sequence of amino acids from the N-terminus to the C-terminus because this is the convention.

d. The side chains, in order from N- to C-terminus, are: —CH$_2$C$_6$H$_4$OH, —H, —H, —CH$_2$C$_6$H$_5$, —CH$_2$CH(CH$_3$)$_2$.

e.

Pentapeptide structure with peptide bonds indicated by arrows.

Peptide bonds

f. The peptide is chiral because at least one of the amino acids from which it is derived are chiral.

g. Endorphins are the body's natural pain killers. Both leu- and met-enkephaline are pentapeptides, but leu-enkephalin has the amino acid leucine as the C-terminal amino acid, whereas met-enkephalin has the amino acid methionine as the C-terminal amino acid.

11 Yes, they are different compounds; they are structural isomers because they have a different arrangement of atoms and bonds, yet their molecular formulas are the same.

12

Tripeptide structure: N-terminus ... Cys — Gly — Val ... C-terminus

13 a. The N-terminal amino acid is cysteine.

b. The C-terminal amino acid is glycine.

c. This is a nonapeptide, derived from nine amino acids.

d. This peptide contains eight peptide bonds, joining the nine amino acids.

e. The two cysteines have a thiol functional group (—SH) in their side chain.

14 a. This appears to be a membrane protein because it is in the cell membrane.

b. Nonpolar, because the inside of the cell membrane is nonpolar.

c. Polar, because water, a polar molecule, is outside the membrane.

15 Fibrous proteins are insoluble in water and have a nonspherical shape. Globular proteins are soluble in water and have a spherical shape.

16 Fibrous proteins

17 Enzymes are typically <u>globular</u> proteins.

18 Both are fibrous proteins, but elastins are more flexible, whereas collagens are more rigid.

19 Hemoglobin has four *N*-termini, one for each of the four polypeptide chains.

20 Primary, secondary, tertiary, and quaternary protein structure

21 The native conformation of a protein is the three-dimensional structure of the protein required for it to perform its function.

22 **a.** Phenylalanine, Phe:

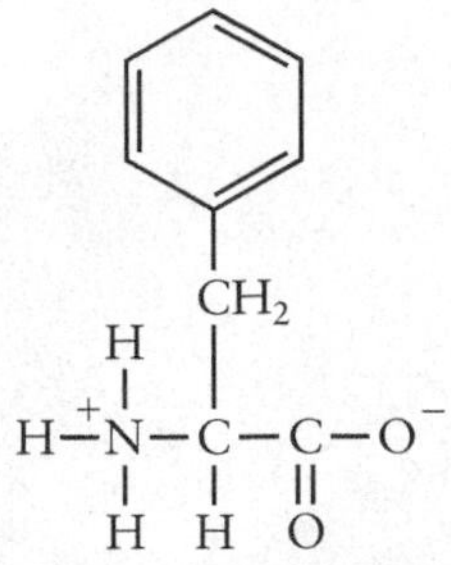

(structure of phenylalanine)

CH₂

H—⁺N—C—C—O⁻

H   H   O

b. At amino acid 508, counting from the *N*-terminus, there is a missing Phe.

c. You would expect cystic fibrosis to be a genetic disease because it is a protein disorder and proteins are constructed based on information encoded in our DNA. Since DNA is the genetic material we inherit from our biological parents, CF is a genetic disorder.

d. A membrane protein, because it functions as an ion channel that allows chloride ions to cross the cell membrane, so it must be located in the cell membrane.

23 Insulin contains three sections of α-helix and no β-sheets. The red ribbon represents a type of secondary structure known as an α-helix.

24 Partial charges arise from polar bonds, which occur between atoms with different electronegativity values. The most polar bonds are seen in N—H and O—H bonds, which are the hydrogen donors in hydrogen bonding (dashed red line).

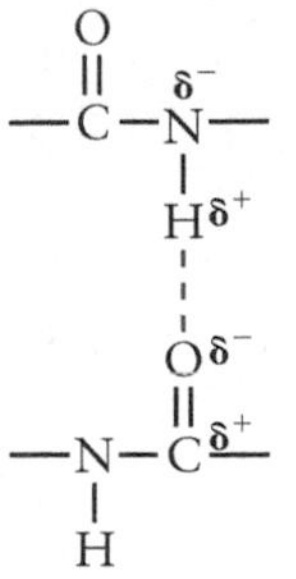

O
‖
—C—N—
|
H δ⁺
⋮
O δ⁻
‖
—N—C δ⁺
|
H

25 **a.** Above and below the sheets
b. Outward from the helix and perpendicular to the length of the helix

26 A ribbon diagram is a way to depict the polypeptide backbone(s) of a protein by showing secondary structure as colored ribbons: coiled for an α-helix and straight arrows for β-pleated sheets. Sections with no secondary structure are shown as colored string.

27 Side chains are not involved in secondary structure; hydrogen bonding occurs between amides in the polypeptide backbone only.

28 Wide arrows in a ribbon diagram represent β-pleated sheets, a type of secondary structure. The arrows point toward the *C*-terminus of the peptide. Since the arrows are pointing in opposite directions, these β-sheets are antiparallel.

29 The products of reduction are the two thiols shown below.

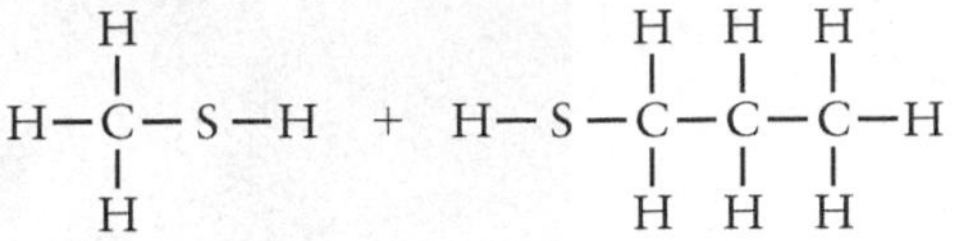

H   H H H
| | | |
H—C—S—H   +   H—S—C—C—C—H
| | | |
H   H H H

30 **a.** Hydrogen bonding between the —O—H of a phenol and an amide
b. Disulfide bridge formed between two cysteines
c. Dispersion forces between hydrocarbon side chains
d. Salt bridge between the conjugate base of Asp or Glu and the conjugate acid of His

31 Tertiary protein architecture is shown because it involves electrostatic attractions and disulfide bridges between the side chains, not the polypeptide backbone, of a *single* polypeptide.

32 **a.** Salt bridge   **b.** Hydrogen bonding
c. Dispersion forces   **d.** Disulfide bridge
e. Dispersion forces

33 Primary protein architecture

34 Quaternary structure refers to electrostatic interactions and disulfide bonds between the side chains of more than one polypeptide chain, in a protein with two or more polypeptide chains.

35 Ways to denature a protein include detergents, change in pH, change in temperature, mechanical agitation, and addition of certain metals. The conformation of a protein changes when it is denatured. In some cases denaturation is reversible.

36 Primary protein structure is unaffected by denaturation because denaturing agents do not affect the peptide bonds of a protein.

37 The induced-fit model of enzyme-substrate binding describes how the enzyme, E, and the substrate, S, bind to form an enzyme substrate complex, ES. The E and S are only approximately complementary in shape, but a conformational change in the enzyme upon binding creates a better fit between substrate and enzyme, hence, an "induced fit."

38 An enzyme is a globular protein. Enzymes are catalysts for biochemical reactions, which are otherwise too slow to sustain the processes required for living.

39 The active site (or binding site) is a cleft or pocket on the enzyme that binds the substrate. It is roughly complementary in shape to the substrate and has amino acid side chains that bind the substrate.

40 The ES complex stands for enzyme-substrate complex, which is the enzyme with its substrate bound to the active site.

41 According to the graph in Figure 13-29a, *pepsin* has its maximum catalytic activity (top of the curve = highest enzyme activity) at pH = 2 (dashed line).

42 For **a–d**, the rate of the reaction decreases in each case; because these are all examples of conditions not optimal for enzymes in terms of pH and temperature.

43 Competitive inhibitor

44 A noncompetitive inhibitor

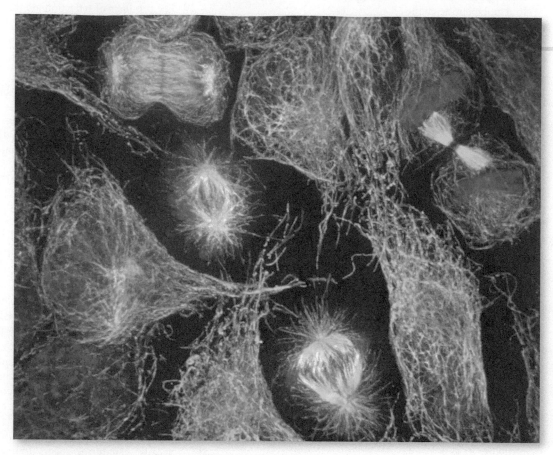

Human cervical cancer cells during cell division. Cancer cells are abnormal, rapidly dividing cells. DNA is fluorescently labeled in blue. [Jennifer C. Waters/Science Source]

14 Nucleotides and Nucleic Acids

Gene Mutations and Disease

Most diseases are caused by or impacted in some way by our genes. Some of these diseases—about 4,000 that we know of—are hereditary diseases caused by a specific gene. Well known examples include sickle cell anemia and cystic fibrosis. Other diseases, such as some types of cancer, are caused by environmental factors that alter the chemical structure of genes in a particular cell type. Environmental factors include carcinogens (cancer causing compounds), radiation, and viral infections. In most cases however, diseases are caused by a combination of certain genes and environmental factors.

Today, we have the technology to test for the presence of altered genes that are known to predispose an individual to certain diseases. For example, the *breast cancer genes BRCA 1 and 2*, discovered in 1994, are indicators of a woman's predisposition for developing a type of breast cancer. A woman who tests positive for these genes has an 85 percent chance that she will develop breast cancer before the age of 65. In this chapter we examine nucleic acids—the chemistry behind genes—and we will see why they play such a prominent role in disease.

A gene is a segment of DNA containing the information for constructing a protein, the subject of the previous chapter. DNA is an information-carrying biomolecule known as a nucleic acid. The term *nucleic* originates from the fact that DNA is found in the *nucleus* of cells. Your DNA contains all the genes (instructions) required for building every protein you have ever needed or will ever need in your lifetime. Thus, your DNA determines every one of your physical traits, from the color of your eyes to your basal metabolism. And now we are learning that our DNA even determines whether we have a predisposition toward developing certain diseases. Almost every cell in your body contains the same DNA, which is subtly different from the DNA of every other person. Indeed, this is why a forensic scientist can analyze the DNA from any bodily fluid or tissue to identify who it came from.

What is the connection between genes and disease? Genes that predispose a person to disease often contain mutations. *A mutation is a permanent alteration of the chemical structure of a gene, which in turn produces an altered protein.* Mutations are either acquired or inherited. *Inherited* mutations are genes that you are born with; received from your biological mother or biological father. *Acquired* mutations are alterations to your genes that take place sometime during your lifetime. Acquired mutations affect only one cell type, so the mutation is not found in the DNA of other cells. Cervical cancer, for example, can occur from an acquired mutation in cervical cells; other cells in the body remain normal.

Environmental factors can trigger mutations. For example, certain forms of cervical cancer are believed to be caused by a persistent infection by the *human papillomavirus (HPV)*, one of the most common sexually transmitted diseases. Some strains of the *HPV* virus cause a mutation in genes responsible for the production of proteins that control cell replication, resulting in uncontrolled growth of cervical cells. Since cancer is the uncontrolled proliferation of cells, these abnormal cervical cells, shown in the opening photo of the chapter, can develop into cervical cancer if subsequent changes, influenced by yet other genes, also occur.

In this chapter you will learn how the cell uses the chemical information encoded within its DNA to build proteins, a process that requires RNA, the other important nucleic acid in the cell. ●

"The ability to store and transmit genetic information from one generation to the next is a fundamental condition for life." —Albert Lehninger

The ability to transmit genetic *information* is a fundamental condition for life. The biomolecules responsible for storing and transmitting genetic information are the nucleic acids *deoxyribonucleic acid* (DNA) and *ribonucleic acid* (RNA).

In some past biology class you most likely learned that every human being starts life as a fertilized egg (a zygote), a single cell containing DNA. This single cell grows and divides ultimately becoming a human being with one trillion cells, each with the same DNA as the zygote. Every protein, including those that determine whether the zygote will develop into a male or a female is encoded within its DNA. It is therefore essential for life that DNA be able to *replicate* itself—a process that occurs prior to cell division and described in section 14.2.

The primary function of DNA, however, is to serve as a blueprint for assembling all the proteins required by the individual throughout his or her lifetime. Encoded within a segment of the DNA, known as a **gene**, is the information for assembling the sequence of amino acids for a specific protein. In section 14.3 we will discuss how information from DNA (the gene) is transferred to mRNA and then used to assemble amino acids into a protein.

In order to understand DNA replication, and the transfer of information from DNA to RNA and protein synthesis, we must first consider the basic chemical structure of a nucleotide, the monomer units of the polynucleotides DNA and RNA.

> In the common medical procedure known as an *amniocentesis*, amniotic fluid surrounding the developing fetus is collected and analyzed, often for genetic abnormalities. Since all cells from the fetus contain the same DNA, the sex of the child is also revealed in this test.

■ 14.1 The Chemical Structure of a Nucleotide and Polynucleotides

Nucleic acids are polymers (polynucleotides) constructed from **nucleotides**, much like proteins are polymers (polypeptides) constructed from amino acids. However, there are 20 different amino acids from which to build a protein, whereas there are only four different nucleotides from which to build either DNA or RNA.

Chemically, a nucleotide contains three parts:

(1) A monosaccharide (sugar)
(2) A phosphate
(3) A nitrogenous base

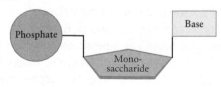

The Monosaccharide

The monosaccharide components of DNA and RNA are slightly different: D-ribose is the sugar in RNA (**Figure 14-1a**), whereas 2-deoxy-2-D-ribose is

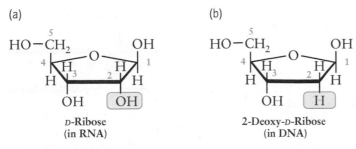

Figure 14-1 Chemical structure of (a) D-ribose, found in RNA, and (b) 2-deoxy-D-ribose, found in DNA. The only structural difference is highlighted in the blue boxes.

the sugar in DNA (Figure 14-1b). Both monosaccharides are furanoses and contain five carbons, except that 2-deoxy-2-D-ribose lacks a hydroxyl group at C(2) as highlighted in blue in Figure 14-1b. In a nucleotide, the carbon atoms of the monosaccharide are numbered starting at the anomeric carbon as 1′ (pronounced "one prime") and proceeding clockwise around the ring to the 5′ carbon, situated above the ring in the usual arrangement for a D-sugar.

At the anomeric carbon, the monosaccharide contains a β-N-glycosidic linkage to a nitrogenous base, which means the nitrogenous base is situated above the ring (β), on the same side of the furanose ring as the 5′ carbon, and a covalent bond joins the 1′ carbon of the sugar to a nitrogen (N) atom in the nitrogenous base.

> The prime symbol (′) is used to identify the carbon atoms in the *monosaccharide* component of a nucleotide (1′, 2′, 3′, . . .) to distinguish them from the atoms in the *nitrogenous base* (1, 2, 3, . . .).

The Nitrogenous Base

A nucleotide contains a **nitrogenous base** derived from either pyrimidine (Figure 14-2) or purine (Figure 14-3): planar (flat), aromatic rings containing two or four nitrogen atoms in the ring, respectively. They are called *nitrogenous* because the rings contain *nitrogen* and *bases* because they are like amines, which act as *bases* in aqueous solution. DNA nucleotides contain one of the

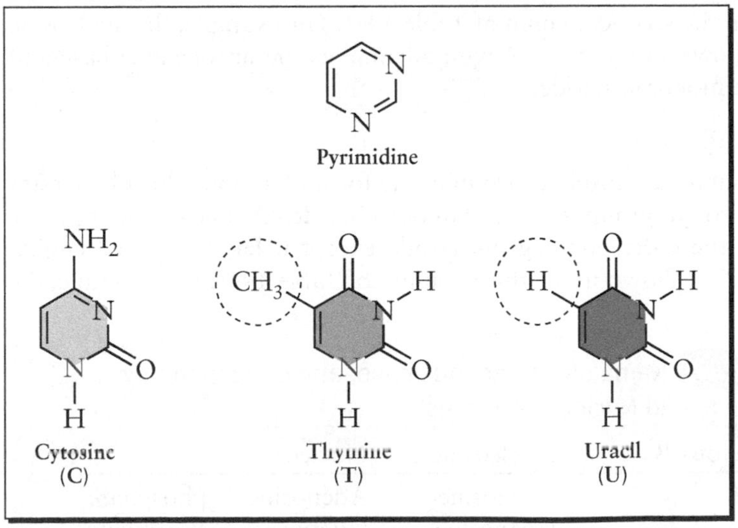

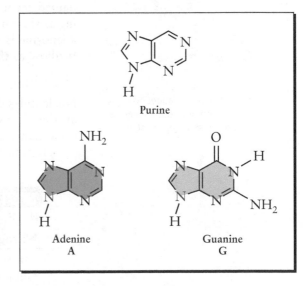

Figure 14-2 The nitrogenous bases derived from pyrimidine: cytosine (C), thymine (T), and uracil (U). DNA does not contain uracil. RNA does not contain thymine. The only difference in structure between thymine and uracil is shown in the dashed circles.

Figure 14-3 The nitrogenous bases derived from purine: adenine (A) and guanine (G).

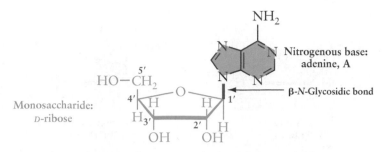

Figure 14-4 The RNA nucleoside adenosine showing the nitrogenous base in a β-*N*-glycosidic linkage to D-ribose.

following four nitrogenous bases, shown next to their common one letter abbreviations:

- (A) adenine,
- (G) guanine,
- (C) cytosine, and
- (T) thymine

RNA nucleotides contain one of the following four nitrogenous bases:

- (A) adenine,
- (G) guanine,
- (C) cytosine, and
- (U) uracil

As you can see, nucleotides in both RNA and DNA contain A, G, and C, but DNA contains thymine, whereas RNA contains uracil. The structure of thymine is slightly different from the structure of uracil: thymine has a —CH$_3$ group where uracil has an —H atom (Figure 14-2). In a nucleotide, the nitrogenous base is covalently bonded to the anomeric carbon of the sugar, in a β-glycosidic linkage to the ring nitrogen atom that is highlighted in red in Figures 14-2 and 14-3.

A nucleo*tide* without its phosphate group is known as a nucleo*side*, and named according to the identity of its nitrogenous base, by changing the ending as shown in the second column of Table 14-1. For example, the nucleoside adenosine is shown in Figure 14-4 with adenine as the nitrogenous base and D-ribose as the monosaccharide.

The Phosphate Group

Nucleotides contain a phosphate group in the form of a monophosphate ester at the 5′ hydroxyl group of the monosaccharide. A nucleotide has the same name as the corresponding nucleoside except it has the added ending "5′ phosphate," as shown in the third column of Table 14-1. For example, the

> The prefix "deoxy" is added to the nucleoside and nucleotide name if 2-deoxyribose instead of ribose is the sugar.

TABLE 14-1 Name of Nitrogenous base, and corresponding Nucleoside and Nucleotide Names

Nitrogenous Base	Nucleoside	Nucleotide
Adenine	Adenosine	Adenosine 5′ phosphate
Guanine	Guanosine	Guanosine 5′ phosphate
Thymine	Thymidine	Thymidine 5′ phosphate
Cytosine	Cytidine	Cytidine 5′ phosphate
Uracil	Uridine	Uridine 5′ phosphate

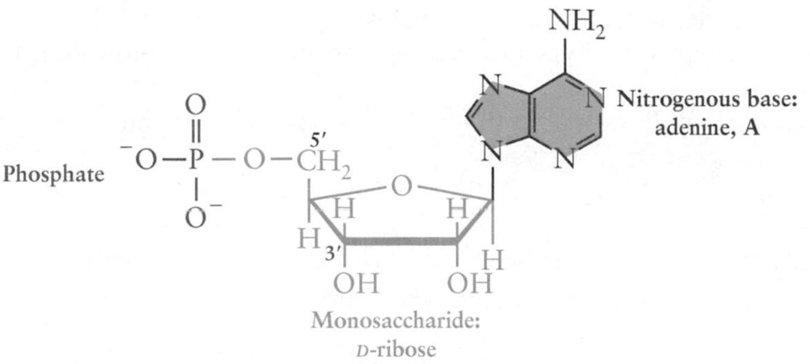

Figure 14-5 The RNA nucleotide adenosine 5′ monophosphate (AMP), showing the three basic parts of a nucleotide: the nitrogenous base, the monosaccharide, and the phosphate group.

nucleotide adenosine 5′ monophosphate (AMP) is a monophosphate ester with the structure shown in **Figure 14-5**. Since nucleotides are organic derivatives of phosphoric acid, H_3PO_4, the term *acid* appears in the terms deoxyribonucleic *acid* (DNA) and ribonucleic *acid* (RNA).

Nucleotide triphosphates (NTPs) are common high energy forms of a nucleotide, as a result of the high energy phosphoanhydride bonds that join three phosphate groups, a concept first introduced in chapter 7. The potential energy in phosphoanhydride bonds is used to drive the reaction that joins a single nucleotide to a growing polynucleotide, as for example during the replication of DNA and the building of RNA polynucleotides. The 5′ phosphate is involved in the reaction that joins one nucleotide to another described in the next section and seen throughout this chapter.

WORKED EXERCISES | Nucleotides

14-1 Which of the following monosaccharides is present in the nucleotides that make up RNA?

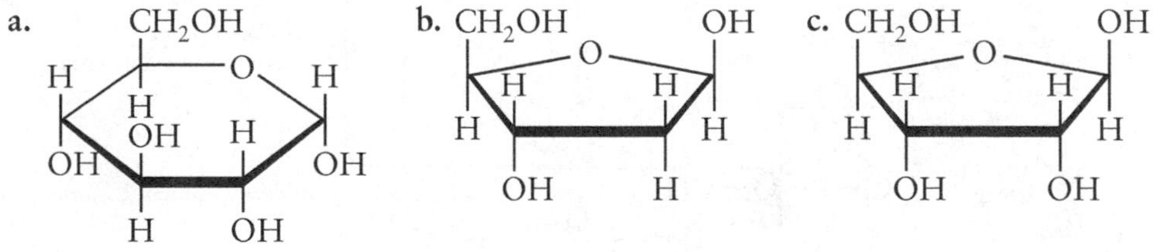

14-2 For the nucleotide shown below, circle and label the following parts and answer the questions below.

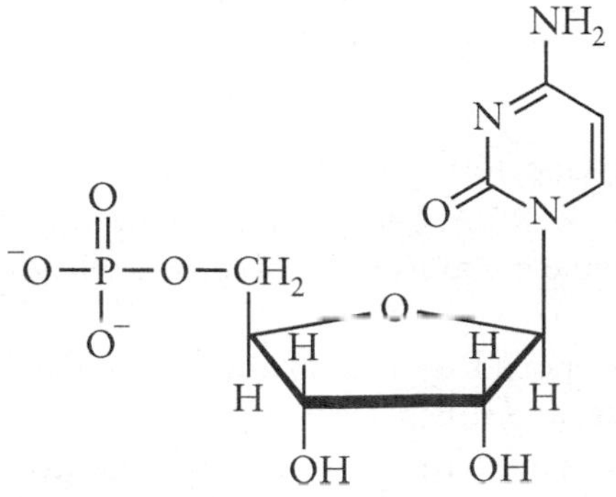

a. The nitrogenous base. What is the name of this base? Is it derived from a purine or a pyrimidine?

b. The monosaccharide. What is the name of this monosaccharide?

c. The phosphate group. What type of functional group joins the phosphate group to the sugar?

d. The anomeric carbon. How would this carbon be identified by number? Is the N-glycosidic linkage α or β? Explain.

e. The 3′ carbon

f. The 5′ carbon

g. Would this particular nucleotide be found in DNA or RNA? Explain.

14-3 Name the four nitrogenous bases found in DNA and give their one letter abbreviations.

14-4 Which nitrogenous base is not found in RNA? Which nitrogenous base is not found in DNA? What is the structural difference between these two nitrogenous bases?

14-5 Is the ring in a nitrogenous base aromatic or not? What does this mean?

14-6 Would you expect a nitrogenous base to be affected by a significant decrease in pH?

Solutions

14-1 The monosaccharide (c), D-ribose, is the monosaccharide component of RNA nucleotides.

14-2 a. See below in black.

b. See below in blue. D-ribose.

c. See below in pink. A monophosphate ester joins the phosphate to the 5′ hydroxyl group of D-ribose.

d. The anomeric carbon is 1′. The N-glycosidic linkage is β because the nitrogenous base is above the ring—and on the same side as the C(5) CH_2OH group.

e. See below in pink.

f. See below in pink.

g. This nucleotide would be found in RNA because the monosaccharide component is D-ribose rather than 2-deoxy-D-ribose.

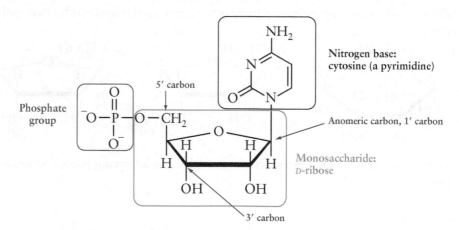

14-3 Adenine, A; guanine, G; cytosine, C; and thymine, T

14-4 Thymine is not found in RNA. Uracil is not found in DNA. The difference in structure is that thymine contains a —CH_3 group where uracil contains a hydrogen (—H) atom.

14-5 The ring in the nitrogenous base is aromatic, which means electrons are delocalized around the entire ring, and it is particularly stable.

14-6 Since nitrogenous bases are basic, they react with protons in acidic solution to form their conjugate acid form, so yes, they would be affected by a decrease in pH.

PRACTICE EXERCISES

1 An amino acid is to a peptide as a _____ is to a nucleic acid.

2 Which of the following monosaccharides is found in DNA and what is it called?

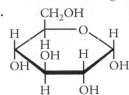

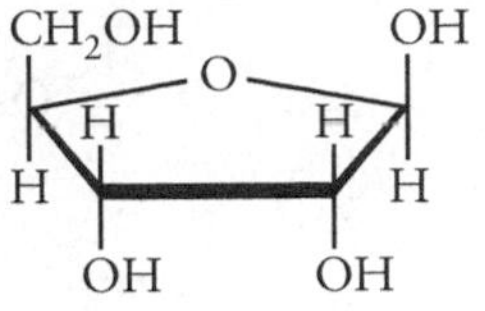

3 For the nucleotide shown below, circle and label the following parts and answer the questions below.

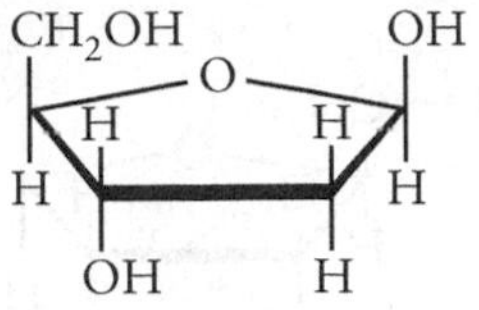

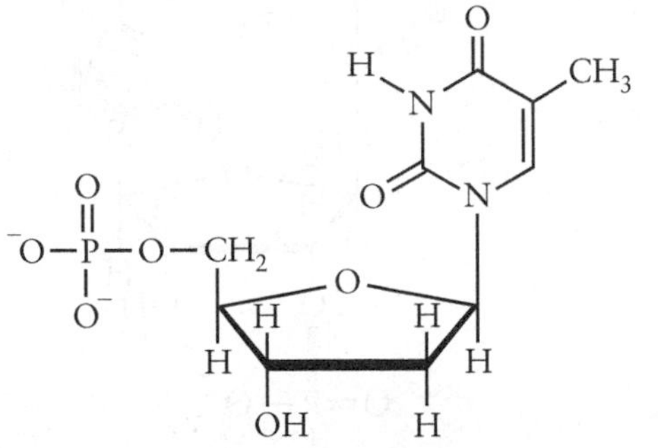

a. The nitrogenous base. What is the name of this base? Is it derived from a purine or a pyrimidine?
b. The monosaccharide. What is the name of this monosaccharide?
c. The phosphate
d. The anomeric carbon
e. The 5′ carbon
f. The 3′ carbon
g. Would this nucleotide be found in DNA or RNA? Explain.

4 Name the four nitrogenous bases found in RNA and give their one letter abbreviations.

5 Which nitrogenous base is not found in DNA?

6 Three parts of a nucleotide are shown below.

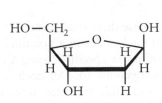

a. Construct a nucleotide from these three molecules. Remember to remove the H atom on the —N-H of the nitrogenous base when forming the covalent bond to the sugar.
b. Label the 3′ and the 5′ carbon atoms in your nucleotide.
c. What nitrogenous base does this nucleotide contain?
d. What monosaccharide does this nucleotide contain?
e. Would this nucleotide be found in RNA or DNA? How can you tell?

Constructing Polynucleotides from Nucleotides

RNA and DNA are linear **poly**nucleotides formed when many nucleotides bond together in a head-to-tail fashion. A covalent bond joins the 3′ hydroxyl group of one nucleotide to the 5′ phosphate group of another nucleotide, forming a phosphate diester, as shown in Figure 14-6 for the RNA trinucleotide ACG.

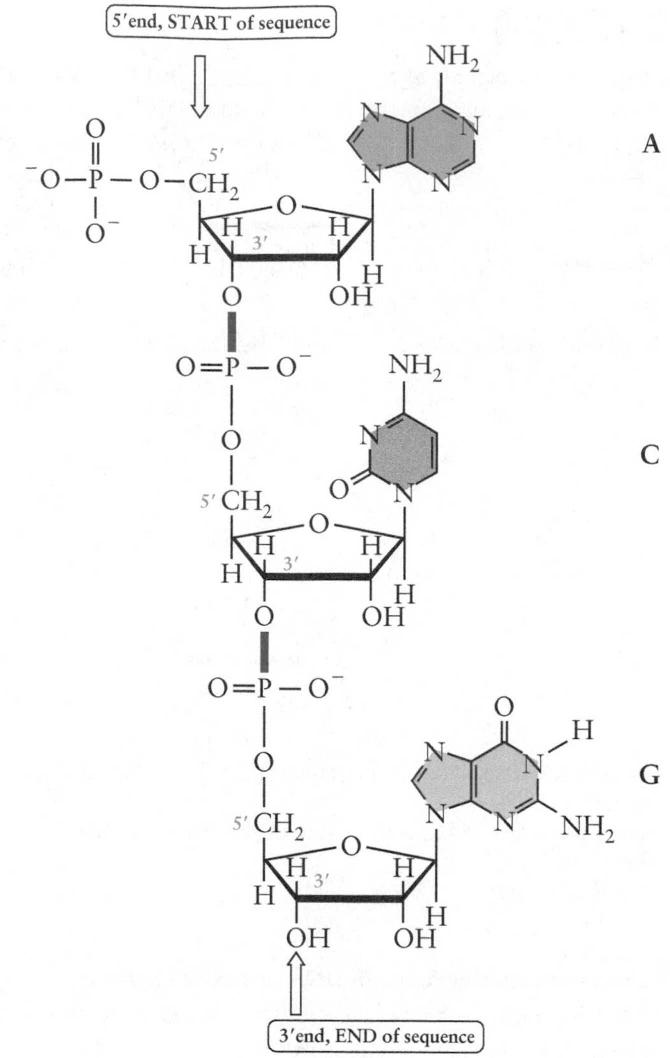

Figure 14-6 The chemical structure of the trinucleotide ACG, showing the nucleotide sequence from the 5' end at the top of the figure to the 3' end at the bottom of the figure. The covalent bonds shown in red join individual nucleotides.

The two phosphate ester bonds that join these three nucleotides are highlighted in red. Since the only difference between nucleotides in a polynucleotide is the nitrogenous base, a sequence of nucleotides is specified using the one letter abbreviation that corresponds to the nitrogenous base of each nucleotide. A lowercase "d" is inserted in front of the nucleotide sequence to indicate a DNA polynucleotide, otherwise it is understood to be RNA.

Every nucleotide in a polynucleotide is joined to two other nucleotides: at the 5' phosphate group and at the 3' hydroxyl group, with the exception of the terminal nucleotides, one of which has a free 5' phosphate group and the other, a free 3' hydroxyl group. The convention when writing a sequence of nucleotides is to list the nucleotides from the free 5' end to the free 3' end (5' → 3'). Hence, the nucleotide sequence shown in Figure 14-6 is abbreviated ACG. The designation dACG would signify the same trinucleotide sequence but with 2-deoxy-D-ribose as the monosaccharide rather than D-ribose.

The reaction that joins a nucleotide to a growing polynucleotide chain is much like an esterification reaction; but instead of an alcohol and a carboxylic acid reacting to form an ester, an alcohol and a phosphate react to form a phosphate ester. Since this type of reaction requires energy, nucleotide triphosphates (NTPs and dNTPs), are used when joining a nucleotide to the polynucleotide. In the cell, the reaction always takes place between the 5′ phosphate group of the incoming nucleotide triphosphate and the 3′ hydroxyl group of the growing polynucleotide (5′ → 3′), as shown in the example below for a DNA polynucleotide.

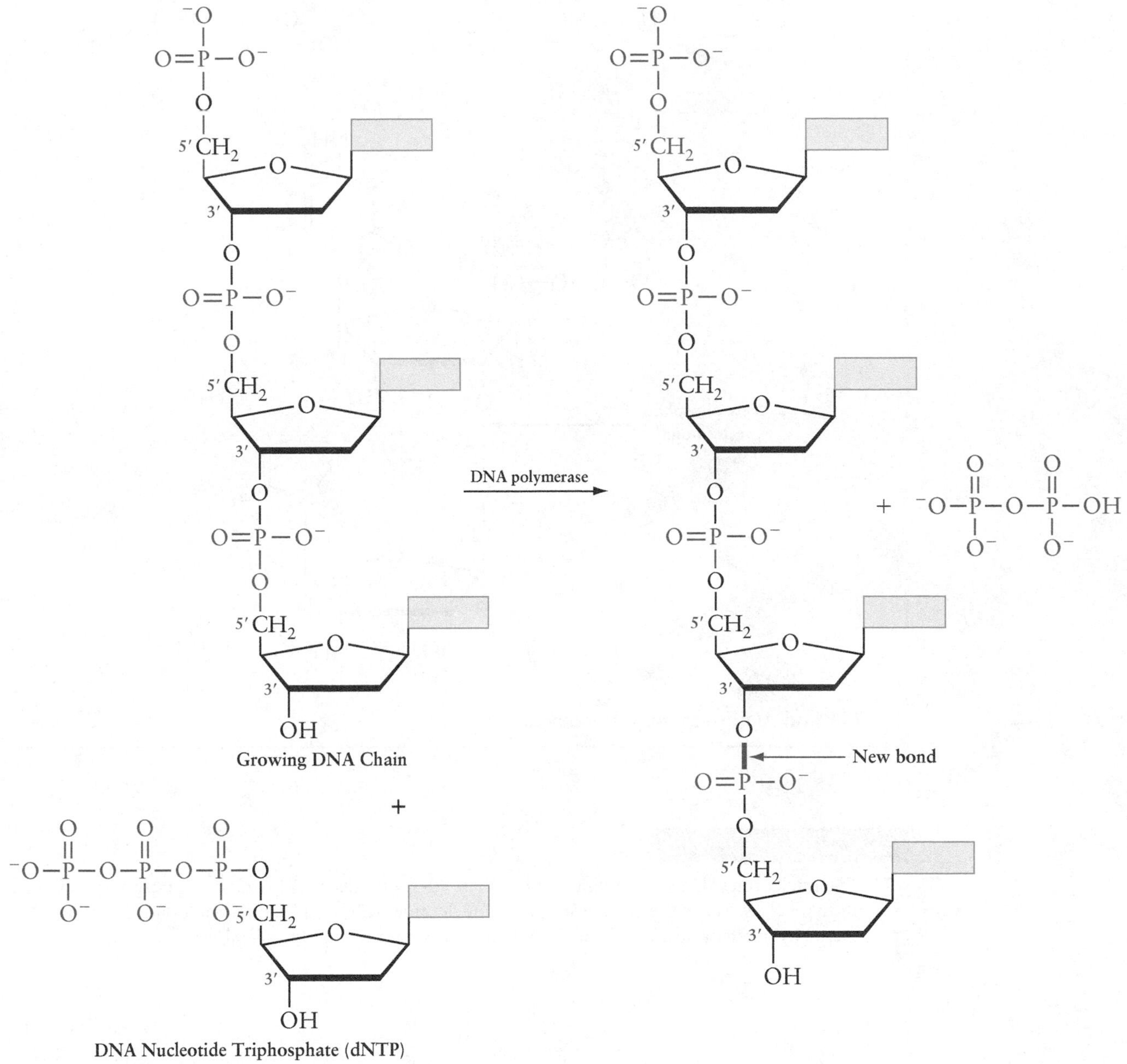

Growing DNA Chain

+

DNA Nucleotide Triphosphate (dNTP)

DNA polymerase

New bond

14-7 The nucleotides cytidine 5′ phosphate (C) and adenosine 5′ phosphate (A) are shown below. Write the structure of the dinucleotide CA and identify the bond that joins these two nucleotides. Label the 5′ end and the 3′ end of the dinucleotide. Is the dinucleotide AC the same or different from the dinucleotide CA?

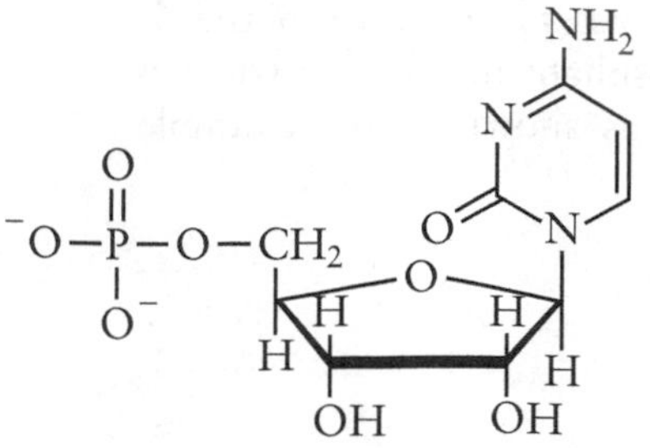

Solution

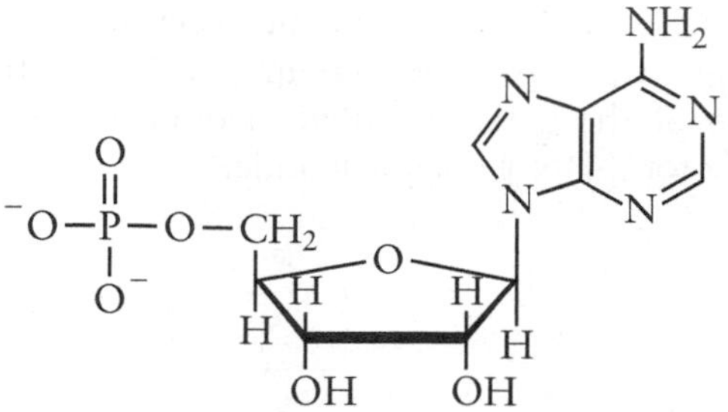

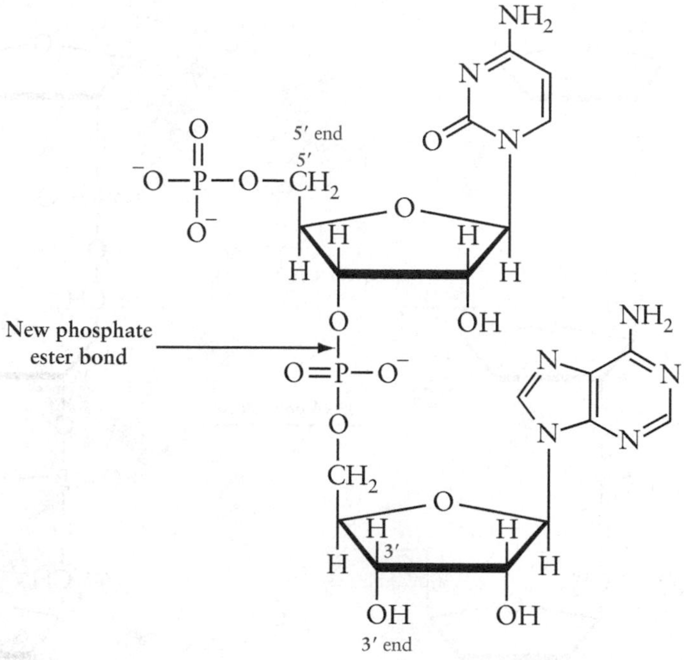

New phosphate
ester bond

CA and AC are structural isomers.

PRACTICE EXERCISES

7   The nucleotides deoxyguanidine 5′ phosphate (dG) and deoxythymidine 5′ phosphate (dT) are
    shown below. Write the structure of the dinucleotide dGT, and identify the bond that joins
    these two nucleotides. Label the 5′ end and the 3′ end of the dinucleotide.

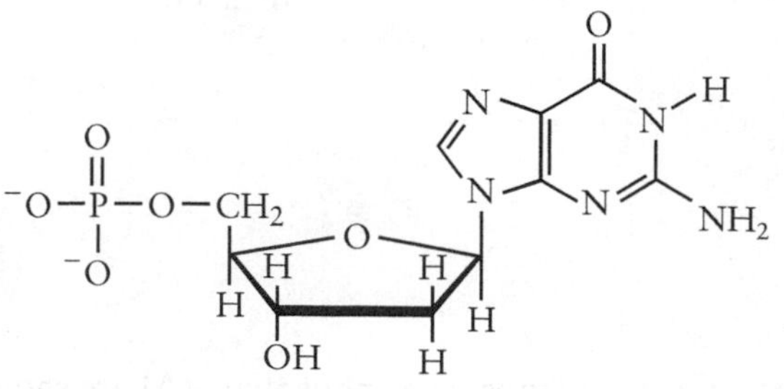

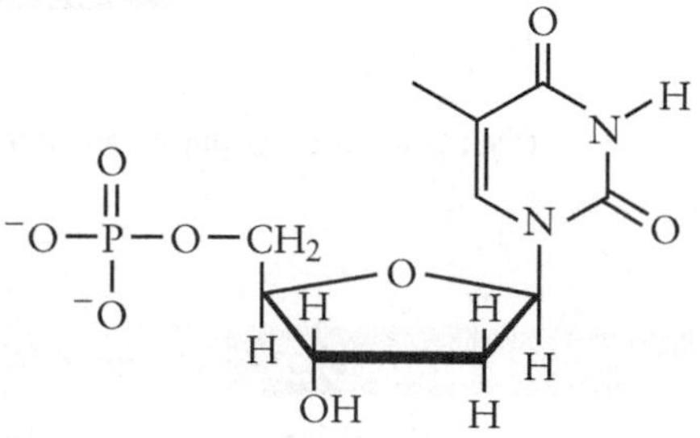

8   The structure of the well-known nucleotide triphosphate ATP is shown below. Why is a
    nucleotide triphosphate needed to add a nucleotide to a growing polynucleotide, even though
    a monophosphate ester is produced?

ATP
adenosine triphosphate

14.2 DNA Structure and Replication

"A structure this pretty just had to exist" —James Watson

For life to exist, an organism or a cell must be able to replicate itself, a process that begins with the replication of DNA. DNA is a polynucleotide constructed from millions of nucleotides. In 1953, James Watson, Francis Crick, and Maurice Wilkins determined the three-dimensional structure of DNA, for which they were later awarded the Nobel Prize. Using molecular models that they created, Watson and Crick were able to propose a structure of DNA that was consistent with the x-ray data collected by Rosalind Franklin and other experimental data available at the time (Figure 14-7).

Double Helix Structure of DNA

DNA is composed of two polynucleotides, each a linear sequence of millions of nucleotides, often referred to as "strands." The two strands are twisted around each other in the form of a right-handed double helix, as shown in Figure 14-8. As a right-handed helix, DNA is also chiral. Since the work of Watson, Crick, Wilkins, and Franklin, several other forms of DNA have also been identified.

The sugar-phosphate backbone of each strand forms the outer lengthwise portion of the double helix, represented by the two brown ribbons in Figure 14-8. The two strands of a DNA double helix are arranged *antiparallel*, which means that their 5′ to 3′ ends are oriented in opposite directions. In other words, one strand has its 3′ end opposite the 5′ end of the other strand.

More than a year before Watson and Crick proposed the structure of DNA, Rosalind Franklin hypothesized that DNA had a helical structure,

Figure 14-7 Rosalind Franklin (left), a British x-ray crystallographer, produced the x-ray images of DNA that were crucial to the determination of the structure of DNA by James Watson and Francis Crick, shown at right standing next to their model of DNA [Left: Science Source. Right: A. Barrington Brown/Science Source]

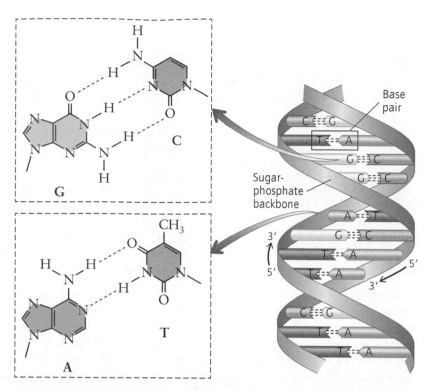

Figure 14-8 The double helix structure of DNA showing the two sugar-phosphate backbones as brown ribbons running the length of the double helix and the base pairs shown as horizontal bars perpendicular to the sugar-phosphate backbone. Dashed lines between the nitrogenous bases on each strand of DNA represent hydrogen bonds: two for A and T, and three for C and G, as seen in the area of detail.

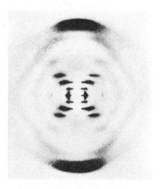

Figure 14-9 X-ray image of DNA, obtained by Rosalind Franklin in 1953. The remarkable symmetry of the diffraction pattern is indicative of a helical structure. [Science Source]

based on the highly symmetrical x-ray images she collected, as seen in the now famous image shown in **Figure 14-9**. Symmetry is the hallmark of a helical structure.

Watson and Crick reasoned that the hydrophilic monosaccharide and phosphate groups of the sugar-phosphate backbone would be on the outside of the helix, where they can interact with the polar aqueous environment. They proposed that the relatively nonpolar nitrogenous bases were on the interior of the helix, and *perpendicular* to the sugar-phosphate backbone, shown as horizontal colored bars in Figure 14-8.

A nitrogenous base on one strand forms two or three hydrogen bonds to a complementary nitrogenous base on the opposite strand, creating what is known as a **base pair**, as shown in the area of detail in Figure 14-8. *The base pairs in double-stranded DNA are complementary, which means that each nitrogenous base interacts with one specific base across from it on the other strand*. The base pairings in DNA are:

- A–T
- G–C

Thus, where there is an adenine (A) on one strand, the complementary base located across from it on the other strand is thymine (T), and vice versa; and where there is a guanine (G) on one strand, the complementary base on the other strand is cytosine (C), and vice versa. Base pairings have two hydrogen bonds between adenine and thymine, and three hydrogen bonds between guanine and cytosine, as shown in Figure 14-8.

The two strands of DNA are held together in a double helix by the hydrogen bonds between base pairs as well as **base stacking**, a stabilizing interaction that occurs between the delocalized electrons of the aromatic rings of the nitrogenous bases above and below. Although these electrostatic interactions are weaker than covalent bonds (chapter 3), their cumulative strength is significant. Yet, because hydrogen bonding and base stacking are noncovalent forces of attraction, they are more easily disrupted, which is necessary when the two strands of DNA need to be separated, as during replication and transcription when the information encoded in a strand of DNA is copied.

WORKED EXERCISES DNA Double Helix

14-8 Below are some sequences of nucleotides located on a segment of one strand of DNA, cited from the 5′ to the 3′ end. Indicate the complementary sequence of nucleotides that would appear on the adjacent DNA strand from the 3′ to the 5′ end.

   a. 5′ dAGTCCG 3′   **b.** 5′ dCCTTGA 3′

14-9 What part of the DNA double helix has the following properties?

   a. Is hydrophobic?
   b. Stabilizes the helix by base stacking?
   c. Is hydrophilic and interacts with the surrounding aqueous environment?
   d. Stabilizes the helix by hydrogen bonding?

14-10 Are the two strands of a DNA double helix *parallel* or *antiparallel*? What does this mean?

Solutions

14-8 The complementary sequences of nucleotides from the 3′ to the 5′ end across from the given strand are shown in red:

   a. 5′ dAGTCCG 3′
       3′ dTCAGGC 5′
   b. 5′ dCCTTGA 3′
       3′ dGGAACT 5′

14-9 **a.** The nitrogenous base pairs
   b. The nitrogenous bases
   c. The sugar phosphate backbone
   d. The base pairs C–G, G–C, T–A, and A–T

14-10 The two strands of DNA are arranged antiparallel, which means the 5′ end of one strand is opposite the 3′ end of the other strand.

PRACTICE EXERCISES

9 Below are sequences of nucleotides located on a segment of one strand of DNA cited from the 5′ to the 3′ end. Indicate the complementary sequence of nucleotides on the adjacent DNA strand from the 3′ to the 5′ end.

   a. 5′ dTTGGCA 3′   **b.** 5′ dATGCCA 3′

10 Name the part of the DNA double helix that fits the description below:

   a. It is on the interior of the double helix.
   b. It is hydrophilic.
   c. They form hydrogen bonds.
   d. They stabilize the double helix by base stacking.

11 What characteristic of the x-ray image of DNA indicates that it has a helical structure?

12 What type of interactions hold the two strands of DNA together? What is the advantage of this type of interaction over covalent bonding?

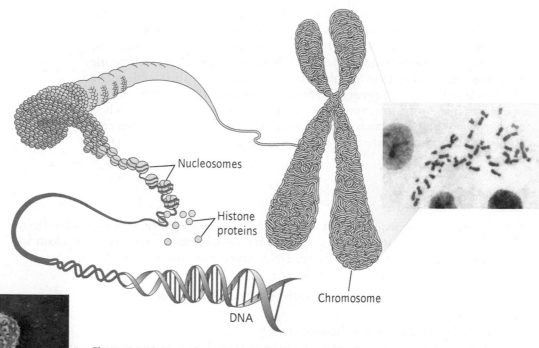

Figure 14-10 DNA, shown in brown, coils around histones, shown in green, to form nucleosomes (brown and green). The nucleosomes coil upon themselves to create larger fibers which form the familiar X-shaped structure of a chromosome. Chromosomes are large enough to be visible under a light microscope. [Michael Abbey/Science Source]

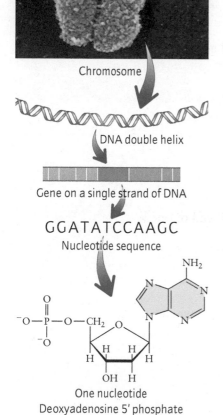

Figure 14-11 Chromosomes contain double-stranded DNA. Genes are segments within the DNA that contain a sequence of nucleotides that code for a particular protein. [Biophoto Associates/ Science Source]

Genes and the Human Genome

DNA, together with specialized proteins, coils into highly compact structures known as **chromosomes**. If stretched out end to end, the DNA from all your chromosomes would span more than a meter in length! However, DNA is not stretched out, but coiled, as illustrated in Figure 14-10. A DNA double helix is coiled around a core of proteins, known as **histones**, creating compact structures known as **nucleosomes**. Since histones have a net positive charge, they facilitate the packing of negatively charged DNA into these compact structures. Nucleosomes coil further to create even larger fibrous structures, which create the familiar X-shape of a chromosome. A DNA double helix is about 2 nm wide, while a chromosome is about 1,400 nm wide and visible under a light microscope, as seen in the micrograph in Figure 14-10.

The complete sequence of nucleotides in your DNA, distributed across 46 chromosomes, is known as your **genome**. Every organism has a genome, ranging in size from the genome of a simple bacterium that might contain 600,000 base pairs, to the human genome, which contains about 3 billion base pairs. The human genome is 99.9 percent the same in all people and almost half of the proteins made by a human cell are similar to those of other organisms.

Only about 2 percent of the entire human genome contains sequences of nucleotides that code for a protein. The sections of DNA that code for a protein are known as genes (Figure 14-11). *A **gene** is a segment of DNA—a sequence of nucleotides—that contain the instructions for making a protein.* The average human gene contains about 3,000 base pairs, and we have approximately 30,000 genes, although the role of many of these genes is still unknown. In the next section we describe how a cell produces a protein from the instructions encoded in the nucleotide sequence of a particular gene.

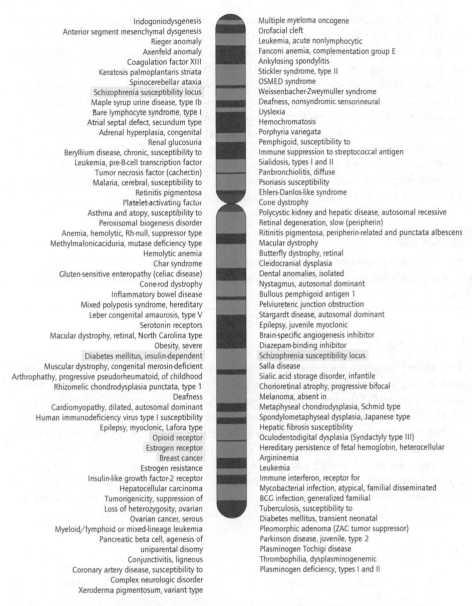

Iridogoniodysgenesis	Multiple myeloma oncogene
Anterior segment mesenchymal dysgenesis	Orofacial cleft
Rieger anomaly	Leukemia, acute nonlymphocytic
Axenfeld anomaly	Fanconi anemia, complementation group E
Coagulation factor XIII	Ankylosing spondylitis
Keratosis palmoplantaris striata	Stickler syndrome, type II
Spinocerebellar ataxia	OSMED syndrome
Schizophrenia susceptibility locus	Weissenbacher-Zweymuller syndrome
Maple syrup urine disease, type Ib	Deafness, nonsyndromic sensorineural
Bare lymphocyte syndrome, type I	Dyslexia
Atrial septal defect, secundum type	Hemochromatosis
Adrenal hyperplasia, congenital	Porphyria variegata
Renal glucosuria	Pemphigoid, susceptibility to
Beryllium disease, chronic, susceptibility to	Immune suppression to streptococcal antigen
Leukemia, pre-B-cell transcription factor	Sialidosis, types I and II
Tumor necrosis factor (cachectin)	Panbronchiolitis, diffuse
Malaria, cerebral, susceptibility to	Psoriasis susceptibility
Retinitis pigmentosa	Ehlers-Danlos-like syndrome
Platelet-activating factor	Cone dystrophy
Asthma and atopy, susceptibility to	Polycystic kidney and hepatic disease, autosomal recessive
Peroxisomal biogenesis disorder	Retinal degeneration, slow (peripherin)
Anemia, hemolytic, Rh-null, suppressor type	Ritinitis pigmentosa, peripherin-related and punctata albescens
Methylmalonicaciduria, mutase deficiency type	Macular dystrophy
Hemolytic anemia	Butterfly dystrophy, retinal
Char syndrome	Cleidocranial dysplasia
Gluten-sensitive enteropathy (celiac disease)	Dental anomalies, isolated
Cone-rod dystrophy	Nystagmus, autosomal dominant
Inflammatory bowel disease	Bullous pemphigoid antigen 1
Mixed polyposis syndrome, hereditary	Pelviureteric junction obstruction
Leber congenital amaurosis, type V	Stargardt disease, autosomal dominant
Serotonin receptors	Epilepsy, juvenile myoclonic
Macular dystrophy, retinal, North Carolina type	Brain-specific angiogenesis inhibitor
Obesity, severe	Diazepam-binding inhibitor
Diabetes mellitus, insulin-dependent	Schizophrenia susceptibility locus
Muscular dystrophy, congenital merosin-deficient	Salla disease
Arthrophathy, progressive pseudorheumatoid, of childhood	Sialic acid storage disorder, infantile
Rhizomelic chondrodysplasia punctata, type 1	Chorioretinal atrophy, progressive bifocal
Deafness	Melanoma, absent in
Cardiomyopathy, dilated, autosomal dominant	Metaphyseal chondrodysplasia, Schmid type
Human immunodeficiency virus type I susceptibility	Spondylometaphyseal dysplasia, Japanese type
Epilepsy, myoclonic, Lafora type	Hepatic fibrosis susceptibility
Opioid receptor	Oculodentodigital dysplasia (Syndactyly type III)
Estrogen receptor	Hereditary persistence of fetal hemoglobin, heterocellular
Breast cancer	Argininemia
Estrogen resistance	Leukemia
Insulin-like growth factor-2 receptor	Immune interferon, receptor for
Hepatocellular carcinoma	Mycobacterial infection, atypical, familial disseminated
Tumorigenicity, suppression of	BCG infection, generalized familial
Loss of heterozygosity, ovarian	Tuberculosis, susceptibility to
Ovarian cancer, serous	Diabetes mellitus, transient neonatal
Myeloid/lymphoid or mixed-lineage leukemia	Pleomorphic adenoma (ZAC tumor suppressor)
Pancreatic beta cell, agenesis of	Parkinson disease, juvenile, type 2
uniparental disomy	Plasminogen Tochigi disease
Conjunctivitis, ligneous	Thrombophilia, dysplasminogenemic
Coronary artery disease, susceptibility to	Plasminogen deficiency, types I and II
Complex neurologic disorder	
Xeroderma pigmentosum, variant type	

Figure 14-12 Genes located on chromosome 6 that code for proteins associated with the indicated traits and disorders. The Human Genome Project has mapped all 46 chromosomes. [From the Human Genome Project Information Web site, U.S. Department of Energy.]

In 2003, a 13-year international effort known as the Human Genome Project was completed. The Human Genome Project mapped out the nucleotide sequence and arrangement of all the genes in the human genome. For example, a simplified map of chromosome 6 is shown in **Figure 14-12**. Chromosome 6 has 170,000,000 base pairs, and some of the traits and disorders associated with genes on this chromosome are shown at their approximate location on the chromosome. For example, a gene linked to breast cancer, a gene linked to Parkinson's disease, as well as the genes for the opioid and estrogen receptors are located on this chromosome. All the other chromosomes have been similarly mapped.

In the opening story, we learned that a genetic disorder is caused by an altered gene, which codes for a protein that cannot perform its function properly. Since most diseases have some genetic component, knowing where a particular gene or set of genes lies on the genome allows us to better understand genetic diseases and, ultimately, closer to finding treatment options.

Mitochondria contain their own DNA, coding for 37 genes essential for mitochondrial function. Mitochondrial DNA is inherited from your biological mother.

Genetic modification, GM, refers to DNA technologies that alter the genetic makeup of an organism such as animals, plants, or bacteria. Combining genes from different organisms is known as recombinant DNA technology, creating genetically modified (GM), genetically engineered, or transgenic products. GM products include medicines, vaccines, foods, and food ingredients. In 2006, 252 million acres of transgenic crops were planted in 22 countries. Most of these crops were herbicide and insect resistant soybeans, corn, cotton, canola and alfalfa. Soon bananas that produce human vaccines against infectious diseases such as hepatitis B will be available. GM technology has its benefits, but like any new technology, also poses potential risks. Controversies surrounding GM foods and crops focus on human and environmental safety, which you will hear much more about in the coming years.

PRACTICE EXERCISES

13 What is a gene? What is a genome? How many base pairs are there in the human genome?

14 What is the charge on a histone? Explain how the charge on a histone facilitates the formation of nucleosomes.

DNA Replication

The ability to reproduce requires that an organism or a cell be able to replicate itself, a process that begins with DNA replication. DNA replication requires several different enzymes and begins with the unwinding of a section of the DNA double helix to expose the two strands of DNA, referred to as **parent strands**. Free nucleotides in solution assemble along the exposed parent strands, forming base pairs. The important enzyme *DNA polymerase* catalyzes the formation of phosphate ester linkages between the 5′ phosphate group of a nucleotide and the 3′ hydroxyl group on the growing nucleotide chain. Both parent strands serve as the template for a new **daughter strand**, complementary to the parent strand, as illustrated in Figure 14-13.

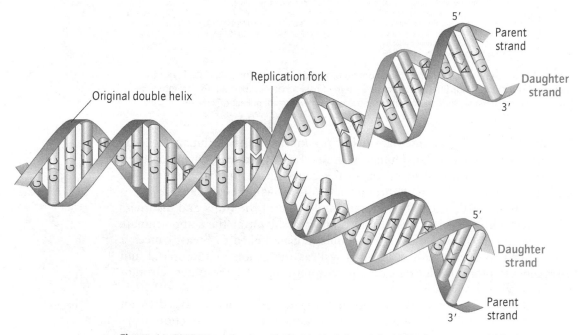

Figure 14-13 DNA replication. Each strand of the original DNA (parent strands) becomes a template for the new strand (daughter strands).

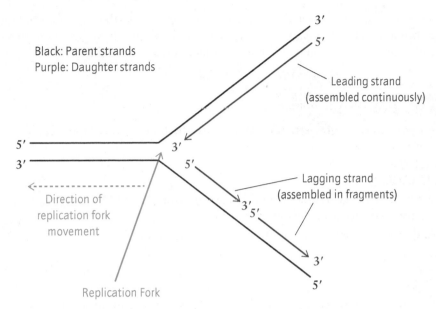

Black: Parent strands
Purple: Daughter strands

Leading strand
(assembled continuously)

Direction of
replication fork
movement

Lagging strand
(assembled in fragments)

Replication Fork

Figure 14-14 Leading strand and lagging strand synthesis always occurs from the 5′ to the 3′ end in DNA replication.

At the end of the replication process, two new double-stranded DNA molecules, identical to the original DNA, are produced. Each new double helix contains one strand from the original double helix and one new daughter strand.

Both strands of DNA are copied simultaneously and always in the 5′ → 3′ direction. Since the two strands are antiparallel, this means that the two daughter strands are synthesized in opposite directions. *An important characteristic of DNA polymerase is that it can only synthesize a nucleic acid in one direction: 5′ → 3′.* Replication begins where the two strands of the double helix have been separated, known as a **replication fork**, illustrated in Figure 14-14. Since *DNA polymerase* can only work in the 5′ → 3′ direction, only one strand, known as the **leading strand**, can be synthesized continuously from the 5′ to the 3′ end. The other strand, known as the **lagging strand**, which is also synthesized from the 5′ to the 3′ end, must therefore be synthesized *discontinuously*, forming polynucleotide fragments that are then later joined.

As the leading strand approaches the replication fork, the enzyme *helicase* further unwinds the two DNA parent strands, so the replication fork moves in the same direction as the synthesis of the leading strand. Hence, the leading strand is formed continuously.

In contrast, synthesis of the lagging strand is discontinuous because it is synthesized in the direction opposite the movement of the replication fork. As a result, new sections of parent strand are constantly being exposed that have not been replicated. *DNA polymerase* begins to assemble another lagging strand fragment from the site of the new replication fork to the start of the previous fragment. In this way, the lagging strand is built in fragments of 100–200 base pairs. The enzyme *DNA ligase* then catalyzes the reaction that joins adjacent fragments to form the lagging strand.

DNA replication occurs at an astonishing rate—about 100 nucleotides per second! *DNA polymerase* is not only fast, but accurate, inserting the wrong nucleotide less than once every 10,000 nucleotides. *DNA polymerase* also proofreads the daughter strands for mistakes. When errors are detected, it signals other enzymes to replace and repair incorrectly placed nucleotides.

After proofreading and repair, the error rate drops to less than one in a billion nucleotides. Natural mistakes in replication do occur, however, and these mistakes account for some of the mutations that exist in our DNA.

DNA Replication

14-11 Consider a portion of double-stranded DNA with the following complementary sequence of base pairs.

5′ dAACCTTGG 3′
3′ dTTGGAACC 5′

Write the sequence of nucleotides found in each new replicated DNA segment. Label the parent strands and the daughter strands.

14-12 What role do the following enzymes play in DNA replication?
   a. *DNA polymerase*
   b. *Helicase*
   c. *DNA ligase*

14-13 For the statements below, indicate whether it applies to the leading strand or to the lagging strand, or both.
   a. _____ Is synthesized in the 5′ → 3′ direction
   b. _____ Is synthesized continuously
   c. _____ Requires *DNA ligase* to join fragments
   d. _____ Is a daughter strand
   e. _____ Is synthesized in the same direction as the movement of the replication fork

Solutions

14-11 Parent strands   5′ dAACCTTGG 3′   3′ dTTGGAACC 5′
        Daughter strands   3′ dTTGGAACC 5′   5′ dAACCTTGG 3′

14-12 a. *DNA polymerase* joins nucleotides in the 5′ to 3′ direction, by forming phosphate esters and proofreads for errors in the nucleotide sequence of the daughter strands.
        b. *Helicase* unwinds the DNA double helix at the replication fork exposing a section of the two parent strands of DNA.
        c. *DNA ligase* joins adjacent lagging strand fragments to create one of the daughter strands.

14-13 a. Both leading and lagging strands
        b. Leading strand
        c. Lagging strand
        d. Both leading and lagging strands
        e. Leading strand

15 Consider a portion of double-stranded DNA with the following nucleotide sequence:

5′ dGACCTAGCCC 3′
3′ dCTGGATCGGG 5′

Write the sequence of nucleotides found in each new DNA daughter strand after replication, and label the parent and daughter strands.

16 In which direction does *DNA polymerase* always build a polynucleotide?

17 Why is the lagging strand synthesized in fragments?

18 Which of the following illustrations accurately represents the direction in which the leading strand and the lagging strand are synthesized during DNA replication:

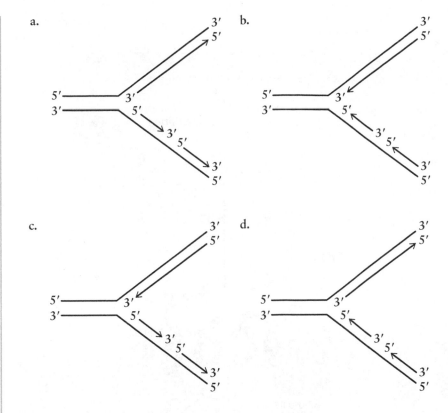

19 For the correct illustration in exercise 18, label the following:

   a. Leading strand
   b. Lagging strand
   c. Replication fork
   d. Direction of movement of replication fork
   e. Parent strands
   f. Daughter strands

■ 14.3 The Role of DNA and RNA in Protein Synthesis

The primary function of DNA is as a repository of genetic information necessary for building proteins. The basic flow of information is from DNA to messenger RNA to protein, known as the **central dogma** of molecular biology. In this section we consider these steps in detail.

Protein synthesis requires both DNA *and* RNA. RNAs, the other important nucleic acids of the cell, contain the monosaccharide D-ribose rather than 2-deoxy-D-ribose, and the nitrogenous base uracil instead of thymine.

RNA is found in three forms, each with a different role in protein synthesis:

• ribosomal RNA (rRNA),

• messenger RNA (mRNA), and

• transfer RNA (tRNA).

Amino acids are assembled into proteins at *ribosomes*, structures located throughout the cytoplasm of the cell. Ribosomes are the small dark brown dots seen in the micrograph of a cell in Figure 14-15. Ribosomes are composed of both ribosomal RNA (rRNA) and proteins. Ribosomal RNA is composed of polynucleotides containing over 1000 nucleotides.

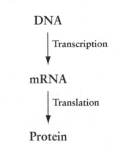

DNA
↓ Transcription
mRNA
↓ Translation
Protein

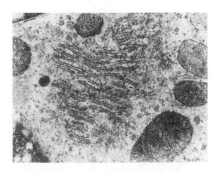

Figure 14-15 This micrograph of an animal cell shows the ribosomes in dark brown and the mitochondria in purple. [D Spector/Photolibrary/Getty]

Figure 14-16 Overview of the flow of genetic information: transcription of DNA (in the nucleus) to mRNA; and translation from mRNA to proteins at the ribosomes (in the cytosol).

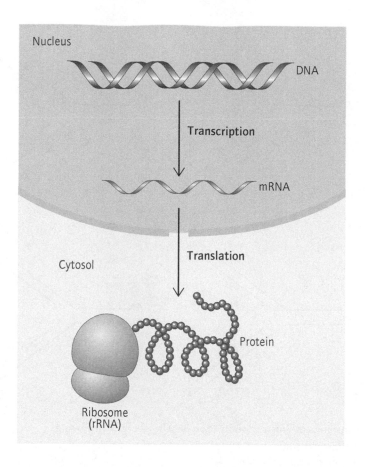

Although ribosomes are located outside the nucleus, DNA never leaves the nucleus. Instead, messenger RNA (mRNA) carries the information encoded in a segment of DNA from the nucleus to the ribosomes, where protein synthesis takes place, a process that takes place in two steps, as illustrated in Figure 14-16:

- **Transcription:** In the nucleus of the cell, a segment of one strand of DNA (a gene) is copied as a complementary single-stranded messenger RNA (mRNA).
- **Translation:** At the ribosome, mRNA serves as a template for assembling tRNAs containing the requisite amino acids, which are joined by peptide bonds to form a polypeptide.

Transcription: DNA to mRNA

When a particular protein is required by the cell, the appropriate gene is expressed, and the protein is synthesized. **Gene expression** is a complex process that is carefully regulated by the cell. Indeed, the expression of different proteins in different cell types is one way that cells are differentiated. A brain cell, for example, will express different genes, thus producing different proteins, than a muscle cell.

Gene expression begins with transcription. During transcription a nucleotide sequence—the gene—on a section of one strand of DNA, known as the **template strand** or (+) strand, is copied as a complementary single-stranded mRNA transcript, as illustrated in Figure 14-17. The DNA strand that is not copied is known as the **coding strand** or (−) strand because it has the same nucleotide sequence as the mRNA transcript.

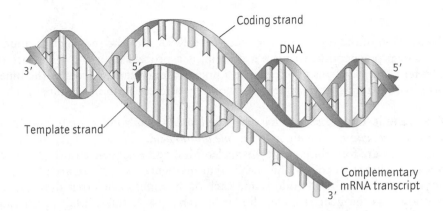

Figure 14-17 Transcription, showing a segment of DNA unraveled to expose the nucleotide sequence on the template strand so that *RNA polymerase II* can join complementary nucleotides that have assembled along the template strand (shown in red). mRNA synthesis occurs from the 5′ to the 3′ end of the mRNA transcript.

As with DNA replication, the mRNA transcript is synthesized in the 5′ → 3′ direction by reading the DNA template strand in the 3′ → 5′ direction. The nucleotide sequence of the mRNA transcript is *complementary* to the nucleotide sequence on the DNA template strand. For example, if the nucleotide sequence on the template strand of DNA is 3′ dGATCAT 5′, the mRNA transcript will have the complementary sequence 5′ CUAGUA 3′. Note that an A on DNA is copied as a U on mRNA because RNA contains uracil (U) instead of thymine (T).

In our cells, *RNA polymerase II* catalyzes the reaction between the 5′ phosphate group on an incoming nucleotide and the hydroxyl group at the 3′ end of the growing mRNA chain. *RNA polymerase II* is an enzyme that assembles nucleotides at a rate of 50 nucleotides per second. This reaction is similar to the reaction catalyzed by *DNA polymerase*, except the nucleotides contain D-ribose as a sugar and the nitrogenous base uracil (U) rather than thymine (T).

Unlike bacteria our genes do not consist of one continuous nucleotide sequence that codes for a protein, but instead our genes contain intervening sections of hundreds to thousands of noncoding nucleotides. Thus, the initial product of transcription is a **pre-mRNA** polynucleotide that must be cut and spliced into the final mRNA before translation and protein synthesis can occur.

WORKED EXERCISES | Transcription

14-14 For a DNA template strand containing the sequence 3′ dAATTGGCC 5′, what is the sequence of nucleotides from the 5′ to the 3′ end in the mRNA transcript after transcription?

14-15 What is the role of *RNA polymerase II*?

14-16 Where in the cell does transcription occur?

Solutions

14-14 A complementary set of nucleotides would be found on mRNA, but U rather than T is found in RNA: 5′ dUUAACCGG 3′.

14-15 *RNA polymerase II* is the enzyme in human cells that catalyzes the synthesis of mRNA from nucleotides that are complementary to the template strand of DNA.

14-16 Transcription occurs in the nucleus of the cell.

PRACTICE EXERCISES

20 What are the three types of RNA? Which type of RNA makes up part of the structure of a ribosome? What other type of biomolecule makes-up a ribosome?

21 Describe the process of transcription.

22 What happens when a gene is expressed?

Information flows from DNA to RNA to proteins.

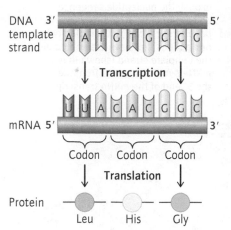

Figure 14-18 The flow of information from DNA to proteins—the central dogma. Transcription of DNA produces a complementary sequence of nucleotides on mRNA. Every codon (three bases) on mRNA codes for one amino acid.

Translation

Translation of the nucleotide sequence on mRNA into the sequence of amino acids of a polypeptide, to form a protein, is a process known as translation. The term *translation* is used because a nucleotide sequence serves as the blueprint for an amino acid sequence—two very different monomer units.

The Genetic Code *Every three non-overlapping nucleotides on an mRNA, known as a* **codon,** *uniquely specifies one amino acid, as shown in* Figure 14-18. Since there are four different RNA nucleotides, and a sequence of three nucleotides constitutes a codon, this means that mathematically there are $(4^3) = 64$ possible codons. The 64 codons and each of the amino acids that they specify is known as the **genetic code,** shown in Table 14-2. Interestingly, all living organisms have the same genetic code.

The codon AUG specifies methionine and also the start of an amino acid sequence. Three of the 64 codons are *stop codons* (UAA, UAG, UGA), which signal the end of the polypeptide. The remaining 61 codons code for the 20 natural amino acids. Since there are more codons than there are amino acids, most amino acids are specified by more than one codon. For example, the two codons CAU and CAC both specify the amino acid histidine (Table 14-2). Generally, when multiple codons specify the same amino acid, the codons differ in the third nucleotide of the codon. For example, the two codons that specify histidine both have CA as the first two nucleotides, but differ in the third nucleotide, U versus C. This redundancy of the genetic code is important because it minimizes the effect of transcription errors. In other words, if CAU

TABLE 14-2 The Genetic Code

	Second nucleotide				
	U	C	A	G	
U	UUU / UUC → Phe UUA / UUG → Leu	UCU / UCC / UCA / UCG → Ser	UAU / UAC → Tyr UAA STOP / UAG STOP	UGU / UGC → Cys UGA STOP / UGG Trp	U C A G
C	CUU / CUC / CUA / CUG → Leu	CCU / CCC / CCA / CCG → Pro	CAU / CAC → His CAA / CAG → Gln	CGU / CGC / CGA / CGG → Arg	U C A G
A	AUU / AUC → Ile AUA Met AUG START	ACU / ACC / ACA / ACG → Thr	AAU / AAC → Asn AAA / AAG → Lys	AGU / AGC → Ser AGA / AGG → Arg	U C A G
G	GUU / GUC / GUA / GUG → Val	GCU / GCC / GCA / GCG → Ala	GAU / GAC → Asp GAA / GAG → Glu	GGU / GGC / GGA / GGG → Gly	U C A G

First nucleotide (left) · Third nucleotide (Wobble position) (right)

incorrectly appeared where a CAC should appear in the mRNA, the amino acid histidine would still be specified. Note that while there is redundancy in the genetic code, a codon always specifies only one amino acid. In other words, if you look at a sequence of codons in an mRNA, you will be able to predict the amino acid sequence produced from that sequence of codons; however, if you look at an amino acid sequence in a polypeptide, you will *not* be able to predict the sequence of nucleotides that coded for that amino acid sequence.

Ribosomes and tRNA Translation occurs at the ribosomes. Ribosomes are enormous structures that consist of two subunits, both composed of rRNA and various proteins, one larger than the other. During protein synthesis, the two subunits lock together with the messenger RNA trapped in the space between them. The larger subunit contains the active site, where peptide bonds are made. During translation, the ribosome moves along the mRNA in the $5' \rightarrow 3'$ direction reading one codon (three nucleotides) at a time.

Assembling amino acids in sequence, according to the codons specified by mRNA, is the job of transfer RNA (tRNA). Transfer RNAs (tRNAs) are smaller nucleic acids composed of 75–85 nucleotides, with a characteristic cloverleaf shape, containing sections of both double-stranded and single-stranded loops, as illustrated in Figure 14-19. The two important regions on a tRNA are:

- the anticodon loop (shown in red in Figure 14-19), a sequence of three nucleotides, that varies for the different tRNAs, and
- the 3' end, which forms a covalent bond (an ester) to the specific amino acid associated with the anticodon of that tRNA.

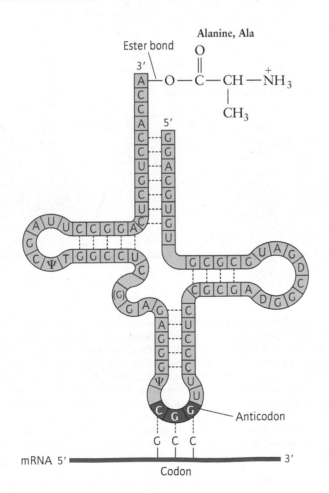

Figure 14-19 The overall structure of a tRNA molecule. The anticodon loop (in red) is usually complementary to the codon on the mRNA. The amino acid is attached in an ester linkage at the 3' end of the tRNA. Here the anticodon CGG is base paired to the mRNA codon GCC, which codes for alanine, the amino acid at the 3' end of this tRNA molecule.

During translation, three base pairs are formed between an mRNA codon and the complementary tRNA **anticodon**. For example, a 5′ GCC 3′ codon on mRNA binds a tRNA with the anticodon 3′ CGG 5′, carrying the amino acid alanine at its 3′ end, as shown in Figure 14-19. Note that mRNA codons are always written in the 5′ → 3′ direction and tRNA anticodons are written in the 3′ → 5′ direction. A tRNA carrying an amino acid has its 3′ hydroxyl group in an ester linkage to the carboxylic acid end of the amino acid.

Although there are 61 mRNA codons that specify an amino acid, most cells contain less than 45 tRNAs. This is because tRNAs can base pair to more than one mRNA codon. The explanation for this is known as the **wobble hypothesis**, which allows nonstandard base pairings in the third position of an mRNA codon, if it codes for the same amino acid. For example, the mRNA codons 5′ CAA 3′ and 5′ CAG 3′ both code for the amino acid glutamine, according to the genetic code. A tRNA molecule with the anticodon 3′ GUU 5′ can base pair with either of these mRNA codons because they both code for glutamine: the 5′ CAA 3′ to 3′ GUU 5′ pairing is the standard pairing, while the 5′ CAG 3′ to 3′ GUU 5′ pairing contains the nonstandard pairing G to U in the third base—the wobble position.

Assembling the amino acids on the ribosome Base pairings between the anticodon on a tRNA molecule and a codon on the mRNA brings an amino acid to the ribosome, in proximity to the growing polypeptide chain, as illustrated for a tRNA carrying isoleucine in Figure 14-20a. Next, a reaction (acyl transfer) occurs between the ester of the growing polypeptide chain on the

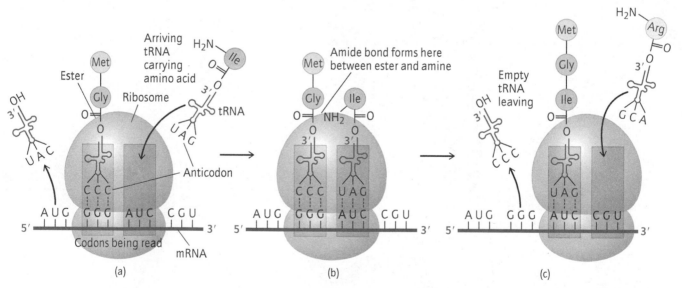

Figure 14-20 Translation, showing the ribosome in orange, the mRNA in red, and the tRNAs as cover-leaf shaped structures with amino acids represented as colored spheres. Transcription occurs from the 5′ to the 3′ end of mRNA: (a) A codon (AUC) on mRNA is read and the anticodon (UAG) on a tRNA carrying isoleucine base pairs with the codon through hydrogen bonding. The previous tRNA is still bound to the mRNA with the polypeptide chain attached to its 3′ hydroxyl as an ester. (b) A peptide bond is formed between the ester of the growing polypeptide chain and the free amine of the new amino acid, causing the polypeptide chain to grow by one amino acid, isoleucine, and move to the last new tRNA (c) The previous tRNA, no longer containing an amino acid, leaves the ribosome and the ribosome shifts to the next codon on mRNA. The process is then repeated until a stop codon is reached.

previous tRNA and the free amine of the amino acid on the most recently recruited tRNA, to form a peptide bond, an amide (see Figure 14-20b). In the reaction, the growing polypeptide chain moves to the most recently recruited tRNA. The previous tRNA molecule, with its amino acid no longer attached, containing a free 3′ hydroxyl group, diffuses away from the ribosome to find another amino acid (another glycine) as shown in Figure 14-20c. The most recently recruited tRNA, still bound to the ribosome, now contains the growing polypeptide chain at its 3′ end.

Note also how the ribosome has shifted by one codon in the 5′ → 3′ direction on the mRNA (Figure 14-20b to Figure 14-20c). The next codon on mRNA is then read, another tRNA carrying the appropriate amino acid is recruited, and the process repeats itself until the entire polypeptide has been assembled and a stop codon has been reached.

Following translation, the polypeptide must still undergo additional modifications to become a functional protein: folding, forming disulfide bonds, and electrostatic interactions with other polypeptides to form quaternary structures. However, the primary structure of the polypeptide has been assembled at the ribosome.

WORKED EXERCISES Transcription and Translation

14-17 Using the genetic code shown in Table 14-2, determine the amino acid specified by the following codons on mRNA:
   a. AAG
   b. UGC

14-18 What is the anticodon on the tRNAs carrying the amino acids given in question 14-17? Assume standard base pairings.

14-19 Consulting Table 14-2, list all the mRNA codons that code for threonine. What is the biological benefit of having more than one codon for a given amino acid?

14-20 Mathematically, why are there 64 possible codons?

14-21 For the following mRNA sequences, what is the amino acid sequence formed? Consult Table 14-2.
   a. 5′ AGUCCGUAC 3′
   b. 5′ AAUUGCUUC 3′

Solutions

14-17 a. The codon AAG codes for lysine.
   b. The codon UGC codes for cysteine.

14-18 a. The anticodon for the codon AAG is UUC.
   b. The anticodon for the codon UGC is ACG.

14-19 The mRNA codons that code for threonine are ACU, ACC, ACA, and ACG. The benefit of this redundancy is that if there is a transcription error in the third nucleotide of the codon, the correct amino acid, in this case threonine, is still specified.

14-20 There are 64 codons because there are four different RNA nucleotides (C, G, A, and U) and a codon consists of three nucleotides, so statistically there are $4^3 = 4 \times 4 \times 4 = 64$ different ways to arrange a sequence of three nucleotides.

14-21 a. Ser-Pro-Tyr
   b. Asn-Cys-Phe

PRACTICE EXERCISES

23 Using Table 14-2, determine the amino acid specified by the following mRNA codons:
  a. GGG
  b. AAA

24 Using Table 14-2, specify all the codons that code for the following amino acids:
  a. Cysteine
  b. Leucine

25 For the following mRNA sequences, indicate the anticodons on the four tRNA molecules involved in building the polypeptide. Assume standard base pairings. Indicate the order in which the tRNA molecules are recruited and the amino acid that each will be carrying. What is the amino acid sequence formed? You may consult Table 14-2.
  a. 5′ GGUACUAUUUAU 3′
  b. 5′ GGUGACCGACAU 3′

26 What are the three stop codons? What is a stop codon?

Genetic Mutations

In the opening story you were introduced to genetic mutations. *A genetic mutation is any permanent chemical change to one or more nucleotides in a gene (DNA) that affects the primary structure of a protein.* Two basic types of mutations include:

- **Substitutions:** One nucleotide in the gene has been substituted by another;
- **Frameshift:** One nucleotide is added to or deleted from the gene.

Both types of mutations have the potential to change the amino acid sequence as a result of transcription and translation. Consider the normal 13 nucleotide sequence in a gene shown in **Figure 14-21a**, which after transcription

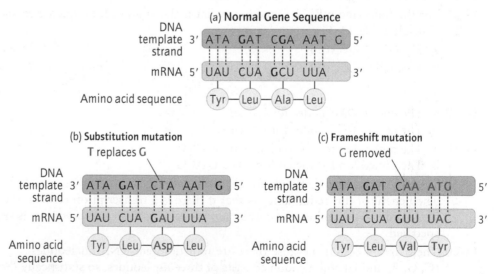

Figure 14-21 Mutations arise when a change in the nucleotide sequence on a gene leads to changes in the amino acid sequence. (a) Normal DNA segment, corresponding mRNA, and sequence of four amino acids. (b) Substitution mutation: The nucleotide T replaces G, causing a change in the third codon in mRNA, and thus the third amino acid. (c) Frame shift mutation: The nucleotide G is deleted, causing a change in the two codons in mRNA, thus causing a change in the last two amino acids.

produces an mRNA with the four codons shown. Referring to the genetic code in Table 14-2, translation yields the sequence Tyr-Leu-Ala-Leu. A *substitution* mutation of T for **G** at one nucleotide as indicated in Figure 14-21b, yields the amino acid sequence Tyr-Leu-Asp-Leu instead, a sequence containing one different amino acid. If instead, the nucleotide G is *deleted* as shown in Figure 14-21c, then a frameshift mutation occurs producing the sequence Tyr-Leu-Val-Tyr, a sequence with different amino acids after the deletion.

The effects of DNA mutations are minimized by the redundancy of the genetic code, as more than one codon exists for every amino acids. For example, if a DNA mutation produces the codon CCA instead of CCC, the mutation has no effect on the polypeptide produced because both these codons code for the same amino acid, proline. The codons CCU and CCG also code for proline. In other words, any possible nucleotide in the third position following CC codes for proline.

While we have seen examples of genetic mutations that have serious health consequences (see opening story), sometimes there are beneficial genetic mutations. One example of a beneficial genetic mutation is tetrachromatic vision. Most people have trichromatic vision, three cone receptors (retinal photopigments) for red, green, and blue. For the sake of comparison, color blind men (8% of the population) have dichromatic vision: cone receptors for only blue and red, or blue and green. In what is believed to be a relatively modern mutation, some women have a mutation on their X-chromosome that gives them tetrachromatic vision: four cone receptors that give them a heighted sense of color and the ability to discern subtle color differences—super color vision. While the average trichromat can discern one million different color hues, a tetrachromat can discern 100 million different color hues.

PRACTICE EXERCISES

27 Indicate whether the following normal mRNA sequence would produce the same or a different amino acid sequence if the mutations listed below occurred (consult Table 14-2). If so, indicate the new amino acid sequence.

DNA template strand: 3′ dTAATGA 5′

mRNA: 5′ AUUACU 3′

Amino acids: Ile-Thr

a. First T on DNA is replaced by G.

b. Third nucleotide on DNA is replaced by G.

c. Third nucleotide on DNA is deleted and seventh nucleotide is G.

d. First nucleotide on DNA is deleted and seventh nucleotide is A.

The year 2013 marked the 60th anniversary of the discovery of the structure of DNA, a macromolecule containing the information, in the form of a nucleotide sequence, for assembling amino acids into proteins. In the decades since this discovery, the entire human genome has been mapped and DNA technologies continue to emerge at a remarkable rate, transforming the landscape of medicine and other fields such as forensic science. The twenty-first century is an exciting time to be working in the healthcare professions, as you will be supremely positioned to observe advances unfold in this rapidly advancing and exciting field.

Chemistry in Medicine | HIV and Viral Nucleic Acids

The *h*uman *i*mmunodeficiency *v*irus, **HIV**, causes *a*cquired *i*mmuno*d*eficiency *s*yndrome, AIDS, a condition in which the immune system can no longer fight off life-threatening opportunistic infections, such as pneumonia. The Joint United Nations Program on HIV/AIDS estimates that 40 million people around the world are infected with the HIV virus, including 1.7 million people in the United States.

A virus is a nucleic acid—RNA or DNA— encapsulated in a protein coat. In fact, viruses are classified according to the type of nucleic acid they contain: RNA or DNA; single-stranded or double-stranded; and for single-stranded nucleic acids, template strand (+) or coding strand (−), giving rise to the seven major types of viruses.

The HIV virus contains two single-stranded (−)-RNAs. Since a virus lacks the full set of enzymes and molecular structures needed to replicate its genome and to produce viral proteins, it uses the enzymes and molecular structures of the host cell. The HIV virus targets T-lymphocytes, a type of white blood cell that plays a central role in the immune system. The virus inserts its viral DNA into the host's DNA, so that when the host cell's DNA undergoes transcription and translation, viral proteins are also produced. Eventually the viral infection destroys the host cell, but not until after many more new viruses have been produced by the host that can infect other healthy T-lymphocytes.

The life cycle of the HIV virus is illustrated in Figure 14-22. The first step is known as *fusion*, wherein the virus attaches itself to specific receptors located on the surface of T-lymphocytes (step 1). The virus then injects its genomic RNA and some key enzymes into the T-lymphocyte (step 2).

Before it can be incorporated into the host's double-stranded *DNA*, single-stranded genomic viral *RNA* must be transcribed into a single strand of viral *DNA*, a process known as *reverse transcription*, since it proceeds in the opposite direction of normal transcription. The single strand of viral DNA produced then serves as a template for creating a second strand, forming viral double-stranded DNA (step 3). It is because of this step that the HIV virus is classified as a **retrovirus**, a class of viruses that transcribes their genomic RNA into viral DNA.

The virus uses its own enzyme *HIV reverse transcriptase* to catalyze reverse transcription. *HIV Reverse transcriptase* is an error-prone enzyme, causing many mutations in the viral DNA produced, thus constantly changing the viral proteins produced. These are beneficial mutations for the virus, as it makes it difficult for the host's immune system to identify viral proteins and mount an immune response. Efforts to develop a vaccine for HIV have been thwarted because of the rapidly mutating viral DNA.

After double-stranded viral DNA is formed in the cytoplasm it moves into the nucleus of the host cell, where it is incorporated into the host's DNA (step 4), because they are simply polynucleotides just like the

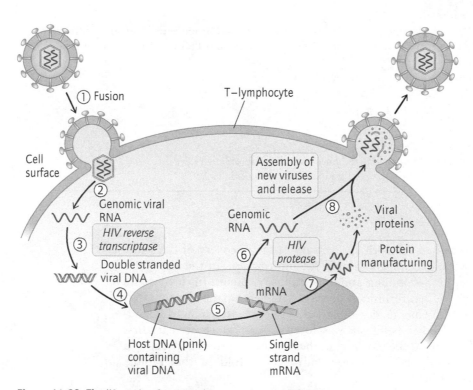

Figure 14-22 The life cycle of an HIV virus as it infects a T-lymphocyte.

host's DNA. Once incorporated into the host's DNA, viral genes lie dormant until genes in the host's DNA are expressed, and protein synthesis begins. Ironically, gene expression in T-cells occurs when the T-cell is needed to fight off an infection. Thus, as the T-cells work to fight off an infection they are also producing viral mRNA during DNA transcription (step 5). Some of the viral mRNA produced is hydrolyzed by the viral enzyme *HIV protease*, to produce more genomic RNA (step 6) while some viral mRNA is translated into proteins, such as the key viral enzymes (step 7). In the final step, genomic RNA and these viral proteins are assembled into many new viruses and released from the host cell (step 8).

Without drug intervention, the T-cell count of an HIV infected person eventually drops to a level where the immune system can no longer mount an effective immune response. The infected individual, therefore, eventually succumbs to common infections. This is the stage at which an HIV infection becomes AIDS. The time period between HIV infection and AIDS ranges from several months to many years, and depends on many factors.

HIV drug therapies to date have focused on developing inhibitors of key enzymes involved in the life cycle of the virus, namely *HIV reverse transcriptase* and *HIV protease*. More recently fusion inhibitors, which interfere with the first step in the life cycle of the virus, have been added to the list of drugs available to treat HIV infection. Since the HIV virus rapidly mutates, it eventually becomes resistant to any one type of drug therapy. Therefore, combination drug therapies are typically used.

As with other diseases you have learned about, understanding the chemistry is the first step in developing a treatment. Although there is no cure for HIV/AIDS, understanding nucleic acids has made it possible for scientists to develop targeted drug therapies that have allowed HIV-positive individuals to live much longer and healthier lives.

Summary

The Chemical Structure of a Nucleotide and Polynucleotides

- A nucleotide contains three parts: a monosaccharide, a phosphate, and a nitrogenous base.
- RNA contains D-ribose as its monosaccharide, and DNA contains 2-deoxy-D-ribose. 2-Deoxy-D-ribose lacks an OH group on the C(2) carbon of D-ribose.
- At the anomeric carbon, C(1), the monosaccharide contains a β-N-glycosidic linkage to the nitrogen atom of a nitrogenous base.
- The four nitrogenous bases found in DNA are adenine, guanine, cytosine, and thymine. The four nitrogenous bases found in RNA are adenine, guanine, cytosine, and uracil. DNA does not contain uracil, and RNA does not contain thymine.
- The phosphate group in a nucleotide is in a phosphate ester linkage to the 5′ alcohol of the monosaccharide.
- Nucleotide triphosphates of deoxyribose (dNTPs) are high energy forms of a nucleotide that contain two phosphoanhydride bonds, which are used to drive the reaction that joins a nucleotide to a polynucleotide.
- RNA and DNA are linear polynucleotides joined between the 5′ phosphate group of one nucleotide and the 3′ hydroxyl group of another nucleotide as a monophosphate ester.
- The only difference on each sugar along the sugar-phosphate backbone of RNA and DNA is the nitrogenous base. Thus, nucleic acids are identified by the one letter abbreviation of their nitrogenous base. A sequence of nucleotides in a polynucleotide is listed from the 5′ end to the 3′ end by convention.

DNA Structure and Replication

- The ability to reproduce is a fundamental condition for life.
- DNA is a polynucleotide constructed from millions of nucleotides.
- DNA consists of two strands of nucleic acids, which are twisted around each other into a right-handed double helix.

- The two strands of DNA are antiparallel with the 5′ end of one strand opposite the 3′ end of the other strand.
- The monosaccharide and phosphate groups of the DNA backbone are hydrophilic and therefore located on the outside of helix.
- A nitrogenous base on one DNA strand hydrogen bonds to a complementary nitrogenous base on the opposite strand to form a base pair.
- The complementary base pairs in DNA are A–T and C–G.
- The two strands of DNA are held together by hydrogen bonds between the base pairs as well as base stacking, the energetically favorable interaction between the aromatic rings of the nitrogenous bases above and below.
- DNA is coiled around a core of positively charged proteins known as histones, forming compact structures known as nucleosomes.
- Nucleosomes coil upon themselves further to form a chromosome.
- Chromosomes, located in the nucleus of the cell, are highly ordered compact structures containing DNA and specialized proteins.
- An organism's genome is the complete sequence of nucleotides in their DNA. The human genome, distributed over 46 chromosomes, contains about 3 billion base pairs.
- A gene is a segment of DNA that contains the instructions for making a protein.
- DNA replication starts with the unraveling of DNA to expose two parent strands, which each serve as a template for a new daughter strand.
- *DNA polymerase* synthesizes a nucleic acid in only one direction: 5′ → 3′. In other words, the 5′ phosphate of an incoming nucleotide is joined to the 3′ hydroxyl group of the growing polynucleotide chain.
- Because the two strands of DNA are antiparallel, and synthesis occurs only in the 5′ → 3′ direction, the two daughter strands are synthesized in opposite directions.
- The leading strand is synthesized continuously from the 5′ → 3′ end toward the replication fork, and in the same direction as the movement of the replication fork, as the DNA double strand unwinds.
- The lagging strand is synthesized discontinuously, in fragments, from the 5′ → 3′ end, in the opposite direction of the movement of the replication fork. The lagging strand fragments are then joined by *DNA ligase*.

The Role of DNA and RNA in Protein Synthesis

- The central dogma of molecular biology states that information flows from DNA to mRNA to protein.
- The three major forms of RNA are ribosomal RNA (rRNA), messenger RNA (mRNA), and transfer RNA (tRNA).
- During transcription a nucleotide sequence, known as a gene, on the template strand of DNA, is copied as a complementary single-stranded mRNA. *RNA polymerase II* catalyzes the synthesis of mRNA in the 5′ → 3′ direction.
- Codons, groups of three non-overlapping nucleotides on an mRNA, code for one of the 20 natural amino acids.
- The 64 codons and each of the amino acids they specify is known as the genetic code.
- Translation is the process that occurs on ribosomes in which the nucleotide sequence in an mRNA is read, and amino acids are assembled into a polypeptide.

- tRNAs are smaller nucleic acids consisting of an anticodon loop at one end. The other end of the tRNA can carry the corresponding amino acid at its 3′ end, joined as an ester between the hydroxyl group of tRNA and the carboxylic acid of the amino acid.
- Translation occurs on ribosomes, where base pairs are formed between an mRNA codon and the complementary anticodon of a tRNA carrying the appropriate amino acid.
- Most cells have less than 45 tRNAs because nonstandard base parings are allowed in the third position of an mRNA codon if it codes for the same amino acid, known as the wobble hypothesis.
- A peptide bond is formed during translation when an acyl transfer reaction moves the growing polypeptide chain onto the amino acid on the most recently recruited tRNA. The ribosome then shifts in the 5′ → 3′ direction to read the next codon and the process is repeated.
- A genetic mutation occurs when one or more nucleotides in a gene are chemically changed in a way that alters the primary structure of the protein.

Key Words

Anticodon A three nucleotide sequence in the anticodon loop of a tRNA that base pairs to the complementary codon on mRNA during translation.

Base pair A nitrogenous base on one polynucleotide hydrogen bonding to the complementary nitrogenous base on another polynucleotide. The complementary base pairs in DNA are A–T and G–C, and in RNA, the pairs are A–U and G–C.

Base stacking The stabilizing interaction resulting from the delocalized electrons on aromatic rings in the nitrogenous bases that hold two strands of DNA together.

Central dogma A reference to the basic flow of information from DNA to mRNA to amino acid to form proteins.

Chromosome The highly compact structures containing DNA, the genome, and specialized proteins. Found in the nucleus of our cells.

Coding strand The DNA strand that does not serve as a template during transcription. It is named as such because it has the same nucleotide sequence as the mRNA produced, except it has T instead of U.

Codon The three non-overlapping nucleotides on an mRNA that code for one amino acid.

Daughter strand A reference to the two new strands of DNA synthesized during DNA replication, as distinguished from the two parent strands that were part of the original DNA.

Frameshift mutation When one nucleotide is added to or deleted from a gene causing a change in the primary structure of a protein.

Gene A nucleotide sequence within DNA that contains the instructions for making a protein.

Genetic code The amino acids, start codon, and three stop codons specified by each of the 64 nucleotide codons of an mRNA.

Gene expression When a protein is required by the cell and transcription and translation ensues.

Genetic mutation Any permanent chemical change that occurs at one or more nucleotides in a gene that leads to a change in the primary structure of a protein.

Genome The complete DNA sequence of an organism.

Histone The positively charged proteins that facilitate the tight packing of DNA (negatively charged) into nucleosomes.

Lagging strand The daughter strand that is synthesized in fragments (in the direction opposite of the movement of the replication fork) during DNA replication by DNA polymerase in the 5′ → 3′ direction.

Leading strand The daughter strand that is synthesized continuously (in the same direction as the movement of the replication fork) during DNA replication by DNA polymerase in the 5′ → 3′ direction.

Nitrogenous base One of the three components of a nucleotide containing an aromatic ring and two or four ring nitrogens. The nitrogenous bases found in DNA are adenine (A), guanine (G), cytosine (C), and thymine (T). The nitrogenous bases found in RNA are adenine (A), guanine (G), cytosine (C), and uracil (U).

Nucleosome Compact structures containing DNA coiled around histone proteins.

Nucleotide A molecule that contains a nitrogenous base, a monosaccharide, and a phosphate group. Nucleotides are the monomer units of the polynucleotides RNA and DNA.

Nucleotide triphosphate (NTP) A nucleotide containing three phosphate groups joined by two high energy phosphoanhydride bonds, which supply the energy for joining a nucleotide to a growing polynucleotide during DNA synthesis, for example.

Parent strands A reference to the two original strands of the DNA double helix that are each copied during replication, as distinguished from the daughter strands.

Replication fork The site where two strands of DNA separate, exposing the two parent strands for replication.

Retrovirus A virus that transcribes RNA to DNA, the opposite of normal transcription (DNA to RNA) before it can be incorporated into the host's DNA.

Ribosomes Complex cellular structures located in the cytoplasm where ribosomal RNA (rRNA), together with many proteins, assemble amino acids into polypeptides by reading mRNA.

Substitution mutation When one nucleotide in a gene is substituted for another nucleotide leading to a change in the primary structure of the protein.

Template strand The DNA strand that is copied, as a complementary sequence of nucleotides in mRNA, during transcription. Note the complementary base for A is U rather than T in mRNA.

Transcription The process in which a complementary mRNA is synthesized from the template strand of a segment of DNA (the gene).

Translation The process of synthesizing a polypeptide from amino acids whose sequence is specified by the nucleotide sequence in an mRNA, using tRNAs carrying the individual amino acids.

Wobble hypothesis The explanation for why there are nonstandard base pairings observed in the third base pair, during translation, between a mRNA codon and a tRNA anticodon, provided it codes for the same amino acid.

Additional Exercises

Gene Mutations and Disease

28 What is DNA?

29 Why can some of our genes predispose us to certain diseases?

30 What is a mutation?

31 How is an acquired mutation different from an inherited mutation?

32 If a person has an acquired mutation in one type of cell, are other cell types affected?

33 What environmental factors can trigger mutations?

34 What two genes are known to predispose a woman to breast cancer?

35 What two molecules are responsible for storing and transmitting genetic information?

36 How has *HPV* been linked to cervical cancer?

The Chemical Structure of a Nucleotide and Polynucleotides

37 What are the three basic parts of a nucleotide?

38 Which of the monosaccharides shown below is part of RNA, and which is part of DNA? What is the difference between these two monosaccharides?

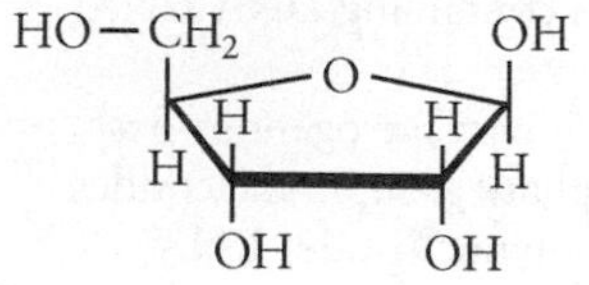

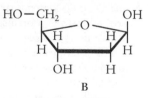

A   B

39 In a nucleotide, which carbon atom of the monosaccharide has a covalent bond to the nitrogenous base? Which nitrogen atom on the nitrogenous base is covalently bonded to the monosaccharide? Why is this bond called a β-*N*-glycosidic linkage?

40 Name the nitrogenous bases (and their one letter abbreviations) that are found in DNA.

41 Name the nitrogenous bases (and their one letter abbreviations) that are found in RNA.

42 Which carbon atom of the monosaccharide contains the hydroxyl group that forms a bond to the phosphate in a nucleotide?

43 For the nucleotide shown, circle and label the following.

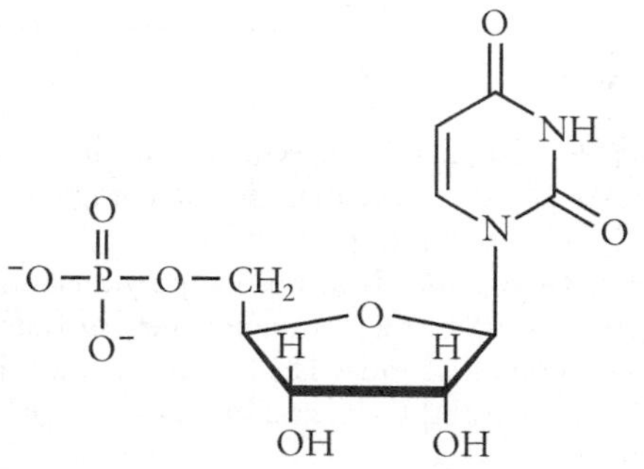

a. The nitrogenous base. What is the name of this nitrogenous base?
b. The monosaccharide. What is the name of this monosaccharide?
c. The phosphate group
d. The 5′ carbon
e. The anomeric carbon
f. The 3′ carbon
g. Would this nucleotide be found in RNA or DNA? Explain.
h. Provide the complete name of this nucleotide.

44 For the nucleotide shown, circle and label the following.

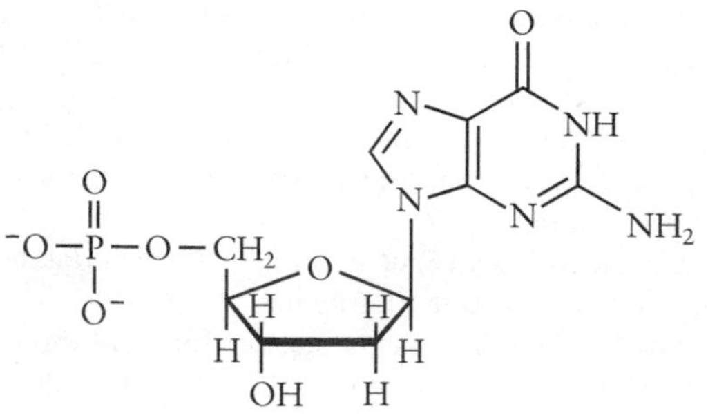

a. The nitrogenous base. What is the name of this nitrogenous base?
b. The monosaccharide. What is the name of this monosaccharide?
c. The phosphate group

d. The 5′ carbon

e. The anomeric carbon

f. The 3′ carbon

g. Would this nucleotide be found in RNA or DNA? Explain.

h. Provide the complete name of this nucleotide.

45 Write the structure of the adenine-containing nucleotide that you would find in DNA.

46 Write the structure of the guanine-containing nucleotide that you would find in RNA.

47 Construct a nucleotide from the three basic parts of a nucleotide shown below. Label the 3′ and the 5′ hydroxyl groups. Would this nucleotide be found in RNA or DNA? Explain. Name this nucleotide.

48 Construct a nucleotide from the three basic parts of a nucleotide shown below. Label the 3′ and the 5′ hydroxyl groups. Would this nucleotide be found in RNA or DNA? Explain. Name this nucleotide.

49 When two nucleotides react to form a dinucleotide, what two functional groups are involved? What new functional group is produced? Which carbon atom in the monosaccharide contains the hydroxyl group involved in this reaction?

50 Draw the structure of dTAC.

51 Draw the structure of GUC.

52 What is a nucleotide triphosphate, and what purpose does it serve in DNA synthesis?

53 What is the difference between a nucleotide and a nucleoside?

54 Explain how a molecule as large as DNA can fit inside the nucleus of a cell.

55 Is the charge on DNA positive, negative, or neutral? Is the charge on histones positive, negative, or neutral?

56 How are DNA, histones, nucleosomes, and chromosomes related?

57 How many chromosomes are found in most human cells?

58 What is your genome? Approximately how many base pairs does your genome contain?

59 Name two diseases associated with a protein that is coded for by a gene located on chromosome 6.

60 What is genetic modification, GM? What are some advantages of genetic modification of plants?

DNA Structure and Replication

61 Describe the three-dimensional shape of DNA.

62 How are the two strands of the DNA double helix arranged: parallel or antiparallel?

63 Is DNA a right-handed or a left-handed double helix? Is DNA chiral?

64 What was the convincing experimental evidence that suggested DNA had a helical structure? Who collected this data?

65 What parts of a polynucleotide are located on the exterior of the DNA double helix and why?

66 What parts of a polynucleotide are located on the interior of the DNA double helix and why?

67 List two stabilizing factors that hold the two strands of DNA together in a double helix.

68 What parts of a DNA double helix are involved in base stacking?

69 Is base stacking a stabilizing or destabilizing interaction?

70 What type of electrostatic attraction exists between base pairs?

71 How many hydrogen bonds are found in an A–T base pair?

72 How many hydrogen bonds are found in a C–G base pair?

73 For the sequence of nucleotides dTATCGC on a segment of one strand of DNA, indicate the sequence of nucleotides that would be found on the DNA strand across from it. How many hydrogen bonds in total hold the base pairs together in this segment of DNA?

74 For the sequence of nucleotides dCGATAG on a segment of one strand of DNA, indicate the sequence of nucleotides that would be found on the DNA strand across from it. How many hydrogen bonds hold the base pairs together in this segment of DNA?

75 For each sequence of nucleotides on a segment of one strand of DNA shown below, indicate the sequence of nucleotides that would be found on the other DNA strand directly across from it.

a. dCTAGGC   **c.** dTTGGAA

b. dACTGAA   **d.** dGGTACT

76 For each sequence of nucleotides on a segment of one strand of DNA shown below, indicate the sequence of nucleotides that would be found on the other DNA strand directly across from it.

a. dTATGCC   **c.** dCCTATT

b. dAACCTG   **d.** dGTATCC

77 Why is DNA replication a critical function for life?

78 Is the replication of DNA an anabolic or a catabolic biochemical process?

79 What is the function of *DNA polymerase*?

80 Where in the cell does DNA replication occur?

81 What is the function of the enzyme *helicase*?

82 Consider a portion of double-stranded DNA with the following base pairing sequence:

5′ dGGTACGCTT 3′

3′ dCCATGCGAA 5′

Write the sequence of nucleotides found in each new DNA after replication, and label the original parent strands and the two new daughter strands.

83 Consider a portion of double-stranded DNA with the following base pairing sequence:

5′ dCATTAAGCCG 3′

3′ dGTAATTCGGC 5′

Write the sequence of nucleotides found in each new DNA after replication, and label the original parent strands and the two new daughter strands.

84 What kinds of electrostatic attractions are formed between the base pairs of the parent and daughter strands of DNA?

85 What kind of reaction does the enzyme *DNA polymerase* catalyze?

86 What is the function of the enzyme *DNA ligase*?

87 Find the errors in the following sequences of parent-daughter strands and replace the incorrect nucleotides with the correct ones.

 a. 5′ dATTCCGTA 3′   parent strand
   3′ dCAAGGTAT 5′   daughter strand
 b. 5′ dGGGCCCTTTAA 3′ parent strand
   3′ dCCCAGGAAGTT 5′ daughter strand

88 Find the errors in the following sequences of parent-daughter strands, and replace the incorrect nucleotides with the correct ones.

 a. 5′ dCGTACTGGA 3′   parent strand
   3′ dCCATGAACT 5′   daughter strand
 b. 5′ dGAGTATCT 3′   parent strand
   3′ dCTCCTCGA 5′   daughter strand

89 In which direction does *DNA polymerase* always synthesize a polynucleotide?

90 Are the two daughter strands synthesized simultaneously during DNA replication? In which direction are the daughter strands synthesized?

91 How is the synthesis of the two daughter strands different? What are these two strands called and why?

92 What is a replication fork?

93 Does the lagging strand or the leading strand move in the same direction as the movement of the replication fork? Explain.

94 Which daughter strand is formed discontinuously? What happens to the lagging strand fragments after they are formed?

The Role of DNA and RNA in Protein Synthesis

95 What is the central dogma of molecular biology?

96 What is a gene?

97 What are the three major types of RNA?

98 Where in the cell does transcription occur? What nucleic acids are involved?

99 Where in the cell does translation occur? What nucleic acids are involved?

100 Which DNA strand is copied during transcription?

101 Does RNA contain D-ribose or 2-deoxy-D-ribose? Does it contain the nitrogenous base T or U?

102 Where in the cell are amino acids assembled into proteins?

103 What type of biomolecules make up a ribosome?

104 How does the genetic information get from DNA, located in the nucleus, to the ribosome, located in the cytosol?

105 What are the two basic processes involved in synthesizing a protein from a gene?

106 Explain what "gene expression" means.

107 What type of reaction does *RNA polymerase II* catalyze?

108 For the following nucleotide sequences found on a DNA template strand, write the sequence of nucleotides produced on mRNA during transcription.

 a. 3′ dATAGGCCTTA 5′
 b. 3′ dTTAACCGGAA 5′
 c. 3′ dCGATCGATCG 5′

109 For the following sequences found on a DNA template strand, write the sequence of nucleotides produced on mRNA during transcription.

 a. 3′ dCCGGAATATA 5′
 b. 3′ dAAGGCCAATT 5′
 c. 3′ dGTACACGTCG 5′

110 What is a codon?

111 What is the genetic code?

112 What are the two important regions of a tRNA molecule?

113 What are the steps involved in translation, the process of building a protein from mRNA?

114 There are 61 codons that specify an amino acid but fewer than 45 tRNAs carrying amino acids, so more than one mRNA codon must be able to base pair with the same tRNA. What must be true for two different codons to base pair with the same tRNA? What is the name of the hypothesis that explains this phenomenon?

115 What is a genetic mutation?

116 What two types of changes in the nucleotide sequence of a gene can lead to a mutation?

117 Using the genetic code (Table 14-2), what amino acid will a tRNA molecule containing the anticodon AAA be carrying?

118 Using the genetic code (Figure 14-2), what amino acid will a tRNA molecule containing the anticodon GAC be carrying?

119 During translation, what happens when a stop codon is reached?

120 There is one start codon. What is it and for what amino acid does it code?

121 List all the codons for leucine. Why is there more than one codon for leucine? What is the benefit of having more than one codon for the same amino acid?

122 Can a codon code for more than one amino acid?

123 Can an amino acid be specified by more than one codon?

124 Can a tRNA anticodon base pair with more than one codon?

125 Write the structure of an amino acid, and show the reaction between an amino acid and the 3' hydroxyl group of a tRNA. Which functional group on the amino acid is involved? What new functional group is produced?

126 What are all the codons for alanine?

127 For the following nucleotide sequences on mRNA, indicate the three amino acid sequence formed?
   a. 5' AAUAGUGUG 3'   c. 5' CACCGGUGG 3'
   b. 5' CCCUUUGGG 3'   d. 5' UCCUUAGCA 3'

128 For the following nucleotide sequences on mRNA, indicate the three amino acid sequence formed.
   a. 5' GUUGCUCGU 3'   c. 5' AGUAACUCG 3'
   b. 5' CUACGCGGU 3'   d. 5' UAUGAUACC 3'

129 Given the following normal DNA and corresponding mRNA sequence, what amino acid sequence would be formed? Indicate whether the following mutations would produce the same or a different amino acid sequence (assume all reading occurs from left to right):
   DNA   3' dGGTGCT 5'
   mRNA   5' CCACGA 3'
   a. The third nucleotide on DNA is replaced by C.
   b. The sixth nucleotide on DNA is replaced by A.
   c. The second nucleotide on DNA is deleted.
   d. The fourth nucleotide on DNA is deleted.

130 Given the following normal DNA and corresponding mRNA sequences, what amino acid sequence would be formed? Indicate whether the following mutations would produce the same or a different amino acid sequence (assume all reading occurs from left to right):
   DNA   3' dAGTAAA 5'
   mRNA   5' UCAUUU 3'
   a. The third nucleotide on DNA is replaced by C.
   b. The sixth nucleotide on DNA is replaced by T.
   c. The second nucleotide on DNA is deleted.
   d. The fourth nucleotide on DNA is deleted.

131 A sequence of nucleotides on mRNA will code for only one sequence of amino acids. If you were given a sequence of amino acids in a protein, could you predict the unique sequence of nucleotides on mRNA that coded for that protein? Explain your answer.

Chemistry in Medicine: HIV and Viral Nucleic Acids

132 What distinguishes HIV infection from AIDS?

133 What is a virus?

134 How are viruses classified? What type of genome does the HIV virus have?

135 What are the two key enzymes that HIV requires during its life cycle?

136 Why do viruses need a host cell to replicate their genetic material and synthesize viral proteins?

137 What type of cell does HIV infect?

138 What is the first step in HIV infection?

139 How is the single-stranded RNA of an HIV virus converted into double-stranded viral DNA? What key viral enzyme is used in this process?

140 What is a retrovirus? Why is HIV classified as a retrovirus?

141 Why have efforts to develop a vaccine for HIV been unsuccessful?

142 What happens when viral DNA is incorporated into the host cell's DNA?

143 When does gene expression occur in T-cells?

144 How does viral DNA get transcribed and translated to produce viral enzymes and other viral proteins needed by the HIV virus?

145 What happens to the viral RNA that is produced by the host cell?

146 Which steps in the life cycle of the HIV virus have HIV drug therapies targeted?

147 Why is a combination of drug therapies typically needed to treat HIV infection?

Answers to Practice Exercises

1 An amino acid is to a peptide as a <u>nucleotide</u> is to a nucleic acid.

2 b. 2-deoxy-D-ribose

3

a. Nitrogenous base: thymine (a pyrimidine)

c. Phosphate

b. Monosaccharide: 2-deoxy-D-ribose

d. Anomeric carbon

g. This nucleotide would be found in DNA because the sugar is 2-deoxy-D-ribose, and the nitrogenous base, thymine is only found in DNA.

4 cytosine, C; guanine, G; adenine, A; uracil, U

5 Uracil, U, is not found in DNA.

6 a.-b. See below.
   c. The nitrogenous base is adenine.
   d. The monosaccharide is 2-deoxy-D-ribose.
   e. This nucleotide would be found in DNA because it contains 2-deoxy-D-ribose as its monosaccharide component.

7

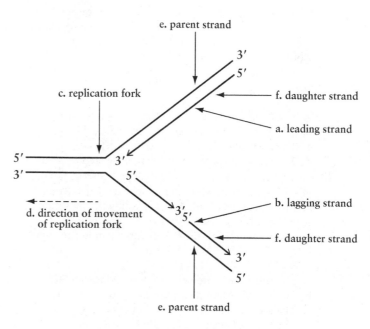

8 A nucleotide triphosphate contains the high energy phosphoanhydride bonds that supply the energy needed for the reaction that forms the new phosphate ester bond.

9 a. 3′ dAACCGT 5′
　 b. 3′ dTACGGT 5′

10 a. nitrogenous base pairs
　 b. sugar-phosphate backbone
　 c. nitrogenous base pairs
　 d. nitrogenous base pairs

11 It is highly symmetrical.

12 Hydrogen bonding and base stacking. The advantage of these non-covalent interactions is that it allows the two strands of DNA to be separated when the strands need to be read, such as during DNA replication and DNA transcription.

13 A gene is a sequence of nucleotides on DNA that codes for a particular protein. A genome is the entire DNA sequence of an organism. There are 3 billion base pairs in the human genome.

14 The charge on a histone is positive, which is attracted to the negative charge on DNA, making it easier to form the compact structure of a nucleosome.

15

parent 1:	5′ dGACCTAGCCC 3′
daughter 1:	3′ dCTGGATCGGG 5′
parent 2:	3′ dCTGGATCGGG 5′
daughter 2:	5′ dGACCTAGCCC 3′

16 *DNA polymerase* always synthesizes a polynucleotide in the 5′ → 3′ direction.

17 The lagging strand is synthesized in fragments because the two strands of DNA are antiparallel and DNA polymerase can only build nucleotides in the 5′ → 3′ direction. So, the lagging strand begins at the replication fork and works away from the replication fork as the replication fork moves in the opposite direction, exposing more of the parent strand. Thus, the next fragment is built from the new site of the replication fork to the start of the previous fragment.

18 c.

19.

20 rRNA, mRNA, and tRNA. rRNA makes up part of the ribosome. A ribosome is composed of nucleic acids and proteins.

21 Transcription is the process in which a gene, a segment of one strand of DNA, is used as a template for synthesizing a single strand of mRNA with a complementary sequence of nucleotides.

22 A protein is produced, by transcription of a gene to form mRNA, followed by translation of mRNA to form a polypeptide.

23 a. GGG: glycine
　 b. AAA: lysine

24 a. cysteine: UGU and UGC
　 b. leucine: UUA, UUG, CUU, CUC, CUA, CUG

25

a. mRNA codons:	5′ GGU	ACU	AUU	UAU 3′
tRNA anticodons:	3′ CCA 5′	3′ UGA 5′	3′ UAA 5′	3′ AUA 5′
amino acids:	Gly - Thr - Ile - Tyr			
b. mRNA codons:	5′ GGU	GAC	CGA	CAU 3′
tRNA anticodons:	3′ CCA 5′	3′ CUG 5′	3′ GCU 5′	3′ GUA 5′
amino acids:	Gly - Asp - Arg - His			

26 The three stop codons are UGA, UAA, and UAG. A stop codon signifies the end of a gene.

27 a. DNA: 3′ d G A A T G A 5′

     mRNA: 5′   C U U A C U 3′

     amino acids: Leu-Thr; different amino acid sequence (one amino acid changed)

  b. DNA: 3′ d T A G T G A 5′

     mRNA: 5′   A U C A C U 3′

     amino acids: Ile-Thr; same amino acid sequence

  c. DNA: 3′ d T A/T G A G 5′

     mRNA: 5′   A U **A C U C** 3′

     amino acids: Met-Leu; different amino acid sequence (both amino acids changed)

  d. DNA: 3′ d/A A T G A A 5′

     mRNA: 5′   U U A C U U 3′

     amino acids: Leu-Leu; different amino acid sequence (both amino acids changed)

When exercising, fatty acids are converted into carbon dioxide and water through biochemical pathways that produce ATP, the energy currency of the cell. [iStockphoto/Thinkstock]

15 | Energy and Metabolism

Exercise and Weight Loss

There are millions of people who wrestle with the problem of excess weight and how to lose that weight. Why is it so hard for some people to lose weight? Isn't it simply a matter of balancing calories consumed (energy in) with calories burned (energy out)? In this chapter we consider energy and metabolism, which will reveal some of the complexities involved in losing weight.

We know that diets that call for a severe reduction in calories usually do not work over the long term. The problem with a drastic reduction in calories is that brain cells need a constant supply of glucose, yet there are no major biochemical pathways that produce glucose from stored fat (glycerol from triglyceride hydrolysis being the one exception). Once the body's limited supply of glycogen has all been converted to glucose, if glucose is not supplied through the diet, the body begins to convert protein in muscle tissue to glucose. Plummeting glucose levels also signal the brain that it needs glucose, creating hunger sensations, and making it even more difficult to stick to a weight loss plan.

Effective diets generally recommend a moderate intake of calories combined with exercise. The United States Department of Agriculture (USDA) recently updated the familiar "food pyramid" with "MyPlate," depicting the five food groups recommended for a healthy diet as portions on a plate (Figure 15-1). Modifications of "MyPlate" have also been publicized. These recommendations include exercise as a key component to a healthy diet, as well as recommending healthy oils and emphasizing drinking water rather than dairy with every meal. How does exercise fit into the weight loss equation? To lose weight a person needs to metabolize triglycerides stored in fat cells. Metabolism of triglycerides occurs in muscle

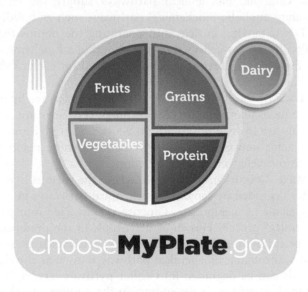

Figure 15-1 The United States Department of Agriculture's (USDA) most recent replacement for the food pyramid: "MyPlate," illustrating the daily proportions of the five food groups recommended for a healthy diet. [Courtesy of USDA, www.ChooseMyPlate.gov]

and liver cells, where fatty acids are converted to carbon dioxide, water, and energy, through a series of biochemical pathways. Hence, the best way to convert stored fat into energy is to place higher energy demands on the body, that is, exercise. Consumption of fewer calories, combined with regular exercise, while maintaining steady blood glucose levels, causes cells to oxidize fatty acids, literally "burning fat." These fat burning catabolic processes occur in organelles of the cell known as the mitochondria, the energy factories of the cell. Indeed, studies have shown that extremely fit endurance athletes have more mitochondria in their muscle cells than the average person.

In addition to the caloric content of food, the type of food a person consumes when dieting is also important. Nutritious foods that are high in B vitamins provide the body with a steady supply of the starting materials needed to produce the coenzymes required for these fat burning biochemical pathways.

Although weight loss involves hormones, brain chemicals, and many other complex processes, the biochemistry of metabolism explains part of the weight loss equation. In this chapter you will learn the biochemical pathways involved in the catabolism of carbohydrates and fats as well as the central catabolic pathways that lead to the production of ATP: the citric acid cycle and oxidative phosphorylation. ●

The cells in your body are constantly carrying out chemical reactions that produce important cellular intermediates. Equally important is the transfer of energy that occurs in these reactions, a central theme in biochemistry known as **bioenergetics**. Metabolism includes all of the reactions of the cell, both the pathways that build molecules (anabolic pathways) and the pathways that break down molecules (catabolic pathways). In this chapter we focus on the catabolic pathways that break down carbohydrates and triglycerides, the compounds in our food that under normal circumstances supply the body with most of its energy.

■ 15.1 Energy and Metabolism: An Overview

Extracting energy from our food and storing it in the form of ATP is the primary function of **cellular respiration**, the biochemical pathways that convert carbohydrates, fats, and sometimes dietary proteins, into CO_2, water, and energy. These catabolic biochemical pathways supply the energy for the energy-demanding anabolic biochemical pathways such as building proteins, synthesizing DNA, moving muscle, and operating the pumps that transport ions and molecules across cell membranes. Catabolic biochemical pathways are generally divided into three stages, as illustrated for each of the food biomolecules in Figure 15-2.

Stage 1: Each of the three major food biomolecules are hydrolyzed into their respective molecular building blocks: polysaccharides are hydrolyzed to glucose; proteins are hydrolyzed to amino acids; and triglycerides are hydrolyzed to fatty acids and glycerol. The products of these reactions are then delivered to various cells via the bloodstream.

Stage 2: Stage 2 occurs inside the cell, where the various hydrolysis products are further degraded by separate biochemical pathways into acetyl CoA, the central molecule of metabolism. Glucose is converted into pyruvate by a 10-step biochemical process known as **glycolysis**. In the presence of oxygen, pyruvate can be further oxidized to acetyl CoA. Fatty acids are also degraded into acetyl CoA, through a biochemical pathway known as β-**oxidation**.

Amino acids can be degraded into either pyruvate, acetyl CoA, or intermediates of the citric acid cycle depending on the number of carbon atoms in the amino acid. However, amino acids are not a primary source of energy, except in the case of illness or starvation. Instead, amino acids are used to

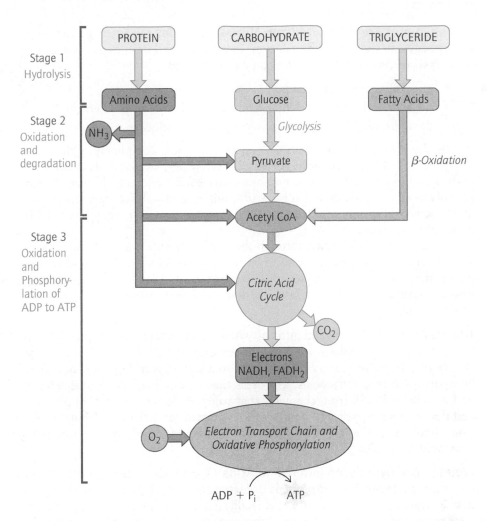

Figure 15-2 An overview of the three stages of catabolism for the three food biomolecules—proteins, carbohydrates, and triglycerides—showing the key intermediates and metabolic pathways leading to the phosphorylation of ADP to ATP.

make proteins and other nitrogen-containing compounds. In this chapter, therefore, we focus primarily on the catabolic fate of glucose and fatty acids.

Stage 3: In the final stage of catabolism, acetyl CoA enters the **citric acid cycle**, an 8-step biochemical pathway that oxidizes acetyl CoA to carbon dioxide (CO_2) while reducing the coenzymes NAD^+ and FAD to NADH and $FADH_2$.

The electrons carried by NADH and $FADH_2$, which are produced in stage 2 and the citric acid cycle, are transferred to the **electron transport chain**, a series of oxidation–reduction reactions concluding with the reduction of oxygen to water. The energy transferred in the electron transport chain drives the phosphorylation of ADP to ATP, a biochemical process known as **oxidative phosphorylation**.

Stage 1: Hydrolysis of Carbohydrates, Fats, and Proteins

In the first stage of catabolism, also known as digestion, the large biomolecules that make up our food react with water in hydrolysis reactions that break them down into their smaller components, as illustrated in Figure 15-2 and described below. Recall from chapter 10 that in one type of *hydrolysis* reaction water breaks the carbonyl carbon-heteroatom bond of a carboxylic acid derivative (esters, thioesters, and amides) to produce a carboxylic acid and an alcohol, thiol, or amine, respectively.

Carbohydrates Are Hydrolyzed into Monosaccharides Digestion of carbohydrates begins in the mouth. The enzyme *amylase,* found in saliva, catalyzes the hydrolysis of starch into the following carbohydrates:

- glucose, a monosaccharide,
- maltose, a disaccharide composed of two glucose monomers, and
- dextrins, oligosaccharides composed of 3–12 glucose molecules.

In the small intestine, maltose and dextrins are completely hydrolyzed into glucose, while the other important disaccharides, lactose and sucrose, are hydrolyzed into their respective monosaccharides. Recall from chapter 11 that hydrolysis of complex carbohydrates (di-, oligo-, and polysaccharides) occurs at the glycosidic linkages. Complete hydrolysis of starch, therefore, yields many glucose molecules.

Monosaccharides diffuse through the small intestine and enter the bloodstream. Glucose, soluble in water, is distributed to cells via the bloodstream, hence, the reference "blood sugar." The monosaccharides fructose and galactose are converted into glucose in the liver and then released back into the bloodstream.

Triglycerides Are Hydrolyzed into Fatty Acids and Glycerol In chapter 12 you learned that triglycerides, stored in adipocytes, serve as the body's long-term supply of energy. To extract the energy stored in triglycerides, they must first be hydrolyzed into fatty acids, a process that occurs in skeletal muscle cells and liver cells. Recall from chapter 7 that triglycerides are triesters of glycerol and that esters are hydrolyzed into carboxylic acids and alcohols (chapter 10). Thus, hydrolysis of a triglyceride produces one molecule of glycerol and three fatty acid molecules.

Proteins Are Hydrolyzed into Amino Acids Hydrolysis of proteins occurs in the stomach. Hydrolysis of proteins occurs at the peptide bonds (amides) producing amino acids, a process that is facilitated by the low pH of the stomach. Generally, amino acids are not metabolized to supply energy, except in special circumstances such as starvation.

ATP and Energy Exchange

Since the main purpose of catabolic reactions is to extract energy from the covalent bonds in glucose, fatty acids, and sometimes amino acids, let us first consider how this energy is transferred and stored. In a biological cell, the energy released in catabolic reactions is not transferred as heat, but rather as chemical potential energy, in the form of high energy bonds, the most important of which are the phosphoanhydride bonds of nucleotide triphosphates, such as adenosine triphosphate (ATP).

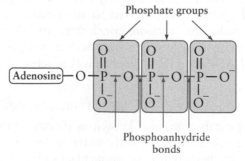

Adenosine triphosphate (ATP)

Hydrolysis of ATP to form adenosine *di*phosphate (ADP) and inorganic phosphate (P_i) breaks one of these high energy phosphoanhydride bonds, releasing 30.5 kJ/mol (7.3 kcal/mol) of energy. As a convention, we often show energy *released* as a negative value—a downhill reaction—and energy *absorbed* as a positive value—an uphill reaction. It is estimated that about 1.5 million ATP molecules are hydrolyzed every second in a cell that is doing cellular work.

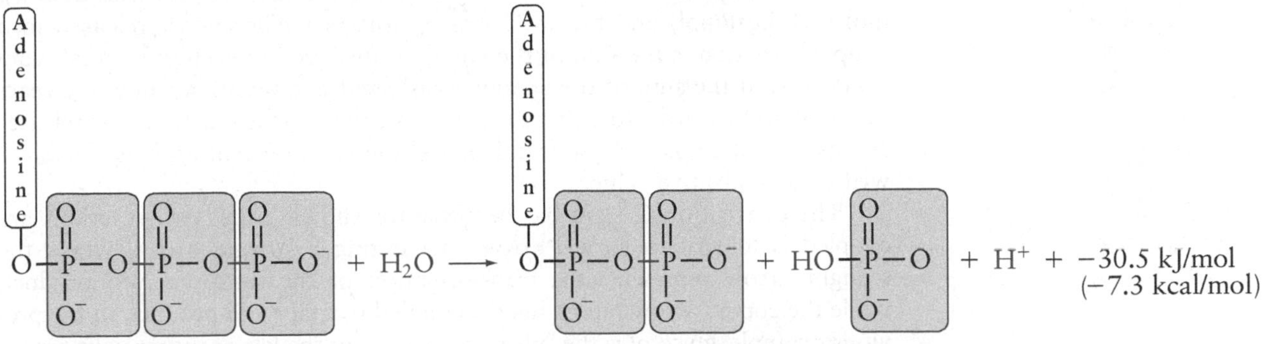

Adenosine triphosphate (ATP)   Adenosine diphosphate (ADP)   Inorganic phosphate, P_i

Hydrolysis of ATP can also occur at both phosphoanhydride bonds to form adenine *mono*phosphate (AMP) and *two* molecules of inorganic phosphate, releasing twice as much energy: 61 kJ/mol (14.6 kcal/mol):

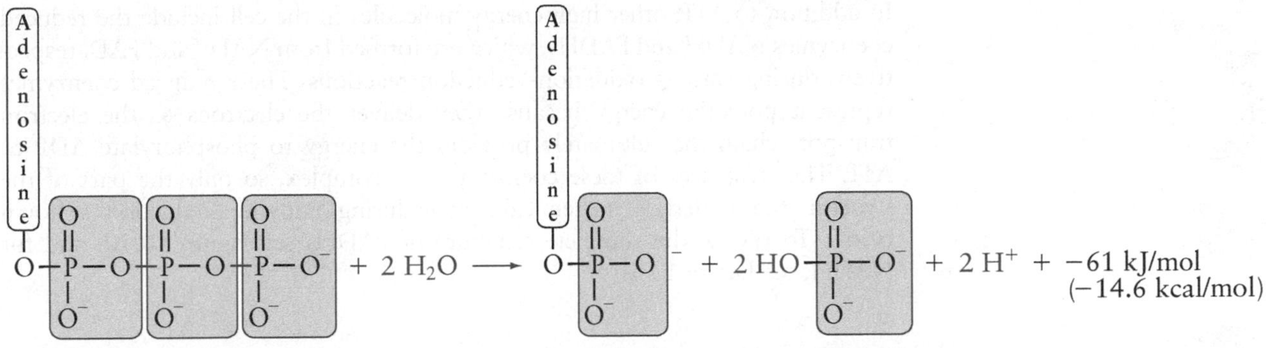

Adenosine triphosphate (ATP)   Adenosine monophosphate (AMP)   Inorganic phosphate, P_i

By definition, the reverse of these reactions requires an *input* of energy of the same amount. For example, the reaction ADP + P_i requires 30.5 kJ/mol (7.3 kcal/mol) of energy. Thus, our cells store energy as ATP by phosphorylating ADP, and then utilize the potential energy stored in ATP by hydrolyzing it back to ADP and P_i. For this reason, ATP is often referred to as the "energy currency" of the cell.

Biochemical reactions that require an input of energy do not happen on their own, just as balls never roll uphill on their own. However, an uphill reaction can be "catapulted" forward if it is linked to a more energetic downhill reaction. In biochemistry, reactions that require an input of energy (uphill reactions) are **coupled** to reactions that release energy (downhill reactions), such as, but not limited to, ATP hydrolysis. For example, consider the first reaction shown below, the reaction of glycerol ($C_3H_8O_3$) and inorganic phosphate, $HOPO_3^{2-}$ (P_i) to produce glycerol-3-phosphate ($C_3H_7O_3PO_3$) and water in the construction of cell membranes:

Reaction

$$P_i + C_3H_8O_3 \longrightarrow C_3H_7O_3PO_3 + H_2O \qquad +9.2 \text{ kJ } (+2.2 \text{ kcal/mol})$$

$$\underline{H_2O + ATP \longrightarrow ADP + P_i} \qquad \underline{-30.5 \text{ kJ } (-7.3 \text{ kcal/mol})}$$

$$\qquad\qquad\qquad ATP \quad ADP \qquad\qquad -21.3 \text{ kJ } (-5.1 \text{ kcal/mol})$$

$$C_3H_8O_3 \overset{\textstyle\curvearrowright}{\underset{\textstyle\curvearrowleft}{}} C_3H_7O_3PO_3$$

This first reaction requires 9.2 kJ/mol (2.2 kcal/mol), and therefore it will not proceed on its own. However, when it is coupled to the second reaction, the hydrolysis of ATP to form P_i and ADP, a reaction that *releases* 30.5 kJ/mole (7.3 kcal/mol), the coupled reaction (third, and net reaction) releases 21.3 kJ/mol (5.1 kcal/mol) and therefore can go forward. The energy released in a coupled reaction is the sum of the energy transferred in each individual reaction. Thus, if the sum of the energy transferred is a negative value, the reaction will go forward. Note that in such a calculation, the reaction that releases energy is written as a negative value, and the reaction that absorbs energy is written as a positive value.

The convention used in biochemistry for showing that two reactions are coupled is to write a curved arrow intersecting a straight arrow, where the straight arrow represents the transformation of the substrate into product, while the curved arrow represents the coupled reactant and product. In the previous example, glycerol is the substrate, placed on the left of the straight arrow, and glycerol-3-phosphate is the product, placed on the right side of the arrow; while ATP is placed on the left-side of the curved arrow intersecting the straight arrow, and ADP is placed on the right-side of the curved arrow. The intersecting arrows indicate a coupled reaction, as shown in the final reaction above.

Coenzymes: NADH, $FADH_2$, and Acetyl CoA

In addition to ATP, other high energy molecules in the cell include the reduced coenzymes NADH and $FADH_2$, which are formed from NAD^+ and FAD, respectively, during various oxidation–reduction reactions. These reduced coenzymes represent potential energy because they deliver the electrons to the electron transport chain that ultimately provides the energy to phosphorylate ADP to ATP. The structures of these coenzymes are complex, so only the part of the structure that undergoes a chemical change during oxidation–reduction is shown below. To review the complete structure of $FADH_2$ see Figure 10-10, and for NADH, see Figure 10-1.

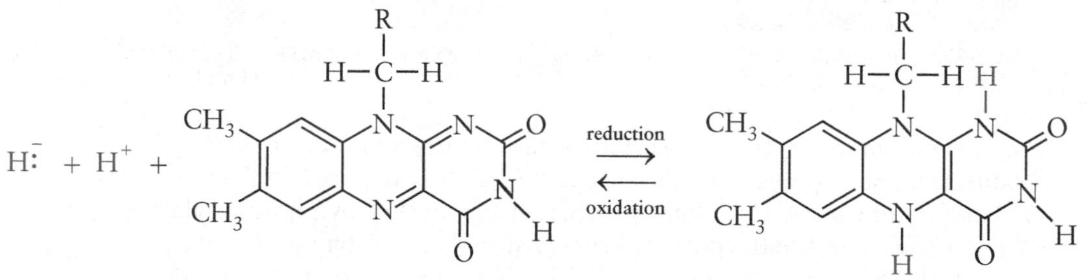

O
‖
R—C—

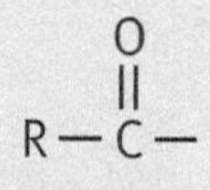

Acyl group

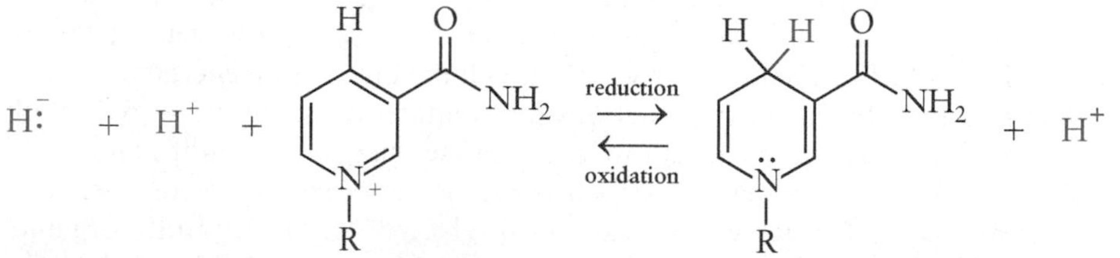

H   O
| ‖
H—C—C—
|
H

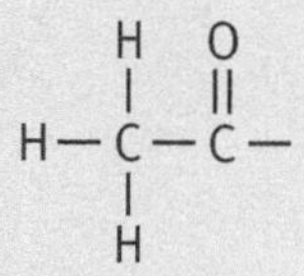

Acetyl group

Recall that an acyl group is a carbonyl group with its attached R group, so an acetyl group is a type of acyl group in which the —R group is —CH_3.

Acetyl coenzyme A, abbreviated acetyl CoA, is one of the most important intermediates of metabolism because it is central to so many biochemical pathways. Indeed, it is the point of convergence for all three of the catabolic pathways that break down the food biomolecules (see Figure 15-2). Coenzyme A is as a biological acyl carrier molecule, involved in transferring acyl groups in various biochemical pathways.

High energy thioester bond that joins
    the acyl group to coenzyme A

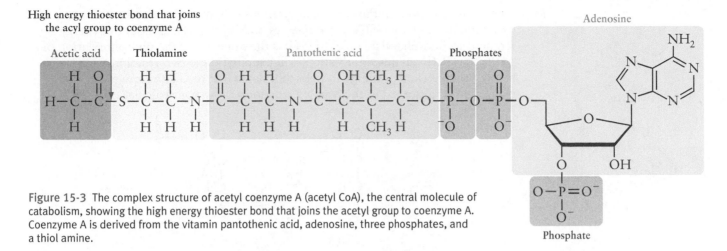

Figure 15-3 The complex structure of acetyl coenzyme A (acetyl CoA), the central molecule of catabolism, showing the high energy thioester bond that joins the acetyl group to coenzyme A. Coenzyme A is derived from the vitamin pantothenic acid, adenosine, three phosphates, and a thiol amine.

The complex structure of acetyl CoA is shown in Figure 15-3. Acetyl CoA is derived from the nucleoside adenosine (chapter 14), the vitamin pantothenic acid (vitamin B_5), the thiol amine $HSCH_2CH_2NH_2$, and acetic acid (CH_3CO_2H). A high energy thioester bond joins the two-carbon acetyl group to the thiol of coenzyme A.

In the next sections we consider in detail the catabolic pathways that convert glucose (section 15.2) and fatty acids (section 15.5) into acetyl CoA. We examine what happens to acetyl CoA in the citric acid cycle (section 15.3) and how oxidative phosphorylation (section 15.4) uses the energy from the electron transport chain to phosphorylate ADP to ATP.

WORKED EXERCISES | Coenzymes and Energy Transfer

15-1 The partial structures of the coenzymes NAD^+ and NADH are shown in the half reaction below. Identify each structure as either NAD^+ or NADH. Is the half reaction shown an oxidation or a reduction? Why is this coenzyme classified as an electron carrier? Why is it classified as a high energy molecule?

15-2 Identify the high energy bonds in each of the molecules, below. How many kilojoules or kilocalories are released when ATP is hydrolyzed to ADP and P_i?

ATP

ADP

15-3 One of the reactions in the citric acid cycle is the hydrolysis of succinyl CoA to succinate and CoA—SH, which releases 33.8 kJ/mol. The amount of energy required to phosphorylate GDP to GTP is the same as for the phosphorylation of ADP to ATP. Given this information, will coupling these two reactions as shown below provide enough energy to phosphorylate GDP to GTP? How much energy is released or absorbed in the coupled reaction? Show your calculations.

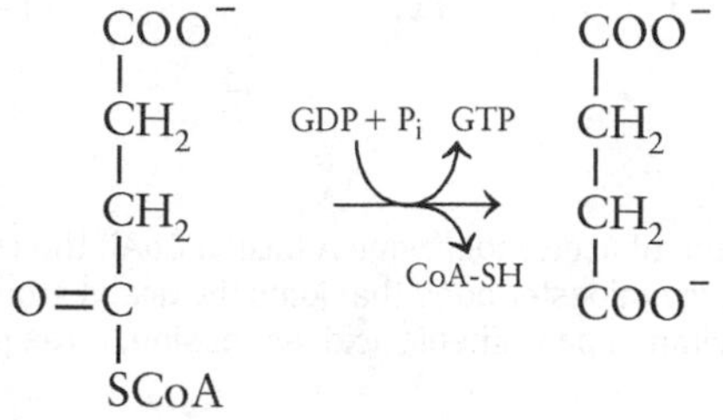

Succinyl CoA   Succinate

15-4 The structures of acetyl CoA, succinyl CoA, and fatty acyl CoA are shown below. Locate the sulfur atom in each molecule, and indicate what type of functional group the sulfur atom is part of. In what way are these molecules similar? Why do they all have "CoA" in their name? How are these molecules different?

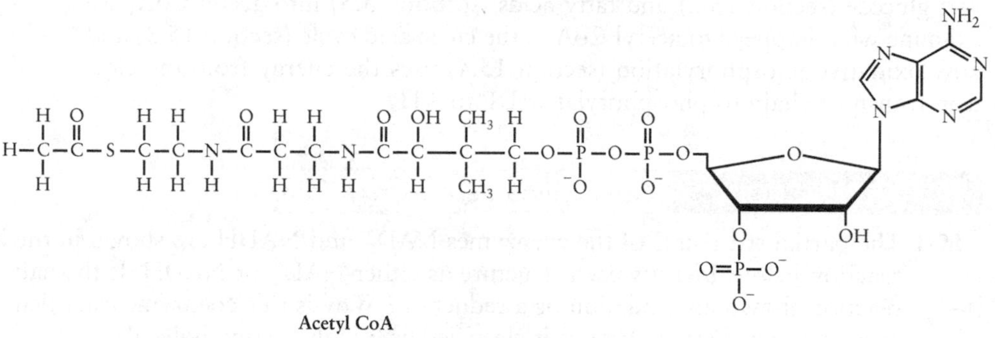

Acetyl CoA

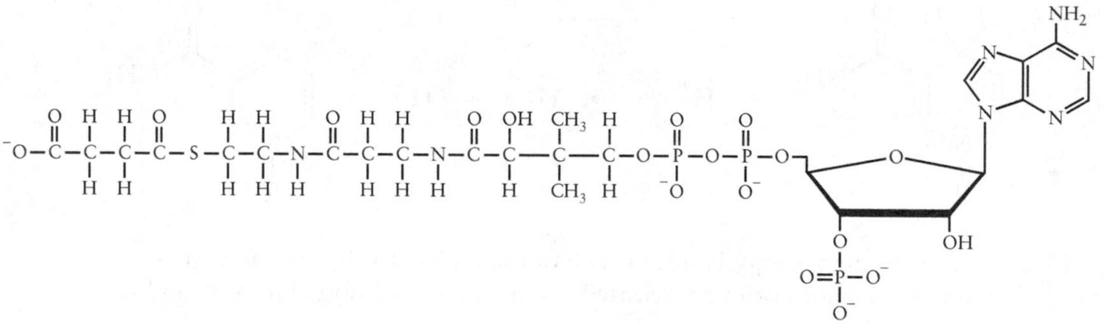

Succinyl CoA

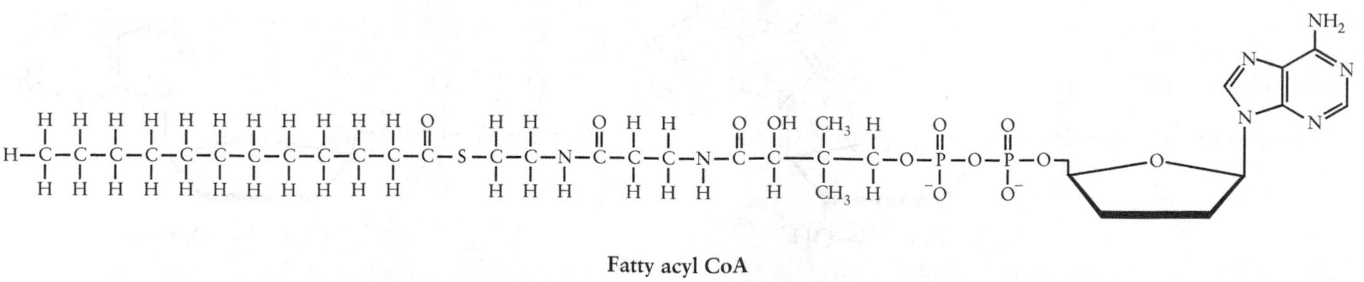

Fatty acyl CoA

Solutions

15-1 The structure on the left is NADH, and the structure on the right is NAD^+, which you can distinguish by the extra H in NADH. The half reaction shown is an oxidation because electrons are lost in going from NADH to NAD^+ (fewer hydrogen atoms in the product than the reactant). NADH is an electron carrier because it contains two more electrons than NAD^+. NADH is referred to as a high energy molecule because it has more potential energy than NAD^+, and can be used to phosphorylate ADP to ATP during oxidative phosphorylation in the last stage of catabolism.

15-2 Hydrolysis of ATP to ADP and P_i releases 30.5 kJ/mol (7.3 kcal/mol). The red arrows point to the high energy phosphoanhydride bonds.

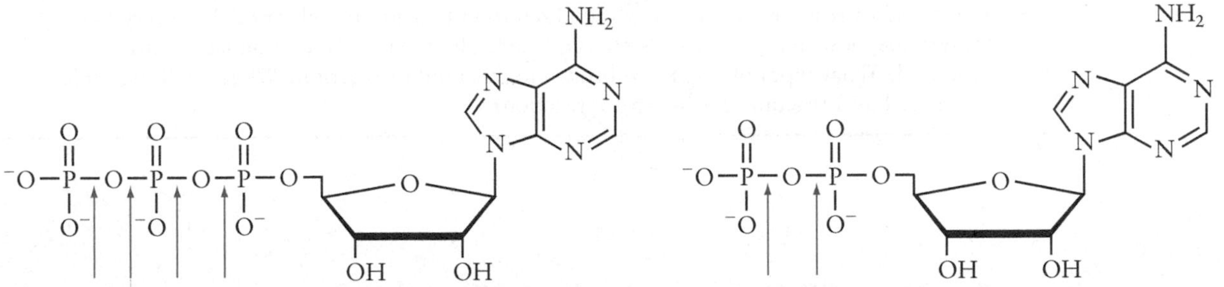

ATP

ADP

15-3 Coupling these two reactions will provide the energy required to phosphorylate GDP, because a net 3.3 kJ/mol is released, so the reaction will go forward:

$$H_2O + \text{Succinyl CoA} \longrightarrow \text{Succinate} + \text{CoA-SH} \qquad -33.8 \text{ kJ/mol}$$
$$P_i + \text{GDP} \longrightarrow \text{GTP} + H_2O \qquad +30.5 \text{ kJ/mol}$$

$$\text{Succinyl CoA} \longrightarrow \text{Succinate} + \text{CoA-SH} \qquad -3.3 \text{ kJ/mol}$$

15-4 The sulfur atom in each of these molecules is part of a thioester functional group. Each of these molecules contains a thioester formed between the thiol of coenzyme A and a carboxylic acid. They all have CoA in their name because they are thioesters of coenzyme A. These molecules differ in the —R group of the acyl group that is carried by coenzyme A. In acetyl CoA the acyl group is the two-carbon acetyl group. In succinyl CoA the acyl group is the four-carbon succinic acid group. In fatty acyl CoA the acyl group is the 12-carbon fatty acyl group shown. The acyl groups are the part of the molecule shown left of the sulfur atoms in the structures shown in exercise 15-4.

PRACTICE EXERCISES

1. Indicate the product(s) formed when each of the following biomolecules is hydrolyzed during the first stage of catabolism:
   a. Triglycerides
   b. Carbohydrates
   c. Proteins
2. What is the primary function of amino acids in the cell?

3   For each of the following reactions, indicate how much energy is absorbed or released by circling either the + or the − and entering the correct value:

   a. $ATP + H_2O \rightarrow ADP + P_i$ ± _____ kJ/mol
   b. $ATP + 2\ H_2O \rightarrow AMP + 2\ P_i$ ± _____ kJ/mol

4   Which two biomolecules are the primary sources of energy for the body?

5   What biochemical intermediate is common to the catabolism of glucose, fatty acids, and some amino acids?

6   What kind of reaction converts carbohydrates into glucose? What type of bond is broken in this type of reaction? What small molecule is required as a reactant in this type of reaction?

7   What kind of reaction converts proteins into amino acids? What type of bond is broken in this kind of reaction? What small molecule is required as a reactant in this type of reaction?

8   What kind of reaction converts triglycerides into fatty acids and glycerol? For every one triglyceride, how many fatty acids are produced? How many glycerol molecules are produced? What types of bonds are broken in this kind of reaction? What small molecule is required as a reactant in this type of reaction?

■ 15.2 Carbohydrate Catabolism: Glycolysis and Pyruvate Metabolism

You have seen that carbohydrates serve as the primary source of energy for many cells. In this section we examine glycolysis, the biochemical pathway that converts glucose into pyruvate in the second stage of carbohydrate catabolism. We will then consider the catabolic fate of pyruvate, the product of glycolysis.

Glycolysis

Glycolysis is the only energy producing metabolic pathway for some cells, including red blood cells, brain cells, and sperm cells. Glycolysis is also the primary energy producing pathway for muscle cells. *Glycolysis is an anaerobic biochemical pathway, which means that it does not require oxygen.* For this reason many microorganisms that live in an anaerobic environment use glycolysis to produce energy. These include the microbes that cause tetanus, botulism, and gas gangrene. Indeed, the earliest living organisms, before there was any molecular oxygen on the planet, are believed to have produced their energy through glycolysis.

Glycolysis occurs in the cytosol, the fluid part of the cell outside the nucleus (Figure 15-4). Glycolysis is a 10-step biochemical pathway that converts glucose, a molecule containing six carbon atoms, into two molecules of pyruvate, a molecule containing three carbon atoms.

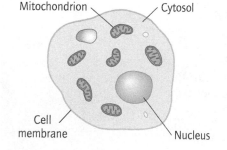

Figure 15-4 The basic parts of a human cell showing the nucleus, the mitochondria, and the cell membrane. The fluid part shown in yellow is the cytosol, where glycolysis takes place.

$$\begin{array}{c} {}^{1}CHO \\ H{-}\!\!\!{}_{2}{-}OH \\ HO{-}\!\!\!{}_{3}{-}H \\ H{-}\!\!\!{}_{4}{-}OH \\ H{-}\!\!\!{}_{5}{-}OH \\ {}_{6}CH_2OH \end{array} \quad \xrightarrow[\text{10 steps}]{\text{Glycolysis}} \quad 2 \;\; \begin{array}{c} O{=}\!\!{}^{1}C{-}O^{-} \\ {}_{2}C{=}O \\ {}_{3}CH_3 \end{array}$$

D-glucose   **Pyruvate**

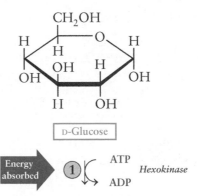

D-Glucose

Figure 15-5 The 10 steps of glycolysis showing the structure of the organic intermediates, the type of reactions involved, and highlighting the energy consuming and producing steps.

For every glucose molecule that enters glycolysis, two ADP are phosphorylated to two ATP, two NAD^+ are reduced to two $NADH + 2H^+$, and two molecules of pyruvate are produced.

$$Glucose + 2\,ADP + 2\,P_i + 2\,NAD^+ \rightarrow 2\,Pyruvate + 2\,ATP + 2\,NADH + 2\,H^+$$

Figure 15-5 shows the 10-step sequence of biochemical reactions that constitute glycolysis, including the structure of each organic intermediate. Every intermediate in glycolysis contains one or more phosphate groups in its structure, indicated as a (P). The -2 charge on these monophosphate ester groups prevents the intermediates of glycolysis from diffusing out of the cell, since polyatomic ions cannot diffuse through the cell membrane.

In step 4, the six-carbon molecule fructose-1,6-bisphosphate is split into two three-carbon structural isomers: glyceraldehyde-3-phosphate and dihydroxy-acetone-3-phosphate. In the next step, step 5, the latter (dihydroxyacetone-3-phosphate) is converted into the former (glyceraldehyde-3-phosphate), in an isomerization reaction. Thus, in the first half of glycolysis, one molecule of glucose is split into two molecules of glyceraldehyde-3-phosphate:

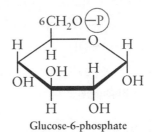

Glucose-6-phosphate

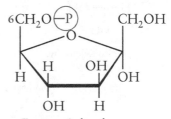

Fructose-6-phosphate

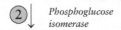

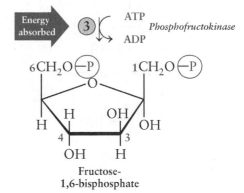

Fructose-1,6-bisphosphate

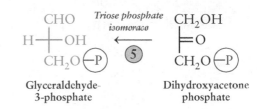

Glyceraldehyde-3-phosphate   Dihydroxyacetone phosphate

Figure 15-5 (*continued*)

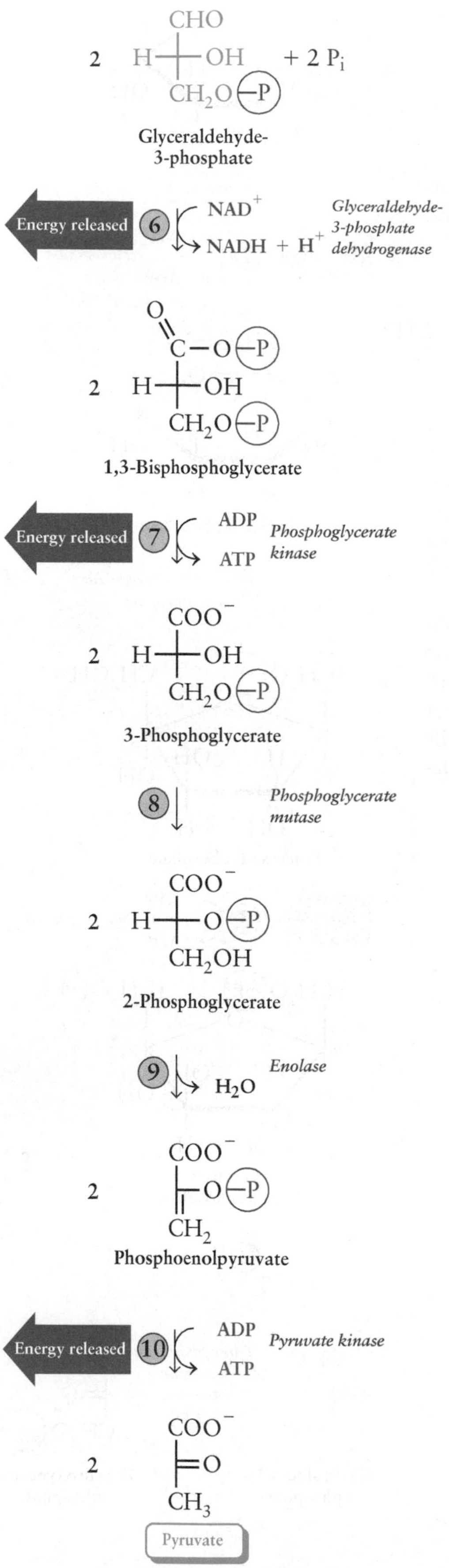

2   CHO
    H——OH   + 2 Pᵢ
    CH₂O—Ⓟ

Glyceraldehyde-
3-phosphate

Energy released ⑥ NAD⁺ → NADH + H⁺   *Glyceraldehyde-3-phosphate dehydrogenase*

2   O‖C—O—Ⓟ
    H——OH
    CH₂O—Ⓟ

1,3-Bisphosphoglycerate

Energy released ⑦ ADP → ATP   *Phosphoglycerate kinase*

2   COO⁻
    H——OH
    CH₂O—Ⓟ

3-Phosphoglycerate

⑧   *Phosphoglycerate mutase*

2   COO⁻
    H——O—Ⓟ
    CH₂OH

2-Phosphoglycerate

⑨ → H₂O   *Enolase*

2   COO⁻
    C—O—Ⓟ
    CH₂

Phosphoenolpyruvate

Energy released ⑩ ADP → ATP   *Pyruvate kinase*

2   COO⁻
    C=O
    CH₃

Pyruvate

In the second half of glycolysis, the two molecules of glyceraldehyde-3-phosphate are converted into two molecules of pyruvate. Thus, every step in the second half of glycolysis is doubled, hence the coefficient "2" appears in each of the steps 6–10.

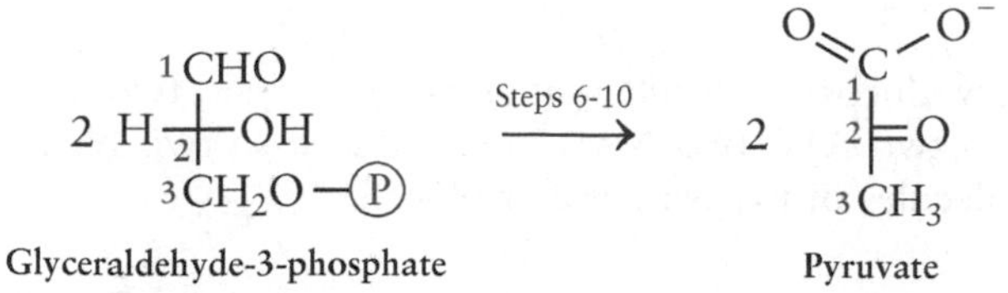

2   ¹CHO
    H——²OH   →[Steps 6-10]   2   ¹C—O⁻ O
    ³CH₂O—Ⓟ   ²C=O
                                   ³CH₃

Glyceraldehyde-3-phosphate   Pyruvate

Each step that involves energy transfer is highlighted in Figure 15-5. Notice how the first half of glycolysis actually requires an input of energy: two ATP are required for every one glucose molecule that undergoes glycolysis (steps 1 and 3). However, in the second half of glycolysis, four ADP are phosphorylated to four ATP, two for each glyceraldehyde-3-phosphate that is converted into pyruvate (steps 7 and 10). By subtracting the two ATP used in the first half of glycolysis, we calculate the net output of ATP in glycolysis as 4 − 2 = 2 ATP per glucose. In addition, two molecules of NAD⁺ are reduced to NADH during glycolysis (step 6). In the next section we consider what happens to the two pyruvate molecules that are produced in glycolysis.

The Metabolism of Pyruvate

In human cells the pyruvate produced from glycolysis can undergo one of two possible catabolic fates, depending on whether or not oxygen is present.

- Aerobic conditions: pyruvate is oxidized to acetyl CoA and carbon dioxide. Also, NAD^+ is reduced to $NADH + H^+$.
- Anaerobic conditions: pyruvate is reduced to lactic acid. Also, $NADH + H^+$ is oxidized to NAD^+.

Oxidation of Pyruvate to Acetyl CoA In the presence of oxygen, pyruvate is converted into acetyl CoA and CO_2. This reaction requires coenzyme A as an acyl carrier molecule and NAD^+ as an oxidizing agent. Thus, the three-carbon pyruvate molecule is converted into carbon dioxide (one carbon) and a two-carbon acetyl group, carried by coenzyme A (acetyl CoA). During the oxidation of pyruvate, electrons are transferred to NAD^+, reducing it to $NADH + H^+$. Although oxygen does not appear in the reaction, it is required later in order to regenerate NAD^+ from NADH, an oxidation.

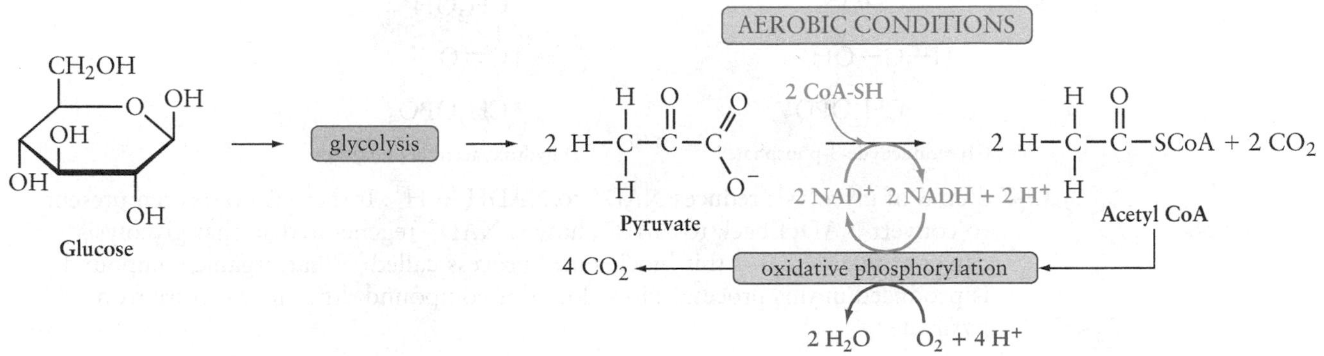

Reduction of Pyruvate to Lactate When oxygen levels in the cell are low, as they are during strenuous exercise, anaerobic conditions exist. Under anaerobic conditions, pyruvate is reduced to lactate and $NADH + H^+$ is oxidized to NAD^+, in a process known as **lactic acid fermentation:**

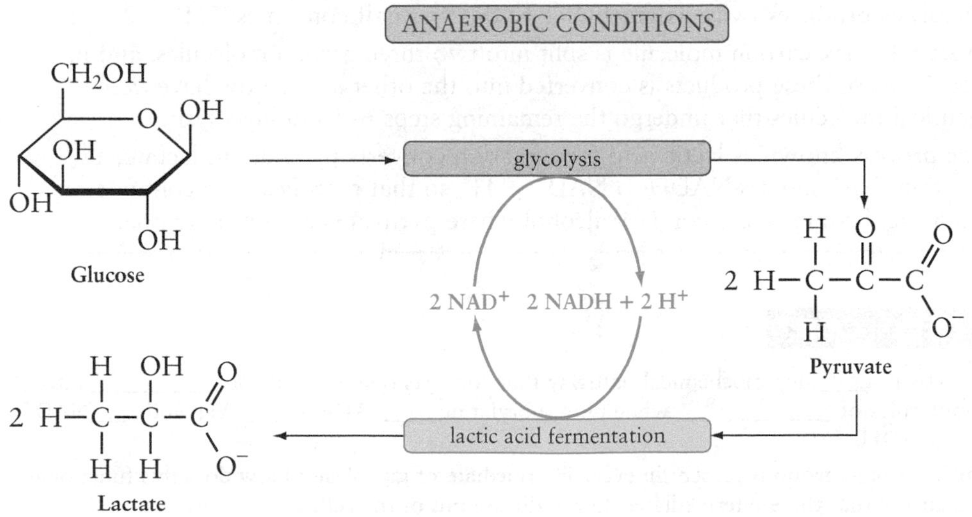

Lactic acid fermentation ensures that NAD^+ again becomes available to the cell so that glycolysis can continue (step 6 of glycolysis requires NAD^+), the only anaerobic biochemical process that produces ATP. Although the citric acid cycle and oxidative phosphorylation produce much more energy than glycolysis, those biochemical pathways require oxygen.

Lactic acid fermentation causes lactic acid to build up in muscles, causing the familiar sensation of sore muscles that typically follows strenuous exercise, particularly physical activities that involve intense quick bursts of energy such as sprinting. Once the energy demanding activity is over, lactate is transported to the liver where it is converted back into pyruvate once oxygen becomes available again.

WORKED EXERCISES   Glycolysis and Pyruvate Catabolism

15-5 Which steps in glycolysis involve ATP? Which steps involve ADP phosphorylation, and which steps involve ATP hydrolysis? How do you arrive at a net yield of two ATP for every pass through glycolysis?

15-6 Step 4 of glycolysis produces glyceraldehyde-3-phosphate and dihydroxyacetone phosphate, shown below. In step 5, one of these molecules is converted into the other. How does step 5 cause all subsequent steps of glycolysis to include a "2" for a coefficient in both the reactants and the products?

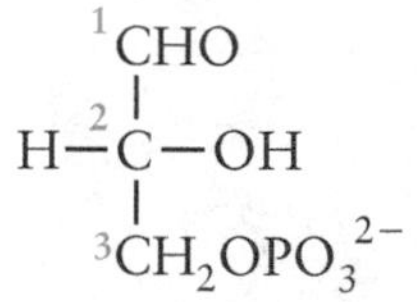

$$^1CHO$$
$$H-^2C-OH$$
$$^3CH_2OPO_3^{2-}$$
Glyceraldehyde-3-phosphate

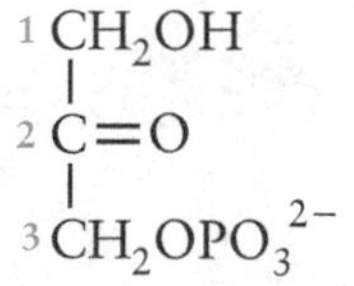

$$^1CH_2OH$$
$$^2C=O$$
$$^3CH_2OPO_3^{2-}$$
Dihydroxyacetone phosphate

15-7 Step 6 of glycolysis reduces NAD^+ to $NADH + H^+$. If there is no oxygen present to convert NADH back to NAD^+, how is NAD^+ regenerated so that glycolysis can continue? What is this biochemical process called? What organic compound is produced in this process? How does this compound differ in structure from pyruvate?

Solutions

15-5 In Figure 15-5 look for the curved arrows showing ADP and ATP: steps 1, 3, 7, and 10. ADP is phosphorylated in the second half of glycolysis: steps 7 and 10. ATP is hydrolyzed in the first half of glycolysis: steps 1 and 3. Since one molecule undergoes steps 1–5, and two molecules undergo steps 6–10, the second half of glycolysis produces twice as much ATP as the first half consumes: $2(2) - 2 = 2$.

15-6 In step 4, a six-carbon molecule is split into two three-carbon molecules, and in step 5 one of these products is converted into the other so we now have *two* identical molecules that undergo the remaining steps 6–10 of glycolysis.

15-7 The process known as lactic acid fermentation converts pyruvate to lactate, a reaction that oxidizes NADH to $NAD^+ + H^+$ so that glycolysis can continue. Lactic acid contains a secondary alcohol where pyruvate contains a ketone.

PRACTICE EXERCISES

9   Glycolysis is a ___ step biochemical pathway that converts one molecule of _____ into ___ molecules of _____ while phosphorylating _____ADP to _____ATP and _____NAD^+ to _____NADH.

10 What functional group is present in every intermediate of glycolysis? How does this functional group ensure that these intermediates do not diffuse out of the cell?

11 Is glycolysis an *anaerobic* or an *aerobic* process? Define these terms.

12 Which two steps of glycolysis require an input of energy in the form of ATP? Which two steps of glycolysis produce ATP? What is the net output of energy in glycolysis, in terms of the direct phosphorylation of ADP to ATP?

13 What is lactic acid fermentation? Where in the cell (cytosol, nucleus, or mitochondria) and under what conditions (anaerobic or aerobic conditions) is lactic acid produced in a human cell?

14 Answer the following questions about glycolysis:
   a. Where in the cell does it take place?
   b. What is the main product of glycolysis, and how many of these molecules are produced?
   c. Is energy produced or consumed during the first half of glycolysis?
   d. Is energy produced or consumed during the second half of glycolysis?

15.3 The Citric Acid Cycle

The last stage of catabolism for glucose, fatty acids, and some amino acids begins with acetyl CoA entering the **citric acid cycle,** an 8-step biochemical pathway that takes place in the mitochondria, the energy producing organelles of the cell. The citric acid cycle is an aerobic pathway that produces much more energy than glycolysis, in the form of the coenzymes NADH and $FADH_2$, as well as the nucleotide guanosine $5'$ triphosphate, GTP. GTP is a nucleotide similar to ATP, except the nitrogenous base is guanine instead of adenine.

For every acetyl CoA molecule that enters the citric acid cycle, one GDP is phosphorylated to GTP; three NAD^+ are reduced to three NADH; one FAD is reduced to $FADH_2$; and two molecules of CO_2 and one molecule of coenzyme A is produced.

$$\text{Acetyl CoA} + 3\ NAD^+ + FAD + GDP + P_i \rightarrow 2\ CO_2 + \text{CoA—SH} + 3\ NADH + 3\ H^+ + FADH_2 + GTP$$

The acetyl group carried by coenzyme A (acetyl CoA) is completely oxidized to two carbon dioxide (CO_2) molecules. As part of this process, three molecules of NAD^+ are reduced to NADH and one molecule of FAD is reduced to $FADH_2$. These coenzymes deposit their electrons into the electron transport chain where they provide the energy to phosphorylate ADP to ATP during oxidative phosphorylation. NADH provides the energy to phosphorylate approximately 2.5 ADP to 2.5 ATP, and $FADH_2$ provides the energy to phosphorylate approximately 1.5 ADP to 1.5 ATP. In addition, one molecule of GDP is directly phosphorylated to GTP in the citric acid cycle.

The citric acid cycle is written as a circular pathway because in the *first* step acetyl CoA reacts with oxaloacetate, a molecule that is also the product of the *last* step. To gain a better understanding of what happens to the two carbon acetyl group of acetyl CoA when it reacts with the four carbon oxaloacetate molecule, consider only the carbon atoms in the citric acid cycle, as illustrated in Figure 15-6, where each carbon is represented as a sphere. Carbon atoms

$$
\begin{array}{c}
COO^- \\
| \\
O = C \\
| \\
CH_2 \\
| \\
COO^-
\end{array}
$$

Oxaloacetate

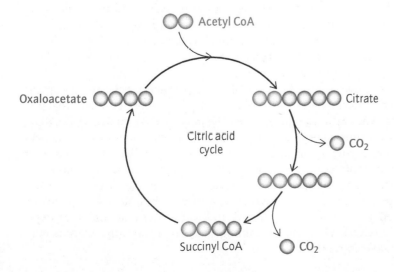

Figure 15-6 An overview of the citric acid cycle following the carbon atoms derived from acetyl CoA, shown in blue, and oxaloacetate, shown in red. Two molecules of carbon dioxide (red) are expelled, and oxaloacetate is regenerated.

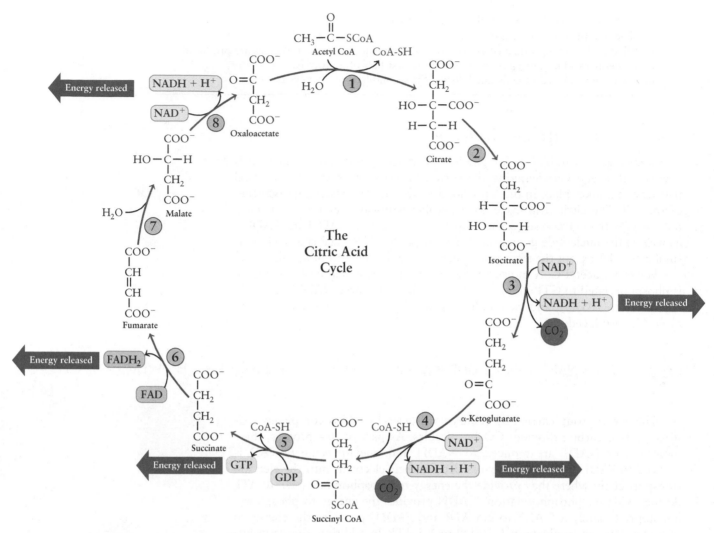

Figure 15-7 The eight steps of the citric acid cycle showing the structure of the organic intermediates and highlighting the energy producing steps.

originating from acetyl CoA are shown as blue spheres, and carbon atoms originating from oxaloacetate are shown as red spheres. In Figure 15-6 we see that it is not actually the two carbon atoms from acetyl CoA that are degraded to carbon dioxide, CO_2, but instead, two of the carbon atoms from oxaloacetate. The two molecules of carbon dioxide are expelled in the first half of the citric acid cycle causing the six carbon-containing citric acid molecule to be shortened to five and then four carbons. In the second half of the citric acid cycle, this four-carbon intermediate, succinyl CoA, is converted to oxaloacetate, also a four-carbon molecule. Hence, the two carbon atoms of the acetyl group become part of the oxaloacetate product.

Figure 15-7 shows the eight steps of the citric acid cycle, including the structure of each intermediate. A key step, step 1, is a condensation reaction that forms a carbon-carbon bond between the acetyl group of acetyl CoA and oxaloacetate, to produce citrate, a molecule with six-carbon atoms, and coenzyme A, the acyl carrier molecule. The biochemical pathway is named after citric acid, the product of this first step. The six-carbon citric acid molecule, shown as citrate at physiological pH, is shortened to the five-carbon molecule, α-ketoglutarate, and then to succinyl CoA, a four-carbon acyl group carried by coenzyme A. These two steps (steps 3 and 4) are the reactions that expel carbon dioxide (CO_2). Over the remaining four steps, succinyl CoA is converted into oxaloacetate.

The citric acid cycle is also known as the Kreb's cycle and the tricarboxylic acid cycle.

The energy producing steps in the citric acid cycle, highlighted in Figure 15-7, are the four oxidation–reduction steps 3, 4, 6, and 8, and the phosphorylation of GDP to GTP in step 5. Steps 3, 4, and 8 each reduce a molecule of NAD^+ to NADH. Step 6 reduces FAD to $FADH_2$. Thus, every acetyl CoA molecule that goes through the citric acid cycle provides the energy to phosphorylate 10 ADP to 10 ATP, either by direct phosphorylation or during oxidative phosphorylation, calculated as follows:

3 NADH	$\times$	2.5	=	7.5	ATP
1 $FADH_2$	$\times$	1.5	=	1.5	ATP
1 GTP	$\times$	1	=	1	ATP
Total			=	10	ATP

In summary, one acetyl CoA molecule that enters the citric acid cycle provides the energy to phosphorylate 10 ADP to 10 ATP.

WORKED EXERCISES   The Citric Acid Cycle

15-8   Referring to Figure 15-6, are the two carbon atoms of the acetyl group in acetyl CoA the same two carbon atoms that appear as expelled carbon dioxide? If not, where are they found at the end of the citric acid cycle? In which steps of the citric acid cycle is carbon dioxide released? Which steps produce CO_2 and energy?

15-9   A condensation is a type of reaction where two molecules are joined, often by forming a carbon-carbon bond. Which reaction in the citric acid cycle is a condensation reaction? What two molecules are joined, and how many carbon atoms do they each contain? What product is formed, and how many carbon atoms does it contain?

15-10   The coenzyme NAD^+ is involved in oxidation reactions involving C—O bonds. How many molecules of NAD^+ are reduced to NADH in the citric acid cycle? How many of these are formed in the first half of the citric acid cycle? How many of these are formed in the second half of the citric acid cycle? How many of these are formed in reactions that involve the expulsion of CO_2?

15-11   How many ADP are phosphorylated to ATP from one NADH during oxidative phosphorylation? How many ADP are phosphorylated to ATP from one $FADH_2$ during oxidative phosphorylation? How many direct phosphorylation reactions occur in the citric acid cycle?

Solution

15-8   No, the carbon atoms in the acetyl group end up as part of oxaloacetate, the final product formed in step 8 of the citric acid cycle. A molecule of carbon dioxide is released in both steps 3 and 4, reactions that both also produce energy in the form of NADH.

15-9   The first step of the citric acid cycle is a condensation reaction between the two-carbon acetyl group of acetyl CoA and the four-carbon oxaloacetate molecule. The product of this condensation reaction is the six-carbon citrate molecule.

15-10   Three molecules of NAD^+ are reduced to NADH + H^+ in the citric acid cycle (steps 3, 4, and 8). Two of these reduced coenzymes are formed in the first half of the citric acid cycle, and the third is produced in the second half of the cycle. Two NADH are formed in the two CO_2 producing steps 3 and 4.

15-11   Approximately 2.5 ADP can be phosphorylated to 2.5 ATP when one NADH is oxidized to NAD^+ in the electron transport chain and oxidative phosphorylation. Approximately 1.5 ADP can be phosphorylated to 1.5 ATP when one $FADH_2$ is oxidized to FAD^+ in the electron transport chain and oxidative phosphorylation. There is one direct phosphorylation in the citric acid cycle: GDP + P_i → GTP.

PRACTICE EXERCISES

15 Referring to Figure 15-6, what happens to the four carbon atoms of oxaloacetate that enter the citric acid cycle? Do any of these carbon atoms end up as carbon dioxide? How many carbon atoms does the acetyl group in acetyl CoA contain?

16 The coenzyme FAD is involved in oxidation and reduction reactions involving C—C bonds. How many molecules of FAD are reduced to $FADH_2$ in the citric acid cycle? In which step(s) of the citric acid cycle is this coenzyme involved? How does the substrate appear to have changed in this reaction? How many ADP can be phosphorylated to ATP from one $FADH_2$ during oxidative phosphorylation?

17 Hydrolysis of the thioester in step 5 provides the energy to phosphorylate GDP to GTP—a coupled reaction. How many ATP is equivalent to one GTP? Complete the reaction below.

$$GTP + ADP \rightarrow GDP + \underline{\hspace{1cm}}.$$

18 In the last half of the citric acid cycle, succinyl CoA is converted to oxaloacetate. What is the structural difference between these two molecules?

19 In the second half of the citric acid cycle, there are two oxidation–reduction steps. Which steps are they and how does the substrate change in these two steps?

20 Based on the structure of citrate, explain why another name for the citric acid cycle is the tricarboxylic acid (TCA) cycle?

21 How many molecules of ADP are phosphorylated into ATP as a result of one acetyl CoA going through the citric acid cycle, assuming that all the NADH and $FADH_2$ are used to phosphorylate ADP to ATP?

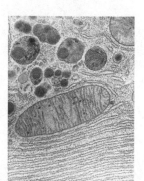

Keith R. Porter/Science Source

(a)

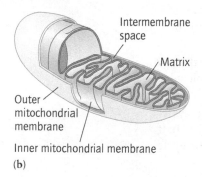

Intermembrane space

Matrix

Outer mitochondrial membrane

Inner mitochondrial membrane

(b)

Figure 15-8 A mitochondrion: (a) micrograph (b) illustration showing the four key parts of a mitochondrion, two membranes, and two aqueous regions.

15.4 The Electron Transport Chain and Oxidative Phosphorylation

The final phase of catabolism for glucose, fatty acids, and some amino acids occurs when the electrons from NADH and $FADH_2$ generated in the citric acid cycle, glycolysis, and pyruvate oxidation, are transferred to the electron transport chain. This transfer of electrons regenerates NAD^+ and FAD, the oxidized forms of these coenzymes, making them available again to shuttle electrons.

The electron transport chain consists of a sequence of electron transfer reactions (oxidation–reduction reactions) within specialized proteins and small carrier molecules, beginning with the oxidation of NADH to NAD^+ or $FADH_2$ to FAD and ending with the reduction of oxygen to water. These oxidation–reduction steps collectively supply the energy needed to phosphorylate ADP to ATP in a process known as oxidative phosphorylation. To understand oxidative phosphorylation and the electron transport chain, we first need to understand the basic structure of a mitochondrion, the organelles of the cell where the citric acid cycle, the electron transport chain, and oxidative phosphorylation take place.

Where It All Happens: The Mitochondria

With the exception of glycolysis, the energy producing pathways of catabolism occur in the **mitochondria** (pronounced mahy-tuh-**kon'**-dree-uh), organelles of approximately one micrometer in size. An organelle is an enclosed structure within the cell that performs a specific function. The structure of a mitochondrion is like a large wrinkled bag inside a smaller unwrinkled bag, creating two compartments: one between the two bags and one inside the wrinkled bag. The parts of a mitochondrion are illustrated in Figure 15-8 and include the:

- outer mitochondrial membrane (the outer bag),
- inner mitochondrial membrane (the inner bag),
- intermembrane space (the space between the bags), and
- matrix (inside the wrinkled bag).

The **outer mitochondrial membrane** is a phospholipid bilayer containing proteins that make it permeable to most molecules, ions, and small proteins. ATP and ADP can readily pass through the outer mitochondrial membrane.

In contrast, the **inner mitochondrial membrane** is permeable only to oxygen, carbon dioxide, and water. The proteins involved in the electron transport chain are embedded within or located along the inner mitochondrial membrane. The key enzyme *ATP synthase*, which catalyzes the phosphorylation of ADP to ATP, lies in the inner mitochondrial membrane, with a significant portion of this large enzyme extending into the matrix.

The inner mitochondrial membrane is a folded membrane that separates the interior of a mitochondrion into two separate aqueous compartments: the **intermembrane space**, located between the outer mitochondrial membrane and the inner mitochondrial membrane; and the **matrix**, the space enclosed by the inner mitochondrial membrane. The matrix contains all the enzymes of the citric acid cycle, as well as dissolved oxygen. The deep folds of the inner mitochondrial membrane expose most of the contents of the matrix to the inner mitochondrial membrane.

In chapter 8 you learned that dialysis of solute through a membrane proceeds in the direction that equalizes the concentration of solute. Since ions, including protons (H^+), are unable to diffuse across the inner mitochondrial membrane, the concentration of protons in the matrix is lower (higher pH; less acidic) than the concentration of protons in the intermembrane space (lower pH; more acidic), creating a different concentration of protons on either side of the membrane known as a **proton gradient**.

The Electron Transport Chain

We have seen that electrons are captured in the form of the reduced coenzymes NADH and $FADH_2$ during glycolysis, pyruvate oxidation, and the citric acid cycle. In this section we examine what happens to these electrons after NADH and $FADH_2$ transfer them to the electron transport chain, where they undergo a series of oxidation–reduction reactions, ending with the reduction of molecular oxygen to water:

$$O_2 + 4\,e^- + 4\,H^+ \rightarrow 2\,H_2O$$

Oxygen, the final destination of the electrons transferred through the electron transport chain, is essential if oxidative phosphorylation is to occur. Indeed, it is oxygen that makes the citric acid cycle and oxidative phosphorylation aerobic processes.

The electron transport chain consists of four large multienzyme protein complexes and two mobile electron carrier molecules, illustrated in the schematic in Figure 15-9. The four multienzyme complexes are known as **Complexes I, II, III, and IV**, shown in green, and the mobile electron carrier molecules are **coenzyme Q** and **cytochrome c**, shown in blue.

The protein Complexes I–IV are membrane proteins found in the inner mitochondrial membrane. They span part of or the entire inner mitochondrial membrane, extending from the matrix to the intermembrane space. Complex II, which extends into the matrix, receives electrons from $FADH_2$, which you will recall was produced during the citric acid cycle. Electrons from NADH enter earlier in the electron transport chain, at Complex I. Both Complexes I and II transfer their electrons to the membrane carrier molecule coenzyme Q, thus reducing coenzyme Q. Coenzyme Q, abbreviated CoQ, is a hydrophobic molecule that moves laterally within the inner mitochondrial membrane transporting electrons from Complex I to Complex III and from Complex II to Complex III. Electrons from Complex III are then transported by the electron

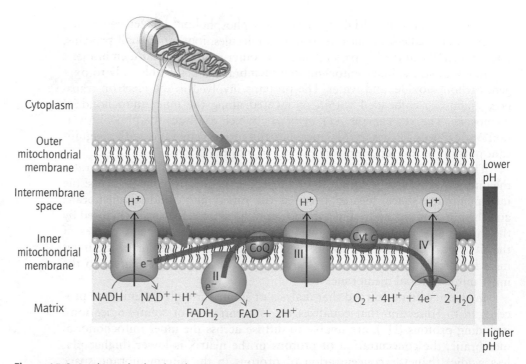

Figure 15-9 A schematic showing the components of the electron transport chain embedded in the inner mitochondrial membrane: the four multienzyme protein complexes I–IV (shown in green), and the two mobile electron carriers CoQ and Cyt C (shown in blue). The brown arrow follows the path of the electrons delivered from NADH and $FADH_2$. Black arrows represent proton pumps, which transport protons from the matrix (higher pH) to the intermembrane space (lower pH), across the inner mitochondrial membrane, and against the proton gradient.

carrier molecule cytochrome c, abbreviated Cyt C, to Complex IV, where the final reduction of oxygen to water occurs.

The Proton Motive Force As electrons are transferred through the protein complexes of the electron transport chain, energy is released that is used to pump protons from the matrix across the inner mitochondrial membrane and into the intermembrane space, against the proton gradient (uphill). Proton pumps are represented as black arrows in Figure 15-9. The intermembrane space therefore has a higher concentration of protons (lower pH, higher H^+ concentration) than the matrix (higher pH, lower H^+ concentration). Since the natural direction of diffusion is from an area of higher solute concentration to an area of lower solute concentration (see chapter 8 and dialysis), energy is required to pump protons in the opposite direction—against the proton gradient. Imagine that every time a proton is pumped from the matrix into the intermembrane space it is like compressing a spring tighter and tighter. In summary, the energy needed to pump protons against the proton gradient (to compress the spring) comes from the energy released in the oxidation–reduction steps within the protein complexes of the electron transport chain—basically, the two processes are coupled.

A proton gradient, like water above a dam or a compressed spring, represents potential energy. The accumulation of protons in the intermembrane space represents both chemical and electrical potential energy, since protons are charged ($+1$). The charge differential between these two spaces creates an electrical potential much like a battery. The chemical and electrical potential energy generated in a mitochondrion is known as the **proton motive force**, which ultimately provides the energy for phosphorylating ADP to ATP, described in the next section.

The proton pumps in these complexes (Complexes I, III, and IV) are proteins with one face oriented toward the matrix side of the membrane where

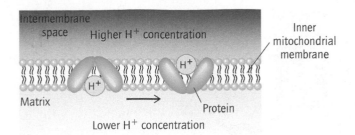

Intermembrane
space   Higher H$^+$ concentration

Inner
mitochondrial
membrane

H$^+$

H$^+$

Matrix

Lower H$^+$ concentration

Protein

Figure 15-10 Schematic showing how a proton pump transports a proton from the matrix to the intermembrane space. A conformational change in the protein exposes one face of the protein to the matrix and then to the intermembrane space.

they can bind a proton at an acidic side chain of the protein. Energy released in the oxidation–reduction reactions within the complex is used to make a conformational change within the protein that then exposes the proton to the other side of the membrane. The proton is then released into the intermembrane space, as illustrated in Figure 15-10.

Oxidation–Reduction Reactions of the Electron Transport Chain How are electrons relayed through the electron transport chain? Electrons are transferred between atoms or molecules within a protein complex or electron carrier molecule in a series of oxidation–reduction reactions. These redox reactions occur at metal atom centers (copper and iron) and in organic cofactors such as coenzyme Q, as illustrated in the reduction of coenzyme Q shown below.

$$2e^- + 2H^+ \qquad 2e^- + 2H^+$$

$$\boxed{\text{Reduction}} \qquad \boxed{\text{Oxidation}}$$

CoQ

Coenzyme Q, also known as CoQ, ubiquinone, or CoQ-10, is found in a wide variety of foods, in particular organ meats such as heart, liver, and kidney, as well as sardines, and peanuts.

Atoms and molecules have an intrinsic affinity for electrons, a physical property of a substance that determines the direction of electron flow in an oxidation–reduction reaction. Atoms and molecules with a greater electron affinity represent substances with lower potential energy. Indeed, electrons cannot flow in the opposite direction without an input of energy. In the electron transport chain, electrons are passed to atom centers with increasingly greater electron affinity, as shown in Figure 15-12, where the y-axis represents potential energy. As you can see, the reduced coenzymes NADH and FADH$_2$ have a lower electron affinity—higher potential energy—while molecular oxygen (O$_2$) has the highest electron affinity—lowest potential energy.

Complex I has the lowest electron affinity (highest potential energy) of the four complexes in the electron transport chain. NADH transfers electrons to Complex I. Electrons from FADH$_2$ enter later in the electron transport chain, at Complex II, which is lower in potential energy than Complex I. Consequently, no protons are pumped through Complex II. Therefore, FADH$_2$ ultimately supplies less energy than NADH, often measured in the number of

CoQ-10 is available as a supplement (Figure 15-11), believed by some to help prevent heart failure, cancer, MS, and periodontal disease. However, there is some debate about whether there are really any beneficial effects associated with taking CoQ-10 supplements.

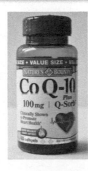

Figure 15-11 CoQ, also known as ubiquinone or CoQ-10, is available as a supplement. [Catherine Bausinger]

Figure 15-12 Graph showing the relative potential energy of the components of the electron transport chain involved in relaying electrons through oxidation–reduction reactions.

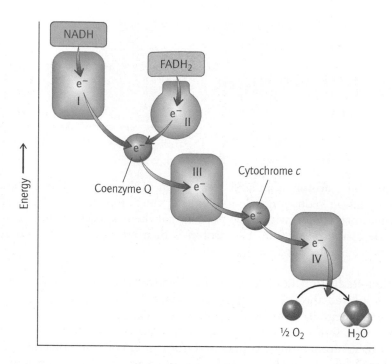

ADP that are phosphorylated to ATP, which is estimated at 2.5 per NADH and 1.5 per FADH₂:

$$2.5 \text{ ATP per NADH}$$
$$1.5 \text{ ATP per FADH}_2$$

As you can see, oxygen is the final destination of the electrons we extract from our food, and the energy released in the process is used to establish a proton gradient in the mitochondria. In the next section, we will see how this proton gradient is used to drive the phosphorylation of ADP to ATP, the primary energy carrier molecule of the cell.

WORKED EXERCISES   The Electron Transport Chain

15-12 Why do protons in the intermembrane space not simply diffuse into the matrix, equalizing the concentration of protons on both sides of the membrane? What is the "proton motive force?"

15-13 Where is the pH higher, in the matrix or the intermembrane space? Explain how this difference in pH arises.

15-14 What are the mobile electron carriers in the electron transport chain? Between which complexes do they carry electrons? Where in the mitochondrion are they located?

15-15 Fill in the blanks for the following chain of events that describes the path of the electrons through the electron transport chain, starting with NADH:

NADH → Complex ___ → coenzyme ___ → Complex ___ → cytochrome ___ → Complex ___ → O₂.

15-16 Which of the four protein complexes does not pump protons from the matrix into the intermembrane space?

15-17 How many valence electrons does one O_2 molecule contain? How many electrons do two H_2O molecules contain? What is the difference in the number of electrons? Where do the electrons that reduce oxygen to water in Complex IV come from? Write the balanced equation, including electrons, for the reduction of oxygen to water.

Solutions

15-12 Protons cannot diffuse through the inner mitochondrial membrane because it is a selectively permeable membrane, and ions cannot pass through a selectively permeable membrane because they have a charge. The proton motive force refers to the chemical and electrical potential energy in the proton gradient established between the matrix and the intermembrane space.

15-13 The pH is lower (higher proton concentration) in the intermembrane space than in the matrix as a result of protons being pumped through Complexes I, III, and IV from the matrix to the intermembrane space.

15-14 Coenzyme Q and cytochrome c are the mobile electron carriers of the electron transport chain. These electron carriers are hydrophobic molecules that are found in but not bound to the inner mitochondrial membrane, and move laterally within the membrane carrying electrons between protein complexes. CoQ carries electrons from Complex I to Complex III and from Complex II to Complex III. Cytochrome c carries electrons from Complex III to Complex IV.

15-15 NADH → Complex I → Coenzyme Q → Complex III → cytochrome c → Complex IV → O_2.

15-16 Complex II does not pump protons from the matrix into the intermembrane space.

15-17 One O_2 molecule has 12 valence electrons and one water molecule has 8 electrons, so two water molecules have 16 electrons total. The difference $16 - 12 = 4$ electrons is supplied by the electron transport chain. The balanced equation is

$$O_2 + 4\,H^+ + 4\,e^- \rightarrow 2\,H_2O$$

PRACTICE EXERCISES

22 Fill in the blank with the appropriate part of the mitochondrion described: *inner mitochondrial membrane, outer mitochondrial membrane, matrix,* or *intermembrane space.*

  a. The _____ is a selectively permeable membrane through which protons cannot pass. Only oxygen, water, and carbon dioxide can pass through this membrane.

  b. The _____ contains the proteins involved in the electron transport chain.

  c. The _____ has a higher pH than the _____.

  d. The _____ and _____ consist of an aqueous medium.

  e. The _____ and _____ consist of a lipid medium.

23 Which membrane in a mitochondrion is permeable to ions?

  a. Outer mitochondrial membrane

  b. Inner mitochondrial membrane

  c. Neither membrane

  d. Both the outer and inner mitochondrial membrane

24 How many multienzyme complexes are part of the electron transport chain?

25 Fill in the blanks for the following chain of events that describe the path of the electrons from $FADH_2$ through the electron transport chain:

$FADH_2$ → Complex ___ → coenzyme ___ → Complex ___ → cytochrome ___ → Complex ___ → O_2.

26 What critical role does oxygen play in the electron transport chain?

27 If a person were deprived of oxygen, how might the electron transport chain be affected?

28 In which part of a mitochondrion is the proton concentration higher: the matrix or the intermembrane space? How do protons get across the inner mitochondrial membrane?

Phosphorylation of ADP to ATP

We have seen that the electron transport chain generates a proton gradient between the matrix and the intermembrane space. The potential energy of this proton motive force is used to phosphorylate ADP to ATP, a process known as

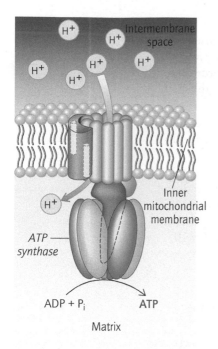

Figure 15-13 An illustration of *ATP synthase*, the enzyme that catalyzes the reaction of ADP + Pᵢ to make ATP, using the energy created by the proton motive force. The enzyme is embedded in the inner mitochondrial membrane and projects well into the matrix. The arrow shows the ion channel that allows protons to flow from the intermembrane space to the matrix, generating the energy to turn the "motor" part of the enzyme for phosphorylation.

oxidative phosphorylation. You saw in section 15.1 that the phosphorylation of ADP requires 30.5 kJ/mol (7.3 kcal/mol):

$$ADP + P_i \rightarrow ATP + H_2O \qquad 30.5 \text{ kJ/mol (7.3 kcal/mol)}$$

Within the inner mitochondrial membrane and extending into the matrix is a remarkable enzyme complex known as *ATP synthase,* also referred to as Complex V. Paul Boyer, of UCLA, and John Walker, an English scientist, were awarded the Nobel Prize in chemistry in 1997 for their discovery that *ATP synthase uses the potential energy of the proton motive force to phosphorylate ADP to ATP—an important coupled biochemical process known as oxidative phosphorylation.*

ATP synthase is a multienzyme protein complex that looks like a cylindrically shaped motor on a stick. The "stick" is embedded in the inner mitochondrial membrane while the cylinder extends well into the matrix, as illustrated in Figure 15-13. When the concentration of protons in the intermembrane space reaches a certain level, and ADP and Pᵢ are both bound to *ATP synthase,* an ion channel within the "stick" opens that allows protons from the intermembrane space to flow freely into the matrix as illustrated in Figure 15-13. Since this proton flow is from a region of higher proton concentration to a region of lower proton concentration (in the direction of the proton gradient), energy is released that drives the rotation of the "motor," much like water flowing over a dam rotates a turbine to generate electricity. Furthermore, rotation of the "motor" brings the two reactants, ADP and Pᵢ, together so that a phosphorylation reaction can occur. Then, a second turn of the "motor" releases the product, ATP, from the enzyme.

ATP Production from the Complete Oxidation of Glucose

Let us calculate the total amount of ATP produced when one molecule of glucose is completely oxidized to six carbon dioxide molecules through the catabolic processes of glycolysis, aerobic oxidation of pyruvate, the citric acid cycle, and oxidative phosphorylation:

Glycolysis (section 15.2):

Glucose + 2ADP + 2 P_i + 2NAD⁺ → 2 Pyruvate + 2ATP + 2NADH + 2 H_2O + 2H⁺

Pyruvate oxidation (section 15.2):

2 Pyruvate + 2 NAD⁺ + 2CoA-SH → 2 Acetyl CoA + 2NADH + 2CO₂

Citric acid cycle (section 15.3):

2 Acetyl CoA + 4H₂O + 6NAD⁺ + 2FAD + 2ADP + 2Pᵢ → 4CO₂ + 6NADH + 2FADH₂ + 2 CoA-SH + 2ATP + 4 H⁺

Net:

Glucose + 10NAD⁺ + 2FAD + 2 H₂O + 4ADP + 4Pᵢ → 6CO₂ + 10NADH + 6H⁺ + 2FADH₂ + 4ATP

Net ATP Energy from one molecule of glucose:

Glycolysis: 2 ATP

Citric acid cycle: 2 ATP

Oxidative phosphorylation: (10 NADH × 2.5) + (2 FADH₂ × 1.5) = 28 ATP

Total: 32 ATP per glucose

Therefore, the usable amount of ATP produced in the complete oxidation of one molecule of glucose is approximately 32 ATP.

The ATP produced both directly and during oxidative phosphorylation is available for anabolic pathways to occur that allow mechanical work such as muscle contraction, active transport of molecules and ions through membranes, as well as the anabolic processes that build proteins, DNA, and cell membranes. In this text we have focused primarily on catabolic process, but equally important are the various anabolic processes that are performed by the cell, which use some of the same metabolic intermediates seen in these catabolic pathways. In anabolic pathways, biochemical reactions are coupled to the hydrolysis of ATP to ADP and P_i or AMP and $2P_i$, providing the energy for these reactions to go forward. Hence the common reference to ATP as the "energy currency of the cell."

WORKED EXERCISES | Oxidative Phosphorylation and Counting ATP

15-18 What is the function of the proton channel in *ATP synthase*, and why do protons not simply diffuse across the inner mitochondrial membrane?

15-19 How many ATP are produced during the complete oxidation of one glucose molecule? How much CO_2 is produced?

15-20 In which biochemical pathways is ADP directly phosphorylated to ATP rather than through oxidative phosphorylation, and how many ATP are produced per molecule of glucose?

15-21 For each of the following biochemical pathways, indicate how many molecules of NADH are produced per molecule of glucose that enters glycolysis.

    a. Glycolysis: _____

    b. Pyruvate oxidation to acetyl CoA: ___

    c. Citric acid cycle: ____

15-22 The yield of ATP during oxidative phosphorylation is _____ per NADH and _____ per $FADH_2$.

Solutions

15-18 The proton channel in *ATP synthase* provides a passageway for protons to cross the inner mitochondrial membrane. A proton channel is necessary because the inner mitochondrial membrane is not permeable to ions.

15-19 32 ATP are produced for every one glucose molecule that undergoes complete catabolism to six molecules of CO_2.

15-20 For one molecule of glucose, glycolysis produces a net two ATP directly, and the citric acid cycle produces two ATP directly by phosphorylating two GDP to two GTP, the equivalent of two ATP.

15-21 **a.** Glycolysis: <u>2</u>

    b. Pyruvate oxidation to acetyl CoA: <u>2</u>

    c. Citric acid cycle: <u>6</u>

15-22 The yield of ATP during oxidative phosphorylation is <u>2.5</u> per NADH and <u>1.5</u> per $FADH_2$.

PRACTICE EXERCISES

29 What important chemical reaction is driven by the flow of protons from the intermembrane space back into the matrix through the ion channel in *ATP Synthase*?

30 How many ADP molecules are phosphorylated to ATP for every $FADH_2$ molecule that transfers electrons to the electron transport chain? What is this value for NADH? Why is it not the same as the value for $FADH_2$?

31 For each of the following biochemical pathways, indicate how many molecules of $FADH_2$ are produced per molecule of glucose:
   a. Glycolysis: _____
   b. Pyruvate oxidation to acetyl CoA: ___
   c. Citric acid cycle: ____

32 For each of the following biochemical pathways, indicate how many molecules of carbon dioxide are produced in one round of the indicated biochemical pathway:
   a. Glycolysis: _____
   b. Pyruvate oxidation to acetyl CoA: ___
   c. Citric acid cycle: ____

■ 15.5 Fatty Acid Catabolism

In this section we examine stage 2 of triglyceride catabolism: the conversion of fatty acids into acetyl CoA in a biochemical process known as β-oxidation. The third stage of catabolism for triglycerides is the same as it is for glucose: the citric acid cycle and oxidative phosphorylation. Thus, once we learn how fatty acids are converted to acetyl CoA, we can also calculate the amount of ATP formed when a fatty acid is completely oxidized to carbon dioxide molecules.

Stage 2: β-Oxidation of Fatty Acids

The primary function of fatty acids is to provide energy for cells over the long term. Fatty acid catabolism provides 80 percent of the energy needs for heart and liver cells. β-oxidation takes place in the mitochondria, the same energy producing organelles where the citric acid cycle and oxidative phosphorylation takes place. Before β-oxidation can take place, however, a fatty acid molecule must be activated to fatty acyl CoA, a reaction that takes place in the cytosol of muscle and liver cells and requires ATP (an uphill step).

Fatty Acyl CoA In chapter 12 you learned that fatty acids are long hydrocarbon chains containing a carboxylic acid at one end. Natural fatty acids contain an even number of carbon atoms. In the cytosol, a fatty acid reacts with coenzyme A to form the thioester of the fatty acid, known as **fatty acyl CoA**. Fatty acyl CoA is similar to acetyl CoA, however, acetyl CoA has a two-carbon acetyl group whereas fatty acyl CoA has more carbon atoms in the acyl group. The reaction between coenzyme A and a fatty acid to form fatty acyl CoA is coupled to ATP hydrolysis to produce adenosine monophosphate (AMP) and two P_i (inorganic phosphate), a reaction that breaks two high energy phosphoanhydride bonds:

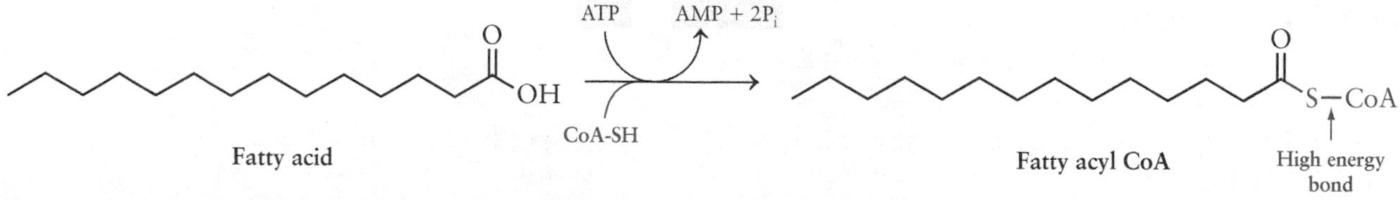

Fatty acid   Fatty acyl CoA   High energy bond

> Note that carbon atoms in a chain are often identified in relation to their proximity to the carbonyl group as shown:

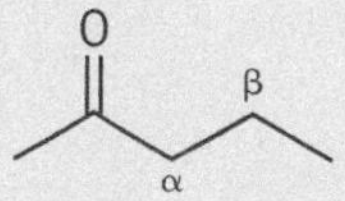

The activated fatty acid, fatty acyl CoA, is then transported into the mitochondrion where β-oxidation takes place.

β-Oxidation is a 4-step biochemical pathway that converts fatty acyl CoA into acetyl CoA and a new fatty acyl CoA molecule with two fewer carbons. In this biochemical pathway, one FAD is reduced to $FADH_2$, and one NAD^+ is reduced to NADH as shown in **Figure 15-14**. A molecule of coenzyme A is also required in the last step.

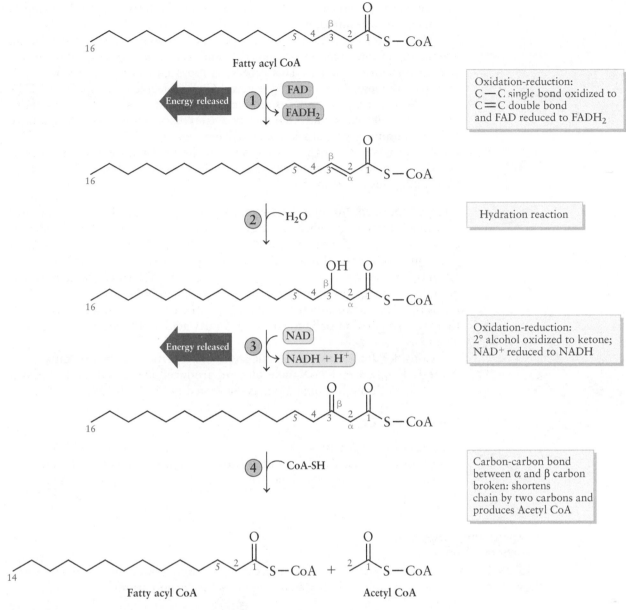

Figure 15-14 The four steps of β-oxidation including the structure of the organic intermediates, describing the type of reactions, and highlighting the energy producing steps.

As you can see from Figure 15-14, two of the four steps in this biochemical pathway are oxidation–reduction reactions that involve the oxidation of a β-carbon, which is where the biochemical pathway gets its name. The four steps of β-oxidation accomplish the following chemical transformations:

Step 1: Oxidation and FADH₂ produced. In the first step, the carbon-carbon single bond between the α- and the β-carbon atoms of fatty acyl CoA is oxidized to a carbon-carbon double bond, and FAD is reduced to FADH₂.

Step 2: Hydration. A water molecule, in the form of an OH group and an H atom, replace the double bond in what is known as a hydration reaction, introducing a hydroxyl group at the β-position and a hydrogen atom at the α-position, resulting in a product that is an alcohol.

Step 3: Oxidation and NADH produced. The hydroxyl group introduced in step 2 is oxidized to a ketone as NAD⁺ is reduced to NADH.

Step 4: Carbon-carbon bond broken. The single bond between the α-carbon and the β-carbon breaks, producing acetyl CoA and a new fatty acyl CoA molecule that is two carbon atoms shorter than the fatty acyl CoA in step 1. Coenzyme A is required, which forms a thioester to the new fatty acid.

The new fatty acyl CoA molecule then enters another round of β-oxidation, producing another acetyl CoA molecule and a new fatty acyl CoA molecule, two carbon atoms shorter. This process is repeated multiple times and with each pass through β-oxidation one acetyl CoA molecule and a fatty acyl CoA molecule two carbons shorter is produced. In the last pass, a four-carbon fatty acyl CoA molecule is converted into two acetyl CoA molecules. Thus, a fatty acid containing x carbons is converted into $x/2$ acetyl CoA molecules. The number of passes through β-oxidation is one fewer than the total number of acetyl CoA molecules produced: $(x/2) - 1$, because in the last pass two acetyl CoA molecules are formed.

Energy Produced During β-Oxidation Most of the energy obtained from fatty acid catabolism occurs when the acetyl CoA molecules produced in β-oxidation are oxidized in the citric acid cycle: a total of 10 ATP per acetyl CoA. In addition, each pass through β-oxidation produces one NADH and one FADH$_2$ for a total of $(1 \times 2.5) + (1 \times 1.5) = 4$ ATP. To calculate the total amount of energy produced from the complete catabolism of a fatty acid, in terms of ATP, we can sum these values and remember to subtract the two ATP that were required initially to convert the fatty acid and coenzyme A into fatty acyl CoA.

Consider for example palmitic acid, a 16-carbon ($x = 16$) fatty acid. A total of eight acetyl CoA molecules are produced ($16/2 = 8$) after 7 passes ($8 - 1 = 7$) through β-oxidation, as illustrated in Figure 15-15.

Palmitic acid: C$_{16}$

7 Cycles of β-oxidation	$\times$	4 =	28 ATP
8 Acetyl CoA molecules through the citric acid cycle	$\times$	10 =	80 ATP
Activation of 1 palmitic acid molecule		=	−2 ATP
Total			**= 106 ATP**

WORKED EXERCISE   β-Oxidation

15-23 Calculate the total number of ATP molecules produced from complete oxidation of stearic acid, a saturated fatty acid containing 18 carbon atoms. Show your calculation.

Solution

15-23 Begin by determining how many acetyl CoA molecules can be produced from the fatty acid, by dividing the total number of carbon atoms in the fatty acid by 2. Since stearic acid contains 18 carbon atoms, it can form $18/2 = 9$ acetyl CoA molecules. Determine the number of passes through β-oxidation by subtracting 1 from the number of acetyl CoA molecules formed: $9 - 1 = 8$. Determine the amount of ATP produced when these nine acetyl CoA molecules go through the citric acid cycle by multiplying by 10. Next, determine the number of ATP produced in each pass through β-oxidation by multiplying by 4. Add these values and then subtract two for the initial activation step:

Stearic Acid: C$_{18}$

9 Acetyl CoA molecules through citric acid cycle	$\times$	10 =	90 ATP
8 β-oxidation steps	$\times$	4 =	32 ATP
Activation of 1 palmitic acid molecule		=	−2 ATP
Total			**120 ATP**

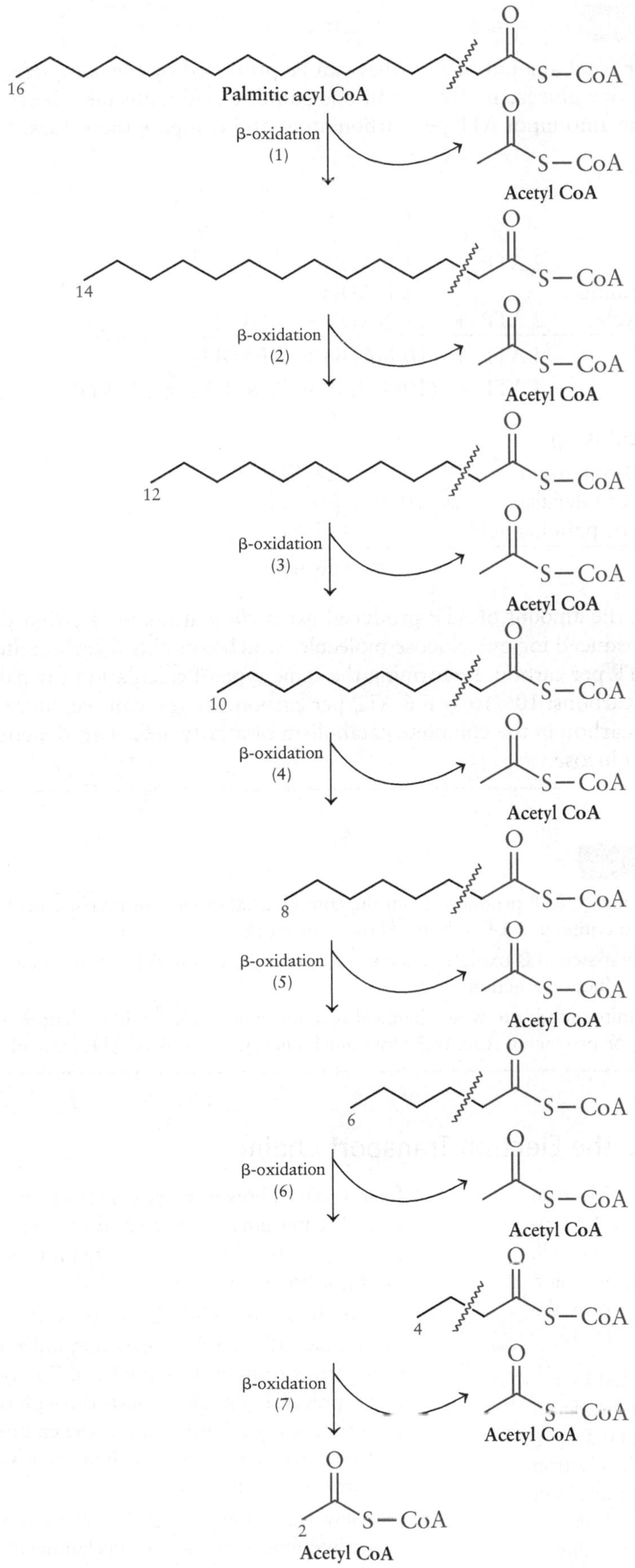

Figure 15-15 The catabolism of palmitic acid, a fatty acid containing 16 carbon atoms, showing the seven successive cycles of β-oxidation to produce eight acetyl CoA molecules.

WORKED EXERCISE | Compare the Amount of ATP Produced

15-24 Calculate the total amount of ATP that can be produced by the complete oxidation of one glucose molecule and one palmitic acid molecule. Next, determine the amount of ATP per carbon atom and compare the values.

Solution
15-24

1 Glucose:

Glycolysis:	2 ATP +	2 NADH
Pyruvate oxidation:		2 NADH
Citric acid cycle:	2 ATP +	6 NADH + 2 FADH₂

Total: **4 ATP + 10 NADH + 2 FADH₂**

$$4\ \text{ATP} + (10 \times 2.5) + (2 \times 1.5) = 32\ \text{ATP}$$

1 Palmitic acid (C₁₆)

7 Cycles of β-oxidation	× 4 =	28 ATP
8 Acetyl CoA molecules	× 10 =	80 ATP
1 Activation of palmitic acid	=	−2 ATP

Total   **= 106 ATP**

If we consider the amount of ATP produced *per carbon* atom, we see that there are 32 ATP produced for one glucose molecule, which contains 6 carbon atoms: 32/6 = 5.3 ATP per carbon. Performing the same type of calculation for palmitic acid, with 16 carbons: 106/16 = 6.6 ATP per carbon. As you can see, more ATP is produced per carbon in the complete catabolism of a fatty acid than the complete catabolism of glucose.

PRACTICE EXERCISE

33 Calculate the amount of ATP produced from the complete catabolism of myristic acid, a saturated fatty acid containing 14 carbons. Show your work.

34 Explain why the first step in β-oxidation is considered an oxidation. What substance is reduced in this oxidation–reduction reaction?

35 Before entering a mitochondrion, what chemical reaction must a fatty acid undergo? Is this an energy consuming or producing reaction? How much energy, in terms of ATP, is involved?

Chemistry in Medicine Cyanide and the Electron Transport Chain

Most compounds that inhibit enzymes of the electron transport chain are lethal because they prevent ATP synthesis, the main energy producing pathway for the cell. An inhibitor of an enzyme in the electron transport chain will block all electron transfers downstream from the inhibitor bound enzyme, as illustrated in **Figure 15-16.**

- Rotenone and Amytal: Rotenone, produced by a plant, has been used as an insecticide and pesticide. Amytal is a sedative. Rotenone and Amytal are competitive inhibitors of an enzyme in the electron transport chain that is part of Complex I; therefore, these toxins prevent the transport of electrons delivered to Complex I by NADH. These toxins do not interfere with the processing of electrons

from FADH₂, however, because electrons from FADH₂ enter later in the electron transport chain, at Complex II, which is downstream from the inhibitor bound enzyme.

- Antimycin, an antibiotic, interferes with an enzyme in Complex III, thereby preventing utilization of electrons from both NADH and FADH₂, and hence preventing ATP synthesis through oxidative phosphorylation. Antimycin, however, does not prevent direct synthesis of ATP in glycolysis and the citric acid cycle.

- Oligomycin, an antibiotic used as an antifungal agent, interferes with the proton channel in *ATP synthase* preventing phosphorylation of ADP to ATP.

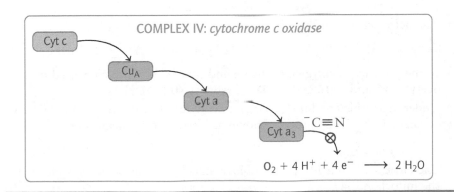

Figure 15-16 The effect of various enzyme inhibitors on the electron transport chain.

- Cyanide poisoning interferes with the final oxidation–reduction step in Complex IV, shown in **Figure 15-17** when electrons are transferred from an iron atom in the enzyme *cytochrome a₃* to oxygen, O_2, the final destination of the electrons in the electron transport chain. Cyanide, CN^-, poisoning is fatal if not treated immediately because it is an *irreversible* inhibitor of *cytochrome a₃*. Therefore, even though the victim may be breathing sufficient oxygen, he or she will be unable to utilize the oxygen because electrons are never transferred from *cytochrome a₃* to oxygen. Consequently, protons are no longer pumped into the inner membrane space, so *ATP Synthase* no longer is supplied with the proton motive force that drives the phosphorylation of ADP to ATP. Without ATP, the cell lacks the energy to drive the processes required for living. In response, cells shifts to anaerobic pathways and lactic acid accumulates. Signs of cyanide poisoning include headache, lethargy, confusion, nausea and vomiting, and convulsions.

Death resulting from cyanide poisoning occurs from respiratory arrest. A victim of cyanide poisoning may become incapacitated within seconds and can die within minutes. A fatal dose of cyanide can be as low as 1.5 mg/kg.

Cyanide is a polyatomic ion that readily diffuses into tissues and binds to target enzymes within seconds. Cyanide can be eliminated from the body by treatment with thiosulfate, a polyatomic ion that reacts with cyanide in the liver to form thiocyanate, a reaction that is catalyzed by the enzyme *rhodanese*:

Rhodanese

$$S_2O_3^- \quad + \quad CN^- \quad \rightarrow \quad SCN^- \quad + \quad SO_3^-$$

Thiosulfate   cyanide   thiocyanate   sulfite

Thiocyanate is then excreted through the kidneys.

Another detoxification treatment for cyanide poisoning is the administration of hydroxycobalamine, the active ingredient in cyanide antidote kits, approved by the FDA in 2006. Cyanide has a strong affinity for the central cobalt atom in hydroxycobalamine forming cyanocobalamine, a form of Vitamin B-12, which is readily excreted through the kidneys.

We have seen many examples of how chemistry allows us to understand biochemical processes at the atomic level, including metabolism, allowing us to understand many of the complex issues relating to health and disease. Often this knowledge leads to finding treatment options for various conditions and diseases, as we saw here for cyanide poisoning.

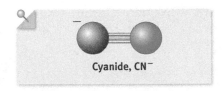

Cyanide, CN^-

COMPLEX IV: *cytochrome c oxidase*

Cyt c → Cu$_A$ → Cyt a → Cyt a₃ — $^-C\equiv N$ ⊗

$$O_2 + 4 H^+ + 4 e^- \longrightarrow 2 H_2O$$

Figure 15-17 Cyanide poisoning affects the last oxidation–reduction reaction of the electron transport chain when oxygen is reduced to water in Complex IV.

Summary

Energy and Metabolism: An Overview

- Extracting energy from our food and storing it in the form of ATP is the function of cellular respiration.
- Cellular respiration is the overall biochemical process that converts carbohydrates, fats, and sometimes dietary proteins, into CO_2, water, and energy.
- In stage 1 of catabolism, the three food biomolecules are hydrolyzed into their respective molecular building blocks: polysaccharides are hydrolyzed to glucose; proteins are hydrolyzed to amino acids; and triglycerides are hydrolyzed to fatty acids and glycerol.
- In stage 2 of catabolism, glucose is converted to two pyruvate molecules, and in the presence of oxygen, pyruvate is oxidized to acetyl CoA.
- In stage 2 of catabolism, fatty acids are converted into acetyl CoA.
- Although amino acids can be metabolized for energy, their primary function is for building proteins and other nitrogen containing compounds.
- The final stage of catabolism is the same for all three food biomolecules and begins with the citric acid cycle and concludes with oxidative phosphorylation.
- In a biological cell, energy released in catabolic reactions is stored as chemical potential energy such as the phosphoanhydride bonds of nucleotide triphosphates like adenosine triphosphate (ATP).
- Hydrolysis of ATP to form adenosine diphosphate (ADP) and inorganic phosphate releases 30.5 kJ/mol (7.3 kcal/mol) of energy.
- Hydrolysis of ATP to form adenine monophosphate (AMP) and two molecules of inorganic phosphate releases 61 kJ/mol (14.6 kcal/mol).
- In biochemistry, reactions that require an input of energy are coupled to reactions that release energy so that they can go forward.
- Other high energy molecules in the cell include the reduced coenzymes NADH and $FADH_2$, which are formed during various oxidation–reduction reactions.
- Acetyl coenzyme A is a central molecule of metabolism for both anabolic and catabolic pathways.
- One of the main functions of coenzyme A is to transfer acyl groups in biochemical reactions.

Carbohydrate Catabolism: Glycolysis and Pyruvate Metabolism

- Glycolysis is an anaerobic pathway, which means that it does not require oxygen.
- Glycolysis is a 10-step biochemical pathway that occurs in the cytosol and converts glucose into two molecules of pyruvate.
- The net reaction for glycolysis is:

$$\text{Glucose} + 2\,\text{ADP} + 2\,P_i + 2\,\text{NAD}^+ \rightarrow 2\,\text{Pyruvate} + 2\,\text{ATP} + 2\,\text{NADH} + 2\,H^+$$

- In the presence of oxygen, pyruvate and coenzyme A are converted into acetyl CoA and CO_2, and NAD^+ is reduced to NADH.
- Under anaerobic conditions, pyruvate is reduced to lactate, and NADH is oxidized to NAD^+, a process known as lactic acid fermentation.

The Citric Acid Cycle

- The citric acid cycle is an 8-step biochemical pathway that takes place in the mitochondria.

- The net reaction that occurs in the citric acid cycle is:

  Acetyl CoA + 3 NAD$^+$ + FAD + GDP + P$_i$ →
  2 CO$_2$ + CoA-SH + 3 NADH + 3 H$^+$ + FADH$_2$ + GTP

- The citric acid cycle is circular in appearance because in the *first* step acetyl CoA reacts with a molecule of oxaloacetate, which is also produced in the *last* step of the biochemical pathway.

- Most of the energy released in the citric acid cycle is in the form of electrons carried by three NADH molecules and one FADH$_2$ molecule. The electrons from these reduced coenzymes eventually enter the electron transport chain, where they are used to drive the phosphorylation of ADP to ATP.

- The citric acid cycle also directly phosphorylates one GDP to GTP, which is the equivalent of one ADP to ATP.

- NADH provides the energy to phosphorylate 2.5 ADP to 2.5 ATP, and FADH$_2$ provides the energy to phosphorylate 1.5 ADP to 1.5 ATP, during oxidative phosphorylation.

- One acetyl CoA molecule that enters the citric acid cycle provides the energy to phosphorylate the equivalent of 10 ATP.

- The first half of the citric acid cycle generates citric acid and then expels two CO$_2$ molecules, producing succinyl CoA. The second half of the citric acid cycle regenerates oxaloacetate from succinyl CoA.

The Electron Transport Chain and Oxidative Phosphorylation

- The electron transport chain is a sequence of electron transfers—oxidation–reduction reactions—within proteins and small carrier molecules, beginning with NADH or FADH$_2$ and ending with the reduction of oxygen to water: O$_2$ + 4 e$^-$ + 4 H$^+$ → 2 H$_2$O

- The oxidation–reduction steps of the electron transport chain supply the energy needed to phosphorylate ADP to ATP in a process known as oxidative phosphorylation.

- β-oxidation of fatty acids, the citric acid cycle, and oxidative phosphorylation occur in the mitochondria.

- The key parts of a mitochondrion are the outer membrane, the intermembrane space, the inner membrane, and the matrix.

- The electron transport chain consists of four large membrane-bound proteins (Complexes I through IV) and two electron carrier molecules, coenzyme Q and cytochrome c.

- Electrons are transferred from atom centers and molecules with lower electron affinity to those with higher electron affinity, with oxygen serving as the final destination of electrons.

- As electrons pass through Complexes I, III, and IV, protons are simultaneously pumped from the matrix to the intermembrane space.

- The chemical and electrical potential energy generated in the mitochondria, as a result of the proton gradient between the matrix and the intermembrane space, is known as the proton motive force, which provides the energy for phosphorylating ADP to ATP.

- *ATP synthase* is a multienzyme protein complex that catalyzes the phosphorylation of ADP to form ATP, using the energy of the proton motive force.

- The net ATP produced in the complete oxidation of glucose to carbon dioxide is:

  (10 NADH × 2.5) + (2 FADH$_2$ × 1.5) + 4 ATP = 32 ATP

Fatty Acid Catabolism

- A fatty acid must first be activated to fatty acyl CoA before β-oxidation can take place, a reaction that takes place in the cytosol.

- The reaction between coenzyme A and a fatty acid to form fatty acyl CoA is coupled to ATP hydrolysis to produce adenosine monophosphate (AMP) and two P_i (inorganic phosphate).
- β-oxidation is a 4-step biochemical pathway that produces acetyl CoA and a new fatty acyl CoA with two fewer carbons than the starting fatty acyl CoA.
- β-oxidation reduces one FAD to $FADH_2$ and one NAD^+ to NADH.
- A fatty acid containing x carbons is converted into $x/2$ acetyl CoA molecules. The number of passes through β-oxidation is one fewer than the number of acetyl CoA molecules produced: $(x/2) - 1$, because in the last pass two acetyl CoA molecules are formed.
- Most of the energy obtained from fatty acid catabolism occurs when the acetyl CoA molecules produced in β-oxidation are oxidized in the citric acid cycle: a total of 10 ATP per acetyl CoA.
- Each pass through β-oxidation produces one NADH and one $FADH_2$ for a total of $(1 \times 2.5) + (1 \times 1.5) = 4$ ATP during oxidative phosphorylation.

Key Words

Acetyl CoA The central molecule of metabolism, characterized by a thioester, and derived from the vitamin pantothenic acid, is used to transfer acyl groups in biochemical reactions.

Aerobic biochemical pathway A biochemical pathway that requires oxygen, such as pyruvate oxidation, the citric acid cycle and oxidative phosphorylation.

Anaerobic biochemical pathway A biochemical pathway that does not require oxygen, such as glycolysis and lactic acid fermentation.

ATP Synthase An enzyme complex in the inner mitochondrial membrane that uses the potential energy of the proton motive force to phosphorylate ADP to ATP. A channel created within *ATP synthase* allows protons to flow from the intermembrane space to the matrix, driving the motor that brings ADP and P_i together so that it can react and form ATP.

β-Oxidation The biochemical pathway that converts fatty acids into many acetyl CoA molecules during the second stage of triglyceride catabolism, and in the process reduces one FAD to $FADH_2$ and one NAD^+ to NADH.

Bioenergetics The study of energy transfer in living cells.

Cellular respiration The oxidation of carbohydrates, proteins, and triglycerides to produce energy in the form of ATP.

Citric acid cycle The aerobic catabolic pathway in the third stage of catabolism that converts acetyl CoA into two molecules of carbon dioxide. The process reduces three molecules of NAD^+ to NADH and one molecule of FAD to $FADH_2$, and directly phosphorylates one GDP to GTP.

Coenzyme Q A hydrophobic electron carrier molecule that moves laterally within the inner mitochondrial membrane shuttling electrons from Complex I to Complex III and from Complex II to Complex III in the electron transport chain.

Complexes I–IV Four large enzyme complexes that lie within the inner mitochondrial membrane where the oxidation–reduction reactions of the electron transport chain occur.

Coupled reaction An energy-consuming reaction linked to an energy producing reaction so that the former can go forward.

Cytochrome c An electron carrier molecule in the inner mitochondrial membrane that shuttles electrons from Complex III to Complex IV in the electron transport chain.

Electron transport chain A series of protein complexes and electron carrier molecules in the inner mitochondrial membrane that relay electrons from NADH and $FADH_2$ to oxygen via a series of oxidation–reduction reactions in the last stage of catabolism. As electrons are relayed through the electron transport chain, protons are pumped from the matrix to the intermembrane space, by a proton pump in complexes I, III, and IV, creating a proton gradient.

Fatty acyl CoA The activated form of a fatty acid required for β-oxidation.

Glycolysis A 10-step biochemical pathway that converts glucose into two molecules of pyruvate in the second stage of carbohydrate catabolism, and in the process directly phosphorylates two ADP to ATP and reduces two NAD^+ to two $NADH + H^+$.

Inner mitochondrial membrane A selectively permeable membrane that separates the matrix from the intermembrane space in a mitochondrion.

Intermembrane space The aqueous region of a mitochondrion located between the outer mitochondrial membrane and the inner mitochondrial membrane. It has a higher proton concentration than the matrix.

Lactic acid fermentation The conversion of pyruvate to lactate, with simultaneous oxidation of NADH to NAD^+, under anaerobic conditions.

Matrix The aqueous region in a mitochondrion within the inner mitochondrial membrane. It has a lower proton concentration than the intermembrane space.

Mitochondria Organelles within the cell where the energy producing pathways of β-oxidation, the citric acid cycle, and oxidative phosphorylation occur.

Outer mitochondrial membrane The outer permeable membrane of a mitochondrion; it encloses and defines the mitochondrion.

Oxidative phosphorylation The biochemical process in the third stage of catabolism that uses the proton motive force to phosphorylate ADP to ATP using *ATP synthase*.

Proton gradient A difference in proton concentration between two regions. In a mitochondrion there is a proton gradient between the intermembrane space and the matrix.

Proton motive force The chemical and electrical potential energy produced as a result of a proton gradient between the matrix and the intermembrane space, which is used to drive the phosphorylation of ADP to ATP.

Additional Exercises

Exercise and Weight Loss

36 Is there a biochemical pathway that produces glucose from fat?

37 Where does metabolism of triglycerides take place in the body?

38 Why is exercise an important component for weight loss?

39 Why are B vitamins essential for metabolism?

40 Define bioenergetics?

Energy and Metabolism: An Overview

41 Do catabolic or anabolic biochemical pathways produce energy overall?

42 What are polysaccharides, proteins, and triglycerides each hydrolyzed to when digested in the first stage of catabolism?

43 What is the central molecule of metabolism? What is "central" about this molecule?

44 What biochemical pathway converts glucose to pyruvate? What biochemical pathway converts fatty acids to acetyl CoA?

45 What biochemical pathways constitute the third stage of catabolism?

46 Where in the body are carbohydrates digested?

47 Where in the body are maltose and dextrins hydrolyzed? What is the product of these hydrolysis reactions?

48 What bonds are broken in the hydrolysis of complex carbohydrates?

49 Where in the body are fructose and galactose converted into glucose?

50 Where in the body are triglycerides hydrolyzed?

51 What compounds are produced when triglycerides are hydrolyzed?

52 Where in the body are proteins hydrolyzed?

53 What bonds are broken in the hydrolysis of proteins?

54 What is the primary function of amino acids hydrolyzed from dietary proteins?

55 How is the energy that is released in catabolic reactions most commonly stored?

56 Which bonds in adenosine triphosphate, ATP, are high energy bonds?

57 What is the other product from the hydrolysis of ATP to ADP? How many phosphoanhydride bonds are broken? How much energy is released?

58 What is the other product from the hydrolysis of ATP to AMP? How many phosphoanhydride bonds are broken? How much energy is released?

59 How much energy is required to phosphorylate ADP to ATP?

60 How do biochemical reactions that require energy occur?

61 What is a coupled reaction?

62 In muscle tissue the coupled reaction shown below is used to produce ATP. Energy is released in the coupled reaction.

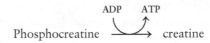

Phosphocreatine ⟶ creatine

a. What does the curved arrow mean? What does the straight arrow mean? Rewrite the reaction in the traditional way with reactants and products and one arrow.
b. Hydrolysis of phosphocreatine releases 43.1 kJ/mol (10.3 kcal/mol). How much energy is released in the coupled reaction shown? Show your calculation.

63 In the last step of glycolysis, phosphoenolpyruvate is converted to pyruvate. This reaction is coupled to the phosphorylation of ADP to ATP. The coupled reaction is shown below. Energy is released in the coupled reaction.

Phosphoenolpyruvate ⟶ pyruvate

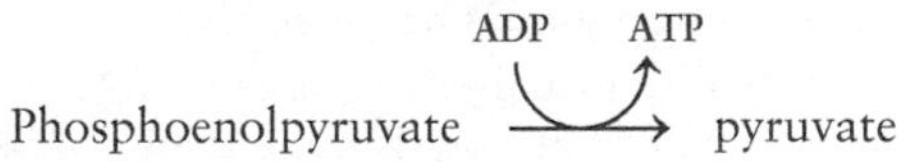

a. Which reaction provided the energy to drive the coupled reaction?
b. What does the straight arrow indicate? What does the curved arrow mean? Rewrite the reaction in the traditional way with reactants and products and one arrow.
c. Hydrolysis of phosphoenolpyruvate releases 61.9 kJ/mol (14.8 kcal/mol). How much energy is released in the coupled reaction shown? Show your calculation.

64 How do NADH and FADH$_2$ represent potential energy? How is the potential energy in these reduced enzymes ultimately used?

65 What catabolic biochemical pathways produce acetyl CoA?

66 The structure of acetyl coenzyme A is shown below. Identify the thioester, and indicate which part of the molecule represents the "acetyl" group? What part of the molecule is derived from coenzyme A? What part of the molecule is derived from the vitamin pantothenic acid? Which part of the molecule is derived from a nucleotide?

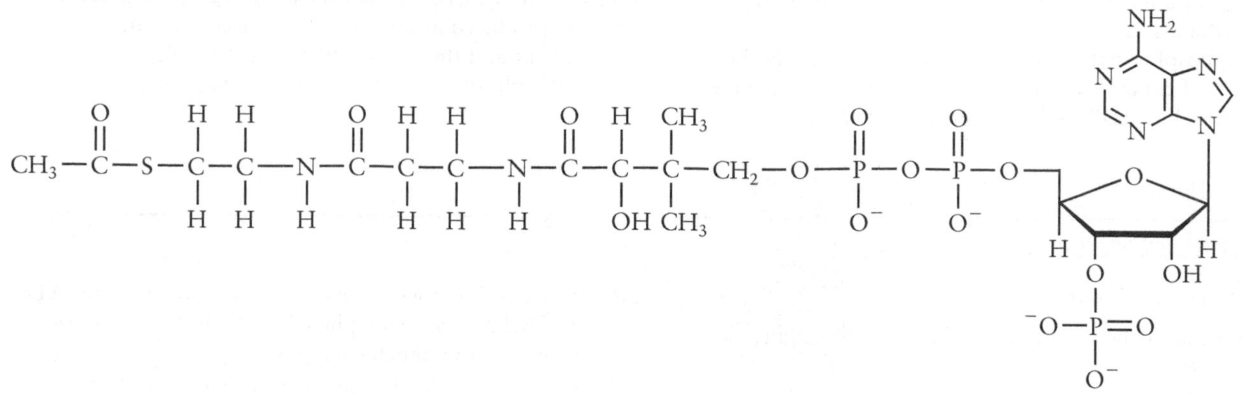

Carbohydrate Catabolism: Glycolysis and Pyruvate Metabolism

67　Glycolysis is a process that converts _____ into _____. In what stage of carbohydrate catabolism does glycolysis take place?

68　What kinds of cells use glycolysis as their sole source of metabolic energy?

69　Does glycolysis require oxygen? Is it an aerobic or anaerobic biochemical pathway?

70　How do microbes that cause tetanus, botulism, and gas gangrene produce energy?

71　Where in the cell does glycolysis take place?

72　How many molecules of ATP, NADH, and pyruvate are produced for every molecule of glucose that enters glycolysis?

73　Every intermediate in glycolysis contains a _____ functional group. State one purpose of this functional group.

74　Which step of glycolysis converts a larger molecule into two smaller molecules? This step is the origin of the *lysis* in gly*colysis*. What does "lysis" mean?

75　The first half of glycolysis requires energy. How does the second half of glycolysis more than compensate for this requirement?

76　How many ADP are directly phosphorylated during glycolysis?

77　Is any NADH or $FADH_2$ produced during glycolysis? If so, how many molecules per glucose?

78　What is the catabolic fate of pyruvate under aerobic conditions? What is the catabolic fate of pyruvate under anaerobic conditions?

79　Write the reaction that shows the formation of acetyl CoA from pyruvate. Identify the carbon atoms in acetyl CoA that come from pyruvate.

80　In the conversion of pyruvate to acetyl CoA, what coenzyme is oxidized or reduced?

81　What small organic molecule is expelled in the aerobic conversion of pyruvate to acetyl CoA? What is the role of coenzyme A in this reaction?

82　Why is oxygen required to convert pyruvate to acetyl CoA?

83　At the end of a marathon, do anaerobic or aerobic conditions exist in the marathoner's muscle cells?

84　How does the body produce ATP under anaerobic conditions?

85　When and why does lactic acid build up in muscles? What is the fate of lactate once a person has stopped their strenuous physical activity?

86　Consider the following reaction:

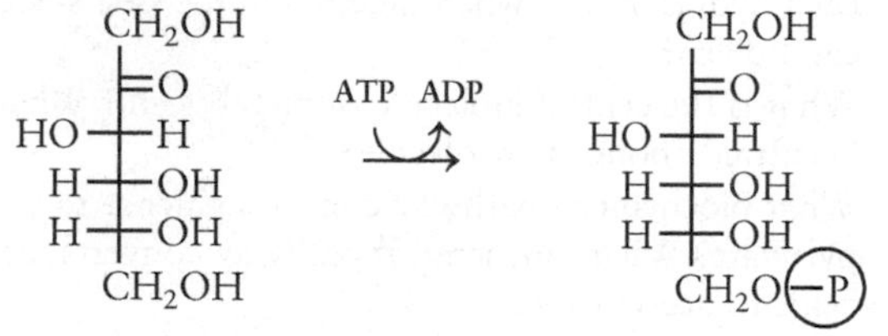

Fructose　　　　　　　　Fructose-6-phosphate

Speculate why the reaction that converts fructose to fructose-6-phosphate is coupled to the reaction that converts ATP to ADP?

The Citric Acid Cycle

87　During what stage of catabolism does the citric acid cycle take place?

88　In what part of the cell does the citric acid cycle take place?

89　Is the citric acid cycle an aerobic or anaerobic pathway?

90　Which biochemical pathway produces more energy: glycolysis or the citric acid cycle?

91　How many molecules of GTP, NADH, $FADH_2$, and CO_2 are produced for every one molecule of acetyl CoA that enters the citric acid cycle?

92　How many ADP molecules can be phosphorylated to ATP from the electrons released by one NADH molecule during oxidative phosphorylation?

93　How many ADP molecules can be phosphorylated to ATP from the electrons released by one $FADH_2$ molecule during oxidative phosphorylation?

94　When written, why does the citric acid cycle have a circular appearance?

95 The structure of citric acid is shown below.

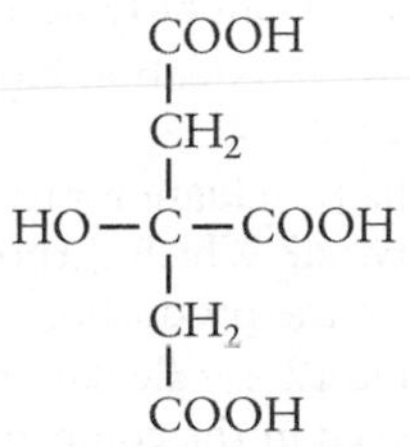

$$
\begin{array}{c}
\text{COOH} \\
|\\
\text{CH}_2 \\
|\\
\text{HO}-\text{C}-\text{COOH} \\
|\\
\text{CH}_2 \\
|\\
\text{COOH}
\end{array}
$$

a. What structure would this compound have at physiological pH?
b. How many carbon atoms does citric acid contain and where do they come from in the citric acid cycle?
c. In which step of the citric acid cycle is citrate produced?

96 The structure of α-ketoglutarate is shown below. How many carbon atoms does it contain? Does it contain fewer, greater, or the same number of carbons as citrate? Explain.

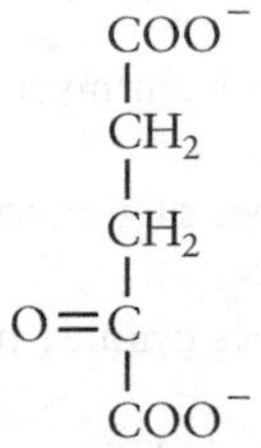

$$
\begin{array}{c}
\text{COO}^- \\
|\\
\text{CH}_2 \\
|\\
\text{CH}_2 \\
|\\
\text{O}=\text{C} \\
|\\
\text{COO}^-
\end{array}
$$

97 Halfway through the citric acid cycle, how many carbon atoms have been lost from citrate? What molecules are expelled during the first half of the citric acid cycle in order to remove these carbon atoms from citrate? How many molecules of NADH are produced in the first half of the citric acid cycle?

98 How many ATP molecules are produced from one acetyl CoA molecule going through the citric acid cycle?

99 The citric acid cycle produces energy in the form of NADH and FADH$_2$. Explain how this energy is used.

100 Step 6 of the citric acid cycle is shown below.

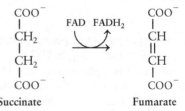

Succinate   Fumarate

a. How does the substrate structure change in this step?
b. Is succinate oxidized or reduced?
c. Is the coenzyme oxidized or reduced?

101 Step 8, the final step of the citric acid cycle, is shown below.

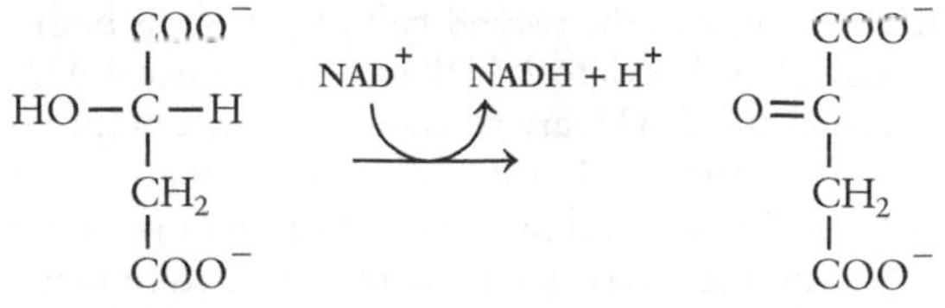

$$
\begin{array}{ccc}
\begin{array}{c}
\text{COO}^- \\
|\\
\text{HO}-\text{C}-\text{H} \\
|\\
\text{CH}_2 \\
|\\
\text{COO}^-
\end{array}
&
\xrightarrow{\text{NAD}^+ \quad \text{NADH} + \text{H}^+}
&
\begin{array}{c}
\text{COO}^- \\
|\\
\text{O}=\text{C} \\
|\\
\text{CH}_2 \\
|\\
\text{COO}^-
\end{array}
\\
\text{Malate} & & \text{Oxaloacetate}
\end{array}
$$

a. How has the substrate changed in this step?
b. How is the product of this reaction related to a reactant in the first step of the citric acid cycle?
c. Is malate oxidized or reduced?
d. Is the coenzyme oxidized or reduced?

The Electron Transport Chain and Oxidative Phosphorylation

102 In what stage of catabolism does the electron transport chain and oxidative phosphorylation occur?

103 Where do the electrons that enter the electron transport chain come from?

104 In what cellular organelles does the electron transport chain and oxidative phosphorylation occur?

105 What biochemical pathways occur in the mitochondria?

106 Can ions readily pass through the outer mitochondrial membrane?

107 Can ions readily pass through the inner mitochondrial membrane?

108 Where in the mitochondrion is a proton gradient formed? Provide an illustration labeling the relevant parts of the mitochondrion. Where is the concentration of protons lowest?

109 Which has a higher pH, the intermembrane space or the matrix of a mitochondrion?

110 List two ways in which coenzyme Q and cytochrome c are different from Complexes I–IV?

111 In Complex I of the electron transport chain, protons are pumped from the matrix to the intermembrane space through the inner mitochondrial membrane via an ion channel. Explain why this process requires an input of energy to occur. What is the source of the energy for this process?

112 What is the proton motive force?

113 Is there a charge difference between the matrix and the intermembrane space?

114 What type of reactions are responsible for relaying electrons through the electron transport chain?

115 Does the affinity for electrons increase or decrease as electrons are transferred through the electron transport chain?

116 Which has the greatest affinity for electrons: NADH, FADH$_2$, or molecular oxygen?

117 What process provides the energy to pump protons against the proton gradient from the matrix to the intermembrane space?

118 Which complex has a lower affinity for electrons: Complex I or Complex II?

119 Which coenzyme supplies more energy during oxidative phosphorylation: FADH$_2$ or NADH? Explain your answer. Which coenzyme delivers its electrons earlier in the electron transport chain?

120 Indicate which mobile electron carrier molecule delivers electrons between the following complexes:
a. from Complex I to Complex III
b. from Complex II to Complex III
c. from Complex III to Complex IV

121 Are electrons ever passed from Complex I to Complex II?

122 Which of the following Complexes in the electron transport chain pump protons from the matrix to the intermembrane space as electrons are passed through the complex:
a. Complex I
b. Complex II
c. Complex III
d. Complex IV

123 Write the reaction that produces water in Complex IV. Is oxygen reduced or oxidized in this reaction?

124 What important enzyme catalyzes the reaction between ADP and P_i during oxidative phosphorylation? What is the product of this reaction?

125 Where does *ATP synthase* obtain the energy needed to phosphorylate ADP?

126 Explain why energy is released when protons flow from the intermembrane space into the matrix. What is this energy used for?

127 How many ATP molecules are produced in the complete oxidation of one molecule of glucose? Show your calculation.

128 What are some examples of anabolic pathways in biochemistry?

129 What processes supply anabolic processes with energy?

Fatty Acid Catabolism

130 What is the primary function of fatty acids?

131 What cells in the body use the energy provided by fatty acid catabolism?

132 Where in the cell does β-oxidation occur?

133 Where in the cell is a fatty acid activated to fatty acyl CoA?

134 Does the reaction that forms fatty acyl CoA require energy or produce energy? Write the overall reaction.

135 In the 4-step β-oxidation process, how many carbon atoms are removed from fatty acyl CoA?

136 Which steps in the β-oxidation biochemical pathway are oxidation steps?

137 Which step in the β-oxidation process breaks a carbon-carbon bond? Which carbon-carbon bond is broken? What are the products?

138 Referring to Table 12-1, calculate the number of ATP molecules produced in the complete oxidation of oleic acid.

139 Referring to Table 12-1, calculate the number of ATP molecules produced in the complete oxidation of linolenic acid.

Chemistry in Medicine: Cyanide and the Electron Transport Chain

140 Why are most compounds that inhibit enzymes of the electron transport chain lethal?

141 Does rotenone interfere with the transport of electrons from NADH or $FADH_2$? Explain your answer.

142 Where and how does antimycin block the electron transport chain?

143 Where and how does oligomycin block the electron transport chain?

144 Where and how does cyanide block the electron transport chain?

145 Why can cyanide be lethal even if the victim is breathing sufficient oxygen?

146 What causes death in cyanide poisoning?

147 What are some symptoms of cyanide poisoning?

148 What are two antidotes for cyanide poisoning? How do they work?

Answers to Practice Exercises

1 a. fatty acids and glycerol
   b. monosaccharides
   c. amino acids

2 The primary function of amino acids is to build proteins and other nitrogen containing compounds.

3 a. +30.5
   b. −61

4 Carbohydrates and triglycerides

5 Acetyl CoA

6 A hydrolysis reaction; glycosidic bonds are broken; water is a reactant.

7 A hydrolysis reaction; peptide bonds (amide bonds) are broken; water is a reactant.

8 A hydrolysis reaction; three fatty acids and one glycerol molecule are produced; ester bonds are broken; water is a reactant.

9 Glycolysis is a __10__ step biochemical pathway that converts one molecule of __glucose__ into __2__ molecules of __pyruvate__ while phosphorylating __2__ ADP to __2__ ATP and __2__ NAD^+ to __2__ NADH.

10 Each intermediate has a phosphate ester. The −2 charge on a phosphate ester prevents the intermediate from diffusing out of the cell because charged species can't cross the cell membrane.

11 Glycolysis is an anaerobic process, which means that it does not require oxygen. An aerobic process requires oxygen while an anaerobic process does not require oxygen.

12 Steps 1 and 3, in the first half of glycolysis require ATP, steps 7 and 10 in the second half of glycolysis both directly phosphorylate 2 ADP to 2 ATP. Since 4 ATP are produced and 2 ATP are consumed, the net output of ATP in glycolysis is 4 − 2 = 2 ATP.

13 Lactic acid fermentation is the reduction of pyruvate, formed during glycolysis, to lactate, a reaction that oxidizes NADH to NAD^+. Lactic acid fermentation

occurs in the cytosol under anaerobic conditions. It occurs in muscle cells when oxygen is scarce, such as during intense exercise.

14 a. cytosol
b. two pyruvate molecules
c. Energy in the form of two ATP is consumed in the first half of glycolysis.
d. Energy in the form of four ATP is produced in the second half of glycolysis.

15 The four atoms in oxaloacetate become part of the molecule of citrate produced in step 1. Two of the carbon atoms from oxaloacetate end up expelled as CO_2 in steps 3 and 4 of the citric acid cycle. There are two carbon atoms in the acetyl group of acetyl CoA.

16 One molecule of FAD is reduced to $FADH_2$ in step 6 of the citric acid cycle. A carbon-carbon single bond in the substrate has been oxidized to a carbon-carbon double bond. For every $FADH_2$ that transfers electrons to the electron transport chain, approximately 1.5 ADP can be phosphorylated to 1.5 ATP during oxidative phosphorylation.

17 One GTP is equivalent to one ATP:
$GTP + ADP \rightarrow GDP + \underline{ATP}$

18 Oxaloacetate has a carboxylic acid where succinyl CoA has a thioester, and a CH_2 appears in succinyl CoA where a ketone appears in oxaloacetate.

19 In step 6, a C—C single bond is oxidized to a C=C double bond, and in step 8 a secondary alcohol is oxidized to a ketone.

20 Citric acid is a molecule containing three carboxylic acid functional groups, hence the "*tri*carboxylic acid" cycle.

21 Ten ADP are phosphorylated into 10 ATP as a result of one pass through the citric acid cycle.

22 a. The inner mitochondrial membrane is a selectively permeable membrane through which protons cannot pass. Only oxygen, water and carbon dioxide can pass through this membrane.
b. The inner mitochondrial membrane contains the proteins involved in the electron transport chain.
c. The matrix has a higher pH than the intermembrane space.
d. The intermembrane space and the matrix consist of an aqueous medium.
e. The inner mitochondrial membrane and outer mitochondrial membrane consist of a lipid medium.

23 a. The outer mitochondrial membrane is permeable to ions.

24 Four multienzyme complexes make up the electron transport chain.

25 $FADH_2 \rightarrow$ Complex __II__ $\rightarrow$ coenzyme __Q__ $\rightarrow$ Complex __III__ $\rightarrow$ cytochrome __c__ $\rightarrow$ Complex __IV__ $\rightarrow O_2$.

26 Oxygen is the final electron acceptor in the electron transport chain, whereupon it is reduced to water. In the process, NADH and $FADH_2$ are oxidized back to NAD^+ and FAD.

27 Without oxygen, the electron transport chain cannot accept electrons, so energy stops being produced, except through glycolysis, an anaerobic pathway.

28 The concentration of protons is higher in the intermembrane space than in the matrix. Protons are pumped from the matrix to the intermembrane space by the proton pumps in Complexes I, III, and IV, as electrons pass through those complexes in the electron transport chain.

29 $ADP + P_i \rightarrow ATP$

30 For $FADH_2$, 1.5 ATP and for NADH, 2.5 ATP. The value is lower for $FADH_2$ because electrons from $FADH_2$ are delivered later in the electron transport chain—at Complex II, a point lower in potential energy, so fewer protons are pumped into the intermembrane space as a result.

31 a. 0   b. 0   c. 2

32 a. 0   b. 1   c. 2

33 1 myristic acid (a fatty acid with 14 carbons)

6 cycles of β-oxidation	× 4 =	24 ATP
7 acetyl CoA molecules	× 10 =	70 ATP
Activation of 1 myristic acid	=	−2 ATP
Total	=	92 ATP

34 The first step of β-oxidation is considered an oxidation because a carbon-carbon single bond is oxidized to a carbon-carbon double bond, a loss of two electrons. The coenzyme FAD is reduced to $FADH_2$; it accepts the two electrons.

35 A fatty acid first has to be activated to fatty acyl CoA, a reaction that converts ATP into AMP and two P_i, hence, the equivalent of two ATP. The reaction is an energy consuming reaction.

APPENDIX A | Basic Math Review with Guidance for Using TI-30XA Scientific Calculator

The study of chemistry requires the use of math. This appendix is written as a reference for how to perform several common math operations.

Arithmetic

The Four Operations

Addition

- Example: There are 5 oranges in one basket, 2 oranges in another basket, and 3 oranges in a third basket. The total number of oranges is 10.
- $5 + 2 + 3 = 10$
- The **sum** is the result of adding. In the example above, 10 is the sum.

Subtraction

- Example: There are 5 apples in a basket, your sister eats 3 apples from the basket, leaving 2 apples.
- $5 - 3 = 2$
- The **difference** is the result of subtracting. In the example above, 2 is the difference.

Multiplication: Repeated addition

- There are 3 groups of 5 yellow lemons. The total number of yellow lemons is 15.
- This expression would normally be written $3 \times 5 = 15$.
- There are several ways to write multiplication. The four most common are:
  - Multiplication symbol $3 \times 5 = 15$
  - Multiplication dot $3 \cdot 5 = 15$
  - Back-to-back parenthesis $(3)(5) = 15$
  - Factor next to parenthesis $3(5) = 15$
- The **product** is the result of multiplying. In the example above, 15 is the product.
- The **factors** are the numbers involved in multiplication. In the example above, 3 and 5 are both factors.

Division: Partitioning into groups of equal amounts

- A bunch of 12 grapes is split into 2 equal groups. Now there are 6 grapes in each group.
- $12 \div 2 = 6$.
- There are several ways to write division. The three most common are:
  - Division symbol $12 \div 2 = 6$
  - Forward slash $12/2 = 6$
  - Fraction form $\dfrac{12}{2} = 6$
- In fraction form, the **numerator** is the number above the fraction bar. The numerator is the part of the whole. The **denominator** is the total number of parts in the whole. The denominator is the number below the fraction bar.
- The **quotient** is the result of dividing. In the example above, 6 is the quotient because it is the number of equally partitioned groups that are created.
- The **dividend** is the amount being partitioned.
- The **divisor** is the quantity in each group.

- To compute an ordinary arithmetic problem:
- Type the first number, press the operational symbol, type the second number, and press the = sign.
  - The TI-30XA, like all basic and scientific calculators, has four basic operation keys: +, −, ×, ÷. They are green buttons on the right side of the calculator.
  - The TI-30XA represents the equality symbol as an = sign. Other calculators may represent the equality symbol with "ANS."

Rounding Numbers

When performing calculations using measured values, the final result must be reported using the correct number of **significant figures** (see chapter 1). In order to correctly report the final result, you will often need to round a calculated number to a precise number of significant figures. *A calculator will not round for you!* Correctly rounding a number is as important as accurately calculating the final result.

To round a number:

1. First, identify the place value of the last significant figure.
2. Then, look one place value to the right of the significant figure; this is the first **non-significant figure**.

   a. If the first non-significant figure is greater than or equal to 5, increase the last significant figure by 1 and truncate (remove) all further place values. This is referred to as rounding up.
   b. If the first non-significant figure is 4 or less, leave the significant figure alone and truncate all further place values to the right.

Example 1: Rounding the boldfaced digit to the right of the decimal.

3.12472	There are four significant figures (the last significant figure is in the thousandths place).
3.124<u>7</u>2	Look at the first non-significant figure (<u>7</u>). Since <u>7</u> is greater than or equal to 5, round up the 4 in the thousandths place to a **5**, and truncate all remaining places to the right.
3.125	3.125 is the correctly rounded final result. Notice that both the 7 and 2 were truncated.

Example 2: Rounding the boldfaced digit to the right of the decimal.

0.08310156	There are three significant figures (the last significant figure is in the ten-thousandths place).
0.0831<u>0</u>156	Look at the first non-significant figure (<u>0</u>). Since <u>0</u> is less than 5, do nothing to the **1** in the ten-thousandths place and truncate the 0 and all remaining places to the right.
0.0831	0.0831 is the correctly rounded final result with three significant figures. Notice that 0 and 1 were truncated.

Example 3: Rounding place values to the left of the decimal.

415,678.87	You have determined there are three significant figures (the last significant figure is in the thousandths place).
415,<u>6</u>78.87	Look at the first non-significant figure (<u>6</u>). Since <u>6</u> is greater than or equal to 5, increase the 5 to a 6 and change the non-significant figures to zeros.
416,000	416,000 is the correctly rounded final result with three significant figures. Note that when you are rounding place values higher than the ones place, you cannot truncate the place values because it changes the value of the number. In other words, 416,000 is not the same as 416!

• *Caution:* Calculators do not round numbers. This must be done by the user.

Fraction Operations

The **fraction bar** $\left(\dfrac{n}{d}\right)$ is equivalent to a division sign.

 • Scientific calculators do not automatically perform fraction addition, subtraction, or multiplication; however, they do perform division operations.
- It is highly recommended that the user set up the fraction operation by writing it down before inputting values into a calculator.

Addition of Fractions

- Example: $\dfrac{1}{3} + \dfrac{1}{7} = \dfrac{10}{21}$
- To add and subtract fractions you must first find the least common denominator. The **least common denominator** is the lowest common multiple both denominators share.
  - In the example $\dfrac{1}{3} + \dfrac{1}{7}$ the least common denominator is 21 because it is the lowest multiple of 3 and 7. In other words, 21 is the lowest number 3 and 7 can divide into equally. The number 14 is a lower multiple of 7, but it is not a multiple of 3.
  - In the example, 7 was the scale factor for $\dfrac{1}{3}$ and 3 was the scale factor for $\dfrac{1}{7}$.
- Multiply both the numerator and denominator by their **scale factor,** the number you multiply the denominator by to increase it to the least common multiple.
  - $\dfrac{1}{3} \times \dfrac{7}{7} = \dfrac{7}{21}$ and $\dfrac{1}{7} \times \dfrac{3}{3} = \dfrac{3}{21}$
- Once both fractions share a common denominator, add their numerators.
  - $\dfrac{7}{21} + \dfrac{3}{21} = \dfrac{10}{21}$

Subtraction of Fractions

- The process for subtracting fractions is similar to the process for adding fractions. Instead, once both fractions have the same denominators, you subtract the numerators.
- Example: 1/3 − 1/7 = 10/21
- To add and subtract fractions you must first find the **least common denominator.** The least common denominator is the least common multiple both denominators share.
- Multiply both the numerator and denominator by their **scale factor** (the number you multiply the denominator by to increase it to the least common multiple).
- When both fractions share a common denominator, subtract their numerators.
  - $\dfrac{7}{21} - \dfrac{3}{21} = \dfrac{4}{21}$

Multiplication of Fractions

- Example: $\dfrac{1}{2} \cdot \dfrac{5}{6} = \dfrac{5}{12}$
  - The way to describe the fraction operation in this example is to say "half of five sixths."

- To multiply fractions simply multiply both numerators and multiply both denominators.
  - In the example above, the product of both numerators 1 and 5 is 5. The product of both denominators 2 and 6 is 12.

Division of Fractions

- Example: $\dfrac{1}{2} \div \dfrac{5}{6} = \dfrac{6}{10}$

- To divide fractions multiply the first term by the reciprocal of the second term. The **reciprocal** is the inverse of the term, or more commonly stated as the numerator and denominator switching places.
  - The reciprocal of $\dfrac{5}{6}$ is $\dfrac{6}{5}$.
- $\dfrac{1}{2} \cdot \dfrac{6}{5} = \dfrac{6}{10}$

Simplification of Fractions

- When the numerator and denominator have a common factor they can be simplified. Simplification is a process used to make a seemingly complicated fraction more recognizable.

- Example: $\dfrac{12}{24} = \dfrac{1}{2}$

- To simplify a fraction divide the numerator and denominator by the **greatest common factor (GCF)**
  - The numerator, 12, and denominator, 24, both have 2, 4, 6, and 12 as a common factor. Since 12 is the largest of these factors, it is the greatest common factor, and this is the number the numerator and denominator each is divided by.
  - $12 \div 12 = 1$ and $24 \div 12 = 2$

Equivalent Forms: Decimals, Fractions, and Percentages

A fraction, decimal, or percent are different ways of representing the same number. All fractions have a decimal and percent equivalent and vice versa. Since fractions can also be considered unsolved division problems, doing the division will return an equivalent decimal or whole number (for fractions like $\dfrac{10}{5} = 2$). The resulting decimal is equivalent to the original fraction.

All rational numbers (numbers that can be written as a ratio of two whole numbers) either repeat or terminate.

- Example: $\dfrac{1}{3} = 1 \div 3 = 0.333333333$

  - $\dfrac{1}{3}$ is a fraction with a decimal equivalent that repeats forever. A calculator will approximate $\dfrac{1}{3}$ to 0.333333333 although technically the 3s repeat forever. It is more precise to use the fraction form of a rational number than a repeating decimal.

Other rational numbers will result in decimals that terminate.

- Example: $\dfrac{1}{4} = 1 \div 4 = 0.25$

 - To convert $\dfrac{1}{4}$ into a decimal value:
  - Type: 1, press ÷, type 4, press =
    - The fraction bar is the mathematical equivalent to a division operation.

- To convert 0.25 to a fraction:
- Type: 0.25, press 2nd, press ←
  - The TI-30XA can convert decimals to fractions. It is the secondary function "F<>D" above the ← on the bottom left of the calculator.

A percent is a ratio of something out of 100. For example 50% is equivalent to $\frac{50}{100}$ or $\frac{1}{2}$. To convert a decimal to a percent, multiply the decimal by 100.

- $0.5 = 0.5 \times 100 = 50\%$

- To convert the decimal 0.5 to a percent:
- Type: 0.5, press ×, press 100, press =
  - Always use the decimal form of a percent when using a calculator. 50 times a number is not equal to 0.5 times the same number.

Exponents

The exponent operation represents repeated multiplication. It is a number multiplied by itself a certain number of times.

The **base** is written normally whereas exponents are written as a superscript.

- For example, if we are multiplying 4 by itself 3 times, we write: $4^3 = 4 \cdot 4 \cdot 4 = 64$
- 4 is the **base**, ^{3} is the **exponent**.

When referring to exponents it is common practice to say a number "raised to the power x." The example above would be stated as "four raised to the third power" or "four raised to the power three."

- When a base is raised to the second power it is said to be "squared."
- When a base is raised to the third power it is said to be "cubed."

- To evaluate $4^3 =$
- Type: 4, press y^x, type 3, press =
  - The TI-30XA has a dedicated exponent function key, y^x. It is located above the division operator.
  - The x^2 key is dedicated for squaring.
  - The secondary function of the 1 key, x^3, is dedicated for cubing.

Exponent Laws

- Any base raised to the first power equals the base.
  - $x^1 = x$
- Any base raised to the power zero equals 1.
  - $x^0 = 1$
- When identical bases with different exponents are multiplied, the result is the base raised to the sum of the two exponents.
  - $x^n \cdot x^m = x^{n+m}$
  - $4^2 \cdot 4^5 = 4^7$
- When similar bases with different exponents are divided, the result is the base raised to the difference of the two exponents.
  - $\frac{x^n}{x^m} = x^{n-m}$
  - $\frac{5^6}{5^3} = 5^3$

Logarithms and Inverse Logarithms

Logarithms and exponents are very closely related. Logarithms are commonly used to determine the exponent and the solution; they answer the question "how many times do I have to multiply the base by itself in order to obtain the solution." Simply put, "if I know the base and the solution, what is the exponent?"

- Example: $\log_4(16) = x$
  - This problem is read as "the logarithm of 16 with base 4 equals x."
    - $x = 2$ because 4 *raised to the second power* equals 16.
- Example: $\log(100) = x$
  - In problems such as this where there is no base written, it is understood to be 10.
    - $x = 2$ because 10 to the second power is 100.

- To evaluate $\log(100) =$
- Type: 100, press LOG
  - The TI-30XA LOG function assumes that the base of the logarithm is 10. Use this key only when the base is 10.

Inverse Logarithms

Inverse logarithms are used to determine the solution when you know the base and the exponent. They are also used to algebraically cancel a logarithm.

- Example: $\log_{10} x = 4.2$, which is the algebraic equivalent of $10^{4.2} = x$
- $x = 15,848.932$
  - In this example we are interested in finding the value of x. Algebra allows us to use the **inverse logarithm** (10^x) to both sides so we can isolate and find the value of x.

- To solve $\log_{10} x = 4.2$ or $10^{4.2} = x$
- Type: 4.2, press *2nd*, press LOG
  - The TI-30XA is equipped with the inverse logarithm function. It is the secondary function written in yellow as 10^x above the LOG key.

Scientific Notation

Scientific notation (also known as exponential notation) is used to express very large or very small numbers that would be inconvenient to write in standard form. Very large and very small numbers are frequently encountered in chemistry. For example, the number of atoms in a mole is such a large number that it would take over a line of text to write out in standard form. It is much more convenient to instead express these large and small numbers with many repeating zeros by using powers of 10.

- Example of a large number: $1.24 \times 10^7 = 1.24(10 \cdot 10 \cdot 10 \cdot 10 \cdot 10 \cdot 10 \cdot 10)$

$$= 1.24(10,000,000)$$

$$= 12,400,000$$

- Example of a small number: $5.006 \times 10^{-5} = 5.006\left(\dfrac{1}{10 \cdot 10 \cdot 10 \cdot 10 \cdot 10}\right)$

$$= 5.006(0.00001)$$

$$= 0.00005006$$

- To evaluate 5.006×10^{-5}
- Type: 5.006, press EE, press $\pm$, type 5, press $=$
  - The TI-30XA has a function key for negating the value of an input. It is located on the bottom right corner adjacent to the equal sign and looks like a + sign swapping places with a − sign.
  - The TI-30XA has a function key for scientific notation. It is "EE," located above the 7. "EE" stands for engineering exponent. The EE key multiplies the input by 10 raised to whatever power the user specifies. Therefore, it is very important that you don't press EE and "× 10", otherwise your answer will be 10 times greater than it should be.

Scientific notation has two parts: the **coefficient** and the **exponent.** The coefficient is the factor (1.24) that multiplies the exponent (10^7). The coefficient must be a number greater than or equal to 1 and less than 10.

A shortcut for interpreting scientific notation is to think of the exponent as the number of place values you should move the decimal in the coefficient to the right or to the left. If the exponent is positive, move the decimal to the right; if it is negative, move the decimal to the left.

- Example: $2.4 \times 10^3 = 2,400$. Notice that the decimal point moved to the right three places, from between the 2 and the 4 to after the second zero.
- Example: $3.05 \times 10^{-6} = 0.00000305$. Notice that the decimal point moved left six places.

Order of Operations

When presented with an equation that includes several different operations, the order in which you perform the operations is important. A different order will give a different answer! The order of operations is a standard procedure to ensure that a series of math operations is completed correctly.

The mnemonic device commonly used to remember the order of operations is **PEMDAS,** "**P**lease **E**xcuse **M**y **D**ear **A**unt **S**ally." This acronym stands for:

Parenthesis or other grouping symbols like brackets [ ]

Exponents

Multiplication and Division

Addition and Subtraction

- Any operations inside of parenthesis must be evaluated first. Note that parentheses are implied around the numerator and the denominator of a fraction.
- Exponents are evaluated once operations inside parentheses have been evaluated.
- Multiplication and division are evaluated after parenthesis and exponents.
  - Multiplication and division are evaluated from left to right. In other words, if you have the expression $10 \div 2 \times 3$ you must do $10 \div 2$ first because it is to the left of 2×3
    - $10 \div 2 \times 3 = 5 \times 3 = 15$
- Addition and subtraction are evaluated last.
  - Addition and subtraction are evaluated from left to right. In other words, if you have the problem $5 - 3 + 2$ you must do $5 - 3$ first.
    - $5 - 3 + 2 = 2 + 2 = 4$

Example:

$\dfrac{(5^3 + 4) + 3 - 2}{10 \div 10 \times 5}$ More than one operation is present in this example so the order of operations must be followed.

$(5^3 + 4) + 3 - 2$ Evaluate the numerator before the denominator. The operations in the numerator must be treated as if they were grouped together by parenthesis. Evaluate the operations in the numerator that are in parentheses first. Five to the third power is evaluated first because it is inside the parenthesis and exponents are evaluated before addition: $5^3 = 125$. Then, add 4: $125 + 4 = 129$. Next, perform addition and subtraction: add 3 and subtract 2 in that order because the addition is to the left of the subtraction. $129 + 3 = 132$; $132 - 2 = 130$.

$$\frac{130}{10 \div 10 \times 5}$$

Rewrite the expression with the simplified numerator.

$$\frac{130}{(10 \div 10 \times 5)}$$

The operations in the denominator must be treated as if they were grouped together by parenthesis. Then, execute the order of operations within the parenthesis. Division must be carried out first because it is to the left of the multiplication: $(10 \div 10 = 1)$. Then, multiply by 5: $1 \times 5 = 5$.

$$\frac{130}{5}$$

Rewrite the simplified expression.

26

The final step is to simplify the fraction. Since a fraction bar symbolizes division, divide.

- *Caution:* Calculators do not automatically carry out the order of operations so you must input the operations in the correct order.
- A calculator has keys for both open and closed parentheses. In the TI-30XA the parentheses keys are located above 8 and 9.
- To simplify $\dfrac{(5^3 + 4) + 3 - 2}{10 \div 10 \times 5}$
- Type in (5, press y^x, type 3 + 4). Once the parentheses have been closed, type in $+3 - 2$, and press = You have now evaluated the numerator.
- To continue, type in $\div (10 \div 10 \times 5)$, then press =

Solving Equations with One Unknown Variable

One goal of algebra is to isolate and solve for an unknown variable in an equation; it is one of the most important skills in math.

Variables and Constants

- A **variable** is a number or quantity that is represented by a letter or symbol. Variables have two important characteristics. The first is that a variable at one time may "hold" the place of a number. $3 + x = 4$ is a good example of the first characteristic. In this example it is clear that the variable, represented by the letter x, has a fixed value of 1. The term *variable* is also used to define a quantity that can vary (change). For example, let's say we are studying the relationship between two quantities: shoe size and height, where shoe size is the independent (x) variable. In general, we will find that the larger the individual's shoe size is, the taller they will be. In this case, a person's shoe size is represented by the variable x, but since there are many people whose data are being analyzed, the value of x changes (varies) depending on the shoe size of the individual.
- A **constant**, in contrast to a variable, is a fixed number or quantity that is represented by a number, letter, or symbol. For example, π is a constant because $\pi \approx 3.14$. In the equation $E = mc^2$, c is actually a constant because it represents the speed of light.

In chemistry you will often be given an equation with a single variable or multiple variables, and you will need to algebraically manipulate the equation in order to find the value of one of the variables. This is called "isolating" the variable. To isolate a variable, it is imperative that the three basic rules of solving an algebraic equation are followed.

1. Reverse Order of Operations

2. Inverse Operation to Undo

3. Balance Equation

The **Reverse Order of Operations** explains the order we should follow to isolate the desired variable. Like the name implies, we isolate the variable by undoing the operations around it while following the order of operations in reverse. Addition and subtraction should always be undone first, followed by multiplication and division, followed last by exponents and roots.

The **Inverse Operation to Undo** informs us of how to undo an operation that is being done to the variable we are trying to isolate. Addition undoes subtraction, multiplication undoes division, and exponents undo roots.

The **Balance Equation** principle reminds us that when solving an algebraic equation, whatever operation is done to one side *must* be done to the other side to maintain the equality. This is because an equals sign in an equation should be thought of as a balance. In other words, if we add 4 to the left side of the equation and not to the right, then we no longer have an equality. This concept is known as **Balancing the Equation**.

Example 1:

$5x - 2 = 8$	To isolate x on one side of the equal sign we must undo the subtraction of 2 and the multiplication by 5 to the variable x. **Reverse Order of Operations** states that subtraction must be undone first, then multiplication.
$5x - 2 + 2 = 8 + 2$	To undo the subtraction of 2, do the **inverse** operation: add 2. Keep the equation **balanced** by also adding 2 to the other side.
$5x = 10$	Rewrite the equation with both sides simplified.
$\dfrac{5x}{5} = \dfrac{10}{5}$	The inverse operation of multiplying by 5 is to divide by 5. Keep the equation balanced by also dividing the other side by 5.
$x = 2$	After simplifying both sides of the equal sign, the variable x is successfully isolated—it is by itself on one side of the equality.

In Example 1 the variable x was arranged with several other numerical terms. Many situations that involve algebra will have more than one variable where the goal is to isolate one variable in terms of the other variables. Boyle's law relating pressure and volume of gas is a good example of this.

Example 2:

$P_1V_1 = P_2V_2$	In Boyle's law four different variables represent properties of a gas in initial and final states. P_1 = Pressure of gas in initial condition V_1 = Volume of gas in initial condition P_2 = Pressure of gas in final condition V_2 = Volume of gas in final condition In this example we describe how to solve for V_2 (isolate V_2).
$\dfrac{P_1V_1}{P_2} = \dfrac{P_2V_2}{P_2}$	To undo the multiplication of V_2 by P_2, do the inverse operation: divide by P_2 on both sides of the equal sign.
$\dfrac{P_1V_1}{P_2} = V_2$	Simplify the equation and rewrite. P_2 divided by P_2 equals 1, so the P_2 terms cancel each other on the right side of the equality. V_2 has successfully been isolated. If the values of P_1, V_1, and P_2 are known, the volume in the final condition (V_2) can now be determined.

APPENDIX B | Answers to Odd-Numbered Additional Exercises

Chapter 1

45 A bone density scan measures the area and the mass of a section of bone. It estimates the third dimension to get the volume of the bone.

47 normal

49 A person with osteoporosis has less bone density than a person with osteopenia.

51 gases and liquids

53 temperature

55 **a.** heat **b.** work **c.** heat **d.** work

57 potential energy

59 **a.** kinetic energy **b.** potential energy
c. potential energy **d.** potential energy

61 The faster moving molecules have more kinetic energy.

63 **a.** Molecules in the solid state have the least amount of kinetic energy.

65 steam (The water molecules are in the gas phase.)

67 **a.** physical change **b.** chemical change
c. physical change **d.** chemical change

69 **a.** A chromium atom is on the atomic scale. (It is too small to be seen.) **b.** The human body is on the macroscopic scale. (You can see it.) **c.** A grain of sand is on the macroscopic scale. (You can see it.) **d.** A virus is on the atomic scale. (It is too small to be seen.)

71 *femto* (There are 10^{15} in a base unit.), *milli* (There are 10^3 in a base unit.) *centi* (There are 10^2 in a base unit.), and *deci* (There are 10 in a base unit.)

73 **a.** 7.6×10^{-6} **b.** 1×10^{-3} **c.** 1×10^4
d. 1.4×10^3

75 **a.** 42,000 **b.** 13,700,000 **c.** 0.013

77 6.3×10^{-3} is the smaller number.

79 **a.** 10^3 **b.** 10^2 **c.** 2.5×10^{-3} **d.** the same number
e. 0.00078

81 **a.** 10 m **b.** the same length **c.** the same length
d. 1 nm

83 **a.** 1 g **b.** 1 μg **c.** the same mass **d.** 10 μg

85 **c.** 150 μg is equivalent to 1.5×10^{-7} kg

87 3.5 mL

89 **a.** 0.25 cm^3 **b.** 80 L **c.** the same volume
d. 0.05 μL

91 2 cm

93 calorie and joule

95 **a.** true **b.** false **c.** true **d.** false

97 The balance is precise and accurate.

99 **a.** three **b.** one **c.** three **d.** three

101 **a.** exact number **b.** not exact **c.** not exact

103 **a.** 1.8 **b.** 4.3 **c.** 28

105 **b.** 79.88

107 **a.** 10.4 There are three significant figures.
b. 18 There are two significant figures.

109 **a.** 33,910.
b. 0.073
c. 3,390 mL

111 **a.** 6,000 mL
b. 20. km

113 $1 \text{ m} = 10^{-3} \text{ km}$

115 0.200 g

117 120.0 sec

119 331 lb

121 0.025 m

123 **b.** 1.05 in

125 60 mL

127 **a.** 0.989 cal
b. 8.65×10^3 cal
c. 14.0 cal
d. 87.7 cal

129 36 mg/dose

131 **a.** A brick has the greater density. It has more mass than a loaf of bread for the same volume.
b. A bowling ball has the greater density. It has more mass than a soccer ball, and they both have the same volume.
c. The bucket of concrete is denser. It has more mass than the bucket of water.
d. Normal bone is denser than a bone from a woman with osteoporosis.
e. A high-density lipoprotein has more density than a low-density lipoprotein.

133 1.9 g/mL

135 A liquid with a density greater than 1.0 g/mL will sink in water.

137 205 g

139 1.025 Yes, the specific gravity is within the normal range.

141 1.014 g/mL The specific gravity of the urine sample is within the normal range.

143 The freezing point of water is 0 °C and 32 °F.

145 **b.** in the refrigerator

147 293 kelvin and 68 °F You are wearing summer clothes.

149 284 K and 51 °F You are wearing winter clothes.

151 Yes, you have a fever.

153 195 K and −110 °F

155 **a.** For water the amount of heat is 290 cal.
b. For copper the amount of heat is 27 cal.
c. For sand the amount of heat is 47 cal.
Water requires the greatest input of heat to warm 13.4 g by 22 °C. Water has the largest heat capacity.

157 27 g

159 21.9 °C

161 0.2 cal/g °C

163 Glucose is the most important source of energy for the body. It is a sugar.

165 A person must consume the same amount of Calories that they expend in daily living to maintain their weight.

167 Glucose is a form of potential energy. The energy is stored in the chemical bonds.

Chapter 2

53 Anemia is a condition that occurs when there is a shortage of red blood cells. Red blood cells transport oxygen to cells throughout the body.

55 Iron deficiency anemia can be caused by poor absorption of iron, lack of iron in the diet, chronic blood loss, or pregnancy. Also, some medications reduce the body's ability to absorb iron.

57 atom

59 An element is a substance that is composed of only one type of atom. A compound is a substance that is composed of two or more different types of atoms held together by chemical bonds. A compound can be broken down by chemical means into its elements.

61 A proton has a +1 charge; a neutron has no charge; and an electron has a −1 charge.

63 The proton and the neutron have more mass than the electrons, so the nucleus is denser.

65 the electron

67 **a.** 23 protons and electrons **b.** 16 electrons and protons **c.** 12 protons and electrons

69 the number of protons in an element

71 The mass of a proton is 1 amu and 1.66×10^{-24} g.

73 $^{16}_{8}O$, $^{17}_{8}O$, $^{18}_{8}O$

75

Isotope	Mass Number	Atomic Number	Number of Protons	Number of Neutrons
Sulfur-32	32	16	16	16
Sulfur-33	33	16	16	17
Sulfur-34	34	16	16	18
Sulfur-36	36	16	16	20

77 The lightest isotope is sulfur-32. It has the smallest mass number.

79 **a.** iron-54 **b.** iron-58 **c.** iron-56. **d.** iron-54. **e.** Iron-56 is present in the greatest amount, but there are three other isotopes of iron that contribute to the mass of iron.

81 **a.** boron, atomic number 5
b. magnesium, atomic number 12
c. osmium, atomic number 76
d. silver, atomic number 47
e. mercury, atomic number 80
f. americium, atomic number 95
g. cesium, atomic number 55

83 **a.** beryllium, Be **b.** manganese, Mn **c.** palladium, Pd **d.** thorium, Th **e.** seaborgium, Sg

85 scandium and vanadium

87 A family or group of elements has elements in the same column in the periodic table.

89 a period

91 **a.** Group 1A, alkali metals **b.** Group 4A **c.** Group 8B **d.** Group 3A **e.** Group 8A, noble gases

93 **a.** nonmetal **b.** metalloid **c.** nonmetal

95 **a.** potassium, K **b.** radon, Rn

97 The building block elements make up the structure of the majority of compounds found in living organisms. They are carbon, 4 valence electrons; hydrogen, 1 valence electron; nitrogen, 5 valence electrons; oxygen, 6 valence electrons; phosphorus, 5 valence electrons; and sulfur, 6 valence electrons.

99 nonmetals

101 Iron carries oxygen through the bloodstream and delivers it to cells throughout the body. Iron is part of the structure of hemoglobin.

103 Zinc is important in the immune system. Good sources of zinc are oysters, breakfast cereal, beef, pork, chicken, yogurt, baked beans, and nuts.

105 A helium atom has the smaller diameter because it has fewer electrons. As the number of electrons increases, the outermost electrons spend more of their time in larger orbitals that extend farther from the nucleus, thereby increasing the diameter of the atom.

107 The group number equals the number of valence electrons for the elements in that group.

109

Element	Symbol	Atomic Number	Group Number	Number of $n = 1$ Electrons	Number of $n = 2$ Electrons	Number of $n = 3$ Electrons
Oxygen	O	8	6A	2	6	0
Beryllium	Be	4	2A	2	2	0
Argon	Ar	18	8A	2	8	8
Fluorine	F	9	7A	2	7	0

111 group 4A

113 the ground state

115 Boron has three valence electrons in the $n = 2$ level, while aluminum has three valence electrons in the $n = 3$ level.

117 2 electrons

119 The magnitude of the charge is equal to the difference between the number of protons and electrons in the ion.

121 A cation has a positive charge, while an anion has a negative charge. Metals form cations.

123 a. Ca^{2+} b. Sn^{2+} and Sn^{4+} c. N^{3-} d. Ag^+

125 Carbon achieves stability by sharing electrons with other nonmetal atoms.

127 Cu^+ has one more electron than Cu^{2+}. Both ions have the same number of protons.

129 a. magnesium ion, 12 protons, 10 electrons. There are two more protons than electrons.   b. iron ion, 26 protons, 24 electrons. There are two more protons than electrons.   c. chloride, 17 protons, 18 electrons. There is one more electron than protons.   d. fluoride, 9 protons, 10 electrons. There is one more electron than protons.   e. oxide, 8 protons, 10 electrons. There are 2 more electrons than protons.

131 Radioisotopes are unstable isotopes of an element. They have an imbalance in the ratio of neutrons to protons in the nucleus.

133 Background radiation is radiation that comes from natural sources such as cosmic rays from the Sun or radon gas from the ground.

135 Electromagnetic radiation or high-energy particles are released when a nucleus undergoes radioactive decay.

137 All types of electromagnetic radiation travel at the speed of light.

139 a. radio wave   b. x-ray   c. visible

141 a. γ-rays are more damaging to biological tissue because they are higher in energy.   b. Ultraviolet radiation is more damaging to biological tissue because they are higher in energy.

143 $^{213}_{83}Bi \rightarrow\ ^{209}_{81}Tl +\ ^4_2\alpha$

145 platinum-192

147 a. $^{26}_{11}Na \rightarrow\ ^{26}_{12}Mg +\ ^{\ 0}_{-1}\beta$ b. $^{210}_{86}Rn \rightarrow\ ^{206}_{84}Po +\ ^4_2\alpha$

149 four half-lives

151 It has a very short half-life and won't be left in the body for a long time.

153 33 hours

155 80.0 g (initial amount), 40.0 g (one half-life), 20.0 g (two half-lives), 10.0 g (three half-lives), 5.0 g (four half-lives), 2.5 g (five half-lives), 1.25 g (six half-lives), 0.625 g (seven half-lives), 0.313 g (eight half-lives) More than five half-lives have elapsed when 3 percent of the material remains.

157 Ionizing radiation damages tissue because it has sufficient energy to dislodge a valence electron from an atom, forming a cation. It can cause mutations in DNA, the genetic material of our cells.

159 Ionizing radiation damages tissue because it has sufficient energy to dislodge a valence electron from an atom, forming a cation.

161 X-rays have more penetrating power than radio waves. Therefore, when you go to the dentist you need to wear a lead apron.

163 The lead apron protects your neck and chest from the penetrating power of the x-rays.

165 a. β particles   b. γ-rays   c. γ-rays

167 α particles, β particles, and x-rays

169 The bequerel and curie measure the same property of radiation. They measure the rate of radiation emissions from a sample. The abbreviations are Bq for bequerel and Ci for curie. A millicurie would be abbreviated with mCi.

171 An absorbed dose measures the energy of radiation absorbed per mass of tissue, but it does not take into account the penetrating power of the radiation. The effective dose takes into account both the penetrating power of radiation and the amount of energy to give a biological effect. The units for absorbed dose are the gray and the rad. The effective dose is measured in sieverts and rem.

173 An LD_{50} indicates a level of exposure that would result in death in 50% of the population in 30 days.

175 the CT scan

177 a. 10 mCi   b. γ-rays

179 The quarter is higher in density than the tissue in the esophagus. The quarter absorbed more x-rays and is lighter in color than the tissue of the esophagus.

181 The MRI is best for imaging soft tissue areas of the body. The technique uses low-energy radio waves that do not pose a risk of biological damage. An MRI is not ideal for imaging denser tissues in the body, such as bones and joints.

Chapter 3

33 adenosine, the brain to prepare for sleep

35 dopamine

37 There are ionic compounds and covalent compounds. Ionic compounds have ionic bonds, and covalent compounds have covalent bonds.

39 An ionic bond is a strong force of attraction. Electrostatic attraction is another name for this kind of attraction.

41 A salt is an ionic compound.

43 A monoatomic ion is formed from a single atom, while a polyatomic ion is formed from a molecule.

45 Na^+, monoatomic ion

47 a. polyatomic   b. polyatomic   c. monoatomic   d. monoatomic   e. polyatomic

49 a. Na^+, sodium cation, charge +1; O^{2-}, oxide, charge −2   b. Cu^{2+}, cupric ion, charge +2; O^{2-}, oxide, charge −2   c. Ag^+, silver cation, charge +1; Cl^-, chloride ion, charge −1   d. Zn^{2+}, zinc ion, charge +2; S^{2-}, sulfide, charge −2   e. Ga^{3+}, gallium ion, charge +3; As^{3-}, arsenide ion, charge −3

51 a. Ba^{2+}, charge +2; SO_4^{2-}, charge −2   b. Li^+, charge +1; CO_3^{2-}, charge −2   c. Mg^{2+}, charge +2; SO_4^{2-}, charge −2   d. K^+, charge +1; $H_2PO_4^-$, charge −1   e. Na^+, charge +1; NO_2^-, charge −1

53 **a.** LiI: lithium (cation), iodide (anion)
   b. RbF: rubidium (cation), fluoride (anion)
   c. $CaBr_2$: calcium (cation), bromide (anion)
   d. BaI_2: barium (cation), iodide (anion)
   e. FeS: iron (cation), sulfide (anion)
   f. Al_2O_3: aluminum (cation), oxide (anion)

55 **a.** strontium oxide **b.** potassium iodide **c.** sodium iodide **d.** lithium fluoride **e.** gallium oxide

57 **b.** potassium hydrogen phosphate

59 **a.** Na_3PO_4 **b.** NH_4Cl **c.** $Mg(OH)_2$ **d.** $NaHCO_3$

61 $Pb_3(PO_4)_2$

63 A molecule is a discrete entity composed of two or more nonmetal atoms held together by covalent bonds.

65 Ionic compounds are solids at room temperature, and it takes more energy to melt or vaporize an ionic compound than a covalent compound. Covalent compounds can be found in all three states of matter at room temperature.

67 **a.** covalent **b.** ionic **c.** ionic

69 **c.** two nonmetals

71 When two atoms share electrons, the atoms achieve a full valence energy level. The shared electrons are in an orbital that encompasses both nuclei.

73 The octet rule states that when writing a Lewis dot structure, the valence electrons are arranged so that each atom in period 2 and higher is surrounded by eight electrons.

75 **a.** ·Ċ· **b.** H· **c.** ·Ö· **d.** ·P̈·

77 **a.** The carbon atoms contain four pairs of bonding electrons and zero pairs of nonbonding electrons. **b.** The nitrogen atom has three pairs of bonding electrons and one pair of nonbonding electrons. **c.** Yes, every atom has an octet.
   d. There are single and double bonds in this molecule.

79 :B̈r—B̈r: This molecule is an element.

81 Selenium would have two bonding pairs and two nonbonding pairs. It is in the same family/group as oxygen and sulfur, so it has the same number of valence electrons.

83 **a.** 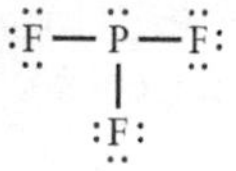 :F̈—P̈—F̈:
        |
      :F̈:
   b. 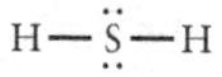 H—S̈—H
   c. 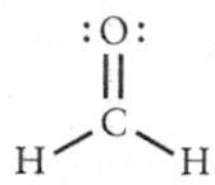
      :O:
      ‖
      C
   H   H
   d. 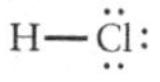 H—C̈l:
   e. 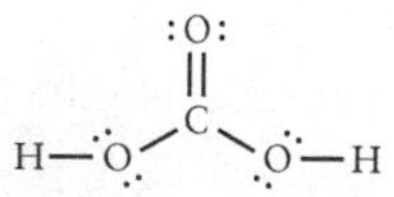
       :O:
       ‖
       C
   H—Ö Ö—H

85
     H
     |
  H—C—C≡N:
     |
     H

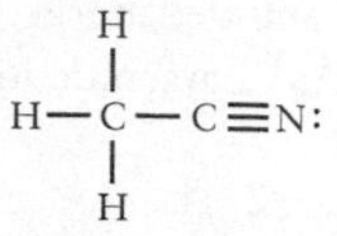

87 sulfur trioxide

89 **a.** nitrogen oxide **b.** nitrogen dioxide **c.** nitrous oxide **d.** dinitrogen trioxide **e.** dinitrogen tetraoxide **f.** dinitrogen pentaoxide

91 **a.** 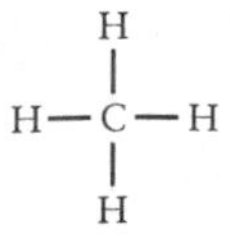
     H   The model shown is a space-filling model.
     |
  H—C—H
     |
     H
   b. H—C≡N: The model shown is a ball-and-stick model. The bond angle is 180°.

93 linear

95 The electron geometry for four electron groups around a central atom is tetrahedral. The possible molecular shapes are tetrahedral, trigonal pyramidal, and bent. The bond angles for the three molecular shapes are 109.5°.

97 **a.** trigonal pyramidal

99 The bond angles in a bent shape can be either 120° or 109.5°. The bond angles of 120° come from a trigonal planar electron geometry, while the bond angles of 109.5° come from a tetrahedral electron geometry.

101 **a.** 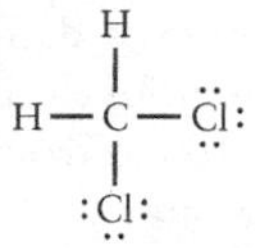
     H
     |
  H—C—C̈l:
     |
    :C̈l:

The electron geometry is tetrahedral. The molecular shape is tetrahedral, and the bond angles are 109.5°.
   b. 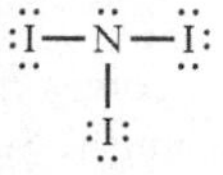 :Ï—N—Ï:
         |
       :Ï:

The electron geometry is tetrahedral. The molecular shape is trigonal pyramidal. The bond angles are 109.5°.

103 **c.** 180°, linear

105 **a.** The molecular shape is linear. The bond angle around the carbon atom is 180°. **b.** The molecular shape is tetrahedral. The bond angles around the phosphorus atom are 109.5°.

107 The electron geometry of this molecule is trigonal planar; there are three bonding groups and no nonbonding groups. The O—N—O bond angle is 120°.

109 **a.** 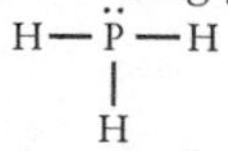 H—P̈—H
      |
      H

The molecular shape is trigonal pyramidal; there are three bonding groups and one nonbonding group. The bond angles are 109.5°.
   b. 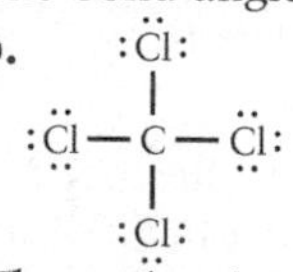
       :C̈l:
       |
   :C̈l—C—C̈l:
       |
      :C̈l:

The molecular shape is tetrahedral; there are four bonding groups and no nonbonding groups. The bond angles are 109.5°.
   c. 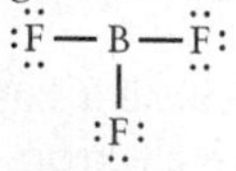 :F̈—B—F̈:
         |
       :F̈:

The molecular shape is trigonal planar; there are three bonding groups and no nonbonding groups. The bond angles are 120°.
   d. :F̈—S̈—F̈: The molecular shape is bent. There are two bonding groups and two nonbonding groups. The bond angles are 109.5°.
   e. Ö=C=S̈ The molecular shape is linear; there are two bonding groups. The bond angle is 180°.

111 a.

b. The bond angle around each carbon atom is 109.5°. There are four bonding groups around each carbon atom. **c.** The bond angle around the oxygen atom is 109.5°. There are two bonding groups and two nonbonding groups around the oxygen atom. **d.** The molecular shape at each carbon center is tetrahedral. **e.** The molecular shape at the oxygen atom is bent.

113 The molecular shape is trigonal planar when there are three bonding groups and no nonbonding groups surrounding a central atom. The bond angles are 120°. The molecular shape is trigonal pyramidal when there are three bonding groups and one nonbonding group surrounding a central atom. The bond angles are 109.5°

115 Based on VSEPR, the bond angles should be 109.5° because each atom is tetrahedral. Based on geometry, the angles within an equilateral triangle are 60°. This molecule is not commonly found in nature because the bond angles predicted by VSEPR and defined by geometry do not match.

117 1. three groups of electrons
2. bonding electrons
3. no
4. trigonal planar
5. No, they are not. The carbon on the left is tetrahedral; it has four bonding groups around it. The carbon on the right is trigonal planar; it has three bonding groups around it.
6. 120°

119 a. oxygen
b. oxygen
c. fluorine

121 increases

123 The halogens are the most electronegative. They are the farthest to the right in the periodic table.

125 If the electronegativities of the two atoms in a covalent bond are similar, then the bond will be nonpolar. If the electronegativities are different, then the covalent bond will be a polar covalent bond.

127 a. Water is polar.

b. Ethanol is polar.

129 a. I_2 is nonpolar. The two atoms in the covalent bond are the same.
b. CH_4 is nonpolar. Carbon and hydrogen have similar electronegativities.

131 A covalent bonding force occurs between two atoms within the same molecule and is much stronger than an intermolecular force of attraction. An intermolecular force of attraction occurs between atoms on different molecules.

133 The three types of intermolecular forces of attraction are: dispersion forces, dipole-dipole interactions, and hydrogen bonds. The strongest forces of attraction are hydrogen bonds. The weakest forces of attraction are dispersion forces.

135 dispersion forces

137 Dipole-dipole interactions are due to the interaction between permanent dipoles, while dispersion forces are due to the interaction between temporary dipoles.

139 H—O, H—N, and H—F bonds

141 a. C_5H_{12} is nonpolar, so it should only exert dispersion forces.
b. Acetone is polar, so it should exert dispersion forces and dipole-dipole interactions. The dipole-dipole interactions are stronger.
c. Water is polar and has O—H bonds, so it should exert hydrogen bonding, as well as dispersion forces, and dipole-dipole interactions. The hydrogen bonding is the strongest intermolecular force.

143 hydrogen bonding

145 Estrogen binds to the estrogen receptor and activates several genes. This gene activation also stimulates the proliferation of breast cancer cells.

147 The estrogen receptor is a large protein molecule. Estradiol fits the estrogen binding site perfectly because it has a complimentary shape to the binding site—the cavity within the receptor.

149 When Tamoxifen binds to the receptor, it changes the shape of the receptor, preventing gene activation from occurring.

Chapter 4

43 b and c

45 acetaldehyde

47 *Alcohol dehydrogenase* catalyzes the conversion of ethanol into acetaldehyde.

49 Alcohol contains "empty" calories because it contains no nutrients.

51 The formula mass of a compound is the sum of the mass of all the ions in a formula unit.

53 430.52 amu/molecule

55 120.38 amu/formula unit

57 Avogadro's number is the number of items in one mole.

59 1×10^{100} eggs

61 58.69 g

63 a. 558.59 g/mol **b.** 369.37 g/mol

65 a. 78.00 g/mol

67 One mole of feathers has the smaller mass. They have an equal number.

69 one mole of zinc

71 The molar mass of a compound is the same as the numerical value as its formula mass except that the units are g/mol rather than amu/formula unit.

73 1.44×10^{-5} mol

75 2.41×10^{-4} mol

77 a. 34 g **b.** 17 g **c.** 340 g **d.** 23.1 g

79 2.2×10^{-4} mol calcium

81 3.33 g sodium

83 1.29×10^{21} molecules

85 2.5×10^{21} formula units

87 2.4×10^{-8} mol lead

1.4×10^{16} atoms

89 a., b., c., and d.

91 The abbreviation (s) indicates that the substance is in the solid state. The abbreviation (g) indicates that the substance is in the gaseous state. The abbreviation (aq) indicates that the substance is in the aqueous state.

93 Matter can be neither created nor destroyed in a chemical reaction. It is an application of the law of conservation of matter.

95 a. $2 C_6H_{14} (l) + 19 O_2 (g) \rightarrow 12 CO_2 (g) + 14 H_2O (l)$

b. $C_2H_6O (g) + 3 O_2 (g) \rightarrow 2 CO_2 (g) + 3 H_2O (l)$

c. $2 C_3H_6 (g) + 9 O_2 (g) \rightarrow 6 CO_2 (g) + 6 H_2O (l)$

97 c. For every 12 moles of iron that react, 6 moles of iron (III) oxide are produced.

E4.5 $2 Al (s) + 3 CuO (s) \rightarrow Al_2O_3 (s) + 3 Cu (s)$

8.39 g of copper

99 Bioenergetics is the study of energy transfer in biological cells.

101 The difference in potential energy between reactants and products can be measured and is known as the change in enthalpy of the reaction, ΔH.

103 In an exothermic reaction, the products are lower in energy than the reactants. If the reaction is reversed, the products become the reactants and the reactants become the products. In the reverse reaction the products are higher in energy than the reactants; therefore it is an endothermic reaction.

105 a. exothermic

b. exothermic

c. endothermic

d. endothermic

107 a calorimeter

109 a. 860 Cal   b. 120 Cal   c. 110 Cal

d. 190 Cal   e. 120 Cal

111 800 Cal

113 7.0×10^3 Cal

115 anabolic reactions and catabolic reactions

117 a. and d.

119 $C_5H_{12} + 8 O_2 \rightarrow 5 CO_2 + 6 H_2O +$ heat

121 Calorimetry is used to determine the number of calories (amount of energy) in foods. The amount of energy in a food is measured as heat energy. This heat energy is what fuels our bodies so we can walk, talk, move, etc.

123 The activation energy is the amount of energy that must be attained by the reactants in order for the reaction to occur. If the reactants do not have the required activation energy, they will only bounce off each other without reacting.

125 a. It illustrates an endothermic reaction. The products are higher in energy than the reactants.

b.

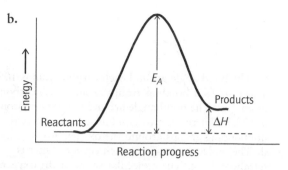

c. In the presence of a catalyst, the reaction curve will look like this.

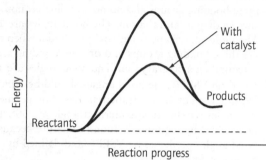

127 A catalyst does not affect the value of ΔH. A catalyst affects the value of E_A by making it smaller.

129 An enzyme contains an active site, where the reactant molecule(s) bind. The enzyme reduces the freedom of motion available to the reactant(s), thereby forcing the reactants into a spatial orientation conducive to reaction and lowering the activation energy for the reaction.

131 In the cell, chemical reactions occur at normal body temperature and at a relatively constant concentration. Therefore, to increase the rate of a biochemical reaction, enzymes are used. Enzymes reduce the freedom of motion available to reactants; they lower the activation energy by forcing reactants into a spatial orientation conducive for reaction.

133

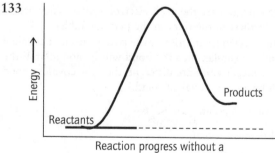

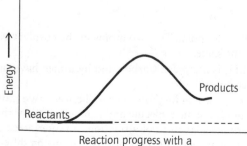

The reaction without a catalyst has the higher activation energy. The catalyst makes it easier for the reaction to occur by lowering the activation energy.

135 In a reversible reaction, the products combine to form the reactants. At the end of the reaction, both reactants and products are present. A reversible reaction is indicated by two half arrows pointing in opposite directions.

137 **a.** constant

139 **a.** acetic acid, water, acetate, and hydronium ion
b. constant
c. If more acetate is added to the reaction, the reaction will shift to the left in order to consume the excess acetate present.
d. If more acetic acid is added, the reaction will shift to the right in order to consume the excess acetic acid present.

141 **a.** Adding more reactant (H_2CO_3 or H_2O) or removing product (HCO_3^- or H_3O^+) will shift the reaction to the right.
b. Removing reactant (H_2CO_3 or H_2O) or adding more product (HCO_3^- or H_3O^+) will shift the reaction to the left.

143 A person on a mechanical respirator cannot be overfed or underfed. If they are underfed, they will be malnourished which can lead to coma and death. If they are overfed, oxygen consumption increases and the ventilator and lungs must work harder.

145 Direct calorimetry places an individual in a human calorimeter and measures the heat radiated from them. Indirect calorimetry measures a patient's oxygen uptake. Direct calorimetry is the gold standard. Indirect calorimetry is the more practical method since the patient only needs to use a spirometer to measure the oxygen uptake, rather than a human calorimeter that requires considerable cost, time, and engineering skills.

Chapter 5

51 air and seawater

53 Confusion, weakness, headache, itching, and joint pains are some of the symptoms of the bends. When the diver is ascending back to the surface, the pressure on the diver decreases and additional nitrogen molecules come out of solution. If a diver ascends too quickly, the dissolved nitrogen gas will diffuse out of the blood too quickly, forming bubbles in the bloodstream.

55 **a.** chemical reaction   **b.** change of state   **c.** change of state   **d.** chemical change

57 freezing, condensation, and deposition

59 **a.** melting (solid → liquid)   **b.** sublimation (solid → gas)   **c.** freezing (liquid → solid)   **d.** vaporization (liquid → gas)   **e.** condensation (gas → liquid)   **f.** vaporization (liquid → gas)   **g.** freezing (liquid → solid)

61 Vaporization, going from a liquid to a gas, makes mercury dangerous.

63 At the melting point or boiling point, the added energy is needed to disrupt all the intermolecular forces of attraction present.

65 Ethanol has the ability to hydrogen bond, which is the strongest intermolecular force of attraction. It takes more energy to disrupt these stronger intermolecular forces of attraction. Carbon dioxide does not have the ability to hydrogen bond.

67 Steam causes burns because of the change of state that occurs when steam comes in contact with your skin. Steam condenses when it comes in contact with the skin, which requires heat to be removed from the steam by an amount equivalent to the heat of vaporization of water. The heat is removed from the steam and transferred to your skin. Additional heat is removed as the liquid cools from 100 °C to 37 °C.

69 Pressure is defined as the force per unit area exerted by gas particles colliding against the walls of the container.

71 phenol

73 The boiling point of water should be lower at the base camp for Mt. Everest because the atmospheric pressure is lower there.

75 At 14,000 ft above sea level, there are fewer air molecules around us than at sea level. At 14,000 feet below sea level, not only is the whole atmosphere pressing down on us, but so is 14,000 ft of water.

77 35 mmHg and 0.046 atm

79 No, the number of breaths given by the mechanical ventilator does not need to be adjusted. The arterial pressure is 43 mmHg.

81 At the lower elevation in Denver, the atmospheric pressure is higher compared to the mountains. The pressure of a gas is inversely proportional to the volume of the gas. The the pressure increases as you come down from the mountains, so the volume of gas in the water bottle decreases.

83 pressure = 0.32 atm There is less pressure; therefore, the child went up in altitude with the balloon.

85 Upon inhalation, the pressure of the lungs *decreases* as the volume of the lungs *increases*.

87 Boyle's law states the pressure of a gas is inversely proportional to the volume of a gas; thus, if one goes up, the other will go down. Upon inhalation, the pressure of the lungs decreases as the volume of the lungs increases. Upon exhalation, the pressure of the lungs increases as the volume of the lungs decreases.

89 Charles' law states that temperature and volume are directly proportional to each other. As you heat the cake in the oven, the carbon dioxide in the cake heats up; therefore, the volume of the carbon dioxide increases, and the cake rises.

91 0.080 mol

93 0.19 mol

95 9.4 L

97 1.9 mol

99 334 K

101 0.500 mol

103 0.683 mol

105 4 L

107 7.6×10^{-3} mol

109 3.38 atm

111 As the patient breathes out, the partial pressure of the carbon dioxide in the bag increases.

113 A glass of soda should have more bubbles in it on the beach since the atmospheric pressure is higher than in the mountains. Henry's law states that the higher the pressure is above a liquid, the higher the concentration of the gas is in the liquid.

115 $P_{nitrogen} = 2.45$ atm

The percentage of nitrogen in the tank is 64 percent. The percentage of nitrogen in the air is 77 percent. The percentage of nitrogen in the tank is less than the percentage of nitrogen in the air.

117 A patient would need to use hyberbaric oxygen therapy if they had the bends, carbon monoxide poisoning, diabetic wounds, or an infection of necrotizing fasciitis.

119 Hemoglobin binds oxygen. Before oxygen can diffuse out of the blood, it first must be released from hemoglobin. Thus, oxygen diffuses out of the blood much more slowly than nitrogen.

121 HBOT significantly reduces the amount of time needed to drive out the CO present in the blood.

123 a. 0.059 mL

   b. 0.026 mL

Chapter 6

37 "Good" fats are unsaturated hydrocarbons that contain one or more carbon-carbon double bond and the carbon-carbon double bond creates a kink in the overall shape of the molecule. "Bad" fats have either a long saturated hydrocarbon chain that does not contain a carbon-carbon double bond or are a *trans* fat.

39 polyunsaturated fats

41 Catalytic hydrogenation reactions convert carbon-carbon double bonds into carbon-carbon single bonds. The food industry uses this method to produce products that have a longer shelf life and a consistency that consumers prefer.

43 Heteroatoms are nonmetal elements (other than hydrogen and carbon), such as oxygen, nitrogen, phosphorus, and sulfur, that form covalent bonds with carbon.

45 a. false **b.** true **c.** true **d.** false

47 C_3H_8 and C_4H_{10} are saturated hydrocarbons. C_5H_{10} and C_2H_2 are unsaturated hydrocarbons.

49 a., c., and **d.** Hydrocarbons are hydrophobic. They are insoluble in water, soluble in other hydrocarbons, and have low boiling points.

51 Water has a higher boiling point than methane. Water molecules form hydrogen bonds with other water molecules. Hydrogen bonds are stronger than the dispersion forces holding the methane molecules together, so heat is needed to separate water molecules and form a gas.

53 The three parts of an IUPAC name are the prefix, the root, and the suffix.

55 c. hexane

57 a. butane **b.** but-1-ene **c.** but-1-yne

59 Alkanes are hydrocarbons that contain only carbon-carbon single bonds and carbon-hydrogen bonds. A cycloalkane is an alkane whose chain of carbon atoms is joined in a way that forms a ring structure. Cycloalkanes are unsaturated hydrocarbons because they contain fewer than the maximum number of hydrogen atoms for a given number of carbon atoms, n.

61 Every carbon atom in an alkane must have a tetrahedral geometry; therefore, the overall shape of the molecule takes on a zigzag appearance when the chain has three or more carbon atoms. An alkane can have a non-zigzag appearance, but the non-zigzag conformations are less stable.

63 a. branched-chain alkane

b. branched-chain alkane

c. straight-chain alkane

65 a.

b.

c.

d.

e.

67 Oxygen is found in menthol.

69 a.

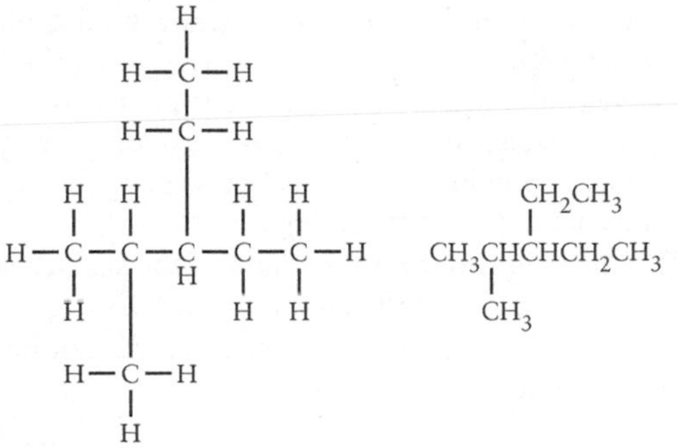

```
         H
         |
     H—C—H
         |
     H—C—H
         |
  H  H  H     H  H
  |  |  |     |  |
H—C—C—C——C—C—H
  |  |  |     |  |
  H  H  H     H  H
         |
     H—C—H
         |
         H
```

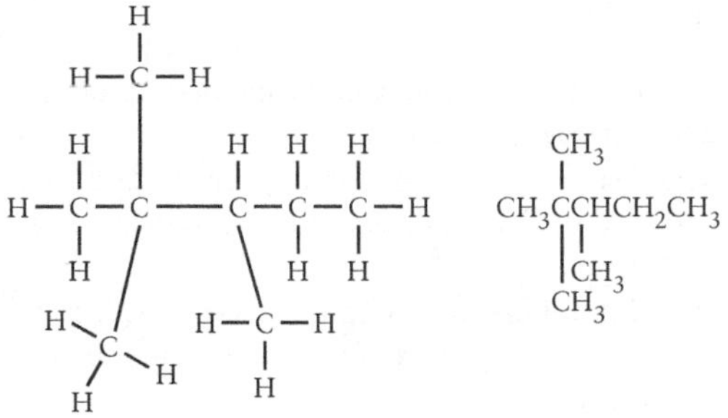

$$CH_3CHCHCH_2CH_3$$
with branches CH_2CH_3 and CH_3

b.

```
         H
         |
     H—C—H
         |
  H  H     H  H  H
  |  |     |  |  |
H—C—C————C—C—C—H
  |  |     |  |  |
  H  H     H  H  H
     |     |
   H—C—H   H—C—H
   /   \     |
  H     H    H
```

$$CH_3CCHCH_2CH_3$$
with branches CH_3 and CH_3

c.

```
            H
            |
        H—C—H
            |
        H—C—H
            |
  H  H  H     H
  |  |  |     |
H—C—C—C———C—C—H
  |  |  |     |  |
  H  H  |     H  H
      H—C—H
        |
      H—C—H
        |
        H
```

$$CH_3CH_2CHCHCH_3$$
with branches CH_2CH_3 and CH_2CH_3

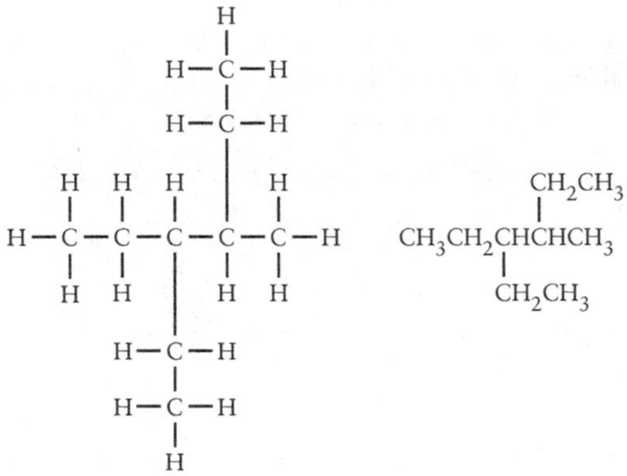

71 a.

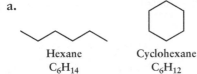

Hexane　　　　　Cyclohexane
C_6H_{14}　　　　　C_6H_{12}

The molecular formulas have the same number of carbon atoms (6), but different numbers of hydrogen atoms (14 in hexane and 12 in cyclohexane). Hexane is a straight-chain alkane; in cyclohexane the end carbons are bonded together to form a ring.

b.

Propane　　　Cyclopropane
C_3H_8　　　　C_3H_6

The molecular formulas have the same number of carbon atoms (6), but different numbers of hydrogen atoms (8 in propane and 6 in cyclopropane). Propane is a straight-chain alkane; in cyclopropane the end carbons are bonded together to form a ring.

73 a. cyclooctane
b. cycloheptane
c. cyclopentane

75 Three carbons in a ring ▷. Five carbons in a ring

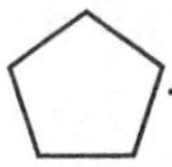

⬠. The ring with five carbons in a ring is more

commonly found in nature. The three-carbon ring is strained; the bond angles are forced to be 60° when they should be 109.5°.

77 Yes, muscone contains a cycloalkane. There are 15 carbon atoms in the ring. Oxygen is the heteroatom in muscone.

79

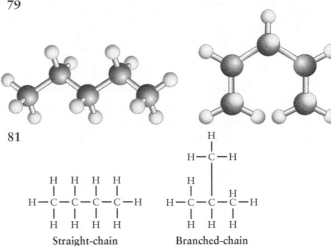

81

```
  H  H  H  H
  |  |  |  |
H—C—C—C—C—H
  |  |  |  |
  H  H  H  H
```

Straight-chain
butane

```
         H
         |
     H—C—H
         |
  H  |     H
  |  |     |
H—C—C————C—H
  |  |     |
  H  H     H
     |
     H
```

Branched-chain

83 The structure on the left has five carbons in the main chain. The structure on the right has three carbons in the main chain and two CH_3 groups branching off the middle carbon. All structural isomers have the same chemical formula.

85 The molecular shape of a carbon atom in a triple bond of an alkyne is linear. The bond angle is 180°.

87 a. simple alkene
b. diene
c. conjugated polyene
d. simple alkene

89 a. The two molecules shown are the same molecule.
b.
　　trans　　　　*cis*
c.
　　cis　　　　　*trans*

91 The double bonds are *cis* double bonds. All the groups are on the same side of the double bond. The molecule shown is the natural *cis* form of the fatty acid.

93

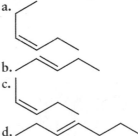

95 a. hept-1-yne　**b.** hex-3-yne　**c.** hex-2-yne
97 a. 2,2,5-trimethyloctane
b. 3,4-dimethylheptane
c. 3,4-diethylheptane
d. 4-ethyl-5-methyl-6-propylnonane
e. 3-ethyloct-4-yne
f. 2-methyldec-5-yne

99

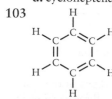

Hexane   2-Methylpentane   3-Methylpentane

2,2-Dimethylbutane   2,3-Dimethylbutane

If two compounds end up having the same name, then they were conformational isomers.

101 a. 2,3-dimethylpent-2-ene
b. 3-ethylhex-2-ene
c. cyclopentene
d. cycloheptene

103

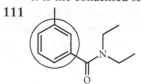

The H—C—C bond angles are 120° at every carbon in the molecule.

105 Delocalization minimizes the electron-electron repulsions so that benzene is more stable and less likely to undergo chemical reactions.

107 a. true   **b.** true   **c.** true   **d.** true

109 All except **c.** are acceptable ways to represent benzene:
a. shows the delocalization of the electrons in benzene;
b. shows the delocalization of the electrons in benzene;
d. shows that benzene has alternating carbon-carbon single bonds and carbon-carbon double bonds;
e. is the condensed structural formula for benzene.

111

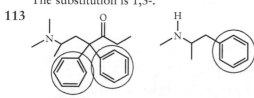

DEET

The substitution is 1,3-.

113

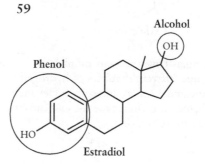

Methadone   Methamphetamine

They both have a nitrogen atom attached to a carbon atom, and an aromatic ring and a methyl group attached to the carbon backbone.

115 a. *trans*   **b.** *cis*   **c.** Yes, the molecule would have a different shape if this double bond is isomerized.

The change in shape of the protein-retinal complex initiates a nerve impulse that travels along the optic nerve to the brain. This impulse is interpreted as a visual image.   **d.** Double bond (b) undergoes isomerization in the chemical process of vision. The energy comes from light.

117 The protein rhodopsin is found in the surface of the cell membranes of the rods and cones.

119 Rhodopsin is the *cis* isomer and bathrhodopsin is the *trans* isomer.

Chapter 7

45 Functional groups in a molecule determine its chemical and physical properties.

47 The distribution of morphine, heroin, and hydrocodone can lead to dependence and, with repeated use, addiction.

49 When opioids bind to opioid receptors, it initiates a sequence of biological events that leads to reduced pain sensations.

51 endorphins and opioids

53 An ether has an oxygen atom bonded to two carbon atoms. An alcohol has oxygen atom bonded to a hydrogen atom and a carbon atom.

55 a. primary alcohol
b. secondary alcohol

c. tertiary alcohol

57 The molecular shape around the oxygen is bent tetrahedral. The C—O—H bond angle is 109°.

59

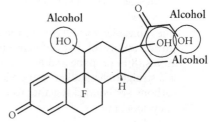

Estradiol

Betamethasone

The phenol is an alcohol attached to an aromatic ring. They both have a three six-membered rings and a five-membered ring fused together.

61

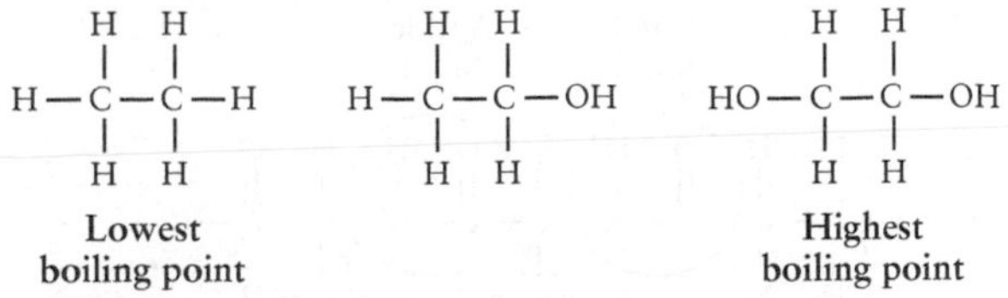

H–C–C–H (H and H on top, H and H on bottom)

Lowest boiling point

H–C–C–OH

HO–C–C–OH

Highest boiling point

63 Alcohol

(HO) Ether (O)

Ether (O) ONO₂

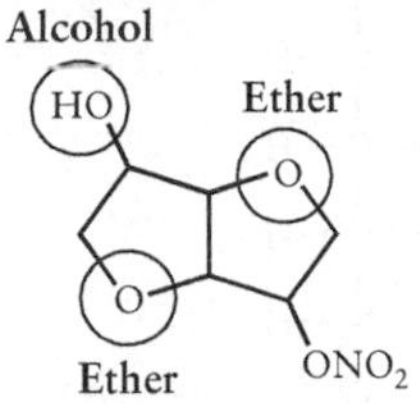

65. a. 1-ethoxyethane
b. 2-methoxypentane
c. 3-ethoxypentane

67 a. ether **b.** primary alcohol **c.** ether **d.** tertiary alcohol **e.** ether **f.** secondary alcohol **g.** primary alcohol

69 a. butan-2-amine, primary amine **b.** butan-1-amine, primary amine **c.** N-methylbutan-1-amine, secondary amine

71 c. is a primary amine. Only structure **c.** is capable of hydrogen bonding.

73 a-b.

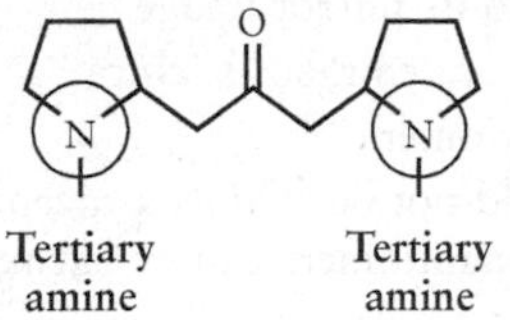

Tertiary amine   Tertiary amine

c. trigonal pyramidal
d. 109.5°
e. It is an alkaloid because it contains a nitrogen atom, and it is produced by a plant.

75 a.

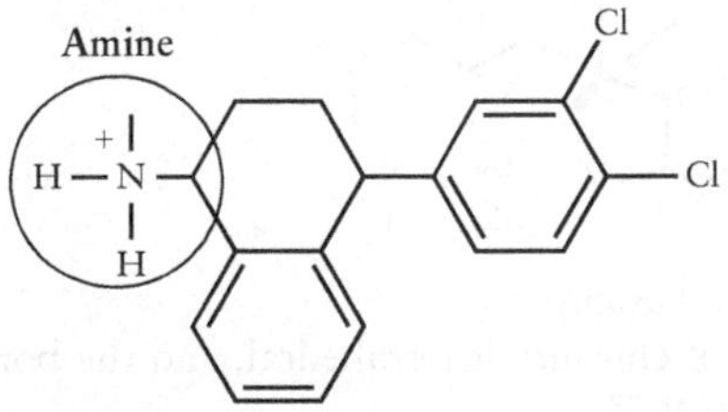

Amine

$H–\overset{+}{N}–$ Cl   Cl

b. ionic form
c. secondary amine

77 a. aldehyde; the IUPAC name is propanal.
b. ketone; the IUPAC name is pentan-3-one.
c. aldehyde; the IUPAC name is pentanal.
d. ketone; the IUPAC name is hexan-2-one.
e. aldehyde; the IUPAC name is 3-methyl pentanal.

79 a.

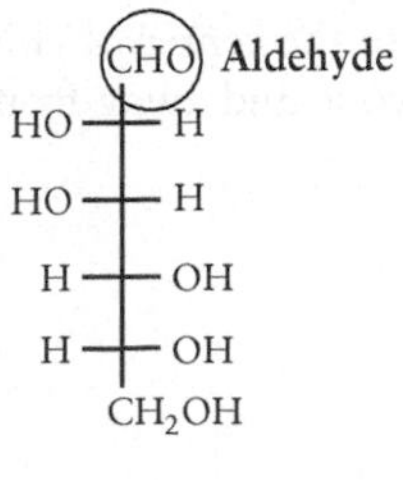

(CHO) Aldehyde
HO — H
HO — H
H — OH
H — OH
CH₂OH

b. The molecular shape around the carbonyl carbon is trigonal planar.

c.

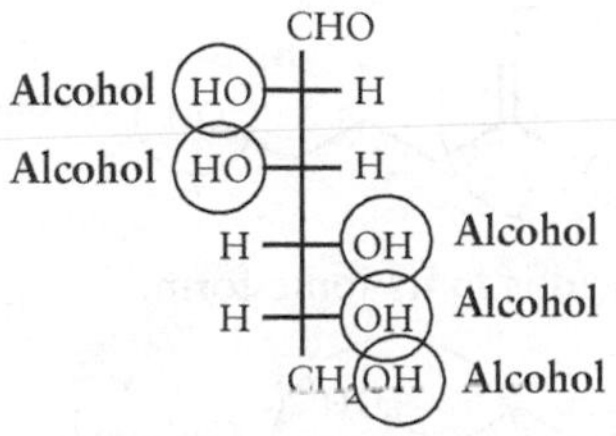

CHO
Alcohol (HO) — H
Alcohol (HO) — H
H — (OH) Alcohol
H — (OH) Alcohol
CH₂(OH) Alcohol

d. Mannose is a carbohydrate.

81 Propan-2-ol has the higher boiling point. Propan-2-ol has hydrogen bonding and dipole-dipole interactions; while propan-2-one only has dipole-dipole interactions.

OH   O

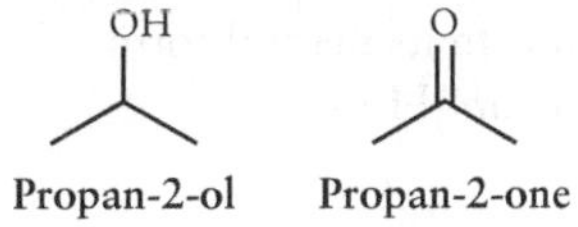

Propan-2-ol   Propan-2-one

83 a.

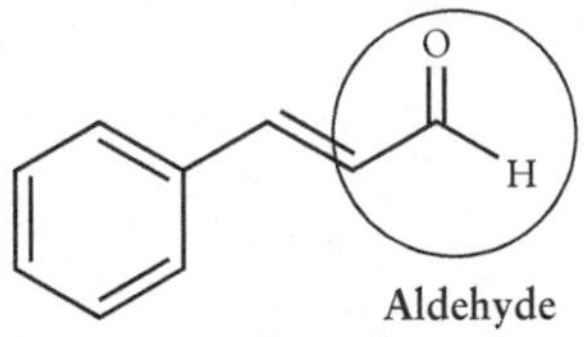

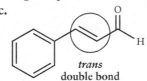

Aldehyde

b. trigonal planar

c.

trans
double bond

85 a. pentanoic acid
b. hexanoic acid
c. propanoic acid
d. 3-methylpentanoic acid

87 a.

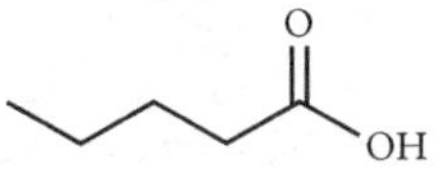

O
OH

b.

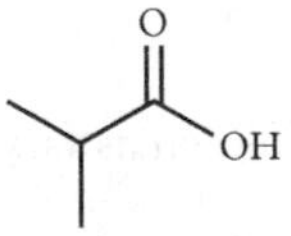

O
OH

c.

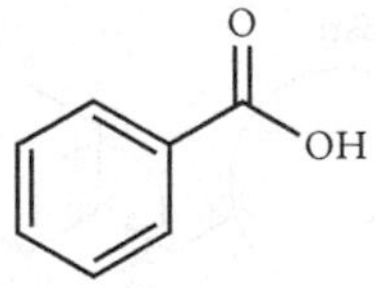

O
OH

89

O
HO — H

O
OH

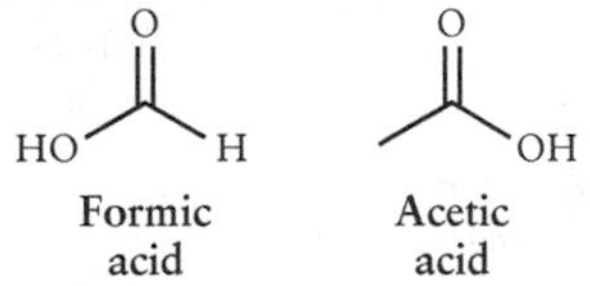

Formic acid   Acetic acid

91 a.

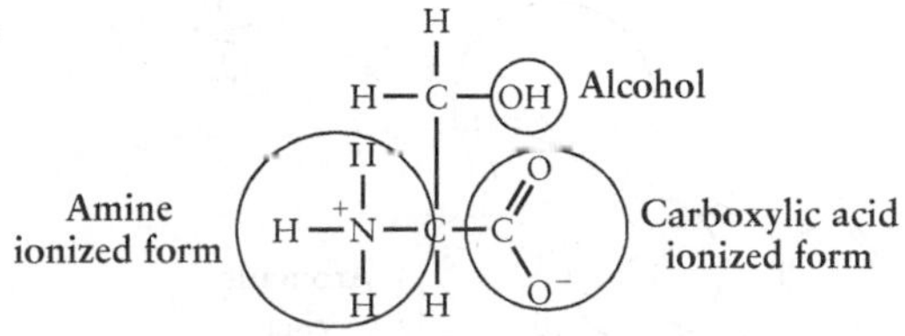

H
H–C–(OH) Alcohol

Amine ionized form $H–\overset{+}{N}–C–C$ Carboxylic acid ionized form
O⁻

b. The amine is in its ionic form.
c. The carboxylic acid is in its ionic form.
d. The overall charge on serine is zero.

93 a.

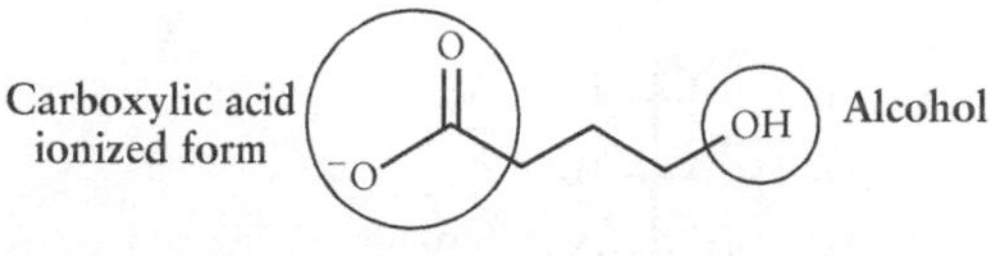

Carboxylic acid ionized form

Alcohol

b. The carboxylic acid is in its ionic form.

95 a. and b.

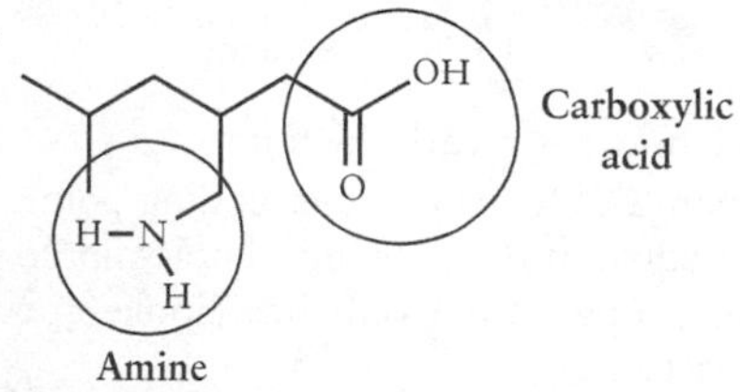

Carboxylic acid

Amine

c. The carboxylic acid is in its neutral form.
d. The amine is in its neutral form.

97 a.

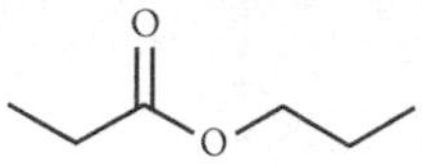

b.

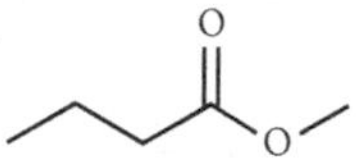

c.

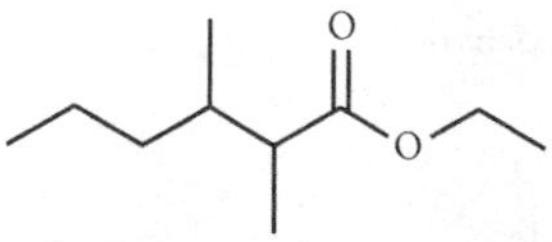

99 butyl ethanoate

101

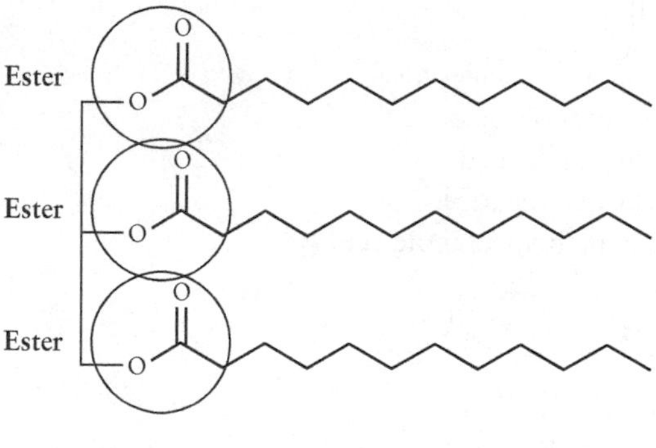

Ester

Ester

Ester

Triglycerides are more commonly known as fats and oils.

103

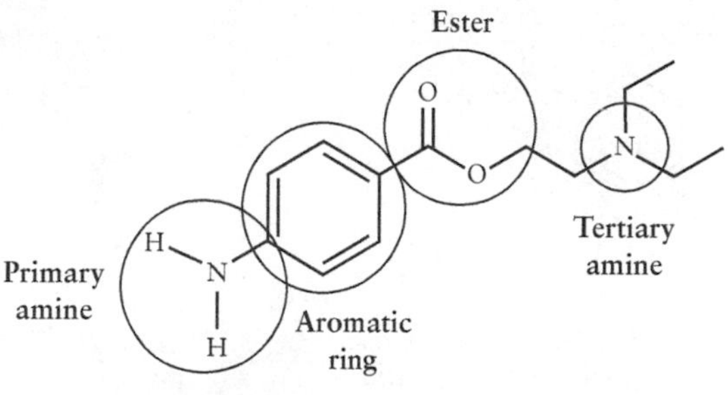

Ester

Primary amine

Tertiary amine

Aromatic ring

105

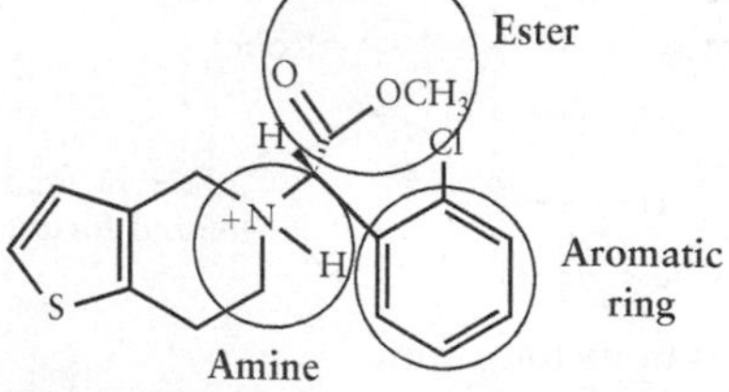

Ester

Amine

Aromatic ring

The amine is in its ionic form.

107 b. and **c.** are amides.

109

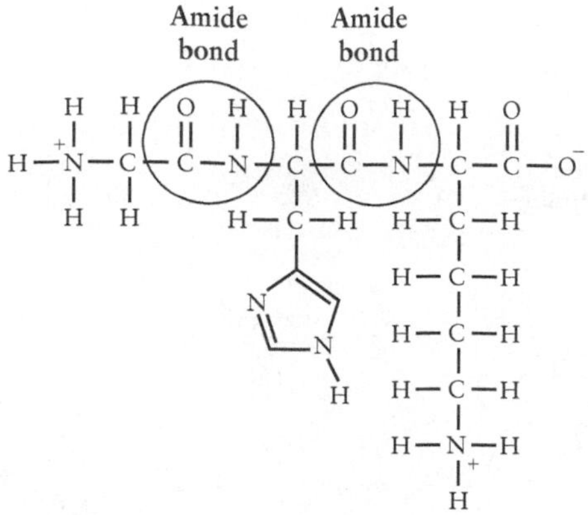

Amide bond   Amide bond

111

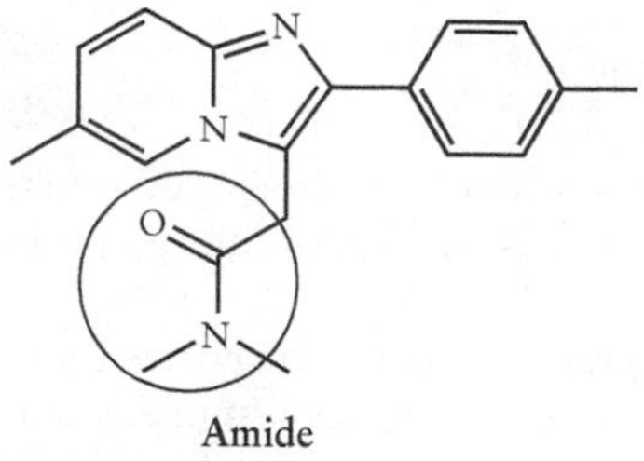

Amide

113 esters and amides

115 An achiral molecule is superimposable on its mirror image; it is identical to its mirror image.

117 a. achiral   **b.** chiral   **c.** chiral   **d.** chiral

119 a. identical   **b.** enantiomers

121 No, ethambutol should not be sold as a racemic mixture. One of the enantiomers causes harm (blindness).

123

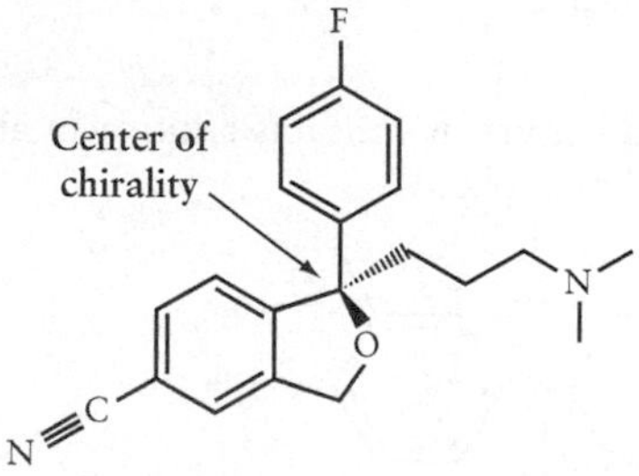

Center of chirality

Lexapro

The center of chirality is tetrahedral, and the bond angles are 109.5°.

125

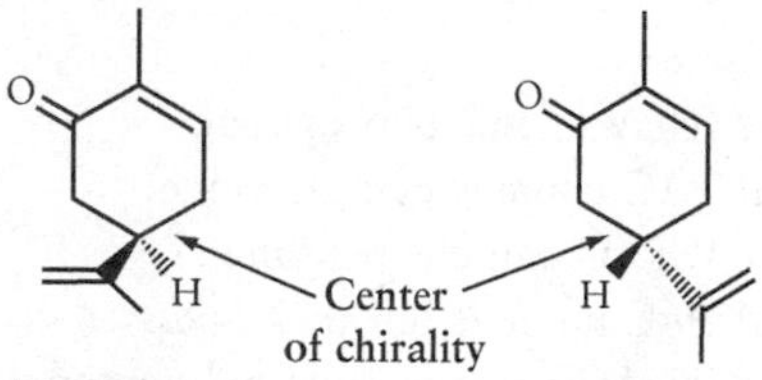

Center of chirality

R-carvone   s-carvone

The hydrogen atom attached to the center of chirality projects towards you in *s*-carvone and away from you in *R*-carvone.

127

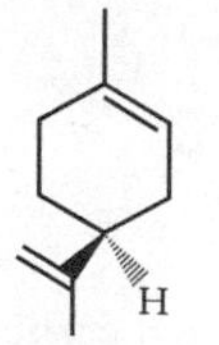

s-limonene

The receptors for smell are a chiral environment. When the two enantiomers are placed in a chiral environment, they exhibit different physical properties, such as smell.

129

Organic   Inorganic   Organic

The organic compounds contain carbon atoms; the inorganic ones do not.

131 **a.** Phosphoanhydride bonds

b. The cell stores potential energy by forming phosphoanhydride bonds.

c. The compound is a phosphate diester. There are two phosphate groups joined together with phosphoanhydride bonds.

133 A person who has schizophrenia has an excess amount of dopamine in the brain; a person who has Parkinson's disease has a decreased amount of dopamine in the brain.

135 neurotransmitters; amine functional group

137 Dopamine cannot pass through the blood-brain barrier to reach the brain where it is needed.

139 Parkinson's disease is the loss of dopamine producing neurons along the neuronal pathway controlling movement.

141 Atypical antipsychotics are the medications that do not affect movement.

Chapter 8

45 The kidneys filter blood to remove waste products (ions and small molecules) and water.

47 It filters out ions, small molecules, and water.

49 The millions of tubules present in the kidneys are selectively permeable membranes that filter the blood. The waste and surplus water exits the kidneys through the ureters to the bladder, where it is stored until it is eliminated.

51 Renal failure is when the kidneys stop working altogether. It may be acute or chronic.

53 Chronic renal failure can result from diabetes, high blood pressure, or certain hereditary factors.

55 A kidney transplant is the other alternative for the patient.

57 **a.** false   **b.** false   **c.** true   **d.** true

59 **a.** homogeneous   **b.** homogeneous   **c.** heterogeneous   **d.** homogenous   **e.** homogenous   **f.** homogeneous

61 In colloidal dispersions, the particles are known as colloids. The major component is called the medium.

63 colloidal dispersions

65 Aggregates are many small molecules clumped together.

67 No, the kidneys do not remove small proteins. Proteins are too big and cannot pass through the selectively permeable membrane of the kidney.

69 Yes, the kidneys remove urea. Urea is a small molecule that can pass through the selectively permeable membrane of the kidney.

71 **a.** solution   **b.** colloid   **c.** solution and colloid   **d.** solution   **e.** solution   **f.** colloid

73 A centrifuge can be used to separate particles out of a suspension.

75 The medication is insoluble in all acceptable media, and some people have trouble swallowing capsules or tablets.

77 the solute and the solvent

79 **a.** Glucose is the solute. It is a solid. The solvent is water.   **b.** Carbon dioxide is the solute and is a gas. The solvent is water.   **c.** Iodine is the solute. It is a solid. The solvent is ethanol.   **d.** Magnesium sulfate is the solute. It is a solid. The solvent is water.

81 The solubility of a solute in a solvent depends on the polarity of the solute and the solvent. Polar compounds will dissolve in polar solvents and nonpolar compounds dissolve in nonpolar solvents.

83 Hydrocarbons are not soluble in water. The hydrocarbons form a separate layer that avoids the water molecules, the hydrophobic effect.

85 **b.**, **c.**, and **d.** would dissolve in water. They are polar molecules, and water is a polar molecule. **a.** and **e.** would not dissolve; they are nonpolar molecules.

87 Propan-2-ol and acetic acid are polar solvents. CCl_4, carbon dioxide, and octane are nonpolar.

89 **a.** $CaCl_2(s) \rightarrow Ca^{2+}(aq) + 2\ Cl^-(aq)$
b. $NH_4Cl(s) \rightarrow NH_4^+(aq) + Cl^-(aq)$
c. $K_3PO_4(s) \rightarrow 3\ K^+(aq) + PO_4^{3-}(aq)$
d. $NaCl(s) \rightarrow Na^+(aq) + Cl^-(aq)$

91 The particles of acetaminophen will settle to the bottom. The bottle needs to be shaken before use so that the particles are distributed throughout the medium.

93 Gout is caused by saturated solutions of calcium salts such as calcium phosphate and calcium oxalate. When the calcium salts precipitate, they form crystals in joints.

95 **b.** and **c.** NaCl and K_3PO_4 are ionic compounds and form electrolytes when dissolved in water.

97 **a.** false   **b.** true   **c.** true   **d.** false

99 In the colorimetric technique, a colored dye is added, and the intensity of the color of the solution determines the concentration. The darker solutions contain more solute and are more concentrated.

101 3.9×10^{-3} g

103 8.0 mL

105 400 ppb

107 4.50 g

109 1.09×10^3 mL

111 3.06×10^3 mL

113 3.31×10^{-5} mol

115 1.3 M

117 4.5 L

119 1.8×10^{-3} eq/L

121 0.3 eq/L SO_4^{2-} and 0.3 eq/L Mg^{2+}

123 0.005 mol/L calcium ion and 0.003 mol/L phosphate ion

125 0.54 M

127 1.2 M

129 7.5 mL

131 0.007 mL/hr

133 12 mg/hr

135 68 mL/hr

137 The nature of the membrane and the charge, polarity, and size of the solute determine whether or not a solute can pass through the membrane.

139 Simple diffusion is the spontaneous movement of a molecule or ion from a region of higher concentration to a region of lower concentration.

141 Dialysis is the movement of solute particles across a selectively permeable membrane.

143 A hypertonic solution is the solution with the higher solute concentration. A hypotonic solution is the solution with the lower solute concentration. Isotonic solutions have equal solute concentrations.

145 The pressure is greater than the osmotic pressure, and water is forced to diffuse against the concentration gradient.

147 **a.** The solution with the higher concentration is solution A. The water will flow from solution B to solution A. The volume of solution A will increase.
b. The solution with the higher concentration is solution B. The question asks about water flow (i.e., osmosis), so the identity of the solutes does not matter, just the total concentration of the solutes. The water will flow from solution A to solution B. The volume of solution B will increase.
c. The solutions are isotonic. The water will not flow between the solutions. The volumes will remain the same.

149 In osmosis, water flows from a hypotonic solution to a hypertonic solution.

151 No, you should not expect dialysis to occur. With isotonic solutions, there is no difference in concentration between the two solutions and no driving force for the solutions to move.

153 If you apply pressure to seawater separated from fresh water, the water will flow from the seawater (the hypertonic solution) to the fresh water (the hypotonic solution). This process is called reverse osmosis. If pressure is not applied, the water will flow from the fresh water (higher water concentration) to the seawater (lower water concentration).

155 plasma and blood cells

157 Plasma contains serum and fibrinogen and other clotting factors. Serum is 90 percent water and does not contain fibrinogen and other clotting factors.

159 Some small molecules found in blood are glucose, amino acids, creatinine, and urea. Some electrolytes found in blood are Na^+, K^+, Ca^{2+}, Cl^-, and HCO_3^-.

161 The glucose level (85 mg/dL) is within the normal range. The patient does not have diabetes.

Chapter 9

33 Ethylene glycol is metabolized into oxalic acid and glyoxylic acid. Oxalic acid contains a carboxylic acid functional group rather than an alcohol functional group.

35

37 Veterinarians can use drugs, such as other alcohols that counteract the affects of oxalic and glyoxlic acids. In humans, hemodialysis can be performed to eliminate the unmetabolized ethylene glycol and the toxic acids from the blood.

39 Acids have a sour taste, can neutralize bases, and turn blue litmus paper red. Bases have a bitter taste, feel slippery, neutralize acids, and turn red litmus paper blue.

41 hydronium ion

43 donates, accepts

45

Carbonic acid + Water ⇌ Hydronium ion + Bicarbonate ion

Water acts as a base.

47

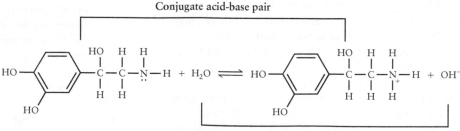

Water acts as an acid.

49 $Ca(OH)_2(s) + H_2O(l) \rightarrow Ca^{2+}(aq) + 2\ OH^-(aq)$
Calcium hydroxide is a base because it produces hydroxide ions.

51 b., d., and e. are bases.

53 a. H_3O^+ b. H_2O c. $CH_3CH_2NH_2CH_3^+$
d. $CH_3CH_2CH_2CH_2COOH$ e. HCO_3H

55 If a concentrated solution of a strong acid is spilled on the skin, it will cause a serious chemical burn. HCl is present in the body.

57 Conjugate acid-base pair

$HI + H_2O \longrightarrow I^- + H_3O^+$

Hydrogen iodide is a strong acid because it completely dissociates into hydronium ion and iodide in water.

59 $Ba(OH)_2(s) + H_2O(l) \rightarrow Ba^{2+}(aq) + 2\ OH^-(aq)$
It is considered a dissolution rather than a reaction because the ionic bonds in barium hydroxide are replaced by ion dipole interactions that occur when the ions are surrounded by water.

61 Weak acids do not completely dissociate in water; they undergo a reversible reaction. At equilibrium, the weak acid, its conjugate base, and hydronium ion are present. Strong acids completely dissociate, only the conjugate base and hydronium ion are present in solution.

63 a. dissociation of a strong base
b. dissociation of a strong acid
c. dissociation of a weak base

65 a. $HCl + H_2O \rightarrow H_3O^+ + Cl^-$
The acid is a strong acid. The reaction is not reversible.
HCl, acid; H_2O, base; H_3O^+, conjugate acid; Cl^-, conjugate base
b. $HNO_3 + H_2O \rightarrow H_3O^+ + NO_3^-$
The acid is strong acid. The reaction is not reversible.
HNO_3, acid; H_2O, base; H_3O^+, conjugate acid; NO_3^-, conjugate base
c.

The acid is a weak acid. The reaction is reversible.
d. $H_2CO_3 + H_2O \rightleftharpoons H_3O^+ + HCO_3$
The acid is a weak acid. The reaction is reversible.
H_2CO_3, acid; H_2O, base; H_3O^+, conjugate acid; HCO_3^-, conjugate base

67 a.

69 a. Propanoic acid, water, propanoate, and hydronium ion are present at equilibrium.
b. At equilibrium, the reactants are favored because propanoic acid is a weak acid. Weak acids do not fully dissociate at equilibrium.
c. The concentrations of propanoic acid and propanoate are constant at equilibrium.
d. The two opposing arrows indicate that both the forward and reverse reactions occur simultaneously.
e.

71

a. Trichloroacetic acid, water, trichloroacetate, and hydronium ion are present at equilibrium.
b. The concentration of trichloroacetic acid and trichloroacetate are constant at equilibrium.
c. The reaction will shift toward the reactants.
d. The reaction will shift toward the left towards the reactants.

73

a. The reaction will shift to the left. Le Châtelier's principle predicts that the reaction will shift to the left to consume the added hydroxide until equilibrium is reached.
b. The reaction will shift to the left. Le Châtelier's principle predicts that the reaction will shift to the left to consume the excess products until equilibrium is reached.

75 pH indicator paper can be used to approximate the pH of a solution. pH indicator paper changes color according to the concentration of the hydronium ions in solution. Red litmus paper will turn blue in the presence of a base.

77 **a.** basic, pH > 7
   b. acidic, pH < 7
   c. basic, pH > 7
   d. acidic, pH < 7
   e. basic, pH > 7

79 When a condition causes excessive amounts of acid or base to enter the blood, the buffer capacity is exceeded. Acidosis occurs when the blood pH falls below its normal range. Alkalosis occurs when blood pH rises above its normal range.

81

$[H_3O^+]$	$[OH^-]$	Is the solution acidic, neutral, or basic?
1.0×10^{-3}	1.0×10^{-11}	Acidic pH = 3
1.0×10^{-12}	1.0×10^{-2}	Basic pH = 12
1.0×10^{-7}	1.0×10^{-7}	Neutral pH = 7
1.0×10^{-5}	1.0×10^{-9}	Acidic pH = 5
1.0×10^{-9}	1.0×10^{-5}	Basic pH = 9

83 pH = 3.5
   Apple juice is acidic.
   $[OH^-] = 3.1 \times 10^{-11} M$

85 pH = 6.5
   Milk is acidic.
   $[OH^-] = 3.1 \times 10^{-8} M$

87 Physiological pH is the pH of arterial blood. It ranges between 7.35 and 7.45. It is slightly basic compared to the pH of water (pH = 7.0).

89 At physiological pH, amines are in their ionized form. The charge on the ion keeps it inside the cell. It also enables them to bind to enzymes and chemically react.

91 Arterial blood pH is measured with a blood gas analyzer. The pH of urine is measured with dipsticks containing dyes made specifically for testing urine.

93 **a.** $HBr(aq) + NaOH(aq) \rightarrow H_2O(l) + NaBr(aq)$
   b. $3\ HBr(aq) + Al(OH)_3(aq) \rightarrow 3\ H_2O(l) + AlBr_3(aq)$
   c. $2\ HBr(aq) + Sr(OH)_2(aq) \rightarrow 2\ H_2O(l) + SrBr_2(aq)$

95 **a.** $HNO_3(aq) + KOH(aq) \rightarrow H_2O(l) + KNO_3(aq)$
   b. $2\ HNO_3(aq) + Mg(OH)_2(aq) \rightarrow 2\ H_2O(l) + Mg(NO_3)_2(aq)$
   c. $3\ HNO_3(aq) + Al(OH)_3(aq) \rightarrow 3\ H_2O(l) + Al(NO_3)_3(aq)$

97 Antacid tablets supply the base for the neutralization reaction with stomach acid, HCl.

99

The salt is charged and therefore soluble in water.

101 A buffer is a solution that resists changes in pH by neutralizing any added acid or base.

103 When a base, OH^-, is added to a buffer, the weak acid component of the buffer, HA, neutralizes the added base, OH^-, to form water and A^-, producing more of the buffer, but not more OH^-.

105 The amount of added acid or base that a buffer can neutralize is known as the buffer capacity. The greater the concentration is of the weak acid and the conjugate base of the buffer system, the greater is its buffer capacity.

107 When a small amount of base enters the bloodstream, carbonic acid reacts with the OH^- ion to form bicarbonate ion and water. These products do not increase the concentration of H_3O^+ and therefore do not change the pH.

109 **a.** The concentration of carbon dioxide decreases when a person hyperventilates. Carbon dioxide is removed faster than it is produced.
   b. As the concentration of carbon dioxide decreases, the H_3O^+ concentration decreases and the pH rises.

111 When sodium hydrogen carbonate is removed from the body, there is an excess of carbonic acid in the blood. Therefore, the pH will decrease. The patient should be given the hydrogen carbonate portion to regulate the pH.

113 In type I diabetes, the pancreas has stopped producing insulin, so diabetics need to replace it.

115 Protein drugs cannot be taken orally because the acidic environment of the gastrointestinal tract will degrade the drugs.

117 A protein does not keep its original shape or its cellular function in the stomach.

119 Enteric coatings are stable in acidic environment; therefore, the enteric coating controls where in the digestive tract the medication is absorbed.

E9.7 $pK_a = 2.12$

E9.9 Phosphoric acid is the strongest acid of the three. Citric acid is the weakest acid of the three.

E9.11 The strongest acid is oxalic acid. The weakest acid is caffeic acid.

Chapter 10

25 Vitamins allow cells to perform the metabolic reactions necessary for living.

27 Riboflavin is essential to the metabolism of carbohydrates, proteins, and fats.

29 If a newborn has jaundice, the newborn will need light therapy to treat it. The light therapy degrades riboflavin.

31 Water-soluble vitamins are readily eliminated through the urine and need to be replenished on a daily basis. Most natural foods are rich in vitamins; there are also supplements that contain vitamins.

33 The functional groups are the reactive sites in a molecule.

35 **a.** **b.**
   c. **d.**

37 A substrate is a particular reactant for a reaction catalyzed by an enzyme.

39 When an organic compound undergoes combustion, carbon dioxide and water are formed as products.

41 The reactant that undergoes loss of electrons in an oxidation–reduction reaction is said to undergo *oxidation*.

43 a. oxidation:

Zn: $\longrightarrow$ Zn^{2+} + 2e$^-$

reduction

2 H$^+$ + 2 e$^-$ $\longrightarrow$ H—H

Cl$^-$ is a spectator ion.

b. oxidation

Na: $\longrightarrow$ Na$^+$ + e$^-$

reduction

:$\ddot{\text{Cl}}$—$\ddot{\text{Cl}}$: + 2e$^-$ $\longrightarrow$ 2 :$\ddot{\text{Cl}}$:$^-$

45 oxidation

47 reduction

49 a.

$$HO-\overset{O}{\overset{\|}{C}}-\overset{H}{\overset{|}{\underset{H}{C}}}-\overset{H}{\overset{|}{\underset{H}{C}}}-\overset{H}{\overset{|}{\underset{H}{C}}}-H \xrightarrow{[H]} H-\overset{OH}{\overset{|}{C}}-\overset{H}{\overset{|}{\underset{H}{C}}}-\overset{H}{\overset{|}{\underset{H}{C}}}-\overset{H}{\overset{|}{\underset{H}{C}}}-H$$

Carboxylic acid   Primary alcohol

The reaction represents a reduction because there is a decrease in the number of oxygen atoms in the product.

b.

$$H-\overset{H}{\overset{|}{\underset{H}{C}}}-\overset{H}{\overset{|}{\underset{H}{C}}}-\overset{H}{\overset{|}{\underset{H}{C}}}-\overset{H}{\overset{|}{\underset{H}{C}}}-\overset{H}{\overset{|}{\underset{H}{C}}}-H + 8 O_2 \xrightarrow{[O]} 5 CO_2 + 6 H_2O$$

The combustion reaction is an oxidation reaction because the carbon atoms have a decrease in the number of hydrogen atoms attached and an increase in the number of oxygen atoms attached.

c. (structure) $\xrightarrow{[H]}$ (structure)

The reaction represents a reduction because there is an increase in the number of hydrogen atoms in the product.

d.

$$HO-\overset{H}{\overset{|}{\underset{H}{C}}}-\overset{H}{\overset{|}{\underset{H}{C}}}-\overset{H}{\overset{|}{\underset{H}{C}}}-\overset{H}{\overset{|}{\underset{H}{C}}}-\overset{H}{\overset{|}{\underset{H}{C}}}-H \xrightarrow{[O]} H-\overset{O}{\overset{\|}{C}}-\overset{H}{\overset{|}{\underset{H}{C}}}-\overset{H}{\overset{|}{\underset{H}{C}}}-\overset{H}{\overset{|}{\underset{H}{C}}}-\overset{H}{\overset{|}{\underset{H}{C}}}-H$$

The reaction represents an oxidation because there is a decrease in the number of hydrogen atoms in the product.

51

$$HO-\overset{O}{\overset{\|}{C}}-\cdots(\text{unsaturated chain})\cdots \xrightarrow[Pt]{H_2(g)} HO-\overset{O}{\overset{\|}{C}}-\cdots(\text{saturated chain})$$

This reaction is a reduction reaction. The product has an increase in the number of hydrogen atoms.

53 An unsaturated fat is more likely to be a liquid.

55 a.

$$H-\overset{H}{\overset{|}{\underset{H}{C}}}-\overset{H}{\overset{|}{\underset{H}{C}}}-\overset{H}{\overset{|}{\underset{H}{C}}}-\overset{O}{\overset{\|}{C}}-H \xrightarrow{[O]} H-\overset{H}{\overset{|}{\underset{H}{C}}}-\overset{H}{\overset{|}{\underset{H}{C}}}-\overset{H}{\overset{|}{\underset{H}{C}}}-\overset{O}{\overset{\|}{C}}-OH$$

b.

(cyclohexyl aldehyde) $\xrightarrow[H]{[O]}$ (cyclohexyl carboxylic acid, —C(=O)—OH)

c.

(carboxylic acid) $\xrightarrow{[H]}$ (aldehyde) or (primary alcohol, —OH)

d.

(cyclopentyl carboxylic acid) $\xrightarrow{[H]}$ (cyclopentyl aldehyde) or (cyclopentyl alcohol)

57 An enzyme is chemically unchanged in the reaction; a coenzyme does undergo a chemical change in the reaction.

59 a. $C_3H_8 + 5 O_2 \rightarrow 3 CO_2 + 4 H_2O$

b.

$$CH_3-\overset{H}{\underset{H}{C}}\!\!=\!\!\overset{}{\underset{}{C}}\!\!<^{CH_3}_{CH_3} + H_2 \xrightarrow{catalyst} CH_3-\overset{H}{\overset{|}{\underset{H}{C}}}-\overset{H}{\overset{|}{C}}<^{CH_3}_{CH_3}$$

c.

$$H-\overset{H}{\overset{|}{\underset{H}{C}}}-\overset{H}{\overset{|}{\underset{H}{C}}}-\overset{OH}{\overset{|}{\underset{H}{C}}}-\overset{H}{\overset{|}{\underset{H}{C}}}-H \xrightarrow{NAD^+ \rightarrow NADH + H+} H-\overset{H}{\overset{|}{\underset{H}{C}}}-\overset{H}{\overset{|}{\underset{H}{C}}}-\overset{O}{\overset{\|}{C}}-\overset{H}{\overset{|}{\underset{H}{C}}}-H$$

d. $FADH_2 + NAD^+ \rightarrow FAD + NADH + H^+ + 2 e^-$

61 a.

$$H_3C-\overset{H}{\overset{|}{\underset{CH_3}{C}}}-\overset{OH}{\overset{|}{\underset{H}{C}}}-CH_3 \xrightarrow{[O]} H_3C-\overset{H}{\overset{|}{\underset{CH_3}{C}}}-\overset{O}{\overset{\|}{C}}-CH_3$$

b.

(acyl—S—CoA) $\xrightarrow{[O]}$ (α,β-unsaturated acyl—S—CoA)

c. $FADH_2$

d.

(benzaldehyde, —C(=O)—H) $\xrightarrow{[O]}$ (benzoic acid, —C(=O)—OH)

63 d. False, a free radical does not contain an octet of electrons.

65 *Transferase* enzymes are the major class of enzymes that catalyze acyl transfer reactions in biochemistry.

67 a.

$$H-\underset{\underset{H}{|}}{\overset{\overset{H}{|}}{C}}-O-H \;+\; \text{(acid)} \quad \xrightarrow{\text{catalyst}}$$

$$\text{(ester)} \; O-\underset{\underset{H}{|}}{\overset{\overset{H}{|}}{C}}-H \;+\; H_2O$$

b.

$$\text{(acid)} \;+\; \text{(OH)} \quad \xrightarrow{\text{catalyst}} \quad \text{(ester)} \;+\; H_2O$$

c.

$$\text{(phenol OH)} \;+\; \text{(acid OH)} \quad \xrightarrow{\text{catalyst}}$$

$$\text{(phenyl ester)} \;+\; H_2O$$

d.

$$\text{(propanoic acid OH)} \;+\; \text{(thiol SH)} \quad \xrightarrow{\text{catalyst}} \quad \text{(thioester S)}$$

69

$$\underset{\text{Acetic acid}}{\text{(O, OH)}} \;+\; \underset{\text{Propan-1-ol}}{\text{(OH)}} \quad \xrightarrow{\text{catalyst}} \quad \text{(ester O)}$$

The ester smells like pear.

71 Three fatty acid molecules react with one glycerol molecule to produce a fat molecule. An esterification reaction produces the fat molecule.

73 a.

$$\text{(acid OH, O)} \;+\; \text{(amine N–H)} \quad \xrightarrow{\text{catalyst}} \quad \text{(amide N, O)}$$

b.

$$\text{(acid OH, O)} \;+\; \text{(amine N)} \quad \xrightarrow{\text{catalyst}} \quad \text{(amide N, O)}$$

c.

$$\text{(acid OH, O)} \;+\; \text{(amine N–H)} \quad \xrightarrow{\text{catalyst}} \quad \text{(amide N, O)}$$

75

$$\underset{\text{Glycine}}{H_2N-\underset{\underset{H}{|}}{\overset{\overset{H}{|}}{C}}-\overset{\overset{O}{\|}}{C}-OH} \;+\; \underset{\text{Cysteine}}{H_2N-\underset{\underset{\underset{\underset{SH}{|}}{CH_2}}{|}}{\overset{\overset{H}{|}}{C}}-\overset{\overset{O}{\|}}{C}-OH} \quad \xrightarrow{\text{enzyme}}$$

$$H_2N-\underset{\underset{H}{|}}{\overset{\overset{H}{|}}{C}}-\overset{\overset{O}{\|}}{C}-N-\underset{\underset{\underset{\underset{SH}{|}}{CH_2}}{|}}{\overset{\overset{H}{|}}{C}}-\overset{\overset{O}{\|}}{C}-OH \quad \text{or} \quad H_2N-\underset{\underset{\underset{\underset{SH}{|}}{CH_2}}{|}}{\overset{\overset{H}{|}}{C}}-\overset{\overset{O}{\|}}{C}-N-\underset{\underset{H}{|}}{\overset{\overset{H}{|}}{C}}-\overset{\overset{O}{\|}}{C}-OH$$

77 Two hydrolysis reactions that occur during digestion are the metabolism of proteins into amino acids and of fats into fatty acids and glycerol.

79 a.

$$H-\underset{\underset{H}{|}}{\overset{\overset{H}{|}}{C}}-\overset{\overset{O}{\|}}{C}-O-\underset{\underset{H}{|}}{\overset{\overset{H}{|}}{C}}-\underset{\underset{H}{|}}{\overset{\overset{H}{|}}{C}}-H \quad \xrightarrow{H_2O}$$

$$H-\underset{\underset{H}{|}}{\overset{\overset{H}{|}}{C}}-\overset{\overset{O}{\|}}{C}-OH \;+\; HO-\underset{\underset{H}{|}}{\overset{\overset{H}{|}}{C}}-\underset{\underset{H}{|}}{\overset{\overset{H}{|}}{C}}-H$$

b.

$$H-\underset{\underset{H}{|}}{\overset{\overset{H}{|}}{C}}-\underset{\underset{H}{|}}{\overset{\overset{CH_3}{|}}{C}}-S-\underset{\underset{H}{|}}{\overset{\overset{O}{\|}}{C}}-\underset{\underset{H}{|}}{\overset{\overset{H}{|}}{C}}-\underset{\underset{H}{|}}{\overset{\overset{H}{|}}{C}}-H \quad \xrightarrow{H_2O}$$

$$H-\underset{\underset{H}{|}}{\overset{\overset{H}{|}}{C}}-\underset{\underset{H}{|}}{\overset{\overset{CH_3}{|}}{C}}-SH \;+\; HO-\overset{\overset{O}{\|}}{C}-\underset{\underset{H}{|}}{\overset{\overset{H}{|}}{C}}-\underset{\underset{H}{|}}{\overset{\overset{H}{|}}{C}}-H$$

c.

$$H-\underset{\underset{H}{|}}{\overset{\overset{H}{|}}{C}}-\underset{\underset{H}{|}}{\overset{\overset{H}{|}}{C}}-\underset{\underset{H}{|}}{N}-\overset{\overset{O}{\|}}{C}-\underset{\underset{H}{|}}{\overset{\overset{H}{|}}{C}}-H \quad \xrightarrow[\text{catalyst}]{H_2O}$$

$$H-\underset{\underset{H}{|}}{\overset{\overset{H}{|}}{C}}-\underset{\underset{H}{|}}{\overset{\overset{H}{|}}{C}}-\underset{\underset{H}{|}}{\overset{\overset{H}{|}}{N^+}}-H \;+\; {}^-O-\overset{\overset{O}{\|}}{C}-\underset{\underset{H}{|}}{\overset{\overset{H}{|}}{C}}-H$$

81 a. The acyl group being transferred contains six carbon atoms.

b. The bond between the carbon atom and the sulfur atom in the thioester is broken.

c. The acyl group is being transferred from a sulfur atom to an oxygen atom.

d. The reaction is classified as a hydrolysis reaction because water is a reactant in the reaction.

83 The acidic environment of the stomach and hydrolase enzymes aid in the digestion of proteins. Hydrolysis of amide bonds occurs during the digestion of proteins, which produces individual amino acids.

85

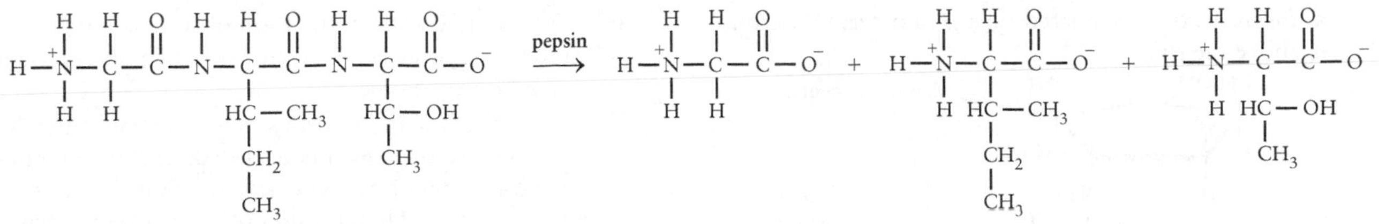

87 The digestion of fats occurs in the small intestine. *Lipase* enzymes catalyze the hydrolysis of fats.

89 Lyase enzymes catalyzes hydration and dehydration reactions.

91 **a.** hydration

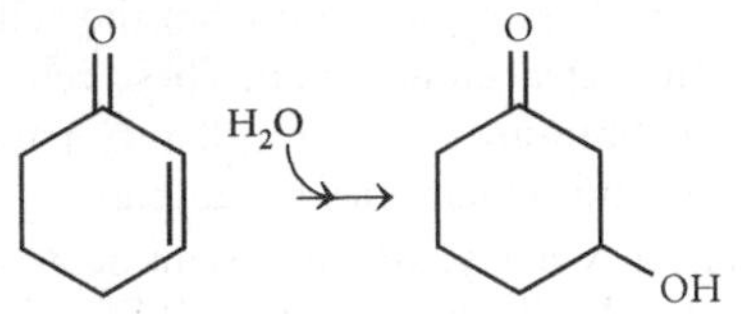

b. dehydration

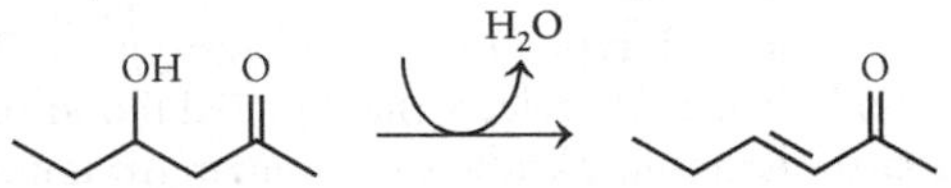

c. hydration

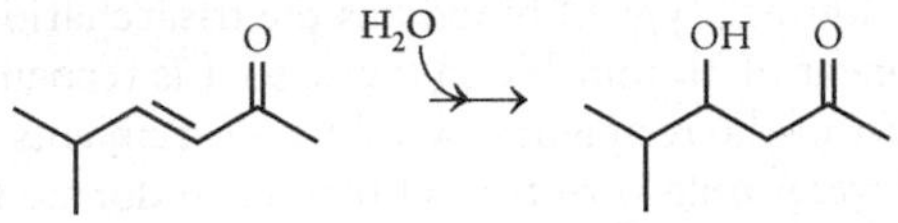

93 **a.**

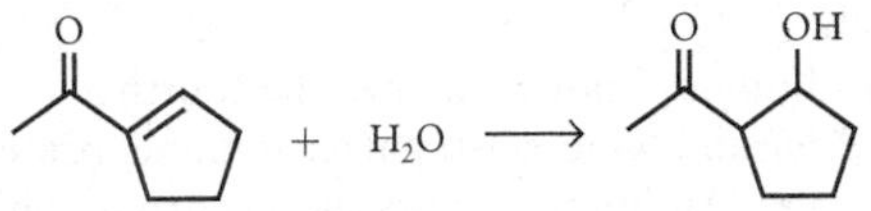

b.

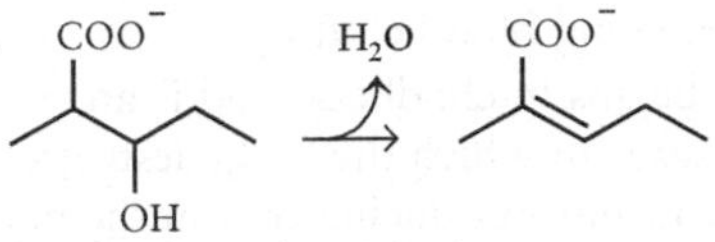

95 **a.** hydrolysis   **b.** dehydration reaction

97 PKU is caused by a deficiency in the enzyme *phenylalanine hydroxylase (PAH)*.

99 BH_4, tetrahydrobiopterin, is the coenzyme required by *PAH*.

101 The treatment for PKU is managed through a strict diet that eliminates foods that include phenylalanine, such as beef, chicken, fish, and dairy products. They must also take tyrosine supplements because they cannot synthesize this amino acid.

Chapter 11

21 Anaerobic bacteria present in the mouth metabolize glucose into lactic acid.

23 The metabolism of glucose into lactic acid by anaerobic bacteria in the mouth is the root cause of dental caries. Lactic acid dissolves tooth enamel.

25 Lactic acid is acidic and dissolves tooth enamel.

27 carbohydrates, lipids, proteins, and nucleic acids

29 When we eat carbohydrates, the potential energy stored in the covalent bonds of the carbohydrates is converted into energy when they are metabolized.

31 the mouth and the small intestine

33 Once glucose enters a cell; the cell uses cellular respiration to metabolize glucose into carbon dioxide, water, and energy.

35 Carbohydrates provide molecular building blocks for RNA and DNA. They are important in cellular recognition.

37 When the structure of carbohydrates were first determined, scientists noted that their composition fit the molecular formula $(CH_2O)_n$. They hypothesized that carbohydrates might be hydrates of carbon.

39 Monosaccharides cannot be hydrolyzed into simpler carbohydrates. Disaccharides can be hydrolyzed into two monosaccharides. Polysaccharides are biological polymers of monosaccharides.

41 Monosaccharides are called simple sugars because they cannot be hydrolyzed into simpler carbohydrates.

43 fructose

45 Most monosaccharides are chiral. They contain one or more centers of chirality.

47 The common monosaccharides contain three to six carbon atoms.

49 **a.** Arabinose contains five carbon atoms, while glucose contains six carbon atoms.

b. These sugars contain an aldehyde as part of their structure as indicated by the CHO at the top of the Fischer projection.

c. The OH group that is farthest away from the CHO group is pointing to the right.

d. L-arabinose

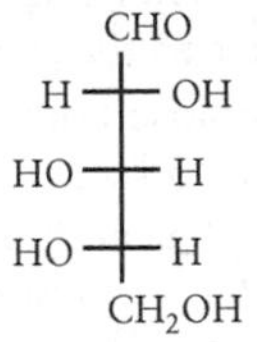

51 Diastereomers are nonsuperimposable, non-mirror image stereoisomers. Diastereomers have a different configuration at one or more centers of chirality, but not at every center of chirality. Enantiomers have a different configuration at every center of chirality.

53 In a Haworth structure it is the carbon that is bonded to two oxygen atoms.

55 The α-anomer and the β-anomer are rapidly interconverting via the open-chain form of the monosaccharide, creating an equilibrium mixture of all three forms of the monosaccharide.

57 a. furanose, five-membered ring, β-anomer; OH group is above the ring.

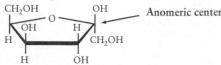
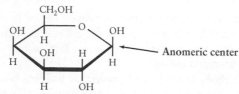

b. pyranose, six-membered ring, β-anomer; OH group is above the ring.

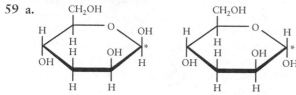

59 a.

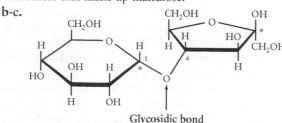

b. The sugar on the left is the β-anomer; the OH group on the anomeric center is above the ring. The sugar on the right is the α-anomer; the OH group on the anomeric center is below the ring.

c. These two structures are diastereomers.

d.

$$
\begin{array}{c}
\text{CHO} \\
\text{HO}-\!\!\!-\text{H} \\
\text{H}-\!\!\!-\text{OH} \\
\text{H}-\!\!\!-\text{OH} \\
\text{H}-\!\!\!-\text{OH} \\
\text{CH}_2\text{OH}
\end{array}
$$

e. One anomer ring opens up to the open-chain form and then recloses to the other anomer ring form.

61 A disaccharide is a carbohydrate that when hydrolyzed yields two monosaccharides, either identical or different monosaccharides. In a disaccharide, the anomeric center at a glycosidic linkage is a stable functional group so it does not open and close in aqueous solutions and cannot change configurations from the α-anomer to the β-anomer.

63 Maltose and cellobiose both consist of two glucose monomers connected by a $1 \rightarrow 4$ glycosidic bond. Maltose, however, has an $\alpha(1 \rightarrow 4)$ bond, whereas cellobiose has a $\beta(1 \rightarrow 4)$ bond.

65 a. Maltulose is a disaccharide. There are two monomers that make up maltulose.

b-c.

Glycosidic bond

d. Maltulose has an α-glycosidic linkage. $\alpha(1 \rightarrow 4)$

e. The hydrolysis of maltulose produces two different monosaccharides.

67 Plants store energy in the form of starch.

69 Starch is hydrolyzed into glucose, which then provides energy for the body.

71 Cellulose plays an important role as dietary fiber. It adds bulk to the contents of the colon and facilitates the absorption of nutrients and water. It does not provide energy. Humans do not have enzymes that can hydrolyze $\beta(1 \rightarrow 4)$ linkages that are found in cellulose.

73 Glycogen, amylose, amylopectin all have $\alpha(1 \rightarrow 4)$ glycosidic linkages.

75 Most cell markers are oligosaccharides that identify a cell like a fingerprint identifies a person. Oligosaccharide markers project out from the cell membrane into the extracellular fluid. These cell markers allow the immune system to identify which cells are host cells and which ones are foreign.

77 Type A blood has a N-acetyl-D-galactosamine as a marker on red blood cells, whereas type B blood has a D-galactose marker.

79 When people with type O blood receive type A, B, or AB blood, their body rejects the donated blood because the fourth monosaccharide is unfamiliar. An immune response occurs, which can lead to life-threatening blood clotting. Type O blood has the trisaccharide component of all four blood types, so it is recognized as a part of the recipient's blood by all recipients. Therefore, people with type O blood can donate to everybody.

81 The girl's body did not recognize the fourth monosaccharide, N-acetyl-D-galactosamine, present in type A blood. An immune response occurred, which lead to life-threatening blood clotting.

83 Type I diabetes begins in childhood and is an autoimmune disease in which the body destroys its own β cells, the insulin-producing cells in the pancreas. There is not enough insulin present in the body. Type II diabetes generally appears in adulthood. Insulin is produced, but cells do not respond properly to the insulin present.

85 β cells produce insulin. They are found in the pancreas.

87 Insulin is a protein. If it were taken orally, it would be rendered useless by the hydrolysis of its peptide bonds in the stomach.

Chapter 12

35 the lipid class of biomolecules

37 The increase in muscle growth is the anabolic effect of steroids. It begins when the steroids bind to the androgen receptor and initiate a cascade of biochemical events which result in protein production that increases muscle growth.

39 Some side effects of anabolic steroid use are infertility, gynecomastia, testicular atrophy, and premature baldness. Prolonged steroid use can lead to liver damage, prostate cancer, and cardiovascular disease.

41 Anabolic steroids treat muscle wasting diseases such as AIDS; they aid in the recovery from burns and are an effective treatment for hypogonadism.

43 The extensive hydrocarbon component of lipids makes them insoluble in water.

45 A fatty acid is a long straight-chain hydrocarbon with a carboxylic acid at one end. They all contain the carboxylic acid.

47 A monounsaturated fatty acid contains one carbon-carbon double bond. A polyunsaturated fatty acid contains more than one carbon-carbon double bond.

49 a. unsaturated fatty acid **b.** saturated fatty acids **c.** unsaturated fatty acid

51 Stearic acid has more intermolecular dispersion forces because it has more surface area to interact with other stearic acid molecules. Linoleic acid has two *cis* double bonds, which create bends in the overall shape of the molecule and reduces the number of contact points between fatty acid molecules, thereby causing fewer dispersion forces.

53 A *cis* double bond will cause a bend or kink in the structure, reducing the number of contact points and creating fewer dispersion forces, thus lowering the melting point.

55 If a *trans* double bond is present in the fatty acid, it is able to adopt a linear zigzag shape with more surface area and more dispersion forces and have a higher melting point. A *cis* double bond will cause a bend or kink in the structure, reducing the number of contact points and creating fewer dispersion forces, thus lowering the melting point.

57 A saturated fatty acid has stronger intermolecular forces of attraction and a higher melting point, therefore, causing it to be a solid at room temperature.

59 flaxseed walnuts, salmon, and canola oil

61 Oleic acid is a Δ^9 fatty acid and ω-9 fatty acid.

63 omega-3

65 a. EPA is a $\Delta^{5,8,11,14,17}$ fatty acid **b.** EPA is a ω-3 fatty acid. **c.** It should be a liquid at room temperature. It has five *cis* double bonds present. A *cis* double bond will cause a bend or kink in the structure, reducing the number of contact points and creating fewer dispersion forces, thus lowering the melting point.

67 Triglycerides are stored in fat cells, known as adipocytes. They are metabolized in muscle and liver cells.

69 Saturated fats and *trans* fats are unhealthy. They have been linked to cardiovascular disease, diabetes, and cancer.

71 A triglyceride is insoluble in water and soluble in organic solvents and therefore classified as a lipid. It is insoluble because the fatty acids give it an extensive hydrocarbon structure.

73 Triglycerides derived from saturated fatty acids are classified as fats. They are solids at room temperature because they are more uniform in shape and can pack more tightly, which creates more intermolecular forces of attraction which leads to a higher melting point. Fats are found in meat, whole milk, butter, cheese, coconut oil, and palm oil.

75 Coconut oil should be a solid at room temperature because it is mostly saturated fats, which pack together tightly.

77 Olive oil is considered a healthy fat because it consists mostly of unsaturated fats.

79 safflower oil

81 olive oil

83 a. an unsaturated fat
   b. liquid at room temperature
   c. It has three ester groups and four carbon-carbon double bonds.
   d.-e.

   f. The carbon-carbon double bonds are *cis* double bonds. It is a healthy fat.

85

87 Triglycerides provide long term energy storage.

89 Emulsification turns the globules of fat into a colloid of finely dispersed microscopic fat droplets suspended in water. This process brings the water-insoluble triglycerides in contact with the water soluble lipase enzymes.

91 Triglycerides are too large to cross the cell membrane of the intestinal mucosa. The triglycerides are hydrolyzed into fatty acids, glycerol, and partially hydrolyzed triglycerides to cross the intestinal mucosa. Once inside the intestinal mucosa, the fatty acids and glycerol are esterified to reform the triglycerides and packaged into chylomicrons, which will carry them through the bloodstream.

93 hydrophobic

95 Triglycerides are found on the inside of chylomicron. The interior of a chylomicron is hydrophobic. Triglycerides are also hydrophobic.

97 phospholipids and glycolipids

99 It amphipathic because it is both polar (the carboxylate end) and nonpolar (the hydrocarbon end).

101 Three fatty acids are needed to form a triglyceride. Two fatty acids are needed to form a glycerophospholipid. One fatty acid is needed to form a sphingolipid.

103 Sphingomyelins and sphingolipids are derived from sphingosine.

105 A glycolipid has a carbohydrate linked to the hydroxyl group farthest from the hydrocarbon chain, rather than an amino alcohol and a phosphate group. This difference is primarily in the polar head.

107 a. It is a phospholipid.
b. It is a glycerophospholipid. It has a glycerol backbone.

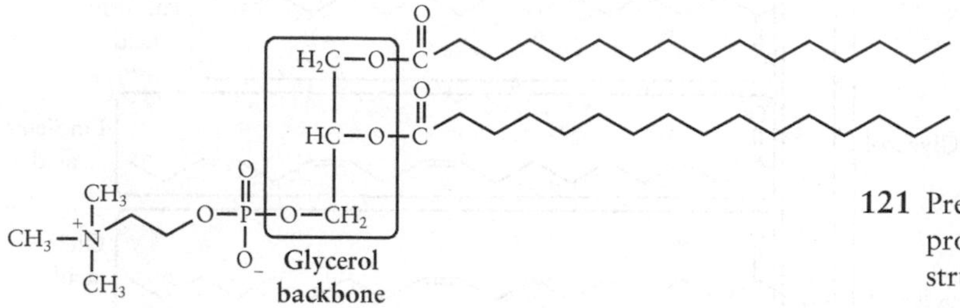

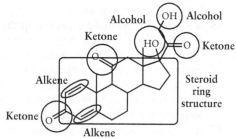

c. The fatty acid derivatives are from palmitic acid. The fatty acid components are saturated. An ester group connects the fatty acids to the backbone.
d. The lipid contains a phosphate.
e. The choline group and the phosphate group are the polar head group.
f.

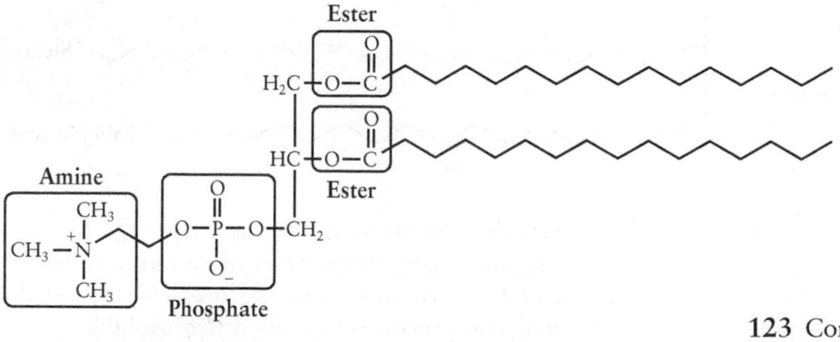

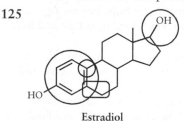

109 The phospholipid and glycolipid molecules interact through intermolecular forces of attraction which creates a flexible fluid-like assembly. The membrane molecules are not covalently bonded to each other.

111 Small nonpolar molecules such as oxygen and carbon dioxide pass freely through the cell membrane.

113 The saturated phospholipids pack more closely than the unsaturated phospholipids; therefore, they are less fluid and more rigid.

115 Steroids are structurally different from triglycerides, phospholipids, and glycolipids because they are not derived from fatty acids.

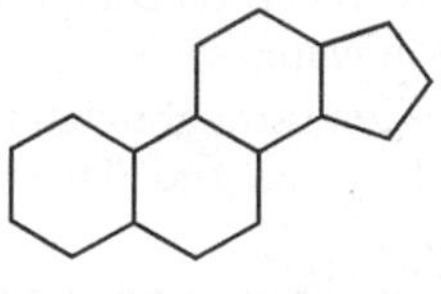

117 Cholesterol helps maintain the integrity of the cell membrane. It is also the biological precursor of all the other important steroids in the cell, including steroid hormones, vitamin D, and bile salts.

119 Cholate has a nonpolar end (the steroid ring system) and a polar end (the ionized carboxylic acid). The steroid ring system is part of the structure.

121 Prednisone is a glucocortoid. Glucocortoids are produced in the adrenal cortex. It has the steroid ring structure in the molecule.

123 Cortisol is released in response to stress. When it is released it causes an increase in blood glucose levels, and it facilitates fat and protein metabolism.

125

Estradiol

The six-membered ring attached to an OH group has been converted into an aromatic ring. The double bond in the adjacent six-membered ring has been removed. The hydrocarbon chain attached to the five-membered ring has been converted into an OH group. The methyl group between rings A and B has been removed.

127 Progesterone, testosterone, and estradiol are all produced from cholesterol. All three hormones have the steroid ring structure. Progesterone and testosterone both have a ketone and a double bond in ring A. Testosterone and estradiol both have an alcohol on ring D. Progesterone has a ketone attached to ring D, while the other two do not. Estradiol has a phenol group for ring A, while the other two do not.

129 Cell membranes have two layers, while lipoproteins have only one layer and are spherical in shape.

131 Chylomicrons have the largest diameter.

133 High density lipoproteins are the densest.

135 HDLs are considered "good." They remove cholesterol from the bloodstream and carry it back to the liver.

137 The three major classes of eicosanoids are prostaglandins, thromboxanes, and leukotrienes.

139 Arachidonic acid is an unsaturated fat. The double bonds are all *cis* double bonds.

141 Prostaglandin H_2 is formed by the action of *cyclooxygenase* on arachidonic acid.

143 *Phospholipase* A_2 catalyzes the release of arachidonic acid from glycerophospholipids. This enzyme catalyzes a hydrolysis reaction.

145 NSAIDs inhibit *cyclooxygenase*, the key enzyme involved in the formation of prostaglandin H_2 and the formation of prostaglandins and thromboxanes.

Chapter 13

45 Sickle-cell anemia is an inherited blood disorder. People who have the disease have crescent (sickle) shaped red blood cells. This shape affects the cells' ability to transport oxygen.

47 The symptoms of sickle-cell anemia are severe pain, fatigue, and lethargy.

49 The β-chains of an individual with sickle-cell anemia have a valine as the sixth amino acid rather than a glutamic acid.

51 Treatment for sickle-cell anemia includes: antibiotics taken for the first 5 years of life, periodic blood transfusions, and treatment with hydroxyurea.

53 a. Tyr   **b.** Cys

55. a. pH = 1   **b.** physiological pH

c. pH = 10

57 a. pH = 1; charge +1
b. physiological pH; charge 0
c. pH =10; charge −1

59 a. The zwitterion, the ionized amine form, water and hydronium ion are all present at equilibrium.
b. If more acid is added, the equilibrium will shift to the left.

61 a. The amino acids with nonpolar side chains are glycine, alanine, valine, leucine, isoleucine, proline, tryptophan, phenylalanine, and methionine.   **b.** The amino acids with polar basic side chains are lysine, arginine, and histidine.   **c.** The amino acids with polar acidic side chains are aspartic acid and glutamic acid.   **d.** The amino acids with polar, neutral side chains are serine, threonine, tyrosine, cysteine, glutamine, and asparagine.

63 a. pH = 10   **b.** physiological pH

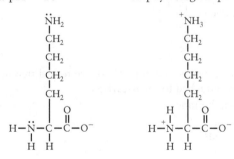

65 Essential amino acids are amino acids that are not synthesized by the body and must be supplied through the diet.

67 a. Aspartic acid and glutamic acid have carboxylic acids in their side chains.   **b.** Glutamine and asparagine both have amides in their side chains.

69

Cysteine contains a thiol functional group.

71

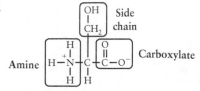

a. The amine is in its ionized form.

b. One of the oxygen atoms in the carboxylate group is missing a hydrogen atom and is negatively charged.

c. This amino acid is serine.

d. The side chain is polar. It contains an alcohol group.

e. At pH = 12 serine has the structure shown below. The net charge is –2.

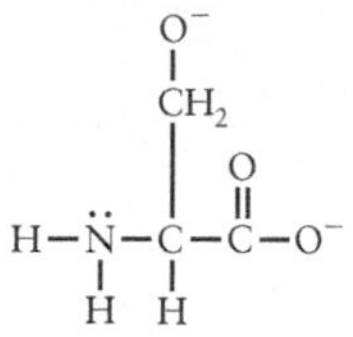

$$\text{O}^-$$
$$|$$
$$\text{CH}_2$$
$$|\qquad \text{O}$$
$$\qquad \parallel$$
$$\text{H}-\ddot{\text{N}}-\text{C}-\text{C}-\text{O}^-$$
$$\qquad |\quad |$$
$$\qquad \text{H}\ \ \text{H}$$

73

Amino acid	Nonpolar	Polar	Acidic	Basic	Neutral	Essential amino acid
Phe	x					x
Asn		x			x	
Met	x				x	x

75 Vegetarians need to eat a meal with a complementary mixture of plant proteins, such as grains with legumes.

77 a. methionine

b. It is L-methionine. The —NH₃⁺ group is pointing to the left in the Fischer projection.

c.
$$\text{COO}^-$$
$$\text{H}_3\overset{+}{\text{N}}-\!\!\!\!-\!\!\text{H}$$
$$|$$
$$\text{CH}_2$$
α-carbon
$$\text{CH}_2$$
$$|$$
$$\text{S}$$
$$|$$
$$\text{CH}_3$$

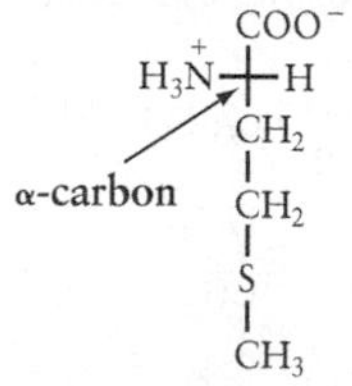

d. The molecule is chiral. There are four different groups attached to the α-carbon.

e. D-methionine

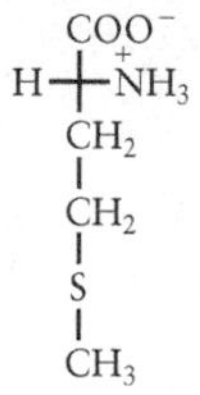

$$\text{COO}^-$$
$$\text{H}-\!\!\!\!-\!\!\overset{+}{\text{NH}_3}$$
$$|$$
$$\text{CH}_2$$
$$|$$
$$\text{CH}_2$$
$$|$$
$$\text{S}$$
$$|$$
$$\text{CH}_3$$

79
$$\text{COO}^-$$
$$\text{H}_3\overset{+}{\text{N}}-\!\!\!\!-\!\!\text{H}$$
$$|$$
$$\text{CH}$$
α-carbon
$$\text{CH}_3\ \ \text{OH}$$

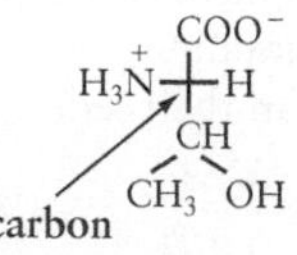

a. Yes, this amino is found in nature.

b. See structure.

c. If the —NH₃⁺ group is on the left, it is an L amino acid. If the —NH₃⁺ group is on the right, it is an D amino acid.

d. D-threonine and L-threonine have the same four groups attached to the α-carbon, but the connectivity and spatial arrangement of the four groups is different.

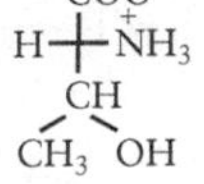

$$\text{COO}^-$$
$$\text{H}-\!\!\!\!-\!\!\overset{+}{\text{NH}_3}$$
$$|$$
$$\text{CH}$$
$$\text{CH}_3\ \ \text{OH}$$

81 a. A hydrogen atom, a carboxylate group, a —NH₃⁺ group, and a CH₂CH₂COO⁻ are attached to the α-carbon.

b. The carboxylate group and the —NH₃⁺ group are present in all amino acids.

c. The CH₂CH₂COO⁻ is unique to glutamic acid.

d. They are enantiomers. L-glutamic acid is found in nature.

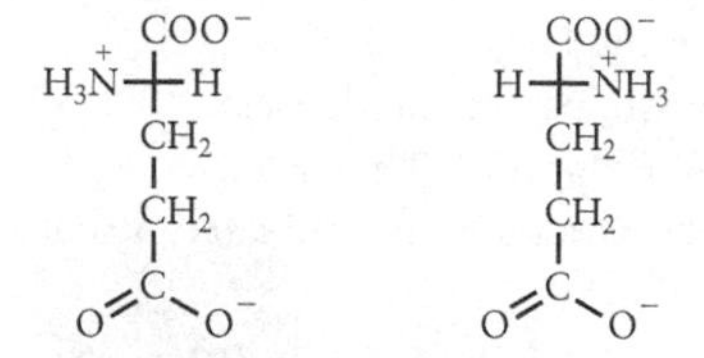

$$\text{COO}^-$$
$$\text{H}_3\overset{+}{\text{N}}-\!\!\!\!-\!\!\text{H}$$
$$|$$
$$\text{CH}_2$$
$$|$$
$$\text{CH}_2$$
$$|$$
$$\text{O}=\!\!\overset{\text{C}}{}\!\!-\text{O}^-$$
L-glutamic acid

$$\text{COO}^-$$
$$\text{H}-\!\!\!\!-\!\!\overset{+}{\text{NH}_3}$$
$$|$$
$$\text{CH}_2$$
$$|$$
$$\text{CH}_2$$
$$|$$
$$\text{O}=\!\!\overset{\text{C}}{}\!\!-\text{O}^-$$
D-glutamic acid

83 A dipeptide contains two amino acids linked together by one amide (peptide) bond. A tripeptide contains three amino acids linked together with two amide (peptide) bonds. A polypeptide has more than 12 amino acids linked together by peptide bonds.

85 a.

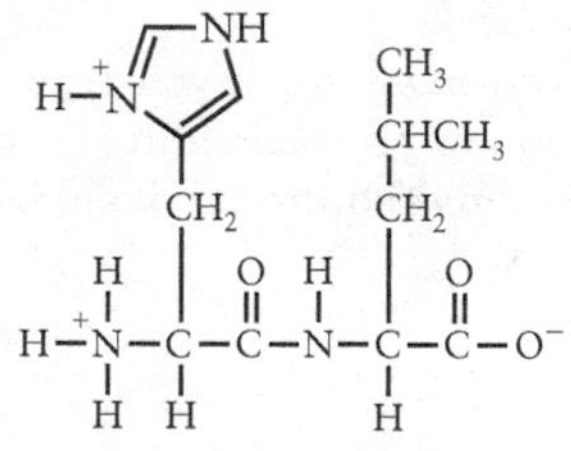

His-Leu

b.

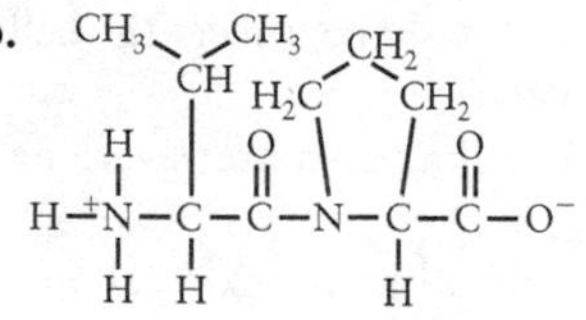

Val-Pro

c.

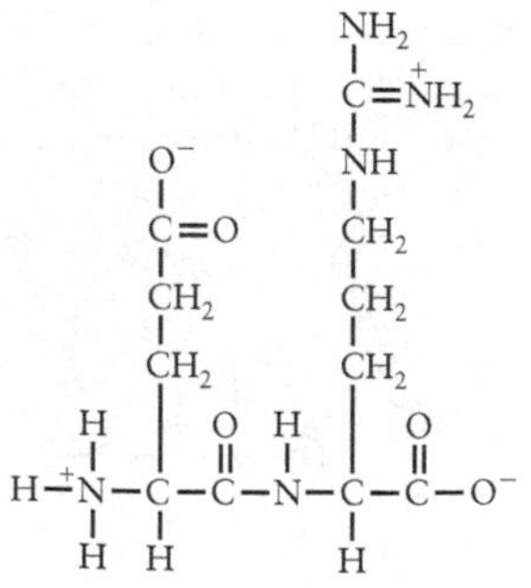

Glu-Arg

d.

```
      NH2        CH3
      |          |
      C=O        CH2
      |          |
      CH2        CHCH3
   H  |  O  H  |  O
H–⁺N–C–C–N–C–C–O⁻
   H  H     H
```

Asn-Ile

87

```
                          OH


                                        SH
           CH2        CH3        CH2
        H  |  O  H  |  O  H  |  O
N-terminus H–⁺N–C–C–N–C–C–N–C–C–O⁻ C-terminus
           H  H     H     H
```

89 a. Gln is the N-terminus.
    b. Asp is the C-terminus.
    c. This pentapeptide contains five amino acids.
    d. -e.

```
   NH2        OH                 NH2        OH
   |                             |          |
   C=O                           C=O        C=O
   |                             |          |
   CH2        CH2        CH3      CH2        CH2
   |          |          |        |          |
   CH2        CH2        |        CH2        CH2
   H  |  O  H  |  O  H  |  O  H  |  O  H  |  O
H–⁺N–C–C–N–C–C–N–C–C–N–C–C–N–C–C–O⁻
   H  H     H     H     H     H
```

91 a. Bradykinin contains nine amino acids.
    b. The amino acid sequence for bradykinin is Arg-Pro-Pro-Gly-Phe-Ser-Pro-Phe-Arg.

93 The native conformation of a protein is its unique three-dimensional shape which is necessary for the protein to be able to perform its specific biological function. Intermolecular interactions, such as hydrogen bonding, electrostatic interactions such as salt bridges, dispersion forces and disulfide bridges stabilize the protein in its native conformation.

95 In humans, hair and fingernails are rich in keratins. In animals, feathers, horns, claws, and hooves are rich in keratins.

97 Collagens are found in skin, tendons, bones, cartilage, and ligaments. Collagens provide rigidity to the tissue.

99 Membrane proteins are found in part or entirely embedded in the cell membrane.

101 the sequence of amino acids that make up the protein

103 The two most common forms of secondary structure are the α-helix and the β-pleated sheet. The secondary structure is formed by hydrogen bonding between different amino acids in the protein. The amide bonds in the polypeptide backbone are involved in the secondary structure. The side chains of a polypeptide are not involved in the secondary structure.

105 A prosthetic group is an essential non-peptide containing organic molecules or metal ions that

form ionic or covalent bonds to the polypeptide. A prosthetic group is part of the tertiary structure.

107 Proteins fold because it lowers the potential energy of the protein and its surrounding environment. A folded protein and its surrounding environment are lower in potential energy than the unfolded protein and its surrounding environment.

109

```
      H  H                      H  H        H  H
      |  |           [O]        |  |        |  |
2  H–C–C–S–H  ———→  H–C–C–S–S–C–C–H
      |  |                      |  |        |  |
      H  H                      H  H        H  H
```

111 aspartic acid, glutamic acid, lysine, arginine, and histidine

113 If the protein is an aqueous environment, the nonpolar side chains are most likely to be found on the interior of the protein. The polar side chains are most likely to be found on the exterior where "like may interact with like."

115 b., d., and e. isoleucine, valine, and proline

117 disulfide bridges, salt bridges, hydrogen bonding, and dispersion forces

119 a. a salt bridge
    b. hydrogen bonding
    c. dispersion forces
    d. disulfide bridge

121 Heat, pH changes, mechanical agitation, detergents, and some metals may denature a protein. When a protein is denatured it loses its secondary, tertiary, and quaternary structure. Since the denatured protein has lost its shape, it has also lost its function. The primary structure of the protein is not affected by denaturation.

123

```
                                    ⁺NH3
                                    |
              H                     CH2
            N–                      |
   O        |                       CH2
    \\      |                       |
     C–OH                           CH2
    /                               |
   CH2        CH2        CH2
   H  |  O  H  |  O  H  |  O
H–⁺N–C–C–N–C–C–N–C–C–OH
   H  H     H     H
```

125 Without enzymes, biological reactions are too slow for life.

127 energy of activation (E_A)

129 When a substrate binds to the active site, the enzyme changes its conformation to improve the "fit" between the substrate and the enzyme. This model is called the induced-fit model.

131 E + S ⇌ ES ⟶ E + P
    E is the enzyme, S is the substrate, ES is the enzyme-substrate complex, and P is the product.

133 The substrate is also one particular reactant.

135 An enzyme is not chemically changed during an enzyme-catalyzed reaction. The substrate is chemically changed to the product during an enzyme-catalyzed reaction.

137 Physiological pH (pH 7.3) is the optimal pH, and body temperature (37 °C) is the optimal temperature.

139 Enzyme inhibitors are compounds that bind to an enzyme and prevent the enzyme from performing its function.

141 Noncompetitive inhibitors bind at a location on the enzyme other than the active site. The binding of a noncompetitive inhibitor causes the shape of the enzyme to change in such a way that the active site can no longer bind to the substrate.

143 Cholesterol inhibits is own synthesis when it binds to the enzyme *HMG-CoA Reductase*. This type of inhibition is called feedback inhibition.

145 Hypertension is chronically elevated blood pressure.

147 Sodium ions cause water to be retained in the circulatory system through osmosis, which increases blood pressure.

149

Bond broken by
angiotensin converting enzyme

151 Captopril binds the zinc atom that is a cofactor at the active site of *ACE*. It is a competitive inhibitor since it prevents angiotensin I, the natural substrate, from binding to ACE.

Chapter 14

29 If the genes contain mutations, they can predispose a person to disease.

31 Acquired mutations are alterations to the genes that take place during a person's lifetime, while inherited mutations are the genes that a person is born with.

33 Carcinogens, radiation, and viral infections can trigger mutations.

35 DNA and RNA

37 A nucleotide consists of a monosaccharide, a phosphate, and a nitrogenous base.

39 The anomeric carbon in the monosaccharide has a covalent bond to the nitrogenous base. The nitrogen atom in the nitrogenous base is bonded to the monosaccharide. The nitrogenous base is situated above the ring (β), and a covalent bond joins the 1'carbon of the sugar to a nitrogen atom on the nitrogenous (N) base.

41 Adenine (A), guanine (G), cytosine (C), and uracil (U) are found in RNA.

43 a.–f.

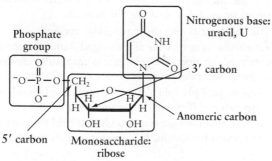

Phosphate group

Nitrogenous base: uracil, U

3' carbon

Anomeric carbon

5' carbon

Monosaccharide: ribose

g. The nucleotide would be found in RNA. The second carbon atom in the monosaccharide has an OH group.

h. uridine 5' monophosphate

45

Phosphate group

Nitrogenous base: adenine, A

Monosaccharide: deoxyribose

47

Hydroxyl group

Hydroxyl group

Guanosine 5' monophosphate This nucleotide would be found in RNA. The second carbon atom in the monosaccharide has an OH group.

49 The 3' hydroxyl group of one nucleotide and the 5' phosphate group of the other nucleotide react to form a phosphate diester.

51

G

U

C

53 A nucleotide contains a nitrogenous base, a monosaccharide, and a phosphate group. A nucleoside has a nitrogenous base and a monosaccharide but no phosphate group.

55 DNA is negatively charged, while the histones are positively charged.

57 Most human cells contain 46 chromosomes.

59 leukemia and epilepsy

61 The three-dimensional shape of DNA is a double helix.

63 DNA is a right-handed double helix. It is chiral.

65 The monosaccharide and phosphate groups are located on the outside of the three-dimensional DNA structure. These groups are hydrophilic and can interact with the polar aqueous environment.

67 Hydrogen bonding and base stacking.

69 Base stacking is a stabilizing interaction.

71 two hydrogen bonds

73 The complementary sequence is dATAGCG. Each A–T pair has two hydrogen bonds, and each C–G pair has three hydrogen bonds: (3 A–T pairs × 2 hydrogen bonds) + (3 C–G pairs × 3 hydrogen bonds) = 15 hydrogen bonds.

75 a. dGATCCG
   b. dTGACTT
   c. dAACCTT
   d. dCCATGA

77 A single cell grows and divides to become a human being with one trillion cells. Before the first cell can divide, DNA must be able to replicate itself.

79 *DNA polymerase* catalyzes the formation of a new daughter strand of DNA from a parent strand. The enzyme also proofreads the daughter strand for mistakes.

81 The DNA parent strands are unwound ahead of the leading strand by the enzyme *helicase*.

83
Parent: 5′ dCATTAAGCCG 3′   3′ dGTAATTCGGC 5′
Daughter: 3′ dGTAATTCGGC 5′   5′ dCATTAAGCCG 3′

85 *DNA polymerase* catalyzes the formation of phosphate ester linkages between the 5′ phosphate group of a nucleotide and the 3′ hydroxyl group on the growing nucleotide chain.

87 a. 5′ dATTCCGTA 3′: parent strand
     3′ dCAAGGTAT 5′: daughter strand
     (Errors in daughter strand are bolded.)
     3′ dTAAGGCAT 5′: corrected daughter strand
   b. 5′ dGGGCCCTTTAA 3′: parent strand
     3′ dCCCAGGAAGTT: 5′ daughter strand
     (Errors in daughter strand are bolded.)
     3′ dCCCGGGAAATT 5′ corrected daughter strand

89 *DNA polymerase* only works in the 5′ → 3′ direction.

91 *DNA polymerase* can only work in one direction, so only one strand, known as the leading strand, can be synthesized continuously from the 5′ to 3′ end. The lagging strand, which is also synthesized in the 5′ to 3′ direction, must be synthesized discontinuously.

93 The leading strand moves in the same direction as the replication fork. The enzyme *helicase* unwinds the two DNA parent strands ahead of the leading strand so the replication fork moves in the same direction as the synthesis of the leading strand.

95 The central dogma of molecular biology is that the basic flow of information is from DNA to messenger RNA to protein.

97 ribosomal RNA (rRNA), messenger RNA (mRNA), and transfer RNA (tRNA)

99 Translation occurs at the ribosome. mRNA and tRNA are involved.

101 RNA contains the monosaccharide D-ribose. RNA contains the nitrogenous base U, uracil.

103 Ribosomes are composed of rRNA and various proteins.

105 Translation and transcription are involved in synthesizing a protein from a gene.

107 *RNA polymerase II* catalyzes the reaction between the 5′ phosphate group on an incoming nucleotide and the hydroxyl group at the 3′ end of the growing mRNA chain.

109 a. 3′ dCCGGAATATA 5′:   DNA strand
       5′ GGCCUUAUAU 3′:   mRNA strand
    b. 3′ dAAGGCCAATT 5′:   DNA strand
       5′ UUCCGGUUAA 3′:   mRNA strand
    c. 3′ dGTACACGTCG 5′:   DNA strand
       5′ CAUGUGCAGC 3′:   mRNA strand

111 The genetic code is the 64 codons and each of the amino acids they specify.

113 Translation begins when the mRNA codon forms base pairs with the complimentary anticodon of a tRNA molecule carrying the correct amino acid at its other end. Another matching tRNA molecule arrives and a peptide bond is formed between the amino acids on adjacent tRNA molecules.

115 A genetic mutation is any permanent chemical change to one or more nucleotides in a gene (DNA) that affects the primary structure of a protein.

117 The anticodon AAA will correspond to the codon UUU; therefore, the amino acid will be phenylalanine.

119 No more amino acids will be added to the growing peptide chain when a stop codon is reached.

121 CUU, CUC, CUA, and CUG all code for leucine. The effects of a genetic mutation are minimized by the presence of more than one codon for leucine.

123 Yes, an amino acid can be specified by more than one codon.

125

3′ hydroxyl on tRNA   Amino acid   Ester

The carboxylic acid of the amino acid is involved. An ester is formed.

127 a. mRNA: 5′ AAUAGUGUG 3′
       amino acids: Asn Ser Val
       The amino acid sequence would be Asn-Ser-Val.
    b. mRNA: 5′ CCCUUUGGG 3′
       amino acids: Pro Phe Gly
       The amino acid sequence would be Pro-Phe-Gly.
    c. mRNA: 5′ CACCGGUGG 3′
       amino acids: His Arg Trp
       The amino acid sequence would be His-Arg-Trp.
    d. mRNA: 5′ UCCUUAGCA 3′
       amino acids: Ser Leu Ala
       The amino acid sequence would be Ser-Leu-Ala.

129 DNA template strand 3' dGGTGCT 5'
mRNA: 5' CCACGA 3'
amino acid sequence: Pro-Arg

  a. DNA mutation: 3' dGGCGCT 5'
    mRNA: 5' CCGCGA 3'
    amino acid sequence: Pro-Arg
    The same amino acid sequence is produced.

  b. DNA mutation: 3' dGGTGCA 5'
    mRNA: 5' CCACGU 3'
    amino acid sequence: Pro-Arg
    The same amino acid sequence is produced.

  c. DNA mutation: 3' dGTGCA 5'
    mRNA: 5' CACGU 3'
    amino acid sequence: His-
    The amino acid sequence is different.

  d. DNA mutation: 3' dGGTCT 5'
    mRNA: 5' CCAGA 3'
    amino acid sequence: Pro-

The amino acid sequence will be different. If the next nucleotide in the mRNA sequence is U or C, the amino acid sequence will be Pro-Asp. If the next nucleotide in the mRNA sequence is A or G, the amino acid sequence will be Pro-Glu.

131 No, you could not predict the unique sequence of nucleotides on mRNA that coded that protein. The amino acids have more than one sequence of mRNA that codes for them; therefore, you cannot tell which nucleotide was used in mRNA.

133 A virus is a nucleic acid RNA or DNA encapsulated in a protein coat.

135 The HIV virus uses *reverse transcriptase* and *protease* in its life cycle.

137 T-lymphocytes

139 Reverse transcription allows a single strand of DNA to be transcribed in the opposite direction of normal transcription; therefore, the single strand of DNA can serve as the template for the second strand. *Reverse transcriptase* is the enzyme that catalyzes this process.

141 HIV is a rapidly mutating viral DNA.

143 Gene expression in T-cells occurs when the T-cell is needed to fight off an infection.

145 Some of the viral mRNA produced is hydrolyzed by the viral enzyme HIV *protease* to produce genomic RNA, and some viral mRNA is transcribed into proteins, such as the key viral enzymes.

147 Since the HIV virus rapidly mutates, it eventually becomes resistant to any one type of drug therapy. Therefore, a combination of drug therapies is needed.

Chapter 15

37 The metabolism of triglycerides takes place in muscle and liver cells.

39 B vitamins provide the body with a steady supply of the starting materials needed to produce the coenzymes required for the fat burning biochemical pathways.

41 Catabolic processes produce energy overall.

43 Acetyl CoA is the central molecule of metabolism. It is produced and used in many biochemical pathways.

45 The citric acid cycle, the electron transport chain, and oxidative phosphorylation are the biochemical pathways of the third stage of catabolism.

47 Maltose and dextrins are hydrolyzed into glucose in the small intestine. They are hydrolyzed into glucose.

49 Fructose and galactose are hydrolyzed into glucose in the liver.

51 The hydrolysis of triglycerides produces glycerol and fatty acids.

53 Peptide bonds are broken in the hydrolysis of amino acids.

55 The energy released from catabolism is stored in phosphoanhydride bonds of nucleotide triphosphates.

57 The products are ADP and inorganic phosphate. One phosphoanhydride bond is broken and 30.5 kJ/mol of energy is released.

59 30.5 kJ/mol is required to phosphorylate ADP to ATP.

61 In coupled reactions, one of the reactants in one of the reactions is a product of the other reaction. The energy transferred in each individual reaction is added together.

63 a. The conversion of phosphoenolpyruvate to pyruvate provided the energy for the coupled reaction.
b. The curved arrow represents the coupled reaction of the phosphorylation of ADP to ATP. The straight arrow represents the transformation of the substrate (phosphoenolpyruvate) into product (pyruvate).
Phosphoenolpyruvate → pyruvate
ADP → ATP

c.

Reaction	
phosphoenolpyruvate ⟶ pyruvate	−61.9 kJ/mol
ADP ⟶ ATP	+30.5 kJ/mol

phosphoenolpyruvate ⤳ pyruvate (ADP ATP)　−31.4 kJ/mol

The coupled reaction releases 31.4 kJ/mol of energy.

65 The biochemical pathways that break down food (carbohydrates, fats, and proteins) produce acetyl CoA.

67 Glycolysis converts glucose into pyruvate. It takes place in the second stage of catabolism.

69 No, it does not require oxygen. It is an anaerobic pathway.

71 Glycolysis occurs in the cytosol, the fluid part of the cell outside the nucleus.

73 Every intermediate in glycolysis contains a phosphate group. The charge on the phosphate groups prevents the intermediates of glycolysis from diffusing out of the cell, since polyatomic ions cannot diffuse through the cell membrane.

75 The formation of pyruvate, ATP, and NADH in the second half of glycolysis more than compensates for the energy used in the first half.

77 Two molecules of NADH are produced from one molecule of glucose during glycolysis.

79

81 Carbon dioxide is expelled in aerobic conversion of pyruvate to acetyl CoA. Coenzyme A is the carrier of the two-carbon acetyl group.

83 Anaerobic conditions exist in a marathoner's muscle cells at the end of a marathon.

85 Lactic acid builds up in muscles through lactic acid fermentation. Lactic acid fermentation occurs when oxygen levels are low, as they are during strenuous exercise. Lactate is formed when pyruvate is reduced and NADH is oxidized to NAD^+. NAD^+ is used in glycolysis to produce ATP. Once the energy demanding activity is over, lactate is transported to the liver where it is converted back into pyruvate.

87 The citric acid cycle occurs during the third and last stage of catabolism.

89 The citric acid cycle is an aerobic pathway.

91 For every molecule of acetyl CoA that enters the citric acid cycle, one GTP, three NADH, one $FADH_2$, and two molecules of CO_2 are produced.

93 About 1.5 ADP molecules can be phosphorylated to ATP from the electrons released by one $FADH_2$ molecule during oxidative phosphorylation.

95 a. At physiological pH the structure of citric acid is:

$$\begin{array}{c} COO^- \\ | \\ CH_2 \\ | \\ HO-C-COO^- \\ | \\ CH_2 \\ | \\ COO^- \end{array}$$

b. There are six carbon atoms in citric acid. Two of the carbon atoms come from acetyl CoA and four of the carbon atoms come from oxaloacetate.
c. It is produced in the first step.

97 Halfway through the citric acid cycle, two carbon atoms have been lost from citrate. Two molecules of CO_2 are expelled to remove these two carbon atoms from citrate. Two molecules of NADH are produced in the first half of the citric acid cycle.

99 The electrons from the NADH and $FADH_2$ produced in the citric acid cycle are used in oxidative phosphorylation to produce ATP.

101 a. Oxaloacetate has a ketone in place of the alcohol in malate. b. Oxaloacetate is one of the reactants in the first step of the citric acid cycle. c. Malate is oxidized. d. NAD^+ is reduced to NADH.

103 The electrons in the electron transport chain come from the NADH and $FADH_2$ produced in the citric acid cycle.

105 The energy producing pathways of catabolism, except glycolysis, occur in the mitochondria.

107 No, ions cannot pass through the inner mitochondrial membrane.

109 The matrix has a higher pH.

111 The protons are moving against the proton gradient. The energy for this process comes from NADH.

113 Yes, there is a difference in charge between the matrix and the intermembrane space.

115 In the electron transport chain the affinity for electrons increases as the electrons are transferred though the electron transport chain.

117 The energy needed to pump protons against the proton gradient comes from the energy released in the oxidation-reduction steps within the protein complexes of the electron transport chain.

119 NADH supplies more energy (as measured in the number of ADP that are phosphorylated to ATP). NADH provides energy to phosphorylate 2.5 ADP, while $FADH_2$ provides energy to phosphorylate 1.5 ADP. NADH transfers its electrons earlier in the electron transport chain.

121 No, electrons are not passed between Complexes I and II.

123 $O_2 + 4 H^+ + 4 e^- \rightarrow 2 H_2O$ Oxygen is reduced to water in Complex IV.

125 *ATP synthase* uses the potential energy of the proton motive force to phosphorylate ADP to ATP.

127 Glycolysis: 2 ATP citric acid cycle: 2 ATP oxidative phosphorylation: 10 NADH (2.5) + 2 $FADH_2$ (1.5) = 28 ATP
2 + 2 + 28 = 32 ATP produced from 1 molecule of glucose

129 The formation of ATP produced directly and during oxidative phosphorylation supply the energy for anabolic processes.

131 Heart and liver cells use the energy provided by fatty acid catabolism.

133 A fatty acid is activated to fatty acyl CoA in the cytosol of muscle and liver cells.

135 Two carbon atoms are removed in the 4-step β-oxidation process.

137 Step 4 breaks the carbon-carbon bond. The single bond between the α-carbon and the β-carbon breaks. This reaction produces acetyl CoA and a new fatty acyl CoA molecule that is two carbon atoms shorter than the fatty acyl CoA at the start of β-oxidation.

139 Linolenic acid: C-18

8 cycles of β-oxidation	× 4 =	32 ATP
9 acetyl CoA molecules through the citric acid cycle	× 10 =	90 ATP
Activation of 1 linolenic acid molecule	=	−2 ATP
Total		120 ATP

141 Rotenone interferes with the transport of electrons from NADH because rotenone is a competitive inhibitor of Complex I where the electrons from NADH are delivered.

143 Oligomycin interferes with the proton channel in *ATP synthase* during phosphorylation of ADP to ATP.

145 Cyanide poisoning is lethal because cyanide is an irreversible inhibitor of *cytochrome a_3*. The victim cannot utilize oxygen because electrons are never transferred from *cytochrome a_3* to oxygen. Consequently, protons are no longer pumped into the inner membrane space, so *ATP synthase* is no longer supplied with the proton motive force that drives the phosphorylation of ADP to ATP. Without ATP, cells lack the energy to drive the processes required for living.

147 Symptoms of cyanide poisoning include headache, lethargy, confusion, nausea, vomiting, and convulsions.

GLOSSARY

α-anomer The ring form of a monosaccharide wherein the —OH group on the anomeric center is below the ring in a Haworth projection.

α-carbon The tetrahedral carbon in an amino acid to which the amine, the carboxylic acid, a hydrogen atom, and the R group are attached. It is also a center of chirality in all the natural amino acids except glycine.

α-helix One of the two major types of protein secondary structure, characterized by a helical shape formed as a result of hydrogen bonding between amides in the polypeptide backbone 3.5 amino acids apart.

α particle A slow-moving, high-energy particle consisting of two protons and two neutrons emitted as a result of nuclear decay. Its nuclear symbol is $_2^4\alpha$ or $_2^4\text{He}$.

Absolute temperature scale A temperature scale, like the kelvin scale, where zero represents a temperature in which particles have no kinetic energy.

Absorbed dose The energy of radiation absorbed per mass of tissue.

Accuracy An indicator of how close repeated measurements are to the "true" value.

Acetyl CoA The central molecule of metabolism, characterized by a thioester, and derived from the vitamin pantothenic acid, used to transfer acyl groups in biochemical reactions.

Acetyl group A carbonyl group and its attached —CH_3 group.

Achiral An object or a molecule that is not chiral. It is superimposable on its mirror image and so identical to its mirror image.

Acid According to the Arrhenius definition, a substance that produces hydrogen ions, H^+, in solution. According to the Brønsted-Lowry definition, a proton donor.

Acid-base homeostasis Maintenance of the proper pH balance in the blood and other bodily fluids.

Acid dissociation constant, K_a The product of the concentrations of products divided by the concentration of reactants for the reaction of an acid with water at equilibrium:

$$K_a = \frac{[H_3O^+] \times [A^-]}{[HA]}$$

Acidosis A serious medical condition characterized by an arterial blood pH below 7.35.

Actinides The elements Ac through Lr in period 7, offset from the main body of the periodic table.

Activation energy, E_A The minimum amount of energy that must be attained by the reactant(s) in order for a chemical reaction to occur.

Active site The pocket or cleft within an enzyme where the substrate binds and a biochemical reaction takes place. Also known as a binding site.

Acyl group A carbonyl group and its attached —R group.

Acyl group transfer reactions A type of reaction that transfers the acyl group from a carboxylic acid to the heteroatom of an alcohol, thiol, or amine to produce a carboxylic acid derivative: ester, thioester, or amide, respectively. Water is also a product.

Adipocytes Fat cells.

Aerobic biochemical pathway A biochemical pathway that requires oxygen, such as pyruvate oxidation, the citric acid cycle and oxidative phosphorylation.

Aggregates Many small molecules or ions clumped together. They constitute the particles of a suspension.

Alcohol A functional group derived from water where one of the H atoms has been replaced with an —R group: R—O—H. The carbon atom bearing the hydroxyl group is an alkyl carbon.

Aldehyde A carbonyl containing functional group characterized by a hydrogen atom bonded to the carbonyl carbon. Also includes formaldehyde which contains two —H atoms bonded to the carbonyl carbon.

Alkali metals The main group metal elements in group 1A.

Alkaline A solution with a pH greater than 7; a basic solution.

Alkaline earth metals The main group metal elements in group 2A.

Alkaloid A compound containing an amine that is found in nature.

Alkalosis A serious medical condition characterized by an arterial blood pH above 7.45.

Alkane A hydrocarbon containing only carbon-carbon single bonds and C—H bonds.

Alkene A hydrocarbon containing one or more carbon-carbon double bonds.

Alkyne A hydrocarbon containing one or more carbon-carbon triple bonds.

Amidation reaction The reaction of a carboxylic acid and an amine, in the presence of a catalyst, that produces an amide. Amidation reactions join amino acids to form the peptide bonds of proteins.

Amide A carbonyl compound with a nitrogen atom bonded to the carbonyl carbon. The nitrogen atom has two additional bonds to either —H or —R.

Amine A functional group derived from ammonia in which one, two, or all three of the hydrogen atoms have been replaced by R groups: RNH_2, R_2NH, or R_3N. There is no carbonyl group bonded to the nitrogen.

Amino acid The molecules from which all proteins are formed. They contain both an amine functional group and a carboxylic acid functional group bonded to the same tetrahedral carbon, the α-carbon. Each also has a unique R group that defines it.

Amphipathic Molecules that have both a polar and a nonpolar region within the same molecule. They include membrane lipids, cholesterol, and lipoproteins.

Amphoteric molecule A molecule, such as water, that can act as either an acid or a base.

Amylopectin The major constituent of starch; composed of primarily $\alpha(1 \rightarrow 4)$ linkages. It is a branched polysaccharide arising from some $\alpha(1 \rightarrow 6)$ linkages.

Amylose The minor constituent of starch; it is an unbranched polysaccharide with $\alpha(1 \rightarrow 4)$ linkages. It has an overall helical shape.

Anabolic pathways Biochemical reactions that convert smaller molecules into larger molecules such as proteins and DNA. Anabolic reactions consume energy overall.

Anaerobic biochemical pathway A biochemical pathway that does not require oxygen, such as glycolysis and lactic acid fermentation.

Analgesic A substance that reduces pain.

Anion A negatively charged ion resulting from a nonmetal atom gaining the electrons required to fill its valence energy level and obtain an electron arrangement like that of the noble gas in the same period.

Anomeric center The carbon atom in the ring form of a monosaccharide that is bonded to two oxygen atoms: the ring oxygen atom and a hydroxyl group (—OH). It is derived from the carbonyl carbon of the open-chain form of the monosaccharide.

Anticodon A three nucleotide sequence in the anticodon loop of a tRNA that base pairs to the complementary codon on mRNA during translation.

Antioxidants Substances found in fruits, vegetables, and tea that prevent the damaging effects of oxidation by reducing harmful oxidizing agents in the cell, or preventing their formation.

Aqueous solution A homogeneous solution in which water is the solvent.

Aromatic hydrocarbon A six-membered ring written as alternating double and single bonds, although the double bond electrons are actually distributed evenly over all six carbon atoms in the ring.

Artificial radioisotopes Man-made radioactive isotopes.

Atmospheric pressure The pressure exerted by the atmosphere (air) at any given place on the earth as a result of gravity. Atmospheric pressure is 1 atm at sea level.

Atomic mass unit A unit of mass for subatomic particles. Protons and neutrons have a mass of 1 amu.

Atomic number The number of protons in an atom; the atomic number defines the element.

Atomic scale Matter that is too small to be seen by a light microscope. The scale of atoms.

Atomic symbol The one- or two-letter symbol used to identify an element.

Atom The smallest intact particle of matter. All matter is made up of atoms.

ATP Synthase An enzyme complex in the inner mitochondrial membrane that uses the potential energy of the proton motive force to phosphorylate ADP to ATP. A channel created within *ATP synthase* allows protons to flow from the intermembrane space to the matrix, driving the motor that brings ADP and P_i together so that it can react and form ATP.

Autoionization of water The reaction that occurs in pure water between two water molecules to produce an equal and small concentration of hydronium ions (H_3O^+) and hydroxide ions (OH^-).

Average atomic mass A weighted average of the mass of all the isotopes of an element based on the natural abundance of each isotope.

Avogadro's law The direct relationship between the number of moles and the volume of a gas if the temperature and the pressure are constant.

Avogadro's number The number 6.02×10^{23}, which represents the number of particles in 1 mole of any substance.

β-anomer The ring form of a monosaccharide wherein the —OH group on the anomeric center is above the ring in a Haworth projection.

β-oxidation The biochemical pathway that converts fatty acids into many acetyl CoA molecules during the second stage of triglyceride catabolism, and in each cycle reduces one FAD to $FADH_2$ and one NAD^+ to NADH.

β particle A high-energy electron emitted as a result of nuclear decay. Its nuclear symbol is $_{-1}^{0}\beta$.

β-pleated sheet A type of protein secondary structure in which two or more sections of parallel or antiparallel polypeptides fold in a pattern that looks like a pleated skirt. Hydrogen bonding between amides on adjacent strands stabilize this type of secondary structure.

Background radiation Radiation emitted from natural sources such as the sun and the earth.

Balanced equation A chemical equation in which the coefficients indicate there are an equal number of each type of atom on both sides of the equation.

Ball-and-stick model A model of a molecule that represents atoms as colored spheres and covalent bonds as sticks. It is a tool for visualizing the shape of a molecule.

Barometer A device used to measure the atmospheric pressure.

Barometric pressure See *atmospheric pressure.*

Base According to the Arrhenius definition, a substance that produces hydroxide ions, OH^-, in solution. According to the Brønsted-Lowry definition, a proton acceptor.

Base pair A nitrogenous base on one polynucleotide hydrogen bonding to the complementary nitrogenous base on another polynucleotide. The complementary base pairs in DNA are A–T and G–C, and in RNA, the pairs are A–U and G–C.

Base stacking The stabilizing interaction resulting from the delocalized electrons on aromatic rings in the nitrogenous bases that hold two strands of DNA together.

Base unit In the metric system, a base unit is the standard unit of measurement for a particular quantity, such as the gram for mass, the meter for length, the liter for volume, and the calorie for energy.

Becquerel (Bq) A unit of measurement indicating the number of radioactive emissions from a sample.

Bent shape A molecular shape that arises either from a tetrahedral electron geometry when there are two nonbonding pairs and two bonding groups of electrons or from a trigonal planar electron geometry when there is one nonbonding pair of electrons and two bonding groups. The bond angle is 109.5° and 120°, respectively.

Benzene The simplest aromatic hydrocarbon, C_6H_6.

Bile salts Amphipathic molecules derived from cholesterol that act as emulsifying agents during the digestion of lipids.

Binary compound A covalent compound composed of only two different types of atoms.

Binding site See active site.

Biochemical pathway A sequence of chemical reactions that occurs in a biological cell wherein the product of one reaction becomes the reactant in the next reaction.

Bioenergetics The study of energy transfer in living cells.

Blood plasma Whole blood that has had the blood cells removed through centrifugation.

Blood serum Blood plasma that has had the fibrinogen and other clotting factors removed.

Bond angle The angle generated by a central atom and any two atoms bonded to the central atom.

Bond dipole The charge separation in a polar covalent bond, created when two atoms have different electronegativities (When EN >0.5 and <2.0).

Bond length The distance between two nuclei in a covalent bond.

Boyle's law The inverse relationship between the pressure and the volume of a gas if the temperature and number of moles are constant and temperature is given in kelvins.

Branched-chain isomer A hydrocarbon in which there are one or more carbon atoms bonded to more than two carbons, creating branches in the chain.

Buffer A solution that resists changes in pH upon addition of a small amount of acid or base. A buffer is composed of a weak acid and its conjugate base, or a weak base and its conjugate acid.

Buffer capacity The amount of acid or base that can be added to a buffer before the pH will begin to change significantly. Buffer capacity is directly related to the amount of buffer components in solution.

C-terminus The terminal end of a polypeptide that contains the free carboxylate group.

Calorie A nutritional calorie, equivalent to 1,000 calories, or 1 kcal.

calorie The amount of heat energy required to raise the temperature of one gram of water by 1 °C.

Calorimeter An apparatus used to measure the change in enthalpy, ΔH, for a chemical reaction. It can be used to measure the Caloric content of foods.

Calorimetry The experimental technique that uses a calorimeter to measure enthalpy changes in chemical reactions.

Carbohydrate A type of biomolecule that includes monosaccharides—simple sugars disaccharides, and polysaccharides. Carbohydrates are a source of energy for cells.

Carbonyl group A carbon-oxygen double bond, C=O.

Carboxylic acid A carbonyl containing functional group that contains an —OH group bonded directly to the carbonyl carbon.

Carboxylic acid derivative Compounds that can be prepared from carboxylic acids, such as esters and amides.

Catabolic pathways Sequences of reactions that convert large molecules, such as carbohydrates, proteins, and fats, into smaller molecules. Catabolic pathways release energy overall.

Catalyst A substance that increases the rate of a reaction by lowering the activation energy for the reaction. A catalyst is not consumed in the reaction but facilitates the reaction.

Catalytic hydrogenation A reduction carried out using hydrogen gas (H_2) in the presence of a metal catalyst (Pd, Pt). Catalytic hydrogenation is used to convert alkenes to alkanes.

Cation A positively charged ion resulting from the loss of valence electrons from a metal.

Cell membrane A selectively permeable barrier that separates the environment of the inside of the cell from the environment of the outside of the cell. Also known as a plasma membrane. Consists of a lipid bilayer composed of phospholipids and glycolipids.

Cellular respiration The combustion reactions cells perform to convert food into energy. The oxidation of carbohydrates, proteins, and triglycerides to produce energy in the form of ATP.

Cellulose A polysaccharide that provides structure for a plant, formed of glucose monomer units in $\beta(1 \rightarrow 4)$ linkages. Cannot be digested by humans but can be digested by ruminants. Serves as dietary fiber.

Center of chirality A tetrahedral carbon atom with bonds to four different groups or atoms. A molecule with one center of chirality is chiral. A molecule with two or more centers of chirality is chiral if it is nonsuperimposable on its mirror image.

Central dogma A reference to the basic flow of information from DNA to mRNA to amino acid to form proteins.

Centrifuge A standard piece of laboratory equipment used to separate the particles in a suspension from the medium. A centrifuge spins test tubes at high speed, 5,000 rpm, causing the suspended particles to collect at the bottom—the pellet.

Cerebroside A glycolipid containing a monosaccharide.

Change in enthalpy, ΔH The heat energy transferred in a chemical reaction under certain defined conditions.

Change of state The process of going from one state of matter (s, l, g) to another state of matter.

Charles's law The direct relationship between the temperature and the volume of a gas if the pressure and number of moles are constant and temperature is given in kelvins.

Chemical change A process where the composition of the substance changes. Also known as a chemical reaction.

Chemical equation A symbolic representation of a balanced chemical reaction that includes a reaction arrow, the formulas of the reactants and products, and whole-number coefficients.

Chemical kinetics The study of reaction rates—how fast reactants are converted into products.

Chemical reaction The transformation of one or more substances into one or more products. Chemical bonds are broken and new ones are formed in a chemical reaction. However, atoms are unchanged.

Chiral An object or a molecule that is nonsuperimposable on its mirror image.

Chromosome The highly compact structures containing DNA, the genome, and specialized proteins found in the nucleus of our cells.

Chylomicrons The least dense and largest of the lipoproteins, one micrometer in diameter, that carry triglycerides from the cells lining the small intestine (intestinal mucosa) to their destination through the lymph system and bloodstream.

Cis An alkene geometric isomer in which the two large groups on the carbon-carbon double bond are on the same side of the double bond.

Citric acid cycle The aerobic catabolic pathway in the third stage of catabolism that converts acetyl CoA into two molecules of carbon dioxide. The process reduces three molecules of NAD^+ to NADH and one molecule of FAD to $FADH_2$, and directly phosphorylates one GDP to GTP.

Coding strand The DNA strand that does not serve as a template during transcription. It is named as such because it has the same nucleotide sequence as the mRNA produced, except it contains T instead of U.

Codon The three non-overlapping nucleotides on an mRNA that code for one amino acid.

Coefficient The whole numbers placed in front of the element symbol, molecular formula, or formula unit of each reactant and product in a chemical equation. Coefficients represent the molar ratio of each of the reactants and products. The coefficient 1 is assumed and not written in.

Coenzyme A nonprotein organic molecule derived from a vitamin required by some enzymes to achieve their catalytic activity.

Coenzyme Q A hydrophobic electron carrier molecule that moves laterally within the inner mitochondrial membrane shuttling electrons from Complex I to Complex III and from Complex II to Complex III in the electron transport chain.

Cofactor A metal ion or organic molecule essential for certain enzymes to achieve their catalytic effect. A type of prosthetic group in the tertiary structure of a protein.

Collagens A type of fibrous protein that serves as the main structural proteins in the body, providing rigidity to connective tissue.

Colloid A particle with a diameter between 1 nm and 1,000 nm dispersed in a medium. Colloids include large molecules such as proteins and polysaccharides.

Colloidal dispersion A homogeneous mixture containing colloids—large molecules or aggregates—with a diameter between 1 nm and 1,000 nm, and evenly distributed throughout the medium. Colloidal dispersions have an opaque appearance.

Combustion reaction An exothermic reaction in which a substance containing C, H, and O reacts with oxygen (O_2) to produce carbon dioxide (CO_2) and water (H_2O).

Common Names Names that have been used traditionally in Non IUPAC, such as the names of carbonyl compounds containing one or two carbon atoms.

Competitive inhibitor A substance that competes with the substrate for the active site of an enzyme and prevents the substrate from being converted into product.

Complexes I–IV Four large enzyme complexes that lie within the inner mitochondrial membrane where the oxidation–reduction reactions of the electron transport chain occur.

Compound A substance composed of two or more different atoms held together by chemical bonds.

Concentration A quantitative measure of the amount of solute per total solution (solute + solvent). Some common concentration units are mg/dL, % m/v, and moles/L.

Concentration gradient Regions with different concentrations often separated by a membrane.

Condensation A change of state from the gas phase to the liquid phase. Requires removal of heat energy.

Condensed structure A shorthand notation for writing the chemical structure of a molecule such that each carbon atom and its attached hydrogen atoms are written as a group: C, CH, CH_2, or CH_3. Bonds are omitted except at branch points.

Conformations Different rotational forms of the same molecule resulting from free rotation about carbon-carbon single bonds.

Conjugate acid A base after it has accepted a hydrogen ion, H^+. It is neutral or positively charged.

Conjugate acid-base pair Molecules and ions that are related by the transfer of a proton: HA/A^- or B/BH^+.

Conjugate base An acid after it has donated a hydrogen ion, H^+. It is neutral or negatively charged.

Conjugated Two or more alternating carbon-carbon double and single bonds.

Conversion A mathematical expression equating two different units.

Conversion factor A mathematical relationship between two units expressed as a ratio or fraction.

Coupled reaction An energy-consuming reaction linked to an energy producing reaction so that the former can go forward.

Covalent bond Electrons shared between two nonmetal atoms in a molecule.

Covalent compound A molecule composed of more than one type of nonmetal atom held together by covalent bonds.

Crenation The shrinkage of red blood cells when immersed in a hypertonic solution as water exits the cell, across the cell membrane, by osmosis.

Curie (Ci) A unit of measurement indicating the number of radioactive emissions from a sample.

Cycloalkane An alkane whose chain of carbon atoms forms a ring structure.

Cytochrome c An electron carrier molecule in the inner mitochondrial membrane that shuttles electrons from Complex III to Complex IV in the electron transport chain.

D-sugar The enantiomeric form of a monosaccharide produced in nature and distinguished in a Fischer projection by the last center of chirality having the —OH group on the right. In a Haworth projection, a D-sugar has the last CH_2OH group, at the far left, above the ring.

Dalton's law of partial pressures The observation that for a mixture of gases each gas exerts a pressure independent of the other gases and each gas will behave as if it alone occupied the total volume:

$$P_{total} = P_1 + P_2 + P_3 + \ldots P_n$$

Daughter nuclide The new isotope formed after radioactive decay.

Daughter strand A reference to the two new strands of DNA synthesized during DNA replication, as distinguished from the two parent strands that were part of the original DNA.

Dehydration reaction The elimination of the elements of water (H and OH) from an alcohol to produce an alkene. It is the reverse of a hydration reaction. Produces preferentially the Zaitsev product.

Delta system A naming system that indicates the location of each carbon-carbon double bond in an unsaturated fatty acid. Numbering starts with the carboxylic acid, and gives the location of the first carbon in each double bond as superscripts, separated by commas, following the Greek letter delta.

Denaturation The disruption of secondary, tertiary, and quaternary structure in a protein so that it can no longer perform its function. Denaturation is caused by heat, pH changes, mechanical agitation, detergents, and certain metals.

Density A physical property of a substance defined as its mass per unit volume: $d = m/V$.

Deposition A change of state from the gas phase directly to the solid state. Requires the removal of heat energy.

Dialysis Diffusion of certain solutes from a region of higher solute concentration to a region of lower solute concentration across a selectively permeable membrane. Also, a life-support treatment that performs the functions the kidneys can no longer perform.

Diastereomers Nonsuperimposable non-mirror image stereoisomers. Monosaccharide stereoisomers that differ at one or more, but not all, centers of chirality, and therefore have different names.

Diene An alkene with two carbon-carbon double bonds.

Diffusion The movement of particles through a solvent as a result of kinetic energy.

Dilute A solution with less solute dissolved in the solvent.

Dilution The process of preparing a dilute solution from a more concentrated solution.

Dimensional analysis A method for converting between units and solving problems that contain units. Conversion factors are used to cancel units.

Diol A compound containing two hydroxyl groups. A type of alcohol.

Dipeptide A peptide derived from two amino acids in an amidation reaction between the carboxylic acid of one amino acid and the amine of the other amino acid.

Dipole arrow An arrow symbol, with a hatch mark on the end opposite the arrow, that indicates the direction of a bond dipole. The head of the arrow points toward the more electronegative atom.

Disaccharide A type of carbohydrate derived from two monosaccharides.

Dissolution The process of dissolving a solute in a solvent—a physical process. On the macroscopic scale, the solute appears to disappear in the solvent. On the atomic level, the solvent surrounds each of the solute particles: individual ions in the case of an ionic compound, and individual molecules in the case of a covalent compound.

Disulfide bridge The disulfide bond, R—S—S—R, a covalent bond formed upon oxidation of two thiol functional groups, such as the side chains in two cysteines.

Effective dose A measure of radiation that includes both the penetrating power and the amount of energy of a particular type of radiation. It conveys the actual biological effect of the radiation.

Eicosanoids Lipids that function as signaling molecules and initiate physiological events, such as the inflammatory response to trauma and infection.

Elastins A type of fibrous protein with elastic fibers that provide flexibility to skin, ligaments, and other tissues that require elasticity.

Electrolyte An aqueous solution containing dissolved ions. The solution can conduct electricity because of the charged ions in solution.

Electromagnetic radiation A form of electromagnetic energy that travels through space as a wave at the speed of light.

Electromagnetic spectrum The entire range of wavelengths or frequencies ranging from γ-rays to radio waves.

Electron A subatomic particle of the atom with a negative charge and negligible mass, found in an electron orbital.

Electron density diagram A space-filling model color coded to show the relative amount of charge on different atoms in a molecule. Red is used for δ^- and blue for δ^+.

Electronegativity A measure of an atom's ability to attract electrons toward itself in a molecule.

Electron geometry The relative position of the bonding and nonbonding electrons around a central atom that determines the molecular shape of the molecule.

Electron orbital A region of space describing where the electron is most likely to be found.

Electron transport chain A series of protein complexes and electron carrier molecules in the inner mitochondrial membrane that relay electrons from NADH and $FADH_2$ to oxygen via a series of oxidation–reduction reactions in the last stage of catabolism. As electrons are relayed through the electron transport chain, protons are pumped from the matrix to the intermembrane space, by a proton pump in complexes I, III, and IV, creating a proton gradient.

Electrostatic attraction The force of attraction between oppositely charged entities such as a proton and an electron or a cation and an anion.

Element A substance composed of only one type of atom, which cannot be broken down into a simpler form of matter. There are 114 elements that have been identified and named.

Emulsification The process whereby fat globules are turned into a colloid of finely dispersed microscopic fat droplets suspended in water.

Enantiomers A pair of nonsuperimposable mirror-image stereoisomers.

Endorphin An endogenous substance that binds to the opioid receptors in the brain, reducing the sensation of pain.

Endothermic reaction A reaction that absorbs heat from the surroundings and therefore has a positive change in enthalpy, $\Delta H > 0$, because the products are higher in energy than the reactants.

Energy The ability to do work or transfer heat, where work is the act of moving an object.

Energy diagram A diagram that depicts the energy changes during the course of a chemical reaction. Energy appears on the y-axis, and the pathway from reactants to products is shown along the x-axis.

English system A system of measure used in the United States for nonscientific everyday applications.

Enthalpy of fusion, ΔH_{fus} See heat of fusion.

Enthalpy of vaporization, ΔH_{vap} See heat of vaporization.

Enzyme A biological catalyst. An enzyme is typically a large globular protein containing a pocket or groove where the reactants bind to the enzyme.

Enzyme-substrate complex, ES The complex formed when the substrate is bound to its enzyme.

Equilibrium When the forward and reverse rates of reaction in a reversible reaction are equal, so the concentration of reactants and products is constant.

Equivalents per liter (eq/L) A unit representing the moles of charge in every liter. Calculated by multiplying the molar concentration of an ion by the numerical value of the charge on the ion.

Essential amino acids Amino acids that must be supplied through the diet because they cannot be synthesized in the body. They include histidine, isoleucine, leucine, lysine, methionine, phenylalanine, threonine, tryptophan, and valine.

Essential fatty acids Fatty acids the body cannot synthesize and therefore must be included in the diet. Examples are linoleic acid and linolenic acid.

Ester A carbonyl containing functional group with an —OR′ group bonded to the carbonyl carbon.

Esterification reaction The reaction of a carboxylic acid and an alcohol, in the presence of a catalyst, that produces an ester. Esterification reactions join fatty acids and glycerol to make fats.

Ether A functional group derived from water where both of the —H atoms have been replaced with —R groups: R—O—R′. The carbon atom bonded to the oxygen can be alkyl or aromatic.

Evaporation A change of state from the liquid to the gas phase at any temperature. It occurs when atoms and/or molecules have enough kinetic energy to leave the surface of a liquid and enter the gas phase.

Evaporative cooling The cooling of the surroundings that occurs when a liquid evaporates (liquid to gas phase change), as a result of heat being transferred from the surroundings to the liquid to achieve the change of state.

Exact number Numbers with an infinite number of significant figures. They are numbers obtained by counting and also apply to definitions.

Excited state A nucleus with excess energy as after undergoing radioactive decay.

Exothermic reaction A reaction that releases energy to the surroundings and has a negative change in enthalpy, $\Delta H < 0$, because the products are lower in energy than the reactants.

Exponential decay A rate of decay expressed by the equation: $N = (\frac{1}{2})^n$, where the number of half-lives, n, is an exponent in the equation. After every half-life, the sample diminishes to half its mass.

FAD/FADH$_2$ Flavin adenine dinucleotide. A coenzyme used in oxidation–reduction reactions of carbon-carbon bonds. Derived from the B vitamin riboflavin.

Family Elements in the same column of the periodic table; also known as a group.

Fats Triglycerides (triacylglycerols) derived from three fatty acids and one molecule of glycerol to form a triester. They provide a long-term supply of energy for the body.

Fatty acid A long unbranched hydrocarbon chain containing an even number of carbons with a carboxylic acid (RCO_2H) at one end. Unsaturated fatty acids also contain one or more carbon-carbon double bonds.

Fatty acyl CoA The activated form of a fatty acid required for β-oxidation.

Feedback inhibition A type of noncompetitive inhibition where the product of one step in a biochemical pathway is a competitive inhibitor of a step earlier in the biochemical pathway, thereby regulating its own production.

Fibrous protein Strong and water insoluble proteins that generally have a structural role in nature. Includes keratins, elastins, and collagens.

Fischer projection A way to represent monosaccharides and other molecules with multiple centers of chirality. Points where horizontal and vertical lines intersect represent centers of chirality.

Flow rate The volume of solution per unit time in which an IV solution is administered.

Formula mass The mass of one formula unit of an ionic compound in units of amu.

Formula unit The lowest whole-number ratio of cations to anions, shown as subscripts, following the atomic symbols used to represent an ionic compound.

Frameshift mutation When one nucleotide is added to or deleted from a gene causing a change in the primary structure of a protein.

Free radical A molecule, atom, or ion that contains an odd number of valence electrons. Free radicals are unstable and reactive species because they contain an atom without an octet. In biochemistry, some free radicals are responsible for cell damage.

Freezing A change of state from liquid to solid. Requires the removal of heat energy.

Freezing point See *melting point*.

Frequency The number of times a full wave cycle passes a given point; used to describe a type of electromagnetic radiation. It is inversely related to the wavelength of light and related to the energy of the light.

Functional group A group of atoms and bonds that react in a characteristic way and give a molecule its physical properties. Most functional groups contain a heteroatom such as O, N, or P. Hydrocarbon functional groups include alkenes, alkynes, and aromatic hydrocarbons. Alkanes are not functional groups.

Furanose ring The five-membered ring form of a monosaccharide containing one ring oxygen, shown with the oxygen at the top in a Haworth projection, by convention.

γ radiation A form of electromagnetic radiation having the shortest wavelengths and greatest energy. Accompanies radioactive decay.

Gamma knife A type of noninvasive radiosurgery in which carefully focused γ radiation is used to destroy tumors or cancer cells.

Gas A state of matter that on the macroscopic scale has neither a definite shape nor a definite volume and takes on the shape of its container. Particles in the gas phase are moving rapidly and in random motion, with mainly empty space between them. The kinetic energy of particles is higher in this state than in the other states.

Geiger counter An inexpensive instrument used in the field to detect all forms of radiation.

Gene A nucleotide sequence within DNA that contains the instructions for making a protein.

Gene expression When a protein is required by the cell and transcription and translation ensues.

Genetic code The amino acids, start codon, and three stop codons specified by each of the 64 nucleotide codons of mRNA.

Genetic mutation Any permanent chemical change that occurs at one or more nucleotides in a gene that leads to a change in the primary structure of a protein.

Genome The complete DNA sequence of an organism.

Geometric isomers Compounds with the same chemical formula and the same connectivity of atoms but a different spatial orientation as a result of the restricted rotation about a carbon-carbon double bond.

Globular proteins Water-soluble polypeptides that fold into an overall spherical shape. Includes most enzymes.

Glucocorticoids A class of steroid hormones produced from cholesterol that regulate the metabolism of carbohydrates, proteins, and lipids.

Glycerophospholipid A lipid derived from glycerol with a phosphate diester linkage to glycerol, and one of three amino alcohols.

Glycolipid A membrane lipid containing a carbohydrate as the polar head. Glycolipids in mammals have a sphingosine backbone.

Glycolysis A 10-step biochemical pathway that converts glucose into two molecules of pyruvate in the second stage of carbohydrate catabolism, and in the process directly phosphorylates two ADP to ATP and reduces two NAD^+ to two $NADH + H^+$.

Glycosidic bond The covalent bond that joins two monosaccharides. It always occurs between the anomeric carbon of the first sugar and one of the hydroxyl group oxygen atoms of the second sugar.

Glycosidic linkage See glycosidic bond.

Glycosphingolipid A membrane lipid with a sphingosine backbone and a carbohydrate as a polar head.

Gray (Gy) A unit of radiation measurement indicating the energy of the radiation absorbed per mass of tissue.

Ground state The lowest energy state of an atom wherein the electrons occupy the lowest allowable energy levels.

Group A column in the periodic table; also known as a family.

Half-life The time that it takes a sample of a radioisotope to decay to one-half of its mass.

Halogens The elements in group 7A.

Haworth projection The conventional way to represent a furanose and a pyranose ring that emphasizes the configuration of the hydroxyl groups, by showing them either above or below the ring.

Heat Kinetic energy (molecular motion) that is transferred from a hot object to a cooler object due to a difference in kinetic energy.

Heating curve A graphical way to show how the temperature of a substance changes as energy is added at a constant rate, where a horizontal plateau occurs as energy is used to effect a phase change rather than an increase in kinetic energy.

Heat of fusion, ΔH_{fus} The amount of energy that must be added to melt a solid or removed to freeze a liquid at its melting point.

Heat of vaporization, ΔH_{vap} The amount of energy that must be added to vaporize a liquid or removed to condense a gas at its boiling point.

Hemolysis The swelling and bursting of red blood cells when immersed in a hypotonic solution as a result of water entering the cells through the cell membrane by osmosis.

Henry's law The observation that the number of gas molecules dissolved in a liquid is directly proportional to the partial pressure of the gas: $P = kC$.

Hertz (Hz) A unit of frequency that equals one wave cycle per second.

Heteroatom An atom in an organic molecule that is not carbon or hydrogen.

Heterogeneous mixture A mixture whose components are unevenly distributed throughout the mixture, such as a suspension or a saturated solution.

Histone The positively charged proteins that facilitate the tight packing of DNA (negatively charged) into nucleosomes.

Homeostasis A balanced system.

Homogeneous mixture A mixture whose components are evenly distributed throughout the mixture, such as a solution or a colloidal dispersion.

Hydration reaction The addition of the elements of water (H and OH) to an alkene to form an alcohol and converting the double bond into a single bond. It is the reverse of a dehydration reaction. Follows Markovnikov's rule when the alkene is unsymmetrical.

Hydrocarbon An organic compound composed of exclusively carbon and hydrogen atoms.

Hydrogen bonding The strongest type of intermolecular forces of attraction between molecules. It occurs in molecules that contain an H—F, H—O, or H—N bond, the three strongest bond dipoles.

Hydrolysis reaction A reaction that uses water to split a carboxylic acid derivative (ester, thioester, or amide) into a carboxylic acid and an alcohol, thiol, or amine.

Hydronium ion A water molecule with a bond to a proton, H_3O^+, produced when an acid is added to water.

Hydrophilic A molecule or part of a molecule that is attracted to water because it is polar.

Hydrophobic A molecule or part of a molecule that is nonpolar and therefore does not dissolve in water.

Hydrophobic effect The tendency for nonpolar molecules to separate from water so as to avoid mixing with polar water molecules.

Hydroxyl group The —OH group in alcohols, phenols, and carboxylic acids.

Hypertonic solution The solution with the higher solute concentration.

Hypotonic solution The solution with the lower solute concentration.

Ideal gas law A combination of Boyle's, Charles's, and Avogadro's laws: $PV = nRT$.

Immunosuppressant A substance that reduces the activity of the immune system.

Induced-fit model The current model of how an enzyme and its substrate bind that involves a conformational change within the enzyme upon binding substrate, which creates a more complementary fit between the enzyme and substrate.

Inner core electrons The electrons in energy levels lower than the energy level containing the valence electrons.

Inner mitochondrial membrane A selectively permeable membrane that separates the matrix from the intermembrane space in a mitochondrion.

Inorganic compound A compound that does not contain carbon.

Inorganic phosphate Monohydrogen phosphate ion, HPO_4^{2-}, abbreviated P_i. The most abundant form of phosphoric acid in the cell.

Intermembrane space The aqueous region of a mitochondrion located between the outer mitochondrial membrane and the inner mitochondrial membrane. It has a higher proton concentration than the matrix.

Intestinal mucosa Cells lining the small intestine.

Intravenous (IV) The administration of a solution through a vein.

Ion A positive or negatively charged atom. The charge results from an unequal number of protons and electrons due to the loss or gain of electrons.

Ion-dipole The electrostatic interaction between an ion and the opposite partial charge on a polar molecule, such as water.

Ionic bond The strong electrostatic attraction that holds anions and cations together, such as in the lattice structure of an ionic compound.

Ionic compound A compound composed of ions held together by electrostatic attractions.

Ionization A process whereby a metal atom loses electrons to become a cation.

Ionizing radiation High-energy radiation with sufficient energy to dislodge an orbital electron from an atom or a molecule, creating a cation.

Ion-product constant, K_w The constant 1.0×10^{-14}, which represents the product of the hydronium ion concentration and the hydroxide ion concentration in an aqueous solution at 25 °C.

Isotonic solutions Solutions that have equal solute concentrations.

Isotopes Atoms having the same number of protons but a different number of neutrons. Isotopes are distinguished by their mass number.

Joule A unit of measure for energy.

Kelvin scale The absolute temperature scale that assigns a temperature of zero to the theoretical condition in which molecular motion has stopped.

Keratins Fibrous proteins that provide structure to hair, feathers, and other tissues.

Ketone A carbonyl containing functional group in which two R groups (carbon atoms) are bonded to the carbonyl carbon.

Kinetic energy The energy of motion; the energy a substance possesses as a result of the motion of its particles.

Kinetic molecular view A model of the arrangement and motions of particles from the perspective of the particles.

Kinetics The study of reaction rates: how fast reactions occur.

L-sugar The enantiomeric form of a monosaccharide not produced in nature and distinguished in a Fischer projection by the last center of chirality having the —OH group on the left.

Lactic acid fermentation The conversion of pyruvate to lactate, with simultaneous oxidation of NADH to NAD^+, under anaerobic conditions.

Lagging strand The daughter strand that is synthesized in fragments (in the direction opposite of the movement of the replication fork) during DNA replication by DNA polymerase in the $5' \rightarrow 3'$ direction.

Lanthanides The elements La through Lu in period 6, seen offset from the main body of the periodic table.

Lattice A three-dimensional array of cations and anions that is an ionic compound. Cations are surrounded by anions and anions by cations.

Law of conservation of mass The universal observation that matter can be neither created nor destroyed.

Leading strand The daughter strand that is synthesized continuously (in the same direction as the movement of the replication fork) during DNA replication by DNA polymerase in the $5' \rightarrow 3'$ direction.

Le Châtelier's principle An observation that governs reversible reactions at equilibrium when a disturbance is introduced. A reversible reaction at equilibrium responds to a disturbance by shifting to the right or the left in such a way as to counteract the disturbance.

Length A measurement of distance. The base unit of length in the metric system is the meter. In the English system, the common units of length are the mile, the yard, the foot, and the inch.

Lethal dose (LD_x) The level of radiation exposure that would result in death in x percent of the population exposed in 30 days.

Lewis dot structure The structure of a molecule showing how valence electrons are distributed around atoms and shared between atoms in a molecule. Constructed by combining Lewis dot symbols. Covalent bonds are drawn as lines, representing shared pairs of electrons, and nonbonding electrons are drawn as pairs of dots.

Lewis dot symbol A way of representing an atom and its valence electrons by writing the atomic symbol surrounded by its valence electrons represented as dots.

Like-dissolves-like A general rule of thumb that predicts that polar solutes will dissolve in polar solvents and nonpolar solutes will dissolve in nonpolar solvents.

Linear An electron geometry and molecular shape resembling a straight line, formed by two groups of electrons surrounding a central atom with a 180° bond angle.

Lipids Biological compounds insoluble in water and soluble in organic solvents. They contain a significant amount of hydrocarbon in their structure.

Lipoproteins Spherical monolayer of phospholipids and proteins assembled so that their polar heads are on the exterior and their hydrophobic tails are on the interior. They function to transport triglycerides and cholesterol through the lymph system and bloodstream.

Liquid A state of matter that occupies a definite volume but does not have a definite shape. Particles in the liquid state are interacting and have more kinetic energy than in the solid phase but less than in the gas phase.

% m/v A unit of concentration that represents the grams of solute per 100 mL of solution. Thus a 1% solution contains 1 g of solute per 100 mL of solution.

Macronutrients Elements that must be supplied through the diet in a quantity greater than 100 mg/day.

Macroscopic scale Matter that you can see with the naked eye.

Main group elements The elements in groups 1A through 8A.

Markovnikov's rule The observation that the major product formed in a hydration reaction of an unsymmetrical alkene has the OH group adding to the double-bond carbon with fewer hydrogens and the H atom adding to the double-bond carbon with more hydrogens.

Mass A measure of the amount of matter, measured on a calibrated balance or scale. The metric base unit of mass is the gram. The common English units of mass are the pound (lb) and the ounce (oz).

Mass number The sum of the number of protons and neutrons in an isotope.

Matrix The aqueous region in a mitochondrion enclosed within the inner mitochondrial membrane. It has a lower proton concentration than the intermembrane space.

Matter Anything that has mass and occupies volume.

Medium The component of a colloidal dispersion that is present in the greater quantity. Analogous to the solvent in a solution.

Melting A change of state from solid to liquid.

Melting point The temperature at which a substance undergoes a change of state between the liquid and solid phases. Same as the freezing point.

Membrane A barrier between two environments such as the cell membrane.

Membrane proteins Proteins that are located in part of or entirely within the cell membrane, a hydrophobic environment.

Metabolic acidosis A medical condition resulting when blood pH falls below normal as a result of a metabolic disorder.

Metabolic alkalosis A medical condition resulting when blood pH rises above normal as a result of a metabolic disorder.

Metabolism The biochemical pathways of catabolism and anabolism.

Metalloids The elements with properties similar to both metals and nonmetals found on each side of the dark zigzag line (excluding aluminum, which is a metal) running diagonally on the right side of the periodic table, separating metals from nonmetals.

Metals The elements on the left side of the periodic table (left of the dark zigzag line).

Metastable An isotope whose nucleus is in an excited state.

Metric prefix When placed preceding a metric base unit, it signifies a multiple or fraction of 10. For example, *milli* is the multiplier 1,000.

Metric system The most common system of measurement used in science and medicine and in most of the world.

Micronutrients Elements that must be obtained through the diet in quantities of less than 100 mg/day.

Microscopic scale Matter that can be seen through a light microscope.

Mineralocorticoids A class of steroid hormones derived from cholesterol, which regulate ion balance in tissues.

Mirror image The image seen when a molecule or object is held up to a mirror.

Mitochondria Organelles within the cell where the energy producing pathways of β-oxidation, the citric acid cycle, and oxidative phosphorylation occur.

Mixture A combination of two or more elements and/or compounds in any proportion.

Molar concentration A unit of concentration represented by the moles of solute per liter of solution: mol/L, abbreviated *M*. Also referred to as molarity.

Molarity See *molar concentration.*

Molar mass The mass of one mole of any element or compound in units of g/mol.

Molar volume of a gas The volume occupied by one mole of any gas at STP, which is 22.4 L.

Mole A counting unit that represents the number 6.02×10^{23}, usually applied to atoms, ions, or molecules.

Molecular dipole A charge separation in a molecule created by bond dipoles together with a molecular shape that doesn't allow cancellation of the bond dipoles.

Molecular mass The mass of one molecule in units of amu calculated from the sum of the individual atomic masses of the atoms in the molecular formula.

Molecular shape The geometry of a molecule based on the relative positions of the atoms in the molecule. Molecular shape is determined from the electron geometry and the relative number of bonding and nonbonding groups around the central atom.

Molecule Two or more nonmetal atoms held together by covalent bonds.

Monatomic ion: An ion formed from an atom, such as Na^+ and Cl^-.

Monomer The repeating structural unit in a polymer.

Monosaccharide A carbohydrate that cannot be hydrolyzed into a simpler carbohydrate. The monomer unit of disaccharides and polysaccharides. Also referred to as a simple sugar. Composed of many hydroxyl groups and either an aldehyde or a ketone. Glucose and galactose are common monosaccharides.

Monounsaturated fatty acid A fatty acid with one carbon-carbon double bond.

Myelin sheath The sphingophospholipids (sphingomyelins) that insulate nerve fibers.

N-terminus The terminal end of a polypeptide that contains the free amine.

NAD$^+$/NADH Nicotinamide adenine dinucleotide. A coenzyme seen in oxidation–reduction reactions of carbon-oxygen bonds. Derived from the B vitamin niacin.

Native conformation The three-dimensional shape of a functional protein.

Net reaction Represents the actual species reacting as in a neutralization reaction. For example: $H^+ + OH^- \rightarrow H_2O$.

Neutralization reaction The chemical reaction between an acid and a base to form a neutral compound like water and a salt, or water, a salt, and CO_2.

Neutron A subatomic particle with no charge and a mass of approximately 1 amu; located in the nucleus of the atom.

Nitrogenous base One of the three components of a nucleotide containing an aromatic ring and two or four ring nitrogens. The nitrogenous bases found in DNA are adenine (A), guanine (G), cytosine (C), and thymine (T). The nitrogenous bases

found in RNA are adenine (A), guanine (G), cytosine (C), and uracil (U).

Noble gases The elements in group 8A. They are uniquely stable because they have a full valence energy level.

Nonbonding electrons Valence electrons on an atom in a molecule that belong solely to that atom and are not shared; represented as a pair of dots in a Lewis dot structure.

Noncompetitive inhibitor A substance that binds to an enzyme at a location other than the active site preventing the substrate from being converted to product.

Nonelectrolyte A solution composed of neutral solute molecules or an insoluble ionic solid, which therefore does not conduct electricity.

Nonmetals Elements on the right side of the periodic table (right of the dark zigzag line).

Nonpolar covalent bond A bond formed between two atoms with the same or comparable electronegativity (EN <0.5).

Nonpolar molecule A molecule with an even distribution of electrons; no separation of charge. It is the result of either all nonpolar bonds in the molecule or polar bonds that cancel because of a symmetrical molecular shape.

Nonsuperimposable When two objects or molecules are overlain and do not match up at every point.

Normal boiling point The temperature at which a pure liquid changes to the gas phase or a gas changes to the liquid phase at atmospheric pressure.

Nuclear equation A common way to represent radioactive decay that shows the parent nuclide on the left side of the arrow and the daughter nuclide and the particle of radiation on the right side of the arrow. The sum of the superscripts on both sides of the arrow must be equal. The sum of the subscripts on both sides of the arrow must be equal.

Nuclear medicine The use of radioisotopes in the diagnosis and treatment of disease.

Nucleosome Compact structures containing DNA coiled around histone proteins.

Nucleotide A molecule that contains a nitrogenous base, a monosaccharide, and a phosphate group. Nucleotides are the monomer units of the polynucleotides RNA and DNA.

Nucleotide triphosphate (NTP) A nucleotide containing three phosphate groups joined by two high energy phosphoanhydride bonds, which supply the energy for joining a nucleotide to a growing polynucleotide during DNA synthesis, for example.

Nucleus The small, dense center of the atom that contains its protons and neutrons.

Octet rule The tendency for most atoms in a molecule to share electrons—form covalent bonds—so that they have eight valence electrons.

Oils Triglycerides derived from unsaturated fatty acids. They tend to be liquids at room temperature (25 °C) because they have lower melting points, due to the *cis* double bonds in their hydrocarbon chains. They are derived from vegetable sources, with the exception of tropical oils.

Omega system A naming system that indicates the location of carbon-carbon double bonds in an unsaturated fatty acid.

Numbering starts at the end of the chain opposite the carboxylic acid, and gives the location of the first carbon with a double-bond following the Greek letter omega.

Opioids Morphine and its derivatives that have analgesic properties.

Oral medication Medications given by mouth in solution or suspension form.

Organic chemistry The branch of chemistry devoted to the study of carbon-containing compounds and their chemical reactions.

Organic compound A compound containing carbon atoms.

Osmosis Simple diffusion of water from a region of lower solute concentration (higher water concentration) to a region of higher solute concentration (lower water concentration) across a selectively permeable membrane.

Osmotic pressure The minimum pressure that must be applied to the hypertonic solution to prevent osmosis.

Outer mitochondrial membrane The outer permeable membrane of a mitochondrion; it encloses and defines the mitochondrion.

Oxidation The reactant that loses electrons in an oxidation–reduction reaction. Observed in organic reactions as a product with fewer hydrogen atoms and or more oxygen atoms than the substrate.

Oxidation–reduction reaction A reaction characterized by the transfer of electrons from one reactant to another.

Oxidative phosphorylation The biochemical process, in the third stage of catabolism, that uses the proton motive force to phosphorylate ADP to ATP using *ATP synthase*.

Oxidizing agent The reactant in an oxidation–reduction reaction that gets reduced; so called because it oxidizes the substrate.

Parent nuclide A radioisotope that underdoes radioactive decay.

Parent strands A reference to the two original strands of the DNA double helix that are each copied during replication, as distinguished from the daughter strands.

Partial pressure The pressure exerted by one gas in a mixture of gases.

Pellet The precipitated particles of a suspension after centrifugation.

Penetrating power A measure of the extent to which a particular type of radiation passes through matter.

Peptide A molecule composed of two or more amino acids joined by peptide bonds.

Peptide bond The carbon-nitrogen bond of an amide, formed between the carboxylic acid of one amino acid and the amine of another amino acid.

Percentage Parts per 100, as for example grams of solute in 100 mL of solution.

Period A row in the periodic table.

Periodicity The repeating trends in chemical and physical properties of the elements, reflected in the layout of the periodic table.

Periodic table of the elements The table showing the 114 elements by their atomic symbol, atomic number, and atomic mass, displayed in characteristic rows and columns.

Permeability The ability of substances to pass through a barrier, such as a membrane.

pH pH $= -\log[H_3O^+]$. A way to express the concentration of hydronium ion that avoids using scientific notation.

pH indicator paper Special dye-coated papers that can be used to determine the approximate pH of a solution by a visible color change.

Pharmaceutical A drug used for therapeutic purposes. Most are organic compounds.

Phases of matter *See* states of matter.

Phenol A functional group composed of an aromatic ring bonded to a hydroxyl group.

Phosphate ester A functional group derived from hydrogen phosphate in which the hydrogen has been replaced by an —R group: $R—O—PO_3^{2-}$.

Phosphoanhydride bond The P—O—P bonds found in, diphosphate, and triphosphate esters. They are high-energy bonds because of the proximity of the negative charges in the phosphate groups.

Phospholipid The most common type of membrane lipid. They contain a phosphate group at the terminal end of either a sphingosine or a glycerol backbone.

Photosynthesis The biochemical pathway used by plants to synthesize carbohydrates from carbon dioxide and water, using sunlight as the source of energy.

Physical change A process that does not affect the composition of the substance. Changes of state and dissolving are physical changes.

Physiological pH The normal pH of arterial blood, which is 7.35–7.45.

Polar covalent bond A covalent bond that contains a bond dipole. A polar covalent bond is formed when two atoms with significantly different electronegativities share electrons (EN >0.5 and <2.0).

Polar molecule A molecule that has a separation of charge. A polar molecule is formed when one or more bond dipoles in the molecule do not cancel, creating a charge separation within the molecule.

Polyatomic ion: An ion formed from the loss or gain of electrons from a molecule, creating a positive or negative charge.

Polyene A molecule with more than two carbon-carbon double bonds.

Polymer A large molecule composed of repeating monomer units.

Polyols Compounds containing many hydroxyl groups.

Polypeptide A biological polymer derived from many amino acids joined by peptide bonds. Polypeptides with more than 50 amino acids are classified as proteins. Many proteins contain more than one polypeptide chain.

Polypeptide backbone The repeating N—C_α—C=O atoms common to all polypeptides. A reference to everything in the polypeptide except the side chains.

Polysaccharide A carbohydrate that can be hydrolyzed into many monosaccharides. Glycogen, starch, and cellulose are polysaccharides that are hydrolyzed into thousands of glucose molecules.

Polyunsaturated fatty acid A fatty acid with more than one carbon-carbon double bond.

Potential energy Stored energy. The energy a substance possesses as a result of its position or composition.

ppb Parts per billion: typically refers to grams of a solute in 10^9 mL of solution.

ppm Parts per million: typically refers to grams of a solute in 10^6 mL of solution.

Precipitate The insoluble solid that eventually settles at the bottom of the container of a heterogeneous mixture or saturated solution.

Precision An indication of how close repeated measurements are to each other.

Pressure A measure of the amount of force applied over a given area: $P = force/area$.

Primary structure The amino acid sequence of the polypeptide(s) that makeup a protein.

Prion A proteinaceous infection agent; an infectious abnormal protein. It is believed to be the causative agent in mad cow disease.

Products The compounds or elements that are produced in a chemical reaction. They are shown to the right of the arrow in a chemical equation.

Prosthetic groups Non-peptide organic molecules and metal ions that are strongly bound to a protein and essential to its function.

Protein Large biomolecules formed from amino acids, containing many amide functional groups. One of the four major biomolecules. It can be either a large polypeptide or multiple polypeptides, which fold into a unique three-dimensional shape.

Proton Subatomic particle with a positive charge and a mass of 1 amu; located in the nucleus of the atom.

Proton gradient A difference in proton concentration between two regions. In a mitochondrion there is a proton gradient between the intermembrane space and the matrix.

Proton motive force The chemical and electrical potential energy produced as a result of a proton gradient between the matrix and the intermembrane space, which is used to drive the phosphorylation of ADP to ATP.

Pyranose ring The six-membered ring form of a monosaccharide containing one ring oxygen, shown with the oxygen at the top right in a Haworth projection, by convention.

Quantum mechanics A theory developed in the early twentieth century that explains phenomena observed on the atomic scale, such as the energy of an electron is quantized: it can only exist in certain fixed energy levels.

Quaternary structure Created by electrostatic interactions between multiple polypeptides in a protein containing more than one polypeptide.

Racemic mixture A 50:50 mixture of enantiomers.

Rad A common unit of absorbed dose of radiation.

Radiation sickness Illness resulting from acute exposure to radiation.

Radiation The energy released by a radioisotope.

Radioactive decay The release of radiation by an unstable nucleus in order to become a more stable nucleus.

Radioisotope A radioactive isotope, unstable due to an imbalance in the number of neutrons and protons or because it contains too many neutrons and protons in the nucleus.

Reactants The compounds or elements that are combined in a chemical reaction. They are shown to the left of the arrow in a chemical equation.

Reaction rate The change in the concentration of reactants or products over time.

Redox An abbreviation for oxidation–reduction.

Reducing agent The substance in an oxidation–reduction reaction that gets oxidized; so called because it reduces the substrate.

Reduction The reactant that gains electrons in an oxidation–reduction reaction. Observed in organic compounds as a product with more hydrogen atoms and or fewer oxygen atoms than the substrate.

Regioselective reaction A reaction that produces one structural isomer preferentially over another structural isomer in a chemical reaction, as seen with Markovnikov's rule and Zaitsev's rule.

Rem A common unit of effective dose of radiation.

Replication fork The site where two strands of DNA separate, exposing the two parent strands for replication.

Respiratory acidosis A condition in which blood pH falls below normal as a result of weak breathing and ineffective removal of CO_2 from the blood.

Respiratory alkalosis A condition in which blood pH rises above normal because CO_2 is eliminated too quickly as a result of hyperventilating or breathing too fast.

Retrovirus A virus that transcribes RNA to DNA, the opposite of normal transcription (DNA to RNA) before it can be incorporated into the host's DNA.

Reverse osmosis The application of pressure greater than the osmotic pressure to a hypertonic solution, so water crosses a membrane against the concentration gradient.

Reversible reaction A reaction in which reactants are converted into products and products are simultaneously converted into reactants. A reversible reaction is represented in a chemical equation by two opposing half-headed reaction arrows.

Ribbon drawings A diagram of a protein wherein the polypeptide backbone is depicted as a flat colored ribbon for the purpose of showing secondary structure and the three-dimensional shape of the protein.

Ribosomes Complex cellular structures located in the cytoplasm where ribosomal RNA (rRNA), together with many proteins, assemble amino acids into polypeptides by reading mRNA.

Salt bridge Ionic bonds formed between charged side chains in a polypeptide or between side chains on different polypeptides.

Saponification reaction The hydrolysis of an ester in the presence of hydroxide ion to produce an alcohol and the conjugate base of a carboxylic acid. If the reactant is a fat, the conjugate base of the carboxylic acids produced is soap.

Saturated hydrocarbon A hydrocarbon with only carbon-carbon single bonds and C—H bonds that fits the formula C_nH_{2n+2}. Does not include cycloalkanes.

Saturated solution A solution with a solute concentration that exceeds the capacity of the solvent to dissolve the solute, resulting in a precipitate.

Secondary structure The sections of α-helix and β-pleated sheet within a protein formed by hydrogen bonding between parts of the amide functional groups in the polypeptide backbone.

Selectively permeable membrane A membrane that allows for the passage of certain molecules and ions across the membrane while preventing the passage of others.

Serum Blood plasma that has had the protein fibrinogen removed, thus preventing the blood from clotting.

Side chain The unique R group bonded to the α-carbon of an amino acid.

Sievert The unit of effective dose when the unit of absorbed dose is the gray.

Significant figures The number of certain digits plus one uncertain digit in a measurement. The number of significant figures reflects the degree of uncertainty in a measurement.

Simple diffusion The spontaneous movement of molecules or ions from a region of higher concentration to a region of lower concentration.

Skeletal line structure A shorthand notation for writing chemical structures in which carbon and hydrogen atom symbols are not written and C—H bonds are omitted. Carbon-carbon bonds are written as lines in a zigzag format.

Solid A state of matter that has a definite shape and volume independent of its container. Particles in the solid phase are in a lattice with only vibrational motion, and possessing the least kinetic energy of the three states of matter.

Solute The component(s) in a solution that is(are) present in the lesser amount. The solute is dissolved in the solvent. The solute has a size less than 1 nm in diameter.

Solution A homogeneous mixture that contains one or more solutes, dissolved in the solvent.

Solvent The component in a solution that is present in the greatest amount and dissolves the solute(s). Water is the solvent in an aqueous solution.

Space-filling model A model of a molecule that shows the relative amount of space occupied by the atoms in the molecule.

Specific enzyme inhibitors Molecules that bind to a specific enzyme and thereby prevent the enzyme catalyzed reaction from occurring. Includes competitive and noncompetitive inhibitors.

Specific gravity The ratio of the density of a substance to the density of water at 4 °C. Specific gravity is a unitless quantity.

Specific heat A measure of the amount of heat required to raise the temperature of 1 g of a substance by 1 °C.

Spectator ions An ion dissolved in solution that is not involved in the chemical reaction.

Sphingomyelin A membrane lipid with a sphingosine backbone and a phosphate group at the terminal hydroxyl group of sphingosine. Found in nerve fibers.

Sphingophospholipid A membrane lipid with a sphingosine backbone and a phosphate group on the terminal hydroxyl group of sphingosine. Also known as a sphingomyelin.

Sphingosine A molecule with a long hydrocarbon chain, two hydroxyl groups, and an amine that serves as the backbone of sphingolipids.

Starch A carbohydrate composed of 20 percent amylose and 80 percent amylopectin. It serves as stored energy for a plant and provides humans and animals with their primary source of energy. The structure is composed of primarily $\alpha(1 \rightarrow 4)$ linkages between glucose monomers.

State of matter Matter is found in three states: solid, liquid, and gas.

Stereoisomers Compounds with the same chemical formula and the same chemical structure but a different three-dimensional orientation in space. Most stereoisomers are chiral.

Steroid A lipid containing the characteristic four fused ring system of a steroid. Includes molecules like cholesterol, glucocorticosteroids, mineralocorticoids, and the sex hormones.

Stock solution A pre prepared solution with a known concentration from which less concentrated (dilute) solutions can be prepared.

Stoichiometry Molar ratios between reactants and/or products in a chemical reaction.

STP (standard temperature and pressure) A standard set of reference conditions for a gas: 0 °C and 1 atm.

Straight-chain isomer A hydrocarbon in which every carbon atom is bonded to at most two other carbon atoms, creating a straight chain without branches.

Strong acid An acid that is fully dissociated in water.

Strong base A base that is fully dissociated in water such as hydroxide ion-containing salts of group 1A or 2A metal cations.

Structural isomers Compounds that have the same chemical formula but differ in the connectivity of the atoms. Structural isomers are different compounds that exhibit different physical and chemical properties and have different IUPAC names.

Subatomic particles The parts of an atom: protons, neutrons, and electrons.

Sublimation A change of state directly from solid to gas. Requires the addition of heat energy.

Substituent A carbon branch along the main chain where a hydrogen atom has been substituted with a chain of one or more carbons.

Substitution mutation When one nucleotide in a gene is substituted for another nucleotide leading to a change in the primary structure of the protein.

Substrate The organic reactant that binds to the enzyme and undergoes a chemical change in a biochemical reaction.

Supernatant The part of a suspension that remains after the suspended particles have been removed by centrifugation.

Suspension A heterogeneous mixture containing particles larger than 1,000 nm (1 micrometer), which are unevenly distributed throughout the medium, and will eventually settle.

Temperature A measure of the average kinetic energy of a substance. The three temperature scales are Celsius, Fahrenheit, and kelvin.

Template strand The DNA strand that is copied, as a complementary sequence of nucleotides in mRNA, during transcription. Note the complementary base for A is U rather than T in mRNA.

Tertiary structure The complex folding of a polypeptide beyond its secondary structure, showing the overall three-dimensional shape of a single polypeptide.

Tetrahedral An electron geometry or molecular shape resembling a tetrahedron. The electron groups around the central atom point to the four corners of a tetrahedron, and bond angles are 109.5°. Occurs when there are four bonding groups of electrons around a central atom.

Thioester A functional group similar to an ester, but where sulfur replaces the oxygen bonded to the carbonyl group: RCOSR′. Thioesters are formed in esterification reactions between a carboxylic acid and a thiol. Water is also a product.

Thiol A functional group similar to an alcohol, but sulfur replaces oxygen: R—S—H, as seen in the amino acid cysteine.

Trace minerals Another term for *micronutrients*.

Trans An alkene geometric isomer in which the two large groups on the carbon-carbon double bond are on the opposite side of the double bond.

Transcription The process in which a complementary mRNA is synthesized from the template strand of a segment of DNA (the gene).

Transition metal elements The metals in groups 1B through 8B, positioned between groups 2A and 3A in the periodic table.

Translation The process of synthesizing a polypeptide from amino acids whose sequence is specified by the nucleotide sequence in an mRNA, using tRNAs carrying the individual amino acids.

Triacylglycerol A chemical term for a fat, also known as a triglyceride.

Triglycerides The most abundant type of lipid in the body, also known as triacylglycerols, or more simply, as fats. They serve as the body's long-term energy supply.

Trigonal planar An electron geometry or molecular shape resembling an equilateral triangle. The electron groups around the central atom point to the three corners of the triangle, and bond angles are 120°. Requires three bonding groups of electrons around a central atom.

Trigonal pyramidal shape The shape of a central atom with a tetrahedral electron geometry in which there is one nonbonding pair of electrons and three bonding groups. Bond angles are approximately 109.5°.

Triols Compounds containing three hydroxyl groups—a type of alcohol.

Tripeptide A peptide derived from three amino acids joined by two peptide bonds.

Tube model A model used to represent complex molecules wherein the bonds and atoms appear as part of a tube. Each end of the tube represents the color-coded atoms.

Unit conversion A type of calculation in which a measurement in one unit is converted into the equivalent value in another unit.

Universal donor A person with blood type O, who can donate blood to an individual with any other blood type.

Universal gas constant The proportionality constant, R, in the ideal gas law:

$$R = 0.08206 \frac{L \cdot atm}{mol \cdot K}$$

Universal recipient A person with blood type AB, who can receive blood from a donor with any other blood type.

Unsaturated hydrocarbon A compound containing one or more carbon-carbon double or triple bonds, or a cycloalkane, or an aromatic hydrocarbon. It has fewer than $2n + 2$ hydrogens, where n is the number of carbon atoms.

Unsaturated solution A homogeneous solution in which all the solute particles are dissolved in the solvent (no precipitate).

Valence electrons The outermost electrons of an atom; those with the highest n value; equal in number to the group number.

Valence shell electron pair repulsion theory (VSEPR) A theory used to predict the shapes of simple molecules. VSEPR predicts the electron geometry from the number of electron groups (usually two to four) around the central atom. Groups of electrons are positioned to achieve the maximum distance between them while maintaining the bond to the central atom.

Vaporization A change of state from liquid to gas that occurs at the boiling point of the liquid.

Vapor pressure The pressure of a gas in equilibrium with the liquid phase at a given temperature.

Volatile A liquid with a high vapor pressure, which therefore enters the gas phase more readily.

Volume A unit of measure that describes the three-dimensional space occupied by a substance. The metric base unit of volume is the liter. Common English units of volume include the gallon, quart, cup, tablespoon, and teaspoon.

Wavelength The distance between wave crests used to describe a type of electromagnetic radiation. Inversely related to frequency and energy.

Weak acid An acid that dissociates to a small extent in aqueous solution. At equilibrium the acid, and a small amount of hydronium ions, and the conjugate base are present.

Weak base A base that dissociates to a small extent in aqueous solution. At equilibrium, the base, and a small amount of hydroxide ions, and the conjugate acid are present.

Wobble hypothesis The explanation for why there are nonstandard base pairings observed in the third base pair, during translation, between a mRNA codon and a tRNA anticodon, provided it codes for the same amino acid.

Work The act of moving an object over a distance against an opposing force.

X-ray detector The part of the x-ray instrument that measures the amount of x-ray radiation that has passed through the tissue.

X-rays Electromagnetic radiation with short wavelengths, high frequency, and high energy. A type of ionizing radiation.

Zaitsev's rule The observation that in the dehydration of an alcohol elimination produces the alkene with fewer hydrogen atoms on the double-bond carbon atoms.

INDEX

NOTE: Page numbers followed by "f" indicate figures; those followed by "t" indicate tables